KU-489-462

Pathology of Domestic Animals

FOURTH EDITION   Volume 3

Pathology of Domestic Animals

FOURTH EDITION   Volume 3

EDITED BY

K. V. F. JUBB
School of Veterinary Science
University of Melbourne
Victoria, Australia

PETER C. KENNEDY
Department of Pathology
School of Veterinary Medicine
University of California, Davis
Davis, California, USA

NIGEL PALMER
Veterinary Laboratory Services
Ontario Ministry of Agriculture and Food
Guelph, Ontario, Canada

ACADEMIC PRESS, INC.
Harcourt Brace Jovanovich, Publishers
San Diego   New York   Boston
London   Sydney   Tokyo   Toronto

This book is printed on acid-free paper. ∞

Copyright © 1993, 1985, 1971, 1970, 1963 by ACADEMIC PRESS, INC.
Copyright renewed 1991 by Peter C. Kennedy and K. F. V. Jubb
All Rights Reserved.
No part of this publication may be reproduced or transmitted in any form or by any
means, electronic or mechanical, including photocopy, recording, or any information
storage and retrieval system, without permission in writing from the publisher.

Academic Press, Inc.
1250 Sixth Avenue, San Diego, California 92101-4311

United Kingdom Edition published by
Academic Press Limited
24–28 Oval Road, London NW1 7DX

Library of Congress Cataloging-in-Publication Data

Jubb, K. V. F.
    Pathology of domestic animals / K.V.F. Jubb, P.C. Kennedy, N.C.
Palmer. – 4th ed.
        p.   cm.
    Includes bibliographical references and index.
    ISBN 0-12-391607-0 vol. 3
    1. Veterinary pathology.   I. Kennedy, Peter C. (Peter Carleton),
date.   II. Palmer, Nigel.   III. Title.
SF769.J82   1992
636.089'607–dc20   92-12261
                                                                    CIP

PRINTED IN THE UNITED STATES OF AMERICA
92 93 94 95 96 97   EB   9 8 7 6 5 4 3 2 1

Contents

CHAPTER 1

The Cardiovascular System
WAYNE F. ROBINSON AND M. GRANT MAXIE

CHAPTER 2

The Hematopoietic System

V. E. O. VALLI WITH A CONTRIBUTION BY B. W. PARRY

CHAPTER 3

The Endocrine Glands

CHARLES C. CAPEN

CHAPTER 4

The Female Genital System
PETER C. KENNEDY AND RICHARD B. MILLER

CHAPTER 5

The Male Genital System
P.W. LADDS

Contents of Other Volumes

Volume 1

Volume 2

Contributors

Volume 1

THOMAS J. HULLAND, Department of Pathology, Ontario Veterinary College, University of Guelph, Guelph, Ontario, Canada N1G 2W1.

C. R. HUXTABLE, School of Veterinary Studies, Murdoch University, Murdoch, Western Australia, Australia 6150.

K. V. F. JUBB, School of Veterinary Science, University of Melbourne, Werribee, Victoria, Australia 3030.

NIGEL PALMER, Veterinary Laboratory Services, Ontario Ministry of Agriculture and Food, Guelph, Ontario, Canada N1H 6R8.

DANNY W. SCOTT, Department of Clinical Sciences, New York State College of Veterinary Medicine, Cornell University, Ithaca, New York, USA 14853-6401.

BRIAN P. WILCOCK, Department of Pathology, Ontario Veterinary College, University of Guelph, Guelph, Ontario, Canada N1G 2W1.

JULIE A. YAGER, Department of Pathology, Ontario Veterinary College, University of Guelph, Guelph, Ontario, Canada N1G 2W1.

Volume 2

IAN K. BARKER, Department of Pathology, Ontario Veterinary College, University of Guelph, Guelph, Ontario, Canada N1G 2W1.

D. L. DUNGWORTH, Department of Pathology, School of Veterinary Medicine, University of California, Davis, California, USA 95616.

K. V. F. JUBB, School of Veterinary Science, University of Melbourne, Werribee, Victoria, Australia 3030.

W. ROGER KELLY, Department of Veterinary Pathology, University of Queensland, St. Lucia, Brisbane, Queensland, Australia 4072.

M. GRANT MAXIE, Veterinary Laboratory Services, Ontario Ministry of Agriculture and Food, Guelph, Ontario, Canada N1H 6R8.

NIGEL PALMER, Veterinary Laboratory Services, Ontario Ministry of Agriculture and Food, Guelph, Ontario, Canada N1H 6R8.

JOHN F. PRESCOTT, Department of Veterinary Microbiology and Immunology, Ontario Veterinary College, University of Guelph, Guelph, Ontario, Canada N1G 2W1.

A. A. VAN DREUMEL, Veterinary Laboratory Services, Ontario Ministry of Agriculture and Food, Guelph, Ontario, Canada N1H 6R8.

Volume 3

CHARLES C. CAPEN, Department of Veterinary Pathology, The Ohio State University, Columbus, Ohio, USA 43210-1093.

PETER C. KENNEDY, Department of Pathology, School of Veterinary Medicine, University of California, Davis, California, USA 95616.

P. W. LADDS, Graduate School of Tropical Veterinary Science and Agriculture, James Cook University, Townsville, Queensland, Australia 4811.

M. GRANT MAXIE, Veterinary Laboratory Services, Ontario Ministry of Agriculture and Food, Guelph, Ontario, Canada N1H 6R8.

RICHARD B. MILLER, Department of Pathology, Ontario Veterinary College, University of Guelph, Guelph, Ontario, Canada N1G 2W1.

B. W. PARRY, School of Veterinary Science, University of Melbourne, Werribee, Victoria, Australia 3030.

WAYNE F. ROBINSON, School of Veterinary Studies, Murdoch University, Murdoch, Western Australia, Australia 6150.

V. E. O. VALLI, College of Veterinary Medicine, University of Illinois, Urbana, Illinois, USA 61801.

Preface to the Fourth Edition

Thirty years will have elapsed since the publication of the first edition of "Pathology of Domestic Animals." In that time it has become like a living thing, changing and adapting as veterinary pathology has changed, yet retaining the generic character of the first edition. The fourth edition will be immediately recognizable to the many users of earlier editions, but it is changed in significant ways.

Yesterday's interesting new case becomes tomorrow's new syndrome. Accordingly, we have attempted to provide a more comprehensive textual inclusion of individual disease entities rather than relegate many diseases to the bibliographies as in earlier editions. This change respects and reflects current veterinary pathology practice. Our concern for the economically significant diseases remains and they are discussed in detail. However, recent rapid growth in knowledge has come from applying modern investigative techniques to comparative pathology and the diseases of companion animals.

Library resources are more readily available than when this work was first produced. The bibliographic listings have been reduced and placed closer to their subjects. These sources have been selected not only for intrinsic merit but for the comprehensive bibliographies they provide.

An essential part of the work has been the illustrations, the number and quality of which have been maintained. We are particularly grateful to our many colleagues who have provided photographs of high quality; they are acknowledged in the figure legends. We have to a large extent departed from the plate format for illustrations; this has allowed us to replace some illustrations and to locate others more appropriately in relation to the text.

The bringing together of this fourth edition has involved many people, far too many to name. Acknowledgments are made at the end of several chapters but we, as editors, are grateful to our contributing authors for their splendid and timely contributions, to Dr. Jennifer Anne Charles for meticulous editorial assistance, and to our word-processor operators Yvonne Pritchard and Kay Vincent who never allowed the way to seem too long or too weary. Edward W. Eaton of the University of Guelph prepared most of the new illustrations for this edition, and we thank him for his assistance. It remains a pleasure to work with Academic Press.

Melbourne, Australia
1993

K. V. F. JUBB
PETER C. KENNEDY
NIGEL PALMER

Preface to the Third Edition

Much has been happening in veterinary pathology between editions of this work. We had, from the time of the first edition, an expectation and a hope that as the number of scientists dedicated to this field of study grew, so also would the variety of publications to serve the diversity of interests. This anticipation has only partly been realized. We still have few books that address themselves to diseases of a single domestic species or to a single organ system of domestic animals. The need for a comprehensive treatment of diseases of domestic species, from the viewpoint of the pathologist, remains. The reception of earlier editions and the interest of our colleagues around the world have influenced us to try for the third time to produce a work of some universal usefulness.

The amount of information available on the pathology of animal disease has grown enormously, and the task of integrating so much new information into a coherent statement has grown on an equal or larger scale. Changes have become necessary in this book. This edition introduces the new generation of veterinary pathologists to a literary task that has grown much beyond what the original authors could handle. We take great satisfaction in this growth and in our colleagues' willingness to join us in the project. The contributors are identified with those chapters or parts of chapters for which they have been individually responsible, although this method does understate the contribution and the dedicated commitment of our coauthors to what has been very much a cooperative effort.

We have retained the original style and format. It was established as the medium for what were personal statements by the authors. We hope that we have been able to maintain some of that flavor. Some features of the style and the format have proven to be awkward in use by the busy working pathologist, and in recognition of this we have given attention to subdivisions in the text, to an expansion of tables of content, to details in indexation (including addition of a cumulative index, in Volume 3), and to an expanded selection of illustrations. The wish to preserve the original style has presented to us and to our contributing authors challenges on content and balance. We hope that these have reasonably been met. Inevitably, we have had to make choices in blending the contributions of our contributing authors into a whole. We have had to reduce excellent sections to keep these volumes within reasonable size, and we have expanded other sections for the sake of completeness. Inevitably too, some of our editorial judgments will be imperfect, and responsibility remains with us for deficiencies in the final compilation.

It is not possible adequately to acknowledge the many people who have contributed to this work; most of them will in this prefatory statement remain unnamed.

The contributors, all of whom volunteered effort without which this work could not have been completed, will find that the uses to which these volumes are put in the next few years will be a fuller tribute to their work than can be written here. The support of our many other colleagues in veterinary pathology is perhaps best indicated by their generosity in providing illustrative material. We have brought forward many of the plates or figures from the earlier editions and have added many new ones. Those brought forward or added are acknowledged in the legends, but many more excellent photographs were received than could be used. We are deeply grateful to those colleagues who offered them.

The institutions with which we are individually affiliated have made time and other resources available to us. The several chapters contain acknowledgments for assistance received, but we must here acknowledge Sandra Brown, Jean Middlemiss, and Edward W. Eaton of the University of Guelph for preparing most of the draft manuscript and many illustrations, Denise Heffernan, Lynette Magill, and Frank Oddi of the University of Melbourne for the preparation of final copy and illustrations, and Tammie Goates of that university for editing the bibliographies. We gratefully acknowledge a generous donation from Syntex Agribusiness toward the costs of preparation of the manuscript. We are grateful again to receive the courtesy and cooperation of Academic Press in this shared contribution to the study of animal disease.

Melbourne, Australia   K. V. F. Jubb
1984   Peter C. Kennedy
                                                                          Nigel Palmer

Preface to the Second Edition

The first edition of "Pathology of Domestic Animals" went to press, not without some sense of satisfaction, with a philosophic acceptance of the many imperfections and a tentative hope that any future edition would provide an opportunity to refine our knowledge, understanding, and technique of communication. Alas, imperfections remain, different ones perhaps, inevitable products of the interaction of limited time, limited intellect, and unlimited supplies of scientific data.

We are impressed by the masses of data that weekly flood our libraries and by the short half-life of much of it, by the exponential increase in knowledge and the splintering of disciplines that proliferate therefrom, and by the inability of many disciplines relevant to medical science to be completely self-sustaining. More and more it is evident that the theme of pathology provides the central and connecting link in medical education and practice and the basis on which a multidisciplined structure can be supported. This is a difficult role for pathologists, but one which they will fill, not by virtue of superior intellects and capacious memories, desirable though these attributes may be, but rather by the proper application of the logic of the scientific method.

Therefore, in preparing this second edition we have attempted to incorporate new knowledge on the specific diseases of animals and, more earnestly, to find a theme of organ susceptibility and responsiveness. We do not doubt the validity of the approach even if we are unable as yet to apply it feasibly to all organs and systems. The format of this edition remains the same as for the first and for the same reasons; the logistics of suitable alternatives are too formidable.

Once again we must express our gratitude to the many people who have contributed in some way to the preparation of this edition. Especially, we are indebted to Professor T. J. Hulland of the University of Guelph for revising the chapter on muscle, to Dr. Anne Jabara, University of Melbourne for the section on mammary tumours, and to Dr. N. C. Palmer and Dr. J. S. Wilkinson of the University of Melbourne for material assistance and many fruitful conversations. As always, a heavy burden falls on those who convert our notes to manuscript and arrange the bibliography, a task shared and cheerfully and devotedly performed by Mrs. Sylvia Lewis and Miss Frances Douglas. We hope that we have done justice to those who have contributed illustrative material: Dr. A. Seawright, University of Queensland; Dr. D. Kradel, Pennsylvania State University; Dr. E. Karbe, University of

Zurich; Dr. B. C. Easterday, National Animal Disease Laboratory at Ames; Mr. J. D. J. Harding, Central Veterinary Laboratory, Weybridge; Dr. J. Morgan, University of California; Miss Virginia Osborne, University of Sydney.

Melbourne, Australia   K. V. F. JUBB
October, 1969   PETER C. KENNEDY

Preface to the First Edition

The preface offers the opportunity to an author to present his excuses for having written the book and his justification of the content and mode of presentation. Our reasons for writing "Pathology of Domestic Animals" are as insubstantial but as compelling as those which committed Captain Ahab to the pursuit of Moby Dick, and we offer no excuses. Neither shall we attempt justification because a bad book cannot be justified and a good book is its own justification.

These volumes are based on our experience and on as much of the relevant literature of the world as we have been able to find and evaluate and we offer them to our colleagues and to all students of pathology in the hope that they will contribute to an understanding of animal disease. We anticipate some criticism in offering these as student texts but in doing so we indicate our confidence in teachers of pathology to guide students in the use of such volumes and in the ability of the student to profit from the exposure. Moreover, these volumes represent, it seems to us, a fair assessment of the needs of veterinary students in these times, since we realize as we should, that the knowledge of pathology possessed by most graduating students must serve them for the rest of their lives.

We should have preferred to write at greater length and in more detail of our chosen field, but practicality and economics have dictated that we can present here no more than a précis of the wealth of information that is the gift of our predecessors and contemporaries to the veterinary profession. In compensation, we have appended to each chapter an extensive but selected bibliography by the proper use of which the earnest seeker after further knowledge will be richly rewarded. Many valuable contributions from the old and foreign literature will not be listed in our bibliographies, perhaps because we have failed to appreciate their significance but largely because we have not obtained access to them.

We wish to emphasize to our younger colleagues that there exist vast gaps in our present knowledge, and we hope future work will do much to fill these gaps. For any errors of established fact that appear and for errors of interpretation of published information we tender, with our apologies, a request that they be drawn to our attention. We have not always attempted to distinguish between what we know and what we think we know, and in stating our position on many matters of controversy it is inevitable that we are sometimes in error; but we do prefer to state our positions while reserving our right to change our opinions when necessary.

The aim of the scientific method is to provide understanding, and the ultimate aim in all study of disease is to understand well enough to preserve the organism and prevent the disease. But disease and the temper of the community do not wait upon the languid spirit of most scientific enquiry; in the annals of veterinary science there are many endemic and epidemic diseases concerning which a broad search for understanding is necessarily postponed in the interest of quickly finding a way to avoid the disease or to face and exert some measure of control over it. Such hastily constructed controls are often satisactory but seldom enough, and usually they merely stem the tide while further enquiry can be made and understanding sought. It is from pathology, viewed broadly, that understanding comes and the need is great because there are old diseases still to be contended with, others now in existence but still to be recognized, and new ones to be anticipated. The pathologist is necessarily concerned with all matters pertaining to disease and we would enjoin him to remember this and meet his responsibilities in an age when urgency disturbs the spirit of the Groves of the Academy.

We have departed somewhat from tradition in the arrangement of these volumes. General pathology is well covered in many existing textbooks and we have not taken space for it, but have restricted our discussions to systemic or special pathology. Almost all we have to say on a particular subject or specific disease is said in one place under the organ system in which it appears most appropriate, although we have waived this general rule in an attempt to make the sections devoted to genitalia and special senses self-sufficient. A few diseases which resisted our systemic classification are relegated to an appendix in Volume 2. Detailed tables of contents are included for each volume to indicate the organization of the text and our classification of the diseases of the systems.

Guelph, Ontario   K. V. F. JUBB
January, 1963   PETER C. KENNEDY

CHAPTER 1

The Cardiovascular System

WAYNE F. ROBINSON
Murdoch University, Australia

M. GRANT MAXIE
Ontario Ministry of Agriculture and Food, Canada

THE HEART*

I. General Considerations

The detection of gross or microscopic abnormalities within the heart may suggest, but not require, the presence of clinical signs attributable to them. Major structural abnormalities such as endocarditis, endocardiosis, and ventricular fibrosis, which are usually associated with clinical signs, may be found incidentally postmortem. There may also be relatively minor but critically located lesions that result in a marked disturbance of function. This is exemplified by focal areas of myocardial degeneration or necrosis precipitating the development of dysrhythmias. Indeed, in some animals with marked clinical signs, there may be no observable structural change. Finally, there are lesions that do not impinge on the functional capacity of the heart; these are either part of a systemic disease, such as distemper, or indicate the presence of a primary disease in a remote organ, such as ulcerative endocarditis in renal failure. Under such circumstances, as well as identifying and classifying morphologic changes, it is desirable to assess the functional significance of the lesion.

The pathologic and functional changes observed are intimately related to the biology of the cell affected. For the heart, the biology, and the reaction to injury, may be

conveniently considered under the anatomic units—the myocardium, endocardium, and pericardium.

The myocardium has two major components: the conduction system, and the atrial and ventricular myocytes. The regularity of myocardial contraction is normally governed by the inherent automaticity of the specialized **conduction system.** This is composed of the sinoatrial node, internodal pathways, atrioventricular (AV) node, the AV trunk (bundle of His), the left and right crura (left and right bundle branches), and the cardiac conducting fibers (Purkinje fibers). These modified cardiac muscle cells have differing rates of diastolic depolarization, with the sinoatrial node the most frequent and dominant. The frequency of depolarization of the sinoatrial node is in turn modified by the autonomic nervous system. Atrial and ventricular myocytes do not normally exhibit the property of automaticity. When injured, however, the myocytes may repeatedly depolarize and may become dominant pacemakers. There are diseases that affect the conduction system, but in domestic animals dysrhythmias are usually the result of a more general involvement of the myocardium.

The atrial and ventricular **myocardium** is composed of cells specialized both to depolarize and to contract, and this is reflected in their structure. Perturbations of the electrical and mechanical properties of the myocardium may be readily induced by a variety of conditions. With some, the mechanism is well established; with others, the end result is observed but the mechanism is obscure. A decrement in these properties may or may not be accompa-

---

* This section was contributed by Wayne F. Robinson, Murdoch University, Australia.

1

nied by a morphologic change. In some cases, there may be insufficient time for the structural change to manifest, but with others, particularly those associated with electrolyte disturbances, the changes remain at a functional level. Given time, those that result in a morphologic manifestation include coronary arterial embolism or thrombosis, catecholamine excess, nutritional deficiencies (such as of vitamin E or selenium), and bacterial or viral infections. As with all tissues, there is a limited set of reactions of the myocardium to injury, but the pattern and distribution of any lesions may aid in arriving at a morphologic and etiologic diagnosis. The stage of irreversible damage to a myocyte, at least in ischemia, is determined by structural and functional changes in the mitochondria. It requires only 30 min of ischemia to produce this state, whether or not flow is restored.

Only in neonatal hearts do myocytes have the capacity to regenerate as those of skeletal muscle do. Once a particular myocyte or group of myocytes is lost, there is no replacement. There is progressive scavenging of the necrotic remnants, and replacement by fibrosis. The remaining myocytes have the capacity for compensatory hypertrophy, which is discussed in detail in a later section.

Myocardial injury may be functionally manifest as irregularities in the rate or rhythm of impulse formation and conduction (dysrhythmias) or as a depression in the force of myocardial contractility. Dysrhythmias are usually associated with acute, often focal myocardial injury. Contractility disturbances occur when insufficient numbers of ventricular myocytes remain for effective contraction or when there are adequate numbers of myocytes, but they contract ineffectually. Generalized ineffective myocardial contraction is most commonly seen as a feature of dilated cardiomyopathies.

Diseases affecting the **endocardium** may be congenital, infectious, or degenerative in origin, and the valvular endocardium is chiefly involved. **Valves** are relatively simple anatomic structures. Covered by endothelium on all faces, the interior of the valve is a thin layer of collagen and elastic fibers with a scattering of stellate cells that produce chondroitin sulfate and hyaluronic acid. The AV valves are attached to the papillary muscles by chordae tendineae, which are composed of dense collagen and covered by endothelium. It is the endothelium that is of significance in infectious states, and the form and function of the interior of the valve that are significant in congenital and degenerative disease.

The propensity of thrombi to form on the free margins of valves is well known, but the factors leading to it are not. It may be related to the continual movement and the resulting apposition of the surfaces of the free margins of the valves. There are also ill-defined factors, such as intercurrent disease or an increased work load, that promote the development of thrombi. The lack of an internal blood supply to the valve may also be a contributing factor and, in common with other vascular structures, injured endothelial cells release inflammatory mediators such as prostaglandins and other substances such as adenosine

diphosphate, which are potent stimulators of platelet aggregation. Once the endothelium is removed, the exposed collagen also stimulates the aggregation of platelets. The presence of the initial thrombus leads to further clotting. In contrast to myocytes, endothelial cells have a great capacity for regeneration and for covering any breach in their continuity. If the original insult is removed, which is not often, the valve heals by fibrosis, but a shrunken, distorted, insufficient, or stenotic valve may remain.

Little is known of the process leading to the abnormal development of valves in the embryo, or the progressive accumulation of glycosaminoglycans in the degenerative disease, endocardiosis. The various theories are discussed under the appropriate headings. However, both lead to deficiencies of valve function.

Valvular abnormality from whatever cause leads to disturbances of blood flow through the heart, either by alterations in the normal unidirectional pattern of flow or by alterations in the impedance of chamber inflow or outflow. These types of abnormalities are usually encompassed under the umbrella of hemodynamic disturbances. Alterations in hemodynamics reflect changes in (1) systolic work loads characterized by changed pressure loading during contraction (afterload), or (2) a changed volume loading during diastole (preload). Most disorders have only a single preload or afterload change imposed. This encompasses all of the valvular disorders, which are either insufficiencies (failure to close) or stenoses (narrowing, failure to open). Some of the congenital heart abnormalities, such as tetralogy of Fallot, have multiple preload and afterload effects. The general rules are (1) all valvular insufficiencies result in increased preload on the ventricles; (2) all semilunar valvular stenoses, outflow tract stenoses, and hypertension place an increased afterload on the ventricles; and (3) all AV valvular stenoses and pericardial disorders place a decreased preload on the ventricles.

Bibliography

Bonagura, J. D. Cardiovascular diseases. *In* "The Cat: Diseases and Clinical Management," R. G. Sherding (ed.), pp. 649–753. New York, Churchill Livingstone, 1989.

Braunwald, E. "Heart Disease. A Textbook of Cardiovascular Medicine." 3rd Ed. Philadelphia, Pennsylvania, W. B. Saunders, 1988.

Darke, P. Cardiac disease syndromes in dogs and cats. *In Pract* **2:** 5–12, 1980.

Else, R. W., and Holmes, J. R. Cardiac pathology in the horse. 1. Gross pathology. *Equine Vet J* **4:** 1–8, 1972.

Else, R. W., and Holmes, J. R. Cardiac pathology in the horse. 2. Microscopic pathology. *Equine Vet J* **4:** 57–62, 1972.

Ettinger, S. J., and Suter, P. F. "Canine Cardiology." Philadelphia, Pennsylvania, W. B. Saunders, 1970.

Fox, P. R. (ed.). "Canine and Feline Cardiology." New York, Churchill Livingstone, 1988.

Hudson, R. E. B. "Cardiovascular Pathology," Vols. 1 and 2. London, Edward Arnold, 1965.

Hudson, R. E. B. "Cardiovascular Pathology," Vol. 3. London, Edward Arnold, 1970.

Hurst, J. W. *et al.* "The Heart," 5th Ed. New York, McGraw-Hill, 1982.

Katz, A. M. Cardiomyopathy of overload: A major determinant of prognosis in congestive heart failure. *N Engl J Med* **322:** 100–110, 1990.

Kunze, R. S., and Wingfield, W. E. Acquired heart disease. *In* "Pathophysiology in Small Animal Surgery," M.J. Bojrab (ed.), pp. 178–195. Philadelphia, Pennsylvania, Lea & Febiger, 1981.

Maher, E. R., and Rush, J. E. Cardiovascular changes in the geriatric dog. *Compend Contin Educ Pract Vet* **12:** 921–923, 1990.

Van Vleet, J. F., and Ferrans, V. J. Myocardial disease of animals. *Am J Pathol* **124:** 98–178, 1986.

Whitney, J. C. Cardiovascular pathology. *J Small Anim Pract* **8:** 459–465, 1967.

II. Heart Failure

The term **heart failure** denotes a situation in which the heart is diseased, all compensatory mechanisms have been exhausted, and characteristic clinical and pathologic signs are present. In this pathophysiologic state, the heart is unable to meet the hemodynamic demands of the animal. However, heart failure is a general term. The addition of **congestive** to heart failure adds more specificity to the condition. **Acute congestive heart failure** refers to a rapid clinical onset of congestive heart failure. Congestive heart failure is characterized by vascular congestion and edema fluid within the interstitium and body cavities. The terms **left-** or **right-sided heart failure** refer to the failure of the left or right ventricular capacity to meet the body's needs and involve the pulmonary or systemic circulation, respectively. **Myocardial failure** refers particularly to the failure of the ventricular myocardium to contract effectively, but it is not a term that is in common use. When the myocardium fails, the more usual term used is congestive heart failure.

Not all cases of heart failure are of the congestive type. In congestive heart failure, the clinical manifestations are more or less constant, but there is a form of heart failure characterized by intermittent weakness and syncope. The cause is usually a substantial change in heart rate or rhythm resulting in a precipitous drop in cardiac output. This type is referred to as **acute heart failure.** Circulatory failure, or shock, is a term used to describe a state which may or may not be the result of heart failure. It is characterized by a drop in effective circulating blood volume. Common causes are acute internal or external hemorrhage, dehydration, or endotoxic shock.

Heart failure with its attendant grammatical modifiers is not an etiologic diagnosis but the end product of numerous causes. As such, it is a syndrome rather than a disease.

Diseases of the heart that lead to failure are those that (1) impose a sustained pressure overload on one or both ventricles, (2) impose a sustained volume overload on one or both ventricles, (3) depress or alter normal contractility of myocardial fibers or lead to loss or replacement of cardiac muscle, and (4) significantly alter the heart's nor-

mal rate and rhythm. As such, they are generally termed hemodynamic, contractility, or arrhythmic disturbances.

Although there are many causes and pathways that lead to an intermittent or permanent lowering of effective cardiac output, there is a limited set of responses to this by the animal. The major compensatory mechanisms include the intrinsic cardiac responses of dilation and hypertrophy, and the systemic response, which includes an increase in heart rate and peripheral resistance, a redistribution of blood flow, venular constriction, and an increase in blood volume. In each case, the compensatory responses are at least temporarily beneficial and are directed toward increasing cardiac output to meet the metabolic needs of the animal. The range within which the compensatory mechanisms result in an increase in cardiac output is wide. Indeed, the increase may be up to five times the basal rate. As cardiac output falls below the requirements of the animal, signs of congestive heart failure appear. These may be intermittent or prolonged, depending on the nature of the defect.

In addition to the intrinsic and systemic responses, there is the added complication of local vascular conditions leading to the accumulation of edema fluid. This is a consequence of an increase in capillary hydrostatic pressure and can involve the systemic or pulmonary veins. Peripheral dependent edema, ascites, hydrothorax, and hydropericardium are seen with right-sided lesions, such as right atrioventricular valvular insufficiency, pulmonic stenosis, or pulmonary hypertension. Pulmonary edema is the predominant finding with left-sided defects, such as left atrioventricular or aortic valvular insufficiency.

A. Intrinsic Cardiac Responses in Heart Failure

1. Cardiac Dilation

The contractile force of the heart can be modified by altering the end-diastolic volume, which within certain limits results in an increase in stroke volume. It is a response to an increased work load in both physiologic and pathologic states. Physical activity of the type observed in the athletic species, or for that matter in any animal, results in an increase in end-diastolic volume. The consequent increased stretching of the myofibers increases contractile force. This is known as the Frank–Starling phenomenon, or heterometric autoregulation. Continued stretch increases contractile force to a limit, after which increased stretch will result in a decrease in tension developed. The limit in most species appears to be a sarcomere length of 2.2–2.4 μm. The same phenomenon is seen in those disease states that produce an increase in diastolic work load on the heart, such as arteriovenous shunts, and in atrioventricular and semilunar valvular insufficiencies.

The physical consequences of dilation are derived from the principle that the pressure developed by a particular level of wall tension is inversely proportional to the radius of the chamber. As a chamber dilates, the expenditure of energy necessary to develop tension in the wall for the required development of intraventricular pressure is increased. Since ~90% of the metabolic energy of contraction is expended in isometric contraction, that is, in raising intraventricular pressure to the level at which the valves will open, the disadvantage of dilation is evident.

2. Cardiac Hypertrophy

Whereas acute volume overload of a chamber is expected to lead to dilation, chronic volume overload is one stimulus to the development of cardiac hypertrophy.

Cardiac hypertrophy is a reversible increase in the mass, but not the number, of myocardial cells. The capacity of the myocyte to undergo division decreases rapidly prior to birth, and little mitotic activity is observed after the first few weeks of life. Cardiac hypertrophy is a feature of any state, including disease states, that gives rise to an increase in either systolic or diastolic work load. In that context, it is considered to be compensatory and is a response to a known, or defined, change in work load. The following discussion will be concerned only with the process of hypertrophy following defined change. The presence of hypertrophy in the absence of an observable increase in work load will be considered to be primary and is discussed under The Cardiomyopathies (Section VII,C of The Heart).

For hypertrophy to occur, there are requirements of time, a healthy myocardium, and an adequacy of nutrition of the myocardium. The search for the biochemical pathway that translates the mechanical stimulus into an increased myocardial mass remains unresolved. The final common pathway to the induction of hypertrophy may be mediated through the cardiac sympathetic nervous system. Prolonged adrenergic stimulation, acting via the myocardial membrane, may lead to increased protein synthesis. In this context, there is increasing interest in the relationship between proto-oncogenes and cardiac hypertrophy. Proto-oncogenes (or cellular oncogenes, c-oncs) are normal cellular genes encoding critical regulatory proteins such as growth factors and growth factor receptors. The α_1-adrenergic receptor is the first, and so far the only, well-documented growth factor receptor for cardiac myocytes. The α_1-receptor in myocytes is coupled to the cytoplasmic transducing proteins phosphoinositide phospholipase C and protein kinase C, for which proto-oncogenes have been suggested. The implication is that cardiac myocyte growth is regulated, like that of other cells, by proto-oncogene products. The distinctions between physiologic and pathologic hypertrophy appear to be quantitative rather than qualitative. In hypertrophy, there is depression of contractility, decreased myosin adenosine triphosphatase (ATPase) activity, and decreased cyclic adenosine monophosphate content.

Hypertrophy of myocardium is associated with reduced contractility of myocardial fibers, irrespective of the circumstances that provoked the hypertrophy. Thus the view of hypertrophy as a compensatory process imposes an apparent paradox, which is resolved only if it is understood that hypertrophy involves an increase in total myocardial mass; that the chamber diameter may not be increased but indeed may be reduced; that the radius of contraction may not be increased; and that hypertrophy of the trabeculae carneae assists ventricular emptying.

In hypertrophy, there is an increase in the size of myocardial fibers. The increase is in length and diameter and is accompanied by an increase in the number of sarcomeres. It appears, however, that although there may be some enlargement of mitochondria, the number of mitochondria and the area of mitochondrial membrane are reduced relative to the volume of myofibrils. These observations suggest that hypertrophic myocardium is at a disadvantage in terms of intracellular energetics—a disadvantage compounded by the ratio of cell volume to unaltered capillary bed.

There are distinctive anatomic patterns of hypertrophy that accompany the increase in work load. **Concentric cardiac hypertrophy,** that is, an increase in mass of the ventricle without accompanying increase in end-diastolic volume, characterizes increased systolic loads such as aortic stenosis, pulmonic stenosis, and pulmonary hypertension in patent ductus arteriosus. There is often a decrease in the volume of the ventricular lumen. An increase in diastolic load, typically produced by atrioventricular or semilunar valvular insufficiencies or by arteriovenous shunts, results in **eccentric hypertrophy,** which is defined as an increase in myocardial mass accompanied by an increased end-diastolic volume. The thickness of the involved ventricular wall is usually no more than normal and may be less.

The gross appearance of hypertrophy depends on the chamber affected and the nature of the insult. In general, hypertrophy of the right side of the heart makes the heart broader at its base; hypertrophy of the left side increases the organ length; bilateral hypertrophy produces a more rounded shape than normal.

In concentric hypertrophy, there is increased thickness of the wall of the affected chamber and a remarkable increase in the size of the papillary muscles and the trabeculae carneae. Although the hypertrophy may emphasize one chamber or another, the whole heart is involved. When the right ventricle is involved, the trabecula septomarginalis (moderator band) may be much thickened (Fig. 1.1A). Extreme hypertrophy of one chamber may encroach on the diastolic capacity of its opposite number (Fig. 1.1B). Microscopically, the myocytes are enlarged, but the increase in the size of fibers is not uniform in the heart.

In eccentric hypertrophy and dilation, the heart tends to be globose in shape and the wall, usually thin, especially with respect to the overall chamber dimensions. The papillary muscles may also be attenuated.

In both types of hypertrophy and dilation, the endocardium may be diffusely opaque as a result of fibrosis, and this endocardial alteration may be the best indication of dilation in the atria, in which dilation and hypertrophy can be difficult to assess.

Cardiac hypertrophy, tachycardia, and dysrhythmias

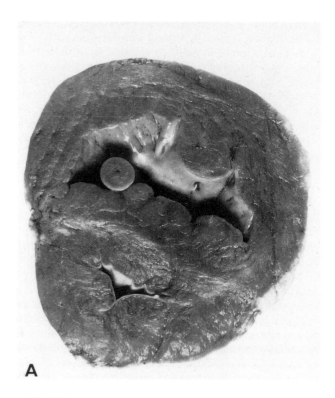

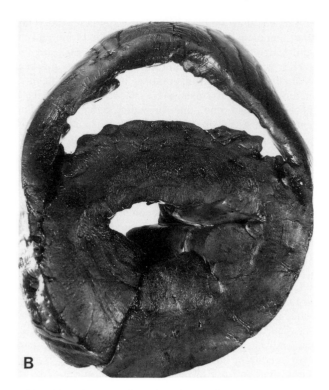

Fig. 1.1A Cross section of cow heart with chronic interstitial pneumonia. The right ventricle is greatly hypertrophied. Note increase in size of trabecula septomarginalis.

Fig. 1.1B Cross section of dog heart with chronic glomerulonephritis. The left ventricle is hypertrophic and encroaching on the diastolic capacity of the right ventricle.

occur in cats with hyperthyroidism (thyrotoxicosis), a condition usually due to thyroid hyperplasia or adenoma. The pathogenesis of ventricular hypertrophy in this disease is not clear but may involve the direct action of thyroid hormones on myocardium, enhanced myocardial adrenergic receptor number or affinity, peripheral vasodilation, and work hypertrophy in response to increased peripheral tissue demands for oxygen and dissipation of heat. The hypertrophy is reversible on return to euthyroidism.

B. Systemic Responses in Heart Failure

The extracardiac clinical signs and gross and microscopic features of heart failure stem from two basic pathophysiologic changes, the accumulation of fluid, and tissue or organ ischemia. Depending on the cause of the heart failure, both effects may be present or, as is more usual, one may predominate.

There is no question that fluid accumulation results from the retention of sodium and water. Additionally, the sodium and water accumulation primarily involves the kidneys. It also involves a hormone, atrial natriuretic factor, released from the heart. The influence of the failing heart on the kidneys stems from its inability to supply them with an adequate flow of blood. Blood flow through different parts of the kidneys depends on the vasomotor tone of blood vessels within the parenchyma. It is consid-

ered that many, if not all, of the intrarenal blood-flow changes in heart failure follow increased activity of the sympathetic nervous system.

The kidneys receive approximately 20% of the output of the left ventricle, almost all of which flows through the renal cortices. One of the earliest changes following a drop in cardiac output is a redistribution of blood flow within the kidney. There is a reduced flow through the outer renal cortex and an increased flow within the outer renal medulla. This results in a readjustment of the filtration fraction, which is the ratio of glomerular filtration rate (GFR) to renal blood flow. Contrary to expectations, there is a *less* than proportionate drop in GFR compared with renal blood flow resulting in an *increased* filtration fraction. As a consequence, proportionally more sodium moves through the glomerular filter, leading to proportionally more sodium being delivered into the proximal convoluted tubule. Because the rate of sodium resorption remains constant, a greater number of sodium ions are resorbed. Also, because of the increased filtration fraction, local plasma osmotic pressure in the efferent arteriole increases, causing greater resorption of sodium and water.

The alterations in renal blood flow in heart failure also increase the activity of the **renin–angiotensin–aldosterone** system, producing more sodium resorption from the distal convoluted tubule. There is also evidence for increased activity for the water-retaining activity of antidiuretic hormone.

A mechanism within the heart also regulates blood volume, complementing the activity of aldosterone and the renin–angiotensin system. Atrial natriuretic peptide (ANP), with natriuretic and diuretic properties, has been isolated and purified from granules in the cells of the atria. In terms of homeostasis, ANP has effects opposite to those of aldosterone, thus providing a balance to fluid regulation. Although plasma ANP is significantly increased in dogs with chronic left AV valvular insufficiency, it is not clear whether the metabolic effects of aldosterone or ANP predominate, but it would appear that the effects of aldosterone override those of ANP.

It should be noted that none of the hormones previously mentioned can produce the edema of congestive heart failure if administered alone. In addition, once a new steady state has been reached, the hormonal state returns to relatively normal limits. Lastly, the mechanisms that are brought into play are not exclusive to the syndrome of heart failure. Any situation that leads to a drop in effective circulating blood volume will activate the sodium- and water-retaining mechanism. The fundamental difference between these states and congestive heart failure is that the total blood volume in heart failure is already more than adequate, but the effective blood volume is much diminished because of the poor cardiac output. The volume changes in heart failure should be viewed as an integrated response by the body to compensate for the inability of the heart to respond to the normal hemodynamic needs of the body.

The expansion of blood volume has both a beneficial and a detrimental effect. By increasing blood volume, venous return is enhanced and, in turn, cardiac output is improved. However, this is to the detriment of the balance between capillary hydrostatic pressure and plasma osmotic pressure. This leads to an increase in the amount of fluid in the interstitial spaces and body cavities.

C. Syndromes of Circulatory Failure

Circulatory failure, the term implying severe systemic consequences, falls into three general types: cardiac syncope, peripheral circulatory failure, and congestive heart failure.

1. Cardiac Syncope

Cardiac syncope is characterized clinically by profound changes in blood pressure and heart rate with bradycardia or tachycardia, both resulting in inadequate output of blood. Both may occur in the presence or absence of organic heart disease. Hypersensitive or hyperactive reflexes, for which the vagus nerve is the efferent limb, may result in reflex inhibition of the heart, manifest as asystole or extreme bradycardia. The sudden deaths that result from acute pleural irritation or the tracheal irritation of aspirated vomitus fall into this group, and obviously there may be no organic heart lesion. In a second form of cardiac syncope, the heart rate is extremely rapid, and the cardiac output, severely reduced or nil. Such may occur in parox-

ysmal tachycardia, atrial flutter or fibrillation, and ventricular fibrillation. Third, in organic heart disease with complete obstruction of impulse conduction from the atrium to the ventricle (complete heart block), syncope may occur if there is sufficient delay before the ventricle assumes an independent rhythm. Finally, cardiac syncope may terminate a syndrome of congestive cardiac failure when the cardiac reserve is depleted and the heart cannot increase its output sufficiently to meet sudden increases in peripheral needs.

2. Peripheral Circulatory Failure

Peripheral circulatory failure is characterized by a reduction in the effective circulating blood volume with insufficient venous return and reduced cardiac output. Acute hemorrhage and shock are examples of this form of circulatory failure.

The combination of compensatory mechanisms, brought into play to maintain cardiac output, are in general successful; however, in a sense, the seeds of destruction are planted. Both the local increase in hydrostatic pressure and the increased sodium and water retention by the kidneys tend to promote development of interstitial edema. Depending on the inciting abnormality, it is usual for one side of the heart to fail before the other, but it must be remembered that the cardiovascular system is a closed circuit and that failure of one side will eventually embarrass the other.

Left-sided heart failure is ushered in by progressive dilation of the left ventricle and atrium, although this progression may be marked by exacerbations and remissions if hypertrophy is given time to develop. The major extracardiac manifestations of left-sided failure arise from the damming back of blood in the lungs and the diminution in the cardiac output. The pulmonary venous congestion is transmitted back to the capillaries of the alveolar wall, and edema fluid accumulates in the interstitial tissue and the alveolar spaces. The consequent reduction in pulmonary vital capacity and impaired gaseous exchange of cardiogenic pulmonary edema result in hypoxic stimulation of the carotid sinus and medullary respiratory centers so that reflex dyspnea occurs. A wheezing bronchial cough is common and is presumed to be due to irritation of the respiratory mucosae by the edema fluid. Cyanosis may be present but is more often the rule in right ventricular failure.

At necropsy the lungs are usually of normal color or slightly cyanotic, but may be light brown in the chronic disease, and are heavy and wet. Stable, white froth is present in the airways, and fluid exudes from the cut surface. There is little evidence of the abundant fluid on microscopic examination, because of its low protein content. The alveoli contain erythrocytes and a scattering of macrophages, some of which contain hemosiderin. It may be necessary in early cases to use a differential stain for iron to confirm the presence of these so-called heart-failure cells. They are more numerous in the chronic disease, and hemosiderin within their cytoplasm may be sufficient to produce a tawny discoloration of the organ.

The major extracardiac manifestations of **right-sided heart failure** depend on increased hydrostatic pressure in the systemic and portal venous systems, and the reduction of flow from the lungs to the left ventricle. Renal complications leading to increased blood volume occur more frequently in right-sided than in left-sided heart failure.

3. Congestive Heart Failure

There is some species difference in the distribution of edema in **congestive heart failure.** In ruminants and the horse, dependent subcutaneous edema is expected; in other species, excess subcutaneous fluid is scant or absent. In the dog, the predominant accumulation of fluid is in the peritoneal cavity; in the cat, in the thorax.

Grossly, the liver is enlarged and congested, and microscopically the sinusoids are dilated with atrophy of the parenchyma about the central veins. In more severe or acute cases, the parenchyma in this location may undergo degeneration or necrosis. It is exceptional for an animal with congestive failure to live long enough for fibrosis and nodularity to occur. Impaired hepatic function is not a significant part of the clinical course, although jaundice may be observed.

Congestion of the stomach and intestine is evident, and this may impair their function, which is manifest usually as diarrhea. The systemic and portal veins are distended, and the spleen may be enlarged and congested in acute failure. However, this latter finding is masked if the animal has been euthanized using barbiturates.

III. Examination of the Heart

The aim of a gross postmortem examination of the heart is to examine four major areas: the pericardium, the myocardium, the mural and valvular endocardium, and the great vessels.

It is preferable to commence by examining the heart and blood vessels *in situ* for abnormalities of size and position. The pericardial sac should also be incised and its contents examined before the thoracic contents are removed. Once removed, the external surface of the heart should be examined, including the pericardium, epicardium, and major vessels.

The accurate assessment of changes in ventricular size and weight is difficult, especially in cases of dilation and eccentric hypertrophy. In such cases, the heart and ventricles should be weighed and the weights compared to body weight. Normal heart weight is between 0.5 and 1.0% of body weight, depending on the species. Tables 1.1 and 1.2 contain reference values for the domestic species.

The right ventricle bears responsibility for systemic circulation in the fetus, and in neonatal hearts, the wall thickness of left and right chambers is about equal; it is not until several months after birth that the mature proportions are attained.

Rigor mortis begins rather earlier in myocardial than in

TABLE 1.1

Reference Values for Heart Weights and Ventricular Ratios in Normal Mature Animals at Necropsy

Species	N	Heart weight/body weight ratio (%)		Left ventricle (free wall and septum)/ right ventricle	
		$\bar{x}$	$\bar{x} \pm 2$ SD	$\bar{x}$	$\bar{x} \pm 2$ SD
Dog	21	0.71	0.43–0.99	3.26	2.39–5.12
Horse	12	0.69	0.41–0.97	3.12	2.43–4.34
Cat	9	0.58	0.28–0.88	3.45	2.94–4.17
Cow	15	0.48	0.30–0.66	3.01	2.43–4.00
Goat	11	0.46	0.26–0.66	3.12	2.50–4.17
Pig	8	0.40	0.32–0.48	2.94	2.38–3.84
Sheep	60	0.26		3.37	2.85–4.13
	8	0.41	0.17–0.65	3.33	2.63–4.54

skeletal musculature and reaches its greater development in the more powerful left ventricle. Rigor should completely express the blood from the left ventricle; rigor of the right ventricle is less efficient, and emptying is incomplete. The presence of some clotted blood in the right ventricle is normal, whereas if present in the left ventricle after a reasonable postmortem interval, it is indicative of incomplete rigor and therefore of severe myocardial degeneration. The presence of unclotted blood in the left ventricle some hours after death is more difficult to interpret. Unclotted blood, usually an indication of death in hypoxia with cascaded fibrinolysis, may flow back into the ventricle when rigor passes.

Blood usually clots slowly after death and permits erythrocytes to sediment. Where blood is present in volume, as in the heart and arterial trunks, this process of sedimentation and subsequent clotting leads to the formation of "currant-jelly" and "chicken-fat" clots, the former containing erythrocytes, and the latter largely devoid of them. Chicken-fat clots are to be expected in horses, which normally have a rapid erythrocyte sedimentation rate, and are in relative excess in anemia. These postmor-

TABLE 1.2

Criteria for Cardiac Hypertrophy in Dogs[a]

LV hypertrophy: $\dfrac{LV+S}{BW} \geq 0.57\%$

RV hypertrophy: $\dfrac{RV}{BW} \geq 0.18\%$

Biventricular hypertrophy: $\dfrac{HW}{BW} \geq 0.94\%$

Ventricular ratios: $\dfrac{LV+S}{RV} = 3.32 \ (2.76–3.88)$

$\qquad$ or $\dfrac{RV}{LV+S} = 0.31 \ (0.26–0.36)$

[a] LV, left ventricle; LV + S, left ventricle plus septum; BW, body weight; RV, right ventricular free wall; HW, heart weight.

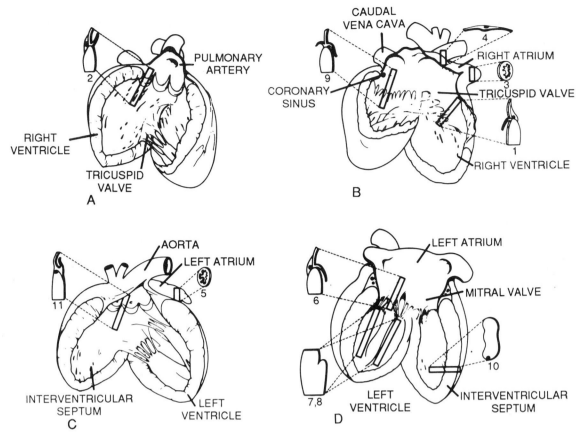

Fig. 1.2 Gross and microscopic examination of the heart. Diagrams A–D illustrate the heart opened as described in the text. The numbers indicate the area and the shape of the tissues removed for histopathology. A. Right ventricular cavity and pulmonary outflow tract. B. Right ventricle and right atrium. C. Left ventricle and aortic outflow tract. D. Left ventricle and left atrium. 1. Right ventricular free wall, atrioventricular valve and atrium. 2. Pulmonic valve, right ventricular outflow tract, and pulmonary artery. 3. Right auricular appendage. 4. Sinoatrial node. 5. Left auricular appendage. 6. Left atrioventricular valve, ventricle and atrium. 7,8. Left ventricular free wall and papillary muscles. 9. Atrioventricular node, right atrioventricular valve and atrium. 10. Interventricular septum. 11. Aortic valve, left aortic outflow tract and aorta. (From "Canine and Feline Cardiology," Philip R. Fox (ed.) and Sanford P. Bishop, Churchill Livingstone, 1988, with permission.)

tem clots are to be distinguished from thrombi; clots are not attached to the endocardium.

Just as a standard technique is used for gross examination, a particular routine should be followed when collecting tissues for microscopy (Fig. 1.2). In addition to the selection of macroscopic lesions, a detailed examination includes samples of the conduction system, each valvular region, each chamber (including endocardial and epicardial surfaces), and great vessels. Areas routinely collected may not be processed, but they are available if required. In routine cases, there are no special requirements for fixation. Stains of particular use are hematoxylin and eosin, Masson's trichrome, phosphotungstic acid hematoxylin, luxol fast blue, and Gomori's aldehyde fuchsin.

The selection of sections to study the specialized conduction system should include the sinus node, the atrioventricular node, the common bundle, left and right crura, and conducting fibers. The sinus node lies subepicardially

at the junction of the cranial vena cava and the right atrium. Sections should include either side of that site, to incorporate tissue in the sulcus terminalis region. The atrioventricular node is obtained by removing a block of tissue from the coronary sinus to the cranial edge of the septal leaflet of the right AV valve. The block should include interatrial septum and dorsal ventricular septum, and will then contain the atrioventricular node, the common bundle, and the left and right crura. The specimen should be serially sectioned into samples 3 mm thick, and all samples should be processed.

Bibliography

Bond, B. R. *et al.* Echocardiographic findings in 103 cats with hyperthyroidism. *J Am Vet Med Assoc* **192:** 1546–1549, 1988.
Isner, J. M. The cardiologist and the pathologist: The interactions of both and the limitations of each. *Hum Pathol* **18:** 441–450, 1987.

Katz, A. M. Congestive heart failure. Role of altered myocardial cellular control. *N Engl J Med* **293**: 1184–1191, 1975.

Liu, S. U. Postmortem examination of the heart. *Vet Clin North Am: Small Anim Pract* **13**: 379–394, 1983.

Maeban, E., and Noretsane, Y. Cell calcium, oncogenes, and hypertrophy. *Hypertension* **15**: 652–658, 1990.

Miller, P. J., and Holmes, J. R. Observations on structure and functions of the equine mitral valve. *Equine Vet J* **16**: 457–460, 1984.

Morgan, H. E., and Baker, K. M. Cardiac hypertrophy. Mechanical, neural, and endocrine dependence. *Circulation* **83**: 13–25, 1991.

Morkin, E., and La Raia, P. J. Biochemical studies on the regulation of myocardial contractility. *N Engl J Med* **290**: 445–451, 1974.

Ostman-Smith, O. Cardiac sympathetic nerves as the final common pathway in the induction of adaptive cardiac hypertrophy. *Clin Sci* **61**: 265–272, 1981.

Peterson, M. E. *et al.* Feline hyperthyroidism: Pretreatment clinical and laboratory evaluation of 131 cases. *J Am Vet Med Assoc* **183**: 103–110, 1983.

Rabinowitz, M., and Zak, R. Mitochondria and cardiac hypertrophy. *Circ Res* **36**: 367–376, 1975.

Robinson, T. F., Factor, S. M., and Sonnenblick, E. H. The heart as a suction pump. *Sci Am* **254 (6)**: 84–91, 1986.

Schwartz, A. *et al.* Abnormal biochemistry in myocardial failure. *Am J Cardiol* **32**: 407–422, 1973.

Simpson, P. C. Protooncogenes and cardiac hypertrophy. *Annu Rev Physiol* **51**: 189–202, 1988.

Steward, A., Allot, P. R., and Mapleson, W. W. Organ weights in the dog. *Res Vet Sci* **19**: 341–342, 1975.

Takemura, N. *et al.* Atrial natriuretic peptide in the dog with mitral regurgitation. *Res Vet Sci* **50**: 86–88, 1991.

Tezuka, F. *et al.* Muscle fiber orientation in the development and regression of right ventricular hypertrophy in pigs. *Jpn J Pathol* **40**: 402–407, 1990.

Wilkmann-Coffelt, J., Parmley, W. W., and Mason, D. T. The cardiac hypertrophy process. Analysis of factors determining pathological vs physiological development. *Circ Res* **45**: 697–707, 1979.

IV. Congenital Abnormalities of the Heart and Large Vessels

In the transition from fetal to neonatal life, substantial adjustments occur within the cardiovascular system: there are alterations in the pressures in cardiac chambers and great vessels, the pattern of blood flow, and the volume of blood flow. Because of these changes, the retention postnatally of fetal vascular shunts, such as the ductus arteriosus, may place an excessive load on the heart in the postnatal period and beyond. There are also congenital heart defects, such as pulmonic stenosis, that compromise the fetus, the newborn, and the adult. It may also be worth stating that it is only those defects that allow adequate *in utero* development and a reasonably successful perinatal life that are recognized, anomalies sufficiently severe to cause death *in utero,* or in the neonatal period, often are not.

As with most diseases, there is a spectrum of change. The variation in severity of a particular lesion may be wide and will necessarily influence whether clinical signs are observed. As such, there is a higher incidence of congenital heart disease than recognized clinically. There is also a group of congenital heart diseases that do not produce clinical signs of heart failure, but which are manifested by upper alimentary dysfunction. Although little is known of the causes of many cardiac malformations in domestic mammals, there is no doubt, especially in dogs, that some are genetically determined. This is demonstrated by the fact that the incidence of defects is higher in purebred then in crossbred populations. However, in some, the mode of inheritance does not conform to simple Mendelian genetics. Both patent ductus arteriosus and tetralogy of Fallot appear to have polygenic inheritance patterns. The full expression of these diseases depends on the inheritance of a number of genes from a number of different loci. Congenital subaortic stenosis in the Newfoundland dog is also genetically determined. It may be either polygenic or a single dominant gene that is variably expressed. Table 1.3 outlines the breed specific predispositions to congenital heart disease in the dog.

There may be other as yet undefined factors that contribute to the development of congenital heart disease in domestic animals. In humans, cardiac and vascular anomalies occur as common features of several syndromes produced by chromosomal abnormalities and virus infections,

TABLE 1.3

Breed Specific Predispositions to Congenital Heart Disease in Dogs

Defect	Breed
Patent ductus arteriosus	Poodle
	Collie
	Pomeranian
	Shetland sheepdog
Pulmonic stenosis	Bulldog
	Fox terrier
	Chihuahua
	Beagle
	Samoyed
	Miniature schnauzer
Subaortic stenosis	German shepherd
	Boxer
	Newfoundland
	German shorthaired pointer
	Golden retriever
Persistent right aortic arch	German shepherd
	Irish setter
Tetralogy of Fallot	Keeshond
	Bulldog
Atrial septal defect	Samoyed
Ventricular septal defect	Bulldog
Tricuspid insufficiency	Great Dane
	Weimaraner
Mitral insufficiency	Bulldog
	Chihuahua
	Great Dane
	Bull terrier

such as rubella. There is little evidence to suggest the presence of similar abnormalities associated with congenital heart disease in domestic animals.

The pattern and incidence of congenital cardiac disease vary with the species examined. In dogs, it appears that patent ductus arteriosus, pulmonic stenosis, subaortic stenosis, and persistent right aortic arch are common. In cattle, atrial and ventricular septal defects and transpositions of the main vessels are most often diagnosed. Subaortic stenosis and endocardial cushion defects are the most common anomalies in pigs. In cats, endocardial cushion defects and congenital mitral insufficiency appear to be common. Congenital cardiovascular disease in horses is rare.

In cases of suspected cardiac abnormality, it is essential to examine the heart and large vessels *in situ* because relations are difficult to trace once the organ is removed. Some animals are born with hearts that, although of normal arrangement, are very small. In the majority of cases of cardiac abnormality, the anomaly is reflected in a gross enlargement of the organ and in an alteration of the size or disposition of the large vessels. Cardiac malformations are extremely variable, and their analysis can be perplexing if it is not remembered that they do follow fairly simple basic patterns that allow, as a first step, the recognition of the primary abnormality, and then, the recognition of the secondary abnormalities which develop as adjustments to allow blood to circulate through the heart. An understanding of the mechanics of abnormal development is necessarily based on an understanding of the normal development of the organ, for which reference should be made to a standard text of embryology.

Variations in the position of the heart are not cardiac malformations but, instead, malformations of adjacent structures. In **ectopia cordis,** the heart may lie in extrathoracic, presternal, or intra-abdominal positions (Fig. 1.3). Dislocations within the thorax are the result of asymmetric

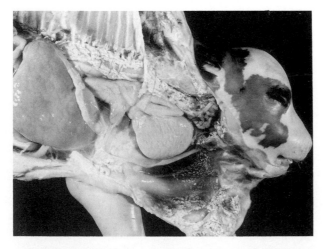

Fig. 1.3 Ectopia cordis. A 7-month bovine fetus with pulmonary hypoplasia, and multiple cardiac and other anomalies. Liver chronically congested.

pressure, as for example in congenital diaphragmatic hernia or pleural effusion. Congenital absence of the pericardium occurs in dogs.

The commonly encountered congenital heart defects may be broadly classified into six categories: (1) failure of closure of fetal or neonatal arteriovenous communication; (2) conotruncal abnormalities; (3) failure of adequate development of the semilunar or atrioventricular valves; (4) endocardial cushion defects; (5) incomplete separation of, or abnormally positioned, vessels; and (6) miscellaneous cardiac anomalies.

A. Failure of Closure of Fetal or Neonatal Arteriovenous Communication

In the development of the heart there are three major arteriovenous communications: between the atria, the ventricles, and the great vessels. Closure of the atrial and ventricular septa occur *in utero,* and the foramen ovale and the ductus arteriosus, in the neonatal period. Failure of closure results in either an atrial septal defect, a ventricular septal defect, or a patent ductus arteriosus.

1. Patent Ductus Arteriosus

Patent ductus arteriosus is one of the more common defects, and it is recorded in all species. In the dog, it has a polygenic inheritance pattern. The ductus develops from the sixth left branchial arch, and functions in the fetus to divert a major portion of blood from the pulmonary artery to the aorta. The flow from venous to arterial side is a consequence of the presence of the high vascular resistance of the fetal pulmonary bed. The structure of the normal ductus differs from that of the adjacent pulmonary artery or aorta. In contrast to the large elastic arteries, the media of the ductus has a dense layer of smooth muscle, which is responsive to a number of compounds. Among the most powerful are epinephrine, norepinephrine, and angiotensin; less effective ones include oxygen, acetylcholine, bradykinin, and indomethacin, a prostaglandin inhibitor. Prostaglandin E_2 keeps the ductus in a dilated state. Of those compounds mentioned, *in vivo* oxygen and acetylcholine are probably the most important in causing constriction.

With the contraction of the medial smooth muscle, the ductus becomes functionally closed in the first few hours after birth. There are exceptions. The ductus in some foals may remain patent for up to 5 days, and a continuous murmur is detectable up to that time. A ductus arteriosus that remains patent beyond this time is considered to be abnormal. Such a ductus is only minimally responsive, or nonresponsive, to oxygen and other compounds that constrict the normal ductus.

The continued patency results in a number of sequelae that are dependent on the size of the ductus and the relationship between the pulmonary and systemic vascular resistance. In uncomplicated cases, the blood flow is from the aorta to the pulmonary artery (left-to-right shunt) during both systole and diastole. Dependent on the ensuing rise of pulmonary arterial pressure, there may be left-to-right flow only during systole; in severe cases the shunt is reversed, and flow from the pulmonary artery to the aorta (right-to-left) occurs. If the ductus is of sufficient size, there is a volume overload on the left ventricle and a pressure overload on the right ventricle, resulting in compensatory hypertrophy of both. There is also left atrial dilation resulting from increased pulmonary blood flow (Fig. 1.4). The ascending aorta and pulmonary artery are dilated, resulting probably from a combination of turbulent flow and altered pressure relationships. The development of congestive heart failure with a large ductus is related not only to the pulmonary vascular resistance, but also to the ability of the left ventricle to handle the volume overload. Experiments in dogs involving aorticopulmonary shunts indicate that a communication of 3 mm or less in diameter may lead to the slow development of left ventricular hypertrophy, but it is otherwise well tolerated. A shunt of 5-mm diameter may lead to pulmonary hyper-

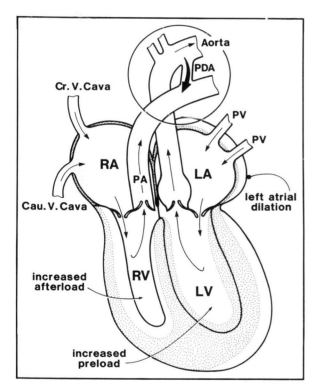

Fig. 1.4 Patent ductus arteriosus (PDA). In most cases blood flow through the PDA occurs during both systole and diastole from the aorta to the pulmonary artery. The right ventricle hypertrophies concentrically because of the increased afterload (increased pulmonary hypertension). The left atrium and ventricle hypertrophy eccentrically because of the increased preload (increased volume of blood returning from the pulmonary circuit). LV, left ventricle; RV, right ventricle; LA, left atrium; RA, right atrium; PA, pulmonary artery; PV, pulmonary veins; Cr.V.Cava, cranial vena cava; Cau.V.Cava, caudal vena cava. (From "Clinicopathologic Principles for Veterinary Medicine." W. F. Robinson and C. R. R. Huxtable (eds.), Cambridge University Press, 1988 with permission).

tension, with degenerative changes in pulmonary vessels and congestive heart failure.

A persistent ductus is, in consequence of turbulent flow, predisposed to thrombosis, referred to as **endocarditis of the ductus.** The anatomic appearance of the patent ductus is varied. In dogs, the severity of the defect increases as the number of genes that determine its presence increases. There is a progression from a ductus diverticulum, which is a blind, funnel-shaped outpouching of the aorta at the site of the ductus, to a ductus that approaches the size of the aorta. The length and the diameter of the patent ductus also vary. In some cases, there is virtually no ductus, but instead, an opening between the aorta and pulmonary artery, which are closely approximated. Also, the histologic appearance of the patent ductus more closely resembles that of the aorta as the gene frequency increases.

2. Atrial Septal Defect

Strictly speaking, the presence of a probe-patent foramen ovale is not an atrial septal defect; it is part of the normal development of a communication between the left and right atria necessary for fetal life. In neonates, and indeed older animals, the foramen ovale may not be anatomically closed, but it is functionally closed (valvular competent) because left atrial pressure exceeds right atrial pressure. In the majority of animals, however, anatomic closure follows functional closure.

True atrial septal defects may result from two phenomena: failure of growth of the septum secundum, and defects of the septum between the right upper pulmonary veins and the cranial vena cava. The latter are termed sinus venosus defects. Those defects associated with the septum primum are discussed under Endocardial Cushion Defects (Section IV,D of The Heart).

In the fetus, separation of the two atria commences with the downgrowth of the septum primum, which moves toward the atrioventricular junction, where the developing endocardial cushions begin to form the atrioventricular valves and separate the ventricles. The septum fuses with the endocardial cushions and begins to fenestrate in the middle. The fenestration is destined to become the ostium secundum. A second septum (septum secundum) develops downward and to the right of the septum primum. With its semilunar edge, the septum secundum and the remains of the septum primum form the foramen ovale. It is the septum primum that moves during fetal life in response to pressure differences between the atria and forms the flap over the foramen ovale.

An atrial septal defect is the result of either failure of fusion of septum primum with the endocardial cushions (ostium primum), an excessively large ostium secundum, or an inadequate development of the septum secundum. The position of ostium primum is low in the interatrial septum, whereas ostium secundum is high in the interatrial septum. The consequence of an atrial septal defect in the neonate is excessive flow from the left to right atrium, with resultant volume overload on the right ventricle and

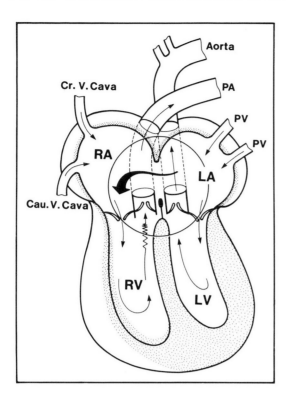

Fig. 1.5 Atrial septal defect. Blood flows through the defect from the left atrium to the right atrium. The right ventricle dilates and hypertrophies under an increased preload. A relative pulmonic stenosis is induced because of the increased volume in the right ventricle. Both atria also dilate following increased volume loads. Reverse shunts may occur with blood flowing from the right to the left atrium. (From "Clinicopathologic Principles for Veterinary Medicine." W. F. Robinson and C. R. R. Huxtable, eds. Cambridge University Press, 1988, with permission.

Fig. 1.6 Ventricular septal defect (VSD). Goat. The defect is just below the aortic valve, and the left ventricle is dilated. (5 cm between arrows.)

elevated central venous pressure (Fig. 1.5). In some cases, following the development of pulmonary hypertension, the flow through the defect is reversed, leading to cyanosis.

3. Ventricular Septal Defect

Ventricular septal defect is one of the most common defects encountered in domestic animals. The position and size of the defect may vary (Fig. 1.6). It may also be single or multiple and may involve the pars membranacea or less commonly the muscular portion of the septum. It may occur as an isolated defect, but it is also seen as part of a number of other defects, such as tetralogy of Fallot, truncus arteriosus communis, and endocardial cushion defects. There appears to be a high incidence of postnatal closure of ventricular septal defects in humans, and a similar phenomenon has been reported in dogs.

The separation of the left and right ventricles is completed by three parts of the embryonic heart: the muscular portion of the septum, the downward growth of the conotruncal ridges, and the membranous portion of the septum derived from the endocardial cushions. Defects can be related to defective development of any of the three parts.

Those most commonly seen in animals involve the membranous portion or the muscular portion of the septum. One may occasionally see subpulmonary ventricular septal defects.

The presence of a septal defect has no deleterious effect on the fetus because left and right ventricular pressures are equal, and there is therefore little flow across the defect. Postnatally, the effects of ventricular septal defects are related to the size of the defect and the level of pulmonic vascular resistance relative to the systemic resistance. There is normally a drop in pulmonary vascular resistance postnatally, leading to a left-to-right shunt. Left ventricular output is maintained by an increase in end-diastolic volume and an augmentation of contractility by the Frank–Starling mechanism. Since the right ventricular pressure equals the left ventricular pressure, the right ventricle is confronted with a large systolic and diastolic load. Both ventricles undergo hypertrophy, the left being more obviously eccentric in nature (Figs. 1.6, 1.7).

B. Conotruncal Abnormalities: Tetralogy of Fallot

There is one relatively common defect that results from anomalous development of the conotruncal septum, namely, tetralogy of Fallot.

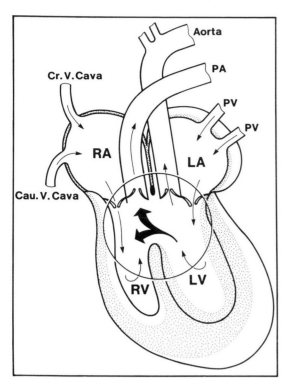

Fig. 1.7 Ventricular septal defect (VSD). Both an increased preload and afterload are imposed on the right ventricle by a VSD. The increased volume of blood returning from the pulmonary circuit also places an increased preload on the left atrium and ventricle. (From "Clinicopathologic Principles for Veterinary Medicine." W. F. Robinson and C. R. R. Huxtable, eds. Cambridge University Press, 1988, with permission.

The three primary developmental anomalies are ventricular septal defect; pulmonic stenosis, and an overriding aorta. These anomalies, accompanied by compensatory hypertrophy of the right ventricle, constitute the tetrad. It is one of the congenital heart diseases that invariably results in clinical signs. Affected animals fatigue easily and are usually cyanotic and polycythemic. The latter is a response to hypoxia. Growth rate is usually retarded.

The association of the three primary anomalies is the result of a defective development of the conotruncal septum. The analysis of affected keeshonds revealed various grades of malformation, determined by the frequency of inheritance of certain genes. The spectrum of anomalies included subclinical defects of the crista supraventricularis; ventricular septal defect; pulmonic stenosis with abnormal nonpatent interventricular septum; and tetralogy of Fallot. As noted, closure of the interventricular septum is contributed to by the conotruncal septum. The flow of blood from the right ventricle into the dextroposed aorta does not depend on the degree of overriding but on the severity of the pulmonic stenosis (Fig. 1.8). The complex is therefore best thought of as a ventricular septal defect accompanied by right ventricular outflow tract obstruction. There have been cases described in humans in

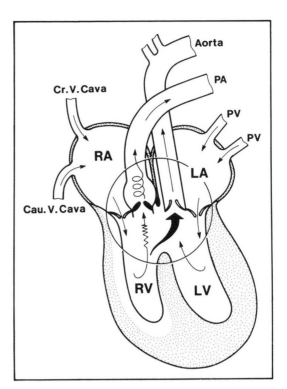

Fig. 1.8 Tetralogy of Fallot. The severity of the pulmonic stenosis determines the predominant direction of blood flow. If severe, a proportion of the blood in the right ventricle flows into the aorta via the VSD. The right ventricle hypertrophies concentrically because of the increased afterload placed on it. (From "Clinicopathologic Principles for Veterinary Medicine," W. F. Robinson and C. R. R. Huxtable, eds. Cambridge University Press, 1988, with permission.

which there is a complete obstruction to right ventricular outflow. The flow of blood to the lungs is then carried by the bronchial arteries.

C. Failure of Adequate Development of Semilunar or Atrioventricular Valves

Failures of adequate development of the semilunar or atrioventricular valves usually result in valvular or subvalvular stenosis. Insufficiencies occur occasionally and especially affect the left atrioventricular valve.

1. Pulmonic Stenosis

Pulmonic stenosis is a relatively common congenital anomaly in dogs, but an unusual finding in other domestic species. It is inherited in beagles, and is suspected to be so in English bulldogs, Chihuahuas, and terrier types. The term pulmonic stenosis encompasses three anatomic variations: supravalvular, valvular, and subvalvular or infundibular stenosis.

Valvular stenosis is probably due to the disordered fusion of the valve cushions and their failure to hollow out properly. The valve, which is then more or less dome shaped, with an irregular central perforation, is sur-

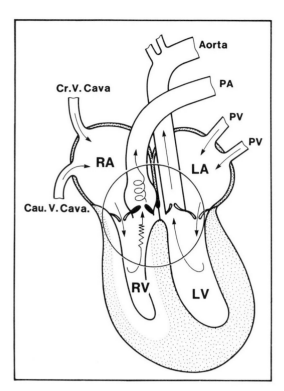

Fig. 1.9 Pulmonic stenosis. The stenosis is almost always valvular and places an increased afterload on the right ventricle which undergoes concentric hypertrophy. There is also poststenotic dilation of the pulmonary artery. (From ''Clinicopathologic Principles for Veterinary Medicine.'' W. F. Robinson and C. R. R. Huxtable, eds. Cambridge University Press, 1988, with permission.)

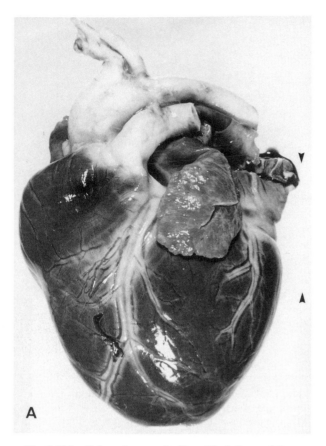

Fig. 1.10A Subaortic stenosis. Dog. The left ventricle occupies most of the ventricular mass, and the ascending aorta is dilated. (5 cm between arrows.)

rounded by three recognizable sinuses of Valsalva, which are small and irregular in form. Subvalvular or infundibular stenosis is produced by a ring of connective tissue encircling the upper portion of the outflow tract of the right ventricle. The third variant is hypertrophy of the crista supraventricularis muscle ridge. With each form, the pulmonary trunk is dilated and thin walled (Fig. 1.9). This is probably due to a combination of turbulent flow and a drop in pressure, creating a venturi effect in the pulmonary artery.

Concentric right ventricular hypertrophy is always present because of the increased systolic pressure generated within the ventricle.

2. Aortic and Subaortic Stenosis

Isolated congenital aortic valve stenosis is distinctly uncommon. Subvalvular aortic stenosis is, in contrast, extraordinarily common, at least in Danish pigs, and is among the more frequently encountered anomalies in dogs. There is a spectrum or gradation in the severity of the anatomic appearance of the subvalvular lesion, ranging from a number of small fibrous plaques on the endocardial surface of the interventricular septum (Fig. 1.10A,B) to a completely encircling fibrous band just below the aortic valve. There may also be involvement of the aortic valve.

The endocardial thickenings are composed of connective tissue in irregular arrangements, with abundant mucinous ground substance. In some cases the change is fibrocartilaginous.

As with pulmonary outflow tract obstruction, there is compensatory concentric hypertrophy of the involved ventricle and poststenotic dilation of the aorta (Fig. 1.11). Also, multifocal myocardial necrosis is a frequent finding. This explains the relatively common clinical finding of sudden death, presumably due to the generation of ventricular dysrhythmias following myocardial necrosis.

The heritability of this anomaly in dogs either is polygenic or involves a major dominant gene with modifiers. There is some question as to whether it is a true congenital disease. In one experimental study, no dog exhibited the typical lesion before 25 days of age. There is also some evidence that the severity of the disease increases with age.

3. Dysplasia of the Tricuspid Valve

Dysplasia of the tricuspid valve appears to be one of the more common defects observed in the cat but is uncommon in other domestic species. The anatomic characteristics of tricuspid dysplasia are focal or diffuse thick-

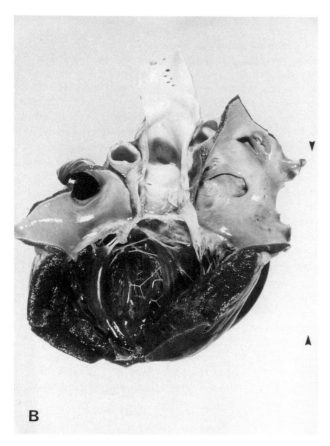

B

Fig. 1.10B Subaortic stenosis. Dog. The opened left ventricle is concentrically hypertrophied, and thick fibrous bands encircle the subaortic area. (10 cm between arrows.)

ening of the leaflets, some of which may be absent, or short chordae tendineae and papillary muscles, and direct fusion of portions of the affected valve with the ventricular wall. There is some overlap between tricuspid dysplasia and Ebstein's anomaly, of which the central feature is downward displacement of the basal portions of the valve. In tricuspid dysplasia, the valve is insufficient, the right atrium is enlarged, and the right ventricle is eccentrically hypertrophied. The anomaly may be associated with malformation of the mitral valve complex or ventricular septal defect.

4. Mitral Valvular Insufficiency

A mixed-frequency holosystolic murmur, with its point of maximal intensity on the left caudal sternal border, is indicative of mitral insufficiency. Commonly heard in old dogs with endocardiosis, it may also be associated with moderate to severe left-sided failure in young dogs and cats. Malformation of the mitral valve complex is probably the most common congenital cardiac anomaly in the cat. Anatomically, there is an enlarged annulus, short thick leaflets, short thickened chordae tendineae, upward malposition of atrophic or hypertrophic papillary muscles, and enlargement of the left atrium and ventricle. There is

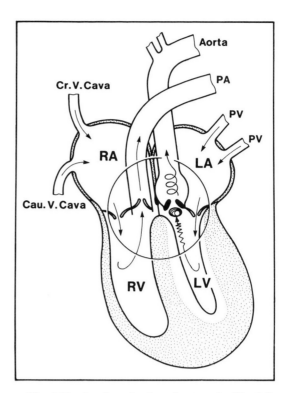

Fig. 1.11 Aortic and subaortic stenosis. The left ventricle hypertrophies because of the increased afterload. The aorta dilates caudal to the stenosis. (From "Clinicopathologic Principles for Veterinary Medicine," W. F. Robinson and C. R. R. Huxtable, eds. Cambridge University Press, 1988, with permission.)

also diffuse endocardial fibrosis. In dogs, the lesion has been most commonly seen in Great Danes, English bull terriers and, to a lesser extent, German shepherds.

Congenital hematomas (hemocysts) on the margins of the atrioventricular valves are common, especially in calves. These are blood-filled cysts lined by an endothelium; they originate in the clefts normally present in the substance of the valves in intrauterine life. These cysts, which may measure up to 1.0 cm in diameter and be multiple, do not usually persist for more than a few months, but occasional ones may enlarge and persist for a year or more, by which time the content is changed to serous fluid.

D. Endocardial Cushion Defects

In the embryo, two masses of loosely organized mesenchyme develop in the narrowed portion of the heart between the atrium and the ventricles. These endocardial cushions of the atrioventricular canal involve both the dorsal and ventral walls. The septum primum develops simultaneously and fuses with the endocardial cushions. Because the endocardial cushions contribute to the development of the atrial and ventricular septa, and to the medial leaflets of the mitral and tricuspid valves, anomalous development may result in a number of anatomic defects. They include an ostium primum defect, a ventricular septal defect with a cleft in the tricuspid valve, or a common atrioventricular canal. Defects of the atrioventricular canal are among the most common defects in the pig. In the largest series examined, 40% of pigs with congenital heart disease had this defect. It is also a frequent finding in cats.

E. Incomplete Separation of, or Abnormally Positioned, Vessels

The defects we are concerned with here are persistence of the right aortic arch, persistent truncus arteriosus, and transposition of the great vessels.

Persistence of right aortic arch is a well-known anomaly in dogs and has been observed in cattle. It is due to persistence of the right fourth aortic arch instead of, as is normal, the left fourth aortic arch. A right aorta descends to the right of the midline, arches over the origin of the right bronchus, and descends either to the left or right of the vertebral column. With the aorta in this position, the ductus arteriosus (ligamentum arteriosum), passing from the aorta to the pulmonary artery, encloses the esophagus and compresses it against the trachea. Obstruction of the esophagus at this point leads to dysphagia and, shortly, to dilation of the cervical portion of the esophagus (Fig. 1.12).

A **double aortic arch** is a variation of the foregoing in which the left arch persists as well as the right. Both arches arise from the ascending aorta. The right arch, which is usually the larger of the two, follows the course previously given. It is joined above the esophagus by the small left arch. The significant end result is constriction of the esophagus.

Congenital aneurysm of the aorta, or of the pulmonary artery, involves, for either vessel, the trunk and arch but may not extend beyond the insertion of the ligamentum arteriosum. Aortic aneurysm may be associated with aneurysm of one or more of the aortic sinuses of Valsalva.

Some of the more severe cardiac anomalies are combinations of defects of the aorta and pulmonary artery. Malposition or **transposition of arterial trunks** is a condition in which the aorta lies in relation to the pulmonary artery such that it receives blood from the right ventricle, the basic defect being a dextropositioning of the aorta. There are four degrees, or types, of this anomaly: in riding or overriding aorta, the aorta straddles the septum, which is defective, and receives blood from both ventricles, and the pulmonary artery leaves the right ventricle; in partial transposition, both vessels leave the right ventricle; in overriding pulmonary artery, the pulmonary artery straddles a defective ventricular septum, and the aorta emerges from the right ventricle; in complete transposition, the aorta emerges from the right ventricle, and the pulmonary artery emerges from the left (Fig. 1.13). It is usual in these transposition complexes for there to be hypoplasia of either the pulmonary or aortic tracts. The first variety of transposition, that of overriding aorta, is the usual one.

Fig. 1.12 Persistent right aortic arch. Dog. Esophagus constricted by aorta, trachea, and ligamentum arteriosum.

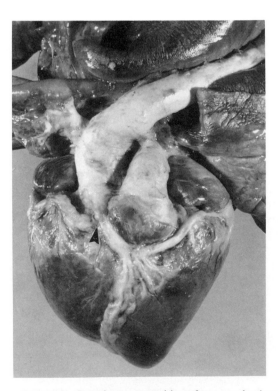

Fig. 1.13 Complete transposition of aorta and pulmonary artery. Foal. Coronary arteries arise from a common trunk (arrowhead).

F. Miscellaneous Cardiac Anomalies

1. Endocardial Fibroelastosis

Diffuse endocardial thickening occurs when any chamber of the heart remains dilated for a prolonged period. That fact has probably led to the confusion that exists about endocardial fibroelastosis. There are reports describing this condition in a number of species, but the primary disease has been well documented only in cats and, to a lesser extent, dogs. Primary endocardial fibroelastosis is characterized by diffuse endocardial thickening by collagen and elastic fibers, and left ventricular hypertrophy and dilation in the absence of any associated cardiac malformation.

The disease in affected Burmese kittens commences with localized endocardial lymphedema. This is detectable microscopically, but not grossly, at 1 day of age. Progressive endocardial collagen and elastin deposition follows and allows macroscopic detection from 20 days of age. The left atrium is dilated, and the left ventricle is hypertrophied and dilated. The endocardial fibrosis progressively involves Purkinje fibers, which exhibit some degenerative change. Although not definitely established, there appears to be little doubt of the inherited nature of the disease.

2. Epithelial Inclusions in the Myocardium

Epithelial tissue is found occasionally in ventricular myocardium in well-defined areas of discoloration or sponginess. The inclusions are present as acinar or ductular structures lined by a simple layer of cuboidal epithelial cells on a basement membrane. The inclusions are possibly of endodermal origin from the foregut and represent displacements arising very early in embryogenesis, when the heart rudiment is adjacent to the developing foregut. An alternative proposal is that they are of mesothelial origin.

Bibliography

Bayly, W. M. *et al.* Multiple congenital heart anomalies in five Arabian foals. *J Am Vet Med Assoc* **181:** 684–689, 1982.

Bolton, G. R., and Liu, S. K. Congenital heart disease of the cat. *Vet Clin North Am* **7:** 341–355, 1977.

Bonagura, J. D. Congenital heart disease. *In* "Contemporary Issues in Small Animal Practice," J. D. Bonagura, (ed.), pp. 1–20. New York, Churchill Livingstone, 1987.

Darke, P. G. G. Valvular incompetence in Cavalier King Charles spaniels. *Vet Rec* **120:** 365–366, 1987.

Dennis, S. M., and Leipold, H. W. Congenital cardiac defects in lambs. *Am J Vet Res* **29:** 2337–2340, 1968.

Edwards, J. N., and Tilley, L. P. Congenital heart defects. *In* "Pathophysiology in Small Animal Surgery," M. J. Bojrab (ed.), pp. 155–177. Philadelphia, Pennsylvania, Lea & Febiger, 1981.

Eyster, G. E. *et al.* Patent ductus arteriosus in the dog: characteristics of occurrence and results of surgery in one hundred consecutive cases. *J Am Vet Med Assoc* **168:** 435–438, 1976.

Friedman, W. F. *et al.* Pharmacologic closure of the patent ductus arteriosus in the premature infant. *N Engl J Med* **295:** 526–529, 1976.

Gopal, T., Leipold, H. W., and Dennis, S. M. Congenital cardiac defects in calves. *Am J Vet Res* **47:** 1120–1121, 1986.

Hagio, M. *et al.* Congenital heart disease in cattle. *Bull Fac Agr, Miyazaki Univ, Japan* **32:** 233–249, 1985.

Hartigan, P. J., Ahern, C. P., and McLoughlin, V. J. Endocardial cushion defects in a litter of malignant hyperthermia-susceptible pigs. *Vet Rec* **106:** 152, 1980.

Heymann, M. A., Rudolph, A. H., and Silverman, N. H. Closure of the ductus arteriosus in premature infants by inhibition of prostaglandin synthesis. *N Engl J Med* **295:** 530–533, 1976.

Hsu, F. S., and Du, S. J. Congenital heart diseases in swine. *Vet Pathol* **19:** 676–686, 1982.

Hunt, G. B. *et al.* A retrospective analysis of congenital cardiac anomalies (1977–1989). *Aust Vet Pract* **20:** 58–63, 1990.

King, J. M., Flint, T. J., and Anderson, W. I. Incomplete subaortic stenotic rings in domestic animals—a newly described congenital anomaly. *Cornell Vet* **78:** 263–271, 1988.

Liu, S. K. Left ventricular false tendons associated with cardiac malfunction in 101 cats. (Abstr.) *Lab Invest* **56:** 44A, 1987.

Liu, S. K., and Brown, B. Malformation of mitral valve complex associated with cardiac dysfunction in 120 cats. (Abstr.) *Lab Invest* **58:** 56A, 1988.

Liu, S. K., and Tilley, L. P. Malformation of the canine mitral valve complex. *J Am Vet Med Assoc* **167:** 465–471, 1975.

Liu, S. K., and Tilley, L. P. Dysplasia of the tricuspid valve in the dog and cat. *J Am Vet Med Assoc* **169:** 623–630, 1976.

Malik, R., and Church, D. B. Congenital mitral insufficiency in bull terriers. *J Small Anim Pract* **29:** 549–557, 1988.

Matic, S. E. Congenital heart disease in the dog. *J Small Anim Pract* **29:** 743–759, 1988.

Nordstoga, N., and Aleksandersen, M. Epithelial inclusions in the bovine myocardium. *Vet Pathol* **25:** 525–526, 1988.

Paasch, L. H., and Zook, B. C. The pathogenesis of endocardial fibroelastosis in Burmese cats. *Lab Invest* **42:** 197–204, 1980.

Patterson, D. F. Hereditary congenital heart defects in dogs. *J Small Anim Pract* **30:** 153–165, 1989.

Pyle, R. L., Patterson, D. F., and Chako, S. The genetics and pathology of discrete subaortic stenosis in the Newfoundland dog. *Am Heart J* **92:** 324–334, 1976.

Rooney, J. R., and Franks, W. C. Congenital cardiac anomalies in horses. *Pathol Vet* **1:** 454–464, 1964.

Sandusky, G. E., and Smith, C. W. Congential cardiac anomalies in calves. *Vet Rec* **108:** 163–165, 1981.

Takeda, T. *et al.* Morphological aspects and morphogenesis of blood cysts on canine cardiac valves. *Vet Pathol* **28:** 16–21, 1991.

van den Ingh, T. S. G. A. M., and van der Linde-Sipman, J. S. Vascular rings in the dog. *J Am Vet Med Assoc* **164:** 939–941, 1974.

Van Nie, C. J. Conduction system in porcine hearts with congenital abnormalities. *Anat Histol Embryol* **9:** 330–336, 1980.

Vitums, A., and Bayly, W. M. Pulmonary atresia with dextroposition of the aorta and ventricular septal defect in three Arabian foals. *Vet Pathol* **19:** 160–168, 1982.

West, H. J. Congenital anomalies of the bovine heart. *Br Vet J* **144:** 123–130, 1988.

V. Pericardium

Primary disease confined to the pericardium is rare, but the close anatomic relationship of the pericardium to the heart, lungs, and pleura sometimes results in the extension of disease processes from the latter organs to the pericardium. The pericardial cavity may communicate with the peritoneal cavity through clefts in the diaphragm; it is usual in such cases to find herniated intestine surrounding the heart. The pericardial sac may be absent congenitally without clinical effect, or it may be partly removed surgically for relief of chronic hemopericardium.

Notwithstanding that the pericardium is not a vital organ, its proper function provides automatic compensations that ensure that end-diastolic transmural pressure is the same for all hydrostatic levels of the ventricle. The role of the pericardium includes prevention of sudden cardiac dilation; maintenance of low transmural pressures; limitation of right ventricular stroke work; hydrostatic compensation for gravitational or inertial forces, and maintenance of cardiac alignment and streamlined intracardiac flow.

A. Noninflammatory Lesions of the Pericardium

1. Hydropericardium

The pericardial sac normally contains a very small quantity of clear, serous fluid. Any excess in the volume of fluid is referred to as hydropericardium. A small increase in volume occurs by transudation after death, and it is soon reddened by postmortem hemolysis.

In hydropericardium, the serosal surfaces remain smooth and glistening, but if the fluid persists for a long time, it may produce slight fibrous thickening and opacity of the pericardial and epicardial surfaces, usually about the base of the ventricles, and small villous proliferations of the serosae, which may rupture and result in a blood-stained effusion. Neoplastic involvement of the pericardium characteristically has this result.

The pericardial fluid is clear or light yellow, without floccules, and with low content of protein. Fluid rich in protein, including fibrinogen, is often found in acute toxemias of clostridial (gas-gangrene group) origin. Clotting occurs soon after exposure to air. In such cases, the effusion is due to direct injury to the endothelium by circulating toxins. Fluid that is copious and gels on exposure also is characteristic of African horse sickness and of heartwater. In pigs with edema disease or mulberry heart disease, fibrin is often formed antemortem. The presence of protein, especially fibrin, in pericardial fluid is an index only of the degree of endothelial injury and increased permeability and is not a precise point of distinction between an exudate and a transudate. Inflammatory exudates can be differentiated from transudates on the basis of the higher content of protein and cells in exudates and histologic evidence of inflammation involving the pericardium and epicardium.

Hydropericardium is often part of generalized anasarca and, as such, is seen in many cachectic illnesses and in congestive heart failure (Fig. 1.14). It may also be the product of local events such as implanted metastases of neoplasms on the pericardium or lymphomatous infiltrations of the myocardium; primary tumors about the base of the heart and in the cranial mediastinum are rather

Fig. 1.14 Hydropericardium. Cat. Pericardium is distended, and pericardial vessels are injected.

common causes of hydropericardium, which is probably the result of venous and lymphatic obstruction. Infectious processes of the pleura, especially granulomatous inflammation, may cause irritation of the pericardium sufficient to provoke a sterile serous effusion. In the latter cases, extension of the infection to the pericardial sac may, in time, occur.

The volume of fluid that accumulates is quite variable and of lesser significance than the rate at which it accumulates. When fluid accumulates rapidly, the pericardium is put under tension, and this is reflected in the pooling of venous blood in the splanchnic and systemic circulations due to impaired ventricular filling. When the fluid accumulates slowly, there is time for stretching and adaptation of the pericardium so that, relative to acute distensions, very large amounts of fluid can accumulate before there is significant impediment to the flow of blood in the large veins and right side of the heart.

2. Hemopericardium

The term hemopericardium is limited to accumulations of pure blood in the pericardial cavity and should not apply to mixtures of blood and serous fluid. If the blood is clotted, it can be assumed that the process is a true hemopericardium. Death usually supervenes rapidly. It is an unusual lesion, except following cardiac puncture or as a spontaneous occurrence in the common hemangiosarcomas of dogs. It is seen in horses in which rupture of the intrapericardial aorta occurs (Fig. 1.15), in atrial rupture in endocardiosis (Fig. 1.16), and in ulcerative atrial endocarditis of uremic dogs, and rarely, following rupture of a coronary artery or cardiac aneurysm. Rupture of the heart, particularly of an atrium, a coronary artery, or the aorta occurs on rare occasions in young, growing swine (see diseases of the arteries, in Section III,C of The Vascular System).

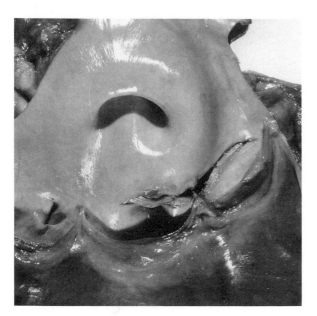

Fig. 1.15 Aortic rupture that caused hemopericardium and sudden death. Horse.

Idiopathic pericardial hemorrhage occurs predominantly in large-breed dogs. The clinical presentation is that of slowly developing right-sided heart failure, with some having accompanying left-sided heart failure. Although the heart sounds are muffled or inaudible on initial presentation, the affected animals have no evidence of dysrhythmias. Subsequently some develop atrial fibrillation. The clinical course can be prolonged. The pericardial effusion is characterized by the presence of both clotted and unclotted blood in the pericardial sac. The clots may with time lose their red color, become rounded and rubbery or impregnated with mineral and persist as the pericardial equivalent of joint mice. The slow onset of clinical signs suggests either slow or intermittent bleeding, allowing distension without tamponade. Indeed, some dogs may make a full clinical recovery following aspiration of the pericardial fluid. The pathogenesis of this condition is not understood. The biopsied pericardium is thickened and sclerotic, and fibrin deposits may be present on both surfaces. Histologically, there is fibrosis and collagenization, capillary proliferation, and microscopic hemorrhage with hemosiderin deposits. Collections of plasma cells and lymphocytes are usual. Papillary proliferations of mesothelium may be a source of continuing hemorrhage.

3. Serous Atrophy of Pericardial Fat

In cachexia of any cause, and paralleling fairly closely the reduction in body weight, there is progressive mobilization of depot fat, which includes that beneath the epicardium. As the lipid vacuoles are reduced in size, they are replaced by proteinaceous fluid and, with a concomitant increase in interstitial fluid, the depots are converted to gray, gelatinous masses that may be flecked by small white foci of fat necrosis (Fig. 1.17).

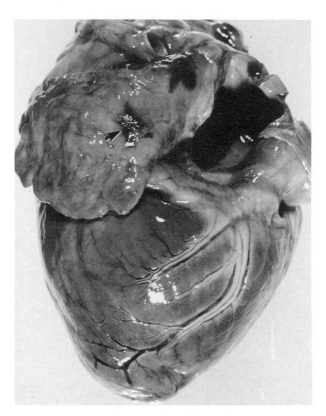

Fig. 1.16 Hemorrhage at site of left atrial rupture (arrow). Dog. Left atrium is dilated following left atrioventricular valve insufficiency.

B. Pericarditis

It is seldom possible by an examination of the lesion itself to decide the cause of pericardial inflammation, although a decision on the cause is often possible when the total pathologic picture in the cadaver is noted.

Fibrinous pericarditis is usually the result of hematogenous infections, but it may arise by lymphatic permeation from an inflammatory process in adjacent tissue. In keeping with this pathogenesis, the microbial causes are varied. In cattle, fibrinous pericarditis, with or without some hemorrhage, is commonly part of sporadic bovine encephalomyelitis, contagious bovine pleuropneumonia, pasteurellosis, blackleg, clostridial hemoglobinuria, and some of the neonatal coliform infections that enter via the navel. In swine it often occurs in Glasser's disease and pasteurellosis, is a common complication of porcine enzootic pneumonia, and is occasionally observed in salmonellosis and in streptococcal infection of piglets. Fibrinous pericarditis in adult sheep is usually part of pasteurellosis; in lambs, it is usually part of pasteurellosis or caused by streptococci. In horses, streptococci are usually present, and the lesion may coexist with polyarthritis. In the cat, pericarditis is rare, but it is seen as part of feline infectious peritonitis.

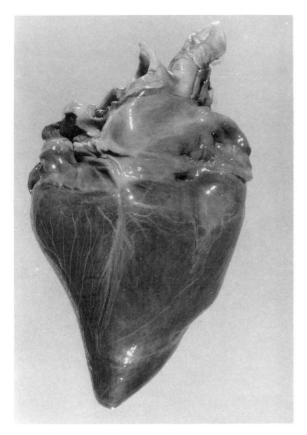

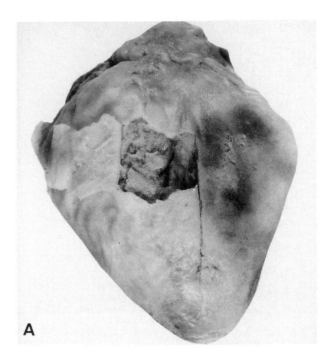

Fig. 1.18A Fibrinous pericarditis. Pig. There is some blood-staining of the exudate, which could be peeled off.

Fig. 1.17 Serous atrophy of epicardial fat. The fat is gelatinous and translucent, and the myocardium is dark.

In fibrinous pericarditis, there is seldom a significant exudation of fluid, so distension of the pericardial sac is not to be expected. The exudation of fibrin usually begins about the base of the heart and extends from there to cover, more or less completely, both the pericardium and epicardium. The fibrin is gray-white, but it may be flecked with blood, or yellow if a large number of leukocytes are added to the exudate. Except for small pools of serum or serous exudate, the pericardium and epicardium are in apposition, and, when at autopsy the visceral and parietal surfaces are separated, the exudate is drawn out into villuslike projections to give an appearance responsible for the names cor villosum, shaggy heart, and bread-and-butter pericarditis (Figs. 1.18A,B).

The completeness of resolution depends on the speed with which the exudate, especially the fibrin, can be removed and the mesothelium, regenerated. In the first instance, this is a function of the amount of fibrin and, in the second, a function of the extent to which the mesothelium has been destroyed. Restorative processes compete, in time, with the processes of organization. Within a week or so, there will be well-formed fibrous tissue in the deepest parts. Immediately superficial to this, there will be young granulation tissue and then a stratum of fibroblasts mixed with leukocytes, and on the

Fig. 1.18B Chronic pericarditis with bread-and-butter exudate. Cow. The reflected pericardium is thickened.

surface there remains the fibrin clot infiltrated with numerous leukocytes.

If the course is prolonged, organization and cicatrization will result in focal or diffuse fibrous adhesions between the pericardial surfaces, with partial or complete obliteration of the sac.

The residual lesions of fibrinous pericarditis occur as more or less distinct variants. Focal, patchy thickenings of the epicardium without adhesions form minimal residue and are usually more distinct on the ventricles than on the atria, where they may be obscured by normal fat. The original scar tissue may undergo fatty metaplasia or be edematous. (The same lesion occurs also covering healed foci of superficial myocarditis.) Focal or diffuse adhesive pericarditis varies in extent from a few "violin strings" across the sac to complete obliteration of the sac. Inclusion cysts lined by mesothelium are often present in fibrous adhesions, and there tends to be excess fluid present in nonobliterated portions of the sac. The pericardium is separable from the epicardium by blunt dissection. As the connective tissue is not dense, there is usually no embarrassment of cardiac function.

Purulent pericarditis almost invariably denotes the presence of pyogenic bacteria, either as primary pathogens or as opportunists in fibrinous pericarditis. It occurs almost solely in cattle as a result of traumatic perforation by a foreign body originating in the reticulum (Fig. 1.19), but it is observed in cats and horses in association with empyema.

The suppurative pericardial fluid may appear as a thin, cloudy exudate, as frank, creamy pus, or as a mixture of pus and masses of fibrin. The color depends on the organisms present, but usually varies from yellow to green, being irregularly dirty gray when putrefactive bacteria are present. The exudate is usually foul smelling. The volume of exudate varies from a thin layer on the serosal surface to 4 liters or more. In the early stages, it can be wiped or washed off to reveal the vascular injection and

granularity of the membranes. Soon, the whole of the epicardium is covered with coagulum; the parietal layer of the pericardium is less severely affected, and separation of the two presents a shaggy appearance of the heart (cor villosum).

Microscopically, the subjacent tissues are densely infiltrated with leukocytes, the reaction extending more deeply than in fibrinous pericarditis, the formation of new blood vessels and connective tissue is prominent, and the overlying coagulated exudate is densely infiltrated with leukocytes.

Resolution probably never occurs. Because of the severity of the inflammatory reaction, healing is by organization, as is characteristic of all suppurative inflammations. Although the course of suppurative pericarditis may be prolonged, death usually occurs before there is complete organization. Organization results in the formation of adhesions that are usually heavy enough to cause constriction of the heart (Fig. 1.20). They cannot be broken down by blunt dissection and may be mineralized. The adhesions obliterate the pericardial cavity, but they may also extend beyond it to involve the mediastinum. The heart, which is confined by scar tissue, may be smaller than normal, or it may be greatly enlarged and hypertrophic, especially when there are also extrapericardial adhesions present.

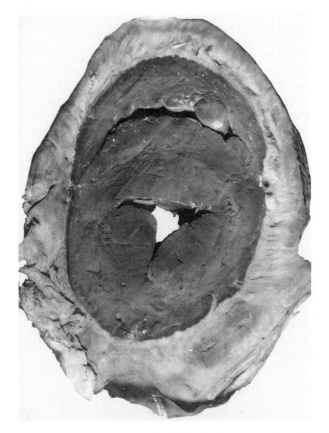

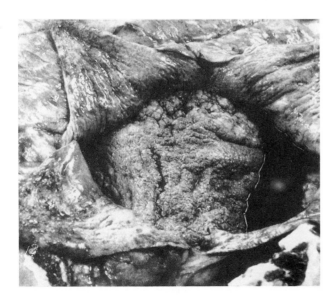

Fig. 1.19 Putrid, traumatic pericarditis. Ox. The pericardium is reflected. Sac contains dirty liquid exudate, and epicardium bears profuse granulation tissue. (Courtesy of P. Olafson.)

Fig. 1.20 Constrictive pericarditis. Ox. Dense scar tissue has obliterated the pericardial sac and constricts the ventricles, the walls of which are hypertrophic.

There is always severe cardiac dysfunction, and death occurs from congestive heart failure.

Bibliography

Berg, R. J., Wingfield, W., and Hooper, P. J. Idiopathic hemorrhagic pericardial effusion in eight dogs. *J Am Vet Med Assoc* **185:** 988–992, 1984.

Freestone, J. F. *et al.* Idiopathic effusive pericarditis with tamponade in the horse. *Equine Vet J* **19:** 38–42, 1987.

Gibbs, C. *et al.* Idiopathic pericardial hemorrhage in dogs: A review of fourteen cases. *J Small Anim Pract* **23:** 483–500, 1982.

Rush, J. E., Keene, B. W., and Fox, P. R. Pericardial disease in the cat: A retrospective evaluation of 66 cases. *J Am Anim Hosp Assoc* **26:** 39–46, 1990.

VI. Endocardium

A. Degenerative Lesions

1. Subendocardial Fibrosis

Subendocardial fibrosis may be diffuse or focal, congenital or acquired (Figs. 1.21A,B). The congenital lesions are discussed under Congenital Abnormalities of the Heart and Large Vessels (Section IV of The Heart). Diffuse subendocardial fibrosis is seen whenever a ventricle or atrium is dilated for a prolonged period. It is probably best exemplified by the dilated cardiomyopathy of large-breed dogs.

Subendocardial fibrosis in localized areas is observed chiefly in the atria and is regarded as a reaction of the endocardium to abnormal jets of blood or to turbulence following congenital or acquired valvular disorders (Fig. 1.22). They are loosely termed jet lesions but are probably areas subject to increased static pressure in turbulent flow.

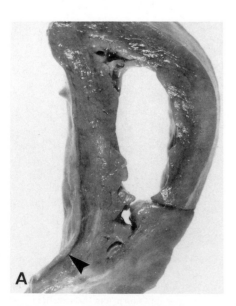

Fig. 1.21A   Subendocardial fibrosis (arrowhead). Calf.

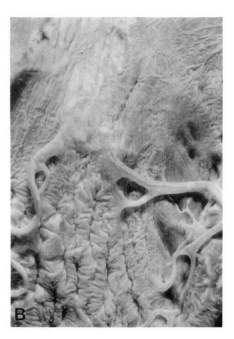

Fig. 1.21B   Endocardial fibrosis. Cow.

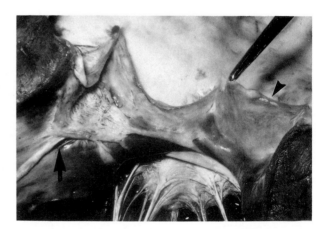

Fig. 1.22   Aortic valve fenestrations (arrowhead) causing endocardial jet lesion (arrow) in left ventricle. Horse.

2. Subendocardial Mineralization

Subendocardial mineralization is associated with a variety of disorders. It occurs commonly in opaque plaques or as small grains in the left atrium, and occasionally in the trunk of the aorta in dogs that have recovered from ulcerative endocarditis of renal insufficiency (see the next section). Prominent white plaques of mineralization also occur beneath the endocardium of the right ventricle in nutritional myopathy of lambs. Subendocardial mineralization in the atria and left ventricle may accompany endocardial fibrosis when these cavities are acutely dilated, either as congenital or acquired defects, and in a variety of prolonged debilitating diseases in cattle (Fig. 1.23). The deposition of calcium and phosphorus may be in degenerate subendocardial muscle, but more commonly it is pre-

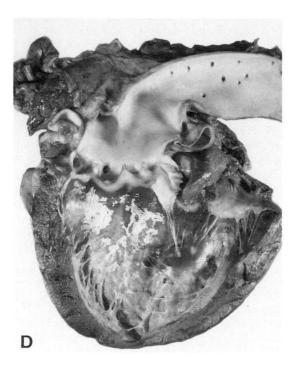

D

Fig. 1.23 White subendocardial plaques of mineralization in the left ventricle. Ox.

cipitated in fibroelastic tissue that is modified by either chronic disease or elevated serum levels of calcium and phosphorus, as in vitamin D intoxication or following the ingestion of plants containing vitamin D analogs (see The Endocrine Glands, Chapter 3 of this volume). Proprietary rodenticides based on calciferols cause this lesion in dogs.

3. Endocardiosis in Dogs

Endocardiosis is the most common cardiovascular lesion in dogs and is encountered most frequently as an incidental finding at autopsy. It is only the more advanced form of the disease that is responsible for the production of clinical signs of left-sided heart failure. The shrunken, distorted atrioventricular valves are seen with greatest frequency in males of particular breeds, such as the poodle, Pomeranian, schnauzer, Chihauhua, Doberman pinscher, fox terrier, Boston terrier, and cocker spaniel. There are significant negative associations for the Labrador retriever and the German shepherd. The prevalence of the disease increases with increasing age, from 5% at younger than 1 year to ~75% at 16 years of age.

The basis of the lesion is a proliferation of loose, fibroblastic tissue in the spongiosa of the valve, accompanied by the deposition of glycosaminoglycan. There is a concomitant degeneration of collagen in the fibrosa of the valve. Alternative, and probably more appropriate, morphologic terms for the condition are mucoid valvular degeneration or myxomatous transformation of the atrioventricular valves.

Endocardiosis affects chiefly the left atrioventricular

valve (Figs. 1.24A,B), the right atrioventricular valve being less severely and less often affected. Only occasionally are the aortic and pulmonary semilunar cusps involved. The atrioventricular cusps are shortened and thickened. The thickening of the leaflet may be more or less uniform with a rounded edge, or prominent nodular thickenings may affect the free margin. The valves are opaque and white, but the surface is smooth and glistening, without any evidence of inflammation. The chordae tendineae may also be thickened and occasionally are ruptured, allowing eversion of the leaflet into the atrium and leading to acute ventricular failure.

For descriptive purposes, it is convenient to grade the gross lesions. The least severe change consists of a few small, discrete nodules at the line of closure of the valve. This progresses to multiple larger nodules that tend to coalesce in the area of contact. There may be irregular

A

Fig. 1.24A Endocardiosis. Dog. Thickened, distorted left atrioventricular valve with focal endocardial fibrosis (jet lesions) in left atrium.

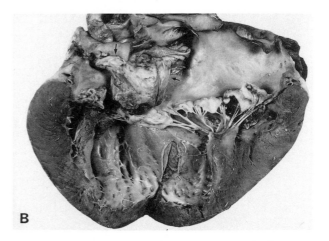

B

Fig. 1.24B Endocardiosis of left A-V valves and atrial thrombosis (arrows). Dog.

areas of opacity in the proximal portions of the leaflet. The nodules may coalesce to form plaquelike deformities in the area of contact. The chordae tendineae are thickened at their points of insertion into the valve, and there are clearly defined areas of opacity on the basal portions of the leaflet. The most severe change is characterized by a gross distortion of the valve by gray-white nodules and elevated plaques. The cusps are contracted, thickened, and irregular. There may also be fixed upward displacement of the free margin of the valve. Chordae tendineae may be ruptured. Care should be taken in the assessment of the change, since both the left and right atrioventricular valve leaflets thicken with increasing age, and the septal leaflet of the right atrioventricular valve is normally distinctly thicker than the free leaflets.

Changes secondary to the atrioventricular valvular insufficiency are dilation of the atria, particularly the left atrium, and dilation of the left ventricle (Fig. 1.25). In a small number of cases, the left ventricle is eccentrically hypertrophied. The left atrial and ventricular endocardium may be diffusely thickened by fibrosis following prolonged dilation. There is, additionally, evidence of regurgitation in the form of focal elevated streaks and plaques of subendocardial fibrosis in the atria (jet lesions). Left atrial tears may also be seen in advanced cases and may lead to hemopericardium (Fig. 1.16).

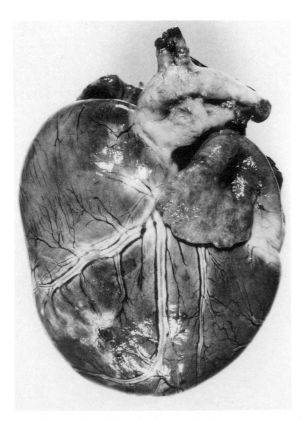

Fig. 1.25 Dilated left ventricle and atrium with focal ventricular fibrosis following endocardiosis. Dog.

Microscopically, the earliest changes are observed on the atrial side of the valves. There is proliferation of the endothelium, an increase in the number of subendothelial fibroblasts and macrophages, and splitting and separation of elastic fibers between the atrialis and spongiosa. However, it is the thickening of the spongiosa and degeneration of the fibrosa that are the most prominent features of endocardiosis. The spongiosa is greatly thickened by the proliferation of loose fibroblastic tissue and the deposition of the proteoglycans, hyaluronic acid and chondroitin sulfate. There is no increase in collagen or elastic tissue in the spongiosa. The changes in the fibrosa are also marked. Collagen bundles become swollen and hyalinized, fragment, and disappear. In advanced cases, only scattered remnants of the fibrosa remain. Similar changes are seen in chordae tendineae. Intramural coronary arteriosclerosis and focal myocardial necrosis and fibrosis are commonly seen in the left ventricular myocardium, especially the papillary muscle, if the ventricle is hypertrophic.

The cause of endocardiosis is not known. It may be a genetically influenced degeneration of connective tissue, particularly collagen. This suggestion is supported by the observation that the most frequently affected breeds are of the chondrodystrophoid type. There is a striking similarity of endocardiosis to the prolapsed mitral valve syndrome of humans, in which there is a deposition of glycosaminoglycans in the spongiosa secondary to an as yet undefined abnormality of collagen metabolism. Mitral valve prolapse occurs in association with connective-tissue disorders such as Marfan's syndrome, Ehlers–Danlos syndrome, osteogenesis imperfecta, and a variety of muscular dystrophies in humans.

B. Endocarditis

Endocarditis is one of the more significant of the endocardial alterations. It is usually bacterial in cause, the exceptions being an occasional parasitic or mycotic lesion. Any portion of the endocardium may become inflamed, but the lesions are usually primary on the valves, from which there may be some encroachment on the adjacent mural endocardium (Fig. 1.26A,B).

Many bacterial species are capable of causing acute **valvular endocarditis,** and the larger the number of cases examined, the wider the variety of pathogens recognized. In cattle, *Actinomyces pyogenes* is probably the most common pathogen, and a primary focus of infection can often be found in some other site, such as a traumatic peritoneal abscess, hepatic abscess, metritis, or mastitis. Streptococci of enteric origin are also of some importance in cattle, although the manner of their systemic invasion is not clear. *Erysipelothrix rhusiopathiae* has caused endocarditis in a variety of animals, but most commonly in the pig; however, streptococci are more common than *E. rhusiopathiae* in acute bacterial endocarditis of swine. Horses seldom develop bacterial endocarditis, but the lesion has been observed in association with *Streptococcus equi, Actinobacillus equuli, Escherichia coli, Pseudomo-*

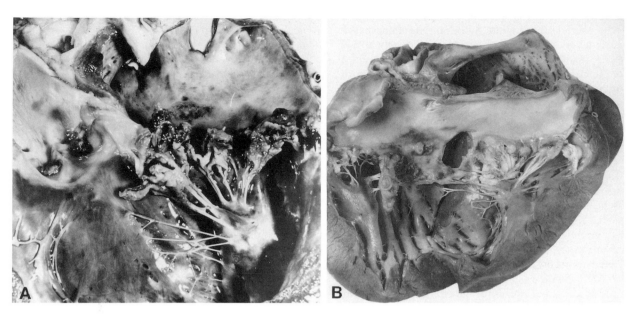

Fig. 1.26 Valvular endocarditis. Dog. (A) Caused by *Escherichia coli*. (B) Showing valvular rupture and extension to ventricular wall.

nas aeruginosa, and *Candida parapsilosis.* Minor outbreaks of endocarditis occur in lambs and are usually caused by *Streptococcus* spp., particularly enterococci; there may be an associated polyarthritis. Endocarditis is seldom observed in dogs but can be associated with a variety of organisms, especially *Streptococcus* spp, *Erysipelothrix rhusiopathiae,* and *E. coli* (Fig. 1.27).

The manner by which bacteria localize on a valve is not clear; there is seldom evidence of some separate underlying disease of the leaflet. In probably all cases of active thrombotic valvular endocarditis, bacteria are present in the lesion. Two factors that must be included in any con-

Fig. 1.27 Aortic valvular endocarditis. Dog. The valve cusp is perforated.

cept of pathogenesis are (1) the tendency for lesions to occur at the lines of apposition of the valve surfaces exposed to the forward flow of blood, and (2) the necessity for sustained or recurrent bacteremia. It is possible that simple debilitation of endothelia in recurrent bacteremia, aggravated locally on the valves by the trauma of apposition, may alter the adhesiveness of the leaflet endothelium sufficiently for bacterial adhesion to occur. It is of significance that valvular endocarditis can be produced in normal animals by a single intravenous injection of bacteria of suitable type and virulence, and that the number of positive results is increased if the hearts of such animals are subjected to increased work loads or the valves are traumatized during the experiments.

It appears that particular strains of *Erysipelothrix rhusiopathiae* can selectively adhere to the valvular endothelium of porcine heart valves *in vitro,* with greatest numbers at the base of chordae tendineae. The phenomenon of selective adherence has been demonstrated for pathogens associated with other organ system disease, notably those organisms causing mastitis and enterotoxigenic colibacillosis. There is also an immunologic cross-reaction between *E. rhusiopathiae* antigens and valvular and myocardial antigens.

The lesion of acute bacterial endocarditis is usually observed as a large vegetation on the affected valve. It is quite exceptional to find the earliest lesions, which consist of irregular ulcerations on the swollen leaflet. The vegetations have a composition similar to that of thrombi with, however, few platelets. Numerous bacterial colonies are present in the vegetations, almost always in pure populations. It is wise to make preliminary identification of the organisms by smears because, although they persist in colonies buried in the vegetations, they may not be cultivable. The bacteria that initiate the lesion are enmeshed in the early layers of thrombi and are buried deeply as new layers are formed on the surface. With continued multiplication, many microscopic colonies are formed and, without appropriate antimicrobial treatment, it is their persistence in the protected environment that is the reason healed bacterial endocarditis is seldom seen. Even though the bacteria in the thrombi may lose their vitality, as indicated by lack of cultivability, they are persistently irritative.

Grossly, the vegetations are yellow-red or yellow-gray and usually covered by a thin clot of blood, which can be easily peeled off. The surface of the vegetation is friable, and small vegetations can be broken off completely to leave a granular, eroded surface on the valve. The ulcerations are especially common in the commissures and at the free margins of the valves, imparting to them a rough, serrated appearance. The primary valvular lesion often extends to the adjacent mural endocardium and, in the cases of aortic valvular endocarditis, to the adjacent aortic intima in the sinuses of Valsalva; in the latter eventuality, the lesion involves an orifice of a coronary artery, predisposing to myocardial embolism. Endocarditis of the atrioventricular valves tends to spread along the chordae tendineae, causing some of them to rupture. Acute swelling and inflammatory change take place in the substance of the valve itself, often with necrosis, which obscures the line of partition between the valve and its surface vegetation.

Organization of the thrombus proceeds from the base by the usual process of granulation, which is likely often to be impeded or destroyed by bacterial growth. Early scarification in the deepest layers separates the thrombus from the underlying myocardium and is frequently accompanied by mineralization.

Valvular endocarditis is commonly fatal, although if there are only slight lesions, resolution may be complete. Clinical signs in animals with endocarditis may be detected only late in the disease with the most common being pyrexia, lameness, and cardiac murmurs. The frequency with which clinical signs are observed is dependent on the husbandry conditions under which the animal is kept. Other sequelae are those resulting from valvular damage or embolism. Portions of the vegetations may become detached and carried as emboli, to become impacted in the vessels of other organs. Such emboli may be bland, that is, consist of thrombotic material only, or septic. Emboli arising in the right heart may produce pulmonary abscesses or pulmonary thrombosis. These latter will not produce pulmonary infarcts in the absence of preexisting pulmonary congestion. Emboli arising in the left heart produce their most obvious effects in the kidney and spleen as septic or aseptic infarcts and embolic glomerulitis, and in the myocardium as myocardial abscesses or interstitial myocarditis. Arthritis is commonly observed in association with endocarditis in the dog. Cerebral embolism is rare in animals.

Unless the valvular lesions are very slight, there is, with healing, some degree of permanent damage, especially if the inflammation is recurrent. Shrinkage or adhesion of leaflets may cause narrowing of the orifice (stenosis), or incompetence as a result of failure of the distorted valves to close the orifice. Stenosis and incompetence may coexist. The usual outcome is congestive heart failure.

Mural endocarditis may be merely an extension to the walls of the cardiac chambers of a process originating on the valves. This is almost invariably true of endocarditis caused by *Actinomyces pyogenes*. Small foci of mural endocarditis may be found adjacent to foci of myocarditis, especially myocardial abscesses (Fig. 1.28). There are some additional occurrences of mural endocarditis worthy of mention.

Parasitic endocarditis caused by the larvae of *Strongylus vulgaris* occurs in horses in the form of caseous and calcareous nodules attached to the endocardium at the apex of the left ventricle and protruding into the cavity. There is, in the same cases, parasitic aortitis of the bulb.

An acute form of **ulcerative mural endocarditis** occurs in cattle dying from **blackleg,** as red thrombotic masses attached to the outer wall of the right ventricle and, less often and less extensively, on the right atrial and valvular endocardium.

Perhaps the most common form of mural endocarditis is that occurring in **renal failure** in dogs. The lesion is confined, within the heart, to the left atrium, but similar lesions occur in the large elastic arteries, being more common in the pulmonary and aortic trunks immediately proximal to the valves, and less numerous in the descending aorta and its major branches (see section on The Vascular System in this chapter). The endocardial

Fig. 1.28 Myocardial abscessation and scarring. There is extension of inflammation to the mural endocardium with thrombus development.

and major arterial lesions are more common in acute than in chronic renal failure, insofar as these syndromes are distinguishable.

The ulcerative atrial lesion, which is identical to the arterial lesion, begins as a swelling of the interstitial spaces of the subendocardium or intima of the arteries, with deposition of glycosaminoglycans. Initially, the endothelium is intact, and when viewed grossly in this stage, the endothelial surface is raised slightly, finely wrinkled, and slightly opaque but still shiny. The lesion may not progress further than this but, instead, heal with some fibrosis, which leaves areas of uneven opacity. More usually, however, there is progression to necrosis involving the cellular elements as well as the collagen and elastic and reticulin fibers. The necrotic tissue ulcerates, and the margin is densely infiltrated by leukocytes. The ulcerations may perforate the wall of the atrium. Thrombi form on the ulcerated surface. Heavy deposits of calcium salts often form in the degenerate tissue. If renal sufficiency is reestablished, the endocardial lesion may heal leaving irregular patches of sclerosis with an intact endothelium, often covering persistent white plaques of mineralization. Other extrarenal lesions of uremia are considered with The Urinary System (Volume 2, Chapter 5).

Bibliography

Bennett, D., and Taylor D. J. Bacterial endocarditis and inflammatory joint disease in the dog. *J Small Anim Pract* **29:** 347–365, 1987.
Bratberg, A. M. Selective adherence of *E. rhusiopathiae* to heart valves of swine investigated in an *in vitro* test. *Acta Vet Scand* **22:** 39–45, 1981.
Bratberg, A. M. Immunological cross-reactions of antigens of *E. rhusiopathiae* and heart tissue from swine. *Acta Vet Scand* **22:** 46–54, 1981.
Buchanan, J. W. Chronic valvular disease (endocardiosis) in dogs. *Adv Vet Sci Comp Med* **21:** 75–104, 1977.
Buergelt, C. D. *et al.* Endocarditis in six horses. *Vet Pathol* **22:** 333–337, 1985.
Calvert, C. A. Valvular bacterial endocarditis in the dog. *J Am Vet Med Assoc* **180:** 1080–1084, 1982.
Darke, P. G. G. Valvular incompetence in Cavalier King Charles spaniels. *Vet Rec* **120:** 365–379, 1987.
Guarda, F., and Negro, M. Pathology of endocardiosis in pigs. *Dtsch Tierärztl Wochenschr* **96:** 377–379, 1989.
Jones, J. E. T. Experimental streptococcal endocarditis in the pig: The development of lesions 3 to 14 days after inoculation. *J Comp Pathol* **91:** 51–62, 1981.
Kogure, K. Pathology of chronic mitral valvular disease in the dog. *Jpn J Vet Sci* **42:** 323–335, 1980.
Murdoch, D. B., and Baker, J. R. Bacterial endocarditis in the dog. *J Small Anim Pract* **18:** 687–699, 1977.
Reef, V. B., and Spencer, P. Echocardiographic evaluation of equine aortic insufficiency. *Am J Vet Res* **48:** 904–909, 1987.
Sanford, S. E. Gross and histopathological findings in unusual lesions caused by *Streptococcus suis* in pigs. I. Cardiac lesions. *Can J Vet Res* **51:** 481–485, 1987.
Whitney, J. C. Observations on the effect of age on the severity of heart valve lesions in the dog. *J Small Anim Pract* **15:** 511–522, 1974.

VII. Myocardium

A. Myocardial Degeneration

Cardiac muscle is structurally similar to skeletal muscle and is subject to the same anatomic types of degeneration. There is, however, a greater liability for cardiac muscle to undergo degeneration as a response to many nonspecific causes, and this is probably related to its continuous activity. Diffuse nonspecific myocardial degeneration occurs in a variety of systemic diseases, especially infectious fevers, anemia, and toxemia.

Hydropic degeneration is characterized by a dull gray appearance and increased friability of the tissue so that it is easily torn. On cut surface, the muscle is smoother than normal, and the outlines of individual muscle bundles, obscured. In **fatty change,** which is more severe than hydropic degeneration, uneven patches of myocardium are pale yellow. The process of fatty change is not uniform, and it is sometimes possible to recognize, beneath the endocardium, alternating bundles of fibers or strips of myocardium more yellow than the remainder (thrushbreast heart). Fatty change occurs in a variety of acute systemic intoxications, especially those of bacterial origin, and it accompanies the dilation of anemia, being best expressed in piglet anemia of iron deficiency.

Atrophy of the heart occurs in chronic wasting diseases and malnutrition. In some cases, the atrophy is accompanied by dark pigmentation, and it is then called brown atrophy. Atrophy can occur without pigmentation, however, and pigmentation can occur without atrophy. Ruminants are principally affected by atrophy, brown or not, and rarely does the atrophic heart produce clinical disturbance. The epicardial fat may be gelatinous, the endocardium, wrinkled, and myocardium, brown and friable. Histologically the muscle fibers are thin, the nuclei small and dark, and masses of brown pigment granules are present within secondary lysosomes at the nuclear poles. The pigment is lipofuscin, an oxidation product of unsaturated lipids.

Mineralization in the myocardium, dystrophic in character, occurs quite commonly and is to be expected whenever there is necrosis of fibers. Calcium salts are selectively deposited in the Purkinje network in organomercurial poisoning in cattle, in which the mineralization is preceded by hyaline necrosis of the Purkinje fibers and is followed by surrounding fibrosis.

B. Myocardial Necrosis

Focal to massive myocardial necrosis, or residual scars thereof, is a common lesion in autopsy material. The causes are quite varied and part of a number of disease syndromes described elsewhere in these volumes.

In infectious disease, the distinction between myocardial necrosis and inflammatory disease in which fiber necrosis is present is somewhat arbitrary. Necrosis predominates in the highly fatal foot-and-mouth disease infections of neonatal lambs, piglets, and calves, whereas the residual lesions in older cattle qualify as myocarditis with a significant inflammatory response. Fulminating infection by canine distemper virus in young puppies is purely degenerative, whereas parvovirus and herpesvirus infections in the same age group contain elements of inflammatory exudate. Of the bacterial diseases, and excepting blackleg, *Haemophilus somnus* appears to be responsible for a residual syndrome of sudden death in cattle with myocardial necrosis or infarction, although in some cases myocardial abscesses may be present.

Severe, often fatal myocardial necrosis is, typically, part of the important vitamin E- and selenium-responsive syndromes of nutritional myopathy in lambs, calves, swine, and horses, of porcine stress syndrome, of equine rhabdomyolysis, and of capture myopathy and other exertional syndromes (Fig. 1.29A,B).

Deficiency disease, additional to those which respond to vitamin E and selenium, may include myocardial necrosis. Myocardial necrosis occurs, although not invariably, in thiamine-deficiency disease of carnivores, always as a single acute episode. Falling disease of cattle in Australia and Florida is a syndrome of sudden death believed to be due to prolonged copper deficiency and consequent depletion of cytochrome oxidase and intracellular respiration. Myocardial scars may accompany acute lesions suggesting repetitive episodes and resembling the lesions of fluoroacetate poisoning in cattle eating the gidgee plant, *Acacia georginae*.

Toxic myodegeneration is common, and some are accompanied by degeneration of skeletal muscle and are described there. Injectable **saccharated-iron** compounds, by virtue of the capacity of iron to generate free radicals in ferric/ferrous translations, cause fatal myocardial necrosis in piglets. Ruminants of early age, that is before the rumen is developed, and pigs are susceptible to the cardiotoxic effects of **gossypol** when cottonseed meal is incorporated in compounded diets as a protein supplement. **Monensin,** an ionophore which is widely used as a coccidiostat for poultry and growth promotant in ruminants, may contaminate compounded feeds for simple-stomached animals and is cardiotoxic in horses and pigs by a mechanism of facilitated transport of cations across cell membranes (see Muscles and Tendons, Volume 1, Chapter 2, for details of toxicity in other species). Myocardial necrosis occurs in dogs ingesting rodenticides that contain **thallium.** These compounds have been widely replaced by warfarin analogs and calciferols.

Plants containing **cardiac glycosides** are of worldwide distribution, but most of them are quite unpalatable and eaten by animals only in extraordinary or inadvertent circumstances. Poisoning is acute, rather than cumulative, and death occurs within a few hours of ingesting the plant. Some cases may live for a few days and exhibit scattered

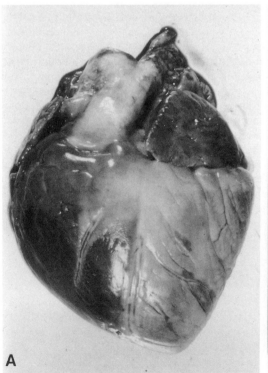

Fig. 1.29 Nutritional myopathy. Calf. (A) Marked pallor of left ventricle due to myodegeneration. (B) Cut surface of left ventricle showing subepicardial and subendocardial myodegeneration.

foci of myocardial necrosis, but usually a diagnosis must be based on circumstantial evidence and the presence of plant parts in the ingesta. The cardiac glycosides are inhibitors of sodium/potassium-ATPase, the sodium pump, which is responsible for maintaining transmembrane ion concentrations and the membrane potential of cardiac muscle fibers. Plants that contain cardiac glycosides and that are responsible for significant mortalities in livestock include *Nerium oleander* (oleander), *Homeria* spp. (the cape tulip), *Thevetia peruviana* (yellow oleander), *Tylecodon* spp., and *Bryophyllum tubiflorum*.

Not all acute cardiotoxicity of botanical source can be ascribed to glycosides. Whereas chronic *Phalaris* toxicity does produce interesting changes in the central nervous system, the acute toxicity is of much greater economic importance as a cause of quick death in animals first exposed to grazing on this pasture species. The acute toxic factor in *Phalaris* appears to be a tryptamine alkaloid, which may interfere with the metabolism and action of serotonin and other catecholamines. The fast-death factors of lupins and algae still need definition. *Lantana camara* causes myocardial necrosis in sheep, which might have an entirely different pathogenesis, possibly dependent on reduced myocardial perfusion as a result of chronic reduction in circulating blood volume. Poisoning by *Cassia occidentalis* (coffee senna) and *Karwinskia humboldtiana* (coyotillo) causes muscular and myocardial necrosis in small ruminants (see Muscles and Tendons, Volume 1, Chapter 2). *Vicia villosa* (hairy vetch) poisoning involves many organs, including the heart. The lesions are unusual for a poisoning in that there is a substantial granulomatous inflammatory response associated with the necrosis (see The Skin and Appendages, Volume 1, Chapter 5).

Ischemic myocardial necrosis occurs in thrombotic disease such as disseminated intravascular coagulation, in the microangiopathy of vitamin E/selenium deficiency, in inflammatory vascular disease such as periarteritis nodosa, and at the capillary level in acute brain injury.

Coronary embolism is a relatively common cause of focal myocardial necrosis, the emboli originating from vegetations of left-sided endocarditis; septic emboli may result in myocardial abscesses. In granulocytopenic diseases, bacterial embolism may result in multiple minute myocardial infarcts, the best examples being seen in bracken fern (*Pteridium aquilinum*) poisoning in cattle. Myocardial infarction may occur in polyarteritis (periarteritis nodosa and allied lesions).

Arteriosclerosis, although common in animals, is very seldom severe enough to be responsible for ischemia of the myocardium. In aged dogs, some degree of hyalinosis of the intramural coronary arteries is frequently seen (Fig. 1.30) and may be associated with, and presumably the cause of, myocardial infarcts (see diseases of arteries in Section III of The Vascular System). Hearts that are dilated or hypertrophic are extraordinarily susceptible to patchy necrosis of apparently random distribution, although larger lesions are more obvious in left ventricle

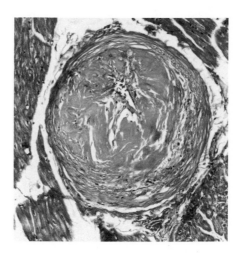

Fig. 1.30 Marked hyalinosis of an intramural coronary artery.

and papillary muscles; it is postulated that shearing forces that develop between adaptable myofibers and inelastic stroma compromise the capillary circulation.

The gross and microscopic appearance of myocardial necrosis is dependent on the interval between the initial insult and death. Gross changes of necrosis are not readily apparent until 12 hr after injury. By 18–24 hr, the affected area, in which calcium salts may be deposited, is paler and gray-brown. The overlying serous membrane is usually normal. Myocardial degeneration, mineralization, and necrosis are most common in the subendocardial myocardium of the left side. This area undergoes the greatest intramyocardial tension during systole and is the most likely to be ischemic. Proliferative or thrombotic lesions completely occluding vessels in these areas are rarely observed even with careful technique. The common small foci of myocardial necrosis may not be visible from the serous surface, and perhaps may be recognizable only microscopically. The necrosis resulting from infarction will, if sufficiently large, involve the serosal surfaces and produce a fibrinohemorrhagic thickening of the epicardium that may neatly overlie and indicate the infarcted area. In animals dying within 24 hr, the epicardial reaction may be the only sign of infarction, the necrotic muscle at this stage not being clearly altered or merely pale. The necrotic area becomes more sharply defined by hyperemia by 2–4 days. By the tenth day, there is beginning replacement of the necrotic zone by fibrous ingrowth. Replacement fibrosis is well established by the end of the sixth week (Fig. 1.31). Loss of muscle substance and replacement by fibrous tissue may lead to the development of aneurysms of the ventricular wall.

Microscopically, lesions are not detectable for the first 6–12 hr. However, ultrastructural changes are observed within 1 hr. Recognition as early as 2 hr may be possible using special stains such as hematoxylin–basic fuchsin–picric acid. Necrosis becomes apparent after 12 hr. Subsequently, neutrophils infiltrate the affected area, and the nuclei of the myocytes become pyknotic. At this

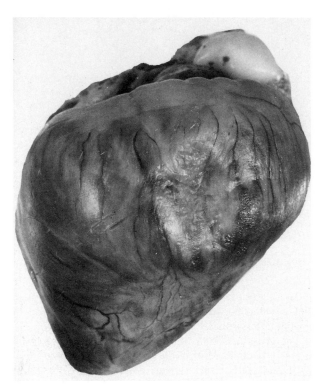

Fig. 1.31 Healed, scarified infarct in wall of right ventricle. Dog.

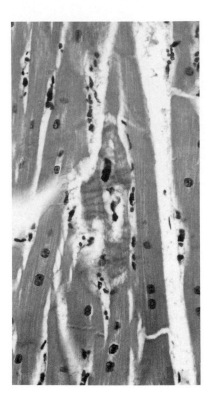

Fig. 1.32 Hypercontracted sarcomeres. Porcine stress syndrome. (Slide courtesy of A. L. Metz.)

stage, macrophages are evident. Granulation tissue appears at the periphery of the lesion toward the end of the first week and usually predominates by the end of the sixth week.

There are variations in the microscopic appearance of myocyte necrosis, and some attempt has been made to associate the differing microscopic patterns with particular causes, but at this stage, few useful conclusions can be drawn. The first variant is **coagulative necrosis,** characterized by cellular swelling, nuclear hyperchromasia, early loss of striations and a granular appearance of the myocyte. This type of change is observed primarily in ischemic states. **Coagulative myocytolysis** is typified by the presence of thick, irregular, eosinophilic bands, with an accompanying loss of striations and slightly staining granular areas in myocytes. The bands are hypercontracted sarcomeres (contraction bands; Fig. 1.32); the light granular areas are mitochondria. Evidence from experimental models suggests that the alteration is the result of the excessive release of endogenous catecholamines. However, this change may be observed in normal myocardium if it is fixed immediately after death. **Colliquative myocytolysis** appears as a loss of striations, and a homogeneous, weakly eosinophilic cytoplasm. The ultrastructural appearance is characterized by intracellular edema, with disintegration of fibrils and other organelles. Waviness of myocardial fibers is claimed by some to be one of the earliest changes observed in myocardial infarction in humans. The affected fibers are not necrotic, but are thinner and undulating

in appearance. It is thought that the waviness is due to irreversible stretching followed by compression of ischemic fibers by the surrounding viable myocardium. The significance of waviness of fibers and, indeed, the morphologic variations of necrosis has yet to be assessed in the domestic species.

Anichkov cells, also known as caterpillar cells, have been observed in myocarditis and a variety of naturally occurring and experimentally induced myocardial necroses. They are also seen in rheumatic fever in humans in association with Aschoff bodies. Anichkov cells appear as large mononuclear cells in which the nuclear chromatin is present in an undulating wavy ribbon with slender processes radiating from it (Fig. 1.33). The origin of these cells is disputed. Suggestions include a fibroblastic, pericytic, endothelial, or myocytic origin. Depending on the theory accepted, the function is that of either a macrophage or an abortive attempt at myocyte regeneration.

1. Fluoroacetate Poisoning

Sodium fluoroacetate (compound 1080) is used extensively in parts of the world as a rodenticide. It is also highly toxic for most domestic mammals. The median lethal dose for most species is about 0.25–1.00 mg per kilogram of body weight. Fluoroacetate is not directly toxic, but becomes so when converted to fluoroacetyl-coenzyme A. This compound is then combined with oxaloacetic acid to form fluorocitrate. The citric acid cycle enzymes *cis*-aconitase and succinic dehydrogenase are

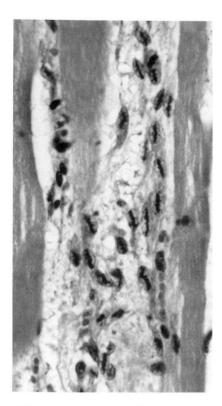

Fig. 1.33 Anichkov cells in interstitium of sheep heart with multifocal myocardial degeneration.

inhibited by fluorocitrate, effectively inhibiting the production of adequate amounts of adenosine triphosphate. Accidental poisoning by fluoroacetate used in rodent control occurs either by direct ingestion of the poison or indirectly, as when dogs eat poisoned rabbits. Fluoroacetate of **plant origin** is responsible for sometimes devastatingly large episodes of poisoning of ruminants. Horses apparently are not affected, perhaps because they do not eat the responsible plants in adequate amounts and perhaps because, like humans, birds, and some native animals, they are considerably more resistant to the poison than are most other species.

Plants known to accumulate fluoroacetate in lethal quantities for ruminants are *Gastrolobium* spp., *Oxylobium* spp., and *Acacia georginae* in Australia, and *Dichapetalum cymosum* in South Africa. Whether fluoroacetate has any role in the physiology of the plants is not known but, in the case of *A. georginae*, there are differences in the development of toxicity in the species, and in *D. cymosum* there are seasonal differences, toxicity being greater in spring and autumn, and especially in young leaves. Young leaves, including sucker growth, and pods and seeds are more toxic than the mature leaves of *A. georginae*.

The syndromes of intoxication vary with the species affected but are chiefly neurologic, as in dogs, which become extremely excited and convulsive, or chiefly cardiac, as in ruminants. Sheep may collapse suddenly and

die within a few minutes, or those less severely affected may develop cardiorespiratory distress and weakness if driven, and shortly die. Some showing these signs, if left undisturbed, may recover to appear normal until again forced to exercise. The syndrome in cattle is the same as that in sheep, death occurring with cardiac failure, fibrillation, cyanosis, dyspnea, and terminal convulsions, and may be precipitated by exercise, excitement, or a large drink of water.

The postmortem findings in ruminants are referable to myocardial injury, which may or may not be conspicuous, depending on the size of the dose and the opportunity for repeated episodes of poisoning. The concentration of fluoroacetate in *D. cymosum* is large, and histologic evidence of acute myocardial injury is seen. In *A. georginae* poisoning, both acute and chronic myocardial changes may be seen. There is some flabbiness of the myocardium, with prominent hemorrhages beneath the cardiac serosae. In acute cases, there are irregular areas of myocardial pallor and mottling sometimes associated with older scars. There is myolysis or hyaline degeneration in the myocardial fibers in multiple foci in the heart, with an intense mononuclear cell response. Loss of sarcoplasm leaves an open meshwork of reticulum and vessels, which eventually condenses and scars.

2. Gousiekte

Gousiekte is a plant poisoning of small ruminants in southern Africa and is of great economic importance in that region. The disease is characterized by acute heart failure 5–8 weeks after the ingestion of certain rubiaceous plants. In natural cases, clinical signs are seldom seen, and animals are usually found dead; the term gousiekte is Afrikaans for quick disease. A number of rubiaceous plants can cause the disease, including *Pachystigma pygmaeum*, *Pachystigma thamnus*, *Fadogia homblei*, *Pavetta harborii*, *Pavetta schumaniana*, and *Pachystigma latifolium*. The toxic principle has not been isolated.

Gross pathologic changes include generalized congestion, ascites, hydropericardium, hydrothorax, and pulmonary edema. Ventricular dilation is an inconstant feature; however, the ventricular walls are thinner and have a tough consistency. In a small proportion of cases, the heart is macroscopically normal. Microscopically there is focal to extensive myofiber loss with replacement fibrosis. The surviving myofibrils may be atrophied and lymphocytic infiltrates of varying intensity may be present.

3. Galenia africana

This bushy plant is toxic to sheep and goats in South Africa. The toxin has not been identified. Its action appears to be cumulative and to result in right-sided failure with congestion of the liver, hepatic fibrosis of periacinar distribution, and ascites. Microscopic examination of myocardium reveals patchy degeneration and necrosis of myofibers and light fibrosis.

4. Myocardial Necrosis Secondary To Neural Injury

The effect of central nervous injury on the myocardium is mediated by the autonomic nervous system, and in particular the sympathetic nervous system. The heart is richly innervated with autonomic fibers, and the myocardium, particularly of the ventricles, has a high concentration of β receptors. Both heart rate and inotropic state are stimulated by catecholamines. In a number of species, the continuous administration of norepinephrine for 1 or 2 weeks results in focal myocardial necrosis. Multifocal myocardial necrosis may develop in dogs with paroxysmal tachycardia due to functional tumors of the adrenal medulla. Also, experimentally induced intracranial hemorrhage produces focal myocardial necrosis, which can be prevented by prior administration of either reserpine, a catecholamine-depleting drug, or propranolol, which blocks β receptors. Focal myocardial necrosis has been demonstrated in dogs, in association with neurologic disease of diverse origin, from external trauma to infectious disease. The microscopic appearance of the lesions is dependent on the time of insult to the time of death. It varies from acute myocardial degeneration and necrosis to almost complete resolution by fibrosis. There are, no doubt, a number of cases with no gross or microscopic lesions where insufficient time elapsed between the onset of clinical signs and death. The findings of multifocal myocardial necrosis on postmortem in any species should always alert the pathologist to examine the central nervous system and adrenal glands.

5. Avocado Poisoning

Sheep and goats may die following the ingestion of fresh leaves from the avocado tree (*Persea americana*). Animals may die suddenly after ingesting comparatively few leaves, whereas others may show few ill effects. In those that die, lesions consist of endocardial hemorrhage, especially of the papillary muscles, and hydropericardium, ascites, hepatic degeneration, and pale kidneys. Cardiac lesions are those of myocyte necrosis and congestion and hemorrhage of variable severity.

Consumption of avocado leaves also results in a depression of milk production in lactating goats. The affected mammary glands are edematous and reddened with clots in the large ducts. The changes in the mammary gland are induced by the Guatemalan but not the Mexican variety of avocado.

6. Doxorubicin (Adriamycin) Cardiotoxicity

Doxorubicin, an antineoplastic agent of the anthracycline antibiotic group, is used in the treatment of lymphosarcoma in dogs; cardiotoxicity is the major factor limiting its use. Acute cardiotoxicity appears to be mediated by peroxidative injury. As well, binding of doxorubicin to nuclear and mitochondrial DNA causes blockage of DNA, RNA, and protein synthesis. Chronic doxorubicin cardiotoxicity, which may be due to decreased protein synthesis as well as altered divalent cation concentration, occurs in dogs given cumulative doses. Congestive heart failure occurs in such dogs. Microscopic myocardial changes consist of myocytic vacuolar degeneration (adria cells), myocytolysis, and myofibril atrophy.

7. Porcine Stress Syndrome

Unexpected death among market-weight pigs occurs particularly when they are subjected to stressful situations. Pietrain, Poland China, and some European strains of Landrace pigs appear to be most commonly affected. Porcine stress syndrome has alternatively been termed malignant hyperthermia, pale, soft, exudative (PSE) pork, back muscle necrosis, and transport death. It is of autosomal recessive character. The pathogenesis of the disease is centered on abnormal intracellular calcium homeostasis exacerbated by excessive catecholamine release. Pigs of the genotype that renders them susceptible to the development of malignant hyperthermia have significantly higher plasma levels of both noradrenaline and adrenaline following exposure to halothane. Ventricular arrhythmias are also more severe than those seen in controls. There is also evidence to show that the sarcoplasmic reticulum calcium release channel–ryanodine receptor protein is the seat of the problem. Sarcoplasmic reticulum from pigs homozygous for the halothane sensitivity gene releases calcium at twice the rate and has a higher binding affinity for ^{3}H ryanodine than does that of normal pigs.

The metabolic basis of malignant hyperthermia may result from a deficiency in inositol 1,4,5-triphosphate phosphatase (Ins $P_3$5-ase) which leads to high intracellular concentrations of inositol 1,4,5 triphosphate (Ins P_3) and of calcium ions. Inositol P_3 mobilizes calcium ions from nonmitochondrial intracellular stores and probably opens calcium channels in the sarcoplasmic reticulum and transverse tubules. As Ins P_3 is mainly metabolized by Ins $P_3$5-ase, a deficiency of this enzyme leads to excessive levels of calcium within the myocyte. Halothane inhibits Ins $P_3$5-ase, and in deficient swine, precipitates malignant hyperthermia.

Affected pigs exhibit muscular tremor, dyspnea, pyrexia, and cutaneous blanching and erythema. Death quickly ensues. Typical postmortem findings include the rapid development of rigor, pulmonary edema, hydropericardium, and splanchnic congestion. In some, there is epicardial and endocardial hemorrhage with irregular areas of pallor in the left ventricular myocardium. The presence or absence of myocardial necrosis in porcine stress syndrome is a major difference between the European and North American descriptions of the disease. Myocardial necrosis is a common finding in the European cases. When present, the necrotic areas are prominent in the inner third of the wall and papillary muscles. The microscopic appearance of the necrotic fibers is characterized by loss of the normal striation pattern, the presence of contraction bands, and fine granulation of the sarcoplasm (Figs. 1.32, 1.34).

The liver is congested, and the stomach may be congested or hemorrhagic with fundic ulceration. The skeletal muscles, especially the biceps femoris and longissimus dorsi, are gray to white and edematous, and the bundles are easily separated. For a complete description of the disease see diseases of muscle (in Muscles and Tendons, Volume 1, Chapter 2).

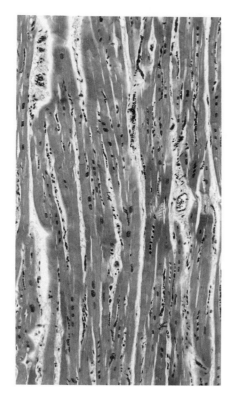

Fig. 1.34 Acute myodegeneration, karyomcgaly, and nuclear rowing. Porcine stress syndrome. (Slide courtesy of A. L. Metz.)

8. Mulberry Heart Disease Of Swine

The name mulberry heart is vaguely suggested by the extensive hemorrhages on the surface of the heart (Fig. 1.35A,B). The disease occurs in pigs only, chiefly those

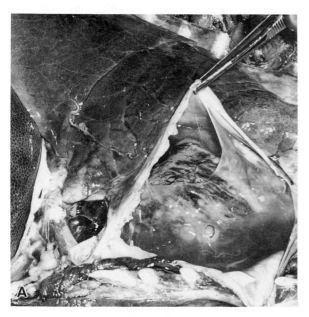

Fig. 1.35A Mulberry heart disease. Pig. Distended pericardial sac and congested lung. Pericardial sac is opened (forceps) to show epicardial hemorrhages.

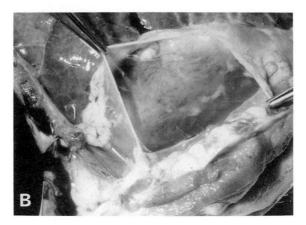

Fig. 1.35B Mulberry heart disease. Pig. Distended pericardial sac and congested lung. Pericardial sac is opened (forceps) to show fibrin clots.

2–4 months of age and in excellent condition, but it has been observed in animals from 3 weeks to 4 years of age. Among old pigs the incidence is sporadic, but in young pigs it occurs in short, snappy outbreaks. In most cases, typically affected animals are found dead. The circumstances of sudden death combined with the gross and microscopic lesions strongly suggest that the affected pigs die of acute congestive heart failure following the development of ventricular dysrhythmias.

The disease can be regularly reproduced by raising weanling piglets on a diet deficient in selenium and vitamin E. Such animals may develop disease within 2 weeks. As discussed in diseases of muscle (in Muscles and Tendons, Volume 1, Chapter 2), selenium is an integral part of the membrane enzyme glutathione peroxidase, which reduces toxic lipid peroxides to hydroxy acids. Vitamin E is an antioxidant and acts synergistically with selenium to protect membranes from high concentrations of lipoperoxides. Field studies, particularly from the Scandinavian countries, have shown that although dietary selenium supplementation markedly reduces the incidence of hepatosis dietetica and nutritional myopathy, the prevalence of mulberry heart disease remains the same. It is also evident that tissue levels of selenium, particularly those in the heart and liver, are within the normal range. Some workers consider that vitamin E deficiency plays a central role in the development of mulberry heart disease. This is not universally accepted, however, since some studies show low tissue levels of vitamin E in affected pigs; others do not. It has also been suggested that the disease is the result of altered vitamin E metabolism and not a consequence of inadequate dietary vitamin E levels. It may be that mulberry heart disease is precipitated following a lack of bioavailability of vitamin E in tissues, or possibly a greater requirement for vitamin E under certain dietary circumstances, such as diets containing large amounts of unsaturated fat.

Animals dead of mulberry heart disease are always in excellent body condition. There may be slight cyanosis of the ears and ventral abdomen, and exophthalmos, the latter due to orbital and palpebral edema. The intermuscular connective tissues are usually wet, and there may be obvious edema, especially in the axillary and inguinal regions and near the xiphoid process. The skeletal muscles are normal.

The principal lesions occur in the thorax. Some 50 ml or more of viscid, straw-colored fluid that clots on exposure to air is invariably present in the pleural cavity. The lungs are edematous and slightly or severely congested, but not severely enough to account for the degree of edema (Fig. 1.36). The pericardium is edematous and opaque and is always acutely distended with fluid transudate. Heavy strands or a lace of fibrin float in the fluid (Fig. 1.35A,B). The fibrin is not fixed to the serous surface and, from where it is in contact with the epicardium, it can easily be lifted off to reveal a clean, glistening, serous membrane. Hemorrhages, linear and ecchymotic, are present beneath the epicardium. They may be few or they may be extensive and involve the epicardium of all chambers, the myocardium, and beneath the endocardium of the papillary muscles and septum (Fig. 1.37). In some, there are multiple pale streaks and patches of necrosis on the epicardial surface that extend into the myocardium. There may be similar lesions on the endocardial surface.

The abdominal cavity contains a small volume of clear transudate and some fine unattached strands of fibrin on the intestine. The stomach usually contains a mass of dry feed, and its fundic mucosa is diffusely congested. The small intestine is empty and dry, its serosal vessels congested. The liver is congested, sometimes markedly so, with edema of the wall of the gallbladder. In some animals

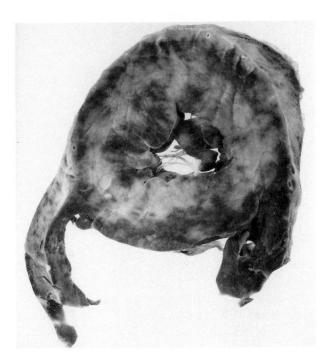

Fig. 1.37 Mulberry heart disease. Pig. Cross section of heart showing myocardial pallor and hemorrhage.

that survive 24 hr or more, there is bilaterally symmetric softening of the white matter forming the cores of the cerebral gyri. The softenings are visible grossly as gray, translucent, depressed areas studded with tiny hemorrhages that sometimes are confluent (Fig. 1.38).

Microscopically in acute deaths, degeneration of myofibers may be minimal or absent. In these cases, there is merely subserosal edema, with some plugging of lymphatics by fibrin, and interstitial hemorrhage. In others, which are probably less acute cases, there are substantial areas

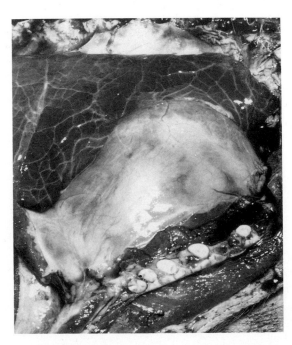

Fig. 1.36 Mulberry heart disease. Pig. Distended pericardial sac and congested lung.

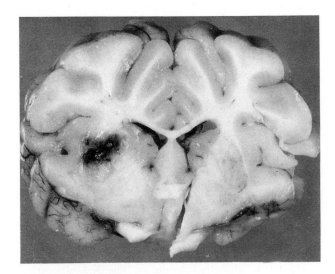

Fig. 1.38 Mulberry heart disease. Pig. Cross section of brain showing hemorrhage in region of internal capsule.

of myocardial necrosis. The extent of mineralization of the necrotic fibers varies, but never appears to be as severe as that seen in lambs or calves with nutritional myopathy.

The second prominent lesion is observed in arterioles of many organs. The change is seen in the heart, kidneys, liver, stomach, intestine, mesentery, skeletal muscle, and skin. There appears to be gradation in the severity of change, from endothelial swelling with increased permeability to necrosis of smooth muscle cells of the media. The basic appearance includes fibrinoid necrosis of the arteriolar walls, formation of hyaline thrombi, disruption of endothelium, and necrosis of smooth muscle cells. Periodic acid–Schiff (PAS) staining of affected arterioles emphasizes the fibrinoid change. The lungs show congestion and edema only. There is congestion and edema of the liver, associated in some cases with central necrosis in the lobules. Proteinaceous fluid produces focal detachments of the retina and intercapillary edema of the glomerular tufts.

The development of the cerebral lesions can be correlated roughly with the duration of the clinical illness. In the majority of cases, being those that die suddenly, there is no microscopic change in the brain, or merely some slight venous congestion and edema of the cortical white matter. In pigs that live for some hours, there is edematous separation of the fiber tracts of the white matter of the frontal cortex, with acute swelling and fragmentation of oligodendroglia. The overlying gray matter is edematous.

In animals that survive for 24 hr or more, it is usual to find severe lysis of the white matter in the cerebrum, which is more or less extensive but usually avoids the internal capsule, corpus medullare, and association fibers (Fig. 1.39). The cores of the gyri of the frontal lobes are most consistently and severely involved, but in some cases all portions of the cerebrum are involved, and in these, there may be similar lesions in the thalamus and brain stem. The vessels, venules especially, in the degenerate areas are severely injured and may be totally necrotic, although more frequently there is endothelial and adventitial swelling and proliferation. Hyaline thrombi are formed in many vessels, and droplets of similar character form in the perivascular spaces. The overlying gray matter is edematous, its vessels, hypertrophic and cuffed by adventitial cells and a few eosinophils. In a few cases, degenerate foci are also present in the cerebellum; these involve the molecular layer as foci of softening (Fig. 1.40) or narrow sheets of hemorrhage.

The postmortem lesions in mulberry heart disease depend on the length of time pigs survive after the initial damage occurs. Most pigs probably die from a ventricular arrhythmia. Those that die immediately no doubt have substantial areas of myofiber death, but there has been insufficient time for the morphologic manifestations of necrosis to appear. A period of 6–12 hr must elapse before myofiber necrosis can be reliably recognized. After that, it becomes increasingly evident and progresses through stages to replacement fibrosis of the necrotic areas.

In the experimentally induced disease, serum enzyme levels indicative of myocardial damage are raised for up to 21 days before death. The extent of cardiac damage is related to the length of time the serum enzymes are raised or, alternatively, to a large increase in a short period.

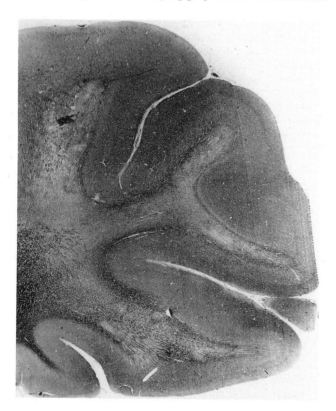

Fig. 1.39 Mulberry heart disease. Pig. Cerebrum. Liquefaction of white matter sparing corpus medullare and U fibers.

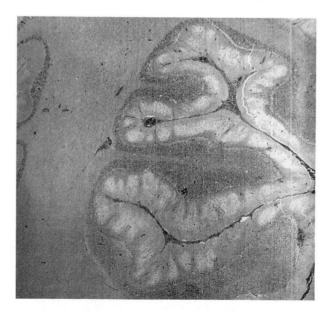

Fig. 1.40 Mulberry heart disease. Pig. Cerebellum. Numerous confluent areas of softening in molecular layer.

The microangiopathy that is sometimes prominent in this disease probably occurs independent of the myofiber necrosis. It is also nonspecific, as the same change is seen in many porcine viral infections and bacterial toxemias. It is, however, a useful indicator of the presence of mulberry-heart disease.

Bibliography

Bradley, R., and Dutfell, S. J. Sudden death and myocardial necrosis in cattle. *J Pathol* **135:** 19–38, 1981.

Combs, A. B., and Acosta, D. Toxic mechanisms of the heart. A review. *Toxicol Pathol* **18:** 583–596, 1990.

Craigmill, A. L. *et al.* Pathological changes in the mammary gland and biochemical changes in milk of the goat following oral dosing with leaf of the avocado (*Persea americana*). *Aust Vet J* **66:** 206–211, 1989.

Cranley, J. J., and McCullagh, K. O. Ischaemic myocardial fibrosis and aortic strongylosis in the horse. *Equine Vet J* **13:** 35–42, 1981.

Doroshow, J. H. Doxorubicin-induced cardiac toxicity. *N Engl J Med* **324:** 843–845, 1991.

Foster, P. S. *et al.* Inositol 1,4,5-triphosphate phosphatase deficiency and malignant hyperpyrexia in swine. *Lancet* **2(8655):** 124–127, 1989.

Fourie, N. *et al.* Clinical pathological changes in gousiekte, a plant-induced cardiotoxicosis of ruminants. *Onderstepoort J Vet Res* **56:** 73–80, 1989.

Gallant, E. M. *et al.* Halothane-sensitivity gene and muscle contractile properties in malignant hyperthermia. *Am J Physiol* **257:** C781–C786, 1989.

Ganote, C. E. Contraction band necrosis and irreversible myocardial injury. *J Mol Cell Cardiol* **15:** 67–73, 1983.

Grant, R. *et al.* Cardiomyopathies caused by januariebos (*Gnidia polycephala*) and avocado (*Persea americana*) leaves. *J South Afr Vet Assoc* **59:** 101, 1988.

Hensen, J. B. *et al.* Myodegeneration in cattle grazing *Cassia* species. *J Am Vet Med Assoc* **147:** 142–145, 1965.

Higgins, R. J. *et al.* Canine distemper virus-associated cardiac necrosis in the dog. *Vet Pathol* **18:** 472–486, 1981.

James, L. F. *et al.* Locoweed (*Oxytropis senicea*) poisoning and congestive heart failure in cattle. *J Am Vet Med Assoc* **189:** 1549–1556, 1986.

Johansson, G., and Jonsson, L. Myocardial cell damage in porcinc stress syndrome. *J Comp Pathol* **87:** 67–74, 1977.

Jönsson, L. *et al.* Cardiac manifestation and blood catecholamine levels during succinylcholine-induced stress of malignant hyperthermia sensitive pigs. *J Vet Med* (cf. vol I, II A), **36:** 772–782, 1989.

Karch, S. B., and Billingham, M. E. Myocardial contraction bands revisited. *Hum Pathol* **17:** 9–13, 1986.

King, J. M., Roth, L., and Haschek, W. M. Myocardial necrosis secondary to neural lesions in domestic animals. *J Am Vet Med Assoc* **180:** 144–148, 1982.

MacIntire, D. K., and Snider, T. G. Cardiac arrhythmias associated with multiple trauma in dogs. *J Am Vet Med Assoc* **184:** 541–545, 1984.

Marple, D. N., and Cassens, R. G. A mechanism for stress susceptibility in swine. *J Anim Sci* **37:** 546–550, 1973.

McCord, J. M. Oxygen-derived free radicals in postischemic tissue injury. *N Engl J Med* **312:** 159–163, 1985.

McEwan, T. Isolation and identification of the toxic principle of *Gastrolobium grandiflorum*. *Nature* **201:** 827, 1964.

McKenzie, R. A. *et al.* The toxicity for cattle of bufadienolide cardiac glycosides from *Bryophyllum tubiflourum* flowers. *Aust Vet J* **66:** 74–376, 1989.

Mickelson, J. R. *et al.* Effects of the halothane-sensitivity gene on sarcoplasmic reticulum function. *Am J Physiol* **257:** C787–C794, 1989.

Moir, D. C., and Masters, H. G. Hepatosis dietetica, nutritional myopathy, mulberry heart disease and associated hepatic selenium levels in pigs. *Aust Vet J* **55:** 360–364, 1979.

Neser, J. A. *et al.* Gossypol poisoning in pigs. *J South Afr Vet Assoc* **59:** 104, 1988.

Newsholme, S. J. Reaction patterns in myocardium in response to injury. *J South Afr Vet Assoc* **53:** 52–59, 1982.

Newsholme, S. J., and Coetzer, J. A. W. Myocardial pathology of domestic ruminants in Southern Africa. *J South Afr Vet Assoc* **55:** 89–96, 1984.

Nielsen, T.K. *et al.* Mulberry heart disease in young pigs without vitamin E and selenium deficiency. *Vet Rec* **124:** 535–537, 1989.

Olson, R. D., and Mushlin, P. S. Doxorubicin cardiotoxicity: Analysis of prevailing hypotheses. *FASEB J* **4:** 3076–3086, 1990.

Pang, V. F. *et al.* Myocardial and pancreatic lesions induced by T-2 toxin, a tricothecene mycotoxin, in swine. *Vet Pathol* **23:** 310–319, 1986.

Prozesky, L. *et al.* A field outbreak in Île-de-France sheep of a cardiotoxicosis caused by the plant *Pachystigma pygmaeum* (Schltr) Robynas (Rubiaceae). *Onderstepoort J Vet Res* **55:** 193–196, 1988.

Reef, V. B., Levitan, C. W., and Spencer, P. A. Factors affecting prognosis and conversion in equine atrial fibrillation. *J Vet Intern Med* **2:** 1–6, 1988.

Rice, D. A., and Kennedy, S. Vitamin E, selenium, and polyunsaturated fatty acid concentrations and glutathione peroxidase activity in tissues from pigs with dietetic microangiopathy (mulberry heart disease). *Am J Vet Res* **50:** 2101–2104, 1989.

Rogers, R. J., Gibson, J., and Reichmann, K. G. The toxicity of *Cassia occidentalis* for cattle. *Aust Vet J* **55:** 408–412, 1979.

Ross, R. D. *et al.* Nutritional myopathy in goats. *Aust Vet J* **66:** 361–363, 1989.

Sanford, S. E. *Streptococcus suis* infections in pigs. *Can Vet J* **28:** 207, 1987.

Sani, Y., Atwell, R. B., and Seawright, A. A. The cardiotoxicity of avocado leaves. *Aust Vet J* **68:** 150–151, 1991.

Schultz, R. A. *et al.* Observations on the clinical, cardiac and histopathological effects of fluoroacetate in sheep. *Onderstepoort J Vet Res* **49:** 237–245, 1982.

Simesen, M. G. *et al.* Clinicopathological findings in young pigs fed different levels of selenium, vitamin E and antioxidant. *Acta Vet Scand* **23:** 295–308, 1982.

Van Der Lugt, J. J., Fourie, N., and Schultz, R. A. *Galenia africana* L. (Aizoaceae) poisoning in sheep. *J South Afr Vet Assoc* **59:** 100, 1988.p

Van Vleet, J. F., and Ferrans, V. J. Myocardial diseases of animals. *Am J Pathol* **124:** 98–178, 1986.

Van Vleet, J. F., and Kennedy, S. Selenium vitamin E deficiency in swine. *Compend Contin Educ Pract Vet* **11:** 662–668, 1989.

Wrogemann, K., and Pena, S. D. J. Mitochondrial calcium overload: A general mechanism for cell necrosis in muscle diseases. *Lancet* **1:** 672–673, 1976.

C. The Cardiomyopathies

The term cardiomyopathy was coined for a group of myocardial diseases in humans which were at the time of unknown or obscure cause. It remains a diagnosis by exclusion. There are, in humans, several criteria that must be satisfied to establish the diagnosis. They are the absence of (1) significant coronary artery disease or anomaly, (2) vascular disease or anomaly, (3) systemic hypertension, past or present, and (4) shunts inside or outside the heart. Features usually present are (1) cardiomegaly, due to generalized dilation, (2) mural thrombosis, usually in the left ventricle, and (3) fibrosis or other lesions in the myocardium indicative of generalized myocardial involvement. Of these features described for humans, the following may be disregarded in domestic animals because of their comparative rarity: coronary arterial disease, systemic hypertension, and thrombosis in the left ventricle.

Since the original definition, the term cardiomyopathy has been used loosely to include varieties of idiopathic dilation or hypertrophy. There has also been some attempt to classify cardiomyopathies into primary (obscure etiology) and secondary (known systemic or etiologic abnormalities) types. Under such classifications, the list of diseases is seemingly endless. It is therefore preferable to restrict use of cardiomyopathy to idiopathic myocardial disease.

Progress has been made in the understanding of the cardiomyopathies, particularly with one of the forms of feline cardiomyopathy, which is almost certainly caused by a deficiency of the amino acid taurine. There is also preliminary evidence that one of the cardiomyopathies in dogs results from a deficiency of carnitine. Thus, these diseases should properly be no longer regarded as cardiomyopathies. Indeed in the case of taurine deficiency in the cat, an alternative term that has been suggested is taurine deficiency myocardial failure (TDMF). We shall continue to refer to them as cardiomyopathies, and discuss them in terms of primary diseases of the cardiac muscle.

Cardiomyopathies generally fall into three categories: (1) hypertrophic, (2) dilated (congestive), and (3) restrictive forms. As the clinical and pathologic patterns of the cardiomyopathies were first described in humans, a brief discussion of the essential features in this species is warranted.

Hypertrophic cardiomyopathy in humans is inherited as an autosomal dominant characteristic, within which there is a high prevalence of subclinical cases. There is asymmetric enlargement of the interventricular septum, which may lead to obstruction of left ventricular outflow. In most cases, a disordered orientation of myofibers in the ventricular septum is evident. There is usually no alteration in the contractile force of the myocardium. It has been postulated that the disorder is a consequence of abnormal myocardial catecholamine activity, probably beginning *in utero*.

Dilated (congestive) **cardiomyopathy** has no such unifying genetic or metabolic basis. It is characterized clinically by a lowered force of contraction and an increased ventricular end-diastolic volume. The final common pathway leads to congestive heart failure. The heart, at postmor-

tem, is dilated and hypertrophied. Evidence of necrosis or fibrosis is nonspecific. The cause is still obscure.

Restrictive cardiomyopathy is characterized by endomyocardial fibrosis with or without eosinophilia of the right and, less frequently, the left ventricle. This results in reduced ventricular compliance and impaired ventricular filling.

Cardiomyopathies occurring in domestic animals have been similarly classified and have been documented in dogs, cats, and to a lesser extent, cattle.

1. Canine Cardiomyopathies

The most commonly recognized cardiomyopathy is the **dilated or congestive form** in young to middle-aged giant and large-breed dogs, such as the St. Bernard, Irish wolfhound, Great Dane, and German shepherd. This type has the clinical features of a sudden onset of various degrees of left- and right-sided heart failure, which is often complicated by atrial fibrillation. There is cardiomegaly, with increased end-diastolic volume and poor contractile function. Soft systolic murmurs, indicative of mitral and tricuspid insufficiency, may be heard on auscultation. The prognosis is poor, with mean survival times of 6–12 months after the onset of the disease.

Postmortem findings in typically affected dogs are those of congestive heart failure. All chambers, particularly the left ventricle, are markedly dilated and may be hypertrophied. The atrioventricular rings are dilated (Fig. 1.41), the endocardium may be opaque due to subendocardial

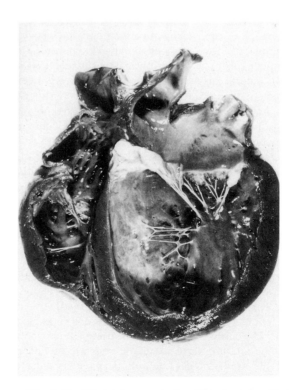

Fig. 1.41 Dilated (congestive) cardiomyopathy. Dog. Left ventricle is dilated, and endocardium is fibrosed.

fibrosis, and there may be atrial thrombosis. There may also be some evidence of infarction in distant organs. The finding, in some cases, of multifocal myocardial degeneration, necrosis, or fibrosis is expected, but unrewarding with regard to possible causes. Ultrastructural examination of myocardium from dogs with dilated cardiomyopathy shows increases in intermyofibrillar spaces, lipofuscin granules, fat droplets and myelin figures, mitochondrial hyperplasia, disruption of myofibrils, and Z-band thickening. These ultrastructural changes are not specific for cardiomyopathy and can be found with other causes of congestive heart failure; however, they are more severe in cardiomyopathy.

There are two further cardiomyopathies in dogs that are variations of those observed in large-breed dogs. The first is in the Doberman pinscher. The postmortem findings are those described for large-breed dogs, with the possible exception of the presence of scattered lymphocyte infiltration of the ventricles. Clinically they are different, in that the dogs exhibit various degrees of ventricular dysrhythmias. Atrial fibrillation is less common than it is in giant breeds. The second is seen in the English cocker spaniel. It appears to be of familial origin, with a high prevalence of subclinical disease. There is electrocardiographic evidence of left or biventricular hypertrophy, and some of these dogs are either found dead or develop acute left ventricular failure. The ventricular enlargement appears at an early age with normal ventricular contractility, which progresses to dilation and poor contractile function. Some of the older dogs have mild endocardiosis as an incidental finding.

Hypertrophic cardiomyopathy is much less common than the dilated form. Associated clinical syndromes include sudden death, death during anesthesia, and congestive heart failure. Disproportionate thickening of the ventricular septum may cause the ratio of the interventricular wall thickness to left ventricular free wall thickness to exceed 1.2 : 1, and there may be histologic evidence of myofiber disarray in the ventricular septum.

The causes and pathogenesis of canine cardiomyopathy are unknown, although as mentioned earlier, carnitine deficiency may play a role. They have, with the exception of the hypertrophic form, the common denominator of lowered contractility and ventricular dilation. There is a familial tendency in the Doberman pinscher and cocker spaniel.

Cardiomyopathy is consistently present in dogs of 3 months or older which are afflicted with inherited **muscular dystrophy** of Duchenne type (see Muscles and Tendons, Volume 1, Chapter 2). Cardiac lesions are relatively late in development and milder than the lesions in skeletal muscles. The cardiac lesions have a striking perivascular distribution and are most prominent beneath the epicardium of the left ventricular free wall, the papillary muscles of the left ventricle, and the right side of the ventricular septum. The initial changes appear to be necrosis and mineralization of muscle, its removal by macrophages and

giant cells, and its replacement by interconnecting bands of fibrous tissue.

2. Feline Cardiomyopathies

Feline cardiomyopathy is probably the most common and well-recognized cardiomyopathy in domestic animals. The condition was recognized as an entity only relatively recently, but a common consequence of cardiomyopathy, iliac thromboembolism, was described more than 50 years ago. There is a wide range in the age of onset of clinical signs, from 7 months to 24 years. Presenting clinical signs include lethargy, anorexia, dyspnea, tachypnea, and occasionally abdominal distension. Murmurs, most often associated with mitral or tricuspid insufficiency, gallop rhythms due to atrial contraction, and arrhythmias of various types are frequently encountered. Approximately one third of cases develop unilateral or bilateral thromboembolic hind-limb ischemia. The originating thrombus begins in the left atrium and is probably a result of turbulence and endocardial injury.

Cardiomyopathies in the cat are a diverse group of morphologic and functional disorders. Early reports (1970) of congestive heart failure in the cat established the existence of an endomyocarditis and an endomyocardial fibrosis accompanied by ventricular hypertrophy or dilation. The feline cardiomyopathies at that time encompassed most described types of myocardial abnormalities, with the hypertrophic and dilated forms predominating. Since that time, further myocardial diseases have been recognized. These include a restrictive cardiomyopathy, the presence of excessive left ventricular trabeculae septomarginalis (moderator bands), and the association of severe ventricular hypertrophy with hyperthyroidism. It is also apparent that these disorders reflect a spectrum of morphologic and functional change, with the presence of intermediate forms. The older classification systems are now becoming obsolete, particularly with the discovery that taurine deficiency accounts for most, if not all, cases of dilated cardiomyopathy in the cat.

Dilated cardiomyopathy is a primary systolic myocardial disorder associated with taurine deficiency. Decreased systolic function results in signs associated with decreased cardiac output, giving rise to bilateral congestive heart failure. End-diastolic ventricular volume and pressure are increased, as is atrial volume and pressure.

On necropsy, all chambers of the heart are enlarged, and the ventricles are dilated and flabby with thinned walls. The endocardium may be pale and slightly thickened by subendocardial fibrosis. The papillary muscles and trabeculae may be atrophied and appear small in relation to the ventricular volume. Often only minor lesions are found on microscopy. The myocardial fibers are moderately hypertrophied with areas of mild diffuse interstitial fibrosis. There may also be focal areas of myofiber loss with replacement fibrosis. In some cases there may be extensive areas of myocardial fibrosis. There is no significant disarray of myofibers. Ultrastructural examination contributes little to the understanding of the disease.

It is now recognized that taurine deficiency is a major contributing cause of dilated cardiomyopathy in the cat. Taurine (2-aminoethane sulfonic acid), an amino acid, was initially implicated when it was shown that a disproportionately high number of cats fed a diet containing marginal amounts of taurine had echocardiographic evidence of myocardial failure. Taurine supplementation restored myocardial function to normal. Moreover, low plasma taurine levels were present in all cats with dilated cardiomyopathy diagnosed at one institution. A substantial majority of these cats recovered, both clinically and electrocardiographically, after treatment with taurine.

A partial explanation for the taurine deficiency is that, when compared with other species, cats have a limited capacity to synthesize and conserve taurine. One of the best-defined functions of taurine is with the conjugation of bile acids and, in most species, bile acids are conjugated to taurine and glycine. If dietary taurine is limiting, then taurine is conserved by using glycine for conjugation. Cats not only have a limited capacity to synthesize taurine but also are unable to utilize glycine for bile acid conjugation.

Although taurine has been shown to be essential for adequate myocardial function, the metabolic basis for its effects are unknown. One suggestion, among others, is that taurine is involved in the modulation of tissue calcium influx through cardiac calcium channels. However, in experimental taurine deficiency, some cats develop disease and some do not. There may be other, as yet unidentified, factors in combination with taurine deficiency that allow the full spectrum of the cardiomyopathy to be expressed.

In contrast to the dilated form, **hypertrophic cardiomyopathy** is a diastolic disorder, resulting from an abnormally stiff ventricular myocardium (decreased compliance) that has normal or near-normal systolic myocardial function, but a diminished capacity to accept diastolic flow from the left atrium. Left ventricular volume is normal or decreased, and the papillary muscles are usually greatly enlarged in relation to the volume of the ventricle. The left atrium is usually dilated.

The pathologic features include an enlarged heart, with marked symmetric hypertrophy of the ventricles in the majority of cases, but especially the left ventricle. The left ventricular lumen is often slitlike. A minority of cases have asymmetric hypertrophy of the left ventricle, with the ventricular septum thicker than the left ventricular free wall. Findings of lesser frequency include atrial thrombi (Fig. 1.42), focal areas of ventricular fibrosis, and fibrosis of the endocardium in the left ventricular outflow tract. Microscopically, myofibers are hypertrophied with vesicular nuclei. Myofiber disarray, characterized by interweaving in the left ventricular free wall and ventricular septum, may be observed. This may also be seen in the right ventricle. Diffuse interstitial fibrosis is also a consistent feature, especially in the inner two thirds of the left ventricular free wall. Scattered focal areas of dense replacement fibrosis are seen in a minority of cats. Focal endocardial thickening by fibrosis occurs commonly. Intramyocardial arteries may show medial hypertrophy, and

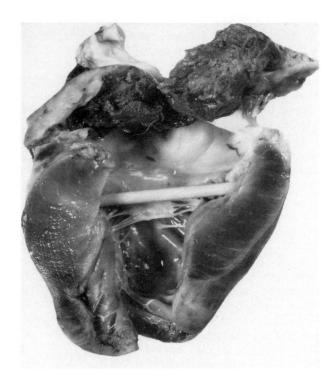

Fig. 1.42 Thrombus in left atrium. Cat. The thrombus is red, recent, and attached in auricle. Aortic embolism caused death.

scattered perivascular accumulations of lymphocytes and monocytes may be present.

Ultrastructurally, the hypertrophied myofibers have large pleomorphic nuclei with undulating surfaces and prominent nucleoli. Dense accumulations of Z-band material are also evident in myocytes, as are lipofuscin granules. Although these changes are seen in normal cat myocytes, the intensity of the changes is greater in cardiomyopathic cases. The cause of feline hypertrophic cardiomyopathy is unknown; it may be heritable in Persian cats.

A concentrically hypertrophied heart also occurs in cats with **hyperthyroidism,** and this should not be confused with hypertrophic cardiomyopathy. The thyroid glands in cases of hyperthyroidism are either unilaterally or bilaterally enlarged, nodular, pink to dark brown, and may contain cysts. Microscopically, they usually exhibit a mixture of hyperplastic areas, adenomatous nodules, and normal follicles (see The Endocrine Glands, Chapter 3 of this volume).

The hearts are symmetrically hypertrophied in most cases; however, some exhibit asymmetric hypertrophy. The left ventricular lumen is usually reduced in size. Affected myofibers are enlarged, but are not in disarray in the great majority of cases.

Excessive moderator bands in the left ventricle bridging the ventricular septum and free wall and entangling the papillary muscles are associated with heart failure in cats. [In the cat, the trabecula septomarginalis (moderator band) is normally found only in the right ventricle.] Although not strictly a cardiomyopathy, it is considered here, as the clinical syndrome occurs in mature cats. In all probability, it is a congenital defect that is manifest only later in life. Affected cats exhibit a range of signs that are mostly referable to left-sided heart failure. They include lethargy, dyspnea, anorexia, pale mucous membranes, and cardiac murmurs. Radiographic findings include cardiomegaly, hydrothorax, and pulmonary edema. Various conduction disturbances also occur.

Gross changes are dominated by the presence of numerous pink-white bands spanning the left ventricular lumen and papillary muscles. Most commonly, the bands connect the cranial and caudal papillary muscles to the ventricular septum. In others, the bands insert between the ventricular septum and the left ventricular free wall.

The left ventricle may be dilated or show increased wall thickness. In neither case is the heart weight significantly different from that of normal cats. The left atrium is enlarged, and pulmonary edema, pleural effusion, hepatic congestion, and aortic thromboembolism may be found. Microscopically, the abnormal moderator bands consist of cardiac conducting fibers and dense, mature collagen covered by endothelium. Loose, fibrous connective tissue is present between the endothelium and the dense collagen, with the variation in size of the moderator bands the result of more or less mature collagen.

Ventricular alterations include myocyte atrophy or hypertrophy, focal fibrosis, and intramural coronary arterial luminal narrowing, intimal thickening, medial hyperplasia, and perivascular fibrosis.

Restrictive cardiomyopathy, endomyocardial fibrosis, is a rare form of cardiomyopathy and is characterized by impaired diastolic ventricular filling because of severe endomyocardial fibrosis. Systolic function is usually normal. The disease occurs predominantly in older male cats, with clinical signs of left-sided or bilateral heart failure. Cardiac murmurs and dysrhythmias are common. Pathologic features include severe endocardial thickening with mural thrombosis. Left atrial enlargement is marked, the result of the restriction of ventricular filling. The left ventricle is thickened, and the lumen volume may be significantly diminished. Microscopically the myocardium and endocardium are replaced by granulation tissue of varying age.

This form of cardiomyopathy is the least well described and, in contrast to the dilated and hypertrophic forms, the criteria for diagnosis are not defined.

3. Bovine Cardiomyopathy

Four syndromes of cardiomyopathy are recorded in cattle. A cardiomyopathy occurs in **Holsteins** in Canada, Australia, and Japan and in Simmental–Red Holstein crossbreds in Switzerland. The unification of each report comes with the knowledge that many of the cases emanated from common sires; an autosomal recessive mode

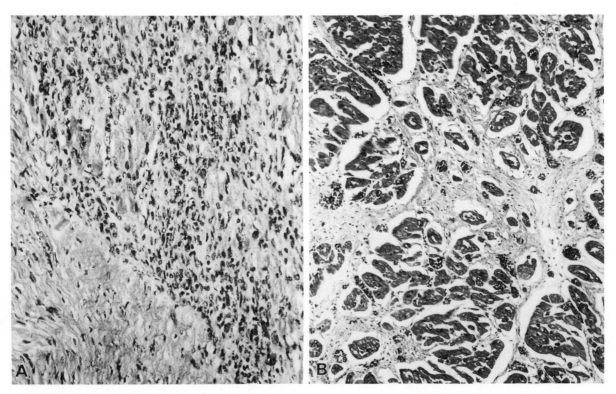

Fig. 1.43 (A) Degeneration and reparative fibrosis of myocardium. Curly-coated polled Hereford, 2 weeks old. (B) Interstitial fibrosis of myocardium. Curly-coated polled Hereford, 1 week old.

of inheritance has been suggested. The cardiomyopathy occurs in young adult to mature animals, with the clinical signs particularly referable to right-sided heart failure. There is no indication of the presence of electrocardiographic abnormalities. Evidence of right-sided heart failure includes dependent subcutaneous edema, hydrothorax, ascites, and severe hepatic congestion. The heart is enlarged with dilation of both ventricles. Microscopically there is a mixture of atrophied and hypertrophied myofibers with vacuolation, and there are islands of interstitial and replacement fibrosis, fatty replacement, and ongoing myofiber necrosis.

A cardiomyopathy associated with a dense curly hair coat in **polled Herefords** has also been described. Affected animals are younger than 6 months of age and are usually found dead. In some cases detected before death, severe ventricular dysrhythmias are evident. Some calves die at 1 week of age, with severe lesions in the ventricles consisting of necrosis, mineralization, and fibrosis of the myocardium (Fig. 1.43A,B). The cardiomyopathy is inherited as an autosomal recessive trait.

Bovine generalized glycogenosis type II (Pompe's disease), is not a primary cardiomyopathy, but occasional cases show clinical signs of left-sided heart failure, and in these the heart is eccentrically hypertrophied. Most affected animals have generalized muscular weakness.

Cardiomyopathy in Japanese Black calves is associated with sudden death or death after a short period of severe dyspnea. Calves usually die at younger than 30 days, although one died at 120 days. At necropsy marked cardiac enlargement with left ventricular dilation, hydropericardium, hydrothorax, ascites, pulmonary edema, and congestion of the liver and spleen are found. Extensive myocardial degeneration and necrosis with fibrosis is present in the left ventricle, particularly affecting the papillary muscles. The right ventricle is similarly but less often involved. The cardiomyopathy is considered to be inherited as an autosomal recessive trait. The clinical and pathologic pattern is reminiscent of that seen in polled Hereford calves with cardiomyopathy.

Bibliography

Abelmann, W. H., and Lorell, B. H. The challenge of cardiomyopathy. *J Am Coll Cardiol* **13:** 1219–1239, 1989.

Bishop, L. Ultrastructural investigations of cardiomyopathy in the dog. *J Comp Pathol* **96:** 685–698, 1986.

Bond, B. R., and Fox, P. R. Advances in feline cardiomyopathy. *Vet Clin North Am: Small Anim Pract* **14:** 1021–1038, 1984.

Bond, B. R. *et al.* Echocardiographic findings in 103 cats with hyperthyroidism. *J Am Vet Med Assoc* **192:** 1546–1549, 1988.

Bradley, R. *et al.* Cardiomyopathy in adult Holstein–Friesian cattle in Britain. *J Comp Pathol* **104:** 100–112, 1991.

Calvert, C. A. Dilated congestive cardiomyopathy in Doberman pinschers. *Compend Contin Educ Pract Vet* **8:** 417–430, 1986.

Gravanis, M. B., and Ansari, A. A. Idiopathic cardiomyopathies. A review of pathologic studies and mechanisms of pathogenesis. *Arch Pathol Lab Med* **111:** 915–929, 1987.

Harpster, N. K. Boxer cardiomyopathy. In "Current Veterinary Therapy VIII. Small Animal Practice." R. W. Kirk (ed.), pp. 329–337. Philadelphia, Pennsylvania, W.B. Saunders, 1983.

Hazlett, M. J. *et al.* A retrospective study of heart disease in Doberman pinscher dogs. *Can Vet J* **24:** 205–210, 1983.

Liu, S. K., and Maron, B. J. Comparison of hypertrophic cardiomyopathy in human, cat and dog. (Abstr.) *Lab Invest* **62:** 58A, 1990.

Liu, S. K., Maron, B. J., and Tilley, L. P. Hypertrophic cardiomyopathy in the dog. *Am J Pathol* **94:** 497–508, 1979.

Liu, S. K., Maron, B. J., and Tilley, L. P. Feline hypertrophic cardiomyopathy. Gross anatomic and quantitative histologic features. *Am J Pathol* **102:** 388–395, 1981.

Liu, S. K., Fox, P. R., and Tilley, L. P. Excessive moderator bands in the left ventricle of 21 cats. *J Am Vet Med Assoc* **180:** 1215–1219, 1982.

Martig, J., and Tschudi, P. Further cases of cardiomyopathy in cattle. *Deutsch Tierärztl Wochenschr* **92:** 363–366, 1985.

McLennan, M. W., and Kelly, W. R. Dilated (congestive) cardiomyopathy in the Friesian heifer. *Aust Vet J* **67:** 75–76, 1990.

Morris, J. G., Rogers, Q. R., and Pacioretty, L. M. Taurine: An essential nutrient for cats. *J Small Anim Pract* **31:** 502–509, 1990.

Pion, P. D. *et al.* Myocardial failure in cats associated with low plasma taurine: A reversible cardiomyopathy. *Science* **237:** 764–768, 1987.

Pion, P. D., Kittleson, M. D., and Rogers, Q. R. Cardiomyopathy in the cat and its relation to taurine deficiency *In* "Current Veterinary Therapy X," R. W. Kirk, (ed.), pp. 251–262. Philadelphia, Pennsylvania, W.B. Saunders, 1989.

Rozengurt, N., and Hayward, A. H. S. Primary myocardial disease of cats in Britain: Pathological findings in twelve cases. *J Small Anim Pract* **25:** 617–626, 1984.

Sandusky, G. E., Jr, Capen, C. C., and Kerr, K. M. Histological and ultrastructural evaluation of cardiac lesions in idiopathic cardiomyopathy in dogs. *Can J Comp Med* **48:** 81–86, 1984.

Sauramura, A. *et al.* Taurine modulates ion influx through cardiac calcium channels. *Cell Calcium* **11:** 251–259, 1990.

Thomas, R. E. Congestive cardiac failure in young cocker spaniels (a form of cardiomyopathy?): Details of eight cases. *J Small Anim Pract* **28:** 265–279, 1987.

Tontis, A. *et al.* Pathology of bovine cardiomyopathy. *Schweiz Arch Tierheilkd* **132:** 105–116, 1990.

Valentine, B. A., Cummings, J. F., and Cooper, B. J. Development of Duchenne-type cardiomyopathy. Morphological studies in a canine model. *Am J Pathol* **135:** 671–678, 1989.

Van Vleet, J. F., Ferrans, V. J., and Weirich, W. E. Pathologic alterations in hypertrophic and congestive cardiomyopathy of cats. *Am J Vet Res* **41:** 2037–2048, 1980.

Van Vleet, J. F., Ferrans, V. J., and Weirich, W. E. Pathologic alterations in congestive cardiomyopathy of dogs. *Am J Vet Res* **42:** 416–424, 1981.

Watanabe, S. *et al.* Evidence for a new lethal gene causing cardiomyopathy in Japanese black calves. *J Hered* **70:** 255–258, 1979.

Whittington, R. J., and Cook, R. W. Cardiomyopathy and woolly haircoat syndrome of polled Hereford cattle: Electrocardiographic findings in affected and unaffected calves. *Aust Vet J* **65:** 341–344, 1988.

D. Hemorrhage of the Heart and Its Membranes

Petechial and larger hemorrhages beneath the epicardium are so common in horses that die naturally that they are almost to be regarded as normal. The incidence of hemorrhages in sheep and cattle is somewhat lower; they are seldom seen in dogs and cats. Subepicardial hemorrhages are common in instances of asphyxia or death in anoxia, and in many acute infectious fevers. Larger ecchymotic hemorrhages that may involve most of the epicardium occur in the hemorrhagic diatheses. Subendocardial hemorrhages have the same pathogenesis as those beneath the epicardium but are much less common. Ecchymotic hemorrhages beneath the endocardium of the left ventricle occur in instances of acute cerebral injury and are useful as a diagnostic indication of clostridial enterotoxemia in lambs and calves, being present in a considerable proportion of cases. Small interstitial capillary hemorrhages within the myocardium accompany those of the subserosa. Deep myocardial hemorrhages that do not extend to the serosa are commonly present in pigs that die of asphyxia: they are distributed in the wall of the right ventricle in the vicinity of the descending coronary grooves. A remarkable degree of myocardial hemorrhage occurs in mulberry heart disease of swine (see Section VII,C,8 of The Heart).

E. Myocarditis

Myocarditis is a common lesion found in a wide variety of systemic diseases, but it is rarely primary. It occurs by direct extension from inflammatory lesions of the endocardium and pericardium, and hematogenously in many infectious diseases; a number of specific parasitic infestations also produce myocarditis. The inflammation is often spotty and therefore often missed even when looked for.

Pyogenic bacteria may originate from any other suppurative focus in the body or, as in the case of *Listeria monocytogenes* and *Actinobacillus equuli,* produce foci of acute inflammation that may develop into abscesses with inevitable destruction of the involved tissue (Fig. 1.44). In these, it is often possible to demonstrate bacterial emboli. *Clostridium chauvoei* may produce myocarditis essentially similar in type to the lesion in skeletal muscle. Tuberculous myocarditis is rare and, when present, the reaction is typical of the species. Necrobacillary myocarditis occurs in cattle secondary to extensive necrobacillosis in other tissues; the reaction is typical for the organism. *Haemophilus somnus* often causes abscesses in the myocardium of feedlot cattle. Protozoan infections include *Trypanosoma cruzi* (Fig. 1.45), *Toxoplasma gondii, Neospora,* and the common *Sarcocystis.*

Apart from these, and perhaps some other forms of specific myocarditis, inflammatory changes of less specific type occur in many infectious diseases and are not detectable by the naked eye. The basic pattern is usually the same, the reaction being centered on the interstitial and perivascular connective tissues, with edema and infiltration of lymphocytes, plasma cells, macrophages, and some eosinophils. Neutrophil leukocytes are few. The changes in the myocardial fibers vary widely in extent and in the severity of degeneration. The cause may sometimes be apparent in the histologic nature of the lesion, as, for example, in malignant catarrhal fever of cattle, in which

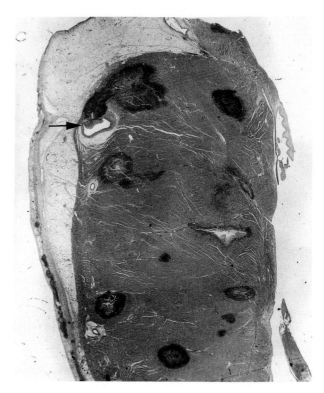

Fig. 1.44 Embolic suppurative myocarditis, *Actinomyces pyogenes.* Lamb. Multiple abscesses, one of which has encroached on and caused thrombosis in a coronary artery (arrow).

the inflammation is centered on the blood vessels, and in toxoplasmosis, in which the fiber degeneration predominates over the inflammatory reaction, and the organism is identifiable. In most cases of interstitial myocarditis, however, diagnosis of the cause can only be made by reference to the total pathologic picture. Only infrequently can death be attributed to the myocardial lesions, and then only when considerable degeneration of the myocardial fibers has occurred, as in foot-and-mouth disease of young animals.

An **eosinophilic myocarditis,** so called because the predominant infiltrative cell is the eosinophil, is observed occasionally in cattle, either incidentally or as a cause of sudden death. The focal lesions are yellowish green without distinct margins. It may occur alone or in association with similar lesions in the skeletal muscles. The cause is not known (see diseases of muscle, in Muscles and Tendons, Volume 1, Chapter 2) but in some cases an association with degenerating *Sarcocystis* is clear.

The myocarditis caused by *Trypanosoma cruzi* is described with The Hematopoietic System (Chapter 2 of this volume) and that due to canine parvovirus infection with Alimentary System (Volume 2, Chapter 1). Myocarditis is also seen in other systemic virus diseases such as distemper, canine herpes virus infection, and Aujeszky's disease in piglets and dogs.

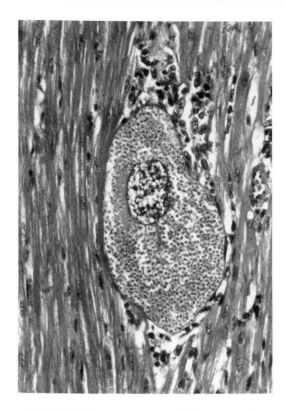

Fig. 1.45 *Trypanosoma cruzi.* Dog. Amastigote-containing pseudocyst in myocardium. Mild, mononuclear cell reaction.

1. Encephalomyocarditis Virus Infection

Encephalomyocarditis virus is a member of the genus *Cardiovirus* and infects humans and animals, principally swine and subhuman primates, although it has been associated with disease in captive elephants in Florida. Rodents in which it appears to be clinically inapparent may act as reservoir hosts, but infection in laboratory rodents commonly produces fatal encephalitis or myocarditis. The source of infection in outbreaks of myocarditis in swine is thought to be feed or water contaminated with virus from rats or other rodents. Ingestion of diseased rodent carcasses is also a likely source of infection. Experimental transmission of disease to swine has been accomplished by feeding the virus, but pigs kept in close contact did not develop disease.

The clinical disease in the peracute form in swine is characterized by sudden death, or death following a brief period of excitation and collapse. In less severe cases, there may be fever, anorexia, and progressive paralysis. The mortality rate is variable. On postmortem, there is hydrothorax, hydropericardium, ascites, pulmonary congestion, and edema. The heart in severe cases shows extreme pallor of the ventricles, with yellow to white foci 2–10 mm in diameter throughout the myocardium. In less severe cases, gross lesions are minimal or absent. Microscopically, the dominant lesions are focal to diffuse myocardial necrosis. Patchy mineralization of necrotic areas

is evident, accompanied by a mononuclear inflammatory reaction that varies greatly in intensity. Survival allows fibrous scarring to develop. There is little evidence of an encephalomyelitis. Encephalomyocarditis virus infection has also been associated with reproductive failure characterized by mummified fetuses and stillbirth. Fetuses that die toward the end of gestation have multifocal myocardial necrosis. Failure of conception and early embryonic deaths are also associated.

Experimentally infected swine may die between 2 and 11 days postinoculation. Virus is present in the feces for 7–9 days following oral administration and may be isolated from all organs, with the myocardium exhibiting the highest titer of virus. The virus titer falls rapidly and may not be found in cases that have been infected for 11 days or more. Experimental inoculation of mice regularly produces both encephalitis and myocarditis. Strains of the virus may be adapted to produce either encephalitis or myocarditis.

2. Parasitic Myocarditis

The parasites that tend to localize in the myocardium are the same as those that have an affinity for skeletal muscle. Parasites without such affinity may also be found in the myocardium in the course of their wanderings, especially if they are migrating in an abnormal host or in unusually large numbers in the normal host.

Of those parasites with an affinity for muscle, the ubiq-

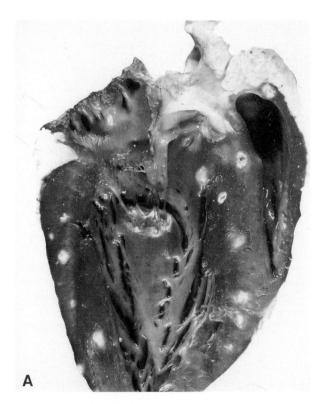

A

Fig. 1.46A Parasitic myocarditis. Lamb. Mineralized foci of degenerate *Cysticercus ovis.*

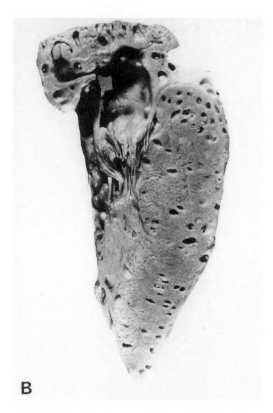

B

Fig. 1.46B *Cysticercus cellulosae.* Pig. Cysts in myocardium.

uitous sarcosporidia are the most common. The prevalence of myocardial sarcocysts is high in ruminants. It increases with age and implies previous acute sarcocystosis in which schizonts are formed in endothelial cells with little effect. The sarcocysts may be found in the cardiac conducting fibers as well as myocardial fibers and normally appear to be harmless. Degenerate cysts are occasionally seen in foci of acute nonsuppurative myocarditis. The cyst content is highly toxic, but whether degeneration or rupture of the cyst is primary or secondary to focal myocardial degeneration of other cause cannot be determined. Sarcosporidia are at least potential causes of focal myocarditis.

The various cysticerci of *Taenia ovis, T. saginata, T. solium,* and hydatid cysts are frequently found in the myocardium, which is one of their sites of predilection (Fig. 1.46A,B). These parasites are described with parasitic diseases of the intestine (see The Alimentary System, Volume 2, Chapter 1), which is the habitat of the definitive stage of each parasite.

Trichinella spiralis is described with diseases of muscle (see Muscles and Tendons, Volume 1, Chapter 2). The larvae of this parasite invade the myocardium but are seldom found there, because they either continue their migration or are destroyed. The myocardial reaction is a severe interstitial myocarditis, with basophilic degeneration and necrosis of fibers in the areas of inflammation.

Bibliography

Acland, H. M., and Littlejohns, I. R. Encephalomyocarditis virus infection of pigs. 1. An outbreak in New South Wales. *Aust Vet J* **51:** 409–415, 1975.

Acland, H. M., and Littlejohns, I. R. Encephalomyocarditis virus. *Aust Vet J* **51:** 416–422, 1975.

Appel, M. J. G. Lyme disease in dogs and cats. *Compend Contin Educ Pract Vet* **12:** 617–626, 1990.

Burke, M. Viral myocarditis. *Histopathology* **17:** 193–200, 1990.

Craighead, J. E. Pathogenicity of the M and E variants of the encephalomyocarditis (EMC) virus. I. Myocardiotropic and neurotropic properties. *Am J Pathol* **48:** 333–342, 1966.

Craighead, J. E. *et al.* Oral infection of swine with the encephalomyocarditis virus. *J Infect Dis* **112:** 205–212, 1963.

Deem, D. A., and Fregin, G. F. Atrial fibrillation in horses: A review of 106 clinical cases, and consideration of prevalence, clinical signs, and prognosis. *J Am Vet Med Assoc* **180:** 261–265, 1982.

Dubey, J. P. *et al.* Neosporosis in cats. *Vet Pathol* **27:** 335–339, 1990.

Gajadhar, A. A., Yates, W. D. G., and Allen, J. R. Association of eosinophilic myositis with an unusual species of *Sarcocystis* in a beef cow. *Can J Vet Res* **51:** 373–378, 1987.

Harris, F. W., and Janzen, E. D. The *Haemophilus somnus* disease complex (hemophilosis). *Can Vet J* **30:** 816–822, 1989.

Higgins, R. J. *et al.* Canine distemper virus-associated cardiac necrosis in the dog. *Vet Pathol* **18:** 472–486, 1981.

Johnson, R. H., and Spradbrow, P. B. Isolation from dogs with severe enteritis of a parvovirus related to feline panleucopenia virus. *Aust Vet J* **55:** 151, 1979.

Links, I. J. *et al.* An association between encephalomyocarditis virus infection and reproductive failure in pigs. *Aust Vet J* **63:** 150–152, 1986.

Love, R. J., and Grewal, A. S. Reproductive failure in pigs caused by encephalomyocarditis virus. *Aust Vet J* **63:** 128–129, 1986.

Meunier, P. C. *et al.* Experimental viral myocarditis: Parvoviral infection of neonatal pups. *Vet Pathol* **21:** 509–515, 1984.

Robinson, W. F., Huxtable, C. R., and Pass, D. A. Canine parvoviral myocarditis: A morphological description of the natural disease. *Vet Pathol* **17:** 282–293, 1980.

Woodruff, J. F. Viral myocarditis. *Am J Pathol* **101:** 428–465, 1980.

VIII. Diseases of the Conduction System

The clinical and electrocardiographic features of primary conduction system disturbances are well documented but comparatively rare. The clinical signs are primarily intermittent collapse, and sometimes sudden death. Electrocardiography may show a variety of abnormalities, such as sinoatrial arrest, second- or third-degree atrioventricular block, atrioventricular dissociation, left or right bundle branch block, Wolff–Parkinson–White syndrome (rarely), and ventricular tachycardia. Many if not most dysrhythmias are secondary to more or less extensive myocardial degeneration and necrosis, inflammation, or neoplasia. In these cases the conduction disturbances should be viewed as secondary phenomena.

Reports on the gross and microscopic examination of hearts with conduction system disturbances are remarkably sparse. This may be due in part to negative findings in some cases, or to time and effort required to adequately examine the conduction system, particularly the common bundle and the left and right bundle branches.

Sudden unexpected death or sudden episodes of viciousness have been associated with atrioventricular, or His, bundle degeneration, particularly in the Doberman pinscher. The predominant finding is focal areas of mineralized cartilage or osseous metaplasia in the central fibrous body in close proximity to conduction tissue. The His bundle is fibrotic, degenerate, and in some cases replaced by fat. In most cases, there is myointimal proliferation in this region, possibly leading to ischemia of the His bundle. A similar phenomenon has been observed in other breeds, including the German shepherd.

Hereditary stenosis of the His bundle has been observed in cases of syncope and sudden death in pug dogs. Electrocardiographically, such dogs have intermittent sinus arrest and paroxysmal second-degree atrioventricular block. The sinus node, the atrioventricular node, and the bundle branches are normal. The paroxysmal atrioventricular block could be related to the lesion in the His bundle, but the sinus pauses are probably the result of excessive parasympathetic activity. Affected dogs are abnormally sensitive to acetylcholine.

Deafness in the Dalmatian dog is associated with an abnormal blotchy coat color, and such animals may die suddenly. A proportion of affected dogs exhibit episodes of sinus pauses. In one the right atrium was fibrotic and atrophied. This was accompanied by marked narrowing of the coronary arteries, including those of the sinus node. These clinical and pathologic findings are similar to those seen in congenital deafness in humans.

Bibliography

Beckett, S. D., Branch, C. E., and Robertson, B. T. Syncopal attacks and sudden death in dogs: Mechanisms and etiologies. *J Am Anim Hosp Assoc* **14:** 378–386, 1978.

Brain, C. M., Kaneene, J. B., and Taylor, R. F. Sudden and unexpected death in horses and ponies: An analysis of 200 cases. *Equine Vet J* **20:** 99–103, 1988.

Castell, S. W., and Bailey, M. Dealing with sudden death in cattle. *Vet Med* **81:** 78–83, 1986.

Hamlin, R. L., Smetzer, D. L., and Breznock, E. M. Sinoatrial syncope in miniature schnauzers. *J Am Vet Med Assoc* **161:** 1022–1028, 1972.

James, T. N. Congenital deafness and cardiac arrhythmias. *Am J Cardiol* **19:** 627–643, 1967.

James, T. N., and Drake, E. H. Sudden death in Doberman pinschers. *Ann Intern Med* **68:** 821–829, 1968.

James, T. N. *et al.* Hereditary stenosis of the His bundle in pug dogs. *Circulation* **52:** 1152–1160, 1975.

Liu, S. K., Tilley, L. P., and Tashjian, R. J. Lesions of the conduction system in the cat with cardiomyopathy. *Recent Adv Cardiac Struct Metab* **10:** 681–693, 1975.

Meierhenry, E. F., and Liu, S. K. Atrioventricular bundle degeneration associated with sudden death in the dog. *J Am Vet Med Assoc* **172:** 1418–1422, 1978.

Robinson, W. F., Thompson, R. R., and Clark, W. T. Sinoatrial arrest associated with primary atrial myocarditis in a dog. *J Small Anim Pract* **22:** 99–107, 1981.

IX. Neoplasms of the Heart

Whereas primary neoplastic disease of the heart is rare, metastases in the myocardium from primary tumors of other organs are common. Two types of neoplasms, hemangiosarcoma and lymphoma, often metastasize to the heart. Primary tumors of the aortic body and of ectopic thyroid and parathyroid may arise in the base of the heart in dogs (Fig. 1.47) (see The Endocrine Glands, Chapter 3 of this volume).

Hemangiosarcoma may be a primary or secondary tumor in the myocardium, and develops most commonly in the right auricle (Fig. 1.48A,B). The tumor is described under Vascular Neoplasms, Section VII of The Vascular System.

Lymphoma involving the heart may be nodular or diffuse (Fig. 1.49A,B). The distinction is arbitrary because there is overlap of the two forms. In the nodular form, there are neoplastic foci, primarily in the atria (particularly the right atrium), which may be up to 5 cm or more in diameter. The nodules are covered by endothelium and are therefore smooth. When beneath the epicardium, they may be difficult to distinguish from epicardial fat. They are prone to central necrosis, and this resembles necrotic adipose tissue. On section, the wall of the atrium is more diffusely thickened and has a pale, fleshlike appearance. Milky fluid exudes from the cut surface if it is squeezed. Globular masses like those beneath the epicardium are also found under the endocardium and projecting into the cardiac cavities. In the diffuse or infiltrating form, the myocardium appears irregularly thickened and gray-

Fig. 1.47 Cardioaortic chemoreceptor tumor. Dog. Tumor is within pericardial sac in adventitia of pulmonary artery (arrow).

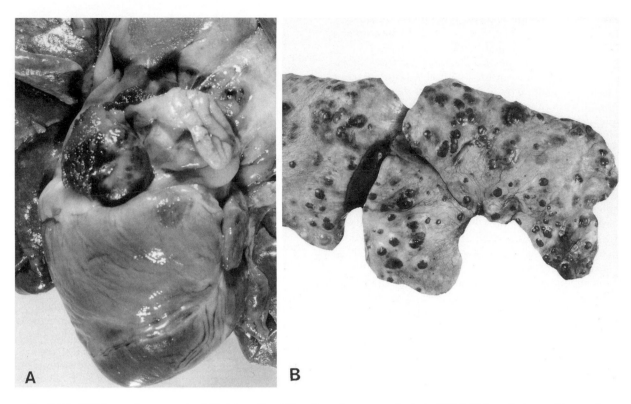

Fig. 1.48 (A) Hemangiosarcoma in right atrium. Dog. There is a pulmonary metastasis. (B) Multiple metastases of hemangiosarcoma to lung. Dog.

white, the ventricular walls being especially involved. These foci or areas of infiltration blend with normal myocardium and, not being discrete, are difficult to distinguish from foci of myocardial degeneration or myocarditis. Such lesions, however, especially if they cause some thickening of the myocardium, are to be first regarded as lymphoma and indicate the need for careful search for primary foci elsewhere. Lymphoid neoplasms are considered in detail with The Hematopoietic System (Chapter 2, Section VI,B of The Leukon, this volume).

There are two tumors of the heart named as if they were neoplasms, but that are developmental abnormalities. The so-called **myxomas** of **heart valves** (true myxo-

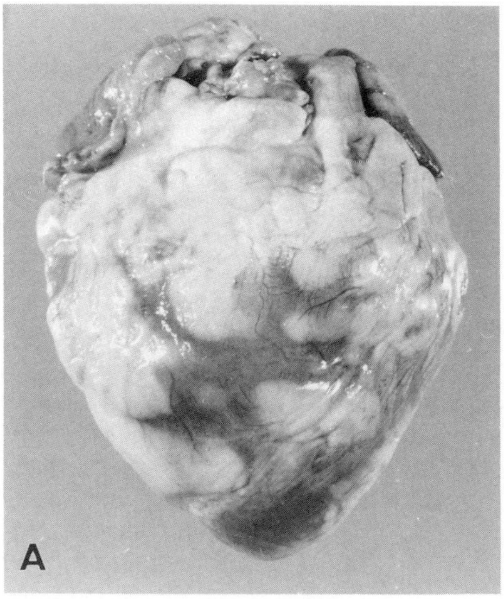

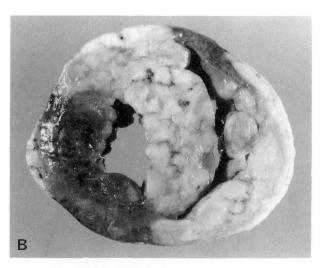

Fig. 1.49A Nodular and confluent lymphomatous infiltration of myocardium. Goat.

Fig. 1.49B Section through heart in Fig. 1.49A.

matous neoplasms do occur) represent a persistence of embryonic myxomatous tissue of which the endocardial cushions are composed. They consist of a loose stroma with few stellate cells, a scattering of multinucleate cells beneath the surface, and an intact endothelial covering. There may be some traumatic hemorrhage in the substance.

Rhabdomyomas are anomalous formations of myocardial fibers detectable as one or more rather discrete but nonencapsulated nodules. The lesion is rare in animals but has been observed in most species. Grossly, the nodules are gray and may be found anywhere in the heart. Those projecting into a chamber are very susceptible to hemorrhage and necrosis and may become cystic (Fig. 1.50). Histologically, they consist of an entanglement of myoblasts, many palisading as long, straplike cells, and others multinucleate with clear striations and myofibrils. Fibrosis can be extensive, and the connective tissues may be impregnated with iron and calcium salts.

Neurofibromas (schwannomas) are isolated or multiple benign neoplasms of peripheral nerves (Fig. 1.51). Cattle are most frequently affected, with the most common sites being peripheral nerves, brachial plexus, autonomic ganglia, intercostal nerves, and cardiac nerves. The disease is usually detected only at slaughter. When the heart is involved, the tumors are single or multiple, round or nodular masses, either on the epicardial surface or within the

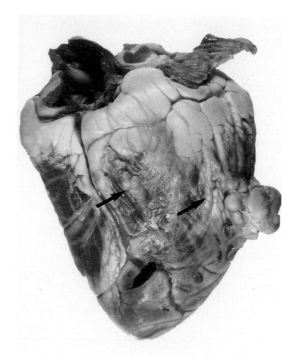

Fig. 1.51 Neurofibroma. Ox. There are corded (arrows) and nodular enlargements of epicardial nerves.

myocardium. The microscopic appearance is of interwoven bundles or whorls of elongated cells (Fig. 1.52). There may be palisading of nuclei and variable amounts

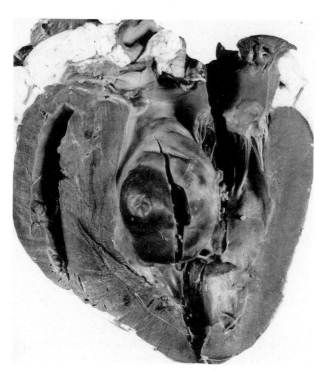

Fig. 1.50 Cystic degeneration of rhabdomyoma in left ventricle. Ox.

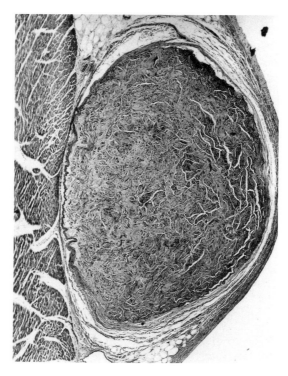

Fig. 1.52 Section through nerve in 1.51.

of collagen present. The principal cell is probably of Schwann or perineural origin.

A **granular cell tumor** or granular cell myoblastoma has been described in the right atrium of a dog. Granular cell tumors are rare in animals but occur in a variety of organs including the lung, brain, meninges, and tongue. The tumor has a distinctive light- and electron-microscopic appearance consisting of spindle-shaped to large polygonal cells containing prominent cytoplasmic granules within lysosomes. The cells are considered to be of neural origin as nerve-specific S100 protein is present within the cytoplasmic granules.

Bibliography

Bradley, R., Wells, G. A. H., and Arbuckle, J. B. R. Ovine and porcine so-called cardiac rhabdomyoma (hamartoma). *J Comp Pathol* **90:** 551–558, 1980.

Canfield, P. J., and Bennett, A. M. Cardiac myxoma in a steer. *Zentralbl Veterinaermed* (A) **26:** 464–467, 1979.

Ikede, B. O., Zubaidy, A., and Gill, C. W. Pericardial mesothelioma with cardiac tamponade in a dog. *Vet Pathol* **17:** 496–500, 1980.

Monlux, A. W., and Davis, C. L. Multiple Schwannomas of cattle (nerve sheath tumors, multiple neurilemmomas, neurofibromatosis). *Am J Vet Res* **14:** 499–509, 1953.

Sanford, S. E., Hoover, D. M., and Miller, R. B. Primary cardiac granular cell tumor in a dog. *Vet Pathol* **21:** 489–494, 1984.

Swartant, M. S. *et al.* Intracardiac tumors in two dogs. *J Am Anim Hosp Assoc* **23:** 533–538, 1987.

THE VASCULAR SYSTEM*

I. General Considerations

The vascular system may be arbitrarily divided into the arterial system, the microcirculation, the venous system, and the lymphatic system. Impaired function of the vascular system can be generalized, as occurs in shock, or may be more localized, as when decreased supply of oxygenated arterial blood to tissue causes ischemia, or impaired venous drainage causes congestion, or decreased lymphatic drainage of tissue fluids causes lymphedema.

The **arterial system** is subdivided into large elastic arteries, medium and small muscular arteries, and arterioles, with gradual transitions between the divisions. These vessels are characterized histologically by walls composed of three layers: the internal, middle, and external tunics, or tunica intima, tunica media, and tunica adventitia. The tunica intima consists of endothelium; subendothelial connective tissue, which contains collagen, elastin, proteoglycan (ground substance), fibroblasts, and smooth muscle cells; and the internal elastic lamina. The tunica media is the thickest layer of the vessel wall, and it consists of concentric layers of smooth muscle cells and elastic fibers. The tunica adventitia consists of the external elastic lamina and a feltwork of elastic and collagen fibers continuous with the surrounding connective tissue.

The thickness of the **tunica intima** decreases as the vessel size decreases, until the subendothelial layer eventually disappears. The internal elastic lamina is not readily distinguishable in elastic arteries because of its continuity with medial elastic tissue, but is very prominent in muscular arteries, and disappears at the level of the smaller arterioles.

The **tunica media** consists of fenestrated elastic laminae in elastic arteries, with smooth muscle cells lying between laminae. Intercellular ground substance is especially prominent in the tunica media of elastic arteries in the horse. The elastic laminae diminish, and smooth muscle predominates as vessels become more distant from the heart. The media of muscular arteries is up to 40 muscle cells thick, whereas the media of arterioles is one to three cells thick. Definitions of the upper limit of arteriolar diameter range from 300 μm if fixed by perfusion to 100 μm if allowed to contract, but there are no sharp distinctions between small arteries and large arterioles. The thickness of the media as seen in routine material is greatest in relation to the luminal diameter in small arteries and arterioles, and it is these vessels, particularly arterioles, which participate actively in blood pressure regulation and also bear the consequences of hypertension. Contraction of muscular arteries and arterioles on an animal's death forces blood from the lumina, and causes longitudinal folding, which appears as scalloping of the internal elastic lamina when seen in cross section.

The external elastic lamina of the **tunica adventitia** is best defined in the large elastic arteries, and becomes indistinct in muscular arteries. An interlacing network of collagen fibers in the adventitia limits expansion of the elastic arteries. Elastic fibers decrease in the adventitia with decreasing vessel size. Vasa vasorum and nervi vasorum supply the adventitia and the outer half of the media. The intima and inner half of the media are avascular and thus more subject to degenerative changes than is the outer media.

The **microcirculation** is the exchange system in which gases, nutrients, and waste products are transferred between the blood and the extravascular tissues. The microcirculation consists of vessels of less than 100 μm diameter: namely, arterioles, terminal arterioles (metarterioles, precapillary sphincter areas), capillaries, postcapillary venules, and venules. This classification overlaps microcirculation and macrocirculation (arteries and veins), but is useful conceptually, as will be seen in disorders such as disseminated intravascular coagulation, which affect primarily the microcirculation. Arterioles open through precapillary sphincters or through metarterioles into capillaries in most tissues, or into sinusoids in the liver. Capillaries are low-pressure tubules, of about 8 μm diameter, composed of endothelial cells surrounded by a basal lamina; the endothelial cells are connected by tight junctions, which allow rapid passage of small molecules such as glucose. The basal lamina is a molecular sieve which pre-

---

* This section was contributed by M. Grant Maxie, Ontario Ministry of Agriculture and Food, Canada.

vents passage of larger protein molecules. Capillary endothelial cells possess few endoplasmic organelles, but often contain pinocytotic vesicles. Capillaries may be continuous, the most common type; fenestrated (60- to 80-μm fenestrae closed by thin diaphragms), as in the gastrointestinal tract and endocrine glands; or porous (without diaphragms closing the pores) as in renal glomeruli. Fenestration of the endothelial cells allows increased permeability. During tissue healing, cords of endothelial cells grow into the tissues by mitosis, develop lumina, and become functional capillaries. Undifferentiated perivascular cells, or pericytes, are ensheathed by glycocalyx, which is continuous with the capillary basal lamina. Mitosis of pericytes is easily stimulated, and they may be transformed into fibroblasts or smooth muscle cells. Sinusoids are less uniform than capillaries and more permeable because there are large openings between the endothelial cells and a discontinuous basal lamina.

Veins are termed small, medium, or large or, alternatively, venules, collecting veins, and great veins; all have large lumina in relation to their wall thickness. Classification of veins by wall characteristics is difficult because the layers in their walls may be absent or difficult to distinguish. Venules of greater than 30 μm diameter have an incomplete muscular media and thin adventitia. Venular endothelium is normally more permeable than that of capillaries and is also more sensitive to vasoactive amines, the action of which can cause leakage. In the lymph node paracortex, postcapillary venules, which are nonmuscular venules with prominent endothelium, are important sites of lymphocytic traffic. The media is of increasing thickness in medium and large veins, and the internal elastic lamina becomes more prominent. However, the adventitia is the thickest layer in veins, whereas the media predominates in arteries. Backflow of blood in veins is prevented by the presence of semilunar valves, which are invaginations of the intima into the venous lumen. Venous valves are not present in venae cavae or hepatic portal vein. Paucity of valves in veins of the head and face may contribute to incidental venous congestion in these areas and to retrograde spread of infections.

Lymphatic capillaries originate in loose connective tissue, and consist of very permeable walls and endothelial cells which may lack, or have a discontinuous, basal lamina. In larger lymphatics, valves are present, the basal lamina is continuous, and walls consist of three ill-defined layers; an internal elastic lamina is usually absent. The thin-walled veins and lymphatics are more susceptible to compression and occlusion, are more often involved by inflammation in adjacent tissue, and are more liable to neoplastic invasion than are the thicker-walled arteries.

Endothelium and smooth muscle, the main components of vessel walls, play important roles in all types of vascular disease. Understanding of these seemingly simple cells and their functions is still incomplete, but a wealth of information about their important properties and reactive capabilities is accumulating. The endothelium is recognized as the largest paracrine organ in the body; it pro-

duces anticoagulants, procoagulants, fibrinolytic agents, prostanoids, connective tissue components, adenosine, and a host of other substances (Table 1.4). Vascular homeostasis is maintained by complex interactions among endothelial cells, leukocytes, and platelets, each of which produce vasodilators and vasoconstrictors and can interact to inhibit or promote thrombosis.

Vascular **endothelial cells,** as well as forming the thromboresistant monolayer lining of the vascular system, mechanically insulate the circulating blood from the highly thrombogenic subendothelial materials. The thromboresistance of the endothelium is normally maintained by a balance between antithrombotic and thrombotic factors; stimulation of excessive prothrombosis can lead to clotting and thrombosis, whereas excessive antithrombosis can lead to ineffective hemostasis and bleeding. Antithrombosis is normally accomplished through (1) binding and inhibition of thrombin via activation by thrombomodulin of protein C and protein S, via accentuation of antithrombin

TABLE 1.4

Endothelial Cell Properties and Functions[a]

Maintenance of permeability barrier
Elaboration of anticoagulant and antithrombotic molecules
   Prostacyclin
   Thrombomodulin
   Plasminogen activator
   Heparinlike molecules
   Adenosine
Elaboration of prothrombotic molecules
   Von Willebrand factor (factor VIIIa)
   Tissue factor
   Plasminogen activator inhibitor
   Platelet-activating factor
   Collagen
   Fibronectin
Modulation of blood flow and vascular reactivity
   Endothelium-derived relaxation factor, EDRF
      Nitric oxide
   Endothelium-derived hyperpolarizing factor
   Endothelium-derived contraction factors, EDCFs
      Endothelin
   Angiotensin-converting enzyme (AI → AII)
   Prostacyclin
   Peptidoleukotrienes
Regulation of inflammation and immunity
   Interleukin-1
   Adhesion molecules
Extracellular matrix production
Regulation of cell growth
Growth stimulators (PDGF, CSF, FGF)
Growth inhibitors (heparin, TGFβ)

[a] PDGF, platelet-derived growth factor; CSF, colony-stimulating factor; FGF, fibroblast growth factor; TGFβ, transforming growth factor-β. Modified, and reprinted with permission, from Cotran, R. S., Kumar, V., Robbins, S. L. "Robbins' Pathologic Basis of Disease," 4th Ed. Philadelphia, Pennsylvania, Saunders, 1989.

III activity by heparinlike molecules, and via the presence of α_2-macroglobulin; (2) inhibition of platelet aggregation through elaboration of prostacyclin (PGI$_2$), a potent inhibitor of platelet aggregation and a vasodilator, and by adenosine diphosphatase (ADPase)-mediated conversion of proaggregating ADP to adenine nucleotide platelet inhibitors; and (3) promotion of fibrinolysis by synthesis of tissue plasminogen activators. Procoagulant activities include (1) the presence of minute amounts of tissue thromboplastin in endothelial cells, further production of which can be stimulated by endotoxin and by cytokines such as interleukin-1 and tumor necrosis factor; (2) synthesis and secretion of von Willebrand factor, needed for adherence of platelets to subendothelial components, and secretion of platelet-activating factor, an activator and aggregator of platelets; and (3) inhibition of fibrinolysis through release of plasminogen activator inhibitors.

In addition to their role in hemostasis, endothelial cells participate in modulation of blood flow and vascular reactivity by production of (1) endothelium-derived relaxing factor (nitric oxide), (2) endothelin, a vasoconstrictor, (3) angiotensin converting enzyme, which converts angiotensin I to angiotensin II, and (4) prostacyclin. Endothelial cells modulate the actions of various vasoconstrictors (catecholamines, serotonin, arginine, vasopressin) and vasodilators (histamine, leukotrienes, adenine nucleotides), and in concert with the endothelium-derived factors can thereby allow blood vessels to rapidly adapt to changes in hemodynamic conditions.

Endothelial cells aid in the regulation of inflammation and immunity through production of interleukin-1 and various adhesion molecules, which moderate adhesion and hence emigration of leukocytes. Endothelial cells also aid in regulation of cell growth by production of growth stimulators (platelet-derived growth factor, colony-stimulating factor, and fibroblast growth factor), and production of growth inhibitors (heparin, transforming growth factor-β). Endothelial cells produce basement membrane, and are capable of contraction. The endothelium is a semipermeable membrane, which can also transport metabolites including proteins through the cytoplasm via pinocytotic vesicles. Weibel–Palade bodies, rod-shaped cytoplasmic organelles, serve as ultrastructural markers of endothelial cells, at least in primates and horses.

Subendothelial connective tissues consist of various types of collagen, elastin, glycosaminoglycans, fibronectin, laminin, and thrombospondin. Of these, fibrillar collagen is the most potent stimulus for platelet adhesion and activation; collagen also activates the intrinsic pathway of coagulation. Fibronectin, or molecular glue, normally functions to stabilize cell-to-cell and cell-to-substrate attachments, but also becomes cross-linked to fibrin and helps to anchor hemostatic plugs.

Mild endothelial injury, as occurs in the course of inflammation, may be apparent histologically as hypertrophy of endothelial cells (reactive cells). As well, increased vascular permeability may lead to insudation of plasma proteins, including fibrinogen, into the subendothelial area, and hence contribute to vascular hyalinosis. Since the intima and inner media depend on diffusion of metabolites from the vascular lumen for their nutrition, such subendothelial deposits may impair the viability of smooth muscle cells and contribute to degeneration, or fibrinoid necrosis, of vessel walls. Thus, the endothelial barrier must be preserved to maintain a normal environment for medial smooth muscle cells. Necrosis of endothelium, which may be caused by a wide variety of agents, leads to exposure of subendothelial collagen, a potent inducer of coagulation and platelet aggregation, and hence thrombosis. Endothelial cells at the edges of denuded areas will proliferate and migrate to repair the endothelial defect or to form an endothelial covering over (endothelialize) a mural thrombus. Severe microvascular damage, as occurs because of lipoperoxidation of endothelial membranes in vitamin E–selenium deficiency in pigs, can result in hemorrhage and organ failure.

Vascular smooth muscle cells are important both as effectors of vasoconstriction and dilation, and as cells capable of synthesizing basement membrane, collagen, elastin, and proteoglycans of the extracellular space. In response to growth factors derived from platelets and monocyte/macrophages, smooth muscle cells of the media migrate through fenestrae of the internal elastic lamina to the subendothelial area where, as myointimal cells, they proliferate and are responsible for organization or collagenization of deposits, including thrombi. Examples of endoarterial organization and arterio-occlusive disease are discussed later in this chapter. In common with other muscle cells, medial smooth muscle cells respond to an increased work load by hypertrophy. Medial hypertrophy is a prominent consequence, as well as a cause, of pulmonary hypertension. Sustained vasoconstriction, as is induced by ergot (*Claviceps purpurea*) or fescue grass (*Festuca arundinacea*) poisoning, can lead to gangrene of the extremities; fescue toxicosis in cattle (summer syndrome) is caused by infestation of the grass by the endophyte *Acremonium coenophialum,* which produces the ergopeptide alkaloids ergovaline, ergosine, and ergonine. Along with the other components of the vessel wall, smooth muscle cells may be involved in degenerative and inflammatory changes, as described subsequently, and contribute to fibrinoid necrosis of the wall.

Normal **aging changes** of vessels include intimal thickening due to proliferation of myointimal cells and accumulation of their products, deterioration of medial elastic fibers, medial fibrosis and loss of medial smooth muscle, and accumulation of ground substance in the media. These slowly progressive changes are usually of little consequence given the large functional reserve of the vascular system, but do lead to loss of resilience of the vessel walls and may cause vessel wall thinning, stretching, and hence increased tortuosity.

Bibliography

Albelda, S. M. Endothelial and epithelial cell adhesion molecules. *Am J Respir Cell Mol Biol* **4:** 195–203, 1991.

Banks W. J. Cardiovascular system. *In* "Applied Veterinary Histology," 2nd Ed. pp. 314–329. Baltimore, Maryland Williams & Wilkins, 1986.

Cotran, R. S., Kumar, V., and Robbins, S. L. "Robbins' Pathologic Basis of Disease," 4th Ed. Philadelphia, Pennsylvania, W.B. Saunders, 1989.

Dellman, H.-D., and Venable, J. H. Cardiovascular system. *In* "Textbook of Veterinary Histology," H.-D. Dellman and E. M. Brown (eds.), 3rd Ed., pp. 145–163. Philadelphia, Pennsylvania, Lea & Febiger, 1987.

Dinerman, J. L., and Mehta, J. L. Endothelial, platelet and leukocyte interactions in ischemic heart disease: Insights into potential mechanisms and their clinical relevance. *J Am Coll Cardiol* **16:** 207–222, 1990.

Fajardo, L. F. The complexity of endothelial cells. A review. *Am J Clin Pathol* **92:** 241–250, 1989.

Hintz, H. F. Ergotism. *Equine Pract* **10:** 6–7, 1988.

Lovenberg, W., and Miller, R. C. Endothelin: A review of its effects and possible mechanisms of action. *Neurochem Res* **15:** 407–417, 1990.

Marin, J., and Sanchez-Ferrer, C. F. Review: Role of endothelium-formed nitric oxide on vascular responses. *Gen Pharmacol* **21:** 575–587, 1990.

Martin, T., and Edwards, W. C. Protecting grazing livestock from tall fescue toxicity. *Vet Med* **81:** 1162–1168, 1986.

Moncada, S. Prostacyclin and arterial wall biology. *Arteriosclerosis* **2:** 193–207, 1982.

Mulvany, M. J., and Aalkjaer, C. Structure and function of small arteries. *Physiol Rev* **70:** 921–961, 1990 (477 refs).

Niederman, C. N. *et al.* Fescue toxicosis—the summer syndrome. *Agri-Pract* **8:** 11–15, 1987.

Peplow, P. V., and Mikhailidis, D. P. Platelet-activating factor (PAF) and its relation to prostaglandins, leukotrienes and other aspects of arachidonate metabolism. *Prostaglandins Leuko Essent Fatty Acids* **41:** 71–82, 1990.

Proctor, R. A. Fibronectin: A brief overview of its structure, function, and physiology. *Rev Infect Dis* **9:** S317–S321, 1987.

Ross, R. The pathogenesis of atherosclerosis—An update. *N Engl J Med* **314:** 488–500 1986.

Rubanyi, G. M., and Vanhoutte, P. M. (eds.). Endothelium-derived contracting factors. Basel, Switzerland, S Karger, 1990. (31 chapters)

Sanchez-Ferrer, C. F., and Marin, J. Endothelium-derived contractile factors. *Gen Pharmacol* **21:** 589–603, 1990.

Schwartz, S. M., Heimark, R. L., and Majesky, M. W. Developmental mechanisms underlying pathology of arteries. *Physiol Rev* **70:** 1177–1209, 1990. (385 refs.)

Silverstein, R. L., Leung, L. L. K., and Nachman, R. L. Thrombospondin: A versatile multifunctional glycoprotein. *Arteriosclerosis* **6:** 245–253, 1986.

II. Vasculitis

Vasculitis, which denotes inflammation of vessels, is a common occurrence in a wide variety of inflammatory diseases, and is of major importance in the pathogenesis of some conditions (Table 1.5). Whereas vascular degeneration occurs in a number of toxic and metabolic conditions, such as mercury poisoning and mulberry heart disease, these have been excluded from the classification because they lack a significant inflammatory cell compo-

TABLE 1.5

Causes of Vasculitis in Domestic Animals

Infectious
  Viral
    Equine viral arteritis
    Equine infectious anemia
    African horse sickness
    Malignant catarrhal fever
    Bovine ephemeral fever
    Bovine virus diarrhea
    Border disease
    Bluetongue
    Maedi–visna (ovine progressive pneumonia)
    Hog cholera
    African swine fever
    Equine viral rhinopneumonitis (encephalomyelitis)
    Feline infectious peritonitis
  Chlamydial
    Sporadic bovine encephalomyelitis
  Rickettsial
    Rickettsia rickettsii
    Erhlichia canis, E. equi
  Bacterial
    Salmonella spp.
    Erysipelothrix rhusiopathiae
    Haemophilus somnus
    Haemophilus suis
    Haemophilus parasuis
    Actinobacillus pleuropneumoniae
    Corynebacterium pseudotuberculosis (lymphangitis)
    Pasteurella haemolytica
  Mycotic
    Mucormycosis
    Aspergillus fumigatus
    Histoplasma farciminosum (lymphangitis)
    Sporothrix schenckii (lymphangitis)
  Protozoal
    Encephalitozoon cuniculi
    Besnoitia besnoiti
  Helminths (medium to large vessels, usually endovasculitis)
    Strongylus vulgaris
    Dirofilaria immitis
    Spirocerca lupi
    Onchocerca spp.
    Elaeophora spp.
    Aelurostrongylus abstrusus
    Angiostrongylus vasorum
    Schistosoma spp. (phlebitis)
    Brugia spp. (lymphangitis)

Noninfectious
  Immune mediated
    Systemic lupus erythematosus
    Rheumatoid arthritis
    Polyarteritis nodosa
    Anaphylactoid purpura
    Staphylococcal hypersensitivity
    Foreign proteins (serum sickness)
  Nonimmune mediated
    Uremia

nent. The terms **vasculitis** and **angiitis** are used interchangeably and may be used in preference to the more specific term **arteritis,** because arteries, capillaries, and veins may be involved simultaneously in some conditions (**arteritis, phlebitis,** and **lymphangitis** are discussed individually in Section III,D of The Vascular System). As well, the wall of the vessel may be obscured or obliterated by the inflammatory changes, precluding definitive classification. The consequences of vasculitis depend on the size, numbers, and types of vessels affected, and on the degree of obstruction caused, as noted under thrombosis.

Vasculitis is characterized by the presence of inflammatory cells within and around blood vessel walls with concomitant vessel wall damage as indicated by fibrin deposition, collagen degeneration, and necrosis of endothelial and smooth muscle cells. A large portion of the fibrillary, homogeneous, or granular eosinophilic material observed by light microscopy and called fibrinoid has been identified as fibrin by electron microscopy; fibrin may be accompanied by immunoglobulins, complement, and platelets. Degeneration of collagen and smooth muscle also contribute to the increased vascular wall eosinophilia. Thrombosis may occur because of endothelial injury and initiation of coagulation. These changes in the blood vessel wall distinguish vasculitis from perivascular infiltration with inflammatory cells. There may be predominance of neutrophils, lymphocytes, or macrophages in vasculitis, and the predominant cell type may be used to classify the vasculitis. The predominant cell type may vary over the course of the vasculitis, as the condition becomes chronic or resolves. Neutrophilic vasculitis may be further subdivided into leukocytoclastic, in which fragmented neutrophil nuclei or nuclear dust are present, and nonleukocytoclastic, in which there is little nuclear dust.

Vasculitis may develop from within the vessel as a result of endothelial injury by infectious agents and/or immune reactions, or may occur by extension from adjacent areas of inflammation and infection, especially in the case of thin-walled vessels. Infectious agents may cause endothelial damage either directly or through the actions of various toxic products, such as endotoxins of Gram-negative bacteria, or exotoxins of bacteria such as *Corynebacterium pseudotuberculosis. Pasteurella haemolytica* endotoxin (lipopolysaccharide, LPS) can directly damage endothelial cells; such damage may be reduced or increased by the activity of neutrophils, depending on the concentration of LPS present. The mechanism of LPS-mediated damage is speculative, and may involve prostaglandins or production of oxygen radicals by endothelial cells. In the absence of LPS, *P. haemolytica* leukotoxin may contribute to endothelial damage. In the case of *Actinobacillus pleuropneumoniae,* rather than endotoxin causing endothelial damage, cytotoxicity is apparently caused by a 104-kDa hemolysin. Many viruses are endotheliotropic, for example, the viruses of equine arteritis, infectious canine hepatitis, canine distemper, African swine fever, and hog cholera.

Endothelial damage and exposure of subendothelial collagen leads to activation of Hageman factor, and hence activation of complement, kinin, and plasmin systems, and to increased vascular permeability and inflammation. Immune reactions appear to be less important primary causes of vasculitis in domestic animals than they are in humans, but they may contribute to vasculitis caused by an infectious agent, as occurs in malignant catarrhal fever.

Type III hypersensitivity reactions, or Arthus reactions, cause necrotizing vasculitis by the deposition of immune complexes in vessel walls, usually in association with the endothelial basement membrane. Complement is fixed by the complexes, and complement fragments attract neutrophils, which release lysosomal enzymes and oxygen radicals (superoxide anion, hydrogen peroxide, hydroxyl radical, hypochlorous acid) in the process of phagocytosing immune complexes, and hence cause necrosis. As the Arthus reaction matures, neutrophils diminish, and mononuclear cells, including plasma cells, predominate. The morphology of immune complex-mediated vasculitis can be modified by the relative concentrations of immune reactants or secondary mediators; increases in antibody or complement concentrations can change an Arthus-like reaction to a Shwartzman-like reaction. Immune complexes can be found in vessel walls in a variety of conditions including systemic lupus erythematosus, Aleutian disease of mink, and in feline infectious peritonitis. However, even though immunoglobulins or immune complexes are found in the walls of affected vessels, this is not definitive proof that the immune reaction caused the vasculitis. Conversely, the absence of immune complexes from a vessel wall lesion does not rule out immune complex vasculitis because immune reactants can disappear rapidly.

Cell-mediated type IV, or delayed hypersensitivity, reactions may be involved in some types of vasculitis in which there is lymphocytic predominance, such as the arteritis of malignant catarrhal fever and of Border disease. In malignant catarrhal fever, the vasculitis is characterized by accumulations of principally lymphocytes and fewer macrophages in the adventitia and intima of affected small and medium arteries and veins of the alimentary tract, eye, kidney, liver, lung, and brain; neutrophils and plasma cells are rare. Lymphocytes may later invade the media, though adventitial infiltration continues to predominate, and there may be medial myocyte necrosis with variable amounts of fibrinoid degeneration and endothelial hyperplasia; thrombosis may be seen in advanced cases. There is a possibility of, but little direct evidence for, direct viral cytolysis in malignant catarrhal fever.

Vasculitis in humans has been associated with a variety of drugs (see hypersensitivity angiitis, Section III,F,1 of The Vascular System). Similar drug hypersensitivities may be expected to occur in domestic animals.

Improvement in the classification of necrotizing vasculitides will require further characterization of causes and of the spectrum of immune-mediated changes in vessels. Until such time, the use of accurate morphologic descriptions of vascular lesions is preferable to placing vasculitides, especially those of known cause, in such ill-defined categories as polyarteritis nodosa. The wide range of lesions observed probably reflects the variety of antigens or other agents involved, the intensity and chronicity of antigenic exposure, the relative contributions of immune complex versus delayed hypersensitivity reactions, the genetic variation among individuals, and the stage of development at which the lesion is observed. Although vasculitides in humans are usually classified into one of seven major divisions, including polyarteritis nodosa and hypersensitivity angiitis, many cases overlap in their morphologic and clinical manifestations, and are simply termed **systemic necrotizing vasculitis.**

Bibliography

Bogman, M. J. T. T. *et al.* The role of complement in the induction of acute antibody-mediated vasculitis of rat skin grafts in the mouse. *Am J Pathol* **109:** 97–106, 1982.

Bratberg, B. Acute vasculitis in pigs: A porcine counterpart to malignant catarrhal fever. *Proc Int Pig Vet Soc Congress Copenhagen, Denmark, Am Assoc Swine Pract,* 1980.

Breider, M. A., Kumar, S., and Corstvet, R. E. Interaction of bovine neutrophils in *Pasteurella haemolytica*-mediated damage to pulmonary endothelial cells. *Vet Immunol Immunopathol* **27:** 337–350, 1991.

Crawford, M. A., and Foil, C. S. Vasculitis: Clinical syndromes in small animals. *Compend Contin Educ Pract Vet* **11:** 400–415, 1989.

Cutlip, R. C. *et al.* Vasculitis associated with ovine progressive pneumonia virus infection in sheep. *Am J Vet Res* **46:** 61–64, 1985.

Easley, J. R. Necrotizing vasculitis: A review. *J Am Anim Hosp Assoc* **15:** 207–211, 1979.

Gardiner, A. C., Zakarian, B., and Barlow, R. M. Periarteritis in experimental border disease of sheep. III. Immunopathological observations. *J Comp Pathol* **90:** 469–474, 1980.

Liggitt, H. D., and DeMartini, J. C. The pathomorphology of malignant catarrhal fever. I. Generalized lymphoid vasculitis. *Vet Pathol* **17:** 58–72, 1980.

Lightfoot, R. W., Jr. The human vasculitis syndromes. *Toxicol Pathol* **17:** 72–76, 1989.

Morris, D. D. Cutaneous vasculitis in horses: 19 cases (1978–1985). *J Am Vet Med Assoc* **191:** 460–464, 1987.

Serebrin, S. *et al.* Endothelial cytotoxicity of *Actinobacillus pleuropneumoniae. Res Vet Sci* **50:** 18–22, 1991.

Van Dellen, A. F. *et al.* Light and electron microscopical studies on canine encephalitozoonosis: cerebral vasculitis. *Onderstepoort J Vet Res* **45:** 165–186, 1978.

Ward, P. A., and Varani, J. Mechanisms of neutrophil-mediated killing of endothelial cells. *J Leukocyte Biol* **48:** 97–102, 1990.

Warren, J. S., and Ward, P. A. Review: Oxidative injury of the vascular endothelium. *Am J Med Sci* **29:** 97–103, 1986.

Whiteley, H. E. *et al.* Ocular lesions of bovine malignant catarrhal fever. *Vet Pathol* **22:** 219–225, 1985.

III. Arteries

A. Congenital Anomalies

Minor variations occur among individuals of a species in the course and distribution of arteries, but these are of little significance except possibly in surgical techniques.

Arteriovenous fistula, the communication of an artery and vein that bypasses the capillary system, is a defect which may arise developmentally or be acquired subsequent to physical trauma, inflammatory necrosis involving adjacent vessels, neoplastic infiltration, or rupture of an arterial aneurysm into a vein. Fistulas occur in dogs, cats, horses, and cattle, but are uncommon. Clinically, fistulas most commonly occur in the extremities and appear as a pulsating, fluctuant swelling in which there is a palpable thrill and a machinerylike bruit on auscultation. In general, arteriovenous fistulas cause decreased peripheral resistance and increased venous return to the right heart. When 20–50% of the cardiac output is shunted through the fistula, high-output cardiac failure can occur. The veins involved are usually thick walled (arterialized), may be dilated and extremely tortuous, and may have intimal sclerosis. Hepatic arteriovenous fistula (congenital hamartoma) occurs in dogs and cats, and can cause portal hypertension, portocaval shunts, and ascites.

Thebesian veins connect the cardiac chambers, between the trabeculae carneae, with the myocardial sinusoids and capillaries. They may also connect with coronary arterioles to form direct connections between the coronary arterioles and the cardiac chambers. Areas of myocardium in which the Thebesian vein system is richly persistent are randomly distributed, and the true nature of the defect may be overlooked because of the tortuous aneurysmal dilation of the affected coronary artery. Injection studies are necessary to demonstrate the extent of the anastomosis.

Bibliography

Bouayad, H. *et al.* Peripheral acquired arteriovenous fistula: A report of four cases and literature review. *J Am Anim Hosp Assoc* **23:** 205–211, 1987.

Hosgood, G. Arteriovenous fistulas: Pathophysiology, diagnosis, and treatment. *Compend Contin Educ Pract Vet* **11:** 625–636, 1989.

Lamb, C. R., and Naylor, J. M. Arteriovenous fistula in the orbit of a calf. *Can Vet J* **26:** 105–107, 1985.

Parks, A. H. *et al.* Lameness in a mare with signs of arteriovenous fistula. *J Am Vet Med Assoc* **194:** 379–380, 1989.

B. Degeneration of Arteries

1. Arteriosclerosis

Arteriosclerosis literally means hardening of the arteries, and is more fully defined as chronic arterial change consisting of hardening, loss of elasticity, and luminal narrowing resulting usually from proliferative and degenerative, rather than inflammatory, changes of the media and intima. The term atherosclerosis is applied to lesions of arteriosclerosis in which degenerative fatty changes also occur. Atherosclerosis is the most common and important type of arteriosclerosis in humans, and the terms can thus be used interchangeably with little loss of meaning when discussing this species. In domestic animals, arteriosclerosis is common, but of little importance, and atherosclerosis is rare.

Arteriosclerosis develops slowly in animals, its extent or incidence being greater in older animals. The disease may become so widespread in individual animals that microscopically normal vessels may be difficult to find. The sclerotic vessels are usually not accompanied by significant disturbances of blood flow, although ischemic change, especially of brain and heart, may be seen. Well-developed lesions may be good sites for thrombus formation in animals when thrombogenic circumstances prevail. Arteriosclerosis uncomplicated by focal degeneration in the sclerotic plaques and with no or minimal lipid deposition is the lesion usually found in older horses, ruminants, and carnivores. There is a predilection for the abdominal aorta and points of arterial branching, but it is also seen in peripheral and pulmonary arteries and in the thoracic aorta. The extent of the lesions is very variable, so that there is no clear differentiation between a permissible aging change and a pathologic process. The nature of the insult (or insults) to the vessel walls which result in sclerosis is not known, although there are numerous possibilities. Hemodynamic factors probably contribute especially to development of plaques about the orifices of arterial branches.

Arteriosclerotic plaques produce a slight thickening and wrinkling of the intima, or more or less distict white, oval or linear elevations, and they can be found microscopically in many animals in which they are not visible to the naked eye. Their presence in peripheral, including coronary and cerebral, vessels, can be correlated with arteriosclerosis of the abdominal aorta.

The initiating lesion(s) and the evolution of plaque formation are not fully elucidated. Microthrombi composed chiefly of platelets are deposited at natural sites of turbulence or of endothelial denudation, and may initate the lesion. Platelets secrete platelet-derived growth factor (PDGF), epidermal growth factor, transforming growth factor-β and other mitogens; PDGF stimulates migration and proliferation of smooth muscle cells. Alternatively, the initial change in the vessel wall may occur as the result of a nondenuding insult to the endothelium; endothelial cells themselves produce several mitogens, including PDGF. Histologically, the internal elastic membrane shows an early distinctive change, being irregular in outline, partially duplicated, and fragmented or discontinuous. Smooth muscle cells migrate into the intima from the media to function as do fibroblasts outside of arteries; these intimal smooth muscle cells produce most of the connective tissue matrix of the intimal plaque, including collagen, elastic tissue, and proteoglycans. In young plaques, or the deeper portions of older ones, the smooth muscle cells (myointimal cells) are numerous and intimately mixed with collagen, elastic fibrils, basement membrane, and proteoglycans as the so-called musculoelastic layer. In large and older plaques, a more superficial elastic hyperplastic layer may be recognized by a more compact arrangement of collagen and elastica with relatively few

cells. This lamination of plaques is not always clearly evident, and its development may depend on the age of the lesion as well as on the segment of vessel involved, whether chiefly muscular or chiefly elastic.

Bibliography

Detweilier, K. D., and Patterson, D. F. The prevalence and types of cardiovascular disease in dogs. *Ann NY Acad Sci* **127:** 481–516, 1965.

Fankhauser, R., Luginbuhl, H., and McGrath, J. T. Cerebrovascular disease in various animal species. *Ann NY Acad Sci* **127:** 817–859, 1965.

Luginbuhl, H. Vascular diseases in animals: Comparative aspects of cerebrovascular anatomy and pathology in different species. *In* "Cerebral Vascular Diseases," New York, Grune & Stratton, 1966.

Maher, E. R., and Rush, J. E. Cardiovascular changes in the geriatric dog. *Compend Contin Educ Pract Vet* **12:** 921–931, 1990.

Marcus, L. C., and Ross, J. R. Microscopic lesions in the hearts of aged horses and mules. *Pathol Vet* **4:** 162–185, 1990.

Nanda, B. S., and Getty, R. Age-related histomorphological changes in the cerebral arteries of domestic pig. *Exp Gerontol* **6:** 453–460, 1971.

Prasad, M. C., Rajya, B. S. and Mohanty, G. C. Caprine arterial diseases. I. Spontaneous aortic lesions. *Exp Mol Pathol* **17:** 14–28, 1972.

Schaub, R. G., Rawlings, C. L., and Keith, J. C. Platelet adhesion and myointimal proliferation in canine pulmonary arteries. *Am J Pathol* **104:** 13–22, 1981.

Whitney, J. C. Some aspects of the pathogenesis of canine arteriosclerosis. *J Small Anim Pract* **17:** 87–97, 1976.

2. Atherosclerosis

Atherosclerosis is of little practical importance in domestic animals, but is primarily of interest for the development of animal models of the human disease. The atherosclerotic susceptibility of animals varies; rabbits, chickens, and pigs are atherosensitive, whereas dogs, cats, cattle, goats, and rats are atheroresistant. Swine and nonhuman primates are usually considered to be the best animal models of human atherosclerosis.

In humans, atherosclerosis and its complications of myocardial infarction, stroke, and peripheral vascular disease are major causes of morbidity and mortality in the Western world. Atherosclerosis affects the large elastic arteries (aorta, iliac). The essential lesion is the atheroma or fibrofatty plaque, which is a focal, raised intimal plaque with a core of lipid (cholesterol and its esters largely) covered by a fibrous cap. Complications of plaques include mineralization, ulceration, superimposed thrombosis, intraplaque hemorrhage, and aneurysmal dilation. Fatty streaks, another type of fatty intimal lesion, are common in the aorta and large arteries of domestic animals, especially ruminants and swine, and humans. The streaks, which are soft, smooth, nonelevated lesions ranging from pinpoint size up to several square centimeters, are rarely visible grossly unless the aorta is stained with a fat stain such as Sudan IV, which stains the lesions bright orange. Streaks and plaques may coexist in the same area, but fatty streaks also occur in areas in which plaques do not occur; thus, the role of fatty streaks in the genesis of atherosclerosis is unresolved.

Multiple pathogenetic influences can contribute to the development of plaques. The various events and theories of development of atherosclerosis have been unified in the response-to-injury hypothesis. Endothelial injury or dysfunction can occur as the result of hyperlipidemia, with the cholesterol present in low-density (LDL) and very-low-density lipoproteins (VLDL) seen as a primary culprit. Additional risk factors are hypertension, smoking, obesity, inactivity, and diabetes mellitus. Immunologic mechanisms, toxins, and viruses may also damage endothelium. Endothelial injury leads to platelet adhesion, monocyte adhesion and infiltration, and smooth muscle migration and proliferation, as discussed in the preceding section, Arteriosclerosis. The feature added in atherosclerosis is insudation of lipid, which is seen as extracellular lipid or as intracellular lipid in foam cells. The lipid-laden foam cells that are characteristic of both fatty streaks and atheromatous plaques may be either altered macrophages or smooth muscle cells; macrophages predominate in streaks, whereas smooth muscle cells predominate in the fibrous cap of plaques. Endothelial cell dysfunction is emerging as an endocrinopathy that may contribute to the effects of atherosclerosis through impaired ability of endothelial cells to produce endothelial-derived relaxing factor and through increased production of endothelin.

Atherosclerosis develops commonly only in the **pig** among domestic species. Fatty streaks and atheromatous plaques develop in the aorta, extramural coronary arteries, cerebral, and iliac arteries, as in humans, and the plaques may cause considerable stenosis of vessels in swine 8–12 years old. The extent of lipid deposition can be influenced by feeding extraordinary diets containing much lipid, including cholesterol, but the atheromas rarely reach the degree of development seen in humans, and they do not lead to occlusive thrombus formation, which apparently requires softening and ulceration of the atheromatous plaque. The initial deposition of lipid occurs in the proliferated smooth muscle cells which show signs of degeneration and which may become foam cells. Macrophages also appear, containing lipid, and fat may be demonstrable extracellularly, presumably released by degenerate cells. Deposits of cholesterol and mineral, not extensive, may be associated with softening of larger atheromatous plaques. However, except in pigs fed high-fat diets, lipid is usually a minor component of the predominantly fibromuscular plaques.

Of the domestic animals, the deposition of cholesterol and other lipids in the arteries in more than trace amounts occurs only in **dogs** (Fig. 1.53), and then almost always in association with hypothyroidism or diabetes mellitus. In dogs, hypercholesterolemia is chiefly related to hypothyroidism, but even in thyroidectomized, cholesterol-fed, hypercholesterolemic experimental dogs, only occasional dogs classed as hyperresponders developed atherosclerosis, perhaps as a result of an increased proportion of

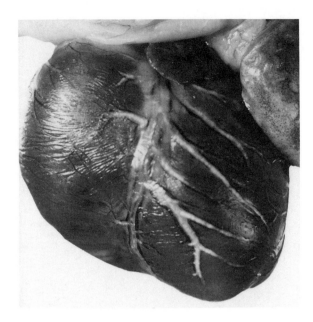

Fig. 1.53 Irregular thickening of coronary arteries. Atherosclerosis. Dog.

cholesterol-rich VLDL. A breed predisposition to atherosclerosis in miniature schnauzers may be the result of idiopathic hyperlipoproteinemia.

The deposition of lipids begins in the middle and outer layers of the media and occurs more extensively perhaps in the small muscular arteries than in the large elastic ones; if present, aortic lesions occur as intimal plaques. The veins are normal. The vascular changes can be severe enough to be seen on gross inspection. The vessels are enlarged, less pliable than normal, and their walls are thickened and ill defined because a hyperplastic adventitia merges irregularly with surrounding tissues. Projecting into the lumens of the vessels on cross section are tiny yellow-brown nodules. In vessels large enough to be opened longitudinally, there are numerous small yellow flat elevations, which may be confluent.

The media may be greatly increased in thickness by the accumulation of lipid, most of which is in foam cells, but some is present in identifiable muscle cells of the media or free in the interstitium as droplets or crystals (Fig. 1.54). The deposition of lipid in the internal layers of the media leads to disruption of the internal elastic lamina and involvement of the intima. The amount of lipid deposited can be, relatively, very great and lead to eccentric enlargement of the vessel. Associated with the lipidosis, and possibly as a response to it, there is progressive proliferation of fibrous connective tissue. Beginning with the formation of fine fibrils in and about the areas of lipid deposition, the fibrosis is progressive and may eventually completely transform the structure of the wall. The connective tissues become hyaline and relatively acellular and may be heavily impregnated with salts of calcium and iron, present as granules, large clumps, or crystalline plaques.

Atheromatosis of the degree described in dogs is rare.

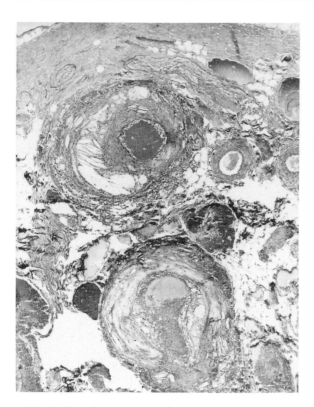

Fig. 1.54 Atherosclerosis of extramural coronary arteries. Dog.

Clinical consequences of severe atherosclerosis appear to be infrequent in dogs, even though the vessels most severely involved are those of the heart, brain, and kidney. Occasionally, however, as in humans, rupture of the atheromatous area into the lumen with thrombosis or even widespread lipid embolism has been observed.

The morphology of atherosclerosis differs in dogs and in humans; lipid is present in the intima but primarily in the media and adventitia of atherosclerotic canine arteries, but is present primarily in the intima in humans. The term xanthomatosis has been applied to the canine condition in the past, but its use is inadvisable as it may cause confusion with the human condition of primary xanthomatosis (type II hyperlipoproteinemia), in which there is premature atherosclerotic coronary artery disease, but also xanthomatous infiltration of the myocardium, dermis, and other organs with an associated granulomatous reaction. A similar condition in other animal species is referred to as xanthomatosis, and is characterized by the presence of lipid-laden foam cells in subcutaneous or intracutaneous accumulations known as xanthomas. A familial hyperlipoproteinemia occurs in lipoprotein lipase-deficient cats and results in the formation of xanthomata in a variety of organs; artherosclerosis is not a feature of this condition.

Bibliography

Armstrong, M. L., and Heistad, D. D. Animal models of atherosclerosis. *Atherosclerosis* **85:** 15–23, 1990.

Chastain, C. B., and Graham, C. L. Xanthomatosis secondary to diabetes mellitus in a dog. *J Am Vet Med Assoc* **172:** 1209–1211, 1978.

DeBowes, L. J. Lipid metabolism and hyperlipoproteinemia in dogs. *Compend Contin Educ Pract Vet* **9:** 727–734, 1987.

Healy, B. Endothelial cell dysfunction: An emerging endocrinopathy linked to coronary disease. *J Am Coll Cardiol* **16:** 357–358, 1990.

Johnstone, A. C. *et al.* The pathology of an inherited hyperlipoproteinaemia of cats. *J Comp Pathol* **102:** 125–137, 1990.

Luginbuhl, H., and Jones, J. E. T. The morphology and morphogenesis of atherosclerosis in aged swine. *Ann NY Acad Sci* **127:** 763–779, 1965.

Maher, E. R., and Rush, J. E. Cardiovascular changes in the geriatric dog. *Compend Contin Educ Pract Vet* **12:** 921–931, 1990.

Munro, J. M., and Cotran, R. S. The pathogenesis of atherosclerosis: Atherogenesis and inflammation. *Lab Invest* **58:** 249–261, 1988.

Rifai, N. Lipoproteins and apolipoproteins. Composition, metabolism, and association with coronary artery disease. *Arch Pathol Lab Med* **110:** 694–701, 1986.

Ross, R. The pathogenesis of atherosclerosis—An update. *N Engl J Med* **314:** 488–500, 1986.

3. Mineralization

Mineralization (calcification) occurs quite frequently in the arteries of animals, either as a dystrophic or metastatic process. **Dystrophic mineralization,** or mineralization of dying or dead tissue, occurs in areas of inflammation, degeneration, and thrombosis, and not necessarily in association with preexisting arteriosclerosis. **Metastatic mineralization** occurs as the result of hypercalcemia and/or hyperphosphatemia. The distinction between dystrophic and metastatic mineralization may be artificial; in vitamin D toxicosis for example, some maintain that degeneration occurs prior to mineralization as a result of direct vitamin D toxicity and that hypercalcemia and mineralization are secondary events. Although the cardiovascular system is particularly susceptible to mineralization, mineralization of other soft tissues, such as the renal tubular and pulmonary alveolar basement membranes, endocardium, myocardium, and gastric mucosa, usually occurs concomitantly with vascular mineralization.

Vitamin D poisoning occurs in a variety of circumstances and leads to mineralization of vessels and other tissues in association with hypercalcemia and/or hyperphosphatemia. The disease can be endemic in herbivores, cattle most often, that have consumed plants (such as *Solanum malacoxylon, S. torvum, Trisetum flavescens, Cestrum diurnum*) that contain 1,25-dihydroxycholecalciferol or a related compound (see The Endocrine Glands, Chapter 3 of this volume). Horses and pigs are occasionally poisoned by inadvertent overdoses of vitamin D in feed. Accidental poisoning of dogs and cats with a cholecalciferol (vitamin D_3) rodenticide is common.

Mincralization occurs in dogs with hypercalcemia due to pseudohyperparathyroidism, in some dogs in associa-

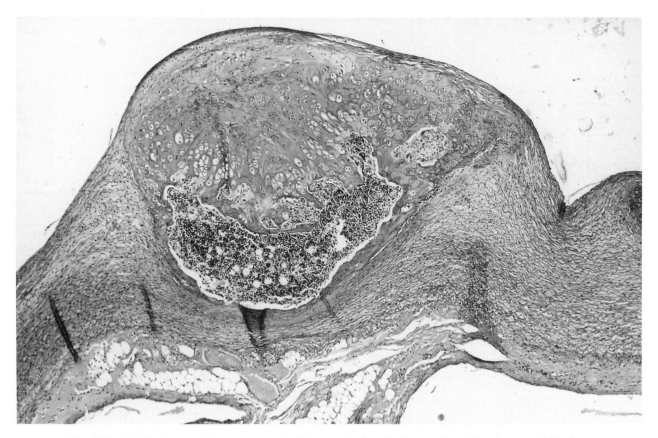

Fig. 1.55 Cartilaginous and osseous metaplasia in arteriosclerosis of aorta. Dog. Note hematopoietic tissue.

tion with chronic renal insufficiency, and in hypomagnese-mic cattle. Mineralization also occurs spontaneously in animals of advanced age and, in cattle especially, in a variety of chronic debilitating diseases such as Johne's disease. In sporadic cases, the causes of mineralization are generally unknown and may be nonspecific.

Calcium and phosphorus are deposited in association with extracellular vesicles adjacent to elastic fibers and, in the smaller vessels, may produce a complete, concentric ring of medial mineralization; this is the domestic animal equivalent of Mönckeberg's arteriosclerosis (medial calcific sclerosis) of humans. Mineralized lesions in the aorta and pulmonary artery are usually in the form of flat plaques and are often very striking grossly; complete encirclement is seldom seen. The buttonlike protuberances which are seen in the aorta about the orifices of smaller branches are not primarily aortic but, instead, concentric lesions in the small efferents that project when the opened vessel undergoes elastic recoil. The plaques in the large vessels are brittle, and the smaller arteries are transformed into rigid tubes which can be fractured easily. Metaplastic ossification may occur (Fig. 1.55).

Intimal bodies commonly occur in small arteries and arterioles of horses, particularly in the submucosa of the intestines, but in other organs as well. Intimal bodies appear in section as multiple irregular mineralized masses covered by endothelium and usually protruding into the arterial lumen (Fig. 1.56) They occur in horses of all ages, are not associated with strongyle infection, and have no apparent functional significance. These

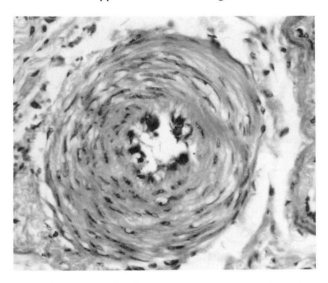

Fig. 1.56 Intimal bodies in submucosal artery of intestine. Horse.

bodies apparently arise from degeneration and mineralization of subendothelial smooth muscle cells and intercellular material.

The deposition of iron salts with calcium (**siderocalcinosis**) is observed commonly in horses, the incidence or severity of the change increasing with age. Vessels of all types are involved, and sites of predilection include the globus pallidus and internal capsule in the brain. Mostly, the deposits are without notable consequence, but secondary malacia may occur. It is also possible that malacia of other cause may exacerbate the cerebrovascular siderosis.

The mineralized nodules in the intima of the aorta in the horse are probably all healed lesions of verminous arteritis.

Bibliography

Bundza, A., and Stevenson, D. A. Arteriosclerosis in seven cattle. *Can Vet J* **28:** 49–51, 1987.

De Olivereira, A. C., Rosenbruch, M., and Schulz, L. C. Intimal asteroid bodies in horses: Light- and electron-microscopic observations. *Vet Pathol* **22:** 226–231, 1985.

Drazner, F. H. Hypercalcemia in the dog and cat. *J Am Vet Med Assoc* **178:** 1252–1256, 1981.

Haggard, D. L., Whitehair, C. K., and Langham, R. F. Tetany associated with magnesium deficiency in suckling beef calves. *J Am Vet Med Assoc* **172:** 495–497, 1978.

Imaizumi, K. *et al.* Morphological changes of the aorta and pulmonary artery in thoroughbred racehorses. *J Comp Pathol* **101:** 1–9, 1989.

Morris, K. M. L. Plant-induced calcinosis: A review. *Vet Hum Toxicol* **24:** 34–48, 1982.

Rosol, T. J., and Capen, C. C. Pathogenesis of humoral hypercalcemia of malignancy. *Domest Anim Endocrinol* **5:** 1–21, 1988.

Thomas, J. B., Hood, J. C., and Gaschk, F. Cholecalciferol rodenticide toxicity in a domestic cat. *Aust Vet J* **67:** 274–275, 1990.

4. Arteriolosclerosis

Arteriolosclerosis describes a heterogeneous group of arteriolar lesions that may be predominantly hyaline or predominantly hyperplastic. The major cause of these changes in vessels in humans is hypertension, but the pathogenesis of the lesions in domestic animals is often not understood. Endothelial damage is likely the primary event in these conditions. Hyaline degeneration—the microscopic appearance of amorphous, brightly eosinophilic material in vessel walls—can occur as the result of increased vascular permeability which allows insudation of various plasma components; it includes the amyloid deposits of renal glomeruli and elsewhere, and the fibrin and necrotic smooth muscle of fibrinoid necrosis. Fibrinoid necrosis may predate and initiate inflammatory cellular infiltration, after which the reaction would be termed arteritis. Thickening of the arteriolar wall may cause luminal narrowing and hence ischemia of the region supplied.

Systemic hypertension, or persistently elevated systemic blood pressure, is an important disease of humans, and occurs as primary or essential hypertension and as secondary hypertension. Both forms of hypertension are typically benign in their clinical course; accelerated or malignant hypertension develops in about 5% of affected individuals. Both primary and secondary hypertension have been reported in dogs and cats; the primary form is rare, and the secondary form is apparently uncommon, though hypertension may often go unrecognized. The diagnosis of hypertension could well become more frequent as noninvasive techniques for measurement of blood pressure, such as Doppler ultrasonic sphygmomanometry, come into common use. Renal disease, likely the most common cause of hypertension in dogs and cats, may be either a cause or an effect of hypertension, thus making difficult the definition of the occurrence of a hypertensive vascular lesion as a primary or a secondary event. Primary hypertensive lesions can lead to decreased renal perfusion, activation of the renin–angiotensin–aldosterone system, and hence more hypertension, but on the other hand, primary chronic renal disease can lead to inadequate excretion of sodium and water, blood volume expansion, and hence secondary hypertension. In either case, hypertension is self-perpetuating as medial hypertrophy and hyalinization of renal arteries lead to more nephrosclerosis, more hypertension, and more pressure-induced vascular damage.

Hypertension which would be classed as secondary is said to occur in more than 60% of dogs with chronic renal disease, and may also occur in association with pheochromocytoma, hyperadrenocorticism, hypo- and hyperthyroidism, and diabetes mellitus. In cats, hypertension occurs in association with chronic renal disease, hyperthyroidism, and chronic anemia. Systemic and local hypertension reportedly accompany acute and chronic laminitis in horses; hypertension is suggested to occur in acute and chronic laminitis in cattle. The hypertension of feline hyperthyroidism is reversible on resolution of the primary condition. The presenting sign in canine and feline hypertension may be blindness due to hypertensive retinopathy; ocular lesions include retinal vascular tortuosity, intraocular hemorrhage, and retinal detachment, the results of retinal arterial degeneration. In the face of hypertension, retinal and choroidal arterioles constrict; sustained vasoconstriction may lead to ischemic necrosis of the arteriolar smooth muscle, increased vascular permeability, hyalization, edema, and hemorrhage. Left ventricular hypertrophy commonly occurs in hypertension in response to increased demand for cardiac output plus increased peripheral vascular resistance.

Pulmonary hypertension—persistent elevation of pulmonary arterial blood pressure—can occur both as the cause and as the result of pulmonary arterial disease (see The Respiratory System, Volume 2, Chapter 6; and, in this chapter, Arterial Hypertrophy, Section III,E, and Dirofilariasis, Section III,F,2,a, both in The Vascular System).

Widespread arterial degeneration commonly occurs in dogs with **uremia,** and is more frequently noted in small muscular arteries and arterioles of the gastric submucosa, tongue, colon, gallbladder, urinary bladder, and kidneys. Less frequently affected are arteries of the small intestine, myocardium, and other organs. The arterial and arteriolar lesions, in concert with degenerative capillary lesions, produce ischemic injury, and contribute to grossly observed lesions of uremia, such as uremic gastric infarction. The arterial lesions of acute renal insufficiency consist of subendothelial deposition of fibrin, disruption of the internal elastic lamina, necrosis of medial smooth muscle, mineralization, and sometimes a neutrophilic reaction. These lesions may be segmental or circumferential in the vessel wall. This arterial reaction is primarily degenerative rather than inflammatory and is similar to **hyaline arteriolosclerosis** of benign nephrosclerosis (patchy ischemic atrophy) in humans. In chronic renal insufficiency, medial hypertrophy and adventitial fibrosis produce thickening and whorling of renal interlobular and intralobular arteries and afferent arterioles. These changes are similar to **hyperplastic arteriolosclerosis** of humans, which is characterized by concentric fibrosis and smooth muscle proliferation in the intima (onion-skinning); these hyperplastic changes, in combination with fibrinoid necrosis of the media of arterioles, are features of malignant hypertension.

Hyalinosis is commonly observed in the splenic arterioles of dogs and pigs, affecting most frequently those of the lymphatic nodules. This change is more severe in dogs with a variety of diseases including diabetes mellitus and chronic renal disease, and thus may be of hypertensive origin. A hypertrophic hyalinization is regularly observed in the uterine and ovarian arteries in pregnant and postpartum animals, the elastic tissue is increased in amount, the intima is markedly thickened and hyaline, and the muscularis may be atrophic.

Hypertrophic hyalization, or **hyalinosis,** is observed in old dogs, affecting principally the **intramural coronary arteries** (Fig. 1.30) and small meningeal and cerebral vessels. The hyalinization is not clearly of clinical significance in the brain but may cause multifocal intramural myocardial infarction (MIMI) and, in concert with valvular endocardiosis, lead to congestive heart failure. The hyaline deposits consist predominantly of fibrin or glycosaminoglycans or, less commonly, amyloid. Other lesions encountered in the intramyocardial arteries include musculoelastic intimal thickenings, intimal cushions, and microthrombi; these lesions occur at sites of turbulent blood flow, namely bifurcations and branchings.

Sclerosis of intramural coronary arteries, predominantly of the left ventricle, occurs in dogs with congenital subaortic stenosis and occurs in the right ventricle of dogs with pulmonic stenosis. Intimal thickening due to fibroelastic and smooth muscle proliferation can result in myocardial infarction and scarring. The intimal proliferation may occur in response to decreased coronary perfusion pressure and increased intramural tension during systole.

Arteriolar hyaline degeneration is seen in swine as part of the pathological picture in meninges in organomercurial poisoning, in various organs in edema disease, and chiefly

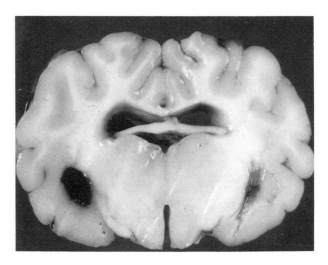

Fig. 1.57 Hemorrhage and malacia in brain. Pig. Caused by necrotizing cerebrospinal angiopathy.

in the heart in hepatosis dietetica and mulberry heart disease. The progression of arterial lesions in **edema disease** is from mural edema, to hyaline degeneration, to medial necrosis, and intramural and perivascular hemorrhage, at which stage acute severe hypertension develops, likely as a result of increased vascular resistance, and hypertension in turn exacerbates vascular damage; adventitial proliferation follows in survivors. Certain strains of *Escherichia coli* produce edema disease toxin (Shiga-like toxin-II variant, SLT-IIv) which preferentially targets vessels of the gastric and colonic submucosa and the cerebellar folia, perhaps because of concentration of a toxin receptor, such as globotetraosylceramide, at these sites. **Cerebrospinal angiopathy** in swine is characterized by lesions in arterioles, primarily in the brainstem, which include subendothelial hyaline deposits, degeneration or fibrinoid necrosis of the media, and perivascular eosinophilic droplets. The vascular lesions cause demyelination and malacia, and hence nervous signs and are probably manifestations of subclinical edema disease (Fig. 1.57).

Bibliography

Anderson, L. J. The follicular arterioles of the spleen in canine interstitial nephritis. *J Pathol Bacteriol* **95:** 55–58, 1968.

Anderson, L., and Bergman, A. Pathology of bovine laminitis especially as regards vascular lesions. *Acta Vet Scand* **21:** 559–566, 1980.

Cheville, N. F. Uremic gastropathy in the dog. *Vet Pathol* **16:** 292–309, 1979.

Davies, T. S., Nielsen, S. W., and Kircher, C. H. The pathology of subacute methylmercurialism in swine. *Cornell Vet* **66:** 32–55, 1976.

Dimski, D. S., and Hawkins, E. C. Canine systemic hypertension. *Compend Contin Educ Pract Vet* **10:** 1152–1158, 1988.

Kamiya, S., and Daigo, M. Relationship between glycosaminoglycans and pregnancy-induced sclerosis in bovine uterine arteries. *Jpn J Vet Sci* **50:** 1055–1059, 1988.

Kobayashi, D. L. *et al.* Hypertension in cats with chronic renal failure or hyperthyroidism. *J Vet Intern Med* **4:** 58–62, 1990.

MacLeod, D. L., Gyles, C. L., and Wilcock, B. P. Reproduction of edema disease of swine with purified Shiga-like toxin-II variant. *Vet Pathol* **28:** 66–73, 1991.

Nakamura, K. *et al.* Perivascular eosinophilic droplets in swine brain induced by *Escherichia coli* toxin. *Can J Vet Res* **50:** 438–440, 1986.

Nielsen, N. O. Edema disease. *In* "Diseases of Swine," A. D. Leman *et al.* (eds.), 6th Ed. pp. 528–540. Ames, Iowa, Iowa State University Press, 1986.

Paulsen, M. E. *et al.* Arterial hypertension in two canine siblings: Ocular and systemic manifestations. *J Am Anim Hosp Assoc* **25:** 287–295, 1989.

Perry, L. A., Dillon, A. R., and Bowers, T. L. Pulmonary hypertension. *Compend Contin Educ Pract Vet* **13:** 226–233, 1991.

Rau, L. Hypertension, endothelium, and cardiovascular risk factors. *Am J Med* **90** (Suppl. 2A): 13S–18S, 1991

Turner, J. L. *et al.* Idiopathic hypertension in a cat with secondary hypertensive retinopathy associated with a high-salt diet. *J Am Anim Hosp Assoc* **26:** 647–651, 1990.

Van Vleet, J. F., Ferrans, V. J., and Ruth, G. R. Ultrastructural alterations in nutritional cardiomyopathy of selenium-vitamin E deficient swine. II. Vascular lesions. *Lab Invest* **37:** 201–211, 1977.

C. Arterial Rupture, Aneurysms

Rupture of arteries as a result of physical trauma is common; spontaneous ruptures are not. The aortic lesions produced by *Spirocerca lupi* in dogs occasionally lead to rupture. Rupture of a uterine artery is a cause of fatal hemorrhage in aged mares at parturition and is associated with low serum copper levels. Mycotic ulceration of the internal carotid or maxillary artery in horses can lead to fatal guttural pouch hemorrhage.

Rupture of the aorta is well known, but certainly uncommon, in the horse. The ruptures occur during periods of excitement and activity, such as racing or in stallions while breeding, and are probably related to increased intra-aortic pressure. A predisposing aortic lesion has not been definitely identified, but fragmentation, degeneration, and mineralization of elastic fibers may contribute to weakening of the wall. Cystic medial necrosis, the presence of pools of ground substance within the elastic media, has been suggested to be an underlying cause, but this change is present in the aortas of many normal horses. Intimal thickening and medial fibrosis of vasa vasora may predispose to aortic medial necrosis and rupture. Tears occur in the ascending aorta at any level from the aortic valvular ring to the level of the brachiocephalic trunk, and range from 0.5- to 1-cm tears in the right coronary sinus to large transverse or three-cornered tears in the cranial wall of the aorta (Fig. 1.15). Hemorrhage may occur into a number of different sites: hemopericardium, the most common sequel to aortic rupture, leads to sudden death; hemorrhage into, and disruption of, the atrioventricular node or bundle of His also leads to sudden death; dissection through the ventricular septum and into the right ventricle (cardioaortic fistula) causes progressive right-heart failure; dissection of blood along the aortic wall may

lead to formation of an aneurysm, which may in turn rupture.

Spontaneous rupture of the pulmonary artery also occurs in horses, but is even less frequent than aortic rupture.

Rupture of the aorta, pulmonary artery, or coronary artery occurs experimentally, though not as a natural event, in copper-deficient swine, and has been extensively investigated because of similarities with Marfan's syndrome, an inherited connective tissue disorder in humans in which dissecting aortic aneurysms occur. Degeneration of elastica appears to be the basis of the vascular lesions of copper deficiency, and is due to a deficiency of a copper-containing enzyme, lysyl oxidase, which is responsible for cross-linking of collagen and elastin. Similarly, feeding β-aminoproprionitrile to pigs, rats, or turkeys causes dissecting aortic aneurysms and skeletal changes, an experimental model known as lathyrism, but is due to inhibition, rather than deficiency, of lysyl oxidase.

An **aneurysm** is a localized abnormal dilation of any vessel. Aneurysms are of most importance when they affect the aorta; they occur most commonly in humans as a complication of atherosclerosis, but are infrequent occurrences in domestic animals other than turkeys. On the basis of their gross appearance, aneurysms may be classified as berry, saccular, fusiform, or dissecting; in a **dissecting aneurysm,** blood enters the wall of an artery through an intimal tear, dissects between medial layers, and creates a cavity within the arterial wall. Dissecting aortic aneurysm (dissecting hematoma, or acute aortic dissection) is one manifestation of aortic rupture in horses, and may precede fatal aortic rupture. Dissecting aneurysms occur in the coronary and renal arteries of young, male, racing greyhounds, possibly as a result of arteriosclerosis and hemodynamic stresses, and can fatally rupture. Dissecting aneurysms may be of consequence by causing stenosis of the aortic lumen, as well as by rupturing. **True aneurysms** are composed of all, or most, of the layers of the vessel wall. Dilation of the cranial mesenteric artery may occur in the course of verminous arteritis in horses; the arterial wall is often greatly thickened, but the dilation could be termed an inflammatory rather than a true aneurysm. **False aneurysms** result from rupture of an artery or aneurysm and are essentially hematomas communicating with an arterial lumen; their walls are formed by the surrounding fibrous tissue.

Bibliography

Bevilacqua, G., Camici, P., and L'Abbate, A. Spontaneous dissecting aneurysm of the aorta in a dog. *Vet Pathol* **18:** 273–275, 1981.

Bjotvedt, G. Spontaneous renal arteriosclerosis in greyhounds. *Canine Pract* **13:** 26–30, 1986.

Coulson, W. F., Linker, A., and Bottcher, E. Lathyrism in swine. *Arch Pathol* **87:** 411–417, 1969.

Fregin, G. F., Hammel, E. P., and Rooney, J. R. Clinico-pathologic conference (pulmonary artery rupture). *J Am Vet Med Assoc* **164:** 813–816, 1974.

Rings, D. M., Constable, P., and Biller, D. S. False carotid aneurysm in a sheep. *J Am Vet Med Assoc* **189:** 799–801, 1986.

Sanger, V. L. *et al.* Cardiovascular disease resembling Marfan's syndrome in a dog. *Indian Vet J* **57:** 703–706, 1980.

Simpson, C. F., Boucek, R. J., and Noble, N. L. Similarity of aortic pathology in Marfan's syndrome, copper deficiency in chicks, and β-aminopropionitrile toxicity in turkeys. *Exp Mol Pathol* **32:** 81–90, 1980.

Smith, K. M., and Barber, S. M. Guttural pouch hemorrhage associated with lesions of the maxillary artery in two horses. *Can Vet J* **25:** 239–242, 1984.

Stowe, H. D. Effects of age and impending parturition upon serum copper of thoroughbred mares. *J Nutr* **95:** 179–183, 1968.

van der Linde-Sipman, J. S. *et al.* Necrosis and rupture of the aorta and pulmonary trunk in four horses. *Vet Pathol* **22:** 51–53, 1985.

D. Arterial Thrombosis and Embolism

The three major predispositions to thrombosis are (1) **injury to endothelium,** e.g., via infectious, toxic, or immunologic mechanisms; (2) **altered blood flow,** as occurs with stasis or turbulence; and (3) **hypercoagulability** of the blood, which may result from increased concentrations of activated procoagulants, increased numbers or stickiness of platelets, or decreased concentrations of inhibitors such as antithrombin III (see The Kidney, Volume 2, Chapter 5). The relative contribution of each of these three features of Virchow's triad may vary. Endothelial injury, by itself, is the major cause of thrombosis, especially in the heart and arteries. Stasis must usually be combined with endothelial damage and/or hypercoagulability to induce thrombosis.

Thrombi are of importance because they occlude the vessel at the site involved (Fig. 1.58), and because pieces of the thrombus break off, giving rise to emboli which may occlude vessels distal to the site of thrombosis. Arterial occlusion is of significance in organs with an end-arterial blood supply, such as the kidney, because of the absence of collateral circulation and the development of infarction. Simultaneous occlusion of a large number of pulmonary arterioles or arteries by thrombi can lead to right heart failure (cor pulmonale) and death. Even in areas with extensive collateral circulation, the development of this circulation may be impaired by release of vasoactive substances such as serotonin and/or prostaglandins from the thrombus.

Specific causes and examples of thrombosis abound: thrombosis is a common complication of arteritis. Specific examples of thrombosis, namely aortic–iliac thrombosis in horses and disseminated intravascular coagulation, are discussed subsequently.

Thromboembolism occurs in a variety of conditions. In verminous arteritis in horses, emboli arise from thrombi in the cranial mesenteric artery and cause intestinal ischemia, and possibly infarction. Saddle thromboemboli of the aortic trifurcation in cats arise from the dilated left atrium in cases of cardiomyopathy and cause posterior

Fig. 1.58 Aortic–iliac thrombosis. Calf. Note laminations in arterial thrombus, organization from intima, and markedly reduced eccentric lumen.

paresis of acute onset; the clinical signs of aortic trifurcation obstruction, referred to as the five p's, are pain, paresis, pulselessness, poikilothermy, and pallor. Arterial thromboembolism occasionally occurs as a result of congenital cardiac defects and the formation of intracardiac thrombi. Emboli commonly arise from the cardiac valves in cases of vegetative bacterial endocarditis. Thrombi that develop in animals undergoing extended anesthesia can result in fatal pulmonary microthromboembolism, a common occurrence in humans, but a condition that is seldom recognized in domestic animals. Pulmonary thromboembolism may well complicate a variety of conditions, such as cardiac or renal disease, physical trauma, disseminated intravascular coagulation, hyperadrenocorticism, and neoplasia, but be underdiagnosed as a cause of morbidity.

Thrombosis is by far the most common antecedent to embolism, but emboli other than thromboemboli occasionally occur. Venous **air embolism** is a rare complication of pneumocystography, laparoscopy, or cryosurgery; an air trap develops in the right ventricle and may cause fatal blockage of circulation. **Foreign bodies,** such as pellets or bullets, may be embolic and occlude major vessels such as the aorta. Larval and adult **parasites,** such as strongyles and heartworms, may be embolic in the arterial system, especially if killed by therapy, and may cause arterial occlusion plus a granulomatous response. Emboli of **neoplastic cells** can result in distant metastases. **Fibrocartilagi-**

nous embolism can result in necrotizing myelopathy in dogs, pigs, and humans. Fractures of bones, especially of those of the pelvis, and trauma of soft tissues can result in **fat embolism;** death may occur due to embolization of the central nervous system. Embolic fat may also originate from rupture of fat-laden hepatocytes.

Bibliography

Anderson, W. I. *et al.* Infarction of the pons and medulla oblongata caused by arteriolar thrombosis in a horse. *Cornell Vet* **80:** 285–289, 1990.

Comp, P. C. Overview of the hypercoagulable states. *Semin Thromb Hemost* **16:** 158–161, 1990.

Daniel, G. B. *et al.* Diagnosis of aortic thromboembolism in two dogs with radionuclide angiography. *Vet Radiol* **31:** 182–185, 1990.

Flanders, J. A. Feline aortic thromboembolism. *Compend Contin Educ Pract Vet* **8:** 473–484, 1986.

Furneaux, R. W. Two cases of fat embolism in the dog. *J Am Anim Hosp Assoc* **10:** 45–47, 1974.

Gilmore, D. R., and deLahunta, A. Necrotizing myelopathy secondary to presumed or confirmed fibrocartilaginous embolism in 24 dogs. *J Am Anim Hosp Assoc* **23:** 373–376, 1987.

Gilroy, B. A., and Anson, L. W. Fatal air embolism during anesthesia for laparoscopy in a dog. *J Am Vet Med Assoc* **190:** 552–554, 1987.

Green, R. A. Clinical implications of antithrombin III deficiency in domestic animals. *Compend Contin Educ Pract Vet* **6:** 537–545, 1984.

Jones, R. S., Payne-Johnson, C. E., and Seymour, C. J. Pulmonary micro-embolism following orthopaedic surgery in a thoroughbred gelding. *Equine Vet J* **20:** 382–384, 1988.

Klein, M. K., Dow, S. W., and Rosychuk, R. A. W. Pulmonary thromboembolism associated with immune-mediated hemolytic anemia in dogs: Ten cases (1982–1987). *J Am Vet Med Assoc* **195:** 246–250, 1989.

Langelier, K. M. Ischemic neuromyopathy associated with steel pellet BB shot aortic obstruction in a cat. *Can Vet J* **23:** 187–189, 1978.

LaRue, M. J., and Murtaugh, R. J. Pulmonary thromboembolism in dogs: 47 cases (1986–1987). *J Am Vet Med Assoc* **197:** 1368–1372, 1990.

Meschter, C. L. Disseminated sweat gland adenocarcinoma with acronecrosis in a cat. *Cornell Vet* **81:** 195–203, 1991.

Pion, P. D. Feline aortic thromboemboli and the potential utility of thrombolytic therapy with tissue plasminogen activator. *Vet Clin North Am: Small Anim Pract* **18:** 79–86, 1988.

Rolfe, D. L. Aortic thromboembolism in a calf. *Can Vet J* **18:** 321–324, 1977.

Spier, S. Arterial thrombosis as the cause of lameness in a foal. *J Am Vet Med Assoc* **187:** 164–165, 1985.

Stuart, B. P. *et al.* Ischemic myopathy associated with systemic dirofilariasis. *J Am Anim Hosp Assoc* **14:** 36–39, 1978.

Swanwick, R. A., and Williams, O. J. Fatal myocardial infarct in a greyhound. *J Small Anim Pract* **23:** 451–455, 1982.

1. Aortic–Iliac Thrombosis in Horses

Aortic–iliac thrombosis causes exercise intolerance and hind-leg lameness (intermittent claudication) in affected horses. The condition is seen most frequently in racing Thoroughbreds and Standardbreds, especially young males. The oldest lesions are located at the aortic quadri-

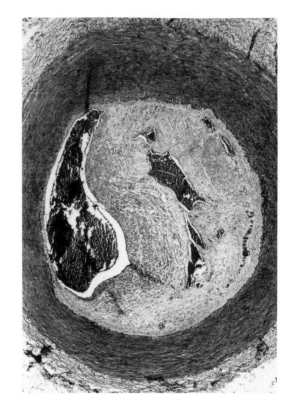

Fig. 1.59 Aortic–iliac thrombosis. Horse. Lumen of artery is almost occluded by an organized, recanalized fibromuscular mass.

furcation and in the distal portions of the femoral and internal iliac arteries, and consist of partially or completely occlusive masses of well-organized and well-vascularized fibrous tissue, occasionally containing hemorrhagic or degenerate areas (Fig. 1.59). Proximal to these organized masses, large unorganized thrombi are often present. The tunica intima is obliterated by the occlusive masses, except for the internal elastic lamina, which usually remains intact. The tunica media is largely unaffected, except for ischemic necrosis of the media in greatly distended arteries or under thick plaques. The pathogenesis of the lesions is unknown. The lesions may result from organization of strongyle-related thromboemboli, or from progressive enlargement and organization of spontaneously developing fibrous intimal plaques.

Bibliography

Edwards, G. B., and Allen, W. E. Aortoiliac thrombosis in two horses: Clinical course of the disease and use of real-time ultrasonography to confirm diagnosis. *Equine Vet J* **20:** 384–387, 1988.

Maxie, M. G., and Physick-Sheard, P. W. Aortic–iliac thrombosis in horses. *Vet Pathol* **22:** 238–249, 1985.

2. Disseminated Intravascular Coagulation

Disseminated intravascular coagulation (DIC) is a common and important intermediary mechanism of disease, but is not a disease in itself. It may be defined as a pathological activation of the coagulation system that leads to generalized intravascular clotting involving particularly arterioles and capillaries. The process may be acute, subacute, or chronic, and may be localized or generalized. Macrothrombosis does not usually develop in this disorder. The terms consumption coagulopathy, defibrination syndrome, and consumptive thrombohemorrhagic disorder are also used because of the massive consumption of coagulation factors which occurs and which may be sufficiently severe that a hemorrhagic diathesis results.

A wide array of agents and conditions will initiate coagulation (Table 1.6), basically either by causing widespread endothelial damage and thus exposing thrombogenic subendothelial collagen, or by directly activating the coagulation cascade via the intrinsic or extrinsic pathway. Exposure of monocytes, macrophages, and endothelial cells to disease agents or mediators will cause expression of tissue thromboplastin on the cell surfaces and activation of the extrinsic pathway of coagulation, and is likely a predominant pathway of DIC. Stasis of blood flow and plasma hyperviscosity are other factors which may contribute.

Many of the agents listed in Table 1.6 may initiate DIC by more than one route. For example, Gram-negative bacterial endotoxin can cause endothelial damage (either

TABLE 1.6

Agents or Conditions Known to Induce Disseminated Intravascular Coagulation in Animals

Bacteria	Gram negative (endotoxin)
	Gram positive
Helminths	*Dirofilaria immitis*
Protozoa	*Theileria* spp.
	Sarcocystis spp.
	Babesia spp.
Rickettsia	*Rickettsia rickettsii*
Viruses	African swine fever
	Hog cholera
	Bluetongue
	Epizootic hemorrhagic disease of deer
	Infectious canine hepatitis
	Feline infectious peritonitis
	Aleutian disease of mink
Neoplasia	Carcinoma
	Leukemia
	Hemangiosarcoma
Other	Aflatoxicosis
	Antigen–antibody complexes (incompatible blood transfusion)
	Gastric dilation (volvulus)
	Heat stroke
	Hyperlipemia in ponies
	Hyperosmolality
	Immunologic endothelial injury
	Ingestion of red maple leaves
	Proteolytic enzymes (pancreatitis, snakebite)
	Shock, vascular stasis, prolonged anesthesia, acidosis
	Tissue necrosis (hepatic, pneumonia, postsurgery, burns)

directly or via mediators such as interleukin-1 and tumor necrosis factor), which exposes subendothelial collagen and leads to activation of platelets and Hageman factor; activation of the complement cascade and hence platelet activation; direct activation of Hageman factor and hence the intrinsic pathway; and leukocyte damage which results in thromboplastin release. Viral infections are postulated to induce DIC by viral- or antiplatelet antibody-mediated platelet aggregation, by causing endothelial damage, and by causing Hageman factor activation by immune complexes. As well, in feline infectious peritonitis, for example, hepatic necrosis impairs macrophage clearance of activated clotting factors and may potentiate the coagulation defect. Another important pathophysiologic event that accompanies activation of coagulation is activation of the kallikrein–kinin system; generation of kinins contributes to increased vascular permeability, hypotension, and shock.

When coagulation is initiated, activated Hageman factor simultaneously converts plasminogen to plasmin. Plasmin cleaves fibrinogen/fibrin to produce fibrinogen/fibrin degradation products, which inhibit thrombin activity, fibrin polymerization, and platelet aggregation. A negative feedback system normally exists between thrombosis (coagulation) and fibrinolysis (anticoagulation), in which thrombin helps convert plasminogen to plasmin to cause fibrinolysis and hence limit thrombosis. In addition, a dynamic equilibrium exists between production of tissue plasminogen activator and production of plasminogen activator inhibitors by endothelial cells.

The clinical manifestations of DIC are protean and vary with the organ(s) in which thrombosis predominates and the severity of a hemorrhagic diathesis. Clinical presentations include shock, organ failure, hemorrhage, or hemolysis depending on the relative activation of thrombin and plasmin and hence the dominance of thrombosis or hemorrhage. Hemorrhage is usually seen only in animals which survive acute episodes with thrombosis and shock, or in those reacting to a moderate procoagulatory stimulus. Perhaps the most common result of acute DIC is shock due to reduced venous return to the right heart, plus systemic and pulmonary arterial hypertension due to thrombosis. Special examples of DIC include

1. the **generalized Shwartzman-like reaction** (GSR) (bilateral hemorrhagic renal cortical necrosis, induced experimentally by timed injections of endotoxin), which is seen as a cause of oliguric renal failure in septicemic or endotoxemic postpartum cows;
2. **hemorrhagic adrenal necrosis** (Waterhouse–Friderichsen syndrome) seen in septicemic calves;
3. **microangiopathic hemolytic anemia** (MHA) (an intravascular fragmentation anemia) seen in septicemic calves;
4. **acrocyanosis and gangrene** seen in the extremities of calves recovering from septicemia; and
5. the **hemolytic–uremic syndrome** (HUS) in which

coagulation is combined with microangiopathic hemolytic anemia and acute renal failure.

Animals that die with DIC may have petechial to ecchymotic hemorrhages on mucosae and serosae and in the skin, and may have variable thrombotic or hemorrhagic involvement of internal organs. Histologically, microthrombi may be associated with congestion, edema, hemorrhage, and necrosis. Microthrombi are usually seen most easily in capillaries of brain, renal glomeruli, adrenals, lungs, and myocardium, but may occur in any organ. They consist primarily of fibrin (hyaline thrombi), platelets (granular thrombi), or of fibrin degradation products (hyaline globules or shock bodies); pulmonary alveolar hyaline membranes are also said to be a form of hyaline microthrombi. Because fibrinolysis is an ongoing process which continues after death, most microthrombi are lysed within 3 hr of death and may not be found if postmortem examination is delayed. With routine hematoxylin-and-eosin staining, microthrombi appear as homogeneous pink masses in dilated small blood vessels. Identification of microthrombi is aided by the use of special stains such as Martius-scarlet-blue, which stains fibrin red, or phosphotungstic acid hematoxylin, which stains fibrin purple. Even with special stains, incompletely polymerized fibrin is difficult to demonstrate. Red cells that impinge on fibrin strands in the microcirculation may fragment; these red cell fragments (schistocytes) may be detected within blood vessels in histologic sections and may thus serve as indicators of DIC.

Bibliography

Bick, R. L. Disseminated intravascular coagulation and related syndromes: A clinical review. *Semin Thromb Hemost* **14:** 299–338, 1988. (354 references).

Feldman, B. F., Thomson, D. B., and O'Neill, S. Plasma fibronectin concentrations in dogs with disseminated intravascular coagulation. *Am J Vet Res* **46:** 1171–1174, 1985.

Gay, C. C. *et al.* Hyperlipemia in ponies. *Aust Vet J* **54:** 459–462, 1978.

Hammer, A. S. *et al.* Hemostatic abnormalities in dogs with hemangiosarcoma. *J Vet Internal Med* **5:** 11–14, 1991.

Hoffmann, R. Adrenal lesions in calves dying from endotoxin shock, with special reference to the Waterhouse–Friderichsen syndrome. *J Comp Pathol* **87:** 231–239, 1977.

Israel, E. *et al.* Microangiopathic hemolytic anemia in a puppy. Grand rounds conference. *J Am Anim Hosp Assoc* **14:** 521–523, 1978.

Kane, K. K. Fibrinolysis—A review. *Ann Clin Lab Med* **14:** 443–449, 1984.

Long, P. H., and Payne, J. W. Red maple-associated pulmonary thrombosis in a horse. *J Am Vet Med Assoc* **184:** 977–978, 1984.

Medleau, L. *et al.* Erythema multiforme and disseminated intravascular coagulation in a dog. *J Am Anim Hosp Assoc* **26:** 643–646, 1990.

Momotani, E. *et al.* Histopathological evaluation of disseminated intravascular coagulation in *Haemophilus somnus* infection in cattle. *J Comp Pathol* **95:** 15–23, 1985.

Morgan, R. V. Acute gastric dilatation–volvulus syndrome. *Compend Contin Educ Pract Vet* **4:** 677–682, 1982.

Morris, C. F. *et al.* Hemolytic uremiclike syndrome in two horses. *J Am Vet Med Assoc* **191:** 1453–1454, 1987.

Morris, D. D., and Beech, J. Disseminated intravascular coagulation in six horses. *J Am Vet Med Assoc* **183:** 1067–1072, 1983.

Muller-Berghaus, G. Pathophysiologic and biochemical events in disseminated intravascular coagulation: Dysregulation of procoagulant and anticoagulant pathways. *Semin Thromb Hemost* **15:** 58–87, 1989, (Review, 344 references).

Nordstoga, K., and Fjolstad, M. The generalized Shwartzman reaction and *Haemophilus* infections in pigs. *Pathol Vet* **4:** 245–253, 1967.

Richardson, S. E. *et al.* The histopathology of the hemolytic uremic syndrome associated with verocytotoxin-producing *Escherichia coli* infections. *Hum Pathol* **19:** 1102–1108, 1988.

Shimamura, K. *et al.* Distribution patterns of microthrombi in disseminated intravascular coagulation. *Arch Pathol Lab Med* **107:** 543–547, 1983.

Slappendel, R. J. Disseminated intravascular coagulation. *Vet Clin North Am: Small Anim Pract* **18:** 169–184, 1988.

E. Arterial Hypertrophy

Hypertrophy of arteries may affect one or all components of the arterial wall. High altitude disease of cattle, described subsequently, is a result of hypoxia-induced pulmonary arterial constriction and hypertrophy, which lead to pulmonary hypertension and eventual right heart failure (cor pulmonale). Various cardiac anomalies cause pulmonary arterial hypertension and hyperperfusion, which result in pulmonary arterial hypertrophy and additional pulmonary hypertension (see the following). Arterial hypertrophy occurs in collateral vessels in response to the extra load they carry after occlusion of the artery which usually supplies an area. A specific form of hypertrophy occurs in the pulmonary arteries of cats and is described subsequently. Systemic arterial involvement in the hypertension associated with chronic renal disease is described in preceding sections.

1. High-Altitude Disease of Cattle

This disease is caused by pulmonary hypertension that results in dilation and hypertrophy of the right ventricle with the ultimate development of cardiac decompensation and congestive cardiac failure. Edematous swelling of the venter, as is typical of congestive heart failure in cattle, is responsible for the synonym brisket-disease. The syndrome in cattle resembles in many respects the chronic mountain sickness of humans resident at high altitude, and represents failure of the cardiorespiratory system to adjust to the hypoxia of higher altitudes. There are species differences in cardiorespiratory sensitivity to hypoxia; lambs appear to be relatively insensitive and humans, moderately so, whereas cattle are hypersensitive and prone to develop hypertensive heart failure at altitudes of about 2500 m and above. There are also variations within species, and some cattle react dramatically to degrees of hypoxia that only mildly inconvenience others. Young cattle are more susceptible than adults, and the morbidity rate is highest in animals exposed to high altitudes for the first time. In animals imported from low altitudes to about 3500 m, the incidence of severe pulmonary hypertension approaches 50%, whereas in herds maintained at such altitudes for many generations, high-altitude disease may not affect more than 2%. The different morbidity rates are probably attributable to natural selection in the resident population.

Cattle exposed to critical altitudes, or experimentally to comparable degrees of hypoxic stimulation, develop a severalfold increase in pulmonary arterial pressure and pulmonary vascular resistance, which may gradually be reduced to near normotensive levels when the animals are returned to low altitudes. A distinctive feature of the pulmonary vasculature in cattle is the well-developed muscular media of the arteries and veins with diameters as small as 20 μm, which implies an unusual potential for vasomotor activity. Hypoxia-induced vasoconstriction leads to work hypertrophy of the muscular media of pulmonary arteries and arterioles, and hence hypertension, and may be compounded by reaction hypertrophy that occurs in response to increased arterial pressure. In cattle affected with high-altitude disease, there is uniformly prominent hypertrophy of the muscular media of pulmonary arteries and arterioles, and usually adventitial proliferation around pulmonary arteries of all sizes. The hypertension may also cause endothelial damage, thrombosis, intimal proliferation, and medial mineralization, especially in elastic pulmonary arteries.

Bibliography

Bisgard, G. E. Pulmonary hypertension in cattle. *Adv Vet Sci Comp Med* **21:** 151–172, 1977.

Dingemans, K. P., and Wagenvoort, C. A. Pulmonary arteries and veins in experimental hypoxia. *Am J Pathol* **93:** 353–368, 1978.

Hull, M. W., and Anderson, C. K. Right ventricular heart failure of Montana cattle. *Cornell Vet* **68:** 199–210, 1978.

2. Cardiac Anomalies and Pulmonary Arterial Hypertension

Increased pulmonary arterial pressure (hypertension) and increased blood flow (hyperperfusion) occur with certain cardiac anomalies in which there is left-to-right shunting of blood, such as a large ventricular septal defect or patent ductus arteriosus. Pulmonary arterial hypertension, which may even lead to reversal of the shunt and cyanosis (Eisenmenger's syndrome), produces pulmonary arterial constriction and hypertrophy, increased pulmonary vascular resistance, and sustained pulmonary hypertension by means of a number of arterial responses. These responses range from persistence of the normal thick muscular wall configuration of fetal life, to lesions that are the acquired response to hypertension (Fig. 1.60A). These acquired lesions include prominent medial hypertrophy, intimal proliferations which may be obliterative, disruption of elastic laminae, adventitial fibrosis, and plexiform lesions (vascular occlusion by myointimal proliferations containing irregular vascular channels). In the early, cellu-

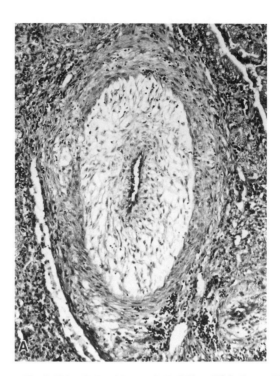

Fig. 1.60A Intimal hyperplasia (idiopathic). Lung. Kitten.

lar phase of the lesion, there is migration of medial smooth muscle cells through the internal elastic lamina and into the intima, possibly in response to a chemical attractant such as endothelium cell-derived growth factor. This is followed by conversion of the smooth muscle cells to myofibroblasts, and production of ground substance, collagen, and elastin, which produces marked intimal thickening (Fig. 1.60B). Aged mature lesions become less cellular, myofibroblasts revert to smooth muscle cells, and thin elastotic septa separate wide vascular channels; these channels may be thrombosed. As the pulmonary vascular lesions progress, the arteries may dilate, and their walls undergo fibrinoid degeneration. Together these lesions constitute **plexogenic pulmonary arteriopathy,** which in its more severe manifestations is an end-stage irreversible arterial condition. It is associated with a poor prognosis, even if the underlying cardiac anomaly can be corrected surgically.

Bibliography

Caslin, A. W. *et al.* The histopathology of 36 cases of plexogenic pulmonary arteriopathy. *Histopathology* **16:** 9–19, 1990.

Jeraj, K. *et al.* Patent ductus arteriosus with pulmonary hypertension in a cat. *J Am Vet Med Assoc* **172:** 1432–1436, 1978.

Nimmo-Wilkie, J. S., and Feldman, E. C. Pulmonary vascular lesions associated with congenital heart defects in three dogs. *J Am Anim Hosp Assoc* **17:** 485–490, 1981.

Pyle, R. L. *et al.* Patent ductus arteriosus with pulmonary hypertension in the dog. *J Am Vet Med Assoc* **178:** 565–571, 1981.

Smith, P. *et al.* The ultrastructure of plexogenic pulmonary arteriopathy. *J Pathol* **160:** 111–121, 1990.

Turk, J. R., Miller, J. B., and Sande, R. D. Plexogenic pulmonary

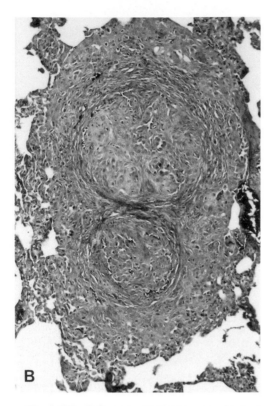

Fig. 1.60B Pulmonary arterial hypertension in dog with cardiac anomaly. Two pulmonary arteries with intimal fibromuscular hyperplasia, medial and adventitial hyperplasia. (Slide, courtesy of J. S. Nimmo-Wilkie.)

arteriopathy in a dog with ventricular septal defect and pulmonary hypertension. *J Am Anim Hosp Assoc* **18:** 608–612, 1982.

3. Medial Hypertrophy of the Pulmonary Arteries of Cats

Pulmonary medial hypertrophy and hyperplasia commonly occur in cats, but are of no clinical significance even when very severe and generalized. The change is seen with equal frequency in germ-free and conventional cats, displays no age, sex, or breed predilection, and appears to be normal anatomic variation in cats. The cat lungworm *Aelurostrongylus abstrusus* has been suggested as an etiologic agent, and although the parasite may produce similar lesions, it is not necessary for the development of the lesion. Pulmonary hypertension does not result from the vascular change; right ventricular pressure does not increase, nor does the right ventricle hypertrophy. The most severely affected vessels may be grossly visible on the cut surface of the lung and through the pleura when the lungs are collapsed, and are even palpable in some cases. The histologic spectrum of arterial changes ranges from mild sporadic or generalized hypertrophy of the tunica media, to marked medial hypertrophy and hyperplasia with concomitant intimal proliferation and severe luminal encroachment. Thrombosis has not been observed, but adhesion of endothelial surfaces across the lumen does occur, providing sinuous channels and an ap-

pearance similar to that of an organized, recanalized thrombus. Frequently, the arterial hypertrophy is accompanied by fibromuscular hyperplasia in the pulmonary parenchyma and around alveolar ducts.

Bibliography

Rawlings, C. A. *et al.* Response of the feline heart to *Aelurostrongylus abstrusus*. *J Am Anim Hosp Assoc* **16:** 573–578, 1980.

F. Arteritis

Arteritis resulting from a large variety of causes—infectious, toxic, immunologic—is a common occurrence, and its pathogenesis has been dealt with in general under vasculitis. Arteritis may be of considerable significance in the pathogenesis of a condition, usually because of the occurrence of thrombosis, ischemia, and infarction (Fig. 1.61). As well as having direct effects due to thrombosis and infarction, arteritis may affect distant organs, as occurs when obliterative pulmonary endoarteritis leads to congestive heart failure in dogs infected with *Dirofilaria immitis*. The term endoarteritis is used in preference to endarteritis to avoid implying that end-arteries are primarily affected.

Arteritis of hematogenous origin occurs in the course of septicemias and bacterial endocarditis. The primary injury may be to the endothelium and intima or it may, presumably when the organisms localize in the vasa vasorum, affect first the adventitia and outer lamellae. Septi-

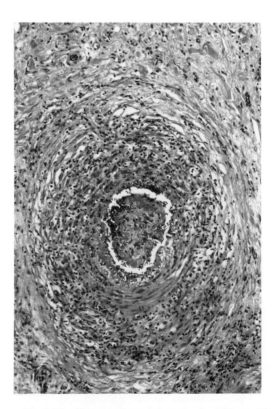

Fig. 1.61 Panarteritis and thrombosis. Kidney. Pig.

cemic salmonellosis and erysipelas, and hog cholera, are often responsible for arteritis in pigs. The results are seen most readily in the skin as erythematous areas of purple discoloration of the ears, perineum, snout, and venter, which, if the animal survives, may become necrotic and slough. Lesions with this basis may also occur in other organs, and indeed are often seen as infarcts in the spleen in erysipelas and hog cholera. In affected portions of the vessels, the endothelium swells and proliferates, and the walls of the vessels undergo acute fibrinoid necrosis. The formation of thrombi in these areas in turn gives rise to infarcts and necrosis.

Arteritis arising by extension of inflammation and infection from adjacent tissues may occur, expecially if the original inflammatory process is suppurative or necrotizing. It is encountered most often in the bacterial pneumonias, adjacent to abscesses, in purulent meningitis, in local lesions of aspergillosis and mucormycosis, and in acute metritis, especially if complicated by necrobacillosis. Fungi of the family Mucorales have a distinct affinity for arteries and produce a necrotizing, thrombotic arteritis.

Arteritis may be found as an incidental background lesion in clinically normal animals. For example, mild idiopathic arteritis and intimal thickening occurs in extramural coronary arteries in 5–10% of young beagle and mongrel dogs, which can complicate the interpretation of lesions in dogs used in arterial toxicity studies. A more severe, febrile syndrome, the beagle pain syndrome, occurs in beagles and other breeds. In this syndrome, necrotizing arteritis affects the coronary and other arteries and results in thrombosis, infarction, and hemorrhage as well as progressive atrophy of temporal and cervical muscles; amyloidosis occurs in some chronically affected dogs. The cause of the syndrome is unknown; immune complex vasculitis is suspected.

An idiopathic vasculopathy occurs in kenneled and racing greyhound dogs, involving skin and occasionally kidneys. Deep, slowly healing cutaneous ulcers occur as the result of fibrinoid arteritis, thrombosis, and hemorrhagic infarction. Renal lesions include acute necrosis of glomeruli and of afferent arterioles, glomerular capillary thrombosis, and acute tubular necrosis. The pathogenesis of the condition is thought to involve a genetic predisposition to an unidentified infectious agent.

1. Polyarteritis Nodosa (Panarteritis or Periarteritis Nodosa)

The term polyarteritis nodosa has been applied to a heterogeneous group of arteritides, which occur sporadically in all species of domestic animals, on the basis of similarities with the human condition in which small and medium-sized arteries undergo severe necrotizing inflammation, often in a nodose pattern and with a predilection for branching points. As all layers of the arterial wall are involved, the lesion is also referred to as panarteritis. The only agent thus far associated with the condition in humans is hepatitis B antigen, which is found in 25–40% of individuals with necrotizing vasculitis; high mortality

occurred in the human disease before the advent of corticosteroid therapy. Polyarteritis nodosa has been reported in dogs having rheumatoid arthritis and systemic lupus erythematosus; separation of the vasculitis component from disease syndromes is of problematic value. Also, as noted under vasculitis, a more appropriate term for idiopathic vasculitides is probably **systemic necrotizing vasculitis** rather than polyarteritis nodosa, which has tended to become a catch-all term.

A wide variety of clinical and pathologic manifestations occurs in polyarteritis nodosa as described in humans and domestic animals. Although occasionally associated with the death of an animal, lesions are also noted in otherwise normal animals at slaughter. Renal, coronary, hepatic, and gastrointestinal vessels are perhaps most commonly involved (Fig. 1.62A,B). Arterial lesions at all stages of development, namely acute, healing, and healed, may occur simultaneously in one individual.

The acute lesions are often typical of immune complex-induced arteritis, and inflammation and necrosis of the arterial wall may be segmental or circumferential. Thrombosis may occur and lead to infarction and hemorrhage. As the reaction progresses, neutrophils are replaced in the media by mononuclear cells, and fibroplasia may result in grossly visible fibrous thickening of the wall. The vascular lumen may be obliterated by organization of a thrombus plus intimal proliferation. The term periarteritis has been used in the past to describe a stage in the vascular reaction in which mononuclear cells surround the vessel, but the term is now little used.

A variant of polyarteritis nodosa, known as **hypersensitivity angiitis,** affects smaller vessels in humans, namely arterioles, capillaries, and venules in skin, mucous membranes, lungs, brain, heart, gastrointestinal tract, kidneys, and muscle. Synonyms include allergic vasculitis, micro-

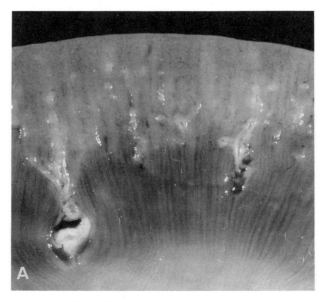

Fig. 1.62A Polyarteritis nodosa. Sheep.

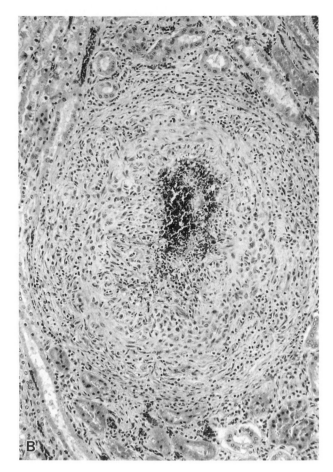

Fig. 1.62B Panarteritis. Kidney. Sheep. Border disease. (Slide, courtesy of J. Orr.)

scopic polyarteritis, and leukocytoclastic vasculitis. Suspected causes include many common drugs (e.g., ampicillin, chloramphenicol, diazepam, levamisole, penicillin, phenylbutazone, potassium iodide, sulfonamides, tetracycline), microbes (β-hemolytic streptococci), heterologous proteins, and tumor antigens. The reaction often occurs 7–10 days after exposure to the stimulus; remission usually follows removal of the agent. Lesions typical of leukocytoclastic vasculitis are seen. Since larger arteries are spared, macroscopic infarcts are uncommon. Hypersensitivity angiitis appears to be the basis of some cases of hemorrhagic purpuric disease in horses; other causes of subcutaneous edema and mucosal hemorrhages, such as equine viral arteritis, equine infectious anemia, and equine ehrlichiosis, must also be considered. Trimethoprim-sulfadiazine and trimethoprim-sulfamethoxazole have caused hypersensitivity vasculitis in dogs.

Bibliography

Albassam, M. A. *et al.* Polyarteritis in a beagle. *J Am Vet Med Assoc* **194**: 1595–1597, 1989.

Arteritis and arterial drug toxicity in the safety assessment of drugs. *Toxicol Pathol* **17**: 65–231, 1989 (17 papers).

Carpenter, J. L. *et al.* Idiopathic cutaneous and renal glomerular vasculopathy of greyhounds. *Vet Pathol* **25**: 401–407, 1988.

Carpenter, J. L., Moore, F. M., and Albert, D. M. Polyarteritis nodosa and rheumatic heart disease in a dog. *J Am Vet Med Assoc* **192**: 929–932, 1988.

Crawford, M. A., and Foil, C. S. Vasculitis: Clinical syndromes in small animals. *Compend Contin Educ Pract Vet* **11**: 400–415, 1989.

Curtis, R., Bell, W. J., and Laing, P. W. Polyarteritis in a cat. *Vet Rec* **105**: 354, 1979.

Elling, F. Nutritionally induced necrotizing glomerulonephritis and polyarteritis nodosa in pigs. *Acta Pathol Microbiol Scand* **87A**: 387–392, 1979.

Filippich, L. J., and Mudie, A. W. Polyarteritis nodosa in a bovine carcass. *Aust Vet J* **48**: 66, 1972.

Hamir, A. N. Polyarteritis nodosa in a sow. *Aust Vet J* **56**: 343–344, 1980.

Helmboldt, C. F., Jungherr, E. L., and Hwang, J. Polyarteritis in sheep. *J Am Vet Med Assoc* **134**: 556–561, 1959.

Kemi, M. *et al.* Histopathology of spontaneous panarteritis in beagle dogs. *Jpn J Vet Sci* **52**: 55–61, 1990.

Landsverk, T., and Bratberg, B. Polyarteritis nodosa associated with sarcocystosis in a lamb. *Acta Vet Scand* **20**: 306–308, 1979.

Morris, D. D. Cutaneous vasculitis in horses: 19 cases (1978–1985). *J Am Vet Med Assoc* **191**: 460–464, 1987.

Rachofsky, M. A. *et al.* Probable hypersensitivity vasculitis in a dog. *J Am Vet Med Assoc* **194**: 1592–1594, 1989.

2. Verminous Arteritis

We are here concerned with those parasites whose natural habitat is the lumen or the wall of arteries. Parasites which, accidentally or as part of their life cycle, migrate in tissues may, depending on the route they take, cause vasculitis in various organs. The chief offenders in this regard are the larvae of the genera *Strongylus* and *Ascaris;* the vascular lesions of *Ascaris* infestations are described under Diseases of the Intestine (Volume 2, Chapter 1). *Angiostrongylus vasorum,* which inhabits the pulmonary arteries of dogs, is described with the lungworms. *Spirocerca lupi,* which passes part of its history in the adventitia of the thoracic aorta, is described in its final habitat, the esophagus.

a. DIROFILARIASIS (HEARTWORM DISEASE). The dog is the only mammal commonly infected by *Dirofilaria immitis* (heartworm) and is the only significant reservoir of infection. In areas in which heartworm infection is enzootic in dogs, a number of other mammals may become infected, including domestic cats and wild felids, wild canids (coyotes, foxes, wolves), sea lions, muskrats, horses and, rarely, humans. If *D. immitis* develops to maturity in hosts other than the dog, microfilaremia is usually low or absent, and heart failure is rare. Pulmonary dirofilariasis is of considerable importance in humans because the spherical, subpleural granulomas produced are usually mistaken radiographically for primary or metastatic lung tumors, leading to thoracotomy and excisional biopsy for diagnosis. *Dirofilaria immitis* has a tropical and subtropical distribution in southern Europe, Asia, Australia, and North and South America, although its range has extended into more northern temperate zones in North America and southern temperate zones in Australia.

Adults of *Dirofilaria repens* occur in subcutaneous connective tissues of dogs and other carnivores (and rarely humans). Similarly, *D. tenuis* is found in the subcutaneous connective tissue of the raccoon in the southern United States. None of these three filariids usually develops to maturity in humans, but they may be found in cysts in various parts of the body, especially in the lungs. *Dirofilaria striata,* a parasite of bobcats (*Lynx rufus*), has been reported in dogs in Florida. The Florida *Dirofilaria* differs slightly from *D. immitis* and *D. striata.*

Adult *D. immitis* worms live in the pulmonary arteries and right heart, but they can develop in various other arteries and veins, and have been reported from a variety of unusual sites, including the eye and brain. Adult males are 12–20 cm long by 0.7–0.9 mm in diameter; females are 25–31 cm by 1.0–1.3 mm. The adult female worms are viviparous, and release microfilariae into the bloodstream; microfilariae are obtained by mosquitoes in blood meals. The first, second, and early third larval stages of *D. immitis* are obligate parasites of mosquitoes, predominantly of the genera *Aedes, Culex,* or *Anopheles*. Larvae develop to the infective stage in about 13 days in the malpighian tubules. They then migrate to the cephalic spaces of the head or to the proboscis of the mosquito and escape into a new host when the mosquito feeds. The infective larvae are ~1.2 mm long when they enter the host, and are ~5 cm long when they reach the right ventricle 3–4 months later. In the meantime, the parasites have wandered and grown in the connective tissues, chiefly those of the subcutis and muscles, finally leaving these to migrate to the heart via the veins. They mature in a further 3 months. Thus the prepatent period is 6–8 months. Adults may live for several years, and microfilariae can survive in a dog for as long as 2.5 years.

The clinical diagnosis of dirofilariasis depends on the detection of microfilariae in the blood, preferably by the microscopic examination of a centrifuged sediment of lysed blood (modified Knott's test), or by a filter technique. Microfilariae of *D. immitis* (290–315 μm long, straight body and tail, tapered head) must be differentiated from those of *Dipetalonema reconditum* (270–290 μm, curved body, blunt head; most have a button-hooked tail), the adults of which are nonpathogenic and are commonly found in the subcutaneous connective tissue of dogs. The average number of circulating microfilariae per adult female *D. immitis* decreases as the number of adult worms increases. Hence the number of circulating microfilariae indicates neither the numbers of adult worms present nor the severity of heartworm disease in an individual dog. Although microfilariae are present in the blood more or less continuously, there is a tendency to periodicity, with maximum numbers occurring in the evening and in the summer, coincident with maximal activity of mosquitoes. Microfilariae are pooled in internal organs, especially the lungs, in the daytime. Fetal pups may be infected by mi-

crofilariae from the bitch and have microfilaremia, but the pups do not harbor adult worms.

Occult dirofilariasis, that is, heartworm infection without microfilaremia, occurs in 10–67% of infected dogs, primarily as a result of destruction of microfilariae by immune mechanisms (immune-mediated occult disease), but also as the result of prepatent infections, single-sex infections, and drug-induced sterility of adult heartworms. Clinical diagnosis of occult dirofilariasis then rests on the findings of the indirect fluorescent antibody test, antigen detection tests, and cardiopulmonary imaging. Circulating antibodies are usually detectable only in dogs without microfilaremia.

The clinical signs of canine dirofilariasis are usually those of cardiovascular dysfunction, such as cough and tiring on exercise, which may progress to congestive heart failure. Heartworm disease is usually seen in dogs which are older than 5 years and have had continuous or multiple infections. There is a rough correlation between the numbers of worms present and the severity of the disturbance, but there are exceptional cases, and severe signs may appear in dogs with only a few worms. Clinically normal dogs may harbor as many as about 30 worms, and clinically affected dogs, 50 or more.

Adult heartworms are normally found in the pulmonary arteries and right ventricle (Fig. 1.63A), but may be found in the right atrium and venae cavae in heavy infestations. Young adult worms usually develop first in small pulmonary arteries and progressively involve more proximal arteries as they grow. Adult worms may be found in the right ventricle if more than about 25 are present, and in the right atrium and venae cavae if there are 50 or more worms. The caudal lobar pulmonary arteries are usually the most heavily infested vessels. Right heart failure develops as a consequence of pulmonary hypertension, which in turn is the result of pulmonary vascular sclerosis. Of secondary importance is mechanical interference of the worms with cardiac blood flow and valve function. The pulmonary sclerosis and hypertension slowly regress following removal of the adult worms.

Heartworm disease is primarily a pulmonary vascular disease, and is characterized by myointimal proliferations in small peripheral arteries initially, and later in the large lobar arteries, especially the right caudal lobar artery (Fig. 1.63B). The fibromuscular intimal proliferations in the larger vessels produce a grossly visible shaggy or roughened appearance, a change pathognomonic of heartworm disease. The vascular reaction is to the immature and mature adult worms, and begins as an endoarteritis with infiltration of leukocytes, primarily eosinophils. The leukocytic response subsides and is replaced by myointimal proliferation, which produces irregular rugose to villous projections enmeshing the worms; the myointimal proliferation occurs at sites of direct contact with worms, is likely due to mechanical irritation and endothelial damage, and is possibly mediated by platelet-derived growth factor. Thrombosis may be associated with either live or dead worms, and thromboembolism, especially following

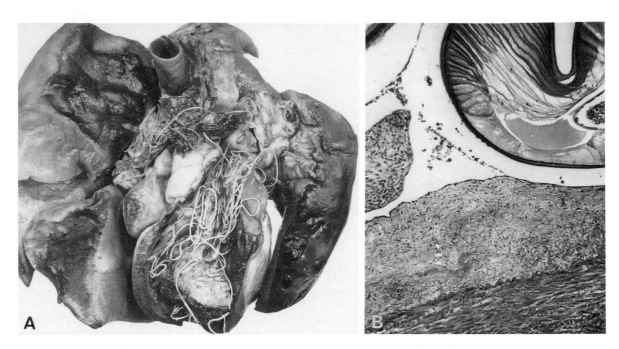

Fig. 1.63 (A) *Dirofilaria immitis* in heart and pulmonary artery. Dog. (Courtesy of P. Olafson.) (B) Sclerosis of pulmonary artery containing *Dirofilaria immitis*. Dog.

adulticide therapy, may further exacerbate pulmonary hypertension; the presence of dead parasites initiates a granulomatous reaction in the vessel wall, which may extend into the pulmonary parenchyma. Thrombi may be organized and recanalized. Pulmonary infarction is an uncommon sequel to thromboembolism.

Pulmonary parenchymal changes which accompany these vascular changes include periarterial granulomatous inflammation, hemosiderosis, diffuse interalveolar fibrosis, proliferation of alveolar epithelium, and fibrous pleural thickenings. Dead worms commonly incite pulmonary granuloma formation. Adulticide therapy causes embolization of worms and resulting thrombosis and granulomatous inflammation; resolution of these lesions is under way by 6 weeks posttreatment. Proliferative intimal lesions will progressively resolve after surgical removal of adult worms; resolution of advanced, fibrotic obstructive arterial lesions is unlikely. Severe pulmonary arterial disease often results in chronic or low-grade DIC, and hence thrombocytopenia and hemoglobinuria. Microfilariae entrapped in the lung in a state of antibody excess are surrounded by neutrophils and eosinophils, producing an eosinophilic pneumonitis. This allergic pneumonitis may be the precursor of pulmonary eosinophilic granulomatosis, which is characterized by 1–8 cm diameter nodules composed of eosinophils, lymphocytes, plasma cells, and macrophages, and devoid of microfilariae, plus peripheral blood eosinophilia.

Additional lesions of heartworm disease include those of right heart failure, such as chronic passive congestion of the liver, and occasionally ascites. Membranoproliferative glomerulonephritis occurs primarily due to glomerular deposition of immune complexes, either those formed in circulating blood in a state of antigen excess, or formed *in situ* in the glomeruli. The intensity of the immune complex deposition varies directly with the intensity and duration of microfilaremia and the antibody response. Physical damage to glomerular endothelium has also been attributed to microfilariae. The result of glomerular disease in dirofilariasis is proteinuria, which is usually mild to moderate in severity. Renal failure and uremia are uncommon. Microfilariae are usually of little consequence to the dog, but when they die they may incite formation of microgranulomas in various organs, such as the lung, liver and kidney.

The **vena caval syndrome** (venae cavae syndrome, liver failure syndrome) is a variant of heartworm disease seen usually in young dogs in which large numbers (100+) of adult worms fill the right atrium and venae cavae, likely as the result of retrograde migration from the pulmonary arteries. The syndrome is characterized by sudden onset of weakness, anorexia, bilirubinuria, hemoglobinuria, and anemia. Shock occurs because of obstruction and decreased venous return. The mass of worms may also interfere with valve function, causing tricuspid regurgitation which, together with pulmonary hypertension, results in decreased left ventricular preload and congestive right ventricular failure. Hepatic congestion is very severe, hepatic lymphatics may be distended, and ascites may occur. Anemia develops due to right-atrial turbulence and shear-induced mechanical hemolysis, with perhaps microangiopathic hemolysis resulting from platelet activation and fibrin formation. Azotemia develops, and death usually occurs in 1–3 days. Phlebosclerosis and thrombosis in the caudal vena cava and hepatic veins are similar to the reactions usually seen in the pulmonary arteries.

Bibliography

Abramowsky, C. R. *et al. Dirofilaria immitis.* 5. Immunopathology of filarial nephropathy in dogs. *Am J Pathol* **104:** 1–12, 1981.

Arnoczky, S. P., Fox, P. R., and Tilley, L. P. Current concepts in heartworm disease. *Semin Vet Med Surg* **2:** 1–84, 1987.

Atkins, C. E., Keene, B. W., and McGuirk, S. M. Pathophysiologic mechanism of cardiac dysfunction in experimentally induced heartworm caval syndrome in dogs: An echocardiographic study. *Am J Vet Res* **49:** 403–410, 1988.

Boreham, P. F. L., and Atwell, R. B. (eds.). '' Dirofilariasis.'' Boca Raton, Florida, CRC Press, 1988.

Buoro, I. B. J, Atwell, R. B., and Heath, T. Angles of branching and the diameters of pulmonary arteries in relation to the distribution of pulmonary lesions in canine dirofilariasis. *Res Vet Sci* **35:** 353–356, 1983.

Calvert, C. A. *et al.* Pulmonary and disseminated eosinophilic granulomatosis in dogs. *J Am Anim Hosp Assoc* **24:** 311–320, 1988.

Grauer, G. F. *et al.* Clinicopathologic and histologic evaluation of *Dirofilaria immitis*-induced nephropathy in dogs. *Am J Trop Med Hyg* **37:** 588–596, 1987.

Ludders, J. W. *et al.* Renal microcirculatory and correlated histologic changes associated with dirofilariasis in dogs. *Am J Vet Res* **49:** 826–830, 1988.

Otto, G. F. (ed.) ''Proceedings of the Heartworm Symposium '89, Charleston, South Carolina.'' American Heartworm Society, Washington, D.C., 1986. (One of a series of the proceedings of the symposia of the American Heartworm Society; extensive bibliography)

Rawlings, C. A. ''Heartworm disease in dogs and cats.'' Philadelphia, Pennsylvania, W.B. Saunders, 1986.

Ro, J. Y. *et al.* Pulmonary dirofilariasis: The great imitator of primary or metastatic lung tumor. *Hum Pathol* **20:** 69–76, 1989.

Sasaki, Y. *et al.* Improvement in pulmonary arterial lesions after heartworm removal using flexible alligator forceps. *Jpn J Vet Sci* **52:** 743–752, 1990.

Sutton, R. H., and Atwell, R. B. Lesions of pulmonary pleura associated with canine heartworm disease. *Vet Pathol* **22:** 637–639, 1985.

b. ONCHOCERCIASIS (PARASITIC AORTITIS). *Onchocerca armillata* is a parasite of the wall of the aorta of cattle, water buffaloes, and goats, in decreasing order of prevalence, in south Asia and equatorial Africa. Several other *Onchocerca* spp. are described with tendons and aponeuroses under Diseases of Muscle (Volume 1, Chapter 2). The life cycle and vectors involved in the transmission of *O. armillata* are unknown; however, black flies (*Simulium*) and midges (*Culicoides*) are the intermediate hosts of other *Onchocerca* spp.

Although virtually all older cattle may be infested in enzootic areas, the lesions produced are apparently not of clinical significance. The preferential site is the arch of the aorta. With chronicity, lesions extend cranially to include the brachiocephalic trunk, costocervical arteries, and brachial arteries, and caudally to include the abdominal aorta to the level of the iliac bifurcation. The intima is corrugated and contains numerous sinuous tunnelings, which end in small nodules in the intima and media; the nodules protrude into the vascular lumen or through the adventitia. The males, which are ~7 cm long, inhabit the nodules along with the anterior end of the female, which bears the vaginal opening, and the microfilariae, which escape into the bloodstream. The body of the female lies in the sinuous tracts, from which the complete specimen cannot be readily dissected, but its length is estimated as >100 cm. Parasitic tunnels in the inner media have a thin fibrous tissue lining without cellular infiltration. There may be acute transmural aortitis with a predominance of eosinophils, but subacute or chronic focal granulomatous inflammation is more frequent and occurs around tunnels containing dead, degenerate, or mineralized worms. The granulomas in the media and adventitia become encapsulated by fibrous tissue to form nodules of 0.5–2.5 cm diameter containing caseous material and intact and/or dead worms, and eventually dry mineralized debris. The granulomatous reaction will resolve with time, and, in very old cattle, the aortic wall becomes thinner, has a corrugated lining with irregular mineralized ridges, and may develop aneurysms up to 3.5 cm in diameter. The suggested causes of the granulomatous aortitis include hypersensitivity to the parasite, foreign-body reaction to degenerate and dead onchocercal worms, and the release of toxic factors from the parasites.

Microfilariae can be found in the blood of infected animals at all times, but there is a definite tendency to nocturnal periodicity; infection can be detected by examination of skin snips for microfilariae. In some bulls with large numbers of circulating microfilariae, repetitive episodes of collapse and tetanic convulsion have been observed and attributed to the microfilariae. Some of the bulls showing convulsions eventually developed acute or recurrent ophthalmitis with ophthalmoscopic evidence of retinal pigmentary degeneration and other changes similar to those seen in ocular onchocerciasis in humans.

Bibliography

Atta El Mannan, A. M. *et al. Onchocerca armillata:* Prevalence and pathology in Sudanese cattle. *Ann Trop Med Parasitol* **78:** 619–625, 1984.

Cheema, A. H., and Ivoghli, B. Bovine onchocerciasis caused by *Onchocerca armillata* and *O. gutturosa.* *Vet Pathol* **15:** 495–505, 1978.

Chodnik, K. S. Histopathology of the aortic lesions in cattle infected with *Onchocerca armillata* (Filariidae). *Ann Trop Med Parasitol* **52:** 145–148, 1958.

Prasad, M. C., Rajya, B. S., and Mohanty, G. C. Pathology of caprine aortic onchocerciasis. *Indian J Anim Sci* **45:** 270–274, 1975.

c. ELAEOPHORIASIS. The genus *Elaeophora* is another of the family Filariidae. Three species of interest are *E. poeli* and *E. schneideri,* which infect ruminants, and *E. bohmi,* a parasite of horses.

Elaeophora poeli occurs frequently in the aorta of cattle and related species in tropical areas of the Far East, but is of little clinical significance. Its life cycle is unknown. These are threadlike worms, the female of which may be up to 30 cm long. The lesions are found in the aorta and can be distinguished from those of *Onchocerca armillata* because the females of *E. poeli* are attached by the head with the body swinging free in the lumen of the vessel (Fig. 1.64). The males become encysted in intimal fibrous nodules, which also contain the head of the female. In light infestations, the lesions are found chiefly on the dorsal wall of the aorta about the openings of the intercostal arteries. Heavy infestations are more diffuse and may provoke a prominent fibrous thickening of the vessel wall.

Elaeophora schneideri is found in the arteries of black-tailed, white-tailed, and mule deer, domestic sheep, elk (wapiti), and moose in mountainous areas of the western and southwestern United States and British Columbia, Canada, and in white-tailed deer in the southeastern United States. Although up to 90% of mule deer are infected in some areas, they are little affected by the parasite, and are the likely reservoir host. Horseflies of the genera *Hybomitra* and *Tabanus* are the intermediate hosts. Adult worms normally inhabit the carotid arteries and their branches, but may be found in many other arteries through the body. Adult females are 6–12 cm long by 0.6–0.9 mm thick, and adult males are 5.5–8.5 cm long by 0.4–0.7 mm thick. Microfilariae are about 280 μm long by 18 μm thick.

In elk, adult worms in the cephalic arteries cause mechanical irritation and hence endothelial damage, intimal

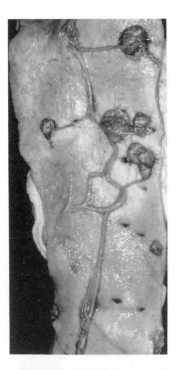

Fig. 1.64 Intimal fibrous nodules in bovine aorta contain males and the heads of females of *Elaeophora poeli*; the body of the female swings free in the aortic lumen. (Courtesy of P. B. Little and *The Canadian Veterinary Journal*.)

proliferation and fibrosis, thrombosis, and embolism, which result in focal ischemia and infarction in the brain, eyes, optic nerves, ears, muzzle, nostrils, and other tissues of the head. Adult worms usually cause only mild intimal sclerosis in deer and sheep, and only rarely do they cause granulomatous verminous thrombosis. Microfilariae can cause multifocal myocardial infarction, myocarditis, and fibrosis in deer. Oral food impactions have been associated with *E. schneideri* in white-tailed deer, but a causal relationship has not been established.

The most important consequence of elaeophoriasis in sheep is filarial dermatosis (sore head), an often severe exudative dermatitis affecting the poll, face, abdomen, and lower parts of the legs, in up to 1% of sheep in enzootic areas. Stomatitis and rhinitis also occur. The lesions are due to microfilariae released by adult females located in arteries supplying the affected areas. Microfilariae can be demonstrated microscopically in the lesions, and they are responsible for intense leukocytic infiltration, chiefly of eosinophils and mononuclear cells. There is hemorrhage, vesiculation, and ulceration of the epidermis with exudation of serum and leukocytes to form crusts. These lesions have the characters of an allergic reaction and are presumably related to the alien nature of the sheep as a host, because they rarely occur in·deer. The skin lesions are intensely pruritic and often lead to severe self-trauma. Lesions often persist for 3–4 years and will regress spontaneously following the death of adults and microfilariae.

Elaeophora bohmi was for some time regarded as one of the genus *Onchocerca*. Approximately 6% of a sample of Austrian horses were found infested, the vessels involved being arteries and veins of the metacarpus, metatarsus, and more distal extremities. The parasites rather selectively involve the media of the vessels, avoiding the intima and adventitia, although the fibrous reaction that develops may cause stenosis of the lumen. The worms are coiled and entwined among the tissue layers, provoking parasitic granulomas with intense eosinophil infiltration and macrophage response. In cases of longer standing, diffuse or nodular fibrous thickenings are visible and palpable in the vessel walls. Microfilariae were found in the blood of 14 of 161 horses (9%) sampled in Iran. The infection appears to be of little significance.

Bibliography

Adcock, J. L., and Hibler, C. P. Vascular and neuro-ophthalmic pathology of elaeophorosis in elk. *Pathol Vet* **6:** 185–213, 1969.

Couvillion, C. E. *et al.* Elaeophoriasis in white-tailed deer: Pathology of the natural disease and its relation to oral food impactions. *J Wildl Dis* **22:** 214–223, 1986.

Hibler, C. P., and Adcock, J. L. Elaeophorosis. *In* "Parasitic Diseases of Wild Mammals," J. W. Davis and R. C. Anderson (eds.). Ames, Iowa, Iowa State Press, 1971.

Jensen, R., and Seghetti, L. Elaeophorosis in sheep. *J Am Vet Med Assoc* **127:** 499–505, 1955.

Kemper, H. E. Filarial dermatosis of sheep. *J Am Vet Med Assoc* **130:** 220–224, 1957. (*E. schneideri*)

Little, P. B. A Malaysian experience with animal disease. *Can Vet J* **20:** 13–21, 1979.

Mirzayans, A., and Maghsoodloo, H. Filarial infection of Equidae in the Tehran area of Iran. *Trop Anim Health Prod* **9**: 19–20, 1977. (*E. bohmi*)

Supperer, R. Filariosen der Pferde in Osterreich. *Wien Tieriarztl Mschr* **40**: 194–220, 1953. (*E. bohmi*)

Worley, D. E., Anderson, C. K., and Greer, K. R. Elaeophorosis in moose from Montana. *J Wildl Dis* **8**: 242–244, 1972. (*E. schneideri*)

d. STRONGYLOSIS IN HORSES.

Horses are commonly infected with both large and small strongyles (see The Alimentary System, Volume 2, Chapter 1), but by far the most damaging to the host is the large strongyle *S. vulgaris*. Whereas the adults of *S. vulgaris* share their large-intestinal habitat with many other strongylids, *S. vulgaris* is the only one to undergo development in the horse's arterial system.

The usual route of migration of *S. vulgaris* is as follows: infective third-stage larvae are ingested, exsheath in the small intestine, penetrate the mucosa and submucosa of the small intestine, cecum, and colon within 3 days of infection, and molt to fourth-stage larvae by day 7. The larval penetration of the gut can cause the formation of large subserosal hemorrhages, called hemomelasma ilei. The fourth-stage larvae, which are ~1 mm long, penetrate submucosal arterioles, migrate in or on the intima of arterioles and arteries, being constrained by the internal elastic lamina from penetrating to the media, and reach the cranial mesenteric artery, the predilection site for further development, between days 11 and 21. After 3–4 months, the larvae molt to the fifth stage (immature adults, 10–18 mm long), return to the wall of the cecum and colon via the intestinal arteries, and are encapsulated in subserosal nodules before returning to the gut lumen as adults when the nodules rupture. The adults require another 6–8 weeks to reach sexual maturity. The prepatent period of *S. vulgaris* is thus ~6–7 months.

Virtually all horses are infected with *Strongylus vulgaris* and have arterial lesions due to larval migration (Fig. 1.65), although the prevalence of *S. vulgaris* infection has been markedly diminished by benzimidazole and ivermectin anthelmintics. A degree of resistance to verminous arteritis is slowly acquired under natural conditions, but horses of all ages remain susceptible to infection. Lesions range from the very common, but insignificant, tortuous intimal tracks, to the less common, but more serious, occlusive thrombotic lesions. Lesions of verminous arteritis are often incorrectly called verminous aneurysms, but any dilation is usually thick-walled due to inflammation and scarring, rather than thinned (Fig. 1.65). Saccular or fusiform aneurysms may occasionally result from weakening of the arterial wall. Lesions occur most frequently in the cranial mesenteric artery and its extension, the ileocecocolic artery. This apparent larval predilection has been suggested to fit a random-walk model in which the larvae sense blood vessel curvature, rather than direction of blood flow, migrate longitudinally along arteries, and are essentially trapped in the cranial mesenteric artery because of its perpendicular connection with the aorta, a border which most of the larvae do not cross. Larvae that spill over into the aorta may be considered as aberrant and lost to their species, in that they are unlikely to return to the cranial mesenteric artery and hence to the gut.

Tortuous intimal tracks consist of fibrin, necrotic debris, and a mixture of inflammatory cells, including eosinophils, in the intima (Fig. 1.66). The tracks become covered by endothelium, the fibrin is removed, and the tracks are converted to fibromusculoelastic thickenings. More extensive larval migration causes more extensive thromboarteritis. Small mural thrombi are formed about the larvae attached to the intima, and there is a reactive leukocytic infiltration, edema, and resultant degeneration of the elastic and muscle fibers of the underlying media. The initial acute inflammation persists as long as the parasite remains but is soon accompanied by a productive connective-tissue response, initiated by smooth muscle cells from the media, with attempted organization of the overlying thrombus and the accumulation of large numbers of mononuclear leukocytes. The proliferating connective tissues of the intima and adventitia may ultimately replace the normal structure of the arterial wall. The affected walls become solid and very thick, and the affected artery may be up to 20 cm in its greatest diameter. The luminal surfaces of active arterial lesions are always rough and covered with layered thrombi in which the parasite is found partially embedded. It is usual to find some worms, both fourth-stage larvae and immature adults, in the thrombi,

Fig. 1.65 Dilation of aorta at origin of mesenteric vessels. Secondary to verminous arteritis. Horse.

Fig. 1.66 Serpiginous tracks due to *Strongylus vulgaris* migration in the aortic intima. Horse.

but the number of worms in any location in any one case varies from a few to several hundred. Following migration of the fifth-stage larvae back to the intestine, or destruction of the larvae by anthelmintics, the intimal lesions resolve as fibrous intimal thickenings, and the luminal diameter will hence increase. The arterial wall thickness also decreases.

The lesions and accompanying worms are most common in the cranial mesenteric artery, although they are also found elsewhere, including the aorta, the renal arteries, the celiac artery and its branches, and the spermatic vessels (Fig. 1.67A,B). They may be in any branch of the cranial mesenteric artery but are concentrated largely in the right division and the colic artery, which is its continuation, and, to a slightly lesser extent, in the cranial branch and its continuation, the right colic artery. The worms are rarely found in the aorta caudal to the origin of the renal vessels; they are relatively common about the origin of the cranial mesenteric and celiac arteries, and less common again in the thoracic aorta. Although this statement on the prevalence of lesions in the thoracic aorta applies to actively thrombotic lesions containing worms, it is probable that older, warty mineralized lesions of the type often seen in the intima of the aortic arch have this origin; they occur frequently in sites where active lesions occur. The lesions in the aortic bulb are located immediately proximal to the aortic valves on the cranial wall of the bulb and in the adjacent cranial aortic and caudal right aortic sinuses, but rarely in the caudal wall and the left sinus (Fig. 1.67A). This localization has been explained on the basis of the curve of the aortic arch and the axis of systolic ejection, which should, theoretically, provide a zone of turbulence and relatively low velocity of blood flow in the cranial segment of the aortic bulb.

The sequelae of verminous arteritis due to *S. vulgaris*

are varied and depend on the location of the arteritis. Involvement of the cranial mesenteric artery is almost constant in horses, but untoward sequelae are much less common, no doubt due to the extensive collateral arterial circulation to the equine intestine, which may mitigate against the effects of frequent thromboembolic episodes. Colic, which is commonly caused in horses by *S. vulgaris* if parasite control is poor, may occur for a number of reasons: thromboembolism and intestinal ischemia or infarction, interference with innervation of the gut due to pressure on abdominal autonomic plexuses, or release of toxic products from degenerating larvae. Fatal acute intestinal infarction occasionally ensues, with ischemia of large areas of the cecum or colon. Thromboembolism of the cecal and colic arteries may also lead to mucosal ulceration and diarrhea, and occasionally death. Hematochezia has been reported in a horse as a consequence of fistulous connections between a verminous cranial mesenteric arterial aneurysm and the cecum and ileum. Cyathostomes (small strongyles) have supplanted *S. vulgaris* as the most important cause of verminous colic. Coronary arterial thrombosis and occlusion, which occurs most commonly in the right coronary artery, can lead to myocardial infarction and death; similar consequences follow the development of verminous aortic valvular endocarditis, a rare event in strongylosis. Embolism of the brachiocephalic trunk can occur and is especially significant when the emboli contain larvae which subsequently migrate in the brain causing encephalomalacia and either chronic incoordination or progressive fatal encephalitis. Renal arterial thromboembolism and verminous arteritis have resulted in renal infarction and, on rare occasions, passage of larvae in the urine. Aortic–iliac thrombosis may be a consequence of strongyle-related thromboembolism, but its pathogenesis remains uncertain. Thrombosis of the celiac artery and embolism of its branches is usually inconsequential.

Bibliography

Aref, S. A random walk model for the migration of *Strongylus vulgaris* in the intestinal arteries of the horse. *Cornell Vet* **72:** 64–75, 1982.

Drudge, J. H., and Lyons, E. T. Large strongyles—recent advances. *Vet Clin North Am: Equine Pract* **2:** 263–280, 1986.

Herd, R. P. The changing world of worms: The rise of the cyathostomes and the decline of *Strongylus vulgaris*. *Compend Contin Educ Pract Vet* **12:** 732–736, 1990.

Kiper, M. L., MacAllister, C., and Qualls, C., Jr. Hematochezia attributable to cranial mesenteric arterial aneurysm with connecting tracts to cecum and ileum in a horse. *J Am Vet Med Assoc* **193:** 1278–1280, 1988.

Klei, T. R. *et al.* Effects of repeated *Strongylus vulgaris* inoculations and concurrent ivermectin treatments on mesenteric arterial lesions in pony foals. *Am J Vet Res* **51:** 654–660, 1990.

Morgan, S. J., and Van Houten, D. S. The ultrastructure of *Strongylus vulgaris*-mediated equine chronic mesenteric arteritis. *Vet Res Commun* **14:** 41–46, 1990.

Morgan, S. J. *et al.* Histology and morphometry of *Strongylus vulgaris*-mediated equine mesenteric arteritis. *J Comp Pathol* **104:** 89–99, 1991.

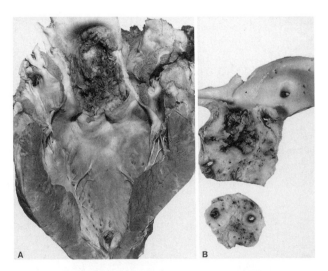

Fig. 1.67 Verminous aortitis (*Strongylus vulgaris*). Horse. (A) Note also interstitial myocarditis and verminous nodule at apex of ventricle. (B) In root of mesenteric vessels. Cross section of mesenteric vessels, below.

Ogbourne, C. P., and Duncan, J. L. "*Strongylus vulgaris* in the horse: Its biology and veterinary importance." 2nd Ed. Farnham Royal, United Kingdom, Commonwealth Agricultural Bureaux, 1985. (Comprehensive review)

Pauli, B., Althaus, S., and Von Tscharner, C. Arterial repair after mechanical injury by migrating fourth-stage larvae of *Strongylus vulgaris* in the horse. (A light and electron microscopic study). *Beitr Pathol* **155:** 357–378, 1975.

Slocombe, J. O. D. *et al. Strongylus vulgaris* in the tunica media of arteries of ponies and treatment with ivermectin. *Can J Vet Res* **51:** 232–235, 1987.

IV. Veins

Venous disorders may be of clinical significance by predisposing to thrombosis and subsequent embolism, or by obstructing venous return and hence causing passive congestion of the affected areas. Obstruction of venous outflow may be followed by the development of collateral venous drainage.

Anomalies of veins occur uncommonly. Arteriovenous fistulas have been discussed with arterial anomalies. A variety of congenital **portosystemic anastomoses** occur in dogs and allow portal blood, with its absorbed toxic substances, such as ammonia, to bypass the liver and cause hepatic encephalopathy. The most common of these defects is persistent ductus venosus. Other portosystemic anastomoses occur extrahepatically between the portal vein and various branches of the caudal vena cava or the azygous vein. As well, portosystemic anastomoses may be acquired due to portal hypertension, which in turn is caused by impedance of blood flow through a diseased liver, an obstructed portal vein, or a kinked intrathoracic caudal vena cava (see The Liver and Biliary System, Volume 2, Chapter 2).

Dilation of a vein, variously termed phlebectasia, varicosity, or aneurysm, occurs but is of little or no importance. It is common as the so-called varicocele of the pampiniform plexus. Dilation may occur because of congenital defects such as venous diaphragms, agenesis, hypoplasia, or ectopia, or may occur secondary to trauma, neoplasia, or surgical intervention. The initial dilation causes insufficiency of the venous valves, and the veins may become dilated, elongated, and tortuous. Thrombosis or sclerosis may be sequelae. The portosystemic anastomoses noted previously usually result from dilation of preexisting microscopic venous anastomoses to produce collateral venous drainage.

Hepatic telangiectasis, peliosis hepatis, and hepatic veno-occlusive disease are described with Diseases of the Liver (Volume 2, Chapter 2).

Rupture of a large vein, usually the result of physical trauma, may result in fatal hemorrhage. Spontaneous idiopathic rupture of the venae cavae or portal vein of horses occurs occasionally; rupture of the great coronary vein of the heart is a rare cause of unexpected death in horses.

Bibliography

Allen, J. R. *et al.* Spontaneous rupture of the great coronary vein in a pony. *Equine Vet J* **19:** 145–147, 1987.

Cornelius, L., and Mahaffey, M. Kinking of the intrathoracic caudal vena cava in five dogs. *J Small Anim Pract* **26:** 67–80, 1985.

Martin, R. A. (ed.). Portosystemic shunts. *Semin Vet Med Surg (Small Anim)* **5:** 75–141, 1990. (Eight articles)

Reimer, J. M. *et al.* Diagnosis and surgical correction of patent ductus venosus in a calf. *J Am Vet Med Assoc* **193:** 1539–1541, 1988.

Valentine, R. W., and Carpenter, J. L. Spleno–mesenteric–renal venous shunt in two dogs. *Vet Pathol* **27:** 58–60, 1990.

A. Phlebothrombosis and Thrombophlebitis

Venous thrombosis (**phlebothrombosis**) results from the usual pathogenetic factors and often leads to inflammation of the vein wall, and hence becomes **thrombophlebitis.** Conversely, phlebitis inevitably leads to thrombosis. Spontaneous thrombosis affects the venae cavae and portal veins occasionally, and the inciting factors are usually unknown. Thrombosis of the renal vein and vena cava occasionally occurs in dogs with the nephrotic syndrome as a result of urinary loss of antithrombin III and resulting hypercoagulability of the blood. Stasis of venous blood results in thrombosis of the femoral tributaries of cattle recumbent for more than 4–6 hrs (downer cows). Venous infarction commonly ensues, and necrosis of the medial muscle masses and sciatic nerves causes permanent posterior paralysis. The thickened, congested, red-black mu-

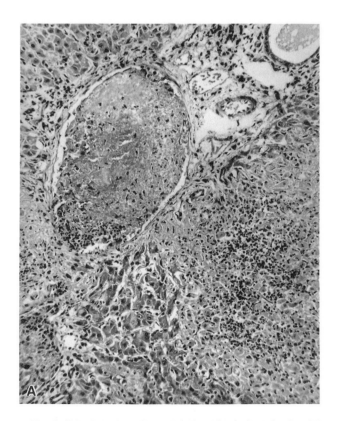

Fig. 1.68A Intrahepatic portal thrombosis in a lamb with omphalophlebitis.

cosa of the gastric fundus in swine with acute septicemia is a venous infarct resulting from thrombosis of veins of the mucosa and submucosa, probably as a sequel to endothelial damage. Gastrointestinal infarction and ulceration can occur in phenylbutazone-intoxicated horses as a result of degeneration of veins and phlebothrombosis. Thrombosis of the jugular veins follows inept or repeated venipuncture, injection of irritant solutions, and the use of indwelling catheters, especially if contaminated with bacteria. Important sequelae to jugular thrombosis are pulmonary embolism, and, if the emboli are septic, valvular endocarditis and pulmonary abscessation.

Endophlebitis may result from the implantation of bacteria carried in the bloodstream as in salmonellosis, and from the injection of irritant solutions. An important form of acute suppurative endophlebitis is the omphalophlebitis of the newborn resulting from infection of the uncicatrized umbilical cord. The resultant bacteremia may result in acute death or give rise to widespread suppurative lesions, often polyarthritis. Hepatic thrombophlebitis is especially common in navel-ill (Fig. 1.68A).

In feline infectious peritonitis, phlebitis occurs as a result of antigen–antibody complex deposition in the walls of small veins and venules, which is followed by complement fixation and neutrophil chemotaxis. Neutrophils cause tissue necrosis, and hence thrombophlebitis, which is followed by the focal pyogranulomatous reaction typical of this disease (Fig. 1.68B).

Phlebitis very commonly occurs by extension from areas of inflammation and importantly provides for the hematogenous dissemination of infection. The entire thickness of the wall of the vein is inflamed, and thrombosis occurs rapidly. If the thrombus is infected, it is likely to soften, disintegrate, and produce emboli. An important example of this pathogenetic sequence is thrombosis of the caudal vena cava in cattle, which occurs as a sequel to rumenitis and hepatic abscessation (Fig. 1.69A). Perivascular hepatic abscesses often involve the vena cava, causing thrombophlebitis and a variety of consequences: pulmonary thromboembolism and arteritis, pulmonary abscessation, or pulmonary arterial aneurysms, which may rupture and cause massive pulmonary hemorrhage (Fig. 1.69B). Valvular endocarditis is an occasional sequel. As well, the wall of the vena cava may be eroded by the

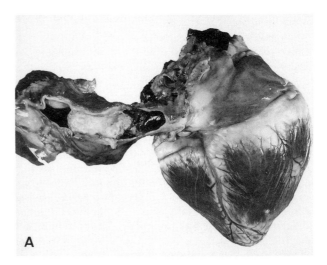

Fig. 1.69A Thrombophlebitis of posterior vena cava originating from hepatic abscess. Ox.

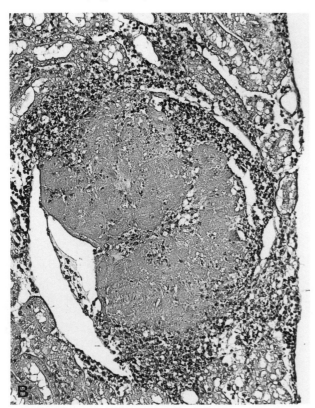

Fig. 1.68B Thrombophlebitis. Renal stellate vein. Cat. Feline infectious peritonitis.

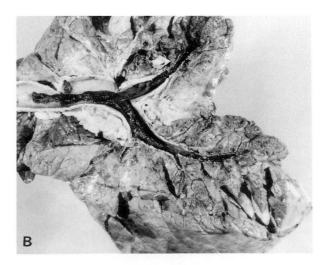

Fig. 1.69B Hemorrhagic tracheal cast and acute pulmonary edema. Cow. Sequel to rupture of hepatic abscess into postcava.

abscess, which ruptures into the lumen causing massive septic embolization and sudden death.

Bibliography

Burrows, C. F. Inadequate skin preparation as a cause of intravenous catheter-related infection in the dog. *J Am Vet Med Assoc* **180:** 747–749, 1982.

Hoskins, J. D., Ochoa, R., and Hawkins, B. J. Portal vein thrombosis in a dog: A case report. *J Am Anim Hosp Assoc* **15:** 497–500, 1979.

Meschter, C. L., Maylin, G. A., and Krook, L. Vascular pathology in phenylbutazone-intoxicated horses. *Cornell Vet* **74:** 282–297, 1984.

Mills, L. L., and Pace, L. W. Caudal vena caval thrombosis in a cow. *J Am Vet Med Assoc* **196:** 1294–1296, 1990.

Morris, D. D. Thrombophlebitis in horses: The contribution of hemostatic dysfunction to pathogenesis. *Compend Contin Educ Pract Vet* **11:** 1386–1394, 1989.

Taylor, A. W. Traumatic duodenitis with subsequent thrombosis of the posterior vena cava in a horse. *Aust Vet J* **46:** 281–283, 1970.

Weiss, R. C., and Scott, F. W. Pathogenesis of feline infectious peritonitis: Pathologic changes and immunofluorescence. *Am J Vet Res* **42:** 2036–2048, 1981.

B. Parasitic Thrombophlebitis

Schistosomes are important venous parasites in tropical Africa and Asia, and will be discussed at length.

Gurltia paralysans, a metastrongyle, causes paralytic disease in domestic cats in South America. The adults lodge in the veins of the thigh and lumbar cord and cause thrombosis and ectasia.

Schistosomiasis (bilharziasis), a snail-borne fluke infection, is prevalent in domestic animals in Asia, Africa, and other tropical and subtropical areas, but does not usually occur as clinical disease; the importance of the disease may increase as irrigation projects enhance the habitat for the intermediate snail host. Damage and clinical effects are the result of localization of permanently coupled adult male and female flukes in blood vessels of the liver, lungs, alimentary and urogenital tracts, and the nasal cavity. Egg-laying continues for many years, and schistosomiasis can be a disease of long chronicity.

Members of the family Schistosomatidae are the blood flukes of mammals and birds; species that parasitize mammals include those of the genera *Schistosoma, Heterobilharzia,* and *Orientobilharzia.* It is rather characteristic of the trematodes, including the blood flukes, that the intermediate stages are selective in their choice of hosts, molluscs, and that the infective and final phases do quite well in a variety of definitive hosts, although there is some host preference. The human schistosome, especially *S. japonicum,* may be discovered in domestic animals, and the mammalian schistosome may be discovered in humans as patent or nonpatent infestations.

Based on egg morphology and intermediate snail hosts, the species which occur in domestic animals may be grouped as follows:

1. *S. haematobium* group—*S. bovis* in the portal and mesenteric veins of ruminants and occasionally horses, camels, and pigs, in southern Europe and tropical areas of Africa and Asia; *S. mattheei* in the portal and mesenteric veins, plus veins of the urogenital tract and stomach, of ruminants in central and southern Africa; *S. curassoni* in ruminants in west Africa; *S. margrebowiei* and *S. leiperi* in wild artiodactyls in central Africa;
2. *S. mansoni* group—*S. rodhaini* in dogs and other carnivores in central Africa;
3. *S. indicum* group, present in India and Southeast Asia—*S. spindale* in mesenteric veins of ruminants and, occasionally, of horses and dogs; *S. nasalis* in nasal mucosal veins of cattle and, occasionally, of goats and horses; *S. indicum* in portal and mesenteric veins of herbivores; *S. incognitum* in swine and dogs; and
4. *S. japonicum* group—*S. japonicum,* human by preference, in all domestic animals in the Far East; *S. mekongi* in dogs and humans in Southeast Asia.

Schistosoma bovis and *S. japonicum* are the most pathogenic of the schistosomes of cattle and sheep. Mixed infections with more than one species of schistosome occur; hybridization may occur in this circumstance.

Other members of the family Schistosomatidae of veterinary significance include *Orientobilharzia turkestanicum* in mesenteric veins of herbivores and cats in the Near East, Russia, China, and Mongolia; *O. dattai* and *O. bomfordi* in artiodactyls in India; and *Heterobilharzia americana* in raccoons, bobcats, and occasionally dogs, in the southern United States.

The life cycles of the schistosomes are comparable to those of other flukes and follow the general pattern described elsewhere for the liver fluke, *Fasciola hepatica.* As a point of difference, the redia phase of *F. hepatica* in the intermediate host is replaced in the schistosome by a generation of daughter sporocysts.

The eggs of the schistosome are laid in the vessels that the adults inhabit. The eggs when laid are immature but contain a miracidium by the time they have worked their way through the overlying mucous membrane to escape to the exterior. The eggs hatch in water, and the miracidia actively penetrate the soft tissues of aquatic snails to which they are physiologically adapted. The fork-tailed cercariae which emerge from the snail are the infective phase, and they enter the definitive host by burrowing through the skin; patent infestations can also develop if the cercariae are ingested with water. They will invade the skin of hosts which are quite alien to them, provoking allergic reactions and dermatitis (swimmer's itch, swamp itch).

Cutaneous lesions develop as a result of penetration of the skin by cercariae. They occur only on those portions of the skin which come in contact with infested water and are seldom observed in animals except under experimental conditions. The cutaneous lesions are tiny petechiated nodules in which there is a sharp leukocytic reaction to the parasite. Passage of a large number of parasites through the lungs will produce pneumonia. Ectopic localization in various organs, such as the kidney, produces small hemorrhages and inflammatory foci.

In the definitive host, the cercariae lose their tails and become schistosomula, which enter venules in the dermis and, thereafter, are conveyed passively in the blood, through the lungs, and into the systemic circulation. Once in the blood there is probably no selective migration; those which are distributed to the vessels of their final habitat develop further, and those which are distributed to other organs die. *Schistosoma nasale* develops in the veins of the nasal mucosa; all other species develop in veins of the abdominal cavity, principally in the portal vessels, until sexual maturity is attained, when they migrate out to the smaller veins of the mesentery (Fig. 1.70). In long-standing infections, *S. mattheei* may reside in veins of the urinary bladder as well as in portomesenteric veins. The prepatent period in the mammalian host ranges from 30 to 77 days.

In clinical cases of schistosomiasis of the portomesenteric distribution, the animals are anemic, and may be severely emaciated in fatal cases. Diarrhea or dysentery occur at the time of patency, persist for a few weeks, and, in the absence of reinfection, may be followed by spontaneous recovery, as the immune response of the host eliminates the parasites and suppresses the egg-laying of survivors. The eggs can be found in the feces together with some mucus and blood. Blood loss leads to anemia and hypoalbuminemia, which may contribute to the presence of fluid in serous cavities. The wall of the intestine, chiefly the large bowel, is thickened by inflammatory and

Fig. 1.70 Schistosomiasis. Lamb. There is an adult worm in a mesenteric vein (arrow).

scar tissue, and may contain small granulomas and lymphoid nodules. Scarring begins in the mucosa and extends to involve all layers, including the serosa. The mucosa may bear polypoid and papillary proliferations. The mesentery is thickened and shortened by fibrous tissue, and the mesenteric nodes are enlarged and fibrotic. The adult parasites, which are about 1–3 cm long, can be found in the mesenteric and portal vessels. There is fibrosis of the liver of variable degree affecting particularly the portal vessels. The liver may be enlarged or small depending on the course and severity of the infestation. On cut surface, the periportal nature of the fibrosis is quite apparent but, with time and more severe infestations, the fibrosis becomes more diffuse, and the surface of the liver may have a hobnailed appearance or be studded by numerous minute granulomas.

Proliferative granulomatous nasal lesions produced by *S. nasalis* result in clinical signs of dyspnea and mouth breathing. Granulomatous lesions are frequently present in the urinary bladder and ureters of cattle with *S. mattheei* infection; lesions in the ureters, and occasionally renal pelvis, are linear, whereas those in the bladder become very extensive and polypoid.

The stage of oviposition and extrusion of eggs through the tissues and mucous membranes is the stage in which most damage is done. The mated schistosomes, in permanent copula, move out into the smallest venous radicles, and the eggs are deposited first in the lumen of the vessel. They adhere to, and are eventually covered by, the endothelium. The eggs of some species are spinous, and this probably facilitates passive migration into the tissue. The migration does appear, however, to be partly active by virtue of secretions elaborated by the developing miracidium, which diffuse through the shell of the egg into the tissues. These secretions may be also in part responsible for inciting the inflammatory reaction to the eggs. Initially most of the eggs are deposited in the lamina propria of the intestine, and break into the lumen with little more than mechanical injury. Rupture of congested mucosal venules accounts for the characteristic dysentery of schistosomiasis.

As the disease progresses and the adults are excluded from smaller venules due to endophlebitis, eggs are laid in veins in deeper tissues, where antigens released by the miracidium induce a delayed hypersensitivity response and formation of small granulomas (pseudotubercles). The granulomas are characterized initially by the infiltration of leukocytes, chiefly eosinophils, and later by the accumulation of mononuclear leukocytes, epithelioid and giant cells, and reactive fibrosis. Microabscesses occasionally form and rupture into the gut lumen. Eggs in submucosal granulomas degenerate and are removed, or may be mineralized and remain for a longer time. Some of the mature or disintegrating eggs may be coated with a bright eosinophilic deposit (the Splendore–Hoeppli reaction), the result of antigen–antibody complex formation. The granulomas gradually regress and are replaced by fibrous tissue. Host cell responses to eggs are primarily the result of host

hypersensitivity to soluble egg antigens, and to a much lesser degree due to physical and chemical stimuli.

As would be expected, not all the eggs move in the proper direction. Some break into lymphatics and are conveyed to the regional nodes to produce the typical granulomas but, more important, a great many of them pass to the liver producing portal phlebitis and granulomatous hepatitis, in concert with the adult flukes. In heavy infestations and possibly by means of venous shunts developed in the liver, many eggs may pass in venous blood to and beyond the lungs, producing granulomatous endoarteritis and adjacent granulomas wherever they lodge. The development of hepatic lesions follows the same course as for those which develop in the intestinal wall. In the liver, the eggs break out of the portal venules to become enclosed by granulomas. Sooner or later, they die and are mineralized, and by that time have provoked an extensive surrounding fibrosis, which may be visible grossly as bilharzial "clay pipestem" portal fibrosis. This course of events occurs similarly in urinary schistosomiasis caused by *S. mattheei,* wherein hematuria accompanies excretion of eggs in the urine, and granulomatous cystitis and ureteritis result from the host's immunologic response to the schistosome eggs trapped in the tissues.

Adult schistosomes in mesenteric and portal veins elicit an eosinophilic endophlebitis, often with irregular intimal proliferation, and may be found within thrombi. Adults may similarly cause reactions in veins of the pancreas and the mesenteric and portal lymph nodes, and in pulmonary arterioles. The dead parasites are removed by a granulomatous reaction, which is often followed by the formation of a grossly visible lymphoid nodule. The lumen of the vein may be occluded, and the venous wall obliterated, by this lymphoid response. There may also be diffuse intimal proliferation and endophlebitis of intrahepatic portal veins, medial hypertrophy, and adventitial inflammation and fibrosis leading to prominent portal fibrosis. Irregular dilated vascular bypasses develop in the portal areas. The adults ingest erythrocytes and regurgitate hematin pigments, which are picked up by the monocyte–macrophage system, especially in the regional lymph nodes and liver; these black iron-porphyrin pigments may be responsible for the gray color of the liver and lungs in severe cases of schistosomiasis.

Bibliography

Bartsch, R. C., and Van Wyk, J. A. Studies on schistosomiasis. 9. Pathology of the bovine urinary tract. *Onderstepoort J Vet Res* **44:** 73–94, 1977.
Fransen, J. *et al.* Pathology of natural infections of *Schistosoma spindale* Montgomery, 1906, in cattle. *J Comp Pathol* **103:** 447–455, 1990.
Hussein, M. F., Bushara, H. O., and Ali, K. E. The pathology of experimental *Schistosoma bovis* infection in sheep. *J Helminthol* **50:** 235–241, 1976.
Lawrence, J. A. The pathology of *Schistosoma mattheei* infection in the ox. 1. Lesions attributable to the eggs. 2. Lesions attributable to the adult parasites. *J Comp Pathol* **88:** 1–14 and 15–29, 1978.
Losos, G. J. "Infectious Tropical Diseases of Domestic Animals." Harlow, England, Longman Scientific & Technical, 1986.
Monrad, J., Christensen, N. O., and Nansen, P. Acquired resistance in goats following a single primary *Schistosoma bovis* infection. *Acta Trop* **48:** 69–77, 1990.
Rollinson, D., and Simpson, A. J. G. (eds.). "The Biology of Schistosome: From Genes to Latrines." San Diego, California, Academic Press, 1987.
Slaughter, J. B., II, Billups, L. H., and Acor, G. K. Canine heterobilharziasis. *Compend Contin Educ Pract Vet* **10:** 606–612, 1988.
Soulsby, E. J. L. "Helminths, Arthropods and Protozoa of Domesticated Animals," 7th Ed. Philadelphia, Pennsylvania, Lea & Febiger, 1982.
Vercruysse, J. *et al.* Clinical pathology of experimental *Schistosoma curassoni* infections in sheep and goats. *Res Vet Sci* **44:** 273–281, 1988.

V. Lymphatics

The richness of the lymphatic plexuses in almost all tissues and their important role as drainage channels for interstitial fluid make for almost inevitable involvement of lymphatics by any inflammation and by the many neoplasms which metastasize via lymph channels. In most instances, the lymphatic lesions are so small as to have no significance, but in some cases the consequence of lymphatic involvement may be the major presenting clinical sign or lesion. Such, for example, is the case when attention is drawn to papillary carcinoma of the canine ovary by severe ascites; disrupted portions of the neoplasm permeate and obstruct the ventral diaphragmatic lymphatics, which are the main efferent channels of the peritoneal cavity.

Lymphedema is defined as swelling of a part of the body by an increased quantity of lymph due to a lymphatic system disorder. The term should not be used to describe edematous swellings due to venostasis, hypoproteinemia, heart failure, or cellulitis. Lymphedema may be classified as primary or secondary. Primary lymphedema, which is usually congenital, and may be hereditary, is due to anomalous development of the lymphatic system. Secondary lymphedema occurs because of obstruction of previously normal lymphatics, due to inflammation, neoplasia, surgery, or trauma. Lymphedema is of significance because it may predispose the affected area, usually a limb, to secondary bacterial infection and poor wound healing. Prolonged lymphedema leads to fibrosis.

A. Congenital Anomalies

There are many degrees of variation of the standard disposition of the large lymphatic vessels, but these are of no importance.

Congenital hereditary lymphedema is a rare condition which has been reported in Ayrshire calves, dogs of various breeds, and Dutch Landrace pigs. The lymphatic abnormality is due to a single autosomal recessive gene in

calves, is inherited as an autosomal dominant trait with variable expressivity in dogs, and appears also to be inherited in pigs.

The central lymphatic system normally forms as outgrowths from the endothelium of primitive veins, and subsequent outgrowths form the peripheral lymphatic system. The primary lymph nodes develop from the primitive central lymphatic anlagen, the jugular and iliac lymph sacs. Secondary and tertiary lymph nodes later form along the peripheral lymphatic system and are the most severely affected nodes in congenital lymphedema. The earlier the stage at which lymphatic malformation occurs, that is, the more central the lesion, the more severe will be the clinical manifestations and the earlier will be their onset.

Grossly, the most severely affected individuals have generalized subcutaneous edema and fluid in serous cavities and are stillborn or must be delivered by cesarean section. The mortality rate of Ayrshire cows with lymphedematous fetuses has been high. Moderate degrees of lymphedema are seen as symmetrical swelling of the head, neck, limbs, and tail. The ears of lymphedematous calves often have edematous accessory lobes. Lesser degrees of lymphedema affect the distal parts of the limbs, especially the hind limbs, and may not develop for several weeks postnatally. Mild regional lymphedema may be of little significance, although chronic subcutaneous edema will lead to fibrosis and may predispose to secondary bacterial infection. Mild lymphedema affecting only the hind limbs will in some cases disappear, presumably due to postnatal development of the peripheral lymphatic system. The peripheral lymph nodes are usually smaller than normal or absent in affected animals. Although obstruction may be demonstrated by lymphangiography, nodal hypoplasia or absence does not necessarily lead to obstruction. Afferent and efferent nodal sinuses may be dilated and may be evident on gross examination. The thoracic duct may be dilated and tortuous in generalized lymphedema.

Histologically, a wide range of lymphatic abnormalities may be seen in the edematous, and possibly fibrotic, subcutis of affected areas. The lymphatics may be aplastic, that is, no lymphatics are found; hypoplastic, in which case lymphatics are too few or too small; or hyperplastic, or varicose, in which case lymphatics are excessive in size and number, and dilation has resulted in valvular incompetence and lymphostasis. Ectatic lymph channels are commonly present around and within hypoplastic peripheral lymph nodes. Central nodes are similarly affected in animals with generalized lymphedema, and will be absent in cases of lymph–vascular agenesis.

Bibliography

Davies, A. P. *et al.* Primary lymphedema in three dogs. *J Am Vet Med Assoc* **174:** 1316–1320, 1979.

Leighton, R. L., and Suter, P. F. Primary lymphedema of the hind limb in the dog. *J Am Vet Med Assoc* **175:** 369–374, 1979.

Mulei, C. M., and Atwell, R. B. Congenital lymphoedema in an Ayrshire–Friesian crossbred female calf. *Aust Vet J* **66:** 227–228, 1989.

van der Patte, S. C. J. The pathogenesis of congenital hereditary lymphedema in the pig. *Lymphology* **11:** 10–21, 1978.

B. Dilation and Rupture of Lymphatics

Dilation of normally developed lymphatic vessels **(lymphangiectasia)** almost invariably results from some form of obstruction and leads to the accumulation of excess interstitial fluid in the drainage area. Causes of such lymphatic obstruction include infiltrating neoplasms, which may obstruct the lymphatic vessel or the drainage node, inflammatory thrombosis and sclerosis of the channels or nodal sinuses, and postsurgical scarring. The vessels become irregularly dilated and tortuous, distal to the obstruction, and the accompanying increase in interstitial fluid, if persistent, will cause an increase in interstitial and subcutaneous fibrous tissue. Epicardial lymphatics in horses are occasionally dilated and tortuous, but this is of no apparent significance (Fig. 1.71).

Intestinal lymphangiectasia is the most common cause of protein-losing enteropathy in dogs, and contributes to intestinal protein loss in cattle with Johne's disease and other enteritides. Dilated lacteals in intestinal villi rupture or leak their contents into the intestinal lumen, and severe hypoproteinemia can result. Leakage of protein from capillaries in the villi may also be an important contribution to protein loss. Chyle may leak from dilated subserosal lymphatics and contribute to the hypoproteinemia-induced ascites.

Chylothorax, the result of leakage or rupture of the

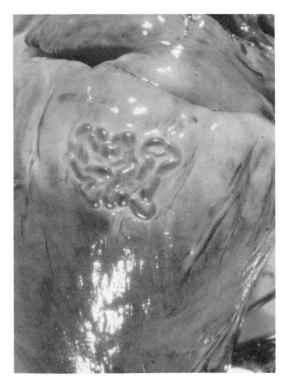

Fig. 1.71 Dilated epicardial lymphatics. Foal.

thoracic duct, occurs infrequently in dogs and cats. The cause of chylothorax is usually not apparent; rupture due to thoracic trauma is often postulated but rarely proven. An injured thoracic duct has limited capacity for spontaneous healing. Reported causes of chylothorax include neoplasia, congenital anomalies, dirofilariasis, and blastomycosis; many cases appear to develop spontaneously. Obstruction of thoracic duct drainage may lead to lymphangiectasia of intrathoracic lymphatics and leakage of chyle. **Chylopericardium** is an unusual accompaniment of chylothorax. The diagnosis of chylothorax is confirmed by analysis of the pleural fluid, which is opalescent to opaque, milky white to yellow, and on standing forms a top layer of cream (chylomicron fat), which is ether-soluble. Microscopically, the fluid contains large numbers of lymphocytes, a few neutrophils, red cells, plasma cells, and fat globules. The site of the rupture is often difficult to locate surgically or at autopsy. Ligation of the thoracic duct leads to the development of alternate lymphaticovenous anastomoses, and cessation of the chylothorax.

Chylous ascites results from rupture of the cisterna chyli, and is a rare event.

Bibliography

Birchard, S. J., and Bilbrey, S. A. Chylothorax associated with dirofilariasis in a cat. *J Am Vet Med Assoc* **197:** 507–509, 1990.

Birchard, S. J., and Fossum, T. W. Chylothorax in the dog and cat. *Vet Clin North Am: Small Anim Pract* **17:** 271–283, 1987.

Fossum, T. W. *et al.* Intestinal lymphangiectasia associated with chylothorax in two dogs. *J Am Vet Med Assoc* **190:** 61–64, 1987.

Fossum, T. W. *et al.* Generalized lymphangiectasis in a dog with subcutaneous chyle and lymphangioma. *J Am Vet Med Assoc* **197:** 231–236, 1990.

Peaston, A. E. *et al.* Combined chylothorax, chylopericardium, and cranial vena cava syndrome in a dog with thymoma. *J Am Vet Med Assoc* **197:** 1354–1356, 1990.

White, S. D. *et al.* Acquired cutaneous lymphangiectasis in a dog. *J Am Vet Med Assoc* **193:** 1093–1094, 1988.

C. Lymphangitis

Lymphangitis with anatomic specificity is a prominent feature of a number of specific diseases. Most of the diseases are discussed elsewhere but warrant brief mention here also. Acute thrombotic lymphangitis, the thrombi containing large numbers of bacilli, with a marked tendency toward necrosis, is a feature of the restricted form of anthrax that is seen in pigs, dogs, and horses. A granulomatous lymphangitis occurs typically in mycobacterial infections; the mesenteric lymphatics stand out like threads beneath the serosa in Johne's disease, and are often beaded in intestinal tuberculosis; they bear the specific microscopic lesions of these infections (Fig. 1.72). The spread of actinobacillosis can often be discerned as long rows of small granulomas from the site of primary infection on the way to the regional nodes. The corded and nodular ulcerating lymphangitis of cutaneous glanders in horses must be distinguished from the specific forms of lymphan-

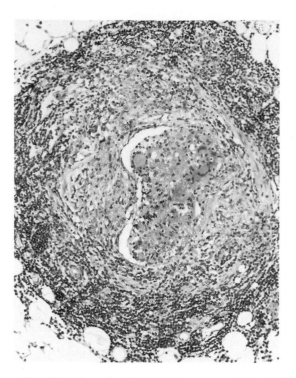

Fig. 1.72 Lymphangitis in ileal mesentery. Johne's disease. Note granulomatous inflammation with giant cells.

gitis discussed subsequently. Lymphangitis is often a prominent development in cutaneous streptothricosis in cattle. Lipogranulomatous lymphangitis has been found in association with intestinal lymphangiectasia in dogs.

Bovine farcy is a chronic granulomatous disease of cattle in the tropics, and is characterized by nodular suppurative dermatitis, lymphangitis, and lymphadenitis. Formerly thought to be caused by *Nocardia farcinica,* most cases of bovine farcy are now thought to be caused by *Mycobacterium farcinogenes* or *Mycobacterium senegalense.*

Sporadic lymphangitis, a nonspecific, nonulcerative lymphangitis affecting the hind legs of horses, is relatively common. Synonyms include Monday-morning disease, bigleg, and weed. The disease is ushered in by acute lameness and severe cording of the lymphatics, although this is usually palpable only where the lymphatics lie near the saphenous vein on the inner aspect of the thigh. The limb swells rapidly, the swelling beginning about the pastern or metatarsus and reaching its zenith in 3–4 days, by which time it has usually extended beyond the stifle. There are transient general disturbances associated with fever. Recovery is usual, although repeated attacks may occur, and some cases are fatal, presumably as a result of septicemia. There is edema of the connective tissues of the distal portion of the limb with very rapid fibroplasia. Above the stifle, the edema is largely confined to the subcutis, this probably depending on the arrangement of fascia there. The lymphatics are dilated, and the larger ones contain

thick pus or a thinner mixture of pus and lymph. There is regional acute lymphadenitis.

The cause and pathogenesis of this form of lymphangitis are unknown. A variety of pyogenic organisms have been recovered from the lesions, but it appears that, in the usual case, the initial acute infection is rapidly overcome. The occurrence of the disease after a period of inactivity, and the experimental production of a similar lesion by inoculation of bacteria into skin of the pasterns in undernourished horses, suggests that "stocking" of the hind limbs, stagnant edema which is common in debilitated horses in periods of inactivity, predisposes to the progress of infection. However, most horses with stocking of the hind limbs do not develop lymphangitis. The entry of infection is probably traumatic.

Bibliography

Meschter, C. L., Rakich, P. M., and Tyler, D. E. Intestinal lymphangiectasia with granulomatous lymphangitis in a dog. *J Am Vet Med Assoc* 190: 427–430, 1987.

Tageldin, M. H. *et al.* Concurrent infection with *Mycobacterium farcinogenes* and *Actinobacillus lignieresi* in slaughtered cattle. *J Comp Pathol* 99: 431–437, 1988.

Tufvesson, G. Lymphangitis in horses. *Nord Vet Med* 4: 529–576, 729–744, 817–860, 1047–1058, 1952. (Nonspecific Monday morning type)

1. Ulcerative Lymphangitis

This is a chronic progressive inflammation of the subcutaneous lymphatics of horses. *Corynebacterium pseudotuberculosis* (*ovis*) is the cause of the classic condition, but similar lesions may be caused by other pyogenic organisms, including streptococci, staphylococci, *Rhodococcus equi*, and *Pseudomonas aeruginosa*. Two biotypes of *C. pseudotuberculosis* are recognized: a nitrate-negative type isolated from sheep and goats, and a nitrate-positive type isolated from horses and cattle. *Corynebacterium pseudotuberculosis* produces a potent phospholipase exotoxin which attacks the sphingomyelin of vascular endothelial cells, and may be important in aiding the spread of the bacteria by acting as a permeability factor. Ulcerative lymphangitis was an important cause of debility in the horse era, but is now sporadic and rather rare.

Some features of the biology of *C. pseudotuberculosis* are discussed with caseous lymphadenitis. In that disease, as in ulcerative lymphangitis, the infection begins in cutaneous wounds. Ulcerative lymphangitis typically begins about the fetlocks of the hind limbs. As a result of lymphangitis, there is diffuse swelling in the leg soon followed by development of dermal nodules. These are abscesses which ulcerate and discharge a thick creamy pus, which may be bloodstained. Only a small area of skin may be sloughed so that, usually, the margins of the ulcer are undermined. The ulcers heal and leave small areas of depilated, depigmented skin. As the primary ulcers heal, new nodules form in adjacent skin, suppurate, ulcerate, and cicatrize, and in this way the disease progresses slowly. As the new nodules develop, the lymphatics be-

tween them become corded and as much as 1–2 cm thick, and fresh abscesses develop along the inflamed vessels. In uncontrolled infections, over a course of many months, much of the skin of the body and neck as well as the limbs may be affected, and this leads to death. Typically, the regional lymph nodes, although moderately enlarged, do not become suppurative or fibrotic. In rare instances, bacteremia results in internal dissemination, especially to the kidneys.

Ulcerative lymphangitis also occurs in cattle, but its course is somewhat different from that in the horse. The initial lesions tend to develop within the dermis or subcutis of the neck or trunk rather than on the distal portions of the limbs. The initial abscesses are large and may be >5 cm in diameter. Extending away from them are thickened lymphatics along which new abscesses develop. The abscesses ulcerate on the skin, and the ulcers persist, oozing a serumlike exudate. There is early lymphadenitis with swelling of the node and its fixation to the surrounding tissue. The inflammation spreads slowly, and there is progressive suppurative lymphadenitis.

The equine biotype of *Corynebacterium pseudotuberculosis* causes deep pectoral and ventral abdominal wall abscesses in horses in the southwestern United States. The transmission of *C. pseudotuberculosis* in these cases is obscure, but may be by arthropod vectors. Ulcerative lymphangitis does not occur; dissemination is by bacteremia. Untoward sequelae include prolonged resolution of the abscesses, multiple and internal abscessation, and abortion.

Bibliography

Abu-Samra, M. T. *et al.* Ulcerative lymphangitis in a horse. *Equine Vet J* 12: 149–150, 1980.

Addo, P. B. A review of epizootic lymphangitis and ulcerative lymphangitis in Nigeria: Misnomer or misdiagnosis? *Bull Anim Health Prod Afr* 28: 103–107, 1980.

Bain, A. M. *Corynebacterium equi* infections in the equine. *Aust Vet J* 39: 116–121, 1963.

Biberstein, E. L., Knight, H. D., and Jang, S. Two biotypes of *Corynebacterium pseudotuberculosis*. *Vet Rec* 89: 691–692, 1971.

Brumbaugh, G. W., and Ekman, T. L. *Corynebacterium pseudotuberculosis* bacteremia in two horses. *J Am Vet Med Assoc* 178: 300–301, 1981.

Mayfield, M. A., and Martin, M. T. *Corynebacterium pseudotuberculosis* in Texas horses. *Southwest Vet* 32: 133–136, 1979.

Miers, K. C., and Ley, W. B. *Corynebacterium pseudotuberculosis* infection in the horse: Study of 117 clinical cases and consideration of its etiopathogenesis. *J Am Vet Med Assoc* 177: 250–253, 1980.

2. Epizootic Lymphangitis

Epizootic lymphangitis, or pseudofarcy, is caused by *Histoplasma farciminosus* and clinically resembles ulcerative lymphangitis and cutaneous glanders (farcy). Epizootic lymphangitis occurs almost exclusively in horses and mules. Equine histoplasmosis is probably a more appropriate name, since *H. farciminosus* infection may be mani-

fest as keratoconjunctivitis or pneumonia in the absence of past or concurrent lymphangitis.

Outbreaks of epizootic lymphangitis have occurred in the past when large numbers of military horses and mules were held in overcrowded facilities. The condition remains enzootic in some Mediterranean countries, Africa, and the Near and Far East. The organism apparently exists as a soil saprophyte, and enters the body through skin wounds of the lower limbs to cause the classic epizootic lymphangitis, but may also invade castration wounds, ruptured abscessed lymph nodes after strangles, and gastric ulcers. *Histoplasma farciminosus* may cause sinusitis, and/or pneumonia via inhalation of contaminated dust, may cause rhinitis by contact with skin lesions, and may be transmitted by flies (*Musca* and *Stomoxys*) to cause keratoconjunctivitis.

The initial cutaneous lesion, or focus of infection in a wound, has a tendency to ulcerate, or it may undergo alternating periods of discharge and closure for some weeks before healing and scarifying. The infection may resolve, but often spreads centripetally in the adjacent tissues and along the lymphatics. In the adjacent tissues, which are swollen, small nodules develop (Fig. 1.73) which are about 1 cm in diameter and, in the course of a few days, these ulcerate to discharge at first a viscid gray exudate and, later, pus. The initial lesions are intradermal and are freely movable over the subcutaneous tissue. Spread of the infection is solely via the lymphatics, and these convey the organisms into the subcutaneous and,

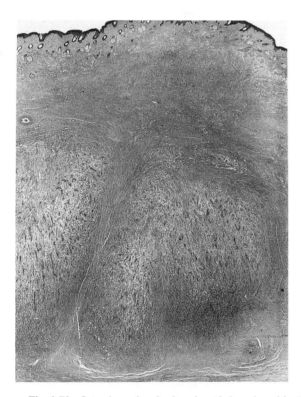

Fig. 1.73 Intradermal nodes in epizootic lymphangitis. (Slide, courtesy of D. R. Cordy.)

occasionally, into the deeper tissues. The inflamed lymphatics are thickened and hard, and along their course new nodules form, ulcerate, discharge, and eventually heal with scarification, although there may be alternating periods of activity and quiescence with the growth of excessive granulation tissue. Sometimes the ulcers continue to enlarge and may coalesce. In this way, the cutaneous infection spreads gradually but irregularly, but is always characterized by intradermal and subcutaneous nodules, ulcers, and irregularly swollen lymphatics. In the early stages, the skin between the lesions remains normal and mobile, but in areas of extensive ulceration, it becomes very thick, indurated, and firmly fused to the underlying tissues. When the thickened skin is incised, it presents the lardaceous appearance of granulation tissue in the horse and contains a number of small, yellow, purulent foci between which course the lymphatics, dilated and filled with pus and serous fluid. The regional lymph nodes are regularly involved and swollen. In the early stages, the swollen nodes contain many small foci of softening, but later the foci coalesce and are heavily encapsulated. The nodes may rupture. The condition is usually chronic, persisting for 3–12 months, and causes considerable debility, but low mortality.

The disease may begin quite frequently on the conjunctiva or nictitating membrane, producing at first a small papule and a serous conjunctival discharge. The papules ulcerate to form flat, buttonlike growths of granulation tissue, the eyelids become severely swollen, and the inflammation extends to the tissues of the forehead.

Nasal infection is usually accompanied by a mucopurulent discharge containing large numbers of the fungus, and may be bloodstained. The lesions begin as yellow, flat papules or nodules on the nasal mucosa, and these soon break down to form craterous granulating ulcers, which bleed easily. The nasal lesions are usually found near the external nares, and they may extend from there on to the muzzle, or they may be found deep in the nasal cavity and in the pharynx. Similar lesions may occur in the nasal sinuses, the larynx, and the bronchi.

The pulmonary lesions may be solid granulomatous areas, or they may be liquefied with puslike contents. When ulcerative lesions are present on the nasal mucosa, there is suppurative regional lymphadenitis. The typical nodules of epizootic lymphangitis and the liquefied foci which form in deep tissues may also be found in the pleura, spleen, liver, testes, tunica vaginalis, and bone marrow.

The diagnosis of epizootic lymphangitis depends on the demonstration of the fungus, which is usually present in very large numbers in the contents of the nodules and fresh discharges from ulcers. The parasitic phase of the fungus is yeastlike, Gram positive, globose or oval in shape, and 2–3 μm in diameter. It occurs extracellularly and intracellularly in macrophages. The fluorescent antibody test is an effective means of diagnosis. In histologic sections, the organism is similar to *H. capsulatum*, consisting of a basophilic nuclear mass and thin capsule which can be demonstrated by stains for polysaccharides.

Bibliography

Ajello, L. Comparative morphology and immunology of members of the genus *Histoplasma*. A review. *Mykosen* **11:** 507–514, 1968.

Fawi, M. T. *Histoplasma farciminosum*, the aetiological agent of equine cryptococcal pneumonia. *Sabouraudia* **9:** 123–125, 1971.

Gabal, M. A., and Khalifa, K. Study on the immune response and serological diagnosis of equine histoplasmosis (epizootic lymphangitis). *Zentralbl Veterinarmed (B)* **30:** 317–321, 1983.

Oppong, E. N. W. Diseases of horses and donkeys in Ghana. *Bull Anim Health Prod Afr* **27:** 47–49, 1979.

Richer, F. J. C. La lymphangite épizootique. Revue générale de la maladie et observations cliniques en Republique du Senegal. These Ecole Nationale Veterinaire d'Alfort, 1977.

Singh, S. Equine cryptococcosis (epizootic lymphangitis). *Indian Vet J* **32:** 260–270, 1955.

Singh, T. Studies on epizootic lymphangitis. I. Modes of infection and transmission of equine histoplasmosis (epizootic lymphangitis). *Indian J Vet Sci* **35:** 102–110, 1965.

Singh, T. Studies on epizootic lymphangitis. Study of clinical cases and experimental transmission. *Indian J Vet Sci* **36:** 45–59, 1966.

Singh, T., and Varmani, B. M. L. Some observations on experimental infection with *Histoplasma farciminosum* (Rivolta) and the morphology of the organism. *Indian J Vet Sci* **37:** 47–57, 1967.

Singh, T., Vermani, B. M. L., and Bhalla, N. P. Studies on epizootic lymphangitis. II. Pathogenesis and histopathology of equine histoplasmosis. *Indian J Vet Sci* **35:** 111–120, 1965.

3. Parasitic Lymphangitis

Filariid worms of the genus *Brugia* (syn. *Wuchereria*) parasitize the lymphatic system of dogs and cats in tropical areas. *Brugia malayi* occurs in cats and primates in India and Malaysia; cats may be reservoir hosts for human infection. *Brugia patei* occurs in dogs and cats in Africa, *B. pahangi* in dogs and cats in Africa and the East, and *B. ceylonensis* in dogs in Sri Lanka. *Brugia* spp. are transmitted by mosquitoes, and their life cycles are similar to those of other filarial worms. Infective larvae enter peripheral lymphatics, migrate to the nearest lymph node, and develop for 2 weeks before migrating down afferent lymphatics, where they mature and produce granulomatous lymphangitis, lymphangiectasis, and lymphadenitis, but do not usually cause lymphedema and elephantiasis as occur in humans due to infection with *B. malayi* or the closely related *Wuchereria bancrofti*. The main importance of *Brugia* spp. lies in the differentiation of their microfilariae from those of *Dirofilaria immitis*, and in their role in human infections.

Bibliography

Denham, D. H., and Rogers, R. Structural and functional studies on the lymphatics of cats infected with *Brugia pahangi*. *Trans R Soc Trop Med Hyg* **69:** 173–176, 1975.

Ewert, A., and Folse, D. Lymphatic filariasis. *Am J Pathol* **115:** 135–137, 1984.

McVay, C. S. *et al*. A comparison of host responses of the Mongolian jird to infections of *Brugia malayi* and *B. pahangi*. *Am J Trop Med Hyg* **43:** 266–273, 1990.

Schacher, J. F., and Sahyoun, P. F. A chronological study of the histopathology of filarial disease in cats and dogs caused by *Brugia pahangi* (Buckley and Edeson, 1956). *Trans R Soc Trop Med Hyg* **61:** 234–243, 1967.

VI. Specific Infectious Diseases of Vessels

A. Equine Viral Arteritis

This disease is caused by an RNA virus of the family Togaviridae (genus *Arterivirus*), which is pathogenic only for horses and is cytopathic in equine kidney culture. Although serologic surveys indicate that infection with the virus is common in North America and Europe, clinical disease is uncommon. Most strains of virus are avirulent, and mortality is rare in natural outbreaks. Long-lasting immunity follows recovery. Transmission of virus occurs primarily by respiratory and venereal routes during the acute phase of infection; venereally infected mares can infect non-immune mares by aerosol transmission. Long-term carrier stallions play an important role in perpetuation and dissemination of the virus; a carrier state has not been demonstrated in mares or foals.

The clinical disease is characterized by fever; variable anorexia and depression; leukopenia; edema of the ventral body wall and limbs, especially the hind limbs; skin rash, most commonly on the neck; serous, later mucopurulent, oculonasal discharge with rhinitis and conjunctivitis; and periorbital/supraorbital edema. Dyspnea, coughing, ataxia, and diarrhea are less frequent. Pregnant mares often abort during or shortly after the febrile period. Abortion is probably due to myometritis in the mare; specific lesions are present occasionally in aborted fetuses, and include necrotizing vasculitis, especially of the placenta, and inflammatory foci in various organs.

The virus is pathogenic to endothelial cells and causes a panvasculitis, that is, inflammation of veins, lymphatics, and arteries. Following initial replication of the virus in macrophages, endothelial cells are invaded beginning 3 days after experimental aerosol infection. As inflammation progresses and neutrophils damage the internal elastic lamina, medial cells are invaded. The most severe edema occurs at days 6 and 7 when phlebitis, lymphangitis, and capillary damage are most pronounced. Arterial necrosis peaks at day 10, when edema has largely disappeared. The vascular lesions will resolve if the horse survives. Antibody appears to play little role in the pathogenesis of the disease, in contrast to the immune-complex component of some other viral vasculitides.

The gross lesions of the disease, in addition to the changes observed clinically, consist principally of hemorrhages and edema. Petechial hemorrhages are found on all serous membranes, in the substance of the lungs and the gastric mucosa. Larger hemorrhages may be present in the adrenals. There is excessive fluid, which contains much protein and some strands of fibrin in all serous cavities.

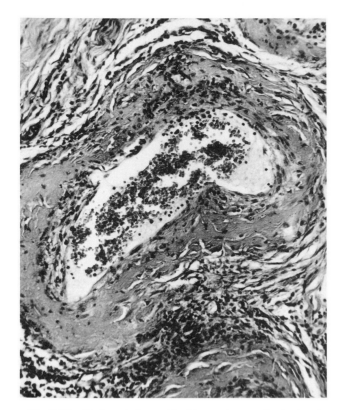

Fig. 1.74 Equine viral arteritis. Horse. Artery of colonic submucosa with fibrinoid necrosis of muscularis and early leukocytic reaction. (Slide, courtesy of J. R. Rooney.)

As much as 10 liters may be present in the pleural and peritoneal cavities. The connective tissues and mesenteries of the body cavities are saturated with edema fluid, and the wall of the gut may be thickened by edema. Enteritis, hemorrhagic or diphtheritic in character and usually more severe in the large than in the small intestine, is regularly present. The lungs are more or less severely edematous.

The characteristic microscopic lesions occur focally or segmentally in the media of small muscular arteries (Fig. 1.74). The muscle cells undergo necrosis and are replaced by a hyaline or fibrinoid material. There is edema of the wall and adventitia and an infiltration of leukocytes, chiefly lymphocytes, the nuclei of which undergo fragmentation with necrosis. The endothelium and intima may have been repaired so that thrombosis is unusual; thrombosis may, however, occur in the intestine and lungs. The arterial lesions occur in many organs but are perhaps most consistently present in the gut and adrenals. Infarction is most common in the intestines, particularly in the cecum and colon. Massive necrosis of lymph nodes also occurs. Extensive necrosis of the adrenals may result from direct viral injury and infarction.

Bibliography

Coignoul, F. L., and Cheville, N. F. Pathology of maternal genital tract, placenta, and fetus in equine viral arteritis. *Vet Pathol* **21**: 333–340, 1984.

Huntington, P. J. *et al.* Equine viral arteritis. *Aust Vet J* **67**: 429–431, 1990. (review)

Huntington, P. J., Forman, A. J., and Ellis, P. M. The occurrence of equine arteritis virus in Australia. *Aust Vet J* **67**: 432–435, 1990.

Johnson, B. *et al.* Arteritis in equine fetuses aborted due to equine viral arteritis. *Vet Pathol* **28**: 248–250, 1991.

B. African Horse Sickness

African horse sickness (AHS) is a disease of solipeds, and occasionally of dogs and camels, caused by an orbivirus that is closely related to bluetongue virus. The disease has been primarily an African one, well established in the south of the continent since the time of European settlement. However, it has spread to and been established in Mediterranean countries, including Spain, and from the Near East to India. Its importance lies in the extreme lethal potential of the infection in susceptible horses and the very wide distribution of the competent insect vectors.

The vectors are primarily gnats of the genus *Culicoides*. Transstadial, but not transovarian, transmission has been shown for the tick *Hyalomma dromedarii* in Egypt, but a natural role for vectors other than *Culicoides* is not clear. Historically, in enzootic areas south of the Sahara, outbreaks of disease occurred in seasons when nocturnal biting insects were active, outbreaks subsiding shortly after the first frosts. It is possible that in climatically suitable areas, transmission might occur throughout the year. Annual recrudescence poses the question of reservoirs for over-winter persistence, but this has not been resolved. Even in districts where AHS occurs annually, the distribution tends to be limited to low-lying areas such as valleys, swamps, and areas of summer rain.

Horse sickness is primarily a disease of the horse and its close relatives, but a wide range of species is susceptible to the virus experimentally, including goats, guinea pigs, mice, ferrets, and rats. The horse is the most susceptible to infection and to illness; the mortality rate in susceptible populations of horses may be as high as 95%. The mortality rate in mules (about 80%) is less than that in horses but greater than that in the Egyptian donkey. The South African donkey and the zebra are resistant. The role of dogs, and other scavenging carnivores, in the disease cycle is not known. The disease is not contagious by contact, but dogs may acquire it from eating infected horseflesh. Infection detected by viral isolation or serologically can be common in dogs in some circumstances, but whether vector transmission occurs is not known.

In many parts of Africa the use of the horse, mule, and donkey is dependent on the control of AHS, but this control has been difficult to attain. The primary obstacle has been the multiplicity of antigenic types of the virus. There are at present nine major antigenic types. These types have a basic antigenic relationship, which may be demonstrated in the horse, but the relationship is not sufficiently close to provide solid immunity among types. Indeed, although all strains investigated can be fitted into one of the major types, the antigenic diversity of strains within a single type is such that very few strains are identical. This is certainly the situation in enzootic areas where a large number of different strains may be isolated during an outbreak. On the other hand, there has been close antigenic relationship between the viruses recovered during the epizootic outbreaks that occurred outside the reservoir-host range where the transmission was from horse to horse. This suggests that, although the virus of African horse sickness is extremely mutable, mutations do not readily occur when the disease is transmitted from horse to horse. Recovered horses are immune to homologous challenge and possess some immunity to heterologous challenge.

Four clinical forms of AHS are described: the pulmonary, cardiac, mixed, and AHS fever forms. Most cases are of the mixed form, perhaps reflecting variations in the susceptibility of the host, the virulence of the virus, or the tropism of virus for different parts of the vascular system.

The peracute or **pulmonary form** of the disease—the most common form in epizootics—has an incubation period of 3–5 days. Fever occurs suddenly and is of short duration, followed by the acute onset of respiratory distress. Death may occur in a few hours from pulmonary edema, and frothy fluid fills the respiratory tract and may flow from the nostrils.

The subacute or **cardiac form** is usually associated with infections which are not fully virulent or are encountered in animals in which infection by heterologous strains has stimulated some immunity. The incubation period is 7–14 days with clinical signs of fever lasting 3–6 days. As the fever subsides, edematous swellings appear in the temporal or supraorbital fossae and the eyelids. The swellings extend to the lips, cheeks, tongue, intermandibular space, and throat. Swelling from subcutaneous fluid may affect the upper neck and sometimes the chest and shoulders. Terminally, petechial hemorrhages usually occur on the conjunctiva and ventral surface of the tongue. The mortality rate may be as high as 50% and is attributed to cardiac failure and pulmonary edema.

As observed, many cases are **mixtures of the pulmonary and cardiac forms.** Initial pulmonary signs of relatively mild degree may be followed by characteristic swelling of the head and death from cardiac failure, or the cardiac form may be followed by sudden onset of dyspnea and pulmonary edema.

The mild clinical form of the disease, **AHS fever,** may not be noticed in outbreaks. It is characterized by remittent fever lasting for a week or so. This is the form of the disease to be expected in animals with immunity to heterologous strains and in resistant species such as donkey and zebra.

The pathogenesis of the signs and lesions is not understood. Virus is found early and develops to highest titer in spleen, lymph nodes, lung, and the large intestine; the titer in myocardium is low. The edema is indicative of increased vascular permeability, but its localized distribution is unusual for the horse and suggests selective injury to the vasculature of some tissues. The nature of the fluid, including the presence of blood cells and fibrin, is suggestive of an inflammatory process. Histologic examination has not revealed structural changes in vessels which would account for the effusions. Myocardial lesions are seen microscopically in the cardiac form of the disease and are less severe in the pulmonary form. These consist of hemorrhage, edema, and focal myocardial necrosis with inflammatory infiltrates and slight proliferation of connective tissue.

The gross lesions observed at autopsy depend largely on the clinical form of disease. In the pulmonary form, the most characteristic changes are edema of the lungs and hydrothorax. In very acute cases alveolar edema and mottled hyperemia of the lungs are seen, whereas in cases with a somewhat more protracted course, extensive interstitial and subpleural edema are also present, but hyperemia is less evident. There may be diffuse alveolitis in some animals (Fig. 1.75A,B). Other lesions commonly observed are periaortic and peritracheal edematous infiltration, diffuse or patchy hyperemia of the glandular fundus of the stomach, hyperemia and petechial hemorrhages in the mucosa and serosa of the small and large intestines, subcapsular hemorrhages in the spleen, and congestion of the renal cortex. All the lymph nodes are enlarged and edematous, especially those in the thoracic and abdominal cavities. The pericardial sac may contain variable amounts of fluid, and numerous petechial hemorrhages occur in the pericardium. Cardiac lesions are usually not conspicuous, but epicardial and endocardial petechiae may sometimes be evident.

In the cardiac form, the most characteristic lesions are the edematous infiltration of the subcutaneous, subfascial, subserous, and intermuscular tissues. Only the head and neck may be involved, but in more severe cases, the edema also involves lower parts of the neck, brisket, and shoulders. Hydropericardium is a constant finding, the pericardial sac often containing 2 liters or more of fluid. The epicardium and endocardium show numerous, almost confluent ecchymotic hemorrhages. These are particularly conspicuous along the course of the coronary vessels, beneath the bicuspid and tricuspid valves and in the papillary muscles adjacent to the insertions of the chordae tendineae. The lungs are usually normal or only slightly engorged, and the thoracic cavity rarely contains an excess of fluid. The lesions in the gastrointestinal tract are generally similar to those found in the pulmonary form, except that submucosal edema tends to be far more pronounced. Ascites is unusual.

In the mixed form, the lesions are a combination of those observed in the pulmonary and cardiac forms. In fact, most of the fatal cases of AHS can be considered to be of the mixed form, with lesions typical of either the pulmonary or the cardiac form predominating.

The infective dose of AHS virus for dogs is large, and high infective titers are not known to be gained except by ingestion of infected horse flesh. The course of the disease in dogs is similar to the pulmonary form in horses, with

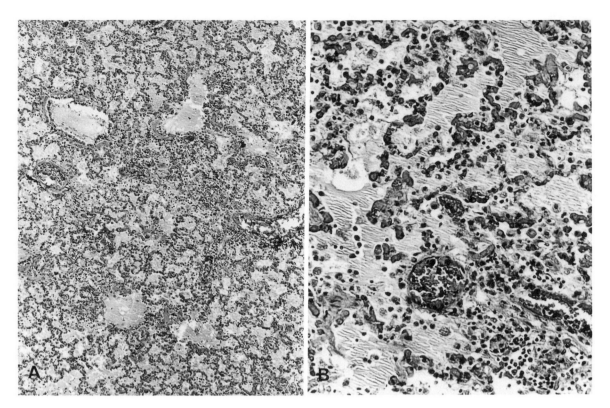

Fig. 1.75 (A) Diffuse pulmonary edema with mononuclear cells in alveoli. African horse sickness. (B) Higher magnification of A.

fever and severe respiratory distress. The mortality in dogs showing respiratory signs is very high. At autopsy the lungs are edematous and may have some superimposed bronchopneumonia. There is much froth in the airways, copious hydrothorax in which the fluid gels on exposure, and saturation of mediastinal tissues. Histologic changes are not helpful in diagnosis or understanding of pathogenesis.

The clinical signs of AHS are usually characteristic, though AHS might be confused with anthrax, equine infectious anemia, equine piroplasmosis, purpura hemorrhagica, equine viral arteritis, equine viral rhinopneumonitis, or surra. Definitive diagnosis requires viral isolation.

Bibliography

Losos, G. J. "Infectious Tropical Diseases of Domestic Animals." Harlow, England, Longman Scientific and Technical, 1986.

Maurer, F. D., and McCully, R. M. African horsesickness—with emphasis on pathology. *Am J Vet Res* **24:** 235–266, 1963.

Mellor, P. S., Hamblin, C., and Graham, S. D. African horse sickness in Saudi Arabia. *Vet Rec* **127:** 41–42, 1990.

Van Rensburg, I. B. J. *et al.* An outbreak of African horse sickness in dogs. *J South Afr Vet Assoc* **52:** 323–325, 1981.

Salama, S. A. *et al.* Isolation and identification of African horsesickness virus from naturally infected dogs in Upper Egypt. *Can J Comp Med* **45:** 392–396, 1981.

C. African Swine Fever

African swine fever (ASF) is an acute-to-chronic, febrile, viral disease of swine, characterized by high fever, cutaneous hyperemia, abortions, edema, and hemorrhage in internal organs, particularly lymph nodes. The African swine fever virus (ASFV) infects only members of the Suidae family and is a harmless companion of the warthog (*Phacochoerus aethiopicus*) and the bush pig (*Potamochoerus porcus*). However, the virus is a constant and major threat to domestic pigs. Originally limited to Africa, the disease has, since the 1960s, appeared in a number of countries, including Spain, Portugal, France, Italy, Malta, Sardinia, Cuba, the Dominican Republic, Haiti, and Brazil, and in 1985–1986, in Belgium and the Netherlands. In an area free of the disease, ASF is typically seen as a peracute or acute disease with a high morbidity and mortality, but as virulence diminishes with time, subacute, chronic, and inapparent forms become increasingly evident. Survivors remain persistently infected.

The causative agent, ASFV, is an enveloped DNA virus currently classified in the genus *Iridovirus,* and is the only one that infects a mammal. It is a relatively resistant virus, which may survive in the environment or in uncooked pork products for prolonged periods. The ASFV apparently replicates in cells of the mononuclear phagocyte system, with a predilection for reticular cells (antigen-processing cells). There are several antigenically different strains of ASFV, which also vary in their cultural characteristics and their ability to hemadsorb. Viral virulence is classified as high, moderate, and mild.

There is no cross-protection conferred by infections by hog cholera virus (HCV) and ASFV. Failure of swine which are immune to HCV to survive challenge with ASFV is the critical test for establishing the diagnosis of ASF in a suspected outbreak. The ability of most isolates of the virus to produce hemadsorption when grown in tissue culture contrasts with the HCV, and this feature allows ready differentiation of most strains of the two viruses.

International spread of ASF is primarily through infected pork products in garbage fed to pigs. Once a focus of infection is established, spread is most likely to occur by direct or indirect contact. In Africa, the transmission of the virus from wild Suidae to domestic pigs is primarily by the argasid tick *Ornithodoros moubata,* a true biological vector and a reservoir of the virus in nature. Additional arthropod vectors exist.

The immunologic response of swine to strains of ASFV is unusual and is incompletely understood. Susceptible animals infected with African field strains develop precipitating and complement-fixing, but rarely neutralizing or protective, antibodies. The animals remain viremic and usually die within 7–10 days. Pigs that recover are resistant to infection with the homologous virus. In the case of infection with highly virulent ASFV, even though 10 days may have elapsed since infection, pigs typically die before development of anti-ASF immunoglobulin, probably because of the demise of antigen-presenting cells in lymph nodes, spleen, and liver.

The peracute form of ASF is characterized by a 1–3 day course of pyrexia, hyperpnea, and cutaneous hyperemia, with morbidity and mortality approaching 100%. The usual clinical course of ASF is acute. Following an incubation period of 4–10 days there is a reduction in appetite, pyrexia, marked cutaneous reddening, dyspnea, and severe leukopenia. Pregnant sows may abort. The clinical signs in the less severe forms of the disease are variable. The signs may be a mild expression of those seen in the acute disease and may easily be confused with other diseases. There are recurring periods of pyrexia, dyspnea, growth retardation, and emaciation. Death may occur during one of the periods of pyrexia. The clinical signs of chronic ASF may be indistinguishable from those seen in hog cholera (HC).

The ASFV tends to affect the same organs and tissues as does the HCV. The anatomic differences tend to be quantitative and dependent on the more fulminating nature of ASF; only splenomegaly and hematomalike visceral lymph nodes are particularly characteristic of ASF (Fig. 1.76). The vascular lesions which can develop rapidly, namely hemorrhage and edema, are apt to be more severe in ASF than in HC, whereas the lesions which develop more slowly, such as infarction, tend to be absent.

The pig dead from acute ASF shows few or no signs of

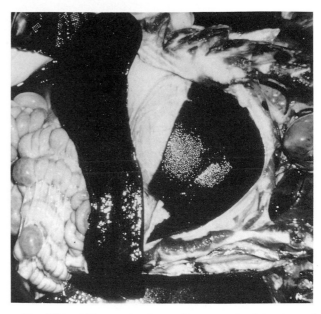

Fig. 1.76 African swine fever. Splenomegaly. (Courtesy of C. C. Brown.)

recent wasting. Gastrosplenic and renal lymph nodes are usually intensely hemorrhagic, and ecchymotic hemorrhages may be present on the serous membranes. The spleen is usually enlarged and may be markedly enlarged and friable; infarction cannot usually be recognized grossly. Pulmonary edema, not common in HC, is present frequently in pigs with ASF. The pulmonary septa are thickened by a yellow gelatinous infiltrate. Pulmonary hemorrhage is also common.

The gallbladder is often edematous, and the vessels of the wall are engorged and conspicuous. Both petechiae and ecchymoses are present on the serosal and mucosal surfaces. In some outbreaks of the disease, there are extensive pancreatic hemorrhages and necrosis. The renal changes are similar to those of HC and consist of subcapsular petechiae (Fig. 1.77).

The stomach often contains ingesta, but the mucosa is apt to be inflamed and eroded. Changes in the small intestine may be absent or consist of segmental areas of congestion and mucosal petechiation. More severe intestinal changes may be found in the large intestine; these may include large areas of hemorrhage, severe congestion, and ulceration. Lesions suggesting button ulcer are very rare in this disease, presumably because few animals survive long enough for their development.

The postmortem appearance of subacute and chronic cases of ASF varies considerably. In the subacute cases that die after 3–4 weeks of illness, there may be lymph node and renal hemorrhage, an enlarged, but not congested spleen, lobular consolidation of cranial lung lobes and mucosal intestinal hemorrhage. Prominent features of the chronic form include fibrinous pericarditis and pleuritis; splenic and lymph node enlargement; lobular consolidation of the lungs, which may progress to necrosis and

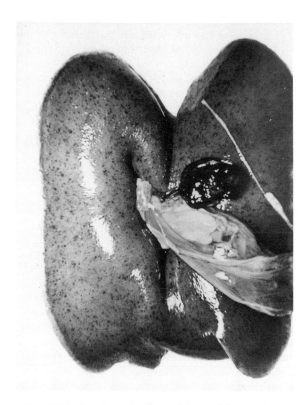

Fig. 1.77 Renal cortical petechiae and hemorrhagic lymph node. African swine fever.

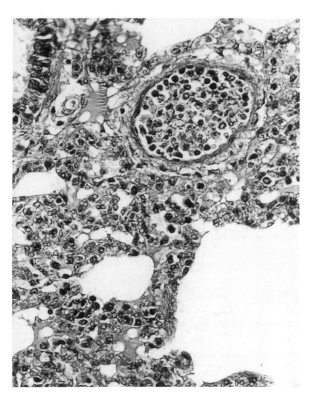

Fig. 1.78 Cellular thrombus in pulmonary vessel, and alveolar thickening. African swine fever.

mineralization of an entire lobe; skin lesions ranging from raised hyperemic plaques to necrotic areas; swollen joints; and arthritis. The meningoencephalomyelitis and periportal hepatitis observed in the acute disease persist in the chronic disease. Lesions in aborted fetuses are inconsistent but include petechiation of placentas, skin, and myocardium, mottled lungs and liver, and anasarca.

Histologically, infection by a highly virulent strain of ASFV causes extensive necrosis of cells of the mononuclear phagocyte system, whereas infection with a moderately virulent strain causes little necrosis. The histologic findings in acute ASF are very similar to those of HC, but there are important differences. A major difference is that ASFV does not infect epithelium. Necrosis with karyorrhexis in lymphoid tissue everywhere is often very obvious in ASF; frank necrosis is quite rare in HC, although mature lymphocytes are apt to be absent in lymphoid tissue. Renal tubular degeneration with amorphous casts in the medulla is common in ASF, but rare in HC. In ASF, necrosis of periportal hepatocytes and infiltrating lymphocytes is common, whereas microscopic hepatic lesions are usually absent in HC. The vascular cuffs in the brain in ASF contain much more necrotic debris than do the lesions in HC. The degeneration of vascular endothelium and the fibrinoid arterial changes are identical. Vascular endothelial damage in the lung may cause thrombosis (Fig. 1.78) of vessels and thickening of alveolar walls. Pigs dead of acute ASF may have glomerular capillary thrombosis;

surviving pigs may develop focal segmental glomerulonephritis.

Since there is no treatment or effective vaccine against ASF, rapid diagnosis is critical to disease control. A diagnosis of ASF is confirmed by detection of ASFV antigen in tissues by direct immunofluorescence, detection of ASF antibody by indirect immunofluorescence or enzyme-linked immunosorbent assay (ELISA), and viral isolation.

Bibliography

Becker, Y. (ed.). "African Swine Fever." Boston, Massachusetts, Martinus Nijhoff Publishing, 1987. (11 chapters)

Edwards, J. F., Dodds, W. J., and Slauson, D. O. Mechanism of thrombocytopenia in African swine fever. *Am J Vet Res* **46:** 2058–2063, 1985.

Losos, G. J. "Infectious Tropical Diseases of Domestic Animals." Harlow, England, Longman Scientific and Technical, 1986.

Martin-Fernandez, J. *et al.* Glomerular pathology in surviving pigs experimentally infected with African swine fever virus. *Histol Histopathol* **6:** 115–121, 1991.

McDaniel, H. A. African swine fever. *In* "Diseases of Swine," 6th Ed., A. D. Leman *et al.* (eds.), pp. 300–309. Ames, Iowa, The Iowa State University Press, 1986.

Mebus, C. A. African swine fever. *Adv Virus Res* **35:** 251–269, 1988.

D. Hog Cholera (Classic Swine Fever)

Hog cholera (HC) is a highly contagious viral disease of swine; it may occur as acute, subacute, chronic, or inapparent syndromes. Acute HC is a disease of high morbidity and mortality caused by a virulent strain of virus; low-virulence virus may cause inapparent disease. Hog cholera is said to have developed *de novo* in Ohio, in the United States, in the early 1800s. It subsequently spread to almost all countries and is also known as swine fever, *Schweinepest, peste du porc,* and *peste svina.* The hog cholera virus (HCV) produces disease only in swine. The disease has been eradicated from Australia, Canada, the United Kingdom, and the United States, but it remains a problem in South America and several European countries.

Hog cholera is caused by a *Pestivirus,* a member of the family Togaviridae. Other members of the genus include bovine virus diarrhea virus (BVDV) and Border disease virus. Hog cholera virus and BVDV are closely related antigenically, and although pestiviruses other than HCV are thought to be harmless to pigs, natural BVDV infection has been reported to cause clinical signs and necropsy lesions indistinguishable from those of chronic HC. Under appropriate circumstances, all three pestiviruses may produce transplacental infection, and hence persistent infections and congenital anomalies. The HCV is an enveloped, single-stranded RNA virus which replicates intracytoplasmically. Most strains do not produce a cytopathic effect in porcine cell cultures. The enveloped nature of the virus makes it sensitive to lipid solvents and to desiccation. However, the virus may persist for prolonged periods in uncooked pork products.

Antigenic variation exists among strains of HCV, and field strains vary widely in virulence. The interaction of a particular strain of virus and the host provides a number of differing disease syndromes. Strains of HCV are usually classified as being of high, moderate, or low virulence, or as being avirulent. Highly virulent strains produce the features of the classic acute disease in pigs of all ages; the morbidity may reach 100%, with a mortality rate of up to 90%. Strains of moderate virulence induce subacute or chronic disease, and pigs may subsequently die or recover. In pigs infected postnatally, strains of low virulence produce few or no signs of disease and induce immunity, but may cause fetal abnormalities. Artificially attenuated strains used as vaccines are avirulent, even for fetuses. Virulence of HCV may be unstable; virulence can increase by one passage through pigs.

Transmission of the disease is usually by direct contact of infected with susceptible pigs. The virus is present in urine, feces, and lacrimal and oronasal secretions of infected pigs. With the less virulent strains, the virus may be excreted in the urine for periods up to 3 months. The major mechanism of spread of virus of low virulence occurs from continuous virus shedding by chronic or persistently infected asymptomatic pigs. Minor modes of transmission include fomites and arthropods.

Hog cholera is characteristically an acute disease of high morbidity and mortality, most animals surviving only to 14 days after showing the first signs of illness, namely anorexia, depression, fever, and leukopenia. Persistent infection (that is, survival of greater than 30 days) is caused by HCV strains of moderate or low virulence and may arbitrarily be divided into chronic and late-onset types. In chronic HC, typical acute disease is followed by clinical improvement, the result of specific antibody formation, but later by immune exhaustion and chronic disease in which the pig is viremic and more susceptible to secondary bacterial infection. Late-onset HC occurs in pigs that are persistently viremic and immunotolerant to HCV as a result of fetal infection by HCV of low virulence. The latest date for the development of a persistent infection appears to be about day 70 of gestation; by day 85 viral antigen may be detected in a few fetuses, but virus cannot be isolated. Signs that develop in viremic, but previously asymptomatic, pigs include anorexia, depression, leukopenia, conjunctivitis, dermatitis, diarrhea, runting, and posterior paresis. The late-onset form of HC is the porcine equivalent of mucosal disease in cattle, in which BVDV infection of fetal calves results in persistently viremic, immunotolerant animals.

The gross and microscopic appearances of HC reflect the effects of the virus on the vascular endothelium, the immune system (lymphoreticular cells and macrophages), and epithelial cells. The effect on the immune system varies with the stage of differentiation; the mature elements degenerate, whereas the undifferentiated cells respond by proliferation. Endothelial changes are primarily degenerative, but some proliferative changes occur. Damage to endothelial and other cells leads to thrombocytopenia, consumption coagulopathy, and in turn disseminated intravascular coagulation and hemorrhage.

The classic or **acute** form of the disease, which affects pigs of all ages, commences by entry of the virus through the mucous membranes with initial replication in the tonsillar epithelium. This is followed by spread to the cervical lymph nodes with viremia within 16 hr of infection. The virus has a propensity to replicate in cells of the immune system, particularly in lymph nodes, bone marrow, and other lymphoid aggregates such as the spleen and Peyer's patches. After 3–4 days the virus invades endothelial cells and epithelial cells, including those of the pharyngeal mucosa, gastrointestinal tract, gallbladder, pancreas, salivary gland, uterus, adrenal, and thyroid. The clinical signs are not specific and do not provide good correlation with pathologic changes, although in the acute disease they may be sufficient for presumptive diagnosis where the disease is endemic.

Superficially, the eyelids are frequently adherent and sticky, the carcass is dehydrated and is soiled by a terminal diarrhea. The irregular erythema which can be seen in the animal when alive is less obvious, but hemorrhages, particularly in unpigmented areas of skin, may be seen (Fig. 1.79). If present they are most numerous on the abdomen and the inner aspects of thighs.

Diagnostic lesions may be sparse in HC, especially if peracute, and it may be impossible to establish the diagnosis on the basis of the gross lesions in a single animal. The lesions most commonly present are hemorrhages in the

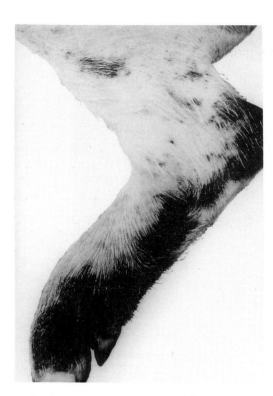

Fig. 1.79 Acute hog cholera. Diffuse and patchy subcutaneous hemorrhages.

Fig. 1.80 Multiple splenic infarcts. Hog cholera. (Courtesy of C. C. Brown.)

periphery of the lymph nodes and renal petechiae. The renal hemorrhages may be very few, and in all cases the kidney capsule should be removed and the surface examined in good light. Hemorrhages in the lymph nodes are more obvious; characteristically, only the periphery of the node is involved. In acute cases this produces a distinctive bright red halo; if the case is more chronic, the halo may be the dirty brown of partially degraded hemoglobin. Less frequently, the hemorrhage in the node is diffuse. The nodal changes are generalized, but nodes most apt to show severe changes are the mandibular, colonic, hepatic, and iliac; changes in the renal lymph nodes are not useful in this context.

Splenic infarction is almost pathognomonic of acute HC. Unfortunately, the tendency for infarcts to develop in the disease varies with the strain of the virus, reports of incidence ranging from 1 to 87%. When present, infarcts occur as single or multiple, dark red, pyramidal blebs 0.5–2 cm in diameter usually along free edges, but occasionally on the flat surface (Fig. 1.80). The spleen in HC is not enlarged, as it is in septicemias, and is reddish brown so that the infarcts contrast quite sharply with the meaty parenchyma. In addition to the hemorrhages of the kidney, petechiae are common in the urinary bladder, the larynx (Fig. 1.81A,B), the gastric mucosa, the lung, and the epicardium, in about that decreasing order of frequency. There is usually a small amount of straw-colored fluid in the pericardial sac. There may be hemorrhagic lobular pneumonia (Fig. 1.82A).

Irregular, but sharply outlined areas of necrosis develop in the tonsils and posterior fauces (Fig. 1.82B). The stomach of an animal dead of HC is empty except for a small amount of watery mucus and ingesta. The fundic mucosa is congested, and mild or severe erosions may be present. Specific lesions do not occur in the small intestine, but the mesentery is usually congested. It is in the colon and cecum that the virus combines with the intestinal flora to produce craterous mucosal defects, button ulcers, that are characteristic of the subacute or chronic stage of the disease. These lesions begin as sharply outlined circular areas of hemorrhage and necrosis (Fig. 1.83). The central area is yellow and dry. As the lesion ages, the rim, which is composed of necrotic epithelium, bacteria, and detritus, is raised above the surrounding mucous membrane, and concentric rings form as if growth were cyclic. The center of lesions sinks, giving a slight diskoid shape. If this necrotic tissue and debris is removed, a deep ulcer is revealed. Growth-arrest lines may be seen in the ribs of chronically sick pigs. Gross lesions are not usually present in the brain, but the brain should be removed because it is there that the histologic changes can be best appreciated.

Within the brain, all areas may be affected, but the lesions are usually most conspicuous in the medulla,

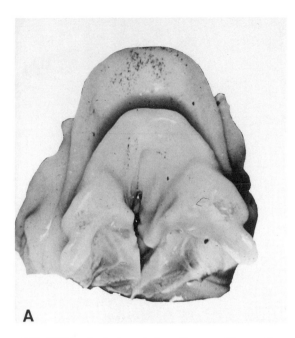

Fig. 1.81A Acute hog cholera. Laryngeal hemorrhages.

pons, midbrain, and thalamus. The response is largely
confined to the vessels and their supporting mesoderm.
Many of the venules have eccentric cuffs, which are
formed in part by transmural migration of monocytes.
Mitotic figures and nuclear chromatin of necrotic cells
are frequently present together in the cuff, as proliferation and necrosis are coexistent. Swelling and degeneration of the endothelial cells occur in the walls of the
smaller vessels (Fig. 1.84). All changes from slight
thickening to necrosis of the wall may be seen. The
lumen is often compromised by these reactions. The
response of the neural parenchyma appears to be entirely
secondary to the vascular lesions; often it is nonexistent
or confined to minimal neuronal degeneration, or it may
consist of glial nodules formed about obstructed or
destroyed capillaries. If severe cerebral edema develops,
there may be widespread damage to the oligodendroglia.
The vessels of the eyes, choroid plexuses, and the
leptomeninges are similarly involved.

Vascular lesions usually are most severe in lymphoid
tissue but may occur anywhere. The lesions vary from
slight thickening of the capillary wall to fibrinoid necrosis
of arterioles. As these changes develop, there is a tendency for extravasations to occur and for tiny thrombi to
form.

Microscopic areas of infarction are common in the
lymph nodes and skin, but only in the spleen, tonsil, gallbladder, and large intestine are these areas usually large
enough to be grossly visible. In some cases, swelling and
paleness of the capillary wall is obvious. The nuclei of
these vessels are enlarged, and the nuclear chromatin is
dispersed as a fine dusting. Less often, the endothelial
proliferation is so marked as to be the most conspicuous

Fig. 1.81B Acute hog cholera. Hemorrhages in mucosa of
urinary bladder.

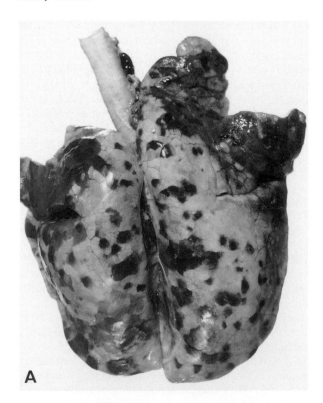

Fig. 1.82A Hemorrhagic pneumonia with lobular distribution. Hog cholera. The cranioventral bronchopneumonia is probably not related.

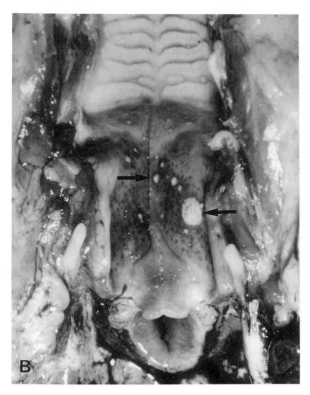

Fig. 1.82B Necrosis of tonsils (arrows). Hog cholera. (Courtesy of C. Brown.)

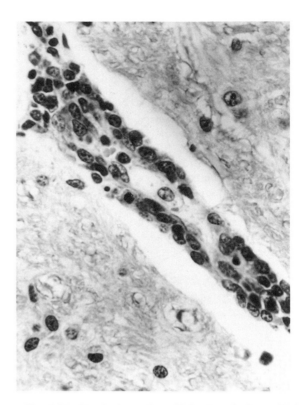

Fig. 1.84 Cerebral venule cuffed primarily by multiplying mononuclear cells. Hog cholera.

Fig. 1.83 Acute hog cholera. Hemorrhagic ulcers in colon.

change. In general, in acute cases the degenerative changes are most prominent, whereas in chronic ones, proliferative changes are more obvious.

In the periphery of lymph nodes (the equivalent of the medulla in other species), lesions vary from slight edema and proliferation of the reticuloendothelial elements to extensive hemorrhage. It is the intermediate stages that are the most common and in which the nature of the lesion is best visualized; in these, there is necrosis of the small vessels and secondary hemorrhages and parenchymal necrosis.

In the spleen, the most pronounced lesions are in the follicular arteries, particularly those at the apex of the pyramidal infarcts. The swelling and hyalinization of these vessels may be so severe as to occlude the lumen. Survival of tissue within the infarcted area will depend on how long the infarction preceded death. Even if the spleen is free of gross and microscopic infarction, vascular changes of the type seen elsewhere can usually be found. They are often associated with perivascular hemorrhage. In addition, there is reticuloendothelial hyperplasia and absence of mature lymphocytes.

The kidney tubules often show nonspecific degenerative changes, and the subcapsular hemorrhages seen grossly can be found, but usually not their origin. Immune complex-mediated glomerulonephritis can develop in the terminal stages of chronic HC.

In utero infections can result in a variety of effects,

including abortion, mummified fetuses, and stillborn piglets. Fetal abnormalities such as ascites, petechiation of skin and other organs, hepatic nodularity, pulmonary hypoplasia, microcephalus, hydrocephalus, cerebellar hypoplasia, and hypomyelinogenesis may also be observed. Persistently infected pigs that survive may be runts and die 2–11 months after birth. These piglets have severe thymic atrophy and pale, swollen lymph nodes; some may have focal colonic mucosal necrosis.

Serologic diagnosis of HC is difficult. Antibodies to HCV and BVDV cross-react in many assays; they may be differentiated by specific virus neutralization tests. Monoclonal antibodies permit differentiation of the pestiviruses in organ sections or in tissue culture. The direct immunofluorescence test for detection of viral antigen in frozen tissue sections is the usual officially accepted test in eradication programs.

Bibliography

Carbrey, E. A. *et al*. Persistent hog cholera infection detected during virulence typing of 135 field isolates. *Am J Vet Res* **41**: 946–949, 1980.

Frey, H. R. *et al*. Experimental transplacental transmission of hog cholera virus in pigs. I. Virological and serological studies. *Zentralbl Veterinaermed (B)* **27**: 154–164, 1980.

Hermanns, W. *et al*. Experimental transplacental transmission of hog cholera virus in pigs. V. Immunopathological findings in newborn pigs. *Zentralbl Veterinaermed (B)* **28**: 669–683, 1981.

Liess, B. "Classical Swine Fever and Related Viral Infections." Boston, Massachusetts, Martinus Nijhoff Publishing, 1988. (12 chapters)

Moennig, V. Pestiviruses: A review. *Vet Microbiol* **23**: 35–54, 1990.

Plateau, E., Vannier, P., and Tillon, J. P. Atypical hog cholera infection: Viral isolation and clinical study of *in utero* transmission. *Am J Vet Res* **41**: 2012–2015, 1980.

Terpstra, C., and Wensvoort, G. Natural infections of pigs with bovine viral diarrhoea virus associated with signs resembling swine fever. *Res Vet Sci* **45**: 137–142, 1988.

van Oirschot, J. T., and Terpstra, C. Hog cholera virus. *In* "Virus Infections of Porcines." M. B. Pensaert, (ed.), pp. 113–130. Amsterdam, Elsevier Science Publishers, 1989.

E. Bovine Ephemeral Fever

Bovine ephemeral fever (BEF), a noncontagious disease of cattle and water buffalo, is characterized by sudden onset of fever, stiffness, and lameness, usually with high morbidity and low mortality, and with rapid spontaneous recovery within 3–4 days of the onset of clinical signs. The cause, the BEF virus, is classified as a rhabdovirus. The disease is endemic in tropical regions of Africa and Asia and occurs as epidemics in subtropical and temperate regions of Africa, Asia, and Australia; it has not been reported from the Western Hemisphere, Europe, New Zealand, or New Guinea. In temperate climates, BEF is a disease of summer and autumn, and ceases abruptly at the first frost; the reservoir for overwintering or between epizootics is probably in wild ruminants. Infection usually

confers lifelong immunity. The likely vectors are various species of *Culicoides* and mosquitoes.

Clinical disease ranges from almost imperceptible clinical signs to death, with considerable individual variation in an outbreak. Mild disease may be exacerbated by environmental stress, e.g., exposure causing dehydration. Disease is milder in young than in mature animals; in lean than in fat animals; in light steers and cows than in bulls; in dry cows than those in heavy lactation. The incubation period is usually 2–4 days. The fever may be biphasic, triphasic, or multiphasic; clinical signs are milder in early than in later phases of fever. Mildly affected animals may have fever, be lame or stiff for a day or two, and then recover completely. Neutrophilia with a left shift and concurrent lymphopenia, increased plasma fibrinogen concentrations, and hypocalcemia occur early in the course of disease. More severely affected animals are anorectic, have oculonasal discharge, may have muscle tremors, stiffness, lameness, and patchy edema of the face or jaw. Rumen motility ceases, and the animal may be constipated. Animals in lateral recumbency are usually still able to rise. Lactation falls by an average of 50%, and does not usually recover to normal during that lactation. Cows in late pregnancy may abort. In very severe cases, there may be temporary or permanent paralysis of all limbs, inability to swallow, and drooling; in 1–4 days, loss of reflexes, coma, and death follow.

Whether the disease is mild or severe, recovery commences quite suddenly and is complete in 95–97% of uncomplicated cases, giving rise to the name ephemeral fever. Although mortality is usually less than 1%, the most productive mature animals are often lost, plus there are losses of milk production, abortion losses, and temporary infertility in bulls. Other sequelae include fatal pulmonary emphysema, pneumonia, mastitis, and locomotor disturbances. The total economic loss in an outbreak can be severe.

The essential gross feature of BEF is serofibrinous polyserositis, which variously affects the synovial, pericardial, thoracic, and peritoneal cavities, and is the result of increased vascular permeability. The serosal membranes themselves are often edematous and may be hemorrhagic; the edema fluid in cavities often contains fibrin clots. Severely affected joints may be surrounded by brown or yellow gelatinous periarticular fluid that also extends along tendon sheaths and fascial planes. There may be patchy pulmonary edema, visceral and parietal pleuritis, epicarditis at the base of the heart, edematous lymph nodes in the febrile stages, and necrosis in some skeletal muscles; myonecrosis is likely the result of recumbency. Histologically, affected tissues are edematous and infiltrated by neutrophils, and may also be hyperemic and hemorrhagic. Vascular changes include endothelial swelling and hyperplasia, hyperplasia of pericytes, fibrinoid necrosis of arterioles, perivascular fibroplasia, and thrombosis of occasional vessels in muscles. Wallerian degeneration has been reported in the upper cervical spinal cord in cases with chronic paralysis, but may be secondary to trauma.

The pathogenesis of BEF is poorly understood; the site of viral replication remains unknown. Neutrophils appear to play a central, but undefined, role; cattle in which neutrophils are experimentally depleted develop viremia but not clinical disease. Interferon-a, interleukin-1, and tissue necrosis factor circulate in high titer during the acute phase of BEF.

The clinical signs observed in an epizootic are usually diagnostic. Serologic diagnosis may be complicated by cross-reaction with antibody to Kimberley virus, an arbovirus which induces production of low levels of serum neutralizing antibody to BEF virus but does not protect against BEF.

Bibliography

Losos, G. J. "Infectious Tropical Diseases of Domestic Animals." Harlow, England, Longman Scientific and Technical, 1986.
St. George, T. D. Bovine ephemeral fever: A review. *Trop Anim Health Prod* **20**: 194–202, 1988.
Uren, M. F. Bovine ephemeral fever. *Aust Vet J* **66**: 233–236, 1989.
Uren, M. F., and Murphy, G. M. Studies on the pathogenesis of bovine ephemeral fever in sentinel cattle. II. Haematological and biochemical data. *Vet Microbiol* **10**: 505–515, 1985.
Young, P. L., and Spradbrow, P. B. Demonstration of vascular permeability changes in cattle infected with bovine ephemeral fever virus. *J Comp Pathol* **102**: 55–62, 1990.

F. Heartwater (Cowdriosis)

Heartwater is a tick-borne rickettsiosis of ruminants that is endemic in sub-Saharan Africa and that has been identified in the Caribbean. The common name, heartwater, is derived from the pericardial effusion which is typically present in small ruminants but sometimes absent in cattle. Heartwater is a major vector-borne disease of ruminants in Africa, third in importance only to East Coast fever and trypanosomiasis, but has a wider distribution than either of these. Mortalities due to heartwater are estimated to be three times greater than those due to babesiosis and anaplasmosis. The causative agent is *Cowdria ruminantium*, which is classified in the order Rickettsiales; *Cowdria* isolates have been shown to cross-react strongly with *Ehrlichia equi* and to a lesser extent with *E. canis* in indirect fluorescent antibody tests. The vectors are three-host ticks of the *Amblyomma* genus, principally *A. variegatum*, *A. hebraeum*, and *A. pomposum*. *Amblyomma maculatum* and *A. cajennense* are suitable vectors present on the American mainland. Transstadial transmission occurs in vector ticks; transovarian transmission is infrequent.

The various strains of *Cowdria ruminantium* are antigenically similar, but they differ considerably in their virulence. The various domestic hosts also differ in their susceptibility. Imported breeds of cattle, sheep, and goats are as a rule much more susceptible than indigenous breeds.

There is also an important age difference in susceptibility. Calves and lambs younger than 3 weeks are highly resistant to heartwater. This is a true age resistance, and it is independent of maternal immunity. Ruminants are most susceptible at about the stage of early maturity, and the mortality then, in the absence of treatment, is expected to be about 60% in cattle and close to 100% in European breeds of sheep, although this will depend on the virulence of the infecting strain.

There also appear to be differences in susceptibility among the African antelopes. The blesbok and black wildebeest are known to be susceptible to infection without showing signs of illness. Wildlife may act as reservoirs of infection, though this is not necessary as the tick itself is a reservoir; infection survives in adult ticks longer than 15 months. Domestic ruminants which have recovered from the disease seldom remain as carriers of the circulating organism for more than 3 weeks after recovery, but it is not clear whether the agent remains in a masked form in tissues. Immunity of variable duration follows recovery, often as premunity; sterile immunity may persist for a year or more after disappearance of the organism. Immunity in natural infections is expected to be reinforced by reinfections.

When a tick feeds on an infected host, organisms apparently first infect and develop within gut cells. Subsequent stages of the organism continue their development in hemolymph and salivary gland, and are transferred to the vertebrate host during tick feeding.

The early development of the parasite is not completely understood, but probably takes place in reticular cells of lymph nodes and spleen in which initial bodies are demonstrable at about the fourth day of infection. Presumably, there is a developmental phase in reticular cells after which the organism is released into the blood and parasitizes cells of vascular endothelium. The organism is demonstrable in capillary endothelium of brain 1–3 days after it is demonstrable in lymph nodes. *Cowdria ruminantium* parasitizes endothelial cells but is also present in the blood; the infection is readily transmissible by inoculation of blood collected during the febrile stage or shortly thereafter. The colonies (groups, clusters, morulae) of microorganisms in endothelial cells lie characteristically in membrane-bound vacuoles in the cytoplasm at the poles of the nuclei. Colonies and organisms are pleomorphic; colonies are formed by binary fission of organisms. Organisms range in size from 0.2 to >2.5 μm and are termed small, medium, or large; each is surrounded by a double membrane, and based on their internal morphology, they may be termed elementary (electron-dense), intermediate, or reticulate bodies; usually only organisms of the same form are found within a vacuole. They can be found in sections and are present in all organs and in vessels of all calibers but are best sought in the capillaries of the cerebral cortex in squash-smear preparations or in imprints from the endothelium of a large vessel such as the jugular vein. Giemsa is the preferred method of staining.

The natural incubation period of heartwater is 2–3

weeks but varies considerably. In the peracute form of the disease, death occurs rapidly in convulsions without warning. The usual clinical syndrome in susceptible animals is acute, with a course of up to 6 days. There is high fever and its attendant signs. As the course advances, nervous signs develop. These are especially prominent in cattle and consist of chewing movements, protrusion of the tongue, twitching of eyelids and blinking, unsteadiness of gait, circling, and a terminal convulsive phase in which there is prominent cutaneous hyperesthesia. Diarrhea is common in cattle but less common in sheep. Mild progressive anemia may develop, as may leukopenia (neutropenia, eosinopenia, and lymphocytosis) and hypoalbuminemia. In the subacute syndrome, the course is prolonged and leads either to slow recovery or to collapse and death. Animals with high natural resistance or partial immunity may show transient fever only.

The gross anatomic changes in heartwater are the same in each ruminant species but generally are less prominent in cattle than in sheep or goats; lesions may be inconspicuous or absent in treated animals. The general condition of the animal is not altered appreciably by the disease. Typical gross lesions include effusion in body cavities, hydropericardium, edema of lungs, brain, mediastinum, and associated lymph nodes, and splenomegaly, though all of the lesions rarely occur simultaneously. The name of the disease is derived from the hydropericardium, but this is sometimes absent. The effusion, which is serous with some fibrinous strands, is more constant and copious in sheep than in cattle. The other serous cavities also contain

excess serous fluid; there may be several liters in the pleural and peritoneal cavities of the ox and up to half a liter in sheep and goats. When the exudation is copious, the mediastinum and retroperitoneal tissues are also edematous. The membranes remain clear and glistening. The lungs are severely edematous, especially in the peracute form. The lymph nodes of the head, neck, and cranial mediastinum especially are swollen and edematous. The spleen is almost invariably much enlarged in sheep and goats, being sometimes up to six times its normal size, but splenomegaly in cattle is only moderate. There is hyperemia with submucosal edema in the lower alimentary tract, but it is most pronounced in the abomasum. Petechial hemorrhages are present in the endocardium, mucosa of stomach and gut, tracheobronchial mucosa, and lymph nodes.

The diagnosis of heartwater depends largely on the demonstration of organisms in endothelial cells in brain either in smears or section (Fig. 1.85A); brain biopsy can be used to establish the diagnosis in the living animal. Lesions in the brain are common (Fig. 1.85B) and do account for the nervous signs which characterize the disease, especially in cattle, and for the failure of animals to recover with chemotherapy.

Gross changes are inconstant on inspection of the surface of the brain. Petechial hemorrhages may be present over the cerebellum, and the choroid plexus may be dull and thickened. On slicing, petechial, ecchymotic, and

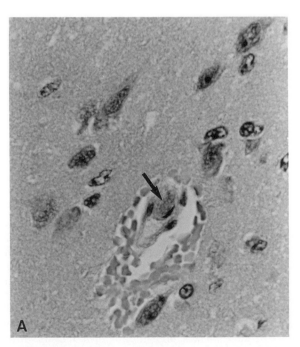

Fig. 1.85A Groups of *Cowdria ruminantium* organisms in endothelial cell of cerebral capillary (arrow). Heartwater. Ox. (Slide courtesy of T. M. Wilson.)

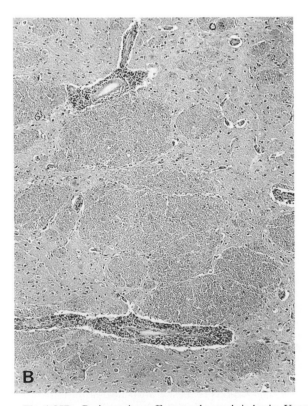

Fig. 1.85B Perivascular cuffs around vessels in brain. Heartwater.

larger hemorrhages may be numerous in all parts of the brain.

The microscopic lesions in the brain are reflections of parasitism and injury to small vessels, especially capillaries and venules. Colonies of *C. ruminantium* are visible in endothelial cytoplasm of untreated cases. There is acute inflammation of the choroid plexus with fibrinous exudate in the stroma, congestion, hemorrhage, and accumulation of inflammatory cells about the vessels. Within the brain, vascular injury leads to perivascular or larger hemorrhages and accumulation of protein droplets in the perivascular space. Focal or larger areas of malacia surround the vessels and are in various stages of development from edematous disruption of white matter with glial activation to microcavitation.

The genesis of the increased capillary permeability seen in heartwater is unclear; the severity of edema is often thought to be out of proportion to the numbers of infecting organisms. Production of a toxin has been suggested, but not proven; *Cowdria* stains negatively with Gram's stain, has a cell wall similar to that of Gram-negative bacteria, and has been suggested to produce endotoxin. Complement and the products of arachidonic acid metabolism may play a role in the generation of vasoactive substances.

Bibliography

Brown, C. C., and Skowronek, A. J. Histologic and immunochemical study of the pathogenesis of heartwater (*Cowdria ruminantium* infection) in goats and mice. *Am J Vet Res* **51:** 1476–1480, 1990.

Camus, E., and Barre, N. "Heartwater—A Review." Paris, Office International des Epizooties, 1988.

Heartwater: Past, present, and future. Proceedings of a workshop held at Berg en Dal, Kruger National Park, Sept. 8–16, 1986. *Onderstepoort J Vet Res* **54:** 1–545, 1987. (61 papers)

Losos, G. J. "Infectious Tropical Diseases of Domestic Animals." Harlow, England, Longman Scientific and Technical, 1986.

Mebus, C. A., and Logan, L. L. Heartwater disease of domestic and wild ruminants. *J Am Vet Med Assoc* **192:** 950–952, 1988.

Prozesky, L., and Du Plessis, J. L. The pathology of heartwater. II. A study of the lung lesions in sheep and goats infected with the Ball₃ strain of *Cowdria ruminantium*. *Onderstepoort J Vet Res* **52:** 81–85, 1985.

Uilenberg, G. Heartwater (*Cowdria ruminantium* infection): Current status. *Adv Vet Sci Comp Med* **27:** 427–480, 1983.

G. Rocky Mountain Spotted Fever

Rocky Mountain spotted fever (RMSF), a febrile exanthema caused by *Rickettsia rickettsii*, is an important rickettsiosis of humans and dogs; *R. rickettsii* or antigenically similar organisms infect a wide variety of mammals and birds. Rocky Mountain spotted fever occurs in dogs throughout the Americas, with most cases in North America occurring during the tick season (April to September). The ticks most commonly responsible for the transmission of RMSF are *Dermacentor variabilis* and *D. andersoni;* these ticks serve as vectors, reservoirs, and natural hosts for *R. rickettsii*. The rickettsiae are maintained in nature in a primary sylvan cycle, involving mostly small rodents. Dogs can develop a high and persistent rickettsemia, and are a potential public health danger; humans may be infected during tick removal by contact with hemolymph of engorged ticks or with tick excreta.

Dogs are usually infected via the bite of infected ticks. The rickettsiae invade and replicate within endothelial cells of small blood vessels; endothelial damage is probably mediated by phospholipase A and a trypsinlike protease, of either rickettsial or host origin. There is little support for production of endotoxin or exotoxin by *R. rickettsii*. Endothelial damage leads to platelet activation and activation of coagulation and fibrinolytic systems; thrombosis in RMSF is thought to arise in reaction to endothelial damage rather than as DIC. Vasculitis in RMSF is a product of endothelial swelling and necrosis with secondary contributions from the immune and phagocytic host responses mounted by lymphocytes and macrophages, but is not primarily caused by immune-complex formation or by cell-mediated mechanisms. The action of various mediators, such as those of the kallikrein–kinin system, in addition to direct endothelial necrosis, leads to increased vascular permeability and hence edema and multifocal hemorrhage. Edema of the medulla oblongata may contribute to cardiorespiratory depression. Fulminant infections may result in peripheral vascular collapse and death during the first week of infection.

Dogs younger than 2 years are predominantly affected. Purebred dogs, especially German shepherds, appear to be more susceptible than are crossbreds. Following an incubation period of a few days, clinical findings in affected dogs include, in decreasing order of frequency, listlessness and depression, high fever, history or actual presence of ticks, anorexia, myalgia/arthralgia, lymphadenomegaly, vestibular deficits, dyspnea, conjunctivitis, paralumbar hyperesthesia, edema of the face and extremities, petechiae or hemorrhagic diathesis, and vomiting and diarrhea. Vasculitis may be visible by means of ophthalmoscopic examination of retinal vessels. Laboratory findings include anemia, leukopenia followed by leukocytosis, thrombocytopenia, prolongation of activated partial thromboplastin time, increased concentration of circulating fibrinogen, and increased concentrations of fibrin/fibrinogen degradation products. Serum biochemistry usually reveals hypoalbuminemia, increased alkaline phosphatase activity, and increased cholesterol concentration. Increased concentrations of aldosterone and antidiuretic hormone may contribute to expansion of plasma and extracellular fluid volumes and hence edema. Death may occur in the acute stages of disease as the result of peripheral vascular collapse, hemorrhagic diathesis, and thrombosis, although fully developed DIC is an unusual event in RMSF in dogs. Necrosis of extremities may occur. Damage to the cardiovascular system, brain, or kidneys may cause death or permanent organ dysfunction; severely affected dogs may die with rapidly progressive meningoencephalitis.

The most reliable serologic test for RMSF is the species-specific microimmunofluorescent method. High antibody titers develop to *R. rickettsii,* but usually decline after 3–5 months. Direct fluorescent antibody staining of a skin biopsy is positive in up to 80% of infected dogs; coccobacillary organisms are present in endothelial cells and vessel walls. Tetracycline therapy is effective, and dogs are immune after recovery.

Lesions seen at necropsy include edema of ears and muzzle; ulcerative glossitis; scrotal dermatitis; petechiation of mucous membranes, caudal abdominal skin, pleura, and gastric wall; hemorrhagic colitis; and hemorrhagic lymphadenopathy. The prominent histologic lesion of RMSF is necrotizing vasculitis of small veins, capillaries, and arterioles, with perivascular accumulations predominantly of lymphocytes and macrophages; this lesion is seen most commonly in skin, testes, alimentary tract, pancreas, kidneys, urinary bladder, myocardium, meninges, retina, and skeletal muscle. Acute meningoencephalitis may be present. There may also be acute splenitis, acute interstitial pneumonia, and multifocal necrosis of myocardium, adrenals, and liver. A moderate degree of glomerulonephritis may be present.

Bibliography

Davidson, M. G. *et al.* Vascular permeability and coagulation during *Rickettsia rickettsii* infection in dogs. *Am J Vet Res* **51:** 165–170, 1990.

Greene, C. E., and Breitschwerdt, E. B. Rocky Mountain spotted fever and Q fever. *In* "Infectious Diseases of the Dog and Cat," 2nd Ed., C. E. Greene (ed.), pp. 419–433. Philadelphia, Pennsylvania, W. B. Saunders, 1990.

Walker, D. H. (ed.). "Biology of Rickettsial Diseases." Boca Raton, Florida, CRC Press, 1988. (2 vols, 18 chapters)

VII. Vascular Neoplasms

Tumors of endothelial origin are classified as benign or malignant, that is angiomas or angiosarcomas, but this distinction is not always easily made. Differentiation of vascular malformations or proliferations from hemangiomas may also be difficult. Hemangiopericytoma is discussed with skin neoplasms. There are rare reports of nonendothelial tumors, e.g., malignant melanoma, chondrosarcoma, arising in the aorta.

The nomenclature of benign vascular tumors or malformations is confusing, and includes hemangioma, hemangioendothelioma, hamartoma, vascular malformation, telangiectasis, and angiokeratoma, as well as variants of those terms. **Hemangioma,** a benign tumor of endothelial cells, is supposedly a true neoplasm capable of independent growth. Some of the "growth" of a hemangioma may be due to congestion, hemorrhage, and thrombosis. **Hemangioendothelioma** describes a benign vascular tumor with hypertrophic endothelial cells forming small hypercellular vascular channels, and is essentially a very cellular capillary hemangioma. Unfortunately, the term hemangioendothelioma has also been used to describe hemangiosarcoma, a clearly malignant vascular tumor. **Hamartoma** is a general term that refers to focal disordered overgrowth of mature tissue that is indigenous to the organ involved. Since vascular tissue is ubiquitous, vascular hamartomas may occur in any site of the body; some authors refer to angiomas as vascular hamartomas. Most hamartomas are present at birth, and their growth is coordinated with that of the surrounding tissue. **Vascular malformation** is obviously a very broad term, which may be included under the term hamartoma, and may include arteriovenous fistula in which thick-walled vessels occur in contrast to the thin-walled vessels of the other benign vascular proliferations. **Telangiectasis** refers to congenital or acquired foci of abnormally dilated capillaries, sinusoids, arterioles, or venules, usually seen as small masses in the skin and mucous membranes. **Angiokeratoma** is a rare benign tumor, usually of the nictitans and conjunctiva of dogs, in which dilated vascular channels are associated with hyperplastic epithelium. Fortunately, these distinctions, though of pathogenetic interest, are of little practical importance, since the lesions described are uniformly benign, and excision, if possible, is curative.

Hemangioma is a benign tumor of endothelial cells and may be classified as capillary or cavernous based on the size of the vascular channels formed. The tumor may be difficult to differentiate from vascular malformations and granulation tissue. Hemangiomas occur most commonly in older dogs, and are seen less frequently in other species. These tumors arise from vascular endothelium and hence may be found in any site in the body, but are most common in the dermis and subcutis, especially of the legs, flank, neck, face, and eyelid. Hemangiomas are usually single, ovoid, red-black masses of 0.5–3 cm diameter, which ooze blood when cut. Histologically, blood-filled vascular spaces are lined by a single layer of well-differentiated endothelium, and may be thrombosed. The vascular spaces are separated by variable amounts of connective tissue stroma. Hemangiomas are not encapsulated, are not invasive, and do not recur after complete surgical excision. **Disseminated cavernous hemangioma** is rare, but has been reported in a calf, and a similar multifocal hemangioma has been reported in a pig. Multiple small tumors are present in skin, subcutis, and internal organs. It has been proposed that this and other similar conditions be included under the syndrome name **juvenile bovine angiomatosis,** to include calves having either solitary, e.g., gingival or spinal canal, or multiple angiomatous lesions. Hemangiomas occur in the urinary bladder of cattle affected with enzootic hematuria, which is discussed with diseases of the urinary system (Volume 2, Chapter 5).

A number of hemangiomalike lesions occur in domestic animals. **Bovine cutaneous angiomatosis** of adult dairy cows is reported from Britain, France, and the United States. Nodular dermal vascular proliferations occur anywhere in the skin, but especially along the back, may have a considerable inflammatory component, and resemble pyogenic granuloma of humans. **Varicose tumor of the scrotum** of dogs is a benign vascular proliferation which resembles cavernous hemangioma when fully developed. Lesions resembling hemangioma occasionally develop in the skin of the scrotum or perineum of boars, or the vulva, mammary glands, or ovaries of sows.

Meningioangiomatosis is a rare condition reported in two dogs and a horse, and characterized by benign focal proliferation of blood vessels and meningothelial cells in the leptomeninges and underlying brain parenchyma.

An unusual intravascular proliferation of endothelial cells resembling **angioendotheliomatosis** in humans has been reported in a cat. Intravascular proliferation of pleomorphic mononuclear cells in a dog has been termed **neoplastic angioendotheliomatosis,** and was thought to be an unusual angiotropic lymphoma. **Lymphomatoid granulomatosis** is a rare pulmonary disorder of middle-aged dogs characterized by infiltration of lung and various other organs by neoplastic mononuclear cells admixed with eosinophils, lymphocytes, and plasma cells; originally thought to be angiocentric, this pulmonary neoplasm is now thought to be a T-cell tumor that is angioinvasive.

Lymphangioma is a rare, benign tumor which consists of lymph channels forming capillary, cavernous, or cystic tumors. Lymphangiomas may occur as congenital malformations (hamartomas) or may develop spontaneously in adults. Cavernous and cystic lymphangiomas may progressively enlarge, dissect along fascial planes, and be difficult or impossible to remove surgically.

Glomangiomas (glomus tumors) of humans, which occur in one of the neuromyoarterial temperature receptors, primarily in the distal parts of the fingers and toes, are characterized by branching vascular channels in a fibrous stroma that contains nests of specialized glomus cells, a type of smooth muscle cell. Glomangiomas have been reported in primates. The glomus jugulare and pulmonale are chemoreceptors, and tumors of these bodies resemble those of other chemoreceptors, namely nests of monomorphic rounded cells with scant cytoplasm separated by thin septa. The tumor is encapsulated and grows expansively. A glomus jugulare tumor has been reported in a dog.

Hemangiosarcoma (malignant hemangioendothelioma), a malignant tumor of endothelial cells, occurs most frequently in old dogs, but is less common than hemangioma. German shepherd dogs (Alsatians) are most commonly affected. Hemangiosarcoma occurs infrequently in horses, cats, cows, and sheep. The tumor may arise in any site of the body. The most common primary sites in dogs are spleen, right atrium, and liver; in cats, spleen, intestines, and subcutaneous tissue; in horses, ocular, cutaneous, and multicentric. Hemangiosarcoma is the most common primary cardiac tumor of dogs, and the tumor usually occurs subepicardially in the wall of the right atrium at the entrance to the auricle near the coronary groove (Fig. 1.48A) or in the auricular appendage. Hemangiosarcoma probably arises *de novo* and not from preexisting hemangiomas. The tumors have a gray to red-black hemorrhagic appearance, and may reach a diameter of 30 cm in the spleen as a result of hemorrhage within the tumor. Large hemorrhagic masses can often be found in the spleens of dogs; some are hematomas rather than either hemangiomas or hemangiosarcomas. The distinction is important because splenic hemangiosarcomas carry a very poor prognosis. However, the distinction can often be difficult because splenic hemangiosarcomas often hemorrhage, and if the resultant hematoma is not sampled carefully, its neoplastic origin will be missed. Following rupture, implants of splenic tissue or tumor may be found on the peritoneum. Hemangiosarcomas typically metastasize widely, especially to the lungs (cannonball metastases) (Fig. 1.48B). Histologically, these tumors consist of vascular spaces lined by elongated, plump, anaplastic endothelial cells. Vascular spaces must be identified in tumors with a highly cellular stroma to differentiate hemangiosarcoma from fibrosarcoma. Hemorrhage and necrosis commonly occur within these tumors, and death may occur due to hemorrhage into the peritoneal cavity, pericardial sac, or brain. When metastases are widespread, the primary tumor site may be difficult to determine, and multicentric origin is a possibility. Poorly differentiated hemangiosarcomas may usually be differentiated from spindle cell sarcomas or other nonendothelial neoplasms by immunohistochemical staining for factor VIII-related antigen, a marker of endothelial cells. *Ulex europaeus* lectin, a marker of neoplastic human endothelial cells, is unreliable as a marker of canine vascular tumors. Ultrastructurally, Weibel–Palade bodies are present in at least some of the typical neoplastic endothelial cells.

Lymphangiosarcoma (malignant lymphangioendothelioma) is an extremely rare tumor in domestic animals. The tumor is histologically similar to hemangiosarcoma, but the irregular vascular channels contain few red cells; it may be diffusely invasive and metastasize widely. Ultrastructurally, lymphangiosarcoma is distinguished from hemangiosarcoma, at least in cats and humans, by the general lack of a basal lamina, few micropinocytotic vesicles, few intercellular junctions, and a discontinuous endothelial cell layer in lymphangiosarcoma.

Bibliography

Anderson, W. I. *et al.* Primary aortic chondrosarcoma in a dog. *Vet Pathol* **25:** 180–181, 1988.

Ando, Y., Yokoiki, Y., and Kadota, K. Ultraviolet-photographical and ultrastructural observations on swine ovarian haemangioma. *Jpn J Vet Sci* **49:** 547–550, 1987.

Augustin-Voss, H. G., Smith, C. A., and Lewis, R. M. Phenotypic characterization of normal and neoplastic endothelial cells by lectin histochemistry. *Vet Pathol* **27:** 103–109, 1990.

Carroll, R. E., and Berman, A. T. Glomus tumors of the hand. *J Bone Joint Surg (Am)* **54A:** 691–703, 1972.

Chan, C. W., and Collins, E. A. Case of angiosarcoma of the nasal passage of the horse—Ultrastructure and differential diagnosis from progressive haematoma. *Equine Vet J* **17:** 214–218, 1985.

Cordy, D. R. Vascular malformations and hemangiomas of the canine spinal cord. *Vet Pathol* **16:** 275–282, 1979.

Dargent, F. J., Fox, L. E., and Anderson, W. I. Neoplastic angioendotheliomatosis in a dog: An angiotropic lymphoma. *Cornell Vet* **78:** 253–262, 1988.

Fitzgerald, S. D., Wolf, D. C., and Carlton, W. W. Eight cases of canine lymphomatoid granulomatosis. *Vet Pathol* **28:** 241–245, 1991.

Freestone, J. F., Williams, M. M., and Norwood, G. Thoracic haemangiosarcoma in a 3-year-old horse. *Aust Vet J* **67:** 269–270, 1990.

George, C., and Summers, B. A. Angiokeratoma: A benign vascular tumour of the dog. *J Small Anim Pract* **31:** 390–392, 1990.

Hubbard, G. B., and Wood, D. H. Glomangiomas in four irradiated *Macaca mulatta*. *Vet Pathol* **21:** 609–610, 1984.

Johnstone, A. C. Congenital vascular tumors in the skin of horses. *J Comp Pathol* **97:** 365–368, 1987.

Ladds, P. W. Vascular hamartomas of the liver of cattle. *Vet Pathol* **20:** 764–767, 1983.

McEntee, M. *et al.* Meningocerebral hemangiomatosis resembling Sturge–Weber disease in a horse. *Acta Neuropathol (Berl)* **74:** 405–410, 1987.

Moore, F. M., and Thornton, G. W. Telangiectasia of Pembroke Welsh corgi dogs. *Vet Pathol* **20:** 203–208, 1983.

Moore, P. F., Hacker, D. V., and Buyukmihci, N. C. Ocular angiosarcoma in the horse: Morphological and immunohistochemical studies. *Vet Pathol* **23:** 240–244, 1986.

Moran, S., Johnson, R. P., and Kreplin, C. M. A. Malignant melanoma involving the aorta in a dog. *Can Vet J* **24:** 148–149, 1983.

Platt, H. Vascular malformations and angiomatous lesions in horses: A review of 10 cases. *Equine Vet J* **19:** 500–504, 1987.

Prymak, C. *et al.* Epidemiologic, clinical, pathologic, and prognostic characteristics of splenic hemangiosarcoma and splenic hematoma in dogs: 217 cases (1985). *J Am Vet Med Assoc* **193:** 706–712, 1988.

Pulley, L. T., and Stannard, A. A. Tumors of the skin and soft tissues. *In* "Tumors in Domestic Animals," J. E. Moulton (ed.), 3rd Ed., pp. 23–87. Berkeley, University of California Press, 1990.

Remedios, A. *et al.* Mediastinal cystic lymphangioma in a dog. *J Am Anim Hosp Assoc* **26:** 161–163, 1990.

Ribas, J. L., Carpenter, J., and Mena, H. Comparison of meningio-angiomatosis in a man and a dog. *Vet Pathol* **27:** 369–371, 1990.

Rossier, Y. *et al.* Pleuroscopic diagnosis of disseminated hemangiosarcoma in a horse. *J Am Vet Med Assoc* **196:** 1639–1640, 1990.

Rothwell, T. L. W. *et al.* Unusual multisystemic vascular lesions in a cat. *Vet Pathol* **22:** 510–512, 1985. (neoplastic angioendotheliosis)

Rudd, R. G. *et al.* Lymphangiosarcoma in dogs. *J Am Anim Hosp Assoc* **25:** 695–698, 1989.

Scavelli, T. D. *et al.* Hemangiosarcoma in the cat: Retrospective evaluation of 31 surgical cases. *J Am Vet Med Assoc* **187:** 817–819, 1985.

Sheikh-Omar, A. R., and Jaafar, M. Ovarian haemangioma in sows. *Vet Rec* **117:** 110, 1985.

Swayne, D. E., Mahaffey, E. A., and Haynes, S. G. Lymphangiosarcoma and haemangiosarcoma in a cat. *J Comp Pathol* **100:** 91–96, 1989.

van der Gaag, I., Vos, J. H., and Goedegebuure, S. A. Lobular capillary haemangiomas in two calves. *J Comp Pathol* **99:** 353–356, 1988.

Van Nes, J. J. *et al.* Glomus jugulare tumor in a dog: A case report. *Tijdschr Diergeneeskd* **103:** 1091–1098, 1978.

von Beust, B. R., Suter, M. M., and Summers, B. A. Factor VIII-related antigen in canine endothelial neoplasms: An immunohistochemical study. *Vet Pathol* **25:** 251–255, 1988.

Waters, D. J. *et al.* Metastatic pattern in dogs with splenic haemangiosarcoma: Clinical implications. *J Small Anim Pract* **29:** 805–814, 1988.

Watson, T. D. G., and Thompson, H. Juvenile bovine angiomatosis: A syndrome of young cattle. *Vet Rec* **127:** 279–282, 1990.

Widmer, W. R., and Carlton, W. W. Persistent hematuria in a dog with renal hemangioma. *J Am Vet Med Assoc* **197:** 237–239, 1990.

CHAPTER 2

The Hematopoietic System

V. E. O. Valli
University of Illinois, U.S.A.
with a contribution by
B. W. Parry
University of Melbourne, Australia

GENERAL CONSIDERATIONS

I. Normal Hematopoiesis

The reticuloendothelial system, as used herein, refers to the hematopoietic system as a whole and acknowledges the ultimate derivation of all fixed and circulating blood cells from a common pluripotential precursor which resides in the bone marrow.

The close relationship between bone and hematopoiesis is a constant finding in mammals and birds. Whereas fetal and splenic hematopoiesis is largely erythroid, the full expression of all cell lines is largely confined to bone marrow. This intimate relationship appears to be dependent on the pattern of blood flow in bone and on a bone–bone marrow portal capillary system. The blood vessels of cortical bone that are derived from periosteal vessels arborize on the endosteal surface, where they form a secondary capillary system which traverses the marrow cavity and drains into the central venous sinus. The radicles from the nutrient artery travel directly to the endosteum and penetrate the cortical bone to return via capillaries to the endosteal network. There is thus anatomic evidence for a preferential microenvironment for hematopoiesis adjacent to endosteal bone which may be generalized to explain the distribution of blood-forming activity in marrow and in ectopic or metaplastic bone.

Marrow capillaries form the conduit for delivery of blood cells to the systemic circulation via the vein which parallels the nutrient artery. All venous flow appears to be centripetal, with periosteal, osteal, and marrow drainage all emptying into the central venous sinus. The marrow capillary endothelium is thin and the cells are without tight junctions. They are incompletely covered on the abluminal surface by a fenestrated basal lamina which is in turn covered to about 60% of the area by cytoplasmic processes of adventitial reticular cells. These reticular cells are believed to be weakly phagocytic but are stromal in origin and distinct from the monocyte–macrophage system. They have no capability for hematopoietic differentiation but are able to form reticulin fibers that cross link between vessels and form the structural support for bone marrow. These marrow stromal cells are hormonally active and essential for the maintenance, growth, and differentiation of the hematopoietic stem cells.

The nonhematopoietic marrow includes nerves and fat cells. The adipocytes have a fatty acid content that differs from that in other sites. Marrow fat is more resistant to lipolysis in response to starvation. The fat cells are probably derived from the perivascular reticular cells. The nerves of bone marrow are both myelinated and unmyelinated and may innervate blood vessels.

In the adults of all species, the hematopoietic marrow is concentrated in the spine, pelvis, sternum, ribs, calvarium, and proximal ends of the limb bones. When increased hematopoiesis is required, there is first lipolysis and myeloid expansion in those areas with residual red marrow. If requirements are further increased, there is hematopoietic conversion of fat in more distal areas. This expansion of activity, although focally irregular, follows a general preference for sites of higher blood flow which are found in close proximity to endosteal surfaces. Thus, regenerative hematopoiesis is most prominent in areas of cancellous bone and in endosteal margins of metaphyses (Fig. 2.1).

Coincident with the remodeling of fetal bone, which provides a marrow cavity, there is colonization by hematopoietic stem cells which arise from the blood islands of the yolk sac. The cellular migration from the blood islands occurs in embryogenesis with seeding into the thymic anlage, the primordial lymph nodes, the liver, and the spleen. At birth, hematopoiesis is largely medullary with some persisting in liver and spleen and to a lesser extent in adrenals and kidney. In the adult, extramedullary hematopoiesis is largely limited to splenic erythropoiesis but may recruit other organs such as kidney and liver in reverting to the embryonic pattern.

The hematopoietic stem cell system consists of a hierarchical lineage with the most primitive cells being capable of differentiation into all blood cell lines and of almost unlimited self-renewal. Growth and differentiation depend on the provision of a suitable inductive microenvironment dependent on an array of **growth factors** and **interleukins** (Fig. 2.2). Commitment to differentiation represents an interplay between inherent genetic capability and the selective hormonal pressure of the microenvironment. Differentiated or committed stem cells can produce only one or two cell types and have limited proliferative capacity. The differentiation pathways result from the interaction of specific receptors for the various growth factors and the proportionate levels of the factors themselves. Thus high levels of a particular factor may up-regulate the production of its specific receptor and thereby recruit additional precursors to the production of a single cell lineage. Growth, differentiation, and maturation are forced by a higher proportion of receptors for the factors acting early in the proliferative cascade on primitive cells and vice versa (Fig. 2.2). These principles apply in leukemic as well as normal states, and manipulation of this interplay will become a feature of antineoplastic chemotherapy. The cloning of growth factors and identification of their genetic origin to loci which are characteristically translocated in leukemias provide a molecular basis for understanding leukemogenesis.

Under normal circumstances, only a few of the primitive cells will be replicating at any time, the remainder being in resting phase. In the event of stem cell depletion by therapy or disease, the small number of active primitive cells is a limiting factor in the rate of recovery; competition for the progeny most in demand (usually neutrophils) will decrease production of other cell lines (usually erythroid). The primitive stem cells resemble medium lymphocytes and cannot be defined on a morphologic basis. Maintenance of the stem cell pool represents a significant proportion of the total marrow production.

Knowledge of relationships between precursor cells is provided by observations on stem cells in *in vitro* cultures. The precursors for the platelet and red cell series are related by genetic capability or by receptor status as are, separately, precursors for the neutrophil and monocyte series. These relationships provide an explanation, determined by demand, for the concurrent increased production of red cells and platelets in some circumstances or of neutrophils and monocytes in other circumstances. Hormonal factors influence the final defensive capability of leukocytes as well as their release from the marrow. Inflammation causes the release of a family of short- and long-acting mediators, of which **tumor necrosis factor** stimulates growth factor production, which increases marrow cell release. Failure of this final step in blood cell production results in **ineffective hematopoiesis** and death of the cells in the marrow. This syndrome is seen most often in the myelodysplasias of the cat and dog where there is the paradoxical concurrence of pancytopenia and hyperplastic marrow. These relationships are set out schematically in Fig. 2.2.

An appreciation of the cell kinetics and production times of blood cells, and the proportions of their progenitors in normal marrow, is essential to the interpretation of

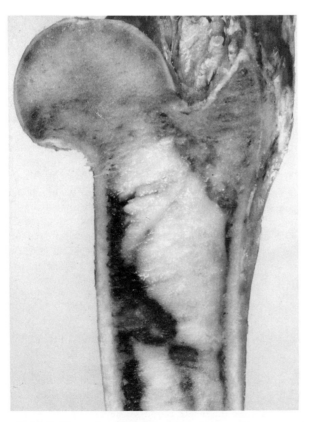

Fig. 2.1 Expansion of erythropoietic marrow adjacent to endosteum. Hemorrhagic anemia. Pig.

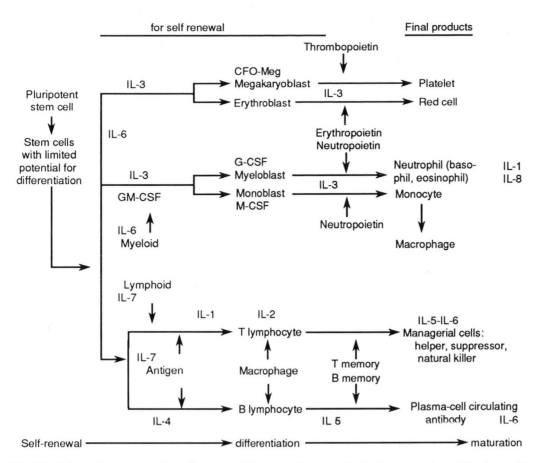

Fig. 2.2 Schematic representation of hematopoietic stem cell system, illustrating reduced capability for proliferation with increased pressure for differentiation. [Revised from D. Metcalfe and M. A. S. Moore, "Hemopoietic Cells," North-Holland, Amsterdam, 1971, and G. M. Keller and R. A. Phillips, *J Cell Physiol* **1**(Suppl): 31–36, 1982.]

this tissue in disease. This information is summarized in Fig. 2.3.

The recognizable blast cells of the four lines (erythroid, granulocytic, megakaryocytic, and monocytic), make up only about 1% of all marrow cells. Six morphologic stages of myeloid and erythroid differentiation are recognized, of which the first three are dividing cells and the last three are undergoing nuclear and cytoplasmic maturation. Very consistently in steady-state conditions, 25% of marrow cells are in the proliferative phase and 75% are in the maturation phase. Deviations from these proportions with an increase of **proliferative phase** cells indicate active regeneration, leukemia, or ineffective hematopoiesis associated with **nuclear maturation defects** such as occurs in folate and vitamin B_{12}-related disorders. Increased proportions of the **maturation phase** cells with pancytopenia indicate ineffective hematopoiesis and myelodysplasia. Late asynchrony limited to the erythroid system concurrent with hypochromic anemia results from the **cytoplasmic maturation defect** of iron deficiency.

The proportions of proliferative phase cells indicate that the promyelocytes and prorubricytes result from a simple doubling of the numbers of precursor blast cells. The proportions of myelocytes and basophilic rubricytes suggest that there is some proliferative activity in these stages, and nuclear labeling shows that metamyelocytes and polychromatic rubricytes can divide to a limited extent in conditions of increased demand. Monocytes have three morphologic stages of differentiation, of which the first two are dividing cells, and the last is a maturation phase. The total production time of the monocyte appears to be only one quarter of that required to produce a neutrophil.

The megakaryoblast is about the size of a prorubricyte and can be identified by binucleation with interphase nuclei and intense cytoplasmic basophilia. Thrombopoietic cells can be divided into two broad groups. The proliferative-phase cells are multinucleated and have minimal, very basophilic cytoplasm. Maturation-phase cells have a multilobulated nucleus and abundant cytoplasm in which the basophilia decreases as the granulation increases over

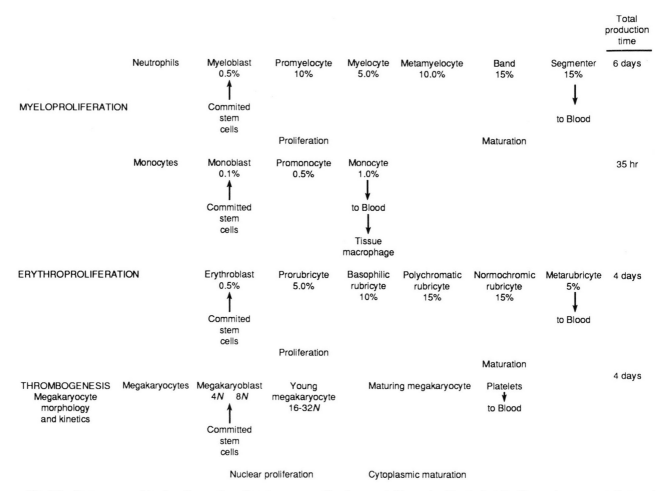

Fig. 2.3 Bone marrow kinetics. Proportion of total marrow cells when myeloid : erythroid ratio is 1.0. All megakaryocyte (all stages) occupy 0.5–1% of marrow cells.

the 4-day production period. Normal or increased nuclear ploidy with inadequate cytoplasmic volume is indicative of **ineffective thrombopoiesis.**

Lymphocytes are present in normal marrow at 3–5% of marrow cells. In inflammatory states, their number apparently increases by several percent, and lymphocyte identification is complicated by an increased proportion of morphologically similar stem cells. Normally, plasma cells are at very low levels in young animals and increase to 1–2% of marrow cells in adults. In chronic inflammatory disease, their numbers may increase to 25% or more of marrow cells. A large increase in plasma cells without cytologic atypia does not support a diagnosis of immunoblastic tumor. Germinal centers are not found in the marrow of normal animals at any age.

Cellularity of marrow is used herein to indicate the proportions of hematopoietic to fatty marrow, and **cell density** to indicate the spatial relationship of hematopoietic cells to each other. It is generally accepted that, excepting focal lesions such as tuberculosis, myeloma, or metastatic carcinoma, red marrow obtained from any site will have the same cellularity and cell differential. In other words, the marrow functions as an organ with proliferative and control mechanisms providing local and whole-body homogeneity. Normal marrow functions at about 50% cellularity. Persistent increased demands for cells may raise this proportion to 100%, in which event the marrow organ will be greatly expanded with conversion of fatty to red marrow in the endosteal marrow of the distal extremities. In normal animals, the hematopoietic cells are densely packed, with their boundaries in close apposition. This close apposition is probably a function of cell–cell interaction in control mechanisms. In chronic disease with failing marrow, such as occurs in equine infectious anemia or bovine trypanosomiasis, the cell density is reduced, and in these circumstances the loss of marrow volume is partially compensated for by sinusoidal dilation. This latter change results in a marrow which appears red grossly but is cytologically hypofunctional.

Marrow iron is an important feature of myeloid evaluation. Stainable iron is rare in normal animals younger than a year and gradually increases with maturity. Hemosiderin is intracellular in macrophages and in this state is only slowly mobilized for erythropoiesis. Thus an animal with substantial marrow iron reserves will respond faster to acute hemorrhage if supplemental iron is provided. A coarse pattern of iron deposition in marrow indicates poor iron utilization, and is the characteristic pattern in the anemia of chronic disease. In hemolytic anemias, there is usually an increase in total marrow iron with abundant coarse and fine hemosiderin indicative of rapid turnover. A hypochromic microcytic anemia with late erythroid asynchrony and scant iron is found in iron deficiency anemia, whereas the same peripheral and marrow picture with increased hemosiderin strongly suggests copper deficiency.

In normal marrow, the fat is firm and opaque on gross examination. There is little change in volume on drying. The fat is completely removed in histologic processing, leaving empty adipocytes. In contrast, in serous atrophy the marrow becomes grossly translucent and pink rather than clearly red or white, and the marrow shrinks rapidly on exposure of the medullary cavity to air, indicating a higher than normal water content. Histologically, the fat cells in serous atrophy contain a thin, pink proteinaceous background. Serous atrophy of marrow fat is a late change in cachexia or starvation and is a significant lesion.

Bibliography

Donahue, R. E. *et al.* Human IL-3 and GM-CSF act synergistically in stimulating hematopoiesis in primates. *Science* **241:** 1820–1823, 1988.

Hartwell, L. H., and Weinert, T. A. Checkpoints: Controls that ensure the order of cell cycle events. *Science* **246:** 629–634, 1989.

Henney, C. S. Interleukin-7: Effects on early events in lymphopoiesis. *Immunol Today* **10:** 170–173, 1989.

Laskey, R. A., Fairman, M. P., and Blow, J. J. S phase of the cell cycle. *Science* **246:** 609–614, 1989.

Le, J., and Vilcek, J. Biology of disease. Interleukin 6: A multifunctional cytokine regulating immune reactions and the acute phase protein response. *Lab Invest* **61:** 588–602, 1989.

McIntosh, J. R., and Koonce, M. P. Mitosis. *Science* **246:** 622–628, 1989.

Morstyn, G., and Burgess, A. W. Hemopoietic growth factors: A review. *Cancer Res* **48:** 5624–5637, 1988.

Murray, A. W., and Kirschner, M. W. Dominoes and clocks: The union of two views of the cell cycle. *Science* **246:** 614–621, 1989.

Nixon, R. K., and Olson, J. P. Diagnostic value of marrow hemosiderin patterns. *Ann Intern Med* **69:** 1249–1254, 1968.

O'Farrell, P. H. *et al.* Directing cell division during development. *Science* **246:** 635–640, 1989.

Pardee, A. B. G_1 events and regulation of cell proliferation. *Science* **246:** 603–608, 1989.

Smith, K. A. Interleukin-2: Inception, impact, and implications. *Science* **240:** 1169–1176, 1988.

Tadmori, W. *et al.* Human B-cell proliferation in response to IL-4 is associated with enhanced production of B cell-derived growth factors. *J Immunol* **142:** 826–832, 1989.

Takatsu, K. *et al.* Interleukin-5, a T cell-derived B-cell differentiation factor also induces cytotoxic T lymphocytes. *Proc Natl Acad Sci USA* **84:** 4234–4238, 1987.

Valli, V. E. *et al.* Evaluation of blood and bone marrow, rat. *In* "Hemopoietic System," T. C. Jones, *et al.* (eds.), pp. 9–26. Monographs on Pathology of Laboratory Animals. Berlin, Springer-Verlag, 1990.

Wallerstein, R. O. Marrow iron. *J Am Med Assoc* **238:** 1661, 1977.

Willems, J. *et al.* Human granulocyte chemotactic peptide (IL-8) as a specific neutrophil degranulator: Comparison with other monokines. *Immunology* **67:** 540–542, 1989.

II. Disorders of Stem Cells

Disorders of stem cell function include errors in the number and/or quality of cells produced. Numerical disorders consist of congenital cytopenias, acquired leukemias, toxic-induced neutropenia, idiopathic aplastic anemia, and pancytopenia. Qualitative disorders include congenital errors in cytoplasmic and nuclear development which result in defective protection against infection, oxygen transport, or clot formation.

A. Congenital Abnormalities in Blood Cell Numbers: Cyclic Hematopoiesis

Deficient cell production is seen in cyclic hematopoiesis of humans and gray collie dogs and in the genetically anemic W/W^v mouse, in which the error is in the restriction of the pluripotent stem-cell pool. In contrast, in the genetically anemic Sl/Sl^d mouse, the stem cell pool is adequate, but marrow stroma is deficient, resulting in an inadequate hematopoietic microenvironment. Whereas not of clinical or diagnostic concern, these latter two diseases have contributed greatly to the understanding of hematopoietic dependence on a functional pluripotent stem cell in an appropriately fertile microenvironment. In this sense, the older concept of hematopoiesis deriving from marrow stroma is vindicated by the array of essential lymphokines which are produced by the supporting structures.

The diseases described in the next two sections are remarkable in that they combine a genetic origin with a dilution of melanin in pigmented hair and hematologic abnormalities.

Cyclic hematopoiesis is a rare abnormality of pluripotent stem cells in gray collies transmitted in an autosomal recessive manner. Affected puppies can be identified by the light silvery gray hair and by microphthalmia. The defect is lethal and death occurs before maturity. There is a trilineage cycling of blood cells with the deficits most apparent in neutrophils which have the shortest peripheral life span. In affected dogs, cycling occurs with a 12- to 14-day periodicity, with severe neutropenia at each period of marrow hypoplasia followed by marked neutrophilia in rebound. The disease appears due to cycling levels of suppressor T lymphocytes, which are in the ascendancy when hematopoiesis is failing and are reduced in recovery. The total leukocyte count of 2–20 × 10^9/liter may not change markedly because, with repeated cycles and the onset of persistent infection, an accumulation of longer-lived monocytes develops which masks the neutropenia. The precursors for megakaryocytes and rubricytes go through similar cycles, but because of their longer peripheral life span, the deficits in these cells are less apparent. The animals are immunologically competent, and lymphoid hyperplasia is an early compensatory mechanism for recurrent neutropenia. Amyloidosis is a common sequel in animals that survive 3 months or longer.

Dogs with cyclic hematopoiesis are difficult to maintain. Disease is recurrent and related to sepsis. Characteristically there are about 3 days of neutropenia followed by normal or increased neutrophil counts for 6–7 days. At periods of neutropenia, there is marked left shift with myelocytes present which have toxic vacuolation and granulation (Fig. 2.4) and Doehle bodies. Apparently cell supply is not the only error and neutrophils have deficits in killing function for a variety of bacteria. Animals may be mildly icteric and there is increased blood urea, especially in those dogs which develop amyloidosis.

The disease tends to vary somewhat with the age of the animal, but the same organ systems are affected at all ages. The animal is usually in poor condition and there are often focal skin lesions with abscessation or cellulitis. Young dogs tend to have bronchopneumonia with purulent exudate, whereas older animals tend to have interstitial pneumonia with focal areas of scarring. There is frequently valvular endocarditis and catarrhal enteritis and often bleeding into the lower intestinal tract. The liver is usually pale and fatty, and the spleen is uniformly enlarged. Lymph nodes are generally enlarged and may be focally abscessed. The bone marrow is uniformly reddened in midfemoral shaft. Focal areas of sepsis may occur virtually anywhere in the body and there is often septic arthritis and epiphysitis with focal hemorrhages and lysis of the growth plates in young animals. Dogs which die in the first few weeks of life tend to succumb to infections of the respiratory and enteric tract, whereas dogs which survive for several months often die of renal failure.

The lesions in the lung, kidney, and gastrointestinal tract are those of chronic and recurrent inflammation usually due to common bacterial invaders. The principal changes occur in the hemic/lymphatic system and consist of a widespread lymphoid atrophy with reduced density of the splenic periarteriolar sheaths (Fig. 2.5) and of the germinal centers. In some dogs, the germinal centers may be markedly enlarged by both perifollicular and interfollicular deposition of amyloid, which may appear as early as 3 months of age. The lymph nodes have normal numbers of germinal centers with poor cell density, and the paracortical areas have reduced area and cell density, and there is a characteristic sinus histiocytosis without medullary cord hyperplasia (Fig. 2.6). The bone marrow varies widely depending on the stage of the cycle at the time of death. Characteristic changes include a progressive in-

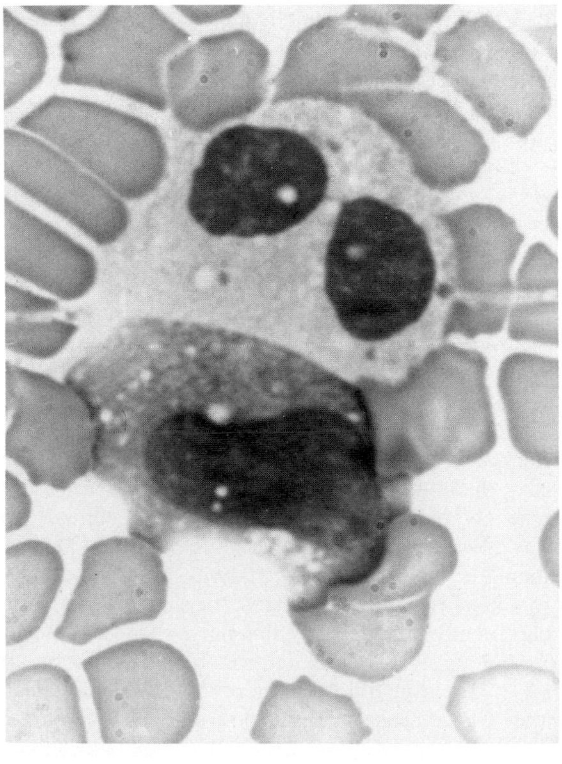

Fig. 2.4 Cyclic hematopoiesis. Dog. Blood smear. Immature monocyte, center. Bilobed neutrophil indicative of premature release from marrow contains toxic granules and vacuolations.

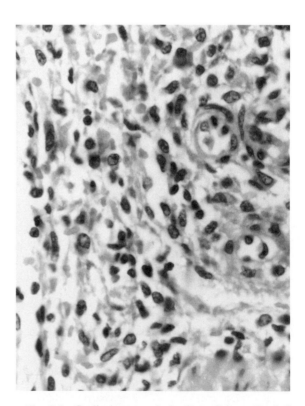

Fig. 2.5 Cyclic hematopoiesis. Dog. Spleen. Depletion of small lymphocytes in periarteriolar sheath.

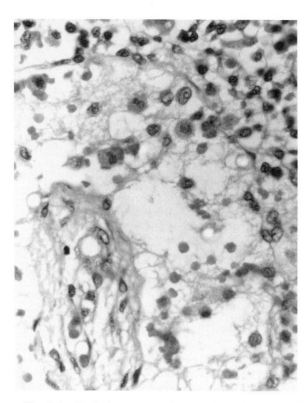

Fig. 2.6 Cyclic hematopoiesis. Dog. Lymph node. Cellular depletion of medullary cords and sinuses.

crease in sinus and hematopoietic area with loss of fat and a generalized increase in connective tissue that becomes more prominent with survival. There is an increase in marrow plasma cells and, at the time of rebound, the marrow granulocyte reserves are increased, whereas at the time of aplasia, there are few young cells present and the marrow is hypocellular. At the stage of rebound, there are many large, round cells with deeply stained cytoplasm which are likely myeloblasts and promyelocytes with a lesser number of erythroid precursors. There is thymic cortical atrophy.

Bibliography

Chusid, M. J., Bujak, J. S., and Dale, D. C. Defective polymorphonuclear leukocyte metabolism and function in canine cyclic neutropenia. *Blood* **46:** 921–930, 1975.
Lothrop, C. D. *et al.* Correction of canine cyclic hematopoiesis with recombinant human granulocyte colony-stimulating factor. *Blood* **72:** 1324–1328, 1988.

B. Congenital Abnormalities in Blood Cell Function

The von Willebrand defect is included under disorders of coagulation (Section I,B of Hemorrhagic Diatheses).

Recurrent or persistent infections in young animals may result from defects in humoral or cell-mediated immunity or in the functions of phagocytes. Selective deficiency of immunoglobulin M (IgM) in Doberman pinschers and of

IgA in beagles and Shar-peis, complement deficiency in Brittany spaniels, T-cell abnormalities in bull terriers, and granulocytopathy in Irish setters are described and the defect is, at least partially, clarified. There are breed-related syndromes in which the defective component is not identified. For example, nasal and systemic aspergillosis is well recognized in German shepherd dogs, as is recurrent and persistent infection in the Weimaraner. The granulocytopathy of Irish setters is associated with a defect in surface adhesive glycoproteins, and that in Weimaraners is associated with deficient production of active radicals in granulocyte lysosomes.

1. Chediak–Higashi Syndrome

The Chediak–Higashi syndrome is a composite disorder of granule formation in cells and is of a simple recessive character in cattle of the Hereford, Brangus, and Japanese Black breeds, cats, mink (Aleutian disease), mice, and humans. The condition may also occur in bison, killer whales, foxes, and white tigers. The disease is manifested clinically by partial albinism or color dilution, high susceptibility to infections, and a hemorrhagic tendency. It is a rare syndrome except in mink, in which all of the favored blue Aleutian strains are affected.

Enlarged cytoplasmic granules, which are the result of fusion of pre-existing granules of normal size, are found in most types of cells which normally contain granules. The partial albinism is ascribed to a clumping of melanin granules in such tissues as eye and skin and their fusion with lysosomes. There are fewer, larger granules than normal also in hepatocytes, renal epithelium, neurons, endothelial cells, and the white cells of blood.

The prolonged bleeding time and hemorrhagic tendency is, at least in part, due to defects in platelets. The clumping response of platelets on exposure to collagen is deficient, and is associated with reduced levels of adenosine diphosphate in the platelets.

Giant lysosomes are present in circulating granulocytes, lymphocytes, and monocytes. Phagocytosis appears to be normal, but bacteriocidal activity is impaired. The impaired capacity of phagocytes to deal with ingested macromolecules or microorganisms appears due to impaired fusion of lysosomes with phagosomes and provides an explanation for the high incidence of recurrent infections.

2. Bovine Granulocytopathy Syndrome

A heritable syndrome of chronic febrile disease characterized by poor growth and early death occurs in Holstein cattle. Clinically there is emaciation, dermatitis, persistent oral ulcers, enlarged lymph nodes, and slow healing of wounds and infections. There is anemia with marked persistent neutrophilic leukocytosis of up to 100,000/μl and mild left shift. Bone marrow has marked myeloid shift with synchronous maturation of all cell lines which appear normal by light and electron microscopy. There is a polyclonal gammopathy with a decreased albumin/globulin ratio.

Pathologically there is fibrosis of lymph nodes and widespread infection of body surfaces. It is significant that neutrophils are present in large numbers in vessels but not in the infected tissues. The neutrophils from these animals have reduced motility, phagocytosis, and intracellular killing. The cases described are all derived from a single sire, apparently as an autosomal recessive trait.

3. Pelger–Huet Anomaly

The Pelger–Huet anomaly is a benign dominant hereditary condition characterized by hyposegmentation of the neutrophil nuclei, which have mature aggregation of chromatin despite their immature shape. The condition occurs in humans, dogs, and rabbits, and a case is reported in the cat. The neutrophils respond normally in functional tests and heterozygotes survive normally. A homozygous form in rabbits is lethal. The main importance of the trait is the need to distinguish it from inflammatory states. Affected animals have a normal total leukocyte count, but the major form of the peripheral neutrophil is at the band stage, with bands comprising 30–70% of total cells. Metamyelocytes are present at much lower frequency and myelocytes are not found in the blood of healthy dogs. An ephemeral form resembling this condition has been described in the dog, possibly resulting from drug idiosyncrasy.

4. Feline Mucopolysaccharidosis

The type I or Hurler–Scheie and type VI or Maroteaux–Lamy forms of this disease occur in cats and the latter is associated with metachromatic granules in the neutrophils. The granules are 0.5–1.0 μm in diameter and are not prominent with routine stains. Typical facial dysmorphia consisting of frontal bossing and depressed nasal bridge is most obvious in the type VI syndrome seen in the Siamese breed (see Bones and Joints, Volume 1, Chapter 1). Affected animals survive to reproductive age with normal mentation, small stature, and cloudy corneas. They are not noted to be unusually susceptible to infection.

5. Sphingomyelinosis

A lysosomal storage disease due to an autosomal recessive trait resulting in a loss of sphingomyelinase occurs in humans and also in the cat, dog, and mouse. Storage in the hematopoietic system results in splenomegaly with psychomotor retardation as a result of accumulation in neurons. The enzyme is usually absent from leukocytes in the form seen in animals, and lymphocytes and monocytes are characteristically vacuolated. Nervous signs predominate (see The Nervous System, Volume 1, Chapter 3).

Bibliography

Ayers, J. R., Leipold, H. W., and Padgett, G. A. Lesions in Brangus cattle with Chediak–Higashi syndrome. *Vet Pathol* **25:** 432–436, 1988.

Berrier, H. H. *et al.* The white-tiger enigma. *Vet Med Small Anim Clin* **70:** 467–472, 1975.

Breitschwerdt, E. B. *et al.* Rhinitis, pneumonia and defective neutrophil function in the Doberman pinscher. *Am J Vet Res* **48:** 1054–1062, 1987.

Bundza, A., Lowden, J. A., and Charlton, K. M. Niemann–Pick disease in a Poodle dog. *Vet Pathol* **16:** 530–538, 1979.

Burns, G. L., Meyers, K. M., and Prieur, D. J. Secondary amyloidosis in a bull with Chediak–Higashi syndrome. *Can J Comp Med* **48:** 113, 1984.

Conto, C. G. *et al. In vitro* immunologic features of Weimaraner dogs with neutrophil abnormalities and recurrent infections. *Vet Immunol Immunopathol* **23:** 103–112, 1989.

Haskins, M. E. *et al.* Mucopolysaccharidosis I. *Comp Pathol Bull* **XIII:** 3, 1981.

Kehrli, M. E. *et al.* Molecular definition of the bovine granulocytopathy syndrome: Identification of deficiency of the Mac-1 (CD-11b/CD18) glycoprotein. *Am J Vet Res* **51:** 1826, 1990.

Kramer, J. W., Davis, W. C., and Prieur, D. J. The Chediak–Higashi syndrome of cats. *Lab Invest* **36:** 554, 1977.

Menard, M., and Meyers, K. M. Storage-pool deficiency in cattle with the Chediak–Higashi syndrome results from an absence of dense granule precursors in their megakaryocytes. *Blood* **72:** 1726, 1988.

Renshaw, H. W., and Davis, W. C. Canine granulocytopathy syndrome: An inherited disorder of leukocyte function. *Am J Pathol* **95:** 731–744, 1979. (Irish setters)

Studdert, V. P. *et al.* Recurrent and persistent infections in related Weimaraner dogs. *Aust Vet J* **61:** 261–263,1984.

Tvedten, H. W. Pelger–Huet anomaly, hereditary hyposegmentation of granulocytes. *Comp Pathol Bull* **XV:** 3, 1983.

THE LEUKON

I. Leukocyte Responses to Peripheral Disease

This section describes changes which occur in the different types of leukocytes. The following section deals with characteristic changes in leukocyte numbers in various species of domestic animals.

A. Neutrophils

Studies of granulokinetics using nuclear labels (tritiated thymidine) show remarkably similar results for humans and those domestic species studied. In the steady state, a total production time of 6 days is followed by a ripening period of 1 day prior to release from marrow. The peripheral circulation time is probably 6–8 hr. Neutrophils do not re-enter the circulation after tissue migration. Band neutrophils are released into the circulation in a bolus or discontinuous fashion.

The fine structure of neutrophils is quite similar for all domestic species. The primary or azurophil granules are peroxidase positive and most prominent in promyelocytes. The secondary or specific granules are peroxidase negative and form the background cytoplasmic density of the mature cells. In toxemic states, the primary granules are visible in the cytoplasm of mature neutrophils as fine azurophilic or pink granules when Romanovsky stains are used. These primary granules are normally present in mature cells and can be demonstrated as artefactual change in aged or overstained neutrophils. Other indications of systemic toxemia include cytoplasmic vacuolations and Doehle body formation. In general, toxic granulation represents the mildest change, followed by vacuolation and Doehle body formation. In severe toxemias, such as those that occur in canine pyometra, bovine mastitis, and equine salmonellosis, all three changes occur together. Doehle bodies are composed of lamellae of rough endoplasmic reticulum persisting from the normal basophilia of immaturity. Since rough endoplasmic reticulum is part of the synthetic mechanism of the precursor cells, it is expected that as toxic changes increase, the specific or secondary granules are decreased. In contrast, animals with well controlled or mild inflammatory disease usually have increased secondary granulation with minimal toxic changes. Presumably the functional capability of these cells varies directly with the level of granule bound enzyme, as evidenced by cytoplasmic density in stained blood films.

There is considerable variation in enzyme content of granules in various species, with the dog and cat neutrophil lacking alkaline phosphatase. Neutrophils function as secretory cells in the inflammatory reaction, with granule exocytosis being mediated by a number of stimuli. The secreted material modulates and enhances the acute inflammatory reaction and appears to stimulate myelopoiesis. Such stimulation may partially explain why in diseases such as enteritis, mastitis, and dermatitis, in which neutrophils are lost from the body, blood neutrophils tend to be fewer than in tissue reactions in which secretory granules are retained. The spilling of enzyme into tissues during phagocytosis may cause tissue injury in acute inflammation. The phagocytic vacuoles of neutrophils have a pH as low as 4.5, and adherence of activated cells to endothelium may cause vascular injury.

Leukocytes are usually quantitated on a numerical basis; however, numbers and functional capability are not synonymous. Neutrophils may be hypofunctional and cytologically appear less dense, in the same manner as red cells may be hypochromic and carry less oxygen. Neutrophils produced under conditions of severe toxemia presumably have decreased function in concert with their cytologic deterioration. In general, the cytologic quality of neutrophil cytoplasm varies directly with leukocyte count and the clinical progress of the animal. Thus animals in extremis tend to have very low leukocyte counts with severe toxic changes, and those that are responding well to their infection have high cell counts, and the density of neutrophil cytoplasm is increased. Many factors may influence the quantitative and qualitative responses of the bone marrow in infectious disease.

Neutrophils complete their maturation under the influence of a number of factors including interleukin-8 (Fig. 2.2). In neonates, free-radical production in neutrophils is below adult levels, which may be a factor in their early susceptibility to infectious disease. Blood cell function is not entirely a separate process, and neutrophils and platelets act synergistically in both coagulation and inflammation. Thus selective cytopenias such as neutropenia or thrombocytopenia will impair the function of cells that are not numerically deficient. Bacterial toxins acting locally induce an inflammatory response, and systemically result in release of mature neutrophils and toxic changes in those in the process of maturation. Chronic infection with bovine virus diarrhea virus decreases the functional reactivity of both neutrophils and lymphocytes. Phagocytosis is depressed in hypercoagulable states, likely due to the release of fibronectin products lysed by plasmin. Leukocyte function is impaired by endogenous toxins as occurs in uremia, and neutrophil enzyme is reduced in iron deficient states. Leukocyte function may be significantly altered by dietary factors. For example, selenium deficiency causes reduced intracellular killing and the 3-omega fatty acids reduce leukotriene B_4 and thus have anti-inflammatory effects. Dexamethasone depresses the functions of both neutrophils and lymphocytes. The effects on neutrophils may be blocked by administration of ascorbic acid.

Antibiotics vary in their ability to enter neutrophils, but those that do, such as tetracycline and chloramphenicol, decrease chemotaxis, and tetracyclines and polymyxin B decrease both phagocytic activity and oxidative metabolism. Finally, the systemic effects of cancer include impairment of phagocyte function.

B. Eosinophils

The production and utilization kinetics of eosinophils are similar to those of neutrophils, but eosinophils respond to different stimuli. Their granulation is cell specific and, to some extent, species specific. They are capable of killing bacteria but are less efficient than neutrophils. Their armamentarium includes the major basic protein of their specific granules, a potent H_2O_2–peroxidase–halide system, and inactivators of the leukotrienes of anaphylaxis. They have receptors for complement 3b, which is probably the means by which they bind to tissue parasites. The association between parasitism and eosinophilia is less than is usually accepted. Eosinophils are late arrivals in the immune response to parasites and their chemotaxis can be explained on the basis of specific lymphokines from prior sensitized T lymphocytes. The increase in mature eosinophils seen in mature sheep with hemonchosis is probably part of the syndrome of iron deficiency and not directly related to parasitism.

In blood films, eosinophils have more cytoplasmic basophilia than neutrophils. This is reflected in their wide spectral separation from other cells in fluorescent-activated cell counters. In diseased animals, eosinophils frequently have more marked toxic changes than neutrophils. These changes consist of reduced granulation, increased diffuse cytoplasmic basophilia, and large vacuolations. The toxic changes are seen most often in diseases involving the gut, lung, and skin, and are not species specific. Eosinophils are most prominent in these three tissues, and this may provide an anatomic basis for the preference. Eosinophils can be considered to be concentrated on body surfaces, with greatest frequency in intestine and lesser numbers in dermis and hemic–lymphatic tissue, lung, and uterus. Significantly, many tissues normally contain none at all. Cows in estrus usually have an eosinophilia of 0.5 to 1.5×10^9/liter. Toxic changes are absent.

The functions of eosinophils are incompletely understood. They are known to interact with homocytotropic antibodies (IgE and IgG) and with mast cells and basophils. The antibody and T lymphocytes provide specificity to the reaction and the IgE on mast cells attracts eosinophils, which modulate the inflammatory reaction. The relative quantities of tissue IgE, extractable histamine, and eosinophils suggest that these components form a system which is most prominent on body surfaces, immunologically mediated, often parasite related and frequently associated with eosinophilia.

When benign eosinophilia is marked and persistent, it can be assumed to be due to immune phenomena. Less clear is the distinction between benign hypereosinophilic syndrome and chronic myelogenous leukemia of the eosinophil type. Whereas both conditions will cause tissue damage, the immaturity and atypia are more prominent in the malignant disease and are usually accompanied by some degree of basophilia. Eosinopenia is characteristic of acute infection. It appears that corticosteroid modulates the level of blood eosinophils, but their production and release may persist despite high levels of steroid if the peripheral stimulus is of sufficient intensity. Eosinophils appear to be specifically directed cells with a short circulation time in disease states. This may explain the apparent increased eosinopoiesis in marrow without eosinophilia and with focal tissue accumulation, such as occurs in eosinophilic myositis. The level of serum cationic protein derived from eosinophils may markedly increase in acute inflammation despite eosinopenia, suggesting that these cells are being produced and are degranulating rapidly but are not obvious in the blood because of rapid transit.

C. Basophils

Basophils arise from bone marrow precursors which can first be recognized at the promyelocyte stage. Their maturation sequence and kinetics are similar to neutrophils with peak peripheral blood arrival of labeled cells 7 days after injection of tritiated thymidine. The circulation time is unknown but is probably short. Newly released cells are specifically directed; they do not have a random and age-dependent removal as do neutrophils. Basophils do not segment as fully as neutrophils, and in this respect resemble eosinophils. Basophils are about a third greater in diameter in blood films than are neutrophils, and retain smooth nuclear boundaries unlike the irregular constrictions of mature neutrophils. These characteristics are of interest because basophils are easily degranulated in slide preparation and may be mistaken for neutrophils.

The relationship between basophils and mast cells appears to be functional, with a reciprocal variation in their relative proportions. Basophils have segmented nuclei with moderate granulation whereas mast cells are mononuclear and heavily granulated. The granules of both cells contain histamine and heparin, but the histamine concentration in mast cells is some 20 times greater than that in basophils. Basophil granules are ultrastructurally distinct, and mast cells contain hydrolytic enzymes and 5-hydroxytryptamine which are not present in basophils.

It appears that basophils have a committed precursor at about the same level as neutrophils and which is probably more primitive than the myeloblast. Mast cells are believed to be derived from an undifferentiated precursor in perivascular connective tissue that may in turn derive from monocytes. In cytologic preparations of mast cell tumors, it is common to find stromal-type cells with bipolar and incomplete cytoplasmic boundaries, which contain only one or two basophilic granules. These may be the stem cells for the bulk of the tumor mass.

Basophils are capable of phagocytosis and are weakly motile. Unlike mast cells, which degranulate by exocytosis and can then regenerate, basophils largely degranulate by diffuse internal lysis and have less synthetic capability. It is likely that basophils of most species have receptors for IgE and degranulate when attached antibody binds with specific antigen. The secretory activity of both basophils and mast cells is dependent on the presence of calcium and magnesium ions. In addition to their specific involvement in hypersensitivity reactions, basophils release heparin during postprandial lipemia which activates the enzyme lipoprotein lipase. Steroid hormones appear to cause a reduction in blood basophils, as well as eosinophils, and basophil counts are reduced in acute infections. Since basophils are rarely found in the blood of normal cats and dogs and are usually $<0.1 \times 10^9$/liter in the blood of other species, a reduction in basophils is seldom detected. Basophilia is seen in hypothyroidism, after protein injection (vaccination), estrogen injection, and accompanying eosinophilia. Basophilia may be seen in immune hemolytic anemia and canine hyperadrenocorticism. Basophilia that accompanies a leukemoid reaction should increase the suspicion of chronic myelogenous leukemia.

D. Monocytes

The blood monocytes derive from a marrow precursor in common with that for neutrophils. There is some switching, probably determined as late as the myeloblast stage. The peak of labeled blood monocytes occurs 2 days after injection of tritiated thymidine and, in contrast to neutrophils, in which the marrow and circulating pools of cells appear equal, the monocyte pool is composed almost entirely of circulating elements. Monocytes leave the peripheral blood with a half time of ~20 hr and are not believed to reenter the circulation.

Monocytes can be distinguished from lymphocytes on routine staining by their more irregular nuclear outline and uniform cytoplasmic density. Lymphocytes usually have a perinuclear halo of reduced cytoplasmic basophilia.

Monocyte chromatin is consistently fine with or without a few large chromocenters, whereas lymphocyte chromatin varies widely with cellular activity. The specific granules of monocyte cytoplasm are not always resolvable by light microscopy. When present they are similar in size and tinctorial properties to toxic granules of neutrophils. Newly released monocytes tend to have more cytoplasmic basophilia and lack vacuolations, whereas the opposite is true of older monocytes. This vacuolation of monocytes is common in animals with inflammatory disease.

Monocytes leave the blood to replenish the entire spectrum of tissue macrophages known specifically by location as macrophages of lung, spleen, marrow, peritoneum, and the lymph node sinuses as well as Kupffer, Langerhans, and giant cells, and possibly the interdigitating reticular cells of germinal centers. As noted earlier, the terms reticulo- and reticulum should not be used with reference to the monocyte–macrophage system. The metallophilic reticular cells of interstitial tissues produce reticulin or collagen and are weakly phagocytic. They are fibroblast derived and distinct from macrophages, which do not produce fibers. Most of the tissue depots for macrophages are at the termination of the marrow–blood–tissue pipeline; however, in some sites, particularly lung and liver, there is local proliferation which may be accelerated in marrow failure.

The activities of the monocyte–macrophage system are almost as diverse as their tissue distribution and include phagocytosis, myeloproliferative control, antigen processing, and production of endogenous pyrogen. The ability of these cells to bind to antigens and lymphocytes is central to these functions. Macrophages have membrane receptors for the Fc portion of immunoglobulin and for C3b of the complement system.

Phagocytosis is considered in three functional steps, consisting of particle attachment, internalization, and ingestion. Attachment is a function of binding capabilities whereas the signal for ingestion may be either immunologically or nonspecifically activated. Ingestion is the energy-dependent step in phagocytosis. The energy is provided by anaerobic glycolysis, and macrophages characteristically function well in anoxic milieu such as occurs in abscesses and granulomas. In contrast, alveolar macrophages are critically dependent on a PO_2 of greater than 25 mm of Hg for phagocytosis. Once ingested, a foreign particle is subjected to an array of lysosomal enzymes by fusion of the primary vesicle with the membrane-bound enzyme to form the phagosomal digestion chamber.

Macrophages produce specific stimulatory molecules, called colony-stimulating factor, which are important in inducing myelopoietic differentiation. It is likely that other cells also produce this factor; however, the critical microenvironment in marrow probably requires a cell–cell interaction in which local macrophages are most effective. Marrow macrophages serve as nurse cells to erythropoietic islands and remove the nuclei (hematogones) from maturing rubricytes.

Monocytes and macrophages are involved in the immune system during antigen recognition and later as effector cells during the cell-mediated response. Administered antigen first becomes associated with the medullary macrophages and cortical dendritic cells of lymph nodes. This process, known as antigen focusing, appears to involve the phagocyte in antigenic degradation and adjuvanting with RNA, followed by transfer to competent B lymphocytes in intimate contact with the phagocyte membrane.

Temperature elevation is part of the host response to inflammation. Leukocyte pyrogen is produced *de novo* by all granulocytes but primarily by macrophages. Most inducers are themselves pyrogens, of which endotoxin is one of the most potent. Pyrogen release from phagocytes accompanies phagocytosis, and like that process is energy dependent. The monocyte system also produces interferon, and appropriately armed macrophages are able to kill tumor cells by mechanisms that are immunologically nonspecific.

In the dog, steroid injection causes a prompt monocytosis which reaches $2-3 \times 10^9$/liter and persists for 8 hr. In the cat, horse, and cow, monocytes decrease in acute stress and return with marrow rebound. A monocytosis of $3-5 \times 10^9$/liter is characteristic of the dog with pyometra and, to a lesser degree, with immune hemolytic anemia. Monocytosis may accompany lymphoma, especially if there are large areas of tumor necrosis. Monocytosis may be seen in muscle bruising and with sterile hematomas.

E. Lymphocytes

Lymphoid precursors arise in the blood islands of the yolk sac and colonize the primary lymphoid organs, the marrow, and thymus, in early embryonic life. Both of these sites remain active in lymphopoiesis, with their progeny circulating to the secondary lymphoid organs constituted by the lymph nodes, spleen, and gut, and lung-associated lymphoid tissue. In general, the primary lymphoid tissues of marrow and thymus remain isolated from the activities which arise during cellular differentiation in the secondary or peripheral lymphoid tissues. Kinetically, the very great bulk of circulating lymphocytes are thymus derived and long-lived (years), whereas a small number of cells with rapid turnover have a shorter life span (days). Lymphocytes more often terminate their sojourn in a particular compartment by division rather than death, and in this sense, the concept of life span differs from that of granulocytes which are end-stage cells with a finite life.

The number of lymphocytes in the peripheral blood is the net result of recirculation from lymph to blood via the thoracic duct and from blood to tissue via the postcapillary venules of lymph node cortex. Lymphopenia results from low transmural traffic across these venules, whereas in lymphoid leukemia they may be dilated and tightly packed with tumor cells. Steroid therapy causes lymphopenia; this change is due partially to lympholysis and partially to redistribution of lymphocytes into tissues. Lymphopenia occurs in chronic renal disease, lymphangiectasia, chylothorax due to rupture of the thoracic duct; in canine parvovirus infection, distemper, and viral hepatitis; in hyperadrenocorticism; and in chronic hog cholera. In all species, persistent lymphopenia indicates a poor prognosis. Lymphocytosis occurs in ruminants responding well to chronic disease. Examples of the latter are found in tuberculosis, brucellosis, trypanosomiasis, and bovine leukemia virus infection. Lymphocytosis occurs in cats as a fear reaction and in dogs and cats with hypoadrenocorticism.

F. Platelets

The blood platelets are produced by megakaryocytes that arise from committed hematopoietic stem cells. There is a common precursor for the erythroid and megakaryocytic systems that differentiates at the same hierarchical level as the precursor for the neutrophil–monocyte system. Evidence for this relationship is found in *in vitro* culture systems where both megakaryocytes and rubricytes may develop under the inductive pressure of erythropoietin. In the fetus, megakaryocytes are found successively in the liver, spleen, and marrow, and in adult mammals, they are found in the marrow, lung, and spleen.

The volume of marrow occupied by megakaryocytes bears a constant relationship to peripheral platelet consumption. Thus with increased platelet utilization there is an increase in megakaryocyte number and volume, giving rise to the concept of megakaryocyte mass. Normally there is also a direct relationship between megakaryocytic ploidy and cytoplasmic volume. Thus, increased output from the system is accomplished by both increased input from the stem cell pool and by increased output per cell as a result of increased nuclear reduplication. This duplication from $2 N$ to 16 to $32 N$ occurs by both endomitosis and mitosis, although the latter tends to be seen only in a stressed system. It has been estimated that about 50 platelets are produced per nuclear lobe or 1600 platelets from a $32 N$ cell. In most species platelets can be produced in 4 days. They have a circulation time of 9 days, which is determined largely by age-dependent utilization and to a lesser extent by random removal at sites of endothelial injury. The cellular production from a hematopoietic system can be said to be effective when the peripheral arrival of new cells is consistent with the volume of marrow precursors. In immune thrombocytopenias there is usually increased peripheral destruction with an appropriate increase in marrow production. Occasionally there is depression of the precursors, and the disease is then greatly complicated by decreased marrow production. A less obvious problem of ineffective thrombopoiesis occurs in humans in pernicious anemia and in calves with trypanosomiasis, and probably in other conditions as well. Ineffective thrombopoiesis is characterized by increased megakaryocytic mass with increased nuclear volume and normal or decreased cytoplasmic volume in the face of a peripheral

thrombocytopenia that is due primarily to decreased production. The rate of thrombopoiesis is probably controlled by regulatory T lymphocytes at the stem cell level and by the α_2 globulin, thrombopoietin, at the level of differentiation and maturation.

The blood platelets are discoid cells generally one third to one half the diameter of homologous red cells. Platelets have mitochondria and are capable of glycolysis, with pathways remarkably similar to those in skeletal muscle. The cell is spongelike and perforated by a microtubular array that probably aids rapid release of granular contents during the secretory phenomenon known as the platelet-release reaction. In stained blood films, platelet granules are pink and may be centralized into a granulomere, leaving a basophilic agranular peripheral hyalomere. Platelets can be quantitated like red cells on the basis of hematocrit and mean volume. Platelets contain fibrinogen and the contractile protein, thrombosthenin, which provides the force for clot retraction. A variety of plasma proteins adhere to the platelet membrane and may function in establishing the electrical properties of platelets. Platelets undergo metabolic aging in the circulation, undergoing progressive lipid peroxidation. This may be the mechanism of senescent removal. Young platelets tend to be larger with greater basophilia, whereas older platelets are less granular and may lack basophilia. Platelets larger than red cells and usually basophilic are called shift platelets. They indicate increased production. Their detection in thrombocytopenias is thus prognostically significant in differentiating increased destruction from decreased production. Normally the blood platelets present a spectrum of size, basophilia, and granulation, which corresponds to cell age and indicates age-dependent removal. In thrombocytopenias with increased turnover, they may all appear young or all appear old if utilization is normal and production has ceased.

The role of the platelet in hemostasis is discussed in a later section (Section I of Hemorrhagic Diatheses). Briefly, the normal contractility of injured vessels is assisted by platelet adhesion, aggregation, release reaction, and contraction, which is associated with a laminar layering of fibrin. Although blood will clot in the absence of platelets, the hemostatic plugs so formed are unstable due to deficient adhesion and contraction. Platelets can be said to support endothelial integrity, and thrombocytopenia results in red cell diapedesis through apparently intact endothelium. Clinically, endothelium appears to remain intact for some hours after thrombocytopenia develops, and similarly, petechiation may persist for some hours after platelet production increases.

Platelets are not equal in their functional capability and may be deficient because of congenital or acquired defects. Congenital thrombopathies are recognized in dogs and cattle, and acquired dysfunction occurs in acute leukemias and with aspirin therapy. In routine hematopathology, the clot-retraction test is a useful measure of the functional capability of the circulating platelets. This variability in function accounts for the variations in the platelet counts at which purpuric hemorrhage occurs. Thus, petechiae may not be seen at a level of 10×10^9/liter if platelets are large and very functional, whereas purpura may be seen at 50×10^9/liter or higher if function is poor or impaired.

Acquired thrombocytopenia was formerly termed idiopathic but is now accepted to be immune in origin unless shown to be otherwise. The spongy nature of platelets makes them susceptible to immune sensitization because of adsorbed drugs, toxins, and virus. Immune-mediated thrombocytopenia of unknown cause occurs in dogs, cats, cattle, and horses and probably in other species. Thrombocytopenia is a constant finding in acute equine infectious anemia and in bovine trypanosomiasis. Low platelet counts due to decreased production occur in cyclic hematopoiesis, pancytopenia, and following estrogen toxicity or marrow irradiation. Concurrent hemolytic anemia and consumptive thrombocytopenia is seen in dogs, and a large spleen in any species may cause deficits in circulating cells. Thrombocytosis of small platelets is characteristic of iron and copper deficiency. Thrombocytosis is transient after splenic removal and occurs in polycythemia vera and in megakaryocytic myelosis, both of which are rare, but are seen most commonly in dogs.

Bibliography

Bielefeldt-Ohmann, H., Baker, P. E., and Babiuk, L. A. Effect of dexamethasone on bovine leukocyte functions and bovine herpesvirus type I replication. *Can J Vet Res* **51:** 350–357, 1987.

Brown, G. B. *et al.* Defective function of leukocytes from cattle persistently infected with bovine viral diarrhea virus, and the influence of recombinant cytokines. *Am J Vet Res* **52:** 381–387, 1991.

Carakostas, M. C., Moore, W. E., and Smith, J. E. Intravascular neutrophilic granulocyte kinetics in horses. *Am J Vet Res* **42:** 623–625, 1981.

Carr, I., and Daems, W. T. (eds.). "The Reticuloendothelial System: A Comprehensive Treatise," Vol. I. New York, Plenum, 1990.

Dvorak, A. M. *et al.* Anaphylactic degranulation of guinea pig basophilic leukocytes. I. Fusion of granule membranes and cytoplasmic vesicles: Formation and resolution of degranulation sacs. *Lab Invest* **42:** 263–276, 1980.

Dvorak, A. M. *et al.* Evidence for a vesicular transport mechanism in guinea pig basophilic leukocytes. *Lab Invest* **42:** 263–276, 1990.

Gallin, J. I., and Fauci, A. S. (eds.). "Advances in Host Defense Mechanisms," Vol. I. New York, Raven Press, 1982.

Gyang, E. O. *et al.* Effects of selenium–vitamin E injection on bovine polymorphonucleated leukocytes phagocytosis and killing of *Staphylococcus aureus*. *Am J Vet Res* **45:** 175–177, 1984.

Hendrick, M. A spectrum of hypereosinophilic syndromes exemplified by six cats with eosinophilic enteritis. *Vet Pathol* **18:** 188–200, 1981.

Henson, P. M. Interactions between neutrophils and platelets. *Lab Invest* **62:** 391–393, 1990.

Jain, N. C. "Schalm's Veterinary Hematology," 4th Ed. Philadelphia, Pennsylvania, Lea & Febiger, 1986.

Kessler, S., and Kuhn, C. Scanning electron microscopy of mast cell degranulation. *Lab Invest* **32:** 71–76, 1975.

McCall, C. E. *et al.* Lysosomal and ultrastructural changes in human "toxic" neutrophils during bacterial infection. *J Exp Med* **129:** 267–293, 1969.

McCall, C. E. *et al.* Human toxic neutrophils III. Metabolic characteristics. *J Infect Dis* **127:** 26–33, 1973.

Movat, H. Z. *et al.* Acute inflammation in Gram-negative infection: Endotoxin, interleukin-l, tumor necrosis factor, and neutrophils. *Fed Proc* **46:** 97–104, 1987.

Paape, M. J., Miller, R. H., and Ziv, G. Pharmacologic enhancement or suppression of phagocytosis by bovine neutrophils. *Am J Vet Res* **52:** 363–366, 1991.

Parmley, R. T. *et al.* Ultrastructural morphology and cytochemistry of iron-deficient polymorphonuclear leukocytes. *Exp Mol Pathol* **44:** 57–66, 1986.

Roth, J. A., and Kaeberle, M. L. Evaluation of bovine polymorphonuclear leukocyte function. *Vet Immunol Immunopathol* **2:** 157–174, 1981.

Roth, J. A., and Kaeberle, M. L. *In vivo* effect of ascorbic acid on neutrophil function in healthy and dexamethasone-treated cattle. *Am J Vet Res* **46:** 2434–2436, 1985.

II. Leukopenia–Leukocytosis in Benign Disease

The hematologic profile is a powerful tool in the assessment of response to disease. Serial blood profiles allow an assessment of an animal's response to a disease process and often are of prognostic value. In general, the **three factors** most important in predicting the magnitude of (neutrophilic) leukocytosis are **animal species,** and **cause** and **localization** of disease. There is marked variation in hematologic reactivity between species of domestic animals, which is related to the normal ratio of neutrophils to lymphocytes in peripheral blood. Thus the dog with a neutrophil : lymphocyte ratio of 2.4 is most reactive, whereas the cow with a ratio of 0.5 is least reactive. The maximal leukocytosis likely to be encountered in each species can be predicted by multiplying the normal neutrophil : lymphocyte ratio times 50 as shown in Table 2.1.

In general, carnivores are more responsive than herbivores, and ruminants are the least reactive of the herbivores. Acute viral infections tend to involve large volumes of tissue and to injure body surfaces (gut, lung, skin) so that not only is the area of chemotaxis large, but also

TABLE 2.1

Leukocyte Responses in Benign Disease

Species	Normal neutrophil : lymphocyte ratio	Maximum expected leukocytosis $\times 10^9$/liter
Dog	2.4	120
Cat	1.5	75
Foal	1.4	70
Horse	1.2	60
Pig	0.7	35
Goat	0.6	30
Sheep	0.5	25
Calf	1.0	50
Cow	0.5	25

leukocytes and their leukopoietic products may be lost from the body. Localized inflammatory processes are likely to be associated with leukocytosis regardless of species and type of agent. The differences between the cow and dog are largely in the rate of acceleration of cell production rather than in their potential for maximal response. The cow is able to mount a sustained neutrophilic leukocytosis like that of the dog with pyometra if given proper conditions of time and containment. These conditions are usually produced by abscessation of lung or liver, which may be either miliary or single and massive. Whereas sustained leukocyte counts in the cow greater than 50×10^9/liter are rare, they do occur. The cow with a neutrophilic leukocytosis of $20-25 \times 10^9$/liter that is sustained for more than a week can confidently be assumed to have abscessation.

In animals, infection frequently causes leukopenia, particularly in ruminants with mastitis and metritis and in carnivores with systemic viral infection, such as infectious canine hepatitis and distemper in the dog and parvovirus infections in the dog and cat. Guidelines for leukocytic interpretation that are valid for all species are that the **absolute increase** in leukocytes indicates **individual** or **species reactivity,** whereas the **proportional increase** in mature and immature neutrophils and the level of toxic changes indicate the **effort of the response.** In all species, a low total leukocyte count with persistent neutrophil immaturity and toxemia is a poor prognostic finding. The return of the leukocyte count to normal, from levels which were either increased or decreased, with reduction in immaturity and cytologic toxicity, is indicative of recovery. The return of eosinophils to the blood is usually prognostically favorable if accompanied by other signs of recovery.

III. Myeloid Reactions

A. Leukemoid Reaction

Leukemoid reactions are abnormalities of peripheral blood which resemble leukemia but are due to other causes. They occur in all species but are most often seen in the cat and dog. The leukemoid reaction is often associated with hemolysis, severe or unusual localized infection, and with metastatic cancer with focal necrosis.

Hematologically there is neutrophilic leukocytosis, which is usually in the range of $50-100 \times 10^9$/liter. Mature neutrophils predominate in the peripheral blood, but there is immaturity to at least the promyelocyte stage. Blast cells are seen occasionally, but these are usually lymphoid and misinterpreted because of the myeloid immaturity. There is usually diffuse cytoplasmic basophilia of mature cells with reduced secondary granulation and variable cytoplasmic vacuolation. Toxic azurophilic granulation and Doehle bodies may be present. There is usually anemia of at least mild degree that is nonresponsive due to the intensity of the myeloid reaction. Mild rubricytosis of maturing cells is irregularly present. There is usually a monocytosis of $2-4 \times 10^9$/liter with mild immaturity and variable cytoplasmic vacuolation. Eosinophils are generally mature and fewer than 1×10^9/liter and may have decreased secondary granulation with vacuolation and increased cytoplasmic basophilia. Basophils are absent.

Aspirated bone marrow is hypercellular, generally greater than 70%, with synchronous myelopoiesis, a marked increase in mature reserves, and fewer than 10% blast cells. There is erythroid atrophy but not phthisis, and hemosiderin is usually coarse and increased.

Animals showing a leukemoid reaction often recover. Those that die show various lesions, depending on the primary cause. Pathologically there is an increased volume of hematopoietic marrow. By definition, the leukemoid reaction must resemble leukemia. The diagnosis of leukemoid reaction tends to be a working clinical designation used until biopsy, autopsy, or recovery resolves the problem.

B. Leukoerythroblastic Reaction

The leukoerythroblastic reaction is characterized by the presence in the peripheral blood of rubricytes as young as basophilic rubricytes and granulocytes as young as myelocytes. It is seen in all species and is most common in the cat and dog, and least common in the horse. Since this reaction is always part of other processes, there are no characteristic clinical signs. It is most apparent in immune hemolytic anemia of animals other than the horse. Immature cells of both systems are likely to appear in the peripheral blood when there is heightened marrow activity. The degree of immaturity is important and is less with reactive marrow than with malignant infiltration.

It is uncommon to have benign erythroid hyperplasia without concurrent myeloid hyperplasia. Thus, responsive immune hemolytic anemia is usually accompanied by a sustained neutrophilia with mild left shift. In contrast, myeloid hyperplasia is regularly accompanied by some degree of erythroid atrophy. Myeloid immaturity is thus expected in increased erythropoiesis, but rubricytes are not expected in myeloid hyperplasia unless there is a considerable level of stress. The leukoerythroblastic reaction is therefore present at some point in all leukemias and is most common in the acute leukemias.

Recognition of the leukoerythroblastic reaction does not constitute a diagnosis but rather a condition of concurrent release of benign immature myeloid and erythroid cells. It needs to be differentiated from malignancy in both the erythroid and myeloid systems and interpreted in the light of relative immaturity and the nature of the concurrent process.

IV. Myeloproliferative Disease

Myeloproliferative disease is a general term for diseases characterized by medullary and extramedullary proliferation of one or more marrow myeloid cell lines, with specific exclusion of the lymphomas and lymphoid leukemias. The FAB (French–American–British) system of classification of the acute leukemias provides for the definition of diseases which are histogenetically separable, but at the same time provides information that relates strongly to biological behavior. This classification has traditionally been based on morphology and rate of progression and divides the acute myeloid leukemias into categories of M1 to M7 and the acute lymphoid leukemias into L1 to L3. A large body of information now adds cytogenetic and immunophenotypic definition to this morphologic classification. Most of the latter data relate to human diseases; however, key information has been derived from work on animals, including birds.

The chronic leukemias, being by definition well differentiated, present little difficulty in recognition and are not considered in the FAB classification. In terms of pathogenetic mechanisms, the chronic leukemias have contributed greatly to understanding of the leukemic process since the description of the 9;22 (Philadelphia) chromosomal translocation characteristic of human chronic myelogenous leukemia. These morphologically detectable cytogenetic alterations have led to studies of the molecular consequences of chromosomal injury.

It appears that the **two crucial steps** in **leukemogenesis** are the clonal acquisition of increased capability for proliferation, and the ability to produce specific growth factors in an autocrine self-driving cell system. Mechanistically, the critical chromosomal alterations place potent promoters near or within cellular genes, thus altering their level of expression. The promoters may be endogenous oncogenes or insertions carried by retroviruses. The avian MH2 virus provides an example of an agent carrying two oncogenes, the v-*myc*, which induces self-renewal, and the v-*mil*, which induces the autocrine production of the growth factor colony-stimulating factor (CSF), together resulting in monocytic leukemia. Those acquired chromosomal alterations with high sensitivity and specificity for lineage-specific malignancies are termed **primary changes,** as in the Philadelphia chromosome. Nonrandom acquired chromosomal changes which occur in the course of the accelerated phase of chronic myelogenous leukemia are termed **secondary changes** since they occur in other conditions and are not useful as tumor markers but are useful for staging and prognosis. Typically in late-stage human chronic myelogenous leukemia, there are further random changes in chromosomes which result from the basic instability of the malignant clone and lack lineage specificity, but are associated with the level of tumor aggression; these are termed **tertiary chromosomal changes.** Tertiary changes are common in solid tumors and for years obscured the specificity of the less common primary and secondary changes. Clonal chromosomal abnormalities are found in as many as 75% of human leukemia–lymphoma and myelodysplasia cases, suggesting that these changes have been inadequately looked for in spontaneous tumors of animals.

The major determination in the diagnosis of acute poorly differentiated or undifferentiated leukemia is the distinction between myeloid and lymphoid lineage. This distinction has traditionally been based on histochemistry (Table 2.2). However, the recognition and characterization of lineage-specific cell surface antigens, has produced a large family of reagent antibodies directed against these antigens, particularly for human and mouse cells. By international concordance, these antigens are grouped by **clusters of differentiation** (CD) in which CD4 and CD8 define T-helper and -suppressor subsets, respectively, and CD13 and TdT, respectively, define poorly differentiated myeloid and lymphoid clones. As might be expected, the simplest techniques suffice in more differentiated malignancies whereas in primitive tumors, all modalities may be required to fully define cell lineage and thus management strategies. Many of these antibodies have broad cross-reactivity with the counterpart antigens on cells of domestic animals, but most do not react in a diagnostically useful manner. Collectively these diagnostic techniques have added greatly to the understanding of leukemias, particularly in the areas of apparent conflict. For example, in myeloid leukemia in blast crisis, in 25% of human cases the tumor cells carry lymphoid markers recapitulating the evolution of the clone from a totipotential stem cell capable of producing both myeloid and lymphoid progeny. Some 30% of human acute lymphoid leukemias in adults have the Philadelphia chromosome and in some cases, the response to treatment varies with the presence of specific cytogenetic alterations, although the tumors may appear morphologically similar.

The myeloproliferative diseases occur mostly in dogs and cats; their recognition in other domestic and wild animals is relatively rare. In dogs, myeloproliferative disease constitutes about 5% of hematopoietic neoplasia. Cats have the greatest proportion and range of myeloproliferative diseases, which constitute 10–15% of hematopoietic neoplasms with a similar proportion of lymphoid leukemias and the rest, lymphomas. Characteristically the acute leukemias have a course of 1–2 months from diagnosis to termination, whereas the chronic leukemias have a natural course of 1–3 years. Acute leukemias of humans and animals, because of their high rate of proliferation, are most responsive to chemotherapy. In contrast, animals with chronic leukemia are often adversely affected by chemotherapy because the mitotic rate of the tumor cells is similar to or less than that of residual normal marrow.

The descriptions provided are of the diseases unaltered by chemotherapy. Treatment of the acute leukemias causes lymphoid atrophy and reduced cellularity of bone marrow. Treatment of the chronic leukemias causes similar changes but with more generalized sclerosis. Chemotherapy tends to reduce the size of the residual tumor cells and to reduce their lineage specific differentiation. It is therefore highly desirable to base histologic and cytologic diagnoses on tissues collected prior to onset of therapy.

Bibliography

Dube, I. D., Carter, R. F., and Pinkerton, P. H. (1990). Chromosome abnormalities in chronic myeloid leukaemia. A model for

TABLE 2.2

Cytochemistry for Acute Myeloid Leukemias[a]

Reaction	Normal cells manifesting reaction	Diagnostic utility
Myeloperoxidase	Neutrophil series	Myeloblastic leukemia without maturation—MI
	Eosinophils (cyanide resistant)	
	Basophils ±	Myeloblastic leukemia with maturation—M2
	Monocytes ±	
Sudan black B	Neutrophil series	M1, M2
	Monocytes ±	
Chloracetate esterase	Neutrophil series	M1, M2
	Monocytes ±	
Nonspecific esterase (α-napthyl acetate substrate or α-napthyl butyrate)	Monocytes (inhibited by sodium fluoride)	Monocyte leukemias
		Monocytic leukemia, differentiated—M5
		Monocytic leukemia, poorly differentiated—M5 (acute monoblastic)
		Myelomonocytic leukemia—??
Periodic acid–Schiff (PAS)		Erythroleukemia—M6
		Acute lymphatic leukemias

[a] With permission of R. D. Brunning, M.D., Department of Pathology, University of Minnesota.

acquired chromosome changes in haematological malignancy. *Tumor Biol* **11**: 3–24, 1990.

Evans, R. J. and Gorman, N. T. Myeloproliferative disease in the dog and cat: Definition, aetiology and classification. *Vet Rec* **121**: 437–443, 1987.

Freedman, A. S., and Nadler, L. M. Cell surface markers in hematologic malignancies. *Semin Oncol* **14**: 193–212, 1987.

Metcalf, D. The roles of stem cell self-renewal and autocrine growth factor production in the biology of myeloid leukemia. *Cancer Res* **49**: 2305–2311, 1989.

A. The Acute Leukemias

1. M1, Myeloblastic Leukemia without Maturation (Acute Myeloblastic Leukemia)

This form of acute myeloblastic leukemia (AML) is characterized by a predominance of blasts in blood and marrow with fewer than 10% having cytoplasmic granulation. At least 5% of malignant blasts stain positively for Sudan black and myeloperoxidase. Acute myeloblastic leukemia (with or without maturation) is largely a disease of young mature dogs and cats with a tendency for increased frequency in males. Swine are occasionally affected, and other domestic animals, very rarely. This variant of acute myeloblastic leukemia is likely underdiagnosed, especially if the few cells with cytoplasmic granules are not recognized, and cytochemistry is not carried out or is inconclusive since lymphoid tumors are much more common and the tumor cells resemble large malignant lymphocytes.

Histologically it is indistinguishable cytologically from acute lymphocytic leukemia (ALL) or large cell lymphoma; thus blood and marrow cytology is essential for accurate diagnosis. Architectural differences between this and lymphoma are usually sufficient to identify the disorder as a primary leukemia, but will not differentiate it from acute lymphocytic leukemia.

2. M2, Myeloblastic Leukemia with Maturation

This condition is similar to the previous form except that maturation proceeds to or past the promyelocyte stage in at least 10% of tumor cells. Approximately half of the tumor cells stain positively with Sudan black and myeloperoxidase. Most animals are presented in good body condition because of epistaxis or melena of acute onset and have mucosal pallor, often with thrombocytopenic petechiation. The liver and spleen are usually not palpably enlarged and the lymph nodes are characteristically of normal size. Acute myeloblastic leukemia usually has a course of 2–3 weeks from the time of diagnosis. Death is due to hemorrhage and sepsis, if not complicated by chemotherapeutic myelosuppression.

Anemia is moderate (80 g/liter) and due to myelophthisis, or marked (20 g/liter) when complicated by thrombocytopenic hemorrhage. Anisocytosis is mild and there are a few poikilocytes, which are more numerous if platelets are deficient. Howell–Jolly bodies are usually present and indicative of reduced splenic function. Hypochromia will be present if the animal survives hemorrhage by a week or more. The erythroid response is mild or absent and usually less than the normal level of 60×10^9 reticulocytes per liter. Rubricytosis is often present to a mild degree, usually less than 1.0×10^9/liter, and with cells as immature as prorubricytes and occasionally rubriblasts in the peripheral blood. These cells are benign, and their release independent of age is indicative of leukemic myelophthisis (leukoerythroblastic reaction).

Thrombocytopenia is generally marked and in the $5–50 \times 10^9$/liter range, with immature platelets rare or absent. The platelets present tend to have uniform basophilia but reduced granulation, and functional deficits accompany low numbers. Cytoplasm shed from the fragile tumor cells rounds up and, with its sparse pink granulation, may be mistaken for platelet material, especially large immature platelets.

The total leukocyte count is highly variable but may be in the normal range and is usually less than 25×10^9/liter, but may rise to 50×10^9/liter near death. There is an absolute neutropenia and usually lymphopenia with myelocytes predominating followed by promyelocytes, metamyelocytes, and myeloblasts. The tumor may be of the neutrophil, eosinophil, or basophil type, but the neutrophil type is most common. Peripheral blood differentiation does not proceed past the metamyelocyte stage, and if more-mature cells are present, they are likely to be the progeny of residual normal marrow. Hypersegmentation may be present in benign neutrophils due to hyposplenism and probably to folate deficiency. If neutropenia has persisted for a week or more, bacteremia is often present and the benign cells have toxic changes.

Cytologically (Fig. 2.7A,B), the leukemic cells in blood have round centrally placed nuclei 2.5 red cells in diameter with a fine uniform or cribriform chromatin pattern. One to three nucleoli, which are round and from one-third to two-thirds the size of a red cell, are visible in the leukemic myeloblasts and promyelocytes. The cytoplasm is moderate in volume and basophilia, and has a smooth outline that is not usually indented by adjacent cells. There is usually sufficient fine azurophilic granulation, most visible in the Golgi clearing, to identify the cells as being of myeloid origin. In addition, the process of recognition by "the company they keep" is important; the mild reniform nuclear indentation in the most mature tumor cells suggests metamyelocyte differentiation. This indentation is much more shallow than the sharp folding characteristic of lymphocytes. Rarely in blood, benign promyelocytes and myelocytes will be recognized by their lower nuclear/cytoplasmic ratio, small nucleoli, and increased granulation. They are part of the process of myelophthisis and the leukoerythroblastic reaction.

Aspirated bone marrow (Fig. 2.8A) has very high cellularity and cell density, with few fat vacuoles. There is virtually complete phthisis (Fig. 2.8B) of the normal elements, resulting in loss of rubricytes, megakaryocytes, and mature granulocytes, and a coarse hemosiderin pattern. Characteristically the marrow cells are about one morphologic stage less mature than those in the peripheral blood. Early metamyelocytes are rare, and myeloblasts and promyelocytes predominate. Mitoses are very common and granulation is usually recognized in metaphase cells when the obscuring basophilia is minimal. The animals are usually presented for examination when 50% or more of the marrow is occupied by tumor, and marrow failure is imminent.

At autopsy there is pallor of membranes and tissues and mild to moderate loss of condition. The lymph nodes are

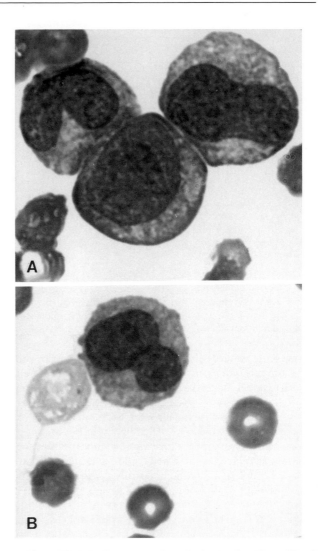

Fig. 2.7 (A) Acute granulocytic leukemia. Dog. Blood. Poorly granulated myeloblasts. Myeloid origin can be deduced from the pattern of maturation which ends at the metamyelocyte stage. (B) Bilobed progeny of malignant myeloid precursors. Cytoplasmic fragment from injured tumor cell, which contains granules, may be confused with a large platelet.

of normal size and often edematous. The liver is of normal size and pale. The spleen is usually mildly and uniformly enlarged and may be focally infarcted. Femoral marrow is uniformly vascular and a glistening pale tan rather than the hard white of normal fat. There are usually foci of hemorrhagic infarction if thrombocytopenia has been severe. The thymus is atrophic. Solid foci of the tumor in parenchymal organs are not found. Secondary lesions of marrow failure include hemorrhage, pneumonia, and cellulitis. Because of the absence of neutrophils, even septic lesions may have minimal cell reaction.

Histologically the neoplastic cells are round with peripheralized chromatin, prominent pink nucleoli, and a moderate amount of amphophilic cytoplasm. There are frequent mitoses and some cells of metamyelocyte type

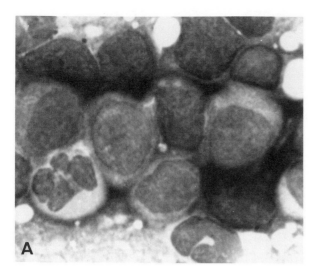

Fig. 2.8A Acute granulocytic leukemia. Bone marrow aspirate. There is early granulocytic asynchrony with phthisis of erythroid and megakaryocytic cells.

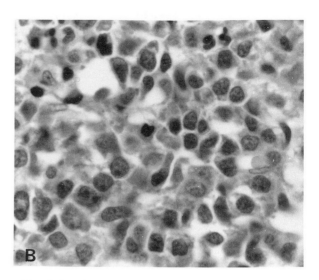

Fig. 2.8B Acute granulocytic leukemia. Marrow histology. Phthisis of normal elements and reduction in mature granulocytes indicates the level of maturation of the malignant myeloid precursors.

are present. Mature cells of the normal series are rare and, if the tumor is of the neutrophil type, most segmented cells will be residual, normal eosinophils. Megakaryocytes and rubricytes are rare or absent.

The spleen has lymphoid atrophy that is most prominent in thymus-dependent areas. The sinus areas contain few erythrocytes, are hypercellular due to tumor infiltration, and have increased, coarse hemosiderin. Benign hematopoiesis, if present, is characteristically paratrabecular and largely limited to thrombopoiesis. There is usually some subendothelial colonization of the large muscular sinuses by tumor cells. The liver usually has periacinar ischemic

degeneration. The tumor has a sinusoidal distribution, and if the leukocyte count is low, tumor cells may not be obvious. Periportal colonization is minimal and absent from many triads; this is an important distinction from the lymphomas. Benign hematopoiesis is similarly inconspicuous, which suggests that there is exhaustion of benign stem cells. Lymph node histology is highly variable and may vary from minimal involvement occurring first in medullary cords (Fig. 2.9A) with moderate lymphoid atrophy to diffuse involvement with capsular colonization (Fig. 2.9B), usually without an increase in overall size. If transfusions have been given, there will be erythrophago-

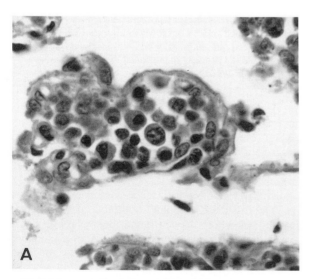

Fig. 2.9A Acute granulocytic leukemia. Medulla of node indicating pattern of tumor colonization in medullary cords. Empty sinuses indicate phthisis of marrow precursors of monocyte–macrophage system.

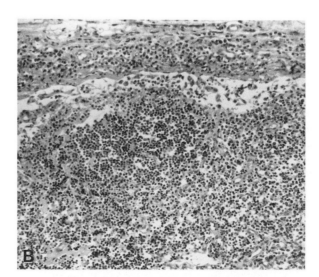

Fig. 2.9B Acute granulocytic leukemia. Lymph node. Cortical and capsular colonization by tumor.

cytosis in the medullary sinuses of the lymph nodes and sinuses of the spleen.

Myeloblastic leukemia with maturation is distinguished from acute lymphoblastic leukemia on the basis of cytoplasmic granulation and a positive reaction of tumor cells to the chloroacetate esterase stain.

Bibliography

Fraser, C. J. *et al.* Acute granulocytic leukemia in cats. *J Am Vet Med Assoc* **164:** 355–359, 1974.

Grindem, C. B. Ultrastructural morphology of leukemic cells in the cat. *Vet Pathol* **22:** 147–155, 1985.

Keller, P. *et al.* Acute myeloblastic leukaemia in a dog. *J Comp Pathol* **95:** 619–632, 1985.

3. M3, Promyelocytic Leukemia

Promyelocytic leukemia (Fig. 2.10) is a rare disease of young mature animals, which likely occurs in all species but is most commonly recognized in dogs and cats and also reported in swine. Clinical signs are similar to those of myeloblastic leukemia but with a greater tendency for bleeding. The course of the disease is usually 1–2 weeks from time of diagnosis. Death may be associated with intracranial hemorrhage.

Promyelocytic leukemia differs from myeloblastic leukemia by a predominance of promyelocytes in both blood and bone marrow. This difference does not extend to

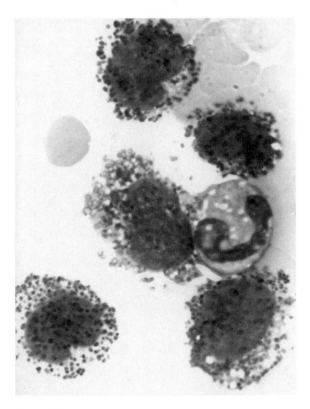

Fig. 2.10 Promyelocytic leukemia, canine blood. Nuclear indentation indicates that this tumor is of basophil type rather than a mast cell leukemia.

greater terminal differentiation, and metamyelocytes are rare. The cells are larger than those of myeloblastic leukemia and the degree of granulation is variable. Most cases described in animals are of the hypergranular type; however, these are the most likely to be recognized. A microgranular form is also described, associated with nuclear folding and a monocytoid appearance. The cytoplasmic granules in both variants of this disease are intensely positive with Sudan black and myeloperoxidase stains. The cytoplasmic granules are procoagulant, and rapid cytoreductive therapy may result in disseminated coagulation. The disease in humans is associated with a specific primary 15;17 chromosomal translocation, which results in the upregulation of the myeloperoxidase gene.

Bibliography

De The, H. *et al.* The t(15;17) translocation of acute promyelocytic leukaemia fuses the retinoic acid receptor α gene to a novel transcribed locus. *Nature* **347:** 558–561, 1990.

Kadota, K. *et al.* Ultrastructure and C-type particles in myeloid leukemia of a pig. *Vet Pathol* **21:** 263–265, 1984.

4. M4, Myelomonocytic Leukemia

Myelomonocytic leukemia is a rare disease recognized in dog, cat, horse, and humans in which there is concurrent neoplasia of the neutrophil and monocytic cell systems. Most commonly the disease consists of concurrent clones of tumor derived from both the monocytic and neutrophil systems, and as such the tumor cells mimic their normal counterparts in both morphology and histochemical reactions. Up to a third of human cases of myelomonocytic leukemia have a single tumor clone which combines the cytochemical characteristics of both cell lines and therefore represents an entity for which there is no normal counterpart. Myelomonocytic leukemia is defined as having at least 20% of both tumor cell lines staining for the neutrophil series (chloroacetate esterase, Sudan black, or myeloperoxidase activity) or for the monocytic series (fluoride-sensitive esterase). The requirement for at least 20% involvement by both tumor lines rules out confusion with benign hyperplasia. The critical point in the diagnosis of this disease is that both types of lineage-specific staining must be found in cells which are independently determined to be malignant.

Animals are presented in good body condition but with a history of weight loss of recent onset accompanied by malaise and anorexia. Clinically there is pallor of mucous membranes, irregular enlargement of cervical nodes, mild splenomegaly, and moderate fever. Anemia is moderate to marked and nonresponsive with some fragmented cells, increased pallor, and Howell–Jolly bodies. Thrombocytopenia is marked and poorly responsive. The leukocyte count is generally between 25 and 50 × 10^9/liter. Morphologically, the leukocytes resemble those of myeloblastic leukemia with some maturation and acute monocytic leukemia occurring in the same animal. Malignancy, as determined by nuclear atypia and immaturity in the peripheral blood, is present in both systems. In general, the degree of esterase staining varies with the degree of cytoplasmic granulation and both cell lines may be positive for peroxidase activity. The course is 4–6 weeks with termination due to sepsis and thrombocytopenic bleeding. Pathologically there are widespread petechial hemorrhages, focal infarctions of liver, kidney, spleen, and marrow, and usually oral ulcerations. Histologically there is widespread tumor infiltration in marrow, splenic sinuses, nodes, portal areas of the liver, and irregularly in other organs including lung. In untreated cases, benign extramedullary hematopoiesis may be found in adrenal glands and nodes. There is prominent phthisis of normal lines in marrow, particularly of megakaryocytes. Myelomonocytic leukemia must be differentiated by cytology and cytochemistry from monocytic leukemia, from which it is indistinguishable histologically.

Bibliography

Brumbaugh, G. W. Myelomonocytic myeloproliferative disease in a horse. *J Am Vet Med Assoc* **180**: 313–316, 1982.

Jain, N. C. *et al*. Clinical–pathological findings and cytochemical characterization of myelomonocytic leukaemia in 5 dogs. *J Comp Pathol* **91**: 17–31, 1981.

Rohrig, K. E. Acute myelomonocytic leukemia in a dog. *J Am Vet Med Assoc* **182**: 137–141, 1983.

5. M5, Monocytic Leukemia

The predominant cell in monocytic leukemia is a monocytic precursor as defined by nonspecific (α-naphthyl butyrate) esterase activity. In contrast to myelomonocytic leukemia, fewer than 20% of cells mark with Sudan black or myeloperoxidase indicating granulocytic origin. Monocytelike myeloblastic leukemia occurs in two variants, which vary in degree of differentiation and in rate of tumor progression. The first variant is characterized by monoblast predominance with little progression to monocytes apparent on routine blood stains, whereas in the second variant, cells with obvious monocytic differentiation predominate in the blood and marrow. The leukocyte count tends to be low or normal in the more acute form and elevated in the differentiated variant.

Monocytic leukemia occurs in young dogs and cats and in mature horses and cattle. The disease in the young tends to be of the first variant described, whereas that in adults is of the second. This description is typical of the disease in dogs. The first form is an acute disease with clinical presentation similar to that of myeloblastic leukemia. Animals are presented because of general signs of lethargy, weight loss, vomition, diarrhea, anorexia, and, more specifically, recurrent epistaxis. Clinically there is pallor that appears worse than the packed-cell volume (PCV) would indicate, and is characteristic of this disease where there is sludging of tumor cells in the microvasculature. The course is generally 2–4 weeks.

Anemia is usually marked (<50 g/liter) and nonresponsive, with pallor in larger red cells. Thrombocytopenia is severe and nonresponsive. The leukocyte count is usually less than 25 × 10⁹/liter at diagnosis but increases to be-

tween 50 and 100 × 10⁹/liter terminally. There is often mild neutrophilia and left shift which is likely due to growth factors produced by the tumor and a field effect of the autocrine pathogenesis driving the tumor cells themselves. Cytologically the tumor is relatively monomorphic, but both monoblasts and promonocytes are present, with the latter predominating in the blood (Fig. 2.11). The monoblasts are large, three to four red cells in diameter with an irregularly indented nucleus whose boundaries are obscured by intense cytoplasmic basophilia. The promonocytes have slightly smaller nuclei less than three red cells in diameter, with convoluted or cerebriform outline, a fine retiform chromatin pattern, and one to two nucleoli of moderate size. The cytoplasm is more basophilic than in young benign monocytes and characteristically of uniform density from the nucleus to the cytoplasmic boundary. This is a useful distinction from the perinuclear clearing seen in lymphocytes. A very fine azurophilic granulation is apparent in the Golgi clearing, and clear vacuolations are frequent and often penetrate the nucleus. Rubricytes may be present, but the anemia is nonresponsive.

Aspirated marrow is highly cellular with relatively complete phthisis of erythroid and megakaryocytic cells. Myelopoiesis is synchronous to the neutrophil stage but relatively reduced in amount. About 75% of marrow cells are undifferentiated blasts, monoblasts, and promonocytes,

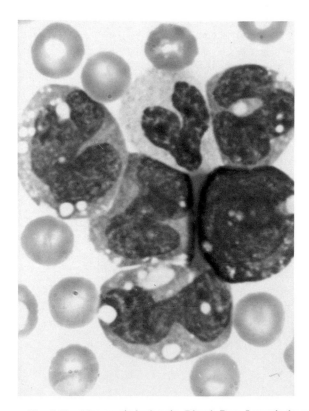

Fig. 2.11 Monocytic leukemia. Blood. Dog. Irregularly convoluted and segmented nuclei with prominent cytoplasmic vacuolations indicate monocytic differentiation. Vacuolation characteristic of monocytes is present in the blastic cell at center right.

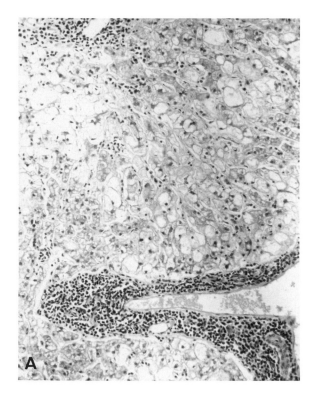

Fig. 2.12A Monocytic leukemia. Dog. Liver. Colonization of periportal (below) and periacinar areas by malignant monocytes.

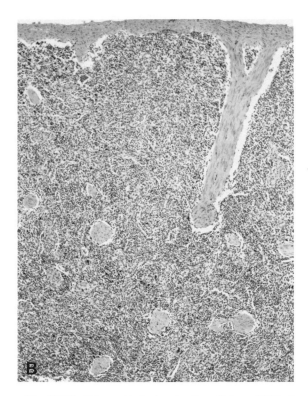

Fig. 2.12B Monocytic leukemia. Dog. Spleen. Follicular atrophy and diffuse sinusoidal colonization.

with nuclear hyperchromicity, prominent nucleoli, and a moderate mitotic rate.

Lesions are similar to those of myeloblastic leukemia, often with focal hemorrhages, but less often with secondary infection. The tumor distribution (Fig. 2.12A,B) is like that of myeloblastic leukemia but with greater dissemination to the urinary system and adrenals. Histologically, the nuclear convolutions are less apparent than on cytologic examination and the tumor cells resemble cleaved lymphocytes with deeply amphophilic cytoplasm. The tumor cells are not regularly phagocytic and this criterion is not helpful in distinguishing this disease from ALL. Definitive diagnosis requires cytochemical staining on blood and marrow aspirates or imprints.

Bibliography

Burkhardt, E., Saldern, F., and Huskamp, B. Monocytic leukemia in a horse. *Vet Pathol* **21:** 394–398, 1984.

Latimer, K. S., and Dykstra, M. J. Acute monocytic leukemia in a dog. *J Am Vet Med Assoc* **184:** 852–855, 1984.

Mackey, L. J., Jarrett, W. F. H., and Wiseman, A. Monocytic leukaemia in a cow. *Res Vet Sci* **13:** 287–289, 1972.

Scott, C. S. *et al.* Diagnostic and prognostic factors in acute monocytic leukaemia: an analysis of 51 cases. *Br J Haematol* **69:** 247–252, 1988.

6. Malignant Histiocytosis

Malignant histiocytosis is the nonleukemic tissue-phase tumor of the mononuclear phagocyte system. It is a rapidly progressive disease characterized by proliferation of markedly atypical histiocytes which are avidly erythrophagocytic. The disease is rare and usually occurs in mature cats and dogs. Clinically there is fever, depression, weight loss, irregular lymphadenopathy, and consistent hepatosplenomegaly. Hematologically there is a rapidly progressive anemia that is initially responsive but within a few weeks progresses to a nonresponsive pancytopenia with marrow failure. Occasionally the disease may be first noticed and diagnosed by an enlarged node that on aspiration yields very large cells with nuclei as large as hepatocytes and nucleoli as large as red cells. The cytoplasm is finely granular and stains positively with α-naphthyl acetate esterase. Erythrophagocytosis is not present in all tumor cells, but hemosiderin in the absence of red cells indicates previous activity. As many as 100 red cells may be identified within a single malignant histiocyte, which indicates that the anemia is initially due to intracellular hemolysis and later to both destruction and impaired production.

Pathologically there is pallor, and hemorrhage may be present if thrombocytopenia has been severe. There is irregular firm enlargement of lymph nodes which are dark on cut surface, and enlargement of spleen and liver, both of which may be focally infiltrated and infarcted. Bone marrow is irregularly reddened, and usually the disease advances so rapidly that fatty areas remain. Histologically the tumor cells have distinct cytoplasmic borders, large round or oval, often paired nuclei, prominent nucleoli, and abundant pale cytoplasm often with ingested red cells and less often platelets and leukocytes. The tumor cells may be discretely packaged and even mildly encapsulated in spleen and liver or in more diffuse and infiltrative distribution, as is more common in nodes and marrow. The diagnosis may need to be confirmed by histochemistry, but the severe atypia and phagocytic tendencies suggest the correct interpretation.

Two histiocytic proliferative diseases are described in Bernese mountain dogs and both are pedigree related. **Systemic histiocytosis** (see The Skin and Appendages, Volume 1, Chapter 5) consistently involves skin and lymph nodes by histiocytes which lack cellular atypia and may not be neoplastic. **Malignant histiocytosis** occurs in older dogs of the breed, mainly males, and is rapidly progressive. Discrete nodular masses, solitary or multiple, are especially frequent in the lungs and the hilar nodes are enlarged. Masses are also common in other lymph nodes, liver, and spleen. The infiltrates are composed of large, pleomorphic mononuclear cells which are individually quite discrete and the tumors are often infiltrated by neutrophils and eosinophils. Large, multinucleate giant cells, many with bizarre mitoses, are present in all tumors. Phagocytosis, by tumor cells, of erythrocytes, neutrophils, and other tumor cells is suggestive of histiocytic origin and immunochemical markers demonstrate lysozyme and α_1-antitrypsin.

Bibliography

Cattoretti, G. *et al.* Malignant histiocytosis. A phenotypic and genotypic investigation. *Am J Pathol* **136:** 1009–1019, 1990.

Moore, P. F., and Rosin, A. Malignant histiocytosis of Bernese mountain dogs. *Vet Pathol* **23:** 1–10, 1986.

Reiner, A. P., and Spivak, J. L. Hematophagic histiocytosis. A report of 23 new patients and a review of the literature. *Medicine* **67:** 369–388, 1988.

Wellman, M. L. *et al.* Malignant histiocytosis in four dogs. *J Am Vet Med Assoc* **187:** 919–921, 1985.

7. M6, Erythremic Myelosis

Erythremic myelosis can be defined as a condition in which nucleated erythroid cells of all stages comprise 50% or more of marrow cells, with at least 10% of cells dysplastic and at least 30% of cells primitive. Erythroleukemia, although rare in animals, is relatively less so in humans. On the other hand, erythremic myelosis, a disease associated with feline leukemia virus infection in the cat, is much more common in animals than in humans, in whom it is exceedingly rare. Erythremic myelosis is a tumor of nucleated erythroid precursors, whereas erythroleukemia is a concurrent tumor of both the erythroid and myeloid (granulocytic) systems. Hematopoietic proliferation in benign cells has two loci of genetic control to which the feedback system of growth factors provides the fine tuning. In malignancies, the system may become deregulated at either level resulting in either a poorly or well-differentiated tumor. In the erythroid system, autonomy at the early

locus results in erythremic myelosis, whereas at the late locus the disease is polycythemia vera with an overproduction of normal-appearing red cells. Within erythremic myelosis, there are variations in the general maturity of the tumor cells such that the more mature variant may initially be mistaken for a benign response to severe anemia, whereas the immature variant resembles the hairy-cell leukemia or reticuloendotheliosis described in cats. These variations are apparent histologically in the accumulation in the spleen and liver of rubricytes that are either relatively mature and well saturated with hemoglobin or are primitive and resemble lymphocytes with round nuclei and an unusual level of anisokaryosis. The malignant erythroid cells mimic their normal counterparts in containing sufficient glycogen to stain positively with the periodic acid–Schiff (PAS) reaction in both cytologic and histologic preparations.

Erythremic myelosis is largely a disease of cats, but it has been reported in a mature Holstein cow. It may appear from 1 year to old age and has a two to one predominance in males. The disease is rare in other domestic animals.

Animals are presented with a history of depression lasting from a few days to 2 months. Anorexia develops and is occasionally accompanied by vomition and diarrhea. With recent onset, the animals are in good condition but later are thin with mild abdominal distension due to splenomegaly and mild hepatomegaly. The lymph nodes are of normal size or small and there is marked mucosal pallor, occasionally with mild icterus. The course of the disease is 1–3 months.

Anemia is severe with hemoglobin levels varying from 20 to 50 g/liter. *Haemobartonella* is often suspected and may occasionally be present but should be recognized as incidental because there is lack of polychromasia and blast cells are present in the blood. Anisocytosis is mild to moderate and the mean corpuscular volume is characteristically increased to the range of 60/fl. Changes in shape are slight and hemoglobin saturation is normal. Howell–Jolly bodies and spherocytes, and occasionally Heinz bodies, are present, but immature or polychromatic red cells average fewer than one per oil immersion field. The total nucleated cell count varies from 20 to 200 × 10⁹/liter and is usually above 50 × 10⁹/liter. Neutrophils are at low normal or deficient levels and there may be a mild left shift. Lymphocytes may be greater than 10×10^9/liter, with higher counts probably reflecting the integrity of extramedullary lymphopoiesis. Rubricytes comprise 90% of the differential count and two-thirds of these are readily recognized polychromatic, normochromic, and metarubricytes which have normal hemoglobin saturation but are larger than normal and have nuclei which appear inappropriately young for the degree of cytoplasmic maturation (early asynchrony). Presumably this larger size accounts for the persistent rise in mean corpuscular volume despite the lack of polychromasia, indicating that many red cells are the progeny of leukemic precursors. By association, the maturing rubricytes can be related to basophilic and prorubricytic precursors, but their significance may be missed if lymphocytosis is present and the numerous leukemic blasts are misinterpreted as part of lymphoproliferation with benign rubricytosis. Rubricytosis in the absence of polychromasia does not indicate a marrow response, and this asynchronous production suggests neoplasia in the erythroid system itself. The blasts (Fig. 2.13) typically have round nuclei with dense cribriform chromatin and a single nucleolus which may be as large as a normal red cell and one-third the nuclear diameter. The cytoplasm is moderate in volume, markedly eccentric, and basophilic. The malignant cells mimic benign rubriblasts in having a few fine azurophilic granules similar to those seen in myeloblasts. The granules differentiate these blasts from the lymphoid series, including the reticuloendotheliosis variant, which these cells otherwise closely resemble. Platelets are generally fewer than 75×10^9/liter, but immature platelets may be numerous. Many of the malignant rubricytes are injured in preparation of blood films and up to 25% of the differential count may be bare nuclei, which in interpretation need to be included with their cohort.

Aspirated marrow is hypercellular and usually without fat or the usual coarse hemosiderin pattern of malignancy. There is an absence of mature granulocytes, and megakaryocytes are dysplastic but at near normal numbers, which may suggest that the leukemogenesis has occurred near the common precursor for the erythroid and platelet systems. There is marked early erythroid asynchrony with 70% of cells having large primitive nuclei, which are described as megaloblastic. Mild dyserythropoiesis is present with some bi- and trinucleation and small satellite nuclei. Mature rubricytes are rare, considering their frequency in the peripheral blood. In general, there is marked immaturity in cells with round nuclei and a high level of cytoplasmic basophilia.

On gross examination all tissues are pale and there may be slight icterus. There is usually loss of condition and occasionally emaciation. The lymph nodes are of normal size or slightly enlarged. Lesions are largely limited to pallor, diffusely reddened femoral marrow, and moderate splenomegaly. The spleen is uniformly enlarged, and fleshy and dry on cut surface, without obvious lymphoid follicles.

Microscopically, the bone marrow has high cellularity with dense cellular packing (Fig. 2.14). Granulocyte reserves are absent. About 90% of the cells appear blastic with the rest much smaller and with apparent cytoplasmic hemoglobin. Megakaryocytes are present, usually in small clusters. The primitive cells have a fine granular chromatin pattern with thickened membranes and a single prominent central nucleolus. Nuclei are round or mildly irregular and the cytoplasm is moderate in volume, faceted against adjacent cells and deeply stained. The spleen has atrophy of thymus-dependent areas and increased sinus stroma with loosely packed, diffusely distributed tumor cells. Benign hematopoiesis is usually inapparent and hemosiderin is reduced. There is some subendothelial colonization of large muscular sinuses. Lymph nodes generally have cortical atrophy with sinus histiocytosis and occasionally contain focal neoplastic invasion. The liver usually has accentuation of zonation due to periacinar ischemia, and increased hemosiderin in Kupffer cells, which suggests increased destruction of red cells. Tumor involvement is usually sinusoidal with dilation of periacinar sinusoids and mild focal investment of triads (Fig. 2.15).

Erythremic myelosis must be distinguished from hemolytic anemia. Cats with immune hemolysis may occasionally have a marked rubricytosis of $50–90 \times 10^9$/liter, but these precursors are accompanied by $300–600 \times 10^9$ reticulocytes/liter. Spherocytosis and rafting of red cells are present in immune hemolysis, the erythroid immaturity is

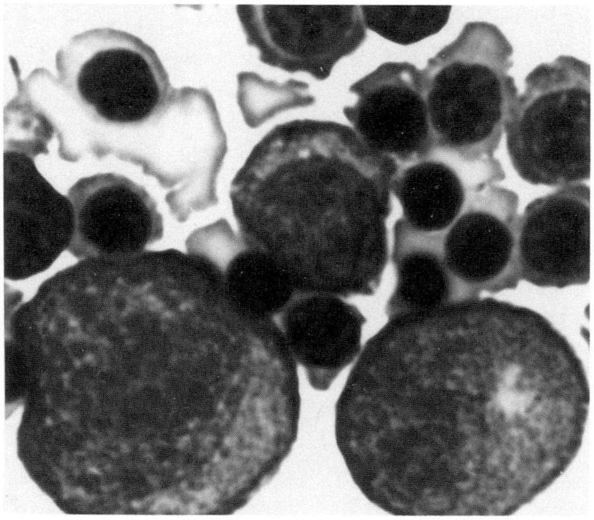

Fig. 2.13 Erythremic myelosis. Cat. Blood. Large malignant blast cells with a cluster of more mature progeny.

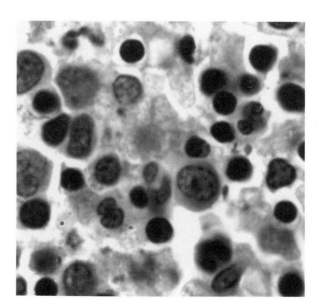

Fig. 2.14 Erythremic myelosis. Cat. Histology of marrow. There is a predominance of maturation phase rubricytes.

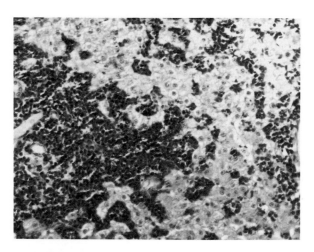

Fig. 2.15 Erythremic myelosis. Cat. Liver with heavy periportal colonization (left) and multiple foci of tumor colonization in sinusoids and around central vein (right).

less severe than in erythremic myelosis, and nuclear atypia is absent.

8. Erythroleukemia

Erythroleukemia is the simultaneous occurrence of neoplasia in both the erythroid and neutrophil precursors (Fig. 2.16). It is a rare disease, primarily of dogs and cats, which is characterized clinically by sudden onset of weakness, depression, and anorexia. Weight loss and signs of marrow failure, including chronic nonresponsive anemia, hemorrhage, and sepsis, follow. Smoldering erythremic changes occur for 1–2 months and are followed by terminal conversion to acute myelogenous leukemia.

The disease usually begins like erythremic myelosis but with less prominent myeloid immaturity. It progresses to a trilineage dysplasia; terminally the myeloblastic compo-

nent is more prominent. Dyserythropoiesis is marked and persistent, the anemia is severe (20 g/liter) and nonresponsive, and mean corpuscular volume is high. Thrombocytopenia is severe. The total nucleated cell count is usually less than 50×10^9/liter with 25–75% rubricytes. In cats, there may be a mild, persistent lymphocytosis. Pathologically the disease resembles erythremic myelosis.

Bibliography

Li, J.-P. *et al.* Activation of cell growth by binding of Friend spleen focus-forming virus gp55 glycoprotein to the erythropoietin receptor. *Nature* **343**: 762–764, 1990.

Maede, Y., and Murata, H. Erythroleukemia in a cat with special reference to the fine structure of primitive cells in its peripheral blood. *Jpn J Vet Sci* **42**: 531–541, 1980.

Shibuya, T., Niho, Y., and Mak, T. W. Erythroleukemia induction by Friend leukemia virus. *J Exp Med* **156**: 398–414, 1982.

Zawidzka, Z. Z., Janzen, E., and Grice, H. C. Erythremic myelosis in a cat: A case resembling diGuglielmo's syndrome in man. *Pathol Vet* **1**: 530–541, 1964.

9. M7, Megakaryoblastic Leukemia

Megakaryoblastic leukemia is a rare disease mainly seen in dogs. It may be underdiagnosed clinically because of confusion with myeloblastic leukemia and acute lymphocytic leukemia (ALL) of the L2 type, and pathologically because of confusion with malignant myelofibrosis. It is of relatively recent definition in human medicine and now is estimated to constitute some 8% of all acute leukemias. The disease affects a wide age range in both humans

Fig. 2.16 Erythroleukemia, canine blood. Concurrent neoplasia of granulocytic and erythrocytic systems. Malignant myeloblast (right) and malignant erythroid precursor (left).

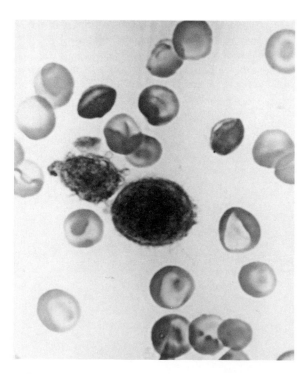

Fig. 2.17 Acute megakaryoblastic leukemia. Dog. Blood. Malignant megakaryoblasts with filiform and blunt cytoplasmic pseudopodia.

and animals. Pathogenetically, the aberrant cytoplasm of the tumor cells is procoagulant, resulting in fibrin deposition and repair by largely reticulin fibrosis. Platelet-derived growth factor (PDGF) may be responsible for the proliferation of benign stroma. The result is myelophthisis and very rapid marrow failure. Clinical signs are typical of acute leukemia, often with fever, subcutaneous hematomas, and gastrointestinal hemorrhage. The animals are usually in good condition, indicative of rapid onset, and typically are presented with hepatosplenomegaly. The course is usually 2–4 weeks.

There is pancytopenia with moderate to severe nonresponsive anemia and thrombocytopenia, and a total leukocyte count around 5×10^9/liter with neutropenia and peripheral blast cells. Tumor nuclei are typically two and a half to three red cells in diameter and round, but irregular constrictions and binucleation may be seen. Chromatin is dense and cribriform and nucleoli are not prominent. Clear vacuoles may penetrate the tumor nuclei and cytoplasm and appear in platelets. Unlike malignant myeloblasts, the cytoplasm is not round and entire but scant and may have one or two blunt pseudopodia (Fig. 2.17). Basophilia is marked and obscures azurophilic granulation, which is variably present and is most easily seen in the cytoplasmic

protrusions. Platelets vary widely in size and are basophilic but poorly granulated. If bleeding is prominent, there will be fragmented red cells.

Aspirated marrow may be highly cellular, but the irregular sclerosis rapidly makes a core biopsy essential to evaluate marrow. There is early loss of erythropoiesis and granulopoiesis and loss of granulocyte reserves. From 70 to 90% of marrow cells are hyperchromatic blasts with irregularities in nuclear size, shape, and number per cell. The presence of nuclear fusion in some cells is highly significant and is characteristic of megakaryoblastic leukemia. The tumor cells mimic their normal counterparts and about a third stain positively with the PAS reaction both cytologically and histologically, depending on their content of cytoplasmic glycogen. Because cytoplasm is scant in the mononuclear blasts, the PAS reaction is most prominent in the larger cells with lobulated nuclei and in these the reaction is highly variable in the same cell. Tumor cells stain negatively with Sudan black and myeloperoxidase and positively with α-naphthyl acetate esterase and for the factor VIII antigen. Monocytic leukemia or malignant histiocytosis is ruled out by the absence of erythrophagocytosis and by lack of reaction in megakaryoblastic leukemia of the monocytic tumors for the factor VIII antigen.

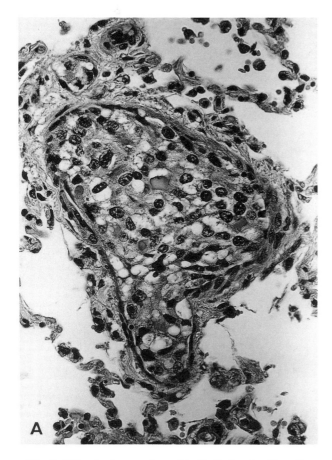

Fig. 2.18A Acute megakaryoblastic leukemia. Dog. Lung, tumor emboli in arteriole. Small homogeneous areas are fibrin.

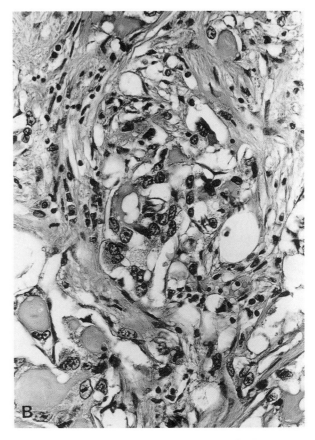

Fig. 2.18B Acute megakaryoblastic leukemia. Dog. Marrow with myelophthisis, myelofibrosis of benign stroma, and colonization by malignant megakaryoblasts.

Acute lymphocytic leukemia is ruled out by the presence of the multinucleated and lobulated nuclei on routine stains of aspirates and sections.

Pathologically there is pallor and irregularly hemorrhage that is more typical of factor deficiency than of thrombocytopenia. The spleen is large and firm, and histologically there is lymphoid atrophy and marked sinus expansion with both blasts and irregular nests of megakaryocytelike cells with marked atypia and occasionally very large nucleoli. Splenic stroma is increased and extramedullary hematopoiesis is absent or minimal. Tumor is present in the hepatic sinusoids and irregularly in nodes. There are remarkable tumor emboli in pulmonary arterioles with tumor cells apparently growing in syncytia (Fig. 2.18A). There is focal positivity for fibrin in these emboli, indicating the propensity of this tumor to cause disseminated coagulation. In marrow, there is loss of fat and irregular phthisis of normal elements (Fig. 2.18B). The tumor cells in the aggregate of single and multilobular nucleated types greatly predominate and their atypia rule out benign disease. On specific staining, there is a marked and variable increase in reticulin but not in collagen.

Bibliography

Bolon, B. *et al.* Megakaryoblastic leukemia in a dog. *Vet Clin Pathol* **18**: 69–72, 1989.

Holscher, M. A. *et al.* Megakaryocytic leukemia in a dog. *Vet Pathol* **15**: 562–565, 1978.

McClellan, J. E., Maddox, J. C., and Innes, D. J. Platelet vacuoles in acute megakaryoblastic leukemia. *Am J Clin Pathol* **92**: 700–702, 1989.

Moscinski, L. C. *et al.* Myeloperoxidase-positive acute megakaryoblastic leukemia. *Am J Clin Pathol* **91**: 607–612, 1989.

Shull, R. M., Denovo, R. C., and McCracken, M. D. Megakaryoblastic leukemia in a dog. *Vet Pathol* **23**: 533–536, 1986.

B. The Chronic Leukemias

1. Chronic Granulocytic Leukemia

Chronic granulocytic leukemia (chronic myelogenous leukemia) is recognized most frequently in mature dogs and cats, and the disease probably occurs in all species. In cats, males are more often affected than females. This disease is less common than the acute leukemias.

In contrast to the acute leukemias, there is a history of weight loss extending over several months with recurrent inappetence and diarrhea, and often several treatments for slow-healing superficial lesions. Clinically there is mucosal pallor, dullness of hair, loss of condition, splenomegaly, and increased heart rate. Lymph nodes are of normal size or slightly enlarged. The course is 1–4 years or longer. Late conversion to an acute leukemia is seen less often than in humans, probably because animals with this disease are usually destroyed when they enter the accelerated phase and prior to the development of a blast crisis.

Anemia is mild to moderate, seems to be less severe in dogs than in other species, and is nonresponsive. A few relatively mature rubricytes are usually present plus a small number of fragmented and large red cells in a normochromic, normocytic or mild hypochromic background. Howell–Jolly bodies are usually present indicating splenic overload. The leukocyte count is usually greater than 50×10^9/liter and often exceeds 200×10^9/liter. The major feature of chronic granulocytic leukemia is that segmented granulocytes are always present in the peripheral blood and usually predominate. Malignancy may occur in any of the three granulocytic lines, but neutrophils are most often affected. Leukemia involving the eosinophil line is seen most often in cats, and basophil tumors are most common in pigs. Secondary granulation is reduced and there may be toxic vacuolation (Fig. 2.19). Hyper- and irregular segmentation are characteristic. There is usually a reasonably synchronous distribution of younger neutrophils to myeloblasts. The latter cells are rare in peripheral blood and usually not found in a differential count of 100 cells. Cytoplasmic basophilia is usually mild and granulation is apparent in promyelocytes and myelocytes. There is usually a mild increase in normal-appearing basophils. Platelet levels may be normal or increased but are usually around 100×10^9/liter and with progression, deficiency becomes limiting to life. Platelet granulation may be normal but is more often reduced, and there are occasional very large platelets.

Aspirated marrow is hypercellular with a myeloid to erythroid ratio of 50 or higher and a marked increase in metamyelocytes and band neutrophils which, with a few segmented cells, occupy more than half of the marrow volume. Erythropoiesis is synchronous and never as deficient as in the acute leukemias. Hemosiderin is normal or reduced, and megakaryocytes are at normal levels for long

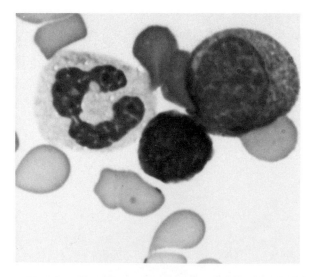

Fig. 2.19 Chronic granulocytic leukemia, feline blood. Malignant promyelocyte (right) with a large poorly granulated neutrophil, probably the progeny of malignant precursors. There is nonresponsive anemia, and hemobartonellae are present in red cells (center). Benign basophilic rubricyte (center), in the absence of polychromasia, indicates myelophthisic erythroblastosis.

periods but become deficient as the disease progresses and becomes more aggressive.

Pathologically there is pallor and usually emaciation. There may be widespread focal hemorrhages, and enteric bleeding is common, as are chronic oral or skin ulcers. The spleen is always enlarged, dry, and fleshy and may be many times normal size. The liver is enlarged to a variable degree, pale with an anemic lobular pattern, and may have focal hemorrhage, infarction, or infiltration. The nodes are irregularly enlarged, often due to septic drainage from secondary disease. There may be focal lesions in other organs, but they are usually minimal. All marrow is uniformly deep red.

Histologically, the bone marrow is solidly cellular with dense cell packing and reduced sinus area. Metamyelocytes and bands predominate with numerous giant metamyelocytes. Erythropoiesis is relatively much reduced but may be near normal on an absolute basis, considering the greatly expanded marrow cavity. Megakaryocytes may be numerous and are hyperlobulated without the usual condensation to a single nuclear mass. They tend to have reduced cytoplasmic volume in proportion to nuclear size (ineffective thrombopoiesis). Primitive cells are numerous, and tend to cluster in laminar fashion one to three cells deep near bony trabeculae. An increase in depth of the primitive cell layer presages onset of the accelerated phase. These proliferating cells have round or oval nuclei with coarsely granular chromatin and prominent single central pink nucleoli. Their cytoplasm is abundant and moderately amphophilic.

Lymph nodes vary depending on whether they are secondarily infected, but tend to have follicular hyperplasia with reduced cell density. Mature tumor cells may fill the peripheral sinus and permeate the capsule, with cells in rows between the collagen layers. Primitive tumor cells variably colonize the medullary cords, whereas the mature cells predominate in sinuses. There is variable colonization of the medullary sinuses. This is most obvious if the tumor is of the eosinophil or basophil type, and if colonization is extensive, it is often accompanied by medullary sclerosis. Benign extramedullary hematopoiesis will also be present in the medullary cords and can be recognized by the presence of megakaryocytes. As the tumor progresses, the benign cells become displaced to the adjacent medullary sinuses.

The splenic capsule is thin and focally bridged. Follicles are present and prominent with low cell density and reduced thymic cuffs. The sinuses are hypercellular, containing focal clusters of primitive cells with prominent cytoplasmic basophilia, and also some mature cells, but fewer than in the marrow. Tumor cells colonize the endothelium of muscular veins. Hemosiderin is not increased and may be deficient if there has been recurrent hemorrhage.

The liver has both focal and sinusoidal involvement, the latter in transit and the former consisting of fixed leukemic infiltrates around both portal triads and the large veins. In periacinar areas, tumor cells replace hepatocytes within the endothelium of the hepatic cords, forming a hematopoietic tissue–sinus relationship similar to that of marrow.

Virtually all other tissues may be focally invaded, especially adrenal medulla, renal cortex around arterioles and, less regularly, the heart and reproductive, pulmonary and enteric systems. Lesions of cachexia are present in advanced cases.

Chronic myeloid leukemia can be differentiated from leukemoid reaction and from hypereosinophilic syndrome by the presence of blasts in the peripheral blood and tissue destruction by the infiltrating tumor cells.

Bibliography

Finlay, D. Eosinophilic leukaemia in the cat: A case report. *Vet Rec* **116:** 567, 1985.

Lewis, M. G. *et al.* Retroviral-associated eosinophilic leukemia in the cat. *Am J Vet Res* **46:** 1066–1070, 1985.

McEwen, S. A., Valli, V. E. O., and Hulland, T. J. Hypereosinophilic syndrome in cats: A report of three cases. *Can J Comp Med* **49:** 248–253, 1985.

Searcy, G. P., and Orr, J. P. Chronic granulocytic leukemia in a horse. *Can Vet J* **22:** 148–151, 1981.

2. Mast Cell Leukemia

Mast cells are a heterogeneous population and are widely distributed in connective tissues. In the mouse, and likely in other species, the mast cell and the basophil share a common precursor. Neoplastic proliferation of mast cells to form solitary or multiple tumors is described with The Skin and Appendages (Volume 1, Chapter 5) and with The Alimentary System (Volume 2, Chapter 1). Systemic mastocytosis, involving primarily the hematopoietic organs in a pattern common to the leukemias, is a rare disease seen in humans and cats. Systemic mastocytosis occurs independent of cutaneous or alimentary mastocytoma. Mastocytemia may or may not be present but tends to be intermittent; detection of mast cells in circulation is best accomplished on smears of buffy coat. The hemogram may be normal or there may be eosinophilia and anemia.

The usual presenting signs of systemic mastocytosis in the cat are vomition and splenomegaly. Gastric and duodenal ulceration accompanies visceral and systemic mast cell neoplasms and is presumed to be caused by high levels of histamine produced by the neoplastic cells. Splenomegaly, which is often massive, is consistently present. The spleen is firm and brownish and there may be irregular white fibrous thickening of the capsule. Microscopically, the architecture of the splenic pulp is effaced by sheets of mast cells in diffuse distribution. The cytoplasm is bland and abundant, and special stains are necessary to demonstrate the specific granules.

Hepatomegaly is often present. The infiltration of mast cells in the liver is diffuse and produces a fine stippled appearance when the triads are mainly infiltrated. Larger accumulations of the malignant cells in irregular foci may be visible as pale foci on the surface. The infiltrations in kidney, bone marrow, and other organs tend to be more discrete than those in the spleen and liver.

Bibliography

Bean-Knudsen, D. E. *et al.* Porcine mast cell leukemia with systemic mastocytosis. *Vet Pathol* **26:** 90–92, 1989.

Galli, S. J. New insights into "the riddle of the mast cells": Microenvironmental regulation of mast cell development and phenotypic heterogeneity. *Lab Invest* **62:** 5–33, 1990.

Parwaresch, M R., Horny, H.-P., and Lennert, K. Tissue mast cells in health and disease. *Pathol Res Pract* **179:** 439–461, 1985.

Garner, F. M., and Lingeman, C. H. Mast cell neoplasms of the domestic cat. *Pathol Vet* **7:** 517–530, 1970.

3. Granulocytic Sarcoma

Granulocytic sarcoma or chloroma is the extramedullary growth of focal granulocytic neoplasms, which may be of neutrophil or eosinophil type (Fig. 2.20A–D). They are initially aleukemic and have variable differentiation which, if minimal, usually leads to classification as lymphoma. They have typical granulocytic (α-naphthyl acetate esterase) staining and the ephemeral greenish hue on exposure to air is presumably due to a high level of myeloperoxidase. The disease in dogs tends to involve lung, gut, and skin but any tissue can be initially involved and skeletal muscle is characteristically affected in cattle. There may be progression to acute myeloblastic leukemia within months of biopsy. Most cases are diagnosed at autopsy. The detection of granulocytic sarcoma in an animal with a previous diagnosis of myeloid leukemia indicates an accelerated phase with rapid progression, and reversion to a less differentiated cell type.

Focal proliferation of eosinophils and precursors will result in fibrosis, presumably as a result of tissue injury by the lysosomal proteins. Granulocytic sarcoma must be differentiated from benign hypereosinophilic syndrome and from eosinophil leukemia. Cytochemical differentiation from lymphoma may be assisted by thin sections at 1–2 μm in which the typical cytoplasmic granulation is most apparent.

Bibliography

Callahan, M. *et al.* Granulocytic sarcoma presenting as pulmonary nodules and lymphadenopathy. *Cancer* **60:** 1902–1904, 1987.

Chan, J. K. C., and Ng, C. S. Diagnosis of granulocytic sarcoma. *Pathology* **19:** 317, 1987.

4. Megakaryocytic Myelosis

This disease (essential thrombocythemia) is the chronic counterpart of acute megakaryoblastic leukemia. It is a rare disease of dogs and cats generally older than 5 years with a clinical course of 1–3 years. Megakaryocytic myelosis is characterized clinically by bleeding and thrombosis of the tips of tail and ears, and hematologically by massive thrombocytosis. There is marked nonresponsive anemia with mild anisocytosis and hypochromia and an occasional oval red cell. The leukocyte count is in the normal range usually with reduced neutrophils and mild left shift, and a mild to moderate basophilia, the latter being easily overlooked, as the basophils are usually poorly granulated. The platelet count may be low or normal but is most often dramatically increased to $2000–4000 \times 10^9$/liter or higher, which imparts a distinct blueness to the stained blood film. At any platelet level, there are marked morphologic changes with wide size variations, frequent vacuolation with pale basophilia, and irregular and reduced granulation (Fig. 2.21). The shift platelet reaction is marked, and functional tests of platelet activity such as aggregation and clot retraction are usually reduced. This thrombasthenia is made evident by the serious hemorrhage that results when even small amounts of aspirin are given to alleviate thrombosis in extremities.

Aspirated marrow is hypercellular with myeloid hyperplasia and early asynchrony involving the megakaryocytic and basophil lines, and erythroid phthisis with loss of mature granulocyte reserves. Hemosiderin has a coarse pattern and is increased in amount. With progression of the disease, aspiration of marrow is prevented by myelofibrosis, and a core biopsy is required. Marrow aspirates are hypercoagulable, and nucleated cells are injured by fibrillar strands and obscured by masses of coherent platelets.

At postmortem most tissues are pale, the spleen is enlarged several times, and is dark with a dry, fleshy cut surface. The liver is symmetrically enlarged, dark to deep brown with rounded and notched borders. The kidneys are of normal shape and size but deep brown. The bone marrow is uniformly dark red and highly cohesive. Lymph nodes are of normal size.

Microscopically the bone marrow is solidly cellular with generalized increase in reticulin and focal reduction in cell packing where sclerosis is most advanced. There is massive megakaryocytic hyperplasia, often with normal cytoplasmic volume, increased basophilia, and hyperchromatic nuclei, but otherwise with little atypia. Isolated tumor cells are irregularly distributed in cortical areas of lymph nodes. Liver cells are atrophic with intrahepatic cholestasis, and sinusoids are two to three times normal width with few red cells and many megakaryocytes. Portal infiltration is mild and irregular. The spleen has lymphoid atrophy with hypercellular sinus areas which resemble bone marrow. Most other tissues are minimally affected.

Bibliography

Michel, R. L., O'Handley, P., and Dade, A. W. Megakaryocytic myelosis in a cat. *J Am Vet Med Assoc* **168:** 1021–1025, 1976.

Thiele, J. *et al.* Histomorphometry of bone marrow biopsies in chronic myeloproliferative disorders with associated thrombocytosis—features of significance for the diagnosis of primary (essential) thrombocythaemia. *Virchows Arch Pathol Anat* **413:** 407–417, 1988.

5. Polycythemia Vera

Polycythemia vera is a rare disease that is seen mainly in dogs and cats but occurs in cattle, and can be induced

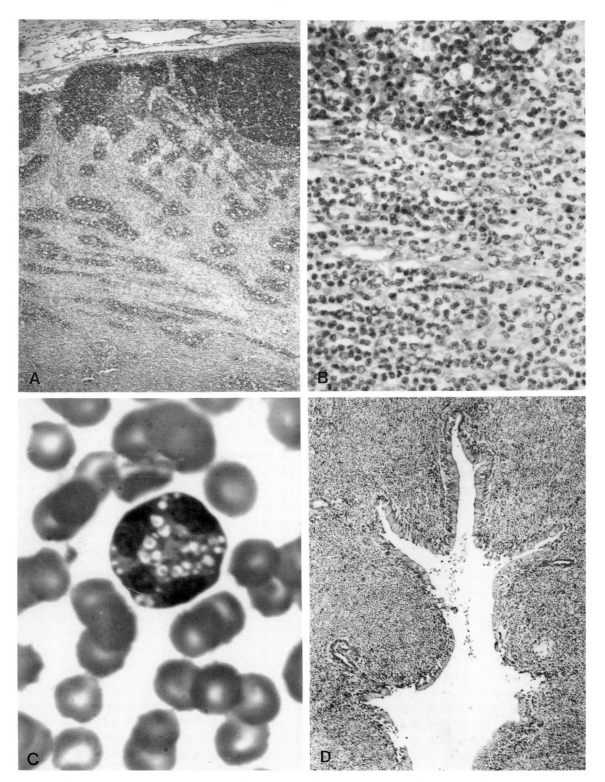

Fig. 2.20 Granulocytic sarcoma. Dog. (A) Lymph node. Follicular and paracortical atrophy. Sinus hypertrophy and colonization by granulocytic tumor. (B) Detail of (A). Medullary sclerosis and infiltration of eosinophils and precursors. (C) Large atypical eosinophil in blood. (D) Lung. Peribronchial infiltration with stenosis.

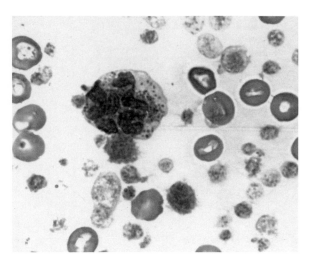

Fig. 2.21 Megakaryocytic myelosis, canine blood. Marked anemia, largely normochromic and normocytic with severe thrombocytosis of platelets, which vary markedly in size, basophilia, and granulation. The cell in center is an atypical basophil.

in mice by the Friend leukemia virus. Mouse studies have shown that anemic and polycythemic strains of this retrovirus interacting with early- and late-acting genetic loci produce either erythremic myelosis or polycythemia vera. In the latter disease, the erythroid stem cells grow in culture without erythropoietin stimulation and are thus an autonomous tumor clone. Growth factors can be isolated from peripheral blood leukocytes that are mitogenic and transforming, and distinct from platelet-derived and epidermal growth factors. The disease in humans is accompanied by leukocytosis and thrombocytosis and by trilineage marrow hyperplasia which progresses to reticulin myelofibrosis, marrow failure, and acute leukemia, usually of the myeloid type. Leukemic transformation appears more common in patients treated with ^{32}P and chlorambucil than in those treated by phlebotomy. Animals with polycythemia vera develop neither thrombocytosis nor apparently myelofibrosis.

Affected animals are presented with reduced exercise tolerance, and often polydipsia and signs of neuromuscular dysfunction including head tremor and posterior paresis. There is mucosal congestion and cyanosis with labored respiration and signs of hemorrhage including epistaxis, hematemesis, hematochezia, and hematuria. Lymph nodes are normal, but occasionally there is splenomegaly.

Arterial thrombosis may be present and may add to the caudal weakness. Males are affected more often than females. The course is usually limited to months because of difficulties in management.

Hematologically there is a marked erythrocytosis with red cells in the 10–15 × 10^{12}/liter range, hemoglobin of 200–250 g/liter, and a hematocrit of 65–80%. The red cells are normochromic and normocytic unless there are thromboses, when poikilocytes will be present. Polychromasia or reticulocytes are usually not present, and the

platelet count is low or normal. About one-third of cases have a mature neutrophilic leukocytosis of up to 30 × 10^9/liter. Normal blood oxygen levels rule out secondary polycythemia. The plasma volume is reduced, but the red cell mass is increased and may be doubled. Plasma iron turnover is greatly increased and red cell life span is essentially normal.

Aspirated marrow has synchronous trilineage hyperplasia with cell proportions near a myeloid : erythroid ratio of 1.0. The cellularity is very high with little remaining fat.

Postmortem findings include generalized congestion and cyanosis, and arterial thrombosis may be present. The spleen is enlarged in about 10% of cases and bone marrow is diffusely reddened. Microscopically there is dilation of hepatic sinusoids with hepatocellular atrophy and hyperpigmentation. The marrow has pancellular hyperplasia with high cellularity and reduced iron. There is splenic sinusoidal distension and congestion compatible with increased blood volume.

Polycythemia vera must be distinguished from secondary polycythemia, which is less rare and usually due to vascular anomalies causing anoxia. Rarely, primary or metastatic tumor in the kidney may also result in an erythropoietin-dependent polycythemia.

Bibliography

Eid, J. *et al.* Intracellular growth factors in polycythemia vera and other myeloproliferative disorders. *Proc Natl Acad Sci USA* **84:** 532–536, 1987.

Kaneko, J. J. *et al.* Iron metabolism in familial polycythemia of Jersey calves. *Am J Vet Res* **29:** 949–952, 1968.

McGrath, C J. Polycythemia vera in dogs. *J Am Vet Med Assoc* **164:** 1117–1122, 1974.

Reed, C. *et al.* Polycythemia vera in a cat. *J Am Vet Med Assoc* **157:** 85–91, 1970.

V. Myelodysplastic Syndromes

The myelodysplastic or dysmyelopoietic syndromes are a diverse group of clonal disorders of hematopoiesis characterized by refractory cytopenias accompanied by qualitative and quantitative dysplastic changes in marrow cells, indicating ineffective blood cell production. About 20% of human cases terminate in acute myeloid leukemia and, in animals, most cases terminate with marrow failure. These diseases were first regarded as maladies of old age; however, it now appears that myeloproliferative disease in the young is preceded by a prodromal phase of myelodysplasia that soon evolves into acute myeloid leukemia. Thus myelodysplasia is a syndrome of clonal instability that tends to evolve into acute myeloid leukemia with a high probability of progression in the young and a low likelihood of progression in the aged. In humans, 80% of cases are characterized by cytogenetic abnormalities, but in animals this aspect is as yet undefined.

Mechanistically, the stem cells in myelodysplastic syndromes are less responsive to granulocyte–macrophage (GM)-CSF than normal, whereas they are hyperresponsive in acute myeloid leukemia. The term myelodysplasia replaces the previous label of preleukemia and is a preferable designation for an animal which might otherwise survive the disease but not the diagnosis. Five variants of ineffective myelopoiesis are recognized: refractory anemia with and without ringed sideroblasts; refractory anemia with excess blasts, and in transition to acute leukemia; and chronic myelomonocytic leukemia. The major utility of this classification is that the first two variants tend to have long survival whereas the other variants do not.

The bone marrow in these syndromes is most often hypercellular, but hypocellular variants do occur. Architectural abnormalities in marrow include topographic distortion, with loss of the normal orientation of myelopoiesis near endosteum and centripetal erythroid islands, to a random distribution of lineage. There is an irregular increase in stroma, which may be diffuse and terminally accompanied by ectasia of sinuses with intravascular hematopoiesis. Plasmacytic cuffing of small vessels is increased and there is increased remodeling of bone.

Most characteristically, there is usually some combination of peripheral blood cytopenia. Monocytosis is usually prominent as are poorly granulated and bilobed neutrophils. There may be trilineage abnormalities in nuclear maturation in the bone marrow, or any combination of cell lines may be involved. Rubricyte alterations include megaloblastoid nuclei with delayed maturation, binucleation, and small satellite nuclei. Myeloid abnormalities include poorly granulated blast cells and myelocytic hyperplasia, whereas megakaryocytes have hypolobulated nuclei with errors in fusion and decreased cytoplasmic volume.

The usual clinical presentation is of a thin lethargic animal with pale mucous membranes, and a history of chronic recurrent infections often involving the respiratory tract. Whereas refractory anemias are relatively common in the cat and dog, the most common type of myelodysplasia in animals is a form of refractory anemia with excess blast cells in the cat.

A. Refractory Anemia with Excess Blast Cells

This is a disease of humans and cats which, in the latter, appears to be a nonleukemic form of refractory anemia associated with feline leukemia virus infection. There is nonresponsive pancytopenia and hyperplastic marrow with excess small myeloblasts. It occurs in young mature cats without apparent sex predisposition. The course is 1–6 months, occasionally followed by recovery. Most animals are killed because of recurrent infections due to marrow failure.

The anemia is moderate to marked, mildly hypochromic and macrocytic with anisocytosis but no polychromasia. The leukocyte count is generally below 5×10^9/liter and there is neutropenia with moderate left shift. The neutrophils have reduced secondary granulation, diffuse basophilia, and toxic vacuolation. There is usually moderate thrombocytopenia with a lot of variation in size, mild uniform basophilia, and poor granulation.

Aspirated marrow is characteristically 60–80% cellular as determined by the proportion of cellular to fatty areas. There is early asynchronous maturation with as many as 10% blast cells, loss of granulocyte reserves, and an apparent maturation arrest at the myelocyte–metamyelocyte level. Most of the blasts are myeloid, but they are poorly granular and may be mistaken for lymphocytes. The erythroid nuclei show marked contrast between large chromocenters and clear areas, and dyserythropoiesis with multinucleation is present to a variable degree. Since the early myeloid hyperplasia is not followed by mature reserves, the myeloid : erythroid ratio remains near 1. The megakaryocytes are normal or decreased but not increased. They have large monocytoid nuclei without normal lobulation. Cytoplasmic basophilia remains high in maturing cells and granulation is decreased. Marrow iron is coarse and increased.

At postmortem there is pallor and emaciation with mild splenomegaly. Lymph nodes are normal or mildly enlarged. Secondary lesions of pleuritis, pneumonia, and lymphadenitis are usually prominent. The bone marrow is uniformly pale, pink, and opaque without serous atrophy and is moderately cohesive.

Histologically the bone marrow has high cellularity and focal absence of fat. Cell density may be variable due to reticulin sclerosis. Cytologically, 60% of the cells present appear primitive with round vesicular nuclei, prominent single nucleoli, and a thin rim of highly basophilic cytoplasm. Erythropoiesis is patchy and few mature rubricytes are present. Dyserythropoiesis is more prominent than on cytological preparations. Megakaryocytes appear immature due to irregular lobulation and low cytoplasmic volume. Marrow germinal centers may be present and numerous. The liver usually has some degree of periacinar ischemic degeneration, but hematopoiesis is minimal or absent. There is splenic follicular hyperplasia and increased sinus histiocytes. There are some paratrabecular megakaryocytes, but hematopoiesis is not prominent. The lymph nodes have follicular hyperplasia with medullary follicles and medullary cord hyperplasia. Neutrophils are not prominent either in sites of inflammation or in their draining nodes because of the neutropenia.

B. Chronic Myelomonocytic Leukemia

Chronic myelomonocytic leukemia (preleukemia syndrome) is largely a disease of dogs older than 5 years.

There is a history of anorexia, lethargy, weight loss, and reduced exercise tolerance. Affected dogs may have a poor hair coat and mucosal pallor, and often oral ulcerations. There is usually mild recurrent pyrexia with poor healing of minor injuries. The course is 6 months to 1–2 years. Some dogs develop acute leukemia and most are killed because of recurrent infections and thrombocytopenic hemorrhage. Complete recovery probably does not occur.

Dogs with chronic myelomonocytic leukemia have a mild to moderate nonresponsive anemia with mild anisochromia, anisocytosis, poikilocytosis, and phthisic rubricytosis. Most animals are thrombocytopenic with fewer than 100×10^9/liter cells that have a relatively uniform basophilia and normal granulation. The leukocyte count is usually about 5×10^9/liter but may be as high as 30×10^9/liter if animals are given steroids. Typically, there is neutropenia with about 1×10^9/liter cells, often with an equal number of cells as immature as myelocytes. Secondary granulation is reduced and there is at least moderate toxemia and often Doehle bodies. Dogs with low numbers of neutrophils and little immaturity tend to have some hypersegmentation. Lymphocytes are low normal. There are characteristically $1-4 \times 10^9$/liter immature monocytes. Some are as young as promonocytes and have nucleoli.

Aspirated marrow is always hypercellular and resembles refractory anemia with excess blast cells (Fig. 2.22A). There is early myeloid hyperplasia and loss or absence of granulocyte reserves, which paradoxically may be concurrent with severe neutropenia. There is at least relative erythroid and megakaryocytic phthisis, and qualitative changes suggest ineffective hematopoiesis (Fig. 2.22B). Marrow iron is coarse and increased.

At postmortem there is anemia with normal lymph nodes and mild splenomegaly. Thrombocytopenic hemorrhage may be present. The microscopic findings are similar to those in refractory anemia with excess blast cells.

Esterase stains may be required to identify lineage (Fig. 2.23) if there is atypia and immaturity in both the neutrophil and monocytic systems (Table 2.2). Chronic myelomonocytic leukemia can be differentiated from acute myelomonocytic leukemia on the basis of a longer course and lack of blast cells in the peripheral blood.

Bibliography

Baker, R. J., and Valli, V. E. O. Dysmyelopoiesis in the cat: A hematological disorder resembling refractory anemia with excess blasts in man. *Can J Vet Res* **50:** 3–6, 1986.

Nand, S., and Godwin, J. E. Hypoplastic myelodysplastic syndrome. *Cancer* **62:** 958–964, 1988.

Testa, N. G. *et al.* Haemopoietic colony formation (BFU-E, GM-CFC) during the development of pure red cell hypoplasia induced in the cat by feline leukaemia virus. *Leukemia Res* **7:** 103, 1983.

Yoshida, Y. Biology of myelodysplastic syndromes. *Int J Cell Cloning* **5:** 356–375, 1987.

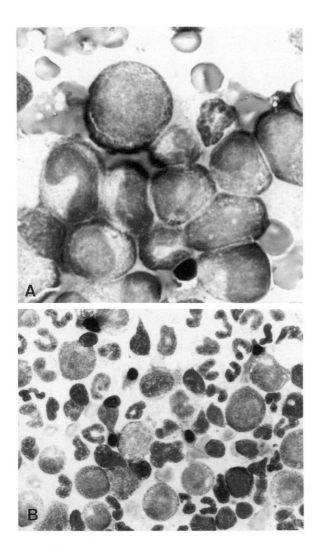

Fig. 2.22 Chronic myelomonocytic leukemia. Dog. (A) Area of early myeloid asynchrony with giant metamyelocytes. (B) Marrow aspirate. Paradoxical myeloid hyperplasia with increased mature granulocyte reserves and moderate erythroid phthisis.

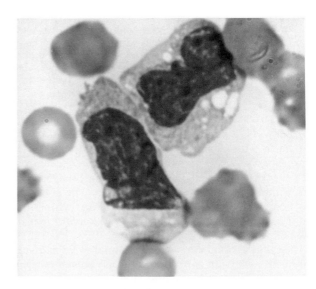

Fig. 2.23 Chronic myelomonocytic leukemia. Dog. Blood. Atypical monocytoid cells which, on esterase staining, appear to be of the neutrophil series.

C. Myeloid Metaplasia with Myelofibrosis

Myeloid metaplasia is a convenient term for those cases characterized by inappropriate myeloid hyperplasia with synchronous maturation and without excess blast cells in which a leukemic diagnosis cannot be justified.

Myeloid metaplasia occurs in dogs and cats. It is nearly always insidious in onset and there is a history of reduced activity and weight loss. There is mucosal pallor and often minor ulcerations. Lameness associated with deep bone pain may be present as in myeloma, and the spleen is usually enlarged. The course is prolonged, generally 6 months to a year or more, and usually ends with euthanasia because of intractable anemia and recurrent infections.

Anemia is a constant finding with variable degrees of hypochromia, anisocytosis, and characteristic teardrop-shaped red cells. Polychromasia is inappropriately mild, but relatively mature rubricytes as well as metamyelocytes are present as part of the leukoerythroblastic reaction. The leukocyte and platelet counts are increased in early cases and decreased later. Hypersegmented neutrophils and toxemic changes are usually present. Platelet size and shape are variable. Marrow aspirates are usually fibrillar and sparsely cellular with numerous bare stromal cell nuclei. The bone marrow is uniformly reddened and shells out of the femoral cavity in a cohesive cylinder.

Microscopically the bone marrow loses its normal tendency for artefactual cracking during histologic processing and is architecturally solid with few and irregular fatty areas. Cell density is reduced as is sinusoidal area. The intercellular interstitium is pale, proteinaceous, and not obviously fibrillar. In some areas, the specialized cells may be arranged in single file rows suggesting containment by reticulin sclerosis. Only a few animals survive until there is obvious collagen birefringence. With reticulin stains, the fiber proliferation is seen to be diffuse and spreading out from its normal perivascular location to invest individual marrow cells (Fig. 2.24). As in myelodysplastic syndromes, there is often venous ectasia with intravascular hematopoiesis. Stromal nuclei are increased, and identifiable by their oval shape and indefinite cell boundaries. By the time collagen is demonstrable, hematopoiesis is absent and the extent of prior myeloid hypertrophy is indicated by hemosiderin-bearing macrophages surviving in poorly cellular areas of collagen matrix. The remaining hematopoiesis is largely granulocytic and megakaryocytic, with little erythropoiesis, and scattered foci of small lymphocytes. Basophilic cells, which are prominent in early aspirates, are present but inconspicuous. The spleen has follicular preservation and occasionally prominent hypertrophy. The sinus areas are solidly cellular and have obvious trilineage hematopoiesis. The liver has both periportal and sinusoidal colonies of hematopoietic cells, and similar foci with varying lineage can be found irregularly in almost all tissues except muscle. The lymph nodes are variable

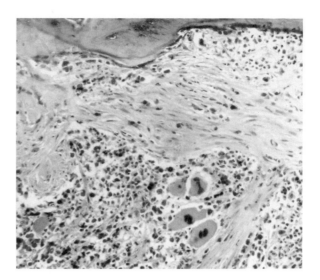

Fig. 2.24 Myelofibrosis in dog marrow causing myelophthisic anemia.

but usually enlarged with thickened capsules, medullary follicles, focal hematopoiesis, and generalized sclerosis. This tendency for fibrogenesis to follow hematopoiesis is also evident in the spleen and liver, and suggests that the prolonged hyperplasia is attended by release of mediators related to the megakaryocytic system and platelet-derived growth factor. Unlike the myelodysplastic syndromes, myeloid metaplasia regularly terminates in myelofibrosis and not in acute leukemia. Myeloid metaplasia must be distinguished from reactive hyperplasia, in which there is a target for the increased cellular output, and from myelodysplasia, in which there is early asynchrony with increased blasts.

Bibliography

Burkhardt, R. et al. Chronic myeloproliferative disorders (CMPD). Pathol Res Pract 179: 131, 1984.

Demory, J. L. et al. Cytogenetic studies and their prognostic significance in agnogenic myeloid metaplasia: A report on 47 cases. Blood 72: 855–859, 1988.

VI. Lymphoproliferative Disease

Lymphoproliferative disease may be defined as neoplastic proliferation of lymphocytes causing lymphoma or lymphocytic leukemia. In general, the lymphoproliferative diseases form a spectrum with lymphoma and leukemia at their extremes, and it is not always possible to determine which came first. The major distinction between leukemia and lymphoma is that in leukemia there are neoplastic cells in bone marrow and in circulation, and in lymphoma there are enlarged peripheral lymph nodes with a relatively normal hematologic picture.

In general, animals with leukemia are presented with some degree of marrow failure characterized by anemia, thrombocytopenia, or neutropenia, which occurs when 50% or more of the bone marrow is involved by tumor. Under these circumstances, the blood and bone marrow are always diagnostic. In contrast, the lymphomas which involve peripheral tissues tend to leave the bone marrow relatively uninvolved and, at the time of diagnosis, there are usually normal hemoglobin, platelet, and leukocyte levels. In 15–20% of animals with lymphoma, it is possible to identify neoplastic cells in the peripheral blood on the basis of cytologic appearance. It is likely that there are always tumor cells in the peripheral blood in lymphomas, even though the total lymphocyte count is not elevated and may be reduced.

A. Lymphocytic Leukemia

1. Acute Lymphocytic Leukemia

In lymphoproliferative disease, malignancy in an ontogenically primitive cell is likely to occur in the bone marrow of a young individual, thus presenting as a leukemia. In contrast, clonal autonomy in a mature lymphocyte is likely to occur in the peripheral tissues in a mature individual with presentation as a lymphoma. Acute leukemia of the lymphoid type (ALL) may be of B or T cell origin. Bone marrow will be heavily involved in both cases but in the case of acute T cell leukemia there is often primary tumor in the thymus with very early and rapid spread to the bone marrow. In contrast to the acute myeloid tumors, which tend to have sufficient morphologic differentiation to permit their histogenesis to be determined, the lymphoid leukemias cannot be classified into B and T cell types on routine stains.

The morphologic characteristics of the acute lymphocytic leukemias are based on nuclear : cytoplasmic ratio, nuclear size and shape, number and size of nucleoli, and cytoplasmic basophilia and vacuolation (Table 2.3). Acute lymphocytic leukemia is also classified immunologically, in terms of degree of differentiation, and cytogenetically, as well as by etiology and prognosis. In human acute lymphocytic leukemia, the disease in children is very predominantly (85%) of the so-called L1 type and a high proportion of these cases attain durable remissions through aggressive therapy. Immunologically, the disease in children is predominantly of early B type with rearrangement of the immunoglobulin gene. In contrast, acute lymphocytic leukemia in adults is predominantly of the L2 type with chromosomal abnormalities occurring in two-thirds of cases, and the response to therapy is much poorer than that achieved in children. The L3 type is the leukemic equivalent of the Burkitt lymphoma cell and is more prone to abdominal and central nervous system spread. Prognosis tends to decline from L1 to L3. Despite their aggressive appearance, the proliferative fraction of these tumors is actually less than that of their normal counterparts and therefore cell cycle-dependent cytoreductive therapy has the potential

to injure the host more than the tumor. In contrast to the acute myeloid leukemias, which compete directly for space with benign stem cells in subendosteal sites, the pattern of lymphoid colonization of marrow is random and therefore the acute lymphocytic leukemias tend not to result in marrow failure as rapidly as the acute tumors of the myeloid system.

Since the pathology of these cell types is ill defined in animals, the diseases are presented as an entity. Acute lymphocytic leukemia in animals is not well classified immunologically and, whereas most are likely to be B cell tumors, many appear to be of the null type.

Acute lymphocytic leukemia occurs in all species. There is a wide age range and apparently some predisposition for intact males. The disease is seen most often in calves and cats younger than a year and in dogs younger than 5 years. It is assumed that most hematopoietic neoplasms in the cat are virus associated but, in the calf younger than a year, malignant lymphoproliferation is not associated with bovine leukemia virus infection.

With the exception of calves, animals are presented with minimal lymph node enlargement. There is some degree of mucosal pallor and splenomegaly, but hemorrhage is less common than with the acute myeloid leukemias. The course is 1–2 months.

Anemia at presentation is usually mild but tends to be more pronounced in cats. Red cells are normochromic, normocytic, without polychromasia, and there is an occasional poikilocyte. Dogs characteristically have Howell–Jolly bodies, indicating splenic involvement. The total leukocyte count is generally between 80 and 100×10^9/liter, but may be low or normal. The differential count is 80–100% lymphocytes, and leukoerythroblastosis is usually not present. Neutrophils are usually mature and often hypersegmented with reduced secondary granulation. Metamyelocytes are occasionally seen, but more immature cells are unusual, as are rubricytes. The leukemic cells may be small and uniform with inapparent nucleoli (L1), or large with irregularly cleft nuclei (L2), or large and of uniform size and shape with prominent nucleoli and highly basophilic cytoplasm, often with vacuolations that stain positively for lipid (L3) (Fig. 2.25A,B). A perinuclear clearing of basophilia with peripheral concentration is characteristic. A type of large granular lymphocyte tumor occurs in the Fischer rat and in dogs and cats and is considered with chronic lymphoid leukemia. The L1–3 cell types are characteristic of the majority of the acute lymphoid leukemias seen in animals. A further type of lymphoid leukemia formerly called **reticuloendotheliosis** and now **hairy-cell leukemia** occurs in humans and is related to infection with the human T-cell lymphotropic virus (HTLV)-II. A counterpart of this disease has been described in cats but appears to be very uncommon. This disease has a very variable course in humans but is generally considered with the chronic leukemias. Since survival in the cat is a matter of weeks after diagnosis, this entity is appropriately considered in the diagnosis of acute lymphocytic

TABLE 2.3

Cytologic Correlates in Acute Lymphoid Leukemia

WHO FAB	Microlymphoblastic L1	Prolymphocytic L2	Lymphoblastic L3
Incidence (%)			
Humans	80	15	5
Animals	10	60	30
Cytologic features			
Cell size	Small predominantly	Large, heterogeneous	Large, homogeneous
Chromatin	Homogeneous	Heterogeneous	Fine and homogeneous
Nucleoli	Absent or small	Absent or small	Prominent
Cytoplasmic volume	Low	Moderate	Moderate
Cytoplasmic basophilia	Moderate	Moderate	Strong
Cytoplasmic vacuolation	Absent	Absent	Prominent
Nuclear shape	Round, rarely cleft	Irregular clefting common	Regularly oval to round

leukemia. Morphologically the tumor cells are characterized by thin cytoplasmic protrusions, which are best seen in wet mounts, and cytochemically by reaction for the tartrate-resistant acid phosphatase (TRAP) stain. Pathologically the tumor has specificity in a broad rim of clear cytoplasm and by the manner in which it dilates, colonizes, and lines hepatic sinusoids, as well as by the portal cuffing more typical of the lymphoid leukemias.

In general, in the acute lymphoid leukemias, the platelets are usually fewer than 100×10^9/liter and often severely deficient. Immature platelets are rare or absent, but there may be confusion with agranular cytoplasm shed from injured lymphocytes.

Aspirated marrow is highly cellular with marked immaturity and trilineage phthisis. The bone marrow lymphocytes are characteristically slightly smaller and more uniform than those in the blood, possibly because of senescence in some of the latter. At postmortem there is usually only mild pallor and loss of condition, except in calves, which may be cachectic and have marked symmetrical lymph node enlargement. Other species have only mild or irregularly enlarged nodes. The cervical and thoracic viscera are usually normal, except as noted for the calf. The liver is markedly enlarged in the calf with irregular pale areas and a lobular pattern on cut surface. In the cat and dog, the liver may be mildly enlarged. It is usually normal in other species. The spleen is moderately and symmetrically enlarged and is dry and fleshy on cut surface, occasionally with focal pale areas. The femoral bone marrow is uniformly reddened in cats and may have foci of residual fat in dogs and horses. Calves have almost solid infiltration of all marrow cavities and there are usually large areas of infarction with yellow areas of necrosis surrounded by hyperemia. This degree of marrow involvement may be present in calves born with the tumor.

Microscopically, the bone marrow has 90–100% cellularity with a solid field of mononuclear cells (Fig. 2.26).

These are usually of medium size with a variable but minor proportion of large and small cells. Nuclei are open and vesicular with central nucleoli, coarse granular chromatin, and irregular thickening of the nuclear membrane. Cytoplasm is moderate in amount and deeply amphophilic. Only an occasional megakaryocyte and rubricyte can be found, and granulocyte reserves are absent. Lymph nodes have intact architecture with follicular atrophy. The cortical cell density is usually low, with tumor cells replacing the small paracortical cells. Postcapillary venules are tightly packed with malignant cells. The spleen has a thinned capsule, follicular atrophy, and solidly cellular sinus areas. There may be foci of necrosis. Benign hematopoiesis is minimal. Subendothelial colonization of large muscular sinuses is well developed. The liver has diffuse sinusoidal and portal colonization. The tumor burden in the sinusoids roughly indicates the level of leukemia. There is irregular sinusoidal dilation, which contributes to the hepatomegaly, and usually some degree of periacinar ischemic degeneration. Foci of tumor may be widespread, but are most common in the kidney, testes, meninges, bowel, and pancreas.

Acute lymphocytic leukemia can be distinguished from acute myeloblastic leukemia by the lack of cytoplasmic granulation and reaction to esterase stains. The most difficult differentiation is from leukemic lymphoma of similar cell type. Generally the leukemias present with marrow failure and minimal lymphadenopathy, whereas the lymphomas have a more normal neutrophil and platelet picture, and may not be anemic unless complicated by enteric hemorrhage. In lymphoma, there are usually localizing signs of peripheral involvement including lymph node enlargement, central nervous system deficits, or melena.

Bibliography

Barcos, M. *et al.* An autopsy study of 1206 acute and chronic leukemias (1958 to 1982). *Cancer* **60:** 827–837, 1987.
Gilmore, C. E., Gilmore, V. H., and Jones, T. C. Reticuloendo-

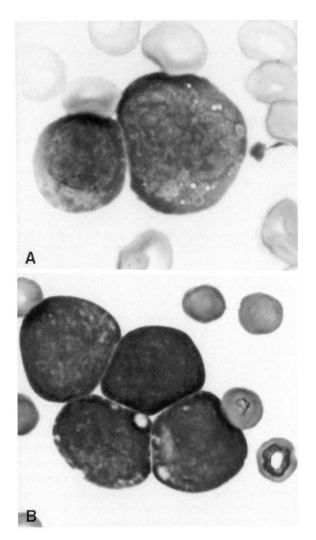

Fig. 2.26 Acute lymphocytic leukemia. Marrow aspirate. Hypercellularity and complete phthisis of normal marrow cells. Fine cytoplasmic vacuolations are characteristic of this cell type.

Fig. 2.25 Acute lymphocytic leukemia. (A) Canine blood. Large granular lymphocyte type. Nuclei are irregularly cleft and indented, and large cytoplasmic lysosomes are apparent in the cell at right. (B) Feline blood. Hyperchromatic medium and large lymphocytes.

theliosis, a myeloproliferative disorder of cats: A comparison with lymphocytic leukemia. *Pathol Vet* **1:** 161–183, 1964.

Leifer, C. E., and Matus, R. E. Lymphoid leukemia in the dog. Acute lymphoblastic leukemia and chronic lymphocytic leukemia. *Vet Clin North Am Small Anim Pract* **15:** 723–739, 1985.

Matus, R. E., Leifer, C. E., and MacEwen, E. G. Acute lymphoblastic leukemia in the dog: A review of 30 cases. *J Am Vet Med Assoc* **183:** 859–862, 1983.

2. Chronic Lymphocytic Leukemia

Chronic lymphocytic leukemia (CLL) is a proliferative disease resulting in the accumulation of a clonal population of B lymphocytes that are morphologically mature but biologically immature. The basic criteria for diagno-

sis are persistent lymphocytosis of mature cells, cell-surface immunoglobulin, and marrow involvement with more than 30% of tumor cells. Immune dysfunction is slowly progressive and hypogammaglobulinemia eventually develops, and a small proportion of cases may also have autoimmune disease, usually evident as immune hemolysis. Factors of prognostic value are the pattern of marrow infiltration, the blood lymphocyte doubling time, and specific karyotypic abnormality. In contrast, staging systems based on enlargement of liver, spleen, and nodes plus degree of lymphocytosis, anemia, and thrombocytopenia are less predictive of progression. Most cases maintain a stable cell type, but in human chronic lymphocytic leukemia a small number of cases evolve into acute lymphocytic leukemia, myeloma, or an aggressive large cell lymphoma termed **Richter's syndrome.** Marrow involvement may be focal or nodular, interstitial, a combination of the two, or diffuse, with this sequence representing the order of progression and decreasingly favorable prognosis. In contrast, it appears that well-differentiated lymphoma of the same cell type as chronic lymphocytic leukemia involves the marrow in a much lower proportion of cases, and predominantly in a focal manner.

Chronic lymphocytic leukemia of the **large granular lymphocyte type** occurs in a high proportion of aged F344 rats and rarely in Sprague–Dawley rats, cats, dogs, and cattle. The large granular cells are derived from a T cytotoxic/suppressor population and tend to colonize liver, spleen, and marrow but not nodes. A further chronic lymphocytic leukemia of the T cell type occurs in the dog, horse, and cow as part of the syndrome of mycosis fungoides, with the leukemic form termed Sézary syndrome. The Sézary cell as it appears in the dermis and in the blood has a markedly convoluted nuclear membrane, which requires ultrathin sectioning for full delineation. In the Sézary syndrome, the prognosis varies inversely with the number of these cerebriform cells which appear in the blood.

This is primarily a disease of cats, cattle, and dogs, usually 8–10 years of age or older. It is an insidiously progressive disease with few localizing signs. The lymph nodes are of normal size, but there is always splenomegaly. There is probably a prodromal period of years and some cases have a late conversion to acute lymphoblastic leukemia. Termination as Richter's syndrome or the conversion to a large-cell lymphoma is reported in animals but is not well documented.

There is a severe lymphocytosis that may range from 50 to 600 × 10⁹/liter. The anemia is moderate and is normochromic with mild anisocytosis and poikilocytosis, occasional Howell–Jolly bodies and no polychromasia. Rubricytes are rarely seen. The lymphocytes constitute 95–100% of the cells seen on a differential count, but a few neutrophils are usually present. The tumor cells are primarily small and medium with an occasional large cell, and 25% or more may be basket cells despite good preparative techniques. The nuclei are round, dense, and coarsely retiform with little internal detail. Cytoplasm is scant and moderately basophilic. In some tumors, the cells have abundant cytoplasm with little basophilia and these may be more slowly progressive. The fact that these cells are malignant emphasizes that cytoplasmic volume and degree of basophilia are not criteria of malignancy in lymphocytes. Nucleoli are present and multiple but are seen easily only in the larger cells, and in the swollen burst nuclei. Platelets have normal morphology, which appears to correlate with normal function, since bleeding is unusual, although the counts are usually below 50 × 10⁹/liter. The aspirated marrow is hypercellular with almost 100% lymphocytes that are

small and have less size variation than those in the blood. Phthisis of normal elements is virtually complete when the disease is well advanced. Marrow iron is not increased.

Pathologically there is mild pallor and usually abundant abdominal fat. The spleen is markedly enlarged, soft and fleshy, deep red, and dry on the cut surface. The liver is at least mildly enlarged and pale brown with a lobular pattern. There may be focal white areas of solid tumor up to 5.0 mm in diameter in spleen, liver, and renal cortices. The nodes are irregularly small or moderately enlarged, but usually are normally mobile and are never markedly enlarged. The femoral bone marrow is uniformly pink and usually friable.

Microscopically the bone marrow has greater than 90% cellularity and presents a solid field of small cells

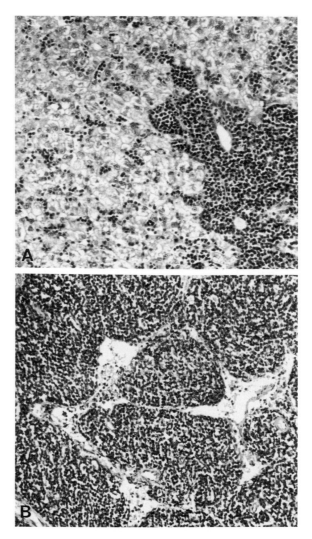

Fig. 2.28 (A) Chronic lymphocytic leukemia. Heavy periportal infiltration of tumor with sinusoidal colonization indicating hyperleukemic state. (B) Chronic lymphocytic leukemia. Lymph node medulla. Note homing of small lymphocytes to medullary cords.

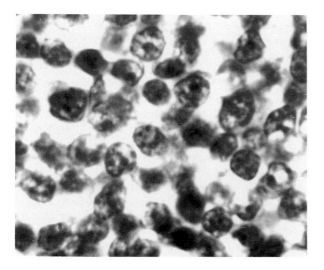

Fig. 2.27 Chronic lymphocytic leukemia. Histology of feline marrow.

with occasional fat vacuoles and megakaryocytes (Fig. 2.27). Cytologically, the tumor nuclei are slightly larger than red cells and have dense chromocenters with irregular incomplete clearing of parachromatin areas and inapparent nucleoli. Cytoplasm is usually scant, and cell packing is uniformly high. The liver has tumor infiltration around virtually all triads (Fig. 2.28A). The sinusoidal tumor cells are less apparent than in acute lymphocytic leukemia despite a higher peripheral blood level. The spleen has a uniform appearance with few residual follicular areas. The sinus areas resemble marrow with diffuse tumor infiltration. Benign hematopoiesis is not obvious either in spleen or liver. The nodes are involved to a variable extent and some are atrophic with sinus histiocytosis. Involved nodes have a thinned but intact capsule with a diffuse cortical infiltration and an occasional residual nodule of benign small cells. Postcapillary venules are dilated and have a high level of luminal and transmural traffic, and may be evident at low magnification as dark foci of tightly packed nuclei. The tumor cells occupy the medullary cords (Fig. 2.28B) and not the sinuses. There is often arteriolar dilation in lung and kidney and many lymphocytes in capillaries. Focal lymphocytic colonization may be found in most tissues, including brain and eye.

Chronic lymphocytic leukemia can be distinguished from leukemic lymphoma of small-cell type by the mild degree of lymph node involvement. The persistent lymphocytosis of benign bovine leukemia virus infection in cattle seldom exceeds 20×10^9/liter, and the cells are more variable in size and shape with a benign homogeneous retiform chromatin pattern.

Bibliography

Agnarsson, B. A. *et al.* The pathology of large granular lymphocyte leukemia. *Hum Pathol* **20:** 643–651, 1989.

Foon, K. A., Rai, K. R., and Gale, R. P. Chronic lymphocytic leukemia: New insights into biology and therapy. *Ann Intern Med* **113:** 525–539, 1990.

Harvey, J. W. *et al.* Well-differentiated lymphocytic leukemia in a dog: Long-term survival without therapy. *Vet Pathol* **18:** 37–47, 1981.

Suster, S., and Rywelin A. A reappraisal of Richter's syndrome. *Cancer* **59:** 1412–1418, 1987.

B. Lymphoma

1. Histologic Interpretation of Lymph Nodes

Lymphoid tissue, like marrow tissue, must be evaluated in the light of its normal architecture and how the component structures are altered in the various types of developmental, degenerative, inflammatory, and hyperplastic processes. The major anatomic structures of the lymph node are the capsule, peripheral sinus, cortex, paracortex and postcapillary venules, medullary cords, and medullary sinuses with associated vascular structures.

A node which has increased rapidly in size will have a **capsule** that is taut and thinned and the **peripheral sinus** will be compressed. If the node has decreased in size, the capsule will be thickened and wavy and the peripheral sinus will open. With chronic inflammation, the capsule becomes thickened, and all fibrous trabeculae, particularly in the medulla, become more prominent.

Germinal centers are normally cortical and subcapsular, and with stimulation may occupy the entire cortex and medulla. These centers have a polarity with a superficial pole that is always directed to the source of antigen whether in node, tonsil, or Peyer's patch. Thus the superficial pole composed of small lymphocytes will be adjacent to the peripheral sinus, whereas the deep pole occupied by large lymphocytes will be deeper in the cortex. The whole is surrounded by a mantle of small lymphocytes so that, three-dimensionally, the germinal center has an architecture like an egg. A section at the edge of a germinal center will therefore appear to be composed of small mantle cells which may be misinterpreted as a lymphoid nodule. Hence the distinction between primary or primordial nodules and secondary or fully developed germinal centers needs to be made on the basis of the overall development of the node and not on an individual focus of lymphocytes.

The germinal centers are thus cytologically heterogeneous and this variation in cell type across the functional areas is an important distinction between benign follicles and follicular lymphomas, which are more homogeneous. The follicular centers contain antigen-focusing **dendritic cells** which are inapparent in normal reactions with a high degree of cellular packing. In stressful conditions that cause lymphoid atrophy, these dendritic cells become more apparent and form the matrix for fibrinoid deposition called **follicular hyalinosis.** There is normally a high rate of mitosis in the deep pole of the germinal centers and a high rate of cell death evident as nuclear debris in tingible body macrophages. The latter large cells provide the clear areas in the densely packed cells of diffuse lymphomas giving the tissue the starry-sky effect on low power examination, and are an indication of rapid cell turnover.

The **paracortex** is composed of small thymus-derived lymphocytes in quiescent nodes, and a mixture of large and small lymphocytes and phages in early response to antigen before the reaction becomes follicular about 10 days after antigen exposure. At this early diffuse stage of cortical hyperplasia, the tissue has a moth-eaten appearance, which has been mistaken for diffuse lymphoma in postvaccinal states.

Within the paracortex and between the germinal centers are the **postcapillary venules,** which are obscured in quiescent nodes and become thick walled with prominent vesicular nuclei in response to antigen stimulation and increased cell traffic, as the lymphocytes from the thoracic duct drainage leave the venous system and reenter the lymphoid tissue. Specific homing receptors on the endothelium and lymphocytes direct the latter to the proper tissue for their level of differentiation and stimulation. The homing tendency is mimicked in differentiated lymphomas, which may tightly and preferentially colonize the medullary cords or intestinal lamina propria. The distinction between **follicular lymphoma** and a pseudonodular proliferation can be made on the location of the postcapillary venules, which are always between follicles in a follicular lesion and within nodules in reactions that are truly diffuse.

The **medullary cords** normally contain small and medium lymphocytes and in chronic follicular activity, they may be solidly occupied by plasma cells, as regularly occurs in nodes associated with the salivary glands and mediastinum in rats, and in the mammary nodes of dairy cattle. The **medullary sinuses** consist of a loose meshwork of reticular fibers, which cross-link the medullary cords and support the vessels arborizing from the node hilus. In quiescent states, the medullary sinuses are relatively empty while, with stimulation and increased cellular drainage from the peripheral sinus, the meshwork becomes filled with macrophages and other cells in transit. The **cellular component** of the peripheral sinus and the medullary sinuses indicates the type of reaction that is occurring in the tissue drained by the node. Thus in septic states there will be neutrophils and some red cells in both areas.

In biopsied nodes that were inadequately isolated prior to extirpation, there will be many red cells in both areas and prominent erythrophagocytosis. Following hemorrhage or transfusion, there will be not only the ingested red cells but also iron pigment. Nodes in the inguinal area of neonatal swine usually contain iron in medullary phages following parenteral iron administration. Metastatic carcinoma cells will appear first in the peripheral sinus and move to colonize the medullary channels.

Lymphoma will distort and efface normal node architecture and in particular will fill and destroy the peripheral sinus. Benign lymphoid proliferation will colonize the capsule and peripheral tissues in hyperplastic states such as equine infectious anemia and trypanosomiasis, but the peripheral sinus will remain intact. In general, the large-cell lymphomas tend to first occupy the cortex of the node and later compress the medullary structures, whereas the small-cell lymphomas home to the medullary cords and compress the adjacent sinuses. Extramedullary hematopoiesis will preferentially colonize the medullary cords as will malignant myeloproliferation, with the benign cells then displaced to the medullary sinuses. The presence or absence of plasma cells is not a useful criterion of lymphoid malignancy since the progeny of non-neoplastic follicles may coexist with lymphoma, and persist even after extensive architectural derangement.

2. Histologic Classification of Lymphoma

A system of tumor classification should be based on clinical relevance and be readily applicable to routinely processed tissue. In a practical sense, it should be based on criteria that relate to biologic behavior and are evident on routine stains, and should not be concerned with details that are not evident in the material at hand, such as B–T differentiation. Operative criteria for the lymphomas include architecture (**follicular** or **diffuse**), cell size (**small** or **large**), nuclear shape (**cleaved** or **entire**), chromatin pattern and distribution, nucleolar size, shape, and number, and cytoplasmic volume, density, and placement. The National Cancer Institute Working Formulation was devised to bring order out of six competing systems which denied oncologists the ability to communicate on specific types of tumors. This system is broadly applicable to animals and has the appeal of providing a comparative bridge on which appropriate models of animal lymphoma can be defined. In addition, it is compatible with measures to further characterize tumors on the basis of cytochemistry, phenotype, and genotype.

The precepts of the Working Formulation and other classifications are that lymphomas of small cells with low mitotic rate have a low rate of progression and thus respond poorly to cell cycle-dependent cytoreductive therapy; lymphomas of larger cells with high proliferative rates are potentially curable diseases, but in the untreated state they rapidly progress and kill the host. Between these extremes there are lymphomas with differentiated characteristics, such as follicular or mantle zone architecture, or tissue localization as in mucosal-associated lymphoid tumor, which render them recognizable as entities with an inherent intermediate rate of progression. The goal of lymphoma classification should be to accurately place a lesion in one of these categories so the oncologist can properly advise the owner, or the researcher can better understand the disease. In applying the Working Formulation system as it is used in the literature and herein, it is essential to recognize that the term **lymphoblastic** is used in a very specific sense to indicate an aggressive small-cell lymphoma with dense uniform chromatin and inapparent nucleoli. In contrast, in veterinary literature the terms blast and lymphoblast are used generically to indicate large-cell lymphomas with prominent nucleoli. This point of distinction is more than academic, since in the dog lymphoblastic lymphoma in the Working Formulation system is an intractable tumor with a tendency to paraneoplastic hypercalcemia, whereas in the generic sense, it refers to a large-cell lymphoma that regularly undergoes remission with appropriate therapy. There is thus a need to use these terms in a context compatible with the large body of information from human oncology.

In a prognostic sense, it appears that in the dog the definition of lymphoma on a cytologic basis into a low-, medium-, or high-grade category is more important than the stage of progression in terms of response to therapy. Thus, the definition of a lesion as a lymphoma without regard to cell type is an inadequate diagnosis. This assertion is supported by extensive studies in the dog and cow which demonstrate that there is a high degree of coherence between cytologic types of lymphomas, regardless of the tissue in which the tumors are found. Further, morphometric studies on human and animal lymphomas demonstrate that the cellular categories of the Working Formulation are in fact morphologic entities that vary widely in criteria that are objectively definable and separable. It is worth noting that cellular categories that are difficult to separate on subjective analysis are also those which are most similar on objective analysis. Thus the lymphomas of diffuse large-cell and small noncleaved types differ principally in size and, in the dog but not the cow, morphometry shows that these entities approach overlap, and therefore may be a continuum with a minority of cases in a gray area where they are not consistently separable by humans or machine. With that disclaimer, the cellular descriptions are given with the confidence that the Working Formulation system represents a catalog of the spectrum of lymphomas that occur in mammals generally, and is fully applicable to domestic and laboratory species of animals. It appears that there is less variation in cellular morphology of lymphomas in birds, reptiles, and fishes than in mammals.

There are a few major factors that must be considered in order for an accurate assessment of a hematologic neoplasm to be made. First, the specimens need to be collected by biopsy or as soon as possible after death. Imprint preparations are very helpful and should always accompany fixed tissue if possible. Extra imprints should be prepared and left unstained, pending the need for cytochemistry or immunophenotyping that can be accomplished only on cytologic preparations. Formalin fixation is adequate, but samples should be given 24 hr of adequate immersion prior to processing. B5 fixative gives superior versatility for immunophenotyping of paraffin-embedded tissues. It may be impossible to adequately classify a lymphoma on well-fixed and processed tissue that is 6–8 μm thick. Optimally, sections should not exceed 4 μm in thickness for adequate assessment of nuclear membranes. Although many cases may be classifiable at 400× magnification, most are not, and maximal resolution is required for a confident assessment. Mitoses are an important aspect of lymphoma interpretation and should be counted, not estimated, since accurate assessment of mitotic rate will prevent confusion between cell types in the low-, medium-, and high-grade categories. Nuclei in metaphase are easily confused with pyknosis at less than ideal resolution. Nuclear size is conveniently estimated in terms of com-

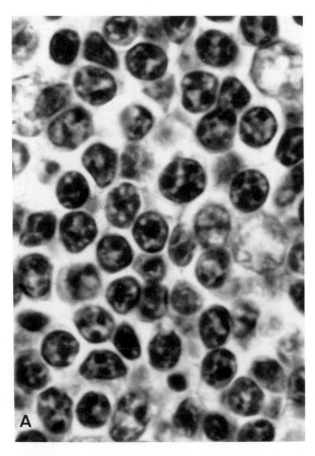

Fig. 2.29A Low-grade (well-differentiated) lymphoma. Cow. Diffuse small lymphocytic (DSL). Compact small round nuclei resembling benign cortical thymocytes.

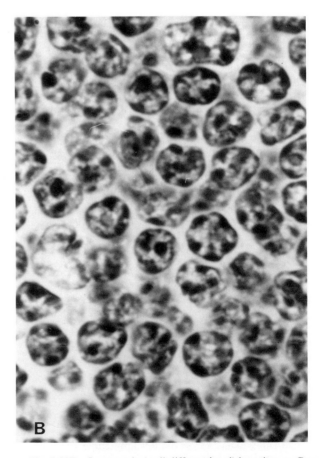

Fig. 2.29B Low-grade (well-differentiated) lymphoma. Cow. Diffuse small lymphocytic intermediate (DSLI). Nuclei are slightly larger than those of DSL. Absent nucleoli and mitoses.

parison with red cells. Finally, the presence and degree of sclerosis should be noted as this component is characteristic of the diffuse mixed lymphomas.

3. Correlation of Cell Type and Topography of Animal Lymphomas

a. Low-Grade Lymphomas The **diffuse small lymphocytic tumors** are the most common of this group and are the peripheral counterpart of the chronic lymphocytic leukemia cell type. The cells resemble those of cortical thymocytes (Fig. 2.29A), nucleoli are not present, and mitoses are rare. Chromatin is aggregated into closely packed, large, dense chromocenters. Lymphomas of diffuse small lymphocyte type occur most often in the gastrointestinal tract. The **diffuse small lymphocytic lymphomas of the intermediate type** (Fig. 2.29B) are composed of cells 25% larger than those previously described, with slightly more internal nuclear detail. They are found in multicentric tumors as well as those of the mediastinum and enteric tract. The **plasmacytoid** (Fig. 2.29C) and **mantle zone types** (Fig. 2.29D) are distributed like diffuse small lymphocytic lymphomas of intermediate type with the plasmacytoid lymphoma occurring in a laminar infiltration sparing the

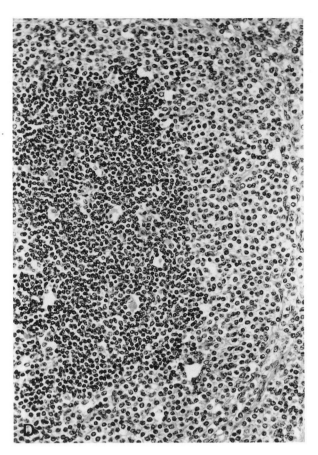

Fig. 2.29D Low-grade (well-differentiated) lymphoma. Mantle zone lymphoma. Tumor cells with compact nuclei centered in abundant pale cytoplasm surround residual foci of benign small lymphocytes in follicular center. Cow.

muscularis in the lamina propria of the small intestine in the horse. This disease is like Mediterranean lymphoma of humans and persists for a long period as a benign infiltrate that resembles inflammation. The **follicular small cleaved** (Fig. 2.30A) and **follicular mixed** (Fig. 2.30B) **tumors** are rarely seen in animals and usually appear as multicentric tumors, possibly because of their homing receptors for B cell areas.

b. Intermediate-Grade Lymphomas The **diffuse small cleaved** and **diffuse mixed lymphomas** are primarily diseases of the enteric tract that may be multicentric when diagnosed (Fig. 2.31A,B). Mitoses are rare in the diffuse small cleaved tumors, and fewer than 1 per field at 1000× is found in diffuse mixed lymphomas. In the cat, the diffuse mixed tumor characteristically appears in the enteric tract as a focal disease of ileocecal junction and associated nodes. There is sparing of the muscularis, indicating a strong homing tendency for the lamina propria. The diffuse mixed lymphoma is usually associated with diffuse fine sclerosis, and in the enteric form, surgical removal has been followed by prolonged treatment-free survival. A diffuse mixed lymphoma with sclerosis occurs in the peri-

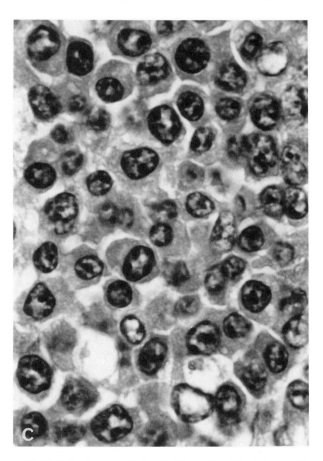

Fig. 2.29C Low-grade (well-differentiated) lymphoma. Cow. Diffuse small lymphocytic, plasmacytoid. Slightly more anisokaryosis than in DSL and with more abundant, deeply stained, eccentric cytoplasm.

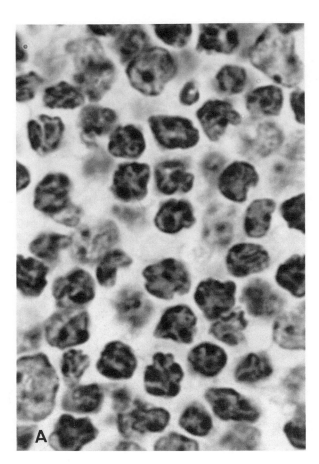

Fig. 2.30A Low-grade (well-differentiated) lymphoma. Lymphoma follicular, small cleaved-cell type (FSC). Area is in center of tumor nodule. Nuclei have compact character and angular outlines. Cow.

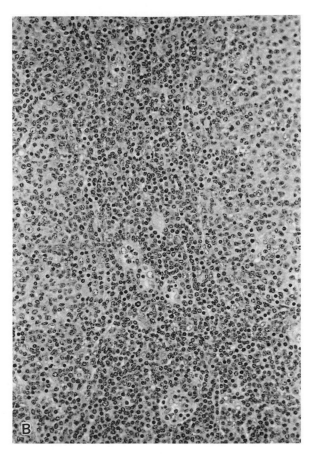

Fig. 2.30B Low-grade (well-differentiated) lymphoma. Follicular lymphoma, mixed-cell type. The postcapillary or high endothelial venules are compressed between enlarging follicles. Cow.

orbital region of the horse and tends to recur locally after removal. The **diffuse large** and **diffuse large cleaved tumors** are multicentric in the dog, cat, and cow and have the largest of the lymphoma cells in these species (Fig. 2.31C,D). In cattle, the diffuse large cleaved tumor is characteristic of the enzootic form of bovine leukosis and is likely clustered in this age group as a result of interaction with the bovine leukemia virus.

c. HIGH-GRADE LYMPHOMAS The **immunoblastic lymphomas** are characterized morphologically by a prominent single central nucleolus (Fig. 2.32) and a high mitotic rate (<13/1000× field). The tumor cells may be large or small in a single case, but retain the same appearance. About one-third of the mediastinal lymphomas of the cat are of the immunoblastic type, and many of the enteric and multicentric lymphomas of the dog and cat are also of this type. The **lymphoblastic lymphomas** (Fig. 2.33A,B) have a mitotic rate only slightly below that of immunoblastic ones and are multicentric in dogs and mediastinal in cats. In dogs, the lymphoblastic tumors are the type most likely to be associated with hypercalcemia. To a lesser extent, the lymphoblastic tumor occurs in the thymic

lymphoma of yearling cattle. Lymphomas of the **small noncleaved type** have a mitotic rate equal to that of the immunoblastic tumors and are multicentric and enteric in the dog, cat, and cow, and tend to cluster in the mediastinal tumors of the latter two species (Table 2.4).

Bibliography

Lardelli, P. Lymphocytic lymphoma of intermediate differentiation. Morphologic and immunophenotypic spectrum and clinical correlations. *Am J Surg Pathol* **14:** 752–763, 1990.

The Non-Hodgkin's Lymphoma Pathologic Classification Project. National Cancer Institute-sponsored study of classifications of non-Hodgkin's lymphomas. Summary and description of a working formulation for clinical usage. *Cancer* **49:** 2112–2135, 1982.

Van Baarlen, J., Schuurman, H-J., and Van Unnik, J. A. M. Multilobated non-Hodgkin's lymphoma. *Cancer* **7:** 1371–1376, 1988.

4. Bovine Lymphoma

The general pathology and classification of lymphoma are discussed in the preceding sections. The following sections deal with the epidemiologic and topographical aspects of lymphoma in several domestic species.

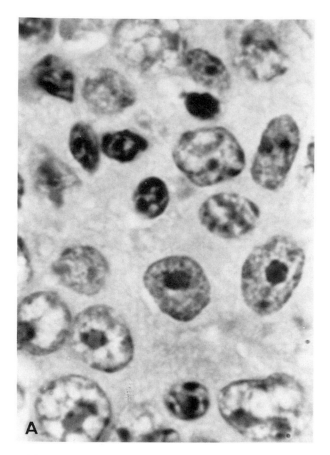

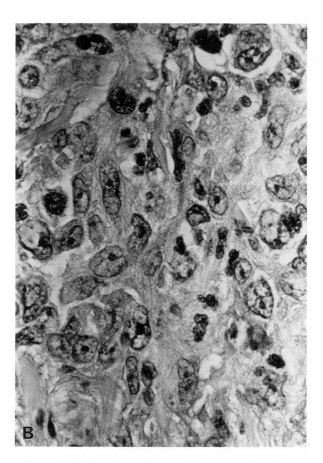

Fig. 2.31A Lymphoma, intermediate grade. Lymph node. Cow. Diffuse mixed-cell type (DM) tumor is composed of small, cleaved, and immunoblastic-type large lymphocytes. Nuclear separation is partially due to background sclerosis.

Fig. 2.31B Lymphoma, intermediate grade. Lymph node. Cow. Diffuse mixed type with more prominent and typical sclerosis.

Lymphoma in adult cattle is very largely the **enzootic bovine leukosis** associated with retroviral infection that is transmitted horizontally by direct contact, including natural breeding and accidental transmission by contaminated needles, ear-tagging and dehorning equipment, and by the use of whole-blood vaccines. Infection, once established, is lifelong and is characterized by the development of circulating antibody, which is also present throughout life. Antibody levels tend to increase with the number of viral antigens recognized, as the animal passes from an asymptomatic carrier to lymphocytosis and to tumorous state (should malignant disease develop). The bovine leukemia virus (BLV) is highly cell associated, and viremia does not occur unless in the period immediately following infection and before antibody is formed. The virus survives in proviral state in bone marrow-derived lymphocytes and monocyte/macrophages, which may be induced to release the C type retrovirus on short-term *in vitro* culture in the absence of antibody. There appears to be some increase in transmission during the summer months, suggesting that biting insects might be vectors. It is, however, possible to maintain infected and noninfected herds on the same premises for prolonged periods with reason-able isolation procedures, which suggests that direct contact is the primary mode of transmission. Some herds tend to have a very high level of infection whereas others have a low level of infection, and the proportions of infected animals tend to remain relatively stable over long periods. There is likely a genetic basis for this phenomenon.

Fewer than 3% of infected animals develop malignant disease, and then only if allowed to survive to the age of peak incidence at 6–8 years. Since viral transmission occurs primarily in closely confined dairy herds, and currently the rate of culling in dairy herds in North America is 20–30% per year, the number of cases of lymphoma appears to be dropping, while the infection rate remains relatively constant. In addition, the development of blood testing for BLV infection has permitted early identification of diseased animals, which are therefore more likely to be found at slaughter than in a clinical setting. Infected cattle which do not develop lymphoma appear to have few other specific problems related to retroviral infection, but may suffer a higher rate of culling for a variety of reasons that may be the result of subtle effects of the host–parasite relationship. The sporadic forms of bovine lymphoma are not associated with BLV infection.

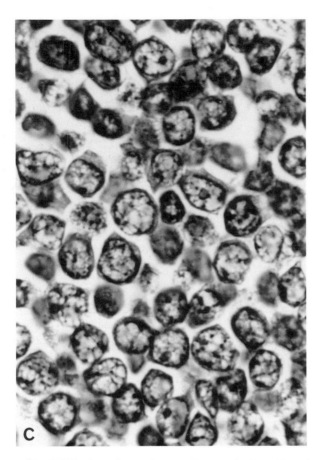

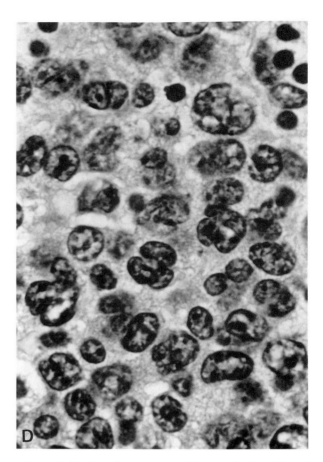

Fig. 2.31C Lymphoma, intermediate grade. Lymph node. Cow. Diffuse large type (DL). Nuclei are round with dispersed chromatin that is branched with irregular thickening of nuclear membranes. Nucleoli irregularly impinge on nuclear membranes.

Fig. 2.31D Lymphoma, intermediate grade. Lymph node. Cow. Diffuse large cleaved cell (DLC). Nuclei are deeply cleft with focal parallelism of reentrant membranes.

Enzootic bovine lymphoma is therefore a disease of mature onset. Hematologically, the animals are normal or nearly so, without anemia or thrombocytopenia. There is a fluctuating lymphocytosis of moderate degree, which is the characteristic response of the cow to infection that is rather well controlled, and also occurs in trypanosomiasis and tuberculosis, and to a lesser extent in brucellosis. In herds in which the disease is endemic, infection is likely acquired in the first year of life. There may be little or no apparent effect of this exposure, but it does serve to perpetuate the virus in the herd. There may be a long period of occult transformation before clinical tumor develops. Cattle with the enzootic form of lymphoma react in lymphocyte stimulation tests against a membrane antigen derived from tumor lymphocytes, which constitutes a diagnostic test for occult tumor. This test remains positive until the later stages of the clinical disease when the animals become anergic. The same membrane antigen when given as a vaccine prevents lymphoma, but not BLV infection, in sheep subsequently challenged with live virus.

Clinically and pathologically, enzootic bovine lymphoma is subdivided on a topographical basis depending on the major organ systems involved. The most common

presentation is an enlarged lymph node characteristically seen in high-producing dairy cows. Enlarged nodes may appear anywhere in the body but are common in the retrobulbar area where they cause unilateral or bilateral proptosis, and in the pharyngeal area where they result in dysphagia and stertorous respiration. The disease is frequently diagnosed on pelvic examination on the basis of enlarged inguinal lymph nodes or involvement of the urogenital system.

The enteric form of the disease is, at least pathologically, part of most cases and is characterized by diffuse thickening of the abomasal submucosa with eventual ulceration leading to hemorrhagic anemia. The latter tendency for gastric and also uterine homing suggests that in many cases the transformed lymphocytes are relatively mature and possess receptors allowing them to accumulate locally as a mucosal-associated lymphoid tumor. There is irregular involvement of the mesenteric nodes associated with the abomasal infiltration and these may at times be large enough to cause obstruction.

Hind limb lameness, which progresses quickly to weakness and recumbency, reflects the spread of tumor from sublumbar nodes to the spinal canal where there is a rela-

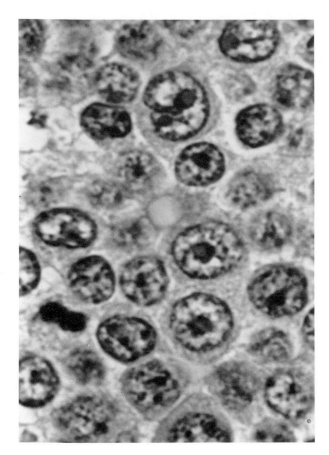

Fig. 2.32 Lymphoma, high grade. Lymph node. Cow. Immunoblastic, large-cell type. Nuclei are round with dispersed chromatin and prominent single central nucleoli. Cytoplasm is typically moderate in amount and densely stained.

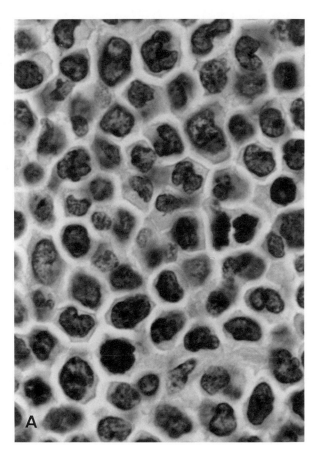

Fig. 2.33A Lymphoma, high grade. Lymph node. Cow. Lymphoblastic convoluted. Nuclei are sharply indented, and chromatin is characteristically densely stained; nucleoli are inapparent in most cells.

tively diffuse infiltration into the perineural fat and heavy involvement of the sheaths of the spinal nerves. Grossly the perineural infiltration has only a slight fleshy pinkness to distinguish it from normal fat, although the distinction is readily made by an imprint of the affected tissues.

Animals occasionally are presented with signs of central venous congestion due to preferential involvement of the right atrium, colonization of the epicardial fat, and irregular and often heavy infiltration into the myocardium that may be accompanied by tamponade.

Almost any organ can be involved, but the liver is involved in most animals, as is the spleen. Some cows die rapidly because of splenomegaly with rupture and abdominal hemorrhage. Rarely, the kidneys are primarily involved in a herd outbreak of lymphoma, again suggesting specific homing, in this case to renal tissue. There is usually cytologic involvement of bone marrow in terminal cases which may become leukemic with very high cell counts, or rarely there is lymphopenia, but myelophthisis as is seen in the lymphocytic leukemia in the calf form is a rare occurrence. The hemolymph nodes are usually involved and occasionally they may be visible clinically as subcutaneous discoid masses 1–2 cm in diameter. The

histologic type of tumor is consistent throughout the body and the duration of the disease is usually a matter of weeks once the disease has been recognized.

The **sporadic forms** of **bovine lymphoma** are not associated with BLV infection. They include the calf type, which is characterized by symmetrical lymphadenopathy and leukemia, the juvenile type with thymic lymphoma in yearlings, and the skin type, which occurs most often in 2-year-old cattle. These three types of bovine lymphomas do not react to membrane antigen from the enzootic form of bovine lymphoma in lymphocyte stimulation tests. As their name suggests, these diseases occur sporadically. It is not clear whether other viruses are involved in their genesis.

The **calf type** of lymphoma may be present at birth but usually develops in the first 6 months of life and typically has symmetrical enlargement of lymph nodes. There is leukemia with leukocyte counts varying from 10 to 100×10^9/liter, a nonresponsive normochromic, normocytic anemia, and thrombocytopenia. Terminally, there is virtually complete myelophthisis and marrow infarction (Fig. 2.34) with neutropenia and thrombocytopenia but without hemorrhage. At autopsy, virtually all nodes are

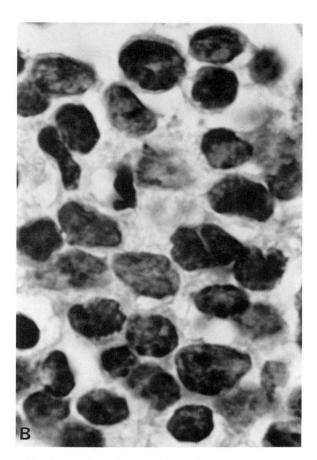

Fig. 2.33B Lymphoma, high grade. Lymph node. Cow. Lymphoblastic convoluted. Different case from Fig. 233A. Comparatively uniform chromatin distribution without parachromatin clearing is typical, as is absent or obscured nucleoli.

enlarged up to 10 times normal size and the kidneys are usually diffusely involved but may not be markedly enlarged. The liver and spleen may be massively involved, or relatively spared. The thymus is usually normal or atrophic, and skeletal muscle and even bone may be focally invaded by tumor.

The **thymic form** of bovine lymphoma is described under Diseases of the Thymus (Section I of Lymphoreticular Tissues in this chapter). Most cases remain aleukemic.

The **skin form** of bovine lymphoma typically occurs in 2- to 3-year-old cattle and is characterized by plaquelike, round, raised lesions which appear on the head, sides, and perineum, become depilated, and are frequently ulcerated. The lesions wax and wane over a period of months with some regressing entirely and new lesions appearing. The disease is remarkable in its indolent nature as compared to the other forms of lymphoma in cattle, and animals allowed to survive may do so for 12–18 months. Ultimately, there is deep organ involvement and the visceral lesions are then indistinguishable from those of the enzootic type of bovine lymphoma. The skin lesions have architectural specificity and are dermotropic and characteristically result in dense infil-

TABLE 2.4

A Modified Working Formulation of Human Non-Hodgkin's Lymphomas

Low Grade	High Grade
A. Diffuse small lymphocytic (DSL), consistent with chronic lymphocytic leukemia DSL, plasmacytoid DSL, intermediate	H. Immunoblastic Plasmacytoid Clear cell Polymorphous
B. Follicular, predominently small cleaved cell	I. Lymphoblastic Convoluted cell Nonconvoluted cell
C. Follicular, mixed, small cleaved, and large noncleaved cell	J. Small noncleaved cell Non-Burkitt Burkitt
Intermediate Grade	**Miscellaneous**
D. Follicular, predominently large cell	K. Composite Mycosis fungoides Histiocytic Extramedullary plasmacytoma Unclassifiable Other
E. Diffuse, small cleaved cell	
F. Diffuse, mixed, small and large cell	
G. Diffuse, large cell Cleaved cell Noncleaved cell	

trations of tumor cells in the papillary dermis. There is focal invasion of the epidermis (Pautrier's microabscess) and in the early stages the disease resembles mycosis fungoides of humans but is more focally distributed, and in the advanced state with leukemia resembles Sézary syndrome. The tumor cells are of moderate size with a very narrow rim of cytoplasm and have cerebriform convolutions of the nuclear membranes. These irregularities of the nuclear outline are not recognizable on examination at 400× magnification. They are best seen in imprints or in buffy coat preparations of blood where the prognosis can be roughly determined by the number of tumor cells in circulation. The skin type of lymphoma is indirectly suggested to be of T cell type, and presumably there are strong homing tendencies driving the colonization of the superficial dermis.

Bibliography

Bundza, A. Sporadic bovine leukosis: A description of eight calves received at Animal Diseases Research Institute from 1974 to 1980. *Can Vet J* **21:** 280–283, 1980.

Ohshima, K.-I. *et al.* Pathologic studies on juvenile bovine leukosis. *Jpn J Vet Sci* **42:** 659–671, 1980.

Parodi, A. L. Pathology of enzootic bovine leukosis. Comparison with the sporadic form. *In* "Enzootic Bovine Leukosis and Bovine Leukemia Virus," A. Burny and M. Mammerickx (eds.), pp. 15–48. Boston, Massachusetts, Martinus Nijhoff Publishing, 1987.

Smith, H. A. The pathology of a malignant lymphoma in cattle. A study of 1113 cases. *Pathol Vet* **2:** 68–94, 1965.

Vernau, W. *et al.* Classification of 1198 cases of bovine lymphoma using the National Cancer Institute Working Formulation for

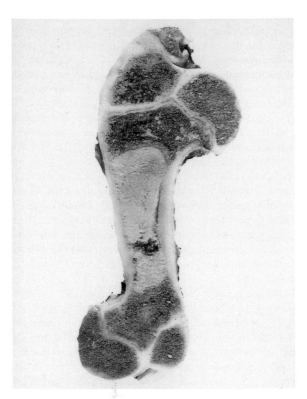

Fig. 2.34 Lymphoma. Medullary infarction in humerus. Calf.

Human Non-Hodgkin's Lymphomas. *Vet Pathol* **29**: 183–195, 1992.

5. Ovine and Caprine Lymphoma

Ovine and caprine lymphomas are widely disseminated tumors similar to the calf or juvenile type of bovine lymphoma, but without leukemia and marrow involvement. A retrovirus indigenous to sheep shares major proteins with the BLV; however, most spontaneous lymphomas of sheep and goats, and virtually all of the those produced experimentally, are caused by BLV. Lymphoma can be produced in sheep in 30–60% of animals 2–3 years after infection with BLV. In sheep, as in cattle, infection with BLV once established is lifelong and the virus persists in integrated provirus in the face of a strong and persistent antibody response. Unlike that in cattle, the host–parasite relationship between BLV and sheep does not produce persistent lymphocytosis but does produce tumor, and the earlier that lambs are infected, the greater will be the proportion of infected animals that develop lymphoma. It appears that, unlike cattle, some sheep after exposure to BLV will develop transient antibody and then are free of infection.

In sheep, BLV infection can be detected by the tests used for diagnosis in cattle. Sheep which develop lymphoma react in lymphocyte-stimulation tests against membrane antigen derived from BLV-associated bovine lymphoma, providing a test for the presence of occult tumor.

Pathologically, there is a typical regular portal involvement of the liver, and there is often hepatomegaly. The spleen is symmetrically enlarged with focal and diffuse involvement, and the heart and kidneys are often very heavily infiltrated. Histologically, the tumors are largely of the small noncleaved type with diffuse architecture. The disease has a very rapid course with most animals dying within 1–2 months of clinical diagnosis.

Bibliography

Kenyon, S. J. *et al.* Induction of lymphosarcoma in sheep by bovine leukemia virus. *J Natl Cancer Inst* **67**: 1157–1163, 1981.
Olson, C. *et al.* Goat lymphosarcoma from bovine leukemia virus. *J Natl Cancer Inst* **67**: 671–675, 1981.
Rohde, W. *et al.* Bovine and ovine leukemia viruses. I. Characterization of viral antigens. *J Virol* **26**: 159–164, 1978.

6. Equine Lymphoma

Lymphoproliferative disease in the horse, like that in the cow, has several characteristic topographical and cytologically distinct modes of presentation as leukemia or lymphoma. In the **leukemic form,** which usually occurs in racing animals in training, there is primary marrow involvement with pancytopenia and severe thrombocytopenia, sudden onset of malaise, subcutaneous edema, and petechial hemorrhages. The bone marrow is diagnostic with myelophthisic destruction of normal cells by lymphocytes of the L2 or L3 type (see Acute Lymphocytic Leukemia, Section VI,A,1 of The Leukon). The clinical course in this form of disease is less than a week. More commonly the cells are more differentiated and typical of chronic lymphocytic leukemia with leukemic lymphocytosis and the course is much longer. A disease like mycosis fungoides of humans occurs in horses and progresses to a form of chronic lymphocytic leukemia with convoluted tumor cells like the human Sézary syndrome. The skin lesions may be minimal and confined to the mucous membranes, which are brick red and do not blanch on compression. Histologically the tumor cells are dermotropic and of medium size with cerebriform nuclear indentations. In humans, the tumor cells have been found to be of T helper phenotype.

The **solid forms** of lymphoid tumor in the horse can be characterized as subcutaneous, alimentary, abdominal, splenic, and thymic/multicentric. **Subcutaneous** lymphoma is remarkable because of the indolent nature of the disease, with some animals surviving for years, and because it occurs very predominantly in females. Regression of the disease has been observed during pregnancy and affected animals may maintain body weight for many months, but are unsightly because of hundreds of subcutaneous masses 1–3 cm in diameter or larger. The lesions cluster on the lower half of the body, particularly on the chest, thorax, flanks, and perineum, sparing the limbs and often involving the face and neck. One of the major reasons for biopsy of these animals is that an otherwise vigorous animal cannot be bridled or saddled without impinging on the masses. Histologically the tumors are truly subcutaneous, likely arise along lymphatics, and rarely involve the deep dermis, which tends not to be fixed over them. Cytologically they are also diverse, but distinctive, and are composed of small cleaved and large-cell types of lymphocytes, interlaced at the cellular level with fine reticulin fibers, and on an architectural level with collagenous bands. They are thus diffuse mixed tumors according to the Working Formulation classification and are characteristically accompanied by diffuse sclerosis. Ultimately there is involvement of regional lymph nodes, but deep organ involvement does not regularly occur in the time that most are allowed to survive. They remain hematologically normal except for mild progressive anemia and hypoproteinemia. A second type of diffuse mixed lymphoma which involves the dermis occurs in horses and characteristically presents as a protruding dorsolateral periorbital lesion 2–3 cm in length, with ulceration. Histologically the lesions are sclerosing, but their most remarkable characteristic is a tendency to recur locally even after careful dissection. In all forms of the diffuse mixed tumor, the mitotic rate is very low and consistent with the indolent rate of progression.

The **alimentary form** of lymphoma occurs in adults that are characteristically thin with reduced appetite and may be colicky, but usually do not have diarrhea. Anemia may be mild or quite severe, normocytic, and with hypoproteinemia. There is usually some degree of neutrophilia with signs of toxemia due to ulceration of the intestinal mucosa, but leukemia is not present. There may be alpha heavy chain paraproteinemia as in Seligman's Mediterranean lymphoma of humans, but this aspect has not been frequently investigated. Some animals with this form of tumor will have hemolytic anemia that may dominate the clinical picture. Pathologically the peripheral and internally palpable lymph nodes are normal and the disease is difficult to diagnose short of an open biopsy. The disease tends to be confined to the upper small bowel, pancreas, and regional nodes, but there may also be involvement of the ileum (Fig. 2.35A,B). There is laminar thickening of the intestinal wall with tight homing of the infiltrate to the muscularis mucosa and submucosa and relative sparing of the muscularis. The tumor cells are uniformly plasmacytoid with round nuclei one and one half red cells in diameter and abundant highly basophilic and eccentric cytoplasm. There is mild nuclear atypia with nucleoli not generally present in mature-appearing plasma cells, hyperchromatic chromatin with an exaggerated heterochromatin clumping, and a low mitotic rate. Cytologically, in imprints, the tumor cells may be mistaken for plasma cell hyperplasia if the extent of the gross lesion has not suggested tumor. There is focal thinning and ulceration of the intestinal epithelium and the tumor cells have the same morphology in the affected nodes, which tend to have thickened capsules and atrophy of normal cells. One is left with the impression that the disease began as benign immune inflammation and progressed to lymphoma, albeit

of well-differentiated type. The disease is an example of a well-ordered, mucosal-associated lymphoid tumor with homing of the tumor cells to the submucosa until very late in the disease.

The **abdominal form** of equine lymphoma is not usually difficult to diagnose clinically because tumor masses involve the large bowel and caudal abdominal nodes. Clinically the animals are thin and have poor appetite, dependent edema, and recurrent diarrhea. The hematologic changes are typical only of chronic disease and inflammation. Pathologically there is widespread tumor in the abdominal cavity and there may be involvement of nodes in the thorax. There is splenomegaly, involvement of the hepatic hilar nodes, irregular thickening of the wall of the small and large intestine, and focal mucosal ulcers. There may be huge tumorous masses involving the intestinal serosa and mesenteric nodes. Histologically the tumors are variable but tend to be diffuse large-cell type with many mitoses.

The **splenic** lymphoma is an isolated lesion that involves the lower part of the organ with a huge mass that may weigh 15 kg or more. This lesion is not usually palpable rectally and the superficial nodes are not enlarged. The animals are presented with responsive hemolytic anemia, and idiopathic immune or infectious equine anemia is usually suspected. There is an upward shift in the mean cell volume (MCV) and increased anisocytosis. Leukemia is not present and the bone marrow is hyperplastic with an erythroid shift and synchronous maturation. The pathogenesis of the anemia appears to involve hypersplenism due to trapping of cells in the normal spleen adjacent to the tumor. Immune hemolysis may contribute as it does in other lymphoid tumors, but this has not been verified. Pathologically the disease usually is limited to the spleen, but occasionally adjacent organs are involved by direct extension. The tumors are of large-cell and immunoblastic type. The major feature of this lesion is the difficulty in clinical diagnosis, and most cases are diagnosed at necropsy.

The **multicentric** lymphomas have a wide age range, but most occur in mature animals. Affected animals are usually in poor condition but continue to eat, and often have low protein levels and dependent edema. Many of these animals have a mediastinal mass as well as irregular enlargement of peripheral nodes, particularly those in the pharyngeal and mandibular areas. Any of the superficial nodes may be involved, but the involvement is not symmetrical. Leukemia is occasionally present, but usually the blood is more typical of cachexia. Pathologically there is widespread tumor in the abdominal cavity, and often there is tumor in the mediastinum with involvement of the heart and even the lung. Cytologically the tumors are not of consistent cell type, but all tend to be of intermediate- or high-grade type, usually small noncleaved, immunoblastic and diffuse large cell. Some of these diseases may be T cell lymphomas arising in the thymus and disseminating widely by the time of diagnosis.

Equine lymphomas do not always observe the rules

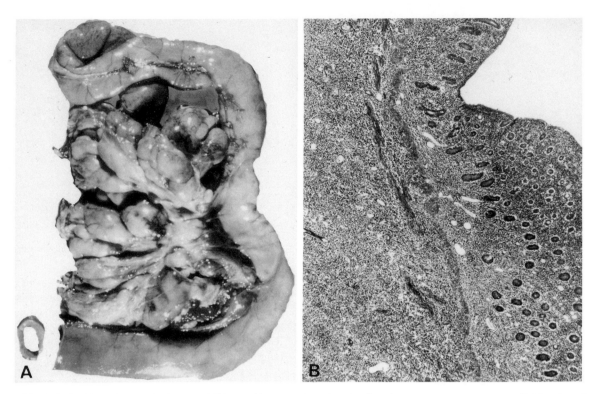

Fig. 2.35A Lymphoma. (A) Horse. Diffuse infiltration of duodenal wall and replacement of pancreas. (B) Section of intestine of (A). The infiltrates may be confined to the mucosa but usually involve full thickness.

described. Lymphoma in the horse, as in other species, may be very variable and primarily involve other tissues such as the central nervous system, and then the signs vary accordingly.

Bibliography

Grindem, C. B. *et al.* Large granular lymphocyte tumor in a horse. *Vet Pathol* **26:** 86–88, 1989.

Gupta, B. N., Keahey, K. K., and Ellis, D. J. Cutaneous involvement of malignant lymphoma in a horse. *Cornell Vet* **62:** 205–215, 1972.

Platt, H. Observations on the pathology of non-alimentary lymphomas in the horse. *J Comp Pathol* **98:** 177–194, 1988.

Rousseaux, C. G., Doige, C. E., and Tuddenham, T. J. Epidural lymphosarcoma with myelomalacia in a seven-year-old Arabian gelding. *Can Vet J* **30:** 751–753, 1989.

Van Den Hoven, R., and Franken, P. Clinical aspects of lymphosarcoma in the horse: A clinical report of 16 cases. *Equine Vet J* **15:** 49–53, 1983.

7. Canine Lymphoma

Hematopoietic tumors in urban dogs occur at an age-adjusted rate of some 30/100,000, with lymphoma constituting 87% of these and the remainder comprising lymphoid and myeloid leukemias. Unlike those in the cat, which have a bimodal pattern of incidence rate, hematologic malignancies in the dog occur with increased frequency throughout life. There is an equal sex incidence in intact animals and a lowered risk for neutered males. Most

lymphomas in dogs are disseminated and of high-grade type; however, there is some clustering of cases based on topographic distribution, with multicentric most common, followed by enteric, cutaneous, and thymic types. The presenting signs vary with the distribution of tumor and tendency, if any, for paraneoplastic changes. There is usually weight loss with some change in dietary habits. The **multicentric** type often has symmetrical lymphadenopathy and tends to be noticed when the animal is still in relatively good condition, as does the cutaneous form since the disease is usually focal and nodular. Dogs with hypercalcemia have polydipsia and polyuria and are usually in a state of collapse, whereas those with primary **enteric** lymphoma (Fig 2.36) have some combination of vomiting and diarrhea, often bloody. There is mucosal ulceration in 50% of enteric cases and these differ in having some degree of anemia accompanied by a moderate to marked neutrophilic leukocytosis. Dogs with **thymic** lymphoma (Fig. 2.37) have reduced exercise tolerance, often dysphagia, and may have tamponade with severe impairment of circulation.

There are few correlations between cell type and topographic distribution of tumor; however, there are more plasmacytoid tumors in the gut, skin, and spleen than in other sites, which suggests that for these areas malignant transformation is preceded by a period, perhaps prolonged, of benign immune hyperplasia. This appears to be particularly likely in the enteric cases in which the diagno-

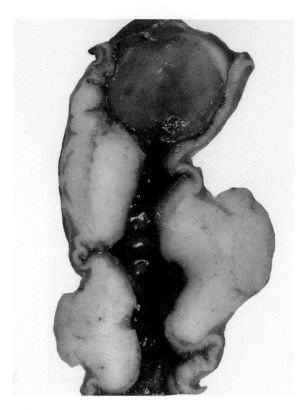

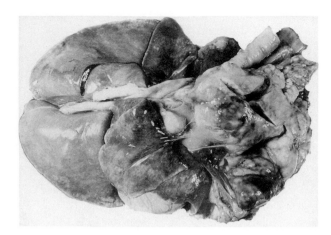

Fig. 2.37 Thymic lymphoma of anterior mediastinum. Dog.

sis of lymphoma is often preceded by changes of lymphocytic/plasmacytic enteritis. There is a clustering of the lymphoblastic type of lymphoma in the mediastinal cases, and T cell phenotype and pseudohyperparathyroidism with hypercalcemia are also associated with mediastinal lymphoma.

The lymph nodes, liver, and, to a lesser extent, the spleen (Fig. 2.38A) are involved in virtually all cases of canine lymphoma and since there are usually lesions that are palpable or identifiable radiographically, most cases are diagnosed by fine needle aspiration biopsy. The lesions may be identified in almost any tissue including tonsils,

Fig. 2.36 Lymphoma. Neoplastic infiltration of pyloric submucosa. Dog.

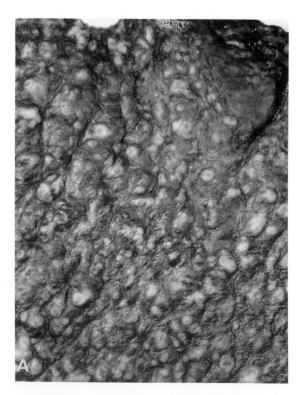

Fig. 2.38A Lymphoma. Neoplastic enlargement of malpighian bodies. Spleen. Dog.

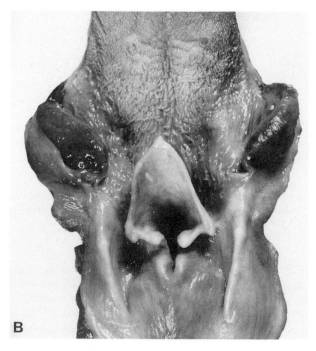

Fig. 2.38B Lymphoma. Bilateral tumor of tonsils. Dog.

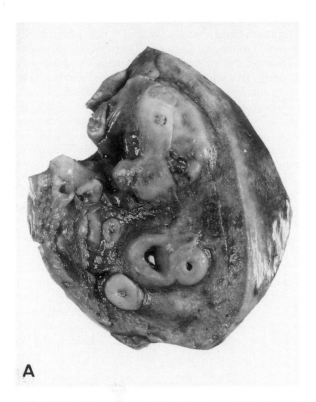

A

Fig. 2.39A Lymphoma of Hodgkin's type infiltrating perivascular and peribronchial lymphatics in lung. Dog.

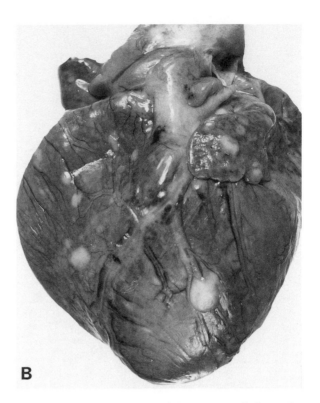

B

Fig. 2.39B Lymphoma. Multiple metastases in heart. Dog.

which are often markedly enlarged (Fig. 2.38B), or in the eye. Involvement of lung (Fig. 2.39A) is uncommon except as pleural implants from tumor in the anterior mediastinum. Secondary deposits in meninges and muscle tend to be diffuse, but those in myocardium tend to be more focal (Fig. 2.39B). About 50% of dogs have some degree of renal involvement and the proportion with impairment of function, due to either or both tumor infiltration or hypercalcemia, is much higher. The bone marrow is involved to some degree in about one-third of cases and the histologic pattern of marrow involvement is a useful parameter in the evaluation of progression.

The low-grade lymphomas constitute only 5% of cases and the follicular tumors, less than 1%. Nearly a third of cases are of intermediate grade and two-thirds are high grade, with immunoblastic and small noncleaved types constituting about 25% each and lymphoblastic, 17%. With the exception of the lymphoblastic tumors, the other high-grade lymphomas usually undergo remission with adequate chemotherapy. Diagnosis needs to be based on untreated cases since therapy alters the cell type to a smaller and resistant cell that resembles small noncleaved morphologically but may be refractory to further treatment.

Cutaneous lymphoma occurs in dogs, cats, cattle, and horses and is considered here because it is seen most frequently in the dog. The disease is composed of a spectrum of entities which require definition in animals for

their biological behavior to be predicted if not understood. This description refers to those lymphomas that arise in the skin in contrast to tumors arising in the deeper structures that may progress to involve the skin.

On a topographic basis, cutaneous lymphoma can be divided into focal and diffuse types. Both types occur in the dog and horse, whereas the disease is most often focal in the cow and diffuse in the cat. In terms of tropism in the skin, the tumor cells may be above or below the basement membrane that defines the dermal epidermal junction. Thus the **dermotropic lymphomas,** as characterized by the plaquelike lesions of 2-year-old cattle, involve the papillary and to a lesser extent the deep or reticular dermis and may focally invade the epidermis to form the Pautrier's microabscess, as described in **mycosis fungoides** of humans (Fig. 2.40). The **epidermotropic lymphomas** cause marked thickening of the epidermal layer, with the tumor cells permeating the corneal layer in a diffuse manner as in **Woringer-Kolopp** disease of humans. The latter disease is rare and appears to occur only in cats and dogs; however, until more specific definition is given for these lesions, their distribution will not be known. In terms of cell type, the cutaneous lymphomas are very predominantly composed of large cells, and those of the mycosis fungoides type have highly convoluted nuclei and are presumed to be thymus derived. Little is known of the phenotype of the epidermotropic lymphomas in animals; however, there is increasing evidence in human pathology that many

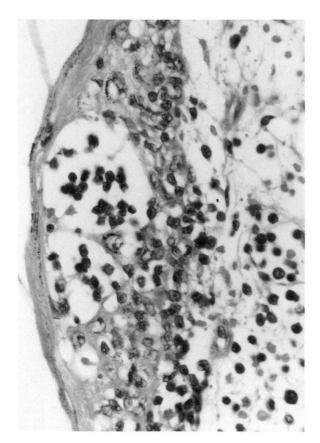

Fig. 2.40 Lymphoma. Cutaneous. Dog. Focal epidermal colonization by small lymphocytes.

cutaneous lymphomas are of bone marrow derivation. Some of these lesions, as in the mycosis fungoides cases of the dog, cow, and horse, progress slowly over as much as a year to involve the parenchymal organs and bone marrow. The more diffuse lesions of the Woringer-Kolopp type cause serious illness due to the extent of the skin involved and are not allowed to progress to death.

Rarely, dogs develop a disease like human **myeloma** with focal lysis of bone, and are usually presented with lameness and nervous deficit. They have hyperproteinemia with monoclonal gammopathy that is in the 8–14g/dl range and may have heavy-chain fragments in the urine. The spleen is often massively enlarged and may contain septic infarcts. The high level of high-molecular-weight plasma protein causes high blood viscosity, and anesthesia in these dogs for marrow or spleen aspiration may result in fatal hypotensive shock. The lesions may be widespread at the time of diagnosis, and qualify for the general term of multiple myeloma, or only a single focus of tumor may be present. The tumor is always locally invasive and readily erodes bone, emphasizing the need for all destructive lesions of bone to be histologically evaluated so as to avoid inappropriate corrective measures such as excision arthroplasty of a femoral head that collapsed due to tumor

lysis. Cytologically the tumor is paradoxical in having more rather than less cytoplasm than benign plasma cells. Nuclei are of moderate size, about two red cells in diameter, with more anisokaryosis than is seen in a plasmacytoma. They are round and hyperchromatic with irregular heterochromatin aggregation and small nucleoli.

Binucleate cells are common. In the classic tumor, the cytoplasm is abundant, centered on the nucleus, and pale staining so that the myeloma may resemble a clear cell tumor at low magnification. There may rarely be dedifferentiation of the myeloma to a more aggressive large-type immunoblastic lymphoma in which the cells are smaller but the nuclei are larger, with prominent central nucleoli. The latter type of lymphoma may on its own be secretory and associated with gammopathy.

Angiotropic large-cell lymphoma is a very unusual form of lymphoma that occurs in humans and dogs and possibly other species. Typically the tumor cells colonize the subintimal area of small muscular arteries of the lung and may be found in the atria and less often in other tissues. Affected animals are mature and may suddenly collapse with circulatory failure while in good body condition and undergo treatment for congestive heart failure. Ultimately there is

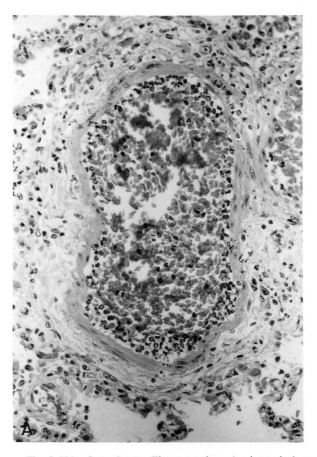

Fig. 2.41A Lymphoma. Tissue tropism. Angiotropic large-cell lymphoma lung. Dog. Laminar subendothelial colonization of dilated artery.

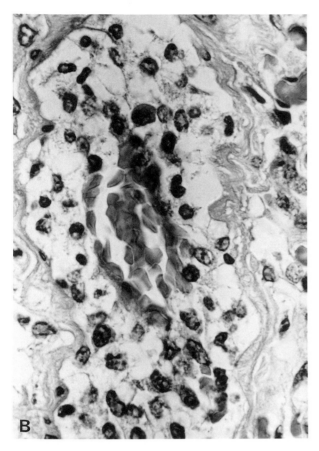

Fig. 2.41B Detail of smaller arteriole of (A). Marked luminal compression due to colonization of subendothelial area by lymphocytes with abundant clear cytoplasm.

pulmonary congestion, flooding, and severe hypoxia and death. Pathologically there is congestive atelectasis of lung but no specific gross changes. Histologically the tumor cells are of the diffuse large or diffuse large cleaved type and tend to form a relatively uniform laminar infiltrate that lifts the endothelium and narrows the lumen (Fig. 2.41A,B). Attention is drawn to these vessels because of the central displacement of red cells in a highly congested lung. Terminal arterioles are less involved and the veins not at all. The animals are not known to be leukemic and marrow is not involved.

Cytogenetic abnormalities are present in at least some canine lymphoproliferative diseases and it appears that cytogenetic techniques may become more basic than histology in the classification of lymphomas on an etiologic and prognostic basis.

Bibliography

Carter, R. F., Valli, V. E. O., and Lumsden, J. H. The cytology, histology and prevalence of cell types in canine lymphoma classified according to the National Cancer Institute working formulation. *Can J Vet Res* **60:** 154–164, 1986.
Couto, C. G. *et al.* Gastrointestinal lymphoma in 20 dogs. A retrospective study. *J Vet Intern Med* **3:** 73–78, 1989.
Greenlee, P. G. *et al.* Lymphomas in dogs. A morphologic, immunologic, and clinical study. *Cancer* **66:** 480–490, 1990.
Grindem, C. B., and Buoen, L. C. Cytogenetic analysis of leukaemic cells in the dog. A report of 10 cases and a review of the literature. *J Comp Pathol* **96:** 623–635, 1986.
Meuten, D. J. *et al.* Hypercalcemia in dogs with lymphosarcoma. Biochemical, ultrastructural, and histomorphometric investigations. *Lab Invest* **49:** 553–562, 1983.
Raskin, R. E., and Krehbiel, J. D. Histopathology of canine bone marrow in malignant lymphoproliferative disorders. *Vet Pathol* **25:** 83–88, 1988.
Rowland, P. H. *et al.* Cutaneous plasmacytomas with amyloid in six dogs. *Vet Pathol* **28:** 125–130, 1991.
Schneider, R. Comparison of age- and sex-specific incidence rate patterns of the leukemia complex in the cat and the dog. *J Natl Cancer Inst* **70:** 971–977, 1983.
Stroup, R. M. *et al.* Angiotropic (intravascular) large cell lymphoma. A clinicopathologic study of seven cases with unique clinical presentations. *Cancer* **66:** 1781–1787, 1990.
Valli, V. E. *et al.* Histocytology of lymphoid tumors in the dog, cat and cow. *Vet Pathol* **18:** 494–512, 1981.
Wilcock, B. P., and Yager, J. A. The behavior of epidermotropic lymphoma in twenty-five dogs. *Can Vet J* **30:** 754–756, 1989.

8. Feline Lymphoma

Feline lymphoma is caused by the horizontally transmitted feline leukemia retrovirus (FeLV). The host–viral relationship in the cat, unlike that in the cow, includes persistent viremia, which is associated with a high incidence of disease; but as in the cow, there may be latent infection with circulating antibody and persistence of provirus in target cells. The presence of a persistent viremic state in the cat likely explains the difference in rate of infection-associated disease in these two species. Like bovine leukemia virus (BLV) the low-molecular-weight envelope proteins (p15E) are very highly immunosuppressive and, in decreasing order of frequency, the disease syndromes associated with infection include leukemia/ lymphoma, refractory anemia, feline infectious peritonitis, diseases of the oral cavity, and a variety of miscellaneous problems. Kittens are susceptible to infection with FeLV, and early infection with high rate of challenge and highly virulent virus favors the formation of myeloid rather than lymphoid tumor. It appears that feline leukemia is preceded by a period of myelodysplasia, which is likely true of leukemia in all species, but is rarely observed outside of experimental infection.

The virus is worldwide in distribution with about 2% of cats infected. In a population of cats at veterinary clinics, as many as 90% of seropositive cats were ill with some form of disease. About 15% of cats infected with FeLV are also infected with feline immunodeficiency virus (FIV); however, the latter agent is not nearly as infectious as FeLV nor as likely to cause disease.

The means by which FeLV induces tumor is unknown and the virus has been found to be polyclonal in some tumors, suggesting a transactivating or indirect role as in BLV. Molecular studies have shown that FeLV isolated from tumors contains a transduced cellular-*myc* gene with the viral-*myc* constituting the entire coding sequence of the cellular gene. Since FeLV is a slow transforming retrovirus lacking an *onc* gene capable of direct leukemogenesis, the ability to acquire cellular oncogenes may be important in its ability to induce malignant transformation. In terms of detection of viral antigen or antiviral antibody, virus-positive cats are associated with the multicentric form of lymphoma, whereas the thymic and alimentary forms of tumor predominate in cats negative for virus. However, in molecular studies, FeLV provirus was found in thymic lymphomas. Thus, viremia is associated with immunosuppression and secondary disease, whereas the latent form of infection (as in the cow) is not. Latent infection is, however, associated with tumor, suggesting that transactivation by integrated FeLV or action of its acquired v-*myc* oncogene is necessary for malignant transformation, but immunosuppression is not. A sterile vaccine based on a viral antigen-producing cell line appears able to produce protective antibody without inducing immunosuppression against FeLV or other feline vaccine viruses.

Since the distribution of lymphoid tumor in cats is so variable, the presenting signs are also variable, but anemia and weight loss are the most consistent clinical signs. Pathologically lymphoma in the cat is grossly like that in the dog except for the more common finding of very advanced mediastinal lymphoma in the cat. In a histologic review of 547 cases of feline lymphoma, there were 67 low-grade, 184 intermediate-grade, and 294 high-grade tumors,

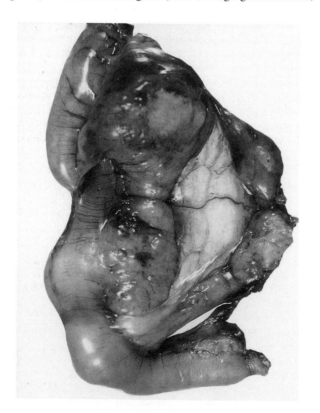

Fig. 2.42 Lymphoma. Primary neoplasia in wall of jejunum. Cat.

plus 2 thymomas. The predominant topographic distribution of tumor was multicentric, 168; gastrointestinal, 150; mediastinal, 120; renal, 33; single enlarged peripheral node, 31; oral cavity, 14; nasal cavity, 10; skin, 10; ocular and periorbital, 8; and 3 had generalized lymphadenopathy. The multicentric lymphomas were composed of a wide range of histologic types. The notable findings of this review were the tight homing of the intestinal lymphomas, particularly of the diffuse mixed type, to the lamina propria, sparing the muscularis and the mucosa until the tumors were very extensive, and suggesting that excision may be an option when the disease is found at a localized stage of progression (Fig. 2.42). In comparison with the dog, immunoblastic lymphoma is more common and constitutes a third of cases in cats, whereas lymphoblastic is much less common, as is hypercalcemia associated with lymphoma. There were 33 lymphomas with primary involvement of the kidney in cats and of these, 16 were of the immunoblastic type. Small noncleaved lymphoma is less frequent in the cat than in the dog or cow.

Bibliography

Baker, J. L., and Scott, D. W. Mycosis fungoides in two cats. *J Am Anim Hosp Assoc* **25:** 97–101, 1989.

Dust, A., Norris, A. M., and Valli, V. E. O. Cutaneous lymphosarcoma with IgG monoclonal gammopathy, serum hyperviscosity and hypercalcemia in a cat. *Can Vet J* **23:** 235–239, 1982.

Grindem, C. B., and Buoen, L. C. Cytogenetic analysis in nine leukaemic cats. *J Comp Pathol* **101:** 21–30, 1989.

Mooney, S. C. *et al.* Treatment and prognostic factors in lymphoma in cats: 103 cases (1977–1981). *J Am Vet Med Assoc* **194:** 696–699, 1986.

Neil, J. C. *et al.* The role of feline leukaemia virus in naturally occurring leukaemias. *Cancer Surv* **6:** 117–137, 1987.

Rojko, J. L. *et al.* Feline lymphomas: Immunological and cytochemical characterization. *Cancer Res* **49:** 345–351, 1989.

Shelton, G. H. *et al.* Feline immunodeficiency virus and feline leukemia virus infections and their relationships to lymphoid malignancies in cats: A retrospective study (1968–1988). *J Acquir Immune Defic Synd* **3:** 623–630, 1990.

Toth, S. R., Onions, D. E., and Jarrett, O. Histopathological and hematological findings in myeloid leukemia induced by a new feline leukemia virus isolate. *Vet Pathol* **23:** 462–470, 1986.

9. Porcine Lymphoma

Lymphoma is the most common tumor in pigs, and in this species there is less diversity, both of topographic distribution and histologic type, compared to other domestic animals. Lymphoma occurs in 25% of animals from herds of Large White pigs bearing an autosomal recessive trait for this condition. The disease may be detectable at 2–3 months of age and average survival is 4–6 months. It has not been possible to mate the homozygous affected animals because of their short survival. The disease appears to be similar in both the sporadic and familial occurrences, with generalized lymph node enlargement, splenomegaly, hepatomegaly, and with particular involvement of the mesenteric nodes, stomach, and intestines (Fig.

Fig. 2.43 Lymphoma. Diffuse infiltration of stomach with pitting ulcers. Pig.

2.43). Discrete multifocal invasion of the renal cortex may be more common in the sporadic disease. There is gradual involvement of the marrow, and many cases die with leukemia, marrow failure, and widespread hemorrhages due to thrombocytopenia. Pigs with the inherited form of lymphoma have a heavy-chain gammopathy with appearance of this fragment in the urine. The abnormal protein is apparently produced by what is therefore a B cell tumor. Histologically, 60% of the tumors are of the diffuse large type, 24%, small noncleaved, and 8%, lymphoblastic (Table 2.5). The histologic inversion of the porcine node has

not attracted an interest in the homing patterns of lymphomas in this species.

Bibliography

Hayashi, M. *et al.* Histopathological classification of malignant lymphomas in slaughtered swine. *J Comp Pathol* **98:** 11–21, 1988.

Imlah, P. *et al.* Serum gamma globulin levels and the detection of IgG heavy chain and light chain in the serum and urine of cases of pig hereditary lymphosarcoma. *Eur J Cancer* **13:** 1337–1349, 1979.

Kadota, K. A case of swine T-cell lymphomas with the Lennert's lesion. *Jpn J Vet Sci* **49:** 913–916, 1987.

Kadota, K., and Nakajima, H. Histological progression of follicular centre cell lymphomas to immunoglobulin-producing tumours in two pigs. *J Comp Pathol* **99:** 145–157, 1988.

10. Lymphomas in Laboratory Animals

Lymphomas are one of the most common tumors in rats and mice on 2-year or lifetime studies. There is lot of diversity of histologic type of tumor in various strains of animals, but the same types of tumors tend to cluster within a given strain. The murine lymphomas have been championed as models of human disease, but unfortunately, rodent lymphomas have not been systematically reported as a function of animal genotype, which is the information required by the pathologist dealing with the lesions in a specific strain of animal. Much of this information is known to workers in the chemical and pharmaceutical industries as a matter of proprietary information. What is published are the results of studies on viral-induced lymphoma models that may be less relevant in application to spontaneous tumors. One of the most studied and inter-

TABLE 2.5

Prevalence of Cell Types in Lymphoma Classified According to NCI Working Formulation

Grade	Cell type	Cattle		Dog		Cat		Horse		Pig		Human	
		n	%	n	%	n	%	n	%	n	%	n	%
Low	Diffuse small lymphocytic (DSL)	22	1.8	14	4.9	12	2.2	1	1.0	1	0.5	41	3.6
	DSL-intermediate	52	4.3	0		28	5.1	11	12.1	2	1.0		
	DSL-plasmacytoid	11	0.9	0		13	2.3*	8	9.0	1	0.5		
	Follicular small cleaved	3	0.2	0	0.0	1	0.2	0	0.0	0	0.0	259	22.5
	Follicular mixed	0	0.0	1	0.4	3	0.5	0	0.0	0	0.0	89	7.7
Intermediate	Diffuse small cleaved	13	1.1	17	5.9	34	6.2	1	1.0	0	0.0	79	6.9
	Diffuse mixed	26	2.2	6	2.1	38	6.9	31	34.2	4	3.0	77	6.7
	Follicular large	1	0.1	1	0.4	1	0.2	0	0.0	0	0.0	44	3.8
	Diffuse large	366	30.5	57	20.0	43	7.9	19	21.1	81	60.0	277	19.7
	Diffuse large cleaved	424	35.4	0	0.0	68	12.5	2	2.1	0	0.0		
High	Immunoblastic	27	2.4	71	24.9	188	34.4	5	5.2	11	8.0	91	7.9
	Lymphoblastic	19	1.6	49	17.2	13	2.4	3	3.3	4	3.0	49	4.2
	Small noncleaved	222	18.5	69	24.2	74	13.5	2	2.2	32	24.0	58	5.0
	Small noncleaved-Burkitt type	9	0.8					1	1.0	0	0.0		
Misc.	Plasmacytoma	3	0.2			19	3.5	5	5.2				
	ALL					10	1.8	1	1.0				
	CLL					12	0.4	2	2.1				
	Thymoma												
Total		1198	100	285	100.0	547	100	89	100	136	100	1064	100

esting spontaneous tumors is the large granular lymphocytic leukemia of the Fischer (F344) rat, which appears to
be a tumor of natural killer (NK) cells and has stimulated
recognition of this tumor in many other species. Lymphomas have been increasingly found in nonhuman primates,
particularly when the primates are infected with immunosuppressive lentivirus. They resemble their etiologic counterparts in humans, are very diverse, and have a tendency
to involve the central nervous system.

Bibliography

Holmberg, C. A. *et al.* Malignant lymphoma in a colony of *Macaca arctoides. Vet Pathol* **22:** 42–45, 1985.
Pattengale, P. K., and Taylor, C. R. Experimental models of
lymphoproliferative disease. The mouse as a model for human
non-Hodgkin's lymphomas and related leukemias. *Am J Pathol* **113:** 237–265, 1983.
Stromberg, P. C. *et al.* Immunologic, biochemical, and ultrastructural characterization of the leukemia cell in F344 rats. *J Natl
Cancer Inst* **71:** 173–181, 1983.
van Berlo, R. J. *et al.* Different homing pattern of isolated mouse
lymphoma cells correlates with a different chromosomal pattern. *Cancer Genet Cytogenet* **38:** 33–42, 1989.

11. Hodgkin's-like Lymphoma

Hodgkin's-like lymphoma (Fig. 2.44) in animals is rare.
Like lymphoma, this disease may occur in all animals, but
has been reported most often in the dog. The diagnosis

of Hodgkin's disease rests on the demonstration of an
acceptable Reed–Sternberg cell in an appropriate background of lymphoproliferation. Few cases meet these criteria. The Reed–Sternberg cell (Fig. 2.45A), which may
be a malignant interdigitating reticular cell, has a precursor
form called a lacunar cell (Fig. 2.45B), which is characteristically seen in the nodular sclerosing type of disease.
The name lacunar is derived from artefactual contraction
of the cytoplasmic boundaries of the reticular cell from
surrounding cells following processing of formalin-fixed
tissue. This artefact facilitates detection of the lacunar and
Reed–Sternberg cells at low magnification.

There is regional lymphadenopathy and the enlarged
nodes tend to be very firm. Diagnosis is based on the
presence of an unequivocal Reed–Sternberg cell in a background of homogeneous or heterogeneous lymphoid proliferation with mild or extensive sclerosis. The disease
tends to be indolent and animals may have large tumor
masses and retain body condition.

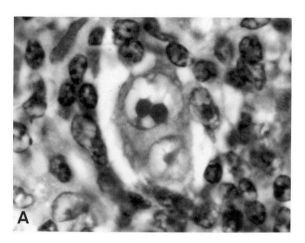

Fig. 2.45A Hodgkin's-like disease. Reed–Sternberg type
cell. Cytoplasmic contraction is characteristic after formalin fixation.

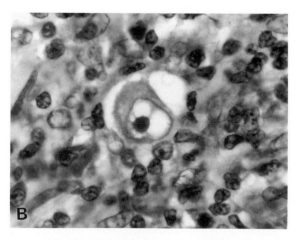

Fig. 2.45B Hodgkin's-like disease. Lacunar cell with typical
cytoplasmic vacuolation.

Fig. 2.44 Hodgkin's-like disease, lymph node. Skunk. Foci
of small lymphocytes in diffuse cellularity which obliterates architecture.

Bibliography

Lukes, R. J. Criteria for involvement of lymph node, bone marrow, spleen and liver in Hodgkin's disease. *Cancer Res* **31:** 1755–1767, 1971.

Smith, D. A., and Barker, I. K. Four cases of Hodgkin's disease in striped skunks (*Mephitis mephitis*). *Vet Pathol* **20:** 223–229, 1983.

THE ERYTHRON

The erythron may be defined as the circulating red cells, their nucleated precursors, and all elements involved in the control of erythroid proliferation. Diseases of the erythron are characterized by too many or too few red cells. Polycythemia, or too many red cells, is rare and usually due to dehydration and plasma volume contraction. Anemia, or too few red cells, is due to decreased red cell production or increased destruction or both.

I. Erythrocyte Physiology

The erythrocyte is specialized to carry out gas transport and in mammals is devoid of nuclei, mitochondria, and ribosomes. Simplistically the erythrocyte can be looked on as a cell membrane surrounding heme, globin, and protective enzymes. The membrane is a bilayer of phospholipid (35%) with interposed molecules of unesterified cholesterol (15%), and glycolipids and protein (50%). The surface proteins make up the receptors and antigens of erythrocytes. On the inner surface the integral membrane proteins bind to a fibrillar cytoskeleton, made up of spectrin and actin, which provides membrane shape, flexibility, and durability. Cell volume is maintained by control of sodium and potassium levels through the action of adenosine triphosphate (ATP)-dependent membrane enzymes. The membrane structural proteins are relatively fixed for the life of the cell, whereas the lipid, particularly cholesterol, is in dynamic interchange with that in plasma. This interchange is under enzyme control and, in the presence of bile salts, cholesterol is gained by the cell, causing increased membrane area and formation of the target cells associated with hepatic dysfunction.

The erythrocyte contents are largely proteins, of which some 95% is hemoglobin and the rest, largely the enzyme systems required for glycolysis, protection of the heme and globin, and ATP production. The loss of the nucleus during maturation is accompanied by the loss of RNA and mitochondria. As a consequence, the mature erythrocyte has also lost the enzymes required for a functional tricarboxylic acid cycle and the cytochromes essential for electron transport. The mature erythrocyte is, therefore, unable to derive phosphate bond energy from the citric acid cycle and is unable to synthesize further heme. The remaining source of energy is through the metabolism of glucose to lactic acid to produce a functional level of ATP. This ATP functions to partition glucose from plasma, maintain homeostatic levels of intracellular sodium and potassium, maintain the biconcave shape, indirectly maintain iron in a ferrous state, and protect the integrity of the globin chains. As a consequence of glycolysis, there is production of the intermediate compound 2,3-diphosphoglycerate (2,3-DPG), which competitively binds to heme to control oxygen affinity and thus offloading. The balance between intracellular synthesis and utilization of glucose is such that glycogen is not formed and therefore erythrocytes are critically dependent on a constant access to plasma glucose.

In circulation, red cells are constantly confronted with oxidant metabolites that are potentially injurious to heme, globin, enzymes, and membranes. The key antioxidant in the red cell system is reduced glutathione. Reduced glutathione may be exhausted by oxidant attack from peroxides produced by macrophages or drugs. This allows denaturation of the globin chains, which form intraerythrocytic precipitates called **Heinz bodies.** In animals the major deficiency affecting glutathione regeneration is an (red cell) age-dependent deficiency of the glucose-6-phosphate dehydrogenase enzyme (G6PD). The G6PD enzyme is the rate-limiting factor in the hexose monophosphate shunt and its deficiency in animals can be acquired by prolonged oxidant exposure, as occurs in kale anemia or phenothiazine toxicity.

In order to function in oxygen transport, hemoglobin iron must be maintained in the reduced or ferrous state. Oxidant attack which reaches the heme iron forms ferric heme or methemoglobin, which is reduced by methemoglobin reductase or the diaphorase system. Some methemoglobin may be reduced by ascorbic acid and glutathione, but the diaphorase system is most important. Both the glutathione and the diaphorase reductase systems are critically dependent on a constant supply of glucose for energy. Unlike the oxidation of globin, which results in hemolysis, the oxidation of heme causes cyanosis. Deficiencies in methemoglobin reductase are inherited in humans and acquired in animals with nitrate poisoning.

Erythropoietin is a constituent of plasma whose action is to increase the total number and volume of circulating red cells. The factor is class specific in that erythropoietin in one species of mammal is active in other mammals but not in birds or amphibians and vice versa. Erythropoietin is a glycoprotein containing 24% carbohydrate and 10% sialic acid and is found in plasma and, at lower concentrations, in urine. Erythropoietin is produced in the kidney.

Tissue anoxia is the specific stimulus for increased erythropoietin production and release. In this respect, arterial oxygen saturation is less basic to erythropoietin release than is tissue saturation, since altered binding affinities may make the blood oxygen unavailable to tissues. It is presumed that oxygen sensors as opposed to flow sensors occur in the kidney, but this is unproven; anoxia in other tissues (liver) may also control erythropoietin release.

Erythropoietin stimulates proliferation of primitive hematopoietic stem cells. It is not clear whether it affects only spontaneously committed erythroid precursors or if it involves undifferentiated stem cells which have receptors for all humoral factors (myeloid, erythroid, lymphoid) and differentiate in response to the specific hormone. There is an increase in the proportion of marrow cells committed to erythropoiesis (myeloid/erythroid ratio is reduced) and with time an increase in hematopoietic marrow volume at the expense first of fat and then of bone. The mitotic interval is shortened and there is shifting of the proliferative phase to include a proportion of the polychromatic rubricytes, thus doubling the output from that proportion of the stem cells. Accelerated hemoglobin synthesis occurs and, in sheep, a switching from synthesis of hemoglobin A to C, suggesting an action at the gene level above that of messenger RNA.

The net result of these changes is increased red cell production with an increase in mean cell volume and a greater proportion of maturation time spent in the peripheral blood. The presence of shift polychromatic macrocytes in the blood is erythropoietin dependent and they give a rough index of erythropoietin level and effect, indicating both increased levels of erythropoietin and a marrow capable and responding to it.

The **metabolism** of **iron** is central to the function of the erythron. Iron is contained in a variety of compounds and compartments within the body. These are conveniently referred to as iron pools. In adults, 75% of body iron is in hemoglobin and there are not sufficient reserves to replace iron lost by hemorrhage, if the loss exceeds a 30% drop in hematocrit. In young and growing animals, there are virtually no reserves, and any iron loss is therefore more serious than in adults.

Free plasma iron like other heavy metals is toxic and a three-tiered system of protein-mediated transport and storage exists. The most labile form of iron is contained in the beta globulin transport protein, **transferrin,** which is produced in the liver and which is normally 40–50% saturated with iron. In iron deficiency, transferrin levels may double, and levels decrease in protein starvation. The level of transferrin in serum is referred to as total iron binding capacity, whereas that portion carrying iron is called serum iron. Transferrin saturation levels of less than 16% are rate limiting to erythropoiesis. This limiting level is easily reached in deficient states since as serum iron drops, total iron binding capacity increases. Each transferrin molecule is able to carry two molecules of iron, of which the first is more easily transferred than the second. The developing rubricytes have some 50,000 specific transferrin receptors which cleave the iron molecules into the erythroid cytoplasm. Although the erythroid marrow receives only 5% of cardiac output, 85% of the iron in transport is retained.

Iron must exist in the ferric state to bind to transferrin. **Ceruloplasmin,** a copper-containing globulin, facilitates iron transport and converts ferrous to ferric iron in the presence of reducing metabolites. Ceruloplasmin decreases in starvation and copper deficiency. It appears to act as an acute-phase protein and increases in late pregnancy, infections, leukemia–lymphoma, and in anemia of ineffective erythropoiesis of iron- or B_{12}-deficient origin. Deficient ceruloplasmin results in impaired iron transport and hypochromic microcytic anemia indistinguishable from that caused by iron deficiency, except that marrow hemosiderin is increased rather than absent. Reticulocytosis following supplementation of deficient animals is delayed for several days while ceruloplasmin is synthesized.

Ferritin forms the second level of iron transport and storage and acts as a buffer between transferrin, where iron turnover is very high, and hemosiderin, where iron turnover is low. Ferritin is an iron protein complex occurring in most cells and plasma, and consists of apoferritin protein subunits and iron. In inflammatory states an acute reacting system continuously scavenges iron from transferrin to ferritin where it is less available to both invading microorganisms and erythropoiesis. This system appears to be the basis for the paradoxical anemia of chronic disease in which body iron stores are increased, but serum iron is rate limiting to erythropoiesis. Serum ferritin levels, unlike serum iron levels, are an accurate measure of body iron stores.

Hemosiderin is the most stable and least available form of storage iron. It is always intracellular, usually in macrophages, and is not found in plasma. Ferritin molecules have been observed in hemosiderin granules, and immunologically, the proteins in both are identical. It appears that hemosiderin is a variable compound consisting of native and denatured ferritin with a total iron content of 25–30% by weight. The iron in hemosiderin is available only after what appears to be prolonged proteolysis. Animals with abundant marrow hemosiderin do not respond maximally to blood-loss anemia unless supplemental iron is given. Iron from all sources, including that derived from hemolysis, hemosiderin, lysis, and oral and parenteral supplementation, appears to be additive in achieving erythropoiesis.

Hemoglobin synthesis begins early in the sequence of erythroid differentiation. The three essential components of hemoglobin are iron, protoporphyrin, and the globin chains. The assembly of these components probably is closely coordinated so that the organic structures are not produced in excess if iron is limiting. Hemoglobin synthesis is under genetic control. The amino acid substitutions of the globin chain seen in human hemoglobinopathies are not found in animals. In animals there is some switching of hemoglobin type as a result of alternate globin chains being produced under conditions of severe anemia and very high erythropoietic levels. This process may represent an acquired mutation which aids survival by substituting a chain with greater efficiency of oxygen delivery in anemic states. It also suggests that erythropoietin influences DNA transcription. Two-thirds of the erythrocytic hemoglobin is synthesized during the rubricyte stages and one-third is completed under the direction of messenger RNA left in the reticulocyte at denucleation.

The initial step in heme synthesis is the intramitochondrial union of glycine and succinyl-coenzyme A (CoA) to form delta-aminolevulinic acid. This step is dependent on the presence of pyridoxal phosphate of pyridoxine (vitamin B_6) origin. The delta-aminolevulinic acid is then transported to the cytosol, where two molecules are combined to form the ringed pyrrole porphobilinogen. Four of these pyrroles are then joined and closed to form the tetrapyrrole intermediates, uroporphyrinogen and coproporphyrinogen, which are then transported back into the mitochondria for the conversion to protoporphyrin and the addition of iron to form heme. In this sequence the porphyrinogens are the true intermediates and the uroporphyrin and coproporphyin are by-products of heme synthesis.

The process of heme synthesis is accelerated by the presence of iron and is slowed by a build-up of the end product hemin (protoporphyrin-iron chloride) probably at the level of delta-aminolevulinic acid synthesis. Lead poisoning blocks the formation of protoporphyrin and causes delta-aminolevulinic acid to circulate in excess; its urinary excretion constitutes a test for lead toxicity. In addition, there is an accumulation of the tetrapyrrols due to inhibition of iron insertion. These compounds are photodynamic and may be detected by fluorescence under ultraviolet examination.

Reticulocytes are newly released erythrocytes which contain RNA and mitochondria. On routine blood stains these cells are polychromatophilic or have diffuse basophilia, whereas on supravital staining with new methylene blue, these organelles condense into punctate basophilia. A reticulocyte can be defined as a cell with three or more focal basophilic aggregates which are readily visible in a single focal plane. They are not found in the peripheral blood of the horse and are rare in the blood of normal cattle, sheep, and goats. Normal dogs and cats have 0.5–1.0% of reticulocytes in peripheral blood and pigs have twice this level.

Reticulocytes are important as a measure of erythrocyte production in all domestic species except the horse. Reticulocytes that are moderately larger (1.5× diameter) than a normal homologous erythrocyte lose their basophilia in about 1 day of circulation. Macroreticulocytes or shift red cells (2.0–2.5× diameter of normal erythrocyte) require 2 or more days to become normochromic, during which time they continue to synthesize hemoglobin. Remodeling of normochromic macrocytes continues for a week or more, during which time there is loss of membrane and hemoglobin until the cell is of characteristic diameter for the species. This distinction is important since reticulocytosis is indicative of present production, whereas macrocytosis indicates previous production and, in the absence of reticulocytosis in an anemic animal, signals recent cessation of red cell production.

Young red cells have unique characteristics which affect their behavior. Because of their larger size, lack of concavity, and greater surface charge, they do not form rouleaux. Because of their reduced hemoglobin saturation, they have lower specific gravity and, in the absence of rouleaux formation, settle more slowly in sedimentation tubes, giving a biphasic reaction. Since they contain mitochondria and are capable of aerobic glycolysis, they are more resistant to crenation and hypotonic lysis.

The term **ineffective erythropoiesis** is used to indicate hematopoietic production that does not result in functional release of new cells. When erythropoiesis is ineffective, there is peripheral blood anisocytosis and poikilocytosis without appropriate reticulocytosis, and a tendency for increased bilirubin as a result of the increased intramedullary erythrocyte catabolism.

Bibliography

Anderson, P. H., Berrett, S., and Patterson, D. S. P. Glutathione peroxidase activity in erythrocytes and muscle of cattle and sheep and its relationship to selenium. *J Comp Pathol* **88:** 181–189, 1978.

Benz, E. J., and Forget, B. G. The biosynthesis of hemoglobin. *Semin Hematol* **11:** 463–523, 1974.

Blunt, M. H. *et al.* Red cell 2,3-diphosphoglycerate and oxygen affinity in newborn goats and sheep. *Proc Soc Exp Biol Med* **138:** 800–803, 1971.

Bunn, H. F. Erythrocyte destruction and hemoglobin catabolism. *Semin Hematol* **9:** 3–17, 1972.

Bunn, H. F., and Kitchen, H. Hemoglobin function in the horse: The role of 2,3-diphosphoglycerate in modifying the oxygen affinity of maternal and fetal blood. *Blood* **42:** 471–479, 1973.

Carter, E. I. *et al.* The kinetics of hematopoiesis in the light horse. I. The lifespan of peripheral blood cells in the normal horse. *Can J Comp Med* **38:** 303–313, 1974.

Dhindsa, D. S., Hoversland, A. S., and Templeton, J. W. Postnatal changes in oxygen affinity and concentrations of 2,3-diphosphoglycerate in dog blood. *Biol Neonate* **20:** 226–235, 1972.

Dunn, C. D. R., and Legendre, A. Humoral regulation of erythropoiesis in cats: Preliminary report. *Am J Vet Res* **40:** 779–781, 1980.

Hawkins, W. B., and Whipple, G. H. The life cycle of the red blood cell in the dog. *Am J Physiol* **122:** 418–427, 1938.

Jain, N. C., and Kono, C. S. Fusiform erythrocytes resembling sickle cells in angora goats; light- and electron-microscopic observations. *Res Vet Sci* **22:** 169–180, 1977.

Jensen, W. N. *et al.* The kinetics of iron metabolism in normal growing swine. *J Exp Med* **103:** 145–159, 1955.

King, M. E., and Mifsud, C. V. J. Postnatal changes in erythrocyte 2,3-diphosphoglycerate in sheep and cattle. *Res Vet Sci* **31:** 37–39, 1981.

Kitchen, H., and Brett, I. Embryonic and fetal hemoglobin in animals. *Ann N Y Acad Sci* **241:** 653–671, 1974.

Lee, C. K. *et al.* A comparison of haemoglobins from the adult and newborn dog. *Res Vet Sci* **15:** 333–337, 1973.

Lewis, I. M., and McLean, J. G. Physiological variations in levels of 2,3-diphosphoglycerate in horse erythrocytes. *Res Vet Sci* **18:** 186–189, 1975.

Valli, V. E. O. *et al.* The kinetics of haematopoiesis in the calf. II. An autoradiographical study of erythropoiesis in normal, anaemic and endotoxin treated calves. *Res Vet Sci* **12:** 551–564, 1971.

II. Anemia

Anemia is a sign of disease that results either from an increased rate of destruction or loss of erythrocytes or from a decreased rate of their production. Anemia is classified here on an etiologic or pathogenetic basis, with the type of anemia inferring some understanding of the underlying cause. The descriptions which follow provide morphologic criteria to assist in the differentiation of the various types of anemias.

The function of hemoglobin is to supply oxygen to the tissues, and the signs and lesions of anemia are largely referable to hypoxia. Depending on the cause, however, other manifestations of the diseases of which anemia is a part are often foremost in appearance. Thus, in anemia of massive hemorrhage, the manifestations relate to hypoxia and sudden reduction in the volume of circulating blood; in hemolytic anemia, the prominent features may be icterus and hemoglobinuria, with fever whether or not the cause is infectious. The rate at which anemia develops will also influence the manifestations: when the development is rapid, signs of anoxia are likely to be predominant; when the development is slow, as in deficiency and aplastic anemia, the condition may be masked by physiologic compensations until some increased requirement, often exercise, places overwhelming demands on the organism.

The appearance of an anemic animal is characterized by pallor, perhaps obscured by icterus. The pallor is due to a reduction in total levels of hemoglobin, and if the disease has developed slowly, to a dilution of the red cell mass by a proportionate increase in the plasma compartment in order to maintain normal blood volume. Pathologically the pallor is accompanied by edema of the lung, and tracheobronchial foam is typical. The heart is dilated and, if the anemia has been chronic, the myocardium tends to be flabby, friable, and pale as a result of fatty degeneration. The proportion of white to red clot is increased, and the liver and kidney are pale and may be fatty. In acute anemia, there is frequently the "nutmeg" appearance of periacinar hepatic necrosis. In acute hemolytic anemias, the gallbladder is distended and the feces are intensely stained with bile pigment. In chronic anemias of deficient hemoglobin production, the gallbladder is empty and the feces are pale. The character of the spleen varies with the cause and course of the disease. In the acute posthemorrhagic anemias, the spleen is contracted; in the acute hemolytic anemias, it is enlarged and pulpy; in the chronic hemolytic and deficiency anemias, it is enlarged and meaty. Since the spleen is a site of both production and destruction of red cells, the splenomegaly of hemolytic diseases may be the product of extramedullary hematopoiesis, or sinus hyperplasia of increased blood destruction, or both. In chronic anemia, the skeletal muscles tend to be pale in proportion to the loss of myohemoglobin. Edema is a frequent accompaniment of anemia, and serous fluid is found in the serosal sacs and in the dependent tissues of the limbs, ventral wall of the body, and throat. Since the full characterization of an anemia may require the examination of both blood and tissues, it is important that diagnostic examinations begin with the live animal. Cytologic preparations of bone marrow, like those of blood, can best be made prior to clotting and therefore both should be collected prior to death.

Bibliography

Jacob, H. S. A pathogenic classification of the anemias. *Med Clin North Am* **50:** 1679–1687, 1966.

A. Congenital Abnormalities of Red Cell Number and Function: Congenital Aplastic/Hypoplastic Anemia

Anemias due to a congenital error in stem cell number (reticular dysgenesis) are rare or lethal defects which are infrequently identified in animals. In contrast, those diseases characterized by normal or increased production of defective red cells with shortened life span are much more prevalent and, because of their potential for dissemination through artificial breeding, they are of much more economic importance. Neonatal llamas suffer a severe anemia associated with slow growth, limb deformity, and hypothyroidism. The disease has not been adequately characterized, but may be a primary defect of red cells or a reflection of systemic errors in metabolism. A disease of polled Hereford calves characterized by anemia and cutaneous abnormalities is described with The Skin and Appendages (Volume 1, Chapter 5).

Bibliography

Smith, B. B. *et al.* Erythrocyte dyscrasia, anemia, and hypothyroidism in chronically underweight llamas. *J Am Vet Med Assoc* **198:** 81–88, 1991.

Steffen, D. J. *et al.* Congenital anemia, dyskeratosis, and progressive alopecia in polled Hereford calves. *Vet Pathol* **28:** 234–240, 1991.

B. Hereditary Hemolytic Disorders of Red Cell Membrane or Enzymes

1. Hereditary Stomatocytosis

Red cells with a rectangular unstained area across their center are termed **stomatocytes** (Fig. 2.46). A hereditary anemia with these membrane changes occurs in chondrodysplastic dwarf Alaskan malamute dogs (see Bones and Joints, Volume 1, Chapter 1). The anemia in these dogs is mild and is characterized by stomatocytosis, macrocytosis, reduced mean corpuscular hemoglobin (MCH), increased osmotic fragility, shortened red cell life span, reticulocytosis with marrow erythroid hyperplasia, and increased iron turnover. The glutathione levels are reduced. The anemia is thus hemolytic and partially compensated. Intracellular sodium and potassium are also abnormal in affected dogs. Changes are most severe in homozygotes, in which both sodium and potassium are elevated. In heterozygotes, only sodium is elevated.

The hemolysis is low grade but intravascular, resulting in reduced serum haptoglobin, deposition of iron in renal epithelium, and hemosiderinuria. The defect is intrinsic to the red cells, since cells of affected donors have a shortened survival in normal dogs, whereas cells of normal donors survive normally in affected dogs. The reduced glutathione level does not render the affected cells more

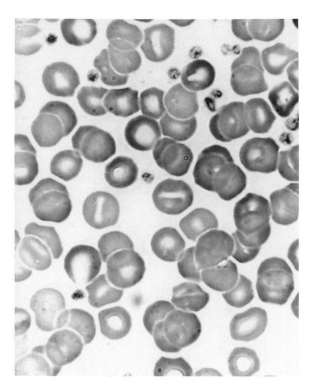

Fig. 2.46 Hereditary stomatocytosis. Central pallor of red cells with rectangular clear areas. Dog.

susceptible to oxidative denaturation of globin (Heinz bodies) because the enzyme present is unusually stable.

2. Familial Anemia in Basenji Dogs

A familial hemolytic anemia found in Basenji dogs is due to **pyruvate kinase deficiency.** The dogs are of normal appearance, size, and weight. They are less active than normal and on clinical examination have pale mucous membranes and often splenomegaly.

The anemia is moderately severe (60 g Hb/liter), macrocytic (MCV 86 fl) with large partially saturated cells [MCH, 23pg, mean corpuscular hemoglobin concentration (MCHC) 27%]. There is a reduction in red cell count $(2-3 \times 10^{12}/\text{liter})$ with a remarkable reticulocytosis of 40%. The young red cells with mitochondria are able to carry out aerobic metabolism and are protected until they mature. There is increased resistance to osmotic lysis for the same reason. There is a marked marrow erythroid hyperplasia with myeloid : erythroid ratios of 0.2–0.4. The apparent (^{51}Cr) red cell life span of normal Basenjis is about 20 days and 2.5–3.0 days in affected dogs. Plasma iron turnover rate is twice normal. Heterozygous carriers are asymptomatic but have higher than normal levels of red cell 2,3-diphosphoglycerate as do the fully affected homozygous animals. The oxygen dissociation curve of affected animals is shifted to the right with reduced hemoglobin oxygen affinity due to the high intracellular levels of 2,3-diphosphoglycerate.

C. Hereditary Disorders of Heme Synthesis: The Porphyrias

Porphyrias are diseases which result from abnormal metabolism of porphyrins, which may be excreted to excess in urine and feces or deposited in various tissues. The pigments are fluorescent unless iron is chelated in them. Fluorescent activation leads to the local production of free radicals in skin and exposed mucous membranes in the condition known as photosensitization.

Porphyrins are basic components of enzymes that contain heme, including hemoglobin, the cytochromes, catalase, and many peroxidases. Porphyrias occur when toxic or inherited metabolic defects limit the activity of one of the enzymes in the synthetic chain.

Bovine protoporphyria affects the Limousin breed of cattle and is caused by a deficiency in the activity of ferrochetalase, the terminal enzyme which catalyzes the insertion of ferrous iron into protoporphyrin IX to form heme. The inheritance is recessive and heterozygotes are asymptomatic. Bovine protoporphyria differs from bovine congenital porphyria in that there is no excretion of porphyrins in the urine, no anemia, and no discoloration of bone or teeth; photodynamic dermatitis is the only consistent clinical manifestation. Pigment is, however, deposited in the liver and heavy concentrations resembling lipofuscin are found in the cytoplasm of hepatocytes, Kupffer cells, endothelium, and portal stroma.

Congenital porphyria occurs in humans, cattle, and cats, and results from the production of the abnormal protoporphyrin isomer I. It occurs also in swine, but the biochemical defect is not known. All cells in the body share this defect but, since the effects on the blood system are most apparent, the disease is usually called erythropoietic porphyria. Only isomer III can be synthesized into heme, although the enzymatic block is never complete, and some normal heme is produced. The abnormal porphyrinogen intermediates accumulate in tissues and are readily oxidized to porphyrins capable of absorbing visible light and emitting visible fluorescence, causing photosensitization (see The Skin and Appendages, Volume 1, Chapter 5). The disease is transmitted as a recessive trait (dominant in pigs) and heterozygotes are unaffected.

The enzyme uroporphyrinogen III cosynthetase is much reduced in affected animals and intermediate in heterozygotes. The disease results from either an absolute or relative deficiency of this enzyme because of overproduction of proximal cosynthetase enzymes. The type I isomer builds up in red cells and tissues and may be detected by fluorescence.

There is a mild to moderate anemia (60–110 g/liter) exacerbated by exposure to light. The anemia is macrocytic normochromic in recovery. Polychromasia is mild and increases with light exposure. Marrow is hyperplastic with increase in late-stage rubricytes. Red cells fluoresce and urine is brown. Red cell protoporphyrin is increased (350–450 mg/liter of red cells, normal being 0.1–0.2 mg/liter red cells). There is reduced ferrochelatase activity, but the level of immunoreactive protein is normal, suggesting a point mutation in the gene coding for the synthesis of this enzyme.

Acquired porphyrias may follow exposure to lindane and related substances (see Bones and Joints, Volume 1, Chapter 1).

Bibliography

Bannerman, R. M., Edwards, J. A., and Pinkerton, P. H. Hereditary disorders of the red cell in animals. *In* "Progress in Hematology," Vol. VIII, E. B. Brown, (ed.), pp. 131–176. San Diego, California, Grune & Stratton, 1973.

Burman, S. L. *et al.* Pyruvate kinase deficiency anaemia in a Basenji dog. *Aust Vet J* **59:** 118–120, 1982.

Giddens, W. E. *et al.* Feline congenital erythropoietic porphyria associated with severe anemia and renal disease: Clinical, morphologic, and biochemical studies. *Am J Pathol* **80:** 367–380, 1975.

Kaneko, J. J. Critical review: Animal models of inherited hematologic disease. *Clin Chim Acta* **165:** 1–19, 1987.

Pinkerton, P. H. *et al.* Hereditary stomatocytosis with hemolytic anemia in the dog. *Blood* **44:** 557–567, 1974.

Straka, J. G. *et al.* Immunochemical studies of ferrochelatase protein: Characterization of the normal and mutant protein in bovine and human protoporphyria. *Am J Hum Genet* **48:** 72–78, 1990.

Troyer, D. L. *et al.* Gross, microscopic and ultrastructural lesions of protoporphyria in Limousin calves. *J Am Vet Med Assoc* **38:** 300–305, 1991.

D. Anemia of Deficient Cell Production

Anemias due to reduction in red cell production are characterized by hypoproliferation that primarily affects cell division as opposed to hemoglobin production. The causes of inadequate cell production include immune phenomena, often drug induced, toxins both exogenous and due to hepatorenal failure, myelophthisis, chronic inflammation, endocrine dysfunction, and nutritional deficiency.

1. Aplastic Pancytopenia

Aplastic pancytopenia occurs in all species but is most often seen in cats and dogs. It may be congenital, but is usually acquired and idiopathic, and may be associated with specific infections or toxic agents. Marrow aplasia may be absolute and rapidly fatal or, more commonly, hypoplastic and chronic. Any of the three blood cell lines may become aplastic, but characteristically, the disease is one either of generalized aplasia resulting in pancytopenia or of pure red cell aplasia (Fig. 2.47) with a hyperplastic myeloid and megakaryocytic marrow. Rarely, diseases characterized by leukopenia are accompanied by marrow myeloid hyperplasia with apparent immune suppression of myeloid maturation. Thus, the apparent marrow hyperactivity is ineffective, and output is deficient. The accepted criteria for aplastic pancytopenia (aplastic anemia) are peripheral blood neutrophils fewer than 0.5×10^9/liter with platelets fewer than 20×10^9/liter, an absence of

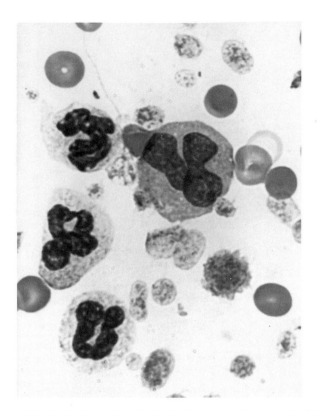

Fig. 2.47 Pure red cell aplasia. Nonresponsive anemia with excess neutrophils, monocytes, platelets. Peripheral blood. Dog.

reticulocytes, and severe marrow hypocellularity in which fewer than 30% of residual cells are of hematopoietic type. Suggested mechanisms for aplasia include decreased numbers of normal stem cells, inhibition of stem cell maturation, abnormal marrow microenvironment, lack of hematopoietic factors, and marrow depression associated with thymoma.

Viruses may cause marrow depression. The most important of these in animals are the canine and feline parvoviruses and feline leukemia virus. **Phenylbutazone** causes aplasia in humans but appears not to have this effect in animals unless there is prolonged exposure at high dosage. All forms of **estrogenic drugs** are potentially myelotoxic in dogs, and disease is characterized by an initial leukocytosis followed by thrombocytopenia, hemorrhage, and aplasia, which is usually irreversible. **Chloramphenicol** causes an irreversible myeloid aplasia in humans as an apparent idiosyncratic reaction. It does not have this idiosyncratic effect in animals, in which marrow suppression is dose related at above-normal rates of administration. **Furazolidine** at high dose rates is acutely neurotoxic in calves and pigs. When administered long term to calves at dose rates used for prophylaxis of infection, furazolidine causes toxic depression of bone marrow manifested acutely as agranulocytopenia with widespread embolic infections, clinically best revealed as discrete necroses and ulcers of mucous membranes.

Animals with aplastic pancytopenia have epistaxis,

bleeding into joints and dependent areas, and are often febrile due to sepsis. Animals with pure red cell aplasia present with mucosal pallor, lethargy, and reduced activity, and are usually afebrile. Pancytopenia may be either moderate and chronic or severe and acute with an absence of neutrophils. Those granulocytes present show toxic changes and little immaturity. Thrombocytopenia is severe and nonregenerative, with all platelets aged, small, and with little basophilia. Anemia is moderate to severe and normochromic normocytic without polychromatic response. Differences in life span and rates of consumption of the cells of blood dictate that the aplastic syndromes are dominated by the purpura of thrombocytopenia and the sepsis of agranulocytosis. Pathologically, hemorrhage and sepsis will predominate in proportion to the relative deficits of platelets and/or neutrophils. If neutropenia is severe, there will be widespread bacterial emboli in lung, liver, and kidney unaccompanied by the edema and cell reaction characteristically seen with a competent marrow.

2. Bracken Fern Poisoning

Bracken fern (*Pteridium aquilinum*) is abundant in many humid grassland areas of the world, and poisoning by it is often the factor which limits the utilization of pastures. There are other species in the bracken fern genus, notably *P. esculentum,* the common bracken, which are toxic, and the diseases produced are identical. *Cheilanthes sieberi,* the mulga or rock fern, contains much higher levels of thiaminase than does bracken and produces the typical hematopoietic syndrome in cattle and polioencephalomalacia in sheep.

Bracken fern poisoning produces a variety of disease syndromes in grazing animals. The acute disease occurs in horses and pigs as a neurologic and cardiac manifestation of thiamine deficiency, in sheep as polioencephalomalacia, and in cattle as aplastic pancytopenia. The chronic syndromes occur in cattle and sheep as enzootic hematuria and bright blindness, and in cattle there is an association with epithelial tumors of the alimentary tract. Some of these syndromes are appropriately discussed in other chapters.

Bracken fern appears to be most poisonous when it is green and growing actively. Hay made from the fern is also toxic, provided that the plants are well dried and retain some of the natural color. The fern is not particularly palatable and the ecologic conditions which are conducive to the ingestion of toxic amounts of the plant are drought resulting in inadequate top growth of improved forage or an excess of lush pasture in which animals seek out coarse growth in order to obtain sufficient roughage. The poisoning is cumulative and the disease may not be evident until after many weeks of continuous exposure, or for as long as 8 weeks after last exposure. Cattle and horses are the species most susceptible; sheep are relatively resistant as well as being fastidious eaters; and pigs, which eat the rhizomes, are occasionally poisoned. The manifestations of poisoning differ in the different species, but depression of hematopoiesis is common to them all. In cattle and sheep, the essential change is irreversible trilineage marrow hypoplasia. This results in **aplastic pancytopenia** and death due to thrombocytopenic hemorrhage complicated by septicemia of neutropenia.

Horses and rats suffer myelosuppression but die of a nervous disease, which is curable by thiamine, before the effects of pancytopenia are fully developed.

Horses whose diet contains 40% or more of bracken fern suffer from inappetence, loss of weight, bradycardia, and nervous signs which progress to recumbency. The illness can be both prevented and treated with thiamine. The syndrome is comparable to that in horses poisoned by *Equisetum* spp., especially *E. arvense,* in which a thiaminase is present as it is in bracken fern and other pteridophytes. As expected in thiamine deficiency, there is, in horses and rats poisoned by bracken fern, a gradual rise in the pyruvate levels in blood. The hematologic changes are not of prognostic significance in bracken poisoning in these species. The hematologic changes are reversed by thiamine, but this effect appears to involve prevention of thiaminase toxicity rather than by correcting a thiamine deficiency, since a simple deficiency in rats does not cause anemia. There are gaps in our knowledge of some importance since the disease in cattle apparently depends on a direct inhibition of a pluripotent stem cell and not a disturbance of the metabolism of thiamine. The principal hematologic change in nonruminants is thrombocytopenia, but there is also a tendency to neutropenia and anemia complicated by hemorrhage. The changes are consistent with what would be expected with inhibition of proliferation of blood cells in a pipeline production, in which the platelets have a 4-day production time and a life span of 9 days, and the neutrophils a 6–7 day production time with a life span of only 6–8 hr. The anemia of hypoproliferation is less significant, unless complicated by bleeding, because of the long life span of the erythrocyte.

The morbidity from bracken poisoning in **cattle** is not high, but the mortality is, and recovery is rare once clinical signs are evident. This is logical since clinical signs are secondary to a complete loss of the pipeline of blood cell formation, and even a return of stem cell function would require survival through a week of thrombocytopenia and neutropenia before the results of renewed production become functional in the peripheral blood. In cattle, thrombocytopenia occurs first and is followed in hours by leukopenia of neutropenia, both likely complicated by increased consumption as hemorrhagic foci remove both from circulation. In contrast to the platelet and neutrophil precursors, some rubricytes persist in the marrow, but the anemia is nevertheless largely nonresponsive. Because of the long red cell life span in cattle (155 days), the anemia is, however, largely due to hemorrhage which is both enteric and generalized. As the disease develops, clot retraction is impaired out of proportion to the reduction in the platelet count, suggesting impaired function (thrombasthenia) as well as deficient numbers. Increased capillary fragility has also been proposed and is likely a consequence of unremitting thrombocytopenia.

The dominant features of the clinical disease are hemorrhages, bacteremia, and fever. The hemorrhages are due to thrombocytopenia, and the bacteremia, due to neutropenia. Bacteremia and bacterial embolism contribute to the genesis of hemorrhage. The latent period of bracken poisoning in cattle is 2 months or more, and even longer in sheep, and the clinical episode develops suddenly and acutely, with high fever, hemorrhages on mucous membranes, and hemorrhages from natural orifices including the urethra. Calves frequently show excessive discharge of mucus, sometimes bloodstained, from the nose and mouth, edema of the throat, and focal, necrotic stomatitis and cheilitis (Fig. 2.48). The course is usually less than 3 days. In the rare cases where recovery has been followed, the return of the leukocyte count to normal may require several months, which suggests continuing toxicity or injury to a very primitive stem cell pool of restricted numbers that requires weeks to replenish.

Pathologically there is hemorrhage in almost all tissues but without any particular pattern (Fig. 2.49). They are particularly numerous, however, in the stomach and intestine, to the mucosal surface of which the blood tends to adhere in clots; in the large intestine, the hemorrhages are more diffuse, and copious free blood may be present in the lumen. Hemorrhages beneath mucous membranes usually lead to overlying ulceration. Hemorrhages in the liver may be sufficiently numerous to give it an odd variegated appearance, especially when it also contains many infarcts

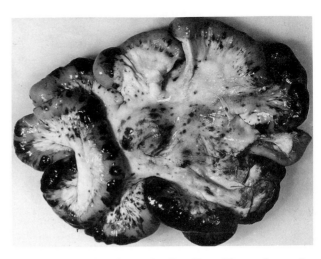

Fig. 2.49 Bracken fern poisoning. Serosal hemorrhages. Intestine. Ox.

(Fig. 2.50A). The hepatic infarcts are spherical or wedge-shaped areas of white or yellow and similar to the lesions of black disease; they may be absent, few, or numerous. The hepatic infarcts, like those in other tissues, are caused by bacterial emboli, which localize in small branches of the portal vein, the sinusoids, and the central veins. The bacteria are ordinary saprophytes in many cases and, if the postmortem interval is prolonged, the focal lesions

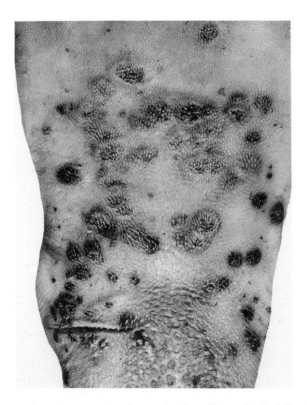

Fig. 2.48 Bracken fern poisoning. Hemorrhagic infarcts. Tongue. Ox.

Fig. 2.50A Bracken fern poisoning. Fatty liver with pale and red infarcts resulting from bacterial embolism. Ox.

expand about the colonies because of local postmortem putrefaction.

The heart has extensive subserosal hemorrhage but, in addition, there are foci of myocardial infarction which are discrete yellow areas up to 1 cm long. There is usually little or no cellular reaction to these embolic bacteria or to the necrosis of tissue in the infarcts (Fig. 2.50B). Infarcts in the kidney are usually red and also caused by bacterial embolism. Extensive hemorrhages are rather consistently present in the walls of excretory passages of the urinary system. The red marrow, which in adults should be firm and fatty, is instead pink and soft, and it dehydrates and shrinks on exposure to air. Marrow cellularity is markedly diminished to the point where either only small islets of erythropoiesis remain, or there is complete aplasia with edematous stroma and dilated sinusoids, and only a few fat cells are present.

Pigs fed rhizomes show only minor signs of toxicity, including depression of growth rate, increases in blood pyruvate, and a reticulocytopenia, all of which can be corrected by administration of thiamine. Neurologic signs such as occur in horses are not observed, but there is myocardial injury of the type observed in experimental thiamine deficiency in pigs. The rhizomes contain substantially more thiaminase than do the fronds.

The experimental production of bracken poisoning in **sheep** is very difficult. Sheep find the fern unpalatable and are slow to develop signs of poisoning. The clinical and

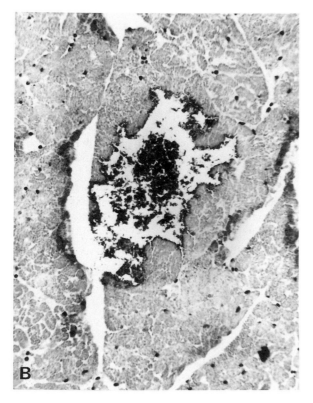

Fig. 2.50B Bracken fern poisoning. Bacterial embolism without inflammatory response. Myocardium. Ox.

pathologic manifestations of the disease are analogous to those in cattle, including the development of pancytopenia.

The previous discussion relates to two toxic factors in bracken fern, one a thiaminase and the other a substance that causes bone marrow depression, especially in ruminants. A nosesquiterpene glucoside, ptaquiloside, extracted from bracken fern, has produced bone marrow depression in a calf, but its significance to the naturally occurring disease is unclear. It appears to be the responsible agent in enzootic hematuria.

Disease in cattle, apparently identical to that produced by bracken fern, has resulted from feeding of soya bean meal which had been extracted with trichlorethylene, and in calves fed diets to which nitrofurans were added for bacteriostatic purposes.

3. Pure Red Cell Aplasia

Pure red cell aplasia is seen in humans and dogs. Pathogenetically there is an immune suppression of differentiated red cell precursors that is T cell driven and mediated by interferon, and there may also be antibody against erythroid nuclei. Therapy, which is likely to be effective if vigorous, is indicative of the mechanism of aplasia and includes immunosuppression by drugs, antithymocyte globulin, and plasmaphoresis. Clinically the disease is likely to occur in a mature dog and may result from exposure to a drug or virus, or may be associated with a lymphoid tumor of the large granular lymphocyte type (see Chronic Lymphocytic Leukemia, Section IV,A,2 of The Leukon). There is weight loss and exercise intolerance, but the appetite is usually retained. There is pallor of membranes but no tendency for bleeding or infection. Hematologically the anemia may be severe with hemoglobin less than 50 g/liter and worse than the red cell numbers might suggest since the red cells, although normochromic and normocytic, become small with age. Whereas reticulocytes are consistently absent, there appears to be stimulation of those lineages able to respond and there is characteristically a mature neutrophilia of $15-25 \times 10^9$/liter and a thrombocytosis of $500-1000 \times 10^9$/liter with a lot of variation in size, basophilia, and granulation such that the larger platelets may interfere with the red cell count (Fig. 2.47). Monocytes are increased, likely as part of the stress reaction, to $2-3 \times 10^9$/liter. The bone marrow has normal cellularity, but the myeloid : erythroid ratio is greatly increased as are megakaryocytes, but rubricytes are rare. Pathologically pallor is the dominant finding, often with pulmonary edema and cardiac dilation. There is ischemic damage to periacinar hepatocytes and proximal renal epithelium. The spleen and nodes are of normal size without extramedullary erythropoiesis, reflecting the generalized suppression of the red cell system.

4. Feline Panleukopenia

Feline panleukopenia (infectious feline enteritis), a parvoviral infection to which all Felidae are susceptible, is described with The Alimentary System (Volume 2, Chapter 1). Clinically the disease may be of such sudden onset that poisoning is suspected. Signs begin 5–6 days after exposure of a susceptible animal and may be mild or fatal. Typically there is prostration with vomiting, diarrhea, and severe dehydration. Fever is variable and may not be present after onset of disease, likely because of the need for functional granulocytes for production of pyrogen. Pathogenetically the disease is radiomimetic and the signs and lesions are referable to destruction of proliferating cells principally of marrow, intestinal epithelium, and spleen, nodes, and thymus. Hematologically the leukocyte counts decrease to less than one-third of normal level on day 5 after experimental infection, remain low for 5 days, and then rise rapidly in cats that survive. In those that die, the leukocytes decline to fewer than 1×10^9/liter by day 7, the final day. This drop is almost entirely due to a reduction in granulocytes, and the timing of both the decline and recovery is related to the production time of the neutrophil, which can be deduced to be 5–6 days, as in the calf and dog. Thus, there is complete cessation of all lines of hematopoiesis shortly after infection and those cells in the postmitotic pool continue to mature and are delivered to the blood in pipeline fashion. After a short period of aplasia, there is recovery of the stem cell pool and an advancing front of early myeloid cells that appears as maturation arrest at days 7–8, when death due to dehydration usually occurs. The intestinal signs do not begin until granulocytes become deficient and, in the germ-free cat, there is a mild decline in total leukocytes but no loss of intestinal epithelium or diarrhea. In recovery, erythropoiesis is delayed until there is a peripheral blood neutrophilic leukocytosis (peak day 13), indicating a shunting of stem cell commitment to the lines essential for survival. The marrow in this disease is extremely labile and changes occur within hours, accounting for the variation seen with disease progression and recovery. Marrow cellularity drops to less than one-third of normal by the third day after inoculation, is less than 20% of normal between days 5 and 7, and returns to about half of normal on day 10. Platelet counts decrease to half the normal level on day 5 and in survivors return to original levels between days 7 and 10 postinfection. Those animals which die have severe aplasia with complete loss of the myeloid and erythroid proliferative phase cells. Survivors have the lowest levels of myeloblasts on day 3 followed by recovery on day 5, whereas the erythroblasts reach their lowest level on day 5 and recovery is delayed until day 10.

Pathologically lymph nodes, spleen, and Peyer's patches have both thymus- and marrow-dependent hypocellularity. Follicles are characteristically large and are occupied by a background of pale dendritic reticular cells made apparent by lysis of the follicular center cells. Fragmented nuclei are usually absent, indicating that lympholysis occurred several days previously. Follicles are rimmed by a narrowed zone of small lymphocytes, and paracortical areas are also hypocellular. All tissues are lacking in neutrophils and there is often extensive erythrophagocytosis in the medullary sinuses of mesenteric nodes. The latter is presumably the result of erythrocyte injury, possibly due to the adherence of luminal toxins during their passage through the capillaries of denuded villi. Striking changes occur in the thymus and consist of complete cortical collapse. The thymus becomes a thin band of faceted lobules composed of hypoplastic medullary areas with collapse of the fascial boundaries around the depopulated cortices. Thymic atrophy is the major lesion in experimental infection of germ-free cats where clinical disease is minimal.

5. Stachybotryotoxicosis

Stachybotryotoxicosis, as a pancytopenic disease, occurs in horses and ruminants in eastern Europe and Russia. The responsible fungus *Stachybotrys alternans* (*atra*) grows on substrates rich in cellulose and contaminates hay and cereal grain. The nature of the marrow depressant toxin is not known but appears to be related to the trichothecenes produced by *Fusarium*. Although best documented as the cause of extensive outbreaks of disease in horses, episodes of comparable magnitude also occur in cattle and sporadic episodes occur in sheep and swine. Other species including humans are susceptible. The syndrome occurs in three stages: the initial local signs of irritation; a period of quiescence in which blood dyscrasia occurs; and a final episode, being the consequence of severe pancytopenia.

Following ingestion of contaminated food, there is drooling, hyperemia of buccal mucosa, and deep fissuring of the lips leading to extensive inflammatory edema of the soft tissues of the buccal cavity. Local lymphadenopathy may persist for several weeks. There is an early leukocytosis, followed by progressive depletion of formed elements of blood, especially platelets and leukocytes, with prolonged prothrombin and clot-retraction times. The terminal stage is fairly brief and manifested as hemorrhagic diathesis and bacterial infection.

The pathological picture in cattle closely resembles that observed in bracken fern poisoning, and that in horses is characterized by profuse hemorrhage and necrosis in many tissues. In sheep, the fleece may be shed at about the time when hemorrhages occur. Necrotic processes in the mucous membranes of the alimentary canal are characteristic, especially of lips, tonsillar fauces, and large intestine. There is more or less severe depletion of hematopoietic and lymphopoietic tissues.

6. Myelophthisic Anemia

Myelophthisic anemia of myelofibrosis is discussed with Myeloproliferative Disease (Section IV of The Leukon) and the displacement of marrow in osteopetrosis and osteodystrophia fibrosa is discussed with Bones and Joints (Volume 1, Chapter 1).

Myelophthisis of leukemia is seen most often in dogs, cats, and cattle. The anemia of leukemia has four mechanisms: stem cell displacement, panmyelosis, immune depression, and cachexia. Marrow failure with consequent loss of all normal cells represents the end stage of acute leukemia. These cell deficits occur earlier and with greater diversity in acute myelogenous leukemia than in acute lymphocytic leukemia. The basic difference between these two diseases appears to depend on the relative distribution of the leukemic stem cells in the bone marrow. Normally the pluripotent stem cells are in highest concentration in the subendosteal marrow, apparently because of the tropism resulting from the bone marrow portal capillary drainage. Leukemic lymphoid cells tend to be distributed randomly throughout the marrow cavity without predisposition for subendosteal hematopoietic areas, and although the marrow may be solidly infiltrated by tumor, marrow failure is a late occurrence. In contrast, in acute myeloid leukemias the malignant stem cells compete directly with their benign counterparts, driving them centripetally, in which location they are stimulated to enter a pattern of differentiation rather than of self-renewal, ultimately leading to depletion of the normal stem cell pool. The anemia in myelogenous leukemia may also be related to the fact that the erythroid cells are themselves the progeny of neoplastic precursors which do not respond to normal demands for proliferation.

In the lymphoid leukemias particularly, there are often immune phenomena associated with the appearance of the neoplastic clone. In chronic lymphocytic leukemia, therapy of the disease may result in the development of immune hemolytic anemia. Similar changes are observed in some lymphomas and the effects may vary from a subtle immune depression of erythropoiesis to destruction of apparently normal peripheral blood erythrocytes.

All animals with leukemia which survive for more than a few weeks show some degree of the cachexia characteristic of cancer, due apparently to the systemic production of tumor necrosis factor by defensive cells. In addition to the wasting of cancer, there is the specific impairment of hemoglobin synthesis as a result of scavenging of serum transport iron, which is part of the biochemical response to chronic disease.

7. Anemia of Uremia

Anemia of uremia occurs in all species but is most often recognized in the dog and cat. Anemia is a characteristic feature of chronic renal failure and urea levels of 35 mmol/liter or higher are almost always associated with a hematocrit of 30% or less. There are four principal mechanisms of anemia in renal failure. These are excessive hemolysis of red cells due to retention of creatinine and guanidinosuccinic acid, toxic depression of erythropoiesis, loss of renal erythropoietin, and blood loss from the kidney. All of these processes may be operative to some degree with variation in their relative importance from case to case, but the major limiting factor in the anemia of uremia in humans and animals appears to be inadequate levels of erythropoietin.

The anemia is normocytic, largely normochromic, and of mild to moderate severity. Polychromasia is rare or absent and there is some poikilocytosis with an occasional hypochromic red cell. Toxic changes in neutrophils are usually prominent. The diagnosis is based on clinical signs and other biochemical evidence of uremia (see The Urinary System, Volume 2, Chapter 5).

8. Anemia of Endocrine Dysfunction

Mild anemia of marrow hypoplasia is associated with loss of function of the pituitary, thyroid, and adrenal glands, and to a lesser extent of the gonads. Anemia related to **pituitary dysfunction** appears to be due both to loss of thyroid-stimulating hormone and adrenocorticotrophin, as both are required to maintain hemoglobin levels. **Hypothyroidism** is associated with normal red cell survival but decreased red cell production, red cell mass, and blood volume. Therapy may temporarily exacerbate the anemia since the plasma volume recovers more rapidly than does the red cell mass. A mild normochromic normocytic anemia occurs in **adrenal insufficiency.** Its true extent is masked by a concomitant reduction in plasma volume. The anemia is corrected by steroid therapy, although the mechanism is unclear. There is often a mild increase in red cell numbers in hyperadrenocorticism: however, it is likely that the effect is erythropoietin mediated and not a direct effect of adrenal steroids on marrow. **Castration** lowers hemoglobin levels 10–20 g/liter in male animals to just above the normal female level. If females are spayed, their hemoglobin levels rise to those of castrated males.

Anemia is rarely of clinical significance in hormonal dysfunction except in the need to distinguish it from anemia of other cause, and the reduction in hematocrit is usually an incidental finding. It is a mild normochromic normocytic anemia without toxemia or polychromatic response.

Bibliography

Adamson, J. W., and Eschbach, J. W. Treatment of the anemia of chronic renal failure with recombinant human erythropoietin. *Annu Rev Med* **41:** 349–360, 1990.

Angel, K. L. *et al.* Myelophthisic pancytopenia in a pony mare. *J Am Vet Med Assoc* **198:** 1039–1042, 1991.

Bartl, R. *et al.* Lymphoproliferations in the bone marrow: Identification and evolution, classification and staging. *J Clin Pathol* **37:** 233–254, 1983.

Carlson, J. H., Scott, F. W., and Duncan, J. R. Feline panleukopenia. I. Pathogenesis in germ-free and specific pathogen-free cats. *Vet Pathol* **14:** 79–88, 1977.

Evans, W. C., Evans, E. T. R., and Hughes, L. E. Studies on bracken poisoning in cattle. Parts I, II, and III. *Br Vet J* **110:** 295–306, 365–380, and 426–444, 1954.

Fenwick, G.R. Bracken (*Pteridium aquilinum*)—Toxic effects and toxic constituents. *J Sci Food Agric* **46:** 147–173, 1988.

Hardy, C. L., and Balducci, L. Review: Hemopoietic alterations of cancer. *Am J Med Sci* **290:** 196–205, 1985.

Hironi, I. *et al.* Reproduction of acute bracken poisoning in a calf with ptaquiloside, a bracken constituent. *Vet Rec* **115:** 375–378, 1984.

Kurtzman, G. *et al.* Pure red-cell aplasia of 10 years' duration due to persistent parvovirus B19 infection and its cure with immunoglobulin therapy. *N Engl J Med* **321:** 519–538, 1989.

Lacey, J. (ed.). "Trichothecenes and Other Mycotoxins." Proc.

Internat. Mycotoxin Symp., Sydney, Australia, John Wiley 1985.

Langheinrich, K. A., and Nielsen, S. W. Histopathology of feline panleukopenia: A report of 65 cases. *J Am Vet Med Assoc* **158**: 863–872, 1971.

Lawrence, J. S. *et al.* Infectious feline agranulocytosis. *Am J Pathol* **16**: 333–354, 1940.

Lund, J. E., and Avolt, M. D. Erythrocyte aplasia in a dog. *J Am Vet Med Assoc* **160**: 1500–1503, 1972.

Mirocha, C. J. Rubratoxin, sterigmatocystin and *Stachybotrys* mycotoxins. *In* "Conference on Mycotoxins in Animal Feeds and Grains Related to Animal Health," Rockville, Maryland, U.S. Department of Commerce National Technical Information Service, PB80-221773, Food and Drug Administration, 1980.

Schneider, D.J. *et al.* A field outbreak of suspected stachybotryotoxicosis in sheep. *J South Afr Vet Assoc* **50**: 73–81, 1979.

Weiss, D. J., and Klausner, J. S. Drug-associated aplastic anemia in dogs: Eight cases (1984–1988). *J Am Vet Med Assoc* **196**: 472–475, 1990.

E. Anemia of Deficient Hemoglobin Production

The anemias of deficient hemoglobin may be subdivided on the basis of pathogenesis into anemias of deficient heme and anemias of deficient globin. In animals, anemias due to insufficient heme are largely a result of iron or copper deficiency or chronic inflammatory disease. In erythropoietic porphyria, there is production of an ineffective isomer of protoporphyrin, and this disease is considered under congenital disorders of red cells (Section II,C of The Erythron). Anemias due to deficient globin (e.g., thalassemia) are not of clinical importance in veterinary medicine; however, true imbalances in the production of the alpha and beta chains of globin have been identified in mouse models of this disease.

1. Iron-Deficiency Anemia

Iron is an essential component of hemoglobin, myoglobin, the cytochrome enzymes of mitochondria and hepatic microsomes, and the metalloflavoproteins dicotinamide adenine dinucleotide, reduced (NADH), and is required as a cofactor for the function of other enzymes. Thus, whereas anemia is the principal sign of iron deficiency, there are other diverse effects which include neutropenia, decreased capacity for killing ingested bacteria, impaired T cell response to mitogens, and decreased capacity for interferon production. Iron deficiency results in alterations in several biogenic amines that are thought to be responsible for irritable behavior and cognitive defects observed in humans and animals. In addition, deficient states result in increased catecholamine levels leading to cardiac hypertrophy in rats and impaired temperature regulation and cold tolerance. A deficiency of body iron stores, as distinguished from iron-deficiency anemia, may develop from either a deficiency of the element in the diet, impaired absorption, or excessive loss. In discussing the pathogenesis of iron deficiency, it is convenient to consider each of these possibilities.

The only natural **diet** of animals which can result in a **deficiency of iron** is milk and therefore deficiency of this pathogenesis is seen only in sucklings. Supplementation of the diet of the dam does not influence the occurrence of iron deficiency in the sucklings. In terms of animal husbandry, the recognition of the need for supplemental iron for neonates has been a relatively recent development which followed serious losses when confinement rearing of swine was adopted early in this century.

Impaired absorption of iron is rarely of significance in animals but does occur in dogs with acquired malabsorption syndromes. Absorption takes place from the stomach and duodenum, and the degree of absorption within limits is geared to demand. About 5% of dietary iron is absorbed in the replete animal and this proportion may increase to near 20% in deficient states. Absorption varies with the form in which iron is ingested. Thus, heme iron is well absorbed, whereas inorganic iron is much less available, but in adequate amounts is adequately absorbed. Iron absorption appears to be regulated as a function of the level of saturation of the serum transport protein transferrin. Transferrin has a high binding coefficient for iron and is normally about 50% saturated. In deficient states, saturation drops below the 16% level, at which point iron is not partitioned to the marrow rubricytes, and iron supply becomes rate limiting to red cell production. The developing enterocytes become imprinted with ferritin, the next level of iron transport, in proportion to body iron stores, and subsequently they absorb iron in proportion to the body's needs. Thus, absorption is partially blocked by high body stores and vice versa.

Free iron in serum in excess of the binding proteins is toxic and promotes bacterial invasion. In well-fed dams and sucklings, these binding proteins rapidly scavenge excess iron given parenterally, so that toxicity or infection does not occur. In debilitated animals, particularly the young, parenteral iron should be withheld until animals are gaining weight and serum proteins can be assumed to be replenished. Thus, teleologically, the physiological anemia of the young is a defensive strategy to deny bacteria access to iron during the period of transition from passive to active immunity. Since sucklings of all species are born with minimal stores, the increased requirements for iron concomitant with rapid growth and expansion of blood volume guarantee that they will develop iron deficiency if not given access to dietary iron or are otherwise supplemented.

Excessive loss of iron is the result of chronic loss of blood and is an important cause of iron deficiency; it is the only cause in adult animals, since normally only very small amounts of iron are excreted and lost.

The signs of iron deficiency are those of chronic anemia, although they may be visibly complicated by the underlying pathologic process. Apart from the effects on blood and those changes which are the result of anemia, lesions specifically attributable to iron deficiency are limited to the recognition of the anemic state in conjunction with an absence of stainable iron in tissues. Body iron stores can be estimated as a function of plasma ferritin where 1 μg of ferritin can be equated with about 120 μg of iron/kg body weight. The **depletion of body iron** can be considered at **three levels of severity** based on the levels of saturation of transferrin, the level of hemoglobin, and of red cell protoporphyrin. At the first stage of depletion, the hemoglobin and transferrin saturation are normal, but ferritin is decreased, and hemosiderin is absent. At the second stage of depletion, both ferritin and transferrin saturation are deficient and red cell production is slowed. Finally, in iron-deficient anemia, there is not only loss of all iron stores, but hemoglobin is decreased, and red cell protoporphyrin is increased as a result of impaired heme synthesis. Rubricyte maturation is coordinated to produce heme and globin in proportional amounts. Lack of iron slows protoporphyrin and globin synthesis, and there is delayed rubricyte maturation which results in an increased proportion of mature rubricytes in the bone marrow (mature asynchrony). The marrow is hyperplastic with an erythroid shift but without peripheral blood reticulocytes. The marrow transit time of the rubricytes is extended beyond the normal 4 days and erythropoiesis is thus termed ineffective.

Hematologically the anemia is classically hypochromic and microcytic in all species and is accompanied by neutropenia with hypersegmentation and marked thrombocytosis of small platelets. Red cells are not only small and poorly saturated but also distorted in shape. The plasma is pale as is the urine because of reduced formation of bilirubin. The serum iron is generally less than 50 μg/dl and the total iron binding capacity is increased generally in the 400–600 μg/dl range. Apparently in a multiply transfused human, the typical anemia can develop in the face of adequate iron by the development of an antibody against the rubricyte transferrin receptor.

Anemia of iron deficiency is common in **piglets** and in **puppies** of the larger breeds; it is probably responsible also for the anemia which is often observed in calves on diets solely of milk, but it is seldom clinically important and probably never fatal. Unless special measures are taken to prevent the disease in piglets, a morbidity of 90% may occur with a mortality of approximately 10–50%. The higher figure for mortality includes those animals which die of secondary infections. As noted, iron deficiency seriously impairs defensive mechanisms and predisposes piglets to a variety of infectious diseases, especially colibacillosis and swine dysentery, as well as spirochetal granulomas, omphalophlebitis, and streptococcal meningitis, pericarditis, and polyarthritis, and other pyogenic processes.

The piglet is particularly predisposed because of its rapid rate of growth. Mortality is higher in the winter than in other seasons, partly because of confinement and partly because of increased susceptibility to cold. The requirement of a young piglet for iron is set at about 7 mg per day; milk supplies only one-half this amount.

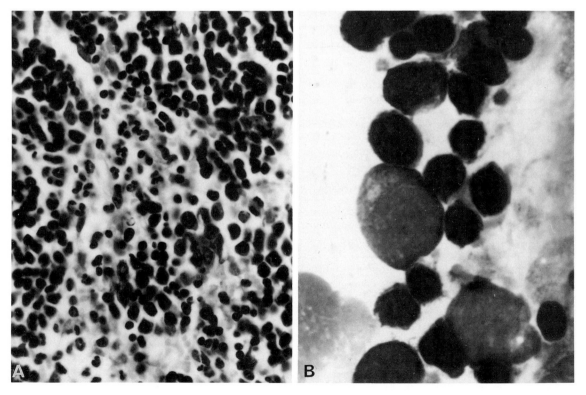

Fig. 2.51 (A) Erythroid hyperplasia. Density of cells indicates late asynchrony. Iron deficiency. Marrow section. Sheep. (B) Iron deficiency. Marrow aspirate. Erythroid hyperplasia with increased rubricytes, late asynchrony. Calf.

The disease is varied in its manifestations and course, but it can be described in three discrete syndromes. The piglets are normal at birth. In the classic syndrome, they become progressively unthrifty after the first week of life and pathologically there is pallor of all tissues, edema of muscles and connective tissues, edema of the lungs, and a thin-walled, flabby heart. In a second syndrome, the piglets often die suddenly in the third week. These piglets are well grown, pale, and appear plump, due as much to edema as to fat. At necropsy, all tissues are extremely pale, wet, and there is pericardial effusion with the heart dilated to more than twice its normal size. The third syndrome is one of secondary infections, and the age incidence is broadened to include pigs of 4–8 or 10 weeks of age and is seen where the piglets are not given parenteral iron and do not begin to feed from a creep feeder until they are 4–5 weeks old. In chronic cases, there will be dyskeratosis with abnormal ringing of horn and hooves or nails.

In all species with iron deficient anemia, the bone marrow has high cellularity and cell density with an erythroid shift and mature asynchrony with increased numbers of megakaryocytes and eosinophils (Fig. 2.51A,B). The pathologic diagnosis of iron-deficient anemia is based on a hypercellular marrow with increased numbers of late-stage rubricytes and an absence of hemosiderin.

Bibliography

Dallman, P. R. Biochemical basis for the manifestations of iron deficiency. *Annu Rev Nutr* **6:** 13–40, 1986.
Gainer, J. H., and Guarnieri, J. Effects of poly I : C in porcine iron-deficient neutropenia. *Cornell Vet* **75:** 454–465, 1985.
Hannan, J. Recent advances in our knowledge of iron-deficiency anemia in piglets. *Vet Rec* **88:** 181–190, 1971.
Kohn, C. W. *et al.* Microcytosis, hypoferremia, hypoferritemia, and hypertransferrinemia in standardbred foals from birth to 4 months of age. *Am J Vet Res* **51:** 1198–1205, 1990.
Larrick, J. W., and Hyman, E. S. Acquired iron-deficiency anemia caused by an antibody against the transferrin receptor. *N Engl J Med* **311:** 214–218, 1984.
Weiser, G., and O'Grady, M. Erythrocyte volume distribution analysis and hematologic changes in dogs with iron deficiency anemia. *Vet Pathol* **20:** 230–241, 1983.

2. Copper-Deficiency Anemia

Copper is a cofactor in a variety of oxidative enzymes of diverse function, and deficiency of the element may affect electron transport (cytochrome oxidase), the absorption of iron and its utilization in hematopoiesis (ceruloplasmin), protection from antioxidants (superoxide dismutase), tyrosine degradation and pigmentation (tyrosinase), neurotransmitter metabolism (dopamine hydroxylase), or cross-linkage of elastin and tropocollagen (lysyl oxidase). Deficiency of the element can therefore be manifested in a variety of ways and in several organ systems. Cytochrome oxidase is a copper-dependent enzyme, and diminished activity of cytochrome oxidase is a sensitive indicator of copper deficiency which apparently limits the synthesis of heme, the prosthetic group of the enzyme.

This association of copper with cytochrome oxidase is probably its fundamental physiological function, but its participation in other enzymologic processes may account for the diverse manifestations of copper deficiency within and between species.

Disturbed hematopoiesis is not the sole or even the most important aspect of copper deficiency, but it is one aspect which is common to all species in which the deficiency is discussed in detail here; the neurologic, vascular, and cutaneous manifestations of copper deficiency are discussed in detail in the chapter on the appropriate system.

Uncomplicated copper deficiency is probably not of widespread occurrence in animals, but in many endemic areas there are chronic diseases with local names, due to a complicated or uncomplicated deficiency of copper. Factors which control the uptake of dietary copper and its excretion are discussed under chronic copper poisoning (Section II,F,5,b of The Erythron). There, reference is made to the ability of high dietary intakes of molybdenum in the presence of sufficient inorganic sulfate to limit hepatic retention of copper by sheep, either by lowering absorption or increasing excretion. There is some evidence for both. The experimental production of a conditioned copper deficiency by appropriate use of molybdenum and sulfate is sometimes successful, sometimes not, which, together with the observation that copper deficiency can occur in ruminants on diets normal with respect to copper, molybdenum, and sulfate, indicates that there are additional unrecognized factors influencing the availability, absorption, excretion, storage, and utilization of copper. There are differences between species, and between breeds of sheep, in the levels of activity of intestinal absorption mechanisms and, probably also, in hepatic excretory mechanisms.

Approximately 25% of ingested copper is absorbed, primarily from the stomach and upper small intestine. Absorption is like that of iron in that it is an energy-dependent, two-step process involving an intraepithelial metallothioneinlike protein that may transfer the copper to the plasma carriers or be shed with the senescent enterocyte. It may be that absorption is more efficient in monogastrics, since deficiency can reliably be reproduced in swine, whereas in sheep, copper deficiency may be more a function of molybdenum toxicity. The absorption is influenced by the chemical form of the copper and by its tendency to form insoluble complexes in the gut. The absorbed copper is transported in blood loosely bound to serum albumin from which it is rapidly distributed to tissues and concentrated principally in liver, kidney, heart, and brain. Copper in blood exists as copper–protein complexes in red cells, loosely bound to serum albumin, and as part of ceruloplasmin, a copper-containing α_2-globulin. Ceruloplasmin is produced in the liver and contains seven copper atoms in each protein molecule, one atom being labile and the remainder, structural. The copper of plasma is much more labile than that of red cells and is a better indicator of fluctuations in copper status than is the copper content of whole blood. Bile is the principal medium of copper excretion.

Copper is present in all cells with principal reserves contained in the liver, although significant amounts are in kidneys, bone marrow, and central nervous system. The latter areas likely represent metabolically functional rather than storage forms. Considerably more copper is found in the liver of the fetus and newborn than in the liver of adults of the same species. An exception is in sheep; ewes, irrespective of their copper status, have much greater hepatic stores than their lambs. In adult ruminants, the levels of hepatic copper vary widely and reflect to some extent the animals' nutritional history with respect to this element. Much smaller, and rather stable, concentrations of copper are found in the liver of nonruminants.

The concentration of copper in the blood of all species ranges close to 1 μg/ml; a decrease of 0.5 μg indicates a significant deficiency; an increase of 0.25 μg or more threatens copper poisoning, provided that the higher levels are not due to molybdenum excess, a condition in which hepatic and other stores of copper are depleted.

Anemia is not invariably present in copper deficiency, but it is probable that if the deficiency is severe enough and prolonged enough, anemia will occur. Several mammalian species, when made deficient in copper, develop anemia, the severity of which depends on the duration of the deficiency as well as on host factors such as age and pregnancy. The morphologic characteristics of the anemia vary with the species, that in rats, pigs, rabbits, and lambs it is microcytic and hypochromic; in cattle and adult sheep, it is macrocytic and hypochromic; and in dogs, normocytic and normochromic.

Unlike iron, where the transport protein transferrin carries the metal in a high-affinity bond with variable saturation, copper circulates in the plasma in an integral bond with the globulin ceruloplasmin. Since copper is required for the synthesis of ceruloplasmin, the protein level drops in copper deficiency. This is unlike iron deficiency in which the total iron-binding capacity rises as the serum iron drops. Ceruloplasmin acts as a specific oxidant for conversion of ferrous to ferric iron in which state it is available for transferrin transport. The protein has been called ferroxidase to indicate this function.

The manner in which copper deficiency leads to anemia is not clear; however, the coincident development of marrow erythroid hyperplasia indicates that, as in iron deficiency, the stem cell system is not impaired but, because of reticulocytopenia, rubricyte maturation must be. The disease is therefore one of ineffective hematopoiesis with hyperplastic marrow but deficient output. The fact that the ineffective asynchrony is not uniformly of the late type in all species, as it is for iron deficiency, and that the red cells are not always hypochromic suggest that the mechanism of development is not solely one of iron inhibition. Rather it appears that the coordination of heme and globin synthesis, that is heme limited in iron deficiency, is more generally affected in copper deficiency, and down-regulation of rubricyte maturation may be mediated by

deficits in both. In iron deficiency, marrow rubricyte mitoses are not prominent but probably do occur, and division without cytoplasmic production is the mechanism of microcytosis. In contrast, in the dog with copper deficiency, the rubricytes are not predominantly late stage, but are generally younger. Mitoses are common and the anemia is normocytic, normochromic. Thus the net effect of copper-deficient anemia in the dog appears to be defective maturation, increased intramedullary cell death and as in iron deficiency, deficient cell production. A shortened survival time of red cells occurs in copper-deficient, and also occurs in iron-deficient pigs, but kinetically the anemia in both is very largely due to reduced production.

Extensive hemosiderin deposits occur in sheep and cattle naturally deficient in copper and in pigs made experimentally deficient, suggesting a defective reutilization of endogenous iron. Copper-deficient swine have impaired ability to transfer iron across the intestinal basement membrane; lowered plasma iron levels; an increase in plasma transferrin; and a lowered rate of turnover of iron from the liver. Thus, although the body is iron replete, there is impaired interorgan transport, and in the pig the anemias of iron or copper deficiency look pathologically similar. There is abundant iron in the latter, however, and the thrombocytosis of iron deficiency is not seen in anemia of copper deficiency.

Anemia of suckling piglets associated with copper deficiency may account for some of the instances in which piglet anemia fails to respond to iron. In copper deficiency, the hemoglobin may fall to 1–2 g/liter and the red cells, to 2×10^{12}/liter. The entire syndrome parallels that of iron deficiency except that it does not respond to iron; a good response follows the administration of copper, but the reticulocytosis is delayed by the 3–8 days required for the synthesis of ceruloplasmin. In contrast, if plasma from a copper-replete animal is given, the response is immediate.

Anemia is an important sign of the enzootic maladies of copper deficiency, but is not invariably present, and in general, develops late in the course of the deficiency. The severity as well as the morphologic features of the anemia are influenced by the severity of copper deficiency, the duration, whether complicated or not, and also by host characters such as age, growth rate, pregnancy, and parasitism.

Bibliography

Blakley, B. R., and Hamilton, D. L. Ceruloplasmin as an indicator of copper status in cattle and sheep. *Can J Comp Med* **49:** 405–408, 1985.

Fields, M., Lewis, C. G., and Lure, M. D. Anemia plays a major role in myocardial hypertrophy of copper deficiency. *Metabolism* **40:** 1–3, 1991.

Lahey, M. E. *et al.* Studies on copper metabolism. II. Hematologic manifestations of copper deficiency in swine. *Blood* **7:** 1053–1074, 1968.

Lee, G. R. *et al.* Iron metabolism in copper-deficient swine. *J Clin Invest* **47:** 2058–2069, 1968.

Roeser, H. P. *et al.* The role of ceruloplasmin in iron metabolism. *J Clin Invest* **49:** 2408–2417, 1970.

Suttle, N. F., and Angus, K. W. Experimental copper deficiency in the calf. *J Comp Pathol* **86:** 595–608, 1976.

Suttle, N. F. The role of comparative pathology in the study of copper and cobalt deficiencies in ruminants. *J Comp Pathol* **99:** 241–258, 1988.

3. Cobalt-Deficiency Anemia

Cobalt deficiency is a disease of ruminants expressed as ill thrift; death, if it occurs, is due to inanition (see ovine white liver disease in The Liver and Biliary System, Volume 2, Chapter 2). The disease is known in many parts of the world, but the distribution, especially of marginal deficiencies, is not well plotted; occurrence is often seasonal and complicated by concurrent deficiencies of copper and phosphorus. The minimal concentration of cobalt in forage consistent with adequate nutrition of ruminants is one part per million on a dry-weight basis, but the only effective test of cobalt deficiency is in the response to oral supplementation. Cobalt deficiency produces a chronic wasting disease to which sheep are slightly more susceptible than cattle and to which horses are immune; lambs and calves are even more susceptible than their respective adults. The disease is characterized by a gradual reduction of appetite for both food and drink, progressive emaciation, and death from starvation, even though food may be plentiful. The European name, licking disease, places emphasis on the allotriophagia which accompanies inappetence in cattle. Abortion, temporary sterility, and decreased production are common in cattle. Pallid mucous membranes are characteristic but, as previously mentioned, anemia may or may not be demonstrable by routine tests and may be masked by blood volume contraction with loss of weight.

In recognized endemic areas, sickness is usually evident after the animals have been at pasture for 4–6 months. Calves and lambs may not survive more than a further 6 months, although adults introduced to deficient pastures may survive for 2 years or more. The postmortem picture is that of starvation.

Cobalt deficiency is a deficiency of vitamin B_{12}. Parenteral treatment of deficient animals with the vitamin will sustain them for long periods, suggesting that hepatic storage is efficient. Pair-feeding studies show that the cachexia of cobalt deficiency is largely due to inanition, although there is apparently some loss of absorptive or metabolic efficiency. Remarkably, the parenteral administration of cobalt as $CoCl_2$ to deficient sheep has no effect on the disease and much of the injected dose is lost in the urine. In contrast, the administration of cobalt orally, or vitamin B_{12} by any route, is curative if enough of either is given. Synthesis of the vitamin occurs in the rumen, and even if there is adequate storage of cobalt in the liver, the disease will progress unless there is enteric cobalt available for microbial metabolism. In horses, the concentration of vitamin B_{12} in the enteric contents increases from the stomach to the rectum. This may explain why horses are refractory to cobalt deficiency and indicates that the supplementation of racing animals is unnecessary.

Energy metabolism in the ruminant depends on acetic, propionic, and butyric acids, and the conversion of these to succinate. The enzyme methylmalonyl-CoA mutase contains vitamin B_{12} and is responsible for the conversion of methylmalonyl-CoA to succinyl-CoA. In cobalt deficiency, there is accumulation of methylmalonic acid in liver and blood, and high levels of propionate and acetate in blood; the latter may be responsible for the inappetence. Vitamin B_{12} also participates via a methyltransferase in the conversion of homocysteine to methionine, an essential amino acid. It is possible that methionine deficiency is responsible for the fatty liver which occurs.

Inadequate ruminal cobalt is permissive of the chronic neurologic syndrome of *Phalaris* poisoning of sheep.

Bibliography

Davies, M. E. The production of vitamin B_{12} in the horse. *Br Vet J* **127:** 34–36, 1971.

Ibbotson, R. N., Allen, S. H., and Gurney, C. W. An abnormality of the bone marrow of sheep fed cobalt-deficient hay-chaff. *Aust J Exp Biol Med Sci* **48:** 161–169, 1970.

Smith, R. M., and Marston, H. R. Some metabolic aspects of vitamin B_{12} deficiency in sheep. *Br J Nutr* **24:** 879–891, 1970.

4. Megaloblastic Anemia

Anemia characterized by reticulocytopenia and marrow erythroid hyperplasia with early asynchrony of rubricytes occurs infrequently in cats and dogs. Vitamin B_{12} is stored in the liver in amounts sufficient to maintain animals for many months without further absorption. On the other hand, there is very little storage of **folic acid,** and signs of deficiency may appear in animals that are anorectic or receive chemotherapy after relatively short periods of disease. This may be due to acceleration of folic acid depletion with the inanition produced by rapidly growing tumors, such as the high-grade lymphomas which cats and dogs are prone to develop. Vitamin B_{12} deficiency occurs rarely in the dog secondary to immune gastric atrophy. It is possible to produce the typical immune destruction of the gastric mucosa in dogs by immunization with homologous gastric juice or mucosal homogenate. A congenital anemia of **giant schnauzers** that is non-responsive but not macrocytic is apparently due to an inability to absorb the vitamin since parenteral vitamin B_{12} reverses the syndrome. Dogs which survive gastric torsion with loss of most or all of the glandular portion of the stomach are at risk of developing vitamin B_{12} deficiency but seem to have sufficient hepatic stores until they die of other causes. A variety of drugs may inhibit the absorption of either folic acid or vitamin B_{12}. Since analyses for these substances are rarely performed, the two deficiencies are not often specifically identified. They will be described together.

Hematologically the first sign of disease due to deficiency of either folate or B_{12} is the presence of large neutrophils with hypersegmented nuclei in an animal with nonresponsive anemia, usually associated with acute leukemia or lymphoma. The marrow precursors of these neutrophils are giant metamyelocytes which have apparently become tetraploid but have not proceeded to division. In the early stages, the rubricytes appear of normal age distribution and the peripheral blood red cells are normochromic, normocytic. With progression, which appears most often in the dog, there is development of marrow erythroid hyperplasia with a predominance of young rubricytes, and macrocytosis in the blood with thrombocytopenia and increased variation in size of platelets. Pathogenetically the disease is an example of an intact and functional stem cell system which is capable of producing a hyperplastic marrow that is ineffective in mature cell production, due to inhibition of DNA synthesis and subsequent cell division. Since divisions are skipped, the few resulting cells are larger than normal.

Bibliography

Bunch, S. E., Easley, J. R., and Cullen, J. M. Hematologic values and plasma and tissue plasma folate concentrations in dogs given phenytoin on a long-term basis. *Am J Vet Res* **51:** 1865–1868, 1990.

Fyfe, J. C. *et al.* Inherited selective malabsorption of vitamin B_{12} in giant schnauzers. *J Am Anim Hosp Assoc* **25:** 533–539, 1989.

Herbert, V. Biology of disease: Megaloblastic anemias. *Lab Invest* **52:** 3–19, 1985.

Thenen, S. W., and Rasmussen, K. M. Megaloblastic erythropoiesis and tissue depletion of folic acid in the cat. *Am J Vet Res* **39:** 1205–1207, 1978.

Waxman, S., Corcino, J. J., and Herbert, V. Drugs, toxins, and dietary amino acids affecting vitamin B_{12} or folic acid absorption or utilization. *Am J Med* **48:** 599–608. 1970.

5. Anemia of Chronic Disorders

In inflammatory states, acute phase reacting proteins scavenge molecular iron from transferrin and tie it up in ferritin and later, if the reaction persists, in hemosiderin. Thus hypoferremia has been considered the primary mechanism in the development of the anemia of chronic disease. However, because the hypochromia is at most mild (MCH less than 31%), while the anemia may be moderate (70-100 g/liter), it appears that anemia is due primarily to a reduction in red cell production rather than inhibition of hemoglobin synthesis. Further, whereas shortened red cell life span can be demonstrated in anemia of chronic disorders, the hemolysis is mild and the red cells survive normally in normal animals, demonstrating that the problem is environmental and not intrinsic to the red cell itself. Finally it is recognized that in primarily hemolytic anemias, the bone marrow is highly productive and can increase output six or more times normal levels, yet in the anemia of chronic disorders, the red cells are normocytic and there is reticulocytopenia out of proportion to the degree of the anemia. Cats with sterile abscessation and anemia respond partially to additional iron by continuous infusion and, more completely, to iron plus cobalt, the latter treatment to increase erythropoietin by tissue anoxia. Thus, pathogenetically, the anemia of chronic disorders is primarily a problem of decreased and deficient production of red cell numbers and, like the anemia of uremia, it may be compensated by increased levels of erythropoietin if iron is available. Although the mechanisms operative in the syndrome of chronic disorders are defined, the only practical therapy is reversal of the primary problem.

Collectively, the anemias associated with inflammatory, degenerative, immune, and neoplastic disease form a large and important group in all species (Fig. 2.52A,B). The common pathway by which erythropoiesis is selectively depressed in anemias of such diverse association is related to tissue injury, and mechanistically to the constant production of tumor necrosis factor or cachectin, a monokine produced by activated macrophages that has both toxic and cytostatic effects on the rubricyte system. Teleologically, this process appears to be a system to deny access to iron to invading microorganisms, with anemia as an acceptable exchange for survival. The anemias of chronic disorders are of moderate severity, largely normochromic, normocytic, and poorly responsive, with associated changes of toxemia in the leukocytes.

Bibliography

Clibon, U. *et al.* Erythropoietin fails to reverse the anemia in mice continuously exposed to tumor necrosis factor-alpha *in vivo.* *Exp Hematol* **18:** 438–441, 1990.

Smith, J. E., and Cipriano, J. E. Inflammation-induced changes in serum iron analytes and ceruloplasmin of Shetland ponies. *Vet Pathol* **24:** 354–356, 1987.

Ward, C. G. Influence of iron on infection. *Am J Surg* **151:** 291–295, 1986.

Weiss, D. J., and Krehbiel, J. D. Studies of the pathogenesis of anemia of inflammation: Erythrocyte survival. *Am J Vet Res* **44:** 1830–1831, 1982.

6. Anemia of Malnutrition

Anemia of malnutrition is most commonly a disease of the young and is seen most frequently in sheep and goats, and less often in kittens and puppies where poor nutrition may be complicated by ecto- and endoparasitism (Fig. 2.53A). Malnutrition in animals, as in humans, is sporadic and worldwide in distribution. The disease is worthy of consideration because of the regular tendency for the anemia to worsen once corrective measures are taken. Hematologically the anemia of malnutrition is largely normochromic, normocytic, but may be hypochromic if complicated by parasitism and blood loss. Kinetically the anemia is primarily due to a reduction in red cell production with minor hemolysis, as in anemia of chronic disorders. Since malnutrition is so often associated with delayed healing and infection, the two syndromes are often coincident and may share a combined development. Malnutrition that is primarily due to **caloric deprivation** is called marasmus. In swine, the anemia is mild, and the response to phlebotomy is vigorous and curative, reflecting a functional stem cell system, adequate iron supply, and normal erythropoietin production.

Protein–calorie deprivation is more common and when severe is called kwashiorkor. It is a syndrome of cachexia characterized by hypoplastic marrow, mild anemia, and reticulocytopenia. Since serum proteins are also deficient,

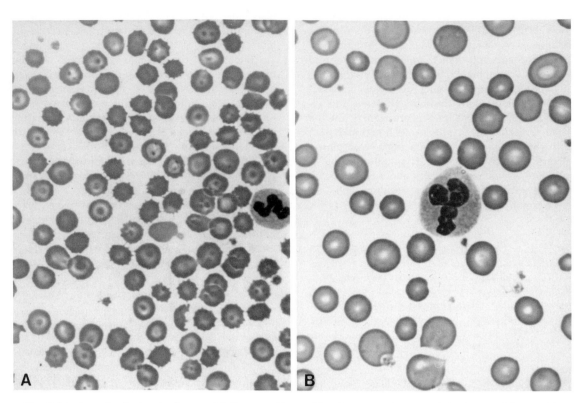

Fig. 2.52 Anemia of chronic disease. (A) Target cells resulting from altered serum cholesterol and lipoprotein in liver disease. Membrane pliability results in acanthocytosis and poikilocytosis. Dog. (B) Microcytosis characteristic of portocaval shunting.

low serum iron is not rate limiting to red cell production because transferrin is also low and the saturation level is adequate for iron delivery. The anemia in these syndromes is partially adaptive and is due to reduced activity and partially to multifactor deficiencies. In the development of starvation, there is first a loss of muscle mass followed by parenchymal atrophy, anemia, and last, a reduction in all serum proteins. Since the latter are the major transport for many minerals, and the major source of amino acids, their loss is most limiting to all aspects of hematopoiesis. In repletion, these elements are replaced in reverse order and the recovery of all tissues is dependent on first increasing the levels of plasma albumin and globulins. As plasma protein levels rise, the plasma volume contraction that developed with loss of oncotic pressure is also reversed. Since this volume expansion is much more rapid than the expansion of the red cell mass, the anemia becomes more severe and may be misinterpreted as therapeutic failure. The pattern of recovery is a rise in level of plasma proteins followed by blood volume expansion, reticulocytosis, a slow rise in hemoglobin, and finally a gain in body weight. The course of these changes takes 3–4 weeks, depending on the level of complicating factors such as specific vitamin deficiencies, parasitism, and other infections. Pathologically starvation is characterized by atrophy of all organs but particularly the thymus, lymph nodes, spleen, and liver. There is pancreatic acinar atrophy and serous atrophy of marrow fat (Fig. 2.53B).

Deficiencies of specific vitamins are capable of causing anemia in animals, but in general, their importance is more in the prevention of related disease. Swine fed **vitamin E**-deficient diets develop a moderately severe nonresponsive anemia characterized by marrow erythroid hyperplasia with multinucleated dyserythropoiesis, increased turnover of serum iron, and an increase in unconjugated bilirubin. Red cell lipids are markedly reduced and red cells are highly susceptible to oxidant attack. The mechanism appears to be at least partially due to inhibition of pyrrole and thus heme synthesis. This mechanism is important in a practical sense since an abundant pool of protoporphyrin is an efficient sink for tying up iron in the synthesis of heme. Piglets which are vitamin E deficient are more susceptible to toxicity when given iron by any route. Premises where other vitamin E- and selenium-related diseases have occurred are likely to have pale piglets that do not respond to repeated iron administration and have the characteristic dyserythropoiesis.

Vitamin B$_6$ deficiency induced in swine results in a microcytic but not hypochromic anemia of moderate severity, characterized by marrow erythroid hyperplasia and hyperferremia, and pathologically by hemosiderosis of marrow, spleen, and liver, and hepatic centrilobular fatty degeneration. Vitamin B$_6$ is required for formation of the precursors of delta-aminolevulinic acid, which then forms the pyrrole ring and ultimately protoporphyrin. Blockage at this early stage of ring synthesis prevents iron utilization

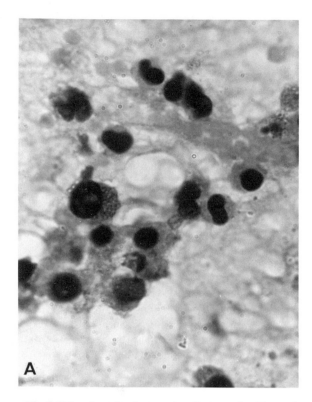

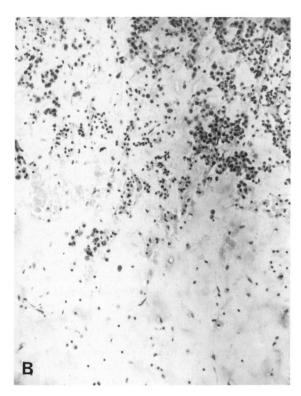

Fig. 2.53A Anemia of starvation. Late erythroid asynchrony and increased proportion of eosinophils. Sheep.

Fig. 2.53B Anemia of starvation. Serous atrophy of marrow with residual cuffs of subendosteal hematopoiesis. Sheep.

and leads to ineffective hematopoiesis. In the cat, in addition to anemia, vitamin B_6 deficiency causes stunting of growth, emaciation, convulsions, and oxalate nephrosis.

Bibliography

Caasi, P. I., Hauswirth, J. W., and Nair, P. O. Biosynthesis of heme in vitamin E deficiency. *Ann N Y Acad Sci* **203:** 93–102, 1972.

Gershoff, S. N. *et al.* Vitamin B_6 deficiency and oxalate nephrocalcinosis in the cat. *Am J Med* **27:** 72–80, 1959.

Lynch, R. E., *et al.* The anemia of vitamin E deficiency in swine: An experimental model of the human congenital dyserythropoietic anemias. *Am J Hematol* **2:** 145–158, 1977.

Warrier, R. P. *et al.* The anemia of malnutrition. *In* "The Malnourished Child," Vol. 19, R. M. Suskind and L. Lewinter-Suskind (eds.), pp. 61–72. New York, Raven Press, 1990.

Wintrobe, M. M. *et al.* Pyridoxine deficiency in swine with particular reference to anemia, epileptiform convulsions and fatty acids. *John Hopkins Hosp Bull* **72:** 1–25, 1943.

F. Hemolytic Anemias

Anemia of hemolysis develops when the rate of destruction exceeds that of production. Since the products of hemoglobin catabolism are readily reutilized, the anemias of hemolysis are characteristically and predominantly highly responsive with reticulocytosis and an upward shift in mean red cell volume (MCV). It is stated that human marrow can increase output sixfold in hemolytic anemia,

and the cat and dog have a similar capacity for regeneration, whereas horses, pigs, and ruminants are somewhat less responsive. The response to anemia is most efficiently measured by the **reticulocyte count,** which should be expressed in absolute numbers and not as a function of the number of leukocytes or peripheral blood rubricytes. Reticulocytes have a maturation time of about 1 day and their level in the blood is always an indicator of current marrow output. With very high levels of erythropoietin, some mitoses are skipped, resulting in the release of even larger shift red cells, which have a maturation time of 2–2.5 days. The presence of a few larger red cells that are not polychromatic in the absence of a reticulocytosis indicates red cell shut-down within the last 1–2 days. Thus a normal dog with 6×10^{12}/liter red cells and 1% reticulocytes has 6×10^{10}/liter new red cells on a steady-state basis. In other terms of expression, there are 60×10^9/liter reticulocytes or about 4 times as many reticulocytes as leukocytes that are normally present. In hemolytic states, the dog may produce as many as 600×10^9/liter reticulocytes or 10 times the normal output. Attempts to correct the reticulocyte count for the number of large shift red cells present is subjective and introduces a misleading concept of precision to a process that is not reproducible. The raw number of reticulocytes expressed in absolute terms is in itself the most accurate measure of marrow erythroid response, just as the neutrophil count is an accurate measure of current myeloid output. In responsive anemias, there is a variable but much lower number of rubricytes accompanying the reticulocytes, whereas in stress erythropoiesis, rubricytes may appear alone. It is important that these nucleated precursors are not considered an indication of a responsive marrow. If their absolute numbers, usually less than 10×10^9/liter, are compared to expected levels of reticulocytes at $300–500 \times 10^9$/liter, it can be seen that this level of contribution is minuscule and far below the normal level of production in a nonanemic animal. Reticulocytes are not released to the blood in the horse, but the upward shift in the MCV can be used as a relative measure of response. Blood loss results in an upward shift of <6 fl, whereas in hemolytic anemia the shift will increase to >12 fl.

The iron economy may be rate limiting in hemolytic anemia, with the contributions from recycling, enteric absorption, and mobilization from stores being additive in their effect. This point is worth emphasis since even animals that have abundant iron stores in hemosiderin will respond more strongly if given additional iron by any route.

Intracellular destruction of red cells is the normal route of removal and occurs in the monocyte–macrophage system in marrow, spleen, liver, and nodes. In hemolytic states, the spleen may become quantitatively more important than marrow. The actual process of lysis is rapid and if erythrophagocytosis is observed, the level of destruction is likely also rapid. This catabolism releases iron from heme to transferrin or to storage as ferritin or hemosiderin, while the protoporphyrin ring is opened and released to the plasma as unconjugated bilirubin. Globin is reutilized via the amino acid pool.

Intravascular destruction of red cells occurs in normal circumstances at a very low rate as a result of high-velocity trauma in small arteries. In some cases of accelerated hemolysis with complement fixation, red cells are lost by intravascular lysis at very rapid rates, resulting in hemoglobinuria. Normally, free plasma hemoglobin is rapidly scavenged from circulation by binding to first haptoglobin, then hemopexin, and finally albumin. Free hemoglobin rapidly dissociates into dimers, which are quickly bound with high affinity to haptoglobin, and are then too large for renal excretion and are removed from circulation by the liver. If haptoglobin is saturated, iron unprotected by the normal cytosol reductase system is oxidized to the ferric state, causing breakage of the heme–globin bond. The free heme is then picked up by hemopexin, which in the bound state is cleared with a half time of 7–8 hr by the liver. Even minor hemolysis will remove haptoglobin from circulation, but reduced levels of hemopexin constitute a sensitive test of intravascular hemolysis. Mild overloading of these systems allows the filtration of heme, which is reabsorbed in the renal proximal tubules. The intracellular accumulation of hemosiderin constitutes a record of previous hemolysis that can be read by detecting the epithelial cells on urinalysis or on histologic examination of the kidney. Severe intravascular hemolysis is a double hazard to the kidney because the toxic and vasoactive effects of the free heme are added to anemia, with resultant ischemic injury to the nephron (see The Urinary System, Volume 2, Chapter 5).

Bibliography

Lumsden, J. H. *et al.* The kinetics of hematopoiesis in the light horse. II. The hematological response to hemorrhagic anemia. *Can J Comp Med* **39:** 324–331, 1975.

Lumsden, J. H. *et al.* The kinetics of hematopoiesis in the light horse. III. The hematological response to hemolytic anemia. *Can J Comp Med* **39:** 332–339, 1975.

Schacter, B. A. Heme catabolism by heme oxygenase: Physiology, regulation, and mechanism of action. *Semin Hematol* **25:** 349–369, 1988.

Weiser, M. G., and Kociba, J. Persistent macrocytosis assessed by erythrocyte subpopulation analysis following erythrocyte regeneration in cats. *Blood* **60:** 295–303, 1982.

1. Immune-Mediated Hemolytic Anemias

The adherence of antibody to the red cell membrane with or without complement fixation causes red cells to become spherical. Their life span is then reduced either by destruction by the monocyte–macrophage system or by intravascular hemolysis. Since the damage is usually to the most mature cells, most hemolytic anemias are highly responsive.

Immune anemias are classified on the basis of mechanism or association. Serologically there are three types: **warm antibody** type with maximal activity at 37°C; **cold antibody** type with maximal activity at 2–4°C; and those **without detectable antibody.** Classification by association is based on identification of an underlying cause and includes primary or idiopathic immune hemolytic anemia and secondary or symptomatic immune hemolytic anemia. These may be isoimmune, drug-induced, associated with infectious disease, associated with other immune disorders (systemic lupus erythematosus), or associated with lymphoid or other malignancy.

In **symptomatic immune anemias,** the cause is identified and is usually adherence of drug or virus to red cells, thereby sensitizing them to the immune system. In **idiopathic immune anemias,** which form the bulk of cases, it is not clear whether the error is due to an actual change in red cell antigenicity or to loss of self-recognition by the immune system.

The detection of antibody on the surface of red cells is performed by means of the **Coombs test.** In this test, an antiglobulin for the intended species (e.g., dog) is prepared in another species (e.g., rabbit, sheep). Antiglobulins are usually prepared against both immunoglobulin and complement for a broader reactivity. About 500 molecules of immunoglobulin G (IgG) must be bound to each red cell to give a positive Coombs or antiglobulin test, yet significant shortening of red cell life span can occur with as few as 24–34 molecules per red cell. Therefore, a negative Coombs test does not rule out clinically significant hemolysis. The test is conducted at 37°C and 4°C to detect warm- and cold-reacting antibodies. In general, the warm-reacting antibodies are IgG or IgA and the cold reacting are IgM. Cases of cold reacting IgG have been described and these may be the cause of agglutination which appears due to complement alone.

The warm-reacting antibodies are termed incomplete and generally do not fix complement on the red cells. The anemias produced are characterized by a slower rate of intracellular hemolysis. In contrast, the cold-reacting antibodies fix complement and may cause acute intravascular hemolysis with hemoglobinuria if extremities are cooled. The IgM antibodies tend to elute from red cells, but if complement is fixed, it remains attached and may cause increased *in vivo* destruction and give a positive Coombs test.

Whereas a negative Coombs test does not rule out immune hemolytic anemia, a positive test may occur in the absence of anemia. Since macrophages do not pick up red cells bound by all classes of IgG, there may be cases which have a positive Coombs test but are asymptomatic.

Five types of immune hemolytic disease occur: (1) erythrocytic type characterized by severe anemia with hyperplastic marrow (most common); (2) hemolytic anemia and immune thrombocytopenia with hyperplastic marrow; (3) pure red cell aplasia with selective immune depression of erythroid precursors; (4) pancytopenic type in which there is immune anemia, neutropenia, and thrombocytopenia with hyperplastic marrow; and (5) the aplastic type, which resembles pancytopenia but with marrow aplasia. Isoimmune hemolytic anemia is recorded in kittens and pups but remains of most importance in foals, calves, and piglets.

a. ISOIMMUNE HEMOLYTIC ANEMIA IN THE FOAL. Mares are isoimmunized naturally by focal placental hemorrhage which allows foal red cells to enter the maternal circulation. Clinical isoerythrolysis rarely occurs in primiparous mares and generally is not seen until the third or fourth foal. It is likely that a considerable amount of fetal blood is required for initial sensitization, but restimulation in subsequent pregnancies may require much less blood of the same genotype, and therefore repeat pregnancies by the same sire carry a greater risk of an affected foal. Blood transfusions and tissue-origin vaccines may complicate the problem of maternal sensitization. Isoimmunization can occur only when the sire and foal possess a blood antigen absent in the mare, usually Aa or Qa. Stallions homozygous for an offending factor will always sire susceptible foals, whereas heterozygous sires will produce susceptible foals in about 50% of matings.

Foals are normal at birth but can absorb dangerous levels of colostral antibody for up to 48 hr after birth and likely longer. Foals which develop clinical disease and are removed from the dam may suffer recurrence of signs if put back on the dam a week or more later. Signs of the disease may occur as early as 8 hr or as late as 120 hr after birth. The severity of the anemia varies greatly depending on the amount and type of isoantibody that is absorbed. Hemolytic antibodies are most damaging, and highest ti-

ters exist in the first milk. Thus, strong foals which nurse vigorously soon after birth may be most severely affected.

Earliest signs are lethargy and reluctance or inability to stand and infrequent suckling. Clinical signs include pallor of mucous membranes with icterus, dyspnea, pounding heart, and hemoglobinuria in severe cases. Mild disease may not be noticed.

There is a severe normochromic anemia that becomes moderately macrocytic (Fig. 2.54A,B). Polychromasia is absent, but an occasional rubricyte may be present. The marrow is hyperplastic with an erythroid shift. The hematocrit may drop below 15%, while the leukocyte count is elevated, usually to 25 $\times$ 10^9/liter or more, and the platelet count is normal or elevated. The serum bilirubin may rise to 20 mg/dl or more, with much of it unconjugated and therefore fat soluble. Transaminase levels are elevated and a continuing rise suggests a poor prognosis. The urine is dark brown. The foal's cells are strongly Coombs positive if the hemolysis is severe. Rouleaux formation, or rafting, of red cells is exaggerated (Fig. 2.54A). The diagnosis is confirmed by demonstration of antibody in maternal serum or milk that will agglutinate foal red cells and usually sire red cells. Foals with isoimmune anemia are usually given high levels of immunosuppressive steroids which, with high levels of iron released from lytic cells and inadequate passive transfer of antibodies in colostrum, renders them extremely susceptible to bacterial infection. In foals thus treated, the appearance of **toxic changes** in blood neutrophils with rising transaminases may be the

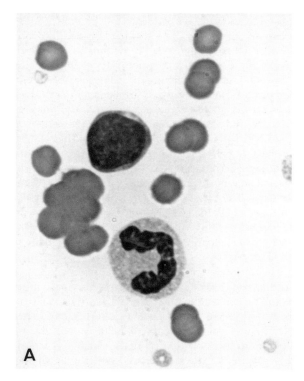

Fig. 2.54A Isoimmune hemolytic anemia. Rafting of red cells. Foal.

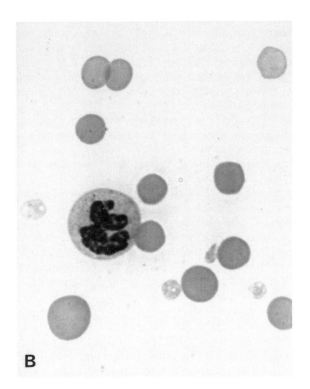

Fig. 2.54B Isoimmune hemolytic anemia. Foal. Spherocytes and macrocytes indicative of defective erythroid response in this species.

first indication of infection since fever and malaise are prevented by the steroids. Prophylactic antibiotics and/or very careful monitoring are indicated. Bovine colostrum has proven effective and may be helpful by stimulating development of the foal's complement system.

Foals which die of isoimmune anemia are usually remarkably icteric and unless they are younger than a week, there is usually some degree of cachexia. The lungs are pale, icteric, and edematous and there is increased pericardial fluid. The liver is turgid, friable, fatty, and bile stained. The spleen is enlarged with rounded borders and may have subcapsular hemorrhage. The splenic parenchyma is friable and bulging on cut surface but does not ooze blood on standing. The kidneys are pale with yellowed papillae and the intestinal tract is usually empty of ingesta and moderately distended with gas. The femoral marrow is diffusely reddened with translucent fat, which rapidly retracts on drying.

The histologic lesions vary with the treatment and period of survival. If repeated transfusions have been given, there will be hemosiderosis of the spleen, liver, marrow, and lung despite the usual absence of storage iron in neonates. There is pulmonary alveolar flooding, often with foci of septic alveolitis. In animals which die acutely, there is ischemic hepatic periacinar degeneration, whereas those surviving 2 weeks or more and receiving steroid therapy have irregular ballooning of hepatocytes with canalicular cholestasis and hyperplasia of highly pigmented

Kupffer cells. Extramedullary hematopoiesis is absent. The spleen has follicular and periarteriolar lymphoid atrophy with germinal centers occupied by dendritic reticular cells with an epithelioid appearance. The sinus areas are densely cellular with increased fixed stromal cells and macrophages. Hemosiderin and erythrophagocytosis are prominent, but hematopoiesis is not. Kidneys have ischemic epithelial degeneration with flattened proximal epithelium that has heavy bile pigmentation and irregularly spaced nuclei. There are often granular and cellular casts in collecting tubules. The lymph nodes and thymus are hypocellular, and node sinuses are usually devoid of macrophages if steroid treatment has been prolonged. The marrow has a reduced amount of fat, and that present undergoes serous atrophy and becomes eosinophilic. In acute cases, there is an erythroid shift and cellular packing is high. Later the erythroid shift remains, but the cell density is reduced and hemosiderin and erythrophagocytosis are prominent. Thrombopoiesis is normal and mature granulocytes are reduced.

b. ISOIMMUNE HEMOLYTIC ANEMIA IN THE CALF   Spontaneous sensitization of cows to fetal red cell antigens is very rare. The calf derives its iron from maternal red cells deposited at the trophoblastic junction, but the reverse cellular exchange does not appear to occur. Vaccines of blood origin for prevention of babesiosis and anaplasmosis may contain red cell antigens which immunize the dams. If the bull has the same red cell antigens as the vaccine donor, then the calves will share these antigens and may develop isoimmune hemolytic anemia when they receive colostrum. Gestation is normal.

The disease may be mild or peracute with signs occurring within hours of birth and severe and terminal dyspnea developing at 1–7 days of age. Severely affected animals are dull and have reduced activity. Jaundice develops in those that survive 1–2 days.

In the peracute form of the disease, there is hemoglobinuria, hypofibrinogenemia, and fibrin degradation products in the blood, and death is rapid. At postmortem a large spongy spleen and hemoglobin discoloration of kidneys are present. There is an excess of bloodstained pleural fluid and the lungs are congested and edematous. Histologically, there are widespread fibrin thrombi, and death is due to disseminated intravascular coagulation.

In the acute form, the packed cell volume drops to 6–7% and there is often hemoglobinuria. Death is expected at ~5 days with anemia and icterus. The marrow is responsive but inadequate with only 1–2% reticulocytes and rubricytosis that exceeds the level of leukocytes. The Coombs' test is positive and the dam's milk agglutinates the calf's red cells. Hemolysis occurs if complement is added. The A,B,C,F–V, and S–U antigen systems are most often the target of maternal antibody.

In the mild form, the hematocrit drops to 18% at a week after birth and rises to 30% at 3 weeks of age. The anemia is normochromic and macrocytic, and recovery is slow.

c. Isoimmune Hemolytic Anemia in Piglets   Isoimmune anemia of piglets was originally attributed to maternal sensitization by a crystal violet-inactivated hog cholera vaccine of blood origin. However, spontaneous transplacental sensitization occurs in swine and the disease is primarily an isoimmune thrombocytopenia with lesser effects on the red cell system. The piglets are normal at birth and disease occurs after suckling. The isoantibodies are usually against the Ea, Ee, Gb, and Kb red cell antigens (see also isoimmune thrombocytopenia, Section III,A of Hemorrhagic Diatheses).

Signs occur between 1 and 4 days after birth and consist of pallor, inactivity, dyspnea, and jaundice. These signs gradually subside. After 10–11 days, widespread petechiation develops, particularly in ventral areas, in association with severe thrombocytopenia. Most piglets die at this time. The first signs are due to red cell and platelet destruction in the peripheral blood. The marrow is at first responsive and the disease is self-correcting for the first week. As antibody continues to be absorbed, there is depression of marrow precursors, and the terminal petechiae are associated with decreased platelet production because of marrow aplasia. The terminal anemia is due both to hemorrhage and hypoproliferation.

The erythrocyte count drops from a normal level of $4.5–5.3 \times 10^{12}$/liter at birth to $1–3 \times 10^{12}$/liter at 4 days of age. As the red cell count drops, anisocytosis and polychromasia develop, along with some circulating rubricytes. The mean corpuscular volume rises from a normal of 70 fl at birth to 100–120 fl at 7–14 days of age and returns to normal in survivors at 4 weeks of age. The neutrophils rise to $9–10 \times 10^9$/liter at 4 days when the red cells are lowest and then decline to normal at $3–4 \times 10^9$/liter. Platelets have a bimodal response and drop from 300×10^9/liter at birth to 100×10^9/liter at day 1, then rise to normal at the first week, followed by severe thrombocytopenia after 10–14 days.

The bone marrow is hypoplastic in piglets which die and megakaryocytes are absent. Red cells are Coombs' positive from days 1 to 7 and are negative by day 14.

d. Drug-Induced Hemolytic Anemia   Many drugs have been associated with Coombs'-positive hemolytic anemia and these same drugs may cause immune thrombocytopenia. The most commonly implicated drugs are quinidine, quinine, para-aminosalicylic acid, phenacetin, penicillin, insecticides, sulfonamides, chlorpromazine, and dipyrine.

There appear to be two types of immune hemolytic anemia associated with drugs. One is caused by an antibody that reacts only with cells exposed to the drug. In the other, the antibody reacts with normal red cells in the absence of the drug. The latter disease thus is indistinguishable from "auto" immune hemolytic anemia except for the history of drug exposure and exacerbation with retreatment. Drug-related hemolysis is usually highly specific, but occasionally closely related drugs cause hemolysis in the same patient. Some drugs require only small doses to cause disease, whereas hemolytic anemia after penicillin therapy is limited to cases with prolonged high-level treatment. In most cases, the antibody eluted from sensitized red cells is drug specific and does not react with normal red cells. Most cases caused by drugs other than penicillin have little IgG on red cells, but complement is usually present.

Since drugs and environmental contaminants are so ubiquitous, and some drugs induce an autoimmune type hemolysis, the distinction between a spontaneous and drug-induced immune anemia is usually not possible.

e. Primary or Idiopathic Immune Hemolytic Anemia   This type of anemia is most common in dogs and cats, less common in cows and horses, and probably occurs in all species.

Clinically there is usually depression of sudden onset in an animal in good body condition. Occasionally the history may indicate previous attacks. Rarely red urine is seen and the animal is examined for suspected urinary infection. There is mild to marked pallor and often mild icterus. When hemolysis and thrombocytopenia are present, there will be mucosal bleeding and often epistaxis, dark stools, and hematuria. Heart rate and respiratory rate are increased in proportion to the degree of anemia and a hemic murmur may be present. The spleen may be palpably enlarged.

Occasionally immune hemolytic anemia in the dog may resemble systemic lupus erythematosus of humans. The canine form of **systemic lupus erythematosus** is characterized by a Coombs-positive hemolytic anemia with thrombocytopenia, polyarthritis, and glomerulonephritis. Rarely these lesions are accompanied by a symmetrical facial dermatosis, serositis with accumulation of fluid in body cavities, and combinations of leukopenia, hepatosplenomegaly, and lymphadenopathy. Antibodies to double-stranded DNA are relatively common as are positive lupus erythematosus preparations. All breeds of dogs may be affected and the disease, as in humans, is most severe in young females. Canine systemic lupus erythematosus is considered also with The Urinary System (Volume 2, Chapter 5), Bones and Joints (Volume 1, Chapter 1), and The Skin and Appendages (Volume 1, Chapter 5).

There is a moderate to severe anemia, which is highly responsive, with 10–20 polychromatic red cells per oil immersion field or $300–400 \times 10^9$ reticulocytes/liter of blood. The anemia is macrocytic and slightly hypochromic with a normal corpuscular hemoglobin concentration. Spherocytosis, leptocytosis, and marked anisocytosis are present (Fig. 2.55A,B,C), usually with rubricytosis of $1–20 \times 10^9$/liter. Rouleaux formation is usually marked, and slide aggregation may be present. The direct Coombs test is positive in the majority of cases but may be negative; a negative result should not eliminate the diagnosis if the other findings are consistent with the disease. Large immature red cells occur in most microscopic fields.

There is an accompanying neutrophilic leukocytosis usually of 30×10^9/liter in the dog and cat, which in-

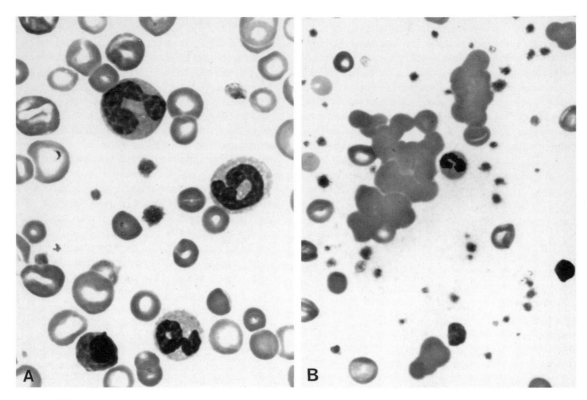

Fig. 2.55A,B Immunohemolytic anemia. Dog. (A) Leptocytes, hypochromic macrocytes, polychromatic red cells, neutrophilia, monocytosis, rubricytosis. (B) Marked immune rafting in butt of film. Polychromatic leptocytes not involved in rouleaux formation.

creases to 40×10^9/liter with initial heavy steroid therapy. There is a mild left shift with eosinopenia, lymphopenia, and a characteristic monocytosis in dog and cow and to a lesser extent in other species. Thrombocytosis is usually present with many large young platelets. The marrow is hypercellular with erythroid shift and normal maturation.

Biochemical icterus may be present and is usually less than 3.0 mg/dl except in the horse, where 10 mg or more may be present. In all species, free and conjugated bilirubin both tend to be markedly elevated. Icterus on presentation usually indicates previous attacks or chronicity with ischemic hepatic damage. Icterus tends to increase with each blood transfusion, especially if the packed cell volume is maintained only for 1–2 days. The erythrocyte sedimentation rate is very rapid.

Lesions usually are prominent. Those described here refer to the dog, but the changes are similar in the cat, calf, and horse. In the horse, immune hemolytic anemia is more often associated with (occult) lymphoma than in other species. Grossly, there is at least mild icterus and in some cases it is marked. The carcass is usually in good condition with some history of weight loss. In dogs, if steroid treatment has been prolonged, there may be abdominal enlargement and ventral hair loss. The lungs are edematous and may contain infarcts. The heart may be normal, or atonic, rounded, and dilated if anemia has led to congestive failure. There is often a prominent yellow

chicken-fat clot. The abdominal viscera are pale except the kidneys, which have dark red-brown cortices if the disease has been prolonged. The liver is usually enlarged, fatty, and friable with distended gallbladder and a nutmeg pattern visible through the capsule. The spleen is enlarged, sometimes up to 6–10 times normal weight. Lymph nodes are usually small if steroid treatment has been given, and red-brown medullary pigmentation due to hemosiderosis is present. The femoral marrow has subendosteal reddening if clinical signs have been present for a month, and diffuse reddening if the disease has persisted 3–4 months or longer. If the immune hemolysis is complicated by thrombocytopenia, hemorrhagic lesions may dominate the gross picture. In the combined immune destruction of erythrocytes and platelets, the lymph nodes are reddened, and may be enlarged, edematous, and cyanotic if draining an area with interstitial hemorrhage. Hemorrhages are most prominent on serosal surfaces, including those of the enteric and respiratory systems, and on the heart and diaphragm.

Histologically the thymus is atrophic, though rarely there is a follicular reaction with plasma cell accumulation or a lymphoid thymoma. There is pulmonary alveolar flooding and, if congestive failure has developed, there are alveolar macrophages with hemosiderin. Some animals have pulmonary thrombosis with focal peripheral infarction. The liver has Kupffer cell hyperplasia of variable cal atrophy and sinus expansion. There is always some degree of sinus histiocytosis and variable accumulation of plasma cells in medullary cords. If hemorrhage has occurred, there is increased hemosiderin in sinus phagocytes and erythrophagocytosis is often very marked. The kidneys have minimal to moderate thickening of glomerular basement membranes and there are occasional hemosiderin granules in mesangial areas. The juxtaglomerular nuclei are prominent with large nucleoli and the proximal epithelium is variably flattened and bile pigmented. Occasionally there are (incidental) loose lymphoid cuffs around corticomedullary arteries, and the kidney, like the lung, liver, and spleen, may have focal thrombotic infarction. The adrenal zona glomerulosa is usually hypertrophic, and the reticularis, reduced.

The femoral bone marrow has 50–60% cellularity or higher in the normally fatty central medullary areas. Cellular packing (density) is high in highly responsive anemias and reduced in animals which die with marrow depression. As in the spleen, hemosiderin is markedly increased and of both fine and coarse character, indicating chronic accumulation and continuing increased turnover. Erythrophagocytosis is often extensive and more easily identified than that in the spleen. The marrow has an erythroid shift and megakaryocytes are increased in numbers and multinuclearity. Animals with concurrent immune thrombocytopenia may have dysthrombopoiesis characterized by reduced megakaryocyte cytoplasmic volume and many pyknotic nuclei. Neutrophil migration through megakaryocytic cytoplasm is often prominent but is apparently without diagnostic value.

The myeloid system usually maintains a picture of early asynchrony due to steroid effects and early release of maturation phase cells. The relative reduction in myelopoiesis is apparently much more than compensated for by the expanded volume of hematopoietic marrow. Erythropoiesis is usually normal, but there may be a mild early asynchrony if there is prominent peripheral blood rubricytosis with release of maturation phase cells. Occasionally there is mild dyserythropoiesis with satellite nuclei (incipient Howell–Jolly bodies), and mitoses may be observed in polychromatic rubricytes. Plasma cells are increased and may be found in cuffs around small arteries. The overall picture is one of increased erythrocyte destruction followed by increased production and, in chronic cases, lymphoid atrophy, macrophage hyperplasia, hemosiderosis, and secondary injury to parenchymal organs.

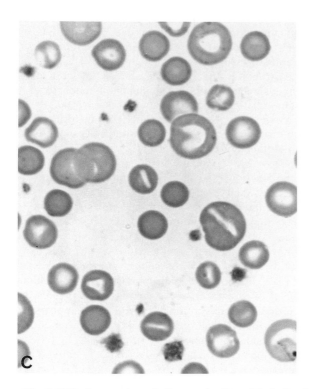

Fig. 2.55C Immunohemolytic anemia. Dog. Marked anisocytosis with spherocytes and macrocytic leptocytes. Howell–Jolly bodies indicate spleen overload.

degree and hemosiderin accumulation may be prominent. There is minimal periportal lymphoid reaction or hematopoiesis, and vacuolar degeneration is present in periacinar and often in midzonal hepatocytes. Inspissated bile plugs form in canaliculi, but the ducts themselves are empty. If changes of circulatory failure are added to the anemia, there is irregular necrosis of periacinar hepatocytes, swelling of sinusoidal endothelium, and separation of hepatic cords. The splenic capsule is thin and muscular trabeculae are narrow and sparse, reflecting splenomegaly and sinus hyperplasia. There is atrophy of follicles and periarteriolar sheaths. In addition, the normal small lymphocytes of the thymus-dependent cuffs are loosely rather than compactly arranged and nuclear size is increased with vesiculation so that these areas are inconspicuous and identifiable by the vessel rather than the lymphoid cuff. Major changes occur in the sinus areas, which have diffuse extramedullary hematopoiesis with prominent erythropoiesis and thrombopoiesis. Hemosiderosis is often marked with focal and diffuse clusters that may exceed three to four red cells in diameter. Erythrophagocytosis is present and often prominent. Plasma cells are increased and all free cells are enmeshed in a stromal network that is markedly hypertrophied and can be identified by the oval reticular nuclei and cellular cleavage planes created by their processes.

The lymph nodes have follicular hyperplasia with bare follicles due to paracortical atrophy. If steroid treatment has been intense and prolonged, there is generalized corti-

Bibliography

Costa, O. Specificities of antinuclear antibodies detected in dogs with systemic lupus erythematosus. *Vet Immunol Immunopathol* **7**: 369–392, 1984.

Dennis, R. A. *et al.* Neonatal immunohemolytic anemia and icterus of calves. *J Am Vet Med Assoc* **156**: 1861–1869, 1970.

Dowsett, K. F., Dimmock, C. K., and Hill, M. W. M. Haemolytic disease in newborn calves. *Aust Vet J* **54**: 65–67, 1978.

Dunn, J. K., Searcy, G. P., and Hirsch, V. M. The diagnostic significance of a positive direct antiglobulin test in anemic cats. *Can J Comp Med* **48**: 349–353, 1984.

Geor, R. J. *et al.* Systemic lupus erythematosus in a filly. *J Am Vet Med Assoc* **11:** 1489–1492, 1990.

Hubler, M. *et al.* Feline neonatal isoerythrolysis in two litters. *J Small Anim Prac* **28:** 833–838, 1987.

Jackson, M. L., and Kruth, S. A. Immune-mediated hemolytic anemia and thrombocytopenia in the dog: A retrospective study of 55 cases diagnosed from 1969 through 1983 at the Western College of Veterinary Medicine. *Can Vet J* **26:** 245–250, 1985.

Jeffcott, L. B. Haemolytic disease of the newborn foal. *Equine Vet J* **4:** 165–170, 1969.

Lavoie, J.-P. *et al.* Complement activity and selected hematologic variables in newborn foals fed bovine colostrum. *Am J Vet Res* **50:** 1532–1536, 1989.

Linklater, K. A., McTaggart, H. S., and Imlah, P. Haemolytic disease of the newborn, thrombocytopenic purpura and neutropenia occurring concurrently in a litter of piglets. *Br Vet J* **129:** 36–46, 1973.

Scott, D. W. *et al.* Autoimmune hemolytic anemia in the cat. *J Am Anim Hosp Assoc* **9:** 530–539, 1973.

Weiser, G., Kohn, C., and Vachon, A. Erythrocyte volume distribution analysis and hematologic changes in two horses with immune-mediated hemolytic anemia. *Vet Pathol* **20:** 424–433, 1983.

Young, L. E. *et al.* Hemolytic disease in newborn dogs. *Blood* **6:** 291–313, 1951.

2. Infectious Hemolytic Anemias

a. EQUINE INFECTIOUS ANEMIA   The equine infectious anemia virus (EIAV) is a member of the Lentivirinae subfamily of retroviruses that infects horses, mules, and donkeys. The disease has a worldwide distribution, and since the development of the agar gel immunodiffusion test, which detects anti-EIAV antibody, has been subject to control measures in many countries. The location of outbreaks in marshy areas led to the common name of swamp fever and their occurrence during the summer led to the demonstration that the virus is arthropod borne. The likely vectors are the stable fly (*Stomoxys calcitrans*), horse flies (*Tabanus* spp.) and *Anopheles* mosquitoes. Transmission is mechanical, not biological, and may also occur accidentally by contaminated needles, syringes, and tattoo equipment or experimentally by parenteral administration of blood or virus. The virus is stable in the environment, but infection appears always to arise by close contact with an infected animal. Infection is lifelong and viremia is cyclic, resulting in a clinical course varying from asymptomatic to acutely fatal, which resembles the idiopathic hemolytic anemias of dogs. The virus infects cells of the monocyte–macrophage system but, in acute viremic states, unintegrated viral DNA is present at levels two orders of magnitude higher in liver than in leukocytes and at lower levels in nodes, marrow, spleen, and kidney. The relationship between the cyclic viremia, viral antigenic variation, and the immune response is complex. The gp90, gp45, and p26 antigens appear to be most apparent to the immune system and, whereas these antigens are bound in early disease, neutralizing antibody is not detected before 2 months of infection. Subsequently there is rapid variation in the glycoprotein envelope antigens that

indicates that the preparation of an effective vaccine will be difficult as for the other retroviridae. The virus has both cis- and trans-acting response elements responding to viral regulatory proteins in controlling gene expression, and shares structural and sequence similarities with human immunodeficiency virus and simian immunodeficiency virus, and functional similarities with visna virus.

It has been suggested that the EIAV has little direct effect on tissues and that the pathogenesis of disease involves the indirect effects of the immune response. When foals with combined immunodeficiency (CID) and lacking functional B and T lymphocytes are infected with EIAV, the typical widespread inflammatory lesions of EIAV infection are not found, suggesting that the lesions in parenchymal organs are a function of the immune system. The foals develop anemia and, although they have a functional complement system, the characteristic fixation of C3 to red cells is not found. Whereas CID foals are not able to clear virus from the blood, the level of viremia in these animals is no higher than that in infected conventional foals. Thus it appears that there may be NK cell or other factors operative in viral control besides antibody.

Following experimental infection of conventional horses, there is a variable incubation period of 5–7 days up to several weeks during which time there is asymptomatic viremia. All signs of disease correspond to the appearance of circulating antibody which is complement fixing but not initially virus neutralizing. C3 appears on erythrocyte membranes, and this probably accounts for macrophage recognition and erythrophagocytosis. Antibodies of the IgG and IgM type are not found fixed to red cells, but it is believed that they have reacted specifically with virus adsorbed to erythrocytes, activated complement, and then been eluted off in the process of analysis. Red cell destruction is thus of the innocent bystander type, with virus playing a similar role to compounds causing drug-induced immune hemolysis. There is shortening of red cell life span to about 20–65% of normal. Hemolysis is largely intracellular, but in acute disease there is erythrocyte fragmentation due to microvascular injury with decreased haptoglobin levels. The marrow is initially highly responsive with an upward shift in MCV and marked anisocytosis. Later the marrow becomes hypoproliferative. Evaluation of the erythroid response requires definition of the stage of the disease, and also the breed affected, since racing breeds tend to react differently from ponies and heavy breeds.

The clinical disease is arbitrarily divided into acute, subacute, and chronic forms, but death in the protracted cases results from acute exacerbation, so the signs are relatively constant. The acute disease is characterized by pyrexia and marked depression, with anorexia, weight loss, and pitting, dependent edema. There are petechial hemorrhages most reliably seen on the ventral surface of the tongue but also present on ocular and vulvar mucosa. The tongue is most reliably examined by rotation *in situ* since forced extraction tends to cause petechiation. Mild icterus develops after a short febrile period and is accompanied by pallor of mild to marked degree. The febrile periods may progress to death in less than a week or regress and recur at irregular intervals. Rarely, an acute episode is followed by many years of an asymptomatic carrier state.

Anemia is severe and sufficient to cause death, with erythrocyte counts as low as 1×10^{12}/liter, hemoglobin of 25–50 g/liter, and hematocrits of 0.08–0.15 liter/liter. There is consistent thrombocytopenia during febrile periods, with resulting petechiae. In the early stages of disease, there is anisocytosis, marked for the horse, with moderate poikilocytosis and, characteristically for the horse, no polychromatic red cells. Nucleated erythrocytes are not present in peripheral blood. In the early stage of an acute attack, there is macrocytosis, sometimes up to an MCV of 60 fl. Later the disease is normochromic, normocytic, and nonresponsive. There is accompanying leukopenia with neutropenia, minimal left shift, lymphopenia, and at least a relative monocytosis. The monocytes often contain ingested red cells (sideroleukocytes) and in buffy-coat smears stain positively for iron. Sideroleukocytes are specific not to equine infectious anemia but to intracellular hemolytic anemia but, prior to diagnosis by antibody, their presence at levels of 4/10,000 leukocytes or higher was considered a positive diagnosis. Sideroleukocytes are present in the blood within 2–3 days of a febrile episode and subside with the temperature. The neutrophils have moderate toxic changes with vacuolation and azurophilic granulation, but not Doehle bodies. The mononuclear cells are large and basophilic and it is difficult to distinguish lymphocytes from young monocytes. The platelets are small and uniformly basophilic with a normal low level of granulation, and immature cells are rarely found. In the acute disease, the aspirated bone marrow is hypercellular with an erythroid shift but later the cellularity and the erythropoiesis are reduced and there is mild dyserythropoiesis and macrophage and plasma cell proliferation.

Icterus is always present in febrile, anemic horses, and bilirubin is usually between 10 and 15 mg/dl; most is unconjugated. Lipemia is occasionally seen in the acute disease. With chronicity (30 days or more), there is a drop in albumin of around 10 g/liter and a corresponding increase in gamma globulin so that the total protein is relatively unchanged but the albumin:globulin ratio is decreased. The serum iron rises in the acute disease and drops with chronicity, whereas the total iron-binding capacity remains normal or mildly elevated.

In horses dead of the acute disease, there is anemia, icterus, and widespread foci of hemorrhage. In animals with protracted disease, there is anemia, emaciation, and edema with serous atrophy of fat which is bilirubin stained. In febrile disease, the thrombocytopenia complicates minor trauma and larger hemorrhages are found subcutaneously and around sites of parasitic migration. There is edema of the ventral abdominal wall and in the suspensory ligaments of the viscera. The spleen is enlarged, turgid,

and fleshy with capsular hemorrhages, a bulging but not oozing cut surface, and inapparent lymphoid follicles. The liver is enlarged, dark, and turgid with a fine lobular pattern and focal capsular hemorrhages. Petechial hemorrhages are present on the renal capsules and in the perirenal tissues. On cut surface there are multiple fine hemorrhages throughout the cortex and medulla (Fig. 2.56). The most significant gross lesions occur in bone marrow where the degree of reddening is in direct proportion to the duration of the disease. In acute cases, the conversion of fat to hematopoiesis in the femur occurs first in proximal cancellous and then subendosteal diaphyseal and distal cancellous areas. The red and yellow areas are initially firm and opaque, often with focal areas of hemorrhagic infarction. In chronic cases, the red conversion may include all of the medullary marrow and is alternately pink and translucent in areas of serous atrophy of fat, and cyanotic where congested sinuses have dilated as adjacent hematopoietic areas atrophy.

Microscopic lesions occur in most tissues, but are most prominent in the heart, lungs, liver, spleen, kidney, marrow, and lymph nodes. Their severity varies with chronic-

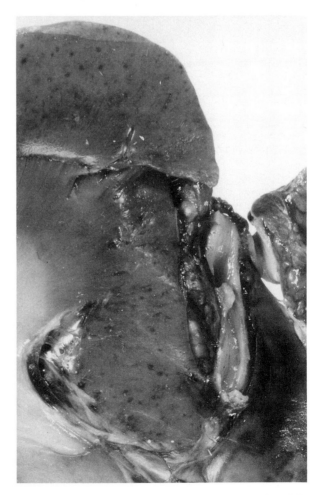

Fig. 2.56 Equine infectious anemia. Sagittal section of kidney. Prominent glomeruli with petechial hemorrhages.

Fig. 2.57 Equine infectious anemia. Interstitial lymphocytic myocarditis in acute disease.

ity and they are described here in the fully developed state. The myocardium has fiber atrophy in chronic cases, and interstitial edema in acute cases, with perivascular lymphocytic aggregations that irregularly permeate the surrounding interstitium (Fig. 2.57). There is mild pulmonary alveolar thickening and an overall appearance of hypercellularity. Occasionally hemosiderin-bearing macrophages are found in alveolar walls and are likely intravascular. The liver presents a spectrum of changes which varies from mild periportal lymphocytic infiltrates to atrophic cords with sinusoidal dilation, Kupffer cell hyperplasia, broad loosely arranged periportal lymphoid infiltrates, and increased interstitial connective tissue. As the disease progresses, hepatic hemosiderosis increases, largely in Kupffer cells. Generalized endothelial prominence, along with sinusoidal Kupffer cells and infiltrates, gives an overall picture of hypercellularity (Fig. 2.58A,B). In subacute cases there is periacinar fatty vacuolation, and in acute cases, hemorrhage and necrosis. In animals in which the disease has been quiescent for some months, the hepatocytes appear normal and the lymphoplasmacytic infiltrates subside, but the sinusoidal hemosiderin-bearing macrophages remain as evidence of previous hemolysis.

Splenic follicles are variably enlarged but hypocellular, often with a bull's-eye appearance due to a sharp distinction between the cells of the follicular center and the mantle and marginal layers, and a sharp transition to con-

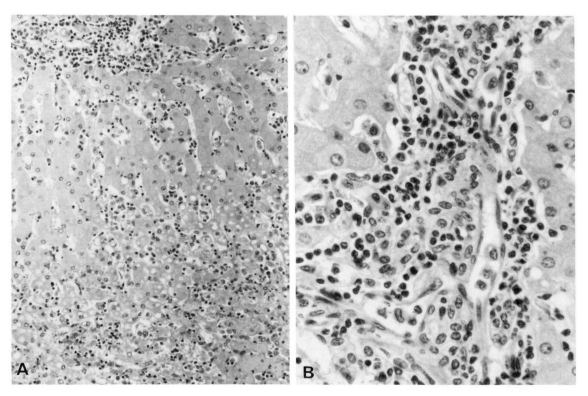

Fig. 2.58 Equine infectious anemia. (A) Portal and sinusoidal infiltrates of lymphocytes and plasma cells, Kupffer cell proliferation and periacinar necrosis in the chronic disease. (B) Detail of portal area of (A).

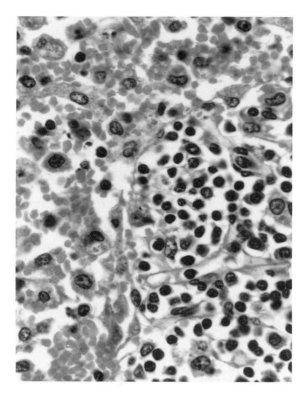

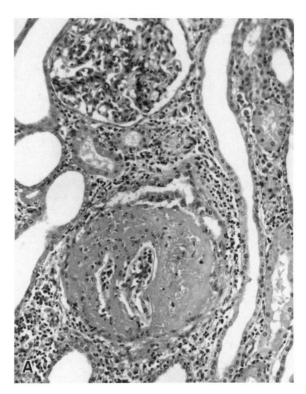

Fig. 2.59 Equine infectious anemia. Lymph node. Medullary cords contain few small lymphocytes and plasma cells. Active sinus histiocytes. Red cells in sinuses derived from thrombocytopenic bleeding.

Fig. 2.60A Equine infectious anemia. Periglomerular interstitial infiltrates of the chronic disease with glomerular hemorrhage from thrombocytopenia in acute exacerbation.

gestion in the surrounding sinusoids. In acute cases, the spleen is congested and, being largely composed of unsupported red cells, fractures on sectioning. Later there is sinus hyperplasia with stromal proliferation. The tissue is then cellular, fleshy, and cohesive. Splenic hematopoiesis is never as prominent as in the dog with idiopathic immune hemolytic anemia, but hemosiderosis with macrophages and plasma cell proliferation is prominent. Lymph nodes are edematous and have medullary hemosiderosis with persistence of follicles and thin moth-eaten paracortical areas. In acute disease, the lymphoid tissue is highly reactive and there may be lymphocytic colonization of perinodal fat. Later there is sclerosis and lymphoid atrophy (Fig. 2.59).

Renal lesions in acute disease are largely hemorrhagic with some glomeruli obscured by erythrocytes and fibrin (Fig. 2.60A). Lymphocytic infiltrates may be intense and separate the tubules in an irregular manner (Fig. 2.60B). There is some degree of epithelial atrophy and mild pigmentation that appears to be both bilirubin and hemosiderin.

The red bone marrow in acute disease has 40% or less of fatty areas and the hematopoietic cells are densely packed. There is an erythroid shift with synchronous maturation (Fig. 2.61) and a reduction in mature granulocytes, although eosinophils are prominent. Megakaryocytes are not increased, which may indicate that the thrombocyto-

penia may be at least partially due to ineffective production. With chronicity and repeated febrile periods, there is a progressive reduction of fat cells until the marrow is converted to hematopoiesis (Fig. 2.62A). At the same time, cellular packing decreases and sinusoidal expansion occurs so that marrow which appears red grossly can be seen histologically to be relatively hypoproliferative. Hemosiderin is prominent as is erythrophagocytosis (Fig. 2.62B). Large macrophages can be found which contain hemosiderin, red cells, and rubricytes. This latter association does not appear to be a nurse cell relationship but rather an extension of the peripheral disease to marrow with destruction of the precursors as well as the progeny. Dyserythropoiesis as evidenced by binucleated interphase rubricytes is present. Megakaryocyte cytoplasmic volume is normal or reduced but not increased, indicating that the response of this lineage is weakened and likely at least partially ineffective. Stromal cells are increased, endothelial nuclei are prominent, and there is a diffuse plasmacytosis. Hematopoiesis may regress from the central femoral cavity and leave fatty tissue with serous atrophy heavily laden with hemosiderin-bearing macrophages that indicate previous sites of production and hemolysis.

The disease is one of acute intracellular hemolysis followed by immunoproliferation and finally by hematopoietic and to a lesser extent immune exhaustion.

Following the development of the agar gel immunodiffusion test, steroid-responsive, idiopathic immune hemo-

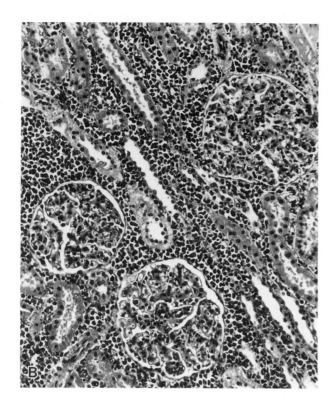

Fig. 2.60B Equine infectious anemia. Diffuse interstitial mononuclear cell infiltrate.

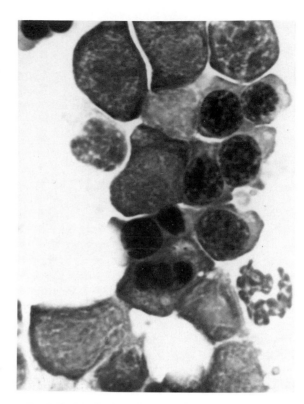

Fig. 2.61 Equine infectious anemia. Marrow aspirate in acute disease. Productive synchronous erythroid hyperplasia.

lytic anemia has been identified in the horse. The lesions resemble those of immune hemolytic anemia of the dog and lack the inflammatory reaction characteristic of equine infectious anemia. Horses with idiopathic immune hemolytic anemia are not usually thrombocytopenic and do not have cyclic fever.

3. Anemia of Blood Cell Parasitism

a. BABESIOSIS The protozoan *Babesia* parasitizes the erythrocytes of a wide range of mammals including humans, and there is normally quite strict host specificity. The disease is transmitted by ticks. Natural transmission is largely transovarial, that is, the adult female becomes infected, and, depending on the babesial and tick species, the ensuing larval or nymphal generation then transmits the infection to a susceptible host. Transstadial transmission occurs in three host ticks, infection being carried through a molt off the host, permitting transmission to the next host encountered. There is no evidence for transmission by biting insects or contaminated instruments (in contrast to the situation with *Anaplasma* infection), but infection is readily transmitted by experimental blood inoculation.

The severity of the disease produced in the mammalian host depends more on the strain and species of infecting *Babesia* than on the number of organisms inoculated. Pathogenic effects are, in most infections, related directly to lysis of red cells by emerging parasites, but other mechanisms, including sludging of infected red cells, hemoglobinuric nephrosis, and release of vasoactive peptides, contribute to the signs and are responsible for death.

Different tick species are vectors in different countries for what are apparently the same species of *Babesia*. Nomenclature is not dealt with here, but rather the pathology and pathogenesis of babesiosis and, in particular, those

Bibliography

Carpenter, S. *et al.* Role of the host immune response in selection of equine infectious anemia virus variants. *J Virol* **61:** 3783–3789, 1987.

Dawson, M. Lentivirus diseases of domesticated animals. *J Comp Pathol* **99:** 401–419, 1988.

Dorn, P. *et al.* Equine infectious anemia virus *tat:* Insights into the structure, function, and evolution of lentivirus trans-activator proteins. *J Virol* **64:** 1616–1624, 1990.

Henson, J. B., and McGuire, T. C. Equine infectious anemia. *Prog Med Virol* **18:** 143–159, 1974.

Messer, N. T., and Arnold, K. Immune-mediated hemolytic anemia in a horse. *J Am Vet Med Assoc* **8:** 1415–1416, 1991.

Perryman, L. E., O'Rourke, K. I., and McGuire, T. C. Immune responses are required to terminate viremia in equine infectious anemia lentivirus infection. *J Virol* **62:** 3073–3076, 1988.

Rice, N. R. *et al.* Viral DNA in horses infected with equine infectious anemia virus. *J Virol* **63:** 5194–5200, 1989.

Rwambo, P. M. *et al.* Equine infectious anemia virus (EIAV): Humoral responses of recipient ponies and antigenic variation during persistent infection. *Arch Virol* **111:** 199–212, 1990.

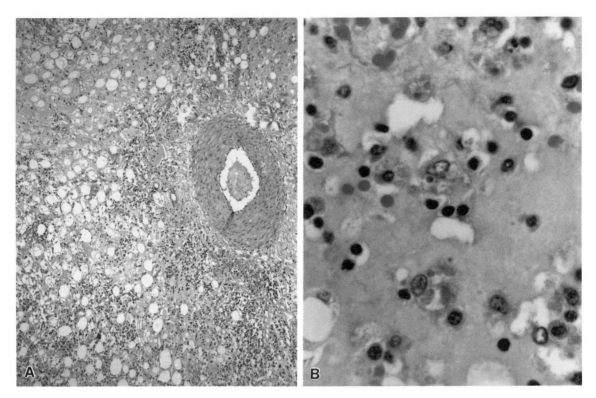

Fig. 2.62 Equine infectious anemia. (A) Central femoral marrow in the chronic disease. Hypertrophy of marrow to occupy entire cavity followed by late reduced hematopoiesis and serous atrophy. (B) Area of serous atrophy in marrow in chronic disease. Macrophages contain hemosiderin indicating earlier hematopoiesis and erythrophagocytosis.

naturally occurring infections which cause the most severe and economically important disease. The severity of many of the anemias caused by parasites of blood cells is enhanced by splenectomy, and this technique is used commonly in pathogenetic studies. Splenectomy is often performed in dogs following abdominal trauma or for treatment of neoplastic disease, and this may activate latent infections by organisms such as *Haemobartonella canis*. In human babesiosis, splenectomy appears to be a major risk factor for infection, primarily with *B. microti*.

i. *Bovine Babesiosis Babesia bovis* (*B. argentina, B. berbera*, and *B. colchica*) is probably the most important cause of the tick fevers of cattle, with local exceptions in some countries. The fact that some *Bos indicus* cattle breeds are relatively resistant to the effects of this parasite has led to the suggestion that the organism evolved in these breeds. Young cattle have pronounced resistance to severe infection. Immunity passively acquired from colostrum in very young calves may persist for up to 9 months of age. Strong acquired immunity follows overt infection, and may persist indefinitely in the absence of latent infection, although the latter is usual. Infection with a single strain of the parasite protects against subsequent infection for at least 4 years. For these reasons, the disease is mostly seen in adult cattle recently introduced from tick-free areas, or in areas of greatly fluctuating tick populations.

Infection by *B. bovis* causes severe febrile illness, which begins about 2 weeks after exposure to ticks. The animal may become extremely ill before severe anemia, parasitemia, or hemoglobinuria are apparent, and sometimes the clinical picture is dominated by nervous derangement, likely associated with the tendency for parasitized red cells to sludge in cerebral capillaries as well as in the microvasculature of the adrenal and kidney of splenectomized animals. Usually, however, the clinical picture is characterized by weakness, fever, hemoglobinuria, and anemia; the latter is more severe in animals surviving more than a week. The parasite is never numerous in circulating blood, rarely being seen in more than 5% of circulating erythrocytes. Parasitized erythrocytes are much more likely to be found in smears of blood squeezed from cut skin capillaries than in routine blood samples from large veins. Recovery is usually fairly rapid in animals which survive the acute phase, although chronic ill thrift has been described. Photosensitization may occur in the convalescent phase, probably due to overload of the phylloerythrin-conjugating sequence by bilirubin derived from hemolysis.

The most typical findings postmortem are those to be expected of an acute intravascular hemolytic crisis. There is anemia, variably severe jaundice, and hemoglobinuria, and the kidneys are deep reddish brown throughout as a result of hemoglobin staining and intense congestion of capillaries. There is capillary congestion of most organs, the spleen is always grossly swollen, soft, and dark, and the liver is also acutely congested and may be heavily stained with bile pigment. The gallbladder is distended with thick, dark viscous bile.

In some acute cases, the lungs are congested and edematous, and the larger airways contain stable foam. Recent hemorrhages are common in the thoracic serosal membranes. The most characteristic macroscopic feature of *B. bovis* infections is a striking, uniform congestion of the gray matter throughout the brain, imparting to it a dramatic, deep pink color (the cerebral flush), which contrasts strongly with the white matter; the latter is often stained a faint yellow by unconjugated bilirubin. In some peracute cases, there is minimal hemoglobinuria or jaundice, but the brain still shows the characteristic capillary congestion from which imprint preparations are usually diagnostic, with many intraerythrocytic organisms present.

Histologically there is thickening and congestion of pulmonary capillaries with hemosiderin in intravascular macrophages. Kidneys and liver have the histologic lesions to be expected of an acute hemolytic disease. There is variably severe hemoglobinuric nephrosis with severe congestion, focal hemorrhage, and hemosiderin in tubular epithelium, and diffuse vacuolar degeneration with interstitial mixed mononuclear cell reaction. There is hepatic periacinar congestion with uninfected red cells and vacuolar degeneration in midzone and periacinar hepatocytes accompanied by canalicular cholestasis. There is hemosiderin accumulation in both hepatocytes and Kupffer cells, and the latter contain both infected and noninfected red cells. Lymphocytes and plasma cells accumulate in portal areas and around central veins. Parasitized erythrocytes may be seen in vessels in all tissues, but they are particularly common in interstitial capillaries in the kidney, in the gray matter of the brain, in the heart, and especially in skeletal muscle; in these locations, nearly every red cell packed into the distended capillaries appears to contain one to two parasites. The organisms are faint blue in routine sections but are best demonstrated in imprint preparations of fresh tissue. In sections, Giemsa's stain demonstrates them as small paired or single spherical bodies with denser staining at one pole. In well-fixed fresh brain of acute cases, there may be edema of the neuropil around the clogged capillaries (Fig. 2.63).

Animals which die with peracute disease have necrosis of lymphocytes in germinal centers of node and spleen, whereas there may be some degree of recovery in animals surviving a week and general depletion of lymphocytes in those dying after protracted illness. Characteristic changes in the medullary region of nodes include sinus histiocytosis with extensive erythrophagocytosis. The histology of the spleen is often unhelpful, the organ being so suffused with intra- and extracorpuscular hemoglobin that evidence of erythrophagocytosis is obscured. There is little extramedullary hematopoiesis in liver or spleen. The bone marrow shows an erythroid shift and, in animals that survive the acute stage, there is a strong reticulocytosis and slow recovery in red cell numbers. There is mild hemosiderin increase, suggesting that marrow is also a site of erythrolysis.

Babesiosis results in metabolic disease more complex than a simple syndrome of intravascular hemolysis. Severely affected animals may die before significant anemia has set in, and in these cattle the intense visceral congestion and pulmonary edema suggest that death may be at least partly due to circulatory shock. This has been supported by studies which have shown that the parasite is the source of proteases which activate plasma kallikrein which, as well as being a hypotensive agent, may activate bradykinin, another potent vasodilatory agent. Rather unexpectedly, in *B. bovis* infections metabolic alkalosis occurs, whereas acidosis is the rule in severe *B. canis* infections. The result of these profound metabolic upsets is a syndrome of circulatory failure, likely due to extensive plugging of microvasculature by sequestered red cells. The apparent anemia is therefore due in considerable measure to the vasodilatory effects of kallikrein and shift of red cells away from large veins. Although plasmin activation occurs, the effects on the coagulation system are unlikely to be a significant factor in the pathogenesis since hemorrhage is not a major part of the morbid picture.

The actual mechanism of penetration of red cells by the merozoites of *Babesia* is as yet unclear, but complement is required, and it appears that the organism causes activation by the alternate pathway, and by C3b achieves adherence and later invasion. The mechanism of hemolysis is also obscure. It may be in large part spherocytic, and complicated by mechanical loss by fragmentation and exiting of parasites. Immune mechanisms do not appear to play a significant role in the early stages of the disease, but they may be a factor in delayed recovery. Osmotic fragility is markedly increased in nonparasitized cells, whereas the fragility of parasitized cells may actually decrease. Parasitized red cells form prominent stellate projections, which appear to be the means by which they adhere to each other and to the endothelium. Fibrin may be involved in some of the sludging phenomena.

Babesia bigemina infection may be transmitted by the same tick as *B. bovis,* and both parasites may cause losses in the same geographic area. *Babesia bigemina,* however, usually causes a much less severe syndrome, notwithstanding the fact that parasitized erythrocytes circulate early (Fig. 2.64), sometimes before clinical signs are apparent, and in much greater numbers than they do in *B. bovis* infection. Like the latter parasite, *B. bigemina* causes erythrocyte destruction, but this appears to be directly proportional to the level of parasitemia, unlike *B. bovis* infection. There are few of the vasoactive and red cell adherence effects, so the disease is not attended by such profound shocklike signs. Thus the anemia in severe cases may be severe before the animal becomes ill and, as in any case of severe anemia, death may occur suddenly as a result of myocardial failure. Fever is not usually as severe as in *B. bovis* infection.

The pathology of *B. bigemina* and *B. bovis* infections

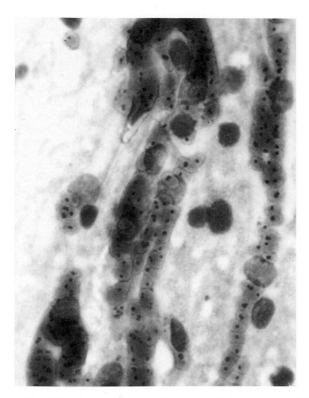

Fig. 2.63 Sludging of parasitized red cells in cerebral capillaries. *Babesia bovis.* Ox. Giemsa. (Courtesy of H. M. D. Hoyte.)

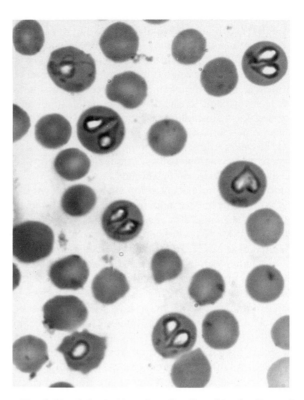

Fig. 2.64 *Babesia bigemina.* Ox. Parasitized cells are large due to presence of parasites and selective parasitism of young cells. Giemsa.

is similar, except that in *B. bigemina* infection there is little capillary congestion of the viscera and none of the cerebral gray matter, the presence or absence of the cerebral flush being the most reliable gross feature for differentiating the two infections.

Pulmonary edema may be more often seen in fatal cases of *B. bigemina* infections, probably because the greater severity of anemia causes terminal left ventricular failure.

Babesia divergens and *B. major* are more or less restricted to northern and western Europe and Asiatic Russia and, for the most part, are of relatively minor importance to the cattle industries there, although significant outbreaks of disease and death may occur. Although *B. divergens* is small and resembles *B. bovis* in size, and *B. major* is as large as *B. bigemina,* the pathology and pathogenesis of the disease produced by these organisms is apparently very similar to that produced by *B. bigemina.* Thus there is intravascular hemolysis with little or no capillary agglutination of erythrocytes, and death is due primarily to severe anemia. Spontaneous splenic rupture has been reported in *B. major* infections, a complication rarely if ever reported in other bovine babesial infections.

Differential diagnosis of bovine babesial infections in countries where more than one species of *Babesia* is present is important and at times difficult, and may be complicated by dual infections and by the additional possibility of *Anaplasma* infection. *Babesia bovis* and *B. bigemina*

may be distinguished from one another in well-prepared blood smears, but erythrocytes parasitized by *B. bovis* are rare in jugular blood, or even in blood from deep skin punctures. They are best demonstrated in smears of blood expressed from a superficial skin scrape; this is most conveniently obtained from the tail-tip in the live animal. *Babesia bigemina*-containing erythrocytes, on the other hand, are quite numerous in circulating blood while clinical signs are severe. The morphology of the parasites will usually be insufficiently preserved in postmortem material to allow distinction, but the preference of *B. bovis* for capillaries of organs such as kidney, heart, and brain will in most cases serve to distinguish the infections. It must be remembered that *Sarcocystis* spp. may be present in smears of myocardial blood. Serologic tests are available for live animals and direct fluorescent antibody techniques may be used for differentiation in postmortem material. It is necessary to demonstrate a very heavy parasite burden in cerebral capillaries if brain smears are to be relied on as evidence of active *B. bovis* infection, as most clinically normal carriers of this organism can be shown to have some parasitized erythrocytes in this site.

Blood smears should always be made from a peripheral site in a dead animal, particularly if fluorescent antibody techniques are available for diagnosis, because such smears may be the only means of distinguishing an *Anaplasma* infection from one of the babesial infections.

Sterile immunity can be obtained with appropriately prepared antigens, though these are not readily available.

ii. *Canine Babesiosis* *Babesia canis* infection is transmitted by different ticks in different countries; it is not known whether this is related to the apparent differences in virulence of the organism. The infection in dogs in Africa is severe and frequently causes death in all ages, whereas in Australia and in the United States of America, the disease is relatively mild in mature animals. The disease is, however, more severe in pups in all countries. As distinct from bovine babesiosis, young dogs appear to be more susceptible than older animals.

Mildly affected animals develop anemia and fever, and are lethargic and have a poor appetite. Icterus and red urine are not seen and they recover after a clinical course of a few days. More severe cases show a wide variety of signs, including severe depression, drooling, vomiting, jaundice and hemoglobinuria, mucosal petechiae and congestion, ulcerative stomatitis, and angioneurotic edema of the head, legs, and body. Anorexia and weight loss may be persistent. A few dogs may show nervous signs that are associated with erythrocyte sludging in cerebral capillaries and cellular infiltration of the meninges (Fig. 2.65). Acidosis is particularly important in this disease.

Pathologically there is staining of tissues both with bilirubin and hemoglobin. The kidneys are dark brown and there is copious thick bile in the gallbladder and splenomegaly. In addition, there is evidence of vascular injury

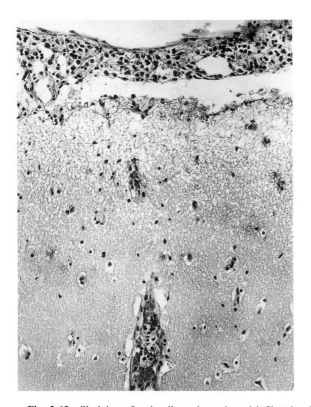

Fig. 2.65 Sludging of red cells and meningeal infiltration in *Babesia canis* infection.

in the form of hemorrhages and edema, which may be severe in the lung. Parasitized erythrocytes may be found plugging capillaries in smears and sections of cerebral cortex, reminiscent of *B. bovis* infection, but the phenomenon is never as severe as in that disease. Disseminated intravascular coagulation is a consistent occurrence in severe *B. canis* infection, and is presumably related to, and is likely to aggravate, the hemolysis and the vascular damage. Microthrombi can be demonstrated in many tissues.

Babesia canis and *B. bovis* infections clearly resemble one another in the severity of the syndromes and the involvement of pathogenetic mechanisms other than simple hemolysis. *Babesia gibsoni* infection is apparently restricted to Asia, with the most severe losses in dogs imported into India. The pathogenesis of this infection seems to be analogous to that of *B. bigemina* in cattle, in that dogs may have heavy parasitemia and quite severe anemia without becoming very ill. The body temperature is not elevated so consistently as in the case of *B. canis* infection, and hemoglobinuria rarely if ever occurs. The pathogenesis of *B. gibsoni* infection appears to be related primarily to the intracellular destruction of erythrocytes, as the spleen becomes very large and there is no evidence of intravascular hemolysis. Death is due to anemia, with the packed cell volumes dropping to less than 10%.

iii. *Equine Babesiosis* Babesiosis in horses is similar to that in dogs in that there are two babesias of equids, one larger than the other, but the larger (*B. caballi*) is usually considered less virulent than the smaller (*B. equi*). *Babesia caballi*, like *B. canis* and *B. bovis*, tends to accumulate in capillaries; thus these species have been referred to as visceral, as distinct from *B. bigemina*, *B. equi*, and *B. gibsoni*, which are more evenly distributed throughout the circulation. There is, however, considerable overlap in virulence between all these species, all of which may cause death.

Babesia caballi infection is often mild and, if severe, the syndrome is characterized by fever, jaundice, and anemia, but hemoglobinuria is unusual. *Babesia equi* infection produces severe anemia, often accompanied by hemoglobinuria, severe depression, lacrimation, and often ventral and periorbital edema and mucosal petechiae. Posterior ataxia is not seen as frequently as in *B. caballi* infection. The lesions in both infections are very similar, and include generalized icterus, anemia, and splenomegaly. More visceral edema and hemoglobinuric nephrosis is expected in *B. equi* infection. Mixed infections are possible, and pathologic differentiation of equine babesiosis from equine infectious anemia may be difficult.

iv. *Babesial Infections of Other Species* These for the most part are mild or clinically inapparent, but severe reactions sometimes occur; the reasons for these are seldom obvious. The course of these diseases in general follows the pattern described for *B. bigemina* infection, in that there is quite severe parasitemia, anemia due to intravascular hemolysis, with hemoglobinuria in acute cases. At autopsy there is splenomegaly and variably severe jaundice and, in some cases (notably *B. trautmanni* infection in pigs), edema and petechiae. *Babesia ovis* may cause cerebral congestion in sheep.

White-tailed deer may be infected by an organism resembling *B. divergens*, and mice may be infected with *B. rodhaini*, *B. hylomysci*, and *B. microti*. The latter infects other rodents as well. *Babesia hylomysci* causes severe hemoglobinuria and nephrosis.

Bibliography

Everitt, J. I. *et al.* Experimental *Babesia bovis* infection in Holstein calves. *Vet Pathol* **23:** 556–562, 1986.

Levy, M. G. *et al.* Studies on the role of complement in the *in vitro* invasion of bovine erythrocytes by *Babesia bovis*. *Rev Elev Med Vet Pays Trop* **39:** 317–322, 1986.

Roher, D. P., Anderson, J. F., and Nielsen, S. W. Experimental babesiosis in coyotes and coydogs. *Am J Vet Res* **46:** 256–262, 1985.

Taylor, W. M. *et al.* Equine piroplasmosis in the United States—A review. *J Am Vet Med Assoc* **7:** 915–919, 1969.

Todorovic, R. A. Bovine babesiosis: Its diagnosis and control. *Am J Vet Res* **35:** 1045–1052, 1974.

Wright, I. G., and Goodger, B. V. Pathogenesis of babesiosis. *In* "Babesiosis of Domestic Animals and Man," M. Ristic, (ed.), pp. 100–113. Boca Raton, Florida, CRC Press, 1988.

b. CYTAUXZOONOSIS An intraerythrocytic piroplasm tentatively classified in the family Theileriidae infects wild ungulate species in Africa, including the kudu, eland, and giraffe, as well as domestic and wild Felidae in North America. Unlike the *Babesia*, which infect only red cells of the mammalian host, the *Cytauxzoon* organisms have a schizogenous phase in intravascular macrophages as well as in red cells. The natural mode of transmission is unknown, but ixodid ticks are believed to be the most likely vectors. The disease is not spread by cage contact but can be experimentally transmitted by fresh blood or by cryopreserved blood or tissue homogenates from infected animals. The clinical signs of cytauxzoonosis in domestic cats include depression, anorexia, pyrexia, dehydration, pallor, and icterus. Whereas the disease is unusual, most reported cases have been fatal, with death occurring within a few days after the onset of clinical signs. The level of parasitemia is low and the piroplasms are commonly found in 1–4% of peripheral blood erythrocytes. Although a much higher level of parasitemia may occur, in some reported cases no parasitized red cells were identified. The sporadic appearance of the disease and fatal course suggest that domestic cats are an unusual host for the organism and that a reservoir host exists in the wildlife population. The bobcat (*Lynx rufus*) given blood from an infected cat develops a persistent parasitemia without signs of clinical illness, suggesting that it may be the reservoir for the acute disease in domestic cats. Remarkably, there appear to be two strains of the organism, which are indigenous to either the eastern or western bobcat. Cats are susceptible to both, and bobcats are susceptible to their nonindigenous strain. This likely explains why cats recovered from the bobcat transmission

are still fully susceptible to other strains from domestic cats. Diagnosis is dependent on detection in red cells of the irregularly pear-shaped organisms. They are often paired and are larger than *Haemobartonella* organisms. There are apparently no pathognomonic gross or histological findings.

Bibliography

Glenn, B. L., Kocan, A. A., and Blouin, E. F. Cytauxzoonosis in bobcats. *J Am Vet Med Assoc* **183:** 1155–1158, 1983.

Kier, A. B., Wrightman, S. R., and Wagner, J. E. Interspecies transmission of *Cytauxzoon felis*. *Am J Vet Res* **43:** 102–105, 1982.

Kier, A. B., Wagner, J. E., and Kinden, D. A. The pathology of experimental cytauxzoonosis. *J Comp Pathol* **97:** 415–432, 1987.

c. ANAPLASMOSIS There are three species of the genus *Anaplasma*, order Rickettsiales: *A. marginale* and *A. centrale* are infectious for cattle; *A. ovis* infects sheep and goats. These parasites are now classified with the Rickettsiales, notwithstanding that the disease produced and its mode of transmission have much in common with babesiosis. Division occurs by binary fission. The organisms are spherical or oval, from 0.3 to 1.0 μm in diameter, and situated in the erythrocytes (Fig. 2.66), usually near the margin (*A. marginale* and *A. ovis*).

Anaplasmosis in **cattle** is probably transmitted under field conditions by *Dermacentor* and other ticks, as well as by *Boophilus microplus*, and may be transmitted me-

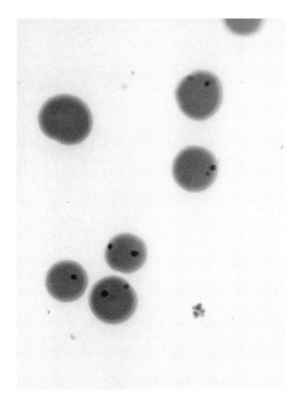

Fig. 2.66 *Anaplasma marginale*. Peripheral blood. Ox.

chanically by bloodsucking flies. Mechanical transmission has resulted from the careless use of hypodermic needles during immunization procedures and bleeding, and by instruments used for dehorning, castration, and ear tagging. The disease can be transmitted by blood from an infected animal, although in the chronic carrier state, the level of parasitism is very low and organisms are difficult to find in the peripheral blood. Unlike those in *Babesia,* the intraerythrocytic antigens are relatively easily separated from red cells and form the basis of an efficient diagnostic test based on complement fixation.

Much of what has already been said of babesiosis is applicable to anaplasmosis. Young animals are susceptible to infection but are relatively resistant to the disease; acquired infections in these are latent and endow the animals with resistance for the duration of the latent infection. Carrier status and immunity are usual following recovery from infection. This has epidemiologic significance in that indigenous animals, wild ruminants as well as cattle, in enzootic areas acquire the infection early in life and, barring accidents, do not become diseased, but represent a source of infection for incoming animals and make eradication of the disease virtually impossible in many areas. Events which upset the balance between the organism and the host's immune system are not common but include hemorrhage and especially other protozoan diseases which cause anemia. As in babesiosis, splenectomy is expected to lead to recrudescence. Acute disease can be regularly produced in splenectomized calves and infection appears to occur preferentially in older red cells.

The natural incubation period of anaplasmosis is 1–3 months or longer and the disease which follows closely resembles babesiosis. One important distinction is that hemoglobinuria does not occur in anaplasmosis as the destruction of erythrocytes occurs intracellularly rather than intravascularly. Whereas the mechanism of hemolysis is apparent, the means by which the phagocytic cells recognize injured or infected red cells is less clear. Serum factors sensitize erythrocytes to phagocytosis by the monocyte–macrophage system and these opsonins increase in the circulation prior to the hemolytic crisis. Since the destruction of parasitized red cells seems to be selective, it is likely that the immune system identifies the erythrocytes containing parasites as abnormal, and these are then removed by splenic and other macrophages. Spherocytosis may be prominent in the acute disease and is likely immune mediated. Nonimmune factors may be operative as well, and the decrease in plasma lipids with parasitemia may be related to the reduction in red cell numbers and may contribute to that reduction by increasing membrane rigidity. Attempts to culture *Anaplasma in vitro* demonstrated that high magnesium levels were required to establish growth and that animals deficient in magnesium are refractory to clinical disease after experimental challenge. It appears that *Anaplasma* and other intracellular parasites require abundant ambient magnesium for their endogenous metabolism.

Animals with acute anaplasmosis have mucosal pallor,

icterus, depression, and occasionally incoordination. The animals resist movement and have a rapid heart rate and increased respiratory rate. Urine may be deep brown but not red. The clinical syndrome in cattle is usually subacute illness with variable fever, anemia, weakness, and ill thrift. The severity of the disease is related to the proportion of the erythrocyte mass destroyed, which may equal 70%.

At a variable period prior to the onset of clinical signs, usually 5–10 days, the *Anaplasma* organisms appear in red cells in the peripheral blood. Since the organisms continue to appear and increase in number, it is apparent that in the acute phase, the parasitic multiplication proceeds more rapidly than removal of parasitized red cells in the monocyte–macrophage system. The erythrocyte count in acute disease drops to as low as 1×10^{12}/liter and the hemoglobin to 30–40 g/liter. The hematocrit tends to stabilize at 0.10–0.15 liter/liter, at which time there is rapid ongoing destruction with a competent marrow response resulting in a macrocytic mildly hypochromic anemia. At this time, as many as 50–60% of the red cells may be parasitized. The strong erythroid response to anaplasmosis, which does not occur in babesiosis, is possibly due to the intracellular rather than intravascular destruction of erythrocytes, thereby avoiding hemoglobinuria and allowing recycling of iron and proteins. Characteristically, anemia is accompanied by a neutrophilic leukocytosis. Chronicity results in apparent recovery or chronic disease with debility and a nonresponsive normochromic normocytic anemia in which the *Anaplasma* bodies are rarely observed in the peripheral blood. Biochemically the bilirubin, most of it unconjugated, is elevated in the acute disease.

There are no pathognomonic gross lesions of anaplasmosis. There is pallor of all tissues and mild icterus, with relatively good body condition in those animals dying acutely and cachexia in chronic cases. The lungs are pale and discolored and may have bullous emphysema if there has been severe terminal dyspnea. There are frequently ecchymotic hemorrhages on the epicardium, and the heart is flabby and dilated. The liver is anemic and icteric, and the gallbladder is usually distended. The spleen is enlarged, turgid, and congested in acute cases and firm, dark red, and fleshy in chronic ones. The enteric tract is unremarkable and the bladder may contain deeply bilirubin-stained urine. The marrow cavity is variably expanded, depending on the stage of the disease, and may have serous atrophy with chronicity.

The tissue changes in anaplasmosis are similar to those in babesiosis, although hemosiderosis is much more marked in anaplasmosis.

Anaplasma centrale produces a natural infection of cattle, but is also employed, being a mild pathogen, as an immunizing agent against *A. marginale* in some areas of endemic infection. It usually produces a mild disease, although sometimes it can be severe, with fever and anemia but no icterus. *Anaplasma centrale,* too, causes a persistent infection.

Anaplasmosis in **sheep and goats** is caused by *A. ovis.*

Cattle are not susceptible to this parasite. On the other hand, sheep and goats develop a latent infection if inoculated with *A. marginale*. In the Mediterranean region, anaplasmosis occurs as a natural disease in sheep and goats and is characterized by progressive anemia and emaciation, but with a relatively low mortality rate. In other regions in which the infection occurs, especially the United States of America and South Africa, the disease is milder unless the resistance of the animal is lowered by intercurrent disease, helminthiasis, or inadequate nutrition. It is possible, however, that in endemic areas ovine anaplasmosis of subclinical severity is responsible for unthriftiness of economic importance. Infected sheep develop complement-fixing antibodies.

Bibliography

Brown, J. E. *et al.* Blood magnesium values in healthy cattle and in cattle affected with anaplasmosis and eperythrozoonosis. *Am J Vet Res* **47:** 158–162, 1986.

Buening, G. M. Cell-mediated immune responses in calves with anaplasmosis. *Am J Vet Res* **34:** 757–763, 1973.

Buening, G. M. Hypolipoidemia and hypergammaglobulinemia associated with experimentally induced anaplasmosis in calves. *Am J Vet Res* **35:** 371–374, 1974.

Hawkins, J. A., Love, J. N., and Hidalgo, R. J. Mechanical transmission of anaplasmosis by tabanids (Diptera: Tabanidae). *Am J Vet Res* **43:** 732–734, 1982.

Kocan, K. M., Venable, J. H., and Brock, W. E. Ultrastructure of anaplasmal inclusions (Pawhuska isolate) and their appendages in intact and hemolyzed erythrocytes and in complement-fixation antigen. *Am J Vet Res* **39:** 1123–1130, 1978.

Swenson, C., and Jacobs, R. Spherocytosis associated with anaplasmosis in two cows. *J Am Vet Med Assoc* **188:** 1061–1063, 1986.

d. EPERYTHROZOONOSIS Parasites of the genus *Eperythrozoon* (family Bartonellaceae, order Rickettsiales) are found on the surface of erythrocytes and occasionally free in the plasma, rather than within the cells, and thus differ from *Anaplasma*. This location of the parasites is typical of the haemobartonellae, but *Eperythrozoon* is distinguished by being predominantly coccoid with multiple organisms in a single red cell and irregularly arranged over the concavity, unlike the Haemobartonellaceae, which tend to have coccoid and rod-shaped forms which typically appear in chains around the red cell periphery. This distinction is somewhat artificial since the round bodies of *Eperythrozoon*, when viewed on the periphery of the red cell membrane, also appear rodlike. The ring-shaped structures frequently observed in Romanovsky-stained blood films are apparently artefactual alterations.

Eperythrozoonosis is, among domestic animals, a disease of sheep, cattle, swine, and llamas, which is characterized by anemia and icterus. Hemoglobinuria does not occur, and in **sheep** the red cells in clinical disease are Coombs' positive, which suggests that the anemia is due to intracellular destruction of parasitized erythrocytes. The disease in sheep has a seasonal incidence, which suggests that infection may be related to changes in the population of *Melophagus* and other bloodsucking insects. Infection can be transmitted artificially by inoculation of blood from a carrier animal, and clinical disease is most easily established in splenectomized recipients.

Eperythrozoon ovis is parasitic in sheep in the form of ring-shaped, bacillary, or coccobacillary forms, from 0.3 to 1.0 μm in diameter, attached to the surface of erythrocytes. Organisms are found free in the plasma, probably being dislodged from the erythrocytes during preparation of the smear. These parasites can cause disease in nonsplenectomized sheep, and relapses occur following splenectomy of carrier animals. The incubation period of the disease is approximately a week, after which parasitemia develops. The parasites multiply rapidly and in the course of a week far outnumber the erythrocytes, after which their relative numbers decline as anemia develops. In severe infections, icterus occurs. The mortality rate is insignificant, but repeated relapses can occur and can be induced by intercurrent disease. The erythrocyte count may drop to 1×10^{12}/liter and there is concurrent monocytosis with lymphocyte counts up to 20×10^9/liter.

Although *E. ovis* is widespread and ordinarily insignificant as a pathogen, a 10% mortality rate has been reported in naturally infected lambs which died with profound anemia. This suggests that the infection, either alone or as a complicating disease, may be responsible for more widespread anemia and unthriftiness than is usually recognized. Summer epizootics of the disease without mortality have been observed in which anemia was profound and affected animals developed edema of the head while grazing by day.

The pathologic picture in clinical or fatal ovine eperythrozoonosis is principally that of anemia and serous effusions, in some cases with modest icterus. The anemia is hemolytic and, although hemoglobinuria is of unusual occurrence, bilirubin levels may be increased severalfold. The bone marrow has subendosteal reddening and shows evidence of regeneration. The kidneys may appear normal or be the bluish color associated with hemoglobin, or more usually a rusty brown color, which is due to large deposits of hemosiderin iron in the proximal tubules. The spleen is enlarged and soft, and the enlargement may persist for some time after clinical recovery. There is lymphoid hyperplasia, and the follicles are very prominent to the naked eye.

Eperythrozoon wenyoni is the **bovine** parasite. It is of wide geographic distribution and in most species resembles *E. ovis* but is host specific. It is not an important infection, and its potential for causing anemia and icterus is realized only during severe intercurrent disease and after splenectomy. The disease was believed to occur in outbreak form in first-calf dairy heifers following calving, and was characterized by edema of the teats and hind limbs believed due to vasculitis. *Eperythrozoon teganodes* is also a parasite of cattle. It is not pathogenic and is distinctive in that it is confined to plasma and not attached to red cells.

Eperythrozoon suis and *E. parvum* are immunologically

distinct, host-specific species in **swine.** The species *E. parvum* is usually coccoid with occasional ring forms less than 0.5 μm in diameter and is benign. The natural disease in swine, which has been known as icteroanemia, is caused by *E. suis,* an organism usually ring shaped and about 0.8 μm in diameter, although in acute disease it may be up to 2.5 μm in diameter. Both parasites are enzootic in South Africa and the United States of America and probably elsewhere. The mode of natural transmission is not known, but probably an arthropod is implicated; the highest incidence of infection occurs during the summer months.

The morbidity in porcine eperythrozoonosis is low, and pigs which acquire the infection remain as carriers. The disease is characterized by fever and anemia, which may be profound, and in cold climates there may be necrosis of extremities. Icterus occurs only in the most severe cases. The usual manifestation of porcine eperythrozoonosis is the sudden death of a few pigs in a herd. Pathologically there is icterus, watery blood, serous effusions, and enlargement of the spleen. In the examination of smears of blood, precipitated stain should not be confused with the parasites. The disease in all species is diagnosed by finding the organism in blood in conjunction with anemia, by serology, and by recombinant techniques.

Bibliography

Henry, S. C. Clinical observations on eperythrozoonosis. *J Am Vet Med Assoc* **174:** 601–603, 1979.

McLaughlin, B. G. *et al.* An *Eperythrozoon*-like parasite in llamas. *J Am Vet Med Assoc* **197:** 1170–1175, 1990.

Oberst, R. D. *et al.* Recombinant DNA probe detecting *Eperythrozoon suis* in swine blood. *Am J Vet Res* **51:** 1760–1764, 1990.

Smith, J. A. *et al. Eperythrozoon wenyoni* infection in dairy cattle. *J Am Vet Med Assoc* **196:** 1244–1250, 1990.

Zachary, J. F., and Basgall, E. J. Erythrocyte membrane alterations associated with the attachment and replication of *Eperythrozoon suis:* A light- and electron-microscopic study. *Vet Pathol* **22:** 164–170, 1985.

e. HAEMOBARTONELLOSIS   The genus *Haemobartonella* is closely related to the genus *Eperythrozoon*. The organisms are attached to the surface of erythrocytes, principally as minute coccoid or bacillary forms which stain blue with Giemsa's stain.

Haemobartonellae are found in cattle, goats, cats, and dogs, and in each species they are capable of causing hemolytic anemia. Geographically they are widespread, but they are unlikely to be seen unless looked for specifically.

The type species is *H. muris,* an ubiquitous parasite of the rat transmitted by lice. Rats harbor latent infections which can be activated by splenectomy, hemolytic poisons, and intercurrent trypanosomiasis; anemia and parasitemia then develop rapidly and can be fatal. *Haemobartonella canis* is the parasite of **dogs.** It is virtually of no pathogenic significance, except that latent infections can be activated by splenectomy. Susceptible dogs deve-

lop acute hemolytic anemia that resembles idiopathic immune-mediated hemolytic anemia except for the presence of the organisms. *Haemobartonella bovis* of **cattle** is of no economic significance, but haemobartonellosis may complicate research on *Babesia* or *Anaplasma* where splenectomy has been carried out and dual infections occur. *Haemobartonella felis* is responsible for the natural disease feline infectious anemia. The organism is ubiquitous, but the uncomplicated disease is unusual. Most often some other disease process has caused a reduction in resistance of the animal, and *Haemobartonella* is an opportunist. One of the most common causes of reduced resistance in the cat is an occult lymphoma.

Haemobartonella felis is presumably spread by fleas and other biting insects and is an important infection in **cats.** Males are at twice the risk of females of developing the clinical disease, and the risk of infection increases with age and in the spring. Direct contact is thought to be important in transmission. The anemia in cats is Coombs' positive and it is likely that either an antiparasitic antibody cross-reacts with normal red cell antigens, or parasitic antigen adsorbs to red cells and sensitizes them to the phagocytic system and selective removal by the spleen and marrow macrophages. The antibody in experimentally infected cats is of the cold-reacting type, which could be a factor in the development of clinical disease, since cats if permitted are regularly nocturnal. The red cell life span is halved in clinical disease.

The most common signs in cats are lethargy and anorexia for a short period of time. Often there is some other history of stress including transport or trauma resulting in sepsis. The anemia is severe and appears suddenly; the erythrocyte count may drop as low as 2×10^9/liter with hemoglobin levels of 20–60 g/liter and hematocrit of 0.06–0.23 liter/liter. Since the disease is characteristically secondary to other processes, the response to the anemia is highly variable, but in the rare uncomplicated case, the anemia is responsive and macrocytic and mildly hypochromic. Diagnosis is dependent on identification of the rod- and ring-shaped organisms which are characteristically arranged in chains around the periphery of the erythrocyte membrane (Fig. 2.67). The leukocyte count is highly variable and there may be leukopenia due to marrow failure, or neutrophilic leukocytosis due to intercurrent sepsis, or leukocytosis of leukemia. Icterus is inconstant and generally mild.

The gross examination is dominated by pallor with or without mild icterus, and usually splenomegaly and mild generalized lymphoid hypertrophy. There is usually reddening of the femoral marrow. Histologically there is splenic sinus hyperplasia and moderate generalized follicular hyperplasia. In uncomplicated cases, the marrow is hypercellular with an erythroid shift. Not infrequently the marrow and nodes are involved with a primary myeloproliferative or lymphoproliferative disorder, which has reduced the efficiency of the monocyte–macrophage system in controlling the level of parasitism.

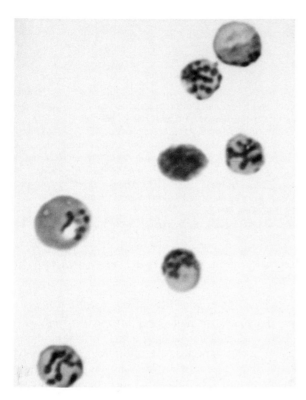

Fig. 2.67 *Haemobartonella felis*. Peripheral blood. Cat.

Bibliography

Bucheler, J., and Giger, U. Cold agglutinins in feline haemobartonellosis. *J Am Vet Med Assoc* **198:** 740–741, 1991.

Hayes, H. M., and Priester, W. A. Feline infectious anemia: Risk by age, sex, and breed; prior disease; seasonal occurrence; mortality. *J Small Anim Pract* **14:** 797–804, 1973.

Maede, Y. Studies on feline haemobartonellosis. IV. Life span of erythrocytes of cats infected with *Haemobartonella felis*. *Jpn J Vet Sci* **37:** 461–464, 1975.

Pryor, W. H., Jr., and Bradbury, R. P. *Haemobartonella canis* infection in research dogs. *Lab Anim Sci* **25:** 566–569, 1975.

Simpson, C. F., and Love, J. N. Fine structure of *Haemobartonella bovis* in blood and liver of splenectomized calves. *Am J Vet Res* **31:** 225–231, 1970.

Wright, I. G. The isolation of *Haemobartonella canis* in association with *Babesia canis* in a splenectomized dog. *Aust Vet J* **47:** 157–159, 1971.

f. EHRLICHIOSIS The Ehrlichieae of the family Rickettsiaeceae includes three species, *Ehrlichia canis, E. bovis,* and *E. ovina,* which produce disease in dogs, cattle, and sheep, respectively. A similar agent has been described (*E. equi*) which causes disease in the horse. The agent infecting the horse causes cytoplasmic morulae in granulocytes, whereas that infecting the dog causes inclusions in lymphocytes and monocytes. It appears that the equine agent occasionally also causes disease in the dog. A tick-borne anemia of dogs caused by *Rickettsia canis* has been known for many years. Subsequently a disease which occurred among introduced breeds in Southeast Asia be-

came known as **canine tropical pancytopenia.** Unlike the disease in horses which is self-limiting, the disease in dogs is characterized by fatal or protracted course in dogs of all ages. Although included here with the infectious hemolytic diseases, canine ehrlichiosis when fully expressed is a symptomatic aplastic pancytopenia with identification of the leukocyte inclusions giving specificity to the disease syndrome.

Ehrlichiosis in **dogs** of the German shepherd breed often produces a severe hemorrhagic syndrome that is less common in other breeds and in experimental infections. The disease is characterized by an incubation period of 10–20 days followed by a febrile phase with variable pyrexia, anorexia, weight loss, depression, and weakness. Some animals die at this acute stage, but most enter a subclinical stage which is of variable duration but generally ends in about 90 days with either epistaxis and generalized hemorrhage or pancytopenia with progressive renal failure.

Typically the most severe hematologic changes occur in platelets, which are depleted by about the 10th day, with slow recovery over the next 30 days, then fall progressively over 2 months or so as marrow aplasia develops. The erythrocyte numbers parallel proportionately the changes in thrombocytes. The leukocyte count fluctuates until 80–90 days postinfection when there is progressive and severe panleukopenia. The anemia is characterized by moderate anisocytosis and polychromasia, with increasing poikilocytosis and loss of polychromasia as marrow failure progresses. There is marked variation of platelet size following the first attack of thrombocytopenia, but signs of production decrease terminally. The neutrophils and lymphocytes, though markedly decreased, retain relatively normal proportions and there is usually eosinopenia and some degree of monocytosis. Diagnosis is based on hematologic changes and the presence of the rickettsial inclusion bodies in the cytoplasm of lymphocytes and monocytes. Typically the morulae of *E. canis* appear rounded and one-third to one-half the diameter of a red cell, with a granular internal structure consisting of basophilic elementary bodies each 0.5–1.0 μm in diameter. In the dog, the inclusions are typically in the cytoplasm of lymphocytes and monocytes, and a predominance of inclusions in the neutrophils associated with a milder disease syndrome suggests infection with *E. equi.* There is an increased erythrocyte sedimentation rate throughout the disease, a progressive reduction in serum albumin with increased gamma globulin, and a much prolonged bleeding time. The late onset of marrow aplasia that becomes limiting as the globulin level peaks suggests that the marrow inhibition may have an immune rather than a direct infectious origin.

Pathologically there is emaciation with subcutaneous and interstitial edema. A consistent finding is epistaxis with petechial hemorrhages on gingiva and conjunctiva. Large subcutaneous hemorrhages are present and there are hemorrhages on the serosa and mucosa of the intestinal tract and on the bladder mucosa. There is mesenteric lymphadenopathy with irregular enlargement of other nodes and medullary discoloration due to hemorrhage. The liver is of normal size and consistency with pallor, focal areas of hemorrhage, and an anemic periacinar pattern. The spleen is of normal size, or slightly enlarged and firm with congestion and obscured splenic corpuscles. Typically hemorrhages are found on the surface of the heart and in the myocardium, and the lungs are edematous and mottled, often with focal areas of hemorrhage associated with sepsis. The postmortem diagnosis is best made by demonstration of morulae in the cytoplasm of pulmonary macrophages obtained from direct imprints of the cut surface of the fresh lung. There are focal subcapsular hemorrhages on the kidney with mild yellowing of the medulla.

Histologically there is lymphoplasmacytic cuffing of veins in the central nervous system and around many vessels in other organs. The liver has panlobular atrophy of cords with sinusoidal dilation, focal plasmacytic cuffing in portal areas, and ischemic periacinar degeneration. There is mild alveolar septal thickening with hypercellularity and mononuclear cuffing around pulmonary vessels. In lymph nodes, the germinal centers are depleted, the paracortical areas are atrophic, and there is marked medullary cord plasmacytosis and sinus histiocytosis, often with prominent erythrophagocytosis. Glomeruli are often ringed with aggregates of lymphocytes and plasma cells, and there are focal interstitial infiltrates of mononuclear cells. The bone marrow is hyperplastic early in the disease, with few fat cells and dense cellular packing, and aplastic terminally with dilated sinusoids and foci of hemorrhagic infarction. The disease in the dog has been called radiomimetic because of the prominent destruction of proliferating cell populations. The prominent plasmacytosis that persists in many organs should, however, alert the diagnostician that these lesions are more inflammatory in nature than those associated with idiopathic or drug-induced marrow aplasia.

Ehrlichiosis in **horses** is similar to that in dogs but less severe. In experimental disease, the incubation period varies from 1 to 9 days with the onset characterized by pyrexia, depression, and complete anorexia. Edema of the ventral midline, limbs, and prepuce persists for 1–2 weeks. There is leukopenia, thrombocytopenia, elevated plasma bilirubin, and granular inclusion bodies, which are circular and 2–3 μm in diameter, in the cytoplasm of the neutrophils. The inclusions in granulocytes parallel the febrile period and are maximal on the third to fourth day and present for up to 15 days after infection. A rise in lymphocyte numbers heralds recovery and a rise in platelets precedes the disappearance of ventral edema.

Gross lesions include petechial hemorrhage, ecchymoses, and edema, particularly in subcutaneous tissues and on the ventral abdominal wall. There is often increased fluid in serous cavities as well as icterus, emaciation, and secondary bacterial infections including bronchopneumonia, arthritis, lymphadenitis, and cellulitis. Histologically there is vasculitis in the areas of edema and hemorrhage.

Potomac horse fever caused by *Ehrlichia risticii* is described with The Alimentary System (Volume 2, Chapter 1).

Bibliography

Buhles, W. C., Huxsoll, D. L., and Hildebrandt, P. K. Tropical canine pancytopenia: Role of aplastic anaemia in the pathogenesis of severe disease. *J Comp Pathol* **85:** 511–521, 1975.

Buhles, W. C., Huxsoll, D. L., and Ristic, M. Tropical canine pancytopenia: Clinical, hematologic, and serologic response of dogs to *Ehrlichia canis* infection, tetracycline therapy, and challenge inoculation. *J Infect Dis* **130:** 357–367, 1974.

Codner, E. C., Roberts, R. E., and Ainsworth, A. G. Atypical findings in 16 cases of canine ehrlichiosis. *J Am Vet Med Assoc* **186:** 166–169, 1974.

Gribble, D. H. Equine ehrlichiosis. *J Am Vet Med Assoc* **155:** 462–469, 1969.

Madewell, B. R., and Gribble, D. H. Infection in two dogs with an agent resembling *Ehrlichia equi*. *J Am Vet Med Assoc* **180:** 512–514, 1982.

4. Anemias of Vascular Parasitism: Trypanosomiasis

Trypanosomes are unicellular organisms, uniform in type but varying in morphologic details of taxonomic importance. Some species are morphologically stable, whereas others are polymorphic, and the differentiation of species requires the familiarity of long acquaintance.

The important trypanosomes to human and animal health are transmitted by arthropods, which carry the metacyclic stages in either the oral or anal tracts. Thus, *Trypanosoma cruzi*, the cause of Chagas' disease in humans and animals in South America, is known as a stercorarian parasite because it is spread by passage in the feces of the reduviid insects (*Triatoma* spp.); after the insects bite (often around the mouth or eye), they defecate and the trypanosomes gain entry to the host through the bite wound, often assisted by the host scratching. The important African trypanosomiases of animals (*T. congolense, T. vivax, T. brucei brucei*) and of humans (*T. brucei rhodesiense* and *T. brucei gambiense*) are called salivarian parasites because of their transmission through the salivary glands of the tsetse fly. Other biting and sucking insects, such as tabanids, stable flies, or fleas, and vampire bats may transmit the trypanosomes mechanically. Infection of the definitive host, the tsetse fly, is probably lifelong. Some researchers feel that the disease resulting from fly transmission is more typical of field cases than that resulting from trypanosomes delivered by hypodermic needle.

Trypanosoma equiperdum of dourine does not require a vector host but is transmissible by direct contact with mucous membranes. It is described with The Female Genital System (Chapter 4 of this volume).

Not all species of *Trypanosoma* of mammals are pathogenic, and among the common nonpathogenic species of veterinary interest are *T. melophagium*, ubiquitous in sheep and transmitted by the ked, *Melophagus ovinus*; *T. theileri*, which is widespread in cattle and transmitted by several types of biting flies, including Tabanidae; *T. theo-*

dori, which occurs in goats in the Middle East, transmitted by the hipposcid fly, *Lipoptena caprina,* and probably synonymous with *T. melophagium;* and *T. lewisi* of rats, which is of much interest as a convenient subject for research on the genus. *Trypanosoma rangeli* of South American cats, dogs, and humans is also apparently nonpathogenic to the vertebrate hosts but differs from these species in that it can be transmitted by inoculation or contamination, the vector being the triatomine, *Rhodnius prolixus.* The species of trypanosomes listed, although widespread in their hosts, are ordinarily sparse in blood and seldom observed unless specifically looked for in thick smears, after splenectomy, or by artificial culture. Occasionally, however, *T. melophagium* and *T. theileri* are found in very large numbers in the blood of cattle and sheep suffering from a primary and usually an immunosuppressive disease. It is not clear whether or not *T. theileri* may become pathogenic in the compromised host; however, its increase in number in association with other diseases suggests that it may be of secondary importance.

In endemic areas, the incidence of infection by pathogenic species varies depending on such factors as the availability of mammalian reservoirs, the density of vector tsetse flies, and the systems of husbandry adopted, but, as a rule, domestic animals and trypanosomes cannot coexist. Trypanosomiasis prohibits domestication of livestock in perhaps one-quarter of the total land surface of Africa. Knowledge of host–parasite relationships is not fully understood, for either the vector or the mammalian hosts. Within a species of the organism, local strains may be distinct in their basic and predominant antigens, but within the host in the course of chronic infections, a series of genetically regulated antigenic variants appears in succession so that the animal exhausts itself without clearing the infection. Successive waves of parasite and antibody dominance account for the characteristic recrudescences of the infection and for the known difficulty in producing an effective vaccine.

The pathogenic trypanosomes differ widely in virulence and are apparently nonpathogenic in mammalian reservoir hosts. Most of them are parasites of a variety of domestic and wild animals, in some of which they produce fulminating disease, in some chronic disease, and in others mild or inapparent infections. Strains within a species of the parasite also differ greatly in virulence, some strains producing rapidly fatal infections, some producing chronic infections with premunition, and some allowing complete recovery and acquired immunity. Much of the knowledge of trypanosomiasis has been gained by studies of *T. brucei,* which conveniently infects dogs, rabbits, rats, and mice and poses little problem of dissemination in developed countries where it has served as a research model. In terms of the overall problem, *T. brucei* is not nearly as important as *T. vivax* and *T. congolense* and the pathogenic trypanosomes of humans. Of the latter organisms, only *T. congolense* is studied with any frequency outside their natural habitats because the other trypanosomes are thought to constitute too great a hazard of dissemination even under conditions of laboratory restraint. Most of the research on *T. vivax* and *T. congolense* has been directed at defining the disease in British breeds of cattle. Efforts are now being made to understand the host–parasite relationship in those indigenous species of animals like the Thomson's gazelle, wildebeest, cape buffalo, and eland, which are heavily exposed and apparently carry the organism, but rarely if ever have clinical disease.

The components of pathogenicity of the trypanosomes are largely unknown. Their effects vary with their tissue tropism. The edema and interstitial reactions caused by *T. brucei* are due to the fact that this parasite resides in perivascular tissues, in contrast to the more serious pathogens of humans and animals, which are obligate intravascular parasites. In many respects, bovine trypanosomiasis is similar to equine infectious anemia in that infection is followed by an asymptomatic period until antibody develops. There is then anemia and generalized compensatory reticuloendothelial hyperplasia, which ultimately results in cachexia and hematopoietic and lymphoid exhaustion.

a. Bovine Trypanosomiasis   In cattle, parasitemia regularly occurs a week after inoculation with 10^5 or more of organisms which have been passed in rats and column separated. By the second week of infection, there is a sharp drop in the red cell count and hemoglobin levels, accompanied by an erythroid shift and macrocytosis, anemia being the predominant aspect of the disease. There is a concurrent thrombocytopenia of moderate degree and hypocomplementemia, with an irregular appearance of immunoglobulin and complement on red cell membranes. Fibrinogen is reduced to about half of normal levels and there is an irregular appearance of fibrin split-products as animals go through febrile periods. Kinetically, the red cell life span is reduced to a half or less of normal, and iron utilization is initially rapidly increased, but slows with cachexia and inhibition of the erythroid response due to the inflammatory changes. The platelet life span is not shortened in infections with *Trypanosoma congolense,* although it may be with *T. vivax.* In the latter, the truncation appears to be due to a proliferation of marrow megakaryocytes which are impaired in their maturation and produce fewer platelets which survive normally, thus constituting an ineffective response. The disease tends to stabilize at about 4–6 weeks in well-fed animals under laboratory conditions, and to proceed to death due to secondary causes in animals which are forced to forage for feed as well as maintain their immune defenses.

The red cells are destroyed in an innocent-bystander fashion by adsorbing toxic and metabolic products of the trypanosomes, as well as suffering direct injury from a parasite-derived phospholipase. There is competition between the stem cells for erythroid and myeloid proliferation in *in vitro* studies, which may explain the clinical finding that animals with severe anemia are more susceptible to secondary infection by bacteria and viruses. A dilutional component to the anemia has been suggested, but no increase in blood volume has been detected. Calves infected during the first week of life have less severe hematologic disease than calves infected at 6 months or older. Some of these animals become cachectic and die, although it appears that infection at an early age results in a more tolerant host–parasite relationship which is less injurious to the host. *Trypanosoma vivax* and *T. congolense* differ in that *T. vivax* tends to circulate relatively freely and uniformly in the vascular system, and, in this respect, the level of parasitemia gives a reasonable estimate of the total parasitic burden. In contrast, *T. congolense* "homes" to the microvasculature of the brain and skeletal muscle, and the paucity of organisms in the peripheral blood makes diagnosis and estimates of total burden difficult. It appears that the tissue tropism of *T. congolense* is a response to developing immunity and is antibody mediated. It is not known why brain and muscle are selectively parasitized, but the phenomenon may be related to trypanosomal energy requirements, since the endothelial cells in these areas tend to have more mitochondria than those in other tissues.

Trypanosoma vivax is more virulent than *T. congolense* and, whereas both cause anemia and cachexia, infection with *T. congolense* regularly results in chronic disease; in a high percentage of animals infected with *T. vivax* there is sudden death due to salmonellosis or other infections. Infected animals are moderately stunted in growth, although this effect is less apparent on body weight than on dressed weight (that part of the carcass which is edible). This disparity is due to the fact that infected animals pool fluid in the intestinal tract and become pot-bellied while failing to deposit muscle and fat. When viewed from behind, infected animals have poorly covered bony prominences, poor "spring of rib," and a deep and pendulous abdomen. The skin is scurfy and rough, particularly about the head, ears, and hindquarters. There is a chronic low-grade pyrexia with intermittent temperature increases, and periodic watery diarrhea with passage of poorly digested ingesta of normal color. There is irregular oculonasal discharge, but no loss of appetite, and the animals continue to eat. There is mild mucosal pallor without petechial hemorrhages and the urine may be dark but never blood tinged. The heart and respiratory rates are not remarkably altered under laboratory conditions but are increased in nonspecific response to anemia under field conditions, and animals may show open mouth-breathing on forced exercise. In the absence of secondary infection, there is little or no depression or abnormal behavior despite high levels of cerebral parasitism (Fig. 2.68A).

There is anemia of moderate degree, generally between 50 and 80 g/liter, which is initially macrocytic and normochromic and later becomes normochromic, normocytic, and poorly responsive (Fig. 2.68B). In the well-developed disease, there are always some hypochromic red cells and at least mild poikilocytosis. In the early stages of infection, the radioiron uptake time is reduced from a normal of about 3 hr to about half that figure, but in chronic disease is only mildly reduced. There is

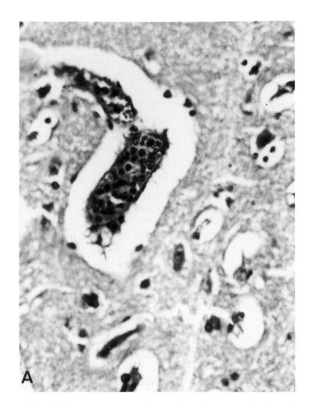

Fig. 2.68A Trypanosomiasis. Ox. Microvascular thrombosis and early poliomalacia. Cerebrum.

increased osmotic fragility of red cells without spherocytosis. The platelets are between 100 and 200 × 10^9/liter and are often lower during intercurrent infection. The leukocyte count is reduced to about half of normal (5–6 × 10^9/liter) due to an absolute reduction in both neutrophils and lymphocytes. Changes are less severe in neonatally infected animals. Biochemical changes consist of a reduction in total protein largely due to reduced albumin, but there are no consistent changes in transaminases or bilirubin. There is a consistent reduction in total serum lipids, cholesterol, and triglyceride, and erythrocyte phospholipid levels are consistently elevated, which may indicate that there is an acquired injury to red cell membranes which contributes to shortened life span. Aspirated marrow is hypercellular with an erythroid shift that drops to 0.4 or lower in animals infected with *T. congolense*. There is a less marked erythroid response in *T. vivax* infection.

There are no pathognomonic gross lesions of trypanosomiasis. Chronically affected animals are cachectic and have a rough hair coat, and there is increased clear fluid in body cavities. There is generalized lymph node enlargement to two to four times normal, and the hemal nodes become prominent in subcutaneous areas and in association with the para-aortic and pelvic nodes, where they may reach 1 cm in diameter or larger. The lungs are heavy and have increased density on palpation, and may show intercurrent anterior bronchopneumonia. The heart is generally flabby with serous atrophy of pericardial fat, and

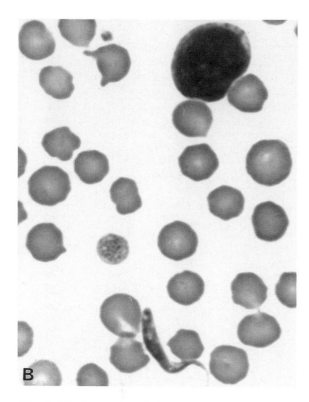

Fig. 2.68B Trypanosomiasis. Ox. Peripheral blood. Moderate poorly responsive anemia, poikilocytosis, and paucity of thrombocytes.

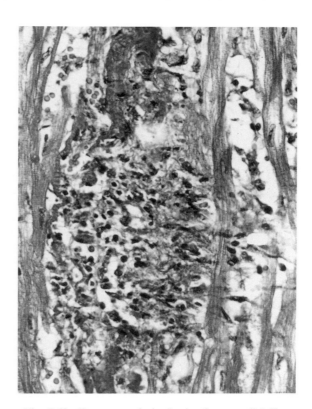

Fig. 2.69 Trypanosomiasis. Lysis of myocardial fibers and myocarditis. Ox.

there may be white foci 1–2 mm in diameter on the epicardium and on the cut surface. The liver and kidneys are symmetrically enlarged and constitute a greater than normal percentage of body weight. The liver is unusually firm and not easily penetrated by digital pressure, and there is a fine lobular pattern visible through the capsule and on cut surface. The renal medullary fat shows serous atrophy. There are fine focal reddened areas 1–2 mm in diameter throughout both layers of the omentum. The enteric tract is unremarkable, except that the small intestine is atonic and contains an excessive amount of normal-appearing fluid content. The most remarkable changes occur in lymph nodes, which have generalized medullary pigmentation, and in bone marrow where there is a regular increase in hematopoietic areas, which may occupy all of the cancellous and diaphyseal fatty areas in animals which have been infected 6 months or more. The spleen is uniformly enlarged and bulges, but does not ooze blood, on cut surface. The lymphoid follicles are generally visible grossly.

Microscopically there is a generalized increase in septal width in the lungs and accumulation of hemosiderin-bearing macrophages. There are multifocal areas of fiber atrophy with sclerosis and lymphoplasmacytic proliferation in the myocardium (Fig. 2.69), most prominently in those animals which have succumbed to the disease. The bone marrow goes through a cycle of changes similar to those in equine infectious anemia. There is early conversion of fatty to hematopoietic marrow with high cell density, erythroid shift, and plasmacytosis, followed by a reduction in cellular packing and dilation of sinusoids, and ultimately by reduction in hematopoiesis and serous atrophy of the remaining tissue. The rubricytes are increased in number and are relatively synchronous in maturation with numerous mature cells, suggesting late asynchrony associated with poor availability of iron. There is a variable degree of hemosiderosis, depending on the duration of the disease, and erythrophagocytosis may be observed in the marrow. There is a reduction in marrow granulocyte reserves and, as a result, a mild early asynchrony, although it is likely that with the expanded volume of hematopoeitic marrow there is a normal volume of granulopoietic tissue. Megakaryocytes are increased in number to about twice the normal level, and the nuclei are increased in diameter, but the cytoplasmic volume is reduced, indicating ineffective thrombopoiesis. There is a mild generalized increase in endothelial and true reticular cells in bone marrow, but myelofibrosis is not severe. Trypanosomes may be found free and phagocytosed in small vessels throughout the body but are most commonly observed in the liver and cerebral cortex. Hepatic changes are constant, and consist of atrophy of hepatocellular cords with sinusoidal dilation and periportal lymphoplasmacytic proliferation. There is increased interstitial stroma, without change in lobular organization, and generalized Kupffer cell hyperplasia.

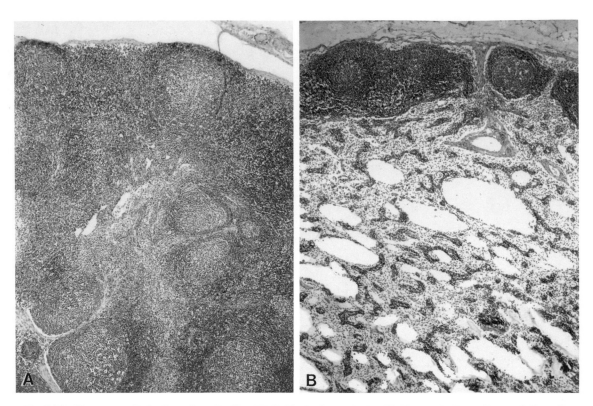

Fig. 2.70 Trypanosomiasis. Lymph node. Ox. (A) Follicular hyperplasia. (B) Paracortical atrophy, medullary sinus hyperplasia, and histiocytosis, dilation of lymphatics in chronic disease.

Changes in the lymphoid tissue are remarkable and constant in all areas of the body. There is first follicular hyperplasia (Fig. 2.70A) with a competent response and the formation of large and densely cellular germinal centers. These changes are accompanied by variable paracortical hyperplasia, and the paracortex has a moth-eaten appearance due to the many large lymphoblasts and tingible body macrophages. At this time the lymph nodes are physically enlarged and there is thinning of the capsules with focal colonization of perinodal fat and a closed peripheral sinus. In chronic disease, the nodes remain enlarged, but this is due, at least in part, to a marked increase in capsular and intranodal stroma with numerous collagenous septa extending from the medulla to the capsule. The medullary follicles remain, but their cellularity is reduced and there is marked atrophy of paracortical areas. Concurrently there is sinus histiocytosis and at least some degree of hemosiderosis without medullary cord hyperplasia. The changes are thus those of continuing stimulation with an initial competent response followed by hyperplasia, then atrophy and sclerosis (Fig. 2.70B).

Concurrent with the changes in lymph nodes, there is remarkable atrophy of the thymus, with the reduction most prominently affecting cortical regions. With the loss of cortical volume, there is a reduction in cell density and an increase in size of nuclei in cortical thymocytes. The splenic lymphoid changes reflect those in other areas of the body, with atrophy of periarteriolar cuffs and the for-

mation of large hypocellular follicles, occasionally with target ringing. The number of follicles is not increased. The most marked splenic change is sinus hyperplasia characterized by a marked increase in fixed cells in the Billroth sinus areas, which contain within their interstices many macrophages and plasma cells, with increased hemosiderin and an irregular appearance of extramedullary thrombopoiesis and erythropoiesis. Renal changes consist of a moderate generalized increase in interstitial connective tissue with mild epithelial atrophy and the appearance of large lymphoid cuffs around arterioles at the corticomedullary junction. Glomeruli are regularly increased in diameter and in cellularity and occasionally contain hemosiderin-laden macrophages. There are focal synechia between the visceral and parietal layers of Bowman's capsule and a mild epithelioid appearance to the central areas of the tuft. The nuclei of the juxta-glomerular cells are unusually prominent, and there is mild pigmentation by both iron and bilirubin in the proximal epithelium. The overall changes in glomeruli are compatible with membranoproliferative glomerulonephritis.

Subtle changes are consistently present in skeletal musculature and consist of an overall hypercellularity with a mild reduction in fiber diameter and increased perivascular lymphocytes. The reddened foci in the omentum consist of focal venous ectasia with mild interstitial sclerosis and perivascular lymphoplasmacytic reaction.

b. TRYPANOSOMIASIS IN OTHER SPECIES   *Trypanosoma*

rhodesiense and *T. gambiense* are human pathogens causing African sleeping sickness and are not important in domestic animals, although *T. gambiense* has produced meningoencephalitis in experimental goats and cats. These trypanosomes are morphologically indistinguishable from each other and from *T. brucei*, from which they are presumed to have evolved.

Trypanosoma brucei brucei is a cause of **nagana** in most domestic species in Africa, but humans are refractory. Transmission is by *Glossina* spp. or mechanical. Equidae, small ruminants, camels, and dogs are very susceptible, the disease is more chronic in cattle, and pigs may recover. Incoordination and spinal paralysis are reported in horses and dogs. Dogs may also develop parasitism and inflammation of the anterior segments of the eye. The neurologic signs are undoubtedly due to invasion of cerebrospinal fluid and to inflammation of meninges, brain, and cord. Diagnosis depends on demonstration of the organism in blood in acute or relapsing febrile cases, or otherwise in lymph nodes by direct smear, inoculation into the very susceptible laboratory mouse, or cultivation.

Trypanosoma evansi was originally shown to be the cause of **surra** in horses and camels, but it is pathogenic for most domestic species and distributed in North Africa, Asia, and Central and South America. Transmission is mechanical, chiefly by Tabanidae, but also by other blood-sucking flies, and apparently by vampire bats in the Americas. The disease in horses and dogs is severe, and probably uniformly fatal in the absence of adequate treatment. Cattle are mildly affected and act as reservoirs, although acute disease may occur in susceptible cattle introduced to endemic areas.

Trypanosoma equinum occurs in South America as the cause of *mal de caderas* of horses, a disease which resembles surra. The parasite is transmitted mechanically by tabanids. It is probably a stable variant of *T. evansi*.

Trypanosoma suis is a West African species transmitted by tsetse flies. It produces a naganalike disease in pigs but is apparently nonpathogenic for other domestic animals.

Trypanosoma simiae was originally isolated from African monkeys. Although its pathogenicity for pigs is unpredictable, it is ordinarily highly virulent and causes death in a few days. Pathogenicity for other domestic species is insignificant. Transmission is cyclical in tsetse flies. Trypanosomes of the species *T. congolense, T. dimorphon, T. vivax,* and *T. uniforme* are causes of nagana in Africa. Most domestic species are susceptible to infection, although dogs are resistant to *T. vivax*. There is considerable variation in the pathogenicity of different strains, especially of *T. congolense*.

Trypanosoma cruzi is the cause of American trypanosomiasis (**Chagas' disease**), which is an uncommon illness of children, although infection of human adults and animals is apparently common. The reservoir hosts are chiefly wild species, but cats, dogs, and pigs can be infected and act as reservoirs. Sheep and goats are susceptible to experimental infections. This trypanosome is of zoological interest because, both in the tissues of the vertebrate host and

in the alimentary canal of the vector, it forms developmental phases resembling the other genera *Leptomonas, Crithidia,* and *Leishmania* of the Trypanosomidae.

Whereas some of the manifestations of *T. cruzi* infection may be attributable to parasitemia with the trypanosomal form, the disease is characterized by invasion of mesodermal tissues, and the growth therein of the leishmanial forms of the organism, which show a preference for cardiac and skeletal muscle. It has been suggested that since the cysts are focal and the interstitial reaction may be widespread and diffuse, there may be an autoimmune component to the myocarditis associated with release of fiber proteins associated with parasitic injury.

Bibliography

Berger, S. L. *et al.* Neurologic manifestations of trypanosomiasis in a dog. *J Am Vet Med Assoc* **198:** 132–134, 1991.

Dargie, J. D. *et al.* Bovine trypanosomiasis: The red cell kinetics of Ndama and Zebu cattle infected with *Trypanosoma congolense. Parasitology* **78:** 271–286, 1978.

Forsberg, C. M. *et al.* The pathogenesis of *Trypanosoma congolense* infection in calves. *Vet Pathol* **16:** 229–242, 1979.

Jenkins, G. C. *et al.* Studies of the anaemia in rabbits infected with *Trypanosoma brucei brucei*. I. Evidence for haemolysis. *J Comp Pathol* **90:** 107–121, 1980.

Kaaya, G. P. *et al.* Inhibition of leukopoiesis by sera from *Trypanosoma congolense*-infected calves: Partial characterization of the inhibitory factor. *Tropenmed Parasitol* **31:** 232–238, 1980.

Mills, J. N. *et al.* The quantitation of *Trypanosoma congolense* in calves. III. A quantitative comparison of trypanosomes in jugular vein and microvasculature and tests of dispersing agents. *Tropenmed Parasitol* **31:** 299–312, 1980.

Murray, M. *et al.* Genetic resistance to African trypanosomiasis. *J Infect Dis* **149:** 311–319, 1984.

Pays, E., and Steinert, M. Control of antigen gene expression in African trypanosomes. *Annu Rev Genet* **22:** 107–126, 1988.

Schmidt, H. The pathogenesis of trypanosomiasis of the CNS. *Virchows Arch* **399:** 333–343, 1983.

Tizard, I. *et al.* Biologically active products from African trypanosomes. *Microbiol Rev* **42:** 661–681, 1978.

Valli, V. E. O., and Forsberg, C. M. The pathogenesis of *Trypanosoma congolense* infection in calves. V. Quantitative histological changes. *Vet Pathol* **16:** 334–368, 1979.

Valli, V. E. O., and Mills, J. N. The quantitation of *Trypanosoma congolense* in calves. 1. Hematological changes. *Tropenmed Parasitol* **31:** 215–231, 1980.

5. *Hemolysis Due to Physical and Chemical Agents*

a. HEINZ BODY HEMOLYTIC ANEMIA   Hemolytic anemia due to denaturation of globin occurs in cattle, horses, and dogs fed **onions,** and in cattle and sheep fed **rape** or **kale.** The same mechanism is operative in the anemia of horses with hemolysis after receiving **phenothiazine** and in cats given **methylene blue** as a urine acidifier. Heinz body anemia also occurs in horses that eat the leaves of red (swamp) maple (**Acer rubrum**). Heinz body hemolytic anemia occurs as an inherited deficiency of the **glucose-6-phosphate dehydrogenase** (G6PD) enzyme in humans and also rarely in sheep and dogs. In inherited deficiency, the level of G6PD may be 70–75% of normal, and low levels are associated with hemolysis on exposure to oxidants that may be dietary, such as **fava beans,** or pharmaceutical, such as the **primaquine** antimalarials. Cats appear particularly susceptible to Heinz body formation and Heinz bodies may be found in as many as 5% of apparently normal cats. The inclusions in cats were formerly called erythrocyte refractile bodies. Unstable hemoglobins resulting from inherited substitutions in the amino acids of the globin chains may have reduced avidity for the heme molecule, the loss of which is followed by globin denaturation and Heinz body formation. The latter pathogenesis is not of importance in domestic animals. Globin injury is involved to some degree in **postparturient hemoglobinuria** in cattle and in **chronic copper poisoning** in sheep. Heinz bodies or intraerythrocytic globin precipitates, when stained supravitally, appear as dense round structures 1–2 μm in diameter at the margin of the red cell.

Oxidizing agents which deplete red cell G6PD, the rate-limiting enzyme of the pentose shunt, diminish the protective biological antioxidant glutathione. Reduced glutathione is a hydrogen donor in the process catalyzed by glutathione peroxidase, a selenium-containing enzyme. **Selenium deficiency** alone may result in Heinz body anemia. Failure to reduce oxidized glutathione results in formation of mixed disulfide linkages between the glutathione and the globin chains. These linked proteins, having lost the elasticity essential to the molecular shape change associated with reversible oxygenation, precipitate within the cell and aggregate to form Heinz bodies. If the oxidizing attack is mild, the Heinz bodies are selectively removed by splenic phagocytes, and the red cells return to the circulation with a slightly reduced volume. If the reaction is severe, the Heinz bodies appear free in plasma (Fig. 2.71), complicate platelet counts, and are removed by the spleen and liver. Hemoglobinemia, hemoglobinuria, and icterus occur in severe cases. Since young red cells have a higher G6PD enzyme level, Heinz body anemia is usually self-limiting and nonfatal, as new red cells added to the circulation are more resistant to episodic toxic exposure. If the exposure to the toxic compound is dietary or continuous, the anemia may be progressive. The anemia may be moderate to severe, normochromic and moderately macrocytic, and responsive, with Heinz bodies free and projecting from the outer edge of red cells. Heinz bodies are unstained with Wright's stain or stain similarly to hemoglobin. With supravital stains such as new methylene blue, they are dark blue against red cells which are unstained. The level of red cell G6PD will be less than 25% of normal.

b. COPPER POISONING Copper poisoning may be acute or chronic, but the syndromes differ markedly, and the terms are misleading because chronic copper poisoning is an acute and catastrophic syndrome. The terms acute and chronic as applied to copper poisoning are not illogical since acute poisoning follows more or less immediately the acquisition of an overdose, whereas the chronic form

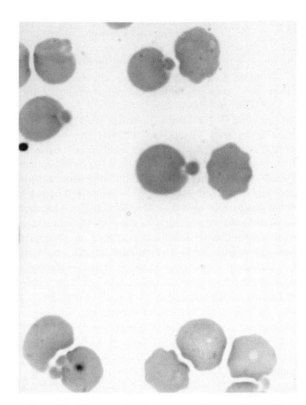

Fig. 2.71 Heinz bodies in peripheral blood. Horse. Phenothiazine toxicity. Aggregates of denatured globin budding from red cell membranes.

represents cumulative poisoning which may, sooner or later, result in acute disease. Chronic copper poisoning is a sporadic and enzootic disease of sheep and a sporadic disease of cattle and swine.

Acute copper poisoning is characterized by a gastroenteritis caused directly by the irritative effects of copper salts. The sole distinguishing feature is the passage of bright bluish-green fluid feces accompanied by severe shock and rapid prostration. Pathologically there is hemorrhagic and edematous abomasitis. The copper salts are obtained by ingestion or by accidental therapeutic, chiefly anthelmintic, overdosage. The toxic dose of $CuSO_4$ for a sheep is ~20 mg/kg of live weight. Oral treatment at an appropriate level may precipitate the chronic syndrome if tissue stores of copper are high.

Chronic copper poisoning as defined here is an acute episode of intravascular hemolysis caused by the release from hepatic storage into the blood of excessive amounts of copper, until then stored by labile binding. The accumulation and release of copper may occur in a variety of ways, some being quite circuitous (see The Liver and Biliary System, Volume 2, Chapter 2).

Copper is a hepatotoxin, and during the period of copper accumulation there are progressive histologic, histochemical and ultrastructural changes in the liver, associated with an increase in plasma of hepatic enzymes. The hemolytic crisis may be precipitated by stress and is associated with release of copper from the liver. Whole blood copper levels rise, initially largely associated with the red cells, and then declining as hemolysis slows. The hemolysis lasts 2–6 days with Heinz body formation, hemoglobinemia, and hemoglobinuria.

The sudden release of copper into the plasma in excess of the capacity of carrier proteins presumably exhausts the antioxidant system of the red cells and results in the denaturation of globin and the formation of Heinz bodies. Coombs'-positive red cells occur in sheep in the recovery stage of chronic copper poisoning. This is presumably the result of autosensitization by circulating red cell stroma and not a prior pathogenetic mechanism.

Hematologically the hemolytic phase is characterized by a sharp increase in hematocrit to 60% or higher in the hours preceding the onset of the hemolytic crisis, and is likely due to dehydration. The hemoglobin drops from a normal level of 100–110 g/liter to 40–50 g/liter in survivors. Since both free and cellular hemoglobin are measured by standard techniques, the massive hemoglobinemia will mask the actual amount of hemoglobin contained within intact erythrocytes. The erythrocyte count drops sharply to 2×10^{12}/liter or less. There is neutrophilic leukocytosis during and following hemolysis. Morphologically, during hemolysis there is marked anisocytosis and poikilocytosis with moderate polychromasia and the appearance of Howell–Jolly bodies and rubricytes, with Heinz bodies budding from up to 15% of erythrocytes. Reticulocytosis heralds the onset of recovery. Serum bilirubin rises after the appearance of hemoglobinuria. The blood urea rises in parallel with the bilirubin and then drops sharply in survivors, but does not return to normal. Levels of ceruloplasmin rise to 2–3 times normal during the hemolytic crisis, while red cell copper increases 15–20 times, plasma copper increases 3–6 times, and whole blood copper increases up to 10 times normal levels.

c. COLD HEMOGLOBINURIA   Hemoglobinuria after the ingestion of cold water occurs in calves and occasionally older cattle, principally as a disease of winter, and is seen in cattle which are housed in warm barns and given water at near-freezing temperature. It occurs also, but rarely, in calves which are maintained in hot barns in the summertime, and it can be readily reproduced by forcing them to drink cold water. The temperature of the water relative to the animals' environmental temperature is important since the disease rarely, if ever, occurs in cattle kept in cooler, open barns. Hemolysis occurs most often in penned animals which are watered at irregular intervals and which therefore tend to drink large volumes when allowed access. The disease is common although rarely fatal. The clinical signs consist of dyspnea with open-mouth breathing, followed by the passage of port-wine colored urine within 1–2 hr of drinking. Hemoglobinuria associated with the ingestion of cold water has also been observed in dogs and swine, but apparently is rare. The mechanism for the hemolysis in calves has been suggested to be an osmotic lysis of red cells caused by the rapid ingestion of water,

but this explanation is unconvincing since the osmotic effects of water should not be temperature dependent. The activity of cold agglutinins has been suggested as a cause of the hemolysis but this seems even more unlikely since they are probably not widespread in cattle. The sporadic and ephemeral nature of the disease has not been conducive to study of the cause of the hemolysis.

d. POSTPARTURIENT HEMOGLOBINURIA   The syndrome of postparturient hemoglobinuria occurs primarily in dairy cows and is typically seen within 2–6 weeks after parturition. The earlier the onset, the more severe the disease. The name, however, is too restrictive because the same disease, although slower in development, occurs before parturition, at other periods of the lactational cycle, and in males. Most cases are, however, postparturient and severe. In North America, postparturient hemoglobinuria is a sporadic disease of multiparous high-producing dairy cattle, and is characterized by intravascular hemolysis, hemoglobinuria, and anemia. The disease is usually seen between the third and sixth lactation and tends to occur in the winter, particularly when preceded by a dry growing season. In New Zealand, the disease occurs in mid and late summer and in younger cows in first and second lactations. Like chronic copper poisoning of sheep, the disease may be precipitated by stress such as forced exercise.

The pathogenesis of erythrocyte destruction leading to intravascular hemolysis is not clear and may vary with geographic distribution. It is generally accepted, however, that the disease is associated with severe hypophosphatemia, and it has been produced when diets deficient in phosphorus are fed for long periods. Severe hypophosphatemia of the order observed with postparturient hemolysis is associated with depletion of red cell ATP and loss of membrane plasticity with formation of spherocytes and rigid, easily fragmented red cells. The latter syndrome is often recognized in humans on hyperalimentation or dialysis, and has been reproduced in dogs. In grazing cattle, an inverse relationship between the incidence of the disease and the fertility of the soil has been observed, and there is a direct relationship to use of beet pulp and cruciferous pastures as forage. In New Zealand, copper deficiency has been suggested as a possible factor in postparturient hemoglobinuria, possibly associated with the use of fertilizers containing molybdenum. Since copper is essential for the red cell superoxide dismutase antioxidant system, it is possible that cruciferous forage and copper deficiency work together to produce a Heinz body type of hemolysis. Whereas the dietary associations are varied, hypophosphatemia appears to be a significant part of the overall disease. The phosphorus levels in the blood, despite hemolysis, range from 0.45 to 0.59 mmol/liter.

The signs of phosphorus deficiency which may accompany hemolytic anemia include severe and shifting lameness, allotriophagia, decreased productivity, lordosis, and retention of long hair. The animals may be emaciated. Usually there are no accompanying signs of phosphorus deficiency, however, and the disease occurs in well-fleshed animals at the peak of their productivity.

It is usual for hemoglobinuria to draw attention to the disease. Hemoglobinuria may not occur in all cases, and weakness with increased respiratory rates and mucosal pallor may be the initiating sign. The disease may progress to death in a day or in 4–5 days. The early mucosal pallor is, with time, replaced by jaundice, and milk production is sharply reduced and the secretion discolored with bile pigment.

The anemia is frequently severe, with erythrocyte counts dropping to 1×10^{12}/liter or less. It is moderately responsive in animals which survive and there is a concurrent neutrophilic leukocytosis. Biochemical changes consist of hypophosphatemia and hyperbilirubinemia. The urine may contain high levels of hemoglobin which rapidly turns dark on conversion to methemoglobin with standing. There is a concurrent albuminuria, characteristic of intravascular hemolysis and hemoglobinuric nephrosis, and the urine forms a stable froth on passage.

Pathologically the principal changes are icterus, pallor, thin watery blood which is dark, and red-to-brown discoloration of the urine. There is fluid accumulation in the serous cavities, and subcutaneous and interfascial edema. There are extensive ecchymotic hemorrhages on the cardiac surfaces and the liver is slightly enlarged, turgid, and pale with an ischemic nutmeg appearance caused by periacinar necrosis. The gallbladder is distended. The spleen is enlarged and congested in peracute cases but, in those which survive several days, it is of normal appearance. In spite of the frequent association of hemoglobinuria and death, it is unusual to find the blue-black pigmentation of kidneys characteristic of chronic copper poisoning in sheep. Icterus is often severe with canalicular cholestasis, which indicates extensive intracellular as well as intravascular hemolysis where the pigment is not lost in urine. Kidneys may be swollen, pale, and soft, without hemoglobin pigmentation. The lungs are edematous and there is frequently interstitial and subcutaneous emphysema associated with terminal dyspnea.

Microscopic changes are characteristic of intravascular and, to a lesser extent, intracellular hemolysis and are not indicative of the cause. There is panlobular hepatocellular swelling with ischemic periacinar coagulation necrosis. If animals survive for some hours with profound anemia, the margin of necrotic tissues within the functional lobules becomes progressively sharpened with a zone of fatty vacuolation separating viable and necrotic hepatocytes. Hypertrophy of Kupffer cells with hemosiderin accumulation becomes distinct in animals which survive for 3–4 days. There is congestion of splenic sinuses. Renal lesions are variable, usually with fatty degeneration of tubular epithelium, bile and hemosiderin pigmentation, and hyaline, granular, and hemoglobin casts in the lower nephron. In more severely affected kidneys, there is a patchy necrosis of the proximal epithelium with desquamation, and interstitial edema.

Postparturient hemoglobinuria closely resembles the syndrome of hemolytic anemia produced in animals which are fed plants of *Brassica* spp. or onions. An unusual amino acid, *S*-methyl cysteine sulfoxide, is produced in these plants and is degraded in the rumen or colon to dimethyl disulfide. The sulfide causes the production of Heinz bodies in the red cells of all exposed species, but the resulting hemolytic anemia is most important in pregnant or parturient animals, especially those which are exposed to diets deficient in phosphorus. An intake by cattle of 10 g/100 kg body weight per day of the amino acid will produce low-grade anemia, and an intake of 15 g/100 kg body weight per day can produce severe fatal anemia.

Bibliography

Beritic, T. Studies on Schmauch bodies. I. The incidence in normal cats (*Felis domestica*) and the morphologic relationship to Heinz bodies. *Blood* 25: 999–1008, 1964.

Greenhalgh, J. F. D., Sharman, G. A. M., and Aitken, J. N. Kale anaemia-1. The toxicity to various species of animal of three types of kale. *Res Vet Sci* 10: 64–72, 1969.

Hickman, M. A., Rogers, Q. R., and Morris, J. G. Effect of diet on Heinz body formation in kittens. *Am J Vet Res* 50: 475–478, 1990.

Ishmael, J., Gopinath, C., and Howell, J. McC. Experimental chronic copper toxicity in sheep. Biochemical and haematological studies during the development of lesions in the liver. *Res Vet Sci* 13: 22–29, 1972.

Jacob, H. S. Hypophosphatemic hemolysis in dogs. *J Lab Clin Med* 84: 643, 1974.

Jacob, H. S., and Amsden, T. Acute hemolytic anemia with rigid red cells in hypophosphatemia. *N Engl J Med* 285: 1446–1450, 1974.

MacWilliams, P. S., Searcy, G. P., and Bellamy, J. E. C. Bovine postparturient hemoglobinuria: A review of the literature. *Can Vet J* 23: 309–312, 1982.

Morris, J. G. *et al.* Selenium deficiency in cattle associated with Heinz bodies and anemia. *Science* 223: 491–492, 1984.

Smith, J. E., Ryer, K., and Wallace, L. Glucose-6-phosphate dehydrogenase deficiency in a dog. *Enzyme* 21: 379–382, 1976.

Smith, R. H. Kale poisoning: The *Brassica* anaemia factor. *Vet Rec* 107: 12–15, 1980.

Tennant, B. *et al.* Acute hemolytic anemia, methemoglobinemia, and Heinz body formation associated with ingestion of red maple leaves by horses. *J Am Vet Med Assoc* 179: 143–150, 1981.

Van Kampen, K. R., James, L. F., and Johnson, A. E. Hemolytic anemia in sheep fed wild onion (*Allium validum*). *J Am Vet Med Assoc* 156: 328–332, 1970.

6. Hemolytic Anemia Due to Mechanical Damage to Red Cells

Microangiopathic or fragmentation hemolysis occurs as a consequence of disseminated intravascular coagulation and, like that syndrome, is secondary to primary disease involving microvascular injury. Microangiopathic hemolytic anemia (MHA) is characterized by severe poikilocytosis with intravascular hemolysis and hemoglobinuria (Fig. 2.72).

The basic mechanism underlying fragmentation hemolysis is the application of high shear forces to red cells

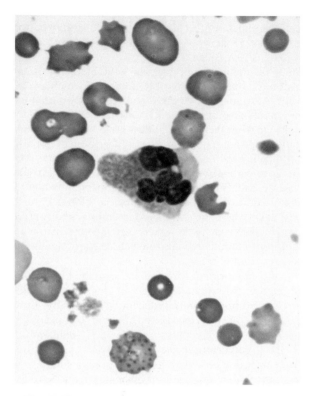

Fig. 2.72 Microangiopathic hemolytic anemia. Peripheral blood. Calf. Hemolytic–uremic syndrome. Nonresponsive anemia, microcytosis, poikilocytosis, and basophilic stippling.

moving at high velocity across a sharp drop in pressure. These conditions are most often met when erythrocytes moving rapidly in small arterioles impinge on intraluminal fibrin strands and are fragmented, usually with a characteristic bite out of the cell periphery. Thus, inflammatory lesions of capillaries and veins do not result in significant fragmentation, since the rate of flow in these vessels is too low. Fragmentation may occur as a result of arteriovenous fistulae when there is high turbulence and flow rates across a small orifice.

Fragmentation anemia occurs most often in association with infectious diseases which cause severe endothelial injury. Mild fragmentation occurs in animals with salmonellosis and with fibrinous pneumonia, but in these cases the degree of hemolysis is usually insignificant compared to the injury to the primary organ. Mild fragmentation also occurs in horses with colic in association with verminous arteritis. Since alterations in red cell shape in the horse are relatively rare, their presence in stained blood films is significant. Fragmentation anemia is seen in association with somatic migration of parasites and is prominent in animals affected by tissue phases of *Strongyloides stercoralis* and in dogs with dirofilariasis. Because of high renal blood flow and juxtaglomerular flow regulation, vascular lesions are prone to develop in the kidney, and fragmentation may occur there in association with a variety of renal diseases. Mild fragmentation occurs in association with

vegetative endocarditis, particularly of the aortic and pulmonary valves, but is rarely seen in septal defects where the openings are usually large and the flow rate, relatively low. In contrast, fragmentation occurs regularly in patent ductus arteriosus where the conduit is small and the pressure drop is high. Fragmentation of erythrocytes occurs during the development of vitamin E–selenium deficiency of swine, whereas marked poikilocytosis is seen in iron deficiency anemia, and in lambs and, to a lesser extent, calves with nutritional muscular dystrophy. Presumably in the latter diseases, the hemolysis is largely due to intrinsic defects in erythrocyte membranes rather than to primary endothelial injury. Mild fragmentation occurs in thrombocytopenia in all species, presumably as a result of widespread endothelial deterioration, some of which involves small arteries. Hemangiomas cause focal coagulation and fragmentation in humans and animals, but the rate of destruction seldom results in anemia or red urine. Fragmentation of red cells may be significant in widely metastatic carcinoma and may be the presenting sign or may herald an accelerated phase of the disease in animals being treated. Hemoglobinuria occurs in event horses on protracted exercise, often in association with exertional myopathy and with a pathogenesis probably similar to that of march hemoglobinuria of humans. Markedly distorted red cells may be seen in hepatic failure in association with severe anemia. The major defect appears to be intrinsic to the red cell membrane as a result of low plasma lipid (Zieve's syndrome). The abnormal shapes and loss of membrane plasticity predispose these cells to fragmentation despite only minor vascular injury.

Bibliography

Hoak, J. C. *et al.* Hemangioma with thrombocytopenia and microangiopathic anemia (Kasabach–Merritt syndrome): An animal model. *J Lab Clin Med* **77:** 941–950, 1971.
Parker, J. C., and Barrett, D. A. Microangiopathic hemolysis and thrombocytopenia related to penicillin drugs. *Arch Intern Med* **127:** 474–477, 1971.
Rebar, A. H. *et al.* Red cell fragmentation in the dog: An editorial review. *Vet Pathol* **18:** 415–426, 1981.
Shull, R. M. *et al.* Spur cell anemia in a dog. *J Am Vet Med Assoc* **173:** 978–982. 1978.

7. Hemolytic Anemia Due to Splenic Hyperfunction

Any condition which causes an enlargement of the spleen results in increased erythrophagocytic activity in that organ. An enlarged spleen (splenomegaly) may be the result of an immunoproliferative reaction due to infection, autoimmune disease, myeloid metaplasia, neoplasia, or simply chronic congestion. Portal hypertension can cause congestive splenomegaly but does so very rarely in animals. The stretching of the splenic reticulum in an enlarged spleen, regardless of the cause, is a specific stimulus to stromal hyperplasia. This increased sinus volume rapidly becomes populated with macrophages and, therefore, splenic enlargement by definition leads to increased phagocytic function. In the short term, conditions which cause splenomegaly with stretching, but not proliferation of the sinus stroma, are reversible. Blood pooled in an enlarged spleen is subjected to areas of slow flow and is biochemically characterized by low glucose, cholesterol, and pH, which collectively cause premature aging and subsequent spherocytosis of erythrocytes, rendering them subject to phagocytic destruction. The clinical disease resulting from hypersplenism may, therefore, be a thrombocytopenic purpura, hemolytic anemia, neutropenia, or any combination of these. If immune factors are added to the congestive pathogenesis, the disease is more severe.

Specific examples of conditions involving splenomegaly and hypersplenism include most if not all of the infectious anemias, but particularly those caused by protozoan agents such as malaria, leishmaniasis, and trypanosomiasis. Occasionally localized disease in the spleen will cause clinical anemia, as in plasmacytoma of the spleen in the dog. The latter condition is not usually a diagnostic problem due to the boundless enthusiasm of surgeons and their ability to palpate the spleen in this species. In the horse, a type of lymphoma occurs as a localized splenic mass that may weigh as much as 35 pounds. The animals are presented with a responsive hemolytic anemia that is usually suspected to be due to equine infectious anemia until proven serologically negative. Since the spleen is not palpable even with this size of mass, a clinical diagnosis depends on recognition of a nonimmunologic hemolytic process.

Bibliography

Keenan, C. M. et al. Visceral leishmaniasis in the German shepherd dog. I. Infection, clinical disease, and clinical pathology. Vet Pathol 21: 74–79, 1984.
Keenan, C. M. et al. Visceral leishmaniasis in the German shepherd dog. II. Pathology. Vet Pathol 21: 80–86, 1984.
Konno, S. Spleen pathology in African swine fever. Cornell Vet 62: 486–506, 1972.
Wadenvik, H., and Kutti, J. The spleen and pooling of blood cells. Eur J Haematol 41: 1–5, 1988.
Weiss, L., and Tavassoli, M. Anatomical hazards to the passage of erythrocytes through the spleen. Semin Hematol 7: 372–380, 1970.

G. Hemorrhagic Anemia

Hemorrhage represents the simplest means of blood loss, although the pathogenesis may be most intricate. (The hemorrhagic diseases are given separate consideration later.) The causes of hemorrhage are very diverse with equally diverse results depending on the cause, the tissues directly involved, and on the rate of blood loss. Bleeding may be acute and the anemia of sudden onset, or continuous with loss of small amounts of blood over prolonged periods. The syndromes representing these two extremes are different, and therefore are discussed separately; however, between the extremes, the various signs form a continuous spectrum. The manifestations also vary depending on whether the hemorrhage is external with the blood lost from the body, or internal with the essential components of hemoglobin available for reutilization. These differences depend largely on the availability of iron and, to a lesser extent, on protein and the essential amino acids (Fig. 2.73A,B).

Acute posthemorrhagic anemia with bleeding to the exterior has a number of possible causes, of which a common one is trauma. Rupture of blood vessels by pathological processes also occurs and is exemplified by massive hemorrhage from the nasal cavities, lungs, and the female reproductive tract. Esophagogastric ulceration of the porcine stomach is a relatively common cause of acute anemia, as are bleeding abomasal ulcers in cattle. Hemorrhage into the stomach and intestines is accompanied by a moderate elevation of nonprotein nitrogen in blood. Parasitism, especially **hemonchosis, ancylostomosis,** and **coccidiosis,** may cause acute posthemorrhagic anemia. Enzootic hematuria of cattle is recognized as a cause of anemia in certain geographic areas.

In acute hemorrhage, immediate examination of the blood may reveal no change since all elements including plasma are proportionately reduced. Very shortly, however, there is an increase in the number of circulating platelets and of leukocytes, particularly neutrophils, which are largely mature. The leukocytosis is initially due to a redistribution of leukocytes from organs such as the spleen, splanchnic vessels, and lung, in which there are normally sequestered a number equal in mass to the cohort normally in circulation. This neutrophilia and thrombocytosis persist as a reflection of increased marrow output directed principally to the production of red cells. As fluid is withdrawn from the tissues and alimentary tract to restore the plasma volume, the anemia becomes evident and it is, for the first 2 days or so, normocytic, after which the arrival of new red cells made macrocytic by the action of increased levels of erythropoietin add to the PCV, MCV, and apparent anisocytosis. Thus the anemia of hemorrhage will be briefly macrocytic, and then scattered hypochromia will be seen, followed by a general reduction in red cell size and saturation as iron becomes limiting. These changes appear to a greater or lesser degree regardless of the level of storage iron, since demand far outstrips mobilization, and the response will be increased incrementally by the addition of iron by both oral and parenteral means. The reticulocyte count, normally at 1% or 60×10^9/liter in the dog and 0% in the cow will rise to $200–300 \times 10^9$/liter depending on the iron supply and whether the blood was lost from the body. In no case will the response equal that for hemolytic anemia. As an example, a typical dog from a group of 36 with enteric bleeding from overdose of sulfa for coccidiosis had a PCV of 15% and a reticulocyte count of 3.2×10^9/liter, and a dog with factor VIII deficiency and an equivalent PCV had a similar level of response. Calves with blood loss anemia are even less responsive than dogs. Horses, which do not release reticulocytes into the circulation, will have an upward shift in MCV of 6 fl in response to removal of half the red cell mass, whereas in response to a similar loss due to Heinz body hemolytic anemia, the MCV rises

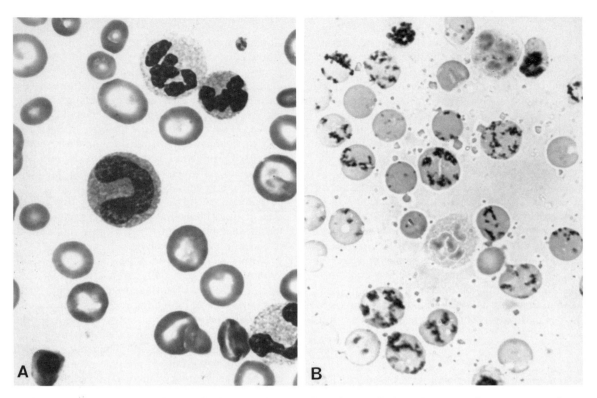

Fig. 2.73A (A) Hemorrhagic anemia. Ox. Duration 1 week. Anisocytosis, hypochromia, and poor response due to rate-limiting effects of iron utilization. (B) Reticulocyte preparation 2 days after parenteral iron administration.

12 fl. The response to anemia caused by a single episode of hemorrhage can be expected to restore a normal PCV in about 30 days, if not complicated by infection.

Following severe hemorrhage, rubricytes appear in the blood. Their number, even when prominent by leukocyte standards, will only be in the 5–10 × 10⁹/liter range and give no indication of response, as can be seen when their numbers are compared to the level of reticulocytes required to raise the PCV. A reticulocytosis which persists longer than 1–2 weeks points to the probability that bleeding is continuing. Significant red cell production can occur without apparent reticulocytosis, particularly in cattle, whose response is always subdued in comparison to dogs.

Marrow hyperplasia is evident grossly by an expansion of the hematopoietic areas along the endosteal surfaces of the long bones, and microscopically by an increase in proportion of hematopoietic to fat cells in red areas to nearly 100% from a normal proportion of about 50%. If iron deficiency is present, which it is likely to be in blood-loss anemia, there will be erythroid hyperplasia with late asynchrony or a predominance of late stage rubricytes.

The immediate causes of chronic, posthemorrhagic anemia are, in general, likely to be enteric, occult, and due to parasitism. Alternatively the loss may be caused by external parasitism, as in heavy flea infestation of kittens and puppies. In the case of ulceration of the gastric squamous mucosa of swine, the loss may occur as recurrent acute episodes or continuous slight bleeding. By arbitrary

but accepted definition, chronic posthemorrhagic anemia is caused not only by loss of blood, but also by inadequate formation, since continued bleeding depletes the body of hematopoietic substrates, principally iron. Under normal circumstances, 75% of the body iron is in the red cell mass so severe hemorrhage represents a significant and irretrievable loss unless supplemental iron is given. Depending on the severity and duration of the hemorrhage, the syndrome may be complicated by hypoproteinemia resulting from chronic loss of plasma proteins. The anemia is hypochromic, microcytic, nonresponsive, and accompanied by a thrombocytosis of small platelets. Since, with deficiency of iron, hemoglobin synthesis lags behind the production of red cells, the hypochromia is more significant than the microcytosis. An indication of hypochromia can be obtained from a stained blood film, in which the erythrocytes will be pale with increased central pallor in dogs, and show pallor in ruminants where it is not normally present. In addition, as part of the syndrome of iron-deficiency anemia, there will be poikilocytosis whose severity will vary directly with the degree of anemia. When the deficiencies are resolved, the response of the bone marrow is evident in a prompt reticulocytosis. When the anemia is excessively prolonged and severe, there may be some degree of extramedullary hematopoiesis by the liver and spleen. The spleen is enlarged and rather meaty and, microscopically, there is increased sinus stroma.

Internal hemorrhage differs somewhat in its course from

external hemorrhage. When acute, the main manifestations are the same but, in addition, there will be some increase in bilirubin even to the level of clinical icterus. Many cases, however, are too rapidly fatal for signs of degradation of the extravasated blood to become manifest. The principal causes of internal hemorrhage of a degree adequate to produce clinical anemia are poisoning by sweet clover, *Melilotus alba*, and spontaneous or traumatic rupture of a viscus, usually the liver or spleen, as often occurs in dogs with hemangiosarcoma. Much of the blood lost into the peritoneal cavity is returned via the lymphatics to the circulation. In the latter disease in the dog, the previously extravasated red cells are apparent on the blood film as a second population of crenated cells in addition to the few acanthocytes characteristic of hemangiosarcoma. In any event of internal hemorrhage, most of the extravasated blood can be reutilized in hematopoiesis after its degradation by the monocyte–macrophage system. Hemosiderosis may therefore be a conspicuous feature in the liver, spleen, bone marrow, and other organs in cases of internal destruction of blood but, as indicated earlier, hemosiderin iron is not readily available for reutilization.

Bibliography

Lumsden, J. H. *et al.* The kinetics of hematopoiesis in the light horse. III. The hematological response to hemolytic anemia. *Can J Comp Med* **39:** 324–331, 1975.

Oehlbeck, L. W. F., Robscheit-Robbins, F. S., and Whipple, G. H. Marrow hyperplasia and hemoglobin reserve in experimental anemia due to bleeding. *J Exp Med* **56:** 425–448, 1932.

Patterson, J. M., and Grenn, H. H. Hemorrhage and death in dogs following the administration of sulfaquinoxaline. *Can Vet J* **16:** 265–268, 1975.

Schnappauf, H. P. *et al.* Erythropoietic response in calves following blood loss. *Am J Vet Res* **28:** 275–278, 1967.

Todd, J. R., and Ross, J. G. Biochemical and haematological changes in the blood of normal sheep following repeated phlebotomy. *Br Vet J* **124:** 353–362, 1968.

H. Disorders of Hemoglobin

Congenital abnormalities of the synthesis of heme and of the molecular structure of hemoglobin are not known in animals, except for porphyria of cattle, pigs, and cats. The porphyrins are photodynamic but otherwise appear to be inert (see The Skin and Appendages, Volume 1, Chapter 5). Acquired disorders of hemoglobin are, however, rather common. Increased protoporphyrins are associated with a number of diseases including lead poisoning, iron deficiency, and the porphyrias. Most hemoglobin disorders result in the formation of methemoglobin. Carboxyhemoglobin formation is rare unless poisoning is intentional.

1. Methemoglobinemia

Methemoglobin is derived from hemoglobin by conversion of iron from the ferrous and functional state to the ferric form, in which it is incapable of transporting oxygen.

A small amount of methemoglobin is normally present in blood, especially of herbivorous animals, and methemoglobin is probably formed at a steady rate but maintained at low concentration by an equilibrium with hemoglobin. Excessive concentrations of methemoglobin in the blood must then be due to increased production or to a defect in the mechanisms which reduce methemoglobin to hemoglobin. Methemoglobin reductase deficiencies have been described in the dog, but in the majority of cases, the formation of methemoglobinemia in animals is due either to increased production of the pigment, under the influence of oxidizing poisons, or to irreversible oxidation of free hemoglobin in hemolytic diseases.

Many drugs and poisons are known to cause the oxidation of intracellular hemoglobin to methemoglobin, but the most important of these are the nitrates, chlorates, phenacetin, and acetanilid.

Nitrate poisoning and nitrite poisoning are often used synonymously, especially in relation to herbivores, in which reduction of nitrate to nitrite may occur in the alimentary tract. Reduction may also occur in forage before ingestion. Nitrate is, however, toxic as the nitrate radical and, although uncommon, poisoning has been observed in dogs and horses following ingestion of excessive amounts of potassium nitrate. Potassium nitrate is an irritant to the alimentary mucosa, and a diuretic, and signs of poisoning are referable to the alimentary and urinary systems, but there are no specific lesions. Excessive dosage results in severe hemolytic anemia and icterus. The reduction of nitrate to nitrite is frequently insignificant in these circumstances and methemoglobinemia does not occur.

When herbivores ingest nitrate, it is reduced in the alimentary tract to nitrite and ultimately to ammonia. Since the amount of ammonia produced is rate limiting, toxicosis by ammonia does not occur. Depending on a number of factors, including the initial concentration of nitrate, the flora of the rumen, and the diet of the animal, the conversion of nitrite to ammonia may be inadequate, in which case nitrite is absorbed and causes methemoglobinemia. The capacity of the rumen microflora to detoxify nitrate, or the nitrite derived from it, is energy dependent and therefore progressively reduced under conditions of poor feeding. The conversion of nitrate to nitrite can and does occur externally as, for example, in hay stacked with a high moisture content.

The sources of nitrite for grazing animals are varied. Agricultural fertilizers which contain large amounts of nitrate or nitrite are frequently responsible for outbreaks of nitrate poisoning. In addition, a number of plants accumulate toxic levels of nitrate under natural conditions. The plants which are most frequently incriminated include the cereal crops, either green or as hay, the leaves of sugar beets and turnips and other Brassicaceae, *Sorghum* spp., and a number of weeds, especially variegated thistle (*Silybum marianum*), redroot pigweed (*Amaranthus retroflexus*), capeweed (*Cryptostemma calandulacium*) and *Tribulus* spp. The nitrate content of these and other plants is extremely variable and is influenced by the species of plant, the availability of the nitrogen of soil, the amount of shading, and the stage of growth of the plant. The application of herbicides may lead to an accumulation of nitrates in plants, especially sugar beet and wheat, and it is probable that a deficiency of other essential nutrients, such as sulfur, may result in metabolic disturbances in the plant and the accumulation of nitrate. Levels of nitrate, expressed as potassium nitrate, of less than 2% in plants are unlikely to result in toxicity. Outbreaks of nitrate poisoning are usually recorded on forage which contains 3–7% of potassium nitrate.

Stagnant waters, especially well water, in areas of high fertility of soils tend to accumulate toxic levels of nitrate, and poisoning has been recorded in cattle and pigs which have ingested such water.

The pathogenesis of nitrate (nitrite) poisoning depends on the oxidation of ferrous to ferric hemoglobin iron. It is likely that there are additional factors, including the oxidation of iron-containing respiratory enzymes and, since nitrates dilate blood vessels, a drop in blood pressure is to be anticipated. The blood vessels which are most sensitive to the dilatory effects of nitrite are those of the head, brain, meninges, and coronary vessels, with the visceral vessels being less sensitive. It is likely that the rapid collapse seen with nitrite poisoning is due to hemodynamic disturbances as well as to the inability of the oxidized hemoglobin to transport oxygen.

Cattle are more susceptible to poisoning than are sheep or horses. Clinically, nitrite poisoning is characterized by weakness, staggering gait, collapse, dyspnea, cyanosis, and rapid pulse. Death occurs in coma. Animals which are less severely affected urinate frequently. Pathologically there is little to see except for cyanosis of mucous membranes and light brown discoloration of the blood and tissues. An excess of bloodstained fluid is commonly present in the pericardial sac. Methemoglobin levels decrease rapidly after death.

Sodium chlorate is used as a herbicide and is occasionally responsible for death of cattle and sheep. It is a powerful but rather slow oxidizing agent, and its principal toxic effect is to cause methemoglobinemia. At high dosage, chlorate also causes intravascular hemolysis so that hemolytic anemia and hemoglobinuria may accompany the methemoglobinemia. When applied to plants, sodium chlorate is progressively converted to sodium chloride so that pastures are most hazardous immediately following spraying. The salt is palatable to livestock, especially in areas of marginal deficiency of sodium chloride, and this no doubt contributes to the likelihood of poisoning.

2. Carboxyhemoglobinemia

Carbon monoxide is formed by incomplete oxidation of combustible carbonaceous materials. It is a common cause of poisoning in humans, uncommon in most animals, even pets, but still occurs in housed pigs. The usual sources of the carbon monoxide are illuminating gas and exhaust fumes from petroleum fuels that accumulate in tightly closed, heated barns. Carbon monoxide has an affinity for hemoglobin which is 200–300 times the affinity of oxygen for hemoglobin, and the conjugated carboxyhemoglobin is incapable of carrying oxygen; the conjugation is, for practical purposes, irreversible. Apart from the bright redness of the blood, which is directly due to carboxyhemoglobin, the signs and lesions of carbon monoxide poisoning are those of anoxia. The anoxia affects the brain most profoundly, and in acute poisoning, the organ is congested and edematous and contains punctate hemorrhages. Degeneration of neurons may be widespread and there is a tendency for foci of softening to develop. Recovery results in residual nervous defects in some cases.

Bibliography

Harvey, J. W., Ling, G. V., and Kaneko, J. J. Methemoglobin reductase deficiency in a dog. *J Am Vet Med Assoc* **164:** 1030–1033, 1974.

Osweiler, G. D. *et al.* Nitrates, nitrites, and related problems. *In* "Clinical and Diagnostic Veterinary Toxicology," 3rd Ed., pp. 460–467. Dubuque, Iowa, Kendall/Hunt Publishing Company, 1985.

LYMPHORETICULAR TISSUES

I. Thymus

The thymus is a composite organ of epithelial and lymphoid tissues, with the epithelial part derived from the third pharyngeal pouch of the foregut endoderm. This epithelial anlagen is colonized by lymphocytes from the blood islands of the primordial yolk sac following migration of the epithelial component to the anterior mediastinal area. The epithelial components are derived bilaterally from the middle series of branchial pouches, which by epithelial outgrowths meet in the midline and are then displaced caudally in the neck.

There are two streams of epithelial migration, with that from the uppermost part of the pharyngeal pouch forming the thymic duct epithelium and later the Hassall's corpuscles, and that from the lower part of the pouch forming the reticular epithelial component of the cortex and medulla of the adult thymic lobule. Embryologically the reticular epithelial component first invades the mediastinal interstitium with irregular, solid, rosettelike buds that enlarge to form the primordial lobules. This reticular epithelium forms loose cuffs around small vessels that persist in adult life and become obvious following lymphoid atrophy. The second component migrates as a system of branching ducts that ramify in the interlobular stroma and penetrate the lobules to form a central cord that broadly communicates between the lobules of a single lobe. In early development, this tubular system has a basement membrane and layered epithelium that becomes solid and cystic and loses the outer membrane in the mature medullary corpuscle. These relationships are pertinent to the histologic interpretation of the thymus in pathologic states, as in dysplasia, atrophy, regeneration, and hyperplasia, these components recapitulate their embryologic relationships. Thus a thymoma with diffuse architecture and medullary predominance can be differentiated from hyperplasia by the presence of the reticular cuffs around the vessels of a thymoma. In terms of induction, it appears that these medullary ducts are essential to colonization of the anlagen by lymphocytes, and are the source of trophic thymic hormones. They are also the source of the cysts lined by ciliated epithelium found in the thymus and anterior mediastinum in many species and which, if large, may be of clinical significance as a cause of dyspnea. A further cell type, that resembles striated muscle fibers, is of indefinite origin, but has importance in the pathogenesis of myasthenia gravis.

In most species, the thymus reaches maximal development about the time of puberty. It can become hyperplastic, and extend from the rami of the mandibles to the base of the heart, in calves given repeated injections of endotoxin. Normally the thymus regresses in adult life, and this regression may be accelerated during severe or chronic illness, but it never entirely disappears. Even a very small organ weighing less than 10 g can be immunologically potent in autoimmune dysfunction.

The thymic cortical lymphocytes are small cells with nuclei slightly larger than erythrocytes, little internal nuclear detail, and minimal cytoplasm. Despite their small size and lack of nucleoli, there is intense lymphopoiesis in the peripheral thymic cortex and a continual migration of cells to the periphery from this area. Most (99%) of this proliferation is ineffective and balanced by a high rate of cell death apparently in a process of immunologic selection. In normal states, the cortical lymphocytes are closely packed so that the cortex appears uniformly dark histologically. In conditions of stress and atrophy, the cortical epithelial cells and phagocytes become prominent and these larger cells with ingested nuclear debris give the cortex a moth-eaten appearance.

The medullary lymphocytes are larger than their cortical precursors and have a more vesicular nucleus with small nucleoli and more cytoplasm. They appear less tightly packed than the cortical cells and, with the higher density of epithelial cells and the greater variability in nuclear size and shape, cause the medullary areas to appear lighter histologically, giving the thymus its characteristic corticomedullary differentiation. In many disease states in which the animal is faring poorly, this corticomedullary boundary is indistinct and blurred by dilution of the cortical cells and an increase in macrophages in both regions. In normal states, the squamous ductal epithelium is obvious in central medullary areas, but the reticular epithelium of the cortex and medulla is not; the latter background of cells is apparent only following atrophy of the normal lymphocyte population.

Physiologically, prothymocytes from bone marrow colonize the outer cortex of the thymus where they proliferate and are selected and trained by the cortical epithelium. In this region, the epithelial cells form large membrane-lined cavities called **caveolae** in which the marrow-derived lymphocytes divide and undergo phenotypic development. On release from the marrow, the prothymocytes have the ability to home selectively to the thymus, but they lack helper, suppressor, or killer functions. In contrast, the peripheral T lymphocytes are characterized phenotypically by CD4 (cluster derived, helper) or CD8 (suppressor) expression, functionally by the MHC determinants they recognize, and by their collective inability to react against normal self-antigens.

Prothymocyte development is believed to occur within the caveolae of the large epithelial nurse cells. Nurse cells tend to express either major histocompatability complex (MHC) II, primarily in the cells of the outer cortex and dendritic cells of the inner cortex, or, in association with the more flattened reticular epithelium of the medulla, both type I and II determinants. This close apposition of epithelium and the young lymphocytes permits the transmission of inductive signals for proliferation and development, or for death of the great majority of the cells produced. The thymic nurse cells themselves produce interleukin-1 (IL-1), which drives lymphoproliferation, and in the course of maturation, the T cells acquire receptor and production capability for interleukin-2 (IL-2), as well as other reactive lymphokines.

The progeny of this system are the recognition and regulation specialists of the adult immune system. They identify antigens presented to them by other cells as having either the MHC I or II determinants, and determine the presence or absence of foreign antigen, which is handled by killing or by stimulation of antibody development. T cell development involves the rearrangement of the T cell receptor genes analogous to the same process by which the B cells gain antibody diversity. Whereas B cells recognize antigen on the basis of molecular shape, T cell recognition is on the basis of peptides. Since antigen is presented to the T cell by antigen-processing phages and dendritic cells, denaturation of antigen is of no consequence since it is the amino acid sequence that is detected as foreign. In nonhematopoietic cells, internal antigen is displayed within a groove in the MHC I molecule on cell membranes so that viral infection or tumor antigen are available for recognition by T lymphocytes. This process of cell recognition is accomplished by intracellular transporter pumps which display all of the proteins (self or foreign) produced within the cell on the cell surface. The diversity of T cell recognition appears to be a function of receptor diversity and permits survival of cells with acceptable recognition of self to avoid holes in antigen detection. Thus most NK (natural killer) cells recognize targets by MHC I as well as foreign antigens, as do the CD8 or suppressor cells, whereas the CD4 or helper cells recognize MHC II and antigen.

Cells exiting the thymus have not only managerial receptors but also homing receptors to guide them to appropriate areas in the peripheral nodes and Peyer's patches. Only 1–3% of thymocytes bear homing receptors, all of which are in the cortex, and all of which are phenotypically mature. Thus it appears that the homing direction is added last with most of the cortical cells homed to nodes and likely medullary cells to Peyer's patches.

It appears that the thymic epithelium, besides making trophic hormone for the immune system, also makes the peptides oxytocin and vasopressin in common with the pituitary, thus indicating a mechanism for the interaction of the nervous, endocrine, and immune systems.

In terms of ontogeny, lymphocytes first appear in the thymus of the fetal lamb at day 43, nodes at day 45, spleen at day 54, and Peyer's patches at day 65. These relationships are altered by fetal infection (see The Female Genital System, Chapter 4 of this volume). In the dog, thymic epithelial anlagen are present by gestational age 23 days, lymphopoiesis begins by day 33, and corticomedullary differentiation with formation of Hassall's corpuscle is first evident at about 38 days.

Bibliography

Bodey, B. *et al.* Development and histogenesis of the thymus in the dog. *Dev Comp Immunol* **11:** 227–238, 1987.

Cottier, H., Kraft, R., and Meister, F. Primary immunodeficiency syndromes and their manifestations in lymph nodes. *In* "Reaction Patterns of the Lymph Node," E. Grundmann and E. Vollmer (eds.). Berlin, Springer-Verlag, 1991.

Geenen, V. *et al.* Neuroendocrinology of the thymus. *Horm Res* **31:** 81–84, 1989.

Liu, S.-K., Patnaik, A. K., and Burk, R. L. Thymic branchial cysts in the dog and cat. *J Am Vet Med Assoc* **182:** 1095–1098, 1983.

Malefijt, R. D. *et al.* T-cell differentiation within thymic nurse cells. *Lab Invest* **55:** 25–34, 1986.

Osburn, B. I. Immune responsiveness of the fetus and neonate. *J Am Vet Med Assoc* **163:** 801 803, 1973.

Schultz, R. D., Dunne, H. W., and Heist, C. E. Ontogeny of the bovine immune response. *Infect Immun* **7:** 981–991, 1972.

Schultz, R. D., Wang, J. T., and Dunne, H. W. Development of the humoral immune response of the pig. *Am J Vet Res* **32:** 1331–1336, 1971.

Shier, K. J. The thymus according to Schambacher: Medullary ducts and reticular epithelium of thymus and thymomas. *Cancer* **48:** 1183–1199, 1981.

Wilson, C. B. The ontogeny of T-lymphocyte maturation and function. *J Pediatr* **118:** S4–S9, 1991.

A. Developmental Diseases of the Thymus

1. Congenital Immunodeficiency

Immunodeficiency is defined as an absence or defect in a class or subclass of the lymphocytes derived either from bone marrow or thymus. Problems not included in this discussion are congenital defects of phagocytic cells or of complement, and the failure passively to transfer immunoglobulin from dam to offspring. Thus, immunodeficiency disease occurs as the result of absence or defect in the bone marrow or thymus-derived lymphocytes, and it may be partial or complete. In individuals, the deficiency may affect bone marrow derived cells with hypogammaglobulinemia, or thymus-dependent cells with lymphopenia and deficient cell-mediated immunity. When the deficiencies involve both B and T derived classes of lymphocytes, they are called combined immunodeficiency states. Since deficiencies of the B and T lymphocytes tend to occur together, they are considered here under the thymus.

2. Agammaglobulinemia

Immunoglobulin deficiencies occur in human males, foals, and probably in other species. The disease in foals resembles X-linked agammaglobulinemia of human infants in both expression and progression. A selective deficiency of immunoglobulin M occurs in foals of quarter horse and Arabian breeding, and both sexes are affected. Immunoglobulin levels vary from undetectable to 10% of normal, and the lymphocyte counts are normal, with normal proportions of B and T derived classes. These animals have a neutrophilic leukocytosis without anemia and are presented for examination at 4–8 months of age with febrile disease involving the respiratory tract.

Bibliography

Banks, K. L., McGuire, T. C., and Jerrells, T. R. Absence of B lymphocytes in a horse with primary agammaglobulinemia. *Clin Immunol Immunopathol* **5:** 282–290, 1976.

McGuire, T. C. *et al.* Agammaglobulinemia in a horse with evidence of functional T lymphocytes. *Am J Vet Res* **37:** 41–46, 1976.

Perryman, L. E., McGuire, T. C., and Hilbert, B. J. Selective immunoglobulin M deficiency in foals. *J Am Vet Med Assoc* **170:** 212–215, 1977.

3. Combined Immunodeficiency

Combined immunodeficiency occurs in humans, foals primarily of the Arabian breed, and in an inbred strain of the C.B-17 mouse. Both sexes are affected and the disease is due to an autosomal recessive trait expressed as an absence of both B and T lymphocyte functions. The syndrome is heterogeneous with a number of entities which differ in severity; however, since the disease is usually fatal in early life, the acronym SCID is used to indicate **severe combined immunodeficiency.** In the fully developed disease, there is extreme susceptibility to infectious agents, and in human infants, vaccination with smallpox and bacillus Calmette–Guérin is uniformly fatal. Affected foals survive about 2–5 months and infants, 2 years, whereas SCID mice if maintained in microisolators will survive 40–80 weeks or about three quarters of a normal life span.

The disease occurs in some 2% of Arabian-bred foals, but the incidence of phenotypically normal heterozygotes will be much higher. Affected foals are normal at birth, but at about 10 days of age develop a range of diseases including pneumonia and diarrhea with which adenoviruses are peculiarly associated. Despite intensive therapy, affected foals die at about 3 months of age. Hematologically, the animals are profoundly lymphopenic. There is hypoplasia of all lymphoid tissue, including thymus, nodes, and spleen, with histologic hypocellularity (Fig.

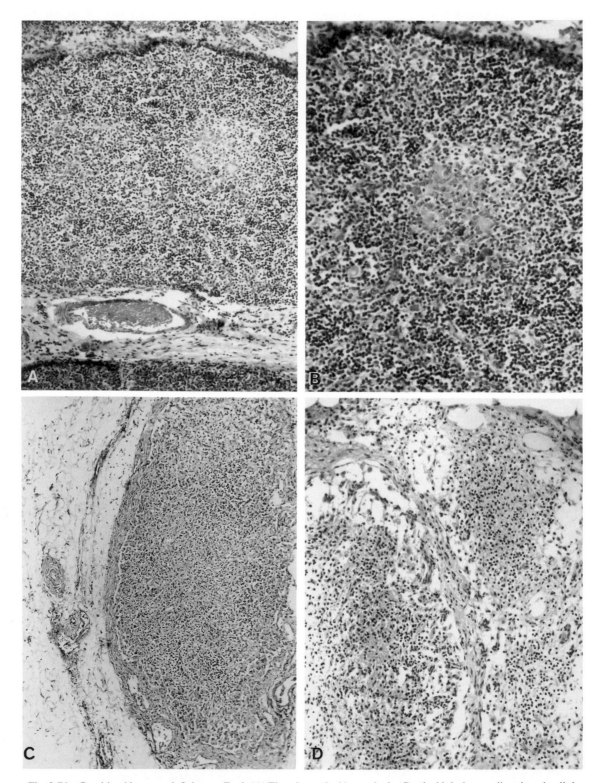

Fig. 2.74 Combined immunodeficiency. Foal. (A) Thymic cortical hypoplasia. Cortical lobules small, reduced cellular density, lack of differentiation between cortex and medulla. (B) Detail of hypoplastic thymus in (A). (C) Hypoplasia of lymph node. (D) Detail of (C) showing absence of follicles and hypocellular paracortex.

2.74A–D). The reduced lymphocyte levels in blood are usually more than compensated by a neutrophilia, so that total leukocyte counts are usually in the normal range. The mechanism of disease appears to be a defect in pro-lymphocytic stem cells at a hierarchical level that involves both bone marrow-derived and thymus-dependent systems. Affected foals develop bilateral nasal discharge, which becomes sufficiently profuse to impair suckling. There is a progressive loss of weight and intermittent pyrexia with coughing, depression, and rough hair coat. Intractable respiratory disease is the most common clinical sign, but diarrhea and swollen joints are also observed. There is a severe lymphopenia and it is suggested that counts of less than 1×10^9/liter are diagnostic for the disease. Lymphopenia is present from birth and is constant throughout life. Since IgM is synthesized by the normal equine fetus, an absence of IgM from serum of foals which have not suckled supports a diagnosis of severe combined immunodeficiency disease. In animals from 1 to 100 days of age, the serum IgM and IgA levels are low and remain between 0.2 and 0.6 g/liter. The IgG, which is maternally derived, progressively drops from 8.0 g/liter shortly after birth to 2.0 g/liter in foals which approach 100 days of age. There is usually mild anemia that increases in severity with resistant and repeated infections, but is often masked by dehydration. Leukocyte counts are highly variable, being primarily dependent on the numbers of neutrophils.

The characteristic lesion associated with severe combined immunodeficiency in foals is bilateral anterior bronchopneumonia in association with small spleen, lymph nodes, and thymus. The latter organ is identified only as a thin mediastinal raphe anterior to the heart. The bone marrow is unusually reactive for the age of the animal, and hematopoiesis persists in cancellous bone and partially occupies the central femoral cavity. Terminally, areas of marrow that appear red grossly may be hypoplastic with dilated sinusoids (Fig. 2.75A).

Histologically, basophilic adenoviral inclusion bodies are usually present in the epithelial nuclei of bronchi, bronchioles (Fig. 2.75B), and pancreatic ducts. The spleen is remarkable in that sinus areas are moderately cellular with macrophages and extramedullary hematopoiesis, whereas the small arterioles are devoid of lymphocytic cuffs. Architecturally, the nodes consist of a stromal framework, which is essentially an unoccupied superstructure. The node capsules are delicate, with the peripheral sinus often heavily colonized by macrophages and granulocytes. Some nodes, particularly from the mediastinal area, may have cortical microabscesses. There is an absence of follicles, and the cortex consists of a reticular framework in which there are small foci of lymphocytes around small arterioles. The postcapillary venules have high vesicular endothelium and are usually well demonstrated because of the lack of perivascular and intramural lymphocytes. The medullary cords are collapsed, and the

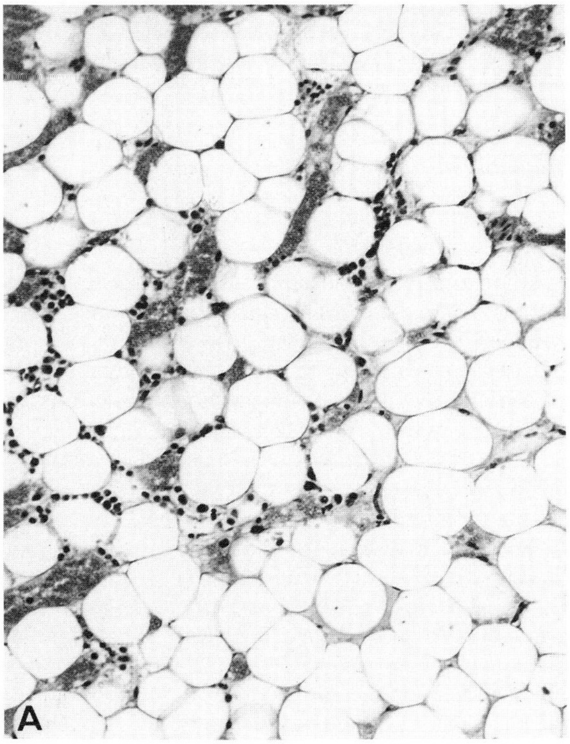

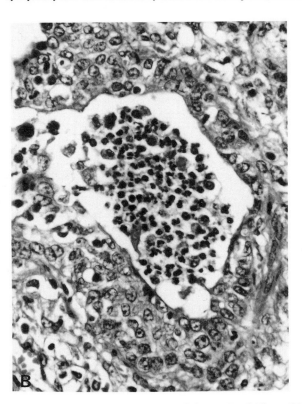

Fig. 2.75A Combined immunodeficiency. Foal. Hypoplasia of bone marrow.

Fig. 2.75B Combined immunodeficiency. Foal. Necrotizing and hyperplastic bronchiolitis. Adenoviral inclusions are present in large, sloughed epithelial cells.

medullary sinuses are dilated and contain neutrophils and macrophages. If thymus can be identified, it consists of small lobules of 1–2 mm diameter surrounded by normal fat. The lobules have normal stromal capsulation and consist almost entirely of medullary structures with central Hassall's corpuscles and a vesicular reticular network with a very low lymphocyte population.

The presence of the **Hassall's corpuscles** is worthy of note since, besides identifying the tissue as thymus, their presence also indicates that there has been successful embryological migration of the epithelial duct system. In human infants, the corpuscles are present only in that variant with adenosine deaminase deficiency; however, this enzyme is not deficient in foals. In contrast, Hassall's corpuscles are not found in SCID mice, including the beige strain that also lacks NK cells. In the latter species, thymus can be distinguished from mediastinal node only by the presence of a peripheral sinus in the latter. Long-term survivors of the SCID mouse strain have a remarkable incidence of some 15% thymic lymphomas. Thymic hormone is not deficient in the SCID syndrome, and possibly the persistent high levels of interleukin-1 acting on a deficient target induce autonomous lymphoproliferation. The cellular defect is not well defined. There is failure of the lymphocytes to mature and, in the mouse, a defective recombinase system may prevent gene rearrangement both for immunoglobulin production and formation of the T cell receptor. A variant of the SCID syndrome in infants is termed the bare lymphocyte syndrome, as the T cells lack human leukocyte antigen (HLA) expression. It appears that some of the variability in these diseases may be due to the level of maternal cells in circulation which are functionally competent by standard tests. The SCID syndrome can be successfully treated in infants and foals by marrow transplant. Graft-versus-host disease follows transplantation of unmatched precursors.

Bibliography

Magnuson, N. S., and Perryman, L. E. Metabolic defects in severe combined immunodeficiency in man and animals. *Comp Biochem Physiol* **83B**: 701–710, 1986.

McChesney, A. E., England, J. J., and Rich, L. J. Adenoviral infection in foals. *J Am Vet Med Assoc* **162**: 545–549, 1973.

McCune, J. M. *et al.* The SCID-hu mouse: Murine model for the analysis of human hematolymphoid differentiation and function. *Science* **241**: 1632–1639, 1988.

McGuire, T. C., Banks, K. L., and Poppie, M. J. Combined immunodeficiency in horses: Characterization of the lymphocyte defect. *Clin Immunol Immunopathol* **3**: 555–566, 1975.

Perryman, L. E. *et al.* Combined immunodeficiency in an Appaloosa foal. *Vet Pathol* **21**: 547–548, 1984.

Rosen, R. S., Cooper, M. D., and Wedgwood, R. J. P. The primary immunodeficiencies. *N Engl J Med* **311**: 235–242 and 300–310, 1984.

B. Acquired Immune Deficiency

Loss of normal immune reactivity occurs in a number of circumstances, most often as a result of therapeutic intervention, either specifically to combat autoimmune disease or indirectly as a result of antineoplastic chemotherapy. An unexpected cause of immune suppression is the transfusion of otherwise compatible allogeneic blood. The mechanism is apparently a blanking of the immune system by the massive introduction of antigen which, despite the beneficial hematologic effects to postsurgical patients, has a number of undesirable immune effects including a reduction in natural killer cell activity. Normal pregnancy induces a mild immune suppression which is usually only of significance in the face of epidemic viral disease, such as occurs with human influenza. The effect of pregnancy on the immune system is readily demonstrated experimentally in animals by alteration of the course of infection with viral, protozoal, and helminthic agents. For example, in Uganda during an outbreak of lumpy skin disease, nonpregnant dairy cattle developed lymphadenopathy and high fevers of short duration followed by complete recovery, while pregnant cattle failed to develop severe lymphadenopathy and fever but subsequently developed skin lesions and protracted disease.

Bibliography

Koller, L. D. *et al.* Immune responses in rats supplemented with selenium. *Clin Exp Immunol* **63**: 570–576, 1986.

Latimer, K. S. *et al.* Effects of cyclosporin A administration in cats. *Vet Immunol Immunopathol* **11**: 161–173, 1986.

Marquet, R. L. *et al.* Modulation of tumor growth by allogeneic blood transfusion. *J Can Res Clin Oncol* **111**: 50–53, 1986.

Mulhern, S. A., and Koller, L. D. Severe or marginal copper deficiency results in a graded reduction in immune status in mice. *J Nutr* **118**: 1041–1047, 1988.

Weinberg, E. D. Pregnancy-associated depression of cell-mediated immunity. *Rev Infect Dis* **6**: 814–831, 1984.

C. Inflammatory Diseases of the Thymus

1. Immune Inflammation of the Thymus: Myasthenia Gravis

Myasthenia gravis is an autoimmune disorder of humans, dogs, and cats, characterized by muscle weakness and reduced exercise tolerance. All of the physiologic features of the disease can be accounted for by reduction in acetylcholine receptors of neuromuscular junctions (see Muscles and Tendons, Volume 1, Chapter 2).

The disease arises as an immune-mediated phenomenon whereby a cellular and antibody reaction is developed against the thymic myoid cells which express acetylcholine receptors. The disease is mediated by the attachment of antibody to peripheral neuromuscular junctions. In dogs the syndrome occurs as an inherited disease of smooth fox terriers, springer spaniels, and Jack Russell terriers, and spontaneously, often associated with thymoma. A possible role of bacteria in the pathogenesis of the spontaneous disease has been suggested because antibody against acetylcholine receptor cross-reacts with proteins found in *Escherichia, Klebsiella,* and *Proteus* species. The thymus in myasthenia gravis of animals has not been well described other than in the recognition of thymoma. In myasthenia gravis of humans, there are three characteristic types of nonneoplastic thymic histology consisting of B cell infiltration in a follicular or diffuse pattern, and thymic atrophy. The follicular type is most common and associated with the greatest density of thymic myoid cells, with the diffuse lesion having an intermediate number, and the atrophic type, the fewest. The myoid cells tend to be associated with Hassall's corpuscles, and the germinal centers, which resemble those of antigen-activated lymph nodes, and diffuse areas of B cell proliferation are centered on the medullary areas with sparing of the cortex, which is often reduced in volume.

The association of myasthenia gravis with thymoma may not be related to the immunologic role of the thymus. Some cases may be expressions of paraneoplastic autoimmunity associated with tumor-derived antigens not necessarily related to the acetylcholine receptors.

Bibliography

Chilosi, M. *et al.* Myasthenia gravis: Immunohistological heterogeneity in microenvironmental organization of hyperplastic and neoplastic thymuses suggesting different mechanisms of tolerance breakdown. *J Neuroimmunol* **11**: 191–204, 1986.

Garlepp, M. J. *et al.* Autoimmunity in spontaneous myasthenia gravis in dogs. *Clin Immunol Immunopathol* **31**: 301–306, 1984.

Stefansson, K. *et al.* Sharing of antigenic determinants between the nicotinic acetylcholine receptor and proteins in *Escherichia coli*, *Proteus vulgaris*, and *Klebsiella pneumoniae*. *N Engl J Med* **312**: 221–225, 1985.

Wilkes, M. K. *et al.* Ultrastructure of motor endplates in canine congenital myasthenia gravis. *J Comp Pathol* **97**: 247–256, 1987.

2. Infectious Inflammation of the Thymus

Inflammation associated with some degree of atrophy of the thymus gland occurs frequently in infectious diseases. Thymic atrophy occurs in dogs with **distemper,** the changes being largely due to net emigration of lymphocytes rather than to lysis. Marked lymphocytolysis with involution does occur in foals aborted due to infection with **equine herpesvirus-1** and in cats with **feline parvovirus.** In calves with **epizootic bovine abortion** and dogs with **salmon poisoning,** there is both lymphocytolysis and infiltration with other inflammatory cells. Chronic **hog cholera** is characterized by progressive lymphoid depletion and death with complete thymic atrophy.

In foals aborted due to **equine herpesvirus-1** infection the thymus is grossly normal or of reduced size with edema. There is lack of distinction between cortical and medullary areas due to loss of cortical cells. Cytologically, there is extensive lymphocytolysis in the subcapsular areas with pyknotic nuclear debris persisting in these areas (Fig. 2.76A,B). The necrotic lesions tend to be patchy, but there is an overall reduction in cell density with edema of the lobules and separation of the lymphocytes by clear fluid. The Hassall's corpuscles may be surrounded by relatively clear acellular spaces, and viral inclusion bodies are often present in reticular epithelial cells. There is, in addition, a mild increase in monocytes and granulocytes in the interlobular connective tissue. The presence of nuclear debris in the outer cortices indicates acute lympholysis. Foals which survive 1–2 days may remove the cellular debris, and the cortices are then characterized by hypocellularity alone.

Cats which succumb to **feline parvovirus** infection have marked thymic atrophy (Fig. 2.77A–D) to the extent that the organ may not be identifiable grossly, and can be examined histologically only by systematically sampling the anterior mediastinal tissues. Microscopically, the thymus is characterized by collapsed lobules with angular facets to the lobes which become flattened and triangular. The changes are uniform in each lobule and consist of complete thymic atrophy (Fig. 2.78A,B) with only a very narrow rim of small lymphocytes 1–2 cells thick beneath the capsule. The Hassall's corpuscles are prominent, and the lobules in effect consist of a condensation of medullary tissue and a prominent epithelial/stromal background which contains few large lymphoblasts. Small lymphocytes typical of cortical cells are virtually absent from the medullary areas.

Cats infected with **feline leukemia virus** suffer thymic cortical atrophy of variable degree and a reduction in functional capability of neutrophils that may persist for a year after viremia is cleared. The p15E envelope protein appears to be the major factor in impaired T cell function, apparently by inhibition of interleukin-2 production and responses. Interferon-γ production is inhibited and NK cell cytolytic function is impaired, but binding to target cells is unaffected. Circulating immune complexes contribute to the impairment of defenses and some improvement in condition can be achieved by their removal. There is a an age-related response to infection, with cats infected at younger than 8 weeks of age surviving about 3 months, and cats 4 months of age at infection surviving about a year. There is an early stage of lymphoid hyperplasia accompanied by weight loss, and followed by lymphoid depletion, lymphopenia, reduced lymphocyte response to lectins, hypogammaglobulinemia, ulcerative stomatitis, diarrhea with necrosis of crypt epithelium, and death. Thymic atrophy occurs in all cats which die, and terminally there is marrow hypoplasia.

The lesions associated with **feline immunodeficiency virus** appear, like the human disease, to be largely those of secondary invaders, and the most specific changes are largely in tests of immune cell number and function. Infected cats maintained in a clean environment survive indefinitely but have reduced numbers of CD4 lymphocytes and reversal of the CD4/CD8 ratio, suggesting specific infection of the helper cell population. Lymphocyte response to lectins is reduced and there is reduced interleukin-2 production by cells from infected cats. The disease process is likely accelerated when FeLV and FIV coexist, which occurs in about 1% of sick cats. Both viruses may be present in cats with lymphoma. It is remarkable how infrequently the thymus is examined in these

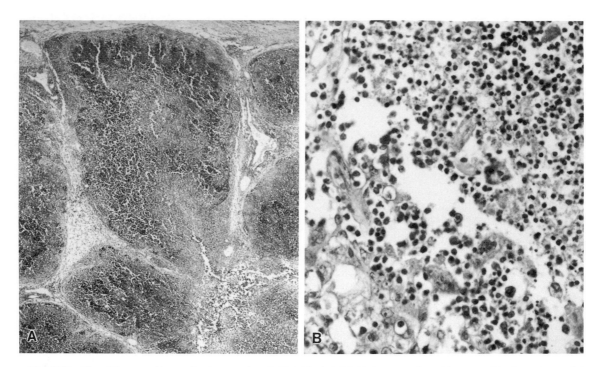

Fig. 2.76   Thymitis caused by equine herpesvirus-1. Fetal foal. (A) Lobules are reduced in size and lack corticomedullary differentiation. The dark areas are viable lymphocytes. (B) Detail of (A) to show extensive cell necrosis.

diseases, even when the focus of the investigation is on immunosuppression.

Bovine immunodeficiency virus (BIV) is an agent of wide distribution and low pathogenicity. In a proportion of infected cattle, there is a moderate lymphocytosis in the $10-15 \times 10^9$/liter range, and mild generalized lymphadenopathy. The disease is very slowly progressive. Specific tests for this virus do not cross-react with bovine leukemia virus or bovine syncytial virus. The BIV agent will infect sheep and rabbits, and they may be appropriate hosts for characterization of the virus and development of detection systems.

Infection with **bovine leukemia virus** (BLV) is lifelong but, unlike infection with feline leukemia virus, there is no viremia after infection is established, the provirus persisting in the genome of B lymphocytes and macrophages. Chronic infection is characterized by the production of an ever-increasing array of antibodies against the various viral antigens, suggesting that the infection is not stable but continues to evolve with the development of lymphocytosis. An increased cull-rate has been shown for BLV-infected dairy cattle, which suggests that, whereas the cattle are outwardly normal, the anti-BLV immune response is at some cost to the animal.

Bibliography

Cheville, N. F., and Mengeling, W. L. The pathogenesis of chronic hog cholera (swine fever). Histologic, immunofluorescent, and electron-microscopic studies. *Lab Invest* **20:** 261–274, 1969.

Cohen, N. D. *et al.* Epizootiologic association between feline immunodeficiency virus infection and feline leukemia virus seropositivity. *J Am Vet Med Assoc* **197:** 220–225, 1990.

Gonda, M. A. *et al.* Development of the bovine immunodeficiency-like virus as a model of lentivirus disease. *Dev Biol Stand* **72:** 97–110, 1990.

Hoover, E. A. *et al.* Experimental transmission and pathogenesis of immunodeficiency syndrome in cats. *Blood* **70:** 1880–1892, 1987.

Hopper, C. D. *et al.* Clinical and laboratory findings in cats infected with feline immunodeficiency virus. *Vet Rec* **125:** 341–346, 1989.

Kimsey, P. B. *et al.* Studies on the pathogenesis of epizootic bovine abortion. *Am J Vet Res* **44:** 1266–1271, 1983.

Novotney, C. *et al.* Lymphocyte population changes in cats naturally infected with feline immunodeficiency virus. *AIDS* **4:** 1213–1218, 1990.

Siebelink, K. H. J. *et al.* Feline immunodeficiency virus (FIV) infection in the cat as a model for HIV infection in man: FIV-induced impairment of immune function. *AIDS Res Hum Retroviruses* **6:** 1373–1378, 1990.

Thurmond, M. C., Maden, C. B., and Carter, R. L. Cull rates of dairy cattle with antibodies to bovine leukemia virus. *Cancer Res* **45:** 1987–1989, 1985.

Van der Maaten, M. J. *et al.* Experimentally-induced infections with bovine immunodeficiency-like virus, a bovine lentivirus. *Dev Biol Stand* **72:** 91–95, 1990.

Whetstone, C. A., Van Der Maaten, M. J., and Black, J. W. Humoral immune response to the bovine immunodeficiency-like virus in experimentally and naturally infected cattle. *J Virol* **64:** 3557–3561, 1990.

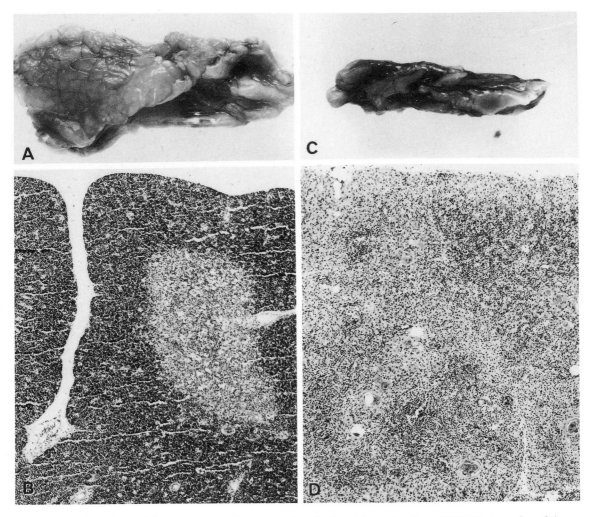

Fig. 2.77 (A,B) Normal thymus, gross and microscopic, of kitten at 8 weeks of age. (C,D) Degeneration of thymus in panleukopenia in kitten at 8 weeks of age.

D. Degenerative Diseases of the Thymus

Thymic involution is part of the normal process of aging and may be slowed by dietary restriction or accelerated by toxic insult. It appears that a high caloric intake promotes early development of the endocrine and immune systems, and that the resulting high level of stimulation to end-organ systems results in an excess of tumor formation in secretory and reproductive organs, and early exhaustion of the immune system. This process may be augmented if a high proportion of the caloric intake is made up of animal fat. When a restricted diet of approximately two thirds of free choice intake is given to normal laboratory rats and mice, their life span is extended, and the level of spontaneous tumors is reduced. In addition, when animals with an inherited tendency for autoimmune disease are subjected to dietary restriction, a tripling of expected life span occurs. Thus unrestricted diet appears to increase the incidence of allergic and autoimmune phenomenon in the young and to contribute to early senescence. It is suggested that these observations are explained by a slowing of the age-associated decline in DNA repair. The improved immune function in old animals maintained on restricted diets may have a similar basis. The recognition that a number of environmental and dietary contaminants may accelerate immune deterioration has resulted in the development of testing protocols specifically designed to assess adverse effects of this nature.

In a general sense, thymic involution can be used as an index of disease severity and duration in young animals in which the organ would normally be much larger.

Bibliography

Licastro, F. *et al.* Effect of dietary restriction upon the age-associated decline of lymphocyte DNA repair activity in mice. *Age* **11:** 48–52, 1988.

Pang, V. F. *et al.* Experimental T-2 toxicosis in swine following inhalation exposure: Effects on pulmonary and systemic immunity, and morphologic changes. *Toxicol Pathol* **15:** 308–319, 1987.

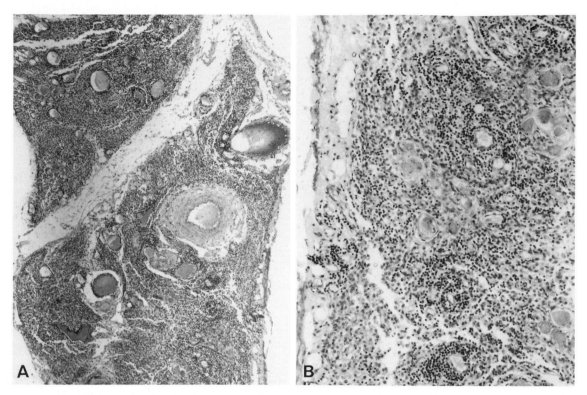

Fig. 2.78 Thymic atrophy. Feline panleukopenia. (A) Lobules reduced in number and size. (B) Proximity of Hassall's corpuscles to capsule indicates degree of cortical atrophy.

van Baarlen, J., Schuurman, H.-J., and Huber, J. Acute thymus involution in infancy and childhood: A reliable marker for duration of acute illness. *Hum Pathol* **19:** 1155–1160, 1988.

Yunis, E. J., and Greenberg, L. J. Immunopathology of aging. *Fed Proc* **33:** 2017–2019, 1974.

E. Hyperplastic and Neoplastic Diseases of the Thymus

1. Thymic Hyperplasia

Hyperplasia of the thymus occurs as an increase in volume of a histologically normal gland in the young, as the development of germinal centers in a gland of small, normal, or increased size, or as a focal area of cortical hyperplasia in an older animal. **Diffuse hyperplasia** may occur in any species but is most likely to be recognized in calves, rabbits, and birds that have been repeatedly immunized. Grossly the glands may fill the anterior mediastinum and extend up the neck in calves. The capsules remain thin and delicate, and the lobulation is distinct. **Follicular hyperplasia** is the result of fully typical B cell germinal centers forming, usually at the corticomedullary junction, in glands that may have a variety of other changes including reactive hyperplasia and atrophy. Follicular hyperplasia is not infrequently found in thymoma, suggesting some period of immune dysfunction preceding development of tumor. In humans, follicular hyperplasia is strongly associated with autoimmune disease, particularly myasthenia gravis, and 85% of patients with thymic lesions

have this change. Follicular development occurs in dogs with immune hemolytic anemia and systemic lupus erythematosus, but the frequency is not documented. Focal hyperplasia is an unusual and incidental finding in older animals, and is most often seen in rats and mice on lifetime studies where a strict protocol requires that the thymus is routinely examined histologically. **Focal hyperplasia** is an appropriate term for lesions characterized by proliferation of the cortical lymphocytes in one or more lobules of the gland, usually with compression of surrounding, often atrophic, lobules suggesting an attempt at regeneration as is seen, for example, in the exocrine pancreas. The lesions are usually lightly encapsulated, often with some laminar infiltration of the perithymic tissues, and significantly with single epithelial cells scattered through a relatively diffuse area of lymphoid proliferation. These masses do not produce clinical signs and are important primarily in their resemblance to early lymphoma.

2. Thymic Neoplasia

Thymoma is a tumor derived from the epithelial components of the thymus that is invested to a greater or lesser degree with benign lymphoid proliferation. It is rare for thymomas to be purely epithelial, and the association of lymphocytes with epithelium, even in metastatic sites, suggests that the latter retains some lymphoid inductive capacity, notwithstanding the malignant state. Thymomas tend to be slow-growing, heavily encapsulated tumors which rarely metastasize. They may be localized to one lobe, or more commonly a remnant of benign thymus may be found compressed to the periphery of the thymic capsule. Thymomas are classified on the basis of cellular composition and the type and degree of atypia of the epithelium, and thus are predominantly lymphocytic, mixed, or predominantly epithelial. The major diagnostic difficulty involves the recognition of epithelium in the lymphocytic type which distinguishes it from the much more common mediastinal lymphoma. The epithelium may be very variable, but it is usually spindle shaped in storiform or rosette pattern, vesicular in diffuse sheets, or squamous. The latter type is most likely to appear cytologically malignant and to metastasize.

In general, lymphoid tumors of the anterior mediastinum in the cat and bovine yearling are aggressive lymphomas, whereas in the goat, sheep, and occasionally in the dog, they are encapsulated masses that may cause clinical signs due to their size but rarely metastasize. In humans and likely in animals, there is rarely neoplasia of the myoid cells to form a rhabdomyosarcomatous mass which is malignant and metastatic.

a. THYMIC LYMPHOMA Thymic lymphomas occur in cats, young cattle, and dogs. In the **cat,** mediastinal lymphomas constituted 22% of 547 cases of lymphoma. Characteristically, the tumors reach a large size with dorsocaudal displacement of the lungs before they are presented for clinical examination. Thymic lymphomas are poorly encapsulated and tend to be traversed by irregular, fine, collagenous septa. It is possible that some of them arise in the anterior mediastinal lymph nodes, but at the time of diagnosis the thymus is obliterated and cannot be dissociated from the tumor. Histologically, two-thirds of feline thymic lymphomas are high-grade tumors of small noncleaved cells and immunoblastic type with a high mitotic rate. They often have a starry-sky appearance due to the presence of many tingible body macrophages associated with a high rate of apoptosis of the tumor cells. There is a wide age distribution of feline thymic lymphoma with 90% occurring relatively evenly between the ages of 1 and 10 years. Fewer than 8% are of the lymphoblastic type, and true small-cell lymphomas of well-differentiated type are even less common. Involvement of the lung, even by direct extension, is unusual, although there may be infiltration into the atria and through the pericardium into the base of the heart and around the origins of the great vessels. Infiltration beneath the sternal pleura is consistently present. Metastases may be widespread, but they are rare and small compared to the mediastinal tumor, suggesting that these lesions spread by local expansion until late in the disease.

Thymic lymphoma in **cattle** characteristically occurs in yearlings of the beef breeds, is not associated with infection with bovine leukemia virus, and does not share tumor-associated antigens with the bovine leukemia virus-induced tumors. Typically, there is swelling, sometimes massive, at the base of the neck and brisket edema. Bloat-

ing and dysphagia associated with esophageal compression are common. The animals are seldom leukemic, although in advanced cases there are usually neoplastic lymphocytes in the peripheral blood without changes in absolute numbers of cells, and tumor cells may be identified in sternal marrow aspirates. The tumors may extend from the rami of the mandibles to the base of the heart with a constriction at the entrance to the thoracic cavity, and may weigh 20 kg or more. They are lightly encapsulated and infiltrate the surrounding tissues, often with smooth, rounded, shiny implants appearing on the pulmonary and parietal pleura. There is compression of the lungs by tumor and pleural fluid. The lungs themselves may be focally invaded but are rarely extensively involved. Invasion of the pericardial sac produces pericardial effusion which contains neoplastic lymphocytes in large numbers. On cut surface, the tumors are gray with irregular yellow areas of infarction which have hyperemic borders, and they tend to have a rather coarse collagenous septation. Histologically thymic lymphomas in cattle have diffuse architecture with a relatively minor stromal component, and most of the septation is limited to rather heavy hyalinized bands which may be several centimeters apart. Cytologically they are high grade, largely of small noncleaved type, and with a disproportionately high number of lymphoblastic types, suggesting T cell derivation.

The thymic lymphomas of other species, including **dogs** and laboratory animals, do not differ markedly from those in the cat, but there is not usually extrathoracic tumor development.

b. LYMPHOID THYMOMA Lymphoid thymomas are seen most frequently in the dog, sheep, and goat, and rarely in the horse. In **dogs and horses,** the tumors are usually clinically apparent with the animals presented for examination because of signs of respiratory and/or cardiovascular impairment. In **goats and sheep,** the lesions are usually found incidentally at autopsy. Lymphoid thymomas are irregularly shaped masses from 5 to 15 cm in diameter, with irregular rounded projections and a firm fibrous encapsulation. Grossly the tumors appear as an exaggeration of normal thymic structure with heavy and irregular fibrous septation. Microscopic cysts are frequently observed, filled with pink proteinaceous material that resembles thyroid colloid. These cysts are branchial duct remnants and have a lining which varies from squamous to columnar and ciliated, and may change abruptly from one form to the other. The corticomedullary distinction is lost, and areas may more closely resemble cortex or medulla depending on the proportion of more pale-staining epithelial cells as compared to lymphocytes; however, the latter always greatly predominate. Hassall's corpuscles are present and irregularly distributed throughout a background of benign but densely packed masses of lymphocytes of medium size. The predominant lymphocytes have nuclei only slightly larger than those of normal cortical thymocytes and are interspersed with large lymphocytes with coarse retiform chromatin and prominent central

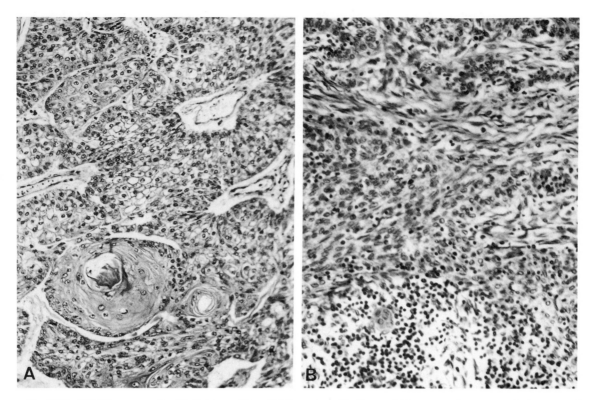

Fig. 2.79   (A) Thymoma of epithelial type. Dog. (B) Lymphoepithelioma of thymus to show stromal character of epithelial component. Ox.

nucleoli. There is a **characteristic cuffing of small veins by reticular epithelial cells,** creating a clear space resembling a lymphatic, or the epithelium may invest the vessel with laminated stellate cells with plump vesicular nuclei in an onion-ring fashion. In dogs and cats, lymphoid thymomas may contain prominent germinal centers, and plasma cells may be present in their vicinity and along the fibrous septa. In these species, follicular activity in thymomas has a high association with autoimmune disease, particularly myasthenia gravis and less often polymyositis. In contrast, germinal centers are not found in thymomas of goats.

c. LYMPHOEPITHELIAL THYMOMA   Tumors of this type are seen most frequently in the goat and sheep as space-occupying lesions. They have firm encapsulation with regular but coarse septation, but with minimal stroma in the pseudolobular areas. The proportion of lymphocytes and spindle cells varies from one area of the tumor to another, with focal predominance of either element. Cystic spaces are frequently present and there may be large areas of hemorrhagic necrosis. Hassall's corpuscles are present singly and in cornified clusters but tend to be less regularly present than in the lymphoid thymomas. Cytologically the lymphocytes are typical of the larger cells of the outer cortex of normal thymus, with compact chromatin and inapparent nuclei. The epithelial cells are characteristically stellate or fusiform (Fig. 2.79A,B), with oval nuclei with sharply peripheralized chromatin and small central

nucleoli. Myoid cells may be absent or occur singly in association with lymphocytes rather than with epithelial areas and may contain striated fibers. The perivascular epithelial cells are more variable than in the lymphoid thymomas and may be present as focal clusters adjacent to vessels, arranged in rosettes rather than as perivascular cuffs. Germinal centers are not usually found.

Rarely **primary germ cell tumors** outwardly resembling seminomas occur in the thymus gland of male humans and animals in the absence of primary tumor in the testicle. Some of these have possibly been described as clear-cell lymphomas if phenotypic identification was not carried out.

Bibliography

Aronsohn, M. G. *et al.* Clinical and pathologic features of thymoma in 15 dogs. *J Am Vet Med Assoc* **184:** 1355–1362, 1984.

Carpenter, J. L., and Holzworth, J. Thymoma in 11 cats. *J Am Vet Med Assoc* **181:** 248–251, 1982.

Hadlow, W. J. High prevalence of thymoma in the dairy goat. Report of seventeen cases. *Vet Pathol* **15:** 153–169, 1978.

Katzin, W. E. *et al.* Immunoglobulin and T-cell receptor genes in thymomas: Genotypic evidence supporting the nonneoplastic nature of the lymphocytic component. *Hum Pathol* **19:** 323–328, 1988.

Marino, M., and Müller-Hermelink, H. K. Thymoma and thymic carcinoma. Relation of thymoma epithelial cells to the cortical and medullary differentiation of thymus. *Virchows Arch (Pathol Anat)* **407:** 119–149, 1985.

Mettler, F., and Hauser, B. Clear cell thymoma in a dog. *J Comp Pathol* **94**: 315–317, 1984.

Nichols, C. R. *et al.* Hematologic malignancies associated with primary mediastinal germ cell tumors. *Ann Intern Med* **102**: 603–609, 1985.

II. Lymph Nodes

A. *Structure and Function of Normal Lymph Nodes*

In the dog, the primordial structures of the largest nodes are present at 35–38 days of gestation, following the formation of the thymus and spleen at 27–28 days. Lymphocytic colonization of the nodes and spleen is prominent at 52–53 days, and of the thymus at 62–65 days. The lymph node anlage appears to arise from periarteriolar mesenchyme. The lymphatic vessels develop prior to the nodes. At the confluence of lymphatics and arterioles, the lymphatics form cuffs enclosing the arterioles and outpouchings which form the capsule of the primordial lymph node, the hilar area being directed toward the arteriole of origin. As the tissue develops, there is a proliferation of reticular cells within the lymphatic enclosure that becomes eccentric to the hilar arteriole, entering the capsule in parallel with the draining efferent lymphatics. The periarteriolar reticular cells form a network which provides an appropriate environment for colonization by thymic and bone marrow-derived lymphocytes.

The lymph nodes, like the spleen, are variably involved in myelopoiesis in the fetal stage of development, and convert to lymphopoiesis shortly after birth. The actual level of lymph node development at birth varies widely with different species. Calves and foals have well-developed lymph nodes at birth and may have germinal center formation if there has been intrauterine infection. In contrast, nestlings at birth have readily recognizable lymph nodes in the neck, mediastinal, and mesenteric areas; the nodes have a loose reticular structure, very low lymphocyte density, and only limited organization into cortex and medulla.

The microcirculation of the mammalian lymph node is closely related to its function, and governs the architectural development of the node. The **afferent lymphatics** enter the capsule and bathe the cortex via the peripheral sinus. Small lymphatic ducts perforate the inner capsule of the peripheral sinus and form spherical microanastomotic networks which correspond in size to the germinal centers. On the medullary aspects of these spherical networks, the lymphatics anastomose into larger ducts which drain into the medullary sinuses and then into the major hilar **efferent lymphatics.** In contrast, the blood vessels both enter (arteriole) and leave (vein) through the hilar region with small vessel arborization in the cortical area. In the outer cortex, the small arteries form meta-arterioles or precapillary arterioles, which perfuse the germinal centers and mantle zone areas. The transition to capillaries occurs at this site, and these vessels then enlarge to form the **postcapillary** or **high endothelial venules** of the paracortex, across which there

is heavy traffic of lymphocytes from the blood to the lymph node parenchyma. These vessels then unite to form progressively larger veins which follow the medullary stroma to form the efferent hilar vein.

The various vascular structures of lymph nodes form characteristic associations with nodal tissue. The lymphatics are identifiable as major ducts entering the peripheral sinus, but the intranodal lymphatics are not obvious, except for the large medullary sinuses and the efferent hilar vessels. The arteries can be identified entering the medullary or hilar area of the node and they follow the fibrous trabeculae to the medullary sinuses. In this area, the arterioles form a regular and obvious network with the medullary cords, such that the arterioles are centrally placed in the sinuses and connected by a network of reticular stroma to the thin limiting membranes of the medullary cords. The smaller arterioles of the paracortex are irregularly visible, as are the terminal arteriolar vessels of the germinal centers and the mantle zone. In the adjacent paracortex, the postcapillary venules are readily identified. These vascular structures of the blood and lymphatic systems through their interactions form **three endothelium-lined compartments, intravascular, intralymphatic,** and **interstitial,** which provide the functional architecture by which the lymph node filters lymph, processes antigen, and returns lymphocytes from blood to tissue.

Functionally, this microcirculatory arrangement directs antigens entering the cortical area in lymphatics to the germinal centers and the outer cortex. The **germinal center** has a superficial pole composed of a mantle of small lymphocytes, a central area of larger lymphocytes of more variable character, and a deep pole also composed of a mantle of small lymphocytes. The processing of antigen occurs at random in lymph node cortex but is always associated with a background of dendritic reticular cells whose origin is unknown but probably is bone marrow. Germinal centers which lie deep within the medulla in cases of follicular hyperplasia (Fig. 2.80A,B) do not have an obvious orientation to the capsule, but it can be assumed that the deep and superficial poles are oriented to the source of antigen and the fine network of lymphatics which drain the peripheral sinus. The paracortical tissue may appear as a solid band encompassing the germinal centers, as is typical of the young. With maturity, the paracortical areas irregularly recede and regenerate in loosely defined, but densely aggregated, **deep cortical units.** These units are also defined by cellular composition and vascular supply as are the germinal centers, but, in contrast to the latter, the deep cortical units are thymus-dependent areas populated by T cells. There are **gaps** in the inner layer of the outer sinus by which the entering lymph can pass directly to the medullary sinuses. Presumably this route is taken by heavy but nonthreatening flow, as occurs when frank blood enters the lymph node during biopsy (red cells appear in the medullary sinuses within minutes of the initiating surgical trauma).

Lymphocytes, dendritic reticular cells, and macrophages cooperate in the immune responses. These cells

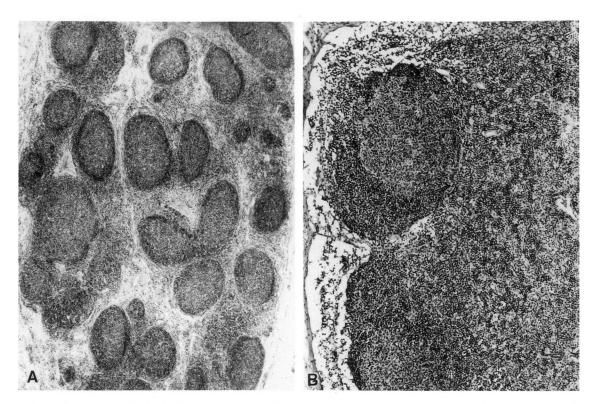

Fig. 2.80 Lymph node. (A) Follicular hyperplasia. Cat. (B) Paracortical hyperplasia with typical starry-sky effect in stimulated node. Dog.

vary in their regional distribution and in their kinetics. The **dendritic reticular cells** occur within the germinal centers and within the deep cortical units or paracortical nodules. Their origin is unknown, but derivation from marrow monocytes has been proposed. The dendritic cells of the germinal centers are long-lived and may retain antigen on their surface membranes for weeks or months in the process of antigen focusing, by which the specificity of a germinal center for a specific antigen is maintained. The dendritic cells of the paracortical nodules are also long-lived and, in superficial nodes of humans, they often contain Birbeck granules characteristic of the Langerhans cells of the skin, suggesting an origin and previous experience in that area. There is functional logic in this arrangement since the follicular macrophages and dendritic cells (B area) are efficient in the digestion and processing of large particulate antigens, whereas those in the paracortex (T area) are more efficient in trapping low-molecular-weight substances such as agents producing contact sensitization in the skin, where the resident reticular cells likely had their initial training. In general, the lymphocytes in the germinal centers are largely B cells, whereas those in the paracortex are largely T cells. The subsets of T cells are also segregated, but not as cleanly, with the suppressor and NK cells confined almost entirely to the paracortex, whereas the helper cells are largely in the paracortex but also in the germinal centers and their small-cell corona.

The process by which lymphocytes exiting the marrow

and thymus are directed to specific tissues is regulated by **homing receptors** on the lymphocytes, which bind to tissue-specific **addressins** on high endothelial or postcapillary venules in the peripheral lymphoid tissues.

The general reactions of nodes to various stimuli are described in the introduction to the lymphomas (see Histologic Interpretation of Lymph Nodes, Section VI,B1 of The Leukon), and that section is pertinent to the following consideration of non-neoplastic disease involving this tissue.

Bibliography

Bryant, B. J., and Shifrine, M. Histogenesis of lymph nodes during development of the dog. *J Reticuloendothelial Soc* **12:** 96–107, 1972.

Fossum, S., and Ford, W. L. The organization of cell populations within lymph nodes: Their origin, life history and functional relationships. *Histopathology* **9:** 469–499, 1985.

Grundmann, E., and Vollmer, E. (eds.). Reaction patterns of the lymph node. Part 1. Cell types and functions. Part 2. Reactions associated with neoplasia and immune-deficient states. "Current Topics in Pathology." Berlin, Springer-Verlag, 1990 and 1991.

Jalkanen, S. (1990). Lymphocyte homing into the gut. *Springer Semin Immunopathol* **12:** 153–164, 1990.

Sainte-Marie, G., Peng, F.-S., and Bélisle, C. Overall architecture and pattern of lymph flow in the rat lymph node. *Am J Anat* **164:** 275–309, 1982.

Spalding, H., and Heath, T. Pathways of lymph flow through

superficial inguinal lymph nodes in the pig. *Anat Rec* **217**: 188–195, 1987.

B. Developmental Diseases of Lymph Nodes

The lesions associated with inherited diseases of the immune system are included in Developmental Diseases of the Thymus (Section I,A of Lymphoreticular Tissues). In severe combined immunodeficiency, the lymph nodes lack both germinal centers and paracortical lymphoid colonization. In severe T cell deficiency, the germinal centers are present, as is antibody, but the paracortical areas are hypoplastic and the spleen lacks periarteriolar lymphoid sheaths. In severe B cell deficiency, as occurs in agammaglobulinemic foals, germinal centers are not formed in node or spleen and plasma cells are not found. The paracortical areas have normal cellularity and cell-mediated immunity is intact. A strain of Black Pied Danish cattle, and probably some other European breeds, carries an autosomal recessive trait for **thymic hypoplasia.** The calves appear normal at birth but by 1–2 months of age, they die of disseminated infection which originates in the skin. These calves are able to produce antibody but lack cell-mediated immune responses. The defect appears to be an inability to absorb zinc and, if this element is provided, they recover full immunologic capability and cutaneous signs of zinc deficiency disappear. Thymic atrophy develops in an inbred strain of Weimaraner dogs which are able to produce antibody but in which cell-mediated responses are depressed.

Bibliography

Brummerstedt, E. *et al.* The effect of zinc on calves with hereditary thymus hypoplasia (lethal tract A46). *Acta Pathol Microbiol Scand* **79**: 686–687, 1971.

Roth, J. A. *et al.* Thymic abnormalities and growth hormone deficiency in dogs. *Am J Vet Res* **41**: 1256–1262, 1980.

C. Degenerative Diseases of Lymph Nodes

Cachectic and **senile atrophy** lead to a moderate reduction in the overall size of body nodes. Senile atrophy is seldom of note in large domestic animals and is seen with frequency only in dogs, cats, and primates. Atrophy due to cachexia is, on the other hand, often encountered in old sheep and goats with dental attrition. Grossly, senile lymph nodes are small with edematous fascia, and characteristically have dark brown pigmentation of medullary areas that may extend in decreasing concentration to the subcapsular sinus. The aging changes are largely limited to a mild thickening of the capsule and medullary trabeculae, with a reduction in density of germinal centers. Nodal atrophy associated with cachexia is much more often encountered, and is characterized by a retention of the overall architecture with a marked reduction in cell density. Characteristically, the germinal centers are small with a hypocellular mantle, and paracortical areas are sparsely populated. In old animals with cachexia, there are numerous pigment-bearing macrophages in medullary sinuses and a thin proteinaceous fluid may be present in the medullary cords and sinuses.

Hemorrhages originating within a node or in tissues drained by the node may be visible in the cortical or medullary sinuses. Those red cells delivered to the node through the afferent lymphatics tend to pass through the cortex to the medullary sinuses where they are ingested by macrophages. The breakdown of erythrocytes within lymph nodes results in accumulation of hemosiderin, which is always most prominent in medullary areas. In cattle and sheep which die with acute respiratory distress, the cervical nodes are frequently red-black owing to the blood contained in them, much of which comes from the tracheal mucosa where extravasations begin in the lymphoid nodules.

Anthracosis is a regular finding in the bronchial nodes of dogs which live in industrial areas and is of lesser frequency and degree in cats. The carbon is retained solely in the phagocytes, principally those of the medullary cords, and therefore only the medulla is black. The pigment is inert and of no consequence. An unidentified black pigment, in appearance like melanin, is present in the hepatic nodes of sheep and cattle when they have, or have had, hepatic distomiasis. The same pigment occurs in bile ducts infested by *Fasciola hepatica,* and, in the case of *F. magna,* wherever the fluke has wandered.

In **emphysema** of lymph nodes, the gas is confined to the sinuses. It affects the mesenteric nodes of swine in association with intestinal emphysema. Emphysema of the bronchial nodes commonly accompanies interstitial pulmonary emphysema of cattle. The lymph nodes are puffy and light. The sinuses are distended and their endothelium is lined by large macrophages and even giant cells, the mobilized cells occurring in small clusters of spotty distribution on the walls of the sinuses.

Lymphoid atrophy, with or without architectural alteration, occurs in a variety of circumstances. Lymphoid atrophy associated with marked dilation of medullary sinuses, termed **vascular sinus transformation,** results from blockage of the efferent lymphatics, and is exacerbated by the development of fibrosis if the venous drainage is also obstructed. These conditions may follow surgical intervention for lymphatic cannulation or occasionally develop in animals that have been recumbent for prolonged periods and suffered concurrent venous infarction of muscle. Complete or focal infarction of node occurs with lymphoma and may be the first sign of disease. Infarction is most likely to occur with large-cell lymphomas and, although vascular obstruction appears to be the cause, thrombosis is seldom observed. Infarction with lymphoma is most often seen in the calf form of sporadic bovine lymphoma.

D. Inflammatory Diseases of Lymph Nodes

Lymphadenitis results when an infectious agent is present in the lymph node, as distinguished from **benign**

lymphoid hyperplasia, in which the node is immunologically reactive but free of local invasion. Lymphadenitis may result from drainage to the node of the products of a distant inflammatory process, which then progresses to involve the node directly. It may be possible to identify the infectious agent responsible for the lymphadenitis or to determine the type of invading agent from examination of the changes in the node (Fig. 2.81A,B). Lymphadenitis may be acute or chronic, suppurative, caseous, or granulomatous. These more or less specific types of lymphadenitis are discussed under specific diseases.

Grossly in **acute lymphadenitis,** the lymph nodes are enlarged, soft, locally mobile, and hyperemic to a variable degree. The capsule is taut and thinned due to the influx of cells, and on the cut surface the parenchyma bulges and exudes blood and lymph, but mainly the latter. Thus, if imprint preparations are made, it is necessary to repeatedly remove the exuding lymph from the cut surface in order to obtain enough adhesion for cellular exfoliation. There may be focal discoloration due to previous pigmentation or to infarction.

Histologically, acute lymphadenitis is characterized by marked hyperemia with unusual prominence of small vessels in cortical areas. Neutrophils are often present, both through drainage from the subcapsular sinus and by migration from the postcapillary venules. Some useful distinction can be made between granulocytes arriving from these two sources. In general, neutrophils in the peripheral and medullary sinuses indicate inflammation in the tissue area drained and are not necessarily indicative of local sepsis. On the other hand, foci of neutrophils in cortical areas are of hematogenous origin and almost always are reactions to bacterial colonization, as is likely to occur in the mesenteric node in bacterial enteritis. When the inflammation is due to one of the pyogenic organisms, abscessation is likely. Necrosis in lymph nodes is an attribute of some acute infections such as salmonellosis and toxoplasmosis (Fig. 2.82).

In **chronic lymphadenitis,** hyperemia and edema are irregularly present, and the infected nodes are large and firm and may be fixed to local tissues if there has been cellulitis. The capsule is thickened as are the internal trabeculae, and with prolonged inflammation, the node becomes dry and hard. Changes of this type are characteristically present in the supramammary lymph nodes of cows with brucellosis of long standing and to a lesser extent in animals with chronic or recurrent bacterial mastitis. In the latter cases, the marked proliferation of the collagenous septa of the medulla may completely displace the medullary cords.

Lymphadenitis may be expressed by changes which are largely degenerative and affect the architecture of the tissue. Thus in some acute viral infections, there is rapid lysis of lymphocytes, as in canine **parvovirus** infection (Fig. 2.83) and **rinderpest. In salmon poisoning** of dogs, the nodes may be much enlarged with follicular hyperpla-

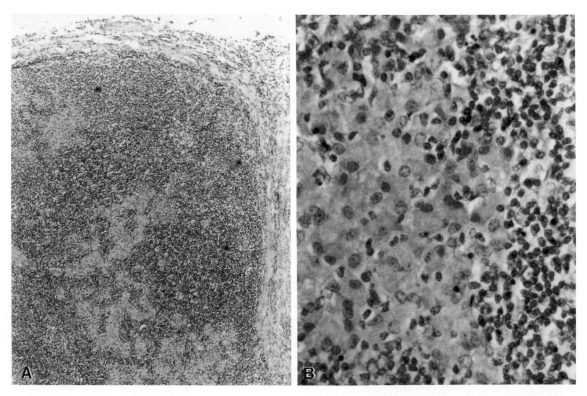

Fig. 2.81 Lymph node. Johne's disease. Goat. (A) Capsule thickened and colonized by inflammatory cells, peripheral sinus narrowed, paracortical hyperplasia. (B) Epithelioid macrophages of sinuses of (A).

Fig. 2.82 Lymph node. Acute lymphadenitis with subcapsular necrosis. Toxoplasmosis. Cat.

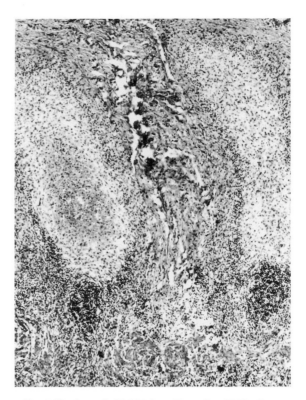

Fig. 2.83 Lymphoid follicles of intestine. Follicular necrosis in parvovirus enteritis. Dog.

sia and accumulation of macrophages (Fig. 2.84A). In some chronic infections, particularly with the viruses causing immunodeficiency, the nodes are atrophic with wrinkled capsule and depletion of lymphocytes, and sinus histiocytosis if the production of monocytes by the bone marrow remains competent (Fig. 2.84B).

E. Parasitic Infestations of Lymph Nodes

The lymph nodes are not a final habitat for any nematode, but some species traverse them as a matter of course and some lodge there accidentally. The larvae of lungworms travel by lymphatics from the intestine to the lungs but cause little injury in the course of these migrations. The first-stage larvae of those lungworms which reside in the alveolar spaces, *Muellerius* spp. and *Protostrongylus* spp. particularly, often find their way to the bronchial nodes and there produce a granulomatous lymphadenitis. The nodes become enlarged, hard, and irregular and have many firm granulomas and encapsulated caseocalcareous nodules. The larvae of *Strongylus* spp. in horses occasionally produce a hemorrhagic lymphadenitis of the abdominal lymph nodes which, when secondarily infected, may suppurate. Larval trematodes, probably *F. hepatica,* are often found in the mesenteric nodes of ruminants, as are larvae of *Oesophagostomum columbianum* in sheep. Nodules develop in the cortex or medulla, as well as in capsular tissues. They are 2–5 mm in size and contain yellowish-green pus, which in older lesions becomes caseous and calcified. The pus is largely the debris of eosinophils and it is surrounded by giant and epithelioid cells and a well-defined capsule. A very similar lesion is produced by the larvae of *Linguatula serrata.* The larvae of this parasite (see Diseases of the Nasal Cavity, Volume 2, Chapter 6) are, in some countries, very common in the mesenteric nodes of sheep and cattle. Viable larvae can be found in normal nodes, but usually the nodes show focal or diffuse hyperplasia and edema, with the larvae lying in cavities which contain a milky fluid resembling chyle. Apparently, the larvae die in the nodes and become encysted in abscesses, which may calcify. In active cutaneous infestations by *Demodex* spp. in dogs, the parasite is often found in the peripheral lymph nodes.

F. Benign Hyperplasia of Lymph Nodes

Lymphadenopathy is defined as a regional or generalized lymph node enlargement of unknown or unspecified cause. **Lymphadenosis,** while not commonly used, more correctly defines generalized lymphoid hypertrophy which the former term is used to indicate. Local enlargement usually reflects a pathologic process limited to the drainage area, particularly inflammatory or neoplastic disease. Generalized enlargement of lymph nodes is a more serious finding, since the possibility of primary tumor is much more likely. Infectious causes of generalized lymph node

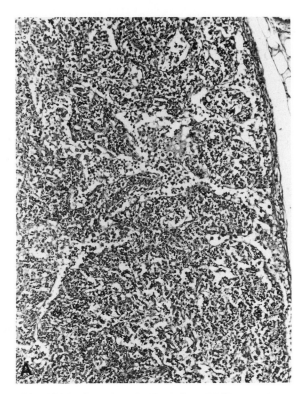

Fig. 2.84A Lymph node. Depletion of follicles and paracortex and proliferation of sinus histiocytes. Salmon poisoning. Dog.

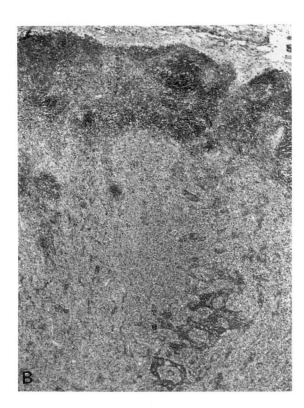

Fig. 2.84B Medullary sinus hyperplasia in bovine virus diarrhea. Ox.

enlargement include tuberculosis, brucellosis, and protozoal infections. A generalized lymphadenopathy produced by an extraordinary degree of paracortical blastogenesis is typical of the early phases of malignant catarrhal fever and theileriosis in cattle. There may be a mild generalized lymph node enlargement in adrenal atrophy, and lymphoid hyperplasia to some degree is transient in most young animals. Endocarditis or abscessation may cause generalized lymphadenopathy, as will lymphoma. In general, in acute septic inflammation the lymph nodes are normally mobile within the subcutaneous tissues, but may become fixed with chronic sepsis. Similarly, lymphadenopathy due to lymphoma usually does not cause fixation of the glands in the early stages, but there is usually perinodal colonization with loss of mobility in advanced stages.

The cytologic picture in benign lymphadenopathy is variable, with small and large lymphocytes, with cleaved and noncleaved nuclei, and macrophages, plasma cells, and neutrophils. Unless the stimulation is less than 10 days old, this mixture of cells should be organized architecturally with evidence of follicular development. In contrast, in primary tumor there tends to be a monomorphic or bimorphic cell population and, in animals in which follicular lymphomas are rare, finding a monomorphic cell population with diffuse architecture suggests lymphoma. In primary tumor, macrophages may be present, and numerous if there is a high mitotic rate in the tumor and a high rate of cell death. In contrast, plasma cells and neutrophils tend to be less numerous than in lymphadeni-

tis, although the presence of plasma cells in itself is no guarantee of a benign reaction. **The most important criteria for distinguishing lymphoma from benign hyperplasia are the integrity of normal architecture and the uniformity of cell and chromatin type.** Architecturally the peripheral sinus is preserved, even in reactive states where there is perinodal colonization by benign lymphocytes. In contrast, the outer sinus is regularly encompassed and destroyed by even low-grade lymphoma.

In terms of the chromatin pattern, **benign and reactive lymphocytes have less internal nuclear detail** than their malignant counterparts and will have a variable amount of chromatin contained within large chromocenters irregularly distributed throughout the nucleus. In inflammatory disease, the chromatin tends to have a fine gradation from light to dark areas, which includes lightly stained but not clear areas. The nucleoli characteristically have a uniform rimming of aggregated chromatin which makes them appear unusually prominent. In lymphoma, large chromocenters are few, and the chromatin pattern tends to be monotonously similar, most often with a retiform or branched pattern with heavy chromatin bands and irregular clear areas, as occurs in the small noncleaved and large-cell tumors. The chromatin aggregations on nucleoli are irregular and discontinuous. Less commonly, tumor cells have cribriform or uniform chromatin patterns without chromatin aggregation as occurs in lymphoblastic tumors. There is more variation in nucleolar size, shape, and number in malignant nuclei than in benign nuclei.

One of the most difficult distinctions is between **diffuse hyperplasia** and lymphoma, and in these cases the decision must be based on both the architectural changes and the character of the cells interpreted in terms of the history and clinical signs. In the early, marked immune response, there will be diffuse parafollicular hyperplasia with effacement of pre-existing follicles. The paracortex has a moth-eaten appearance due to the mixture of cells present which includes many large macrophages and pale dendritic reticular cells. Lymphomas of the diffuse mixed type have in some proportion small cleaved and large cells in every field, with the latter population predominating. There are **no lymphomas** with **three or more lymphocyte types regularly present,** and this fact tends to assist in correctly interpreting the early diffuse lesions. If a firm decision cannot be made, it is fully justifiable to request a repeat biopsy in 10 days, at which time a diffuse benign reaction will have developed a follicular pattern.

In **follicular hyperplasia,** even with extension of follicles to the medullary areas, there will be preservation of at least a narrow band of the paracortex between the follicles. In contrast, in follicular lymphomas the follicles tend to appear bare as a result of paracortical atrophy. Follicular lymphomas have a narrowed peripheral mantle of small lymphocytes, and the lighter and darker areas of the nodules are composed of cells which are cytologically similar but differ in density, whereas in benign follicular hyperplasia there is a marked variation in cell type across the diameter of the follicle.

Some of the more difficult distinctions between benign hyperplasia and primary malignancy of lymph nodes must be made in hyperimmune diseases. Eosinophilic myositis causes a remarkable alteration in lymph nodes with generalized sclerosis and thickening of the capsule, which is irregularly colonized by small lymphocytes and eosinophils in linear arrays, and marked thickening of the cortical and medullary trabeculae. There is diffuse follicular hyperplasia with paracortical atrophy and medullary cord hyperplasia with prominent plasma cell accumulation (Fig. 2.85A,B). The germinal centers vary markedly in size and shape and have accentuation of the large- and small-cell layers, and the sinuses contain a dense matrix of epithelioid macrophages and eosinophils. Eosinophils are also prominent throughout the collagenous trabeculae. Brucellosis and trypanosomiasis present no problem in distinction from cancer and result in a regular picture of follicular hyperplasia followed by paracortical atrophy, with generalized sclerosis and medullary cord hyperplasia.

An unusual type of follicular hyperplasia occurs in old sheep and goats that are in poor condition and may have been destroyed due to debility. There is regular capsular thickening and increased prominence of medullary trabeculae. The follicles may be very large, angular, and faceted in an irregular fashion. Paracortical atrophy is marked, and there is medullary cord hyperplasia and empty medullary sinuses. The most marked changes affect the follicular

vessels, which are large, tortuous arterioles with hyalinized media and often mineralized endothelium. The follicular centers include large epithelioid macrophages and often a very marked tingible body reaction. This reaction resembles **angiofollicular lymph node hyperplasia of the hyaline vascular type** as it is described in humans.

A nodular reaction of the paracortical areas occurs in lymph nodes draining malignant tumors as well as other conditions, and appears to represent an exaggerated thymus-dependent response. The nodules may be multiple and coexist with follicular atrophy, raising concern that they represent a multifocal diffuse lymphoma. The nodules represent deep cortical units. They are centered on networks of dendritic reticular cells and are densely cellular, but with the cytologic variation in cell type characteristic of paracortex. These paracortical nodules, unlike germinal centers, are poorly defined from the surrounding paracortex, which tends to be less densely cellular.

A distinctive peripheral node hyperplasia of young cats is illustrative of the difficulty in separating strong reactive hyperplasia and neoplasia. It is usual in these cases, which range in age of onset from about 6 months to 2 years, for all peripheral nodes to be enlarged. The nodes are firm with smooth capsules and yellow-tan to white on cut surface. The architecture is severely distorted with loss of discernible trabeculae and of subcapsular, trabecular, and medullary sinuses. Cortical lymphoid follicles may be many and large, and in some areas, they are encroached on and partially obliterated by the expanded paracortex. The paracortex comprises a mixture of lymphocytes, blast cells, histiocytes, and plasma cells with abundant mitoses and starry-sky appearance. The paracortical expansion may involve the capsule and adjacent adventitial tissue. High-endothelial venules are prominent in the paracortex.

The cause of this lymphadenopathy in cats is not known. Recovery is expected after a course of variable duration. A similar lymphadenopathy can be caused by feline leukemia virus infection, as an unusual response to that virus. Argyrophilic intracellular bacteria are described in some of the spontaneous cases of lymphadenopathy, but their nature and role remain to be determined; of interest is the morphologic similarity of these organisms to those described in humans affected by cat-scratch disease.

Bibliography

Ellis, J. A., and DeMartini, J. C. Immunomorphologic and morphometric changes in pulmonary lymph nodes of sheep with progressive pneumonia. *Vet Pathol* **22:** 32–41, 1985.

Hernandez, D. *et al.* Association of sinus histiocytosis with massive lymphadenopathy and idiopathic hypereosinophilic syndrome. *Histol Histopathol* **2:** 239–242, 1987.

Kirkpatrick, C. E. *et al.* Argyrophilic intracellular bacteria in some cats with idiopathic peripheral lymphadenopathy. *J Comp Pathol* **101:** 341–349, 1989.

Koo, C. H. *et al.* Atypical lymphoplasmacytic and immunoblastic proliferation in lymph nodes of patients with autoimmune disease (autoimmune disease-associated lymphadenopathy). *Medicine* **63:** 274–290, 1984.

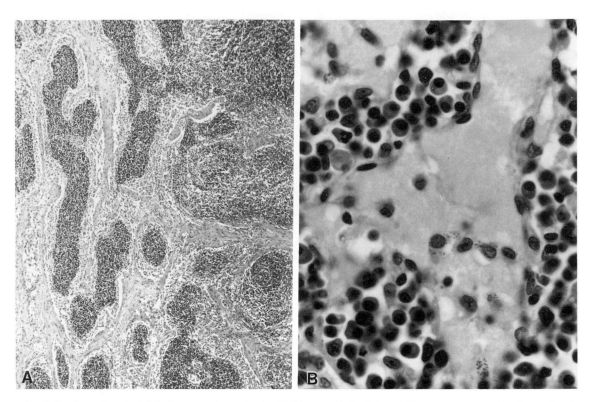

Fig. 2.85 Lymph node. Medullary cord hyperplasia. (A) Characteristic of chronic immune response. Cords are sharply defined by high density of small lymphocytes and mature plasma cells. (B) The sinus macrophages contain small granules of hemosiderin. Leishmaniasis. Dog.

Moore, F. M. *et al*. Distinctive peripheral lymph node hyperplasia of young cats. *Vet Pathol* **23**: 386–391, 1986.

Perrone, T., de Wolfe-Peeters, C., and Frizzera G. Inflammatory pseudotumor of lymph nodes. A distinctive pattern of nodal reaction. *Am J Surg Pathol* **12**: 351–361, 1988.

Suster, S., Hilsenbeck, S., and Rywlin, A. M. Reactive histiocytic hyperplasia with hemophagocytosis in hematopoietic organs: A reevaluation of the benign hemophagocytic proliferations. *Hum Pathol* **19**: 705–712, 1988.

Weisenburger, D. D. *et al*. Multicentric angiofollicular lymph node hyperplasia. *Hum Pathol* **16**: 162–172, 1985.

G. Neoplastic Metastatic Diseases of Lymph Nodes

The general reaction of lymph nodes to nonlymphoid cancer is atrophy of greater or lesser degree due to the systemic effects of cancer, apparently mediated by cachectin or tumor necrosis factor. If the tumor causes inflammatory drainage to local nodes, these will have changes of reactive hyperplasia and lymphadenitis, as occurs in mesenteric nodes draining an area of ulceration due to carcinoma. It is significant that the specific reaction of local nodes to local tumor is usually little more substantive than that of nodes distant from the primary site of malignancy.

In general the response of local nodes to adjacent cancer is **sinus histiocytosis,** and it has been felt that the strength of this reaction is as important to the prognosis in a given case as the characteristics of the tumor itself. It is therefore part of the nature of the malignant process that the nodes are not more reactive and that the malignant clone was permitted to survive and spread. Direct evidence of this process may be seen in dogs with mammary carcinoma, in which the inguinal and axillary nodes are usually buried in fat and hard to find unless enlarged due to growth of metastatic disease. Normally, these glands are thin, oval, and discoid, and only 3–4 mm in thickness, and are identified by medullary pigmentation. Histologically the capsule is lax and undulating, and the peripheral sinus, wide and poorly cellular. There is little follicular activity, paracortical atrophy with a few small deep cortical units which do not alter the depth of the gland, and relatively empty medullary cords and sinuses. If metastases are present, they will consist of single cells and small clusters in the peripheral and medullary sinuses. The tumor cells appear to pass through the cortex to the medulla through the cortical gaps, and largely avoid processing through the cortical system of antigen surveillance. The sinus macrophages do not appear to be specifically attracted to the tumor cells, even when they invade the cortical parenchyma. Indirect evidence for the mechanism of this restrained lymphoid response has been found in that nodes draining an area containing melanoma (one of the more highly antigenic tumors) have an increased level of suppressor cell activity, which presumably damps down the local defensive responses.

There are minor variations in this reaction with specific tumor systems. Highly secretory adenocarcinomas may have more reaction to their secretion than to the tumor cells themselves. Metastatic mast cells cause the same reaction in nodes as in primary sites but tend not to incite a strong cellular and stromal reaction even when highly granulated. As noted earlier, the medullary cords of lymph nodes form a tissue–vascular boundary similar to marrow, and extramedullary hematopoiesis will colonize these sites.

Similarly, in myeloid leukemias the tumor cells colonize the medullary cords and, as in marrow, the benign cells are displaced, in this case to the adjacent medullary sinuses. In the acute myeloid leukemias, the primitive tumor cells have round nuclei and resemble the surrounding residual lymphocytes, making recognition of the metastatic disease difficult with oversight stains. Diagnosis is aided by the identification of megakaryocytes in either the cords or sinuses since, if conditions in the marrow are conducive to extramedullary colonization, it is likely that both malignant and benign stem cells are displaced and that the benign system will have trilineage progeny.

The most difficult metastatic tumor to recognize in a lymph node is a tumor of lymphoid origin. *In situ* lymphoma is rarely observed, but appears occasionally in nodes collected from abattoirs, suggesting that the animal affected was culled for poor performance even though the tumor was undetected. These cases provide a unique opportunity to observe the effects of lymphocyte homing in practical application. Lymphoid tumors represent clones of cells "frozen" in some state of maturation, either primitive for the high-grade tumors, or relatively mature for the low grade lymphomas. The **lymphocyte homing receptors** vary in their expression directly with the degree of lymphocyte maturity. There is, however, marked heterogeneity of homing receptor expression within a given histologic category of lymphoma. Thus those large-cell lymphomas which have well-developed homing receptors tend to be widespread at diagnosis because the tumor cells are able to bind to **addressins** on the **high endothelial venules** of lymphoid tissue throughout the body and subsequently invade locally. In contrast, animals with large-cell lymphomas possessing poorly developed homing receptors tend to be presented with localized disease, since although the tumor cells may be in circulation, they are not able to bind to endothelium in a site conducive to colonization. As would be expected, these interactions are complex and, while some primitive lymphomas such as those in the mediastinum may become very large locally without widespread dissemination, others such as the follicular lymphomas (Fig. 2.86) which tend to be clinically indolent have no, or low, levels of homing receptor but disseminate widely. The recognition of these receptors provides a logical mechanism for the observation that some lymphomas such as the mucosal oriented tumors [mucosa-associated lymphoid tissue (MALT), bronchus-associated lymphoid tissue (BALT)] remain localized for long periods and may be best handled by surgi-

Fig. 2.86 Follicular lymphoma. Dog.

cal removal. In cattle, the effects of these homing tendencies are likely demonstrated by the plasmacytomas that convert whole nodes to medullary cords and by the large-cell lymphomas where the tumor cells are recognizable by nuclear cleavage and are confined to the medullary sinuses, whereas the benign large cells occupy the cords.

III. Spleen and Hemolymph Nodes

The spleen can be defined as a hematopoietic organ which **filters blood** through a **sinusoidal system.** In contrast, the lymph nodes are part of the same organ system with capability for filtering lymph through sinusoids. The vascular system of the spleen consists of arteries, veins, and efferent but no afferent lymphatics. As a consequence, all antigen reaches the spleen through the blood, which determines that primary sensitization and antibody production take place in the spleen only if the first contact with antigen occurs by hematogenous sensitization. On subsequent challenge, the germinal centers of the spleen play an important role in the humoral anamnestic response, both by local production of antibody and by the provision of committed B-memory cells to the peripheral lymphoid organ. The clearance of bacteria from the blood by the spleen predominates only if opsonization is deficient. In other circumstances, the clearing appears to occur mainly in hepatic Kupffer cells. Thus, splenectomy increases susceptibility to a number of blood-borne infections including septicemia and protozoal infections such as malaria, anaplasmosis, and hemobartonellosis.

The circulation of the spleen is central to its function and consists of arborizing arterioles which form the **penicilliary radicles.** They arise largely at right angles to the larger vessels and form the nutritive and antigen sources for the germinal centers. Histologically, the penicilliary vessels are seldom observed, but the eccentric relationship of the germinal centers to the larger arterioles is clearly apparent. The **periarteriolar sheath** of small lymphocytes invests the eccentric germinal center to provide a **mantle zone** of small lymphocytes similar to that surrounding germinal centers in lymph nodes. The penicilliary arterioles end abruptly on leaving the germinal center, and are ensheathed by a few plump reticular cells, which form a potentially **contractile ellipsoid.** Under normal circumstances, the spleen functions largely as a closed system with the blood from the penicilliary vessels entering more or less directly into large adjacent venous sinuses that drain into the splenic veins. Thus, normally almost all of the blood delivered to the spleen passes directly to the venous system, whereas as little as 3% perfuses the sinus areas where sorting and processing by the macrophages is carried out to remove effete or injured red cells. The flow rate through the spleen is relatively large and even with this low sorting fraction, all of the blood passes through the sinusoidal system once per day. With increased arterial flow, or decreased direct drainage to veins, the spleen functions as an open vascular system in which the penicilliary arterioles discharge largely into the Billroth sinuses, and thus a much greater fraction of the arterial flow passes through the sinusoidal filter.

The structure of the **sinus walls** is central to the filtering function. They are composed of elongated endothelial cells which are irregularly supported by encircling reticular fibers like the staves of a barrel. Adjacent endothelial cells are not joined by junctions and they therefore form a latticework through which blood cells readily permeate to pass from the functional Billroth sinus or filtering area back into the venous or systemic circulation. The Billroth sinus areas of the spleen consist of a loose reticular stroma in which the macrophages, lymphocytes, and plasma cells are enmeshed. This system allows for slow flow and maximal opportunity for recognition of altered cells and their removal by the splenic macrophages. Under normal circumstances, there appears to be a constant relationship between splenic volume and body weight, as is the case for liver and kidney.

On a functional histogenetic basis, the small lymphocytes immediately surrounding the larger arterioles are thymus derived, whereas those of the germinal center are bone marrow derived. The germinal center characteristically has three layers of cells consisting of the **germinal center** itself, a **corona** of medium-sized lymphocytes immediately peripheral to the germinal center which are also bone marrow derived, and peripheral to the corona an investment of small **mantle zone** lymphocytes which is continuous with the periarteriolar lymphatic sheath. This

outer mantle layer of the germinal center appears to be composed of both B and T lymphocytes and, in contrast to the cells of the germinal centers and lymphoid sheaths which are long-term residents, the mantle zone lymphocytes are in dynamic interchange with those in the blood and have a turnover time as short as 5–6 hr.

Hemal nodes occur in ruminants and a similar structure occurs in rats. They are enclosed by a fibromuscular capsule and have a general architecture similar to that of lymph nodes. Hemal nodes have a blood vascular circulation and afferent and efferent lymphatics, although the former are apparently much diminished in comparison to lymph nodes. The arteries entering through the hilar area arborize peripherally and are followed closely by veins which apparently communicate through a trabecular structure similar to that of the spleen. Hemal nodes have germinal centers, but the paracortical areas are much reduced in comparison to those of lymph nodes. Unlike lymph nodes, where the subcapsular sinus is poorly cellular, hemolymph nodes contain red cells in density similar to that of peripheral blood. Red cells are not present in the afferent lymphatics and very few are present in the efferent lymphatics so the interposition of the vascular and lymphatic circulations is rather efficient in the removal of erythrocytes from the exiting lymph. Macrophages occupy the trabecular areas and phagocytosis occurs as in the spleen. Myelopoiesis does not occur in hemal nodes either under normal circumstances or under conditions of hematopoietic stress.

Bibliography

Hayes, T. G. Development of ellipsoids in the spleen of the dog. *Am J Vet Res* **29**: 1245–1250. 1968.

Jønsson, V. Comparison and definition of spleen and lymph nodes: A phylogenetic analysis. *J Theor Biol* **117**: 691–699, 1985.

Nopajaroonsri, C., Luk, S. C., and Simon, G. T. The structure of the hemolymph node—A light-, transmission-, and scanning electron-microscopic study. *J Ultrastruct Res* **48**: 325–341, 1974.

Schmidt, E. E., MacDonald, I. C., and Groom, A. C. Circulatory pathways in the sinusal spleen of the dog, studied by scanning electron-microscopy of microcorrosion casts. *J Morphol* **178**: 111–123, 1983.

Yang, T. J., and Gawlak, S. L. Lymphoid organ weights and organ: body-weight ratios of growing beagles. *Lab Anim* **23**: 143–146, 1989.

A. Developmental Diseases of the Spleen

The spleen may be congenitally absent. This occurs regularly in a strain of inbred mice, and rarely in humans and outbred animals when it usually occurs in conjunction with multiple anomalies. In nude mice and rats with thymic deficiency, there is loss of the periarteriolar lymphoid sheath areas of the spleen. The least rare anomaly is the presence of accessory spleens, one or more, usually present in the gastrosplenic omentum. Acquired accessory spleens are implants of splenic parenchyma following traumatic rupture of the main organ (see splenosis in Section III,C of Lymphoreticular Tissues).

Duplication of the spleen is occasionally observed in normal swine and as one of multiple visceral defects in nonviable calves and lambs. Ectopic pancreatic tissue, either exocrine or endocrine, is occasionally observed in spleens which are otherwise developmentally normal.

B. Degenerative Diseases of the Spleen

Senile atrophy affects the spleen, as it does other lymphoid tissues, and is seen particularly in old dogs and old horses. Atrophy accompanies starvation and is seen in wasting disease. Microscopically the capsule is contracted and thickened, and there is lymphoid atrophy, the severity of which is an indication of the duration and severity of the atrophic process. The sinus areas appear fibrous due to condensation of the sinusoids, and lack of blood and resident hematopoietic cells.

In hyperimmune states, there are characteristic changes in the germinal centers of animals that are severely stressed. Lympholysis occurs and the germinal centers appear epithelioid and hypocellular. If the cell loss is acute, there will be prominent tingible body macrophages, as in foals aborted due to equine herpesvirus type 1 infection. If the reaction has persisted for more than a day, the cell debris is gone, and only the dendritic cells remain. A variety of vascular changes occur consisting of hyaline changes in small arterioles which may proceed to mineralization.

Old dogs often have yellowish encrustations along the splenic margins, or even covering most of the capsule (Fig. 2.87). Microscopically, there is condensation and thickening of the capsule, trabeculae, and perivascular tissue. The yellow or brownish encrustations are known as **siderotic nodules** or **Gamna–Gandy bodies** and represent deposits of iron and calcium in connective tissue, usually the trabeculae. The salts become encrusted on connective tissue and elastic fibers, which are swollen, refractile, and stain with hematoxylin. The encrusted fibers occasionally are misinterpreted as fungal hyphae. In association with ceroid, these nodules likely represent the residual effects from areas of hemorrhage. Their deposition in the capsular margins and internally suggests that a dystrophic or storage phenomenon may be more likely.

Amyloidosis of the spleen (Fig. 2.88) occurs as part of generalized amyloidosis, but the deposits in the spleen may not be detectable grossly or histologically without appropriate stains. Amyloidosis of an obvious degree is not common and when present, it involves the germinal centers and may spare the sinus areas completely. This distribution of amyloid gives the organ the sago-spleen appearance, in which the enlarged and waxy follicles protrude from the cut surface. The spleen is not enlarged. The deposition of amyloid occurs first in the small arteries of the lymphoid nodules and minor depositions are common.

Histiocytes of spleen and lymph nodes are enlarged in

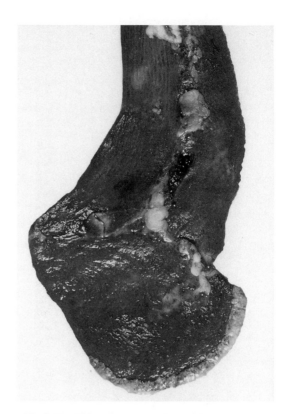

Fig. 2.87 Siderotic encrustations of tip and capsule of spleen. Dog.

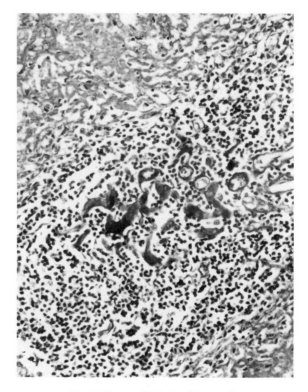

Fig. 2.88 Amyloidosis. Spleen. Cat.

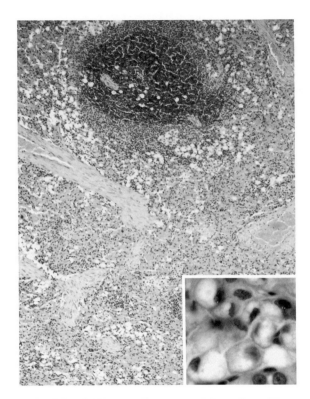

Fig. 2.89 Sphingomyelin storage. Spleen. Dog. Clusters of large macrophages with clear cytoplasm in both follicles and pulp. (Inset) storage cells.

most **storage diseases** (Fig. 2.89) (see The Nervous System, Volume 1, Chapter 3).

Hemosiderin is a storage form of iron and the only splenic pigment of importance. The amount and form of storage iron varies greatly in normal animals with age and species. The pigment is usually present only in macrophages, but in long-standing iron accumulation it may be encrusted on fibers of connective tissue. The amount of hemosiderin is significant only when it is in excess to the point of causing tissue injury and fibrosis, and justifies the name hemosiderosis. Iron is absorbed in excess in any anemia where it is sufficiently available. Increased iron is one of the splenic changes in the hemolytic anemias. The patterns of iron storage in the spleen and marrow can yield information about the type of anemia. In nonresponsive anemias, the iron is very largely aggregated into large coarse granules within the phagocytes, which themselves are inapparent. In responsive hemolytic anemias where there is rapid iron turnover, there is a spectrum of iron deposition from barely stainable, diffusely or focally distributed ferritin, to coarse hemosiderin. In the anemias of chronic disease, there is increased coarse splenic iron due to continued scavenging of the transferrin system by the acute phase reacting proteins.

Bibliography

Dorling, P. R. Lysosomal storage diseases in animals. *In* "Lysosomes in Biology and Pathology," J. T. Dingle, R. T. Dean,

and W. Sly (eds.), pp. 347–349. New York, Elsevier Science Publishers, 1984.

van Houte, A. J. *et al.* The periarteriolar lymphocyte sheath in immunodeficiency. T- or B-lymphocyte area? *Am J Clin Pathol* **92**: 318–322, 1990.

C. Rupture of the Spleen

A normal spleen can be ruptured by the level of trauma that cats and dogs suffer regularly in automobile collisions. The result is more or less severe loss of blood into the abdomen.

There is usually a concurrent rise in transaminases, indicating hepatic laceration as an additional source of abdominal hemorrhage. The spleen may be fully divided into two or more parts, or merely ruptured and healed, the lines of healing producing notches and fissures. Pathologic rupture may occur in any species with only minor trauma if the spleen is enlarged and the capsule thinned. The latter pathogenesis may occur with splenomegaly of any cause including congestive, hemolytic, infectious, or neoplastic types (Fig. 2.90). Lymphoma is a common cause of splenic enlargement leading to fatal internal hemorrhage. Following rupture of the splenic capsule and spilling of the sinus cells into the abdominal cavity, there may be widespread seeding of splenic explants onto the serosal surfaces. This

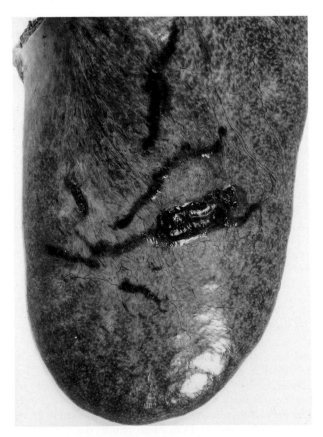

Fig. 2.90 Spontaneous rupture of swollen spleen. Lymphoma. Ox.

is called **splenosis.** These accessory spleens are functional and there may be hundreds on the omentum with a few on the peritoneum. They appear grossly like hemal lymph nodes and histologically like normal spleen. Following splenectomy, any splenic implants present will increase in size and may give some protection against septicemic disease.

D. Torsion of the Spleen

Torsion affecting only the spleen occurs in pigs, dogs, and humans and rarely in horses, and torsion of the spleen and stomach occurs in dogs. When the whole spleen is twisted on its mesentery, there is severe congestion and hemorrhagic infarction due to occlusion of the vein and ultimately blockage of the artery. When the distal portion is twisted, only the vein tends to be occluded, causing hemorrhagic infarction and congestion. Splenic torsion in the dog occurs in large breeds and is characterized by enlargement of the upper abdomen with painful guarding on palpation, and the passage of dark brown urine. Hematologically, there is a characteristic brown discoloration of plasma due to leaching of blood pigments through the capsule as it becomes necrotic. Associated signs consist of the postsplenectomy state with many distorted red cells, hypersegmented neutrophils, and the presence of Howell–Jolly bodies and **red cell pits,** which are unusual in the peripheral blood of the dog with a normally functional spleen. The spleen, on surgical removal, may weigh as much as 5 kg and contain a large proportion of the original red cell mass. As is the case in humans who survive splenectomy, there is risk of infection by organisms normally controlled by the spleen. In the dog, hemobartonellosis and severe hemolytic anemia are the most likely adverse sequelae.

Splenic torsion in pigs usually occurs in sows, and often several are affected in the same herd over a period of a few days. Circumstantial evidence suggests that a change from once- to twice-daily feeding will prevent further losses, but other factors must be involved. Affected animals usually die acutely, and in white-skinned pigs, cutaneous pallor suggests esophagogastric ulceration. Grossly the spleen is blue-black, massively and uniformly enlarged, and twisted 180° about its long axis.

E. Cysts

Splenic cysts are unusual. Occasionally, *Cysticercus tenuicollis* and hydatids are found there. Pseudocysts may result from cystic degeneration of hematomas. Neoplastic cysts are hemangiomas or hemangiosarcomas. The latter are relatively common in the dog and have a rather characteristic history of repeated bouts of malaise and weakness, with apparent recovery over 2–3 days. Severe anemia accounts for the weakness. Pallor may be noticed and results from rupture of the tumor with abdominal hemorrhage. The recovery over a few days is too rapid for new red cell production and indicates autotransfusion of the

extravasated blood. The hematologic picture is also typical with distorted red cells, even in the absence of abdominal hemorrhage, due to the effects of abnormal vascular walls and arteriovenous shunts. After hemorrhage, the red cell distortion is much worse and the formerly extravasated red cells appear tattered and effete, and the plasma is brown due to extensive hemolysis. There are usually Howell–Jolly bodies in the blood along with hypersegmented neutrophils. If abdominocentesis is performed, the presence of an occasional rubricyte is indicative of hemorrhage from the spleen rather than from another abdominal site.

F. Circulatory Diseases of the Spleen

Active hyperemia is common in acute systemic infections and also occurs in some acute bacterial intoxications, such as clostridial enterotoxemia of calves, erysipelas of swine, and *Streptococcus pneumoniae* infection in neonatal calves and goats. Passive congestion of the spleen may arise as a result of disturbances in the systemic and portal circulation, and is a feature of some of the acute hemolytic anemias. The most common cause of passive congestion is the use of barbiturate drugs. Central (cardiac or pulmonary) venous congestion does not usually cause congestion of the spleen in animals, partly because they usually do not live long enough, and partly because the splenic capsule is not elastic as in sheep, or is highly muscular as in the dog and horse and not easily distended. Portal obstruction, with chronic hepatic fibrosis and portal obstruction, causes significant splenic congestion in humans, but this sequence of events is rare in animals.

An acutely congested spleen is enlarged, moderately turgid, and cyanotic with the capsule bluish black. The normal architecture is not discernible on the cut surface and the pulp is red-black and progressively exudes blood. Microscopically the splenic sinuses are dilated with packed red cells, the germinal centers are widely separated, and trabeculae are thinned. The extent of acute congestion may make histologic examination difficult in the horse because the tissue, being largely clotted blood, fragments on sectioning. Splenic samples from horses destroyed with barbiturate should be taken from a thin marginal area where the parenchyma is supported on two sides by capsule. With chronic congestion, the organ becomes firm and the capsule and trabeculae thickened. There is lymphoid atrophy and sinus proliferation, with a marked increase in fixed cells in the red pulp and heavy sinus colonization with macrophages and hemosiderin.

Thrombosis of the splenic vein has been observed in association with traumatic reticulitis and portal thrombosis in cattle, and with splenic abscesses in horses. Thrombosis of both arteries and veins is seen in hypercoagulable states such as immune hemolytic anemias, purpura, and hemorrhagic pancreatitis. Spleens which are enlarged for any reason are prone to thrombosis and infarction, which is likely to occur in lymphoma and in leukemia due to both the thrombocytopenia and the procoagulant content of the hypogranular promyelocytic leukemia. In acute hog cholera, there is widespread endothelial swelling and proliferation with fibrinoid thrombosis of splenic arterioles in 50% of cases, resulting in raised dark infarcts 0.2–2.0 cm in diameter in the splenic capsule. The spleen may suffer septic infarction in the hyperviscosity syndrome associated with myeloma and gammopathy. The spleen is enlarged due to infiltration by tumor cells and flow is further compromised by the hyperviscosity of blood. The septic component is likely due to loss of normal antibody and displacement of phagocytes by the myeloma cells.

Embolism with infarction is not uncommon. The emboli are usually derived from valvular vegetations. The outcome has the usual dependence on whether the emboli are septic or bland.

Bibliography

Berendt, H. L., Mant, M. J., and Jewell, L. D. Periarterial fibrosis in the spleen in idiopathic thrombocytopenic purpura. *Arch Pathol Lab Med* **110:** 1152–1154, 1986.

Holdsworth, R. J. Regeneration of the spleen and splenic autotransplantation. *Br J Surg* **78:** 270–278, 1991.

Maxie, M. G. *et al.* Splenic torsion in three Great Danes. *Can Vet J* **11:** 249–255, 1970.

Obwolo, M. J. Primary splenic haemangiopericytoma in a German shepherd dog. *J Comp Pathol* **96:** 285–288, 1986.

O'Keefe, J. H. *et al.* Thromboembolic splenic infarction. *Mayo Clin Proc* **61:** 967–972, 1986.

Tumen, H. J. Hypersplenism and portal hypertension. *Ann N Y Acad Sci* **170:** 332–334, 1970.

Warner, E. D., Hoak, J. C., and Fry, G. L. Hemangioma, thrombocytopenia, and anemia. The Kasabach–Merritt syndrome in an animal model. *Arch Pathol* **91:** 523–528, 1971.

Weinstein, M. J., Carpenter, J. L., and Mehlhaff Schunk, C. J. Reports of retrospective studies. Nonangiogenic and nonlymphomatous sarcomas of the canine spleen: 57 cases (1975–1987). *J Am Vet Med Assoc* **195:** 784–788, 1989.

Zago, M. A. *et al.* Red cell pits appear preferentially in old cells after splenectomy. *Acta Hematol* **76:** 54–56, 1986.

G. Inflammatory Diseases of the Spleen

Systemic inflammations cause a regular and fairly predictable pattern of response in the spleen. In septicemia and following injection of endotoxin or gram-negative bacterins, there is rapid accumulation of neutrophils in the splenic mantle zone and surrounding area of sinus. There is bacterial destruction and processing of antigen in this area followed by a predictable pattern of centripetal migration, the foreign material moving in specialized macrophages (metallophils) successively into the mantle zone and then to the lymphocyte corona, reaching the germinal center in 10–12 hr. In contrast, immune complexes appear to move by flow and not by cellular transport. Overwhelming infections result in lympholysis of the follicular center cells, with the production of nuclear debris and the exposure of the underlying dendritic reticular cells, whose cytoplasm becomes more intensely stained in disease to attain the tinctorial properties and density of epithelioid macro-

phages. The nuclear fragments are rapidly removed so that the follicular center appears empty, with the nuclei of the dendritic cells relatively widely spaced and surrounded by their densely pink cytoplasm. These epithelioid germinal centers are seen in a variety of acute toxemic diseases in young animals. In older animals, the changes occurring in splenic germinal centers are often more severe and longlasting, and consist of a depopulation of the follicular center cells and transudation of plasma proteins into the germinal centers to form a fairly persistent coagulum recognized as **intrafollicular hyalinosis.** These fibrinous exudates may be removed, with restitution of the follicular center cells or, if there is extensive vascular damage, they may become mineralized and result in involution of the germinal center.

Septicemia and bacteremia result in microbial deposition in the Billroth sinus areas, where the abundance of phagocytic cells usually leads to their rapid destruction. Occasionally small foci of bacterial colonization develop in the sinus areas of spleens, primarily in young animals with severe gram-negative infection of the enteric tract. In septicemic splenitis, there is splenic enlargement, with acute congestion and degeneration of the lymphoid follicles and hypercellularity of sinus areas. In very acute diseases such as anthrax in cattle, the reaction is almost solely vascular. In the less severe septicemias such as erysipelas in swine, there is moderate reactive hyperplasia, which may be more developed than that seen in acute fulminant gram-negative septicemias such as salmonellosis. Marked hyperplastic changes of the sinus areas are a feature of some diseases such as malaria, trypanosomiasis, equine infectious anemia, and malignant catarrhal fever, although involution characterizes some infections.

Abscesses of the spleen may be miliary or large and focal, but both types are uncommon. Abscessation may be due to various organisms including *Actinomyces pyogenes,* but some commonly localize preferentially in the spleen (see Specific Infections of the Lymphoid Tissues, Section IV of the Lymphoreticular Tissues, in this chapter). Purulent splenitis may develop by local extension in cattle from penetrating wounds of the reticulum and, in horses, by extension from the colon due to strongyle migration.

H. Benign Hyperplastic Diseases of the Spleen

1. Nodular Hyperplasia

Nodular hyperplasia is frequently observed in the spleens of old dogs, occasionally in old bulls, and rarely in other species. Most nodules are up to 2 cm in diameter and project hemispherically above the capsule, but may be 5 cm or more in diameter. On the cut surface, the nodules vary from gray to pink to a variegated red and white, with yellow necrotic areas in the larger ones. The variegated pattern is due to persistent islands of sinus in a diffuse field of lymphocytic proliferation. Architecturally the nodules are unencapsulated but compress adjacent trabeculae, and irregularly meld with surrounding sinus areas, which have a much higher density of red cells. Lacking germinal centers, the nodules consist of monomorphic medium or large lymphocytes with moderate cytoplasmic volume and small nucleoli. The nuclei are round with thin nuclear boundaries, and are more open and less hyperchromatic than are lymphoma cells. There may be many mitoses, and the lesions are judged to be benign more on a knowledge of behavior than cytology. There is no evidence to suggest that these lesions ever progress to lymphoma.

In contrast, **splenic plasmacytoma** is a rare condition of dogs with gross features indistinguishable from nodular hyperplasia. Cytologically the cells are uniform and of moderate size but with abundant, usually eccentric cytoplasm that stains deeply. Nucleoli are small but prominent and central, and there is mild heterochromatin aggregation. Mitoses are less numerous than in the surrounding normal tissue and there may be some degree of gammopathy. Plasmacytoma may be indolent for long periods but may progress to an immunoblastic type of lymphoma within a year of biopsy.

2. Hematopoietic Alterations

Extramedullary hematopoiesis results when there is the hormonal induction for increased cell production and pluripotential hematopoietic stem cells available. These cells normally circulate in very low numbers and, in preparation for extramedullary hematopoiesis, they return to embryonic sites of colonization, sparing the germinal centers and periarteriolar lymphoid sheaths. Extramedullary hematopoiesis is characteristically trilineage, but one cell line may predominate, such as the myeloid system in canine pyometra or the erythroid system in hemolytic anemia (Fig. 2.91). Megakaryocytes are the most obvious hematopoietic precursor and characteristically lie adjacent to the smooth muscle trabeculae when stimulation is mild, but may become diffusely distributed in the sinus areas when stimuli are pronounced and prolonged. Splenic erythropoiesis is coincident with increased splenic red cell destruction in immune hemolytic anemias.

In contrast, **myeloid metaplasia** is a condition in which there is marrow myeloid hyperplasia without an apparent target for the increased cell production (see Myelodysplastic Syndromes, Section V of The Leukon). The disease tends to run a course of prolonged hyperplasia followed by myelofibrosis and displacement of the still functional stem cell system to embryonic sites of production. Grossly the spleen is uniformly enlarged with turgid capsule and rounded borders. The parenchyma is dry, and the light pink of hematopoietic marrow rather than dark and cyanotic as with congestion. In myeloid metaplasia of the spleen, there is myeloid predominance, but all cell lines are present to some degree. In the absence of a normal marrow environment, **dyspoiesis** occurs. It is characterized by multinucleated rubricytes, incipient Howell–Jolly bodies, giant and donut forms of metamyelocytes, and

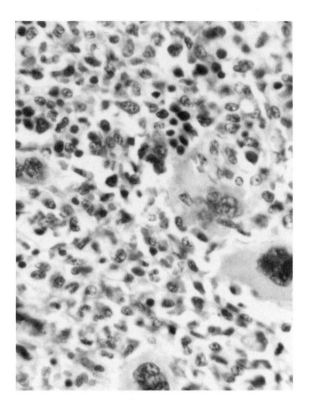

Fig. 2.91 Extramedullary thrombopoiesis and erythropoiesis in pyometra. Spleen. Dog.

hypofusion and segmentation of megakaryocyte nuclei. Myeloid metaplasia is accompanied by immune hyperplasia, hemosiderosis, and splenic fibrosis. This is preceded by follicular hyperplasia of the spleen and nodes, and later hematopoietic colonization with gradual reduction in the lymphoid volume in the spleen, and eventual atrophy as the stem cell system becomes increasingly committed to myeloproliferation. Small infarcts develop and are likely the result of loss of plasticity, and fibrous impairment of vascular structures.

Rarely the spleen contains a **myelolipoma,** which stands out grossly as a focal pale area on the cut surface, and histologically, because the empty fat cells contrast sharply with the sinus parenchyma. Myelolipoma is composed of outwardly normal marrow with about 50% hematopoietic and the rest, large typical fat cells.

Lymphoid hyperplasia of the spleen occurs by expansion of the periarteriolar lymphoid sheaths so that histologically the germinal centers are joined by an irregular branching pattern of densely packed columns of small lymphocytes. The germinal centers in hyperimmune states increase in both size and density, and the accentuation of the mantle, corona, and follicular center cells is increased, often resulting in a target appearance. Since sinus and follicular hyperplasia of the spleen tend to occur together, their concurrence results in an overall increase in splenic weight.

I. Splenomegaly

The hematologic effects of an enlarged spleen include pan or selective cytopenia but almost always anemia and thrombocytopenia (see Anemia Due to Splenic Hyperfunction, Section II,F,7 of The Erythron). **Splenomegaly** is an enlarged spleen, whereas **hypersplenism** describes a spleen that is overactive in cell destruction. Usually the two coincide, but whereas the enlarged spleen is always hyperactive, the hyperactive spleen may not always be enlarged. Since the spleen is almost always involved to some degree in both red cell production and destruction, the decision on whether to remove the spleen in hemolytic anemia may require scintillation counting of labeled red cells to determine whether splenectomy would be helpful. In concurrent hemolytic anemia and thrombocytopenia, as it occurs in dogs, immediate splenectomy is indicated to prevent fatal intracranial hemorrhage. The apparent decoupling of splenic size and function in cell destruction is mediated in large measure by the presence or absence of opsonizing antibody, which greatly influences the chances of a given cell reentering the vascular system after being shunted into the splenic sinus system.

The degree to which splenic size departs from normal is difficult to assess, but weight is always a useful parameter, particularly in animals which die without exposure to barbiturate anesthetics. Splenomegaly is common, and the interpretation of splenic size at autopsy can be gained by reference to the whole carcass and to the cut surface of the organ. The following is an arbitrary list of diseases to be considered when splenomegaly is present. The absence of splenomegaly does not eliminate any of them, but makes them unlikely. Focal multiple lesions do not as a rule cause splenomegaly, although hemangiomas of the canine spleen and metastatic melanomas of the equine spleen are exceptions.

Cattle and sheep
    Anthrax
    Salmonellosis
    Lymphoma–leukemia
    Babesiosis, trypanosomiasis, anaplasmosis
    Congestive splenomegaly
    Acute septic processes with bacteremia and toxemia
Horses
    Equine infectious anemia
    Lymphoma–leukemia
    Metastatic melanoma
    Isoimmune hemolytic anemia
    Salmonellosis
    Anthrax
Pigs
    Torsion
    Congestive splenomegaly
    Salmonellosis
    Erysipelas
    Lymphoma–leukemia
    Isoimmune hemolytic anemia
    Protozoan diseases of blood
    African swine fever

Dogs and cats
    Barbiturates
    Lymphoma–leukemia
    Acquired hemolytic anemias
    Histoplasmosis
    Amyloidosis
    Leishmaniasis
    Hemangiomas and hemangiosarcomas
    Torsion (dogs)
    Mast cell tumor (cats)

Splenic enlargement may occur in storage diseases of which there are many (see The Nervous System, Volume 1, Chapter 3). Sphingomyelinosis of dogs is an appropriate example (Fig. 2.89).

Bibliography

Berge, T. Extramedullary haemopoiesis and metastases in the spleen. *Acta Pathol Microbiol Scand* **82:** 507–513, 1974.

Cooper, J. H., Haq, B. M., and Bagnell, H. Intrafollicular hyalinosis and arterial hyalinosis of the spleen: Histochemical and immunofluorescence studies. *J Pathol* **98:** 193–199, 1969.

Douay, L. *et al.* Blood and spleen haematopoiesis in patients with myelofibrosis. *Leukemia Res* **11:** 725–730, 1987.

Hayes, M. M. *et al.* Splenic pathology in immune thrombocytopenia. *J Clin Pathol* **38:** 985–988, 1985.

Laman, J. D. *et al.* Mechanism of follicular trapping: Localization of immune complexes and cell remnants after elimination and repopulation of different spleen cell populations. *Immunology* **71:** 57–62, 1990.

Lawhorne, T. W., Jr., and Zuidema, G. D. Splenic abscess. *Surgery* **79:** 686–689, 1976.

Meis, J. M. *et al.* Solitary plasmacytomas of bone and extramedullary plasmacytomas. A clinicopathologic and immunohistochemical study. *Cancer* **59:** 1475–1485, 1987.

Miyakawa, K. *et al.* Localization in the rat spleen of carbon-laden macrophages introduced into the splenic artery: A subpopulation of macrophages entering the white pulp. *Anat Rec* **227:** 464–474, 1990.

Valli, V. E. O., and Forsberg, C. M. The pathogenesis of *Trypanosoma congolense* infection in calves. V. Quantitative histological changes. *Vet Pathol* **16:** 334–368, 1979.

Wiernik, P. H. *et al.* Inflammatory pseudotumor of spleen. *Cancer* **66:** 597–600. 1990.

J. Neoplastic Diseases of the Spleen

The vascular tumors of the spleen (Fig. 2.92) are described with The Circulatory System (Chapter 1 of this volume), and leukemic diseases involving the spleen are described with the Myeloproliferative and Lymphoproliferative Diseases (Sections IV and VI of The Leukon, in this chapter). The spleen is expected to be involved in systemic malignancies of the lymphoid system, but this is not always the case or apparent. In a review of lymphomas (and assuming that when lymphoma was suspected, the spleen would always be examined), there was splenic involvement in only 41 of 72 cases (57%) in dogs, 35 of 81 cases (43%) in cats, and 27 of 90 cases (30%) in cattle. Although the spleen may not be invaded in lymphoma, it is

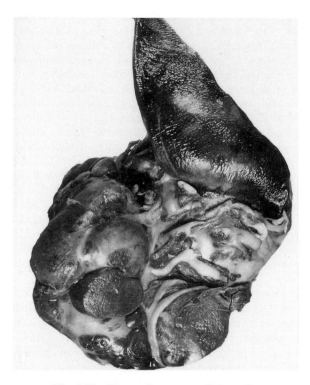

Fig. 2.92 Hemangiosarcoma. Spleen. Dog.

much more likely to be involved in lymphoid and myeloid leukemia. In all situations, but particularly in lymphoma, the size of the organ or the microscopic structure cannot be taken in itself as evidence for the presence or absence of tumor. Careful cytologic examination must be carried out to determine whether a suspect cell population is like that of unequivocal tumor elsewhere.

The evidence for well-defined tissue homing by lymphocytes, added to recognition of B–T subsets, facilitates the interpretation of lymphoid changes. Thus the loss of specific trophic factors, or the displacement of stem cells from a cellular compartment, explain the atrophy of lymphoid, particularly thymus dependent, areas that are seen in cancer generally. More than 90% of peripheral lymphocytes are T cells, and under normal circumstances they make up most of the cellularity of the sinus areas; consequently in lymphopenia, the sinus areas of the spleen appear hypocellular and vice versa. In polyclonal B cell hyperplasia, as occurs in cattle with trypanosomiasis and the persistent lymphocytosis stage of bovine leukemia virus infection, there is enlargement of the mantle zone areas around the germinal centers.

In cats and dogs with lymphoma, there may be unequivocal involvement of lymph nodes, portal colonization of the liver, and atrophy of the periarteriolar lymphoid sheaths and germinal centers in the spleen. Splenic involvement is typically a late phenomenon in thymic lymphoma of all species.

When the spleen is symmetrically enlarged to a mild or greater degree, the histologic infiltration is usually diffuse,

as occurs in leukemia, and occupies all of the sinus areas. There is mild atrophy of the germinal centers and periarteriolar lymphoid sheaths. A diagnosis of lymphoma in the spleen may be difficult when there is a prominent follicular pattern, and the most reliable criterion of malignancy is subendothelial colonization of large muscular sinuses by lymphocytes. Usually, when this change is found, lymphocytes of similar morphologic type will be identified in other areas, particularly the liver. In follicular lymphoma of large-cell type (a rare disease), the tumor is multifocal and centered on germinal follicles, as can be demonstrated by identification of residual benign small cells in the center of the enlarging tumor foci. The associated atrophic changes are more apparent in nodes than in spleen where there is a complete loss of the small lymphocytes of the corona and mantle zone areas. In the horse and dog, lymphoma may be focal in the spleen, and in these cases there is more likely to be atrophy of lymphoid tissue in the benign area as the tumor increases in size. When the spleen is involved in adult cattle with lymphoma, there is usually malignant colonization of the germinal centers, probably reflecting the bone marrow origin of the cells involved. A similar pattern of involvement occurs in cats and dogs, and there is usually a dense homogeneous cluster of lymphocytes two to three times the diameter of a normal follicle, including its investment of corona and mantle cells. In contrast to follicular lymphoma, these are focal aggregates of diffuse tumor, and again their homing pattern can be demonstrated by identification of residual benign cells in some of their centers.

Metastatic neoplastic disease of the spleen is uncommon. It is not likely that migration of tumor cells to the spleen is unusual, but rather that the functional efficiency of the sinus macrophages usually prevents colonization. Metastatic carcinoma involving the spleen and liver usually arises in the pancreas or reproductive system. Whereas neoplasia arising in the liver does move in a retrograde fashion in veins to the spleen, metastases are isolated, and most are probably of arterial origin. Carcinomatous implants originating from other abdominal viscera can mimic peritoneal tuberculosis (Fig. 2.93).

The diffuse colonization of splenic sinuses with mast cells, as seen in the cat, is described under Mast Cell Leukemia (Section IV,B,2 of The Leukon).

Bibliography

Berg, T. Splenic metastases. Frequencies and patterns. *Acta Pathol Microbiol Scand* **82:** 499–506, 1974.

Falk, S., and Stutte, J. H. Hamartomas of the spleen: A study of 20 biopsy cases. *Histopathology* **14:** 603–612, 1989.

Fidler, I. J. General concepts of tumor metastasis in the dog and cat. *J Am Anim Hosp Assoc* **12:** 374–380, 1976.

Palutke, M. *et al.* B lymphocytic lymphoma (large cell) of possible splenic marginal zone origin presenting with prominent splenomegaly and unusual cordal red pulp distribution. *Cancer* **62:** 593–600, 1988.

IV. Specific Infections of the Lymphoid Tissues

The strategic location of the lymph nodes, and blood flow through the spleen, determine their role in resistance

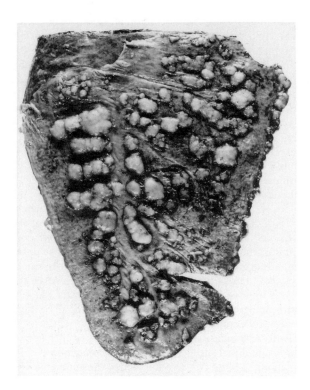

Fig. 2.93 Implants of squamous carcinoma on capsule. Spleen. Ox.

to infections and make their involvement, either locally or generally, directly or indirectly, inevitable in any infectious disease. For the most part, the lesions produced in the nodes and spleen are not specific for the cause. On the other hand, there are some infectious diseases in which these tissues are chiefly or consistently involved, and they are the subject of this section.

A. Caseous Lymphadenitis

Caseous lymphadenitis is a suppurative infection of the lymph nodes primarily of sheep and goats caused by *Corynebacterium pseudotuberculosis (ovis)*. The disease occurs in sheep wherever they are raised, but horses, camels, deer, mules, and rarely cattle and humans may be affected. Cattle can develop a pathologic syndrome which resembles that in sheep, but they do so quite rarely, and the infection generally remains localized to one to two regional nodes draining an infected surface wound or a segment of intestine. *Corynebacterium pseudotuberculosis* is also the cause of ulcerative lymphangitis in cattle and horses, and of pectoral abscesses in horses.

Although the ovine and equine strains are culturally, antigenically and toxigenically similar, there may be undetermined differences in virulence factors since in some areas in which caseous lymphadenitis is endemic, either ulcerative lymphangitis or pectoral abscesses, or both, are unknown.

The disease in **goats** can be more severe than that in

sheep, with the most frequent lesions being in the lymph nodes of the head and neck. The lesions in goats closely resemble those of pseudoglanders or melioidosis (*Pseudomonas pseudomallei*), and it is important to distinguish these diseases (Fig. 2.94). Female goats and intact males have more extensive lesions than do castrated males. In sheep and goats, there is slow spread of the disease with the prevalence of abscesses increasing as a function of age. Caseous lymphadenitis is widespread in goats and the disease differs from that in sheep in the distribution of the abscesses. In goats the mandibular lymph nodes, followed by the parotid nodes, are most often involved, suggesting that the organism is acquired through the buccal mucosa, as well as through skin wounds. Superficial abscesses of the jaw and neck region are common and infection is probably acquired from feed stalls or collars carrying the bacterium. In only a small proportion of goats does infection spread to involve the mesenteric or mediastinal nodes grossly.

Caseous lymphadenitis in **sheep** almost always follows a wound infection, usually a shear wound. The organism can penetrate the intact skin of freshly shorn sheep, however, and may be transmitted by dipping fluids. Docking and castration wounds and the umbilicus are of minor importance. Occasionally, the infection may be acquired by ingestion, as indicated by confinement of the lesions to the nodes draining the buccal cavity. Less commonly the organism is inhaled producing lung abscesses. Parasitic

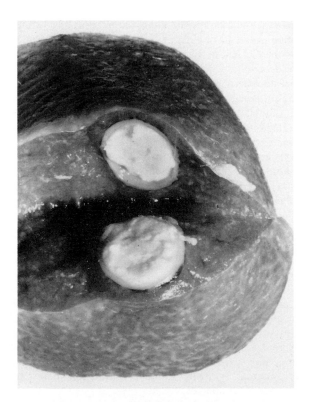

Fig. 2.94 Abscess of melioidosis. Spleen. Goat. (Courtesy of W. R. Kelly.)

wounds of the lower alimentary tract are not a portal of infection.

Corynebacterium pseudotuberculosis is capable of brief survival in the environment, permitting indirect infection from heavily contaminated ground in yards or feeding areas. It can occasionally be obtained from the feces of sheep, possibly as an alimentary passenger, and it is probable that the organism is normally a parasite of the alimentary tract. By this view, infection is transmitted by contamination of wounds with pus from discharging lesions or, more indirectly, with soil contaminated by exudates or feces.

The pathogenicity of *C. pseudotuberculosis* is related to the production of heat-labile toxin, which causes increased vascular permeability and which may be identical to the hemolytic phospholipase which the organism produces, and a surface lipid which is leukotoxic. The toxin is lethal for many animal species and on intravenous injection causes hemolytic anemia and icterus. The natural infection of sheep, even generalized, however, gives scant clinical evidence of intoxication, though the exotoxin facilitates the spread of infection from the primary site. Vaccination with the exotoxin provides a degree of protective immunity. The leukotoxic surface lipid allows persistence within inactivated macrophages in which the organism is effectively a **facultative intracellular parasite.** The acquisition of host resistance may also be in part due to the selection of a population of macrophages resistant to and capable of destroying the toxic surface lipids of the bacteria.

The sequence of events in progressive caseous lymphadenitis is infection of a superficial wound, spread of infection to the local lymph nodes which suppurate, and then lymphogenous and hematogenous extension to produce abscesses in internal organs. The progression is slow and may reach the bloodstream only in older animals, whereas in young animals the disease tends to be confined to the superficial lymph nodes, of which the precapsular and precrural are most often affected.

Spontaneous recovery from the primary infection of cutaneous wounds, even when they suppurate, is more common than is progression of the infection to the lymph node, as judged by results in the experimental disease in which only about 20% of sheep develop abscesses in the nodes. Even when abscesses develop in the nodes, the primary cutaneous lesion heals, but secondary cutaneous lesions may develop as fistulae when the nodal abscess ruptures to the exterior. This occurs more often in goats than in sheep. As a general rule, once the infection gains the nodes it is persistent, although a few infections may be cleared when the node ruptures and discharges onto the skin.

The initial lesion in lymphoid tissues is a diffuse lymphadenitis, which is probably the result of the soluble exotoxin. When the organism reaches the nodes, multiple microscopic abscesses form in the cortex. Eosinophils are a prominent part of the reaction and give a green color to the pus. These foci rapidly coalesce and the central areas

caseate to form a structureless mass which contains fragments of nuclear material and discrete clumps of bacteria. The abscesses are rapidly encapsulated, and when this occurs the acute reaction in the surrounding tissues subsides, but the abscesses continue to enlarge. With enlargement, there is progressive necrosis and reformation of the capsule, which gives the lesion a very characteristic structure of concentric lamellations (Fig. 2.95); these are particularly prominent when calcareous granules are deposited in successive layers at the margin of the expanding lesion. In old lesions, the contents lose their greenish color, become inspissated, and resemble putty. The lamellation is specific to the organism, not the organ involved. The nodal lesions often attain a diameter of 4–5 cm and exceptionally they may reach 15 cm. The larger lesions in superficial nodes cause pressure atrophy and depilation of the overlying skin; they frequently rupture to discharge chronically through a narrow fistula.

Caseous lymphadenitis is rarely fatal, and indeed it seldom even causes debilitation. Its economic importance is due to regulations concerning the trading of carcasses that show evidence of the disease. When fatalities do occur they are caused principally by large pleuropulmonary abscesses. Rarely, *Corynebacterium pseudotuberculosis* causes small outbreaks of polyarthritis in lambs, but they recover spontaneously.

The mature lesion of caseous lymphadenitis is an encapsulated abscess with pus of a distinct greenish color and

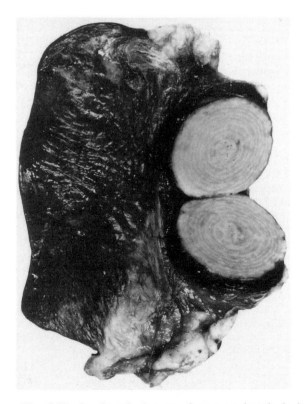

Fig. 2.95 Laminated abscess of caseous lymphadenitis. Spleen. Sheep.

of caseous or caseopurulent consistency. The initial cutaneous lesion may not be evident, having resolved, but it is noted that occasional subcutaneous abscesses, principally of the face and belly, do occur without relation to known aggregations of lymphoid tissue. Mastitis, as occasionally observed in sheep and often observed in goats, probably represents an extension from a wound of the overlying skin; when the mastitis is early and acute it is diffuse and suppurative, and when chronic, it is localized to encapsulated abscesses. The disease is easily diagnosed by fine-needle aspiration of the focal lesions, which allows both cytopathology and bacterial culture.

Spread from the lymph nodes produces lesions in the lungs, and these are rather common with advancing age. They also occur in young lambs, in which the progression is more rapid than in adults. The pulmonary lesions may consist of extensive bronchopneumonia, when abscesses rupture into bronchi in which there are soft caseopurulent foci, or there may be discrete nodules of varying size and number. In cases of bronchopneumonia, and overlying those nodules which are immediately subserosal, there is a pleuritis, often with adhesions. When the adhesions are few and localized adjacent to the nodules, the remaining pleural cavity may be normal. When the adhesions are more diffuse, there is a large amount of serous fluid in the cavities and a thin layer of fibrin on the pleura. The nodular lesions in the lung are similar to those in lymph nodes and have a narrow zone of bronchopneumonia outside the capsule. With time, the pulmonary nodules become sharply circumscribed, encapsulated, subpleural abscesses. The pulmonary lesions are associated with characteristic lesions in bronchial lymph nodes, which may be much enlarged. Dissemination of the infection from the lungs to other viscera is uncommon. Metastases are occasionally observed in the renal cortex as discrete abscesses or descending pyelonephritis. Other viscera, chiefly the liver and spleen, may contain solitary abscesses of the typical form.

Bibliography

Ayers, J. L. Caseous lymphadenitis in goats and sheep: A review of diagnosis, pathogenesis, and immunity. *J Am Vet Med Assoc* **171:** 1251–1254, 1977.

Hsu, T.-Y. Caseous lymphadenitis in small ruminants: Clinical, pathological, and immunological responses to *Corynebacterium pseudotuberculosis* and to fractions and toxins from the micro-organism. *Diss Abstr Int (B)* **45:** 1396, 1984.

Maki, L. R. *et al.* Diagnosis of *Corynebacterium pseudotuberculosis* infections in sheep, using an enzyme-linked immunosorbent assay. *Am J Vet Res* **46:** 212–214, 1985.

Miers, K. C., and Ley, W. B. *Corynebacterium pseudotuberculosis* infection in the horse: Study of 117 clinical cases and consideration of etiopathogenesis. *J Am Vet Med Assoc* **177:** 250–253, 1980.

B. Anthrax

Anthrax is caused by *Bacillus anthracis,* a large, gram-positive, spore-forming organism which is highly pathogenic for most herbivorous animals and humans, whereas carnivorous birds and reptiles are resistant. Domestic animals are susceptible to *B. anthracis* in the decreasing order goats, sheep, cattle, horses, pigs, and dogs. Farmed mink are highly susceptible. In ruminants, the disease is usually brief and septicemic; in horses, pigs, and dogs, it is frequently localized to the throat or intestine and may be fatal before invasion of the blood occurs.

When the disease is septicemic, as it usually is in herbivores, the blood and tissues of the animal swarm with vegetative organisms which, when exposed to air or oxygen, form spores of most remarkable durability. It is the combination of these two factors, the number of organisms and the resistance of spores, which is of paramount importance in the epidemiology of the disease.

Bacillus anthracis probably has limited capacity for growth in the external environment, due in part to antagonistic soil bacteria. Growth may occur in alkaline soils with much decaying vegetable matter, and alternate periods of rain and drought and temperatures in excess of 15.5°C may also facilitate growth. The spores are known to remain viable in soil for at least 15 years, and probably much longer, since they have been noted to retain their vitality and virulence for 50 years in the laboratory. Natural survival of vegetative organisms is rather short, equivalent to the short duration of an infection; vegetative organisms do not survive in a carcass but are rapidly killed by putrefactive bacteria. In the terminal stages of the disease, large numbers of bacilli are excreted in all natural excretions, as well as pathological exudates, and these organisms sporulate and perpetuate infections. As a general rule, the spores are very resistant to methods of disinfection with the exception of chemical disinfectants which are oxidizing agents. Spores on skin have even survived tanning processes to become a hazard for humans.

Reliable details on how spontaneous infections are initiated in animals are not available, but the disease in wild herbivores often follows construction requiring excavation, such as fencing and excavation of dug-outs for watering livestock. It is accepted that dogs and pigs acquire the infection as a result of eating an animal which had anthrax, and deaths in humans have occurred after eating inadequately cooked meat from a goat dead of anthrax. Anthrax in pigs has been traced to the ingestion of bone meal which was not sufficiently sterilized. Vegetative bacilli are unlikely to cause the disease since they are rapidly destroyed in the acid medium of the stomach. Cattle and sheep are presumed to obtain the infection by ingestion of contaminated food and water, entry through mucous membranes possibly being aided by local trauma. Cutaneous infection is rare in cattle but has been reported in India, and pulmonary anthrax resulting from the inhalation of spore-laden dust can occur. Infection through the skin is occasionally seen in sheep and may be assisted by grass seed infestation. Ingestion is an important mode of infection in horses and dogs, as indicated by the common occurrence of lesions in the throat. It is also thought that infection can be transmitted to horses by bloodsucking insects.

Intestinal anthrax in pigs probably reflects infection by ingestion.

The pathogenesis of anthrax is an initial lymphangitis and lymphadenitis (Fig. 2.96), which develop into septicemia. This sequence is especially well illustrated in the pulmonary form of the experimental disease; spores which are inhaled are ingested by cells lining alveoli and transported in them to the tracheobronchial nodes, in which vegetation and true initiation of the infection occur. Spread to the blood is via lymphatics as well as by lymphovenous connections within lymph nodes, and numerous bacilli spread in the lymph from node to node as the filtering mechanism of each is successively swamped. The bacilli which enter the blood are taken up in other parts of the reticuloendothelial system, especially the spleen, to establish secondary centers of infection and proliferation.

There is notably little response on the part of a susceptible animal to the local establishment of anthrax infection. Physiological disturbances, clinical signs, and death depend on the development of a massive septicemia. In immune animals, e.g., guinea pigs, infection is followed by a period of 2–4 hr in which the bacilli proliferate; the area then becomes infiltrated by leukocytes and the bacilli fragment with little or no phagocytosis of intact organisms. The lysis of bacilli, of which the first sign is loss of capsule and staining properties followed by fragmentation, is apparently due to anthracidal substances in plasma or, more likely, liberated from leukocytes. Anthrax has always been regarded as dependent on invasiveness rather than

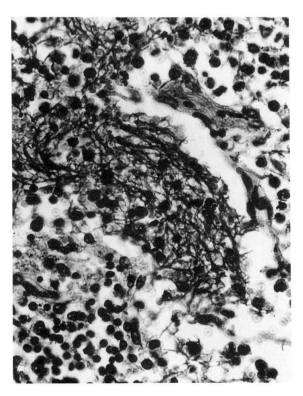

Fig. 2.96 Anthrax bacilli in peripheral sinus. Lymph node. Ox.

toxigenicity, but it appears that antitoxic immunity does, in some manner, inhibit invasiveness.

The vegetative cells produce a small array of toxins. The organisms themselves and the capsular material are virtually nontoxic, although the capsular material may act as a spreading factor and inhibit the activity of leukocytes. The activity of toxins in septicemic anthrax is well illustrated by the demonstration that, once the degree of bacteremia passes a certain threshold which is about 0.3% of the usual maximum, death will occur even though all bacilli may have been destroyed by antibiotic therapy. The toxin consists of three complementary components designated factors I, II and III, or **edema factor, protective antigen,** and **lethal factor,** respectively. Edema factor is an adenylate cyclase that increases cyclic adenosine monophosphate (AMP) after activation by calmodulin. Protective antigen is likely a receptor binding protein that appears to be essential for the biological effects of edema and lethal factors. Lethal factor is a central nervous system (CNS) depressant, but its major effects may be elsewhere. The toxins have been assessed only in experimental laboratory animals, between species and strains of which there are considerable differences in sensitivity. The three toxins are serologically distinct and, since there is autostimulation, it is not surprising that they do not produce lesions when injected separately. The **combined effects** of the three toxins are injury and inactivation of phagocytes, increased capillary permeability, anticomplementary activity, and impairment of coagulation.

Immunity to anthrax appears to depend on the neutralization of toxin. Antibodies against the bacterial cells and capsules are useless, as is evident from the experience that completely avirulent or dead bacilli are not immunogenic; in fact, the most potent vaccines are of bacilli which proliferate and cause an acute local inflammation but which do not invade the blood. The antigen which provokes antitoxic immunity is the protective antigen previously discussed; it is presently regarded as being a nontoxic degradation product (toxoid) of the very labile toxin found in the blood of affected animals. The vaccine of Pasteur apparently resulted from the loss of a temperature-sensitive plasmid that encoded for the protective antigen.

When an animal is suspected of having died of anthrax, the organisms should be detected in smears of blood or local exudate. All bacilli in internal organs are likely to be destroyed in 48 hr or less by putrefaction. Hence, it is best to obtain blood for diagnostic purposes from close to the coronet or the tip of the tail, places which are likely to be involved last by putrefactive processes which destroy the vegetative bacilli. *Bacillus anthracis* occurs in blood in pairs or in short chains of three to four cells. They are large truncate organisms which are easily observed. They are differentiated from putrefactive bacteria by their distinct capsule which stains pink with old methylene blue, and by having square ends when these are apposed. The free ends of *B. anthracis* are often rounded as in *Clostridium* spp. The organism can be cultured readily from putrefied exudates if advantage is taken of its aerobic requirements and the heat resistance of the anthrax spores, to separate it from nonsporulating organisms and sporulating anaerobes. The organism can also be purified in guinea pigs by applying the inoculum to scarified skin. In nonsepticemic anthrax, such as occurs in swine and dogs, it is best to look for the organism in the local exudates and affected lymph nodes. For the diagnosis of anthrax, when other methods have failed or the tissues are old and dry, the Ascoli agar-gel precipitin test is useful but not completely specific.

Bovine anthrax is usually septicemic, and sudden death is usually the first indication of its presence in a herd. Even when cattle are observed closely, they may be dead within 1 hr of showing signs of illness, although some will show general signs of illness for about 24 hr before death. The signs of illness vary with the route of entry and when, as usually happens, entry is by inhalation or ingestion with no area of localization, the animals are depressed and listless. On examination there is high fever, increased heart and respiratory rates, and congested and terminally cyanotic mucosae, which show evidence of bleeding. Animals which survive for a day may have dysentery, abortion, edematous swellings of the perineum, throat, and abdominal wall, and bloodstained milk. Although infection may be virtually synonymous with disease and death in goats and sheep, it is not necessarily so in cattle, in which species some infected animals probably recover. This is suggested by the recovery of some animals which are experimentally inoculated and by the occurrence of transient febrile reactions attributable to no other causes, in herds in which fatal anthrax is occurring.

The carcass of an animal dead of this disease putrefies quickly and becomes very rapidly distended with putrefactive gases, and blood exudes from the natural orifices. These changes are, of course, not diagnostic, but when they are observed in an animal which has died suddenly in an area in which anthrax is endemic, or has at any time occurred, the examination of smears of blood should always precede autopsy. Septicemia in anthrax is a terminal event, and smears of blood may not be helpful when prepared more than a few hours before death.

The morbid picture of the disease in cattle is characterized by splenomegaly, multiple hemorrhages, and edematous effusions in connective tissues. A very large soft spleen is the most significant lesion, and very rarely is it absent. Splenomegaly occurs in other diseases of cattle, but rarely is it as large in association with sudden death. In anthrax the spleen is soft, sometimes it ruptures spontaneously, and when it is incised, the pulp exudes very thick black-red blood which brightens in color on exposure to air. Smears and sections of the spleen reveal very large numbers of bacilli if the carcass is fresh but, when decomposition is advanced, they are destroyed by putrefactive changes. In some cases, splenomegaly is the only lesion. The histology of the spleen is not revealing. The sinus areas are distended with sludged red cells, and the lymphoid follicles are widely separated and hypocellular, but numerous leukocytes and bacilli in chains are present.

It is typical of septicemic anthrax that the organisms are always intravascular.

The blood is thick and dark, frequently described as tarry, and either it is not clotted or the clots are very soft and friable, compatible with the effects of the combined toxins in inhibiting the clotting system. There are likely nonspecific effects present such as the disseminated coagulopathy and fibrinogenolysis that accompany hyperthermia. Small hemorrhages are common in the mucous and serous membranes and in the subcutaneous connective tissues. Loose connective tissues in any location may be infiltrated with gelatinous fluid, and accumulations of such fluid in serous cavities are stained with blood. There is congestion, swelling, and degeneration of parenchymatous organs. The myocardium is dull and flabby.

Cattle are moderately resistant to *B. anthracis* so that local lesions may occur at the site of entry. The local lesions are usually in the small intestine and take the form of ulcerative hemorrhagic enteritis, but acute inflammation in the abomasum and large intestine may also occur. The most severe lesions may be over the lymphoid tissues of the intestine, or extend for a considerable distance from these. The mucous membrane is intensely red and at a greater distance is sprinkled with small hemorrhages. The contents of the intestine are then deeply stained with blood. Superficial necrosis and ulceration occur in some areas of most intense hyperemia. The corresponding mesentery, up to the regional nodes, is infiltrated with gelatinous fluid as a result of acute lymphangitis, and the fluid may be stained with blood. The regional nodes have the appearance of the spleen. They are enlarged, red-black, and on cut surface are moist and shiny. The vessels are intensely congested, and hemorrhage extends into the peripheral sinuses and cortex. Bacilli are numerous and leukocytes are present, but there is no necrosis.

In some cases in which the organisms gain entry through the oropharynx, there is hemorrhagic lymphadenitis of the glands of the throat and edema of the connective tissues in these regions. The occasional case of pulmonary anthrax in cattle is characterized by acute congestion and consolidation of a portion of the lung with larger areas of interstitial edema, edema of the mediastinum, and regional hemorrhagic lymphadenitis. The pulmonary lesion is exudative but lacks the full gamut of inflammatory change unless another cause of pneumonia is superimposed.

Sheep are more susceptible to *B. anthracis* than are cattle, and local lesions do not occur except in the unusual instances of percutaneous infection, in which the lesion may take the form of a spreading edema from the outset or initially appear as hard circumscribed nodules. The disease in sheep takes the same course as that in cattle except that it is even more rapid. Splenomegaly is not so prominent in sheep as in cattle, likely because of the greater level of collagen in the splenic capsule of sheep. The parenchyma is, however, dark and soft. Edematous effusions do not occur in sheep.

Clinical signs of anthrax in **horses** may last for several days and are characterized by colic or by large edematous swellings. The swellings, which can be very extensive, occur on the ventral part of the abdomen and thorax, the legs, in the perineal region, and about the external genitalia. Dysentery may accompany the acute colic. When ingestion is the route of infection in horses, the primary lesion may be in the throat or the intestine, and death may occur from the local reaction and without septicemia. The intestinal lesions are similar to those described for cattle and the pharyngeal lesions are similar to those subsequently described for swine. When septicemia occurs, the morbid changes are the same as in cattle, including very prominent enlargement and congestion of the spleen.

Pigs are relatively resistant to anthrax. They acquire the infection from eating infected flesh and the infection remains localized to the throat or intestine. Since septicemia is exceptional, splenomegaly is not a prominent part of the gross picture. The characteristic sign is swelling of the pharyngeal region and neck. Some pigs have diarrhea and dysentery, but it is unusual to have intestinal localization without pharyngeal localization. Anthrax without illness has been observed in swine; in such cases the disease is limited to isolated mesenteric or pharyngeal nodes.

The local lesion of anthrax in swine is a typical carbuncle at the point of entry, with acute regional lymphadenitis and lymphangitis. Some bacilli no doubt reach the blood, but they do not establish septicemia. Bacteria may localize in liver, spleen, or kidney to produce a metastatic carbuncle but, usually, only individual nodes near the site of entry are involved. The lymphadenitis may be diffuse or focal, but in both cases it is, in the initial stages, hemorrhagic. An intense leukocytic infiltration occurs, all cells within the affected portions of the node die, and the focus becomes encapsulated. With necrosis, the affected tissue changes from a brick-red to a gray friable mass which can be easily shelled-out when the gland is incised.

In primary intestinal anthrax in pigs, the initial lesion is a focal or multifocal hemorrhagic enteritis, with a central zone of diphtheresis which eventually ulcerates. The adjacent serosa and mesentery are thickened with edema fluid and yellowish, with foci and streaks of hemorrhage; they are the site of focal hemorrhagic necrosis due to acute necrotizing vasculitis and lymphangitis. These mesenteric lesions extend only as far as the regional nodes, which show the type of lymphadenitis characteristic of anthrax in swine.

Dogs are reputedly quite resistant to *B. anthracis,* but a number of outbreaks have been observed in kennels in which the dogs have inadvertently been fed meat from an animal which has died of the disease. Anthrax in dogs may pursue a peracute course to sudden death, it may be of the pharyngeal type in which extensive edema develops in the face, head, and neck, or it may be of the intestinal type with signs of acute gastroenteritis. Anthrax may occur in mink with high mortality after feeding fresh meat from infected animals.

Bibliography

Gill, I. J. Antibiotic therapy in the control of an outbreak of anthrax in dairy cows (correspondence). *Aust Vet J* **58:** 214–215, 1982.

Gleiser, C. A. Pathology of anthrax infection in animal hosts. *Fed Proc* **26:** 1518–1521, 1967.

Leppla, S. H. Anthrax toxin edema factor: A bacterial adenylate cyclase that increases cyclic AMP concentrations in eukaryotic cells. *Proc Natl Acad Sci USA* **79:** 7–9, 1982.

Mikesell, P. *et al.* Evidence for plasmid-mediated toxin production in *Bacillus anthracis. Infect Immun* **39:** 371–376, 1983.

Prokupek, K., Dvorak, R., and Polaecnk, R. Immunogenic effect of anthrax protective antigen. *Vet Med* **26:** 279–290, 1981.

Timoney, J. F. *et al.* The genus *Bacillus. In* "Hagan and Bruner's Microbiology and Infectious Diseases of Domestic Animals," 8th Ed., p. 206. Ithaca, New York, Comstock Publishing Associates, 1988.

C. Streptococcal Adenitis in Swine

Jowl abscess is a cervical adenitis caused by *Streptococcus porcinus,* and like strangles of horses, the deep infection follows colonization of the oral cavity and likely the tonsils. The disease has become less common with the introduction of better hygiene, age-matched groups, and feeding equipment designed to avoid sharp projections. The organism has eight serotypes, based on carbohydrate antigens, of which type 4 is most frequently isolated from cervical lymphadenitis of swine. Baby pigs have passive immunity and transmission occurs through direct contact with infected animals which can shed organisms for months, and also from contaminated environment. Invasion of the nasopharynx is followed by mild fever and leukocytosis which resolve, and by lymph node enlargement in 2 weeks, which less often progresses to open drainage. The mandibular nodes are most often involved, followed by the retropharyngeal and parotid nodes. The disease is not often of clinical importance, and affected animals grow well, but losses result from condemnation at slaughter. Bacterins are less effective than live oral culture in prevention.

The abscesses are usually multiple and measure 1–10 cm in diameter. The pus is typically without odor, greenish in color, and creamy in consistency. *Streptococcus porcinus,* a beta hemolytic *Streptococcus* which belongs to Lancefield's group E, can be isolated regularly from the pus.

Rhodococcus equi is often present in the submaxillary lymph nodes of swine and has been proposed as the cause of lesions typical of tuberculosis in these nodes. Evidence suggests, however, that *R. equi* is not responsible for the tuberclelike lesions, which are thought to be caused by *Mycobacterium* spp. of one or other variety. *Rhodococcus equi* has been isolated from lymphadenitis in cattle.

Bibliography

Armstrong, C. H., Boehm, P. N., and Ellis, R. P. Experimental transmission of streptococcal lymphadenitis (jowl abscess) of swine. *Am J Vet Res* **31:** 823–829, 1970.

Miller, R. B., and Olsen, L. D. Frequency of joint abscesses in feeder and market swine exposed to group E streptococci and nursing pigs. *Am J Vet Res* **44**: 945–948, 1983.

Thal, E., and Rutqvist, L. The pathogenicity of *Corynebacterium equi* for pigs and small laboratory animals. *Nord Vet Med* **11**: 298–304, 1959.

Wessman, G. E., Wood, R. L., and Nord, N. A. Identification of three new serotypes of group E streptococci isolated from swine. *Cornell Vet* **73**: 307–313, 1983.

D. Streptococcal Adenitis in Dogs

Streptococcal adenitis of dogs is similar to the disease in swine. It occurs in minor endemics in kennels and is characterized by pharyngitis, fever, conjunctival discharge, and enlargement of the submaxillary nodes. The illness is transient in most cases, but in approximately 10% of dogs, the course is prolonged for 2 weeks or more, and in these, there is a tendency for the nodes to suppurate. The abscesses may be encapsulated and sterilized, or they may fistulate onto the skin and heal. Dysphagia, and occasionally asphyxia, occur in puppies. Typically, the *Streptococcus* belongs to Lancefield's group G and, coincident with the pharyngeal infection, bitches may have inflammation of the genital tract caused by the same organism.

Cervical adenitis caused by the group G organism may occur in kittens.

Bibliography

Ettinger, S. J. "Textbook of Veterinary Internal Medicine," 3rd Ed., Vol. 2. Philadelphia, Pennsylvania, W. B. Saunders, 1989.

Muller, G. H., Kirk, R. W., and Scott, D. W. "Small Animal Dermatology." 3rd Ed., Philadelphia, Pennsylvania, W. B. Saunders, 1983.

Swindel, M. M. *et al.* Pathogenesis of contagious streptococcal lymphadenitis in cats. *J Am Vet Med Assoc* **179**: 1208, 1981.

E. Tularemia

Tularemia (deer fly fever) is caused by *Francisella tularensis*. The disease was first described in Tulare county of California and is still more common in the western United States of America than elsewhere. The organism is a tiny, gram-negative, very pleomorphic coccobacillus which is a strict aerobe and shares many cultural and epidemiologic features with *Yersinia pestis,* the cause of bubonic plague. *Francisella tularensis* is found worldwide except for Australia and Antarctica, but the biovar tularensis is found only in North America. *Francisella tularensis* infects a wide range of species, including most domestic animals, humans, and wild rodents, and it is in these last two that the disease is most often fatal.

The organism is abundant in nature as an infection of many species of rodents, and it is from these, either directly or by the mediation of insect vectors, that humans and domestic animals acquire the infection. The organism is able to penetrate intact skin and mucous membranes,

but it is also infective by ingestion, inhalation, and inoculation by biting insects and ticks.

Tularemia in humans is a severe systemic disease, with various manifestations depending on dissemination or localization. The disease in rodents is recognized by the presence of miliary white foci 2 mm or more in diameter in the liver, spleen, and lymph nodes. They are indistinguishable grossly from the lesions caused by *Yersinia pseudotuberculosis*. Histologically the lesions are characterized by very focal but complete necrosis. Neutrophils and pus may be present early, and macrophages accumulate, but in slightly older and larger lesions there is total coagulative necrosis with a granularity that resembles caseation (Fig. 2.97). A very few fibroblasts and macrophages produce a sharp, narrow margin for the lesion. The bacteria can be demonstrated quite readily in the lesions, in clumps in macrophages and especially as dead but still distinguishable ones in the center of the focus. The lesions in lymph nodes are often larger than those in the liver and may be readily visible grossly as wedge-shaped areas of cortical necrosis demarcated by a narrow zone of intense reactive hyperemia. The affected nodes are palpably enlarged in the living animal and may discharge a thin reddish pus onto the skin. The lymphadenitis may be generalized or restricted to the nodes draining the site of infection which, if visible, is an ulcerated papule.

Francisella tularensis is probably common as an infection in animals, but the disease tularemia is not. It is probable, too, that the infection may remain latent in do-

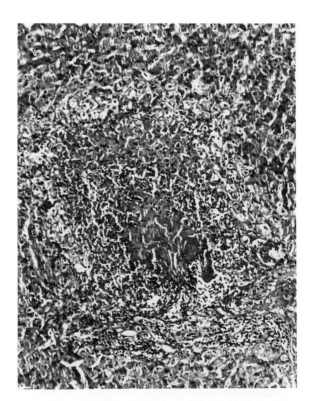

Fig. 2.97 Pyogenic granuloma in tularemia. Liver. Sheep.

mestic animals for long periods without causing ill health and with or without focal lesions, as has been observed in dogs. As a facultative intracellular parasite, it may persist for years as a latent infection, for which effective immunity is cell mediated.

Naturally acquired, latent (subclinical) infections with *F. tularensis* have been observed in dogs and cats but, as a general rule, these species are resistant to the disease. Fatalities due to tularemia have been observed in foals and sheep in association with heavy infestations by ticks, and it is probable that the disease is endemic in sheep in areas where the reservoir rodents and ticks, especially *Dermacentor andersoni* and *Amblyomma americanum* abound. The disease in foals is characterized by a systemic and febrile illness with, at autopsy, enlargement of liver, spleen, and kidneys, and the presence of the typical small necrotic foci of this disease. Tularemia in sheep is also associated with a heavy infestation of ticks; recovery can occur if the ticks are removed before the illness is far advanced. Affected sheep have high fever, stiffness of gait, depression, diarrhea, and an increased respiratory rate. Lesions may be confined to the superficial lymph nodes or show the more classical localization. Pneumonia, probably of other cause, and anemia caused by the ticks are frequently present.

Bibliography

Bell, J. F. *et al.* Enigmatic resistance of sheep (*Ovis aries*) to infection by virulent *Francisella tularensis. Can J Comp Med* **42:** 310–315, 1978.

Calhoun, E. L. Natural occurrence of tularemia in the Lone Star tick, *Amblyomma americanum* (Linn.), and in dogs in Arkansas. *Am J Trop Med Hyg* **3:** 360–366, 1954.

Koskela, P., and Herva, E. Cell-mediated and humoral immunity induced by a live *Francisella tularensis* vaccine. *Infect Immun* **36:** 983–989, 1982.

F. Pseudotuberculosis

Yersinia pseudotuberculosis occurs worldwide in wild rodents and birds and is widespread in nature, recoveries having been made from soil, milk, and feed. The organism regularly produces disease, often in epidemic proportions, only in rodents and birds. Sporadic infections with this organism and occasional outbreaks of disease occur in domestic species, laboratory colonies, and zoos. Cats, because of their contact with rodents and birds, are the domestic species most apt to be secondarily involved by outbreaks of the disease in its natural hosts. Losses of serious proportion, however, occur in sheep, which are exposed to large numbers of organisms during outbreaks of the disease in rodents during cold weather. The ovine disease is known as pyemic hepatitis.

The general pattern of this disease is the same in all species, the variation which occurs with respect to severity and duration being related to species susceptibility, and numbers and virulence of the organisms. *Yersinia pseudotuberculosis* is a facultative intracellular parasite

which explains the latent carrier state and the need for strong cell-mediated immunity for protection from infection. The route of transmission is by ingestion and, in susceptible animals, organisms enter the body through the intestine. Small necrotic foci develop in the Peyer's patches of the ileum and colon and extend as lymphangitis to the regional nodes. The organism becomes septicemic and may kill susceptible rodents at this stage; more typically, and in all domestic species, caseonecrotic foci form in the mesenteric nodes, spleen, and liver, often in association with fibrino-hemorrhagic inflammation in the small intestine (Fig. 2.98A, B). The hepatic foci, which are the most obvious, are 1–10 mm in diameter, whitish, and have no or scant tendency to encapsulation or softening. They are interspersed with irregular areas of parenchymal collapse, which probably result from vasculitis and thrombosis (Fig. 2.99A). Microscopically, there is necrosis, with bacterial colonies and fragmented leukocytes surrounded by macrophages. Giant cells are absent even from later contracting granulomas (Fig. 2.99B). The mesenteric nodes and spleen contain similar foci, and are enlarged by lymphoid and histiocytic hyperplasia. The mesenteric nodes in the cat may be 2–4 cm in diameter and can be grossly confused with intestinal toxoplasmosis or lymphomatosis.

Yersiniosis has emerged as a significant cause of disease in farmed ruminants, including deer. Affected deer may be found moribund or dead, but animals under observation

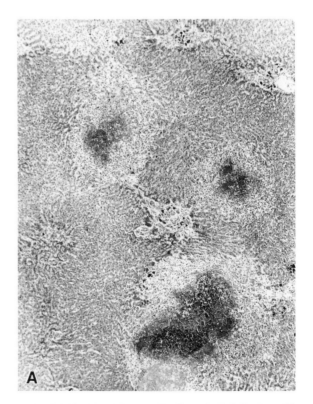

Fig. 2.98A *Yersinia pseudotuberculosis* infection. Sheep. Multiple pyogenic granulomas in liver.

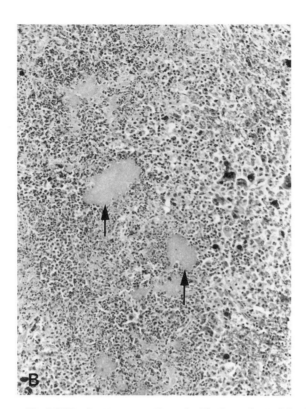

Fig. 2.98B Acute suppurative splenitis. Large bacterial colonies (arrows).

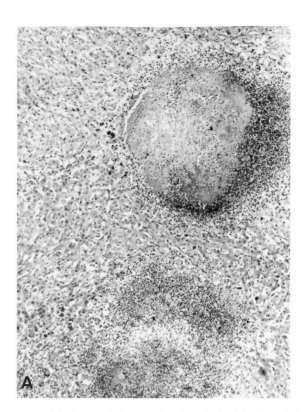

Fig. 2.99A *Yersinia pseudotuberculosis* infection. Liver. Sheep. Septic phlebitis, sublobular vein.

are systemically ill with profuse diarrhea. The incidence of disease in younger age groups rivals that of salmonellosis. In addition to the mesenteric lymphadenitis and hepatitis, enteritis is consistently present and characteristic in its histologic expression. Numerous bacterial colonies are present in the lamina propria, associated with multiple suppurative foci or a diffuse suppurative enteritis.

In view of the extraordinarily wide host range of *Y. pseudotuberculosis* in natural infections, it is surprising that the infection is not recorded more often. This may be due in part to a preference of the organism for growth at room temperature rather than at 37°C. Although susceptible laboratory species, such as guinea pigs and mice, are readily infected experimentally, there are few pathogenetic studies in domestic species. The frequency of natural infection in cats contrasts with the difficulty in establishing experimental infection in this species, which suggests the need for predisposing alteration in the alimentary mucosa.

Bibliography

Gemski, P. *et al.* Presence of a virulence-associated plasmid in *Yersinia pseudotuberculosis*. *Infect Immun* **28:** 1044–1047, 1980.

Harcourt-Brown, N. H. *Yersinia pseudotuberculosis* infection in birds. *Vet Rec* **102:** 315, 1978.

Jerrett, I. V., Slee, K. J., and Robertson, B. I. Yersiniosis in farmed deer. *Aust Vet J* **67:** 212–214, 1990.

Oburolo, M. J. A review of yersiniosis (*Yersinia pseudotuberculosis*) infection. *Vet Bull* **46:** 167–171, 1976.

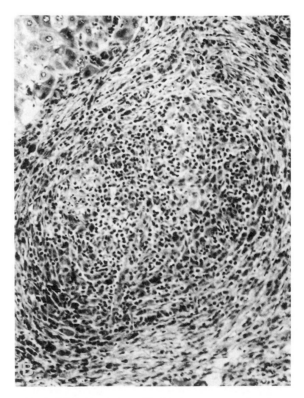

Fig. 2.99B Yersiniosis. Residual granuloma of protracted infection.

Slee, K. J., Brightling, P., and Seiler, R. J. Enteritis in cattle due to *Yersinia pseudotuberculosis* infection. *Aust Vet J* **65:** 271–275, 1988.

Timoney, J. F. *et al.* The Enterobacteriaceae: The non-lactose-fermenters. *In* "Hagan and Bruner's Microbiology and Infectious Diseases of Domestic Animals," 8th Ed., p. 89. Ithaca, New York, Comstock Publishing Associates, 1988.

G. Histoplasmosis

Histoplasmosis is caused by *Histoplasma capsulatum,* an intracellular parasite of the monocyte–macrophage system, and is characterized by diffuse involvement of the reticuloendothelial system. *Histoplasma capsulatum* is a dimorphic fungus. The parasitic phase is yeastlike, whereas the nonparasitic phase produces an abundant mycelium. The mycelium produces two types of spores: a small, smooth, globose microconidium 2–3 μm in diameter, and a large, thick-walled, globose macroconidium or chlamydospore, which bears prominent surface projections. The chlamydospores are an identifying feature of the fungus and also the initiators of the parasitic phase.

Histoplasmosis has a worldwide distribution and is frequently diagnosed in the Mississippi, Ohio, and St. Lawrence River valley areas of North America. It is a largely noncontagious disease of humans, dogs, cats, swine, cattle, horses, and wild animals, and it is suspected that the endemic distribution of the disease is related to factors which favor the persistence and growth of the organism in the soil. The ubiquitous nature of the organism, its intracellular habitat, and its low infectivity suggest that cell-mediated immunity may be a major factor in the pathogenesis of disease.

It is generally agreed that infection is obtained from the soil, either by ingestion or by inhalation of dust, and a distinct association has been repeatedly demonstrated between *Histoplasma capsulatum* and places where pigeons congregate, with abandoned chicken runs and houses, and with bat caves. Chickens themselves are not infected because their high body temperature is not conducive to growth of the organism.

The disease in animals is seen most frequently in dogs and less often in cats. Transmission from dog to dog has been established, but it is not known whether dogs can transmit the disease to humans. When histoplasmosis becomes apparent clinically, the infection is disseminated, and the disease is then progressive and always fatal. Only a minority of infections become disseminated and cause disease, whereas the majority occur without signs or lesions. A high incidence of infection in endemic areas has been confirmed by cultural means, either by direct culture of organs or by first passing infected tissues through mice, which are highly susceptible to *H. capsulatum*. Positive cultures can be obtained from about 50% of unrelated cats and dogs.

Transmission is by inhalation of air-borne conidia followed by phagocytosis in the lung and then transport in phages to other areas of the body. This route accounts for the high incidence of infection in the bronchial and cervical lymph nodes. Since intestinal lesions are usually present, it is felt that ingestion might provide a direct route of infection. However, in disseminated disease, the intestine could be secondarily infected like many other tissues.

The course of latent infections in animals is not known, but the evidence suggests that such infections can persist for many months and possibly are permanent. Infection has been known to persist, without causing ill health, for 2 years in guinea pigs, 6 months in other laboratory rodents, and 4 months in kittens.

Animals with advanced disease due to *Histoplasma capsulatum* show emaciation, persistent diarrhea, pyrexia, and enlargement of liver, spleen, and lymph nodes. In a particular animal, the clinical signs are most commonly related to either the respiratory or the gastrointestinal tract.

Animals with histoplasmosis most consistently have a mild nonresponsive, normochromic, normocytic anemia. The leukocytic changes vary with the reserves of the animal and the stage of the disease. If the animal is in good body condition, the bone marrow is competent, and there is neutrophilic leukocytosis with some degree of monocytosis. Later with debilitation, the leukocyte count drops to normal levels or lower with left shift and marked toxic changes, including the appearance of Doehle bodies. There is usually lymphopenia and eosinopenia. Since the organisms are readily recognized on microscopic examination, fine-needle aspiration of liver, spleen, enlarged lymph nodes, marrow, or skin provides adequate material for diagnosis in appropriate cases. Occasionally the organism may be diagnosed in fine-needle aspirates of lung or bronchoalveolar lavage fluids, or lavage of the turbinates in animals with sinusitis. Cytologically the organisms are present in the cytoplasm of macrophages, unless the cells have been injured, and are somewhat larger than when seen in histologic preparations.

Pathologically, the pulmonary lesions of histoplasmosis may be in the form of grayish, rounded nodules of 1–2 cm diameter and with a distinct tendency to become confluent, or there may be a diffuse increase in the consistency of the lungs. When the intestine is involved, as it frequently is, the lesions are located chiefly in the lower part of the small intestine. The mucosa is the site of nodular thickenings or of corrugations similar to those seen in Johne's disease of cattle. The thickenings are due to infiltration of lymphocytes, plasma cells, and macrophages in the lamina propria and submucosa, and they may extend also through the wall to the subserosa, giving the gut a thickened pipestem appearance. When the thickening is extreme, ischemic ulcerations occur.

The lymph nodes are greatly enlarged but are discrete and without adhesions. There may be no indication of normal architecture on the cut surface, with the uniformity resembling lymphoma, except that the nodes are firm and dry. Histologically there are coalescing granulomas with histiocytosis, and cortical replacement by the reaction.

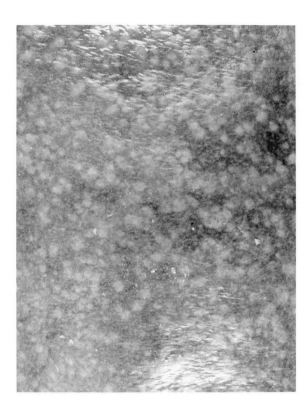

Fig. 2.100 Miliary epithelioid granulomas. Capsular surface of enlarged spleen. Histoplasmosis. Dog.

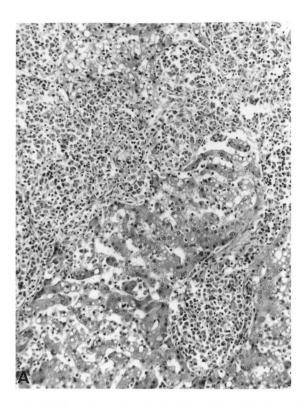

Fig. 2.101A Macrophage infiltration of liver. Histoplasmosis. Dog.

The spleen is enlarged (Fig. 2.100), sometimes to several times its normal size, gray and firm. There is marked sinus expansion and filling by fixed cells of stromal origin and by colonization with macrophages, many of which contain the ingested organisms. There is lymphoid atrophy varying in degree with the stage of debility of the animal. The liver is uniformly enlarged, firm, and gray. The discoloration is diffuse and related to capsular thickening without focal lesions. The infiltrating cells collect in miliary foci in the portal triads and sinusoids, causing extensive displacement and atrophy of the parenchyma (Fig. 2.101A). The adrenal glands are often involved, either the medulla, the cortex, or both, in histoplasmosis which is allowed to run its full course (Fig. 2.101B). Scant normal tissues may then remain in these organs.

Since *Histoplasma capsulatum* is an intracellular parasite of macrophages and causes diffuse involvement of the reticuloendothelial system, focal discrete lesions are not to be expected. The enlargement of organs is due to extensive proliferation and infiltration with monocytes and epithelioid macrophages in whose cytoplasm many of the typical yeasts in small or large numbers are found (Fig. 2.101B). The yeast bodies are 2–4 μm in diameter in sections, and when stained with hematoxylin and eosin, they appear as basophilic dots surrounded by a clear halo. The halo is part of the yeast's cell wall and can be demonstrated by stains for bound glycogen, the organism then appearing as a ring. Such stains are very helpful in distinguishing

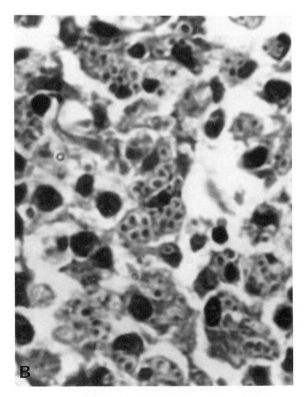

Fig. 2.101B Intracellular *Histoplasma capsulatum*. Adrenal cortex. Dog.

these organisms from cellular debris, but are not necessary when the yeasts are plentiful. Necrosis of tissues is not often present in histoplasmosis, although central caseation without calcification may occur in large, dense accumulations of macrophages. Giant cells are seldom seen and there is only mild fibrosis.

The diagnosis of occult histoplasmosis can be made only by isolation of the organism from tissues. Responsibility of this organism for focal nonprogressive lesions can be accepted only when the fungus can be demonstrated histologically in the lesion, and this is rarely possible even in animals which are sensitive to histoplasmin.

The organs mentioned are those most consistently and obviously involved in histoplasmosis. In the fatal disseminated disease, lesions of similar type may be found in other organs and tissues. There is seldom any difficulty in demonstrating organisms in sections, and clinically they are readily cultured. Culture should not be undertaken without precautions to prevent the inhalation of chlamydospores. Biopsy of enlarged lymph nodes and aspiration biopsy of bone marrow are useful for diagnostic purposes. In some cases, the organisms can be demonstrated in circulating monocytes, but they are never plentiful, so examination is best performed on buffy coat.

Bibliography

Clinkenbeard, K. D., Cowell, R. L., and Tyler, R. D. Disseminated histoplasmosis in cats: 12 cases (1981–1986). *J Am Vet Med Assoc* **190:** 1445–1448, 1987.

Farrell, R. L., and Cole, C. R. Experimental canine histoplasmosis with acute fatal and chronic recovered courses. *Am J Pathol* **53:** 425–445, 1968.

Rowley, D. A., Haberman, R. T., and Emmons, C. W. Histoplasmosis: Pathological studies of fifty cats and fifty dogs from Loudoun County, Virginia. *J Infect Dis* **95:** 98–108, 1954.

H. Leishmaniasis

The genus *Leishmania* is of protozoan parasites which are classified with the family Trypanosomidae. Species of the genus are parasites of humans, dogs, and other mammals. Species of sandflies (*Lutzomia*) are the intermediate hosts and are essential for the maintenance of virulence, whereas other bloodsucking insects, including members of the genera *Phlebotomus* and *Stomoxys* and the tick *Rhipicephalus*, may act as mechanical vectors.

The protozoa proliferate by binary fission in the midgut of the female sandfly and assume a leptomonad form, leaf shaped with a single flagellum arising at the anterior pole. Infection of mammals occurs when the sandfly sucks blood, and in the mammalian host the protozoa assume the leishmanial form as rounded cells of ~2.0 μm diameter with a vesicular nucleus and small kinetoplast but no flagellum. They multiply by binary fission and as leishmanial forms, the Leishman–Donovan bodies, are intracellular parasites of macrophages.

Leishmaniasis is important as a disease of humans, and wherever it occurs in humans, it may also occur in dogs.

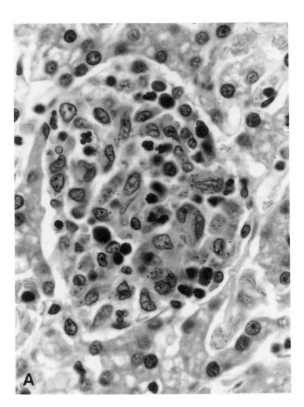

Fig. 2.102A Leishmaniasis. Dog. Epithelioid granuloma. Liver.

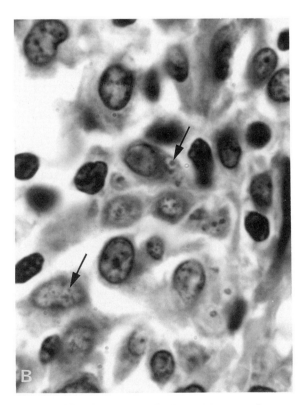

Fig. 2.102B Sinus area of spleen with macrophages, some containing *Leishmania* (arrows). Dog.

Leishmaniasis actually includes three diseases: cutaneous leishmaniasis (oriental sore) is caused by *L. tropica* and occurs in countries about the Mediterranean Sea; mucocutaneous leishmaniasis *(espundia)* is caused by *L. braziliensis* and occurs in Central America; visceral leishmaniasis *(kala-azar)* is caused by *L. donovani* and is endemic in parts of Europe, Africa, and Asia. The species of *Leishmania* are not well distinguished. Speciation depends on degrees of cross-immunity, different clinical manifestations, geographic location, and reservoir hosts. In some regions, such as India, the human population may be the reservoir, but in other regions dogs, cats, and other carnivores including rats probably serve as reservoir hosts.

Infections of animals with *Leishmania* spp. are probably quite common in endemic areas, but disease appears not to be. Latency of the infections, especially in the case of *L. donovani,* is also probable. Both cutaneous and visceral forms of the disease have been described in dogs, as well as cases of visceral disease in which the organisms are diffusely present in the dermis. Leishmaniasis is a disease of the monocyte–macrophage system and the visceral disease mimics histoplasmosis. The protozoa are not cytopathogenic in the usual sense, and destruction of the host macrophages appears to be purely a mechanical consequence of proliferation of the protozoa in the cytoplasm.

The **cutaneous lesions** take the form of chronic ulcers which develop from inflammatory papules at the site of the insect bite. The organisms are inoculated by the biting insect and are soon ingested in histiocytes. Rapid proliferation of the protozoa disrupts the phagocytes, and the released organisms are ingested by further phagocytes to repeat the process. Lymphocytes and plasma cells surround the lesion and neutrophils are attracted to the debris. When the inflammation extends to the overlying epithelium, ulceration occurs. Numerous parasites are present within the macrophages and some are free in the tissue (Fig. 2.102).

The clinical signs of **visceral leishmaniasis** in the dog are of chronic debility and often recurrent oculonasal discharge, with some crusting of the nose, and recurrent diarrhea. There may be mild enlargement of lymph nodes and the spleen is always enlarged to some extent in the visceral form, but at any particular time may not attract clinical attention.

Pathologically there are often focal scurfy skin lesions but generally, in visceral leishmaniasis, the presentation is of a mature dog in poor body condition with a rough hair coat. The oral and cervical viscera are normal and the lungs generally have mild tan mottling but are otherwise normal, as is the heart. The liver contains numerous granulomas (Fig. 2.102A) and is symmetrically enlarged and dark brown. The spleen is two to three times or more normal size with symmetrical enlargement and is dark brown to black on capsular surface. There is mild irregular enlargement of lymph nodes with no other significant changes in the abdominal cavity, except the kidneys are darker than normal. The bone marrow is uniformly red-

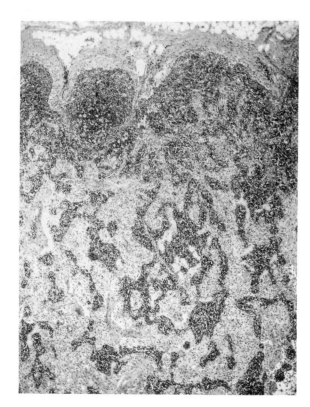

Fig. 2.103 Leishmaniasis. Dog. Lymph node. Follicular and paracortical atrophy of chronic disease. Sinus histiocytosis and medullary cord hyperplasia.

dened in midfemoral shaft in well-developed cases, but the fat is generally of normal character.

The lesions initially are of hemic–lymphatic hypertrophy (Fig. 2.102B) with macrophage proliferation and focal granulomas. The splenic follicles are hyperplastic, often with follicular hyalinosis. In advanced cases, there is atrophy of nodes and spleen, and the sinus areas may be diffusely occupied by large macrophages heavily laden with intracytoplasmic organisms and numerous plasma cells (Fig. 2.103). The lesions in bone marrow may be focal, but consist of clusters of epithelioid macrophages with phagocytosed organisms (Fig. 2.104A,B). The renal changes are variable and may consist of interstitial scarring with some parasitic involvement; however, the effects of hyperproteinemia and immune complexes appear to injure the kidney indirectly. Animals with leishmaniasis and amyloidosis have been described, and there may be concurrent changes due to both processes. Diagnosis is based on the demonstration of the organism in cytologic (Fig. 2.104A) or in histologic preparation, the former being easily achieved by aspiration of marrow, node, spleen, or liver.

Bibliography

Corbeil, L. B. *et al.* Canine visceral leishmaniasis with amyloidosis: An immunopathological case study. *Clin Immunol Immunopathol* **6:** 165–173, 1976.

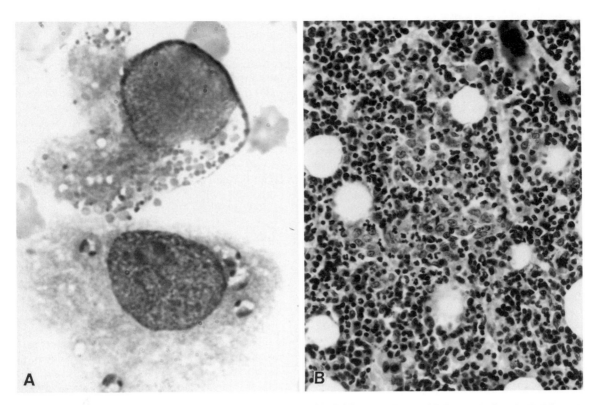

Fig. 2.104 Leishmaniasis. Dog. (A) Marrow aspirate. Epithelioid macrophage with intracytoplasmic *Leishmania*. Ruptured eosinophil at top. (B) Marrow histology. Hyperplasia with dense cellularity, increased granulocytes and macrophages.

McElrath, M. J., Murray, H. W., and Cohn, Z. A. The dynamics of granuloma formation in experimental visceral leishmaniasis. *J Exp Med* **167**: 1927–1937, 1988.

Petersen, E. A. *et al*. Specific inhibition of lymphocyte-proliferation responses by adherent suppressor cells in diffuse cutaneous leishmaniasis. *N Engl J Med* **306**: 387–392, 1982.

Tryphonas, L. *et al*. Visceral leishmaniasis in a dog: Clinical, hematological and pathological observations. *Can J Comp Med* **41**: 1–12, 1977.

I. Theileriosis

Theileria are protozoan parasites of the family Babesiidae. Speciation within the genus is not precise and the system in use depends on geographic distribution, vector, mode of pathogenesis, morphology, and frequency of schizonts in lymphocytes and of piroplasmic stages in red cells, as well as on cross-immunity and host specificity. The species at present recognized as pathogenic for domestic animals include *T. parva*, the cause of **East Coast fever** of Africa; *T. lawrenci*, the cause of **corridor disease** of southern Africa; *T. annulata*, which causes tropical theileriosis in the Middle East; and *T. mutans* (*sergenti*) of very wide distribution. These species are parasites of cattle. *Theileria hirci* and *T. ovis* are parasites of sheep in North Africa, the Middle East, and southern parts of Eurasia. The *Theileria* spp. are found in red cells and reproduce by schizogony in lymphocytes. Related genera are the *Gonderi*, which reproduce by fission in red cells and by schizogony in lymphocytes, and the *Cytauxzoon*, which reproduce by fission in red cells and by schizogony primarily in macrophages.

The *Theileria* are spread by biological tick vectors, principally of the *Rhipicephalus* and *Hyalomma* genera. Larval and nymphal ticks ingest parasites in the erythrocytes of infected hosts. After engorgement, they detach from the host animal and molt through to the next instar. During this period, the parasite migrates to the acinar cells of the salivary gland of the tick. At this stage, the parasite in the salivary gland of the unfed tick is not infective for cattle. Once the infected tick has started to feed, the parasites multiply in the salivary gland and maximal infective stages for cattle are excreted in the saliva between 3 and 5 days after commencement of blood feeding.

This description relates particularly to *T. parva* and the acute form of the disease East Coast fever. **East Coast fever** is encountered when susceptible animals are introduced to endemic areas and subjected to challenge by large numbers of infected ticks. The number of adult ticks provides an index of infection pressure, but the outcome of exposure is determined also by the susceptibility of the cattle and the virulence of the infecting strain of parasite. Thus, theilerial disease can vary from a benign or nonclinical infection to one of acute course with high morbidity and mortality.

There is no evidence that different breeds of cattle vary in their initial susceptibility to *T. parva* or that, antigenic differences between strains notwithstanding, the parasites are capable of antigenic drift in the course of infection, although this might be expected. The components of virulence of *Theileria* organisms are incompletely understood. The sporozoites, of tick origin, are required to invade and infect host lymphocytes, and this capability is one component of virulence. Infected lymphocytes, are transformed *in vitro* into lymphoblastic cells, and this appears to parallel the *in vivo* event as the main indicator of pathogenicity. The macroschizonts develop in the cytoplasm of the transformed cells and then divide synchronously with the host lymphocytes to infect their daughter cells. The proliferation so induced causes the clinically recognizable polylymphadenopathy. Whereas East Coast fever, and corridor disease, are ultimately lymphocytolytic, and virulence is probably associated with macroschizogony, the pathogenesis of disease associated with some other species of the genus may be more intimately related to fission of the piroplasmic forms in erythrocytes.

Most breeds of cattle are highly susceptible to *Theileria parva* and with a high level of challenge, mortality approaches 100%. The main vector is the tick *Rhipicephalus appendiculatus*. In enzootic areas, 40% of the calf crop may die of this disease if ticks are uncontrolled. The severity of the disease is directly proportional to the number of organisms inoculated into the animals by the ticks. From 4 to 20 days after infection by the transmission of the organisms with an infected tick, macroschizonts are found in lymphoblasts of the lymph nodes draining the site of infection. These schizonts increase in number throughout the lymph nodes by an order of magnitude each 10 days, until virtually all of the lymphocytes are parasitized. The macroschizonts may be identified in cytoplasm of infected lymphocytes using blood stains and are known as **Koch's blue bodies** (Fig. 2.105) which are considered diagnostic for the disease. From the tenth day of lymph node infection, increasing numbers of macroschizonts enter the microschizont stage, in which the host cells are destroyed and the micromerozoites are released to invade the erythrocytes. The piroplasms, which are the intraerythrocytic stage of the parasite, are considered to be infective for ticks and first appear on day 12 after infection. They increase rapidly in number until death, at which point a high proportion of the erythrocytes are parasitized.

Enlargement of lymph nodes is a typical sign of East Coast fever disease affecting all nodes, but the enlargement of other than the local nodes is not always apparent, and enlarged nodes shrink to about normal size terminally. The node of particular interest is that local to the site of infection which, with *Rhipicephalus appendiculatus*, is about the ears.

The end of the incubation period of ~2 weeks is heralded by high fever, followed in a couple of days by red cell parasitemia. Drooling, lacrimation, depression, diarrhea, and particularly progressive and prominent enlargement of the superficial lymph nodes give the disease a close clinical resemblance to malignant catarrhal fever. Severe pulmonary edema with dyspnea is common and terminal in many animals, whereas others become prostrate and comatose. The course of the disease is ~1 month and the mortality is ~95%. During the acute phase of the disease, more than 60% of lymphocytes may contain Koch's bodies. The acute disease appears to be caused by massive lympholysis and progressive anemia. An outstanding feature of East Coast fever is leukopenia, which is progressive from the onset of fever, and, terminally, very few leukocytes may be seen in blood smears. The total white count is seldom above 2×10^9/liter and often only a fraction of this, with the cells remaining primarily neutrophils and lymphocytes. There is brief initial stimulation of myelopoiesis, but by the third day of disease, there is neutropenia with immaturity and toxemia. The disease is progressive with loss of precursors, and ultimately there is a trilineage depression with accumulation of hemosiderin in macrophages. Terminally the bone marrow is hypoplastic, and the remaining cells consist of large blastic parasitized lymphocytes and atypical erythroblasts.

Gross lesions in the acute disease resemble those of malignant catarrhal fever. Of principal interest is the enlargement of lymphoid tissues, including Peyer's patches. On cut section the lymph nodes are diffusely discolored with a reddish-brown cortex containing focal hemorrhages, and a dark reddish-brown medullary area. Serous effusion and gelatinous or hemorrhagic edema of connective tissues is common. The spleen is enlarged in the acute disease, but in cases with a prolonged course, it may be shrunken and straplike. There is ulcerative abomasitis, likely nonspecific. The so-called infarcts of the liver and kidney are actually proliferative foci of perivascular lymphocytes. These foci, which project slightly, produce a mottling of small grayish-white patches visible on the surface of the liver and kidney. The lungs are congested and edematous with increased texture on palpation, and increased weight. Small hemorrhages associated with foci of hyaline degeneration occur in the muscles, and petechiae are commonly present under the tongue and in the vulva. Erosive or catarrhal enteritis overlies lymphocytic hyperplasia and infiltration of the gut mucosa (Fig. 2.106).

Histologically in the early stages of infection, there is diffuse lymphoid hyperplasia (Fig. 2.107A), but in animals which have died with East Coast fever, there is widespread lympholysis with hemorrhage and fibrinous exudate throughout the cortical areas of nodes. Lympholysis is prominent in germinal centers and there is a general loss of small lymphocytes with those that remain, appearing large and blastic (Fig. 2.107B). There is hepatic periacinar and, to a lesser extent, periportal lymphocytic infiltration and, in addition, there are focal infiltrations of the hepatic capsule, which give rise to the raised foci seen grossly. In some cases, there is periacinar hepatocellular necrosis and an irregular canalicular cholestasis with foci of inspissated bile. There is early splenic lymphoid hypertrophy, which is later followed by lympholysis. The germinal centers remain prominent and are surrounded by areas of hemor-

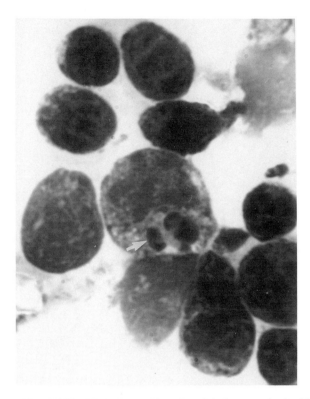

Fig. 2.105 Theileriosis. Ox. Imprint from node in Fig. 2.107(A). Large lymphocytes with Koch's blue body (arrow).

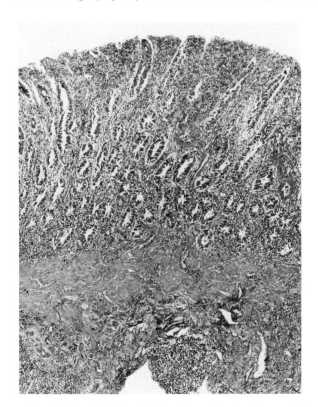

Fig. 2.106 Theileriosis. Ox. Mucosal and submucosal lymphoid infiltration with mucosal erosion. Small intestine.

rhage. The hypocellular follicular centers are usually occupied by a fibrinous or hyaline exudate similar to that seen in lymph nodes. The kidneys are remarkably congested with focal hemorrhage, and there is interstitial infiltration with lymphocytes. The lymphocytic infiltration is prominent around vessels, and often around the parietal layer of Bowman's capsule. There is a variable level of parenchymal necrosis with the formation of hyaline casts and brown pigmentation of the remaining epithelium. The pulmonary changes are characteristic and consist of lymphocytic infiltration of the septa and interstitial tissues, resulting in a widespread severe interstitial alveolitis (Fig. 2.108). The bone marrow is hypocellular with early asynchrony of the granulocytic system and a less severely affected erythroid system accompanied by proliferation of large lymphocytes similar to those infiltrating other tissues.

An aberrant expression of the disease is described as **turning sickness** although it is not clear whether the responsible organism is *T. parva*, *T. mutans*, or another. The condition occurs occasionally in partially resistant cattle reexposed to heavy infestations of infective ticks. Parasitized lymphocytes localize by embolism or sequestration in cerebrospinal vessels (Fig. 2.109) and produce hemorrhagic infarcts. The infarcts are of usual character, but muddy discoloration of tissues and meninges with hemosiderin is suggestive of repetitive minor episodes of infarction. Infarcts of varying age may be found in other organs, especially kidney and spleen, but are small and easily overlooked. Koch's bodies may be rare or not demonstrable in peripheral blood or lymphoid organs but are numerous in the clumped cells in vessels in the infarcts.

Theileria lawrenci is the cause of **corridor disease** of calves, so named because it occurred in cattle moved into an area called the Corridor lying between game reserves in Kwazuia. *Rhipicephalus* ticks collected in the area were able to infect cattle, but ticks feeding on infected cattle were not able to transmit the infection further to other cattle. Thus, cattle were not infective for ticks. Whereas the disease in cattle closely resembled East Coast fever clinically and pathologically, there were differences in parasitic behavior. In cattle affected with corridor disease, less than 5% of red cells in the blood and less than 5% of lymphocytes in smears of lymph nodes contained the parasite, whereas in East Coast fever, the corresponding figures would be about 50 and 80%. Cross-immunity, however, exists between *T. parva* and *T. lawrenci*. *Theileria lawrenci* is endemic in buffalo, which may be carriers; some strains are transmissible from cattle to cattle by ticks and, in these transmissions, *T. lawrenci* comes to behave like and to assume the morphologic characters, particularly with respect to macroschizonts, of *T. parva*, of which it is probably a variant.

Theileria mutans (*sergenti*) is the cause of **Tzaneen disease** of cattle, usually a mild and symptomless infection. Sheep can be infected artificially, the infections remaining latent and persisting at least for some weeks; the organisms in sheep do not appear in smears of blood. *Theileria*

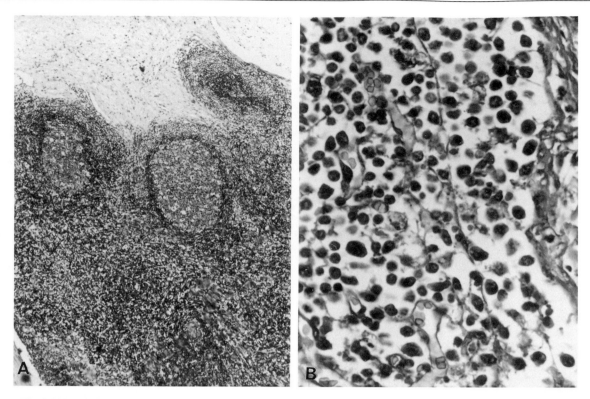

Fig. 2.107   Theileriosis. Ox. (A) Early follicular hyperplasia. Reduced paracortical density with moth-eaten appearance due to tingible body macrophages. (B) Lymphoblastoid transformation in early disease.

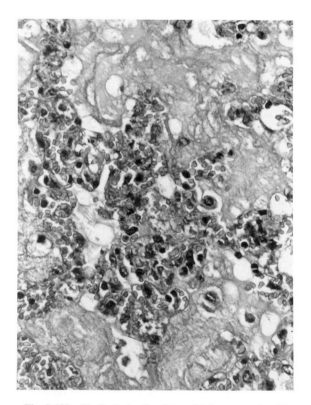

Fig. 2.108   Theileriosis. Ox. Interstitial pneumonia with massive proteinaceous effusion.

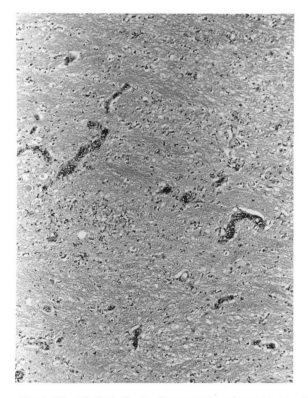

Fig. 2.109   Theileriosis. Ox. Sequestration of parasitized cells in cerebral vessels in turning sickness.

mutans is widely distributed in tropical and subtropical countries and is transmitted naturally by ticks of the genera *Rhipicephalus, Amblyomma,* and *Haemaphysalis,* and possibly by ticks of other genera as well. The importance of the parasite lies in the need to differentiate it from *T. parva.* Cross-immunity trials are necessary for certainty, but it is useful that *T. mutans* can be readily transmitted by the inoculation of even small quantities of blood, and in this infection schizonts are rare or impossible to find. The phases in erythrocytes are rod shaped and ring shaped, in about equal numbers.

There are occasional reports of fatal *T. mutans* infections in adult cattle manifested as hemolytic anemia, icterus, and hemoglobinuria. The lesions in lymphoid tissue are similar in type, but less severe than in East Coast fever. There is very heavy parasitism of erythrocytes, but schizonts are not observed. The pathogenesis of hemolysis is unclear, but there is some evidence that it is associated with the activities of macroschizonts in lymphocytes rather than with piroplasmic forms in erythrocytes.

Theileria annulata (*T. dispar*) is the cause of **tropical theileriosis** in the Middle East. The vectors are ticks of the genus *Hyalomma.* Infection with *T. annulata* is persistent and latent, relatively benign, and often with no parasitemia detectable in smears of blood. It is, however, readily transmissible by inoculations of blood. Occasionally, acute cases do occur and these tend to be fatal, with the disease closely resembling East Coast fever. The parasite, both within erythrocytes and forming Koch's bodies, becomes very numerous in the blood during acute attacks of the disease. The majority of the parasites in the red blood cells are ring or oval shaped, and only a few are bacilliform.

Theileria ovis is the ovine equivalent of *Th. mutans;* it is benign. *Theileria hirci* is somewhat more virulent and is also pathogenic for goats, producing a disease closely resembling East Coast fever.

Bibliography

Baldwin C. L., *et al.* Comparative analysis of infection and transformation of lymphocytes from African buffalo and Boran cattle with *Theileria parva* subsp. *parva* and *T. parva* subsp. *lawrencei. Infect Immun* **53**: 186–191, 1986.

Irvin, A. D., Cunningham, M. P., and Young, A. S. (eds.). "Advances in the Control of Theileriosis." The Hague, The Netherlands, Martinus Nijhoff, 1981.

Irvin, A. D., and Mwamachi, D. M. Clinical and diagnostic features of East Coast fever (*Theileria parva*) infection of cattle. *Vet Rec* **113**: 192–198, 1983.

Ole-Moi Yoi, O. K. *Theileria parva:* An intracellular protozoan parasite that induces reversible lymphocyte transformation. *Exp Parasitol* **69**: 204–210, 1989.

J. Jembrana Disease

An enzootic disease of Indonesian cattle is concentrated on the island of Bali, especially among the indigenous *Bos sondaicus* syn. *Bos javanicus.* The cause is still uncertain, though an agent capable of passing through filter pores of 200 μm is associated with the condition. A rickettsia-like organism or a virus have both been suspected as the cause. The natural mode of transmission is unknown, but an arthropod vector may not be an essential link in the epidemiology. Cattle of European breeds and sheep are resistant to the disease, although subclinical carrier status may persist for many months.

The disease can be transmitted with blood or tissue suspensions from a clinical case when, following an incubation period of 5–10 days, there is transient fever, dullness, and anorexia. Mild oculonasal discharge, pallor of the mucosae, and diarrhea with bloodstained feces are common. Superficial lymph nodes are visibly enlarged, whereas examination of the blood during the febrile period usually reveals anemia, leukopenia, thrombocytopenia, and in some cases which may be fatal, an elevated blood urea level. Basophilic cytoplasmic inclusions are occasionally seen in circulating mononuclear cells. Mortality rates may reach 15% or more.

At autopsy, lesions may not be remarkable, but several are constant and of diagnostic value. Enlargement of lymph nodes with slight edema and blurred corticomedullary junctions are widespread but most marked when regional to other affected organs (e.g., lungs and liver). During the acute phase there is splenomegaly, and on incision, the surface shows small grayish-red foci. Scattered petechiae or ecchymoses occur irregularly in the serosae of the heart, gastrointestinal tract, kidneys, and below the endocardium. The myocardium and liver contain grayish foci 1–3 mm in size. Pulmonary lesions are subtle but can be detected in the cranial lobes, which are firmer than normal, bluish-red, finely mottled, and slightly edematous. Adjacent lobules show mild overinflation. Shallow erosions may occur in the gastrointestinal tract and posterior tongue where an occasional ulcer is also possible. The kidneys are consistently pale and have small numbers of grayish foci.

The general picture may be complicated by concurrent infections, which have included pasteurella pneumonia and helminths of the liver, pancreas, or alimentary tract.

Histologically, three phases can be recognized in the acute disease process of about 6 weeks. During the first week there is a general response by the lymphoreticular system, but in the second phase, up to the fifth week approximately, there is an intense nonfollicular proliferation of reticular and large lymphoid cells. A similar infiltrative process occurs in the liver, kidneys, adrenal medulla, and choroid plexus systems of the brain, although the central nervous tissue shows little change. In the pulmonary cranial lobes alveolar cells are enlarged, pleomorphic, and proliferative alongside infiltrating mononuclear cells. In the third infectious phase, the follicular lymphoid system is reactivated and plasma cells appear progressively. The sequence is suggestive of a transient immunosuppressive disease. Lesions regress, but traces of change can be seen up to 60 days after infection.

Pleomorphic basophilic cytoplasmic inclusions are present in all affected tissues. By light microscopy, clumps of minute coccoid bodies about 0.5 μm, and larger, pleomorphic intravacuolar structures occur in reticular and large lymphoid cells, macrophages including Kupffer cells, pulmonary alveolar cells, and occasionally in vascular endothelium.

Bibliography

Dharma, D. M. N. *et al.* Studies on experimental Jembrana disease in Bali cattle. III. Pathology. *J Comp Pathol* **105**: 397–414, 1991.

K. Tick-Borne Fever

Tick-borne fever is a disease of cattle, sheep, and goats that is borne by the vector tick *Ixodes ricinus* and occurs in Great Britain, Ireland, the Netherlands, Norway, and Finland. The causative agent is *Ehrlichia phagocytophila, Rickettsia phagocytophila, Cytoecetes phagocytophila,* or *Cytoecetes bovis.* The infection is transmitted stage to stage in *I. ricinus,* but not hereditarily, and other blood-sucking arthropods may also transmit the disease. Similar diseases have been reported in sheep and cattle in North Africa. The agent in cattle is identified as *Ehrlichia bovis* and is thought to be transmitted by ticks of the genus *Hyalomma,* whereas the infection in sheep is caused by *Ehrlichia ovina* transmitted by *Rhipicephalus bursa.*

Tick-borne fever is a transient and mild illness which causes some loss of condition. The disease is, however, of some significance because it appears to predispose sheep to cerebrospinal inflammation in louping ill and to staphylococcal pyemia (tick pyemia), the virus of louping ill being conveyed by the ticks and *Staphylococcus aureus* being first established as an infection in the tick bites. Tick-borne fever also appears to predispose animals to pasteurellosis.

Clinically, adult ruminants are more susceptible to tick-borne fever than are the young. After incubation for 2–6 days, or a few days longer in cattle, there is a sudden onset of moderate fever. The fever lasts 2–3 days if the agent is of relatively low virulence and subsides quickly. If more virulent organisms are involved, the febrile reaction may be prolonged for nearly 2 weeks, the decline to normal temperature being gradual. Ewes in late pregnancy may abort. In terms of laboratory diagnosis, with the onset of fever, the organism is detectable in circulating granulocytes and large lymphocytes. The most useful stains are Giemsa and polychrome methylene blue. As the fever declines, the organisms disappear from smears and there is a progressive and quite severe neutropenia for a few days, after which the neutrophil count returns to normal. The organisms are no longer visible in smears unless there is a recrudescence with fever, but they do persist in the blood in some animals for a year or more. During fever, up to 95% of the neutrophils may contain parasites. The parasites are restricted to the cytoplasm and are of varied morphology, the variations probably representing developmental stages of the parasite. They may be round, rod shaped, or irregular masses measuring 0.75–3 μm in diameter present either singly or in clusters; ring-shaped bodies 2–3.5 μm in size breaking up into irregular fragments; round or oval morulae of 2.5–3.5 μm in diameter containing many small granules; or in occasional cells, there are small masses of about 0.5 μm in diameter.

Persistence of infection after the febrile period is associated with a state of labile premunition, with resistance to reinfection being demonstrable at ~5 weeks after the primary infection. Reactions to reinfection are usually much milder than those to the initial infection, but the situation is complicated by the immunologic differences and differences in virulence of strains of the organism. It is probable that in endemic areas, sheep remain infected for life, reinfection occurring at each season of tick activity.

L. Bovine Petechial Fever

Bovine petechial fever or **Ondiri disease** of Kenya and Tanzania is very similar to tick-borne fever in that the causative agent is also *Ehrlichia*-like and parasitizes circulating neutrophils. The agent is most often identified as *Cytoecetes ondiri.* The disease has not been demonstrated to be tick transmitted despite considerable efforts to do so. An arthropod vector is suspected, but unidentified. Several species of wild ruminants have been shown to be susceptible to experimental infection, and there is evidence suggesting that the bushbuck is a reservoir host of the infection.

Experimental transmission has been possible only by injecting blood, spleen, or lymph node taken from affected animals, but transmission by these means is possible for only a limited number of passages. The disease occurs naturally only in restricted areas of Kenya at altitudes above 5000 feet and there appear to be some differences in breed susceptibility, with Borans susceptible but Ayrshires and Herefords resistant to experimental infection.

Clinically, the disease is often mild and, unless cattle are being closely observed, signs may not be noticed. After an incubation period of 5–14 days, the affected animals become febrile, and within 1–2 days fine petechiae develop on the tongue, conjunctiva, and the mucosa of the vulva. The petechiae may persist for only 1–2 or up to 10 days. Occasionally the conjunctival lesions are severe, and the conjunctiva may become edematous and prolapse. Epistaxis and other hemorrhagic tendencies develop occasionally. Diagnostically, early in the course of the disease, affected animals have a leukopenia, due to a decrease in the numbers of circulating lymphocytes and monocytes, and thrombocytopenia of greater or lesser degree accounting for the variation in persistence of the petechiation. There follows in ~1 week a rebound leukocytosis produced by an increase in immature mononuclear cells. The organisms are found in neutrophils and lymphocytes, but are not common, and are not demonstrable in every case.

Pathologically, the lymph nodes of affected animals are enlarged and congested. The petechial hemorrhages that are seen clinically are also found in the internal organs.

The lungs are usually normal in experimental cases, but in severe field cases, death is often due to hemorrhage and edema of the lungs. The folds of the abomasum are edematous and congested, and the contents of the ileum and colon are tarry. The gallbladder is usually distended and the wall, thickened by gelatinous edema. The histologic appearance is nonspecific, and the agent cannot be recognized in routine sections.

Bibliography

Plowright, W. Some notes on bovine petechial fever (Ondiri disease) at Muguga, Kenya. *Bull Epizoot Dis Afr* **10:** 499–505, 1962.

Snodgrass, D. R. Clinical response and apparent breed resistance in bovine petechial fever. *Trop Anim Health Proc* **7:** 213–218, 1975.

HEMORRHAGIC DIATHESES*

Hemostasis involves the interaction of the damaged vascular wall with circulating platelets and coagulation proteins. Conceptually, it can be separated into primary and secondary phases, the former involving local vasoconstriction and platelets, and the latter involving coagulation factors. These phases occur simultaneously, however, and are interrelated. The events are set out schematically in Fig. 2.110.

I. Platelets and Hemostasis

Platelets do not normally interact with healthy vascular endothelium. This is partly because of the negative charge on cell membranes and partly because of prostacyclin, a potent local antagonist of platelet aggregation, which is synthesized and released by endothelial cells. Injury to these cells impairs prostacyclin production and release, thus allowing platelet adhesion and associated changes of platelet activation, aggregation, and release of cytoplasmic constituents.

Exposure of platelets to abnormal vascular endothelium or to extravascular tissue causes them to become activated. This is associated with loss of their normal discoid shape, the development of pseudopodia, and the centralization of cytoplasmic granules. Poorly defined biochemical changes induce alterations in the platelet membrane which make the cell sticky and facilitate its adhesion to the damaged surfaces. von Willebrand factor, previously called factor VIII-related antigen, is an important confactor for such adhesion. von Willebrand factor is synthesized by megakaryocytes and vascular endothelial cells.

Platelet aggregation refers to platelet–platelet interactions, in contrast to platelet–surface interactions. The initial step involves aggregating agents with specific platelet-

* This section was contributed by B. W. Parry, University of Melbourne, Australia.

membrane receptors. Thrombin, produced by activation of the coagulation cascade, and adenosine diphosphate (ADP), released from platelets, are two particularly important stimulants of platelet aggregation. Significant species differences exist in the qualitative and quantitative responses of platelets to aggregating agents.

The platelet-release reaction is an active calcium-dependent process. It is highly selective, depending on the nature and strength of the initiating stimulus. Stimulators include ADP, surface activators such as collagen, and calcium-mobilizing agents. Alterations in platelet endoperoxide/thromboxane A concentrations, and decreases in cyclic adenosine monophosphate (cAMP) are particularly important in mobilizing calcium within the platelet and in the contraction process which is necessary for the secretion of alpha granules, dense granules, and lysosomal vesicles.

The contents of the alpha granules include thromboglobulin, heparin-neutralizing activity (platelet factor 4), platelet-derived growth factor, and thrombospondin; these are more readily released than the contents of the dense-body granules, namely ADP, serotonin, and calcium. Only intense stimulation causes release of platelet lysosomal enzymes from their vesicular storage sites.

The importance of platelets in primary hemostasis is best illustrated by the occurrence of excessive bleeding in animals with thrombocytopenia, or in those with thrombocytopathy (defective adherance, aggregation and/or release). Platelets play an important role in the maintenance of vascular integrity. Weakened endothelial cells may be replaced by the incorporation of a platelet into the endothelial cell, thus preventing the extravasation of red cells. In animals with insufficient functional platelets, this vascular integrity is lost and red cells accumulate in the extravascular space, where they appear as petechial hemorrhages.

Platelets also contribute to secondary hemostasis. When activated, they make available a membrane-bound phospholipid, platelet factor 3, which functions as a cofactor in the activation of factors X and II (prothrombin). Platelets can also activate factor XII, in the presence of ADP, and factor XI in the presence of collagen. The physiologic significance of these latter interactions is not completely understood.

A. Laboratory Diagnosis of Platelet Disorders

Thrombocytopenias are the most common platelet abnormalities. These are readily diagnosed antemortem by a platelet count. Care must be taken to prevent artefactual decreases in platelet numbers as a result of sample collection and handling, especially in ruminants and cats. Thus the venipuncture should be carefully performed to minimize the release of tissue factor and subsequent thrombin production, since thrombin is a potent platelet-aggregating agent. The use of a strong calcium chelator, such as ethylenediaminetetraacetic acid (EDTA), as the anticoagulant will help in this regard.

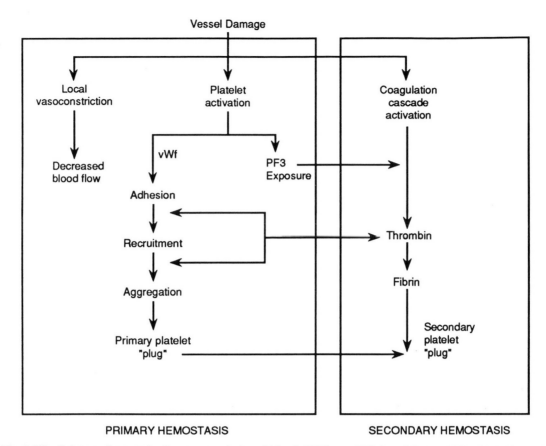

Fig. 2.110 Schema of events leading to coagulation of blood. VWf, von Willebrand factor; PF3, platelet factor 3.

Tests of platelet function are more difficult to perform *in vitro* and generally require sophisticated instrumentatiion. Furthermore, because platelets are quite fragile *ex vivo*, samples must be processed expeditiously. Nevertheless there are two screening tests which can be used to assess platelet function, namely platelet retraction *in vitro* and the buccal mucosal bleeding time.

The buccal mucosal bleeding time is performed on the labial mucosa. A standard superficial incision is made, and the duration of bleeding is recorded. Bleeding is prolonged by severe thrombocytopenia, von Willebrand's disease and abnormalities of platelet function.

B. von Willebrand Factor

von Willebrand factor is a plasma glycoprotein involved in platelet aggregation and in the adhesion of platelets to exposed subendothelial collagen. It is produced by vascular endothelial cells and megakaryocytes and circulates in the blood as macromolecules of various sizes, complexed with factor VIII. The von Willebrand factor macromolecules (multimers) may be large, intermediate, or small in size, depending on the number of polymerized subunits that they contain. The larger multimers are more function-

ally active. Normal animals have all three sizes present in the blood.

von Willebrand factor may be assessed by antigenic or functional assays. One-dimensional immunoelectrophoresis is the most common antigenic assay at present. These assays do not separate the different multimers. The latter can be achieved by two-dimensional immunoelectrophoresis or multimeric analysis. The functional activity of von Willebrand factor can be subjectively assessed by the buccal mucosal bleeding time (see the preceding). However, the latter is not sensitive to mild deficiencies. *In vitro*, functional activity can be assessed by a number of platelet agglutination and/or aggregation tests. These include ristocetin, botrocetin, and collagen-binding assays. The results of the different tests do not always correlate in animals with von Willebrand's disease.

II. Coagulation Factors in Hemostasis

A. Coagulation Cascade

Secondary hemostasis involves the orderly activation and interaction of circulating procoagulant proteins (Fig.

2.111). The culmination of this coagulation cascade is the formation of an insoluble meshwork of fibrin, which stabilizes the primary platelet plug and prevents its rapid dissolution.

Most clotting factors are referred to by Roman numerals. The common exceptions are fibrinogen (factor I), prothrombin (factor II), tissue factor (factor III), and calcium (factor IV). There is no factor VI. Two factors, prekallikrein and high-molecular-weight kininogen, do not have Roman numerals assigned. Tissue factor is a cell-bound lipoprotein, found throughout the body. All other proteins in the coagulation cascade circulate in the blood as inactive precursors. In the healthy animal, they are converted to their active form only when blood is exposed to foreign material, including subendothelial and intracellular constituents. The active form of a clotting factor, produced by proteolytic cleavage usually by another factor preceeding it in the cascade, is designated by the suffix "a." The notable exceptions are fibrin (the active form

of fibrinogen), thrombin (the active form of prothrombin), and kallikrein (the active form of prekallikrein). Four of the coagulation factors are really cofactors rather than proteolytically active substances. These are high-molecular-weight kininogen, tissue factor, and factors VIII and V.

Almost all of the coagulation proteins are synthesized in the liver. The exception is factor VIII, which is thought to be synthesized mainly by vascular endothelial cells. There are no significant reserves of coagulation factors. Consequently, normal hepatic function is necessary for normal hemostasis. The *in vivo* half-life of the coagulation factors varies from 4 to 6 hr for factor VII to 3–6 days for fibrinogen.

A simplified schematic diagram (Fig. 2.111) illustrates the coagulation sequence that results in an insoluble fibrin clot. When blood contacts a foreign surface, such as a damaged vessel wall, prekallikrein is activated to kallikrein and factor XII to factor XIIa. In the presence of high molecular weight kininogen, these factors also activate each other. Factor XIIa also activates factor XI. In the presence of calcium, factor XIa activates factor IX. Factor IXa is only a weak proteolytic enzyme, but when complexed with platelet factor 3, factor VIII and calcium ions, it acquires potent proteolytic activity and rapidly activates factor X. In an analogous fashion, factor Xa combined with factor V, platelet factor 3, and calcium activates prothrombin to form thrombin. Thrombin selectively cleaves negatively charged polypeptides from the ends of fibrinogen, permitting the modified proteins to form a loosely bound network of fibrin strands, which are stabilized into an insoluble polymeric structure by covalent bonds that form under the catalytic action of factor XIIIa. (The latter is also activated by thrombin). This sequence can produce an insoluble fibrin clot within minutes of the activation of factor XII. The process can be accelerated approximately 10-fold when tissue factor is released into the circulation from damaged tissue cells. The chief role of tissue factor is the direct activation of factor X in the presence of factor VIIa and calcium. Hence, fibrin formation can occur within seconds of the release of tissue factor into the circulation. Tissue factor in the presence of factor VIIa and calcium can also effectively activate factor IX, although the physiological significance of this remains to be determined.

As the fibrin strands polymerize, the cellular elements of blood are entrapped in the clot. This close proximity of blood cells, especially platelets, and plasma proteins is essential for the maintenance of normal homeostatic balance. Once the fibrin clot has formed, and bleeding is stopped, tissue repair at the site of damage can occur. To prevent permanent blockage of blood vessels, the fibrin clot must be removed. This is a two-step process involving contraction of the clot, mediated by platelets, and lysis and degradation of fibrin.

Platelets contain a contractile protein, thrombosthenin, which, in the presence of calcium and adenosine triphosphate (ATP), contracts in a manner similar to that of acto-

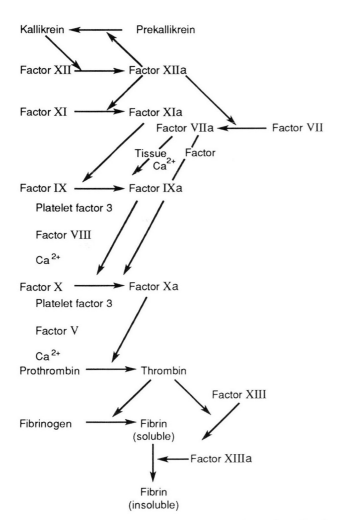

Fig. 2.111 Sequence of events resulting in formation of insoluble fibrin blood clot.

myosin of muscle. The fibrin clot shrinks and serum is extruded into the circulation.

Fibrinolysis is mediated by plasmin which is produced by the activation of the circulating proenzyme, plasminogen. Plasminogen is incorporated in the clot during its formation along with various plasminogen activators, which are released from blood vessels and tissues. Plasmin cleaves the fibrin strands into small fragments, fibrin degradation products (FDP). As the fibrin network is degraded, platelets, protein fragments, and cellular debris enter the circulation for subsequent removal by the phagocytic system. Activated coagulation factors are degraded, and thereby kept in balance, by various proteolytic enzymes such as α_1-antitrypsin and antithrombin III. Inactivated products are removed by hepatocytes and macrophages.

B. Laboratory Diagnosis of Disorders of the Coagulation Factors

The coagulation cascade can be conveniently partitioned into three sections *in vitro*. The intrinsic pathway commences with contact activation of prekallikrein and/or factor XII. The extrinsic pathway commences with the addition of tissue factor. Both pathways converge at factor X into the common pathway. The intrinsic and common pathways are usually assessed by the activated partial thromboplastin time (APTT). The extrinsic and common pathways are assessed by the prothrombin time (PT). Other tests are available, but not routinely performed. They include the activated coagulation time (ACT), which assesses the intrinsic and common pathways, but is not as sensitive as the APTT to partial factor deficiencies; the Russell's viper venom time (RVVT), which assesses the common pathway; and the thrombin time (TT), which assesses fibrinogen alone.

Animals with a prolonged APTT and normal PT may have deficiencies of factors VIII, IX, XI, XII, prekallikrein, or high-molecular-weight kininogen. Factor VIII deficiency is most likely. Deficiencies of factors XII, prekallikrein, and high-molecular-weight kininogen are often asymptomatic; indeed, factor XII is absent in reptiles, birds, and various cetaceans.

Animals with prolonged PT and normal APTT must have a deficiency of factor VII. Animals with prolonged APTT and PT may have a deficiency of one or more factors in the common pathway. They may also have deficiencies in both the intrinsic and extrinsic pathways.

III. Disorders of Primary Hemostasis

The pattern of hemorrhage in disorders of primary hemostasis typically involves the mucous membranes and microvasculature and is of petechial or ecchymotic type. Such disorders include defects of platelets, including thrombocytopenia and thrombocytopathy; von Willebrand's disease; and congenital and acquired defects of blood vessels, which are, however, rare in animals. Hemorrhage may occur spontaneously or following trauma. Common sites include nose, mouth, and the urogenital and gastrointestinal tracts.

A. Thrombocytopenias

Acquired thrombocytopenias are the most common cause of hemorrhagic diathesis in animals. They may be primary disorders or secondary to other diseases. The number of platelets in circulation greatly exceeds normal requirements, and a substantial reduction in numbers, perhaps to as few as $24–40 \times 10^9$/liter, is necessary before bleeding tendencies are detected; spontaneous hemorrhages are unlikely until platelet numbers are reduced to $\sim 10 \times 10^9$ liter.

These generalizations are modified by several factors including the cause of the thrombocytopenia. If accelerated thrombopoiesis occurs in the bone marrow in response to a thrombocytopenia, large platelets (macrothrombocytes) are produced. These are more hemostatically active than their normal-sized counterparts; consequently, more severe thrombocytopenias can be tolerated by the animal before spontaneous hemorrhage is manifest.

Thrombocytopenia may result from increased destruction or utilization of platelets, decreased production of platelets in the bone marrow, and sequestration of circulating platelets in the spleen.

Thrombocytopenia resulting from increased destruction or utilization usually induces megakaryocytic hyperplasia in the bone marrow and often increased proportions of macrothrombocytes in the peripheral blood. In contrast, thrombocytopenia associated with depressed thrombopoiesis is revealed by megakaryocytic hypoplasia in the bone marrow and usually a population of normal-sized to small platelets in the peripheral blood. Other cell series are also often affected in the latter cases and a concurrent leukopenia (neutropenia) and nonregenerative anemia are common accompaniments. The spleen of resting healthy animals contains a considerable pool of platelets which are in equilibrium with the peripheral blood. Splenomegaly may increase the numbers of platelets in this organ, thus producing a modest thrombocytopenia. This will not be severe enough in itself to cause a hemorrhagic tendency.

Excessive destruction and/or consumption of platelets are important causes of thrombocytopenia. They may be immune mediated or nonimmune mediated. The former include immunologic reactions which involve autoantibodies, alloantibodies, isoantibodies, and drug-associated antibodies. The latter usually result from disseminated or localized intravascular coagulation.

Autoimmune thrombocytopenia usually occurs alone (idiopathic thrombocytopenic purpura). It may also develop in conjunction with autoimmune hemolytic anemia or systemic lupus erythematosus. It is most common in dogs, especially smaller breeds, and in horses. It more often affects females. Antibody may be demonstrated on the surface of the platelets.

Antiplatelet antibodies (alloantibodies) can be produced if multiple transfusions are given. They may also occur if a gravid female is exposed to foreign platelet antigens during pregnancy or at parturition. Subsequent progeny may suffer from an isoimmune thrombocytopenia if they have the same platelet antigens as the earlier offspring.

Such **isoimmune thrombocytopenia** is most common in piglets. Affected piglets ingest colostrum with a high titer of the offending antibody. They typically develop a biphasic thrombocytopenia. The initial bout occurs 1–3 days after birth as a result of platelet destruction by maternal antiplatelet antibody. Those piglets which survive often suffer a secondary and more persistent thrombocytopenia at 10–14 days of age because of damage to megakaryocytes and their precursors in the bone marrow. Affected piglets may be very pale and exhibit petechial and ecchymotic hemorrhages in the skin and internal organs. The disease is often fatal.

Drug-induced platelet destruction may be the result of a drug–antibody complex combining with specific immunologic receptors of the platelet membrane, or of the action of an antibody reacting to a drug–platelet complex. Based on reports in humans, many drugs used in veterinary medicine have the potential to induce an immunologic thrombocytopenia. They include antibiotics, such as penicillin, streptomycin, erythromycin, and the sulfonamides; anticonvulsant drugs, such as diazepam and phenobarbital; cardiovascular drugs, such as digoxin, hydralazine, and quinidine; nonsteroidal antiinflammatory drugs, such as acetaminophen, aspirin, and phenylbutazone; and miscellaneous drugs, such as heparin, lidocaine, and thiazide diuretics.

Excessive nonimmune-mediated consumption of platelets may also cause thrombocytopenia and is a feature of disseminated (or localized) intravascular coagulation (see The Cardiovascular System, Chapter 1 of this volume). Such consumptive coagulopathies may result from exposure of platelets to extensive areas of damaged or denuded vascular endothelium or foreign surfaces. Concurrent excessive production of thrombin in the coagulation cascade exacerbates utilization of the platelets and may deplete the blood of coagulation factors, thereby enhancing the tendency for hemorrhage at other sites in the body. Furthermore, excessive production of FDP also occurs because of increased fibrinolytic activity in response to increased intravascular coagulation. High concentrations of FDP have an anticoagulant effect by interfering with platelet function and with the proteolytic activity of thrombin. Various viruses, endotoxins, and antigen–antibody complexes can cause such a consumptive thrombocytopenia (through vascular injury and/or direct platelet injury), for example, distemper in dogs, enterotoxemia in sheep, enzootic pneumonia in calves, and infectious anemia in horses.

Megakaryocytic hypoplasia is a common cause of thrombocytopenia. Such depressed thrombopoiesis is a feature of some systemic infections, such as canine ehrlichiosis and some cases of feline leukemia virus infection, and of

idiopathic aplastic anemia, myelophthisis, and myelotoxicity. There is often concurrent leukopenia and nonregenerative anemia; the important diseases are discussed earlier in this chapter.

Bibliography

Byars, T. D., and Green, C. E. Idiopathic thrombocytopenic purpura in the horse. *J Am Vet Med Assoc* **180**: 1422–1424, 1982.

Feldman, B. F., Thomason, K. J., and Jain, N. C. Quantitative platelet disorders. *Vet Clin North Am: Small Anim Pract* **18**: 35–49, 1988.

Grinden, C. B. *et al.* Epidemiologic survey of thrombocytopenia in dogs: A report of 987 cases. *Vet Clin Pathol* **20**: 38–43, 1991.

Handagama, P., and Feldman, B. F. Thrombocytopenia and drugs. *Vet Clin North Am: Small Anim Pract* **18**: 51–65, 1988.

B. Thrombocytopathies

Inherited defects of platelet function are uncommon to rare in domestic animals. They are reported in several breeds of dogs, especially basset hounds, in Simmental cattle, and in animals affected by the Chediak–Higashi syndrome.

The thrombocytopathies can be broadly divided into two major groups: in thrombasthenialike disorders, platelet function is impaired by a defect in platelet-membrane receptors; in secretory disorders, there is a deficiency of secretory products (storage-pool disease) or an inability of the platelet to undergo secretion (release defect). All of the reported defects are autosomal traits. The heterozygotes may be mildly affected but are usually asymptomatic. In homozygotes, clinical signs are similar to those of thrombocytopenia. They vary from mild to severe and are exacerbated by surgery, trauma, and stress. These diseases are characterized by a prolonged bleeding time without thrombocytopenia and with normal coagulation screening tests. Clot retraction is often, but not always, abnormal. Poor platelet granulation may be evident on microscopic examination of platelets from cases with storage-pool defects.

Acquired defects of platelet function are probably quite common, but may often be subclinical, unless the animal has a preexisting coagulopathy, such as von Willebrand's disease. Acquired thrombocytopathies may be drug-induced or secondary to some other disease. Offending **drugs** include nonsteroidal antiinflammatory agents; caffeine; dextrans; estrogen; some antibiotics such as penicillin, ampicillin, pentamycin, and the sulfonamides; glycerol guaiacolate; halothane; heparin; lidocaine; phenothiazines; propranolol; procaine; and theophylline.

Diseases which may be associated with decreased platelet function include disseminated intravascular coagulation; dysproteinemias such as multiple myeloma and macroglobulinemia; myeloproliferative diseases; systemic lupus erythematosus; immune-mediated thrombocytopenia; and chronic liver and renal disease. Abnormal platelet function in dysproteinemias may result from paraproteins coating the platelet membrane. In chronic liver disease, the platelet dysfunction may be related to decreased amounts of a platelet membrane receptor or increased amounts of fibrin degradation products. In chronic renal disease, the platelet dysfunction may be related to the accumulation of toxic metabolites, such as guanidinosuccinic acid, and an imbalance between prostacyclin and thromboxane synthesis by vascular endothelial cells.

Bibliography

Catalfamo, J. L., and Dodds, W. J. Hereditary and acquired thrombopathies. *Vet Clin North Am: Small Anim Pract* **18:** 185–193, 1988.

Davenport, D. J., Breitschwerdt, E. B., and Carakostas, M. C. Platelet disorders in the dog and cat. Part 1. Physiology and pathogenesis. *Compend Contin Educ Pract Vet* **4:** 762–773, 1982.

Dodds, W. J. Familial canine thrombocytopathy. *Throm Diath Haemorrh* **26:** 241–248, 1967.

Helfand, S. C. Platelets and neoplasia. *Vet Clin North Am: Small Anim Pract* **18:** 131–156, 1988.

Johnstone, I. B. Comparative effects of phenylbutazone, naproxen, and flunixine meglumine on equine platelet aggregation and platelet factor 3 availability *in vitro. Can J Comp Med* **47:** 172–179, 1983.

Johnstone, I. B., and Lotz, F. An inherited platelet function defect in basset hounds. *Can Vet J* **20:** 211–215, 1979.

Meyers, K. M. Pathobiology of animal platelets. *Adv Vet Sci Comp Med* **30:** 131–165, 1985.

Meyers, K. M. *et al.* Storage-pool deficiency in platelets from Chediak–Higashi cattle. *Am J Physiol* **237:** R239–248, 1979.

Meyers, K. M. *et al.* Ultrastructure of resting and activated storage pool-deficient platelets from animals with Chediak–Higashi syndrome. *Am J Pathol* **106:** 364–377, 1982.

Raymond, S. L., and Dodds, W. J. Characterization of the fawn-hooded rat as a model for hemostatic studies. *Thromb Diath Haemorrh* **33:** 361–369, 1975.

Searcy, G. P., and Petrie, L. Clinical and laboratory findings of a bleeding disorder in eight Simmental cattle. *Can Vet J* **31:** 101–103, 1990.

C. von Willebrand's Disease

von Willebrand's disease is a heterogeneous group of disorders involving defective primary hemostasis. The number of platelets and their functional activity, as assessed by *in vitro* tests, are normal. von Willebrand's disease usually results from an absolute deficiency of von Willebrand factor, with all multimers decreased in concentration in the blood. It has been described in over 50 breeds of dog and crossbreeds. In North America it is very common in Doberman pinschers; however, there is also a reasonably high prevalence in German shepherd dogs, golden retrievers, miniature schnauzers, Scottish terriers, and Pembroke Welsh corgis. The disease is also probably relatively common in Doberman pinschers in other countries. von Willebrand's disease has also been described in Poland China and Yorkshire–Hampshire swine, in rabbits, a cat, and a horse.

Clinical signs of von Willebrand's disease vary markedly from almost asymptomatic to very severe. In general, particularly within a species or particular breed, the severity of the hemorrhagic tendency is related to the magnitude of the deficiency of von Willebrand factor, which can vary from only slightly subnormal to very low or is even absent in some cases.

von Willebrand's disease is thought to be inherited as an autosomal dominant trait with variable penetrance in most breeds of dogs. It is inherited as an autosomal recessive trait in Scottish terriers, Chesapeake Bay retrievers, swine, and rabbits.

von Willebrand's disease is usually characterized by a normal platelet count and decreased von Willebrand factor antigenic activity, with all multimers being deficient. Factor VIII activity is normal to somewhat decreased. The German short-haired pointer is an exception, with the smaller multimers present, but the large multimers being deficient. Antigenic activity is thus only moderately decreased, while functional activity is markedly decreased.

Bibliography

Brooks, M. *et al.* Bleeding disorder (von Willebrand's disease) in a quarter horse. *J Am Vet Med Assoc* **198:** 114–116, 1991.

Kraus, K. H., and Johnson, G. S. von Willebrand's disease in dogs. *In* "Current Veterinary Therapy X. Small Animal Practice," R. W. Kirk (ed.), pp. 446–451. Philadelphia, Pennsylvania, Saunders, 1989.

Waters, D. C. *et al.* Expression of von Willebrand factor in plasma and platelets of cats. *Am J Vet Res* **50:** 201–204, 1989.

D. Purpura Hemorrhagica

Purpura hemorrhagica is an extensive hemorrhagic diapedesis which results from increased capillary permeability, which may be secondary to an allergic response. It occurs most frequently in horses but is also noted in pigs and dogs. Purpura hemorrhagica often develops following an infectious or destructive disease process such as strangles or other forms of respiratory disease. The pathogenesis is thought to be related to the accumulation in the circulation of soluble immune complexes formed as a result of antigen excess. Such complexes are poorly cleared by the reticuloendothelial system and cause endothelial injury.

The disease is characterized by extensive petechial and ecchymotic hemorrhages in the skin and mucous membranes. In horses, localized or generalized edema is a common clinical feature of the disorder, with the edematous fluid often being blood tinged. Extensive pulmonary edema may be fatal. Usually there is no evidence of thrombocytopenia or of a coagulation defect to account for the bleeding tendency.

Bibliography

Blood, D. C. *et al.* "Veterinary Medicine," 6th Ed. London, Baillière, 1983.

Reef, V. B. Vasculitis. *In* "Current Therapy in Equine Medicine 2," N. E. Robinson (ed.), pp. 312–314. Philadelphia, Pennsylvania, Saunders, 1987.

IV. Disorders of Secondary Hemostasis

The pattern of hemorrhage in disorders of secondary hemostasis is typically into body cavities or tissues and from wounds. Subcutaneous hematomas and hemorrhage into urogenital and gastrointestinal tracts are common.

A. Inherited Disorders

Factor VIII deficiency, also called hemophilia A and classical hemophilia, is the most common inherited disorder of coagulation proteins. It is a sex-linked recessive trait which most frequently occurs in dogs, being documented in more than 20 breeds. It is especially common in the German shepherd breed. Hemophilia A has also been described in cats, horses, and Hereford cattle. Clinically affected animals are usually hemizygous males. Heterozygous females are carriers. Homozygous females will be clinically affected; this is uncommon but can occur when an affected male is mated to a carrier female, since some mildly affected males do reach sexual maturity.

The extent of hemorrhage varies from slight to severe depending on the degree of factor VIII deficiency, the size of the animal, and the nature and level of physical activity.

Bouts of hemorrhage are often precipitated by stresses, including minor surgery, vaccination, estrus, and the like. In dogs, the first clinical evidence of hemophilia A may be excessive gingival bleeding as the permanent teeth erupt. Other common clinical signs include recurrent or shifting lameness, because of intra-articular hemorrhage, particularly into the stifle joint, and subcutaneous and intramuscular hematomas, probably following trauma. Neurologic deficits may develop because of pressure on peripheral nerves, such as the brachial or sciatic nerves or following hematomyelia. Recurrent hematuria and melena and protracted estral or postpartum hemorrhage may also occur. Fatalities may occur at birth because of umbilical hemorrhage; however, this is probably uncommon. In older animals, fatalities are usually associated with massive internal or external hemorrhage, often following trauma. This includes bleeding into the thoracic and abdominal cavities and subarachnoid space.

Hemophilia A in cats is often a mild disease and may not be recognized until surgery or severe trauma provokes serious prolonged hemorrhage. In contrast, hemophilia A in horses usually causes severe clinical signs very early in life.

Hemophilia A is characterized by a normal platelet count, normal PT, and prolonged APTT. Factor VIII activity is decreased, whereas von Willebrand factor activity is normal to increased. The buccal mucosal bleeding time is normal.

Other causes of prolonged APTT and normal PT are deficiencies of factors IX, XI, XII, prekallikrein, and high-molecular-weight kininogen. Deficiencies of the last three factors are uncommon to rare. **Factor XII deficiency** has been described in miniature poodles and mixed-breed cats. It is asymptomatic. **Prekallikrein deficiency** has been diagnosed in Belgian horses, miniature horses, and a poodle. It may be asymptomatic or cause a mild hemorrhagic tendency.

Factor IX deficiency, also called hemophilia B, is an X-linked resessive trait, clinically identical to hemophilia A. It has been described in several pedigreed breeds of dogs and is the second most common inherited disorder of the coagulation cascade in British shorthair cats.

Factor XI deficiency, or hemophilia C, is an autosomal trait. It has been described in Holstein–Friesian cattle, various pedigreed dogs, and mixed-breed cats. It is a recessive trait in cattle and probably a dominant trait in dogs. Homozygous affected cattle have negligible factor XI activity, yet clinical signs are comparatively mild. Dehorning and calving may be uneventful or, at worst, produce protracted low-grade hemorrhage or large hematomas, respectively.

Factor VII deficiency causes a prolonged PT but normal APTT. It has been diagnosed in beagles, Alaskan malamutes, and miniature schnauzers. It is an autosomal trait. In beagles, it is considered to be inherited in an autosomal manner. It is phenotypically recessive but genotypically codominant. In beagles and miniature schnauzers, it is often asymptomatic. Affected animals may exhibit increased bruising following trauma and prolonged vaginal bleeding after whelping. Clinical signs are more obvious in Alaskan malamutes, with superficial hematomas and hemarthroses more commonly observed.

A prolonged APTT and PT suggests a disorder of the common pathway. Inherited disorders which may cause this are deficiencies of factors X and V, prothrombin, and fibrinogen. Congenital factor V deficiency has not been described in domestic animals. Congenital deficiencies of the other factors are uncommon to rare. They generally produce a severe hemorrhagic tendency. **Factor X deficiency** has been described in the American cocker spaniel; **prothrombin deficiency,** in the boxer and cocker spaniel; and **fibrinogen deficiency,** in the Dürrbach and Saint Bernard dogs and in the Saanen goat. A **dysfibrinogenemia** has been documented in the borzoi (Russian wolfhound) and collie breeds. Clinical signs in the latter were milder than those for hypofibrinogenemia.

Prolonged APTT and PT also occur in animals with deficiencies of several clotting factors. These are usually acquired disorders (see the following); however, such a congenital coagulopathy has been described in Devon rex cats. The vitamin K-dependent factors (II, VII, IX, and X) were deficient. Clinical signs varied from asymptomatic to fatal internal hemorrhage. The deficiencies responded to vitamin K supplementation.

Bibliography

Breukink, H. J. *et al.* Congenital afibrinogenemia in goats. *Zentralbl Veterinarmed (A)* **19:** 661–676, 1972.

Dodds, W. J. Second international registry of animal models of thrombosis and hemorrhagic diseases. *Inst Lab Anim Resources News* **24:** R1–R50, 1981.

Gentry, P. A., and Brush, P. J. Factor XI deficiency in Canadian Holsteins. *Can Vet J* **28:** 110, 1987.

Geor, R. J. *et al.* Prekallikrein deficiency in a family of Belgian horses. *J Am Vet Med Assoc* **197:** 741–743, 1990.

Healy, P. J. *et al.* Haemophilia in Hereford cattle: Factor VIII deficiency. *Aust Vet J* **61:** 132–133, 1984.

Johnstone, I. B., Morton, J. C., and Allen, D. G. Factor VIII deficiency in a cat. *Can Vet J* **28:** 671–673, 1987.

Maddison, J. E. *et al.* Vitamin K-dependent multifactor coagulopathy in Devon rex cats. *J Am Vet Med Assoc* **197:** 1495–1497, 1990.

Parry, B. W. *et al.* Haemophilia A in German shepherd dogs. *Aust Vet J* **65:** 276–279, 1988.

Randolph, J. F., Center, S. A., and Dodds, W. J. Factor XII deficiency and von Willebrand's disease in a family of miniature poodle dogs. *Cornell Vet* **76:** 3–10, 1986.

Spurling, N. W. The haematology of the dog. *In* "Comparative Clinical Haematology," R. K. Archer and L. B. Jeffcott (eds.), pp. 365–440. Oxford, Blackwell Scientific Publications.

Spurling, N. W. Hereditary disorders of hemostasis in dogs: A critical review of the literature. *Vet Bull* **50:** 157–173, 1980.

B. Acquired Disorders

1. Poisoning by Vitamin K Antagonists

Factors II (prothrombin), VII, IX, and X require vitamin K for their production as functional molecules by the liver. Ingestion of vitamin K antagonists such as coumarin and its derivatives and sulfaquinoxaline results in the production of nonfunctional forms of these factors, referred to as proteins induced by vitamin K antagonism (PIVKAs). Factor VII has the shortest biological half-life, ~4–6 hr, and is therefore the first of the vitamin K-dependent factors to decrease in the blood. Consequently, early in the course of such poisonings, the PT may be prolonged while the APTT is still normal. This finding may actually precede the onset of clinical signs. The activities of factors IX and X, with half-lives of ~14–18 hr, are next to decrease, followed by prothrombin with a half-life of ~40 hrs. Depletion of these factors prolongs both the PT and the APTT, often quite markedly. Vitamin K antagonists do not decrease fibrinogen concentration and do not interfere with platelet numbers or function.

The interval between ingestion of vitamin K antagonists and the occurrence of hemorrhage and its severity vary with the relative toxicity of the antagonist, the amount ingested, and the rate of metabolism of the antagonist. They are also influenced by the amount of vitamin K stored in the body and by its bioavailability. Nevertheless, hemorrhage usually occurs within 3–5 days after exposure. Current formulations of the potent coumarin-derived rodenticides may cause marked depletion of vitamin K-dependent clotting factors, the half-life of which being shorter than that of the antagonist.

The type of hemorrhage which occurs is typical for factor deficiencies. It may be very acute in onset or observed only after physical trauma. External hemorrhage may be expressed as epistaxis, melena, or hematuria. Internal hemorrhage occurs into muscle and body cavities, including joints, and beneath skin and mucosae. Hemorrhage in vital organs such as pericardial sac or meninges may be rapidly fatal, or death may occur in hypovolemic shock.

There are no clear sites of predilection for spontaneous hemorrhage except for the thymus and anterior mediastinum in dogs poisoned by rodenticides.

Poisoning of cattle by **moldy sweet clover** was one of the first acquired coagulopathies documented in domestic animals. Sweet clover (*Melilotus alba*) contains coumarin. When it spoils, in hay or silage, the latter is converted to dicumarol. Cattle, pigs, and rabbits appear most susceptible to such poisoning, whereas sheep and horses seem more resistant. Young calves are much more susceptible to dicumarol than mature animals, probably because they have lower body reserves of vitamin K. Newborn calves may bleed to death while their dams are free of clinical signs and exhibit only a prolonged PT.

Subsequent to the discovery of dicumarol, various derivatives of coumarin have been synthesized for use as **anticoagulant rodenticides.** These include warfarin, fumarin, pindone, diphacinone, brodifacoum, and bromadiolone. These agents often cause hemorrhagic diatheses in dogs and occasionally do so in cats, horses, and swine. Sulfaquinoxaline is another potent vitamin K antagonist. It is used as a coccidiostat in livestock and poultry, and sometimes as an active ingredient of rodenticides. Its ingestion may cause hemorrhage in dogs.

2. Hemostatic Defects in Hepatobiliary Disease

Acute hepatic injury, especially that which involves parenchymal necrosis and damage to sinusoidal endothelium, may result in disseminated intravascular coagulation and a hemorrhagic diathesis. The intravascular coagulation is often prominent in adjacent hepatic sinusoids and probably involves both activation of clotting factors and inadequate clearance of activated factors by the injured liver. There may also be reduced hepatic synthesis of clotting factors and their circulating inhibitors, such as antithrombin III and antiplasmin. Lack of antithrombin III, combined with increased amounts of activated clotting factors, encourages fibrin formation and can actually promote thrombosis. Fibrinolytic activity, mediated by plasmin, will produce increased amounts of fibrin-degradation products. If antiplasmin activity is reduced, this fibrinolysis may proceed unchecked, producing greater amounts of fibrin-degradation products. The latter can exert an anticoagulant effect by interfering with platelet function and with the proteolytic activity of thrombin.

Chronic hepatic disease, in which the cell mass is reduced by 70% or more, may result in inadequate synthesis of clotting factors and their inhibitors. This reflects the general reduction in hepatic protein synthesis. Nevertheless, the activity of circulating clotting factors is usually only moderately reduced. Whereas it may be sufficient to prolong the PT and/or the APTT, it seldom causes hemorrhage unless the hemostatic mechanism is further challenged by trauma or surgery. In chronic hepatic disease, retention of bile and interference with the enterohepatic circulation of bile salts may result in malabsorption of fat-soluble vitamins, of which vitamin K is one. A deficiency of vitamin K leads to depletion of vitamin K-dependent factors II, (prothrombin), VII, IX, and X. This will exacerbate the generally decreased rate of protein synthesis noted earlier. Consequently, hemorrhagic diathesis, disseminated intravascular coagulation, and/or thrombosis may result. Thrombocytopenia may also occur in hepatic failure, and there may be qualitative changes in the platelets. The causes of platelet inadequacy are multiple and include excessive utilization, bone marrow depression, sequestration of platelets in splenomegaly, and abnormal metabolism of vitamin B_{12} and folic acid.

3. Hemostatic Abnormalities Associated with Neoplasia

Hemorrhage or thrombosis caused by hemostatic abnormalities may be frequent complications of neoplastic disease. Thrombocytopenia may occur because of myelophthisis, chemotherapy, or the development of localized or disseminated intravascular coagulation. Malignant cells may release trypsinlike substances that act as potent activators of the clotting mechanism. In addition, tumor cells can release fibrinolytic and platelet activators. In these tumor-induced consumptive conditions, qualitative and quantitative platelet abnormalities may coexist with coagulation abnormalities. Platelet dysfunction may develop as the result of the coating of platelet membranes by fibrindegradation products or by immunoglobulins or abnormal proteins produced by the tumor.

Metastasizing tumor cells within the bloodstream are often surrounded by deposits of platelets, fibrin, and other blood elements. The incorporation of these constituents within a tumor embolus may help to protect neoplastic cells from detection by clearance mechanisms and facilitate thrombosis and adhesion to the endothelial wall. The promotion of platelet aggregation and fibrin production by malignant cells and the thrombosis of small vessels may be important in enhancing penetration of the endothelium and the establishment of metastases.

Bibliography

Badylak, S. F. Coagulation disorders and liver disease. *Vet Clin North Am: Small Anim Pract* 18: 87–93, 1988.

Gentry, P. A., and Cooper, M. L. Effect of intravenous administration of T-2 toxin on blood coagulation in calves. *Am J Vet Res* 44: 741–746, 1983.

McDonald, G. K. Moldy sweetclover poisoning in a horse. *Can Vet J* 21: 250–251, 1980.

Murphy, J. J., and Gerken, D. F. The anticoagulant rodenticides. *In* "Current Veterinary Therapy X. Small Animal Practice," R. W. Kirk (ed.), pp. 143–146. Philadelphia, Pennsylvania, Saunders, 1989.

O'Keefe, D. A., and Cuoto, C. G. Coagulation abnormalities associated with neoplasia. *Vet Clin North Am: Small Anim Pract* 18: 157–168, 1988.

Slappendel, R. J. Disseminated intravascular coagulation. *In* "Current Veterinary Therapy X. Small Animal Practice," R. W. Kirk (ed.), pp. 451–457. Philadelphia, Pennsylvania, Saunders, 1989.

ACKNOWLEDGMENTS

The staff in the Word Processing Center typed the references and carried out a great deal of editing. Slides for photography were prepared by the technical staff in the histopathology laboratories of the Colleges of Veterinary Medicine at Guelph and Urbana.

CHAPTER 3

The Endocrine Glands

CHARLES C. CAPEN
The Ohio State University, U.S.A.

I. General Considerations

Endocrine glands are collections of specialized cells that synthesize and store their secretions, and release them directly into the bloodstream. They are sensing and signaling devices located in the extracellular fluid compartment and are capable of responding to changes in the internal and external environments to coordinate a multiplicity of activities that maintain homeostasis.

Secretions of endocrine glands are either peptide, steroid, catecholamine, or iodothyronine hormones that are transported by the blood to influence the functional activity of target cells elsewhere in the body. Other populations of cells are concerned with degradation of hormone after it has exerted its physiologic function. This is accomplished by peptidases on the cell surface, uptake by cells and degradation by lysosomal enzymes, or conjugation with glucuronide or sulfate, and excretion in the bile or urine.

Endocrine glands are small compared with other organs, widely distributed in the body, and connected with one another by the bloodstream. They are richly supplied with blood, and there is a close anatomic relationship between endocrine cells and the capillary network. To facilitate rapid transport of raw materials and secretory products between the blood and endocrine cells, the peripheral cytoplasmic extensions of capillary endothelial cells have fenestrae covered by a single membrane.

A. Types of Hormones

1. Polypeptide Hormones

The primary site of action for polypeptide hormones is the plasma membrane of target cells. Receptor proteins for the hormone are present on the outer surface of the plasma membrane. These hormones are water soluble, have a short half-life in blood (usually measured in minutes), and lack specific plasma-binding proteins.

The receptors for polypeptide hormones on the plasma membrane of target cells perform two key functions. First, they recognize the active hormone from among many other proteins to which the cell is exposed. The concentration of hormone in extracellular fluids often is much lower than that of other proteins. The hormone binds to the receptor site and forms a reversible hormone–receptor complex. Second, the receptor conveys the message of the bound hormone from the outside to the inside of the target cell. The magnitude of this transmembrane signal depends on the concentration of hormone to which the target cell is exposed, the affinity of the receptor for the hormone, and the concentration of receptors on the target cell.

There appears to be a single common intracellular pathway for many different polypeptide hormones. It begins with the activation of an enzyme, adenylate cyclase, in the plasma membrane of target cells. Cyclic adenosine monophosphate (cAMP) is formed intracellularly from adenosine triphosphate (ATP) and activates cAMP-dependent protein kinases. These protein kinases activate or inactivate a variety of enzymes by phosphorylating them, using adenosine triphosphate as a source of phosphate. The intracellular pathway for each polypeptide hormone subsequently branches into a multiplicity of pathways leading to a variety of effects on any given target cell.

Cells that produce polypeptide hormones have a well-developed endoplasmic reticulum with many attached ribosomes that assemble hormone, and a prominent Golgi apparatus for packaging hormone into granules for in-

267

tracellular storage and transport. Secretory granules are unique to polypeptide hormone- and catecholamine-secreting endocrine cells and provide a mechanism for intracellular storage of substantial amounts of preformed active hormone. These membrane-limited granules are macromolecular aggregations of active hormone, often associated with a specific binding protein. When the cell receives a signal for hormone secretion, secretory granules are directed to the periphery of the endocrine cell, probably by the contraction of microfilaments. The limiting membrane of the granule then fuses with the plasma membrane of the cell. The hormone-containing granule core is extruded into the extracellular perivascular space by emiocytosis or exocytosis. The granule core is fragmented subsequently, and hormone is rapidly transported through capillary fenestrae into the circulation. Hormone synthesized in excess of requirement is degraded by fusion of the hormone-containing granules with lysosomes, a process termed crinophagy.

2. Steroid Hormones

Steroid hormone-secreting cells are characterized by large lipid bodies in the cytoplasm that contain cholesterol and other precursor molecules. The lipid bodies are in close proximity to an extensive tubular network of smooth endoplasmic reticulum and large mitochondria, which contain the hydroxylase and dehydrogenase enzyme systems. These enzyme systems function to attach various side chains to the basic steroid nucleus. Steroid-producing cells lack secretory granules and do not store significant amounts of preformed hormone. They are dependent on continued biosynthesis to maintain the normal secretory rate for a particular hormone.

Steroid hormones having a basic nucleus of three cyclohexane rings and one pentane ring account for ~15% of mammalian hormones. The primary site of their action is the nucleus of target cells. Receptors are proteins that bind the hormone in the cytoplasm and nucleus of target cells. The steroid hormones are lipid soluble, which facilitates their transport through the cell membrane. They have a long half-life in blood (typically measured in hours) and reversibly bind to high-affinity, specific binding proteins in plasma.

After steroid hormones are within target cells and bound to cytoplasmic receptors, the hormone–receptor complex is translocated to the nucleus where the hormones bind to receptors in the nuclear chromatin. The interaction of steroid hormones with the genetic information results in increased transcription of messenger ribonucleic acid (mRNA), which directs new protein synthesis by specific target cells.

3. Catecholamine and Iodothyronine Hormones

This chemical group of hormones are tryosine derivatives. They account for ~5% of mammalian hormones and include the catecholamines (epinephrine, norepinephrine) secreted by the adrenal medulla, and iodothyronines (thyroxine, triiodothyronine) produced by follicular cells of the thyroid gland. Catecholamines share similar mechanisms of action with polypeptide hormones, whereas iodothyronines more closely resemble the steroid hormones.

B. Proliferative Lesions in Endocrine Glands

1. Characteristics

Neoplasms derived from polypeptide hormone-secreting endocrine cells usually consist of one predominant cell type and may be associated with the secretion of one polypeptide hormone. However, there is evidence from immunocytochemical and electron-microscopic investigations that some endocrine tumors are composed of more than one type of neoplastic cell and are capable of synthesizing multiple hormones.

Neoplasms derived from steroid hormone-secreting endocrine cells are characterized by having in their cytoplasm large lipid bodies that contain cholesterol and other precursor molecules for hormone synthesis. The lipid bodies are in close proximity to an extensive tubular network of smooth endoplasmic reticulum and large mitochondria. Steroid hormone-producing cells lack secretory granules and are unable to store significant amounts of preformed hormone.

The histopathologic separation between nodular hyperplasia, adenoma, and carcinoma often is more difficult in endocrine glands than in most other organs of the body. For many endocrine glands (especially thyroid C cells, secretory cells of the adrenal medulla, and specific tropic hormone-secreting cells of the adenohypophysis), there is a continuous spectrum of proliferative lesions from diffuse or focal hyperplasia to adenomas derived from a specific population of cells.

It is a common feature of endocrine glands that prolonged stimulation of a population of secretory cells predisposes to the subsequent development of a higher than expected incidence of tumors. Long-term stimulation leads to the development of clones of cells within the hyperplastic endocrine gland that grow more rapidly than the rest and are more susceptible to neoplastic transformation when exposed to the right combination of promoting carcinogens.

Nodular hyperplasia usually appears as multiple small foci, in one or both (for paired) endocrine gland(s), that are well demarcated but not encapsulated from adjacent normal cells. Cells composing an area of nodular hyperplasia closely resemble the cell of origin; however, the cytoplasmic area may be slightly enlarged and the nucleus more hyperchromatic than in the normal endocrine cell.

Adenomas are solitary nodules, in one (or occasionally both for paired) endocrine gland(s), that are larger than the multiple foci of nodular hyperplasia. They are sharply demarcated from the adjacent normal glandular parenchyma, often by a thin, partial to complete, fibrous capsule. The adjacent parenchyma is compressed to varying degrees depending on the size of the adenoma. Cells composing an adenoma closely resemble the cell of origin morphologically and in their architectural pattern of arrangement; however, there often are histologic differences such as multiple layers of cells lining follicles or vascular trabeculae and solid clusters of secretory cells subdivided into packets by a fine fibrovascular stroma.

Carcinomas usually are larger than adenomas and produce enlargement in one (or occasionally both, for paired) endocrine gland(s). Distinguishing adenoma from carcinoma of an endocrine gland often is more difficult than in other tissues. Histopathologic features that are suggestive of malignancy in an endocrine tumor include intraglandular invasion, invasion into and through the capsule of the gland with establishment of secondary foci of growth in the periglandular fibrous and adipose connective tissues, formation of tumor cell thrombi within vessels (especially muscular walled), and particularly the establishment of metastases at distant sites. The spread of neoplastic cells subendothelially in highly vascular benign endocrine tumors should not be mistaken for vascular invasion. Malignant endocrine cells often are more pleomorphic (including oval or spindle shaped) than normal, but nuclear pleomorphism is not a consistent criterion to distinguish adenomas and carcinomas. Mitotic figures may be frequent in malignant endocrine cells, but the significance of this criterion can vary considerably with the degree of background stimulation of the endocrine gland.

2. Functional Aspects of Endocrine Tumors

Many neoplasms derived from endocrine glands are functionally active, secrete an excessive amount of hormone either continuously or episodically, and result in clinical syndromes of hormone excess. Examples that occur in animals include the hypoglycemia of beta cell neoplasms of the pancreatic islets in dogs, hyperthyroidism associated with adenomas and carcinomas derived from thyroid follicular cells in cats and dogs, hypercalcitoninism in bulls and other animal species with thyroid C-cell tumors, and hyperadrenocorticism either associated with adrenocorticotropin (ACTH)-secreting pituitary corticotroph adenomas or neoplasms derived from the adrenal cortex (zona fasciculata) in dogs.

Measurement of hormone levels in serum or plasma in the basal, suppressed, or stimulated state and, in some instances, hormonal metabolites in the urine over a 24-hr period of excretion are essential to confirm that an endocrine tumor is releasing hormone at an abnormally elevated rate. An endocrine tumor can be interpreted as being endocrinologically active if either the rim of normal tissue around the tumor, or the opposite of paired endocrine glands, undergoes atrophy due to negative feedback inhibition by the elevated hormone levels or an altered blood constituent. In response to the autonomous secretion of hormone by the tumor, these non-neoplastic secretory cells become smaller than normal, especially in the cytoplasmic area, and eventually the number of cells is decreased. Functional pituitary neoplasms secreting an excess of a particularly tropic hormone (e.g., adrenocorticotropin) will be associated with hypertrophy and hyper-

plasia of target cells in the adrenal cortex (zonae fasciculata and reticularis).

C. Mechanisms of Endocrine Disease

1. Primary Hyperfunction of an Endocrine Gland

In primary hyperfunction of an endocrine gland, cells, often neoplastic and derived from the gland, synthesize and secrete hormone at an autonomous rate in excess of the body's ability to utilize and subsequently degrade the hormone, thereby resulting in functional disturbances of hormone excess. These include hyperfunction of parathyroid chief cells, thyroid C (parafollicular) cells, beta cells of the pancreatic islets, secretory cells of the adrenal medulla, and follicular cells of the thyroid, among many others.

The autonomous secretion of parathyroid hormone results in progressive and generalized demineralization of the skeleton, leading to hypercalcemia, which results in soft-tissue mineralization and the development of renal calculi. The accelerated osteoclastic resorption of bone results in marked thinning and osteoclastic tunneling of cortical bone, and predisposes bone to the development of pathologic fractures.

The autonomous hypersecretion of thyroxine and triiodothyronine is a common endocrinopathy in cats. This hypersecretion is associated with a spectrum of proliferative lesions of follicular cells. Although the majority of the functional thyroid lesions are adenomas derived from follicular cells, other distinctly multifocal lesions in one or both thyroids are more consistent with multinodular hyperplasia. Functional thyroid lesions in cats should be considered potentially malignant because a few are adenocarcinomas, which may metastasize to regional lymph nodes. The functional disturbances of hyperactivity, weight loss despite increased appetite, hyperthermia, and tachycardia reflect long-term stimulation of multiple populations of target cells by the abnormally elevated blood levels of thyroid hormones.

2. Secondary Hyperfunction of an Endocrine Gland

In secondary hyperfunction of an endocrine gland, a lesion in one organ (e.g., adenohypophysis) secretes an excess of a tropic hormone that leads to long-term stimulation and hypersecretion of hormone by a target organ. An example of this pathogenic mechanism in animals is the adrenocorticotropic hormone (ACTH)-secreting tumor derived from pituitary corticotrophs in the pars distalis and intermedia of dogs (see Fig. 3.6). The clinical signs and lesions in the animal primarily are the result of the elevated blood cortisol levels associated with the ACTH-stimulated hypertrophy and hyperplasia of the zonae fasciculata and reticularis of the adrenal cortex. The syndrome of cortisol excess in dogs is characterized by progressive alopecia, hyperpigmentation, and muscle wasting caused by the protein catabolic effects of glucocorticoids.

Some dogs, particularly poodles, have a similar marked

adrenocortical enlargement and clinical evidence of corti-sol excess, but without a tumor in the pituitary gland. These dogs may have a change in their set point to the negative feedback signal (e.g., blood cortisol). This can be caused by an abnormal accumulation of certain neu-rotransmitter substances (e.g., serotonin) near neuro-secretory neurons in the hypothalamus that secrete cor-ticotropic hormone-releasing factor. The end result is corticotroph hyperplasia, elevated ACTH levels in the blood, and long-term stimulation of the adrenal cortex, which results in hyperplasia of the zona fasciculata and zona reticularis (see Fig. 3.41B) and the clinical syndrome of cortisol excess.

3. Primary Hypofunction of an Endocrine Gland

In primary hypofunction of an endocrine gland, hor-mone secretion is subnormal because of extensive destruc-tion of secretory cells by a disease process, the failure of an endocrine gland to develop properly, or the result of a specific biochemical defect in the synthetic pathway of a hormone.

Immune-mediated injury causes hypofunction of sev-eral endocrine glands in animals, including the parathyroid gland, adrenal cortex, and thyroid gland. Thyroiditis caused by this mechanism is characterized by marked infiltration of lymphocytes and plasmacytes, and deposi-tion of electron-dense immune complexes along the follic-ular basement membranes with progressive destruction of secretory parenchyma.

Failure of development also results in primary hypo-function of an endocrine gland. An example of this mecha-nism in animals is the failure of oropharyngeal ectoderm to differentiate into tropic hormone-secreting cells of the adenohypophysis in dogs, resulting in pituitary dwarfism and a failure to attain somatic maturation (see Fig. 3.5). A large, multicompartmented cyst is present on the ventral aspect of the brain in the pituitary region of these dogs. The cyst compresses the normally developed neurohypo-physis and results in disturbances of water metabolism.

Another form of primary hypofunction is a failure of hormone synthesis caused by a genetically determined defect in a biosynthetic pathway or lack of a specific en-zyme. One of the best-documented examples of this condi-tion in animals is vitamin D-dependent rickets in pigs. It is caused by a lack of 25-hydroxycholecalciferol-1-alpha-hydroxylase in the proximal convoluted tubules of the kidney. This enzyme is needed to synthesize the hormonal form of vitamin D. Blood calcium and phosphorus levels progressively decrease because of the subnormal ability of the pig to convert 25-hydroxycholecalciferol to the bio-logically active, hormonal form (1,25-dihydroxycholecal-ciferol) in the kidney. The lowered blood concentrations of calcium and phosphorus lead to a failure of mineraliza-tion of osteoid and persistence of cartilage in the physes, leading to severe deformities in the skeleton (see Bones and Joints, Volume 1, Chapter 1).

Congenital dyshormonogenetic goiter in sheep, goats, and cattle is another example of primary hypofunction

caused by failure of hormone synthesis. The low blood thyroxine and triiodothyronine levels and clinical evi-dence of severe hypothyroidism in these animals are due to an inability of follicular cells to synthesize thyroglobu-lin. The molecular defect has been shown to be a defective processing of the primary transcripts for thyroglobulin mRNA and aberrant transport of mRNA from the nucleus to ribosomes. This results in subnormal amounts of thyro-globulin mRNA in follicular cells, particularly mRNA which is attached to membranes of the endoplasmic reticu-lum in the cytoplasm.

Thyroglobulin is one of the major components of colloid in the lumen of thyroid follicles. This high-molecular-weight glycoprotein is synthesized on ribosomes of the rough endoplasmic reticulum in follicular cells. Thyro-globulin is packaged into apical granules that are secreted into the follicular lumen to serve as an extracellular matrix for the step-wise iodination of the tyrosyl residues incor-porated into its structure, resulting in the formation of thyroxine and triiodothyronine. These animals do not have a defect in the ability of the thyroid to concentrate ^{131}I; however, there is a greatly reduced ability to iodinate tyrosyl residues and form thyroid hormones. The subnor-mal blood levels of thyroxine and triiodothyronine are detected by the hypothalamus and adenohypophysis, re-sulting in an increased secretion of thyrotropin and intense hyperplasia of follicular cells (see Fig. 3.27A).

4. Secondary Hypofunction of an Endocrine Gland

In secondary hypofunction of an endocrine gland, a destructive lesion in one organ, such as the pituitary, interferes with the secretion of tropic hormone. This re-sults in hypofunction of the target endocrine glands. Large, endocrinologically inactive, pituitary neoplasms in adult dogs, cats, and other animal species may interfere with the secretion of the multiple pituitary tropic hor-mones and result in clinically detectable hypofunction of the adrenal cortex, follicular cells of the thyroid, and go-nads (see Fig. 3.8).

The adrenal cortex of an animal with a large pituitary neoplasm of this type has marked atrophy and degenera-tion of the ACTH-dependent inner zones; however, the aldosterone-secreting zona glomerulosa, which is not un-der direct ACTH control, remains intact. Thyroid function may be subnormal because of a lack of thyrotropin and atrophy of follicular cells, but the calcitonin-secreting C cells function normally since they are not controlled by pituitary tropic hormones.

5. Endocrine Hyperactivity Secondary to Diseases of Other Organs

An example of endocrine hyperactivity secondary to diseases of other organs is the hyperparathyroidism that develops secondary to chronic renal failure or nutritional imbalances. In the renal form, the retention of phosphorus early and subsequent progressive destruction of cells in the proximal convoluted tubules interferes with the meta-bolic activation of vitamin D by the 1-alpha-hydroxylase in the kidney. This is the rate-limiting step in the metabolic activation of vitamin D and is tightly controlled by parathyroid hormone and several other factors. The impaired intestinal absorption of calcium results in the development of progressive hypocalcemia, which leads to long-term parathyroid stimulation, and subsequent development of generalized demineralization of the skeleton. Many bones, particularly those of the head, become weakened and susceptible to fractures.

Nutritional hyperparathyroidism develops in animals fed abnormal diets that are low in calcium, high in phosphorus or oxalate, or (for certain primates) deficient in cholecalciferol. Unsupplemented all-meat diets fed to carnivores fail to supply the daily requirements for calcium, leading to progressive hypocalcemia that stimulates the parathyroid gland to increased activity. The normal kidneys in these animals respond to the increased parathyroid hormone secretion by increasing phosphorus excretion and lowering blood phosphorus levels. After a carnivore is fed this imbalanced diet for several weeks to months, its skeleton becomes severely demineralized and predisposed to fractures. The cortices of long bones are thin, and the medullary cavity widened because of intense osteoclastic resorption of bone that is stimulated indirectly by the increased secretion of parathyroid hormone (see Bones and Joints, Volume 1, Chapter 1).

6. Hypersecretion of Hormones or Hormonelike Substances by Nonendocrine Tumors

The hypersecretion by nonendocrine tumors of hormonelike substances that are similar chemically and/or biologically to the native hormone secreted by the endocrine gland occurs animals and humans. Most of these substances secreted by nonendocrine tumors are peptides. Steroids and iodothyronines require more complex biosynthetic pathways and do not appear to be secreted by nonendocrine tumors.

An example of this mechanism in animals is the adenocarcinoma derived from apocrine glands of the anal sac in dogs. This tumor produces a hormone, parathyroid hormone-related protein (PTHrP), that stimulates osteoclasts indirectly. The resulting accelerated mobilization of calcium from bone leads to the development of persistent hypercalcemia, even though the animal's parathyroid glands are smaller than normal and composed of inactive and atrophic chief cells.

This neoplasm develops predominantly in elderly female dogs and is composed of solid and glandular areas. The columnar tumor cells which line the acini often have distinctive cytoplasmic apices which project into the lumen. Small membrane-limited granules, resembling secretion granules of polypeptide hormone-secreting endocrine cells, are present in the cytoplasm. Circulating levels of immunoreactive parathyroid hormone in hypercalcemic dogs with apocrine adenocarcinomas do not differ significantly from the levels in control dogs with or without other tumors, but blood levels of PTHrP are greatly elevated.

Following surgical removal of the apocrine adenocarcinoma, the serum calcium and phosphorus levels return to normal, immunoreactive parathyroid hormone levels increase rapidly, and the 1,25-dihydroxycholecalciferol level decreases.

7. Endocrine Dysfunction Due to Failure of Target Cell Response

Endocrine dysfunction due to a failure of target cell response has been recognized coincident with a more complete understanding of how hormones convey their biologic message. Steroid and iodothyronine hormones penetrate the cell membrane, bind to cytosolic receptors, and are transported to the nucleus, where they interact with the genetic information in the cell to increase new protein synthesis. Polypeptide and catecholamine hormones bind to receptors on the surface of target cells and activate a membrane-bound enzyme that generates an intracellular messenger (cAMP), which elicits the physiologic response.

Failure of target cells to respond to hormone may be due to a lack of adenylate cyclase in the cell membrane or to an alteration in hormone receptors on the cell surface. Hormone is secreted in normal or increased amounts by cells of the endocrine gland. Certain forms of insulin resistance associated with obesity in animals and humans result from a decrease in the number of receptors on the surface of target cells. This develops in response to the chronic increased insulin secretion stimulated by the hyperglycemia resulting from the excessive food intake. Secretory cells in the corresponding endocrine gland (for example, pancreatic islets) undergo compensatory hypertrophy and hyperplasia in an attempt to secrete additional hormone. The normal pancreatic islets contain predominantly granulated beta cells, whereas the beta cells in the enlarged islets from an obese diabetic animal are markedly hyperplastic and depleted of insulin-containing secretory granules.

8. Endocrine Dysfunction Resulting from Abnormal Degradation of Hormone

In endocrine dysfunction resulting from abnormal degradation of hormone, the secretion of hormone by an endocrine gland is normal, but blood levels are elevated persistently. A decreased rate of degradation simulates a state of hypersecretion. The syndrome of feminization due to hyperestrogenism, associated with cirrhosis and decreased hepatic degradation of estrogens in men, is an example of this pathogenic mechanism.

Chronic renal disease in animals may be associated with subnormal, normal, or elevated blood concentrations of calcium. The hypercalcemia associated with certain forms of renal disease may be related, in part, to diminished degradation of parathyroid hormone along with decreased urinary excretion of calcium by the diseased kidney. Biologically active parathyroid hormone is degraded in the kidney either by peptidases on the surface of tubular cells or by lysosomal enzymes after uptake of the hormone from the glomerular filtrate.

Another example of abnormal degradation of hormone is the induction of hepatic microsomal enzymes, such as T_4-uridine diphosphate (UDP)-glucuronyl transferase, by long-term administration of various chemicals and drugs. This causes a disruption of thyroid hormone economy. The increased microsomal enzyme activity results in an increased excretion of conjugated thyroxine in the bile, subnormal circulating T_4 levels, and a compensatory increased secretion of thyroid-stimulating hormone (TSH) by the pituitary gland. The long-term stimulation of thyroid follicular cells by TSH in a sensitive animal species, such as the laboratory rat, predisposes to the development of an increased incidence of focal hyperplasia and adenomas.

9. Iatrogenic Syndromes of Hormone Excess

In iatrogenic syndromes of hormone excess, the administration of hormone, either directly or indirectly, influences the activity of target cells and results in clinical disturbances. The daily administration of potent preparations of adrenal corticosteroids at inappropriate dosages, for prolonged periods in the symptomatic treatment of various diseases, will reproduce most of the functional disturbances associated with endogenous hypersecretion of cortisol. These disturbances include muscle weakness, marked hair loss, and mineral deposition in the skin associated with cortisol excess. The elevated blood levels of exogenous cortisol result in marked atrophy of the adrenal cortex, particularly the ACTH-dependent zone fasciculata and reticularis.

The administration of certain progestagens will indirectly result in a syndrome of hormone excess. The injection of medroxyprogesterone acetate for the prevention of estrus in dogs stimulates an increased secretion of growth hormone by pituitary somatotrophs, resulting in many of the clinical manifestations of acromegaly. The excessive skin folds, expansion of interdigital spaces, and abdominal enlargement in dogs with iatrogenic acromegaly are related to the protein anabolic effects of growth hormone on connective tissues.

Bibliography

Aurbach, G. D. Inherited disorders of hormone resistance. *In* "Animal Models of Inherited Metabolic Diseases," R. J. Desnick *et al.* (eds.), pp. 353–368. New York, Alan R. Liss, 1982.

Capen, C. C., and Martin, S. L. Calcium metabolism and disorders of parathyroid glands. *Vet Clin North Am* 7: 513–548, 1977.

Chan, L., and O'Malley, B. W. Mechanism of action of the sex steroid hormones. *N Engl J Med* 294: 1322–1328, 1976.

Cox, R. I. The endocrinologic changes of gestation and parturition in the sheep. *Adv Vet Sci Comp Med* 19: 287–305, 1975.

Haussler, M. R. *et al.* Assay of 1,25-dihydroxyvitamin D and other vitamin D metabolites in serum: Application to animals and humans. *In* "Vitamin D—Basic Research and Its Clinical Application," A. W. Normal *et al.* (eds.), pp. 189–196. Berlin, de Gruyter, 1979.

Liggens, G. C., and Kennedy, P. C. Effects of electrocoagulation of the foetal lamb hypophysis on growth and development. *J Endocrinol* 40: 371–381, 1968.

Roth, J. Polypeptide hormone receptors. *In* "Membrane Receptors for Viruses, Antigens and Antibodies, Polypeptide Hormones, and Small Molecules." New York, Raven Press, 1976.

Stewart, A. E. *et al.* Quantitative bone histomorphometry in humoral hypercalcemia of malignancy: Uncoupling of bone cell activity. *J Clin Endocrinol Metab* 55: 219–227, 1982.

Winkler, I., Grave, C., and Harmeyer, J. Pseudo vitamin D-deficiency rickets in pigs: *In vitro* measurements of renal 25-hydroxycholecal-ciferol-1-hydroxylase activity. *Zentralbl Veterinaermed* [A] 29: 81–88, 1982.

II. Pituitary Gland

The pituitary gland in an adult animal is completely separated from the oral cavity. It is situated in the sella turcica, a concavity of the sphenoid bone, and enveloped by an extension of dura mater. The pituitary gland (hypophysis) is subdivided anatomically into the adenohypophysis (anterior lobe) and neurohypophysis (posterior lobe).

A. Development, Structure, and Function

1. Embryologic Development of the Pituitary Gland

The pituitary gland develops embryologically from a dorsal evagination of oropharyngeal ectoderm (Rathke's pouch) (Fig. 3.1) and a ventral downgrowth of diencephalic neuroectoderm as shown schematically in the diagram. The point of fusion of the two primordia develops

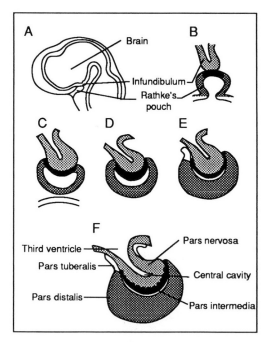

Fig. 3.1 Embryologic development of the pituitary gland. (From C. D. Turner, "General Endocrinology," 2nd Ed. Philadelphia and London, W.B. Saunders, 1955.)

into the pars intermedia. The pars distalis undergoes extensive proliferation to form the major part of the adenohypophysis and is responsible for the secretion of the multiple tropic hormones.

The pituitary gland has two preformed cavities, the residual lumen of Rathke's pouch (cleft) and the infundibular recess of the third ventricle (central cavity). Separation of the developing adenohypophysis from the oropharynx is completed by formation of the sphenoid bone.

2. Structure of the Adenohypophysis

The adenohypophysis consists of three portions, the pars distalis, pars tuberalis, and pars intermedia. In many species the adenohypophysis completely surrounds the pars nervosa of the neurohypophyseal system. The pars distalis is the largest of the three parts of the adenohypophysis and contains the populations of endocrine cells that secrete the pituitary tropic hormones. The secretory cells are supplied with abundant capillaries that have fenestrae in their peripheral cytoplasmic extensions and are supported by the cytoplasmic processes of stellate (follicular or sustentacular) cells.

The pars tuberalis consists of dorsal projections of cells along the infundibular stalk. It functions primarily as a scaffold for the capillary network of the hypophyseal portal system during its course from the median eminence to the pars distalis. The pars intermedia forms the junction between the pars distalis and pars nervosa. It lines the residual lumen of Rathke's pouch and contains two populations of cells. In the dog, one of these cell types synthesizes and secretes adrenocorticotropic hormone (ACTH), similar to corticotrophs in the pars distalis.

3. Functional Cytology of the Adenohypophysis

A specific population of endocrine cells is present in the pars distalis (and in the pars intermedia of dogs for ACTH) that synthesize and secrete each of the pituitary tropic hormones. Secretory cells in the adenohypophysis are subdivided into acidophils, basophils, and chromophobes, based on the staining of their secretory granules with pH-dependent histochemical stains.

Acidophils are further subdivided functionally into somatotrophs and luteotrophs that secrete growth hormone (GH, somatotropin) and luteotropic hormone (LTH, prolactin), respectively. Their granules contain hormones that are simple proteins that stain with orange G, azocarmine, or erythrosin. Basophils include both gonadotrophs that secrete luteinizing hormone (LH) and follicle-stimulating hormone (FSH), and thyrotrophs that secrete thyroid-stimulating or thyrotropic hormone (TSH). Secretory granules of basophils contain glycoproteins that react with the periodic acid–Schiff (PAS) reagent. Chromophobes are pituitary cells that do not have obvious cytoplasmic secretory granules on light-microscopic evaluation. They include the pituitary cells concerned with the synthesis of ACTH and melanocyte-stimulating hormone (MSH), nonsecretory follicular (stellate) cells, degranulated chromophils (acidophils and basophils) in the actively synthe-

sizing phase of the secretory cycle, and undifferentiated stem embryonic cells of the adenohypophysis.

Specific immunocytochemical staining of the adenohypophysis in dogs has demonstrated that ACTH- and MSH-staining cells are polyhedral to round, sparsely granulated, and most numerous in the ventrocentral and cranial portions of the pars distalis. They are less frequent in the dorsal and caudal regions of the pars distalis and throughout the pars tuberalis. In the pars intermedia of dogs, most cells demonstrate immunoreactivity to either ACTH, or alpha-MSH.

Thyrotrophs are large polyhedral cells situated singly or in small groups ventrocentrally in the paramedian plane in the pars distalis of dogs. Gonadotrophs (cells reacting with antisera to human FSH and/or bovine LH) are oval to polyhedral and distributed singly in the pars distalis, particularly in the dorsocranial region and in the caudal extensions along the pars intermedia.

Immunoreactive prolactin cells occur in small groups of large polygonal cells with prominent granules in the ventrocentral and cranial portions of the canine pars distalis. A diffuse increase in this population of cells occurs in females near parturition. Growth hormone-secreting cells are present singly along capillaries in the dorsal region of the pars distalis near the pars intermedia. They are small, round to oval, and have numerous cytoplasmic granules. Somatotrophs frequently undergo diffuse hyperplasia and hypertrophy in old dogs, especially females with mammary dysplasia or neoplasia.

4. Hypothalamic Control of the Adenohypophysis

Each population of endocrine cells in the pars distalis is under the control of a corresponding releasing hormone from the hypothalamus (Fig. 3.2). These releasing hormones are small peptides synthesized via neurosecretion by neurons in the hypothalamus. They are transported by axonal processes to the median eminence where they are released into capillaries and conveyed by the hypophyseal portal system to specific endocrine cells in the adenohypophysis, where each stimulates the rapid release of secretory granules containing a specific preformed tropic hormone.

There are separate hypothalamic releasing hormones that regulate the rate of secretion of each tropic hormone secreted by the adenohypophysis. For most pituitary tropic hormones, negative feedback control is accomplished by the blood concentration of the hormone produced by the target endocrine gland, e.g., thyroid gland, adrenal cortex, ovary, or testis. The hormone produced by the endocrine glands exerts negative feedback control either on the neurosecretory neurons in the hypothalamus that synthesize the corresponding releasing hormone or on tropic hormone-secreting cells in the adenohypophysis or at both sites. However, GH, LTH, and MSH do not act on target endocrine organs to stimulate secretion of a hormone. Negative feedback control of these three pituitary hormones is effected by production of a corresponding release-inhibiting hormone (factor) by neurons in the

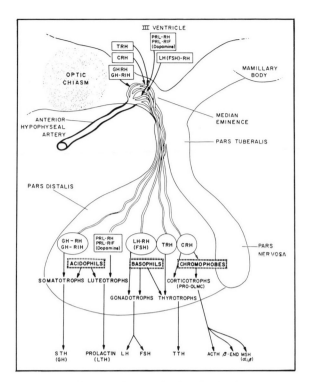

Fig. 3.2 Separate hypothalamic releasing hormones control the release of each trophic hormone by the adenohypophysis. Tropic hormones that do not act on a target endocrine organ to release a final endocrine product (e.g., somatotropin and prolactin) also are regulated by hypothalamic release-inhibiting hormone (RIH) or factor (dopamine for prolactin).

hypothalamus. The relative local concentrations of the specific releasing hormone and release-inhibiting hormone appear to govern the rate of release of GH, LTH, and MSH from the adenohypophysis. Growth hormone release-inhibiting hormone (somatostatin) that controls the release of somatotropin also is produced by endocrine cells in the pancreatic islets. Dopamine is the principal release-inhibiting hormone for prolactin.

5. Structure and Function of the Neurohypophysis

The neurohypophysis has three anatomic subdivisions. The pars nervosa (posterior lobe) represents the distal component of the neurohypophyseal system. It is composed of numerous capillaries that are supported by modified glial cells (pituicytes). The capillaries in the pars nervosa are termination sites for the nonmyelinated axonal processes of neurosecretory neurons in the hypothalamus. Secretion granules that contain the neurohypophyseal hormones, oxytocin and antidiuretic hormone (ADH or vasopressin), are synthesized in hypothalamic neurons but are released into the bloodstream in the pars nervosa. The infundibular stalk joins the pars nervosa to the overlying hypothalamus and is composed of nonmyelinated axonal processes from neurosecretory neurons.

Neurosecretory neurons in the hypothalamus receive neural input from higher centers and translate this into

endocrine output in the form of hormonal secretion. In addition to the usual structural features of neurons, they contain prominent lamellar arrays of rough endoplasmic reticulum, large Golgi apparatuses, and numerous membrane-limited neurosecretory granules in the cell body and axonal process (Fig. 3.3).

The neurosecretory neurons concerned with hormone synthesis are segregated into anatomically defined regions, termed nuclei, in the hypothalamus. The supraoptic nucleus is concerned primarily with the synthesis of ADH, whereas oxytocin is produced predominantly by neurons in the paraventricular nucleus.

Antidiuretic hormone and oxytocin are nonapeptides synthesized by neurons situated in either the supraoptic or the paraventricular nucleus. Antidiuretic hormone and its corresponding neurophysin appear to be synthesized as part of a common larger biosynthetic precursor molecule, termed propressophysin. The hormones are packaged into membrane-limited neurosecretary granules with a corresponding binding protein (neurophysin) and transported to the pars nervosa by axonal processes of the neurosecretory neurons. These axons terminate on fenestrated capillaries in the pars nervosa and release ADH or oxytocin into the circulation.

Antidiuretic hormone is transported by the bloodstream to its site of action in the kidney, where it binds to specific receptors on epithelial cells in the distal part of the nephron and collecting ducts. The overall effect of ADH on the kidney is to increase the active renal tubular reabsorption of water from the glomerular filtrate. The hormone (ADH)–receptor complex activates the membrane-bound enzyme, adenylate cyclase, resulting in the intracellular formation of cAMP from ATP. The intracellular accumulation of cAMP activates protein kinases (involved in the phosphorylation of proteins in the luminal membrane of distal tubular cells) that increase their permeability to water.

6. Blood Supply to the Pituitary Gland

The neurohypophysis in most animals is supplied directly by the posterior (inferior) hypophyseal arteries that branch from the internal carotid arteries. Branches of the anterior (superior) hypophyseal arteries originate from the internal carotid arteries and form the posterior communicating arteries of the circle of Willis. Arteriolar branches penetrate the pars tuberalis, lose their muscular coat, and form a capillary plexus near the median eminence. These vessels subsequently drain into hypophyseal portal veins that supply the pars distalis and transport the hypothalamic releasing hormones to the pituitary. A small artery that arises from the posterior hypophyseal artery may provide a minor blood supply to the adenohypophysis.

B. Diseases of the Pituitary Gland

1. Hypophyseal Changes Associated with Alterations in Target Organs

In response to surgical removal or disease processes in a target endocrine organ of the pituitary, there is a

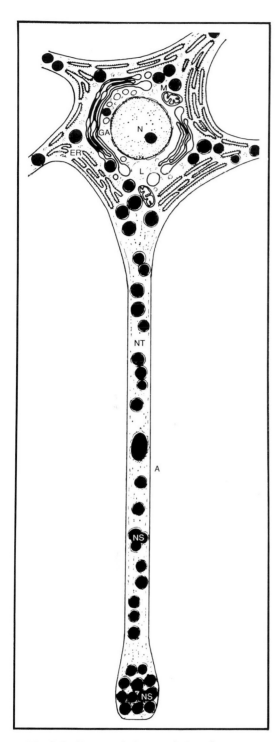

Fig. 3.3 Structural characteristics of a neurosecretory neuron. Nerve cell body (N, nucleus) has dendritic and axonal (A) processes, arrays of rough endoplasmic reticulum (ER), a prominent Golgi apparatus (GA), and large mitochondria (M). Hormone-containing, membrane-limited neurosecretory granules (NS) are transported along axon to site of release.

progressive decline in the circulating concentration of hormone produced by the target organ. This reduction in circulating hormone level is detected by the hypothalamic–adenohypophyseal system and results in structural changes in the specific tropic hormone-secreting cell population in the pars distalis.

The initial reaction is a rapid release of storage granules containing preformed tropic hormone from one population of endocrine cells in the pars distalis. For example, only thyrotropic basophils degranulate following thyroidectomy or thyroid disease; corticotropic chromophobes degranulate after adrenalectomy; and gonadotropic basophils release their secretory granules in response to gonadectomy. After an interval of several days following surgical ablation or extensive destruction of a target endocrine organ, the corresponding degranulated tropic hormone-secreting cell in the pars distalis undergoes hypertrophy with expansion of the cytoplasmic area in response to the sustained increased demand for the particular tropic hormone. The abundant cytoplasm contains extensive organelles concerned with hormonal synthesis (rough endoplasmic reticulum) and packaging into secretory granules (Golgi apparatus) plus large mitochondria.

If the demand for tropic hormone secretion is sustained for days or weeks, one specific population of endocrine cells in the pars distalis undergoes hyperplasia. Groups of hyperplastic tropic hormone-secreting cells are present, scattered as small nests throughout an otherwise normal pars distalis. In response to long-term (weeks to months) stimulation, the hypertrophied cytoplasm of the cells becomes vacuolated due to the extensive distension of profiles of rough endoplasmic reticulum with a finely granular, moderately electron-dense material. The multiple small cytoplasmic vacuoles coalesce subsequently to form a large vacuole that may displace the nucleus eccentrically, characteristic of cells in the pituitary gland which degranulate following thyroidectomy, adrenalectomy, or gonadectomy.

2. Pituitary Cysts

a. CYSTS OF THE CRANIOPHARYNGEAL DUCT   Cysts may develop from remnants of the distal craniopharyngeal duct, which normally disappears by birth in most animal species. The cysts are lined by cuboidal to columnar, often ciliated, epithelium and contain mucin. In dogs, especially of the brachycephalic breeds, cysts from these remnants are often found at the periphery of the pars tuberalis and pars distalis. Cystic remnants of the craniopharyngeal duct in one survey were found in 53% of dogs of several breeds.

Craniopharyngeal duct cysts occasionally become large enough to exert pressure on the infundibular stalk and hypophyseal portal system, median eminence, or pars distalis. Structures adjacent to the cysts atrophy to varying degrees owing to compression and interference with the blood supply. Disruption of a large cyst with escape of the proteinaceous contents into adjacent tissues may incite an intense, local inflammation with subsequent fibrosis that may interfere with pituitary function. Clinical signs can include visual difficulties due to pressure on the optic chiasm, diabetes insipidus, obesity, and hypofunction of the adenohypophysis (gonadal atrophy, decreased basal metabolic rate, and hypoglycemia).

b. Cysts Derived from the Pharyngeal Hypophysis

The proximal portion of the adenohypophyseal anlage may persist in the dorsal aspect of the oral cavity in adults as undifferentiated remnants of cells along the craniopharyngeal canal or as differentiated cells similar to those of the definitive adenohypophysis. These remnants, called the pharyngeal hypophysis, have been described in dogs, cats, other animal species, and humans. The pharyngeal hypophysis is physically separated from the sellar adenohypophysis in dogs, but in cats these structures may be continuous because of persistence of the craniopharyngeal canal.

The pharyngeal hypophysis is seen most often in brachycephalic breeds of dogs. It is a tubular structure lined by ciliated columnar epithelium, located on the midline of the nasopharynx, and is frequently continuous with a multilocular cyst that is lined by ciliated cuboidal or columnar epithelium. The cyst contains colloid material and cellular debris. A mass of differentiated acidophilic, basophilic, and chromophobic cells similar to those of the sellar adenohypophysis usually extends from the cyst wall.

Cysts (up to several centimeters in diameter) may be derived from the oropharyngeal end of the craniopharyngeal duct and project as a space-occupying mass into the nasopharynx in dogs. The predominant clinical sign may be related to respiratory distress because of ventral displacement of the soft palate and occlusion of the posterior nares. The cyst wall may be hard on palpation because of the presence of partially mineralized woven bone. The contents of the cyst are often yellow-gray and caseous due to the accumulation of keratin and desquamated epithelial cells from the cyst lining. The squamous epithelial lining of the cyst is derived from metaplasia of the remnants of the primitive oropharyngeal epithelium.

c. Cysts Resulting from a Failure of Differentiation of Oropharyngeal Ectoderm of Rathke's Pouch

Pituitary dwarfism in dogs usually is associated with a failure of the oropharyngeal ectoderm of Rathke's pouch to differentiate into tropic hormone-secreting cells of the pars distalis. This results in a progressively enlarging, multiloculated cyst in the sella turcica and an absence of the adenohypophysis. The cyst is lined by pseudostratified, often ciliated, columnar epithelium with interspersed mucin-secreting goblet cells. The mucin-filled cysts eventually occupy the entire pituitary area in the sella turcica and severely compress the pars nervosa and infundibular stalk (Fig. 3.4A,B). In the pituitary region, few differentiated, tropic hormone-secreting chromophils immunocytochemically stain for the specific tropic hormones. An occasional small nest or rosette of poorly dif-

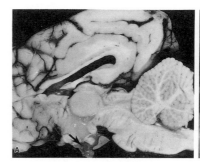
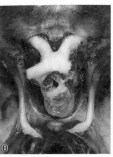

Fig. 3.4 Failure of oropharyngeal ectoderm to differentiate into tropic hormone-secreting cell of the adenohypophysis. German shepherd. Dwarfism. (A) Sagittal section. Pars nervosa compressed by mucin-filled cyst (arrow). (B) Dorsal view of cyst.

ferentiated epithelial cells is interspersed between the multiloculated cysts, but the cytoplasm of these cells is usually devoid of hormone-containing secretory granules.

Juvenile panhypopituitarism occurs most frequently in German shepherd dogs, but it has also been reported in other breeds, such as the spitz, toy pinscher, and Carelian bear dogs from Denmark. The dwarf pups appear normal or are indistinguishable from littermates at birth and until ~2 months of age. Subsequently the slower growth rate than the littermates, retention of puppy hair, and lack of primary guard hairs gradually become evident in dwarf pups. German shepherd dogs with pituitary dwarfism appear coyote- or foxlike due to their small size and soft woolly coat (Fig. 3.5). A bilaterally symmetrical alopecia develops gradually and often progresses to complete alopecia except for the head and tufts of hair on the legs. There is progressive hyperpigmentation of the skin until it is uniformly brown-black over most of the body. Adult German shepherd dogs with panhypopituitarism vary in size from as small as 2 kg up to nearly half normal size,

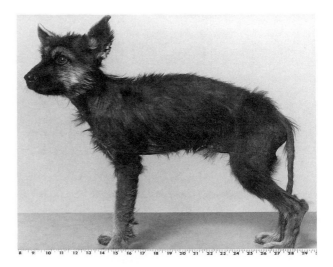

Fig. 3.5 Panhypopituitarism (pituitary dwarfism). German shepherd, 1 year of age.

depending on whether the failure of formation of the adenohypophysis is nearly complete or only partial.

Permanent dentition is delayed or completely absent. Closure of physes is delayed as long as 4 years, depending on the severity of hormonal insufficiency. There are few trabeculae in the primary and secondary spongiosa of the metaphysis of long bones, and osteoblasts are decreased in dwarf pups when compared with those of normal littermates. The external genitalia usually remain infantile. The testes and penis are small, mineralization of the os penis is delayed or incomplete, and the penile sheath is flaccid. In females the ovarian cortex is hypoplastic, and estrus, irregular or absent.

The shortened life span in dogs with pituitary dwarfism results not only from the panhypopituitarism but also from the resulting secondary endocrine dysfunction, such as hypothyroidism and hypoadrenocorticism. The increase in blood thyroxine and cortisol levels in response to challenge by exogenous thyrotropin and adrenocorticotropin are subnormal, owing to hypoplasia of thyroid follicular cells and zonae fasciculata and reticularis of the adrenal cortex. The variation in severity and onset of the lesions in pituitary dwarfism appears to be related to the degree that the oropharyngeal epithelium fails to differentiate and the rapidity with which the mucin-filled cysts enlarge and exert pressure on adjacent structures.

Panhypopituitarism in German shepherd dogs often occurs in littermates and related litters, suggesting a simple autosomal recessive mode of inheritance. The activity of somatomedin (a cartilage growth-promoting peptide whose production in the liver and plasma activity is controlled by somatotropin) is low in dwarf dogs. Intermediate somatomedin activity is present in the phenotypically normal ancestors suspected to be heterozygous carriers.

Useful diagnostic aids for pituitary dwarfs include comparison of height with littermates, radiographs of open epiphyseal lines, thyroid function tests, and skin biopsy. Cutaneous lesions include hyperkeratosis, follicular keratosis, hyperpigmentation, adnexal atrophy and loss of elastin fibers, and the loose network of collagen fibers in the dermis. Hair shafts are absent, and hair follicles are primarily in the telogen (resting) phase of the growth cycle. Assays for somatomedin (a non-species-specific, somatotropin-dependent peptide) provide an indirect measurement of circulating growth hormone activity in dogs suspected to be heterozygous carriers. Basal levels of circulating canine growth hormone are detectable but low (normal range, 1.75 ± 0.17 ng/ml) in pituitary dwarfs and fail to increase following a provocative test for growth hormone secretion provided by clonidine injection as in normal dogs. Insulin hypersensitivity has been demonstrated in pituitary dwarf dogs, probably due to a change in insulin receptor numbers or affinity of binding in response to the low growth hormone levels. Dwarf dogs develop more profound hypoglycemia following an insulin injection than do normal dogs. Their response is similar to that in experimentally hypophysectomized dogs.

Cysts associated with pituitary dwarfism are distinct morphologically from the cysts that develop following the abnormal accumulation of colloid in the residual lumen of Rathke's pouch. The normally developed pars distalis and pars nervosa are compressed to varying degrees by the abnormal accumulation of colloid in a normal cavity of the pituitary.

C. Neoplastic Diseases of the Pituitary

1. Corticotroph (ACTH-Secreting) Adenoma of the Adenohypophysis

Functional tumors arising in the pituitary gland may be derived from corticotroph (ACTH-secreting) cells in either the pars distalis or the pars intermedia. They cause a clinical syndrome of cortisol excess (Cushing's disease). These neoplasms are encountered most frequently in dogs and infrequently in other animal species. In horses pituitary tumors arise almost exclusively from the pars intermedia. Corticotroph adenomas develop in adult to aged dogs and have been reported in a number of breeds. Boxers, Boston terriers, and dachshunds appear to have a higher incidence of ACTH-secreting pituitary tumors than do other breeds. The spectrum of clinical manifestations and lesions that develop is primarily the result of long-term overproduction of cortisol by hyperplastic adrenal cortices. These changes are the result of the combined gluconeogenic, lipolytic, protein catabolic, and antiinflammatory actions of glucocorticoid hormones on many organ systems of the body.

The pituitary gland is consistently enlarged in dogs with corticotroph adenomas (Fig. 3.6). Neither the occurrence nor the severity of functional disturbances appears to be

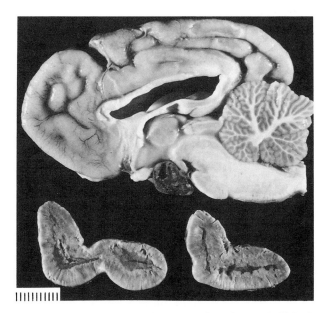

Fig. 3.6 Corticotroph (ACTH-secreting; chromophobic) adenoma of the pituitary with bilateral cortical hyperplasia of adrenals. Dog. Scale in millimeters.

directly related to the size of the neoplasm. Small corticotroph adenomas are as likely to be endocrinologically active as are larger neoplasms. The larger adenomas are often firmly attached to the base of the sella turcica but without evidence of erosion of the sphenoid bone. In the dog the diaphragma sellae is incomplete. Therefore, the line of least resistance favors dorsal expansion of the gradually enlarging pituitary mass and invagination into the infundibular cavity, dilation of the infundibular recess and the third ventricle, with eventual compression or replacement of the hypothalamus, and possible extension of the tumor into the thalamus.

Dorsal expansion of larger corticotroph adenomas results in either a broad-based indentation and compression of the overlying hypothalamus or extension into and replacement of the parenchyma of the hypothalamus and occasionally the thalamus. In the larger neoplasms there are often focal areas of hemorrhage, necrosis, mineralization, and liquefaction. Growth of the pituitary tumor along the basilar aspects of the brain may result in incorporation of the second, third, and fourth cranial nerves, leading to disturbances of their function.

Bilateral enlargement of the adrenal glands occurs in dogs with functional corticotroph adenomas (Fig. 3.6). This enlargement often is striking and is due to increased cortical parenchyma, primarily in the zonae fasciculata and reticularis. Nodules of yellow-orange cortical tissue often are found outside the capsule in the periadrenal fat, as well as extending down into the adrenal medulla. The corticomedullary junction is irregular, and the medulla is compressed.

Pituitary corticotroph adenomas are composed of well-differentiated, large or small, chromophobic cells supported by fine connective tissue septa. They can be divided into sinusoidal and diffuse types on the basis of the predominant pattern of cellular architecture. The cytoplasm of the tumor cells usually is devoid of secretory granules detectable by routine histochemical procedures used for pituitary cytology. However, pituitary corticotroph adenomas arising in both the pars distalis and the pars intermedia associated with the syndrome of cortisol excess are composed of polyhedral cells that immunocytochemically stain selectively for ACTH and MSH. Nodules of focal hyperplasia and microadenomas, composed of similar ACTH–MSH cells, often are present in the adenohypophysis of older dogs.

Pituitary adenomas arising in both the pars distalis and pars intermedia have positive immunocytochemical staining for ACTH, beta-lipotrophin, and beta-endorphin. Immunocytochemical studies have demonstrated that ACTH- and MSH-staining cells are polyhedral to round, sparsely granulated, and most numerous in the ventrocentral and cranial portions of the pars distalis in normal dogs. Corticotrophs are less numerous in the dorsal and caudal regions of the pars distalis and throughout the pars tuberalis.

Although remnants of the pars distalis can be identified near the periphery of corticotroph adenomas, the demarcation between the neoplasm and pars distalis often is not distinct. The pars distalis is either partly replaced by the neoplasm or severely compressed. The pars nervosa and infundibular stalk are either infiltrated and disrupted by tumor cells or completely incorporated within the larger neoplasms.

Cells constituting functional corticotroph adenomas have definite ultrastructural evidence of secretory activity. Organelles concerned with protein synthesis (endoplasmic reticulum) and packaging of secretory products (Golgi apparatus) are well developed in tumor cells. Hormone-containing secretory granules can be demonstrated by electron microscopy in functional corticotroph adenomas of dogs. This is in contrast to the absence of demonstrable secretory granules within neoplastic cells observed with light microscopy. The granules vary in number from cell to cell, are spherical, and are surrounded by a delicate limiting membrane. They are small (mean diameter, 170 nm), electron dense, often situated peripherally in the cell, and may have a prominent submembranous space.

A number of distinctive clinical and functional alterations develop in dogs with corticotroph (ACTH-secreting) adenomas, resulting in the syndrome of hyperadrenocorticism. Centripetal redistribution of adipose tissue leads to prominent fat pads on the dorsal midline of the neck, giving the neck and shoulders a thick appearance. Appetite and intake of food often are increased, either as a direct result of the hypercortisolism or involvement of hypothalamic appetite control centers by a large pituitary tumor. The muscles of the extremities and abdomen are weakened and atrophied. The loss of abdominal muscles and muscles of the abaxial skeleton results in gradual abdominal enlargement (pot belly), lordosis, muscle trembling, and a straight-legged skeletal-braced posture in order to support the body's weight. Profound atrophy of the temporal muscles may result in concave indentations and readily palpable prominences of underlying skull bones. Hepatomegaly due to increased fat and glycogen deposition plus vacuolation of hepatocytes produces a distended, often pendulous, abdomen. Functional alterations of the skin, muscle, lung, and other tissues are the direct effects of elevated blood cortisol levels. They are similar to those resulting from a cortisol-secreting primary adrenal cortical tumor and will be described elsewhere in this chapter.

2. Adenoma of the Pars Intermedia

Adenomas derived from cells of the pars intermedia are the most common type of pituitary tumor in horses, the second most common type in dogs, and rare in other species. They develop in older horses, with females affected more frequently than males. Nonbrachycephalic breeds of dogs develop adenomas in the pars intermedia more often than do brachycephalic breeds.

Adenomas of the pars intermedia in **dogs** result in only a moderate enlargement of the pituitary gland. The pars distalis is readily identifiable and sharply demarcated from the anterior margin of the neoplasm. The tumor may extend across the residual hypophyseal lumen and result in compression atrophy but usually does not invade the parenchyma of the pars distalis (Fig. 3.7A). The posterior lobe is incorporated within the tumor, but the infundibular stalk is intact. Degenerative changes within the neoplasm are minimal.

Adenomas of the pars intermedia in dogs appear to arise from the lining epithelium of the residual hypophyseal lumen covering the pars nervosa. They often are relatively small and more strictly localized than corticotroph adenomas arising in the pars distalis of dogs. Adenomas of the pars intermedia are sharply demarcated from the pars distalis, usually by an incomplete layer of condensed reticulum and focal accumulations of lymphocytes, but are not encapsulated. The histologic appearance is different from that of corticotroph adenomas of the pars distalis in that there are numerous large colloid-filled follicles interspersed between nests of large chromophobic cells (Fig. 3.7B). The follicles are lined by simple columnar epithelium, which is partly ciliated, and contain interspersed mucin-secreting goblet cells. The follicular colloid is densely eosinophilic and PAS positive. The nests of cells between the follicles are primarily chromophobic, but occasional cells contain secretory granules of simple protein (acidophilic) or mucoprotein (basophilic). Endocrinologically active (ACTH-secreting) adenomas of the pars intermedia in dogs have prominent groups of corticotrophs with an abundant eosinophilic cytoplasm and more widely scattered follicles. Dense bands of fibrous connective tissue are occasionally interspersed between the follicles and nests of chromophobic cells, particularly in the endocrinologically inactive adenomas of the pars intermedia. Mitotic figures are observed infrequently. The neoplastic cells compress and often invade the pars nervosa and infundibular stalk.

Adenomas of the pars intermedia in dogs either are endocrinologically inactive and associated with varying degrees of hypopituitarism and diabetes insipidus or endocrinologically active and secrete excessive ACTH, leading to bilateral adrenal cortical hyperplasia and a syndrome of cortisol excess. The clinical signs in the dogs with functional adenomas are similar to those described for corticotroph adenomas of the pars distalis.

Two cell populations have been identified in the pars intermedia of normal dogs by immunocytochemistry. The predominant cell type (A cell) stains strongly for α-MSH as in the pars intermedia of other species. A second cell type (B cell) in the canine pars intermedia stains intensely for ACTH but not for α-MSH. This second cell population accounts for the high bioactive ACTH concentration found in the pars intermedia of dogs and appears to be the cell of origin of corticotroph adenomas of the pars intermedia in dogs with cortisol excess.

Adenomas of the pars intermedia in **horses** can be large tumors that extend out of the sella turcica and severely compress the overlying hypothalamus (Fig. 3.7C). The adenomas are yellow to white, multinodular, and incorpo-

rate the pars nervosa. On sectioning of the pituitary mass, the pars distalis usually can be identified as a compressed subcapsular rim of tissue on the anterior margin. A sharp line of demarcation remains between the neoplasm and the atrophic pars distalis.

Adenomas of the pars intermedia in horses are partly encapsulated and sharply delineated from the compressed parenchyma of the pars distalis. The tumors are subdivided into nodules or compartments by fine septa of connective tissue that contain numerous capillaries and rare inflammatory cells. Tumor cells are large, cylindrical, spindle shaped or polyhedral, with an oval hyperchromatic nucleus (Fig. 3.7D). The histologic pattern is often reminiscent of the prominent pars intermedia of normal horses. Occasionally, cuboidal cells form follicular structures which contain dense eosinophilic colloid. In other areas the spindle-shaped cells may assume a more sarcomatous pattern and palisade around vessels. The cytoplasm is lightly eosinophilic and granular.

Electron microscopy of the large cells composing adenomas of the pars intermedia in horses reveals that the rough endoplasmic reticulum and Golgi apparatus are particularly well developed, suggesting they are synthesizing and packaging considerable amounts of protein (pro-opiolipomelanocortin for secretion). The neoplastic cells contain in their cytoplasm numerous membrane-limited secretory granules (with a mean diameter of ~300 nm) that are surrounded by a closely applied limiting membrane.

The clinical syndrome associated with tumors of the pars intermedia in horses is characterized by polyuria, polydipsia, increased appetite, muscle weakness, somnolence, intermittent hyperpyrexia, and generalized hyperhidrosis. The affected horses often develop a striking hirsutism because of a failure of the seasonal shedding of hair. The hair over most of the trunk and extremities is long (up to 9–10 cm), abnormally thick, wavy, and often matted together.

Horses with larger tumors may have hyperglycemia (insulin resistant) and glycosuria, probably the result of a down-regulation of insulin receptors on target cells induced by the chronic excessive intake of food and hyperinsulinemia. The disturbances in carbohydrate metabolism, ravenous appetite, hirsutism, and hyperhidrosis are considered to be primarily a reflection of deranged hypothalamic function caused by the large pituitary tumors. Adenomas of the pars intermedia in horses often extend out of the sella turcica and expand dorsally because of the incomplete diaphragma sellae and severely compress the overlying hypothalamus. The hypothalamus is the primary center for homeostatic regulation of body temperature, appetite, and cyclic shedding of hair.

Plasma cortisol and immunoreactive adrenocorticotropin levels are either normal or only modestly elevated in horses with adenomas of the pars intermedia. The cortisol levels lack the normal diurnal rhythm and are not suppressed by either high or low doses of dexamethasone.

Tumor tissue and plasma from horses with adenomas of the pars intermedia contain high concentrations of im-

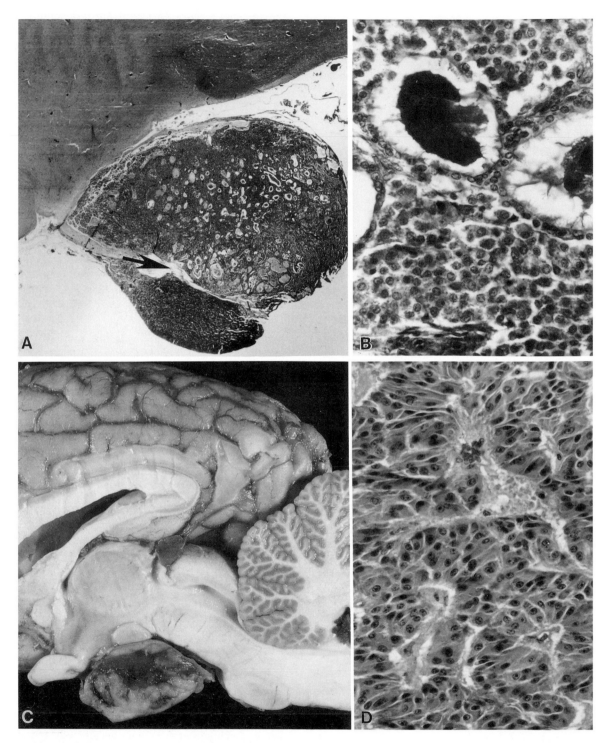

Fig. 3.7 (A and B) Corticotroph adenoma of pars intermedia. Pituitary, dog. (A) Neoplasm demarcated from pars distalis. Arrow indicates the residual hypophyseal lumen. (B) Nests of chromophobic cells interspersed between follicles. (C and D) Adenoma of pars intermedia. Horse. (C) Dorsal extension of tumor compressing hypothalmus and optic nerves with invasion of pars nervosa. (D) Spindle-shaped tumor cells arranged in cords.

munoreactive peptides [corticotropinlike intermediate lobe peptide (CLIP)], α and β melanocyte-stimulating hormones (α- and β-MSH), and beta endorphin (beta END) derived from proopiolipomelanocortin (pro-OLMC) and processed in the pars intermedia. This biosynthetic precursor is a high-molecular-mass (31,000–37,000 daltons) glycoprotein that undergoes different post-translational processing in the pars distalis and pars intermedia. In the normal pars distalis pro-OLMC is processed to ACTH (4,500 daltons), β-lipotropin (β-LPH) and γ-LPH, whereas in the normal pars intermedia, the same precursor molecule is cleaved into α-MSH, CLIP (that contains amino acids 18–39 of the ACTH molecule), β-MSH, and β-END. Plasma cortisol strongly inhibits ACTH secretion by the pars distalis but has a much lesser effect on peptides secreted by the pars intermedia, which are under tonic dopaminergic inhibitory control.

The modest elevation of plasma immunoreactive adrenocorticotropin appears to be due to the different processing of pro-OLMC in tumors derived from cells of the pars intermedia. This may explain the normal or slightly elevated blood cortisol levels and normal or mildly hyperplastic adrenal cortices observed in horses with adenomas of the pars intermedia. The plasma and tumor levels of pars intermedia-derived peptides (CLIP, α- and β-MSH, α-END) are disproportionately elevated (40 times or more) compared to those of ACTH, apparently as the result of selective post-translational processing of pro-OLMC in a manner similar to that of the normal pars intermedia. Extracts of adenomas derived from pars intermedia also contain immunoreactive peptides of a larger molecular mass than those present in the normal pituitary tissue. The smallest (38,500 daltons) of these peptides could represent prepro-OLMC with an attached signal or leader sequence of ~26 amino acid residues. The larger peptides (47,000 and 63,000 daltons) may be derived from improper intranuclear processing of pro-OLMC mRNA with retention of additional coding nucleotide sequences.

Horses with pituitary adenomas derived from pars intermedia develop a clinical syndrome that is associated with the autonomous production of excess proopiomelanocortin(POMC)-derived peptides. Immunocytochemical evaluation of adenomas of the pars intermedia reveals a diffuse moderate to strong staining for POMC, α-melanocyte-stimulating hormone (α-MSH), and β-endorphin (β-END). Although many of the functional disturbances in horses with pituitary adenomas (e.g., diabetes insipidus, polyphagia, hyperpyrexia, hyperhidrosis, and hirsutism) appear to be the result of hypothalamic or neurohypophyseal dysfunction, other signs (e.g., docility and diminished responsiveness to painful stimuli) may be related to the elevated plasma and cerebrospinal levels of β-END. Although adrenocorticotropic hormone (ACTH) is demonstrable in adenomas of the pars intermedia, the staining intensity is patchy and considerably weaker than that of POMC, α-MSH, and β-END. These findings are in accord with biochemical studies that report markedly elevated concentrations of immunoreactive POMC and POMC-derived peptides including α- and β-MSH, corticotropinlike intermediate lobe peptide, and β-END in adenomas and plasma of affected horses relative to ACTH. The overall processing of peptides in adenomas in the pars intermedia appears to be similar to that in the normal equine pars intermedia.

Corticotrophs in the pars distalis of horses have strong immunostaining for ACTH, whereas only a few cells stain for α-MSH. These immunocytochemical findings illustrate the differences between adenomas of the pars intermedia in horses and corticotroph adenomas of the pars distalis (also pars intermedia in dogs) that result in the classic Cushing's disease in humans and dogs. Corticotroph adenomas associated with Cushing's disease are characterized by strong immunostaining for ACTH and weak to moderate immunostaining for α-MSH.

3. Hormonally Inactive Chromophobe Adenoma of the Pars Distalis

Nonfunctional pituitary tumors occur in dogs, cats, laboratory rodents, and parakeets, but are uncommon in other species. Although chromophobe adenomas appear to be hormonally inactive, they may cause significant functional disturbances and clinical signs by compression of adjacent portions of the pituitary gland and dorsal extension into the overlying brain.

Nonfunctional pituitary adenomas result in clinical disturbances by either interference with secretion of pituitary tropic hormones and diminished target organ function or dysfunction of the central nervous system. Affected animals often are depressed, have incoordination and other disturbances of balance, are weak, and may collapse with exercise. In long-standing cases there may be evidence of blindness, with dilated and fixed pupils due to compression and disruption of optic nerves by dorsal extension of the pituitary tumor (Fig. 3.8).

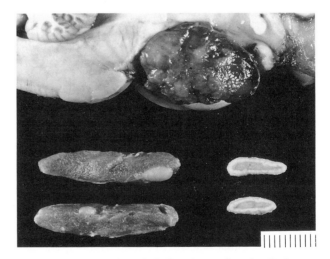

Fig. 3.8 Destruction of pituitary by nonfunctional adenoma with atrophy of adrenal cortex and thyroid follicular cells. Accumulation of colloid in involuted follicles prevents marked reduction in gland size. Scale in millimeters.

Animals with hormonally inactive pituitary adenomas often have progressive loss of weight (pituitary cachexia) with muscle atrophy due to a lack of the protein anabolic effects of growth hormone. Compression of the cells that secrete gonadotropic hormones or the corresponding hypothalamic releasing hormone(s) results in atrophy of the gonads, resulting in decreased libido or anestrus. The affected animals appear dehydrated, as evidenced by a lusterless dry coat, and they consume increased amounts of water.

Hormonally inactive pituitary adenomas often attain considerable size before they cause obvious signs or kill the animal. The proliferating tumor cells incorporate the remaining structures of the adenohypophysis and infundibular stalk. The neoplasms attach firmly to the base of the sella turcica, but cause no erosion of the sphenoid bone. The incomplete diaphragma sellae permits dorsal growth of the adenoma along lines of least resistance. The entire hypothalamus may become compressed and replaced by the tumor.

The adrenal glands in animals with large nonfunctional pituitary adenomas are small and consist primarily of medullary tissue surrounded by a narrow zone of cortex. The adrenal cortex appears as a thin yellow-brown rim composed of a moderately thickened capsule and secretory cells of the outer zona glomerulosa, which are not predominantly under the control of ACTH. The zonae fasciculata and reticularis are severely atrophied compared with these zones in normal adrenal glands. Thyroid glands in animals with large pituitary adenomas are often smaller than normal, though less atrophied than the adrenal cortex. Most of the atrophic thyroid follicles are large, lined by a flattened cuboidal epithelium, and have few endocytotic vacuoles near the interface between the colloid and luminal aspect of the follicular cells. The thyroid lesion is due to lack of TSH-induced endocytosis of colloid. Seminiferous tubules of the testes are small and show little evidence of active spermatogenesis.

The cells composing nonfunctional pituitary adenomas are cuboidal to polyhedral and either arranged in diffuse sheets or subdivided into small packets by fine connective tissue septa. Special histochemical techniques for pituitary cytology fail to demonstrate specific secretory granules within the cytoplasm of tumor cells. The histogenesis of nonfunctional chromophobe adenomas often is difficult to define precisely, but they appear to be derived from less differentiated pituitary cells that neither store nor secrete a specific hypophyseal tropic hormone.

4. Acidophil Adenoma of the Pars Distalis

Neoplasms derived from granulated acidophils are uncommon in all domestic animal species but are common in old animals of many strains of rats. Acidophil adenomas and adenocarcinomas have been reported in the cat, dog, and sheep.

Acidophil adenomas enlarge the pituitary gland and indent the overlying hypothalamus or extend into the overlying brain to varying degrees. The enlarged hypophysis is composed of irregular columns of acidophils interspersed between numerous large blood-filled sinusoids. The fibrous stroma is sparse. Although the degree of cytoplasmic granulation of acidophils varies from cell to cell, the predominating type of neoplastic acidophil usually contains many secretory granules. The nuclei of the densely granulated acidophils are small, oval, and hyperchromatic. Sparsely granulated (chromophobic) cells are interspersed between the densely granulated acidophils. Their cytoplasm is more abundant and lightly eosinophilic but contains only an occasional secretory granule. Secretory granules of the acidophils are bright red when stained with acid fuchsin–aniline blue and with Crossman's modification of Mallory's trichrome. Orange-G stains the granules an intense yellowish orange, but they are PAS negative.

Colloid-containing follicles lined by follicular cells are found occasionally within acidophil adenomas in **dogs**. The colloid is intensely PAS positive. Numerous sinusoids are distended with erythrocytes, detached neoplastic cells, and large masses of fibrin. The pars nervosa and infundibular stalk are infiltrated at the periphery by neoplastic cells, compressed, and partly replaced by fibrous astrocytes.

Electron-microscopic evaluation reveals that two types of acidophils occur within pituitary acidophil adenomas. The predominating type of acidophil is smaller and contains many secretory granules. The granules are spherical to oval, uniformly electron dense, finely granular, and surrounded by a delicate limiting membrane. The submembranous space of the granule is narrow. The mean diameter of mature secretory granules in the neoplastic acidophils is 420 nm (range, 320–600 nm). The plasma membranes of adjacent cells are relatively straight with uncomplicated interdigitations and are connected by an occasional desmosome. The Golgi apparatus is comparatively small and associated with few prosecretory granules. The rough endoplasmic reticulum is composed of small, flattened membranous sacs with attached ribosomes. A few mitochondria are randomly distributed throughout the cytoplasm. Acidophils of this type are interpreted to be in the storage phase of their secretory cycle.

The less common type of neoplastic acidophil has a greater cytoplasmic and nuclear area but contains numerous organelles and few mature secretory granules. The rough endoplasmic reticulum is extensive and consists of aggregates of lamellar arrays of granular membranes. The Golgi apparatus is prominent and associated with numerous small prosecretory granules. Mitochondria are observed more often in the cytoplasm of this type of acidophil. These hypertrophied acidophils are considered to be secretorily active cells. Cells with varying amounts of organellar development and numbers of mature secretory granules are present between the extremes of storage and actively synthesizing acidophils.

In the pars distalis of normal dogs, immunoreactive prolactin cells occur as small groups of large polygonal cells with prominent granules in the ventrocentral and cranial portion. A diffuse increase in this population of cells occurs in female dogs near parturition. In comparison, growth hormone-secreting acidophils are present singly along capillaries in the dorsal region of the pars distalis near the pars intermedia. They are small, round to oval, and have fine cytoplasmic granules. Somatotrophs frequently undergo hypertrophy and diffuse hyperplasia in old dogs, especially females with mammary dysplasia or neoplasia.

In none of the reported acidophil tumors of dogs has conclusive evidence been presented that there were functional disturbances resulting from a hypersecretion of somatotropin or prolactin. However, acidophil adenomas in dogs have been associated with thickened cranial bones and metahypophyseal diabetes with fewer than normal pancreatic islets.

Acidophil adenomas in **cats** have been associated with clinical signs of diabetes mellitus with degranulation of the pancreatic islets and vacuolar changes in beta cells. The persistent elevation of serum somatotropin in cats also is associated with proliferation of joint cartilage, leading to degenerative arthropathy, and with kidney disease due to periglomerular fibrosis and mesangial proliferation. Acidophil adenomas in **sheep** may attain considerable size and remain confined to the sella turcica because sheep have a complete diaphragma sellae separating the pituitary fossa from the brain. The remaining adeno- and neurohypophysis are compressed severely, and the sella turcica is enlarged and deepened because of pressure-induced osteolysis.

Although **growth hormone-secreting acidophils** are one of the major cell types in the adenohypophysis, the development of functional adenomas derived from this population of acidophils is infrequent in animals. Acromegaly is a disease characterized by an overgrowth of connective tissue, increased appositional growth of bone, coarsening of facial features, and enlargement of viscera due to a chronic excessive secretion of growth hormone (somatotropin).

Stimulation of growth hormone-secreting acidophils in the adenohypophysis by progestational agents also occurs under both experimental and naturally occurring conditions in dogs. An increased number of somatotrophs have been observed after the administration of progesterone and cyproterone acetate. Diabetes mellitus has been reported in dogs following the administration of medroxyprogesterone acetate and megestrol acetate. The stimulation of growth hormone release in dogs by progestogens differs from the situation in humans, in whom high prolactin levels result from treatment with these drugs.

Changes in plasma growth hormone, prolactin, cortisol, and progesterone levels in beagles following intramuscular injection of 75 mg/kg medroxyprogesterone acetate (6α-methyl-17-acetoxyprogesterone) every 3 months for 17 months have been correlated with the development of acromegalic features and mammary nodules. Medroxyprogesterone acetate increased mean growth hormone lev-

els, incidence of acromegalylike changes, and frequency of palpable mammary nodules in beagles under controlled conditions.

Growth hormone levels were elevated (2.5 ng/ml and above) in all female beagles with acromegalic features compared with those of placebo controls and dogs receiving crystalline progesterone implants. All dogs with elevated growth hormone levels had multiple mammary nodules. The elevation in growth hormone levels and development of acromegalic features was greater in older than in younger dogs. Initial elevations of growth hormone occurred after 8 months of medroxyprogesterone acetate treatment. Serum prolactin levels were not changed by either medroxyprogesterone acetate or crystalline progesterone implants, but serum cortisol levels were suppressed significantly compared with those in controls. The latter most likely was the result of medroxyprogesterone-induced suppression of pituitary ACTH secretion and corresponding decrease in cortisol synthesis by the adrenal cortex. Elevated growth hormone levels also are reported in dogs with spontaneous mammary tumors, and somatotrophs show cytologic evidence of increased secretory activity.

5. Craniopharyngioma

Craniopharyngioma is a benign tumor which is derived from epithelial remnants of the oropharyngeal ectoderm of the craniopharyngeal duct (Rathke's pouch). It occurs in animals younger than those with other types of pituitary neoplasm and is present in either a suprasellar or infrasellar location. It is one cause of panhypopituitarism and dwarfism in young dogs due to a subnormal secretion of somatotropin and other tropic hormones beginning at an early age, prior to closure of the growth plates.

Craniopharyngiomas often are large and grow along the ventral aspect of the brain where they can incorporate several cranial nerves. In addition, they extend dorsally into the hypothalamus and thalamus. The clinical signs resulting from this type of pituitary tumor often are a combination of (1) lack of secretion of pituitary tropic hormones resulting in tropic atrophy and subnormal function of the adrenal cortex and thyroid, gonadal atrophy, and failure to attain somatic maturation due to a lack of growth hormone secretion; (2) disturbances in water metabolism (polyuria, polydipsia, low urine specific gravity and osmolality) from an interference with the release and synthesis of antidiuretic hormone by the large tumor; (3) deficits in cranial nerve function; and (4) central nervous system dysfunction due to extension into the overlying brain.

Craniopharyngiomas have alternating solid and cystic areas. The solid areas are composed of nests of epithelial cells (cuboidal, columnar, or squamous) with focal areas of mineralization. The cystic spaces are lined by either columnar or squamous cells and contain keratin debris and colloid.

6. Basophil Adenoma of the Pars Distalis

Tumors composed of granulated basophils are among the most uncommon pituitary tumors in all animal species.

Cushing's disease in human beings was initially attributed to a hypersecretion of ACTH by small basophilic adenomas in the pars distalis. Current evidence suggests they are a possible cause in a small percentage of patients with Cushing's disease. Several of the early reports on corticotropin-secreting pituitary tumors in dogs with hyperadrenocorticism reflected this concept and considered them to be basophil adenomas. Corticotroph (chromophobic) adenomas of the pars distalis and pars intermedia are responsible for the great majority of cases of Cushing's-like disease in dogs.

Basophil adenomas in humans may secrete thyrotropin or thyrotropic hormone (TTH), resulting in bilateral enlargement of both thyroid lobes (goiter). Serum thyroxine, triiodothyronine, and TTH are elevated and may be responsive to thyrotropin-releasing hormone (TRH) from the hypothalamus. The neoplastic cells contain small secretory granules (diameter less than 150 nm) with prominent rough endoplasmic reticulum and Golgi apparatus, characteristic of pituitary thyrotrophs.

7. Pituitary Chromophobe Carcinoma

Pituitary chromophobe carcinomas are uncommon compared with pituitary adenomas but have been seen in older dogs and cows. They usually are endocrinologically inactive but may result in significant functional disturbances by destruction of the pars distalis and neurohypophyseal system, leading to panhypopituitarism and diabetes insipidus.

Pituitary carcinomas are large and extensively invade the overlying brain, along the ventral aspect of skull, and the sphenoid bone of the sella turcica, inducing resorption of the bone. Metastases may occur to regional lymph nodes or to distant sites, such as the spleen or liver.

Malignant tumors of pituitary chromophobes are highly cellular and often have large areas of hemorrhage and necrosis. Giant cells, nuclear pleomorphism, and mitotic figures are encountered more frequently than in chromophobe adenomas.

8. Tumors Metastatic to the Pituitary Gland

The pituitary gland occasionally is either partially or completely destroyed by metastatic tumors from distant sites. Examples include malignant lymphoma in cattle and dogs; malignant melanoma in horses and dogs; and transmissible venereal tumor and adenocarcinoma of the mammary gland in dogs. In addition, the pituitary may be compressed or destroyed by local infiltration from osteosarcomas of the sphenoid bone, ependymomas arising in the infundibular recess of the third ventricle, by abscesses (Fig. 3.9A), and meningiomas (Fig. 3.9B) and gliomas (infundibuloma) of the infundibular stalk.

D. Diseases of the Neurohypophysis

Diabetes insipidus is a disorder in which inadequate antidiuretic hormone (ADH) is produced or target cells in the kidney lack the biochemical machinery (i.e., adenylate

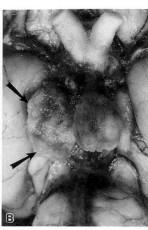

Fig. 3.9 (A) Pituitary abscess. Calf. Midsagittal section. (B) Meningioma on ventral aspect of brain (arrows) exerted pressure on the hypothalamic–hypophyseal portal system. Dog.

cyclase) necessary to respond to the secretion of normal or elevated circulating levels of hormone. The hypophyseal form of diabetes insipidus develops as a result of compression and destruction of the pars nervosa, infundibular stalk, or supraoptic nucleus in the hypothalamus.

The lesions responsible for the disruption of ADH synthesis or secretion in hypophyseal diabetes insipidus include large pituitary neoplasms (endocrinologically active or inactive), dorsally expanding cysts or inflammatory granulomas, and traumatic injury to the skull with hemorrhage and glial proliferation in the neurohypophyseal system. The posterior lobe, infundibular stalk, and hypothalamus are compressed or disrupted by neoplastic cells. This interrupts the nonmyelinated axons that transport ADH from its site of production, primarily in the supraoptic nucleus of the hypothalamus, to the site of release in the capillary plexus of the pars nervosa. Compression of neurosecretory neurons in the supraoptic nucleus of the hypothalamus by the dorsally expanding neoplasm may also result in decreased ADH synthesis. Axons in the compressed pars nervosa with hypophyseal diabetes insipidus associated with pituitary neoplasms are depleted of ADH-containing, dense neurosecretory granules, unlike those in normal animals (Fig. 3.10A,B).

Sporadic cases of hypophyseal diabetes insipidus may be the result of an inherited biochemical defect in the synthesis of ADH and its corresponding neurophysin I, as has been described in the Brattleboro strain of rat. In the nephrogenic form of diabetes insipidus, blood levels of ADH are normal or elevated, but target cells in the distal nephron and collecting ducts are unable to respond because of a lack of adenylate cyclase in the plasma membrane.

Animals with diabetes insipidus excrete large volumes of hypotonic urine, which in turn obliges them to take in equally large amounts of water to prevent hyperosmolality of body fluids and dehydration. Urine osmolality is decreased below normal plasma osmolality (~300 mOsm/

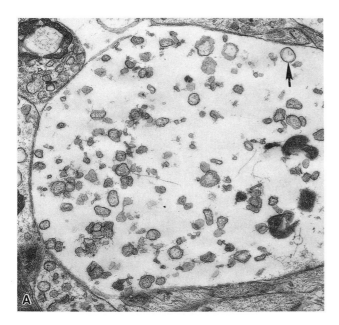

Fig. 3.10A Cross section of axonal process in pars nervosa of dog with hypophyseal diabetes insipidus associated with a pituitary adenoma. Axonal swelling contains few neurosecretory granules with dense cores but occasional irregularly shaped, empty vesicles (arrows).

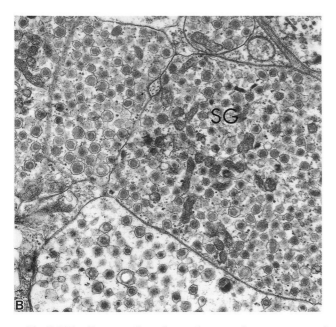

Fig. 3.10B Cross section of axonal process in pars nervosa of normal dog illustrating numerous membrane-bound, antidiuretic hormone-containing neurosecretory granules (SG).

kg) in both hypophyseal and nephrogenic forms of diabetes insipidus. In response to water deprivation, urine osmolality remains below that of plasma in both forms, in contrast to what is observed in normal animals. The elevation of urine osmolality above that of plasma in response to exogenous ADH in the hypophyseal form, but not in nephrogenic diabetes insipidus, allows separation of these two forms of the disease.

Bibliography

Alexander, J. E. Anomaly of craniopharyngeal duct and hypophysis. *Can Vet J* **3:** 83, 1962.

Allens, G. S. *et al.* Pituitary dwarfism in German shepherd dogs. *J Small Anim Pract* **19:** 711–727, 1978.

Andresen, E., and Willeberg, P. Pituitary dwarfism in German shepherd dogs: Additional evidence of simple autosomal recessive inheritance. *Nord Vet Med* **28:** 481–486, 1976.

Baker, E. Congenital hypoplasia of the pituitary and pancreas glands in the dog. *J Am Vet Med Assoc* **126:** 468, 1955.

Binns, W. *et al.* A congenital cyclopian-type malformation in lambs induced by maternal ingestion of a range plant, *Veratrum californicum. Am J Vet Res* **24:** 1164–1175, 1963.

Binns, W. *et al.* Effects of teratogenic agents in range plants. *Cancer Res* **28:** 2323–2326, 1968.

Capen, C. C. Tumors of the endocrine glands. *In* "Tumors in Domestic Animals," 3rd Ed. J. E. Moulton (ed.), pp. 553–639. Berkeley and Los Angeles, University of California Press, 1990.

Capen, C. C., and Koestner, A. Functional chromophobe adenomas of the canine adenohypophysis. An ultrastructural evaluation of a neoplasm of pituitary corticotrophs. *Vet Pathol* **4:** 326–347, 1967.

Capen, C. C., Martin, S. L., and Koestner, A. Neoplasms in the adenohypophysis of dogs. A clinical and pathologic study. *Vet Pathol* **4:** 301–325, 1967.

Cassel, S. E. Ovarian imbalance in a German shepherd dwarf. *Vet Med Small Anim Clin* **73:** 162–163, 1978.

Clarkson, T. B., Netsky, M. G., and de la Torre, E. Chromophobe adenoma in a dog: Angiographic and anatomic study. *J Neuropathol Exp Neurol* **18:** 558–562, 1959.

Dammrich, K. Ein polymorphzelliges basophiles adenom der hypophyse beim hund. *Berl Munch Tierarztl Wochenschr* **24:** 109–113, 1959.

Dousa, T. P. Cellular action of antidiuretic hormone in nephrogenic diabetes insipidus. *Mayo Clin Proc* **49:** 188–199, 1974.

El Etreby, M. F., and Dubois, M. P. The utility of antisera to different synthetic adrenocorticotrophins (ACTH) and melanotrophins (MSH) for immunocytochemical staining of the dog pituitary gland. *Histochemistry* **66:** 245–260, 1980.

El Etreby, M. F. *et al.* The role of the pituitary gland in spontaneous canine mammary tumorigensis. *Vet Pathol* **17:** 2–16, 1980.

El Etreby, M. F. *et al.* Functional morphology of spontaneous hyperplastic and neoplastic lesions in the canine pituitary gland. *Vet Pathol* **17:** 109–122, 1980.

Engel, F. L., and Kahana, L. Cushing's syndrome with malignant corticotrophin-producing tumor. *Am J Med* **34:** 726–734, 1963.

Farrow, B. R. H. Chromophobe adenoma of the pituitary in a dog. *Vet Rec* **84:** 609–610, 1969.

Feldman, E. C. Effect of functional adrenocortical tumors on plasma cortisol and corticotropin concentrations in dogs. *J Am Vet Med Assoc* **178:** 823–826, 1981.

Feldman, E. C., Bohannon, N. V., and Tyrrell, J. B. Plasma adrenocorticotropin levels in normal dogs. *Am J Vet Res* **38:** 1643–1634, 1977.

Gilbert, G. J., and Willey, E. N. Pituitary chromophobe adenoma in the bulldog. *J Am Vet Med Assoc* **154:** 1071–1074, 1969.

Green, R. A., and Farrow, C. S. Diabetes insipidus in a cat. *J Am Vet Med Assoc* **164:** 524–526, 1974.

Gribble, D. H. The endocrine system. In "Equine Medicine and Surgery," E. J. Catcott and J. R. Smithcors (eds.), pp. 433–457. Wheaton, Illinois, American Veterinary Publications, 1972.

Halmi, N. S. et al. Pituitary intermediate lobe in dog: Two cell types and high bioactive adrenocorticotropin content. Science 211: 72–74, 1981.

Hansel, W., Concannon, P. W., and McEntee, K. Plasma hormone profiles and pathological observations in medroxyprogesterone acetate-treated beagle bitches. In "Pharmacology of Steroid Contraceptive Drugs," S. Garattini and H. W. Berendes (eds.), pp. 145–161. New York, Raven Press, 1977.

Harris, W. H., and Heaney, R. P. Effect of growth hormone on skeletal mass in adult dogs. Nature 223: 403–404, 1969.

Hart, M. M., Reagan, R. L., and Adamson, R. H. The effect of isomers of DDD on the ACTH-induced steroid output, histology, and ultrastructure of the dog adrenal cortex. Toxicol Appl Pharmacol 24: 101–113, 1973.

Hottendorf, G. H., Nielsen, S. W., and Liberman, L. L. Acidophil adenoma of the pituitary gland and other neoplasms in a boxer. J Am Vet Med Assoc 148: 1046–1050, 1966.

Howe, A. The mammalian pars intermedia: A review of its structure and function. J Endocrinol 59: 385–409, 1973.

Jensen, E. C. Hypopituitarism associated with cystic Rathke's cleft in a dog. J Am Vet Med Assoc 135: 572–575, 1959.

King, J. M., Kavanaugh, J. F., and Bentinck-Smith, J. Diabetes mellitus with pituitary neoplasms in a horse and a dog. Cornell Vet 52: 133–145, 1962.

Koestner, A., and Capen, C.C. Ultrastructural evaluation of the canine hypothalamic–neurohypophyseal system in diabetes insipidus associated with pituitary neoplasms. Vet Pathol 4: 513–536, 1967.

Kovacs, K., and Horvath, E. "Tumors of the Pituitary Gland. Atlas of Tumor Pathology," Fascicle 21 Second Series. Washington, D.C., Armed Forces Institute of Pathology, 1986.

Kovacs, K., and Horvath, E. Pathology of pituitary tumors. Endocrinol Metab Clin North Am 16: 529–550, 1987.

Lage, A. L. Nephrogenic diabetes insipidus in a dog. J Am Vet Med Assoc 163: 251–253, 1973.

Lichtensteiger, C. A., Wortman, J. A., and Eigenmann, J. E. Functional pituitary acidophil adenoma in a cat with diabetes mellitus and acromegalic features. Vet Pathol 23: 518–521, 1986.

Loeb, W. F., Capen, C. C., and Johnson, L. E. Adenomas of the pars intermedia associated with hyperglycemia and glycosuria in two horses. Cornell Vet 56: 623–639, 1966.

Lubberink, A. A. M. E. et al. Hyperfunction of the adrenal cortex: A review. Aust Vet J 57: 504–509, 1971.

Lund-Larsen, T. R., and Grondalen, J. Ateliotic dwarfism in the German shepherd dog: Low somatomedin activity associated with apparently normal pituitary function (2 cases) and with panadenopituitary dysfunction (1 case). Acta Vet Scand 17: 293–306, 1976.

McGrath, P. The pharyngeal hypophysis in some laboratory animals. J Anat 117: 95–115, 1974.

Millington, W. et al. Equine Cushing's disease: Differential regulation of β-endorphin processing in tumors of the intermediate pituitary. Endocrinology 123: 1598–1604, 1988.

Moore, J. N. et al. A case of pituitary adenocorticotropin-dependent Cushing's syndrome in a horse. Endocrinology 104: 546–582, 1979.

Muller, G. H. Pituitary dwarfism: Cutaneous manifestations of an endocrine disorder. Vet Clin North Am 9: 41–48, 1979.

Muller, G. H., and Jones, S. R. Pituitary dwarfism and alopecia

in a German shepherd with cystic Rathke's cleft. J Am Anim Hosp Assoc 9: 567–572, 1973.

Muller-Peddinghaus, R. et al. Hypophysärer Zwergwuchs beim Deutschen Schaferhund. Vet Pathol 17: 406–421, 1980.

Nelson, D. H. et al. ACTH-producing tumor of the pituitary gland. N Engl J Med 259: 161–164, 1958.

Nicholas, F. Pituitary dwarfism in German shepherd dogs: A genetic analysis of some Australian data. J Small Anim Pract 19: 167–174, 1978.

Orth, D. N., and Nicholson, W. E. Bioactive and immunoreactive adrenocorticotropin in normal equine pituitary and in pituitary tumors of horses with Cushing's disease. Endocrinology 111: 559–563, 1982.

Orth, D. N. et al. Equine Cushing's disease: Plasma immunoreactive proopiolipomelanocortin peptide and cortisol levels basally and in response to diagnostic tests. Endocrinology 110: 1430–1441, 1982.

Peterson, M. E. Pathophysiology of canine pituitary-dependent hyperadrenocorticism (canine Cushing's disease). Front Horm Res 17: 37–47, 1987.

Peterson, M. E. et al. Immunocytochemical study of the hypophysis in 25 dogs with pituitary-dependent hyperadrenocorticism. Acta Endocrinol (Copenh) 101: 15–22, 1982.

Peterson, M. E. et al. Plasma immunoreactive proopiomelanocortin peptides and cortisol in normal dogs and dogs with Addison's disease and Cushing's syndrome: Basal concentrations. Endocrinology 119: 720–730, 1986.

Peterson, M. E. et al. Acromegaly in 14 cats. J Vet Intern Med 4: 192–201, 1990.

Rao, R. R., and Bhat, N. G. Incidence of cysts in pars distalis of mongrel dogs. Indian Vet J 48: 128–133, 1971.

Ricci, V., and Russolo, M. Immunocytological observations on the localization of ACTH in the hypophysis of the dog. Acta Anat (Basel) 84: 10–18, 1973.

Richards, M. A. Polydipsia in the dog. The differential diagnosis of polyuric syndromes in the dog. J Small Anim Pract 10: 651–667, 1970.

Rijnberk, A., der Kinderen, P. J., and Thijssen, J. H. H. Canine Cushing's syndrome. Zentralbl Veterinaermed [A] 16: 13–28, 1969.

Rijnberk, A. et al. Acromegaly associated with transient overproduction of growth hormone in a dog. J Am Vet Med Assoc 177: 534–557, 1980.

Rogers, W. A. et al. Partial deficiency of antidiuretic hormone in a cat. J Am Vet Med Assoc 170: 545–547, 1977.

Saunders, L. Z., and Rickard, C. G. Craniopharyngioma in a dog with apparent adiposogenital syndrome and diabetes insipidus. Cornell Vet 42: 490–495, 1952.

Schally, A. V. Aspects of hypothalamic regulation of the pituitary gland: Its implications for the control of reproductive processes. Science 202: 18–28, 1978.

Schecter, R. D. et al. Treatment of Cushing's syndrome in the dog with an adrenocorticolytic agent (o,p′DDD). J Am Vet Med Assoc 162: 629–639, 1973.

Siegel, E. T., Kelly, D. F., and Berg, P. Cushing's syndrome in the dog. J Am Vet Med Assoc 157: 2081–2089, 1970.

Spaar, F. W., and Wille, J. Zur verleichenden Pathologie der Hypophysenadenoma der Tiere. Zentralbl Veterinaermed 6: 925–944, 1959.

Van Wyk, J. J. et al. The somatomedins: A family of insulin-like hormones under growth hormone control. Recent Prog Horm Res 30: 259–318, 1974.

White, E. G. A suprasellar tumor in a dog. J Pathol Bacteriol 47: 323–326, 1938.

Willeberg, P., Kastrup, K. W., and Andresen, E. Pituitary dwarfism in German shepherd dogs: Studies on somatomedin activity. *Nord Vet Med* **27**: 448–454, 1975.

Wilson, M. G. *et al.* Proopiolipomelanocortin peptides in normal pituitary, pituitary tumor, and plasma of normal and Cushing's horses. *Endocrinology* **110**: 941–954, 1982.

III. Parathyroid Glands and Calcium-Regulating Hormones

Calcium ion plays a key role in many fundamental biologic processes including muscle contraction, blood coagulation, enzyme activity, neural excitability, hormone release, and membrane permeability, in addition to being an essential structural component of the skeleton. To maintain a constant concentration of calcium, despite variations in intake and excretion, endocrine control mechanisms have evolved that primarily consist of the interactions of three major hormones. Although the direct roles of parathyroid hormone, calcitonin, and vitamin D frequently are emphasized in the control of blood calcium, other hormones such as adrenal corticosteroids, estrogen, thyroxine, somatotropin, and glucagon also contribute to the maintenance of calcium homeostasis under certain conditions.

A. Parathyroid Glands and Parathyroid Hormone

Parathyroids are of entodermal origin and derived from the third and fourth pharyngeal pouches in close association with the primordia of the thymus (Fig. 3.11). The entodermal bud that forms the thyroid gland arises on the midline at the level of the first pharyngeal pouch. This gives rise to the thyroglossal duct, which migrates caudally. The proliferation of cell cords at the distal end of the thyroglossal duct forms the follicles of each thyroid lobe. The area at the base of the tongue marking the origin on the thyroid gland is referred to as the foramen cecum linguae in postnatal life. Calcitonin-secreting C cells of neural crest origin reach the postnatal thyroid gland by migrating into the ultimobranchial body. This last pharyngeal pouch moves caudally in mammals to fuse with the

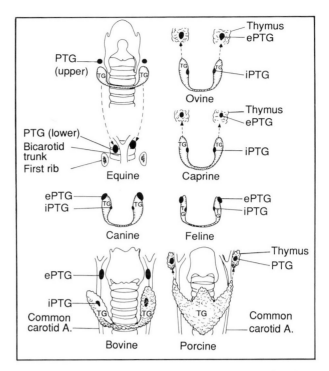

Fig. 3.12 Anatomic location of parathyroids and thyroids in various species. A, artery; e, external and, i, internal PTG, parathyroid gland; TG, thyroid gland. (Modified from H. Grau and H.-D. Dellmann. Über tierartliche Unterschiede der Epithelkörperschen Haussäugetiere. *Z Mikrosk Anat Forsch* **64**:192–214, 1958.)

primordia of the thyroid gland and distribute C cells into each thyroid lobe. Anatomic location of parathyroid glands in relation to thyroids and related structures in several species of animals is shown in Fig. 3.12.

The parathyroid glands contain a single basic type of secretory cell, the chief cell, which is concerned with the elaboration of a single hormone. The chief cells are in different stages of secretory activity, and in certain species are in transition to oxyphil cells. Chief cells interpreted to be in an inactive (resting or involuted) stage of their secretory cycle predominate in the parathyroid glands un-

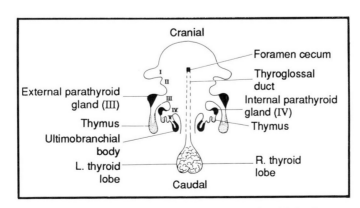

Fig. 3.11 Embryology of thyroid and parathyroid glands.

der normal conditions. Inactive chief cells are cuboidal and have uncomplicated interdigitations between contiguous cells. The relatively electron-transparent cytoplasm contains poorly developed organelles, and secretory granules are sparse. The cytoplasm often has either numerous lipid bodies and lipofuscin granules or aggregations of glycogen. Chief cells in the active stage of the secretory cycle are in the minority in the parathyroid glands of most species. The cytoplasm of active chief cells has an increased electron density due to the close proximity of organelles and secretory granules, increased density of the cytoplasmic matrix, and loss of glycogen particles and lipid bodies.

The second cell type in the parathyroid glands of certain species, including humans, is the oxyphil cell. These cells are absent in rats, chickens, and many species of lower animals. Oxyphil cells occur either singly or in small groups interspersed between chief cells. They are larger than chief cells, and their abundant cytoplasm is filled with numerous large, often bizarre-shaped, mitochondria. Glycogen particles and free ribosomes are interspersed between the mitochondria. Granular endoplasmic reticulum, Golgi apparatuses, and secretory granules are poorly developed in oxyphil cells of normal parathyroid glands, suggesting that oxyphil cells do not have an active role in the biosynthesis of parathyroid hormone. Associated with the marked increase in numbers of mitochondria, oxyphil cells have a higher oxidative and hydrolytic enzyme activity than do chief cells.

Cells are observed with cytoplasmic characteristics intermediate between those of chief and oxyphil cells. These transitional oxyphil cells have numerous mitochondria, but other organelles are present, including rough endoplasmic reticulum, Golgi apparatuses, and secretory granules. The significance of oxyphil cells in the pathophysiology of the parathyroid glands is not completely understood. They are not altered in response to either short-term hypocalcemia or hypercalcemia in animals, but both oxyphil cells and transitional forms may be increased in response to long-term stimulation of human parathyroid glands. Therefore, oxyphil cells do not appear to be degenerate chief cells, as previously thought, but rather are derived from chief cells as the result of aging or some other metabolic derangement.

The production of parathyroid hormone involves the biosynthesis of a large precursor on ribosomes of the rough endoplasmic reticulum in chief cells. This initial translation product is preproparathyroid hormone (preproPTH). It is composed of 115 amino acids and contains a hydrophobic signal sequence of 25 amino acid residues that facilitates the penetration and subsequent vectorial discharge of the nascent peptide into the cisternal space of the rough endoplasmic reticulum. PreproPTH is rapidly converted (within 1 min or less) to proPTH by the proteolytic cleavage of 25 amino acids from the N-terminal end of the molecule. The intermediate precursor, proPTH, is composed of 90 amino acids and moves within the membranous channels of the rough endoplasmic reticulum

to the Golgi apparatus. Enzymes within membranes of the Golgi apparatus cleave a hexapeptide from the N-terminal (biologically active) end of the molecule, forming active parathyroid hormone. Active PTH is packaged into membrane-limited, macromolecular aggregates in the Golgi apparatus for subsequent storage in chief cells. Under certain conditions of increased demand, PTH may be released directly from chief cells without being packaged into secretion granules.

Biologically active parathyroid hormone secreted by chief cells is a straight-chain polypeptide consisting of 84 amino acid residues with a molecular weight of 9,500. Although the principal form of active parathyroid hormone secreted from chief cells is a straight-chain peptide of 84 amino acids, the molecule is rapidly cleaved into amino- and carboxy-terminal fragments in the peripheral circulation and especially in the liver. The purpose of this fragmentation is uncertain since the biologically active amino-terminal fragment is no more active than the entire (1–84) PTH molecule. The plasma half-life of the N-terminal fragment is considerably shorter than that of the biologically inactive carboxy-terminal fragment of parathyroid hormone. The C-terminal and other portions of the PTH molecule are degraded primarily in the kidney and tend to accumulate with chronic renal disease.

Chief cells synthesize and secrete another major protein termed parathyroid secretory protein (I) or chromogranin A. It has a higher molecular weight than PTH, being composed of 430–448 amino acids, and is co-stored and secreted with parathyroid hormone. A similar molecule has been found in secretory granules of a wide variety of peptide hormone-secreting cells and in neurotransmitter secretory vesicles. An internal region of the parathyroid secretory protein or chromogranin A molecule is identical in sequence to pancreatstatin, a C-terminal amidated peptide that inhibits glucose-stimulated insulin secretion. This 49-amino-acid proteolytic cleavage product (240–280 amino acids) of parathyroid secretory protein inhibits low calcium-stimulated secretion of parathyroid hormone and chromogranin A from parathyroid cells. These findings suggest that chromogranin A-derived peptides may act locally in an autocrine manner to inhibit the secretion of active hormone by endocrine cells, such as those of the parathyroid gland.

Secretory cells in the parathyroids of most animals store relatively small amounts of preformed hormone but are capable of responding to minor fluctuations in calcium ion concentration by rapidly altering the rate of hormonal secretion, and more slowly by altering the rate of hormonal synthesis. In contrast to most endocrine organs, which are under complex controls, the parathyroids have a unique feedback control system based primarily on the concentration of calcium (and to a lesser extent of magnesium) ion in blood.

If the blood calcium level is elevated, there is rapid and pronounced reduction in circulating levels of parathyroid hormone. Conversely, if the blood calcium level is lowered, there is an increase in parathyroid hormone levels. The concentration of blood phosphorus has no direct regulatory influence on the synthesis and secretion of the hormone; however, certain disease conditions with hyperphosphatemia are associated clinically with hyperparathyroidism. An elevated blood phosphorus level may lead indirectly to parathyroid stimulation by virtue of its ability to lower blood calcium according to the mass-law equation when the serum is saturated with these two ions. Hyperphosphatemia also suppresses the rate of formation of the biologically active, hormonal form of vitamin D₃ (1,25-dihydroxycholecalciferol) in the kidney, which further contributes to the development of hypocalcemia and parathyroid stimulation.

Magnesium ion has an effect on parathyroid secretory rate similar to that of calcium, but its effect is not equipotent to that of calcium. The more potent effects of calcium ion in the control of parathyroid hormone secretion, together with its preponderance over magnesium in the extracellular fluid, suggest a secondary role for magnesium in parathyroid control.

Calcium ion controls not only the rate of biosynthesis and secretion of parathyroid hormone, but also other metabolic and intracellular degradative processes within chief cells. Increased calcium ion in extracellular fluids rapidly inhibits the uptake of amino acids by chief cells, synthesis of proparathyroid hormone and conversion to parathyroid hormone, and secretion of the stored hormone. The shifting of the percentage of flow of proparathyroid hormone from the degradative pathways to the synthetic route represents a key adaptive response of the parathyroid gland to a low-calcium diet. During periods of long-term calcium restriction, the enhanced synthesis and secretion of parathyroid hormone is accomplished by an increased capacity of the entire pathway in individual chief cells and through hyperplasia of active chief cells. Recently synthesized and processed active parathyroid hormone may be released directly in response to increased demand and bypass the chief cell's storage pool of mature secretory granules in the cytoplasm. Bypass secretion of parathyroid hormone can be stimulated only by a low circulating concentration of calcium ion and not by other secretagogues such as cAMP and beta agonists, epinephrine, norepinephrine, and isoproterenol, for parathyroid hormone. Degradation of mature parathyroid hormone by lysosomal enzymes occurs after prolonged exposure of chief cells to a high-calcium environment.

Parathyroid hormone is the principal hormone involved in the minute-to-minute, fine regulation of blood calcium. It exerts its biologic actions by directly influencing the function of target cells primarily in bone and kidney, and indirectly in the intestine.

The action on bone is to mobilize calcium from skeletal reserves into extracellular fluids. The administration of parathyroid hormone causes an initial decline followed by a sustained increase in circulating levels of calcium. The transitory decrease in blood calcium probably results from a sequestration of calcium phosphate in bone and soft tissues. The subsequent increase in blood calcium results

from an interaction of parathyroid hormone with osteocytes and (indirectly) osteoclasts in bone.

Osteoclasts are primarily responsible for the catabolic action of PTH on bone by increasing resorption. Parathyroid hormone stimulates an increased activity of preformed osteoclasts, but receptors for PTH are not present on osteoclasts; receptors are present, however, on osteoblasts. Isolated osteoclasts respond to PTH only in the presence of osteoblasts. The mechanism by which binding of PTH to osteoblasts results in stimulation of osteoclasts is not completely understood, but appears to include direct effects on the osteoblast as well as release by osteoblasts of secretory products which are capable of stimulating osteoclastic bone resorption (Fig. 3.13).

Binding of PTH to specific receptors on bone cells results in the activation of adenylate cyclase in the plasma membrane. The adenylate cyclase catalyzes the conversion of ATP to cAMP in target cells. The accumulation of cAMP in target cells functions as an intracellular messenger of PTH action in osteoblasts. Parathyroid hormone also induces an increase in cytoplasmic calcium and stimulates phosphatidylinositol turnover in osteoblasts; however, it has not been determined which intracellular messenger is required for induction of bone resorption by PTH stimulation of osteoblasts. The increase in cytosolic calcium is partially dependent on cAMP accumulation. Calcium concentration in the osteoblast may also be increased by the activation of protein kinase C, resulting in the production of inositol triphosphate and subsequent release of calcium from the endoplasmic reticulum. Parathyroid hormone also appears to use a second intracellular messenger in target cells by stimulating inositol triphosphate and diacylglycerol production.

The initial binding of parathyroid hormone to osteoblasts lining bone surfaces causes the cells to contract, thereby exposing the underlying osteoid-covered bone matrix. Osteoblasts secrete latent collagenase and neutral proteases, which results in degradation of the osteoid and exposes the underlying bone mineral, which attracts osteoclasts. Osteoblasts stimulated by PTH also appear to release unknown paracrine factors (possibly cytokines, lymphokines, and prostaglandins, among others), which locally stimulate osteoclastic bone resorption.

The response of bone to parathyroid hormone is biphasic. The immediate effects are the result of increasing the activity of existing osteoclasts and osteocytes present in bone. This depends on the continuous presence of hormone, and results in an increased flow of calcium from deep in bone to bone surfaces through the coordinated action of osteocytes and endosteal lining cells (inactive osteoblasts). This osteocyte–osteoblast pump is concerned with movement of calcium from the bone fluid to the extracellular fluid compartment.

The late effects on bone are potentially greater and are not dependent on the continuous presence of hormone. Osteoclasts are primarily responsible for the long-term action of parathyroid hormone on increasing bone resorption and overall bone remodeling. If the increase in para-

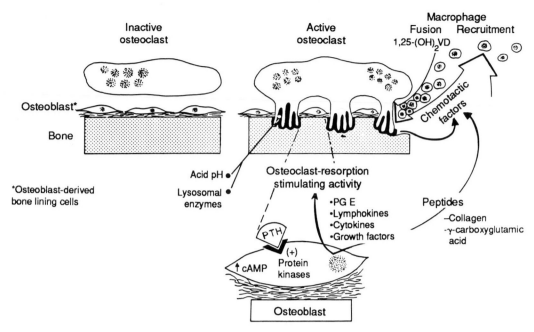

Fig. 3.13 Paracrine control of bone resorption. Specific receptors for PTH are on osteoblasts but not on osteoclasts. In response to PTH, osteoblasts contract and expose the bone surface to osteoclasts and release soluble factors that increase osteoclastic activity.

thyroid hormone is sustained, the active osteoclast pool in bone is increased, and the mineral and organic components (e.g., hydroxyproline) released from bone are released into the extracellular fluid compartment.

A long-term increase in parathyroid hormone secretion may also result in the formation of greater numbers of osteoblasts with a resultant increase in bone formation as well as resorption. However, resorption is usually greater than formation, leading to a net negative skeletal balance.

Parathyroid hormone has a rapid (5–10 min) and direct effect on renal tubular function, leading to decreased reabsorption of phosphorus and phosphaturia. The site where PTH blocks tubular reabsorption of phosphorus has been localized by micropuncture methods to the proximal tubule of the nephron. Parathyroid hormone binds to a receptor on the basolateral aspect of renal epithelial cells. The hormone stimulates adenylate cyclase, increases intracellular cAMP, and inhibits phosphorus reabsorption across the brush border through the actions of protein kinases. The parathyroid hormone receptor has been cloned and sequenced from complementary DNA (cDNA) isolated from kidney cells. The PTH receptor has a striking degree of sequence homology (~56%) with the calcitonin receptor but lacks similarity with other G protein-linked receptors. The receptors for these calcium-regulating hormones belong to a family of G protein-linked receptors with seven transmembrane spanning domains that activate adenylate cyclase.

Although the effects of PTH on the tubular reabsorption of phosphorus have been considered to be of major importance, evidence has accumulated indicating that the ability

of PTH to enhance the reabsorption of calcium is of considerable importance in the maintenance of calcium homeostasis. This effect of PTH on tubular reabsorption of calcium appears to be due to a direct action on the distal convoluted tubule. The biochemical mechanism by which PTH enhances calcium reabsorption is unknown, but it is coupled to increases in intracellular cAMP.

The other important effect of parathyroid hormone is the regulation of the conversion of 25-hydroxycholecalciferol to biologically active (1,25-dihydroxycholecalciferol) and other metabolites of vitamin D.

Under normal conditions parathyroid hormone is secreted continuously from chief cells. In the liver, kidney, peripheral circulation, and at target cells, it is cleaved into a biologically active portion and a biologically inactive portion.

The calcium-mobilizing and phosphaturic activities of parathyroid hormone appear to be mediated through the intracellular accumulation of cyclic 3′,5′-adenosine monophosphate (cAMP) and cytosol calcium in target cells. Binding of parathyroid hormone to specific receptors on target cells results in the activation of adenylate cyclase in the plasma membrane. The adenylate cyclase catalyzes the conversion of ATP and cAMP in target cells. Cyclic 3′,5′ AMP accumulation functions as an intracellular mediator (second messenger) of parathyroid hormone action in target cells, resulting in an increased permeability for calcium ion. The resultant increase in cytosolic calcium content in combination with the cAMP accumulation initiates the synthesis and release of lysosomal enzymes, and activates glycolysis in osteolytic cells with accumulation of acidic products (hydrogen ion, lactate, and citrate) that results eventually in breakdown of both the inorganic and organic phases of bone.

B. Thyroid C Cells and Calcitonin

There is a second calcium-regulating hormone, calcitonin (secreted by the thyroid C cells in response to hypercalcemia), which lowers plasma calcium.

C cells are distinct from follicular cells, which secrete thyroxine and triiodothyronine. They are situated either within the follicular wall between follicular cells or as small groups between follicles. They do not border the follicular colloid directly, and their secretory polarity is oriented toward the interfollicular capillaries.

The concentration of calcium ion in plasma and extracellular fluids is the principal physiologic stimulus for the secretion of calcitonin by C cells. Calcitonin is secreted continuously under conditions of normocalcemia, but the rate of secretion is increased greatly in response to elevations in blood calcium. Magnesium ion has an effect on calcitonin secretion similar to that of calcium, but this effect is observed only under experimental conditions with nonphysiologic levels.

Hyperplasia of C cells occurs in response to long-term hypercalcemia. When the blood calcium is lowered, the stimulus for calcitonin secretion is diminished, and numerous secretory granules accumulate in the cytoplasm of C cells. The storage of large amounts of preformed hormone in C cells and rapid release in response to moderate elevations in blood calcium probably are a reflection of the physiologic role of calcitonin as an emergency hormone to protect against the development of hypercalcemia.

Calcitonin secretion is increased in response to a high-calcium meal, often before a significant rise in plasma calcium can be detected. The cause of this increase in calcitonin secretion could be either a small undetectable rise in plasma ionized calcium or a direct stimulation by the oral calcium load of certain gastrointestinal hormones, which in turn act as secretagogues for calcitonin release from the thyroid gland. Gastrin, pancreozymin, and glucagon all stimulate calcitonin release under experimental conditions. Thus gastrointestinal hormones may be important in triggering the early release of calcitonin to prevent the development of hypercalcemia following ingestion of a high-calcium meal.

Calcitonin exerts its function by interacting with target cells primarily in bone and kidney. The actions of parathyroid hormone and calcitonin are antagonistic to bone resorption but synergistic with decreasing the renal tubular reabsorption of phosphorus. The hypocalcemic effects of calcitonin are primarily the result of decreased entry of calcium from the skeleton into plasma, due to a temporary inhibition of parathyroid hormone-stimulated bone resorption. The hypophosphatemia develops from a direct action of calcitonin, increasing the rate of movement of phosphate out of plasma into soft tissue and bone, as well as from the inhibition of bone resorption. The action of calcitonin is not dependent on vitamin D, since it acts both in vitamin D-deficient animals and following the administration of large doses of vitamin D.

The action of calcitonin on inhibiting bone resorption stimulated by parathyroid hormone and other factors results from blockage of osteoclastic bone resorption. Specific structural alterations are produced by calcitonin in osteoclasts, which have specific receptors for the hormone on their surfaces. Osteoclasts withdraw from resorptive surfaces, and the brush border and transitional zone become atrophic. In addition, there are fewer osteoclasts in bone. Although calcitonin can block bone resorption completely, the inhibition is a transitory effect. Eventually the effects of parathyroid hormone on increasing bone resorption become manifest in the presence of calcitonin.

Both calcitonin and parathyroid hormone decrease renal tubular reabsorption of phosphate, leading to phosphaturia; however, the adenylate cyclase-linked receptors for calcitonin are found in the ascending limb of the loop of Henle and the distal convoluted tubule. In addition, calcitonin results in diuresis of sodium, chloride, and calcium, whereas parathyroid hormone causes renal retention of calcium and hydrogen ions.

Calcitonin and parathyroid hormone, acting in concert, provide a dual negative-feedback control mechanism to maintain the concentration of calcium in extracellular fluids within narrow limits. Parathyroid hormone probably is the major factor involved in the minute-to-minute regulation of blood calcium under normal conditions. In most higher mammals that live in a relatively low calcium–high phosphorus environment, protection against the development of hypocalcemia by parathyroid hormone is probably a life-sustaining function. Calcitonin appears to function more as an emergency hormone to prevent the development of hypercalcemia during the rapid postprandial absorption of calcium and to protect against excessive loss of calcium and phosphorus from the maternal skeleton during pregnancy.

C. Cholecalciferol (Vitamin D)

The third major hormone involved in the regulation of calcium metabolism and skeletal remodeling is cholecalciferol or vitamin D_3. Although this compound has been designated a vitamin for a long time, it can equally be considered a hormone. Cholecalciferol is ingested in small amounts in the diet and can be synthesized in the epidermis from precursor molecules (e.g., 7-dehydrocholesterol) through a previtamin D_3 intermediate form. This reaction is catalyzed by ultraviolet irradiation (wave length 2900–3200 Angström) from the sun. A high-affinity vitamin D-binding protein transports cholecalciferol in the blood.

Vitamin D must be metabolically activated before it can produce its known physiologic functions. Vitamin D_3 from dietary sources is absorbed by facilitated diffusion and bound to an alpha-2-globulin in the blood for transport. Endogenous cholecalciferol synthesized in the skin from

7-dehydrocholesterol also is protein bound for transport to the liver.

In the liver cholecalciferol is converted to 25-hydroxycholecalciferol by a hepatic microsomal enzyme, calciferol-25-hydroxylase, which is associated with the endoplasmic reticulum in hepatocytes.

This first metabolite, 25-hydroxycholecalciferol, is transported to the kidney and undergoes further transformation to a more polar active metabolite, 1,25-dihydroxycholecalciferol. Other metabolites are formed, such as 25,26-dihydroxycholecalciferol and 24,25-dihydroxycholecalciferol. The formation of 1,25-dihydroxycholecalciferol is catalyzed by 25-hydroxycholecalciferol-1-α-hydroxylase in mitochondria, primarily in cells of the proximal convoluted tubule. The conversion of 25-hydroxycholecalciferol to 1,25-dihydroxycholecalciferol is the rate-limiting step in vitamin D metabolism and is the primary reason for the delay between vitamin D administration and expression of its biologic effects.

The control of this final step in the metabolic activation is complex and appears to be regulated in part by the plasma calcium concentration and its influence on the rates of secretion of parathyroid hormone. Parathyroid hormone and conditions that stimulate its secretion increase the transformation of 25-hydroxycholecalciferol to 1,25-dihydroxycholecalciferol. Low blood phosphorus increases the formation of 1,25-dihydroxycholecalciferol, whereas high blood phosphorus suppresses the activity of the 1-α-hydroxylase.

The rates of synthesis of 24,25-dihydroxycholecalciferol and 1,25-dihydroxycholecalciferol appear to be reciprocally related and controlled by similar factors. When 1,25-dihydroxycholecalciferol synthesis increases, the synthesis of 24,25-dihydroxycholecalciferol declines, and vice versa. 24,25-Dihydroxycholecalciferol may play a role in bone formation and egg hatchability in certain animals, and with 1,25-dihydroxycholecalciferol may exert negative feedback control on the parathyroid gland.

Other hormones may increase the activity of renal 1-α-hydroxylase and the formation of 1,25-dihydroxycholecalciferol under certain conditions. Prolactin, estradiol, placental lactogen, and possibly somatotropin enhance 1-α-hydroxylase activity. Increased secretion of these hormones, either alone or in combination, appears to be important in the efficient adaptation to major calcium demands, such as pregnancy, lactation, and growth.

The active metabolites of vitamin D function to increase the absorption of calcium and phosphorus from the intestine. From a functional point of view, vitamin D can be thought to bring about the retention of sufficient mineral ions to ensure mineralization of bone matrix, whereas parathyroid hormone maintains the proper ratio of calcium to phosphate in extracellular fluids.

The major target tissue for 1,25-dihydroxycholecalciferol is the mucosa of the small intestine, where it increases the active transcellular transport of calcium (proximal part) and phosphorus (distal part). Following synthesis in the kidney, 1,25-dihydroxycholecalciferol is transported in a protein-bound form to specific target cells in the intestine and bone. Free 1,25-dihydroxycholecalciferol penetrates the plasma membrane of target cells and initially binds to a cytoplasmic receptor in cells of the intestine. Subsequently, the hormone–receptor complex is transferred to the nucleus, and 1,25-dihydroxycholecalciferol binds to specific receptors in the nuclear chromatin. In the nucleus it stimulates gene expression with increased mRNA formation, which directs the synthesis of vitamin D-dependent proteins.

Intestinal absorptive cells are concerned with the transport of calcium from the lumen to the bloodstream, and in response to 1,25-dihydroxycholecalciferol, they synthesize a specific calcium-binding protein. Calcium-binding protein also has been isolated from small intestine, kidney, parathyroid gland, and the shell gland of laying hens. A vitamin D-dependent calcium-binding protein is present in bone, particularly in the spongiosa and cartilaginous growth plate.

The absorptive capacity of the intestine for calcium is a direct function of the amount of calcium-binding protein present. Vitamin D or low-calcium diets stimulate the synthesis of calcium-binding protein. The physiologic function of calcium-binding protein may be to protect absorptive cells against high cytosolic concentrations of calcium ion during the transcellular transport of calcium from the luminal border to the laterobasilar border of intestinal cells. At the basilar aspect of intestinal absorptive cells, calcium is exchanged for sodium and enters the extracellular fluids.

The active metabolites of cholecalciferol also act on bone. Rickets and osteomalacia develop in young and mature animals respectively, when it is deficient (see Bones and Joints, Volume 1, Chapter 1). Vitamin D also is necessary for osteoclastic resorption and calcium mobilization from bone in adults. Small amounts of active vitamin D metabolite are necessary to permit osteolytic cells to respond to parathyroid hormone (permissive effect) under physiologic conditions. Both 25-hydroxycholecalciferol and 1,25-dihydroxycholecalciferol, and cholecalciferol in pharmacologic doses, stimulate osteoclastic proliferation and the resorption of bone.

Considerably less is known regarding the action of cholecalciferol and its active metabolite on the kidney. The active metabolite(s) of vitamin D appear to stimulate the retention of calcium by increasing renal tubular reabsorption, probably in the distal part of the nephron. The active metabolite may have a direct effect on the parathyroid gland, in addition to its well-characterized action on intestine and bone.

Deficiency of vitamin D results not only from simple dietary lack or inadequate exposure to sunlight but also from a deficiency of the hydroxylase enzymes essential for metabolic activation of precursor molecules (see vitamin D-dependent rickets in Bones and Joints, Volume 1, Chapter 1).

D. Diseases of the Parathyroid Glands

1. Parathyroid and Related Cysts

Small cysts are observed within the parenchyma of the parathyroid or in the immediate vicinity of the gland, frequently in dogs and occasionally in other animal species. Parathyroid cysts are usually multiloculated, lined by a cuboidal to columnar (often ciliated) epithelium, and contain a densely eosinophilic proteinaceous material. Chief cells adjacent to larger cysts may be moderately compressed.

Parathyroid cysts (Kürsteiner's cyst) appear to develop from a persistence and dilatation of remnants of the duct that connects the parathyroid and thymic primordia during embryonic development (Fig. 3.11). Similar cysts may be present in the anterior mediastinum when remnants of the embryonic duct are displaced with the caudal migration of the thymus. They are lined by pseudostratified columnar epithelium and contain a proteinaceous material.

Parathyroid cysts are distinct from midline cysts derived from remnants of the **thyroglossal duct.** The latter are lined by multilayered thyroidogenic epithelium, which often has colloid-containing follicles, and usually are located near the midline, from the base of the tongue caudally into the mediastinum.

Other cystic structures in the thyroid–parathyroid area include ultimobranchial cysts and branchial cysts. **Ultimo-branchial duct cysts** often are present in the parenchyma of the thyroid (near the thyroid hilus in ruminants) and have a keratinizing squamous epithelial lining. They are derived from remnants of the ultimobranchial body (last, usually fifth, pharyngeal pouch) that fuses with the lateral thyroid lobes during embryonic development, and distributes calcitonin-secreting C cells (derived from neural crest) into each thyroid lobe.

Branchial cysts are located lateral to the parathyroid–thyroid area, often near the base of the ear, attached deeply to cervical structures. They are lined by a pseudostratified columnar, partially ciliated, epithelium that is derived from remnants of the second pharyngeal pouch. **Salivary mucoceles** may also be present in the parathyroid–thyroid region. These cysts are lined only with granulation tissue that develops in response to the escape of saliva into the interstitium following disruption of a salivary duct.

2. Degenerative Changes of the Parathyroid Glands

Parathyroid glands of dogs and rats may develop multinucleated syncytial giant cells. They often are more numerous near the periphery of the parathyroid gland, but this varies among parathyroids, and considerable numbers can be present in the more central portions of the gland. The number of syncytial cells varies considerably between parathyroid glands in the same animal but may account for up to half of the parenchyma of the gland. They appear to form by the fusion of the cytoplasmic area of adjacent chief cells. The cytoplasm of syncytial cells is densely eosinophilic and homogeneous, and the plasma membranes between adjacent cells are often indistinct. The nuclei are smaller, and more hyperchromatic and oval, than those in adjacent chief cells.

The mechanism by which syncytial cells form is uncertain, but it does not appear to be related to improper fixation of the parathyroid. They do not usually occur in sufficient numbers to interfere significantly with parathyroid function.

3. Lymphocytic Parathyroiditis and Hypoparathyroidism

In hypoparathyroidism, either subnormal amounts of parathyroid hormone are secreted by pathologic parathyroid glands, or the hormone secreted is unable to interact normally with target cells. Hypoparathyroidism has been recognized in dogs, particularly in smaller breeds such as schnauzers and terriers, but infrequently in other animal species.

Several pathogenic mechanisms can result in inadequate secretion of parathyroid hormone. Idiopathic hypoparathyroidism in adult dogs is associated usually with diffuse lymphocytic parathyroiditis, resulting in extensive degeneration of chief cells and replacement by fibrous connective tissue. In the early stages there is infiltration of the gland with lymphocytes and plasma cells, and nodular regenerative hyperplasia of remaining chief cells. Later, the parathyroid gland is completely replaced by lymphocytes, fibroblasts, and neocapillaries, with only an occasional viable chief cell.

Lymphocytic parathyroiditis may be an immune-mediated disease. Similar lesions are produced in dogs by injection of parathyroid tissue emulsions and by repeated injections of parathyroid emulsions with Freund's adjuvant.

Other possible causes of hypoparathyroidism include invasion and destruction of parathyroids by primary or metastatic neoplasms in the anterior cervical region and trophic atrophy of parathyroids associated with long-term hypercalcemia. The parathyroid glands may be damaged or inadvertently removed during the course of thyroid surgery. If the parathyroid glands or their vascular supply is intact but damaged, there is often regeneration of adequate functional parenchyma and subsequent disappearance of clinical signs. The presence of numerous distemper virus particles in chief cells of the parathyroid gland may contribute to the low blood calcium in certain dogs with this disease. Agenesis of both pairs of parathyroids is a rare cause of congenital hypoparathyroidism in pups.

4. Acute Hypocalcemia near Parturition

Parturient paresis is a complex metabolic disease characterized by the development of severe hypocalcemia and hypophosphatemia near parturition and the initiation of lactation in dairy cattle. The serum calcium falls to less than 50% of normal concentration in spite of an increased secretion of parathyroid hormone.

The parathyroid glands are capable of responding to the acute hypocalcemia by increased development of cellular organelles for hormonal synthesis and by the secretion of parathyroid hormone. Chief cells in the active stage of synthesis and secretion predominate in the parathyroid glands (Fig. 3.14A); however, decreased numbers of storage granules in chief cells suggest that the increased rate of hormone secretion exceeds the rate of synthesis by the gland. There is a significant increase in plasma parathyroid hormone levels in cows with parturient hypocalcemia as compared to those in normal parturient cows.

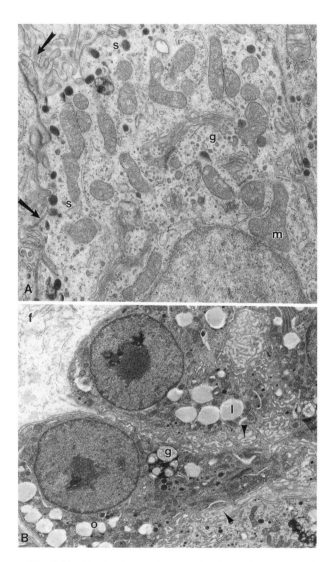

Fig. 3.14 (A) Acute parathyroid stimulation in response to severe hypocalcemia associated with parturition and initiation of the lactational drain for calcium. Parathyroid hormone-containing secretory (s) granules migrate peripherally and fuse with plasma membrane of actively secreting chief cells (arrow). Golgi apparatus (g) is hypertrophied. Many large mitochondria (m) are present. (B) Chronic parathyroid suppression. Cow. High-calcium diet. Irregularly shaped atrophic chief cells are separated by cytoplasmic processes (arrowheads) in widened intercellular spaces. The electron-dense cytoplasm contains many lipid bodies (l), large lipofuscin bodies (g), and occasional secretory granules. Cytoplasmic organelles (o) other than mitochondria are poorly developed. Widened perivascular space contains many collagen fibers (f).

The composition of the prepartal diet is a significant factor in the pathogenesis of parturient hypocalcemia. High-calcium diets increase the incidence of the disease, and low-calcium diets, or prepartal diets supplemented with pharmacologic doses of vitamin D, reduce the incidence of the disease.

Cows fed a high-calcium diet have higher blood calcium levels prepartum, but are less able to maintain serum calcium near the critical time of parturition. Plasma parathyroid hormone levels are lower prepartum than those in cows fed a balanced diet and decline further at 48 hr postpartum. Chief cells in the inactive stage of the secretory cycle predominate in cows fed high-calcium diets (Fig. 3.14B), whereas actively synthesizing chief cells are most numerous in parathyroids of cows fed balanced prepartal diets. In response to the elevated blood calcium in cows fed high-calcium prepartal diets, thyroid stores of calcitonin are diminished, and C cells are partially degranulated and appear to be actively synthesizing more calcitonin. This stimulation of C cells is accompanied by a decrease in bone turnover near parturition. Trabecular bone surfaces are inactive, and few osteoclasts are resorbing bone.

Calcium homeostasis in pregnant cows fed a high-calcium diet appears to be maintained principally by intestinal calcium absorption. This greater reliance on intestinal absorption than on parathyroid hormone-stimulated bone resorption probably is a significant factor in the more frequent development of profound hypocalcemia near parturition. These cows are more susceptible to the decreased calcium available for absorption as a result of the anorexia often associated with the high blood estrogen levels at parturition.

Calcium homeostasis in cows fed balanced or relatively low-calcium diets prepartum appears to be more under the fine control of parathyroid hormone secretion with the approach of parturition. The higher levels of parathyroid hormone secreted during the prepartal period by an expanded population of actively synthesizing chief cells results in a larger pool of active bone-resorbing cells to fulfill the increased needs for calcium mobilization at the critical time near parturition and the initiation of lactation. These cows are less susceptible to the influence of decreased calcium absorption and flow into the extracellular calcium pool resulting from the anorexia and intestinal stasis associated with parturition.

Functional disturbances associated with hypocalcemia in cows are related to paresis, whereas in the bitch they are primarily the result of neuromuscular tetany. The occurrence of either tetany or paresis in response to hypocalcemia appears to be the result of basic physiologic differences in the function of the neuromuscular junction of the cow and the bitch. The release of acetylcholine and transmission of nerve impulses across neuromuscular junctions are blocked by hypocalcemia in cows (but not in the bitch), leading to muscle paresis. The dog appears to have a higher margin of safety in neuromuscular transmission in that the degree to which the end plate potential exceeds the firing threshold is greater than that in the cow. Excitation–secretion coupling is maintained at the neuromuscular junction in the bitch with hypocalcemia, and tetany occurs as a result of spontaneous repetitive firing of motor nerve fibers. Due to the loss of stabilizing membrane-bound calcium, nerve membranes become more permeable to ions and require a stimulus of lesser magnitude to depolarize.

5. Chronic Hyperparathyroidism

a. FOCAL (NODULAR) HYPERPLASIA. Chief cell hyperplasia may affect the parathyroid in a distinctly focal or multifocal distribution. In focal parathyroid hyperplasia, there are single or multiple nodules in one or both glands where there is an increased number of closely packed chief cells, often with an expanded cytoplasmic area. The foci of chief cell hyperplasia are not encapsulated and are poorly demarcated from adjacent parenchyma. Chief cells within the nodules have a relatively uniform composition with a high cytoplasm:nucleus ratio and a slightly more hyperchromatic nucleus than those of adjacent normal chief cells. There may be slight compression of adjacent chief cells around larger areas of focal hyperplasia. Focal chief cell hyperplasia often is difficult to separate from a chief cell adenoma using only morphologic criteria. The presence of multiple nodules of varying sizes and uniform cellularity in one or both parathyroids with minimal compression and no encapsulation is more compatible with focal hyperplasia than with chief cell adenoma.

b. DIFFUSE HYPERPLASIA. Parathyroid hyperplasia, such as that seen with chronic renal failure and long-term dietary imbalances, results in a uniform enlargement of all parathyroid glands. The uniform enlargement of parathyroid glands in diffuse hyperplasia is due to both hypertrophy and hyperplasia of chief cells. There is not a peripheral rim of compressed atrophic parathyroid parenchyma as around a functional adenoma, but rather a uniform population of hyperplastic chief cells extending to the capsule of the gland. The chief cells are packed together often with indistinct cell boundaries. The expanded cytoplasmic area of chronically stimulated chief cells is lightly eosinophilic with occasional distinct vacuoles. A more prominent fibrovascular stroma in some diffusely hyperplastic parathyroids may result in a lobulated appearance. In other hyperplastic parathyroids the chief cells form distinct acinuslike structures.

6. Hyperparathyroidism Secondary to Renal Disease

Secondary hyperparathyroidism as a complication of chronic renal failure is characterized by excessive production of parathyroid hormone in response to chronic hypocalcemia. When the renal disease reaches the point at which there is a significant reduction in the glomerular filtration rate, phosphorus is retained and progressive hyperphosphatemia develops, and it may contribute to parathyroid stimulation by virtue of its ability to lower blood calcium.

Impaired intestinal absorption of calcium due to an ac-

quired defect in vitamin D metabolism probably also plays a significant role in the development of hypocalcemia in chronic renal insufficiency and uremia. Chronic renal disease impairs the production of 1,25-dihydroxycholecalciferol by the kidney and thereby diminishes intestinal calcium transport and increases mobilization of calcium from the skeleton.

Chronic renal insufficiency occurs frequently as a result of several acquired or congenital kidney lesions. All four parathyroid glands undergo marked chief cell hyperplasia (see Fig. 3.18D), and the bones have varying degrees of generalized osteodystrophia fibrosa (see Bones and Joints, Volume 1, Chapter 1). Chief cells in the parathyroid glands of dogs with chronic renal disease are primarily in the actively synthesizing stage of the secretory cycle.

7. Hyperparathyroidism Secondary to Nutritional Imbalances

Nutritional secondary hyperparathyroidism occurs commonly in cats, dogs, certain nonhuman primates, horses, domestic and captive birds, and reptiles. The increased secretion of parathyroid hormone is a compensatory mechanism directed against a disturbance in mineral homeostasis induced by diets with either a low calcium content or an excessive oxalate or phosphorus content with normal or low calcium content. Inadequate vitamin D_3, besides causing rickets, also causes lesions of hyperparathyroidism in some species. The significant result of these imbalances is hypocalcemia, which results in parathyroid stimulation. In response to the diet-induced hypocalcemia, chief cells undergo hypertrophy and eventually hyperplasia (Fig. 3.15A). The expanded cytoplasmic area

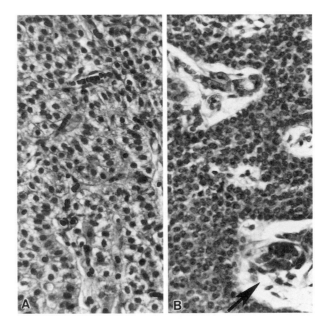

Fig. 3.15 (A) Parathyroid chief cell hyperplasia. Kitten. Low-calcium diet. (B) Chief cells in parathyroid gland of kitten fed a diet that supplied daily requirements for calcium and phosphorus.

is lightly eosinophilic and vacuolated compared with that of chief cells in normal animals (Fig. 3.15B). Perivascular spaces are narrow, and there are few fat cells in the interstitium. The skeletal lesions of nutritional hyperparathyroidism are discussed with Bones and Joints (Volume 1, Chapter 1).

8. Primary Parathyroid Hyperplasia

Primary parathyroid hyperplasia has been described in German shepherd pups associated with hypercalcemia, hypophosphatemia, increased immunoreactive parathyroid hormone, and increased fractional clearance of inorganic phosphate in the urine. Clinical signs included stunted growth, muscular weakness, polyuria, polydipsia, and a diffuse reduction in bone density. Intravenous infusion of calcium failed to suppress the autonomous secretion of parathyroid hormone by the diffuse hyperplasia of chief cells in all parathyroids. Lesions included nodular hyperplasia of thyroid C cells and widespread mineralization of lungs, kidney, and gastric mucosa. The disease was inherited as an autosomal recessive trait.

9. Intoxication by Calcinogenic Plants

Livestock grazing on various calcinogenic plants develop a progressive debilitating disease with widespread soft-tissue mineralization. *Cestrum diurnum* (day-blooming jessamine) in Florida and elsewhere in the southern United States and *Trisetum flavescens* in the Bavarian and Austrian Alps cause calcinosis in horses, cattle, and sheep. *Solanum malacoxylon* in Argentina and Brazil produces the disease *enteque seco* or *espichamento* in cattle, which is characterized by the development of hypercalcemia, hyperphosphatemia, and severe soft-tissue mineralization. Morphologically similar diseases occur in Jamaica (Manchester wasting disease) and Hawaii (*naalehu* disease). Other plants that contain vitamin D-like activity are *Solanum torvum, S. verbascifolium, Dactylis glomerata,* and *Medicago sativa.* Sporadic cases of extensive mineralization occur in cattle throughout the world.

The leaves of these calcinogenic plants contain a substance(s) possessing vitamin D-like biologic activity. The dried leaves of *Solanum malacoxylon* contain a steroid–glycoside conjugate in which the steroidal component is identical to 1,25-dihydroxycholecalciferol. The leaves are not palatable when green, but are eaten readily when dry. The toxin appears to produce disease only in herbivores and then only after ingestion. Intoxication produces rapid wasting and elevations of the calcium and phosphorus levels in the blood. The patterns of these elevations as well as the soft-tissue mineralization are comparable to those produced by intoxication by vitamin D. The leaves of *S. malacoxylon* are remarkably potent, as little as 2 g being capable of producing elevations of calcium and phosphorus in the blood in an adult. The typical wasting is seen when animals are forced to eat large quantities of the leaves. Associated with this wasting, the animals develop kyphosis, and contraction of the tendons and ligaments makes the animals unable completely to

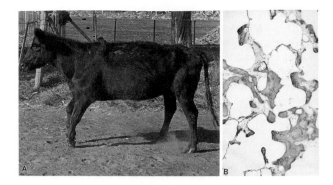

Fig. 3.16 *Enteque seco.* (A) Contraction of tendons and kyphosis in affected heifer. (B) Pulmonary ossification.

straighten their front limbs (Fig. 3.16A). Contraction of the abdominal aponeurosis gives a unique tucked-up line to the abdomen, affected animals having a racehorselike appearance. Many animals develop severe vascular and pulmonary disease (Fig. 3.16B) as a result of irregular or occasional exposure to the plant. These animals cannot be recognized by usual clinical methods, their infirmities mimicking those of advanced age.

The diseases are not restricted to cattle, but occur also, although with lesser frequency and severity, in goats, sheep, and horses. Mature animals are most commonly affected, the syndrome usually producing progressive debility. The earliest signs are those of stiffness and wasting. There is a subtle progression to permanent lameness due to contraction of the tendons. At this stage, the animals are often excitable, tire easily if made to exercise, and may show signs of acute cardiac insufficiency. If they are removed from affected areas at this time, the stiffness and deformity are not lessened, but the wasting is usually stopped, and the animals may gain weight on good pasture. At autopsy, there is a widespread mineralization of the tissues, particularly the aorta, heart, and lungs (Fig.

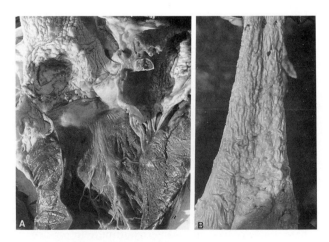

Fig. 3.17 Mineralization of heart (A) and aorta (B) in poisoning by *Solanum malacoxylon.* Ox.

3.17A,B). Mineralized tissues, especially lung, often become ossified (Fig. 3.16B).

Small mineralized blood vessels are encountered in the subcutaneous tissues, and mineralized plaques on the parietal and visceral pleura. Portions of the lungs fail to collapse, and these represent areas of mineralization in which the parenchyma is broken up and emphysematous. The pulmonary mineralization affects the caudal lobes chiefly and, when examined microscopically, is seen to consist of collagenous and irregular mineralized thickenings of alveolar septa. The opened heart often reveals mineralization of the right atrium, but not of the right ventricle. The endocardial mineralization is most severe in the left chambers, both atrium and ventricle, chiefly the basal portion of the latter, involving also the semilunar cusps, the bicuspid, and chordae tendineae, and extending into the aorta. In early cases, the aortic mineralization is restricted to the bulb and zones about the orifices of large branches, but in advanced cases it is diffuse. In these advanced cases many small arteries, especially the coronary and mesenteric arteries, are mineralized severely but irregularly. The arterial mineralization begins in the intima and, with accretion, extends to the media, disrupting the elastic laminae of the vessels. Calcium salts can be found, microscopically, in the mucosa of the abomasal fundus as well as in other fibroelastic tissues such as Glisson's capsule. Mineralization of ligaments and of the fibro-cartilaginous portions of the tendons is probably responsible for the stiffness, notwithstanding the degenerative arthropathy which is usually present.

In chronic cases the bones become extremely dense as new trabeculae are formed in the marrow spaces. These trabeculae appear to develop in atypical basophilic matrix similar to that deposited in vitamin D poisoning (see Bones and Joints, Volume 1, Chapter 1).

Parathyroid chief cells appear initially to accumulate secretory granules after feeding of *S. malacoxylon* and later to undergo involution and atrophy.

E. Neoplasms of the Parathyroid Glands

Functional adenomas and carcinomas of parathyroid glands secrete parathyroid hormone far in excess of normal, causing a syndrome of primary hyperparathyroidism. The normal control mechanism for parathyroid hormone is lost, and hormone secretion is excessive in spite of an increased level of blood calcium. Cells of the renal tubules are very sensitive to alterations in the amount of circulating parathyroid hormone. Initially the excretion of phosphorus and retention of calcium are enhanced. A prolonged increased secretion of parathyroid hormone accelerates osteoclastic bone resorption, and generalized osteodystrophia fibrosa results. The lesion is accentuated in local areas.

Adenomas of parathyroid glands are encountered in older animals, particularly dogs. The number of cases studied in dogs is inadequate to determine any breed or sex predisposition. Tumors of parathyroid chief cells do not appear to be sequelae of long-standing secondary hyperparathyroidism of renal or nutritional origin.

Chief cell adenomas usually result in considerable enlargement of a single parathyroid gland. They are light brown to red and are located either in the cervical region by the thyroids or infrequently within the thoracic cavity near the base of the heart. Parathyroid neoplasms in the precardiac mediastinum are derived from ectopic parathyroid tissue, displaced into the thorax with the expanding thymus during embryonic development. Parathyroid adenomas are sharply demarcated and encapsulated from the adjacent thyroid gland (Fig. 3.18C). Multiple white foci may be seen in the thyroids of dogs with functional parathyroid tumors. These represent areas of C-cell hyperplasia in response to the long-term hypercalcemia. In secondary hyperparathyroidism all four parathyroids are enlarged two to five times normal size (Fig. 3.18D). In comparison, a parathyroid adenoma will enlarge a single gland to a much greater degree; however, the remaining parathyroids will be atrophic. Histopathologic demonstration of a compressed rim of parathyroid parenchyma and a partial to complete fibrous capsule in an enlarged gland points to the diagnosis of adenoma rather than chief cell hyperplasia.

Parathyroid adenomas are composed of closely packed chief cells subdivided into small groups by fine connective tissue septa with many capillaries. The chief cells are cuboidal or polyhedral, and the cytoplasm stains lightly eosinophilic. Occasional oxyphil cells, water-clear cells, and transitional forms may be distributed throughout the adenoma. The oxyphil cells and transitional forms contain well-developed organelles concerned with hormonal synthesis and packaging, as well as secretory granules. This is in contrast to the oxyphil cell of normal parathyroid glands, which has a cytoplasm filled with tightly packed mitochondria, but a poorly developed endoplasmic reticulum and Golgi apparatus. Fat cells and mast cells often are present in the stroma of adenomas.

Adenomas are surrounded by a fine connective tissue capsule and may compress the adjacent thyroid gland. A rim of compressed parathyroid parenchyma usually is present outside the capsule of a small adenoma. The parathyroid tissue in the rim and in other parathyroids of an animal with a functional parathyroid adenoma is chronically suppressed by the elevated serum calcium. The suppressed chief cells have numerous droplets of lipofuscin and lipid. These atrophic chief cells are small, irregular in shape, and have densely eosinophilic cytoplasm.

Parathyroid adenomas usually are slow growing and compress the adjacent thyroid. They are well encapsulated and can be surgically excised without difficulty. Successful removal of a functional parathyroid adenoma results in a rapid decrease in circulating parathyroid hormone levels, since the half-life of parathyroid hormone in plasma is ~20 min. Plasma calcium levels in animals with overt bone disease may decrease rapidly following surgical removal of the parathyroid adenoma and reach subnormal levels within 12–24 hr, resulting in hypocalcemic tetany.

Postoperative hypocalcemia is the result of depressed secretory activity in the remaining parathyroid tissue due

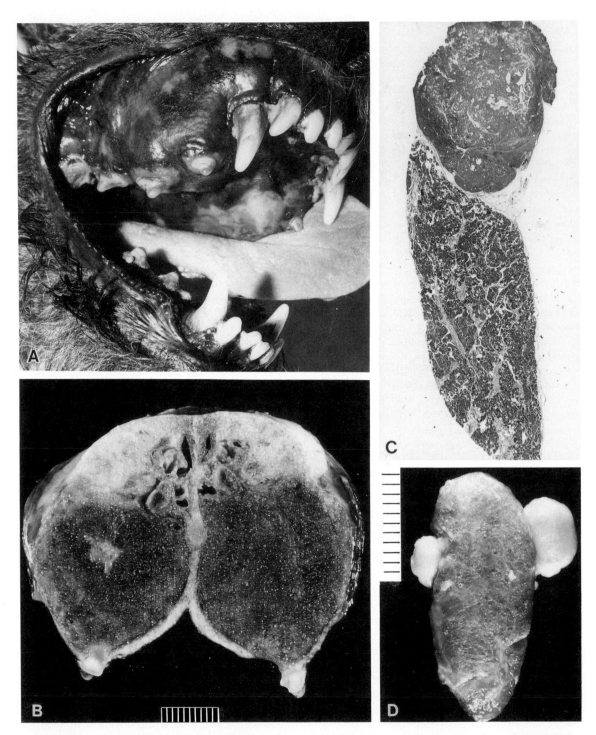

Fig. 3.18 (A and B) Hyperstotic fibrous osteodystrophy, maxilla. Dog. Primary hyperparathyroidism. (A) Bony enlargement of maxilla. (B) Cross section showing encroachment on the nasal cavity by proliferation of woven bone. (C) Chief cell adenoma sharply demarcated from, and compressing the thyroid. Dog. Primary hyperparathyroidism. (D) Chief cell hyperplasia resulting in enlargement of internal and external parathyroids. Dog. Chronic renal failure.

to long-term suppression by the chronic hypercalcemia, decreased bone resorption, and accelerated mineralization or organic matrix formed by the hyperplastic osteoblasts along bone surfaces. Infusion of calcium gluconate, feeding a high-calcium diet, and supplemental vitamin D therapy will correct this postoperative complication.

Parathyroid carcinomas are larger than adenomas, invade the capsule and adjacent structures (e.g., thyroid glands and cervical muscles), and may metastasize to regional lymph nodes and the lung. They are rare in animals.

The clinical disturbances observed with functional parathyroid tumors primarily are the result of a weakening of bones by excessive resorption. Cortical bone is thinned due to increased resorption by osteoclasts stimulated by the autonomous secretion of parathyroid hormone. Lameness due to fractures of long bones may occur after relatively minor physical trauma. Compression fractures of vertebral bodies can exert pressure on the spinal cord and nerves, resulting in motor or sensory dysfunction, or both. Facial hyperostosis due to extensive osteoblastic proliferation and deposition of poorly mineralized osteoid, and loosening or loss of teeth from alveolar sockets has been observed in dogs with primary hyperparathyroidism (Fig. 3.18A,B). Hypercalcemia results in anorexia, vomiting, constipation, depression, polyuria, polydipsia, and generalized muscular weakness due to decreased neuromuscular excitability.

Primary hyperparathyroidism occurs in older animals. They usually have a history of multiple fractures associated with severe generalized skeletal demineralization and normal renal function. Radiographic evaluation reveals areas of subperiosteal cortical resorption, loss of lamina dura around the teeth, soft-tissue mineralization, bone cysts, and a generalized decrease in bone density with multiple fractures in advanced cases. Dogs with primary hyperparathyroidism have greatly elevated blood calcium (13–29 mg/100 ml or more) and low blood phosphorus (4 mg/100 ml or less).

F. Hypercalcemia Associated with Neoplasms of Nonparathyroid Origin

Humoral hypercalcemia of malignancy (HHM) or pseudohyperparathyroidism is one form of cancer-associated hypercalcemia that is induced by the secretion of humoral factors which have effects distant to the site of the neoplasms. Some neoplasms that cause HHM may metastasize to bone (Fig. 3.19A), but the primary mechanism of hypercalcemia involves the distant effects of humoral factors and not the localized effects of bone metastases. Multiple humoral factors have been associated with HHM, including parathyroid hormone, parathyroid hormone-like proteins, cytokines, steroids such as 1,25-dihydroxyvitamin D, and prostaglandins. Many of the humoral factors produced by tumor cells that alter calcium metabolism and bone resorption also can be produced by osteogenic cells, such as osteoblasts.

The clinical syndrome of HHM in human patients and

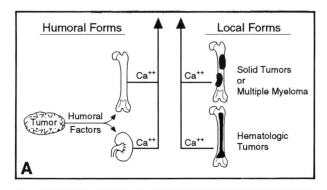

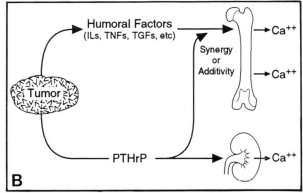

Fig. 3.19 (A) Humoral and local forms of cancer-associated hypercalcemia increase circulating concentrations of calcium by stimulating osteoclastic bone resorption or increasing tubular reabsorption of calcium. (From Rosol and Capen, *Laboratory Investigation* **66:** 1992.) (B) Multiple humoral factors can act additively or synergistically to induce hypercalcemia associated with humoral hypercalcemia of malignancy (HHM). (From Rosol and Capen, *Laboratory Investigation* **66:** 1992.

domestic animals mimics primary hyperparathyroidism, and a term used previously to describe HHM was pseudo-hyperparathyroidism. However, HHM does not mimic hyperparathyroidism in all aspects, as there are distinct differences in bone remodeling and vitamin D metabolism between patients with HHM and patients with primary hyperparathyroidism.

The most consistent feature of HHM in animals is increased osteoclastic bone resorption distant to the site of the neoplasm. Other features include hypercalciuria, decreased fractional excretion of calcium, increased nephrogenous 3′,5′-cAMP, hypophosphatemia, and hyperphosphaturia. The similarities of these clinicopathologic alterations to hyperparathyroidism led to the hypothesis that the neoplasms secrete PTH or PTH-like substances; however, native PTH is not produced by the vast majority of neoplasms associated with HHM. Immunoreactive PTH is normal or low in patients with HHM, and the tumors do not produce PTH mRNA. There are well-validated reports of ectopic production of PTH by some tumors, but these occurrences are rare. Most tumors and serum from patients with HHM do have PTH-like biologic activity, even though the tumors do not contain PTH.

Purification of the PTH-like biologic activity from neoplasms associated with HHM resulted in the discovery of a hormone currently named parathyroid hormone-related protein (PTHrP), parathyroid hormonelike peptide (PLP), or hypercalcemia of malignancy factor. It is likely that PTHrP plays a central role in the pathogenesis of HHM, since it is a consistent feature of HHM and shares most or all of the biologic activities of PTH. Parathyroid hormone-related protein acts alone in some forms of HHM but may act synergistically or additively with other humoral factors in the pathogenesis of hypercalcemia (Fig. 3.19B). It is also likely that some examples of HHM are not associated with PTHrP production by the inciting neoplasms, but rather are due to recognized or unrecognized humoral factors, cytokines, or hormones that are capable of stimulating osteoclastic bone resorption and result in hypercalcemia.

Humoral hypercalcemia of malignancy occurs as a spontaneous disease in domestic and laboratory animals. The dog is most often affected. It occurs in ~25% of dogs with lymphoma (usually thymic or multicentric forms) and the majority of dogs with adenocarcinomas derived from apocrine glands of the anal sac. Lymphoma is a relatively common tumor in older, large-breed dogs, whereas the apocrine adenocarcinoma is less common, and occurs most often in older, female dogs. Humoral hypocalcemia of malignancy also occurs sporadically in dogs, usually in association with other carcinomas. Spontaneous HHM occurs less commonly in other domestic animals but has been reported in cats and rarely in horses.

There are three primary mechanisms by which humoral factors can induce hypercalcemia: (1) stimulation of osteoclastic bone resorption, (2) increase in calcium reabsorption from the kidney, and (3) increase in calcium absorption from the intestine. Increased osteoclastic bone resorption is a fundamental component of the pathogenesis of hypercalcemia and is a consistent feature of HHM. The kidney also plays a central role in the induction of hypercalcemia. Increased osteoclastic bone resorption would increase the flux of calcium into the circulation, but this could be readily excreted by the kidney under normal circumstances. In HHM fractional excretion of calcium often is reduced, which results in decreased excretion and increased tubular reabsorption of calcium. Therefore, bone and renal mechanisms both appear to be responsible for hypercalcemia in HHM. The role of intestinal absorption of calcium in the pathogenesis of hypercalcemia has not been well evaluated.

Most of the humoral factors implicated in the pathogenesis of HHM are capable of stimulating osteoclastic bone resorption *in vivo* and *in vitro*. The humoral factors must be secreted by the neoplasms in large enough quantities to increase circulating concentrations and stimulate bone resorption alone or by cooperative interactions with other humoral factors. None of the humoral factors identified stimulates osteoclast formation or activation directly. Most factors, including PTHrP and cytokines, bind to receptors on osteoblasts which are responsible for coordi-

nating the stimulation of osteoclastic bone resorption. The presence of parathyroid hormonelike activity in tumors associated with HHM and the similarities of HHM to primary hyperparathyroidism led to the isolation and discovery of parathyroid hormone-related protein (PTHrP), initially purified from human tumors associated with HHM. Subsequently, the cDNA and gene for PTHrP were isolated, sequenced, and characterized. The PTHrP gene is complicated and contains at least six exons; two 5'-untranslated regions each with its own promoter, a prepro-hormone coding region, a main coding region, and two additional 3' regions that have coding and noncoding regions. There is alternate splicing of PTHrP mRNA so that three mature peptides can be formed composed of 139, 141, and 173 amino acid residues. In humans the PTHrP gene is located on the short arm of chromosome 12 and is thought to have evolved from the PTH gene on the short arm of chromosome 11 through chromosome duplication. The PTH gene is simpler and has only three exons (5'-untranslated region, preprohormone coding region, and main coding region with a 3'-noncoding region) and encodes for a mature protein of 84 amino acids.

Parathyroid hormone-related protein binds to the well-defined PTH receptors in bone and kidney with similar affinity to PTH. The first 34 N-terminal amino acids of PTHrP contain the PTH-like biologic activity. This is similar to PTH in which the first 34 N-terminal amino acids (of a total of 84) contain most of the biologic activities. There is 70% sequence homology of the first 13 N-terminal amino acids of PTHrP and PTH, but little homology after amino acid 13. The N-terminus is important in the activation of the PTH receptor. Sequence homology in this region is important for the ability of PTHrP to stimulate PTH receptors. The biologic actions of PTHrP in bone and kidney are similar to those of PTH. Both PTHrP and PTH induce their cellular actions in osteoblasts and renal cells by stimulating adenylate cyclase, phospholipase C, and pathways leading to increased intracellular calcium concentration. The efficiency of signal transduction by PTHrP in different cells or membranes may be affected by differences in species specificity for PTHrP. N-Terminal PTHrP stimulates osteoclastic bone resorption *in vivo* and *in vitro* with potency similar to that of PTH.

Parathyroid hormone-related protein plays a central role in the pathogenesis of HHM. It is produced by most tumors and is present in the circulation at increased concentrations in human patients and animals with HHM. The administration of anti-PTHrP antibodies to animal models of HHM decreases serum calcium levels and bone resorption.

Increased circulating concentrations of PTHrP also have been detected in experimental animal models of HHM and in dogs with spontaneous HHM. Dogs with adenocarcinomas derived from apocrine glands of the anal sac have markedly elevated concentrations of PTHrP (10–100 pM), and this suggests that PTHrP plays a primary role in the pathogenesis of the hypercalcemia associated with these tumors. In contrast, dogs with HHM and lymphoma have PTHrP concentrations which range from undetectable to 17 pM. The PTHrP appears not to be the sole humoral factor responsible for the induction of hypercalcemia in dogs with lymphoma, but may interact cooperatively with other humoral factors.

In contrast to parathyroid hormone, which is produced only by the parathyroid gland, PTHrP is produced by many normal tissues including stratified squamous epithelium, endocrine glands (adrenal cortex and medulla, fetal and adult parathyroid glands, adenohypophysis, thyroid), skeletal and smooth muscle, kidney, bone, lactating mammary gland, brain, pancreas, ovary, testicle, myometrium, avian oviduct, and placenta. Present evidence suggests that PTHrP acts as an autocrine or paracrine regulator in these tissues, but its exact function in most normal adult tissues is unknown.

In the fetus, PTHrP appears to function in an endocrine manner as a hormone to control the serum calcium concentration and calcium transport across the placenta. Rat and human fetuses have a wide distribution of PTHrP immunoreactivity, which suggests that PTHrP plays a role in cell growth and differentiation. It is produced by the lactating mammary gland and expression is stimulated by prolactin and results in a high concentration of PTHrP in milk. It may have a systemic role in the lactating female and fetus since suckling increases urinary cAMP and phosphate excretion in rats, and nursing calves have increased circulating levels of immunoreactive and bioactive PTHrP. Neutralization of PTHrP by passive immunization does not alter calcium homeostasis in lactating or neonatal mice.

A syndrome of hypercalcemia of malignancy (pseudo-hyperparathyroidism) in elderly female dogs is associated with perirectal **adenocarcinomas** derived from **apocrine glands** of the **anal sac.** The dogs have persistent hypercalcemia and hypophosphatemia that return to normal following surgical excision of the neoplasm. The hypercalcemia persists following removal of the parathyroid glands.

All tumors have histopathologic features of malignancy and almost all metastasize to iliac and sublumbar lymph nodes. Functional disturbances in dogs with HHM include generalized muscular weakness, anorexia, vomiting, bradycardia, depression, polyuria, and polydipsia. These clinical signs primarily are the result of severe hypercalcemia, and complicate the problems associated with the malignant neoplasm.

Apocrine adenocarcinomas develop as firm masses in the perirectal area, ventrolateral to the anus, in close association with the anal sac, but are not attached to the overlying skin. The tumor arises in the wall of the anal sac and projects as a mass of variable size into its lumen (Fig. 3.20A). It forms glandular acini with projections of apical cytoplasm extending into a lumen (Fig. 3.20B) and is histologically distinct from the more common perianal (circumanal) gland tumor in dogs. Most neoplasms are histologically bimorphic with glandular and solid areas (Fig.

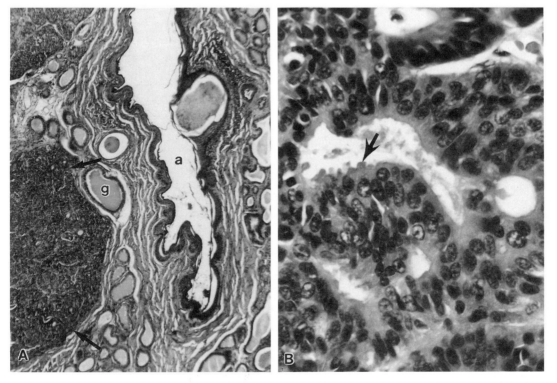

Fig. 3.20A,B Apocrine adenocarcinoma of anal sac. Dog. (Courtesy of *Veterinary Pathology* 1982) (A) Adeno-carcinoma (arrow), normal glands (G), and anal sac (A). (B) Detail of (A). Glandular acini lined by neoplastic cells with apical cytoplasmic projections (arrow).

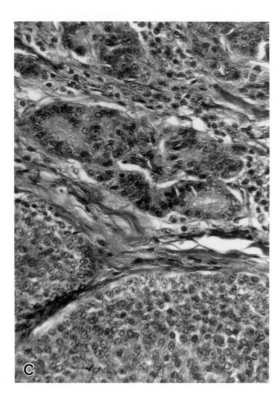

Fig. 3.20C Bimorphic growth pattern of apocrine adenocarcinoma of anal sac. Dog.

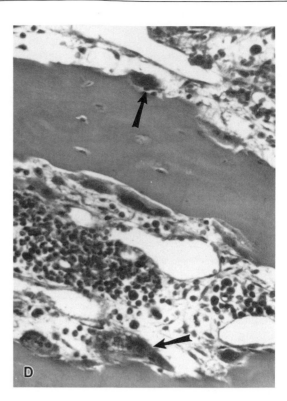

Fig. 3.20D Osteoclastic resorption. There are numerous osteoclasts with excavations (arrows) aligned on trabecular bone surfaces.

3.20C). The solid pattern of arrangement of neoplastic cells is characterized by sheets, microlobules, and packets separated by a thin fibrovascular stroma. Pseudorosettes are common in solid areas adjacent to small blood vessels.

Skeletal demineralization in dogs with HHM is mild in comparison with that from other causes of hypercalcemia and usually undetectable by conventional radiographic methods. Neoplastic cells from the perirectal adenocarcinomas rarely metastasize to bone and cause resorption. Variable numbers of osteoclasts are present on bone surfaces in dogs with marked hypercalcemia (Fig. 3.20D), possibly reflecting different states in the course of the disease and phases of bone-remodeling activity. Osteocytic osteolysis is not detected microscopically, and the cement lines are smooth and linear.

Renal mineralization is detected histologically in ~90% of dogs with pseudohyperparathyroidism associated with apocrine adenocarcinomas of the anal sac, particularly when the Ca × P product expressed as mg/100 ml is 50 or greater. Tubular mineralization is most pronounced near the corticomedullary junction but also is present in cortical and deep medullary tubules, Bowman's capsule, and glomerular tuft. Mineralization is present less frequently in the fundic mucosa of the stomach and endocardium.

The parathyroid glands are small and difficult to locate or not visible macroscopically. Microscopically they are characterized by narrow cords of inactive chief cells with an abundant fibrous connective tissue stroma and widened perivascular spaces. This suggests that apocrine adenocar-

cinomas do not stimulate parathyroid hormone secretion but rather the parathyroids respond to the persistent hypercalcemia by undergoing trophic atrophy. Thyroid C cells respond to the persistent elevation in blood calcium by undergoing diffuse or nodular hyperplasia. There is an inappropriate serum concentration of 1,25-dihydroxycholecalciferol for the degree of hypercalcemia in tumor-bearing dogs, and the rapid decrease following excision of the tumor suggests that the tumor may secrete a substance(s) that alters the activity of the 1-alpha-hydroxylase in the kidney.

Lymphosarcoma is the most common neoplasm associated with hypercalcemia in dogs and cats. Estimates of the prevalence of hypercalcemia in lymphoma dogs vary from 10 to 40%. Peripheral lymph node enlargement may or may not be detected, but there usually is evidence of anterior mediastinal or visceral involvement. Most lymphomas associated with HHM in dogs are of the T-cell subset, whereas lymphomas associated with normal blood calcium are usually of B-cell origin. Most dogs with lymphoma and hypercalcemia have increased circulating concentrations of PTHrP, but the concentrations are not so high as those in dogs with HHM and apocrine adenocarcinomas, and PTHrP does not correlate with serum calcium concentration. It is not completely resolved whether the hypercalcemia develops from the production of humoral substances by neoplastic cells or from physical disruption of trabecular bone due to frequent marrow

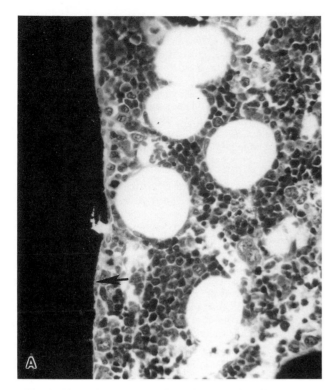

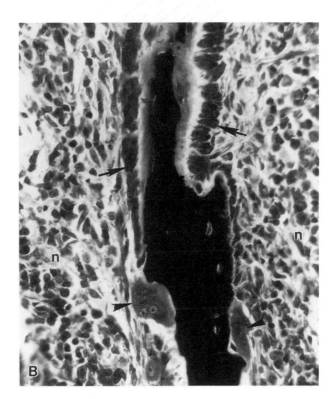

Fig. 3.21A Lumbar vertebrae. Dogs. Von Kossa tetrachrome. (Courtesy of *Laboratory Investigation* 1983.) Normocalcemic dog with smooth bone surface and flattened osteoblasts (arrow).

Fig. 3.21B Hypercalcemic dog with lymphosarcoma of marrow (neoplastic cells, N). Several osteoclasts (arrowheads) present in lacunae. Hyperplastic osteoblasts (arrows) and prominent osteoid seams present adjacent to resorptive surfaces.

involvement, or to both mechanisms. In dogs local production of bone-resorbing factors such as IL-1 or tumor necrosis factors (see Bones and Joints, Volume 1, Chapter 1) probably is important in stimulating calcium release from bone, because only dogs with lymphosarcoma cells in bone marrow have an increase in osteoclastic bone resorption. These dogs often have osteoclasts on trabecular bone surfaces opposite a surface lined by osteoid and large columnar osteoblasts (Fig. 3.21A,B).

G. Evaluation of Parathyroid Function

Parathyroid function may be altered by a wide variety of diseases or by exposure to chemicals that either elevate or lower the blood concentration of calcium (particularly calcium ion). In response to hypocalcemia, chief cells undergo hypertrophy and eventually hyperplasia. On formalin- or Bouin's-fixed sections, the expanded cytoplasmic area is lightly eosinophilic and vacuolated. Perivascular spaces are narrow in a hyperplastic parathyroid, and there are few fat cells in the interstitium. In response to hypercalcemia the cytoplasmic area of chief cells is decreased and more densely eosinophilic, often with a widening of intercellular and pericapillary spaces. If the hypercalcemia is prolonged, there is an overall reduction of glandular parenchyma with increased fibrous or adipose connective tissue in the interstitium. Subtle

differences between affected and normal animals can be best evaluated by morphometric evaluation. The area ratios of parenchyma:interstitium of the glands and cytoplasm:nucleus of chief cells should be assessed.

Ultrastructural evaluation of chief cells is a sensitive means of morphologically evaluating the parathyroid gland. Perfusion of the thyroid–parathyroid area with glutaraldehyde-based fixatives followed by postfixation with osmium tetroxide results in the best retention of structural detail in parathyroids of animals. Morphometric studies can be used to quantitate total cytoplasmic area and area occupied by a particular organelle (e.g., secretory granule).

In response to an acute lowering of blood calcium, a larger percentage of chief cells ultrastructurally will be in the active stage of synthesis and secretion than under steady-state conditions. This is indicated by a peripheral migration of secretory granules and alignment along the plasma membrane, aggregation of the endoplasmic reticulum into lamellar arrays, and enlargement of the Golgi apparatus associated with many small dense granules in the process of formation. Conversely, chief cells in response to hypercalcemia are predominantly in the inactive stage of the secretory cycle, as evaluated by electron microscopy, with dispersed profiles of endoplasmic reticulum, small Golgi complexes with few granules, and often accumulations of either glycogen or lipid (depending on species) in the cytoplasm. Secretory granules accumulate initially in response to an elevation in blood calcium but subsequently decrease due to degradation by lysosomal enzymes by the process of crinophagia. Atrophic chief cells develop in response to sustained and/or more severe hypercalcemia. Their cytoplasm is more electron dense and irregularly shrunken with widened intercellular spaces. Cytoplasmic organelles are poorly developed and may have early degenerative changes suggested by mitochondrial vacuolation with disruption of cristae and distension of endoplasmic reticulum with loss of ribosomes.

Parathyroid hormone in the circulation of animals can be measured by sensitive radioimmunoassays (RIA) or immunoradiometric assays (IRMA). Although the hormone is secreted from chief cells primarily as a straight-chain (1–84 amino acids) peptide, molecular fragments (amino and carboxy terminal) are formed in the periphery (primarily by Kupffer cells in the liver). The immunoheterogenicity created by the multiple circulating fragments of PTH has caused significant problems in the development of sensitive assays in both humans and animals. Since the amino (N-)-terminal end of the molecule (that portion which interacts with the receptor in target cells) is highly conserved between humans and all mammalian species thus far tested, assays directed against this end of PTH are the most sensitive and accurate in assessing parathyroid function. The amino-terminal assay is particularly useful in measuring ongoing or recent functional changes in the parathyroid following exposure to various physiologic perturbations or involvement by disease processes.

Parathyroid hormone assays utilizing antibody generated against the carboxy (C-)-terminal end of the human PTH molecule usually give less consistent results in animals than in human patients. The amino acid sequence of the C-terminal portion of PTH is less well conserved between animal species and humans than the N-terminal region, thereby rendering the antibody less specific and the assay less sensitive. The C-terminal fragment in the circulation is biologically inactive and has a longer plasma half-life than the N-terminal end of the PTH molecule. Therefore, C-terminal assays for PTH in species for which a specific antibody is available tend to give a more integrated evaluation of parathyroid function over time due to the slower turnover rate of this portion of the molecule in the circulation.

Bibliography

Altenahr, E. Ultrastructural pathology of parathyroid glands. *Curr Top Pathol* **56:** 1–54, 1972.

Arnold, S. A. *et al.* Nutritional secondary hyperparathyroidism in the parakeet. *Cornell Vet* **64:** 37–46, 1974.

Austin, L. A., and Heath, H. Calcitonin. Physiology and pathophysiology. *N Engl J Med* **304:** 269–278, 1981.

Ayer, L. M. *et al.* Analysis of parathyroid hormone in bovine parathyroid cysts. *J Bone Miner Res* **4:** 335–340, 1974.

Bilezikian, J. P. Parathyroid hormone-related protein in sickness and in health. *N Engl J Med* **322:** 1151–1153, 1990.

Black, H. E., Capen, C. C., and Arnaud, C. D. Ultrastructure of parathyroid glands and plasma immunoreactive parathyroid

hormone in pregnant cows fed normal and high-calcium diets. *Lab Invest* **29:** 173–185, 1973.

Black, H. E., Capen, C. C., and Young, D. M. Ultimobranchial thyroid neoplasms in bulls. A syndrome resembling medullary thyroid carcinoma in man. *Cancer* **32:** 865–878, 1973.

Black, H. E. *et al.* Effect of a high-calcium prepartal diet on calcium homeostatic mechanisms in thyroid glands, bone, and intestine of cows. *Lab Invest* **29:** 437–448, 1973.

Blum, J. W., Mayer, G. P., and Potts, J. T. Parathyroid hormone response during spontaneous hypocalcemia and induced hypercalcemia in cows. *Endocrinology* **95:** 84–92, 1974.

Boland, R. L. Plants as a source of vitamin D₃ metabolites. *Nutr Rev* **44:** 1–8, 1986.

Burk, R. L., and Schaubhut, C. S., Jr. Spontaneous primary hypoparathyroidism in a dog. *J Am Anim Hosp Assoc* **11:** 784–785, 1975.

Canalis, E., McCarthy, T., and Centrella, M. Growth factors and the regulation of bone remodeling. *J Clin Invest* **81:** 277–281, 1988.

Capen, C. C. Fine structural alterations of parathyroid glands in response to experimental and spontaneous changes of calcium in extracellular fluids. *Am J Med* **50:** 598–611, 1971.

Capen, C. C. Functional and fine structural relationships of parathyroid glands. *Adv Vet Sci Comp Med* **19:** 249–286, 1975.

Capen, C. C. Neoplasms of the parathyroid glands. *In* "Atlas of Tumor Pathology in the Fischer Rat," S. F. Stinson, H. M. Schutter, and G. K. Reznik (eds.), pp. 367–378. Boca Raton, Florida, CRC Press, 1990.

Capen, C. C., and Rosol, T. J. Recent advances in the structure and function of the parathyroid gland in animals and the effects of xenobiotics. *Toxicol Pathol* **17:** 333–345, 1989.

Capen, C. C., and Roth, S. L. Ultrastructural and functional relationships of normal and pathologic parathyroid cells. *Pathobiol Annu* **8:** 129–175, 1973.

Capen, C. C., and Rowland, G. N. Ultrastructural evaluation of the parathyroid glands of young cats with experimental hyperparathyroidism. *Z Mikrosk Anat Forsch* **90:** 495–506, 1968.

Capen, C. C., and Young, D. M. The ultrastructure of the parathyroid gland and thyroid parafollicular cells of cows with parturient paresis and hypocalcemia. *Lab Invest* **17:** 717–737, 1967.

Capen, C. C., Cole, C. R., and Hibbs, J. W. The pathology of hypervitaminosis D in cattle. *Pathol Vet* **3:** 350–378, 1966.

Capen, C. C., Cole, C. R., and Hibbs, J. W. Influence of vitamin D on calcium metabolism and the parathyroid glands of cattle. *Fed Proc* **27:** 142–152, 1968.

Carillo, J. M., Burk, R. L., and Bode, C. Primary hyperparathyroidism in a dog. *J Am Vet Med Assoc* **174:** 67–71, 1979.

Cheville, N. F. Ultrastructure of canine carotid body and aortic body tumors: Comparison with tumors of thyroid and parathyroid origin. *Vet Pathol* **9:** 166–189, 1972.

Cohn, D. V., and MacGregor, R. R. The biosynthesis, intracellular processing, and secretion of parathormone. *Endocr Rev* **2:** 1–26, 1981.

Collins, W. T. *et al.* Ultrastructural evaluation of parathyroid glands and thyroid C cells of cattle fed *Solanum malacoxylon*. *Am J Pathol* **87:** 603–614, 1977.

Conaway, H. H., Diez, L. F., and Raisz, L. G. Effects of prostacyclin and prostaglandin E₁ (PGE₁) on bone resorption in the presence and absence of parathyroid hormone. *Calcif Tissue Int* **38:** 130–134, 1986.

Conrad, H. R., and Hansard, S. L. Effects of massive doses of vitamin D on physiological behavior of calcium in cattle. *J Appl Physiol* **10:** 98–102, 1957.

DeLuca, H. F. The kidney as an endocrine organ for the production of 1,25-dihydroxyvitamin D_3, a calcium-mobilizing hormone. *N Engl J Med* **289:** 359–365, 1973.

Drezner, M., Neelon, F. A., and Lebovitz, H. E. Pseudohypoparathyroidism type II: A possible defect in the reception of the cyclic AMP signal. *N Engl J Med* **289:** 1056–1060, 1973.

Engelman, R. W. *et al.* Hypercalcemia in cats with feline-leukemia-virus-associated leukemia–lymphoma. *Cancer* **56:** 777–781, 1985.

Farfel, Z. *et al.* Defect of receptor-cyclase coupling protein in pseudohypoparathyroidism. *N Engl J Med* **303:** 237–242, 1980.

Fetter, A. W., and Capen, C. C. Ultrastructural evaluation of the parathyroid glands of pigs with naturally occurring atrophic rhinitis. *Vet Pathol* **5:** 481–503, 1968.

Fetter, A. W., and Capen, C. C. The ultrastructure of the parathyroid glands of young pigs. *Acta Anat (Basel)* **75:** 359–372, 1970.

Fischer, J. A. *et al.* Calcium-regulated parathyroid hormone peptidase. *Proc Natl Acad Sci U.S.A.* **69:** 2341–2345, 1972.

Fjalling, M. *et al.* Radiation-induced parathyroid adenomas and thyroid tumors in rats. *Acta Pathol Microbiol Scand* **89:** 425–429, 1982.

Garel, J.-M. Hormonal control of calcium metabolism during the reproductive cycle in mammals. *Physiol Rev* **67:** 1–66, 1987.

Goff, J. P. *et al.* Parathyroid hormone-related peptide content of bovine milk and calf blood assessed by radioimmunoassay and bioassay. *Endocrinology* **129:** 2815–2819, 1991.

Goltzman, D., Hendy, G. N., and Banville, D. Parathyroid hormonelike peptide: Molecular characterization and biological properties. *Trends Endocrinol Metab* **1:** 39–44, 1989.

Goulden, B. E., and MacKenzie, C. P. Suspected primary hyperparathyroidism in the dog. *N Z Vet J* **16:** 131–140, 1968.

Grain, E. G. Hypercalcemia associated with squamous cell carcinoma in a dog. *J Am Vet Med Assoc* **181:** 165–166, 1982.

Gray, R. W. *et al.* 25-Hydroxycholecalciferol-1-hydroxylase: Subcellular location and properties. *J Biol Chem* **247:** 7528–7532, 1972.

Habener, J. R., and Potts, J. T. Biosynthesis of parathyroid hormone. *N Engl J Med* **299:** 580–585, 1978.

Habener, J. F., and Potts, J. T. Subcellular distribution of parathyroid hormone, hormonal precursors, and parathyroid secretory protein. *Endocrinology* **104:** 265–275, 1979.

Harcourt-Webster, J. M., and Truman, R. F. Histochemical study of oxidative and hydrolytic enzymes in the abnormal human parathyroid. *J Pathol* **97:** 687–693, 1969.

Heath, H., Weller, R. E., and Mundy, G. R. Canine lymphosarcoma: A model for study of the hypercalcemia of cancer. *Calcif Tissue Int* **30:** 127–133, 1980.

Holick, M. F. *et al.* Photosynthesis of previtamin D_3 in human skin and the physiologic consequences. *Science* **210:** 203–205, 1980.

Hruska, K. A. *et al.* Metabolism of immunoreactive parathyroid hormone in the dog. The role of the kidney and the effects of chronic renal disease. *J Clin Invest* **56:** 39–48, 1975.

Hruska, K. A. *et al.* Degradation of parathyroid hormone and fragment production by the isolated perfused dog kidney. The effect of glomerular filtration rate and perfusate Ca^{2+} concentrations. *J Clin Invest* **60:** 501–510, 1977.

Hunt, R. D., Garcia, F. G., and Hegsted, D. M. A comparison of vitamin D_2 and D_3 in New World primates. I. Production and regression of osteodystrophia fibrosa. *Lab Anim Care* **16:** 222–234, 1967.

Klausner, J. S. *et al.* Hypercalcemia in two cats with squamous cell carcinomas. *J Am Vet Med Assoc* **196:** 103–105, 1990.

Kornegay, J. N. *et al.* Idiopathic hypocalcemia in four dogs. *J Am Anim Hosp Assoc* **16:** 723–734, 1980.

Krook, L. Spontaneous hyperparathyroidism in the dog. A pathologic–anatomical study. *Acta Pathol Microbiol Scand* **41:** (Suppl. 122) 1–88, 1957.

Krook, L., and Barrett, R. B. Simian bone disease—a secondary hyperparathyroidism. *Cornell Vet* **52:** 459–492, 1962.

Krook, L., and Lowe, J. E. Nutritional secondary hyperparathyroidism in the horse. *Vet Pathol* **1:** (Suppl. 1) 1–98, 1964.

Krook, L. *et al.* Nutritional hypercalcitoninism in bulls. *Cornell Vet* **61:** 625–639, 1971.

Law, F. M. K. *et al.* Parathyroid hormone-related protein in milk and its correlation with bovine milk calcium. *J Endocrinol* **128:** 21–26, 1991.

Lin, H. Y. *et al.* Expression cloning of an adenylate cyclase-coupled calcitonin receptor. *Science* **254:** 1022–1024, 1991.

Mallette, L. E. The parathyroid polyhormones: New concepts in the spectrum of peptide hormone action. *Endocrinol Rev* **12:** 110–117, 1991.

Martin, T. J. Properties of parathyroid hormone-related protein and its role in malignant hypercalcemia. *Q J Med* **76:** 771–786, 1990.

Martin, T. J., and Suva, L. J. Parathyroid hormone-related protein: A novel gene product. *Baillière's Clin Endocrinol Metab* **2:** 1003–1029, 1988.

Matus, R. E., and Weir, E. C. Hypercalcemia of malignancy. *In* "Current Veterinary Therapy," R. W. Kirk (ed.), pp. 988–993. Philadelphia, Pennsylvania, W. B. Saunders, 1989.

Mayer, G. P., Ramberg, C. R., Jr., and Kronfield, D. S. Calcium homeostasis in the cow. *Clin Orthop* **62:** 79–94, 1969.

Meuten, D. J. *et al.* Gastric carcinoma with pseudohyperparathyroidism in a horse. *Cornell Vet* **68:** 179–195, 1978.

Meuten, D. J. *et al.* Hypercalcemia associated with an adenocarcinoma derived from the apocrine glands of the anal sac. *Vet Pathol* **18:** 454–471, 1981.

Meuten, D. J. *et al.* Relationship of serum total calcium to serum albumin and total protein concentrations in dogs. *J Am Vet Med Assoc* **180:** 63–67, 1982.

Meuten, D. J. *et al.* Hypercalcemia of malignancy: Hypercalcemia associated with an adenocarcinoma of the apocrine glands of the anal sac. *Am J Pathol* **108:** 366–370, 1982.

Meuten, D. J. *et al.* Hypercalcemia in dogs with adenocarcinoma derived from apocrine glands of anal sac: Biochemical and histomorphometric investigations. *Lab Invest* **48:** 428–435, 1983.

Meyer, D. J., and Terrell, T. G. Idiopathic hypoparathyroidism in a dog. *J Am Vet Med Assoc* **68:** 858–860, 1976.

Morris, M. L., Teerer, S. M., and Collins, D. R. The effects of the exclusive feeding of an all-meat dog food. *J Am Vet Med Assoc* **158:** 477–488, 1971.

Mundy, G. R. Hypercalcemic factors other than parathyroid hormone-related protein. *Endocrinol Metab Clin North Am* **18:** 795–806, 1989.

Nafe, L. A., Patnaik, A. K., and Lyman, R. Hypercalcemia associated with epidermoid carcinoma in a dog. *J Am Vet Med Assoc* **176:** 1253–1254, 1980.

Nunez, E. A. *et al.* Ultrastructure of the parafollicular (C-) cells and the parathyroid cells in growing dogs on a high-calcium diet. *Lab Invest* **31:** 96–108, 1974.

Oksanen, A. The ultrastructure of the multinucleated cells in canine parathyroid glands. *J Comp Pathol* **90:** 293–301, 1980.

Orloff, J. J. *et al.* Interspecies comparison of renal cortical receptors for parathyroid hormone and parathyroid hormone-related protein. *J Bone Miner Res* **6:** 279–287, 1991.

Osborne, C. A., and Stevens, J. B. Pseudohyperparathyroidism in the dog. *J Am Vet Med Assoc* **162:** 125–135, 1973.

Raisz, L. G., and Martin, T. J. Prostaglandins in bone and mineral metabolism. *Bone Miner Res* **2:** 286–310, 1983.

Rasmussen, H. Parathyroid hormone, calcitonin, and the calciferol. *In* "Textbook of Endocrinology," R. H. Williams (ed.), 5th Ed. Philadelphia, Pennsylvania, W. B. Saunders, 1974.

Rasmussen, H., and Bordier, P. "The Physiological and Cellular Basis of Metabolic Bone Disease." Baltimore, Maryland, Williams & Wilkins, 1974.

Rijnberk, A. *et al.* Pseudohyperparathyroidism associated with perirectal adenocarcinomas in elderly female dogs. *Tijdschr Diergeneeskd* **103:** 1069–1075, 1978.

Rosol, T. J., and Capen, C. C. Pathogenesis of humoral hypercalcemia of malignancy. *Domest Anim Endocrinol* **5:** 1–21, 1988.

Rosol, T. J., and Capen, C. C. Inhibition of *in vitro* bone resorption by a parathyroid hormone receptor antagonist in the canine adenocarcinoma model of humoral hypercalcemia of malignancy. *Endocrinology* **122:** 2098–2102, 1988.

Rosol, T. J. *et al.* Identification of parathyroid hormone-related protein in canine apocrine adenocarcinoma of the anal sac. *Vet Pathol* **27:** 89–95, 1990.

Roth, S. I. The ultrastructure of primary water-clear cell hyperplasia of the parathyroid glands. *Am J Pathol* **61:** 233–248, 1978.

Roth, S. I., and Capen, C. C. Ultrastructural and functional correlations of the parathyroid gland. *Int Rev Exp Pathol* **13:** 161–221, 1974.

Roth, S. I. *et al.* The immunocytochemical localization of parathyroid hormone (PTH) in the bovine. *Fed Proc* **33:** 241, 1974.

Rowland, G. N., Capen, C. C., and Nagode, L. A. Experimental hyperparathyroidism in young cats. *Pathol Vet* **5:** 504–591, 1968.

Rowland, G. N. *et al.* Microradiographic evaluation of bone from cows with experimental hypervitaminosis D, diet-induced hypocalcemia, and naturally occurring parturient paresis. *Calcif Tissue Int* **9:** 179–193, 1972.

Sherding, R. G. *et al.* Primary hypoparathyroidism in the dog. *J Am Vet Med Assoc* **176:** 439–444, 1980.

Sherwood, L. M. *et al.* Production of parathyroid hormone by nonparathyroid tumors. *J Clin Endocrinol Metab* **27:** 140–146, 1967.

Stewart, A. F. *et al.* Biochemical evaluation of patients with cancer-associated hypercalcemia: Evidence for humoral and nonhumoral groups. *N Engl J Med* **303:** 1377–1383, 1980.

Stott, G. H. Dietary influence of the incidence of parturient paresis. *Fed Proc* **27:** 156–161, 1968.

Thompson, K. G. *et al.* Primary hyperparathyroidism in German shepherd dogs: A disorder of probable genetic origin. *Vet Pathol* **21:** 370–376, 1984.

Toverud, S. U. *et al.* Circulating parathyroid hormone. Concentration in normal and vitamin D-deprived rat pups determined with an amino-terminal-specific radioimmunoassay. *Bone Miner Metab* **1:** 145–155, 1986.

Tremblay, G., and Pearse, A. G. E. A cytochemical study of oxidative enzymes in the parathyroid oxyphil cell and their functional significance. *Br J Exp Pathol* **40:** 66–70, 1959.

Vaes, G. Cellular biology and biochemical mechanism of bone resorption. A review of recent developments on the formation, activation, and mode of action of osteoclasts. *Clin Orthop* **231:** 239–271, 1988.

Wallach, J. D., and Flieg, G. M. Nutritional secondary hyperparathyroidism in captive birds. *J Am Vet Med Assoc* **155:** 1046–1051, 1969.

Wallach, J. D., and Hoessle, C. Fibrous osteodystrophy in green iguanas. *J Am Vet Med Assoc* **153:** 863–865, 1968.

Walthall, J. C., and McKenzie, R. A. Osteodystrophia fibrosa in horses at pasture in Queensland: Field and laboratory observations. *Aust Vet J* **52:** 11–16, 1976.

Weir, E. C. *et al.* Humoral hypercalcemia of malignancy in canine lymphosarcoma. *Endocrinology* **122:** 602–608, 1988.

Wilke, R. *et al.* Regulatory hyperparathyroidism in a pig breed with vitamin D-dependency rickets. *Acta Endocrinol* **92:** 295–308, 1979.

Yarrington, J. T. *et al.* Effects of a low-calcium prepartal diet on calcium homeostatic mechanisms in the cow: Morphologic and biochemical studies. *J Nutr* **107:** 2244–2256, 1977.

Young, D. M., and Capen, C. C. Thyrocalcitonin content of thyroid glands from cows with vitamin D-induced hypercalcemia. *Endocrinology* **86:** 1463–1466, 1970.

Zarrin, K. Naturally occurring parafollicular cell carcinoma of the thyroid in dogs. A histological and ultrastructural study. *Vet Pathol* **14:** 556–566, 1977.

IV. Thyroid Gland

A. Development, Structure, and Function

1. Embryologic Development of the Thyroid Gland

The thyroid gland originates as a thickened plate of epithelium in the floor of the pharynx. It is intimately related to the aortic sac in its development, and this association leads to the frequent finding of accessory thyroid parenchyma in mediastinal structures, especially in dogs. This accessory thyroid tissue may undergo neoplastic transformation. Branched cell-cords develop from the pharyngeal plate and migrate dorsolaterally but remain attached to the pharyngeal area by the narrow thyroglossal duct.

A portion of the thyroglossal duct may persist postnatally and form a cyst. Thyroglossal duct cysts are present in the ventral aspect of the anterior cervical region in dogs. Their lining epithelium may undergo neoplastic transformation (see preceding discussion of cervical cysts in parathyroid–thyroid area, Section III,D,1 of this chapter).

Accessory thyroid tissue is common in the dog and may be located anywhere from the larynx to the diaphragm. About 50% of adult dogs have accessory thyroid tissue embedded in the fat on the intrapericardial aorta. These nodules are usually 1–2 mm in dimension and may be multiple. They are lacking in C (parafollicular) cells, which secrete calcitonin, but their follicular structure and function are the same as those of the main thyroid lobes. Attempts to induce hypothyroidism in the dog by a surgical thyroidectomy are not consistently successful because the accessory thyroids readily respond to the prompt increase in endogenous thyrotropin and can undergo hyperplasia sufficient to sustain adequate hormone production.

2. Structure of the Thyroid Gland

The two thyroid lobes in most species are located on the lateral surfaces of the trachea. In pigs, the main lobe of the thyroid is in the midline in the caudal cervical region with dorsolateral projections from each side.

The principal blood supply to each lobe in the dog is from the cranial thyroid artery (a branch of common carotid), and the major venous drainage is via the caudal thyroid vein that enters the internal jugular vein. Lymph drainage from the cranial pole of the thyroid lobes in dogs is to the retropharyngeal lymph nodes. Lymph flow from the caudal aspect of each thyroid lobe is variable, but it often bypasses any lymph nodes before entering the brachiocephalic trunk. Efferent lymphatics usually enter directly into the cervical lymphatic trunk or internal jugular vein. This explains the frequent occurrence of pulmonary metastases from thyroid carcinoma in dogs prior to development of secondary foci in regional lymph nodes. Small efferent lymphatics may pass through the caudal cervical lymph nodes along the ventral surface of the trachea before entering the cranial mediastinum. These vascular arrangements are significant for tumor spread.

The thyroid gland is the largest of the endocrine organs that function exclusively as an endocrine gland. The basic histologic structure of the thyroid is unique for endocrine glands, consisting of follicles of varying size (20–250 μm) that contain colloid produced by the follicular cells. The follicular cells are cuboidal to columnar, and their secretory polarity is directed toward the lumen of the follicles. An extensive network of capillaries provides the follicular cells with an abundant blood supply. Follicular cells have extensive profiles of rough endoplasmic reticulum and a large Golgi apparatus in their cytoplasm for synthesis and packaging substantial amounts of protein that are then transported into the follicular lumen. The interface between the luminal side of follicular cells and the colloid is modified by numerous microvillous projections.

3. Biosynthesis of Thyroid Hormones

The synthesis of thyroid hormones is unique among endocrine glands because the final assembly of hormone occurs extracellularly within the follicular lumen. Essential raw materials, such as iodide, are trapped by follicular cells from plasma, transported rapidly against a concentration gradient to the lumen, and oxidized by a peroxidase in the microvillous membranes to iodine.

The assembly of thyroid hormones within the follicular lumen is made possible by a unique protein (thyroglobulin) synthesized by follicular cells (Fig. 3.22). Thyroglobulin is a high-molecular-weight glycoprotein synthesized in successive subunits on the ribosomes of the endoplasmic reticulum in follicular cells. The constituent amino acids (tryosine and others) and carbohydrates (mannose, fructose, galactose) are derived from the circulation. Recently synthesized thyroglobulin leaving the Golgi apparatus is packaged into apical vesicles and extruded into the follicular lumen.

The amino acid tyrosine, an essential component of thyroid hormones, is incorporated within the molecular structure of thyroglobulin. Iodine is bound to tyrosyl residues in thyroglobulin at the apical surface of follicular cells to form successively monoiodotyrosine and diiodotyrosine. These combine to form the two biologically active iodothyronines (thyroxine, T_4; triiodothyronine, T_3) secreted by the thyroid gland.

4. Thyroid Hormone Secretion

The secretion of thyroid hormones from stores within luminal colloid is initiated by elongation of microvilli on follicular cells and formation of pseudopodia. These elongated cytoplasmic projections are increased by pituitary thyrotropin, extend into the follicular lumen, and indiscriminately phagocytize a portion of adjacent colloid. Colloid droplets within follicular cells fuse with numerous lysosomal bodies that contain proteolytic enzymes.

Triiodothyronine and thyroxine are released from the thyroglobulin molecule and secreted into adjacent capillaries. The iodinated tyrosines simultaneously released from the colloid droplets are deiodinated enzymatically, and the iodide generated is either recycled to the lumen to iodinate new tyrosyl residues or released into the circulation under normal conditions. Thyroxine is rapidly bound in plasma to albumin and three globulin fractions, and triiodothyronine is bound to albumin and one globulin fraction in dogs.

Negative-feedback control of thyroid hormone secretion is accomplished by the coordinated response of the adenohypophysis and certain hypothalamic nuclei to circulating levels of thyroid hormone. A decrease in thyroid hormone concentration in plasma is sensed by groups of neurosecretory neurons in the hypothalamus that synthesize and secrete a small peptide (three amino acids), thyrotropin-releasing hormone (TRH), into the hypophyseal portal circulation.

Thyrotropin (TTH) or thyroid-stimulating hormone (TSH) is conveyed to thyroid follicular cells where it binds to the basal aspect of the cell, activates adenylate cyclase, and increases the rate of biochemical reactions concerned with the biosynthesis and secretion of thyroid hormones. One of the initial responses by follicular cells to thyrotropin is the formation of numerous cytoplasmic pseudopodia, resulting in increased endocytosis of colloid and release of preformed hormone stored within the follicular lumen.

If the secretion of thyrotropin is sustained (hours or days), thyroid follicular cells become more columnar, and follicular lumina become smaller due to increased endocytosis of colloid. Numerous PAS-positive colloid droplets are present in the luminal aspect of the hypertrophied follicular cells.

The converse of what has been just described occurs in response to an increase in circulating thyroid hormone and a corresponding decrease in circulating pituitary thyrotropin. Thyroid follicles become enlarged and distended with colloid due to decreased thyrotropin-mediated endocytosis of colloid. Follicular cells lining the involuted follicles are cuboidal, and there are few endocytotic vacuoles at the interface between the colloid and follicular cells.

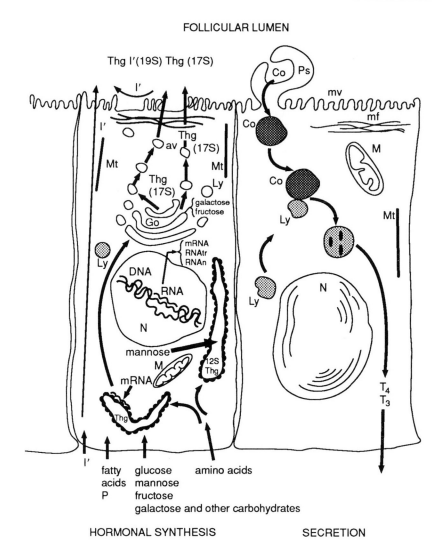

FOLLICULAR LUMEN

Fig. 3.22 Thyroid follicular cells illustrating two-way traffic of materials from capillaries to follicular lumen. Raw materials, such as iodine, are concentrated and rapidly transported into lumen (left side of drawing). Amino acids (e.g., tyrosine) and sugars are assembled by follicular cells into thyroglobulin (thg), packaged into apical vesicles (av) and released into lumen. Iodination of tyrosyl residues occurs within thyroglobulin molecule to form thyroid hormones in the follicular lumen. Elongation of microvilli and endocytosis of colloid by follicular cells occurs in response to TSH stimulation (right side of drawing). Intracellular colloid droplets (Co) fuse with lysosomal bodies (Ly), and active thyroid hormone is enzymatically cleaved from thyroglobulin, and free T_4 and T_3 are released into circulation. (From Bastenie *et al.*, 1975, with permission.)

There are differences in thyroid morphology and function between canine breeds of European origin and the basenji, which originated in Africa. At the same level of iodine intake, thyroidal turnover of iodine in the basenji is two to three times faster than that in European breeds. The corresponding differences in thyroid morphology in the basenji include smaller follicles, more widespread and uniform vacuolation of the colloid, taller follicular epithelium, and ultrastructural features of the follicular cell that more nearly resemble those of the thyrotropin-stimulated gland in European breeds, such as the beagle.

5. Action of Thyroid Hormones

Thyroxine and triiodothyronine, once released into the circulation, act on many different target cells in the body. The overall functions of the hormones are similar though much of the biologic activity may be the result of monodeiodination to 3,5,3'-triiodothyronine prior to interacting with target cells. Under certain conditions (protein starvation, neonatal animals, liver and kidney disease, febrile illness, etc.) thyroxine is preferentially monodeiodinated to 3,3',5'-triiodothyronine (reverse T_3). Since this form is biologically inactive, monodeiodination to form reverse T_3 provides a mechanism to attenuate the metabolic effects of thyroid hormones.

The subcellular mechanism of action of thyroid hormones resembles that for steroids, in that free hormone enters target cells and binds to a cytosol-binding protein. Free triiodothyronine initially binds to receptors on the inner mitochondrial membrane to activate mitochondrial energy metabolism, and subsequently binds to nuclear receptors and increases transcription of the genetic message (mRNA) to facilitate new protein synthesis.

The overall physiologic effects of thyroid hormones are to increase the basal metabolic rate; make more glucose available to meet the elevated metabolic demands by increasing glycolysis, gluconeogenesis, and glucose absorption from the intestine; stimulate new protein synthesis; increase lipid metabolism and conversion of cholesterol into bile acids and other substances; activate lipoprotein lipase, and increase sensitivity of adipose tissue to lipolysis by other hormones; stimulate the heart rate, cardiac output, and blood flow; and increase neural transmission, cerebration, and neuronal development in young animals.

B. Diseases of the Thyroid Gland

1. Developmental Disturbances of the Thyroid Gland

Aplasia and **hypoplasia** may be bilateral in species with paired thyroids, or unilateral, but is uncommon. Bilateral aplasia or hypoplasia of the thyroid in a young animal may result in dwarfism and other manifestations of early-onset hypothyroidism (i.e., subnormal central nervous system development, failure to grow at a normal rate).

Congenital hypothyroidism has been reported with primary thyroid hypoplasia in Scottish deerhound pups. Thyroid glands were smaller than normal and composed of small follicles with little or no colloid (Fig. 3.23A) compared to those of age-matched normal pups (Fig. 3.23B). The pituitary gland was normal macroscopically but contained few granulated acidophils with a high percentage of markedly vacuolated cells with few cytoplasmic granules, similar to thyroidectomy cells (Fig. 3.23C). Serum thyroxine levels were markedly decreased and did not increase in response to exogenous TSH. The affected puppies had retarded growth, fine hair coat, weakness, difficulty in walking, mental depression, and somnolence. Pups with thyroid hypoplasia were smaller than their littermates and had short limbs and a short broad head (Fig. 3.23D). There was an absence of ossification of epiphyses in the axial and appendicular skeleton (see Bones and Joints, Volume 1, Chapter 1).

Thyroglossal duct cysts develop most frequently in dogs and pigs and occasionally other animals by a persistence

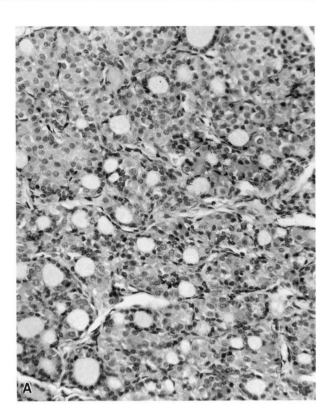

Fig. 3.23A Primary thyroid hypoplasia, Scottish deerhound with non-goitrous congenital hypothyroid gland. (A) Small thyroid follicles that contain little colloid.

of portions of the midline embryologic primordium of the thyroid that migrates caudally from the floor of the primitive pharynx to form the postnatal thyroid lobes. Thyroglossal duct cysts are present in the ventral aspect of the anterior cervical region in dogs as fluctuant masses that may rupture with a fistulous tract to the exterior. Their lining epithelium consists of multiple layers of follicular cells with occasional colloid-containing follicles that may undergo neoplastic transformation and give rise to papillary carcinomas (see Neoplasms of the Thyroid Gland, Section IV,C of this chapter).

Other cysts that develop in the thyroid region that must be differentiated from thyroglossal duct cysts include parathyroid cysts, branchial cysts, ultimobranchial duct cysts, follicular (colloid) cysts, and salivary mucoceles. These are discussed under Parathyroid and Related Cysts (Section III,D,1 of this chapter).

2. Degenerative Changes of the Thyroid Gland

The formation of densely basophilic spherules or granules in thyroid follicles occurs frequently in adult to aged dogs and occasionally in other species. These **corpora amylacealike bodies** appear to represent the precipitation of mineral on aggregated colloid (possibly abnormal chemical structure). **Colloid mineralization** develops in animals with normal blood calcium and phosphorus levels, and does not appear to be consistently associated with other

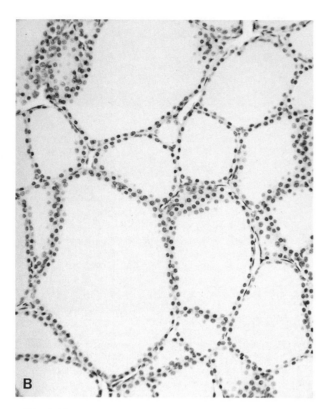

Fig. 3.23B Normal thyroid gland from 8-week-old Scottish deerhound pup with large colloid-filled follicles lined by low cuboidal follicular cells.

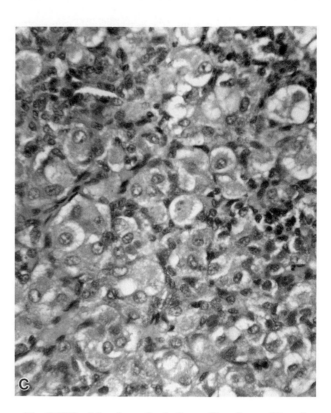

Fig. 3.23C Adenohypophysis from affected pup. Note few granulated acidophils and numerous vacuolated tropic hormone-production cells.

diseases. It is distinctly different from the interstitial mineralization, often associated with interfollicular capillaries, that is associated with chronic renal failure. The mineralization of colloid usually is not associated with any interference in thyroid function, even though it affects many follicles.

The accumulation of **lipofuscin** in follicular cells occurs in many animals and accounts for the progressive darkening (brown-red) of the thyroid with advancing age. The numerous PAS-positive lipofuscin granules accumulate particularly in the luminal aspect of follicular cells but do not appear significantly to interfere with thyroid function.

The deposition of **amyloid** in the interstitium between thyroid follicles has been observed in dogs, cats (especially certain breeds such as Abyssinian), cattle, and occasionally in other animals. Affected thyroids may be slightly enlarged, pale, and firmer than normal. The interstitial accumulation of amyloid may compress adjacent follicles but usually not to a degree that interferes with thyroid function. Amyloid in the thyroid of dogs and cattle almost always is part of a generalized amyloidosis associated with chronic suppurative process elsewhere in the body.

3. Hypofunction of the Thyroid Gland

Hypothyroidism of adult onset is a well-recognized clinical disease in dogs but is encountered only occasionally in other animals. (Congenital hypothyroidism is discussed

Fig. 3.23D The affected pup is smaller, has a shortened, domed head, and a finer haircoat. (A–D courtesy of W. F. Robinson and the *Australian Veterinary Journal*.)

under Hyperplasia of the Thyroid Gland, Section IV,B,7 of this chapter) Although disease may occur in many adult purebred and mixed-breed dogs, certain breeds (golden retrievers, Doberman pinchers, dachshunds, Shetland sheep dogs, Irish setters, miniature schnauzers, cocker spaniels, and Airedales) have been reported to be more commonly affected. Hypothyroidism in dogs usually is the result of primary lesions in the thyroid gland, particularly idiopathic follicular collapse and lymphocytic thyroiditis. Less common causes of hypothyroidism in dogs are bilateral nonfunctional thyroid tumors or severe iodine-deficient goiter. Hypothyroidism secondary to long-standing pituitary or hypothalamic lesions that prevent the release of either thyrotropin or thyrotropin-releasing hormone is less common in dogs. The thyroid gland in secondary hypothyroidism is moderately reduced in size and composed of colloid-distended follicles lined by flattened follicular cells.

a. IDIOPATHIC FOLLICULAR ATROPHY   In follicular atrophy there is a progressive loss of follicular epithelium and replacement by adipose tissue with a minimal inflammatory response. The gland usually is smaller and lighter in color than normal (see Fig. 3.25C). The early lesion of follicular atrophy that is seen in dogs with mild clinical signs of hypothyroidism appears to be confined to one part of the thyroid. The affected part of the thyroid gland contains small follicles lined by tall columnar follicular cells, often with little colloid (Fig. 3.24A). Immediately adjacent thyroid follicles are normal. Individual or small groups of degenerate follicular cells with eosinophilic cytoplasm and pyknotic nuclei are present in the follicular wall, colloid, and interstitium (Fig. 3.24B).

A more advanced form of follicular atrophy is seen in dogs with clinical hypothyroidism and low circulating concentrations of thyroid hormones. The thyroid gland is composed predominantly of adipose tissue with only occasional clusters of small follicles containing vacuolated colloid (Fig. 3.24C). When the thyroid is reduced markedly in size, it consists primarily of small follicles and individual follicular cells with PAS-positive colloid in a microfollicle in the cytoplasm. The mild increase of connective tissue in the interstitium appears to be the result of condensation of the normal stroma. Occasionally, thyroid glands with idiopathic follicular atrophy have either an encapsulated microadenoma or a nonencapsulated area of nodular hyperplasia of follicular cells.

The earliest lesion of idiopathic follicular atrophy observed by electron microscopy is degeneration of individual follicular cells lining thyroid follicles. The advanced stage of follicular atrophy is characterized ultrastructurally by a lack of normal thyroid follicles and the presence of microfollicles in the cytoplasm of individual follicular cells. The remaining hypertrophic follicular cells form small nests and are arranged closely along capillaries. The microfollicles appear to form by invagination of the apical surface of the follicular cells. Long microvilli extend into the colloid, and membrane-limited colloid droplets

are present in the cytoplasm near the microfollicle, suggesting endocytotic activity. Degenerate follicular cells with poorly defined plasma membranes, an irregularly shrunken nucleus, markedly dilated rough endoplasmic reticulum, and swollen mitochondria with disrupted cristae often are scattered in the interstitium. Macrophages with membrane-limited vacuoles containing myelin bodies, other membranous debris, and lysosomes are present near the degenerate follicular cells. The basement membrane may be focally thickened around remaining follicular cells. Small groups of thyroid C cells remain in thyroids with idiopathic follicular atrophy. They appear normal ultrastructurally and have many membrane-limited secretory granules in their cytoplasm.

Idiopathic follicular atrophy appears to be a primary degenerative disease of the thyroid and differs distinctly from the trophic atrophy of follicular cells secondary to diminished secretion of thyrotropin. Under the latter conditions, thyroid follicles are lined by a low cuboidal epithelium and distended by uniformly dense PAS-positive colloid with little evidence of endocytosis. Hypertrophy and hyperplasia of thyrotropic basophils occurs in the pars distalis of dogs with idiopathic follicular collapse and hyperthyroidism.

b. LYMPHOCYTIC THYROIDITIS   Lymphocytic thyroiditis in dogs, obese strain of chickens, nonhuman primates, and Buffalo rats closely resembles Hashimoto's disease in humans. Though the exact mechanism in the dog is not well established, evidence suggests a polygenic pattern of inheritance similar to that observed in the human disease. The immunologic basis for the development of chronic lymphocytic thyroiditis in both humans and dogs appears to be through production of autoantibodies directed against thyroglobulin, a microsomal antigen, and a second colloid antigen. Thyroglobulin autoantibodies have been found in 48% of dogs with hypothyroidism and may be related to the thyroiditis. Laboratory beagles with spontaneous lymphocytic thyroiditis also have circulating thyroid autoantibodies, but the focal thyroiditis usually is not associated with clinical hypothyroidism. Thyroiditis in laboratory beagles is similar serologically to human thyroiditis in that antibodies are present against thyroglobulin, a second colloid antigen, and a microsomal antigen (thyroperoxidase); however, there is not a positive correlation between the occurrence or height of the thyroglobulin antibody titers and the occurrence or severity of thyroiditis.

Thyroid glands of dogs with lymphocytic thyroiditis may be slightly enlarged and tan-white, normal, or reduced in size. Histologic alterations consist of multifocal to diffuse infiltrates of lymphocytes, plasma cells, and macrophages. Lymphoid nodules are present occasionally between follicles. The remaining follicles are small and lined by columnar epithelial cells. Lymphocytes, macrophages, and degenerate follicular cells often are present in the vacuolated colloid (Fig. 3.24D). Thyroid C (parafollicular)

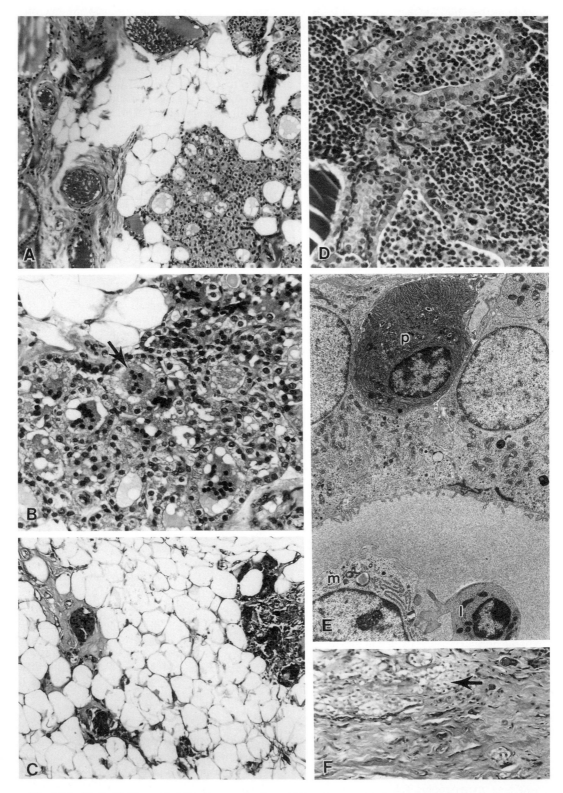

Fig. 3.24 (A, B, C) Idiopathic follicular atrophy. Dog. (A) Early stage. (B) Degenerate follicular cells in follicular wall and colloid (arrow). (C) Late stage. Thyroid consists mainly of adipose tissue, scattered follicles and nests of C cells. (D, E, F) Lymphocytic thyroiditis. Dog. (D) Infiltration of interstitium and follicular lumina. (E) A lymphocyte (l) and macrophage (m) are present in colloid of follicle, and a plasma cell (p) is infiltrating between follicular cells. (Courtesy of the *American Journal of Pathology.*)(F) End-stage thyroiditis with scarring, few follicles, and nests of C cells (arrow).

cells are seen in small nests or nodules between follicles and often are more prominent than those in normal dogs.

Lymphocytic thyroiditis is characterized ultrastructurally by lymphocytes, plasma cells, and macrophages in the interstitium, migrating through the follicular wall, and into the lumen, where they are mixed with colloid and degenerate follicular cells (Fig. 3.24E). Either the basement membrane around thyroid follicles is thick or discrete electron-dense deposits are present between the follicular cells and the basement membrane. These electron-dense deposits are similar to those reported to represent immune complexes in humans with Hashimoto's thyroiditis.

The thyroid lesion of lymphocytic thyroiditis progresses to replacement of the thyroid gland by mature fibrous connective tissue with only a few remaining scattered foci of inflammatory cells (Fig. 3.24F). Thyroid follicles are sparse and widely separated in the end stage of thyroiditis and contain only a small amount of vacuolated colloid, and there is a marked increase in collagen fibers surrounding small follicles or groups of follicular cells.

Many of the follicular lesions with lymphocytic thyroiditis are the result of lymphocytes and plasma cells migrating between thyroid follicular cells. This migration causes separation of adjacent follicular cells from the basement membrane, exfoliation of follicular cells, migration of lymphoid cells into the lumen, and eventual destruction of the follicle. To compensate for the progressive destruction of thyroid follicles by inflammatory cells and the decreased production of thyroid hormones, follicular cells in the remaining follicles often undergo hypertrophy, presumably in response to an increased secretion of thyrotropin.

Oxyphil cells (oncocytes) with an abundant eosinophilic cytoplasm often are seen in thyroids of dogs with long-standing lymphocytic thyroiditis. They contain abundant oxidative enzymes but sparse rough endoplasmic reticulum, suggesting an imbalance between the respiratory and synthesizing activities of the cell. Thyroid hormone synthesis appears to be severely compromised in these modified follicular cells. The occurrence of oxyphilic cells in thyroid glands may be related to aging or overstimulation.

c. Extrathyroidal Lesions and Functional Disturbances in Hypothyroidism   Many functional disturbances associated with hypothyroidism are due to a reduction in basal metabolic rate. A gain in body weight without an associated change in appetite occurs in some hypothyroid dogs. The weight gain varies from slight to striking, and the dog is usually less active.

Lesions in skin and haircoat occur in the majority of hypothyroid dogs. Thyroxine stimulates the anagen or active phase of hair growth, whereas a reduction in blood levels of thyroid hormones favors the telogen or resting phase. Telogen hairs are more easily dislodged from the hair follicles, resulting in thinning of the haircoat and often a bilaterally symmetrical alopecia. Areas affected initially by hair loss are those receiving frictional wear such as

the tail and cervical area. The tail may become almost completely bare in dogs with long-standing hypothyroidism.

Hyperkeratosis is a consistent finding in hypothyroidism and results in an increased scaliness of the skin. It may become severe and occur in circular scaling patches, so that the skin lesion is suggestive of seborrhea. Microscopic examination consistently reveals marked hyperkeratosis, which incudes the external root sheath. The excessive keratin formation and accumulation within hair follicles often results in follicular keratosis that may cause a grossly observable distension of follicles.

Hyperpigmentation occurs in many dogs with hypothyroidism, especially in localized areas of alopecia such as the dorsal aspect of the nose and distal portion of the tail. Increased numbers of melanocytes are present in the basal layer of the epidermis. Changes in the thickness of the epidermis are variable in dogs with hypothyroidism. Epidermal atrophy is reported in ~50% of hypothyroid dogs, normal thickness in ~20%, and a mild to moderate epidermal thickening due to a prominent stratum granulosum with an atrophic stratum spinosum in the remaining 30% of dogs.

Myxedema may develop and produce a characteristic clinical appearance in long-standing or severe hypothyroidism. There is accumulation of mucin (neutral and acid glycosaminoglycans combined with protein) in the dermis and subcutis. This material binds considerable amounts of water and produces marked thickening of the skin. This is most obvious around the face and head where accentuation of the normal skin folds causes a "tragic" appearance (Fig. 3.25A). The eyelids appear thick and drooping, thus contributing to the sad facial expression. The skin feels thick and doughy, but the characteristic pitting observed with other types of edema does not occur with myxedema. Histologically, mucin appears as a blue-purple, granular

Fig. 3.25A Hypothyroidism. Dog. Myxedema with thickened skin folds.

or fibrillar material that disrupts the normal collagen and elastin fibers on skin sections. On skin sections stained with hematoxylin and eosin, mucin is not well visualized, but a disruption of the normal pattern of collagen and elastin fibers can be seen in the dermis.

Abnormalities in **reproduction** are common. Lack of libido and reduction in sperm count may occur in males, whereas abnormal or absent estrous cycles with reduced conception rates may result in females. The spermatogenic epithelium in the testis often is markedly atrophic in long-standing cases of hypothyroidism. Impaired joint function with evidence of pain and joint effusion also can result from severe or prolonged hypothyroidism.

Hypothyroidism in dogs is accompanied by low protein-bound iodine (PBI) levels and decreased ^{131}I uptake by the thyroid gland. The normal blood level of thyroxine in the dog is 1.5–3.6 μg/dl and for triiodothyronine, 48–154 ng/dl. In dogs with hypothyroidism, the thyroxine level determined by radioimmunoassay is usually <1.0 μg/dl and triiodothyronine, <50 ng/dl. The occurrence of mild, non-responsive, normocytic, normochromic anemia is consistent with a diagnosis of hypothyroidism.

The serum cholesterol is elevated significantly. The marked **hypercholesterolemia** with long-standing and severe hypothyroidism results in a variety of secondary lesions including atherosclerosis, hepatomegaly, and glomerular and corneal lipidosis. The decreased rate of lipid metabolism with diminished intestinal excretion of cholesterol and the conversion of lipids into bile acids and other compounds in hypothyroidism result in the hypercholesterolemia. Atherosclerosis of coronary, cerebral, and other vessels may develop in dogs with severe hypothyroidism and long-standing hyperlipidemia (Fig. 3.25B,C). This occasionally results in hemorrhage and ischemic necrosis of the myocardium due to impingement of the vessel lumina by numerous lipid-laden macrophages in the tunica media and adventitia. In dogs with markedly elevated

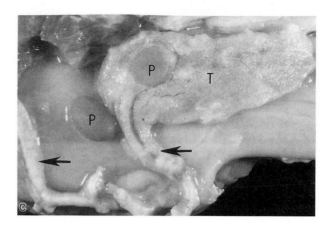

Fig. 3.25C Idiopathic follicular atrophy. Thyroid is reduced in size compared to adjacent parathyroids, and lighter in color. Thyroid arteries (arrows) are prominent due to atherosclerosis.

plasma lipids, renal glomeruli may become plugged with lipid, resulting in progressive renal failure. The lipid-filled glomeruli can be seen macroscopically as yellow-white foci throughout the kidney cortex (Fig. 3.25D). The accumulation of excessive lipid in the liver often results in varying degrees of hepatomegaly with abdominal distension and hepatic failure. Corneal lipidosis is observed occasionally in hypothyroid dogs with hypercholesterolemia. This lesion often is unilateral because the lipid is deposited in corneas that have been previously injured and have a network of "ghost" vessels from which the lipid diffuses into the connective tissue stroma.

4. Evaluation of Thyroid Function

Serum cholesterol concentration is an indirect and variable index of the peripheral action of thyroid hormone.

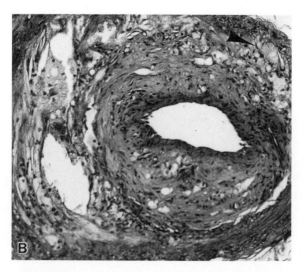

Fig. 3.25B Coronary atherosclerosis with lipid-filled macrophages (arrowhead).

Fig. 3.25D Glomerular lipidosis.

Two-thirds of dogs with spontaneous hypothyroidism have fasting serum cholesterol concentration greater than 300 mg/100 ml. Hypercholesterolemia is as dependent, however, on the composition of the dog's diet as on the severity and duration of hypothyroidism. Cholesterol values tend to be higher in the general dog population that is fed table scraps or home diets than in dogs maintained exclusively on commercially manufactured dry dog foods. Hypercholesterolemia also occurs in some dogs with cortisol excess, a disease which must be considered in the differential diagnosis of hypothyroidism. The measurement of serum cholesterol thus is not a specific and dependable test of thyroid function, but fasting cholesterol values >600 mg/100 ml, which are often observed in hypothyroidism, seldom occur in any other disease in the dog.

The most sensitive and accurate method for evaluation of thyroid function is measurement of blood thyroxine (T_4) and triiodothyronine (T_3) levels by radioimmunoassay. The normal blood level of thyroxine in the dog is 1.5–3.6 μg/dl (mean, 2.48 μg/dl), and for triiodothyronine, 48–154 ng/dl (mean, 95 ng/dl). In dogs with hypothyroidism, the thyroxine level usually is <1.0 μg/dl, and triiodothyronine is <50 ng/dl. When levels are borderline, clearer separation of dogs with hypothyroidism from euthyroid dogs can be made by injecting TSH. In euthyroid dogs the thyroxine level will at least double 8 hr after I.V. or I.M. administration of TSH. In dogs with hypothyroidism, thyroxine and triiodothyronine levels do not change significantly after injection of TSH. Plasma T_4 levels are no more than 0.2 μg/dl greater than control values at 8 hr post-TSH, and plasma T_3 is increased by no more than 10 ng/dl after TSH. The increase in serum T_3 after TSH is more variable in normal dogs than that for T_4, but in dogs with primary hypothyroidism, there is little (10 ng/ml or less) change in serum T_3 at 8 hr after TSH administration.

Histologic examination of a biopsy of the thyroid is a useful and reliable aid in the diagnosis of thyroid disease when either the results of serum assays for T_4 and T_3 are equivocal or a nodule is palpated in the thyroid area. The removal of the caudal one fourth of either lobe of the thyroid for histologic examination is a simple surgical procedure without significant risk, even when the dog is hypothyroid. In most cases of hypothyroidism, there is either marked loss of thyroid follicles with replacement by adipose connective tissue or multifocal lymphocytic thyroiditis, which develops on an immunologic basis.

5. Hyperplasia of the Thyroid Gland

Non-neoplastic and noninflammatory enlargements of the thyroid develop in all domestic mammals, birds, and submammalian vertebrates. The major pathogenetic mechanisms responsible for the development of thyroid hyperplasia include iodine-deficient diets, goitrogenic compounds that interfere with thyroxinogenesis, dietary iodide excess, and genetic enzyme defects in the biosynthesis of thyroid hormones. All of these factors result in inadequate thyroxine synthesis and decreased blood levels of thyroxine and triiodothyronine. This is detected by the hypothalamus and pituitary to increase the secretion of thyrotropin, which results in hypertrophy and hyperplasia of follicular cells.

a. DIFFUSE HYPERPLASTIC AND COLLOID GOITER   Iodine deficiency that resulted in diffuse thyroid hyperplasia was common in many goitrogenic areas throughout the world before the widespread addition of iodized salt to animal diets. Although iodine-deficient goiter still occurs in large areas of the world in domestic animals, the outbreaks are sporadic, and fewer animals are affected. Marginal iodine-deficient diets containing certain goitrogenic compounds result in severe thyroid hyperplasia and clinical evidence of goiter. These substances include thiouracil, sulfonamides, anions of the Hofmeister series, and a number of plants from the family Brassicaceae (see Thyrotoxic Drugs and Chemicals, Section IV,B,7 of this chapter). Young animals born to females on iodine-deficient diets are more likely to develop severe thyroid hyperplasia and have clinical signs of hypothyroidism (Fig. 3.26A).

Although seemingly paradoxical, an excess of iodide in the diet also can result in thyroid hyperplasia in animals and humans. Foals of mares fed dry seaweed containing excessive iodide may develop thyroid hyperplasia and clinically evident goiter. The thyroid glands of the young are exposed to higher blood iodide levels than those of the dam because of concentration of iodide, first by the placenta and then by the mammary gland. High blood iodide interferes with one or more steps of thyroxinogenesis, leading to lowered blood thyroxine levels, and a compensatory increase in pituitary thyrotropin secretion. Excess iodine appears to inhibit the uptake of iodine by the thyroid, blocks the peroxidation of iodide to iodine,

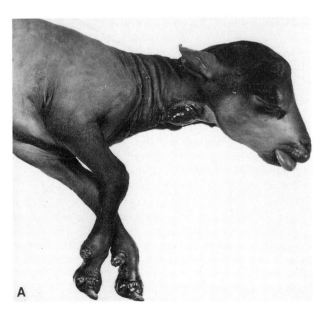

Fig. 3.26A Congenital goiter in iodine deficiency. Lamb. Note large thyroids, alopecia, and swollen tongue.

interferes with the conversion of monoiodotyrosine to di-iodotyrosine, and blocks the release of hormone from the follicle by interfering with proteolysis of colloid.

Iodine deficiency may be conditioned by other antithyroid compounds present in animal feeds and be responsible, in particular situations, for a high incidence of goiter. Prolonged low-level exposure to thiocyanates, which are produced by ruminal degradation of cyanogenetic glucosides from plants as white clover (*Trifolium repens*), couch grasses (*Cynodon* spp.) and linseed meal, and by degradation of glucosinolates of *Brassica* crops, can cause congenital goiter in ruminants. Goitrin (5-vinyl-oxazolidine-2-thione) derived from the glucosinolates of *Brassica* spp. inhibits organification of iodine. *Leucaena leucocephala* and other legumes of the genus are native or cultivated in many subtropical areas and contain the toxic aminoacid, mimosine. The effects of mimosine are discussed with Skin and Appendages (Volume 1, Chapter 5). It is not directly goitrogenic but is converted in the rumen to 3-hydroxy-4(1*H*)-pyridone, which prevents organic binding of iodine by the thyroid.

Goiter in adult animals usually is of little clinical significance and, except for occasional local pressure influences, the general health is not impaired. It does, however, continue to be of significance as a disease of the newborn, although the previous drastic losses in endemic areas are now controlled by the prophylactic use of iodized salt.

Congenital hypothyroidism in domestic animals is almost exclusively associated with hyperplastic goiter, even though the dam may have shown no evidence of thyroid dysfunction. Gestation is significantly prolonged, the larger goiters may cause dystocia, and there is a tendency to retain the fetal placenta. Affected foals show extreme weakness and die within a few days after birth. The thyroids may be only slightly enlarged. Calves seem to be somewhat more resistant to the effects of hypothyroidism and although up to 70–80% may have quite large goiters in endemic areas, the majority survive and thrive. A few are partially or completely hairless, but these are born dead or die soon after birth. Newborn, goitrous pigs, goats, and lambs (Fig. 3.26A) frequently show myxedema and alopecia, and the mortality rate is high, the majority being born dead or dying within a few hours of birth. Enlarged thyroid glands are readily palpable or visible in kids and lambs, but are not apparent in piglets because of the combination of short neck and myxedema. The myxedema of skin, subcutis, and connective tissue in pigs affects especially the enlarged foreparts of the body. The tongue is swollen, and there is edematous swelling of the fauces and larynx, probable contributory factors in death. Asphyxiation may result also from pressure by the enlarged thyroid gland. Young, goitrous animals which are treated and survive usually do not show permanent ill effects.

Congenital hyperplastic goiter with hypothyroidism is not a prominent feature of thyroid disease in domestic carnivores; however, degrees of hyperplasia will be found to parallel the changes in the maternal gland. In puppies,

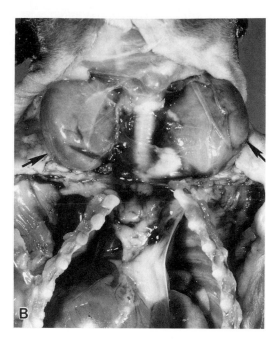

Fig. 3.26B Symmetrical hyperplastic goiter (arrows). Pup.

the thyroid enlargement may be sufficient to cause dystocia or asphyxiation (Fig. 3.26B). In the most severe form, death occurs shortly after birth, and those affected have poorly developed, coarse hair, anemia, and defective bone mineralization. Less severe forms are more common, and affected pups often recover spontaneously. They are characterized by dry, coarse, and sparse hair coat, narrow palpebral fissures, broad skulls, and legs which are relatively thicker and heavier than normal.

Both lobes of the thyroid are uniformly enlarged in young animals with diffuse hyperplastic goiter (Fig. 3.27A). The enlargements may be extensive and result in palpable swellings in the cranial cervical area. The affected lobes are firm and dark red because an extensive interfollicular capillary network develops under the influence of long-term TTH stimulation.

Colloid goiter represents the involutionary phase of diffuse hyperplastic goiter in young adult and adult animals. The markedly hyperplastic follicular cells continue to produce colloid, but endocytosis of colloid is decreased due to diminished pituitary TTH levels in response to the return of blood thyroxine and triiodothyronine to normal. Both thyroid lobes are diffusely enlarged but are more translucent and lighter in color than with hyperplastic goiter (Fig. 3.27B). The differences in macroscopic appearance are the result of reduced vascularity in colloid goiter and development of macrofollicles distended with colloid (Fig. 3.27C).

The changes in diffuse hyperplastic and colloid goiters are consistent throughout the diffusely enlarged thyroid lobes. The follicles are irregular in size and shape in hyperplastic goiter because of varying amounts of lightly eosinophilic and vacuolated colloid in the lumen. Some follicles

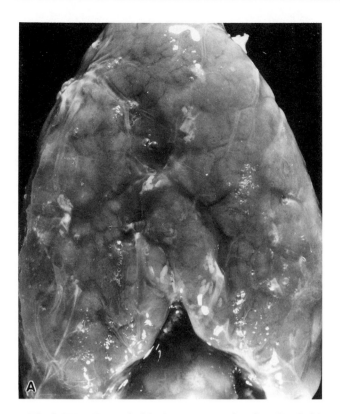

Fig. 3.27A Congenital dyshormonogenetic goiter. Corriedale lamb.

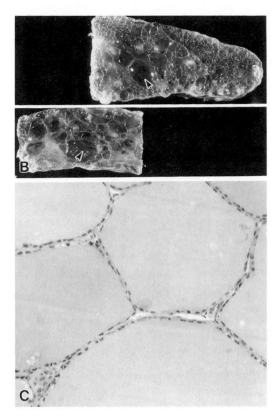

Fig. 3.27B,C Colloid goiter with large follicles (arrows) throughout each lobe. (C) Colloid goiter. Follicles lined by flattened inactive cells.

are collapsed due to lack of colloid (Fig. 3.28). The lining epithelial cells are columnar with a deeply eosinophilic cytoplasm and small hyperchromatic nuclei that are often situated in the basilar part of the cell. The follicles are lined by single or multiple layers of hyperplastic follicular cells that in some follicles may form papillary projections into the lumen. Similar proliferative changes are present in ectopic thyroid parenchyma in the neck or mediastinum.

Colloid goiter may develop either after sufficient amounts of iodide have been added to the diet of animals with iodine-deficient hyperplastic goiter or after the requirements for thyroxine have diminished in an older animal. Blood thyroxine levels return toward normal, and the secretion of TTH by the pituitary gland is correspondingly decreased. Follicles are progressively distended with densely eosinophilic colloid due to diminished TTH-induced endocytosis. The follicular cells lining the macrofollicles are flattened and atrophic. The interface between the colloid and luminal surface of follicular cells is smooth and lacks the characteristic endocytotic vacuoles of actively secreting follicular cells (Fig. 3.27C). Some involuted follicles in colloid goiter have remnants of the papillary projections of follicular cells extending into their lumen. Interfollicular capillaries are less well developed than those with diffuse hyperplastic goiter.

b. NODULAR THYROID HYPERPLASIA Nodular hyperplasia (goiter) in thyroid glands of old horses, cats, and dogs

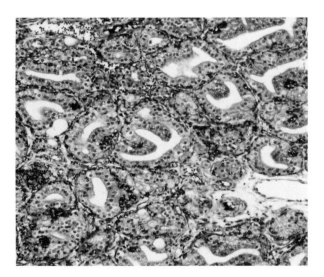

Fig. 3.28 Hypertrophy and hyperplasia of follicular epithelium in congenital goiter.

appears as multiple white to tan nodules of varying size (Fig. 3.29A). The affected lobes are moderately enlarged and irregular in contour. Nodular goiter in most animals is endocrinologically inactive and encountered as an incidental lesion at autopsy; however, functional thyroid ade-

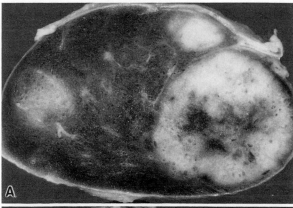

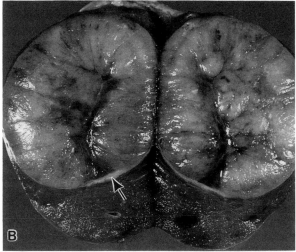

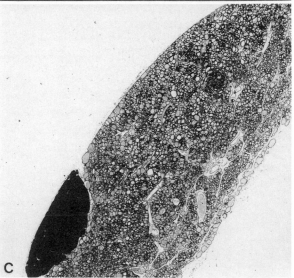

Fig. 3.29 (A) Nodular goiter with multiple hyperplastic nodules. Horse. (B) Solitary thyroid adenoma with prominent fibrous capsule (arrow). Horse. (C) Nodular goiter. Dog. (The dark body, lower left, is parathyroid.)

nomas often develop in glands with multinodular hyperplasia in old cats. In contrast to thyroid adenomas, nodules of hyperplasia are not encapsulated and result in minimal compression of adjacent parenchyma.

Nodular goiter consists of multiple foci of hyperplastic follicular cells that are sharply demarcated but not encapsulated from the adjacent thyroid parenchyma. The microscopic appearance within a nodule is variable. Some hyperplastic cells form small follicles with little or no colloid. Other nodules are formed by larger irregularly shaped follicles lined by one or more layers of columnar cells that form papillary projections into the lumen. Some of the follicles are involuted and filled with densely eosinophilic colloid. These changes appear to be the result of alternating periods of hyperplasia and colloid involution in the thyroid glands of old animals. The areas of nodular hyperplasia may be microscopic (Fig. 3.29C; as in old cats) or grossly visible, causing enlargement of the thyroid (as in old horses; Fig. 3.29A).

There are no absolute morphologic criteria for distinguishing hyperplastic nodules and adenomas derived from follicular cells. As a general rule, hyperplastic nodules are multiple, poorly or not at all encapsulated, variable in their histologic structure, and do not cause compression of adjacent thyroid parenchyma. Adenomas, on the other hand, tend to be solitary, well encapsulated, fairly uniform in histologic structure, and cause compression of the surrounding parenchyma owing to progressive expansile growth (Fig. 3.29B). Endocrinologically active thyroid adenomas result in colloid involution of follicles in the rim of surrounding thyroid due to inhibition of TTH secretion.

c. INHERITED DYSHORMONOGENETIC GOITER   An inability to synthesize and secrete adequate amounts of thyroid hormones beginning before or at birth has been documented in human infants and in several animal species. Congenital dyshormonogenetic goiter is inherited as an autosomal recessive trait in Corriedale, Dorset Horn, Merino, and Romney Marsh sheep, Afrikander cattle, and Saanen dwarf goats. The subnormal growth rate, absence of normal wool development or a rough sparse hair coat, myxedematous swellings of the subcutis, weakness, and sluggish behavior suggest that the affected young are clinically hypothyroid. Most lambs with congenital goiter either die shortly after birth or are highly sensitive to the effects of adverse environmental conditions.

Thyroid glands are symmetrically enlarged at birth because of an intense diffuse hyperplasia of follicular cells (Fig. 3.27A,B). Thyroid follicles are lined by tall columnar cells but often are collapsed because of a lack of colloid resulting from the marked endocytotic activity. The tall columnar follicular cells lining thyroid follicles have extensively dilated profiles of rough endoplasmic reticulum and large mitochondria, but there are relatively few dense granules associated with the Golgi apparatus and few apical vesicles near the luminal plasma membrane. Numerous long microvilli extend into the follicular lumen.

Although thyroidal uptake and turnover of ^{131}I are

greatly increased compared with those of euthyroid controls, circulating thyroxine and triiodothyronine levels are consistently low. The absence of a defect in the iodide-transport mechanism and organification or dehalogenation, together with an absence of normal 19-S thyroglobulin in goitrous thyroids, suggest an impairment in thyroglobulin biosynthesis in animals with inherited congenital goiter. A closely related or similar defect appears operational in the examples of congenital goiter in sheep, cattle, and goats. The protein-bound iodine levels in sheep, cattle, and goats with inherited congenital goiter are markedly elevated. This appears to be the result of iodination of albumin and other plasma proteins by the thyroid gland under long-term thyrotropin stimulation, since hormonal iodide levels are significantly lower than those of in controls. The hypothyroid goats with congenital goiter can be returned to a state of euthyroidism by the addition of iodide (1.0 mg/day) to the diet. Although the goats remain unable to synthesize thyroglobulin, supplementation with excess iodide results in sufficient formation of triiodothyronine and thyroxine in the abnormal iodoproteins to make the animals euthyroid.

The presence of mRNA coding for thyroglobulin has been investigated to elucidate further the molecular basis for the impairment of thyroid hormone biosynthesis in congenital goiter. Although thyroglobulin mRNA sequences are present in the goitrous tissue, their concentration is reduced to one tenth to one fortieth of normal, and the intracellular distribution is abnormal (nuclear, 42% of normal; cytoplasmic, 7%; membrane fraction, 1–2%). The lack of thyroglobulin in these examples of congenital goiter in animals appears to be due to a defect in thyroglobulin mRNA, leading to aberrant processing of primary transcripts or transport of the thyroglobulin mRNA from the nucleus to the endoplasmic reticulum.

6. Thyrotoxic Effects of Drugs and Chemicals

Many of the significant physiologic and pathologic data in the literature on thyroid function and structure have come from studies in animals. Although the basic hypothalamic–pituitary–thyroid axis functions in a similar manner in animals and humans, there are differences between species that are important when extrapolating animal data from toxicity and carcinogenicity studies for human risk assessment.

Long-term perturbations of the pituitary–thyroid axis by various xenobiotics or physiologic alterations (e.g., iodine deficiency, partial thyroidectomy) are more likely to predispose the laboratory animal to a higher incidence of proliferative lesions (e.g., hyperplasia and adenomas of follicular cells) than is the case in the human thyroid. This appears to be particularly true in the male rat, in which there usually are higher circulating levels of TSH than in females. The greater sensitivity of the animal thyroid to derangement by drugs, chemicals, and physiologic perturbations also is related to the shorter plasma half-life of thyroxine (T_4) (12–24 hr) than that in humans (5–9 days),

due, in part, to the considerable differences between species in the transport proteins for T_4.

There also are marked species differences in the sensitivity of the functionally important peroxidase enzyme to inhibition by xenobiotics. Thioamides (e.g., sulfonamides) and other chemicals can selectively inhibit the thyroperoxidase and significantly interfere with the iodination of tyrosyl residues incorporated in the thyroglobulin molecule, thereby disrupting the orderly synthesis of T_4 and T_3. Long-term administration of sulfonamides results in the development of thyroid nodules frequently in the sensitive species (such as the rat, dog, and mice) but not in species resistant (e.g., monkey, guinea pig, chicken, and humans) to the inhibition of peroxidase in follicular cells.

Many drugs and chemicals are goitrogenic in that they disrupt one or more steps in the biosynthesis and secretion of thyroid hormones, resulting in subnormal levels of T_4 and T_3, associated with a compensatory increased secretion of pituitary TSH. When tested in highly sensitive animal species, these compounds often result in follicular cell hyperplasia and, in long-term studies, an increased incidence of thyroid tumors by a secondary mechanism. In the commonly observed secondary mechanism of thyroid oncogenesis in rodents, the initiating factor (e.g., specific xenobiotic or physiologic perturbation) evokes another stimulus (e.g., chronic hypersecretion of TSH) that promotes the development of nodular proliferative lesions (initially hypertrophy, followed by hyperplasia, subsequently adenomas, infrequently carcinomas) derived from follicular cells. Thresholds for a no-effect on the thyroid gland can be established by determining the dose of xenobiotic that fails to elicit an elevation in the circulating level of TSH.

The plasma thyroxine half-life in rats is considerably shorter (12–24 hr) than that in humans (5–9 days). This is related in part to considerable differences between species in the transport proteins for T_4 and T_3. In humans and the monkey, circulating T_4 is bound primarily to thyroxine-binding globulin (TBG), but this high-affinity binding protein is not present in rodents, birds, amphibians, or fish. Triiodothyronine (T_3) is transported bound to TBG and albumin in human beings, monkey, and dog, but only to albumin in mouse, rat, and chicken. In general T_3 is bound less avidly to transport proteins than is T_4, resulting in a faster turnover and shorter plasma half-life in most species.

In the evaluation of potential thyroid toxicity of various xenobiotics in animals, an accurate quantitation of circulating levels of TSH is essential in order to determine whether proliferative lesions of follicular cells are mediated by a chronic hypersecretion of TSH. The immunoassay for TSH is highly species specific with considerable inter-animal and interassay variations. Xenobiotics that disrupt either thyroid hormone synthesis, secretion, or peripheral metabolism often result in prompt increases in circulating TSH levels.

Thyroid-stimulating hormone levels are higher in male than female rats, and castration decreases both the baseline serum TSH and response to thyrotropin-releasing hormone (TRH) injection. Follicular cell height is often greater in male than female rats in response to the greater circulating TSH levels. The administration of exogenous testosterone to castrated male rats restores the TSH level to that of intact rats.

Chemical disruption of thyroid hormone synthesis and secretion in animals may occur at a number of different steps in thyroxinogenesis. These include blockage of iodine uptake, organification defects, blockage of hormone release, drug-induced thyroid pigmentation, inhibition of 5'-deiodinase, and induction of hepatic microsomal enzymes.

a. BLOCKAGE OF IODINE UPTAKE.   The initial step in the biosynthesis of thyroid hormones is the uptake of iodide from the circulation and transport against a gradient across follicular cells to the lumen of the follicle. A number of anions act as competitive inhibitors of iodide transport in the thyroid including perchlorate (ClO_4^-), thiocyanate (SCN^-), and pertechnetate. Thiocyanate is a potent inhibitor of iodide transport and is a competitive substrate for the thyroid peroxidase, but it does not appear to be concentrated in the thyroid. Blockage of the iodide-trapping mechanism has a similar disruptive effect on the thyroid–pituitary axis as does iodine deficiency. The blood levels of T_4 and T_3 decrease, resulting in a compensatory increase in the secretion of TSH by the pituitary gland. The hypertrophy and hyperplasia of follicular cells following sustained exposure results in an increased thyroid weight and the development of goiter.

b. ORGANIFICATION DEFECT   A wide variety of chemicals, drugs, and other xenobiotics affect the second step in thyroid hormone biosynthesis. The step-wise binding of iodide to the tyrosyl residues in thyroglobulin requires oxidation of inorganic iodide (I^-) to molecular (reactive) iodine (I_2) by the thyroid peroxidase present in the luminal aspect (microvillar membranes and apical cytoplasm) of follicular cells and adjacent colloid. Classes of chemicals that inhibit the organification of thyroglobulin include: (1) thionamides (such as thiourea, thiouracil, propylthiouracil, methimazole, carbimazole, and goitrin); (2) aniline derivatives and related compounds (e.g., sulfonamides, *para*-aminobenzoic acid, *para*-aminosalicylic acid, and amphenone); (3) substituted phenols (such as resorcinol, phloroglucinol, and 2,4-dihydroxybenzoic acid), and miscellaneous inhibitors [e.g., aminotriazole, tricyanoaminopropene, antipyrine and its iodinated derivative (iodopyrine)].

These chemicals exert their action by inhibiting the thyroid peroxidase, which results in a disruption of the iodination of tyrosyl residues in thyroglobulin but also of the coupling reaction of iodotyrosines [e.g., monoiodotyrosine (MIT) and diiodotyrosine (DIT)] to form iodothyronines (T_3 and T_4). Propylthiouracil (PTU) has been shown in rats to affect each step in thyroid hormone synthesis beyond iodide transport.

The goitrogenic effects of sulfonamides have been known for ~50 years since the reports of the action of sulfaguanidine on the thyroid. Sulfamethoxazole and trimethoprim exert a potent goitrogenic effect in rats, resulting in marked decreases in circulating T_3 and T_4, a substantial compensatory increase in TSH, and increased thyroid weights due to follicular cell hyperplasia. Dogs also are sensitive to the effects of sulfonamides, resulting in markedly decreased serum T_4 and T_3 levels, hyperplasia of thyrotropic basophils, and increased thyroid weights. In comparison, the thyroids of monkeys and humans are resistant to the development of changes that sulfonamides produced in rats, mice, and dogs. The sensitive species are much more likely to develop follicular cell hyperplasia and thyroid nodules after long-term exposure to sulfonamides than are the resistant species.

c. BLOCKAGE OF THYROID HORMONE RELEASE   Relatively few chemicals selectively inhibit the secretion of thyroid hormone from the thyroid gland. An excess of iodine inhibits secretion of thyroid hormone and occasionally can result in goiter and hypothyroidism in animals and human patients. Several mechanisms have been suggested for this effect of high iodide levels on the thyroid hormone secretion including a decrease in lysosomal protease activity (human glands), inhibition of colloid droplet formation (mice and rats), and inhibition of TSH-mediated increase in cAMP (dog thyroid slices). Rats fed an iodide-excess diet have hypertrophy of the cytoplasmic area of follicular cells with an accumulation of numerous colloid droplets and lysosomal bodies. There is limited evidence ultrastructurally of fusion of the membranes of these organelles and degradation of the colloid necessary for the release of T_4 and T_3 from the thyroglobulin.

Lithium also has a striking inhibitory effect on thyroid hormone release. The widespread use of lithium carbonate in the treatment of manic states occasionally results in the development of goiter with either euthyroidism or occasionally hypothyroidism in human patients.

d. DRUG-INDUCED THYROID PIGMENTATION   The antibiotic minocycline produces a striking black discoloration of the thyroid lobes in laboratory animals and humans, with the formation of brown pigment granules within follicular cells. The pigment granules stain similarly to melanin and are best visualized on thyroid sections stained with the Fontana–Masson procedure. Electron-dense material first accumulates in lysosomelike granules and in the rough endoplasmic reticulum. The pigment appears to be a metabolic derivative of minocycline, and administration of the antibiotic at high dose to rats for extended periods may result in a disruption of thyroid function and the development of goiter. The release of T_4 from perfused thyroids of minocycline-treated rats is significantly decreased, but the follicular cells retain the ability to phagocytose colloid in response to TSH and have numerous colloid droplets in their cytoplasm.

Other chemicals (or their metabolites) selectively localize in the thyroid colloid, resulting in abnormal clumping and increased basophilia to the colloid. Brown to black pigment granules may be present in follicular cells, colloid, and macrophages in the interthyroidal tissues, resulting in a macroscopic darkening of both thyroid lobes. The physicochemically altered colloid in the lumina of thyroid follicles appears to be less capable than normal colloid either of reacting with organic iodine in a step-wise manner to result in the orderly synthesis of iodothyronines or of being phagocytized by follicular cells and enzymatically processed to release active thyroid hormones into the circulation. Serum T_4 and T_3 are decreased, serum TSH levels are increased by an expanded population of pituitary thyrotrophs, and thyroid follicular cells undergo hypertrophy and hyperplasia and eventually develop tumors.

e. INHIBITION OF 5'-DEIODINASE Erythrosine (FD & C Red No. 3) is one the best-characterized chemicals that acts as a 5'-deiodinase inhibitor and results in perturbations of thyroid function. It is a tetraiodinated derivative of fluorescein with iodine accounting for ~58% of the molecular weight of the color, and is a red dye widely used as a color additive in foods, cosmetics, and pharmaceuticals. Amiodarone is another organic iodinated antiarrhythmic compound that disrupts thyroid hormone economy by inhibiting 5'-deiodinase. Iopanoic acid and flavonoids also inhibit the enzyme in hepatocytes.

Erythrosine is an example of a xenobiotic that causes changes in circulating levels of thyroid hormones and morphologic evidence of follicular cell stimulation by producing alterations in the peripheral metabolism of T_4. Inhibition of 5'-deiodinase in the liver and kidney by Red No. 3 explains the lower circulating T_3 levels. The monodeiodination of T_4 by another enzyme (5-deiodinase) to reverse T_3 and inhibition of the 5'-deiodinase [which is necessary to further degrade this inactive iodothyronine to 3,3'-diiodothyronine (T_2)] results in the striking accumulation of serum reverse T_3 (Fig. 3.30). The pituitary, sensing the lowered circulating levels of T_3, compensates by increasing the secretion of TSH, which results in the morphologic evidence of follicular cell stimulation.

Erythrosine does not appear to be a direct-acting thyroid oncogen. Rather, in massive doses (4% of diet over lifetime, beginning *in utero*), it acts through a secondary mechanism to promote the development of benign thyroid tumors in laboratory animals. This action is similar to those of a wide variety of other drugs, chemicals, and physiologic perturbations.

f. INDUCTION OF HEPATIC MICROSOMAL ENZYMES Hepatic microsomal enzymes play an important role in thyroid hormone economy since glucuronidation is the rate-limiting step in the biliary excretion of T_4, and sulfation, for the excretion of T_3 by phenol sulfotransferase. Long-term exposure to many chemicals may induce these enzyme pathways and result in chronic stimulation of the thyroid by disrupting the hypothalamic–pituitary–thyroid axis. The resulting chronic stimulation of the thyroid by increased circulating levels of TSH often increases the risk of developing tumors derived from follicular cells, in chronic studies with such compounds in some species. Xenobiotics that induce liver microsomal enzymes and disrupt thyroid function include central nervous system drugs (e.g., phenobarbital, benzodiazepines); calcium channel blockers (e.g., nicardipine, bepridil); steroids (spironolactone); retinoids; chlorinated hydrocarbons [chlordane, dichlorodiphenyltrichloroethane (DDT), tetrachlorodibenzodioxin (TCDD)], polyhalogenated biphenyls [polychlorinated biphenyl (PCB), polybrominated biphenyl (PBB)]. Most of the hepatic microsomal enzyme inducers have no apparent intrinsic carcinogenic activity and produce little or no mutagenicity or DNA damage.

Phenobarbital has been studied extensively as the prototype for hepatic microsomal inducers that increase a similar spectrum of cytochrome P450 isoenzymes. Polychlorinated biphenyls also induce hepatic microsomal enzymes and disrupt thyroid function, and are commonly used industrial compounds which are widespread environmental contaminants. Their disease-producing capability includes alterations in reproduction, growth, and development. Polychlorinated biphenyls cause a significant reduction in serum levels of thyroid hormones due to alterations in thyroid structure, in addition to the induction of hepatic UDP-glucuronyl transferase and increased secretion of thyroxine-glucuronide in the bile.

In contrast to the previous six categories of indirect-acting thyrotoxic compounds, certain chemicals and irradiation appear to have a direct effect on the thyroid gland, resulting in genetic damage that leads to cell transformation and tumor formation in animals. Examples of thyroid initiators include 2-acetylaminofluorine (2-AAF), *N*-methyl-*N*-nitrosourea (MNU), *N*-bis(2-hydroxypropyl) nitrosamine (DHPN), methylcholanthrene, dichlorobenzidine, and polycyclic hydrocarbons. Chemicals in this group often increase the incidence of both benign and malignant thyroid tumors. Iodine deficiency is a strong promoter of MNU-initiated thyroid tumors in rats.

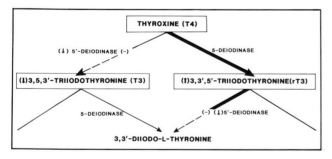

Fig. 3.30 Effects of FD & C Red No. 3 on peripheral metabolism of thyroxine. Inhibition of 5'-deiodinase results in lowered serum T_3 and a compensatory increased secretion of TSH by the pituitary gland, leading to follicular cell stimulation. Preferential monodeiodination of thyroxine by 5'-deiodinase results in markedly elevated blood levels of inactive reverse T_3. (From Capen and Martin, 1989, with permission.)

C. Neoplasms of the Thyroid Gland

Most thyroid tumors are of follicular origin. They are common in dogs and cats, moderately so in horses, and distinctly rare in cattle, sheep, and swine. These tumors are tumors of advanced age, and the low incidence in cattle, sheep, and swine relative to that in dogs, cats, and horses may be partially accounted for by the relatively advanced age the latter may reach.

1. Follicular Cell Adenoma

Adenomas usually are white to tan, small, solid nodules that are well demarcated from the adjacent thyroid parenchyma. The affected thyroid lobe is only moderately enlarged and distorted. A distinct white fibrous capsule of variable thickness separates the adenoma from the compressed parenchyma. Only a single adenoma usually is present in a thyroid lobe (Fig. 3.29B).

Other thyroid adenomas are composed of thin-walled cysts filled with a yellow to red fluid. The external surface is smooth and covered by an extensive network of blood vessels. Small masses of neoplastic tissue remain in the wall and form rugose projections into the cyst lumen. The thyroid parenchyma of the affected lobe may be completely obliterated.

The adenomas are classified into follicular and papillary types. They are sharply demarcated and encapsulated from the adjacent compressed thyroid parenchyma by a fibrous capsule of varying thickness. Adenomas derived from follicular cells that retain the ability to form follicles, according to one of several patterns, are more common by far than papillary adenomas in animals. Each **follicular adenoma** tends to have a consistent growth pattern within itself.

There are several different patterns of growth for follicles similar to those of the normal thyroid. Microfollicular adenomas consist of tumor cells arranged in miniature follicles with small amounts of colloid or an absence of colloid. Macrofollicular adenomas are formed by irregularly shaped large follicles that are greatly distended with colloid and lined by flattened follicular cells. There often is extensive hemorrhage and desquamation of follicular cells into the lumina of the distended follicles.

Cystic adenomas consist of one or two large cavities filled with proteinaceous fluid, necrotic debris, and erythrocytes. Focal accumulations of tumor cells, forming either follicles or solid nests, are present in the capsule of dense fibrous connective tissue. These adenomas may develop by progressive cystic degeneration of one of the several types of follicular adenomas. **Trabecular adenomas** are the most poorly differentiated of the follicular type. The tumor cells are small and are arranged in narrow columns separated by an edematous fibrous stroma. There is little evidence of follicle formation.

Oxyphilic adenomas are composed (predominantly or entirely) of large cells with a densely eosinophilic granular cytoplasm arranged in indistinct follicles with little or no colloid formation. Oxyphilic (Hürthle) cells appear to be metabolically altered follicular cells that accumulate abnormally large numbers of mitochondria in their cytoplasm.

Papillary adenomas of thyroid origin are recognized infrequently in most animal species. Columnar or cuboidal follicular cells are arranged in a single layer about a thin vascular connective tissue stalk. These papillary projections extend into the lumina of cystic spaces of various sizes. The cysts contain desquamated tumor cells, colloid, erythrocytes, and occasionally laminated foci of mineralization resembling psammoma bodies.

2. Hyperthyroidism Associated with Thyroid Tumors

Multinodular hyperplasias and adenomas of the thyroid gland are common in adult to aged cats. Follicular cell adenomas, often developing in a thyroid with multinodular hyperplasia, are encountered more commonly than thyroid carcinomas. Adenomas and carcinomas are most likely to be encountered in aged cats, whereas nodular hyperplasia can occur at any age. The mean age of cats with benign tumors has been reported to be 12.4 years and with thyroid carcinomas as 15.8 years.

A syndrome of hyperthyroidism occurs in aged cats either associated with multinodular hyperplasia, adenomas, or adenocarcinomas derived from follicular cells. The most common clinical sign is weight loss in spite of a normal or increased appetite. Polydipsia and polyuria, increased frequency of defecation, increased volume of stools, and increased activity occur. A common functional disturbance is tachycardia accompanied by premature beats and/or a systolic murmur. Cardiomegaly due to left ventricular hypertrophy may be evident on radiographs or at autopsy.

Thyroid adenomas in cats usually appear as solitary, soft nodules that enlarge and distort the contour of the affected lobe. A thin, fibrous tissue capsule separates the adenoma from the adjacent, often compressed, thyroid parenchyma. Focal areas of necrosis, mineralization, and cystic degeneration are present in larger adenomas. Functional follicular adenomas often develop in thyroids that have multinodular hyperplasia of follicular cells in both lobes.

Hyperthyroidism in cats also occurs in association with bilateral multinodular (adenomatous) hyperplasia. These multiple areas of thyroid hyperplasia usually do not appreciably enlarge the affected lobe(s) (Fig. 3.31A). The hyperplastic nodules are composed of irregularly shaped, colloid-filled follicles lined by cuboidal follicular cells. These multiple nodules of follicular cell hyperplasia may coalesce to form macroscopically observable thyroid adenomas.

Functional thyroid adenomas are composed of cuboidal to columnar follicular cells with occasional papillary infoldings that form follicles containing variable amounts of colloid. The follicles usually are partially collapsed and contain little colloid because of the intense endocytotic activity of neoplastic follicular cells. Long cytoplasmic projections extend from the follicular cells into the lumen

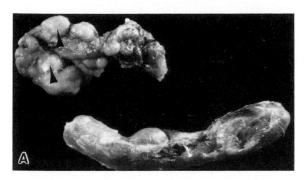

Fig. 3.31A Multinodular follicular cell hyperplasia (arrowheads) in both thyroid lobes. Cat with hyperthyroidism.

to phagocytize colloid (Fig. 3.31B). As a result of the marked endocytotic activity, numerous colloid droplets are present in the apical cytoplasm of follicular cells in close proximity to the many lysosomal bodies.

Functional thyroid adenomas are partially or completely separated from remnants of the adjacent normal thyroid by a fine connective tissue capsule. Surrounding follicles are markedly enlarged and distended by an accumulation of colloid, and their follicular cells are low cuboidal and atrophied, with little evidence of endocytotic activity in response to the elevated levels of thyroid hormones.

Cats with hyperthyroidism usually have markedly ele-

vated serum thyroxine and triiodothyronine levels. Normal feline serum levels of T_4 measured by radioimmunoassay are $\sim$1.5–5.0 μg/dl, and serum T_3 levels are 60–200 ng/dl. The serum T_4 levels in cats with hyperthyroidism range from 5.0 to >50 μg/dl, and serum T_3 levels range from 100 to 1000 ng/dl. Increased serum aspartate aminotransferase, alanine aminotransferase, and alkaline phosphatase levels often occur in hyperthyroid cats.

3. Follicular Cell Carcinoma

Thyroid carcinomas are detected as masses in the ventral cervical region, and they may be unilateral or bilateral. Most thyroid tumors are palpated near the larynx, but larger ones may extend toward the thoracic inlet. The tumor may be firm or soft. Respiratory distress occurs if the tumor encroaches on the larynx or trachea. Carcinomas become fixed in position by extensive local invasion of adjacent structures, whereas adenomas are freely movable under the skin.

Earlier reports have suggested that the incidence of thyroid tumors is highest in areas that are iodine deficient and have many animals with long-standing, diffuse hyperplastic goiter. Most animals with thyroid tumors are adults or aged.

Thyroid carcinomas occur more often than adenomas in dogs, but in cats, adenomas occur more frequently than carcinomas. No sex prevalence has been observed in dogs.

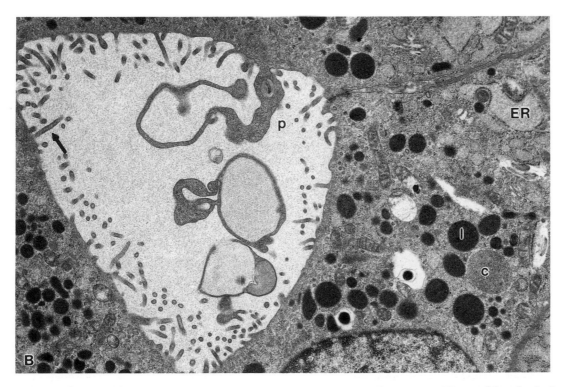

Fig. 3.31B Functional follicular cell adenoma. Cat. Numerous cytoplasmic processes (P) extend from luminal surface of neoplastic follicular cells to engulf colloid by endocytosis. Note lysosomal (l) bodies associated with colloid droplets (c) and long microvilli (arrow) on surfaces bordering colloid. Profiles of rough endoplasmic reticulum (ER) are dilated by finely granular material.

It has been reported that boxers develop thyroid carcinomas more frequently than any other breed of dog. There also are data indicating that beagles, boxers, and golden retrievers have a significantly greater risk for thyroid carcinoma than all other breeds of dogs combined.

Adenocarcinomas of the thyroid are larger than adenomas, are coarsely multinodular, and often have large areas of central hemorrhage and necrosis. Unilateral involvement is about twice as frequent in dogs as is involvement of both thyroid lobes. Carcinomas are poorly encapsulated and invade locally into the wall of the trachea, cervical muscles, esophagus, larynx, nerves, and vessels. Early invasion into branches of the cranial and caudal thyroid veins with the formation of tumor cell thrombi (Fig. 3.32A,C) leads to multiple pulmonary metastases (Fig. 3.32D,E), often before involvement of the retropharyngeal and caudal cervical lymph nodes. Focal white gritty areas of mineralization or bone formation are scattered throughout some tumors. Although adenomas and carcinomas derived from follicular cells usually arise in the neck from the thyroid lobes, they may develop from ectopic thyroid parenchyma in the mediastinum and must be included in the differential diagnosis of heart-base tumors in dogs.

Malignant tumors of thyroid follicular cells generally are more highly cellular and have a greater degree of cellular pleomorphism than do adenomas. On the basis of the predominant histologic pattern of growth, differentiated thyroid adenocarcinomas are subdivided into follicular, papillary, and compact cellular (solid) types. In **dogs** where thyroid carcinomas are seen most frequently, they often have both a follicular and compact cellular growth pattern (Fig. 3.32B), whereas papillary carcinomas are uncommon.

In **follicular adenocarcinomas** the majority of tumor cells are arranged in a recognizable follicular pattern. It is possible to subdivide follicular carcinomas further on the basis of degree of follicle formation, as described for follicular adenomas, but this often is difficult because of the admixture of growth patterns present in any one tumor. Such subdivisions appear to be of little prognostic value. The tumor cells are tall cuboidal to columnar and form follicles of varying size, shape, and colloid content. Mitotic activity in the tumor cells is minimal. The colloid in follicular lumina occasionally is clumped and extensively mineralized.

In **compact cellular carcinoma,** tumor cells form aggregations or solid sheets of cells often separated by a fibrous stroma with little or no attempts at follicle formation and colloid secretion. The polyhedral cells are closely arranged together and have an eosinophilic cytoplasm that is finely granulated or vacuolated. Immunocytochemical and ultrastructural studies have demonstrated that compact cellular (solid) carcinomas are derived from follicular cells of the canine thyroid and not from C cells.

Follicular–compact cellular carcinoma, having approximately equal follicular and compact cellular growth patterns, is the most common type of malignant thyroid tumor in dogs. The follicles formed often are smaller and contain less colloid than those in pure follicular carcinomas. The compactly arranged tumor cells appear to be morphologically and functionally less differentiated than those which form follicles and secrete colloid.

Papillary carcinomas, in which tumor cells form papillae extending into cystic spaces, are uncommon in animals. Single or multiple layers of cuboidal cells surround fibrovascular stalks that project into cystic spaces. Their nuclei are vesicular and pleomorphic with prominent nucleoli. The nuclear vacuoles seen by light microscopy have been shown by electron microscopy to represent cytoplasmic evaginations into the nucleus. Infiltration of tumor cells through the fibrous connective tissue capsule and into adjacent tissues is frequent in thyroid adenocarcinomas.

Undifferentiated thyroid carcinomas lack a characteristic architectural pattern of arrangement of tumor cells. They are an uncommon form of thyroid carcinoma in animals. **Small-cell carcinoma** is one type of undifferentiated thyroid carcinoma. They are composed of highly malignant follicular cells with either a diffuse or a compact pattern of growth. The small tumor cells are uniform in appearance and are closely packed together in clusters separated by a fibrous stroma. The scant cytoplasm is eosinophilic, and the oval nucleus is densely hyperchromatic. Mitotic figures are frequent.

Giant-cell carcinoma is a highly malignant thyroid tumor derived from poorly differentiated follicular cells. The anaplastic tumor cells are large, pleomorphic, and often spindle shaped, making differentiation from a fibrosarcoma difficult. The demonstration of identifiable epithelial structures may require multiple sections from several areas of the tumor. Follicular remnants of transitional forms suggest that giant-cell carcinomas are derived from thyroid follicular cells.

Malignant mixed thyroid tumors have been reported in the dog. The tumors contain both malignant thyroid follicular cells and mesenchymal elements, usually osteogenic or cartilaginous.

Thyroid carcinomas often grow rapidly, invading adjacent structures such as the trachea, esophagus, and larynx, and usually are fixed in position. The earliest and most frequent site of metastasis is the lung (Fig. 3.32D,E) because thyroid carcinomas tend to invade branches of the thyroid vein. Tumor cords may be palpated in the thyroid or jugular veins in some animals with thyroid carcinoma. The retropharyngeal and caudal cervical lymph nodes are less frequent sites of metastasis.

Some thyroid tumors in the dog secrete sufficient thyroid hormone to produce mild clinical signs of hyperthyroidism. It is surprising that hyperthyroidism occurs even with functional tumors since experimental induction of hyperthyroidism in the dog requires daily administration of ~25 times the normal replacement dose of desiccated thyroid or *l*-thyroxine. The clinical signs of hyperthyroidism in dogs with functional thyroid tumors include weight loss, polyphagia, weakness and fatigue, intolerance to heat, and nervousness.

Thyroid carcinomas occur less frequently in **cats** than

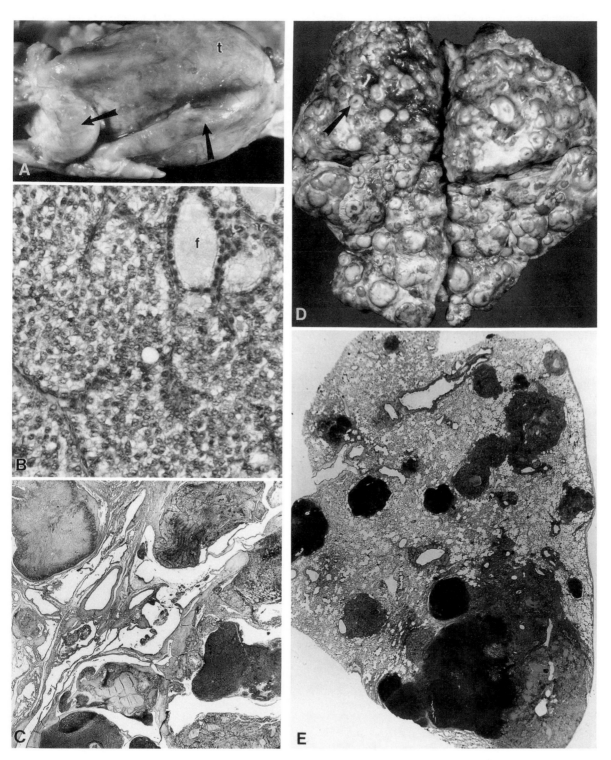

Fig. 3.32 (A) Prominent tumor thrombi distending thyroid veins (arrows) in a dog with an adenocarcinoma of the thyroid (t). (B) Follicular (F)-compact cellular pattern of thyroid adenocarcinoma. Dog. (C) Vascular invasion with tumor thrombi in adenocarcinoma of the thyroid. (D) Extensive pulmonary metastases in lungs from a thyroid adenocarcinoma. Dog. Many nodules have an indented surface (arrow). (E) Miliary metastases of thyroid carcinoma in lung. Dog.

either adenomas or multinodular hyperplasia. They often result in considerable enlargement of one or both thyroid lobes and may invade adjacent structures. Thyroid carcinomas are characterized by the invasion of vessels and connective tissue capsule by neoplastic cells. Metastases to regional lymph nodes (retropharyngeal, mandibular, deep cervical) and distant sites have been reported in <50% of thyroid carcinomas in cats. The well-differentiated thyroid adenocarcinomas are relatively solid and composed of a uniform pattern of small follicles containing little colloid and occasional compact cellular areas. Strands of dense connective tissue with an abundant capillary network and foci of lymphocytes subdivide the neoplastic cells into small lobules.

4. Neoplasms of Thyroglossal Duct Remnants

Tumors arising in cystic remnants of the thyroglossal duct are rare in animals but have been reported in dogs. They appear as well-circumscribed, fluctuant, movable enlargements (~2–4 cm in diameter) on the ventral midline in the cranial cervical region. The clinical history usually indicates a slowly progressive expansion of the cervical mass. On cross section there are multilocular cystic areas containing a translucent proteinaceous fluid alternating with white solid areas.

The thyroid glands appear to be normal in the few cases studied in dogs. These tumors appear to develop from the epithelium of the thyroglossal duct and are not cystic metastases from primary carcinomas in the thyroid gland.

Neoplasms of thyroglossal duct remnants are well-differentiated papillary carcinomas. Multiple papillary outgrowths covered by multiple layers of tall cuboidal to columnar epithelial cells extend from the cyst wall into the lumen (Fig. 3.33). The cyst wall is composed of dense fibrous connective tissue with focal areas of hemorrhage and cholesterol clefts. Aggregations of thyroidogenic epithelium in the form of small follicles and cell cords are present within the fibrous capsule and in surrounding connective tissue. These follicles are lined by a low cuboidal epithelium and contain variable amounts of colloid. Carcinomas of thyroglossal duct remnants appear to be well differentiated and slow growing.

5. Thyroid C (Parafollicular)-Cell Tumors

Tumors derived from C (parafollicular) cells of the thyroid gland are most frequently encountered in adult to aged bulls, certain strains of laboratory rats, and adult to aged horses, but infrequently in other species. A high percentage of aged bulls has been reported to develop C-cell tumors (30%) or hyperplasia of C cells and ultimobranchial derivatives (15–20%). These frequently occurring hyperplastic and neoplastic lesions of C cells have been observed only in bulls and not in cows. The incidence of C-cell tumors increases with advancing age in bulls and is associated with the development of vertebral osteophytes.

The syndrome of C-cell tumors in bulls shares many similarities with medullary thyroid carcinoma in humans. Multiple endocrine tumors, especially bilateral pheochromocytomas and occasionally pituitary adenomas, are detected coincidentally in bulls and humans with C-cell tumors. This may represent a simultaneous neoplastic transformation of multiple endocrine cell populations of neural crest origin in the same individual. A high frequency of thyroid C-cell tumors and pheochromocytomas has been reported in a family of Guernsey bulls, suggesting an autosomal dominant pattern of inheritance. A diffuse or nodular hyperplasia of secretory cells in the adrenal medulla appears to precede the development of pheochromocytoma.

C-cell adenomas appear as discrete, single or multiple, gray to tan nodules in one or both thyroid lobes (Fig. 3.34A). Adenomas are smaller (~1–3 cm in diameter) than carcinomas and are separated from the thyroid parenchyma by a thin, fibrous connective tissue capsule. The adjacent thyroid is compressed but not invaded by neo-

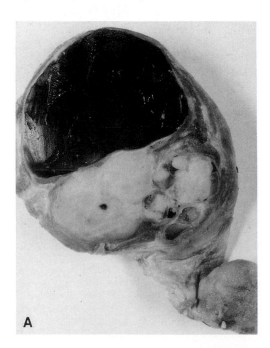

Fig. 3.34A C-cell adenoma with cyst formation. Bull.

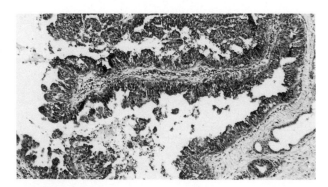

Fig. 3.33 Papillary cystadenocarcinoma derived from thyroglossal duct remnants. Dog.

plastic C cells. In horses C-cell adenomas may result in a palpable enlargement in the anterior cervical region. Larger C-cell adenomas incorporate most of the thyroid lobe, but a rim of dark brown-red thyroid often is present on one side.

Focal and/or nodular hyperplasia of C cells often precedes the development of C-cell neoplasms in animal and humans. Hyperplastic C cells appear normal with an abundant lightly eosinophilic granular cytoplasm (Fig. 3.34B). Nodular hyperplasia of C cells consists of foci less than the size of a colloid-filled follicle. Calcitonin immunoreactivity is localized to the cytoplasm of hyperplastic C cells.

Thyroid C-cell adenomas microscopically are a discrete, expansive mass of cells greater in size than a colloid-distended follicle. They are well circumscribed or partially encapsulated from adjacent follicles that are compressed to varying degrees. C-Cell adenomas may be subdivided into packets of cells by fine or coarse connective tissue septa and capillaries. The neoplastic C cells are well differentiated and have an abundant cytoplasmic area that is lightly eosinophilic or clear. The well-differentiated cells composing C-cell adenomas have numerous membrane-limited secretory granules, circular and lamellar profiles of rough endoplasmic reticulum, and prominent Golgi complexes (Fig. 3.34C).

Thyroid **C-cell carcinomas** result in extensive multinodular enlargements of one or both thyroid lobes. The entire thyroid gland may be incorporated by the proliferating neoplastic tissue. Multiple metastases in anterior cervical lymph nodes usually are large and have areas of necrosis and hemorrhage. Pulmonary metastases are present infrequently and appear as discrete tan nodules throughout all lobes of the lung.

C-Cell carcinomas are more highly cellular, and the tumor cells are more pleomorphic than those of C-cell adenomas. They often have evidence of intrathyroidal and/or extracapsular invasion, occasionally with metastasis to distant sites. The neoplastic cells are polyhedral to spindle shaped with a lightly eosinophilic, finely granular, indistinct cytoplasmic area (Fig. 3.35). The vesicular nuclei are oval or elongate and have more frequent mitotic figures than those in adenomas. Malignant C cells often are subdivided into small groups by fine connective tissue septa, which contain small capillaries.

Thyroid C-cell tumors in bulls, other animals, and humans are firm and in some areas, the stroma consists of dense bands of fibrous connective tissue. In both adenomas and carcinomas there may be deposits of a homogeneous eosinophilic material which stains positively for amyloid. Ultrastructurally, large aggregations of fine amyloid fibrils are observed between the bundles of collagen fibers.

The etiology of the localized amyloid deposition in thyroid C-cell neoplasms is uncertain, but it is not associated with chronic suppurative lesions in other organs. Amyloid production is consistently associated with medullary thyroid carcinoma in humans and also has been reported in certain other endocrine tumors. Chemical differences ex-

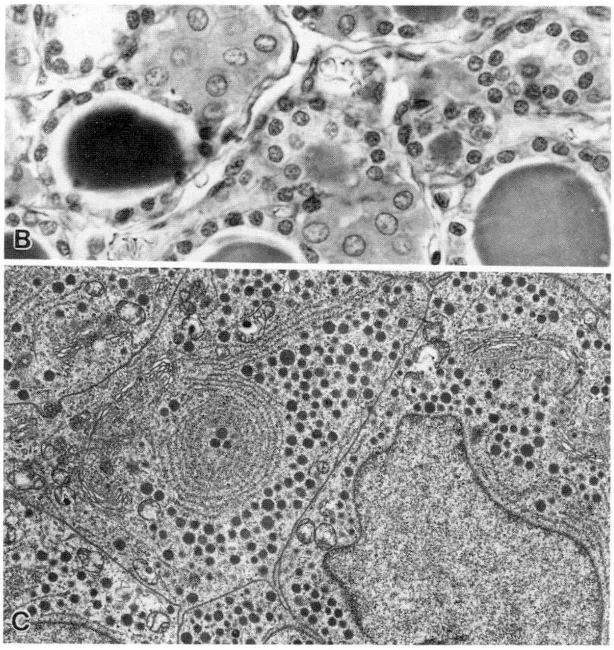

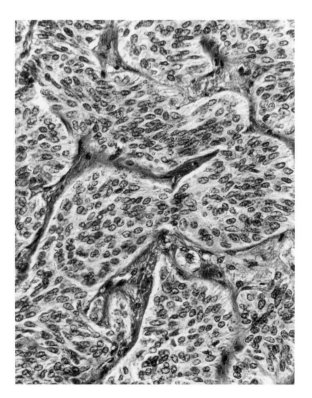

Fig. 3.34B,C (B) Hyperplasia of C cells. Dog. (C) C-cell adenoma composed of well-differentiated cells that contain numerous membrane-limited secretory granules, circular profiles of rough endoplasmic reticulum, and prominent Golgi apparatuses. Horse.

Fig. 3.35 C-cell carcinoma illustrating neuroendocrine pattern. Bull.

ist between amyloid fibrils of immunoglobulin origin and those produced by endocrine tumors.

Bioassay of C-cell adenomas and carcinomas from bulls has demonstrated the presence of calcitonin activity in tumor tissue, and calcitonin has been detected at higher than normal levels in plasma of bulls by immunoassay. Calcium infusion to raise serum calcium increases plasma calcitoninlike activity. Mean serum calcium levels in bulls with calcitonin-secreting thyroid tumors are only slightly lower than those in adult control bulls. The near normal or mild decrease in serum electrolyte values in animals with chronic hypersecretion of calcitonin probably is a result of the low turnover rate of bone and compensatory mechanisms of other endocrine organs.

The etiology of C-cell neoplasms is unknown, but a possible relationship has been suggested between the long-term dietary intake of excessive calcium and the high incidence of these tumors in bulls. Adult bulls frequently ingest 3.5–6.0 times the amount of calcium normally recommended for maintenance. The chronic stimulation of C cells by high levels of calcium absorbed from the digestive tract may be related to the pathogenesis of the neoplasms. A significant decline in the incidence of C-cell tumors has been reported when bulls are switched from a high calcium intake to a reduced calcium intake. Cows do not develop proliferative lesions of C cells under similar dietary conditions, possibly because of the high physiologic requirements for calcium imposed by pregnancy and lactation.

Ultimobranchial tumors in the thyroid glands of bulls often have a more complex histologic structure than the typical C-cell (medullary) carcinoma in human patients and certain strains of laboratory rats. Areas in the tumor composed of differentiated C cells consist of focal accumulations of neoplastic cells with an abundant lightly eosinophilic cytoplasm in the wall of thyroid and ultimobranchial follicles or of larger nodules with a solid histologic structure. This often is accompanied by a multifocal hyperplasia of C cells in other parts of the thyroid lobes and hilus. The neoplastic C cells often are embedded in an increased amount of hyalinized stroma that may contain amyloid. Parts of this thyroid neoplasm in bulls that appear to be derived from less differentiated ultimobranchial remnants consist of folliclelike structures, cysts, and tubules composed of immature small basophilic cells. They closely resemble undifferentiated or stem cells of the normal ultimobranchial body in bulls and other species that possibly develop into both C and follicular cells. Thyroid follicles and cribriform structures with colloidlike material formed by cells resembling differentiated follicular cells often are present in the neoplasms in close association with these more primitive ultimobranchial-derived structures. The heterogeneous histologic structure of ultimobranchial neoplasms in bulls resembles a type of thyroid carcinoma of human patients. This variant, designated as an intermediate type of differentiated carcinoma, has structural and immunocytochemical characteristics of both medullary (C-cell) and follicular carcinomas.

Bibliography

Allison, A. C. Self-tolerance and autoimmunity in the thyroid. *N Engl J Med* **295:** 821–827, 1976.

Anderson, M. P., and Capen, C. C. Diseases of the endocrine system. *In* "Pathology of Laboratory Animals," K. Benirschke *et al.* (eds.), Vol. 1, pp. 423–508. New York, Springer-Verlag, 1978.

Baker, H. J., and Lindsey, J. R. Equine goiter due to excess dietary iodide. *J Am Vet Med Assoc* **153:** 616–630, 1968.

Bastenie, P. A. *et al.* Thyroid auto-immunity disease. *In* "Molecular Pathology," R. A. Good *et al.* (eds.), pp. 234–261. Springfield, Illinois, Charles C. Thomas, 1975.

Belshaw, B. E., and Becker, D. V. Necrosis of follicular cells and discharge of thyroidal iodine induced by administering iodide to iodide-deficient dogs. *J Clin Endocrinol Metab* **36:** 466–474, 1973.

Belshaw, B. E., and Rijnberk, A. Radioimmunoassay of plasma T_4 and T_3 in the diagnosis of primary hypothyroidism in dogs. *J Am Anim Hosp Assoc* **15:** 15–23, 1979.

Bigazzi, P. E., and Rose, N. R. Spontaneous autoimmune thyroiditis in animals as a model of human disease. *Prog Allergy* **19:** 245–274, 1975.

Black, H. E., Capen, C. C., and Young, D. M. Ultimobranchial thyroid neoplasms in bulls. A syndrome resembling medullary thyroid carcinoma in man. *Cancer* **32:** 865–878, 1973.

Buergelt, C.-D. Mixed thyroid tumors in two dogs. *J Am Vet Med Assoc* **152:** 1658–1663, 1968.

Cheville, N. F. Ultrastructure of canine carotid body and aortic body tumors. Comparison with tissues of thyroid and parathyroid origin. *Vet Pathol* **9:** 166–189, 1972.

Collins, W. T., Jr., and Capen, C. C. Ultrastructural and functional alterations in thyroid glands of rats produced by polychlorinated biphenyls compared with the effects of iodide excess and deficiency, and thyrotropin and thyroxine administration. *Virchows Arch B Cell Pathol* **33:** 213–231, 1980.

Crispin, S. M., and Barnett, K. C. Arcus lipoides corneae secondary to hypothyroidism in the Alsatian. *J Small Anim Pract* **19:** 127–142, 1978.

de Vijlder, J. J. M. *et al.* Hereditary congenital goiter with thyroglobulin deficiency in a breed of goats. *Endocrinology* **102:** 1214–1222, 1978.

Ekholm, R. *et al.* Exocytosis of protein into the thyroid follicle lumen: An early effect of TSH. *Endocrinology* **97:** 337–346, 1975.

Falconer, I. R. Studies of the congenitally goitrous sheep: The iodinated compounds of serum and circulating thyroid-stimulating hormone. *Biochem J* **100:** 190–196, 1966.

Gosselin, S. J. *et al.* Biochemical and immunological investigations of hypothyroidism in dogs. *Can J Comp Med* **44:** 158–168, 1980.

Gosselin, S. J., Capen, C. C., and Martin, S. L. Histopathologic and ultrastructural evaluation of thyroid lesions associated with hypothyroidism in dogs. *Vet Pathol* **18:** 299–309, 1981.

Gosselin, S. J. *et al.* Autoimmune lymphocytic thyroiditis in dogs. *Vet Immunol Immunopathol* **3:** 185–201, 1982.

Haley, P. J. *et al.* Thyroid neoplasms in a colony of beagle dogs. *Vet Pathol* **26:** 438–441, 1989.

Hayes, H. M., and Fraumeni, J. F. Canine thyroid neoplasms: Epidemiologic features. *J Natl Cancer Inst* **55:** 931–934, 1975.

High, W. B., Black, H. E., and Capen, C. C. Effects of thyroxine on cortical bone remodeling in adult dogs: A histomorphometric study. *Am J Pathol* **102:** 438–446, 1981.

High, W. B., Capen, C. C., and Black, H. E. Effects of 1,25-

dihydroxy-cholecalciferol, parathyroid hormone, and thyroxine on trabecular bone remodeling in adult dogs: A histomorphometric study. *Am J Pathol* **105**: 1856–1864, 1981.

Hoenig, M. *et al.* Toxic nodular goitre in the cat. *J Small Anim Pract* **23**: 1–12, 1982.

Holzworth, J. *et al.* Hyperthyroidism in the cat: Ten cases. *J Am Vet Med Assoc* **176**: 345–353, 1980.

Johnson, J. A., and Patterson, J. M. Multifocal myxedema and mixed thyroid neoplasm in a dog. *Vet Pathol* **18**: 13–20, 1981.

Joyce, J. R. *et al.* Thyroid carcinoma in a horse. *J Am Vet Med Assoc* **168**: 610–612, 1976.

Karbe, E. Lateral neck cysts in the dog. *Am J Vet Res* **26**: 717–722, 1965.

Karbe, E., and Nielsen, S. W. Branchial cyst in a dog. *J Am Vet Med Assoc* **147**: 637–640, 1965.

Krook, L., Olsson, S., and Rooney, J. R. Thyroid carcinoma in the dog. A case of bone-metastasizing thyroid carcinoma simulating hyperparathyroidism. *Cornell Vet* **50**: 106–114, 1960.

Laurberg, P. Non-parallel variations in the preferential secretion of 3,5,3'-triiodothyronine (T$_4$) and 3,3',5'-triiodothyronine (rT$_3$) from dog thyroid. *Endocrinology* **102**: 757–766, 1978.

Leav, I. *et al.* Adenomas and carcinomas of the canine and feline thyroid. *Am J Pathol* **83**: 61–94, 1976.

Lissitzky, S. *et al.* Congenital goiter with impaired thyroglobulin synthesis. *J Clin Endocrinol Metab* **36**: 17–29, 1973.

Little, J. W., and Rickles, N. H. The histogenesis of the branchial cyst. *Am J Pathol* **50**: 533–547, 1967.

Ljungberg, O., and Nilsson, P.-O. Hyperplastic and neoplastic changes in ultimobranchial remnants and in parafollicular (C) cells in bulls: A histologic and immunohistochemical study. *Vet Pathol* **22**: 95–103, 1985.

Ljungberg, O., Bondeson, L., and Bondeson, A.-G. Differentiated thyroid carcinoma, intermediate type: A new tumor entity with features of follicular and parafollicular cell carcinoma. *Hum Pathol* **15**: 218–228, 1984.

Matsuta, M. Immunohistochemical and electron-microscopic studies on Hashimoto's thyroiditis. *Acta Pathol Jpn* **32**: 41–56, 1982.

Mazzaferri, E. L. *et al.* Papillary thyroid carcinoma: The impact of therapy in 576 patients. *Medicine (Baltimore)* **56**: 171–196, 1977.

Milne, K. L., and Hayes, H. M. Epidemiologic features of canine hypothyroidism. *Cornell Vet* **71**: 3–14, 1981.

Mizejewski, G. J., Baron, J., and Poissant, G. Immunologic investigations of naturally occurring canine thyroiditis. *J Immunol* **107**: 1152–1160, 1971.

Moore, F. M. *et al.* Thyroglobulin and calcitonin immunoreactivity in canine thyroid carcinomas. *Vet Pathol* **21**: 168–173, 1984.

Nesbitt, G. H. *et al.* Canine hypothyroidism: A retrospective study of 108 cases. *J Am Vet Med Assoc* **177**: 1117–1122, 1980.

Nunez, E. A. *et al.* Breed differences and similarities in thyroid function in purebred dogs. *Am J Physiol* **218**: 1337–1341, 1970.

Nunez, E. A., Belshaw, B. E., and Gershon, M. D. A fine structural study of the highly active thyroid follicular cell of the African basenji dog. *Am J Anat* **133**: 463–482, 1972.

Oppenheimer, J. H. Thyroid hormone action at the cellular level. *Science* **203**: 971–979, 1979.

Pammenter, M. *et al.* Afrikander cattle congenital goiter: Characteristics of its morphology and iodoprotein pattern. *Endocrinology* **102**: 954–965, 1978.

Rac, R. *et al.* Congenital goitre in Merino sheep due to an inherited defect in the biosynthesis of thyroid hormone. *Res Vet Sci* **9**: 209–223, 1968.

Reid, C. F. *et al.* Functioning adenocarcinoma of the thyroid gland in a dog with mitral insufficiency. *J Am Vet Radiol Soc* **4**: 36–40, 1963.

Rijnberk, A., and der Kinderen, P. J. Toxic thyroid carcinoma in the dog. *Acta Endocrinol (Suppl)* **138**: 177, 1969.

Rijnberk, A. *et al.* Congenital defect in iodothyronine synthesis: Clinical aspects of iodine metabolism in goats with congenital goitre and hypothyroidism. *Br Vet J* **133**: 495–503, 1977.

Robinson, W. F. *et al.* Congenital hypothyroidism in Scottish deerhound puppies. *Aust Vet J* **65**: 368–389, 1988.

Rogers, W. A., Donovan, E. F., and Kociba, G. J. Lipids and lipoproteins in normal dogs and in dogs with secondary hyperlipoproteinemia. *J Am Vet Med Assoc* **166**: 1092–1100, 1975.

Schiller, I. *et al.* Hypercholesteremia in pet dogs. *Arch Pathol* **77**: 389–392, 1964.

Thake, D. C., Cheville, N. F., and Sharp, R. K. Ectopic thyroid adenomas at the base of the heart of the dog: Ultrastructural identification of dense tubular structures in endoplasmic reticulum. *Vet Pathol* **8**: 421–432, 1971.

van der Velden, M. A., and Meulenaar, H. Medullary thyroid carcinoma in a horse. *Vet Pathol* **23**: 622–624, 1986.

van Herle, A. J., Vassart, G., and Dunmont, J. E. Control of thyroglobulin synthesis and secretion. *N Engl J Med* **301**: 239–249 and 307–314, 1979.

van Voorthuizen, W. F. *et al.* Euthyroidism via iodine supplementation in hereditary congenital goiter with thyroglobulin deficiency. *Endocrinology* **103**: 2105–2111, 1978.

van Voorthuizen, W. F. *et al.* Abnormal cellular localization of thyroglobulin mRNA associated with hereditary congenital goiter and thyroglobulin deficiency. *Proc Natl Acad Sci U.S.A.* **75**: 74–78, 1978.

van Zyl, A. *et al.* Congenital goiter in Afrikander cattle. *In* "Spontaneous Animal Models of Human Disease," E. J. Andrews *et al.* (eds.), Vol. I, p. 108. New York, Academic Press, 1979.

Wallach, S., Carstens, J. B., and Avioli, L. V. Calcitonin, osteoclasts, and bone turnover. *Calcif Tissue Int* **47**: 388–391, 1990.

Young, D. M., and Capen, C. C. Fine structural alterations of thyroid C cells in response to diet-induced hypocalcemia and vitamin D in adult cows. *Virchows Arch B Cell Pathol* **8**: 288–298, 1971.

Young, D. M., Capen, C. C., and Black, H. E. Calcitonin content of ultimobranchial neoplasms from bulls. *Vet Pathol* **8**: 19–27, 1971.

Young, D. M., Capen, C. C., and Black, H. E. Effect of a calcium-deficient diet and vitamin D on thyrocalcitonin in cows. *Endocrinology* **90**: 276–281, 1972.

Zarrin, K. Naturally occurring parafollicular cell carcinoma of the thyroids in dogs. A histological and ultrastructural study. *Vet Pathol* **14**: 556–566, 1977.

V. Adrenal Cortex

The adrenal glands of mammals consist of two distinct parts, which differ not only in morphology and function but also in origin. Because of their close structural relationships, the outer cortex and inner medulla of the adrenal gland usually have been considered parts of one organ (Fig. 3.36). The adrenal cortex develops from celomic epithelium cells that are of mesodermal origin. The chromaffin tissue and sympathetic ganglion cells of the adrenal medulla are derived from ectoderm of the neural crest.

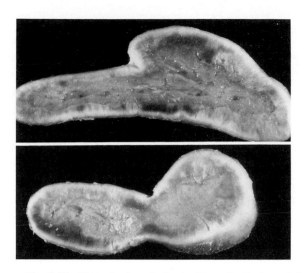

Fig. 3.36 Normal adrenal glands. Dog. Cortical : medullary ratio of approximately 2 : 1.

It is not until relatively late in fetal development that a definitive relationship between the two primordia occurs.

The adrenal glands are richly vascularized, receiving arterial branches either directly from the aorta or from the phrenic, renal, and lumbar arteries. In the capsule, the arteries form a vascular plexus, which eventually supplies the entire adrenal gland through separate channels to the capsule, cortex, and medulla. About the cell columns of the adrenal cortex, a sinusoidal network is formed, which empties into the venous tree at the periphery of the medulla. The larger branches of the venous tree empty into the adrenal vein.

The adrenal cortex classically is subdivided into three layers or zones, although the demarcation between zones often is not distinct. The **zona glomerulosa** (multiformis, arcuata) is composed of columns of cells that have a sigmoid arrangement next to the capsule. It represents ~15% of the cortex and is responsible for the secretion of mineralocorticoid hormones. The secretory cells of the **zona fasciculata** are arranged in long anastomosing cords separated by numerous small capillaries. This middle zone, which forms ~70% of the cortex, is composed of cells that contain abundant cytoplasmic lipid and are responsible for the secretion of the glucocorticoid hormones. The **zona reticularis** accounts for the remaining 15% of the cortex. The secretory cells are arranged in small groups surrounded by capillaries. This inner layer is responsible for the secretion of sex steroids by the adrenal gland. Cortical cells may project for a short distance into the medulla.

Mineralocorticoids are adrenal steroids which have their principal effects on ion transport by epithelial cells, resulting in a loss of potassium and conservation of sodium. The most potent and important naturally occurring mineralocorticoid is aldosterone. The enzymatically controlled electrolyte pumps in epithelial cells of the renal tubule and sweat glands respond to mineralocorticoids by conserving sodium and chloride and by wasting potassium. In the

distal convoluted tubule of the mammalian nephron, a cation-exchange mechanism exists for the resorption of sodium from the glomerular filtrate and secretion of potassium into the lumen. These reactions are accelerated by mineralocorticoids but proceed at a much slower rate in their absence. A lack of secretion of mineralocorticoids (such as in idiopathic adrenal atrophy of dogs) may result in a lethal retention of potassium and loss of sodium.

Glucocorticoid hormones secreted by the adrenal cortex are concerned with the intermediary metabolism of glucose. Cortisol and lesser amounts of corticosterone are the most important naturally occurring glucocorticoids secreted by the adrenal gland. In general, the actions of glucocortocoids on carbohydrate, protein, and lipid metabolism result in sparing of glucose and a tendency to hyperglycemia and increased glucose production. The acute effects of glucocorticoids are observed within 15–30 min before the compensatory effects of insulin become prominent. There is a decrease in glucose uptake in adipose tissue, skin, fibroblasts, and lymphoid tissue followed shortly by increased catabolism in these tissues and muscle. This provides the amino acids for gluconeogenesis, which is increased mainly in the liver. In addition, glucocorticoids decrease lipogenesis and increase lipolysis in adipose tissue, which results in release of glycerol and free fatty acids.

Glucocorticoids also function to suppress inflammatory and immunologic responses and thereby attenuate the associated tissue destruction and fibroplasia. However, under the influence of high levels of glucocorticoids, there is an enhancement of the spread of infections and reduced resistance to a number of bacterial, viral, and fungal diseases. Glucocorticoids may impair the immunologic response at any stage from the initial interaction and processing of antigens by cells of the monocyte–macrophage system through the induction and proliferation of immunocompetent lymphocytes and subsequent antibody production. Inhibition by glucocorticoids of a number of lymphoid cell functions forms part of the basis for suppression of the immunologic response.

Glucocorticoids also decrease the initial inflammatory reaction and its classic manifestations of heat, swelling, and pain. The degree of hyperemia, extravasation, cellular migration, and infiltration at the site of injury is decreased. Especially important are the effects of glucocorticoids on the usual vascular responses of increased permeability, diapedesis, and extravasation. Capillary blood flow is decreased, and there is less endothelial swelling. In addition, a number of phagocytic mechanisms are inhibited by glucocorticoids, and clearance of particulate substances from the blood and lymph is impaired. The accumulation of engulfed antigens in macrophages probably is related in part to the enhanced stability of lysosomal membranes caused by glucocorticoids. There is a diminished capacity of lysosomes to interact with phagocytized material and to release hydrolytic enzymes involved in intracellular digestion.

Glucocorticoids exert a profound negative effect on

wound healing. Dogs receiving high therapeutic levels of adrenal corticosteroids, or patients with hypercortisolism may have wound dehiscence following surgery. The basis mechanism involved is an inhibition of fibroblast proliferation and collagen synthesis leading to a decrease in scar tissue formation.

Secretion of **adrenal sex hormones** by cells of the zona reticularis occurs under normal conditions but in minute amounts that probably are of little physiologic significance. Progesterone, estrogens, and androgens are synthesized by secretory cells of the inner zone of the cortex. Under pathologic conditions, an excessive secretion of adrenal sex steroids infrequently may occur, associated with a neoplasm arising in the zona reticularis. The clinical manifestations of virilism, precocious sexual development, or feminization depend on which steroid is secreted in excess, sex of the patient, and the age of onset.

The **renin–angiotensin system** is the major regulator of aldosterone production by the glomerular zone of the adrenal cortex. Renin is an enzyme secreted into the circulation by cells of the juxtaglomerular apparatus in the kidney. It acts to cleave the plasma globulin, angiotensinogen, to form angiotensin I. The decapeptide is further hydrolyzed to angiotensin II by a converting enzyme. Angiotensin II is both a potent vasoconstrictor and a tropic hormone which stimulates the synthesis and secretion of aldosterone. It is a very labile peptide, which is quickly inactivated in plasma and tissues by angiotensinases.

A number of factors are concerned with the regulation of renin secretion by the kidney. The short loop of negative feedback control is the direct inhibition exerted by circulating levels of angiotensin II. The long loop of this servomechanism is exerted by an indirect feedback inhibition by aldosterone on renin secretion. Renin release and eventually aldosterone secretion are increased by conditions that compromise blood flow and pressure to the kidney, severe dehydration that results in decreased intravascular blood volume, and sodium depletion.

Corticotropin (ACTH) secreted by the adenohypophysis is the principal regulator of adrenal cortical growth and secretory activity, particularly of cells in the zonae fasciculata and reticularis. The adrenal cortex secretes significant quantities of cortisol only in response to ACTH stimulation. Adrenocorticotropic hormone (ACTH) is one of many polypeptide hormones known to exert its action on target cells through the mediation of $3',5'$-cyclic AMP. The tropic hormone attaches to receptors on the plasma membrane of secretory cells in the adrenal cortex, resulting in the activation of adenylate cyclase. This enzyme converts ATP to cyclic AMP, which accumulates in the cytoplasm of target cells in the adrenal cortex and certain extraadrenal tissues. Cyclic AMP serves as the intracellular mediator of ACTH action by stimulating certain key enzymes (e.g., protein kinases) to initiate the biochemical events leading to the biosynthesis of corticosteroid hormones.

Control of the secretion of ACTH by the anterior pituitary gland is governed largely by the hypothalamus through the secretion of **corticotropin-releasing factor.** This peptide is secreted by neurons of the hypothalamus into capillaries which form the hypothalamic–pituitary portal system and convey it to corticotrophs of the pituitary. Corticotropin-releasing factor is thought to act by stimulating cyclic AMP formation within ACTH-secreting cells.

Negative feedback control of ACTH secretion is exerted primarily by the circulating level of cortisol acting on secretory cells in the hypothalamus or anterior pituitary gland or both. When plasma cortisol levels are elevated beyond the normal physiologic range (exogenous administration or cortisol-producing adrenal lesion), ACTH secretion is suppressed, secretory cells in the zonae fasciculata and reticularis decrease the rate of synthesis and release of corticosteroid hormones, and the adrenal cortex undergoes trophic atrophy. Conversely, when cortisol levels are subnormal, there is an increased release of ACTH from the pituitary gland in an attempt to increase cortisol secretion and return blood levels toward normal.

A. Diseases of the Adrenal Cortex

1. Developmental Disturbances of the Adrenal Cortex

The adrenal glands are composites of steroid-secreting cells of mesodermal origin and catecholamine-secreting cells of neural crest origin that become associated anatomically to varying degrees in different animals. The development of one component does not depend on the development of the other, nor does it depend on the presence or absence of the kidney.

Unilateral agenesis of an adrenal gland occurs occasionally in dogs, most frequently on the left side. **Total agenesis** of the adrenal cortex is fatal in all species, but the medulla is not essential for life. Developmental anomalies or diseases that decrease or prevent the production or release of ACTH from the fetal pituitary result in hypoplasia or atrophy of the zonae fasciculata and reticularis but not the zona glomerulosa. **Accessory adrenal cortical** tissue is frequent in many species. It may be found in the capsule of adrenal, in the periadrenal or perirenal adipose tissue, and in the mesorchium. Accessory adrenocortical tissue is seen frequently in the vicinity of the equine testis.

Hypoplasia of the adrenal cortex is associated with maldevelopment of the hypophysis associated with anencephaly, in some cases of cyclopia, and in hypophyseal aplasia. The adrenal cortices are small and histologically have nests of cortical cells without zonal development, similar to the arrangement of the fetal adrenal cortex. The adrenal medulla is normal.

Foci of **hematopoietic cells** are found incidentally in the adrenal glands and usually are not associated with anemia or other evidence of bone marrow depression or extramedullary myelopoiesis. They are observed in cattle as round white foci (up to 3–4 mm in diameter) and histologically are composed predominantly of eosinophils. Foci of lymphopoiesis are seen in the cortex of both sheep and cattle and also may be present in the medulla.

2. Congenital Enzyme Defects of Adrenal Cortical Steroidogenesis

Several genetically determined enzyme defects in the adrenal cortex of human beings interfere with the synthesis of certain cortical steroids and lead to varying degrees of hypoadrenocorticism. For example, a lack of 3-hydroxycorticosteroid dehydrogenase blocks the synthesis of cortisol, aldosterone, and adrenal sex steroids, whereas a deficiency of 18-hydroxy dehydrogenase results in an isolated deficiency of aldosterone. It is uncertain whether any of these congenital enzyme defects occurs in domestic animals.

Rabbits (III VD/J strain) with a congenital (autosomal recessive gene) deficiency of cholesterol-20α-hydroxylase in the adrenal cortex and testis have marked adrenal cortical hyperplasia and feminization at birth. This is a lethal trait and results in death soon after birth. The blockage at an early stage of steroidogenesis in the adrenal cortex interferes with the ability to synthesize cortisol (probably other corticosteroids as well) and leads to a marked accumulation of substrate (e.g., cholesterol) in cortical cells. The marked hyperplasia of the zonae fasciculata and reticularis in the adrenal cortex is a response to the elevated ACTH levels stimulated by the low blood cortisol. The hyperplastic cortical cells are vacuolated and lipid laden because of the accumulation of cholesterol. The lack of cholesterol-20α-hydroxylase in the testis interferes with the synthesis of androgenic steroids. Inadequate androgen levels during development result in a failure of virilization of structures derived from the urogenital sinus and account for the finding of female external genitalia in ~80% of rabbits with the hyperplasia (*ah/ah*) genotype.

3. Degenerative Changes of the Adrenal Cortex

Mineralization of the adrenal glands as extensive deposits of calcium salts occurs frequently in adult cats. A 30% incidence has been reported for cats compared with <6% for dogs. Mineralization also is common in adrenal glands of monkeys. The cause is unknown. The mineral deposits, although often bilateral and extensive, usually are not associated with signs attributable to hypoadrenocorticism. The mineralized adrenals may be detected at autopsy or during routine radiographic evaluation of the abdomen.

Extensively mineralized adrenal glands are coarsely nodular, firm, and mottled with multiple yellow-white foci throughout the cortex and extending into the medulla. They are gritty and difficult to cut. Histologically, there are large areas of necrosis with mineral deposition adjacent to areas of nodular regenerative hyperplasia of remaining viable cortical cells. These hypertrophic cortical cells have abundant lipid in their expanded cytoplasmic area and appear to be able to maintain near-normal blood cortisol levels in response to an apparent increased secretion of ACTH.

Capsular sclerosis affects the adrenal glands of old cows often with ovarian follicular cysts. The connective tissue changes often proceed to collagenous and osseous meta-

plasia in old bulls. There may be an associated moderate reduction in the width of the cortex.

Amyloid deposition in the adrenal glands usually involves the cortex and not the medulla. It occurs in all species and is regularly part of the syndrome of generalized amyloidosis in cattle. Affected adrenal cortices often are widened, and the amyloid deposits may be grossly visible as translucent areas. Amyloid deposition begins around the sinusoids in the inner portions of the zona fasciculata and often is largely confined to this zone. Signs of adrenal cortical insufficiency usually do not develop.

Hemorrhages in the adrenals occur in the newborn of any species. They are presumed to be due to birth trauma. Widespread hemorrhage and early degeneration occurs in the adrenal cortex as part of the exhaustion phase of the stress response. It is seen in wild animals that die suddenly during restraint and horses that die from overexertion (e.g., struggling to free themselves from being trapped in a fence). Toxemia (e.g., intestinal torsion in horses) and septicemia also may injure the endothelial lining of adrenal sinusoids and result in extensive cortical hemorrhage (Fig. 3.37A) or the formation of hematomas.

Telangiectasis of the adrenal cortex occurs in middle-aged or older animals as single or multiple dark foci near the corticomedullary junction. These areas typically are depressed on cut surfaces. Areas of telangiectasis appear to develop subsequent to degeneration and loss of cortical cells and ectasia of adrenal sinusoids in the area. Persisting

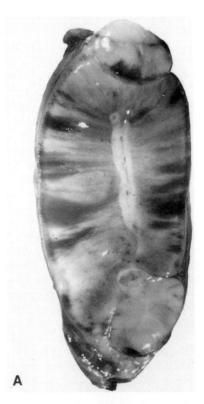

Fig. 3.37A Multiple hemorrhages in the adrenal cortex in a severely stressed horse.

parenchymal cells in these foci may either be hypertrophied and contain much lipid or be small and atrophic.

4. Inflammation of the Adrenal Cortex

Infectious and parasitic agents frequently localize in the adrenal gland and elicit varying degrees of inflammation and necrosis. Focal inflammations usually are suppurative, arising in the course of bacterial septicemia. The adrenal capsule provides an effective barrier against direct invasion by inflammatory processes in adjacent tissue. Gram-negative bacteria, especially coliforms, produce suppurative inflammation with necrosis as part of septicemia in many species (Fig. 3.37B). Emboli often lodge in the adrenal sinusoids to produce focal necrosis and suppurative infection. Tuberculosis of the adrenal is encountered mainly in cattle and humans. *Toxoplasma gondii* localizes in the adrenal cortex of many animals and produces necrosis with reticuloendothelial cell infiltration. Granulomatous inflammation due to *Histoplasma capsulatum, Coccidioides immitis,* or *Cryptococcus neoformans* occasionally occurs in dogs and cats in areas where these fungi are endemic. Multiple distinct granulomas with central areas of necrosis and mineralization may nearly destroy the entire cortex (Fig. 3.38). Small nodules of viable hypertrophic cortical cells are present between the granulomas. The reaction to *Cryptococcus* in the adrenal may vary from no reaction to an extensive infiltration by macrophages.

Several viruses affect the adrenals, and focal lesions arc expected in herpesvirus infections such as Aujeszky's disease. The adrenal lesions are characterized by intra-

Fig. 3.38 Multifocal granulomatous adrenalitis. Dog. There is thinning of the cortices with multiple granulomas (arrow).

nuclear viral inclusion bodies, necrosis, and hemorrhage in the cortex and medulla. Lymphocytes and macrophages infiltrate the necrotic and hemorrhagic foci in the later stages of the infection.

There is good evidence that the high local concentration of anti-inflammatory steroids in the adrenal cortex suppresses local cell-mediated immunity and permits the progressive growth of certain fungi, protozoa, and bacteria.

Inflammatory edema of the adrenal glands occurs in older animals with acute infectious diseases. These conditions would be more likely to cause cortical hemorrhages in young animals. The glands are enlarged due to swelling of the cortex from hyperemia and edema. The cortical cells are swollen, granular, depleted of lipid, and dissociated by the inflammatory edema.

5. Adrenal Diseases Associated with Hypoadrenocorticism: Idiopathic Adrenocortical Atrophy

The most frequently observed lesion in dogs with hypoadrenocorticism is bilateral idiopathic adrenocortical atrophy, in which all layers of the cortex are markedly reduced in thickness (Fig. 3.39A,B). There usually is a deficient production of all classes of corticosteroids (mineralocorticoids, glucocorticoids, and adrenal sex steroids). The adrenal cortex is reduced to one tenth or less its normal thickness and consists primarily of the adrenal capsule. The adrenal medulla is relatively more prominent and, with the capsule, makes up the bulk of the remaining adrenal glands. The pathogenesis of idiopathic adrenocortical atrophy is unknown, but it has been proposed that the lesion is immune mediated. Multiple foci of lymphocytes and plasma cells often are seen interspersed between adrenal sinusoids and groups of fibroblasts in the earlier stages of the disease. The entire three zones of the cortex are nearly absent in dogs that die from untreated hypoadrenocorticism. The capsule appears thickened because of collapse of the adrenal cortex plus fibroblastic proliferation.

No obvious pituitary lesions other than compensatory hyperplasia of corticotrophs have been observed in dogs with idiopathic adrenal cortical atrophy. All zones of the adrenal cortex are involved, including the zona glomeru-

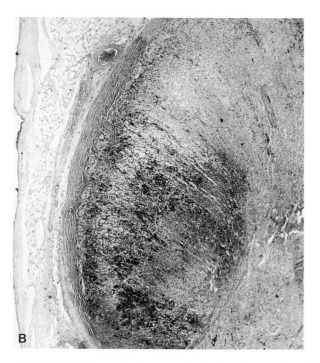

Fig. 3.37B Hemorrhagic necrosis in adrenal cortex in coliform septicemia. Lamb.

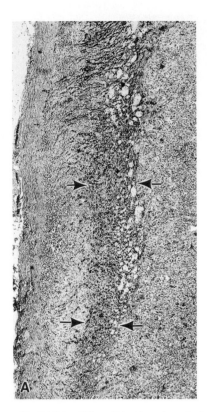

Fig. 3.39A Idiopathic adrenal cortical atrophy. Dog. All three layers of the cortex are reduced in thickness. Medulla lies between arrows.

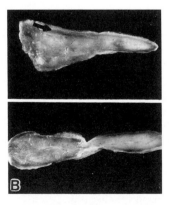

Fig. 3.39B Idiopathic adrenal cortical atrophy. Dog. All three layers of the cortex are reduced in thickness (arrow).

losa that is not under ACTH control. In comparison, trophic atrophy of the adrenal cortex secondary to a destructive pituitary lesion that decreases ACTH secretion is characterized by severe atrophy only of the inner two cortical zones. The outer zona glomerulosa that is not under ACTH control remains intact. These animals do not have the characteristic electrolyte abnormalities present in hypoadrenocorticism, since the secretion of aldosterone remains within normal limits.

Causes of **adrenocortical insufficiency** in the dog include

granulomatous inflammation caused by such diseases as histoplasmosis, blastomycosis, or tuberculosis that destroy the cortex; thrombosis of adrenal vessels with infarction of the gland associated with disseminated intravascular coagulopathies; invasion and destruction of adrenal cortex by metastatic neoplasms; hemorrhage and necrosis with subsequent replacement by fibrous connective tissue; and extensive amyloid deposition along adrenal sinusoids. Functional disturbances of hypoadrenocorticism in chronic adrenal insufficiency produce a distinct and common endocrinopathy of dogs. Many of the functional disturbances are not specific and include recurrent episodes of gastroenteritis, a slowly progressive loss of body condition, and failure to respond appropriately in stressful situations such as minor illnesses or surgery. Less commonly, the dog is presented with a shocklike coma and no history of previous illness. Although hypoadrenocorticism occurs in dogs of any breed or sex and at any age, idiopathic adrenocortical insufficiency occurs more frequently in young adult dogs. This may be related to its suspected immune-mediated pathogenesis.

A reduction in the synthesis and secretion of mineralocorticoids, primarily aldosterone, results in marked alterations of serum potassium, sodium, and chloride levels. Less potassium is excreted by the kidney, resulting in a progressive rise in serum potassium levels and severe hyperkalemia (5.5–9.0 mEq/liter). Less sodium and chloride are reabsorbed from renal tubules, leading to varying degrees of hypernatriuria and hyperchloruria and a corresponding decline in blood levels.

The severe hyperkalemia frequently produces marked cardiovascular disturbances that are reflected by changes in the electrocardiogram. Although the development of clinical signs often is insidious and not readily apparent, the dog frequently is presented with acute circulatory collapse and evidence of renal failure. A progressive decrease in blood volume contributes to hypotension, weakness, and microcardia. Peripheral circulatory collapse may result from the progressive hemoconcentration. Increased excretion of water by the kidney, due to the hyponatremia and hypochloremia, results in progressive dehydration and hemoconcentration. Emesis, diarrhea, and anorexia are common in dogs with hypoadrenocorticism, and they contribute to the animal's deterioration. Weight loss is frequently severe.

A decreased production of glucocorticoids results in several characteristic functional disturbances of hypoadrenocorticism. A failure of gluconeogenesis and increased sensitivity to insulin contribute to development of moderate hypoglycemia. Hyperpigmentation of the skin occurs in some dogs with long-standing adrenocortical insufficiency. This change results from lack of negative feedback on the pituitary gland and increased release of ACTH (and possibly melanocyte-stimulating hormone).

Peripheral lymph nodes may be moderately increased in size due to lymphoid hyperplasia. Histologically, prominent germinal centers are observed in lymph nodes, and substantial numbers of eosinophils may infiltrate the peripheral lymph sinus. There often are increased numbers of eosinophils and lymphocytes in the circulation as a result of the diminished cortisol levels or specificity spillover and binding of ACTH to MSH receptors in melanocytes of the skin.

Hematologic alterations that occur in hypoadrenocorticism include an increase in the packed cell volume and hemoglobin due to dehydration and loss of intravascular fluid volume. In some dogs the numbers of eosinophils and lymphocytes in the blood are not increased above normal, but the differential count is not consistent with the animal's severely stressed appearance. Blood urea nitrogen and creatinine levels begin to rise as renal perfusion and urine output decrease. The serum glucose level ranges from low normal to moderately subnormal in affected dogs, but this usually is not responsible for functional disturbances.

The plasma cortisol levels in dogs with hypoadrenocorticism range from 0.1 to 1.5 μg/dl. Because of the severe atrophy or destruction of the adrenal cortex, there is little or no increase in blood cortisol levels following administration of ACTH.

Carbadox (Mecadox) is a synthetic antibacterial agent and growth promotant belonging to the family of quinoxaline-di-N-oxide compounds. It has been used in young pigs as a growth promotant at 50 ppm in feed and as a preventive or therapeutic agent for diarrhea up to 100–150 ppm. Exposure to 150–200 ppm of carbadox in the feed for 2–3 weeks or 100 ppm for 4–5 weeks can produce a disease in pigs that is characterized by growth retardation and dehydration associated with lesions in the adrenal zona glomerulosa and in the kidney. Prolonged exposure to even 50 ppm can produce disease, the only safe level being 25 ppm or less. The related compounds olaquindox and cyadox have effects similar to those of carbadox, but the latter drug is somewhat less toxic.

The adrenal changes consist of disorganization of the zona glomerulosa with loss of distinction from the zona fasciculata. The cells of the glomerulosa develop hydropic degeneration, and eventually there is atrophy, mild diffuse fibrosis and mononuclear cell infiltrates in the zone, and fibrosis of the capsule, which consistently contains cells with PAS-positive cytoplasmic granules. Depending on duration and severity of exposure, adrenal lesions may persist following removal of the drug. In the kidney there is degeneration and desquamation of epithelium in the collecting tubules of the medulla and in the pelvic epithelium.

Plasma aldosterone and sodium levels are reduced significantly after 2 weeks' exposure to toxic levels of carbadox, and potassium is increased.

6. Hyperplasia of the Adrenal Cortex

Accessory cortical nodules are common in the adrenal glands of adult to aged animals and are found in the capsule, cortex, and medulla (Fig. 3.40A). Many accessory cortical nodules arise as either evaginations of outer cortex into the capsule and surrounding periadrenal fat or invaginations of the cortex into the medulla.

Nodular hyperplasia (measuring from <1.0 to 2.0 cm) also is common in the adrenal as well-defined spherical nodules in the cortex or attached to the capsule. Hyperplastic nodules are usually multiple and bilateral, yellow, and may involve any of the three zones of the cortex. Histologically, the nodules near the capsule resemble the zona glomerulosa, sometimes with areas like the zona fasciculata. The cells either are about the same size as those of the adrenal cortex or hypertrophied. The lipid content in these hyperplastic nodules is not depleted in circumstances which reduce the amount of lipid in the rest of the cortex. Hyperplastic cortical nodules are most common in older horses, dogs, and cats.

Nodular hyperplasia of the zona reticularis may appear as discrete foci of cortical parenchyma extending into the medulla and result in an irregular corticomedullary junction. This adrenal lesion has been seen in animals with functional disturbances suggesting androgen excess (e.g., greater muscle mass, well-developed crest, hypertrophy of clitoris, and involution of mammary gland).

Diffuse cortical hyperplasia results in a uniform, usually bilateral, enlargement of the adrenal cortex. There is a marked hypertrophy and hyperplasia of cells in the zonae fasciculata and reticularis in response to an autonomous hypersecretion of ACTH by a corticotroph adenoma of the pituitary. The hyperplastic cells in the zona fasciculata are vacuolated (lipid-rich) and arranged in straight columns separated by sinusoids. Cells in the outer zona glomerulosa often are compressed by the expanded inner two zones. The corticomedullary junction is irregular in diffuse hyperplasia, and projections of cortical cells often extend into the medulla. Areas of discrete nodular hyperplasia may be present in a diffusely hyperplastic cortex (Fig. 3.40B).

Hyperplasia of **zona glomerulosa** occurs as part of the compensatory mechanism of the body to a variety of factors that stimulate long-term renin release from the juxtaglomerular apparatus of the kidney. Chronic sodium depletion or potassium excess, decreased blood volume to the kidney, or a sustained reduction in systemic blood pressure all may increase aldosterone production by stimulating renin release. Renin results in the formation of angiotensin II, a tropic hormone for the zona glomerulosa of the adrenal cortex.

B. Neoplasms of the Adrenal Cortex

1. Myelolipoma

This is a benign lesion encountered in the adrenals of cattle, nonhuman primates, and infrequently in other animals. It is composed of accumulations of well-differentiated adipose connective tissue cells and hematopoietic tissue, including both myeloid and lymphoid elements. Focal areas of mineralization or bone formation may occur in the tumor. Although the origin of these nodular aggrega-

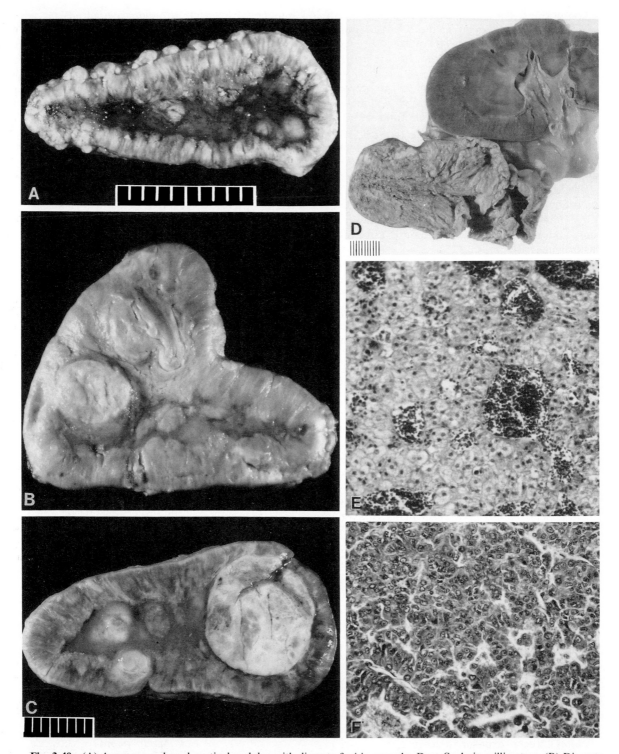

Fig. 3.40 (A) Accessory adrenal cortical nodules with discrete foci in capsule. Dog. Scale in millimeters. (B) Discrete area of nodular hyperplasia in a diffusely hyperplastic adrenal cortex. Dog with corticotroph pituitary adenoma. (C) Cortical adenoma in an adrenal with nodular hyperplasia. Dog. Scale in millimeters. (D) Functional adrenal cortical carcinoma in a dog with cortisol excess. The adrenal gland was completely incorporated within the mass at the cranial pole of the kidney. Scale in millimeters. (E) Adrenal cortical adenoma with hematopoiesis. Dog. (F) Adrenal cortical carcinoma. Dog.

tions of fat, bone, and myeloid cells is uncertain, they appear to develop by metaplastic transformation of cells in the adrenal cortex and are not interpreted to be true neoplasms.

2. Cortical Adenoma

Adenomas of the adrenal cortex are seen most frequently in old dogs (8 years and older) and sporadically in horses, cattle, and sheep. They usually are encountered as incidental findings at autopsy. Castrated male goats are reported to have a higher incidence of cortical adenomas than are intact males.

Cortical adenomas are well-demarcated, usually single, nodules in one adrenal gland, but may be bilateral. Larger cortical adenomas are yellow to red, distort the external contour of the affected gland, and are partially or completely encapsulated. Adjacent cortical parenchyma is compressed, and the tumor may extend into the medulla.

Discrete cortical adenomas often develop in an adrenal with multiple areas of nodular hyperplasia (Fig. 3.40C). Smaller cortical adenomas are more yellow and similar in color to the normal adrenal cortex because of the high lipid content. They are surrounded on all sides by mildly compressed cortex with attempts at encapsulation and may be difficult to distinguish from nodular cortical hyperplasia in old dogs. However, nodular hyperplasia consists of multiple small foci, usually in both adrenals, with no evidence of encapsulation, and often is associated with extracapsular nodules of hyperplastic cortical tissue.

Cortical adenomas are composed of well-differentiated cells that resemble secretory cells of the normal zonae fasciculata or reticularis (Fig. 3.40E). Tumor cells are arranged in broad trabeculae or nests separated by small vascular spaces. The abundant cytoplasmic area of tumor cells is lightly eosinophilic, often vacuolated, and filled with many lipid droplets. Adenomas are partially or completely surrounded by a fibrous connective tissue capsule of varying thickness and a rim of compressed cortical parenchyma. Focal areas of mineralization, hematopoiesis, and accumulations of fat cells may be found in cortical adenomas.

3. Cortical Carcinoma

Adrenal cortical carcinomas occur less often than adenomas and have been reported most often in cattle and older dogs but infrequently in other species. Carcinomas develop in adult to older individuals, and there is no apparent breed or sex prevalence.

Adrenal carcinomas are larger than adenomas and more likely to be bilateral. In dogs they are composed of a variegated, yellow-red, friable tissue that incorporates the affected adrenal gland (Fig. 3.40D). They are often fixed in location because of extensive invasion of surrounding tissues with extension into the posterior vena cava, forming large tumor cell thrombi. Carcinomas may attain considerable size in cattle (up to 10 cm or more in diameter) and may have multiple areas of mineralization or ossification.

Carcinomas are composed of more highly pleomorphic secretory cells than are adenomas that are subdivided into small groups by a fibro-vascular stroma of varying thickness. The architecture of the affected adrenal usually is completely obliterated by the carcinoma. The pattern of growth varies between individual tumors and within the same carcinoma. Tumor cells usually are large and polyhedral with prominent nucleoi and a densely eosinophilic or vacuolated cytoplasm (Fig. 3.40F). Areas of hemorrhage within the tumors are common because of rupture of thin-walled sinusoids.

Carcinomas and adenomas of the adrenal cortex in dogs occasionally are functional and secrete excessive amounts of cortisol. The clinical picture of adrenal cortical carcinoma may be complicated by compression of adjacent organs by the large tumor, invasion into the aorta or posterior vena cava leading to intra-abdominal hemorrhage, and metastases to distant sites.

Functional cortical adenomas and carcinomas are associated with atrophy of the contralateral cortex because of negative-feedback inhibition of the pituitary ACTH secretion by the elevated blood cortisol levels. The atrophic cortex consists primarily of the adrenal capsule and zona glomerulosa. Few secretory cells remain in the zonae fasciculata and reticularis. A similar atrophy may be present in the remnants of compressed adrenal cortex around functional adenomas. The adrenal medulla in the contralateral gland appears expanded and relatively more conspicuous because of the lack of cortical parenchyma.

4. Hypercortisolism

The manifestations and lesions associated with hyperadenocorticism, mainly in dogs, result from a long-term overproduction of cortisol by hyperactive adrenal cortices. As a result of the cortisol excess, affected dogs develop a spectrum of functional disturbances and lesions from the combined gluconeogenic, lipolytic, protein catabolic, and antiinflammatory effects of the glucocorticoid hormones on many organs. The course of the disease is insidious and slowly progressive. Cortisol excess is one of the most frequent endocrinopathies in adult to aged dogs but occurs infrequently in other species of domestic animals.

The elevation in circulating cortisol levels in dogs with hyperadrenocorticism may result from one of several different pathogenetic mechanisms. The most common mechanism is a functional corticotroph (ACTH-secreting) adenoma of the pituitary gland (pars distalis or pars intermedia in dogs), which causes bilateral adrenal cortical hypertrophy and hyperplasia (Fig. 3.6).

Hypercortisolism associated with idiopathic adrenal cortical hyperplasia occurs most frequently in poodles and occasionally in other breeds (Fig. 3.41A). The cortex of both adrenal glands is thickened considerably owing to diffuse or nodular hyperplasia of secretory cells, primarily in the zona fasciculata, and the corticomedullary junction is irregular (Figs. 3.41B and 3.42E).

Functional adrenal neoplasms are an infrequent

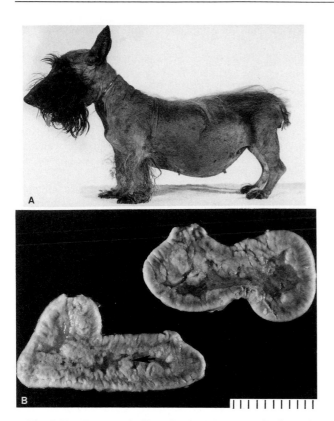

Fig. 3.41 Hypercortisolism. Dog. (A) Hypercortisolism with alopecia extending over most of the body. The skin is thin, wrinkled, and hyperpigmented. Muscle asthenia is evident from the pendulous abdomen and swayed back. (B) Idiopathic adrenocortical hyperplasia results in bilateral cortical thickening and irregularity of corticomedullary junction (arrow). Scale in millimeters.

(10–15% of cases) cause of cortisol excess in the dog. Many of the clinical signs and lesions of naturally occurring hyperadrenocorticism can be induced by the long-term, daily administration of large doses of corticosteroids for the treatment of other diseases.

The appetite and intake of food often are increased as a direct result of either the hypercortisolism or the involvement of hypothalamic appetite centers by a pituitary tumor. The muscles of the extremities and abdomen are weakened and atrophied. The loss in tone of the abdominal muscles and muscles of the abaxial skeleton results in gradual abdominal enlargement, lordosis, muscle trembling, and a straight-legged, skeletal-braced posture (Fig. 3.41A). Profound atrophy of the temporal muscles may result in obvious concave indentations and readily palpable prominences of underlying skull bones. Hepatomegaly due to increased fat and glycogen deposition may contribute to the development of the distended, often pendulous, abdomen. The muscular asthenia and wasting is the result of increased catabolism of structural proteins, combined with diminished protein synthesis under the influence of long-term cortisol excess.

Skin lesions occur in >90% of the dogs with hypercortisolism. The initial changes in the skin often are observed over points of wear such as neck, flanks, behind the ears, and over bony prominences. These initial skin changes spread in a bilaterally symmetric pattern to involve a significant percentage of the body surface. The basic skin lesion is atrophy of the epidermis and pilosebaceous apparatus, combined with loss of collagen and elastin in the dermis and subcutis (Fig. 3.42D). The skin is of fine texture and often paper thin. Most of the hair follicles are inactive and in the telogen phase. Accumulation of keratin in the atrophic hair follicles disrupts the base of attachment of the hair shaft, thereby permitting easy dislodgement from the follicle.

The epidermis is markedly thinned and consists of only one or two cell layers derived from the basilar stratum germinativum. However, the stratum corneum is considerably thickened owing to the accumulation of multiple layers of keratin on the surface (Fig. 3.42).

Cutaneous mineralization is a characteristic lesion in dogs (~40%) with hypercortisolism. Although mineral deposition may occur in any region of the body, the dorsal midline, ventral abdomen, and inguinal region are frequently affected sites. Numerous mineral crystals are deposited along collagen and elastin fibers in the dermis and may protrude through the atrophic and thinned epidermis (Fig. 3.43B).

Fig. 3.42 Skin of dog with hypercortisolism. The skin surface is rough and scaly due to the accumulation of interlapping plates of keratin (K) in the stratum corneum. A few hair shafts (arrow) are emerging from the hair follicles. Scanning electron micrograph.

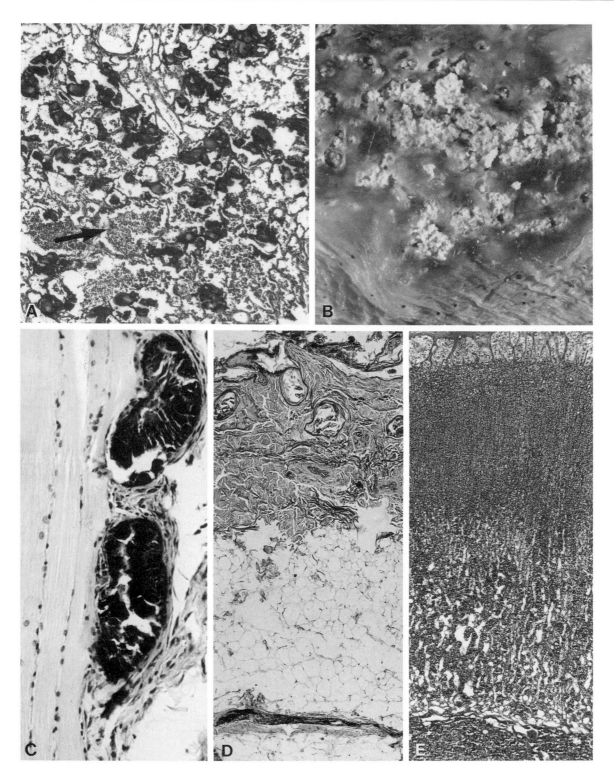

Fig. 3.43 Hypercortisolism. Dog. (A) Pulmonary mineralization and suppurative bronchopneumonia (accumulation of leukocytes, arrow). (B) Mineralization of the skin. Note empty hair follicles. (C) Mineralization of connective tissue between muscle bundles. (D) Atrophy of epidermis with accumulation of layers of keratin and follicular keratosis. (E) Hyperplasia of zonae fasciculata and reticularis in adrenal cortex of a dog with a corticotroph adenoma of the pituitary. Outer zona glomerulosa is approximately normal thickness.

The epidermis remains intact in less severe cases and appears irregularly elevated by the opaque white deposits of mineral. A narrow rim of hyperemia and foreign-body granulomatous inflammation often surrounds the areas of mineralization. The mineral deposits occur in dogs with normal blood calcium and phosphorus levels. The pathogenesis of this manifestation of hypercortisolism most likely is related to the gluconeogenic and protein catabolic action of cortisol, resulting in the rearrangement of the molecular structure of proteins such as collagen and elastin, and the formation of an organic matrix that attracts and binds calcium.

Severe mineralization also occurs in several other tissues. The lungs are most frequently affected (>90%) (Fig. 3.43A), but active skeletal muscles (Fig. 3.43C) and the wall of the stomach may have extensive areas of mineralization. Large, basophilic mineral crystals are deposited in interalveolar and bronchiolar walls with little accompanying inflammation. Pulmonary function may be compromised when mineralization is extensive.

The syndrome of long-term cortisol excess often is complicated by an increased susceptibility to bacterial infections of the skin, urinary tract, conjunctiva, and lung. Multifocal areas of suppurative folliculitis and dermatitis develop near the lip folds and footpads and elsewhere in the skin. Suppurative bronchopneumonia frequently is a serious complication, since it develops in lungs whose function is already partially impaired by extensive mineralization of alveolar walls and terminal bronchioles.

The consistent effects that elevated blood levels of corticosteroids have on the hematopoietic system may help establish a diagnosis of adrenal cortical disease. Involution of lymphatic tissue results in a significant lymphopenia. There is intravascular destruction of eosinophils or sequestration of circulating eosinophils in the lung and spleen, which leads to eosinopenia. The total white blood cell count is elevated with hypercortisolism, and the percentage of neutrophils is increased. An excess of corticosteroids can cause a leukocytosis by stimulating the formation of neutrophils and decreasing the migration of these cells into areas of infection. The nuclei of neutrophils in dogs and cats with naturally occurring or iatrogenic hyperadrenocorticism may become hypersegmented as a reflection of their prolonged life span in the circulation.

Low doses of dexamethasone (15 μg/kg) suppress ACTH production and subsequently plasma cortisol levels in normal dogs but do not suppress production in dogs with pituitary-dependent hyperadrenocorticism or adrenal cortical neoplasms. High doses of dexamethasone (1.0 mg/kg) usually suppress plasma cortisol levels in dogs with pituitary-dependent hyperadrenocorticism but do not significantly influence levels in dogs with adrenal cortical neoplasms. Radioimmunoassays for plasma ACTH in the dog demonstrate a mean concentration of 45.8 pg/ml (range, 17–98 pg/ml). Assays for plasma ACTH are useful in differentiating pituitary-dependent and other causes of adrenal cortical hyperplasia and the syndrome of cortisol excess. Dogs with functional adrenal cortical neoplasms have plasma ACTH concentrations two standard deviations or more below the mean value for normal dogs.

Dogs with the syndrome of cortisol excess caused by diffuse cortical hyperplasia (corticotroph adenoma of pituitary or idiopathic adrenal hyperplasia) are managed medically by the oral administration of the drug *ortho,-para'*2, 2-bis(2-chlorophenyl-4-chlorophenyl)-1, 1-dichlorethane (*o,p'* DDD). The drug is an isomer of the insecticide dichlorodiphenyldichloroethane (DDD) and is selectively cytotoxic, at the recommended dosage, for the zonae fasciculata and reticularis of the canine adrenal cortex. In dogs the cells of these zones are highly susceptible to the cytotoxic effects of *o,p'*DDD. After 5–10 days of daily treatment (50 mg/kg), the secretory cells of the inner two zones are swollen, vacuolated, and undergo coagulation necrosis. After *o,p'*DDD administration, mitochondria are vacuolated, profiles of smooth endoplasmic reticulum are disrupted, and lipid accumulates in secretory cells of the zona fasciculata. Necrosis is followed progressively by collapse of the two inner zones, reduction in cortical thickness, and dilatation of vascular sinusoids. After several months of treatment with *o,p'*DDD, the cortex is markedly thinned and consists of an intact zona glomerulosa surrounded by a fibrous capsule. The levels of plasma cortisol decline progressively in dogs with hyperadrenocorticism treated with *o,p'*DDD. The exaggerated increase in plasma corticosteroids as a response to ACTH stimulation in dogs with adrenal cortical hyperplasia is eliminated after treatment with *o,p'*DDD. The weekly doses of *o,p'*DDD must be continued for the life of the dog in order to prevent recurrence of the functional disturbances and lesions of hypercortisolism. If the intermittent doses of *o,p'*DDD are discontinued, small nests of viable cortical cells which remain in the zonae fasciculata and reticularis will regenerate the pretreatment diffuse hyperplasia, with the return of the functional disturbances and the biochemical alterations. Dose levels of *o,p'*DDD which are effectively cytotoxic to hyperplastic adrenal cortices will not destroy functional adrenal cortical adenomas and carcinomas in the dog. Higher dose levels of the drug are toxic to several parenchymal organs.

VI. Adrenal Medulla

The adrenal medulla has a different embryonic derivation than the cortical portion of the gland. It is derived from neuroectoderm of the neural crest. Secretory cells in the adrenal medulla are involved in biosynthesis of the catecholamine group of hormones.

The main biosynthetic pathway for catecholamines in mammals starts with tyrosine derived either from the diet or formed through hydroxylation of phenylalanine in the liver. Tyrosine is first converted to 1-dihydroxyphenylalanine (Dopa) by tyrosine hydroxylase, and Dopa is then decarboxylated to 1-dihydroxyphenylethylamine (Dopamine) by l-amino acid decarboxylase, which subsequently undergoes beta-hydroxylation (by Dopamine beta-oxidase) to norepinephrine.

In mammals, the medulla is completely surrounded by the adrenal cortex, and venous drainage from the cortex bathes the medullary cells with blood containing the highest concentration of corticosteroids of any fluid in the body. This close anatomic association between the adrenal cortex and medulla in mammals is not fortuitous, because the N-methylating enzyme (phenylethanolamine-N-methyl transferase) that converts norepinephrine to epinephrine is corticosteroid hormone dependent.

Though the adrenal medulla does not appear to be absolutely essential for life, there are a number of physiologic conditions (e.g., sudden hypoglycemia and acute stress) in which release of epinephrine from the medulla is beneficial. Stimulation by preganglionic sympathetic fibers that synapse with secretory cells of the adrenal medulla also increases the output of catecholamines, primarily by increasing tyrosine hydroxylase and Dopamine beta-oxidase activity.

Catecholamines are degraded in peripheral tissues by catechol-O-methyl transferase and monamine oxidase to several metabolites, such as vanillylmandelic acid, metanephrine, and normetanephrine that are excreted in the urine. Measurement of the urinary excretion of these metabolites plus the relatively small amount of free catecholamines excreted in the urine is useful in the diagnosis of certain disorders characterized by overproduction of adrenal medullary hormones.

A. Neoplasms of Medullary Secretory Cells

Pheochromocytomas are the most common tumors in the adrenal medulla of animals. They develop most often in cattle and dogs, and infrequently in other domestic animals. In bulls and humans, pheochromocytomas develop concurrent with calcitonin-secreting C-cell tumors of the thyroid gland. This appears to represent a neoplastic transformation of multiple types of endocrine cells of neuroectodermal origin in the same individual. Malignant pheochromocytoma often is used to designate medullary tumors that invade through the adrenal capsule and into adjacent structures or that metastasize to distant sites.

Pheochromocytomas are tumors of chromaffin cells and are almost always located in the adrenal gland of animals. They can be either unilateral or bilateral. Although size varies considerably, they often are large (10 cm or more in diameter) and incorporate the majority of the affected adrenal. A small remnant of the adrenal gland often can be found at one pole (Fig. 3.44A). Smaller tumors are completely surrounded by a thin compressed rim of adrenal cortex. Large pheochromocytomas are multilobular and variegated light brown to yellow-red because of areas of hemorrhage and necrosis.

A valuable aid in macroscopic diagnosis of pheochromocytoma is the Henle chromoreaction with either potassium dichromate or iodate. Application of Zenker's solution to the flat-cut surface of a freshly resected tumor results in oxidation of catecholamines, forming a dark brown pigment within 20 min (Fig. 3.44B).

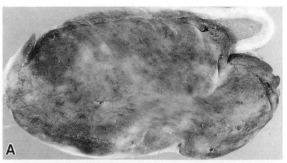
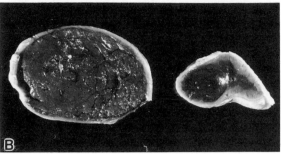

Fig. 3.44 (A) Pheochromocytoma. Dog. (B) Chromaffin-positive reaction in bilateral pheochromocytomas. Ox.

Tumor cells in pheochromocytomas vary from small cuboidal or polyhedral cells to large pleomorphic cells with multiple hyperchromatic nuclei. The cytoplasmic area is lightly eosinophilic, finely granular, and often indistinct because of early onset of autolysis in adrenal medullary tissue. Tumor cells are characteristically subdivided into small lobules by fine connective tissue septa and capillaries (Fig. 3.45).

Pheochromocytomas are composed of either epinephrine-secreting cells, norepinephrine-secreting cells, or both. The principal distinguishing feature of these two

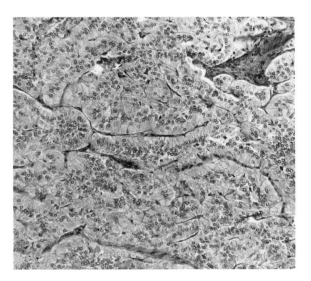

Fig. 3.45 Well-differentiated pheochromocytoma. Bull.

populations of cells is in the fine structure of their secretory granules. Pheochromocytomas, from which norepinephrine is the principal catecholamine extracted, are composed of cells with secretory granules that have an eccentrically situated electron-dense core surrounded by a wide submembranous space (Fig. 3.46A). When epinephrine is the principal catecholamine, the secretion granules in tumor cells have a coarsely granular internal core of lower density and a narrow submembranous space (Fig. 3.46B).

Small pheochromocytomas are well encapsulated and remain confined to the affected adrenal gland. Larger malignant pheochromocytomas may exert pressure on and invade adjacent tissues, particularly the vena cava and aorta. Tumor cells often invade the capsule and penetrate through the wall of the posterior vena cava, forming a large thrombus that partially occludes the venous return from the posterior extremities. Metastases occur in ~50% of pheochromocytomas to the liver, regional lymph nodes, spleen, and lungs in dogs.

Functional pheochromocytomas occur infrequently in animals. Tachycardia, edema, and cardiac hypertrophy observed in several dogs and horses with pheochromocytomas have been attributed to excessive catecholamine secretion. Arteriolar sclerosis and widespread medial hy-

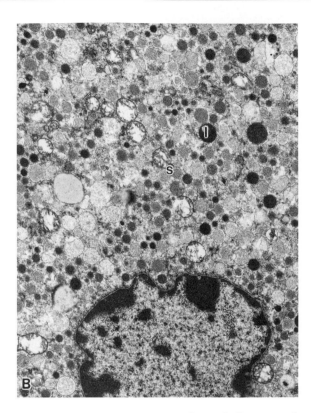

Fig. 3.46B Pheochromocytoma from a bull composed predominantly of cells with characteristics of epinephrine-secreting type of cell of adrenal medulla. Medullary cells contain many secretory granules (S) of low electron density and a narrow submembranous space, occasional lysosomal bodies (l), and few organelles.

perplasia of arterioles have been reported in dogs with pheochromocytomas that were associated with clinical signs suggestive of paroxysmal hypertension. The catecholamine content in pheochromocytomas from bulls with concurrent C-cell tumors of the thyroid gland is higher than that in the normal bovine adrenal medulla. Urinary excretion of vanillylmandelic acid and free unconjugated catecholamines is elevated in bulls with pheochromocytomas.

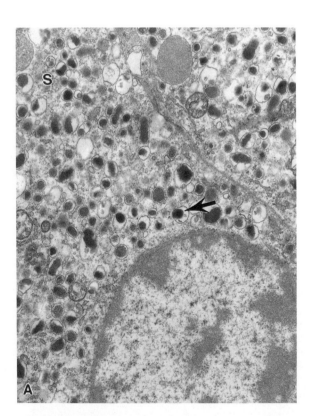

Fig. 3.46A Pheochromocytoma from a bull composed predominantly of cells with ultrastructural characteristics of norepinephrine-secreting cells in the adrenal medulla. Secretory granules (S) have an eccentric electron-dense core surrounded by a prominent submembranous space (arrow).

Fig. 3.47 Bilateral diffuse hyperplasia of adrenal medulla. Bull. Scale in millimeters. (Courtesy of *Veterinary Pathology* 1981.)

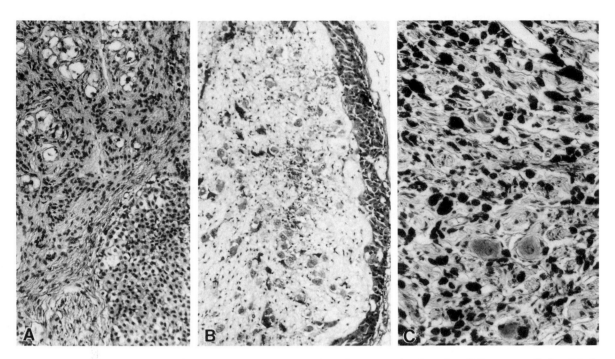

Fig. 3.48 (A) Ganglioneuroma in adrenal medulla. Ox. (B) Ganglioneuroma in adrenal medulla composed of multipolar neurons compressing surrounding cortex. (C) Pigmented ganglioneuroma in adrenal medulla. Ox.

B. Adrenal Medullary Hyperplasia

Diffuse or nodular adrenal medullary hyperplasia appears to precede the development of pheochromocytoma in bulls and humans with C-cell tumors of the thyroid gland. This diffuse proliferation of chromaffin cells is non-encapsulated but compresses the surrounding adrenal cortex (Fig. 3.47). The hyperplastic cells are columnar to cuboidal and have a lightly basophilic cytoplasm. In bulls with prominent diffuse medullary hyperplasia, there are often a few small foci of intense nodular proliferation of medullary cells. Medullary hyperplasia is detected by an increased total adrenal weight, a decrease in cortico-medullary ratio due to an increase in the size and number of medullary cells, and the presence of frequent mitotic figures in the adrenal medulla.

C. Neoplasms of Cells of the Sympathetic Nervous System in Adrenal Medulla

1. Neuroblastoma and Ganglioneuroma

Neuroblastomas arise from primitive neuroectodermal cells, often in younger animals, and form a large intra-abdominal mass. Ganglioneuromas are usually well-differentiated small tumors that have multipolar ganglion cells and neurofibrils. In cattle, they are often pigmented with melanin.

Neuroblastomas are differentiated from pheochromocytomas by being composed of small tumor cells with hyperchromatic nuclei and scant amounts of cytoplasm. They often resemble lymphocytes and tend to form pseudorosettes. Neurofibrils or unmyelinated nerve fibers can be demonstrated in neuroblastomas.

Ganglioneuromas are benign tumors in the medulla composed of multipolar ganglion cells and neurofibrils with a prominent fibrous connective tissue stroma (Fig. 3.48A–C). The surrounding adrenal cortex is severely compressed. Neoplastic cells in medullary tumors occasionally may differentiate in two directions, resulting in adjacent pheochromocytomas and ganglioneuromas in the same adrenal gland.

2. Metastatic Tumors of the Adrenal Medulla

Secondary foci of neoplastic growth in the adrenal glands are common in cases of disseminated neoplasia and originate primarily as emboli. The metastases are usually bilateral, and the medulla is an early site of tumor growth. Direct invasion of the adrenal glands may occur from primary or secondary tumors of contiguous structures.

Involvement of the adrenal glands by lymphosarcoma and mammary carcinoma may be extensive, and both adrenals are likely to be affected. Tumor emboli lodge in sinusoids of the adrenal medulla and may grow to an extent that they compress the surrounding adrenal cortex.

Bibliography

Anderson, M. P., and Capen, C. C. Diseases of the endocrine system. *In* "Pathology of Laboratory Animals," K. Benirschke *et al.* (eds.), Vol. 1, pp. 423–508. New York, Springer-Verlag, 1978.

Anis, J. R. Adrenal cortical failure in the dog. *Vet Med* **55:** 35–38, 1960.

Appleby, E. C., and Sohrabi-Haghdoost, I. Cortical hyperplasia of the adrenal gland in the dog. *Res Vet Sci* **29:** 190–197, 1980.

Bath, M. L., and Hill, F. W. G. Adrenocortical insufficiency in the dog. *Aust Vet J* **54:** 128–132, 1978.

Black, H. E., Rosenblum, I. Y., and Capen, C. C. Chemically induced (streptozotocin-alloxan) diabetes mellitus in the dog: Biochemical and ultrastructural studies. *Am J Pathol* **98:** 295–310, 1980.

Carney, J. A., Sizemore, G. W., and Sheps, S. G. Adrenal medullary disease in multiple endocrine neoplasia, type 2: Pheochromocytoma and its precursors. *Am J Clin Pathol* **66:** 279–290, 1976.

Chastain, C. B., Mitten, R. W., and Kluge, A. P. An ACTH-hyperresponsive adrenal carcinoma in a dog. *J Am Vet Med Assoc* **172:** 586–588, 1978.

Conaway, C. H. Adrenal cortical rests of the ovarian hilus of the patas monkey. *Folia Primatol* **11:** 175–180, 1969.

Cornelius, L. M. Canine distemper presenting as acute adrenocortical insufficiency: A case report. *J Am Anim Hosp Assoc* **10:** 153–157, 1974.

Dahme, E., and Schiefer, B. Intracranielle Geschwulste bei Tieren. *Zentralbl Veterinaermed* **7:** 341–363, 1960.

Deftos, L. J., Bone, H. G., and Parthemore, J. G. Immunohistological studies of medullary thyroid carcinoma and C-cell hyperplasia. *J Clin Endocrinol Metab* **51:** 857–682, 1980.

DeLellis, R. A. *et al.* Adrenal medullary hyperplasia. A morphometric analysis in patients with familial medullary thyroid carcinoma. *Am J Pathol* **83:** 177–196, 1976.

Duckett, W. M. *et al.* Functional pheochromocytoma in a horse. *Compend Contin Ed* **9:** 1118–1121, 1987.

Feldman, E. C. Effect of functional adrenocortical tumors in plasma cortisol and corticotropin concentrations in dogs. *J Am Vet Med Assoc* **178:** 823–826, 1981.

Fox, R. R., and Crary, D. D. Genetics and pathology of hereditary adrenal hyperplasia in the rabbit. *J Hered* **69:** 251–254, 1978.

Frenkel, J. K. Infections involving the adrenal cortex. *In* "The Adrenal Cortex," H. D. Moon (ed.). Paul B. Hoeber, 1961.

Frenkel, J. K. Adrenal infection, necrosis, and hypercorticism. *Am J Pathol* **86:** 749–752, 1977.

Friedman, N. B. The pathology of the adrenal gland in Addison's disease with special reference to adrenocortical contraction. *Endocrinology* **42:** 181–200, 1948.

Greene, C. E. *et al.* Myopathy associated with hyperadrenocorticism in the dog. *J Am Vet Med Assoc* **174:** 1310–1315, 1979.

Hadlow, W. J. Adrenal cortical atrophy in the dog. Report of three cases. *Am J Pathol* **29:** 353–361, 1953.

Head, K. W., and West, G. B. Pathological and pharmacological examination of a canine tumor of the adrenal medulla. *J Comp Pathol* **68:** 167–173, 1958.

Hoenig, E. M. *et al.* The early development and fine structure of allergic adrenalitis. *Lab Invest* **22:** 198–205, 1970.

Howard, E. B., and Nielsen, S. W. Pheochromocytomas associated with hypertensive lesions in dogs. *J Am Vet Med Assoc* **147:** 245–252, 1965.

Howell, J. McC., and Pickering, C. M. Calcium deposits in the adrenal glands of dogs and cats. *J Comp Pathol* **74:** 280–285, 1964.

Hullinger, R. L. Adrenal cortex of the dog (*Canis familiaris*). Histomorphological changes during growth, maturity, and aging. *Zentralbl Veterinaermed [C]* **7:** 1–27, 1978.

Kelly, D. F. Neuroblastoma in the dog. *J Pathol* **116:** 209–212, 1975.

Kelly, D. F., Siegel, E. T., and Berg, P. The adrenal gland in dogs with hyperadrenocorticalism. A pathologic study. *Vet Pathol* **8:** 385–400, 1971.

Kral, F. Epileptiform manifestations of endocrine disturbances associated with adrenal tumors. *J Am Vet Med Assoc* **118:** 235–239, 1951.

Lauper, N. T. *et al.* Pheochromocytoma: Fine structural, biochemical, and clinical observations. *Am J Cardiol* **30:** 197–204, 1972.

Morales, G. A., and Nielsen, S. W. Canine adrenocortical atrophy. Review of literature and a report of two cases. *J Small Anim Pract* **11:** 257–263, 1970.

Mulnix, J. A. Hypoadrenocorticism in the dog. *J Am Anim Hosp Assoc* **7:** 220–241, 1971.

Mulnix, J. A., and Smith, K. W. Hyperadrenocorticism in a dog: A case report. *J Small Anim Pract* **16:** 193–200, 1975.

Nelville, A. M. The nodular adrenal. *Invest Cell Pathol* **1:** 99–111, 1978.

Peterson, M. E., and Feinman, J. M. Hypercalcemia associated with hypoadrenocorticism in sixteen dogs. *J Am Vet Med Assoc* **181:** 802–804, 1982.

Peterson, M. E., Gilbertson, S. R., and Drucker, W. D. Plasma cortisol response to exogenous ACTH in 22 dogs with hyperadrenocorticism caused by adrenocortical neoplasia. *J Am Vet Med Assoc* **180:** 542–544, 1982.

Powers, J. M. *et al.* Adrenal cortical degeneration and regeneration following administration of DDD. *Am J Pathol* **75:** 181–194, 1974.

Power, S. B. *et al.* Accidental carbadox overdosage in pigs in an Irish weaner-producing herd. *Vet Rec* **124:** 367–370, 1989.

Richter, W. R. Tubular adenomata of the adrenal of the goat. *Cornell Vet* **47:** 558–577, 1957.

Rijnberk, A., der Kinderen, P. J., and Thijssen, J. H. H. Canine Cushing's syndrome. *Zentralbl Veterinaermed* **16:** 13–28, 1969.

Ross, M. A., and Innes, J. R. M. Dystrophic calcification in the adrenal glands of monkeys, cats, and dogs. *Arch Pathol* **60:** 655–662, 1955.

Schaer, M., and Chen, C. L. A clinical survey of 48 dogs with adrenocortical hypofunction. *J Am Anim Hosp Assoc* **19:** 443–452, 1983.

Schechter, R. D. *et al.* Treatment of Cushing's syndrome in the dog with an adrenocorticolytic agent (*o,p'*DDD). *J Am Vet Med Assoc* **162:** 629–639, 1973.

Siegel, E. T. Determination of 17-hydroxycorticosteroids in canine urine. *Am J Vet Res* **26:** 1152–1156, 1965.

Siegel, E. T., Kelly, D. F., and Berg, P. Cushing's syndrome in the dog. *J Am Vet Med Assoc* **150:** 760–766, 1967.

Sponenberg, D. P., and McEntee, K. Pheochromocytomas and ultimobranchial (C-cell) neoplasms in the bull: Evidence of autosomal dominant inheritance in the Guernsey breed. *Vet Pathol* **20:** 396–400, 1983.

Twedt, D. C., and Wheeler, S. L. Pheochromocytoma in the dog. *Vet Clin North Am: Small Anim Pract* **14:** 767–782, 1984.

van den Brom, W. E. *et al.* Uptake of [131]I-19-cholesterol by normal and spontaneously hyperfunctioning canine adrenals. *Eur J Nucl Med* **4:** 61–67, 1979.

Van Der Molen, E. J. "Toxicological Pathology of Quinoxaline-di-*N*-oxide Feed Additives for Young Pigs." Utrecht, The Netherlands, Drukkerij Elinkwijk BV, 1989.

West, J. L. Bovine pheochromocytoma: Case report and review of literature. *Am J Vet Res* **36:** 1371–1373, 1975.

White, R. A. S., and Cheyne, I. A. Bone metastases from a

phaeochromocytoma in the dog. *J Small Anim Pract* **18:** 579–584, 1977.

Wright, B. J., and Conner, G. H. Adrenal neoplasms in slaughtered cattle. *Cancer Res* **28:** 251–263, 1968.

Yarrington, J. T., and Capen, C. C. Ultrastructural and biochemical evaluation of adrenal medullary hyperplasia and pheochromocytoma in aged bulls. *Vet Pathol* **18:** 316–325, 1981.

VII. Chemoreceptor Organs

The chemoreceptor organs (nonchromaffin paraganglia) are sensitive to changes in the blood carbon dioxide content, pH, and oxygen tension, thereby aiding in the regulation of respiration and circulation. Carotid and aortic bodies can initiate an increase in the depth, minute volume, and rate of respiration by way of parasympathetic nerves and result in an increased heart rate and elevated arterial blood pressure by way of the sympathetic nervous system. They are composed normally of parenchymal (chemoreceptor, glomus) cells and stellate sustentacular cells. Nerve endings with synaptic vesicles as well as nerve fibers are seen in close association with the chemoreceptor cells.

Chemoreceptor tissue is present at several sites in the body including the carotid body, aortic bodies, nodose ganglion of the vagus nerve, ciliary ganglion in the orbit, pancreas, bodies on the internal jugular vein below the middle ear, and glomus jugulare along the recurrent branch of the glossopharyngeal nerve. Aortic bodies of dogs normally consist of clusters of cells embedded in adventitia at multiple sites including the innominate artery immediately below the origin of the right subclavian artery, on the anterior surface of the aortic arch, beneath the arch between the aorta and pulmonary artery, between the ascending aorta and pulmonary artery near the left coronary artery, and scattered in the wall of the pulmonary artery. Although chemoreceptor tissue appears to be widely distributed in the body, tumors develop principally in the aortic and carotid bodies of animals.

Aortic body tumors are encountered more frequently than neoplasms of the carotid body in animals, but the reverse is true for humans. These tumors develop primarily in dogs, and infrequently in cats and cattle. Brachycephalic breeds of dogs such as the boxer and Boston terrier are highly predisposed to develop tumors of the aortic and carotid bodies.

Aortic body tumors appear most frequently as single masses or as multiple nodules within the pericardial sac near the base of the heart (Fig. 3.49). They vary considerably in size (0.5–12.5 cm) with carcinomas, in general, being larger than adenomas. Solitary small **adenomas** either are attached to the adventitia of the pulmonary artery and ascending aorta or are embedded in the adipose connective tissue between these major vascular trunks. They have a smooth external surface and on cross section are white, mottled with red to brown areas. Larger adenomas may indent the atria or displace the trachea, are multilobular, and partially surround the major arterial trunks at the

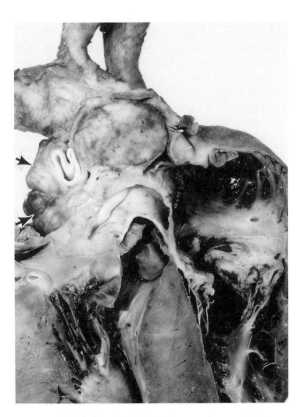

Fig. 3.49 Aortic body tumors (arrows). Dog.

base of the heart. Although the vessels may be completely surrounded by neoplastic tissue, there usually is little evidence of vascular constriction.

Malignant aortic body tumors occur less frequently in dogs than do adenomas. Carcinomas may infiltrate the wall of the pulmonary artery to form papillary projections into the lumen or invade through the wall into the lumen of the atria. Although tumor cells often invade blood vessels, metastases to the lung and liver occur infrequently.

Aortic body tumors in animals tend to be more benign than tumors of the carotid body. They grow slowly by expansion and exert pressure on the vena cava and atria. Aortic body carcinomas may invade locally into the atria, pericardium, and adjacent large thin-walled vessels. When they metastasize, secondary foci of growth are found most frequently in the lung and liver.

Tumors of the aortic bodies in animals are not functional (i.e., they do not secrete excess hormone into the circulation), but as space-occupying lesions may result in a variety of functional disturbances. These include manifestations of cardiac decompensation due to pressure on the atria, vena cava, or both, associated with larger aortic body adenomas and carcinomas. There may be dyspnea, coughing, vomiting, cyanosis, hydrothorax, hydropericardium, ascites, edema of the subcutaneous tissue of the head, neck, and forelimbs, and passive congestion of the liver. The accumulation of serous, often blood-tinged, fluid in the pericardial sac results from the invasion of tumor cells into lymphatics at the base of the heart or the compression of small pericardial veins.

Carotid body tumors arise near the bifurcation of the common carotid artery in the cranial cervical area. They usually appear as a unilateral slow-growing mass and only rarely develop on both sides in the same animal.

Adenomas are 1–4 cm in diameter, are well encapsulated, and have a smooth external surface. The bifurcation of the carotid artery is incorporated in the mass, and tumor cells are firmly adherent to the tunica adventitia. A branch of the glossopharyngeal nerve may be traced into the capsule of the tumor by careful dissection. Adenomas are firm, white with scattered areas of hemorrhage, and extremely vascular. Complete surgical excision or biopsy often is difficult due to the high degree of vascularity and intimate relationship with major arterial trunks in the neck.

Malignant carotid body tumors are larger (up to 12 cm in diameter) and more coarsely multinodular than adenomas. Although carcinomas appear to be encapsulated, tumor cells invade the capsule and penetrate into the walls of adjacent vessels and lymphatics. The external jugular vein and several cranial nerves (in addition to the carotid bifurcation) may be incorporated by the neoplasm. Larger tumors result in extensive dorsal–lateral deviation of the trachea.

Metastases of carotid body tumors occur in ~30% of cases and have been found in the lung, bronchial and mediastinal lymph nodes, liver, pancreas, and kidney. Multicentric neoplastic transformation of chemoreceptor tissue occurs frequently in brachycephalic breeds of dogs. About 65% of the reported cases of carotid body tumors also had aortic body tumors.

The histologic characteristics of chemoreceptor tumors (chemodectomas) are essentially similar whether they are derived from the carotid or aortic body. The neoplastic chemoreceptor cells are subdivided into lobules by prominent branching trabeculae of connective tissue that originate from the fibrous capsule (Fig. 3.50). They are further subdivided into small compartments by fine septa that contain collagen and reticulin fibers plus small capillaries. Tumor cells are commonly aligned along and around small capillaries.

The tumor cells of chemodectomas are discrete, cuboidal to polyhedral, and closely packed together. The cytoplasm is lightly eosinophilic, finely granular, and often vacuolated. Cells composing chemodectomas rapidly undergo autolysis. Cell boundaries become indistinct, and the cytoplasm appears clear if the postmortem interval is prolonged. Chromaffin granules cannot be demonstrated in the cytoplasm of cells composing chemodectomas in contrast to the pheochromocytomas of the adrenal medulla.

Carotid body tumors are very vascular and have numerous muscular arterioles, large thin-walled veins, and an abundant network of capillaries in the connective tissue septa. There are focal areas of hemorrhage from disruption of thin-walled vessels and areas of coagulation necrosis. Several layers of tumor cells may radiate along fine con-

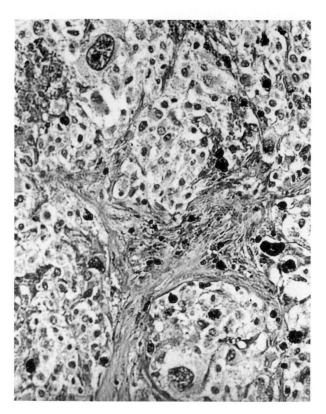

Fig. 3.50 Cellular pleomorphism of chemoreceptor tumor. Dog.

nective tissue septa from the thin-walled vessels. Tumor cells frequently invade blood vessels and lymphatics with the formation of emboli. Carcinomas have evidence of tumor cell invasion through the capsule and into the walls of large muscular arteries and adjacent structures.

Dogs with carotid body tumors usually are presented with a palpable, slowly enlarging mass in the anteriolateral cervical region near the angle of the mandible. Larger neoplasms interfere with swallowing due to pressure on the esophagus and result in circulatory disturbances from compression of the larger veins in the neck. Other functional disturbances may be related to the presence of an aortic body tumor in the same animal.

Carotid body tumors tend to be more malignant than aortic body tumors, and metastases are present in ~30% of the reported cases. Metastases have been found in the lung, bronchial and mediastinal lymph nodes, liver, pancreas, and kidney. The presence of vascular or lymphatic emboli in a biopsy of the primary chemodectoma does not necessarily indicate that metastases are present in distant organs.

The etiology of carotid and aortic body tumors is unknown, but it has been suggested that a genetic predisposition aggravated by chronic hypoxia may account for the higher risk of certain brachycephalic breeds such as the boxer and Boston terrier. Carotid bodies of several mammalian species, including dogs, undergo hyperplasia when subjected to chronic hypoxia by living in a high-altitude environment. Humans living at high altitudes have 10 times the frequency of chemodectomas of those living at sea level.

Adenomas and carcinomas derived from ectopic thyroid tissue account for ~5–10% of **heart-base tumors** in dogs. They often compress or invade structures in the anterior mediastinum near the base of the heart. Areas of ectopic thyroid tumors with a compact cellular (solid) pattern of arrangement are difficult to distinguish histologically from aortic body tumors. Cells composing ectopic thyroid tumors generally are smaller than those in aortic body tumors, with more hyperchromatic nuclei and an eosinophilic cytoplasm.

Neoplastic thyroid follicular cells are not consistently sub-divided into small packets by fine strands of connective tissue. Tumor giant cells are infrequent in ectopic thyroid tumors, and the stroma is less prominent. Multiple sections usually reveal the formation of primitive follicular structures or colloid-containing follicles by neoplastic follicular cells in ectopic thyroid tumors but not in aortic body tumors.

Ectopic thyroid carcinomas arising at the base of the heart in dogs may remain localized and enlarge in the anterior mediastinum or occasionally metastasize to extra-thoracic sites.

Bibliography

Biscoe, T. J., and Stehbens, W.E. Ultrastructure of the carotid body. *J Cell Biol* **30**: 563–578, 1966.

Bloom, F. Structure and histogenesis of tumors of the aortic bodies in dogs, with a consideration of the morphology of the aortic and carotid bodies. *Arch Pathol* **36**: 1–12, 1943.

Cheville, N. F. Ultrastructure of canine carotid body and aortic body tumors: Comparison with tissues of thyroid and parathyroid origin. *Vet Pathol* **9**: 166–189, 1972.

DeKock, L. I., and Dunn, A. E. G. An electron-microscopy study of the carotid body. *Acta Anat* **64**: 163–178, 1966.

Edwards, C. *et al.* The carotid body in animals at high altitude. *J Pathol* **104**: 231–238, 1971.

Godwin, M. C. The early development of the thyroid gland in the dog with special reference to the origin and position of accessory thyroid tissue within the thoracic cavity. *Anat Rec* **66**: 233–251, 1936.

Hayes, H. M. An hypothesis for the aetiology of canine chemo-receptor system neoplasms, based upon an epidemiological study of 73 cases among hospital patients. *J Small Anim Pract* **16**: 337–343, 1975.

Hayes, H. M., and Sass, B. Chemoreceptor neoplasia: A study of the epidemiological features of 357 canine cases. *J Vet Med (A)* **35**: 401–408, 1988.

Hoglund, R. An ultrastructural study of the carotid body of horse and dog. *Z Zellforsch Mikrosk Anat* **76**: 568–576, 1967.

Hubben, K., Patterson, D. F., and Detweiler, D. K. Carotid body tumor in the dog. *J Am Anim Hosp Assoc* **137**: 411–416, 1960.

Johnson, K. H. Aortic body tumors in the dog. *J Am Vet Med Assoc* **152**: 154–160, 1968.

Jubb, K. V., and Kennedy, P. C. Tumors of the nonchromaffin paraganglia in dogs. *Cancer* **10**: 89–99, 1957.

Kameda, Y. The accessory thyroid glands of the dog around the intra-pericardial aorta. *Arch Histol Jpn* **34**: 375–391, 1972.

Kurtz, H. J., and Finco, D. R. Carotid body and aortic body tumors in a dog—a case report. *Am J Vet Res* **30**: 1217–1251, 1969.

Lauper, N. T. *et al.* Pheochromocytoma: Fine structural, biochemical, and clinical observations. *Am J Cardiol* **30**: 197–204, 1972.

LeCompte, P. M. Tumors of the carotid body and related structures (chemoreceptor system). *In* "Atlas of Tumor Pathology," Section IV, Fascicle 16, pp. 1–40. Washington, D.C., Armed Forces Institute of Pathology, 1951.

Misdorp, W. Z., and Elders, R. A. R. Paragangliomen bij mens en hind. *Tijdschr Diergeneeskd* **90**: 205–230, 1965.

Nilsson, A. A case of metastasising tumour of the glomus aorticus in a dog. *Nord Vet Med* **8**: 875–881, 1956.

Pryse-Davies, J., Dawson, I. M. P., and Westbury, G. Some morphological, histochemical, and chemical obervations on chemodectomas and the normal carotid body, including a study of the chromaffin reaction and possible ganglion cell elements. *Cancer* **17**: 185–202, 1964.

Ross, L. L. Electron-microscopic observations of the carotid body of the cat. *J Biophys Biochem Cytol* **6**: 253–262, 1959.

Sander, C. H., and Whitenack, D. I. Canine malignant carotid body tumor. *J Am Vet Med Assoc* **156**: 606–610, 1970.

Scott, T. M. The carotid body tumor in dogs. *J Am Vet Med Assoc* **132**: 413–419, 1958.

Tisseur, H., and Parodi, A. Tumeurs multiples du systeme chemorecepteur chez un chien. *Rec Med Vet* **139**: 99–111, 1963.

Toker, C. Ultrastructure of a chemodectoma. *Cancer* **20**: 271–280, 1967.

Yates, W. D. G., Lester, S. J., and Mills, J. H. L. Chemoreceptor tumors diagnosed at the Western College of Veterinary Medicine 1967–1979. *Can Vet J* **21**: 124–129, 1980.

CHAPTER 4

The Female Genital System

PETER C. KENNEDY
University of California, Davis, U.S.A.

RICHARD B. MILLER
University of Guelph, Canada

I. Anomalies of Development

A. Sex Determination and Differentiation

Disorders of genital development occur in all domestic species, but are common in none. They are caused by abnormalities of genetic or chromosomal origin or inappropriate hormone exposure. The mechanisms involved in the production of many abnormalities are not known and a condition cannot often be classified precisely. Hermaphroditism is a broad grouping. **Hermaphrodites** have ambiguous genitalia with part or all of the genital organs of both sexes present. The intersex is subclassified into true and pseudo, the distinction being based on the presence of both types of gonadal tissue in the true hermaphrodite. The pseudohermaphrodite has only a single type of gonadal tissue and the individual is classified as either a male or female pseudohermaphrodite on the basis of the gonadal tissue present. These terms serve only limited goals of description. The application of sophisticated techniques to the study of abnormal sexual development has elucidated, in some cases, the mechanism by which the anomalies developed, and as well has clarified some of the intricate hormonal and genetic mechanisms by which normal sexual differentiation occurs.

The determination of genetic sex is fixed at the time of fertilization. The genetic sex is then imposed on the undifferentiated gonad. The mechanism by which this occurs is still not entirely clear. It is clear, as it has been for some time, that the Y chromosome is strongly male determining. It is now proposed that the **testis-determining factor** (TDF) of the Y chromosome is a single gene on the

short arm of the Y chromosome close to the pseudoautosomal pairing region. This gene has been called SRY (sex-determining region of the Y). Much of the search for the location of TDF has been carried out in XX human males and mice carrying a mutation that incapacitates the testis-determining gene, but it is hoped that this proposed TDF candidate will be the definitive one because sequences homologous to SRY are located on the Y chromosome of all eutherian mammals tested.

The gene controlling expression of the H-Y antigen, a minor histocompatibility antigen, is also present on the short arm of the Y chromosome. It had been an earlier candidate for TDF because of its near-perfect association with maleness. This association now seems due to the close proximity of the gene on the Y chromosome to SRY.

The germ cells, which will give rise to either oocytes or spermatogonia, originate in the yolk sac and migrate to the still undifferentiated genital ridge. There they colonize the celomic epithelium and the mesenchymal tissue between the mesonephros and the celomic epithelium. The undifferentiated gonad at this early stage, the fourth week of gestation in the dog, the sixth week in cattle, consists of germ cells, mesenchymal cells, and cells of both the mesonephros and the celomic epithelium. The precise contribution that each of the three types makes to the differentiated gonad remains in doubt. Functionally, the cells of the developing gonad are of four types: the germ cells, supporting cells, steroid-producing cells, and unspecialized mesenchyme. Testis determining factor has its first action on the supporting cell lineage to bring about differentiation of Sertoli cells. This conversion is thought to be the triggering mechanism resulting in the other cells

349

following the male pathway. It results in the differentiation of Leydig cells from the steroid-producing cell line, the mitotic arrest in the germ cells, and the organization of the connective tissue into a testicular pattern. The Sertoli cells form cords which enclose the germ cells.

In female differentiation, which occurs in the absence of TDF, the first detectable event is the entry of the germ cells into meiosis, followed by differentiation of the supporting cells into follicle cells which surround the oocytes. The cells of the steroid-producing cell line give rise to the theca cells. The ovary fails to develop in the absence of the germ cells, which suggests that ovarian development depends on gene action expressed within germ cells.

In the developing embryo, the genital tract of both sexes consists of Wolffian and Müllerian ducts and bipotential sinusal and external genital primordia. The Wolffian ducts, the primordia for male accessory organs, originate from the excretory canals of the primitive kidney, the meso-nephros. The female counterparts, the Müllerian or par-amesonephric ducts, arise as longitudinal invaginations of celomic epithelium adjacent to the urogenital ridge. The ducts run parallel to Wolffian ducts, but unlike the Wolffian ducts, open directly into the celomic cavity. The Wolffian ducts remain connected to the mesonephric tissue by a series of tubules. Both duct systems make contact with the urogenital sinus. Femaleness corresponds to an intrinsic program of the primordia.

Female differentiation occurs in the absence of a functional male gonad. Masculine differentiation is imposed on the system at an early stage by testosterone produced by the Leydig cells of the developing testis, and this differentiation prevents further female development. All sexual development would be feminine if not prevented from being so by testicular hormones.

An important feature of the gonadal differentiation is the chronological differences that exist between the sexes. In all mammalian species studied, males differentiate earlier than females. The fetal testis produces two hormones critical for normal male differentiation. The Sertoli cells of the developing testis secrete a Müllerian duct inhibitory factor that brings about Müllerian duct regression. At the same time the Leydig cells of the testis secrete testosterone, which prevents the Wolffian duct regression and brings about its differentiation into vasa deferentia, seminal vesicles, and epididymides. The testosterone is also converted to dihydrotestosterone, and this hormone induces development of penis and scrotum from the bipotential external genital primordia.

1. Freemartinism

The bovine freemartin is a genetic female born co-twin with a male. Twinning occurs in 1–2% of pregnancies in cattle. The twinning is dizygotic in most cases, and anastomoses develop between the placental circulations. This is the critical feature in the sterilization of the female by her male twin. If anastomoses fail to develop, the female is not damaged. The effect of the association on the male is always minimal. In the female, the ovaries

are stunted and frequently contain sterile seminiferous tubules. The derivatives of the Müllerian ducts may be nearly completely absent or lack only communication with the vagina (Fig. 4.1). Various parts of the Wolffian ducts are frequently present. The external genitalia are usually not altered.

Because of the clear association between freemartinism and the presence of interplacental anastomoses, and because the male gonads develop earlier than the female, it was presumed that male sex hormones were carried from the male fetus to the female, sterilizing and modifying the latter. The inadequacy of this explanation is demonstrated by the fact that androgens can masculinize a female fetus, but they cannot inhibit the ovaries or the Müllerian ducts. A critical observation in our understanding of the condition was that embryonic blood of the twins is exchanged and leads to permanent colonization by hematopoietic cells with the result that each twin develops an acquired tolerance for the blood cells of the other. This condition, known as chimerism, appears to be confined to hematopoietic cells. Exchanges of nonhematopoietic cells do not appear to take place in freemartinism. It is now believed that the male fetus sterilizes the female by the partial expression of TDF of the Y chromosome carried by the bloodstream via the placental anastomoses to the female gonad where it brings about an ovarian inhibition. This inhibition is manifest first functionally; estrogen produc-

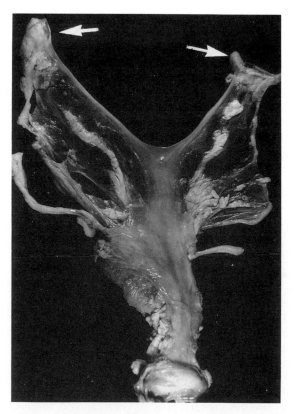

Fig. 4.1 Freemartin. Calf. Rudimentary uterus is attached to gonads (arrows).

tion by the freemartin ovary is reduced and androstenedione is elevated as early as 40 days of gestation. This is well before morphologic evidence of damage can be recognized.

The gonad of a freemartin is usually a cordlike thickening in the cranial border of the ovarian ligament. Varying degrees of modification of the gonad may occur. The rete is well developed, but other ovarian structures may be absent. Alternatively, greatly reduced ovarian tissue can be present, but it lacks the normal germ cell complement. Approximately 50% of freemartin fetuses develop seminiferous cords and varying numbers of Leydig-like steroidogenic cells. The morphologic variation shown in the gonadal development is causally reflected in the variation of the tubular genitalia.

The epididymis is absent and the spermatic cord is poorly developed in cases with the lowest degree of male transformation. With increasing degrees of differentiation toward a testis, the spermatic cord and epididymis become more well developed.

The tubular genitalia, of Müllerian duct origin, may vary from cordlike structures, without any lumina, to well-developed uterine horns with lumina and endometrial glands. Communication with the vagina is always absent, no matter how well developed the uterus may be.

Vestigial seminal vesicles are always present, the vagina is hypoplastic or nonpatent with a complete hymen, the vulva and vestibule are hypoplastic, and the clitoris is enlarged. Many freemartins have a prominent fold of skin on the median plane of the body, extending from a position ventral to the vulva to an area near the umbilicus. The mammary gland fails to undergo normal development.

The allantoic fusion in bovine twin pregnancies occurs at about the 10-mm stage of development. Approximately 90% of females born co-twin with males are freemartins and sterile. Freemartins occur in bovine triplets, quadruplets, and quintuplets as well as in twins.

In sheep, dizygotic twinning is common and freemartinism does occur, but it is much less common, only 1% of heterosexual twins being freemartins in this species. The reason for this is the infrequency with which large-caliber interplacental anastomoses develop. Freemartinism does occurs in goats, complete with blood chimerism, but is rare because, as in the case of sheep, large anastomoses between fetuses are rare. Large anastomoses have been demonstrated between chorionic sacs in swine and freemartinism occurs in that species. However, because of the large litter size, it is difficult to identify the neighbor of intersex piglets, and the condition has not often been proven. Twin ovulation is relatively common in horses. Most twins do not survive to term, but vascular anastomoses with erythrocyte chimerism are frequent in equine twins, although freemartinism has not been convincingly demonstrated.

2. XX Sex Reversal

According to the scheme presented in the previous section, the female role in sexual differentiation is largely passive. The undifferentiated gonad will develop into an ovary unless the testis-determining factor is present, and the undifferentiated tubular genital tract will develop into normal female genitalia unless male hormones are present. Accordingly, ovarian hypoplasia would not be expected to, and does not, lead to abnormalities of sexual differentiation. There are, however, several syndromes of sex reversals in genetic females. Much of what we know about these conditions we have learned from the condition in mice in which an autosomal dominant gene, Sxr, acts like a Y chromosome. Thus XX mice which carry the Sxr gene develop testes and male tubular genitalia. A somewhat similar condition has been observed in dogs, pigs, and goats, although in these species the autosome-associated gene with the Y action functions as a recessive gene.

The condition has been most carefully studied in goats in which the autosomal gene with the Y effect is on the chromosome which controls polling or is closely linked to it. The polled gene is a dominant autosome, a single copy of which produces hornlessness without affecting sex differentiation. In the homozygous condition, however, XX goats are rendered hermaphrodites. They have testes or ovotestes. It has been presumed that the translocation of a subcritical portion of the testis-determining gene to the autosome which contains the gene for polling has produced the recessive mode of sex determination. There is a considerable range of masculinization. The affected individuals may have largely female appearance with only an enlarged clitoris and abdominal testes. Other animals have the appearance of near normal, although sterile, males.

Sex reversal caused by autosomal genes with a Y action has also been reported in dogs. It occurs most often in the American cocker spaniel breed in which several families have been studied in which XX males and XX true hermaphrodites have been described. Some of the true hermaphrodites have functional ovarian tissue and have given birth to pups. The testicular tissue present in the gonads of either the XX males or the XX true hermaphrodites has invariably lacked germ cells. The condition is inherited as an autosomal recessive.

Sex reversal is common in pigs. The mode of inheritance is not well understood, and some cases may have been confused with freemartinism. The masculinizing effect is incomplete. The gonads in affected individuals are often ovotestes and, although the testicular tissue lacks germ cells, ovulation and pregnancy have been reported. This condition can be distinguished from freemartinism by demonstrating that blood chimerism does not exist in affected individuals.

3. XY Sex Reversal in the Mare

Individual examples of domestic animals that had XY karyotypes but were phenotypically female have been reported. In the horse, an inherited XY sex-reversal syndrome has been identified. Pedigree analysis suggests that the trait is inherited as the result of an autosomal sex-limited dominant gene or as a Y chromosomal mutation with variable expression. There is a considerable range of phenotypic expression in this syndrome. Affected individuals vary from phenotypically normal but sterile mares with inactive ovaries and normal tubular genitalia to individuals with streak gonads or ovotestes and severely hypoplastic or aplastic tubular genitalia.

4. Androgen Insensitivity

A syndrome of androgen insensitivity has been recognized in humans, mice, rats, cats, cattle, and horses. This inherited syndrome is most often called testicular feminization and the controlling mutation identified as *Tfm*. Affected individuals are of XY genotype and have testosterone-producing testicles, but possess neither Wolffian nor Müllerian duct-derived tubular genitalia. The external genitalia are female in type, the *Tfm* mutation producing a deficiency of intracellular androgen receptors rendering all cells insensitive to androgens. The Müllerian duct derivatives are absent because their sensitivity to the anti-Müllerian hormone is retained, and anti-Müllerian hormone is produced by the testes of affected animals. The inheritance of the syndrome has been studied most extensively in mice. In this species the controlling gene is X linked. The genes of the X chromosome are strongly conserved, and X linkage in one species makes X linkage in others likely. On this basis, it is presumed that the gene which regulates androgen insensitivity is X linked in all species in which the syndrome occurs. Although an apparent XY sex reversal occurs in this syndrome, it is distinct genetically and anatomically from the XY sex-reversal syndrome previously discussed.

Equine male pseudohermaphrodites of this syndrome have normal-appearing female external genitalia (Fig. 4.2A). The vagina ends in a blind sac, no cervix, uterus, or uterine tubes being present. The gonads are small and clearly testicular, but present in the normal ovarian position (Fig. 4.2B). The seminiferous tubules are small

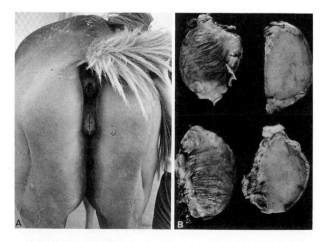

Fig. 4.2 Androgen insensitivity (testicular feminization). Horse. (A) Normal vulva of genetic male (XY) lacking androgen receptors. (B) Hypoplastic testicles of affected animal. Note lack of epididymides.

and lined by inactive Sertoli cells (Fig. 4.2C). Rare spermatogonia can be found, but spermatogenesis does not occur. Well-differentiated Leydig cells are closely applied to the seminiferous tubules (Fig. 4.2D). Horses with the syndrome have no detectable male accessory sex glands. The cattle which have been described have rudimentary seminal vesicles and ampullae. Both cattle and horses which lack androgen receptors have normally formed female mammary glands. Horses are unique among the species in which the syndrome has been reported in that affected individuals show male behavior patterns. This syndrome of male pseudohermaphroditism can be distinguished from others by assaying the cytosol receptors of the androgen target cells. This can be done most conveniently by culturing skin fibroblasts from the genital skin. It is important that labial tissue be used because receptor activity is sharply reduced in nongenital skin.

B. Vulva

The vulva and vestibule are of ectodermal origin, being derived by partition from the urogenital sinus. Failure of proper partitioning allows the cloaca to persist. Hypoplasia of the vulva and vestibule is occasionally observed, sometimes alone, but usually in association with hypoplasia of derivatives of the Müllerian ducts. Hypertrophy of the clitoris occurs in pseudohermaphroditism, and in cows and sows with functional follicular cysts.

C. Ovary

Agenesis of one or both ovaries is occasionally observed in ruminants, swine, and dogs. In bilateral agenesis, the tubular genitalia may be absent as part of the defect or, if present, are infantile or underdeveloped. Supernumerary and accessory ovaries occur very rarely in domestic animals. A supernumerary ovary is a third gonad which is distinctly separate from the normally placed ovaries and apparently arises from a separate anlage. An accessory ovary is located close to a normally placed gonad, is usually connected to it, and appears to develop as a result of splitting of the embryonic gonad.

Vascular hamartomas of the ovary have been described in swine and cattle. Since they are often confused with neoplasms, they are described in the section on ovarian tumors (Section II,D,7 of this chapter).

Severe **ovarian dysgenesis** develops in mares which lack a second X chromosome. These XO individuals have small inactive ovaries which lack germ cells (Fig. 4.3). The tubular genitalia are normally formed but small, and the endometrium is hypoplastic. The external genitalia are small, but not conspicuously so. This syndrome occurs in other domestic species, but possibly not in all because the X deletion has been shown to cause a high mortality in human conceptuses and the XO condition may be lethal in some species. Mares with the XO sex constitution tend to be small, but no somatic anomalies have been identified. A similar syndrome has been seen in mares in which the

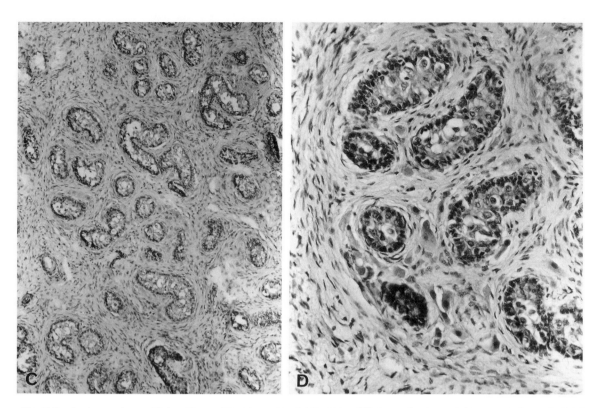

Fig. 4.2 Androgen insensitivity. Horse. Histologic detail of testicles. (C) Hypoplastic seminiferous tubules embedded in stroma in which numerous interstitial (Leydig) cells are present. (D) Higher magnification of (C).

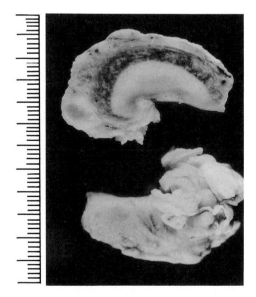

Fig. 4.3 Ovarian dysgenesis. Mare. XO sex chromosome complement.

short arm of the X chromosome is deleted. In both the XO and the X-deletion X syndromes, the germ cells fail to survive and the ovaries become inactive, although some follicles may form in juveniles, and at least one X-deletion X mare has conceived.

Hypoplasia of the ovaries has been studied primarily in cattle, but occurs in other species. It is usually bilateral but varies considerably in its severity and symmetry, so that severe hypoplasia or partial hypoplasia may be applicable to one or both ovaries. In severe hypoplasia, the defective gonad varies in size from a cordlike thickening in the anterior border of the mesovarium to a flat, smooth, firm, bean-shaped structure in the normal position (Fig. 4.4A). There are neither follicles nor luteal scars and, microscopically, the ovary is composed largely of medullary connective tissues and blood vessels. The ovarian cortex may be totally absent, or it may be only partially absent, forming then a thin or incomplete investment for the medulla. The cortical germinative stroma is much deficient in amount and is diluted with collagen. Ova are not detectable and there is no sign of organization into follicles (Fig. 4.4B). In hypoplasia of lesser severity, the ovarian stroma is deficient, but a few primary follicles can be found. Occasionally, a follicle grows to a large size and may become cystic, with luteal tissue in patches in the theca; size, however, is not a criterion of maturity, and the life cycles of follicles in hypoplastic ovaries are not known.

In association with ovarian hypoplasia, there is relative hypoplasia or infantilism of the remainder of the genital tract. Genital infantilism also occurs in young females in association with the nonfunctional ovaries of debility or

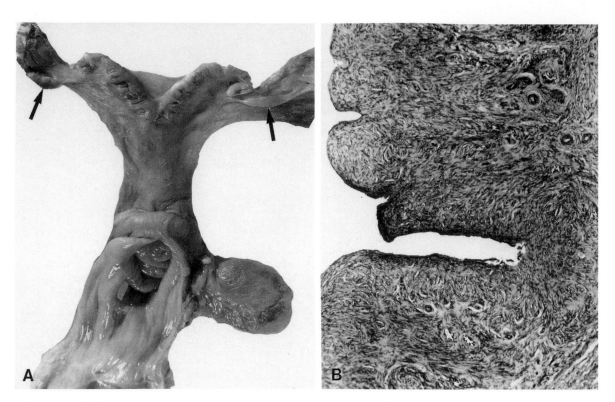

Fig. 4.4 (A) Bilateral ovarian hypoplasia with uterine infantilism. Cow. (Courtesy of C. A. V. Barker). (B) Ovarian hypoplasia. Cow. Paucity of germinal stroma and absence of follicular development.

malnutrition; these cases can be distinguished from those of hypoplasia by the presence of adequate amounts of cortical stroma, the presence of primordial follicles, and the responsiveness of the ovaries to gonadotrophins and improved nutrition. There is, in debility, a failure of release of hypophyseal gonadotrophins so that the follicles do not develop much past the antrum stage.

Ovarian hypoplasia occurs with other chromosomal aberrations such as XXX in mares and cows. The ovaries of female mules are small, and estrous cycles, long and irregular. Very few details of these forms of ovarian hypoplasia have been reported.

1. Ovarian Hypoplasia in Swedish Highland Cattle

Ovarian hypoplasia occurs in Swedish Highland cattle as a genetic defect conditioned by an autosomal recessive gene with incomplete penetrance. Elsewhere, the defect is not known to occur in any definite pattern or particular breed but is sporadic in incidence.

The primary morphologic defect in hypoplastic ovaries in cattle is a deficiency in the number of germ cells in the ovary. There is considerable variation in the degree of expression of the defect and these variations may be correlated with quantitative variations in the number of germ cells present. It is not known whether the deficiency in germ cells reflects a deficiency of their development in the yolk sac, or in their migration to the primitive gonad, or in their persistence and multiplication in the gonad. In

defective cattle, the left ovary is more frequently and severely hypoplastic than the right, although individuals, presumably homozygous for the defect, may have bilateral hypoplasia. Even with unilateral hypoplasia, that is with one competent ovary, this functionally normal organ contains fewer primordial follicles than a normal ovary, and, further, these appear to be inhibited in their development to Graafian follicles. In partial hypoplasia, this inhibited development is in the form of retarded growth of the follicle with a tendency to early luteinization without ovulation. The smaller the number of follicles, the greater is the likelihood of this inhibition.

D. Arrests in the Development of the Müllerian Duct System

These defects are of significance only in cattle and swine. In other species, they are rarely recorded and the degree of inhibited development is less. Thus in the bitch, uterus unicornis, partial fusion of the horns, or reduction in length of one horn are probably the most severe defects recorded and are not associated with sterility. In swine and cattle on the other hand, segmental defects are not uncommon; segmental defects occur in the mare but are rare. In almost all instances, the anomaly is based on the inhibited development of portions of the Müllerian ducts and their differentiated derivatives, or on aberrations of the pattern of fusion of the Müllerian ducts, which produce the cranial part of the vagina, cervix, and uterine fundus.

Segmental aplasia of the Müllerian duct system was a frequent enough occurrence in white Shorthorn cattle that it gave rise to the name white heifer disease for the syndrome, but no breed is exempt from the condition. The arrested development is thought to represent a pleomorphic action of genes which are autosomal recessive. It is likely that separate genes are responsible for abnormalities of right and left horns and that these are linked on the same autosomal chromosome. In the Shorthorn breed, these genes are presumably located close to the gene (or genes) that restricts the normal and basic pigmentation of the Shorthorn and produces a white coat color.

The complicated developmental history of the Müllerian ducts suggests the wide variety of defects that can and do occur. As a rule, the ovaries, uterine fallopian tubes, and cornual apices are present and are normal except for retention of secretion in, and dilation of, the apices of the horns. Sows may provide an exception to the rule because, in this species, hydrosalpinx is relatively common among the gross genital abnormalities; the basis of the hydrosalpinx must be an obstruction to the tube, but whether the obstruction is developmental or acquired has not been determined. If a total or large segmental defect of aplastic type occurs in a salpinx (uterine tube), the basic defect is likely to be gonadal, involving some testicular development due to freemartinism, sex reversal, or chimerism.

The simplest defects involve only the caudal sections of the Müllerian ducts. If these fail to make patent connection with the urogenital sinus, an **imperforate hymen** persists. In association with imperforate hymen, the remainder of the genital tract may be normal, but in time and with the accumulation of secretions (there may be 2 gallons or more), the vagina, cervix, and uterus become distended and atonic (Fig. 4.5). If the hymen is perforated and the distension relieved early, the genital tract may function normally, but prolonged and severe dilation leads to atrophy and permanent loss of tone of the walls. Bacteria may be fortuitously implanted in the uterine contents to produce pyometra.

Lesser degrees of hymeneal persistence are common and usually of no significance. The perforated remnant may consist of a thin fenestrated membrane or dorsoventral bands. In some instances, the bands, especially those of sagittal disposition, may be remnants of the medial walls of the Müllerian ducts. The most severe defect is characterized by hymeneal constriction in association with absence or rudimentary development of the cranial vagina, cervix, or uterus including the horns, together or in any combination (Fig. 4.6A).

Proximal to a segmental defect, the tubular structures become dilated cystic cavities filled with dark reddish-brown mucus which may, if inspissated, form hard rubbery masses (Fig. 4.6B). In the defective segment, there may be no trace of the Müllerian duct or there may be a pencil-thick cord of connective tissue representing it. **Uterus unicornis** is quite common as an expression of

Fig. 4.5 Arrests in development of Müllerian duct system. Segmental aplasia. Cow. Imperforate hymen with distension of vagina and uterus.

segmental aplasia of one duct. The whole horn, excepting its apex, may be absent or the defect may be more limited, usually occurring then at the bifurcation of the uterus (Fig. 4.7A).

In addition to aplastic or hypoplastic defects of the Müllerian ducts, there are common anomalies which result from **failure of proper fusion** of the caudal portions of the ducts. Complete failure of fusion leaves a double vagina, double cervix, and divided fundus; this is rather rare. The more common failures of fusion occur in or adjacent to the cervix. The anterior vagina may be partitioned by a dorsoventral septum in conjunction with a **double cervix.** A dorsoventral band may be across the external os, the cervix and vagina being properly fused. The failure of fusion may involve only part of the cervix, chiefly the caudal part, so that there is one uterine fundus, one internal os, and a bifurcated cervical canal with duplication of the external os. The cervix and uterine fundus may be completely divided, a condition known as **uterus didelphys** (Fig. 4.7B). The normal sequence of fusion involves the uterus last, so that at an early stage of fetal development, uterus didelphis is a normal formation.

The gross anatomic form of the cervix, in terms of length, thickness, number and disposition of rugae, length, direction, and tortuosity of the canal, and so on, is quite variable. The majority of these variations fall within a narrow range that is regarded as normal, but significant

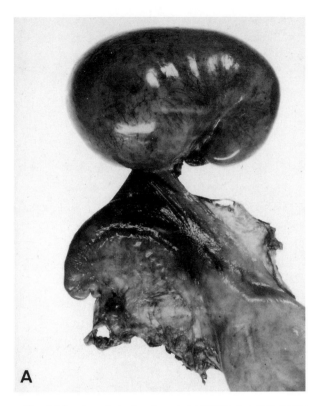

Fig. 4.6A Arrests in development of Müllerian duct system. Segmental aplasia. Cow. Defect in right horn with distension of proximal segment.

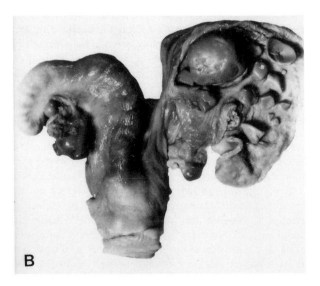

Fig. 4.6B Segmental aplasia. Cow. Defect in right uterine horn with dilation of proximal segment and inspissation of secretions.

departures from normal are not uncommon, especially in heifers. Gross hypertrophy of the cervix is observed occasionally, the structure being two to three times its

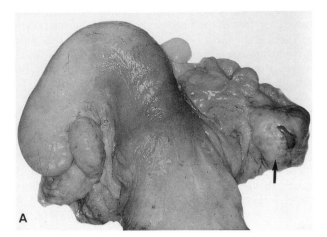

Fig. 4.7A Uterus unicornis. Cow. Normal ovary (arrow).

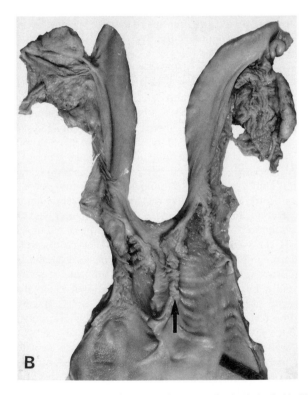

Fig. 4.7B Uterus didelphys. Cow. Uterine body is divided by dorsoventral septum (arrow).

normal length and thickness, and it may present an impenetrable barrier to the passage of sperm or the discharge of uterine secretions. Moderate hypoplasia of the cervix, in which some rugae are absent and the canal widely patent, may militate against functional closure of the canal and predispose to persistent invasion of the uterus by microbes, and chronic endometritis (Fig. 4.8). Occasionally, the cervical canal is irregular in its course or even tortuous and, although this may not interfere with conception fol-

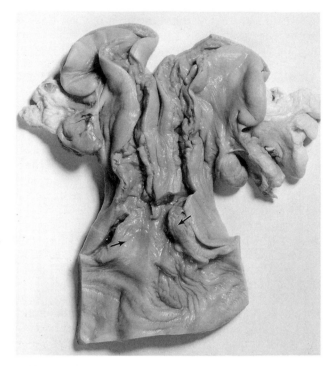

Fig. 4.8 Cervical hypoplasia. Failure of development of cervix (arrows) allows direct communication between vagina and uterus.

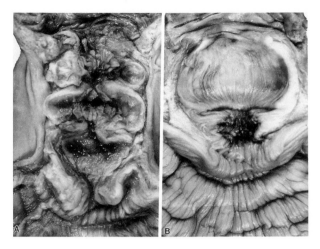

Fig. 4.9 Cervical anomaly. Cow. (A) Dorsal diverticula. (B) Marked dilation of cervix between anterior rugae.

lowing natural breeding, the insertion of an inseminating catheter may be difficult or impossible, and persistent attempts to do so lead to cervical trauma occasionally complicated by abscessation or the formation of traumatic inclusion cysts.

Dilations and diverticula of the cervix have been observed as a cause of infertility in heifers. The malformations occur at the level of the third and fourth rugae, and the cervical canal is usually constricted caudal to the defect (Fig. 4.9A). Individual animals may have cervical dilations or diverticula, whereas others have both lesions (Fig. 4.9B). The dilated areas are usually spherical and located near the internal os. Diverticula are dorsal or dorsolateral, usually single but occasionally double (Fig. 4.9A). The cavities of the defects eventually become filled with tenacious mucus. The cause of the condition has not been established, but it is assumed to be developmental.

Bibliography

Benirschke, K. Hermaphrodites, freemartins, mosaics and chimaeras in animals. *In* "Mechanisms of Sex Differentiation in Animals and Man," C. R. Austin and R. G. Edwards (eds.), pp. 421–463. New York and London, Academic Press, 1981.

Biggers, J. D., and McFeely, R. A. Intersexuality in domestic animals. *Adv Reprod Physiol* **1**: 29–59, 1966.

BonDurant, R. H., McDonald, M. C,. and Trommerhausen-Bowling, A. Probable freemartinism in a goat. *J Am Vet Med Assoc* **177**: 1024–1025, 1980.

Bongo, T. A. *et al*. Foetal membrane fusion and its developmental consequences in goat twins. *Br Vet J* **142**: 59–64, 1986.

Bruere, A. N., and Macnab, J. A cytogenetical investigation of six intersex sheep shown to be freemartins. *Res Vet Sci* **9**: 170–180, 1967.

Clarke, I. J. Prenatal sexual development. *In* "Oxford Reviews of Reproductive Biology," C. A. Finn (ed.), Vol. 4, pp. 100–147. Oxford, Clarendon Press, 1982.

Dunn, H. O. *et al*. Cytogenetic and reproductive studies of bulls born co-twin with freemartins. *J Reprod Fertil* **57**: 21–30, 1979.

Griffin, J. F., and Wilson, J. D. The syndromes of androgen resistance. *N Engl J Med* **302**: 198–209, 1980.

Hare, W. C. D. Intersexuality in the dog. *Can Vet J* **17**: 7–15, 1976.

Hare, W. C. D. Cytogenetics. *In* "Current Therapy in Theriogenology," D. A. Morrow (ed.), 119–155. Philadelphia, Pennsylvania, W. B. Saunders Co., 1980.

Jost, A. *et al*. Studies on sex differentiation in mammals. *Recent Prog Horm Res* **29**: 1–41, 1973.

Kieffer, N. M., Burns, S. J., and Judge, N. G. Male pseudohermaphroditism of the testicular feminizing type of a horse. *Equine Vet J* **8**: 38–41, 1976.

Koopman, P. *et al*. Male development of chromosomally female mice transgenic for *Sry*. *Nature* **351**: 117–121, 1991.

McLaren, A. What makes a man? *Nature* **346**: 216–217, 1990.

Scofield, A. M., Cooper, K. J., and Lamming, G. E. The distribution of embryos in intersex pigs. *J Reprod Fertil* **20**: 161–163, 1969.

Seldon, J. R. *et al*. Genetic basis of XX male syndrome and XX true hermaphroditism: Evidence in the dog. *Science* **201**: 644–646, 1978.

Settergren, I. The ovarian morphology in clinical bovine gonadal hypoplasia with some aspects of its endocrine relations. *Acta Vet Scand* **5**: (Suppl. 1), 1964.

Wachtel, S. S., Basrur, P., and Koo, G. C. Recessive male-determining genes. *Cell* **15**: 279–281, 1978.

Zamboni, L., Bezard, J., and Mauleon, P. The role of the mesonephros in the development of the sheep fetal ovary. *Ann Biol Anim Biochem Biophys* **19**: 1158–1178, 1979.

II. The Ovary

A. Miscellaneous Lesions

Intrafollicular hemorrhage, for unknown reasons, occurs commonly in calves, in follicular cysts in the bitch, and occasionally in atretic follicles of cows. Hemorrhage occurs during ovulation in all species of domestic animals, but is usually minimal and confined largely to the cavity of the collapsed follicle. However, it can be severe and occasionally even lethal in the mare. Hemorrhage into anovulatory follicles occurs in the mare in the autumn at the beginning of the period of seasonal anestrus.

Focal areas of serositis develop adjacent to corpora lutea, indicating that they occur following ovulation. The recently formed strands of tissue, which are referred to as **ovulation tags,** are composed of fibrin, proliferating capillaries, and leukocytes. As the lesion regresses, neutrophils are replaced by lymphocytes, macrophages, and plasma cells, and mesothelial cells cover the tags. Fine bursal adhesions may result, but they are too delicate to interfere with ovulation or the passage of ova into the oviduct. Ovulation tags occur in all species but most frequently in the cow and mare. Scars which form on the surface of the ovary are small and of no consequence.

A significant form of **ovarian hemorrhage** is that which follows manual enucleation of corpora lutea in cattle (Fig. 4.10). The blood loss may vary from 0.5 liter to several liters and cause death. Hemorrhage is more profuse in pregnant cows and those with pyometra than in normal cycling cattle. Relatively little hemorrhage follows manual

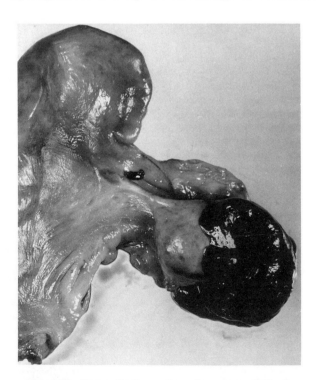

Fig. 4.10 Traumatic hemorrhage from ovary following expression of corpus luteum. Cow.

rupture of follicular cysts, especially if they are thin walled. In some cases, organization of the clot leads to fibrous bursal adhesions which may interfere with the function of the ovary and oviduct. In cases of pyometra and salpingitis, the release of inflammatory detritus into the bursa often results in the formation of extensive **adhesions** between the oviduct and ovary. Obliteration of the bursa by adhesions results in cystic degeneration of follicles. Prominent star-shaped scars develop on the bovine ovary following enucleation of corpora lutea. They can be easily distinguished from ovulation scars on the basis of their size and shape. Remnants of expressed corpora lutea can be found free in the peritoneal cavity or attached to the serosa of a pelvic or abdominal organ. The expressed mass of luteal tissue persists indefinitely as a roughly spherical, flattened body, shaped somewhat like a lima bean. The mass develops a fibrous capsule enclosing the necrotic luteal cells. These necrotic masses of luteal tissue can be differentiated from detached lipomas of similar shape and size by histologic examination.

Postparturient vascular lesions occur frequently in the bovine ovary. The changes consist of intimal proliferation of mucoid tissue in ovarian arteries and hyalinization of the walls of arteries and arterioles. Some of the more severely affected vessels occasionally undergo thrombosis. Degeneration of arterioles also occurs in rapidly regressing corpora lutea associated with abnormally short estrous cycles.

Oophoritis is relatively rare and, when it occurs, it is usually pyogenic (Fig. 4.11). Abscessation of the ovary may follow enucleation of the corpus luteum in cows with pyometra. Serosal granulomas on the ovary occur in bovine peritoneal tuberculosis and in brucellosis, visible macroscopically as small reddish nodules or tags. Similar lesions are usually visible on adjacent serosal surfaces of the genital organs and adnexa. These infective granulomas remain localized to the surface of the ovary and do not penetrate its substance.

B. Ovarian Cysts: Atretic Follicles

The term **parovarian cyst** is used rather loosely in reference to a variety of cystic structures located adjacent to the ovary. Such cysts originate from remnants of the mesonephros. The Wolffian duct (mesonephric duct) and tubules largely degenerate in the female fetus. These embryonic vestiges are homologs of structures associated with the testis. The rete, the efferent ductules, and the epididymis develop from the cranial mesonephros; the female homolog is known as the **epoophoron.** The caudal portion of the mesonephros gives rise to the caudal aberrant ductules. These are collectively called the **paroophoron** in the female and the **paradidymis** in the male. Cysts of the epoophoron and paroophoron have a thin wall of connective tissue and muscle fibers, are lined by low columnar epithelium with clear cytoplasm, and have a basement membrane. Since the tubules are located in the mesovarium close to the ovary, and since their cysts may reach

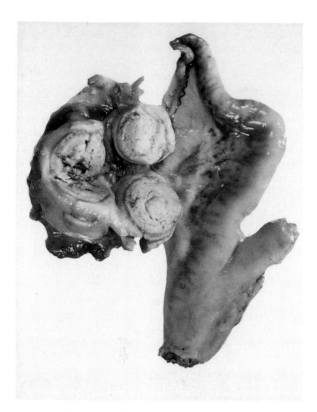

Fig. 4.11 Oophoritis caused by *Corynebacterium pseudotuberculosis*. Ewe.

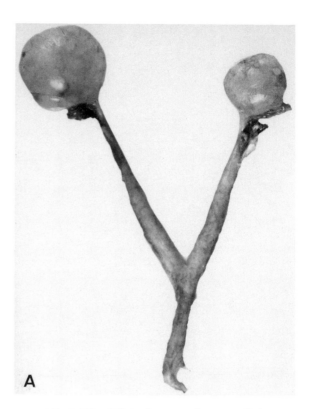

Fig. 4.12A Bilateral paraovarian cysts. Cat.

several centimeters in diameter, they may be confused with true ovarian cysts (Fig. 4.12A). The Wolffian duct has a more prominent muscular wall than the Müllerian duct and has inner longitudinal and outer circular fibers. Wolffian duct cysts are located adjacent to the oviduct.

Müllerian derivatives have epithelial cells similar to those of the oviduct and show a similar response to hormonal stimulation. The epithelium of Müllerian tubules lacks a basement membrane. Cysts of Müllerian origin may be found in the same locations as mesonephric cysts, but the major Müllerian cysts are located near the fimbria of the oviduct. These, including the **hydatid of Morgagni,** are known as accessory oviducts. They occur in all species but most frequently in the mare.

Cystic rete tubules have been observed in the bitch, cat, and cow. Those in the cow seldom attain significant size, but those of the bitch and cat become large enough to be confused with cystic follicles. Cystadenomatous tumors of the ovary may arise focally in these cysts.

Serous inclusion cysts are thought to arise by pinching off of surface epithelial indentations. They are hormonally inactive, but the lining cuboidal epithelium may be hormonally sensitive. Most lie close to the surface, but some are deep in the cortex. They may vary from a few millimeters to two or more centimeters in diameter. They can be distinguished from parovarian cysts by their intraovarian position. They occur in all species, but appear to be unim-

portant in most. In the mare, inclusion cysts develop close to the ovulation fossa and have been called fossa cysts (Fig. 4.12B). Most cause no difficulty, but when large and multiple, they can block ovulation.

The tubules and cords of epithelial cells which bud off the surface of the ovary in the bitch, the so-called subsurface epithelial structures, can also give rise to ovarian cysts. The incidence of these cysts increases with age. They are small, seldom larger than 5 mm in diameter, and can be located anywhere within the ovary, but their relationship to the surface is usually obvious. These cysts are lined by cuboidal epithelium. They do not appear to have any effect on fertility but, since the subsurface epithelial structures are hormonally sensitive, cyst formation may be associated with hormonal dysfunction. The lining of the cysts, like the surface epithelium and the subsurface epithelial structures, expresses cytokeratin, and this allows differentiation from atretic follicular cysts.

Follicular atresia is a normal, but poorly understood, phenomenon and is the destiny of the vast majority of follicles present at birth in all species. It is discussed here because follicular atresia can be pathologic when unnatural influences inhibit maturation, and also because the condition shares features with anovulatory cystic graafian follicles and luteinized cysts. During the estrous cycle of uniparous species, many follicles develop, but only one is supposed to mature and ovulate; the remainder undergo atresia, degenerate, and disappear. In some species which are seasonally anestrous, such as the mare and the ewe,

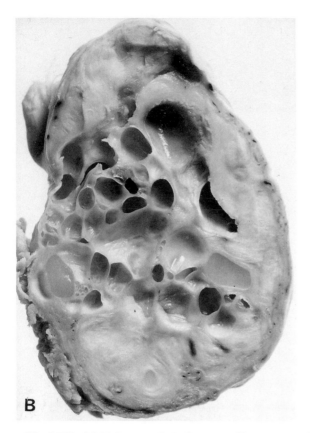

Fig. 4.12B Multiple serous inclusion cysts. The cysts contain coagulated protein. Ovary. Mare. (Courtesy of V. Osborne.)

cyclic follicular growth and atresia without maturation occur throughout the period of anestrus. Some mares develop multiple prominent anovulatory follicular cysts during late anestrus. These anovulatory cysts in the mare do not interfere with the commencement of normal cyclic activity during the breeding season. The development of anovulatory follicles is normal, at least in some species, during pregnancy.

Follicular atresia is pathologic when the degeneration is brought about by the anestrus of debility and inanition. Follicular growth is independent of hypophyseal gonadotrophins up to the stage of antrum formation. Thereafter, growth and maturation depend on stimulation by hypophyseal gonadotrophins in proper proportion and sequence. Debilitated and malnourished animals are capable of synthesizing the gonadotrophins and the ovaries are capable of responding to them. The defect presumably lies in the failure of the hypothalamus to produce or discharge gonadotrophin-releasing hormone or the pituitary's failure to respond to it. The affected follicles may cease to develop at any stage between that of the antrum and that of the finally mature. It is not known how long they may persist before degenerating.

The form of atresia depends mainly on the stage of development of the follicle when degeneration begins. It is always preceded by degeneration of the ovum and its

zona pellucida. In small follicles, the degeneration may be direct; a small cavity forms about the remnant of the zona pellucida, the granulosa cells show nuclear pyknosis and cytoplasmic vacuolation, and they desquamate into the cavity where they undergo complete disintegration. The ovum and zona pellucida are invaded by phagocytes of uncertain origin. In larger follicles, degeneration of the ovum and granulosa epithelium is followed by hypertrophy and ingrowth of theca, the cellular elements of which resemble immature lutein tissue, to enclose the degenerate remnants of the zona pellucida. The final process in atresia is infiltration by connective tissues which, when mature, form a collagenous core.

Atresia of ovarian follicles in the bitch can give rise to anovular cords or islands of persisting granulosa cells. These cords of granulosa cells in which the granulosa cells are elongated and arranged perpendicular to the axis of the cord, suggesting testicular Sertoli cells, are a distinctive feature of the ovary of the bitch.

C. Cystic Ovarian Disease

Anovulatory cystic ovarian disease occurs in most if not all species. Its importance as a disease entity varies greatly between species; in domestic species, it is a serious problem only in cows and sows. Cystic ovarian disease occurs infrequently in the bitch and queen and rarely in the ewe, doe, and mare. It is only in cattle that the condition has received detailed study. Accordingly, the major consideration here is of the bovine disease.

1. Cystic Ovarian Disease in Cows

A well-known feature of anovulatory follicular cysts is their relation to nymphomania, but most ovarian cysts are not associated with signs of persistent estrus. The behavior of cows with cystic ovaries is variable. The majority of cows with ovarian cysts are anestrous. The disease arises from the failure of mature follicles to ovulate. It occurs most often before the first postpartum ovulation. Approximately 45–60% of animals which develop anovulatory follicular cysts will reestablish normal ovarian cycles spontaneously. Cystic follicles also may develop after postpartum ovarian cycles have been established, and these cysts are more likely to persist if effective treatment is not instituted.

The cause of cystic ovarian disease is not understood in any species. The disease in cattle occurs more frequently after parturient or post-parturient disease, and there is evidence that intrauterine infections play a role in the pathogenesis of the disease. There is clearly a genetic predisposition to the disease in certain families. The daughters of cows that have had cystic ovaries have a substantially increased risk of developing the disease as compared to the general population. The disease tends to involve primarily dairy cows, but it can occur in cattle of any breed if they are withheld from breeding for a prolonged period of time. One of the factors that has made the understanding of the disease difficult is the criteria of the disease itself. Follicular cysts in cattle are usually defined as follicles greater than 2.5 cm in diameter that fail to ovulate and may persist. By the time a follicle can meet the criteria to be defined as cystic, the conditions that led to its formation have passed and are not available for study. To offset this difficulty, anovulatory ovarian disease has been produced by a variety of experimental techniques. Unfortunately it is not known whether any, or all, of the experimental manipulations induces anovulatory cystic disease by the mechanisms that operate in the natural disease. The most widely held theory of the origin of cystic ovarian disease has involved aberration of the preovulatory surge of luteinizing hormone, either the absence of the surge or the mistiming of the surge. This hypothesis, or some modification of it, is still the most attractive. Cows which develop cystic ovaries as the result of estrogen and progesterone treatment have an increased mean basal concentration of luteinizing hormone secretion with increase in the frequency and amplitude of pulses, but the characteristic preovulatory luteinizing hormone surge is deficient. This increase in luteinizing hormone secretion is thought to be due to aberrant hypothalamic function altered experimentally by steroids or naturally by ovarian secretion.

The cysts may be single or multiple on one or both ovaries. These cysts may persist, but during the course of the disease additional cysts may be recruited and some cysts undergo atresia. Patches of luteal tissue can be seen grossly in the wall of some of the cysts and can be recognized histologically in about a quarter of them (Fig. 4.13A,B). The walls of the cysts show the same type of degeneration that occurs in normal atresia. Degeneration of the granulosa cells takes place first, the cells undergoing pyknosis and karyorrhexis and sloughing into the cyst lumen. The oocyte also undergoes degeneration. The changes of the theca interna are variable. The theca is partially luteinized in some cases, whereas in others it degenerates and is infiltrated by fibrous tissue. The luteal tissue may occur in patches or form a crescent of variable thickness.

Because the disease is a dynamic one with new cysts being added and cysts undergoing variable luteinization and degeneration, the hormonal consequence is also variable, the peripheral concentrations of luteinizing hormone, estradiol, and progesterone depending on structural and functional features of the cysts.

Cows which bear follicular cysts for long periods ultimately develop permanent anestrus. The cysts are then reduced in size and no longer dominate the contour of the ovary, the pressure within the cysts is reduced, and the walls are appreciably thicker (Fig. 4.14). These atrophic cysts are usually multiple and bilateral; histologically, they do not differ clearly from active follicular cysts.

a. LUTEINIZED CYST This type of cyst develops when ovulation fails to occur and the theca undergoes luteinization. There is no ovulation papilla and the luteal mass is smooth and rounded (Fig. 4.13A). The cavity of the cyst

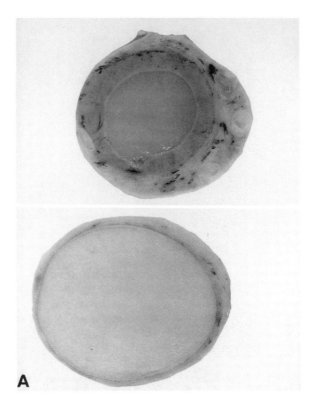

Fig. 4.13A Anovulatory ovarian cysts. Cow. Luteinized cyst above, follicular cyst below.

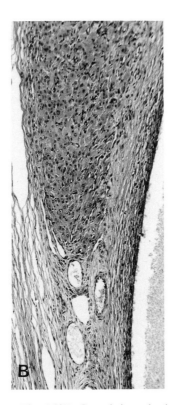

Fig. 4.13B Luteal tissue in theca in ovarian follicular cyst. Cow.

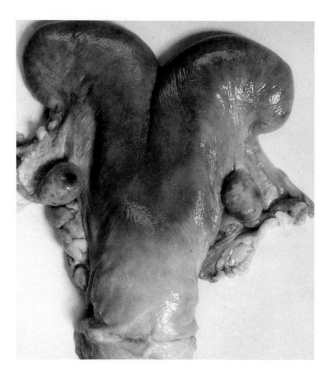

Fig. 4.14 Atrophic follicular cysts with mucometra. Cow. Note dilation and apparent flaccidity of uterus.

is spherical and lined by a layer of fibrous tissue adjacent to the zone of luteinized theca cells. One or more small blebs are usually present on the surface of a normal follicle which is about to ovulate, and luteinization of these produces the ovulation papillae. The presence of papillae is usually, but not invariably, an indication that ovulation has occurred.

Luteinized cysts occur more frequently in cattle and swine than in other species of domestic animals. In cattle, they usually occur as single cysts. Single luteinized cysts may occur in pregnant sows, but multiple cysts are associated with infertility.

The pathogenesis of luteinized cysts is probably the same as that proposed for the cystic follicle, that is, by failure of the hypophysis to release an adequate surge of luteinizing hormone. As previously mentioned, it is not unusual to find some evidence of luteinization in follicular cysts (Fig. 4.13B) and the two conditions are assumed to be expression of different degrees of the same dysfunction.

Anovulatory luteinized cysts should not be confused with cystic corpora lutea. A cystic corpus luteum is a corpus luteum which has formed after ovulation in which a central cavity has persisted in the center of a mass of developing luteal tissue. Cystic corpora lutea are not evidence of ovarian malfunction. They form after ovulation and do not affect the length of the estrous cycle. If the cow has been successfully bred, the cavity will slowly become obliterated. Large central cysts may occasionally persist for 30–40 days after conception. Cystic corpora lutea can be distinguished from luteinized cyst by the

ovulation papilla which distorts the outline of the cyst at the point of ovulation.

b. Extraovarian Lesions of Cystic Ovarian Degeneration in Cows   Changes in other organs develop only if the ovarian cysts persist. Since most cases are responsive to treatment, this is now the exception. It has been shown that in cystic ovarian disease, the hormone limits are within normal levels; however, the toxicity of ovarian hormones depends not on the absolute levels, but on the loss of cyclicity and on persistence.

In established cases of cystic ovarian disease, the uterus is usually altered, the differences in degree being related to the duration of the condition. In association with functionally active follicular cysts, the uterus is enlarged, and the wall is edematous. The cervix is enlarged, the external os patent, and the plicae edematous. The endometrium may appear grossly normal, being smooth, semitransparent, grayish pink and moist. There may be some degree of cystic endometrial hyperplasia detectable by the naked eye as tiny gray-white elevated blisters in the surface of the endometrium, which is overlain by a thin layer of clear viscous mucus. The accumulation of mucus (mucometra) and the development of cystic endometrial hyperplasia progress with time and are most striking in those animals which have been affected long enough for atrophy of the ovarian cysts to have occurred and the animal to have become anestrous. By then, the endometrium may be of the descriptive Swiss-cheese type, and the volume of mucus be 100–1000 ml or even much more. In animals with these atrophic cysts, the endometrium may be atrophic with very few dilated glands and no mucometra, but irrespective of the endometrial changes, the myometrium will be atrophic. In the early stage of the development of the cystic endometrial change when the degree is not clearly in excess, the physiological hyperplasia can be distinguished from the pathologic by the fact that in the former the endometrial activity is uniform and in the latter it is not.

The cervix may be hypertrophic or atrophic as previously described. It usually contains thick, viscid, grayish-white, and cellular mucus. The epithelium undergoes squamous nonkeratinizing metaplasia of mild degree (Fig. 4.15).

The vagina is edematous when the cysts are active, but otherwise the most striking change is the formation of cysts in Gartner's ducts (Fig. 4.16). These Wolffian vestiges which lie in the wall of the vagina, one on each side of its floor, are normally visible only microscopically as continuous or discontinuous ducts lined by a simple pavement or cuboidal epithelium. When chronically stimulated by estrogen, the epithelium becomes squamous and the ducts, cystic and visible, or palpable, as a series of blebs or as tubules up to 1.0 cm in diameter on the floor of the vagina.

The vulva may be edematous when the cysts are active, and the clitoris may be enlarged in long-standing cases. Cystic Gartner's ducts are usually accompanied by cystic

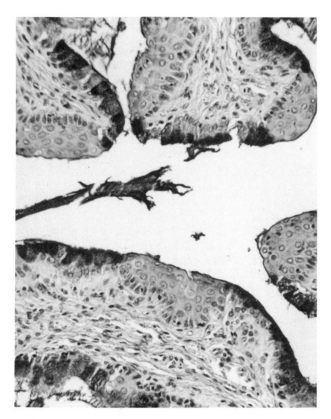

Fig. 4.15 Squamous metaplasia of cervical epithelium in hyperestrogenism. Cow.

Bartholin's glands. Either may become abscessed. Bartholin's glands, one on each side of the floor of the vulva, undergo cyclic secretory changes during the estrous cycle and a squamous epithelium on the ducts is normal in estrus. Exaggeration of this change in chronic estrogenism causes occlusion of the ducts and the formation of retention cysts.

2. Cystic Ovarian Degeneration in Other Species

In **swine,** cystic ovaries are common and are important as a cause of infertility. Failure of a single follicle to ovulate need not interfere with pregnancy, and single large cysts, 2–3 cm in diameter, may be found in pregnant animals. They are presumed to represent mature follicles that did not ovulate, but single anovulatory cysts may be also associated with irregularity of estrous cycle. The association is more clearly manifested with multiple cysts. The cysts vary in type from follicular, which are lined by normal granulosa cells, to luteinized cysts. Multiple luteinized cysts with some cysts up to 5 cm in diameter are a distinctive feature of cystic ovarian disease in swine. The luteal tissue may form a complete rim or be present in patches. The endometrium in cases of multiple luteal cysts shows hyperplasia of progestational type, and in long-standing cases the clitoris is enlarged.

Cystic degeneration of the graafian follicle with the as-

Fig. 4.16 Cystic Gartner's ducts. Cow.

sociated loss of cyclicity and infertility comparable to that of the cow and the sow does not appear to occur in the **mare,** but the matter is not clear because anovulatory follicular cysts do develop during the winter anestrous period, and some of these mares will show irregular signs of estrus. The number and size of these anovulatory cysts vary considerably. There is no evidence that development of these cysts, regardless of their size, is indicative of reproductive disease.

In the **bitch,** anovulatory cysts of both follicular and luteinized types tend to occur in older age groups. It is not clear whether or not the disease represents a single entity or whether it represents a spectrum of related disorders characterized by anovulatory cysts. The incidence of cystic ovarian disease is low in the bitch, but there is some confusion on this point because cystic rete tubules, cysts of subsurface epithelial structures, parovarian cysts, and cystadenomas have been confused with follicular cysts. The cysts may be multiple or single, and the latter are of little clinical significance. **Polycystic ovarian disease** in the bitch is rare, but because it can cause hyperestrogenism it carries a special risk in the dog because of the species' sensitivity to either endogenous or exogenous estrogen which can induce lethal bone marrow suppression. The condition can involve either one or both ovaries, the affected ovary being greatly enlarged by multiple, thin-walled cysts that vary from 1 to 12 mm in diameter. The disease develops in mature animals, which may be either nulliparous or multiparous. If clinical signs occur, they are apt to be associated with hyperestrogenism and include persistent estrus with cornification of the vagina and swelling of the mammary glands or the toxic effects of hyperestrogenism on the bone marrow with resultant pancytopenia. This pancytopenia can result in anemia, thrombocytopenia with bleeding, most often occurring as epistaxis, or granulocytopenia with increased susceptibility to infection. The mechanism of the inhibitory action of estrogen on the hematopoietic tissues in estrogen-sensitive species is not known. Estrogen appears to inhibit the differentiation of pluripotent stem cells while stimulating the differentiation and maturation of committed stem cells.

Most of the cysts are obviously of follicular origin but in various stages of atresia. Degenerating oocytes can be identified in some of the cysts, but the disease is not just one of anovulation because in some cases, hundreds of follicles have been recruited. The granulosa and theca lining of the cysts is attenuated in the larger ones. Patchy areas of luteinization form in the walls of some of the cysts. This luteinization may involve individual cells or sheets of cells and either the theca or the granulosa.

Bibliography

Al-Dahash, S. Y. A., and David, J. S. E. The incidence of ovarian activity, pregnancy and bovine genital abnormalities shown by an abattoir survey. *Vet Rec* **101:** 296–299, 1977.

Al-Dahash, S. Y. A., and David, J. S. E. Histologic examination of ovaries and uteri from cows with cystic ovaries. *Vet Rec* **101:** 342–347, 1977.

Bosu, W. T. K., and Peter, A. T. Evidence for a role of intrauterine infections in the pathogenesis of cystic ovaries in postpartum dairy cows. *Theriogenology* **28:** 725–736, 1987.

Cook, D. L. *et al.* Fate and turnover rate of ovarian follicular cysts in dairy cattle. *J Reprod Fertil* **90:** 37–46, 1990.

Cook, D. L. *et al.* Secretory patterns of LH and FSH during development and hypothalamic and hypophysial characteristics following development of steroid-induced ovarian follicular cysts in dairy cattle. *J Reprod Fertil* **91:** 19–28, 1991.

Dow, C. Ovarian abnormalities in the bitch. *J Comp Pathol* **70:** 59–69, 1960.

Eyestone, W. H., and Ax, R. L. A review of ovarian follicular cysts in cows with comparisons to the condition in women, rats, and rabbits. *Theriogenology* **22:** 109–125, 1984.

Kesler, D. J., and Garverick, H. A. Ovarian cysts in dairy cattle: A review. *J Anim Sci* **55:** 1147–1159, 1982.

McKenzie, B. E., and Kenney, R. M. Histologic features of ovarian follicles of gonadotropin-injected heifers. *Am J Vet Res* **34:** 1033–1040, 1973.

Nalbandov, A. V. Anatomic and endocrine causes of sterility in female swine. *Fertil Steril* **3:** 100–114, 1952.

Wrathal, A. E. Ovarian disorders in the sow. *Vet Bull* **50:** 253–272, 1980.

D. Neoplastic Diseases of the Ovary

1. General Considerations

Tumors arising from tissues that are specifically ovarian can be divided into three broad categories: tumors of the surface celomic epithelium, tumors of the gonadal stroma, and tumors of germ cells. Tumors developing from nongonadal supporting tissues of the ovary can be of the usual variety of fibroblastic, smooth muscle, and vascular tumors, but are, in fact, uncommon in domestic animals.

Tumors of the surface epithelium and its tubular invaginations, the subsurface epithelial structures, give rise to the papillary and cystic adenomas and less frequently, papillary adenocarcinomas. These tumors are common only in the bitch among domestic animals.

Gonadal stromal tumors are tumors of granulosa and theca cells and their luteinized counterparts. They are also referred to as sex cord–stromal tumors to reflect the concern that the granulosa cells may arise from the sex cords of the surface celomic epithelium or the mesonephric tubules. In spite of the histogenetic uncertainty, these tumors must be dealt with as a group because granulosa cells and theca cells often coexist in the same tumor. This tendency to coexistence is dealt with by hyphenation, referring to the tumors as granulosa–theca cell tumors. Included in this group are tumors which resemble testicular tumors of the Sertoli cell and Leydig cell types. Whether these are, in fact, Sertoli cells and Leydig cells or whether the ovarian stromal cells are merely growing in patterns of their testicular homologs is difficult to determine. The question is made more difficult by the fact that this group of tumors frequently produces hormones. The hormones are sometimes appropriate products for the apparent cells of origin, that is, estrogens and progesterone, but many of the ovarian stromal tumors produce androgens. This is particularly true of the ovarian tumors of the mare. It is not useful to attempt to classify ovarian tumors on the basis of hormone production, because, in addition to variations in hormone production from a single class of ovarian tumors, variation in hormone production from a single tumor can occur over time.

The germ cell tumors of the ovary, like their testicular counterparts, arise from germ cells, and are usually divided into two main categories in domestic animals: the dysgerminoma and the teratoma. The dysgerminoma is composed of cells that show morphologic similarity to primordial germ cells and resemble their testicular homolog, the seminoma. In the teratomas, the totipotential germ cells have undergone somatic differentiation, giving rise to two or more germinal layers, with a variety of tissues being present in the tumors. Germ cell tumors are rare in domestic species, except for the testicular seminoma of dogs, and further subdivision of these two categories is probably unwise at this time; however, some germ cell tumors do not fit well into this simple two-category classification. When more extensive series are available for study and analyzed by sophisticated methods, it is likely that the germ cell tumors of animals will show the wide variability that human germ cell tumors display. In addition to the germ cell tumors that show no somatic differentiation (the dysgerminomas and the seminomas) and tumors which have extensive differentiation into tissues of multiple germinal layers (the teratomas), human germ cell tumors have shown varying stages of differentiation, i.e., endodermal sinus tumors (yolk-sac tumors), choriocarcinomas, and embryonal carcinomas, as well as tumors from single germ layers. This is the variability that is to be expected in tumors arising from pluripotential germ cells.

Ovarian tumors occur most frequently in the bitch, the mare, and cow. In the bitch, they are often bilateral, especially those arising from the surface epithelium and the subsurface epithelial structures. In the cow, ewe, and mare, they are usually unilateral and of the gonadal stromal type. In cows, there is a tendency for the tumor to occur in the daughters of affected dams and, in these animals, removal of the affected ovary may be followed by the development of a tumor in the other ovary. Very few ovarian neoplasms have been reported in the feline and porcine species. The tumors of the cat are similar to those of the bitch.

2. Sex Cord–Stromal Tumors

These are the tumors that are derived from the cells of the sex cords or the specialized ovarian stroma. They include granulosa cell tumors and thecomas, either of which may be luteinized.

The **granulosa–theca cell tumor** is the most common tumor in this group. They are generally unilateral, usually nonmalignant tumors in any species. Although they may be observed in young animals, the incidence increases with age. Part of their interest lies in the production of steroids by some of them.

Clinical signs associated with steroidogenesis are frequently associated with the neoplasm in the cow and mare. In mares, in which species the tumors are common and have been studied most carefully, three behavioral patterns have been recognized: anestrus, continuous or intermittent estrus, and male behavior. Surprisingly, testosterone levels in the peripheral plasma are elevated in most cases. Male behavior is usually seen only in those cases in which the testosterone levels are above 100 pg/ml plasma. Some mares with granulosa–theca cell tumors also have elevated estrogen levels, but these elevations are less clearly related to behavioral patterns. Even in cases in which hormone production associated with the tumor is low, atrophy of the opposite ovary usually occurs. The cause of this atrophy has not been determined. It was thought to be the result of testosterone production by the

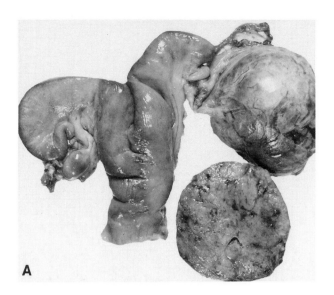

Fig. 4.17A Luteinized granulosa–theca cell tumor. Cow.

tumor, but convincing association between testosterone levels and the ovarian atrophy has not been established, and androgen production by tumors other than the granulosa–theca group has not caused atrophy of the contralateral ovary. Granulosa cells produce the peptide hormone, inhibin, which inhibits release of follicle-stimulating hormone. This hormone has been demonstrated to be elevated in granulosa cell tumors, and it may be that the production of inhibin produces the characteristic ovarian atrophy seen in this disease. Normal function of the atrophic ovary returns usually within a year of removal of the tumor. In bitches with sex cord–stromal tumors, cystic endometrial hyperplasia and pyometra are common.

The surface of the granulosa–theca cell tumor is smooth (Fig. 4.17A) and the cut surface may be solid or cystic (Fig. 4.17B). The solid portions of the tumor are white or yellow, depending on the lipid content. The cells in these tumors resemble their counterparts in normal follicles, but their histologic arrangement is quite varied. Diagnostic, when present, are glandlike or rosette patterns of abortive follicles, some of which may contain a secretory globule resembling an ovum and called a Call–Exner body (Fig. 4.18A). This type of differentiation is frequently observed in the early stages of development of bovine neoplasms but is less common in other species and in large tumors. Instead, the arrangement of cells is usually diffuse with, in some tumors or parts of tumors, pseudoalveolar or cylindromatous patterns, depending on the disposition of the stroma. There is a tendency for some granulosa–theca cell tumors to develop a tubular pattern similar to that of the Sertoli cell tumor of the testis (Fig. 4.18B). The stroma consists of broad irregular bands of dense collagen. Cyst formation and hemorrhage are common. The more thecomatous portions of such tumors may resemble normal theca cells and, at the other extreme, be distinguishable from plump fibroblasts only by the demonstration of su-

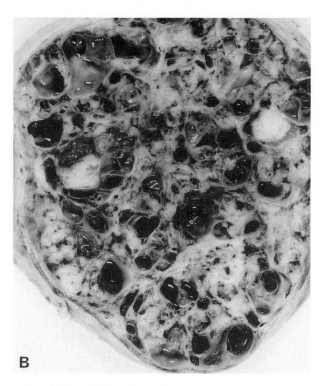

Fig. 4.17B Solid and cystic granulosa–theca cell tumor. Mare.

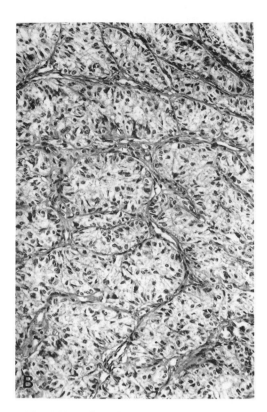

Fig. 4.18B Granulosa–theca cell tumor with tubular pattern reminiscent of Sertoli cell tumor. Bitch.

danophilic droplets in the cytoplasm or by histochemical techniques for steroids.

The cysts which often make up the bulk of the tumor are lined by granulosa cells, which are surrounded by a variable population of thecal cells (Fig. 4.19A,B). In some equine tumors, particularly those associated with high testosterone levels, embedded in the thecal cell layer are large polyhedral eosinophilic cells (Fig. 4.20A). These cells resemble the testosterone-producing Leydig cells and

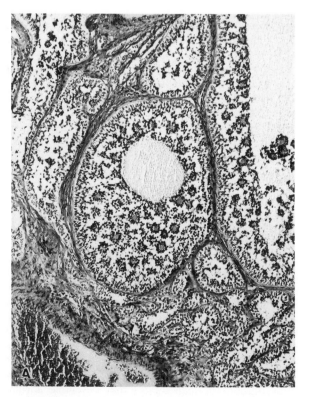

Fig. 4.18A Granulosa–theca cell tumor with rosette formation around Call–Exner bodies. Cow.

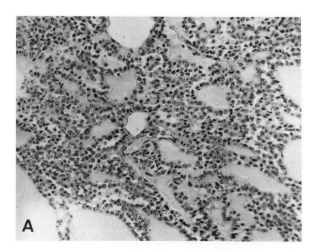

Fig. 4.19A Granulosa–theca cell tumor. Cow.

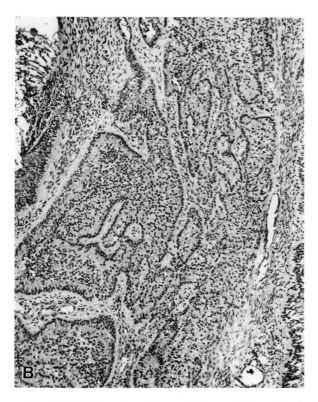

Fig. 4.19B Granulosa–theca cell tumor. Mare. Neoplastic granulosa cells are divided into cords and islands by theca cells.

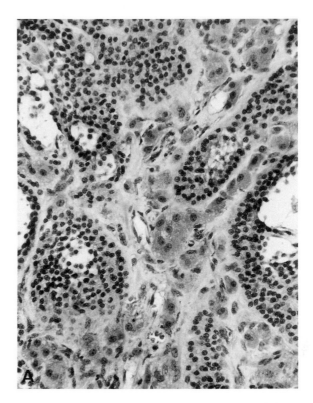

Fig. 4.20A Granulosa–theca cell tumor containing large Leydig-like cells. Mare. These cells are associated with high testosterone levels.

may be the source of the testosterone, but this is at present unproven. The correlation between histologic appearance and behavior pattern is difficult to establish as the tumors are large, and histologic patterns vary from one area to another.

Tumors composed only of theca cells, **thecoma** or **luteoma,** are much less common in all domestic animals than those arising from granulosa cells or containing a mixture of theca and granulosa cells (Fig. 4.20B). The few tumors reported have been firm, white to orange in color, and composed of streaming oval or spindle cells that resembled the cells of the theca interna. The demonstration of lipid in the cytoplasm allows differentiation from fibromas.

3. Tumors of the Surface Epithelium

The papillary cystadenoma and its malignant counterpart have been reported primarily in the bitch. As with the granulosa theca–cell tumor, many of the affected bitches have cystic hyperplasia of the endometrium. It should be emphasized that hyperplasia of the endometrium has been observed in bitches with a variety of ovarian neoplasms. It appears that the tumor, irrespective of type, may occasionally stimulate the ovarian stroma with concomitant production of steroid hormones. Papillary cystadenocarcinomas should be considered in the differential diagnosis of ascites in adult entire bitches. The ascites develops from obstruction of the diaphragmatic lymphatics by permeating tumor fragments, with perhaps a contribution by

secretion by the tumor epithelium. The tumors are frequently multicentric in origin and bilateral. They vary in size from small neoplasms which are not detectable on gross examination to irregular-shaped masses 10 cm or more in diameter. Confinement of the tumor within the rather complete ovarian bursa of the bitch, with consequent compression of the papillae, gives them a typical cauliflowerlike appearance (Fig. 4.21A). Once free of the bursa, the papillary nature becomes obvious (Fig. 4.21B) and peritoneal implantation occurs readily. The presence of such papillae is indicative of malignancy. Cysts of varying size are usually scattered throughout the neoplastic mass (Fig. 4.22A,B). The epithelium varies from low cuboidal to columnar, with stratification in some areas, and mitotic figures are scant even in the malignant form. The tumor appears to develop from the surface and from tubular structures (cortical tubes, subsurface epithelial structures) of the ovarian cortex. Many of the tubules are continuous with the surface epithelium and responsive to estrogenic stimulation. Papillary tumors have been induced in the bitch by prolonged administration of diethylstilbestrol. The experimentally induced tumors regress following withdrawal of hormone treatment. Tumors of the surface epithelium can occur in association with sex cord–stromal tumors, possibly by hormonal induction by the granulosa–theca tumors. The separate identification of the two tumors can be difficult, particularly in areas

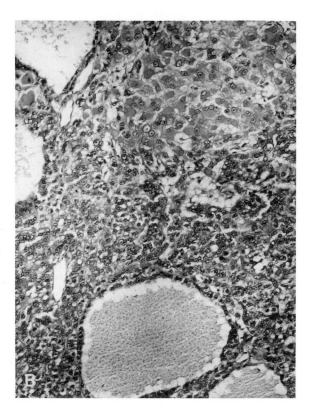

Fig. 4.20B Granulosa–theca cell tumor with areas of luteinization. Cat.

Fig. 4.21B Papillary growth of papillary adenocarcinoma when free of bursa.

Fig. 4.21A Adenocarcinoma of canine ovary. Cauliflower appearance produced by growth within bursa.

where the tumors join, but is aided by immunohistochemical staining. Cytokeratin intermediate filaments are expressed in the surface epithelium and the subsurface epithelial structures. On the basis of limited investigation, it appears that this specificity has been conserved during tumor development, and affords easy differentiation of these from cytokeratin negative tumors of the sex cord–stromal group.

4. Tumors of Germ Cells

The **dysgerminoma** is a rare tumor which is comparable to the more common seminoma of the testicle in gross and microscopic features. It is usually unilateral and has been observed in the bitch, cow, mare, and sow. The tumor is smooth, or nearly so, and relatively soft. The cut surface is gray and may have areas of hemorrhage or yellow patches of necrosis. It is composed of a uniform population of large rounded cells with large central chromatic nuclei. Mitotic figures and giant cells are frequent. The pattern of growth is diffuse; the stroma is always scanty. Just as in the canine testicular seminoma, accumulations of lymphocytes are present within the tumors.

The tumors are uncommon, and so it is difficult to generalize as to age of occurrence and clinical behavior in various species. In the bitch, in which most cases have been reported, the tumor is one of advanced age, and metastases have occurred in 10–20% of the cases.

Ovarian **teratomas** are observed rarely in domestic animals, but have been reported in the bitch, sow, mare, and cow. Most are well differentiated and benign. They usually have solid and cystic areas containing sebaceous material and hair. A wide variety of other tissues is often present including neural tissue, adipose tissue, bone, teeth, and respiratory epithelium (Fig. 4.23). There were several theories on the histogenesis of teratomas, but it is now believed that the benign cystic ovarian teratomas are parthenogenetic tumors that develop from a single germ cell which has completed its first meiotic division, but not its second. The evidence for this has come from a series of

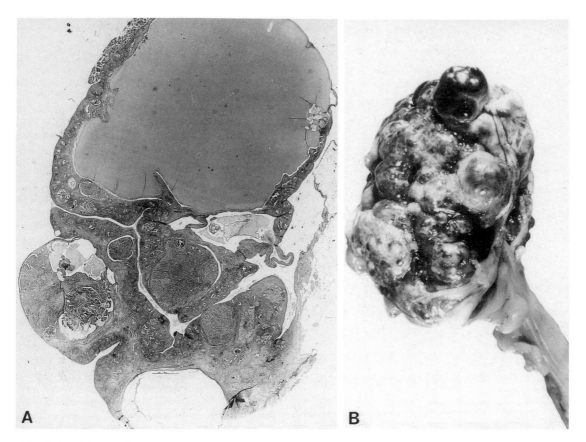

Fig. 4.22 Adenocarcinoma of canine ovary. (A) Papillary cystadenocarcinoma with widespread involvement of ovary and papillary growth on surface and in cysts. (B) Cystadenocarcinoma. The bursa has been removed to expose the tumor. Bitch.

elegant cytogenetic and biochemical studies which have shown that tissues from ovarian teratomas are unique in that they are XX diploid cells, but are homozygous at chromosomal loci for which the host is heterozygous. This lack of heterozygosity is most reasonably explained as being the result of meiotic division.

5. Tumors of Nongonadal Tissues

Ovarian **hemangioma** is rare in all domestic animals except the pig, in which species it is the most common ovarian tumor in mature and aged sows. Ovarian hemangiomas are globular, well circumscribed, and tan to reddish brown. Their surface is smooth and glistening and contains prominent vessels. The tumors arise in the ovarian cortex, and they are occasionally bilateral. The tumor is composed of well-differentiated endothelial cells, which line vascular clefts and spaces. The tumors are often subdivided by bands of connective tissue. Malignant vascular tumors of the ovary have not been reported.

Ovarian tumors of other supporting tissues are very infrequent. **Leiomyomas** developing from the smooth muscle in the mesovarium have been reported in the bitch, queen, and sow, and mesovarial leiomyomas have been induced in rats treated with β-adrenergic receptor stimulants.

6. Metastatic Tumors

Secondary tumors of the ovary are probably not less rare than primary ones, the relatively high incidence being attributable largely to secondary deposits of lymphomas. Mammary carcinomas in the bitch and intestinal carcinomas in the cow may metastasize to the ovary and apparently have an affinity for corpora lutea.

7. Vascular Hamartomas

Vascular hamartomas of the ovary are included with tumors because they may be confused with neoplasms, and because granulosa–theca cell tumors occasionally occur in the same gonad. Hamartomas are tumorlike malformations which are present at birth and cease to grow after the animal reaches maturity, unless the mass is subjected to trauma, infection, or vascular embarrassment. They have been observed in the cow and sow. The mass of malformed vessels may vary in size from those barely visible on gross examination to those weighing several kilograms. The smaller hamartomas are composed principally of mature-appearing, tortuous arteries and veins with relatively little intervascular connective tissue. In the mature individual, the increase in size of the hamartoma is due to thrombosis, with subsequent edema, hemorrhage,

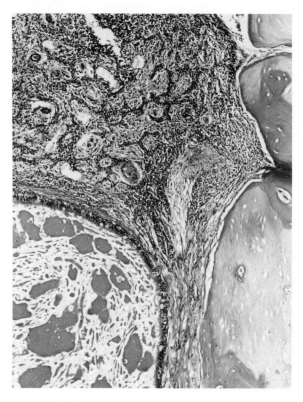

Fig. 4.23 Teratoma. Ovary. A mixture of mature tissues is present. Bitch.

and necrosis, followed by proliferation of fibrous connective tissue. Small hamartomas are clearly demarcated from the adjacent ovarian tissue, but the ovary is gradually replaced as the mass enlarges. The distinction between hemangiomas and vascular hamartomas is a difficult one, and it may be that some of the reported ovarian hemangiomas are in fact vascular hamartomas that have grown as the result of vascular accidents.

Bibliography

Andrew, E. J. *et al.* A histopathologic study of canine and feline ovarian dysgerminomas. *Can J Comp Med* **38:** 85–89, 1974.

Cotchin, E. Spontaneous tumors of the uterus and ovaries in animals. *In* "Pathology of the Female Genital Tract," A. Blaustein, (ed.), pp. 822–861. New York, Springer-Verlag, 1977.

Craig, J. M. The pathology of the female reproductive tract. *Am J Pathol* **94:** 385–437, 1979.

Dehner, L. P. *et al.* Comparative pathology of ovarian neoplasms. III. Germ cell tumors of canine, bovine, feline, rodent and human species. *J Comp Pathol* **80:** 299–306, 1970.

Frazer, G. S., Robertson, J. T., and Boyce, R. W. Teratocarcinoma of the ovary in a mare. *J Am Vet Med Assoc* **193:** 953–955, 1988.

Gerald, P. S. Origin of teratomas. *N Engl J Med* **292:** 103–104, 1975.

Greenlee, P. G., and Patnaik, A. K. Canine ovarian tumors of germ cell origin. *Vet Pathol* **22:** 117–122, 1985.

Hinrichs, K. *et al.* Serous cystadenoma in a normally cyclic mare with high plasma testosterone values. *J Am Vet Med Assoc* **194:** 381–382, 1989.

Hinrichs, K., Watson, E. D., and Kenney, R. M. Granulosa cell tumor in a mare with a functional contralateral ovary. *J Am Vet Med Assoc* **197:** 1037–1038, 1990.

Hsu, F. S. Ovarian hemangioma in swine. *Vet Pathol* **20:** 401–409, 1983.

Jabara, A. G. Canine ovarian tumours following stilboestrol administration. *Aust J Exp Biol Med Sci* **37:** 549–565, 1959.

Lappohn, R. E. *et al.* Inhibin as a marker for granulosa-cell tumors. *N Engl J Med* **321:** 790–793, 1989.

Linder, D., McCaw, B. K., and Hecht, F. Parthenogenic origin of benign ovarian teratomas. *N Engl J Med* **292:** 63–66, 1975.

McCandlish, I. A. P. *et al.* Hormone-producing ovarian tumours in the dog. *Vet Rec* **105:** 9–11, 1979.

Meinecke, B., and Gips, H. Steroid hormone secretory patterns in mares with granulosa cell tumors. *J Vet Med (A)* **34:** 545–560, 1987.

Nielsen, S. W., Misdorp, W., and McEntee, K. Tumours of the ovary. International histological classification of tumours of domestic animals. *WHO Bull* **53:** 215–230, 1976.

Norris, H. J., Garner, F. M., and Taylor, H. B. Comparative pathology of ovarian neoplasms. IV. Gonadal stromal tumours of canine species. *J Comp Pathol* **80:** 399–405, 1970.

O'Shea, J. D. Carcinoma with squamous metaplasia (adenoacanthoma) in the bovine ovary. *J Comp Pathol* **76:** 437–438, 1963.

O'Shea, J. D. Histochemical observations on mucin secretion by subsurface epithelial structures in the canine ovary. *J Morphol* **120:** 347–358, 1966.

O'Shea, J. D., and Jabara, A. G. The histogenesis of canine ovarian tumours induced by stilboestrol administration. *Pathol Vet* **4:** 137–148, 1967.

Patnaik, A. K., and Greenlee, P. G. Canine ovarian neoplasms: A clinicopathologic study of 71 cases including histology of 12 granulosa cell tumors. *Vet Pathol* **24:** 509–514, 1987.

Scully, R. E. Ovarian tumors: A review. *Am J Pathol* **87:** 686–720, 1977.

Whitacre, M. D. Premature lactation in a heifer with a sex cord–stromal tumor. *J Am Vet Med Assoc* **193:** 946–948, 1988.

III. The Uterine (Fallopian) Tubes

Primary lesions in the uterine tubes are uncommon. Hydrosalpinx, pyosalpinx, and salpingitis are the only ones of importance, and these are usually secondary to disease of the uterus or to manual manipulation of the ovary. As regards incidence of salpingeal lesions, the only agreement is that they are much more common than is the diagnosis of them. They are recognized to be important in the cow and sow, but not in other species.

A. Hydrosalpinx

Distension of the uterine tube by fluid follows loss of patency of the lumen (it is possible, too, that a condition akin to paralytic ileus may affect the tube when the surrounding tissues are inflamed), and it does not matter whether the obstruction is to the abdominal ostium or the uterine ostium. The obstruction may have a congenital or inflammatory basis. Congenital anomalies involving the tubes are very rare, except in freemartins, but secondary hydrosalpinx can be associated with obstructions and segmental aplasias of the uterine horns. In the latter instances, the apex of the horn is usually distended with fluid also. The acute septic inflammations are more likely to produce pyosalpinx; the chronic infective inflammations tend to produce loculations and stenosis of the lumen with secondary hydrosalpinx. As almost all infective inflammations are of ascending type from coexisting uterine inflammation, the most severe changes and the obstruction in such cases are usually at or near the uterine end of the tube.

A common form of inflammation which results in hydrosalpinx in cattle is at least in part traumatic in origin, following manual manipulation of the ovary. Uterine irrigations are somewhat less important, but, in either event, it is the adhesion between the tube and adnexa (usually the ovary with partial or complete obliteration of the bursal cavity) which causes obstruction of the abdominal ostium and secondary hydrosalpinx. Squashing of the normal diestrual corpus luteum (there is at best only partial expression of the body) is not difficult but, because of its vascularity, there is always some hemorrhage, varying in extent from very slight to fatal (Fig. 4.10). Small clots, which are usually retained in the bursa, can be completely resorbed. Larger clots must be organized, and this results in the formation of adhesions within the bursa and the fimbriated portion of the tube. With more extensive hemorrhage, adhesions may form to adjacent abdominal viscera and genital adnexa. Expression of pathologically persistent corpora lutea may result in little hemorrhage but a lot of tissue damage, and a more or less proportional degree of organization by fibrosis will result. The important types of persistent corpora lutea occur in association with pathological distension of the uterus. If this latter is pyometra, an additional complication is involved in that, as well as the manual trauma, there is liability to spillage of inflammatory products and extension of infection to traumatically devitalized tissue.

Therapeutic uterine irrigation in cattle, especially if performed in the early postpartum period, has also caused the type of adhesion referred to and follows one of two complications: the irrigation fluid may leak through the tube into the bursa and provoke an inflammatory reaction, the usual types of irrigating fluid being irritative, or the infusion of an excessive volume of fluid causes rupture of the uterus and perimetritis. Uteri vary in the amount of fluid they may retain, but any amount in excess of 120 ml is liable to cause uterine rupture, which occurs on the lesser curvature at the mesometrial attachment with permeation of fluid through or beneath the serosa. Inflammatory thickening is the usual result, and the development of serosal adhesions produces tubal obstruction and dilation.

Hydrosalpinx is so called because the uterine tube is distended, uniformly or irregularly, up to 1.5 cm or so with a clear watery mucus which fluctuates. The tube is also increased in length and tortuosity and is thin walled. Histologically, there is extensive multilocular cyst formation in the mucosa with obliteration of the lumen, and, in some chronically inflamed uterine tubes, mononuclear cell infiltrations of the substantia propria (Fig. 4.24).

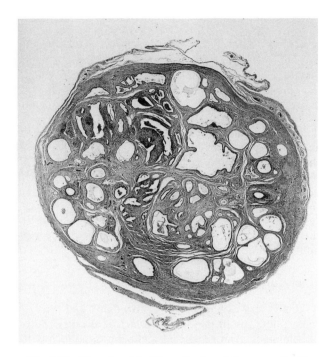

Fig. 4.24 Chronic salpingitis and hydrosalpinx caused by *Brucella suis*. Sow.

B. Salpingitis

Inflammation of the uterine (fallopian) tubes without significant enlargement is the commonest and most important tubal lesion. It is usually bilateral, is usually not detectable macroscopically, and may show any of the usual forms of inflammation, either serous, catarrhal, or fibrinous. In the mildest forms of salpingitis, the mucosa alone is affected and changes of functional significance may be slight enough to be overlooked histologically. Congestion of the mucosal vessels, mononuclear cell infiltration, loss of epithelial cilia, and some desquamation of epithelium may be the only changes detectable. With more severe infections, catarrhal exudate collects in the lumen, the mucosal folds are thickened by cellular infiltration and congestion, and the epithelium is in large part destroyed. Loss of epithelium occurs first in the free edges of the mucosal folds and these denuded areas tend to fuse and adhere to produce intramucosal cysts; alternatively, in chronic catarrhal salpingitis, the mucosa is virtually destroyed and replaced by proliferated connective tissue and cellular infiltrations with more or less occlusion of the lumen.

The uterine tube is a rather simple structure histologically, but even minor inflammatory changes, evidenced by slight congestion or the presence of a few plasma cells, appear to be important because of the readiness with which the epithelial cells desquamate or lose their cilia. The proper function of the living epithelium is necessary for the propulsion of the ovum, for the dissolution of the cumulus oophorus prior to fertilization, and for the maintenance of a luminal environment suitable to survival of the ovum. The salpingeal mucosa has a much lesser capacity for restitution than does the endometrium.

Salpingitis is a common lesion in animals with both *Mycoplasma* and *Ureaplasma* infections. Nonspecific infections causing salpingitis almost invariably do so following spread from the uterus. There is probably 70–75% association between uterine and salpingeal inflammation when diagnosis of the latter is based on histologic evidence. In some cases, there will be a perimetritis with adhesions, pyosalpinx, or bursal abscess. Adhesions of the infundibulum of the mare are very common. The cause of these adhesions is unknown. Some are associated with perimetritis, but most are not. It has been suggested that they may develop as a result of ovulatory hemorrhage. However, this explanation does not easily account for the substantial predominance of adhesions of the right infundibulum.

C. Pyosalpinx

This is less common than hydrosalpinx and typically follows metritis in the same manner as do other forms of salpingitis. The significant anatomic difference is the accumulation of pus in the tube following obstruction of the lumen. The obstruction may be produced by inspissated exudate, inflammatory thickening and fusion of the mucosal folds, or chronic granulation tissue. The length of the tube is usually not uniformly involved by the inflammatory process; rather, there are segments in which the reaction is more acute or more advanced so that the obstruction tends to involve irregular segments with the intervening portions distended with exudate.

The entire thickness of the wall of the duct is infiltrated with neutrophils, lymphocytes, and plasma cells, and the same cells collect in the lumen and in the mucosal cysts formed by adhesions between the denuded epithelial folds. Surviving epithelium may be partly squamous. Eventu-

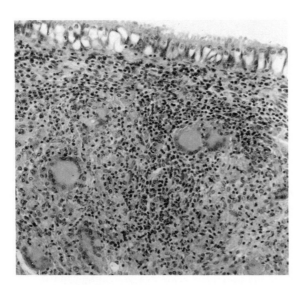

Fig. 4.25 Tuberculous salpingitis. Cow.

ally, the bacteria will be destroyed and the exudate converted to a watery fluid (hydrosalpinx). Frequently accompanying pyosalpinx are the bursal adhesions and local peritonitis described earlier.

Among the organisms which may be found in inflammatory diseases of the uterine tube are streptococci, staphylococci, *Escherichia coli,* and *Actinomyces pyogenes,* and, of these, the latter is the most common and important. *Brucella suis* in swine and *Mycobacterium tuberculosis* in cattle are responsible for specific forms of salpingitis; the lesions are as described for these infections in the uterus (Figs. 4.24, 4.25).

IV. The Uterus

A. Acquired Changes in Position

Of the possible positional changes of the uterus, only three—herniation, prolapse, and torsion—are important. Displacement of the uterus may occur in abdominal and ventral **hernias** in any species, but displacement into the inguinal canal occurs only in the bitch. The sequelae depend on whether or not the herniated uterus becomes incarcerated by a developing pregnancy, pyometra, or traumatic adhesions.

Torsion of the uterus is exceptional in species other than cattle. In almost all cases, such twisted uteri are pregnant, but torsion may occur also with pyometra and hydrometra. The torsion is of the same nature as an intestinal volvulus and occurs about the transverse axis of the organ with the mesovarium as one fixed point. In uniparous species (cow) in which a well-developed intercornual ligament does not permit much independent movement of the horns, the entire organ is involved in the torsion, which is about the mesovarium and vagina or cervix as fixed points. In multiparous species (bitch, cat) with long horns and no intercornual ligament, the torsion will involve part of one horn or the entire horn, the fixed points in the latter instance being the mesovarium and the site of attachment of the horn to the uterine body. There seem to be no rules governing the direction of the twist. Minor degrees of torsion (up to 90 degrees) are fairly common in cows and apparently resolve themselves. The condition becomes of importance only when the torsion is 180 degrees or more and results in dystocia. Any twist in excess of 180 degrees may result also in local circulatory embarrassment. The thinner-walled veins are obstructed before the arteries, and the uterus becomes congested and edematous, with edema of the placenta and death of the fetus. The devitalized uterine wall becomes friable and susceptible to rupture or, if cesarean section is performed, the friability of the wall makes suturing difficult. Death of the fetus may be followed by mummification if the cervix remains closed; if air and infection enter the uterus, the fetus putrefies. In the bitch and cat, transverse rupture of the twisted segment near parturition releases the dead fetuses into the peritoneal cavity (Fig. 4.26). The fetuses which escape into the peritoneal cavity undergo mummification, attach

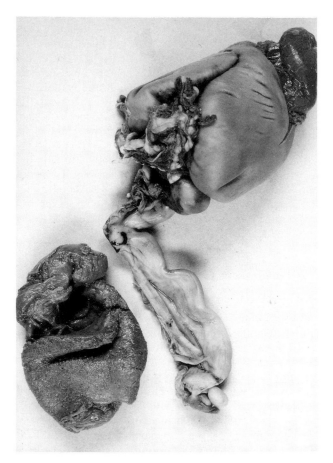

Fig. 4.26 Torsion and rupture of pregnant uterus with release of fetus into abdominal cavity. Bitch.

to the omentum, liver, or intestine, and become covered by a rather thin membrane. Occasionally, a fetus becomes dismembered and fetal bones may be scattered throughout the omentum. In some cases, the mummified fetuses remain in the peritoneal cavity for months or years without causing clinical signs of disease.

Prolapse of the uterus occurs fairly commonly in ruminants, exceptionally in other species. Predisposing causes in the cow are essentially those which cause, or are associated with, uterine hypotony and probably also with dysrhythmia of involutionary contractions. Among the most common associations in the cow are prolonged dystocia relieved by forced traction, retained placenta, and postparturient hypocalcemia. Probably the same sorts of influences operate in ewes and, in addition, uterine prolapse after parturition is a common complication of the hyperestrogenism which results from the ingestion of legumes with a high content of estrogens. Uterine polyps may lead to prolapse of the involved horn in the bitch (Fig. 4.27).

In any species, usually only the previously gravid horn prolapses. In the cow and ewe, the nongravid horn, and sometimes intestine and bladder also, may be present within the everted horn. The pathologic sequelae of pro-

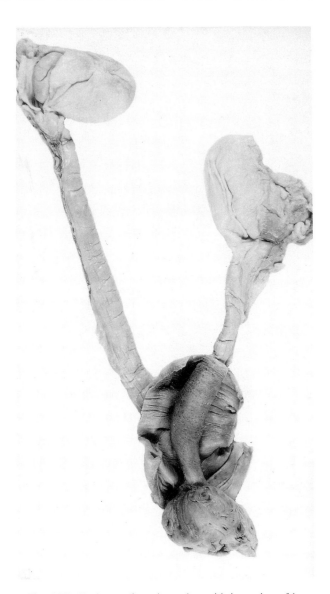

Fig. 4.27 Prolapse of uterine polyp with inversion of horn. Bitch.

lapse are comparable to those of intestinal intussusception with the added factor of trauma. Congestion and edema are followed by hemorrhage, necrosis, and sepsis. Gangrene may supervene.

Rupture of the uterus may occur spontaneously, but is usually a result of obstetrical manipulations. Most ruptures occur in the fundus adjacent to the pelvic brim as irregular tears which may involve the full width of the wall or only the mucosa. Mucosal ruptures are of little consequence. Complete ruptures are often fatal either by virtue of hemorrhage, or spread of uterine inflammation to the peritoneum, or displacement of retained membranes into the abdominal cavity. The majority of ruptures occurs in uteri which are devitalized as a result of torsion or prolonged dystocia.

Rupture may also follow acute distension of the uterus

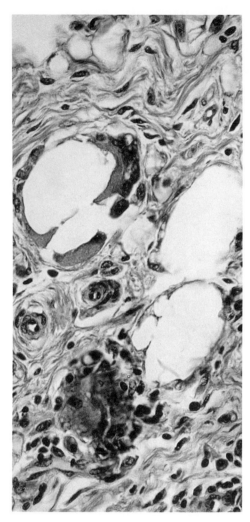

Fig. 4.28 Lipid granuloma in broad ligament following rupture of uterus during irrigation. Cow.

produced by infusion fluids. This is not an uncommon accident. The rupture occurs on the lesser curvature along the line of attachment of the mesometrium, and the irrigating fluids spread into the ligament. Many such fluids have an oily base and a granulomatous inflammatory response to the oil thickens the mesenteries of the uterus, tube, and ovary (Fig. 4.28). Complete sterility is a common sequel to perimetritis.

B. Circulatory Disturbances

Endometrial hyperemia and edema occur normally at estrus and reach the greatest relative development in the bitch in proestrus. The resulting diapedesis and endometrial exfoliation account for the uterine hemorrhage at proestrus in this species. Hemorrhage is common in heifers, less common in older cows, and occurs immediately after estrus. The source of the hemorrhage is the endometrial capillary bed immediately anterior to the cervix. It is

probably an estrogen-withdrawal effect and the nearest thing to menstruation in domestic animals. Punctate hemorrhages occur in the uterine serosa in heifers in estrus. Hemorrhage and hematoma formation in the center and at the periphery of the placental zone are normal in the dog and cat; the breakdown products of the hemorrhage are responsible for the brown and greenish pigmentation in the normal pregnant uterus in these species. A small degree of interplacental hemorrhage in the center of the placentome is normal in the ewe.

Hemorrhage of pathologic import follows torsion and inversion, by obvious mechanisms. Perhaps the most common association of abnormal bleeding is endometrial hyperplasia in the bitch, especially if there is superimposed infection. A less common cause in this species is the uterine fibroleiomyoma (fibroid).

C. Irregularities of Endometrial Growth

1. Atrophy

Atrophy of the endometrium results from loss of trophic ovarian function. Senile atrophy is not important in domestic animals, and they retain their reproductive potential throughout their life span. Atrophy is of course most common after castration, but may reflect hypopituitarism of chronic inanition or wasting disease, or a primary hypophyseal lesion. The endometrium is flat, thin, and grayish in appearance, and in ruminants there are no evident primitive caruncles. The more superficial portions of the endometrium are the more atrophic and, in advanced atrophy, the lining mucosa covers a thin layer of condensed stroma in the depths of which are the inactive glandular remnants, which are sometimes cystic.

The endometrium of those species which have a seasonal period of anestrus undergoes a normal atrophic change. In mares, in which these changes have been studied by endometrial biopsies, the luminal and glandular epithelium may become cuboidal and the glands are straight during the winter anestrous period, but there is considerable individual variation in the degree of atrophy mares develop.

2. Hyperplasia

Endometrial hyperplasia is usually called cystic hyperplasia of the endometrium or cystic hyperplastic endometritis. Both terms are misleading. Cystic hyperplasia, although perhaps the only form of endometrial hyperplasia which is recognizable as indubitably pathologic, is nonetheless the extreme degree of hyperplasia; lesser and earlier noncystic degrees of hyperplasia are also of pathologic importance. Cystic hyperplastic endometritis is really only an inflammatory complication of cystic hyperplasia. In the bitch and cat with endometrial hyperplasia, infectious inflammation is common, but in other species it is less common (Fig. 4.29A,B).

The origin of endometrial hyperplasia can be attributed in some species to excessive and prolonged estrogenic

Fig. 4.29A Cystic endometrial hyperplasia and mucometra associated with multiple corpora lutea. Bitch.

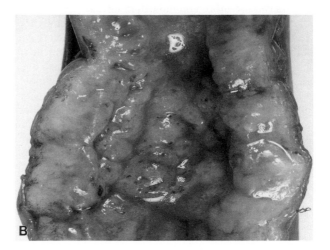

Fig. 4.29B Closeup view of the uterine horn in Fig. 4.29A.

stimulation. Progesterone has been shown to play the major role in the induction of endometrial hyperplasia in the dog and cat, but even here the endometrial response to progesterone depends on estrogen priming. Estrogens act by binding to the estrogen receptors that are present in the endometrial cells and act on these susceptible cells to induce the synthesis of intracellular receptors for progesterone. The progestational effect of conversion of the endometrium to its secretory mode depends on this estrogenic priming, and it is likely that in some cases the disturbances of endometrial growth giving rise to endometrial hyperplasia are to be found in the disturbances of the timing and duration of the priming, but cystic endometrial hyperplasia can be produced in the bitch in the normal

early-diestral uterus by mild trauma (see also Pyometra in the Bitch and Queen, Section IV,F,1 of this chapter). Many cases of cystic hyperplasia of the endometrium have developed following the use of long-acting progestational compounds to delay the onset of estrus in bitches. Cystic endometrial hyperplasia in the cow is associated with ovarian follicular cysts or granulosa cell tumors, both of which can produce prolonged hyperestrogenism. Cystic endometrial hyperplasia is very uncommon in the mare. It has not been associated with granulosa–theca cell tumors in this species.

Ovarian tumors, especially granulosa–theca cell and papillary cystadenocarcinomas, are present in some cases of endometrial hyperplasia of the bitch, but in the majority of clinically recognizable cases, the ovaries contain apparently normal corpora lutea. The hyperplasia develops during the long luteal phase that is normal in the bitch. Often parovarian cysts are present in cases of endometrial hyperplasia.

Noncystic endometrial hyperplasia is not recognizable macroscopically except as an equivocal thickening of the endometrium. The thickening is due to an increase in the size and number of glands which are irregular in their distribution and course, the normal parallel alignment being lost. The stroma is not hyperplastic but is edematous. The glands may show both proliferative and secretory activity. In the bitch, the glands are tortuous and secretory, the mucosal epithelial cells are typically progestational, being hypertrophic with clear cytoplasm, the glands of the basal endometrium are also active, and there is little evidence of the normal partitioning of the endometrium into layers. There is usually some degree of adenomyosis. Cystic hyperplasia, the so-called Swiss-cheese endometrium, is the histologic extreme of the condition and is irreversible. It is probable that endometrial hyperplasia is but an exaggeration of the normal proliferative activity of the endometrium in response to ovarian hormones.

In the bitch and the cat, nondiffuse forms of endometrial

hyperplasia are more common than diffuse, the severely affected areas often being segmental, but structurally they are similar. The cysts are up to 2 mm in diameter, with clear watery contents. In the more severe form in the bitch, there may be localized villous and papillary overgrowths, which histologically are almost solely composed of epithelial tissue with scant connective tissue stroma. Inflammatory lesions are invariably present in advanced cases. In the cow, cystic hyperplasia is typically nonuniform and in long-standing cases is often associated with the presence of excess mucus (mucometra) or fluid (hydrometra) in the uterine lumen.

Endometrial hyperplasia is a significant precancerous lesion in women. In domestic animals, this is not so. Endometrial hyperplasia is very common in the bitch, but uterine cancer is extremely rare.

3. Estrogenic Plants

Pasture legumes as sources of estrogenic activity have claimed attention as a cause of a spectacular syndrome of infertility in sheep, and as a cause of less obvious depression of fertility in **sheep**. Although many estrogenic substances are known to occur in plants, pasture plants which have been found to produce estrogenic effects are mainly some varieties of *Trifolium subterraneum, T. pratense, Medicago sativa,* and *M. truncatula*. The estrogens of the clover species are usually isoflavones; alfalfa and the barrel medic contain coumestans.

Extensive metabolism of phytoestrogens occurs in the rumen. During such metabolism, compounds of very different estrogenic potency from those in the plant can be produced. Among the isoflavones, formononetin, a compound with little estrogenic activity in itself, is the main compound producing histologic effects in sheep after its conversion in the rumen to the potent estrogenic metabolite, equol.

There is variation in estrogenic potency between strains of clover and also with the season or stage of growth, activity being greatest in winter and early spring. Potency is diminished if the clover is allowed to wilt and dry, or if it is made into hay, but the activity is retained if the clover is artificially dried.

Estrogenic pastures grazed during the breeding season may temporarily impair fertility, and ewes so affected may have a very low lambing rate and a high incidence of dystocia and uterine prolapse. Wethers on such pastures develop enlargement of the bulbourethral glands (see The Male Genital System, Chapter 5 of this volume). The fertility of these ewes is returned after removal from the offending pastures. The infertility is in large part attributable to failure of transport of sperm through the cervix due to changes in the cervical mucus. Estral cycles are normal, and normal ovulation apparently occurs. The dystocia that is common is attributed to reduced myometrial tone at parturition; it results in a high rate of maternal and neonatal death. Uterine prolapse occurs in maiden and nonpregnant ewes as well as postpartum ewes; mammary development and lactation occur in nonpregnant ewes.

The principal morphologic changes in the cervices of affected ewes consist in greater glandular development, reduced numbers of goblet cells, and a reduction in the amount of stratified epithelium. These changes are associated with an increased incidence of cervicitis. The endometrium is hyperplastic and macroscopic cysts are often present. Hydrometra and pyometra are occasionally seen. These are the signs and lesions that were described in the original reports of infertility in ewes caused by estrogenic plants, commonly known as clover disease.

The development of varieties of clover low in formononetin has made it possible through changes in management and agronomic measures to control the most severe forms of the disease. But as the original syndrome has been brought under control, a much more insidious form of estrogenic infertility in ewes has emerged as the result of chronic or repeated ingestion of low levels of phytoestrogens. This syndrome causes less severe reproductive losses, but the infertility tends to be permanent. Its anatomic basis consists primarily of subtle cervical changes which also impair sperm transport. The cervical lesion are those of blunting of the cervical folds with a reduction of the number of folds and crypts associated with an increase in the stroma of the lamina propria. There is an increase in coiled tubular glands of the endometrial type in the cervical lamina propria. This change is most severe at the cranial end of the cervix. This glandular induction is the reason the disease has been referred to by the awkward name of estrogen-induced transdifferentiation. The persistent effects of estrogen also bring about modification of sexual behaviour, some ewes developing a more male behavior, clitoromegaly, and fusion of the ventral labia. These changes are mild and the disease can be best identified by careful cervical histologic examination, determining the number of folds and quantifying the area of the lamina propria. Since this is unknown territory for most pathologists, adequate control material is essential.

The effects of estrogenic pastures or of the various isoflavones on **cows** have not been conclusively examined, but there is little doubt from information accrued in Israel and Tasmania that the effects are important, alfalfa (lucerne) and the clovers listed being blamed. The principal features of the syndromes reported are similar to those produced by cystic follicles. The infertility rate is high and is associated with persistent cystic ovaries, aberrations of the estrous cycle with estrus abnormally long, swelling of the vulva independent of estrus and sometimes enlargement of the clitoris, increase in size of the uterus and cervix with cystic endometrial hyperplasia, and enlargement and function of the virgin mammary gland.

Bibliography

Adams, N. R. Pathological changes in the tissues of infertile ewes with clover disease. *J Comp Pathol* **86:** 29–35, 1976.
Adams, N. R. Measurement of histologic changes in the cervix of ewes after prolonged exposure to oestrogenic clover or oestradiol-17β. *Aust Vet J* **63:** 279–282, 1986.
Bellenger, C. R., and Chen, J. C. Effect of megestrol acetate on

the endometrium of the prepubertally ovariectomised kitten. *Res Vet Sci* **48:** 112–118, 1990.

Bennett, D., Morley, F. H. W., and Axelsen, A. Bioassay responses of ewes to legume swards. *Aust J Agric Res* **18:** 495–504, 1967.

Brodey, R. S., and Fidler, I. J. Clinical and pathological findings in bitches treated with progestational compounds. *J Am Vet Med Assoc* **149:** 1406–1415, 1966.

Cox, R. I. Plant estrogens affecting livestock in Australia. *In* "Effects of Poisonous Plants on Livestock," R. F. Keeler, K. R. VanKampen, and L. F. James (eds.), pp. 451–464. New York, Academic Press, 1978.

Francis, C. M., Millington, A. J., and Bailey, E. T. The distribution of oestrogenic isoflavones in the genus *Trifolium. Aust J Agric Res* **18:** 47–54, 1966.

Lightfoot, R. J., and Adams, N. R. Changes in cervical histology in ewes following prolonged grazing on oestrogenic subterranean clover. *J Comp Pathol* **89:** 367–373, 1979.

4. Adenomyosis

This term applies to the presence of endometrial glands and stroma between the muscle bundles of myometrium (Fig. 4.30). In some cases, it is a malformation, and in others it arises by hyperplastic overgrowth of the endometrium. It is not a common lesion in any domestic species but is seen in the bitch with cystic endometrial hyperpla-

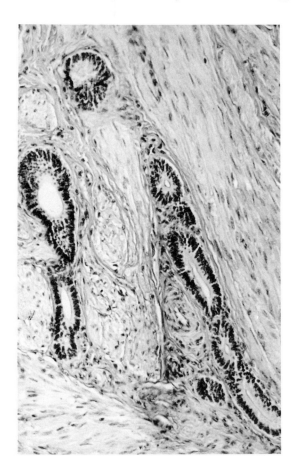

Fig. 4.30 Adenomyosis. Glands of normal configuration in myometrium.

sia. Adenomyosis is observed occasionally in cows as part of the local disarray of segmental aplasia. It may also be present as a malformation of the tips of the uterine horns in cows.

Adenomyosis as seen in domestic animals shares features with endometriosis of menstruating primates. However, there are important distinctions. Endometriosis is a condition in which actively growing endometrial tissues are explanted to aberrant sites within and outside the uterus. One of the aberrant sites can be the myometrium; in this site the two conditions are histologically similar.

5. Endometrial Polyp

This is a rare lesion seen only in the bitch. In contrast to the polypoid form of endometrial hyperplasia, the true polyp contains a substantial connective tissue stroma in addition to glands, and is pedunculated. Polyps may be multiple or isolated, and their shape is molded to the uterine lumen (Fig. 4.27). They tend to cause prolapse of the affected horn, the polyp then observed protruding from the vagina.

D. Hydrometra and Mucometra

The two conditions are considered together as the difference is probably only in physical properties and depends on the degree of hydration of the mucin, which in turn may be related to the relative activity of estrogenic hormone. The accumulation of thin or viscid fluid in the uterus is concurrent with the development of endometrial hyperplasia or is proximal to an obstruction of the lumen of the uterus, cervix, or vagina. In the first instance, the amount of fluid may be several liters, and the greater the volume of fluid, the less viscous it is. Small amounts of mucin give the mucosal surface a gummy stickiness. In cows with cystic ovaries and endometrial hyperplasia, the large volumes of fluid are usually associated with functional cysts of the follicles. When terminal anestrus occurs, apparently much of the fluid is resorbed, leaving a small quantity of tenacious mucus (Fig. 4.31). As de-

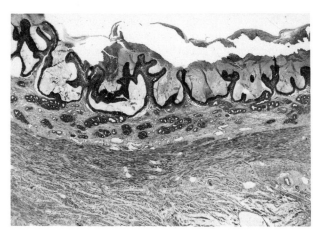

Fig. 4.31 Mucometra. Cat.

scribed earlier, such uteri usually have thin atrophic walls and the ovarian cysts are multiple, small, and thick walled. In the second instance, that of obstruction to the lumen, the volume of fluid depends on the site of the obstruction; in uterus unicornis there may be 500 ml; in imperforate hymen there may be 10 liters or more. The fluid is slightly cloudy and watery but, in some cases of segmental aplasia where the volume of retained secretion is not great, it may be very viscid, ocher-colored, and sometimes inspissated to rubbery masses of mucin and cellular detritus. In these cows, the ovaries are normal. Animals with mucometra are sterile. If affected uteri become infected, an intractable pyometra results. An abnormally long and tortuous cervix may result in a form of mucometra caused by the retention of uterine secretions.

E. Inflammatory Diseases of the Uterus

1. General Considerations

Inflammation limited in extent to the endometrium is termed **endometritis;** involvement of the entire thickness of the wall is **metritis;** of the serosa, **perimetritis;** and of the suspensory ligaments, **parametritis.** The classification is to some extent a useful index of the severity of reaction and of the pathogenesis. The great majority of inflammatory conditions of the uterus do begin in the endometrium and are in some manner associated with the reproductive process. The predisposing factors are to be sought then at either end of the gestation period.

The normal nonpregnant uterus is endowed with a high degree of resistance to infection, and even in the case of the specific genital diseases, brucellosis, trichomoniasis, and *Campylobacter* infection, is incapable of supporting bacterial growth or even the persistence of bacteria for any extended period. Something is known of factors which render the uterus susceptible, at least temporarily, to infection, but little is known of the mechanisms of its resistance. The self-limiting nature of most infections, other than those associated with pyometra, has been long recognized and formed the basis for the recommendation of a period of sexual rest for animals with uterine infections. Even the specific genital infections are self-limited in duration; active infection of a uterus with *Campylobacter fetus* or *Tritrichomonas foetus,* in the absence of pregnancy, survives only for two to three estrual cycles. *Brucella abortus* is an infection of the pregnant uterus and does not persist well in the nonpregnant uterus, although it is apparently capable of persisting in the mammary gland and lymph nodes of infected cows. Uterine resistance varies during the estrous cycle, with susceptibility being the greatest during the luteal phase of the cycle. Several factors may be involved. Uterine motility is increased during estrus and this aids in the physical clearance of microorganisms. Uterine synthesis and secretion of immunoglobulin G (IgG) and IgA vary somewhat during the cycle, but the variations are not great and the timing varies between species. The functional activity of the polymorphonuclear leukocytes migrating into the uterus may also be affected by the stage of the cycle, being more active during estrus than diestrus, but much of the reduced uterine resistance during diestrus and pregnancy is related to the progesterone-induced secretion into the uterine lumen of immunosuppressants which are capable of inhibiting lymphocyte proliferation. It has been well demonstrated that the uterus which is under the influence of progesterone (which includes the pregnant uterus) is very susceptible to many nonspecific bacteria and that a uterus not under the influence of progesterone is remarkably resistant to the same organism even when its expulsion is prevented by ligation of the cervix. This comparison can be carried a step further to show that even greater susceptibility to infection is present at the implantation sites in the pregnant uterus. The uterus of the castrate is resistant to infection but does not clear infections as promptly as does the uterus under the influence of estrogens. These factors of susceptibility and resistance provide some insight into the pathogenesis of postcoital uterine infections and pyometra in the dog and cat.

A different set of influences operate on the puerperium. It is well recognized that uterine infections are likely to follow any abnormal parturition such as an abortion, retained placenta, twin births, dystocia, and traumatic lacerations of the genital canal. A significant proportion of abortions indeed result from specific or nonspecific endometritides, but these are often complicated by the general factors that pertain to all postpartum infections of the uterus. Of the specific abortive agents in the cow, *Campylobacter fetus* and *Tritrichomonas foetus* typically cause early death of the conceptus, which is not complicated by placental retention or metritis. But with other causes, *Brucella abortus* especially, abortion may occur later in pregnancy, at which time the well-developed and manifold interdigitations of the cotyledons favor retention of the placenta and the development of acute metritis. But in these instances much of the uterine disease is the result of secondary invasion by bacteria that gain access to the uterus as the result of an abnormal delivery of a dead or diseased fetus.

Some cases of postpartum metritis are a continuation and exaggeration of a gestational uterine infection, but most puerperal infections of the uterus can be viewed as analogous to wound infections, with the organisms entering via the cervix. Probably all mares have uterine infections by streptococci within 1–3 days after parturition, but these do not persist for more than 2–3 days. In the cow, too, it has been observed that 25–30% are infected in the normal puerperium, but that most recover spontaneously from these infections with *Actinomyces pyogenes, Escherichia coli,* and other bacterial mixtures. Clearly, the outcome will be determined as much by the number and virulence of invading organisms as by the environment within the uterus. The recognized predisposing causes outlined will all be associated with retarded uterine involution, as will most cases of metritis. With the more virulent anaerobic bacteria, such predisposition is not essential.

The period necessary for normal involution varies with the species and is probably determined by the nature of placentation and the degree of epithelial reconstitution necessary. It is well advanced in the mare within 9 days, as judged by the capacity for fertile mating at that time. The conception rate is, however, not high in mares bred at this first postpartum estrus, so that involution is probably not complete. In the cow, on the other hand, involution cannot be considered complete until about 50 days postpartum. In the bitch, involution is occasionally much prolonged, with persistence of decidual remnants deep in the endometrium. The two main ingredients of involution are contraction to obliteration of the uterine cavity and reconstitution of the epithelium; they proceed in parallel. Retarded postpartum contraction is an index of diminished myometrial tonus. This diminished tone may be part of general debility, it may be an exhaustion phenomenon consequent on prolonged dystocia, on excess stretching as with twin births or hydrallantois, or secondary in some obscure manner to already existing inflammation. Whatever the preexisting cause, it is exacerbated by the establishment of postpartum infection. With significantly diminished tone and contractions, there is retention, beyond the normal period, of a patent lumen and retention within the lumen of autolyzing lochia—a mixture of blood, fetal fluid, mucus, and epithelial detritus, often including placental remnants.

Involution of the bovine maternal placenta begins by vasoconstriction in the caruncular stalk with degeneration and dissolution and desquamation of the entire superficial layer of the caruncle, complete in about 10 days. By 40–50 days, epithelial repair is complete, taking its origin from surviving epithelium and the glands. Functional closure of the cervix is present at 24–36 hr after parturition, although it is still readily dilatable. The greatest volume of uterine lochia is present in the first 48 hr after parturition and is rapidly removed, partly by discharge and partially by resorption. Probably all factors predisposing to infection favor the accumulation of an excessive volume of lochia and its retention beyond the normal period and, by so doing, provide an excellent environment for all types of bacterial growth, of which the most devastating are the anaerobes. Their nutrient supply is increased and improved by the presence of decomposing placental remnants and, in some cases, macerated fetuses.

2. Endometritis

In endometritis, it is the endometrium or uterine mucosa which is mainly involved. The term should not be accepted too literally for, except with the very mild forms of infection, there is necessarily some effect of the inflammation on the remainder of the organ. Further, almost all uterine infections begin as an endometritis but may progress so rapidly to any of numerous variations and manifestations that definitions have little worth. The mildest forms are seen postcoitus; they are caused by *Tritrichomonas foetus* and *Campylobacter fetus* or pyogenic cocci and coliforms of low pathogenicity. It is doubtful if these last-named nonspecific infections will produce postcoital endometritis in adult cows as readily as they do in heifers; rather, it seems as if the virgin uterus requires exposure before acquiring resistance to the organisms ordinarily contained in semen.

Uterine infection is common in mares both following foaling and after coitus. The common infecting organisms are β-hemolytic streptococci, *Klebsiella pneumoniae*, *Escherichia coli*, and *Taylorella equigenitalis*, the contagious equine metritis organism. The endometritis is usually mild, but the impact on fertility can be substantial.

There are no gross lesions in this simple form of endometritis. A slight opacity of the normally crystal-clear estrual mucus may be all that is seen. Histologically, the changes are not striking and consist for the most part of a diffuse but light infiltration of inflammatory cells with slight desquamation of the superficial epithelium and no significant vascular changes. Involvement of the glands is minimal. The significance of a few leukocytes in the stroma is always equivocal in cattle; they follow within 2–3 days after parturition and are present during estrus. However, the presence of neutrophils in the stroma of the endometrium is evidence of inflammation in the mare. The best indication of mild endometritis in all species consists of accumulations of plasma cells and foci of lymphocytes in the stroma. Resolution of this type of endometritis may occur with no more residue than a few cystic glands with periglandular fibrosis (Fig. 4.32A,B), although during its course it may be responsible for a series of aborted embryos. If the endometritis is more severe or persists for a longer time, the cumulative damage to the endometrium may render the mare sterile.

Contagious equine metritis is a venereal disease of mares caused by *Taylorella equigenitalis*, a gram-negative, microaerophilic coccobacillus. The disease produced by this organism does not appear to be significantly different from other common infecting organisms of the mare's genitalia. The interest in this disease stems from its apparently abrupt appearance in the horse-breeding establishments in England in 1977 and its rapid spread to the equine studs around the world. Strict control measures seem to have limited its spread and the clinical disease is now rare.

The disease causes temporary infertility in mares and a mucopurulent discharge that lasts 2–3 weeks. The organism can persist in infected mares for several months and recovered mares represent an important reservoir of infection. Stallions transmit the organism by genital contact but do not develop clinical disease.

The temporary infertility which is the clinical hallmark of the disease is the result of mild to moderate inflammation of the endometrium and adjacent structures. After experimental introduction of the organism into the uterus, and presumably after natural infection, the bacteria can be regularly demonstrated in the uterine tubes, endometrium, cervix, and vagina for a 2 to 3-week period, during which time the organism elicits a mild to moderate inflammatory response. The endometrial folds are turgid and swollen and covered by a small amount of a cloudy viscid exudate.

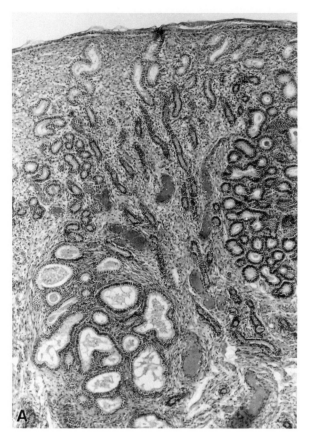

Fig. 4.32A Chronic endometritis with periglandular fibrosis. Mare.

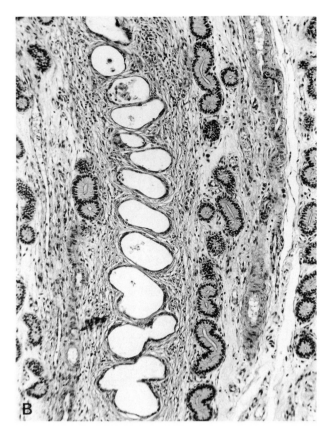

Fig. 4.32B Chronic endometritis. Cow. Dilated, atrophied glands with periglandular fibrosis.

At this stage the endometrium is edematous and the inflammatory infiltrate consists mainly of neutrophils. As the reaction subsides, the edema disappears and plasma cells predominate. Changes in the cervix parallel those of the endometrium. Salpingitis develops in some cases but is a less regular feature of the infection. Infection produces no gross changes in the vagina or vaginal vestibule and histologic changes are mild. No lesions develop in the clitoral fossa or sinus, but the organism can be recovered from these sites after it has disappeared elsewhere.

More severe grades of endometritis are common to the puerperium in cattle. Nothing of significance may be visible on the serosal surface, but the organ is enlarged and flabby, and collapsed rather than firm and contracted. The lumen contains chocolate-colored lochia which is slightly tenacious and often without foul odor. With the admixture of inflammatory exudate and placental detritus, the uterine content becomes progressively dirty grayish yellow. The endometrium is congested and swollen, and the intercotyledonary areas are ragged and tattered with shreds of mucosa free in the lumen. Small hemorrhages are common in the congested mucosa, and there is a prominent leukocytic infiltration massing at the surface involving all mucosal elements, including the glands (Fig. 4.33). Where suppuration and superficial necrosis produce the tattered

mucosa, the surface is comparable to a pyogenic membrane. The remainder of the genital canal may show nothing more than the traumatic lesions incident to parturition. If the uterus is paretic, there may be no discharge in the vagina.

3. Metritis

The distinction drawn here between endometritis and metritis for purposes of description is that in the latter all layers of the uterine wall show evidence of acute inflammation. The uterus is paretic and there may be little or no vaginal discharge. The wall of the uterus is thickened with suffused blood and edema fluid and is very friable. The serosa is dull and finely granular with paint-brush hemorrhages and a thin deposition of fibrin, or the subserosal vessels may be darkly congested. Other than traumatic rupture, perforation with secondary peritonitis is not common except in anaerobic infections; death in untreated cases usually occurs first from toxemia or septicemia.

The secretion may be scant or abundant, is fetid, and is dirty yellow to red-black. The microscopic picture is that of a purulent inflammation. The subserosal connective tissues are edematous and infiltrated with leukocytes, and the same process surrounds the blood vessels of the myometrium and permeates bundles of, and individual, muscle fibers, which themselves undergo granular degeneration.

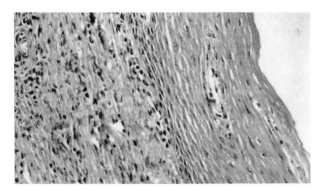

Fig. 4.34 Squamous endometrial metaplasia in chronic endometritis. Cow.

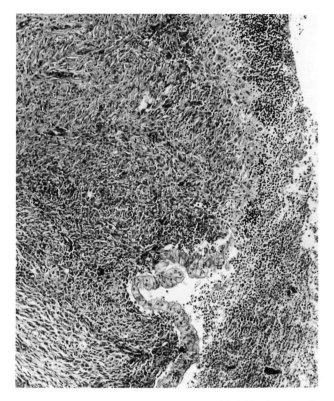

Fig. 4.33 Acute suppurative endometritis following abortion caused by *Listeria monocytogenes*. Cow.

In metritis, as in acute endometritis, the leukocytes mass on the mucosal surface and are associated with extensive hemorrhage, necrosis, and sloughing. Invasion of blood vessels, both arteriolar and venous, aggravates the lesion. Thrombosis may extend to the vessels of mesometrium with the usual sequelae of hemorrhage and infarction.

4. The Sequelae of Metritis and Endometritis

Many of the milder cases of endometritis recover health and fertility spontaneously. Many of the acute cases of metritis are fatal, despite therapy. Among the residual and complicating conditions are chronic endometritis, uterine abscess, parametritis, salpingitis, pyemia, and pyometra. Pyelonephritis (see The Urinary System, Volume 2, Chapter 5) is an occasional complication.

a. CHRONIC ENDOMETRITIS   Recovery from the acute phase of the infection often results in chronic endometrial involvement. With greater or less degrees of endometrial destruction and replacement by granulation scar tissue, the uterus takes on the nature of a fistulous tract. The changes depend on the duration and severity of the inflammation but consist essentially of productive fibrosis and leukocytosis in which lymphocytes and plasma cells predominate. Thickening of the endometrium is by inflammatory tissue; the glands are depleted; those which survive are atrophic, flattened, and attenuated, or cystic due to the periglandular fibrosis. The lining mucosa may

be intact, denuded in places, or show foci of polypoid hyperplasia or squamous metaplasia (Fig. 4.34), as do any chronically irritated mucous membranes. The exudate in the lumen is not copious and may be serous, catarrhal, or frankly purulent. Much of the labile endometrial stroma, especially that of caruncles, may be replaced by useless scar tissue, and dystrophic calcification of necrotic portions of the endometrium may sometimes be extensive enough that the lining of the uterus feels gravelly.

b. UTERINE ABSCESSES   These are not common observations. The localization of an infection to one part of the uterine wall is thought to follow severe metritis or localized traumatic injury to the infected endometrium. Such an abscess may reach 15 cm in diameter and is usually well encapsulated, although there may be some perimetrial adhesion and, in few instances, rupture into the peritoneal cavity or an adjacent hollow viscus.

Uterine abscesses are observed more frequently in cattle than in other species. There appears to be a relationship between the frequency of abscesses and uterine manipulations involving the use of instruments. In cattle, most large abscesses are located in the dorsal wall of the uterine body. This is the area most subject to trauma during the passage of insemination pipettes and uterine catheters. Abscesses which develop following severe metritis are usually small (1–3 cm) and do not have preferential sites.

c. PARAMETRITIS AND PERIMETRITIS   Chronic adhesive peritonitis involving the genital tract does not usually result from septic metritis because the uterine serosa offers an efficient barrier to the spread of infection, and spontaneous rupture of an infected uterus is not common. Few virulent infections spread to the supporting ligaments. Excluding an origin from an extragenital focus, perimetritis and parametritis in cattle usually follow manual manipulation of the ovary, pyosalpinx, obstetrical operation, removal of retained placenta, and uterine irrigations. In each of the latter three circumstances, there may be accidental perforation or rupture of the uterus. The extent of the adhesion may vary from a few fibrous bands to dense

Fig. 4.35 Pyometra with perimetrial and bursal adhesions. Cow.

connective tissue which obscures the contour of the organs and fixes them to adjacent viscera (Fig. 4.35). Abscesses may form in the adhesions of the ovarian bursa and the rectovaginal pouch.

d. SALPINGITIS   See under The Uterine (Fallopian) Tubes (Section III of this chapter).

e. PYEMIA   Acute deaths from metritis usually result from toxemia or septicemia. In cases of longer-standing and pyogenic infection, intermittent bacteremia and septic emboli may result in metastatic infection of other organs and tissues, *Actinomyces pyogenes* being the organism most often incriminated. The tissues most often involved in the metastatic infection are the valvular endocardium, joints, myocardium, and lungs.

F. Pyometra

Pyometra is an acute or chronic suppurative infection of the uterus with accumulation of pus in the uterine lumen. The escape of the pus is usually prevented by a functionally closed cervix, but the discharge may be prevented by an acquired or congenital cervical stenosis, and in mares the gravitational pull of the flaccid, distended uterus over the brim of the pelvis may limit the discharge of pus. Pyometra may occur as a sequel to uterine infections of the types described in the previous sections, but as it is a pathologic entity with a number of factors entering into its pathogenesis, it is considered separately here. Pyometra is an uncommon condition in the ewe and the sow. It is relatively common in the bitch, queen, cow,

and mare, but the circumstances under which the disease develops in these species vary.

1. Pyometra in the Bitch and Queen

Pyometra in the **bitch** is a disease that characteristically affects older animals, especially those that are not bred. The condition most often develops a few weeks after estrus. Affected animals are depressed and anorectic, frequently vomit, and have polyuria and polydipsia, usually accompanied by a vaginal discharge. The pathologic findings vary with the stage of the disease. In less advanced cases, the uterus may be only slightly enlarged with mild endometrial hyperplasia and inflammation. In the more advanced stages, there is a remarkable distension of the uterine horns, which may come to occupy most of the peritoneal cavity (Fig. 4.36). Distension of the cornua may be symmetrical or asymmetrical, uniform or as ampullalike dilatations as in the uterus of midpregnancy. The cervix is completely or almost completely closed as a functional response to the luteal hormones. The serosal surface of the uterus is dark and the vessels are congested and prominent. The wall is friable, and either rupture or perforation with secondary peritonitis is not uncommon. There may be obvious inflammation of the peritoneal serosa and suspensory ligaments, but this is unusual. The nature of the uterine content is variable. In the more severe cases, usually those infected with *Escherichia coli* and *Proteus* spp., the exudate is thick, viscid, tenacious, opaque red-brown, and with a characteristic fetid odor. In other cases, usually those infected with streptococci and staphylococci, the exudate is more typically purulent. The mucosa is irregular in thickness, necrotic, and ulcerated in portions with irregular superficial hemorrhages, and in other portions obviously hyperplastic, dull-white and dry in appearance with small cysts visible in these hyperplastic areas (Fig. 4.37A). Microscopically, the most significant feature is the remarkable endometrial hyperplasia and progestational proliferation in almost all cases (Fig. 4.37B). The cells of such progestational epithelium are enlarged, columnar, vacuolated, and have small pyknotic nuclei. In

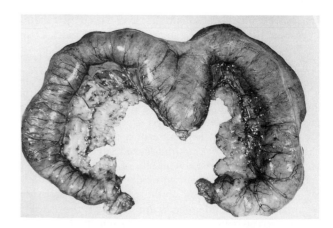

Fig. 4.36 Pyometra. Bitch.

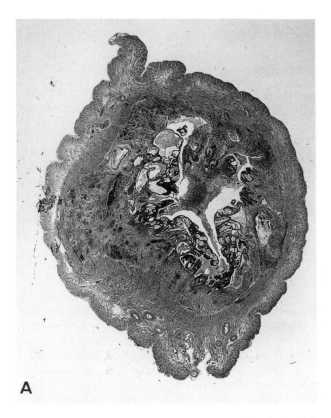

Fig. 4.37A   Cross section of uterus. Bitch. Cystic glandular hyperplasia and pyometra.

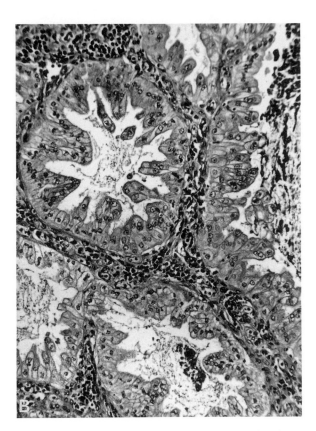

Fig. 4.37B   Glandular hyperplasia with progestational change in the epithelium. Pyometra. Bitch.

some cases, the normal single layer of cells piles up to produce pseudostratification or localized papillary proliferations. Whatever remains of the endometrial lining may show this development, or it may be patchy and alternating with normal epithelium.

These changes are produced by the luteal hormones which inevitably pave the way for the development of this disease. The histologic changes due to infection vary with the bacterial cause and time. Masses of neutrophils collect in the uterine lumen and in the glands, although there is relative sparing of the latter unless they are cystic. The migrating leukocytes collect near the surface and then penetrate the epithelium. In milder cases, there may be few neutrophils in the endometrial stroma, but they are not many when compared with the numerous plasma cells and lymphocytes. There may not be much vascular reaction over and above that of hormonal origin, although perivascular reaction and leukocytosis of lymphatic vessels is almost constant in the myometrium. Sometimes the reaction is much more severe than that just described, and in the endometrial stroma, there are all the exudative phenomena of acute inflammation accompanied by the early reparative response of granulation tissue. The blood vessels are very congested and some show thrombosis, and about others there is diapedesis or larger hemorrhage. The stroma is edematous, and bullae often lift off the overlying epithelium. Numerous neutrophils infiltrate the

stroma and collect in the lumen. There is little formation of fibrin, and rarely are there microabscesses in the mucosa.

It is clear that most cases of pyometra in the bitch are infective inflammations associated with endometrial hyperplasia. It is usually a disease of the diestral (metestrus) period, a time at which corpora lutea are present and progesterone levels are high and, since the condition has been produced both experimentally and accidentally by the administration of progestins, it is clear that progesterone is an essential feature in the pathogenesis.

Since cystic endometrial hyperplasia had been produced in dogs by administration of high levels of progesterone for extended periods of time, and since cystic endometrial hyperplasia was a hallmark of canine pyometra, it was presumed that cystic endometrial hyperplasia induced by abnormal progesterone secretion, or response, preceded and paved the way for the bacterial component. Some doubts concerning the validity of this view were generated when hormonal abnormalities could not be demonstrated in cases of pyometra. Now it has been shown that cystic endometrial hyperplasia can be produced in the normal, early diestral uterus by mild trauma to the endometrium. The investigators scratched the endometrium with a thin wire, but endometrial biopsy has caused the same lesion. More directly to the point, inoculation of appropriate strains of *Escherichia coli* into the uterine lumen, if given 1–2 weeks after estrus, causes a pyometra that has all the features of the natural disease, including cystic endometrial hyperplasia.

It now appears that an appropriate stimulus applied to the progesterone-primed endometrium produces cystic endometrial hyperplasia, and that one such stimulus can be a bacterial infection. Accordingly, rather than pyometra in the bitch occurring as the result of bacterial infection being superimposed on an abnormal endometrium, it appears that the bacterial infection of the progesterone-primed endometrium produces both the pyometra and the cystic endometrial hyperplasia.

The bacterium most commonly present in pyometra in the bitch is *Escherichia coli*. Most of the strains involved are urinary tract pathogens and possess adhesins that facilitate adherence to the urinary tract epithelium. These adhesins appear also to facilitate the colonization of the progesterone-primed endometrium. Urinary tract infections are common in bitches with pyometra, and the strains of bacteria infecting both are often identical. It seems probable that the urinary infection predisposes to pyometra, and that the uterus is infected at the stage that suitable receptors are developed in the endometrium in response to hormone stimulation. This explanation of the bacterial infection can be only a partial one, as not all bitches with pyometra have urinary tract infection, and not all infecting organisms are urinary pathogens. In some cases of pyometra, bacteria cannot be recovered.

The clinical signs and extragenital lesions associated with pyometra in the bitch are due to severe intoxication and probably also intermittent bacteremia. Renal signs and lesions are common. Profound hypotension may reduce renal perfusion and lead to prerenal uremia. This is worsened by a membranoproliferative glomerulonephritis that often develops as a result of immune-complex deposition. The polyuria that is such a common feature of the disease is due to impaired tubular ability to concentrate urine. The mechanisms involved in this are not completely understood, but they are thought to be brought about by *E. coli* antigen and have an immunologic basis. Most animals have a profound leukocytosis, reflected in the bone marrow as an increased myeloid : erythroid ratio. There is myeloid hyperplasia in bone marrow and extramedullary myelopoiesis in liver, spleen, lymph nodes, and adrenals.

Pyometra in the **cat** is roughly comparable to the disease in the dog, but there are differences, some of which are related to the differences in the estrous cycle of the two species. Cats are induced ovulators, and usually if not bred have repeated estrous periods. However, pyometra develops most often 2–5 weeks after estrus and usually after spontaneous ovulation and presumably corpora lutea formation. Unlike the bitch, many queens with pyometra do not have corpora lutea present in their ovaries at death or surgery, but approximately half do, and there is an obvious correlation between the presence of corpora lutea and the development of pyometra in queens. The correlation with endometrial hyperplasia is less clear. Most queens with pyometra do have some degree of cystic endometrial hyperplasia, but this change is very common in the age group affected.

2. Pyometra in the Cow

Pyometra in the cow is also associated with corpus luteum activity in the ovary. However, in contrast to the situation in the bitch in which the high progesterone levels of early diestrum predispose to uterine infection, in the cow uterine disease causes the corpus luteum to persist and maintain a high progesterone level. The retention of the corpus luteum appears to be due to a reduction in or inhibition of the synthesis and release of the luteolytic factor, prostaglandin $F_{2\alpha}$, by the diseased endometrium.

Broadly, there are two periods in which a uterine infection can lead to retention of a corpus luteum with the accompanying hormonal effects which convert an endometritis to a pyometra. They are during the early postpartum period, following dystocia, retained placenta, and metritis, and at varying times after breeding, as a result of venereal infections with early embryonic death. Insemination during the luteal phase of the cycle, or similar operations which can introduce contamination into the uterus during the luteal phase, can mimic venereal infection.

The role of retained luteal tissue in the pathogenesis of pyometra appears clear. The secreted progesterone endows the uterus with a high degree of susceptibility to infection, maintains functional closure of the cervix, and inhibits myometrial contractility.

The amount of pus retained in the uterus of a cow with pyometra varies from a few milliliters to more than several liters and is thick, rather mucinous, and cream or grayish green in color. The cervix has no seal of mucus so, although contracted, there is usually escape of a small amount of pus into the anterior vagina. The wall of the uterus is thick, doughy, and paretic but, in long-standing cases, especially as complications of mucometra, the walls are thin or fibrosed. Pyosalpinx and perimetritis may coexist (Fig. 4.35). There are no significant extragenital lesions. The histologic changes do not differ significantly from those of endometritis of comparable duration and severity. The organisms involved are hemolytic streptococci, staphylococci, coliforms, *Actinomyces pyogenes,* and *Pseudomonas.* The venereally transmitted protozoan, *Tritrichomonas foetus,* can be the cause of pyometra after breeding.

Rarely, the anomalous developments of the uterus that were described earlier can lead to pyometra. In the segmental aplasias and with imperforate hymen, the lumen proximal to the site of obstruction becomes distended with mucus and cellular detritus. These closed cavities provide a satisfactory environment for bacterial growth, but this is not a frequent complication. When it does occur, the source of infecting organisms is probably hematogenous. Pyometras seldom spontaneously resolve and, although the condition is not usually life threatening as it is in bitches, unless cyclic activity can be reinstituted, the condition will persist.

3. Pyometra in the Mare

Pyometra in the mare differs from the disease in the bitch and the cow in several particulars. Whereas some cases develop following difficult parturitions with infections, as they do in cattle, many do not. Remarkably, most mares continue to cycle during the disease, and the hormonal influences which are so marked in the bitch, queen, and cow are much less important in the mare.

In some mares, cervical adhesions and closure may lead to pyometra, but in most instances the purulent material collects without demonstrable cervical closure, and in some cases the cervix is fully dilated. Copious amounts of pus can be discharged under such circumstances, particularly during estrus. Pyometra in the mare seldom leads to evidence of systemic disease, although some mares acquire a mild anemia.

The length of the estrous cycle appears to be related to the severity of the endometrial damage. In rare cases where the endometrial damage is severe, the cycles are prolonged with long luteal phases due to delayed or inadequate prostaglandin $F_{2\alpha}$ release. Mares with less severe endometritis have normal or shortened cycles.

A variety of organisms may be present in pyometra in the mare; *Streptococcus zooepidemicus* is the most common, but *Escherichia coli*, *Actinomyces* spp., *Pasteurella* spp., and *Pseudomonas* are often present.

G. Specific Causative Types of Uterine Infection

To be considered under this heading are three specific infections of the uterus, each of which is responsible for a characteristic anatomical form of metritis. Those infectious diseases in which abortion is a prominent feature are discussed under Diseases of the Pregnant Uterus (Section V of this Chapter).

1. Necrobacillosis

Infection of the vagina and uterus with *Fusobacterium necrophorum* is seen in postparturient sheep and cows as an infection of contamination superimposed on traumatic or inflammatory genital disease. Necrobacillary metritis is usually fatal. The lesions are typical of those produced by this organism in any location and are characterized by dry coagulative necrosis separated by a narrow red zone of intense hyperemia from surrounding viable tissue. An affected uterus is enlarged, and the wall is thick and rigid. There is scant inflammatory exudate in the lumen, although it may contain necrotic placental and caruncular remnants. The mucosa is thickened and folded, and in large patches is necrotic, fragile, and ulcerated with a dark ragged surface. On section, the mucosal thickening is of a yellowish necrotic layer of tissue, and this is separated by a zone of hyperemia from an outer layer of firm gray granulation tissue which replaces the myometrium. Similar lesions are seen in the cervix and vagina. Typically, the masses of bacteria are found facing a wall of leukocytes at the advancing margin of the necrotic zone, which is

itself structureless. Extensive vasculitis with thrombosis is also characteristic of this infection. Thrombi can usually be found in the uterine veins, and they may extend to the vena cava.

2. Tuberculosis

It is estimated that some 20% of cows with generalized tuberculosis and 4% of all tuberculous cows have involvement of the endometrium. There are three routes of infection, namely, hematogenous, via the uterine (fallopian) tubes from the peritoneum, and coital; of these, the last is exceptional. As in tuberculous lesions generally in cattle, there are two anatomic forms of the lesion, miliary tuberculosis and diffuse caseating tuberculosis, although transitional forms do exist. It is generally accepted that the disseminated miliary lesion is of hematogenous origin during the phase of early dissemination.

In miliary tuberculosis, the uterus may appear normal externally. In the early stages, there may be no exudate in the lumen, but later, as the granulomas enlarge and ulcerate, the uterus will contain a yellow purulent exudate. The granulomas are visible as few or many nodules in the mucosa, usually the more superficial portions, and microscopically are of typical tuberculoid structure. A common site is near the bifurcation of the uterus or in the caruncles of the pregnant uterus.

Caseous tuberculosis causes thickening and ridigity of the horns with serofibrinous or purulent fluid in the lumen. The endometrium is thickened, dry, and extensively caseous (Fig. 4.38). There may be a marked exudative reaction with intense leukocytic infiltration. The caseated area is usually demarcated by a zone of epithelioid cells from a margin of connective tissue.

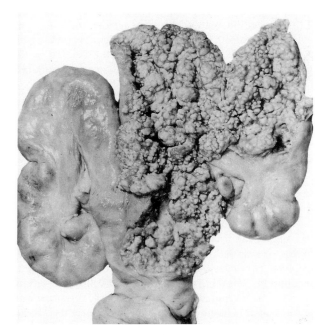

Fig. 4.38 Caseous uterine tuberculosis. Cow.

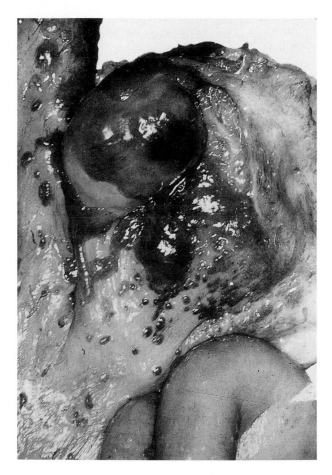

Fig. 4.39 Tuberculous granulation tissue on the ovary and adjacent structures. Cow.

In association with the uterine lesion, there is often involvement of other portions of the genital tract. There may be multiple red granulations on the surface of the ovary and its ligaments (Fig. 4.39) or there may be enlargement of the ovaries with parenchymal tubercles. Tuberculous salpingitis or pyosalpinx is probably the rule when the uterine tubes are the portal of uterine infection. Similar lesions may be found infrequently in the lower genital tract. Infection of cows with the avian strain of the organism also produces a tuberculous metritis and placentitis with multiple small endometrial granulomas.

3. Staphylococcal Granuloma

Staphylococcal infections of the sow's uterus appear to be common as a cause of postpartum metritis and also appear to respond readily to treatment. Granulomatous lesions occur occasionally in sows and cows but are quite uncommon. They have been reported following service by a boar with seminal vesiculitis caused by this organism. The lesion is a large node of fibrous connective tissue enclosing small abscesses in which colonies of the organism are demonstrable.

Bibliography

Acland, H. M., and Kenney, R. M. Lesions of contagious equine metritis in mares. *Vet Pathol* **20:** 330–341, 1983.

Al-Bassam, M. A., Thomson, R. G., and O'Donnell, L. Involution abnormalities in the postpartum uterus of the bitch. *Vet Pathol* **18:** 208–218, 1981.

Asburg, A. C. *et al.* Factors affecting phagocytosis of bacteria by neutrophils in the mare's uterus. *J Reprod Fertil* (Suppl.) **32:** 151–159, 1982.

Austad, R., Blom, A. K., and Borresen, B. Pyometra in the dog. III. Plasma progesterone levels and ovarian morphology. *Nord Vet Med* **31:** 258–262, 1979.

Borresen, B. Pyometra in the dog—A pathophysiological investigation. *Nord Vet Med* **27:** 508–517, 1975.

Carson, R. L. *et al.* The effects of ovarian hormones and ACTH on uterine defense to *Corynebacterium pyogenes* in cows. *Theriogenology* **30:** 91–97, 1988.

Coudert, S. P., and Short, R. V. Prolongation of the functional life of the corpus luteum in sheep with experimental uterine infections. *J Reprod Fertil* **12:** 579–582, 1966.

Dawson, F. L. M. Uterine pathology in bovine infertility. *J Reprod Fertil* **5:** 397–407, 1963.

Dow, C. The cystic hyperplasia–pyometra complex in the bitch. *J Comp Pathol* **69:** 237–250, 1959.

Dow, C. Experimental reproduction of the cystic hyperplasia–pyometra complex in the bitch. *J Pathol Bacteriol* **78:** 267–278, 1959.

Dow, C. The cystic hyperplasia–pyometra complex in the cat. *Vet Rec* **74:** 141–147, 1962.

Fennestad, K. L., Pedersen, P. S., and Moller, T. *Staphylococcus aureus* as a cause of reproductive failure and so-called actinomycosis in swine. *Nord Vet Med* **7:** 929–947, 1955.

Hadley, J. C. The development of cystic endometrial hyperplasia in the bitch following serial uterine biopsies. *J Small Anim Pract* **16:** 249–257, 1974.

Hansen, P. J., and Asbury, A. C. Opsonins of *Streptococcus* in uterine flushings of mares susceptible and resistant to endometritis: control of secretion and partial characterization. *Am J Vet Res* **48:** 646–650, 1987.

Hughes, J. P. *et al.* Pyometra in the mare. *J Reprod Fertil* (Suppl.) **27:** 321–329, 1979.

Kenney, K. J. *et al.* Pyometra in cats: 183 cases (1979–1984). *J Am Vet Med Assoc* **191:** 1130–1132, 1987.

Kenney, R. M. Cyclic and pathologic changes of the mare endometrium as detected by biopsy, with a note on early embryonic death. *J Am Vet Med Assoc* **172:** 241–262, 1978.

Lander Chacin, M. F., Hansen, P. J., and Drost, M. Effects of stage of the estrous cycle and steroid treatment on uterine immunoglobulin content and polymorphonuclear leukocytes in cattle. *Theriogenology* **34:** 1169–1184, 1990.

Nomura, K. Canine pyometra with cystic endometrial hyperplasia experimentally induced by *E. coli* inoculation. *Jpn J Vet Sci* **45:** 237–240, 1983.

Nomura, K., Kawasoe, K., and Shimada, Y. Histological observations of canine cystic endometrial hyperplasia induced by intrauterine scratching. *Jpn J Vet Sci* **52:** 979–983, 1990.

Potter, K., Hancock, D. H., and Gallina, A. M. Clinical and pathologic features of endometrial hyperplasia, pyometra, and endometritis in cats: 79 cases (1980–1985). *J Am Vet Med Assoc* **198:** 1427–1431, 1991.

Sandhold, M., Vasenius, H., and Kivisto, A.-K. Pathogenesis of canine pyometra. *J Am Vet Med Assoc* **167:** 1006–1010, 1975.

Simon, J., and McNutt, S. H. Histopathological alterations of the bovine uterus. 2. Uterine tissue from cows of low fertility. *Am J Vet Res* **18:** 241–245, 1957.

Teunissen, G. H. B. The development of endometritis in the dog and the effect of oestradiol and progesterone on the uterus. *Acta Endocrinol* **9:** 407–420, 1952.

Washburn, S. M. *et al.* Effect of estrogen and progesterone on the phagocytic response of ovariectomized mares infected *in utero* with β-hemolytic streptococci. *Am J Vet Res* **43:** 1367–1370, 1982.

V. Diseases of the Pregnant Uterus

A. General Considerations

There is much variation in the reproductive affairs of the various species. The mechanisms by which gestation is initiated, by which pregnancy is maintained, and by which parturition is triggered have profound effects on manifestations of diseases of the conceptus.

For a pregnancy to proceed in cattle, sheep, goats, swine, and horses, the normal cyclic lysis of the corpus luteum must be avoided. This is achieved by inhibition, on the part of the blastocyst, of synthesis and release of prostaglandin $F_{2\alpha}$ ($PGF_{2\alpha}$) from the endometrium. The cyclic corpus luteum then provides part, or all, of the hormonal support of the pregnancy. Early embryonic death will permit release of $PGF_{2\alpha}$ and lysis of the corpus luteum and subsequent restitution of the estrual cycles in these species. The delay between death of the conceptus and the maternal return to estrus is usually sufficient to allow for complete, or near complete, autolysis and dissolution of the fragile embryo. As a result, the early diseased products of conception are rarely available for study.

Venereal infections cause some of the early embryonic losses but, on the basis of the few studies that have been done and on the experience in other species, there is evidence that chromosomal abnormalities account for most of these early losses. In monotocous species, embryos which die undergo prompt dissolution, the products are resorbed or expelled, and the disease is recognized only as infertility with slightly prolonged estral cycles. In polytocous species the death of a single embryo, or even several, does not interrupt the pregnancy and dead embryos are resorbed or mummified.

In the mare, gonadotrophic hormones are produced in temporary structures in the endometrium, the endometrial cups (Fig. 4.40). These structures form early in pregnancy. The cells that make up the cups are fetal in origin and develop from a band of elongated trophoblast cells of the chorionic girdle. They become detached from the fetal chorion on about day 35 of gestation, and invade the stroma of the endometrium and enlarge and take on the appearance of decidual cells. The cup tissue becomes visible about day 40 of gestation as a horseshoe or ring-shaped band in the endometrium of the pregnant horn. The cups increase in size until day 60 of gestation, at which time their hormone production is at maximum. Hormone levels fall after day 60 and the cups become pale and slough between then and day 100 of gestation. The process of

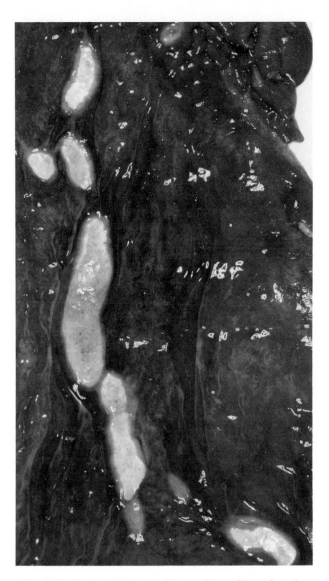

Fig. 4.40 Endometrial cups. Uterus. Mare. Sites of production of gonadotrophic hormones.

degeneration resembles that of graft rejection and is mediated by maternal lymphocytes. The necrotic tissue detaches and lies free between the endometrium and allantochorion.

The function of the potent chorionic gonadotrophin from endometrial cups in mares is unknown. However, if the conceptus dies after day 35 of gestation, at which time the fetal cells have invaded the endometrium, the cups may still form and secrete chorionic gonadotrophin, accessory corpora lutea may be produced, and the mare may fail to return to estrus and instead continue for a variable period in a state of pseudopregnancy.

The corpora lutea in domestic carnivores usually persist during most of the gestation period. They develop after ovulation, spontaneously in the case of the bitch, and induced by coitus in the cat. Once formed their life span appears to be predetermined and the hormone production by them to be independent of continuing pregnancy. As a result, death of the embryos, or death or sickness of the fetuses, shortens the length of the gestation period only slightly. Fetal death before the terminal stage does not usually lead to premature delivery; rather the process of autolysis and resorption or mummification begins, and the products are expelled at or near term. As a consequence, it is very rare to obtain diseased fetuses from these species in a satisfactory state for study. Most studies of fetal diseases in carnivores have been based on experimental investigation.

The corpus luteum is essential for the maintenance of pregnancy during the entire gestation period in swine and goats, as it is in dogs and cats; however, in the sow and the goat, the life span of the corpus luteum is not predetermined and $PGF_{2\alpha}$ will terminate it, and expulsion of dead fetuses can occur before term.

The corpus luteum is essential only during the first half of pregnancy in the cow, mare, and ewe. During this time, the effects of fetal death are unpredictable. The corpus luteum in some cases may undergo lysis, and the dead fetus or fetuses are expelled, usually severely autolyzed; but in some cases the corpus luteum will persist, and the fetus will be resorbed or, if older, and particularly in cattle, mummified. During the last third of gestation in these species, the pregnancy requires hormonal support from the fetus, and fetal death at this stage leads to expulsion within a few days. Even a few days, however, allow time for autolysis to be evident; the fetal tissues are pale, the red cells lysed, and pleural and peritoneal cavities contain hemoglobin-stained fluid.

The timing of parturition is species specific and related to appropriate fetal maturity. How the fetal maturity is sensed, and how it is translated into maternal myometrial contraction appears to vary between species and is unknown for most. Knowledge of fetal control of parturition has been gained largely in sheep, a species in which fetal control of parturition is nearly absolute. In other species, the fetus plays a lesser role or none at all.

Parturition in the ewe is initiated by a sharp rise in cortisol secretion by the adrenal cortex of the fetal lamb. The increased secretion is due partly to increased responsiveness to adrenocorticotropic hormone (corticotropin, ACTH) and partly to increased fetal ACTH production. It has been shown experimentally that either hypothalamic or pituitary destruction or bilateral adrenalectomy in the lamb fetus will prevent its initiation of parturition and indefinitely prolong gestation. Fetal cortisol induces enzyme synthesis in the placenta. These enzymes synthesize estrogen from progesterone thereby increasing placental estrogen production and decreasing progesterone. The change in ratio of estrogen to progesterone stimulates release of $PGF_{2\alpha}$ from the maternal endometrium. Prostaglandin induces compliance in the cervix by changing the physical properties of the collagenous fibers. These alterations are based in part on loss of collagen but more on an increase in glycosaminoglycans and water, which loosen

the matrix in which the collagen fibers are imbedded. The $PGF_{2\alpha}$ also enhances the myometrial response to oxytocin and stimulates uterine contractions. The mechanisms by which parturition is triggered in cattle appear to be similar to those in sheep. Parturition in the corpus luteum-dependent species is initiated by regression of the corpus luteum and a corresponding decrease in circulating progesterone. The details are known only in the goat, in which species the fetal adrenal cortex produces a cortisol surge which causes $PGF_{2\alpha}$ release and luteolysis. It is unlikely that the fetal control of parturition in other corpus luteum-dependent species is as strong as it is in goats.

The mechanisms by which dead fetuses are expelled in cattle and sheep are not known, but it is probable that they share features with normal parturition. They must involve $PGF_{2\alpha}$ release, cervical compliance, and coordinated myometrial contraction. Chronic fetal disease in cattle and sheep results in premature delivery of live but diseased fetuses. Since fetal stress activates the same hypothalamic–pituitary–adrenal endocrine chain used to initiate normal parturition, it is probable that chronic fetal disease operates through the same chain to cause abortion of live fetuses. Equine herpesvirus infection regularly produces abortion of live, diseased, equine fetuses, but very little is known of the mechanisms involved in either these abortions or normal parturition in this species.

B. Embryonic Death

There is little that can be said with profit of the pathology of the ovum, zygote, or early embryo, but a brief consideration is included here for purposes of orientation.

When fertilized ova die, they undergo progressive cytolysis within the zona pellucida, which often remains intact after the cytoplasm and nucleus are dead. The whole structure then disintegrates and is resorbed or discharged at the next estrus. It is rare to recover abnormal ova from the bovine uterine tube during the first 72 hr postestrus. Presumably, in those cases in which estrus occurs after a normal cycle, the ova die most commonly between the fourth and tenth day of the cycle. Most embryonic deaths in cows, however, are associated with prolongation of the interestrual period. In the embryonic stage, the embryonic tissue proper disappears first, and the trophoblast remains for a while before degenerating.

The incidence of **zygotic and embryonic mortality** is remarkably high in the species which have been studied, and probably is in all species. Humans and the domestic species studied suffer a zygotic loss of 15–30%. The causes are presumably diverse and may vary somewhat from species to species. However, the demonstration of abnormal karyotypes in early porcine zygotes and in a high percentage of unselected spontaneous human abortions suggests that chromosomal abnormalities are an important cause of this mortality. The frequency with which chromosomal abnormalities are found varies with the stage of pregnancy; they are most common in the early stages of pregnancy. This is the pattern one would anticipate since some of the most severe anomalies are not compatible with attachment or implantation. Most of these chromosomal abnormalities represent numerical changes such as monosomy, polysomy, polyploidy, and mixoploidy. Only a small percentage are structural. The monosomies and polysomies are the result of nondysjunction during meiosis or early cleavage stages. Triploidy can be caused either by two sperm fertilizing one egg or by the suppression of the second polar body and one sperm fertilizing a diploid egg; both possibilities increase with increased age of the gametes. Tetraploidy occurs as a result of suppression of the first cleavage division of the zygote. Mixoploids develop as a consequence of mitotic error during cleavage division of the zygote. Structural anomalies are caused by spontaneous breakage and reunion. Viruses, drugs, and radiation are known to cause this type of chromosome damage.

C. Fetal Death

A degenerate zygote or early embryo may be resorbed or expelled from the uterus. At a later stage in development, the dead fetus may be mummified, macerated, or aborted; an abortion being defined as the expulsion of a fetus prior to the time of expected viability. A dead fetus delivered within the period of expected viability is arbitrarily referred to as stillborn. It should be appreciated that these different terms might be applied to fetuses suffering from the same basic disease.

In uniparous domestic animals, death of the fetus in late pregnancy is usually followed by abortion. In multiparous species, if most of the fetuses die at the same time, all are likely to be aborted, but it is more usual for one or several dead fetuses to be retained with the remaining viable ones and delivered at parturition. It is often apparent that the deaths have occurred at different ages, the dead fetuses being of different sizes and degrees of mummification or maceration (Fig. 4.41).

D. Mummification of the Fetus

Mummification of a dead fetus is seen occasionally in any, but usually multiparous, species and most commonly in the sow. In the mare, it is typically one of twin fetuses which is mummified. A prerequisite for mummification is that bacterial infection not be present. The fluids are resorbed and the membranes become closely applied to the desiccated fetus. The whole mass becomes brown or black and rather leathery, moist on the surface with sticky mucus, but without odor or exudate. The time required for complete mummificaton will depend to some extent on the size of the fetus. Dehydration is advanced by 7 days in sheep fetuses; however, complete mummification probably requires as long as 6–8 months in the case of a 6-month bovine fetus. All stages may be observed from the earliest of beginning separation of the placenta with hemoglobin staining of the tissues to the latest, when all that remains is a firm and shrunken remnant consisting

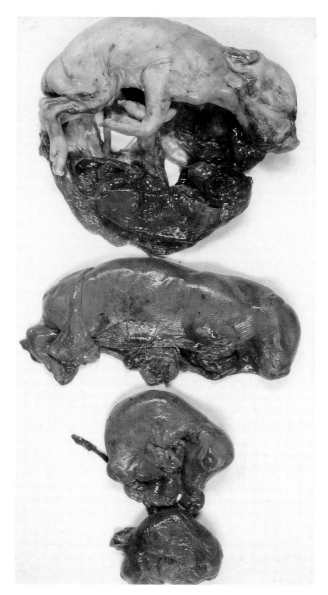

Fig. 4.41 Porcine littermates which have died at differing stages of gestation and are undergoing mummification.

almost wholly of dried skin and bones. Until the time of expulsion, which on occasion occurs spontaneously, or parturition in multipara which still carry some viable fetuses, the cervix remains closed and sealed. In uniparous animals, the mummified fetus may be retained indefinitely. Animals which have had and delivered a mummified fetus usually breed normally on subsequent occasions so there cannot be any serious uterine lesion accompanying the fetal death. Genetic diseases (both inherited and chromosomal abnormalities), viral and protozoan infections, and placental insufficiencies have been proposed as the causes of fetal death leading to mummification, but the cause is rarely firmly established for any particular case. Very likely all the proposed causes and others can produce fetal death at an appropriate time and mummification can result.

The critical features for mummification appear to be the retention of the dead fetus within the uterus through the action of a functional corpus luteum, and a fetal skin mature enough to resist autolysis.

E. Fetal Maceration and Emphysema

Maceration and emphysema of the fetus requires the presence of an infection in the uterus. If the early embryo succumbs to uterine or embryonic infection, maceration is usually followed by resorption within the uterus or expulsion along with a small amount of purulent exudate. This is the usual course of events in venereal infections by *Campylobacter fetus* and *Tritrichomonas foetus* in cows, but is also to be expected with any sort of nonspecific endometrial infection. The causes of infectious death of the conceptus during the period of the fetus are the same as previously described, but the consequences, pyometra or acute endometritis, are usually more severe because of the presence of decomposing fetal remnants. The differences between pyometra and endometritis are as discussed earlier; in pyometra the corpus luteum may persist, the cervix remains sealed, and the uterus withholds its contents. If the corpus luteum and cervical seal break down, there is a purulent discharge of the contents, and except for the presence of fetal remnants, this endometritis does not differ much from that of the puerperium. After about the third month of pregnancy in the cow, complete fetal maceration does not occur. The bones, either more or less complete or only the ossification centers, resist maceration (Fig. 4.42). They may be discharged or be retained in the pus of pyometra indefinitely, often near the cervix. The uterine pus is usually thick and intensely fetid, but this depends on the nature of the infection; in trichomoniasis, the exudate is watery and odorless.

The usual infectious causes of fetal death and endometritis are not potent gas formers; the development of fetal emphysema almost invariably depends on patency of the cervix and on invasion of the uterus and dead fetus by putrefactive organisms from the vagina. There are two common antecedents to emphysema: dystocia at or near term, and incomplete abortion. In incomplete abortion, the cervix is open, but not completely dilated, and the fetus may be delivered into the cervix or anterior vagina, or because of malpresentation, uterine inertia, or inadequate dilation of the cervix, it may be retained in the uterus. The fetus putrefies, becomes distended with foul gas, and crepitates. If the fetus is small, the dam may survive the initial acute episode of fetal emphysema after which maceration occurs in a chronic, foul, purulent metritis. The cervix is not sealed, but because of uterine paresis, the pus is retained in the flaccid, dependent organ.

Advanced uterine lesions accompany the macerated fetus. The uterine wall is thickened, and the reaction within it varies from the acute exudative inflammation of pyometra to more or less complete sclerosis and replacement by granulation tissue in long-standing cases. In these the uterus closes firmly about the bones, and the bones may cause perforation.

Fetal emphysema, at or near term, complicating dystocia is fatal unless treated, and maceration is not an expected sequel. In some cases of uterine torsion, however, the twisted cervix and vagina produce an adequate seal against bacterial invasion of the dead fetus, and mummification, rather than maceration or emphysema, results.

A special form of fetal emphysema occurs in ewes caused by *Clostridium chauvoei*, the organism responsible for blackleg. It affects fetuses near full term, causing acute tympanitic distension of the uterus, typical hemorrhagic and necrotizing lesions of the fetus, and the accumulation of dark thin discharge in small amounts in the uterus. The pathogenesis of this special infection is not entirely clear, but there is usually a relation to rough handling, such as at shearing.

F. Embryonic Death with Persistence of Membranes

This is an uncommon development that can follow embryonic death. In the period of the embryo, the fetal membranes are in volume and mass the greater part of the products of conception, and it is possible for the embryo to die and to be resorbed while the membranes persist and continue to grow. The remaining empty cyst has been referred to as a **cystic placental mole.** Most empty placental cysts correspond in size to fetal membranes at 3–4 months of gestation. There is usually no patent lumen to the cyst. It is filled with a mass of clear gelled fluid together with placental stroma. In those cases with placentary development, the allantoic and amniotic epithelia may be present. The condition may become infected and converted to a pyometra or undergo necrosis and be discharged.

G. Adventitial Placentation

The development of **intercotyledonary placentation** in cattle is a mechanism of compensation for inadequate development of placentomes. The inadequate number of

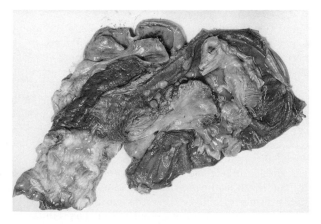

Fig. 4.42 Macerating bovine fetus.

placentomes is primarily endometrial and may be congenital or acquired. There are between 75 and 120 caruncles in the cow and 40 and 125 in the ewe and goat. Not all of them are utilized in a normal single pregnancy. Occasionally, the numbers are much less as a congenital disorder of endometrial organization; more commonly, the reduction in number is acquired by inflammatory destruction of portions of the endometrium. Compensation consists of a great increase in the size of remaining caruncles during pregnancy, many of which may fuse, and by the development of a more primitive villous placentation between the placentomes (Fig. 4.43A,B). The adventitious placenta develops first adjacent to the placentomes; the process may remain localized or involve virtually all of the intercotyledonary placenta, chiefly along the floor of the uterus. In the latter instances, pregnancy is insecure and may not proceed beyond midterm. Hydrallantois is a complication. Adventitial placentation has not been reported in the sheep placenta.

H. Hydramnios and Hydrallantois

Excessive accumulation of fluid in the amniotic and allantoic sacs are infrequent diseases of pregnancy. They occur most often in cattle and are rare in other species. Excess fluid may accumulate in both sacs, but this is the exception. The source, nature, and control of the fluid in

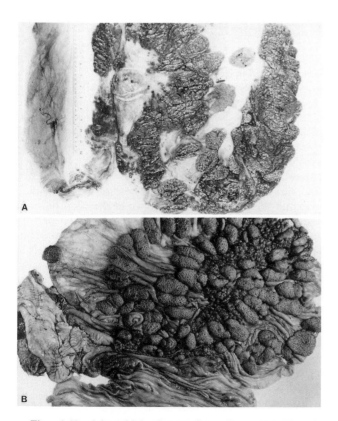

Fig. 4.43 Adventitial placentation. Cow. (A) Placenta. (B) Uterus.

the two sacs are different, and the conditions that give rise to excess fluid in each sac tend to be different. The total amount of fetal fluid increases progressively throughout pregnancy, and the volume at term in cattle is 15–20 liters. This represents a fivefold increase during the last 4 months of gestation. The relative quantity of fluid in each sac varies during the gestation period. During the first and third trimesters, the amount of allantoic fluid is greater. The nature of the fluid in each sac varies during gestation. At term the amniotic fluid is glairy, slightly viscous, and consists largely of fetal salivary secretion, the volume being largely controlled by fetal swallowing. The allantoic fluid is thin and watery and is derived from the fetal kidney via the urachus.

Hydramnios, or **hydrops** of the **amnion,** is usually associated with malformation of the fetus. The malformations may be either inherited or acquired. A variety of inherited diseases can cause hydramnios. There is a high incidence of the condition in bison–cattle hybrids, in association with chondrodysplastic and muscle contracture malformations of bovine and ovine fetuses, as well as in other types of fetal malformations, particularly those involving gross facial anomalies.

Hydrallantois in **cattle** is most often associated with uterine disease with inadequate numbers of caruncles and the development of adventitial placentation. It also occurs with increased frequency in cows bearing twins. The quantity of excess allantoic fluid may be as much as 170 liters, but it is not notably different in quality from normal. The fetal membranes are only slightly thickened, but may be tough and rupture with difficulty. Dystocia with uterine paresis, retention of placenta, and metritis are sequelae of those few cases that do not abort earlier in gestation. The fetuses are usually dead when aborted or delivered at term and small for their age, but may have anasarca and ascites.

The disease occurs but rarely in mares. The conditions which predispose in this species are not known.

I. Amniotic Plaques, Placental Mineralization, and Avascular Chorion

These are normal and their inclusion here is warranted only to avoid confusion to anyone seeing them for the first time. **Amniotic plaques** are foci of squamous epithelium on the internal surface of the amnion. They may or may not be keratinized. They are 2–4 mm or so in diameter, flat, and resemble lesions of the poxes. They are especially concentrated on the umbilical stump where they are taller, cylindrical, or papilliform. They seem to be constantly present on the bovine amnion during the middle trimester, and they do also occur in other species but have not received much attention.

The **deposition of mineral,** visible about the small blood vessels of the placenta as white streaks and spots, occurs in many species from about the end of the first to the middle of the second trimester. The degree of calcium deposition is quite variable and is more extensive in the

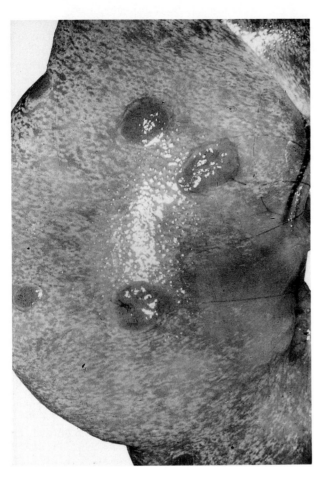

Fig. 4.44 Normal mineralization (calcification). Placenta. Bovine fetus.

allantois than in the amnion (Fig. 4.44). It can be distinguished from dystrophic calcification in that it is not gritty, but resembles milk in appearance and texture. There is no indication of why the calcium is deposited or what, if any, place it has in placental economy. **Metaplastic ossification,** complete with large multinucleated osteoclastlike cells, occurs commonly in the placenta of sows giving birth to normal litters but is seen rarely in other species.

The **avascular chorion** commonly called the necrotic tips is a normal region which is avascular and avillous and occurs at the extremities of the chorioallantoic membrane in porcine, ovine, and bovine placentas. It is almost invariably present from the period of elongation to term. Sometimes the region may extend for several centimeters along the chorion. In appearance, it is clearly demarcated from the vascular chorion, varies from white to brown, and is usually wrinkled.

J. Adenomatous Dysplasia of the Equine Allantois

Adenomatous nodules and plaques develop infrequently in the allantois of the placenta of the mare. The reported cases, and our own, have been associated with

fetal disease, but whether these tumorlike structures represent a hazard to the conceptus or whether they develop as the result of disease of the fetus or placenta is unknown. The nodules or plaques are multiple, tan, and firm and are clustered on the allantois usually near the umbilical stalk. The nodules are solid or cystic ranging in size from 1 to 5 cm in diameter, but adenomatous plaques may be 1–2 cm thick and up to 20 cm in diameter. The overlying chorion can be covered by normal microcotyledonary villi or the surface may be matted or ulcerated by an associated chorionitis.

The affected allantois is thickened by variably sized glandular structures which are lined by cells which are occasionally continuous with the epithelial cells of the allantoic surface. The cells of the glands may be thin and attenuated or tall and columnar. The larger cells have a copious foamy cytoplasm, prominent cell borders, and large vesicular nuclei. Mitotic figures are rare. These hyperplastic glandlike structures are supported by a loose fibrovascular stroma. This stroma is abundant in some areas of the plaque or nodule making up the bulk of the thickened allantoic wall, but in areas where the glands are densely packed, it is thin and inconspicuous. The lumina of the larger cysts contain proteinaceous and cellular debris. The inflammatory reaction in the dysplastic allantois is mild, and even when the chorionitis is severe and ulcerating, the lumina of the cystic glands contain only a modest number of neutrophils.

K. Prolonged Gestation

Syndromes characterized by abnormally long gestation periods occur in cattle and sheep. Some of these syndromes have been only partially defined, but in all cases which have been adequately documented there is evidence of fetal anomaly, either anatomic or functional. The anatomic anomalies usually involve the developing cerebral vesicles leading to holoprosencephaly, the range of expression varying from cyclopia to anencephaly. Individual fetuses are involved, but multiple cases may follow, for example, vaccination of pregnant ewes against Rift Valley fever and Wesselsbron disease. Two of the syndromes in cattle have a genetic basis, the trait being governed by autosomal recessive genes, the defective fetus being homozygous. The first of these, which is most common in the Holstein and Ayrshire breeds, produces a nonviable large calf after a gestation period which is ~2 months longer than normal. The parturition is abnormal; the maternal preparations of relaxation of the pelvic ligaments and filling of the udder are minimal, and assistance is usually necessary. The calves are large, but of nearly normal proportions. They suffer from severe respiratory distress. Death is due to either these respiratory difficulties or uncontrollable hypoglycemia. The adrenals are hypoplastic, and this is thought to be the lethal defect in the syndrome.

In the second type of inherited prolonged gestation, which is seen in Guernsey, Jersey, and Swedish Red and

White breeds, fetal monsters are common. Many of these animals have severe head anomalies of the cyclopian type, but in some the defect is discrete and consists only of the absence of the pituitary. Adenohypophyseal aplasia is common to all of them. These fetuses fail to develop after approximately the seventh month of gestation, and the length of gestation appears to be determined by the viability of the fetus. Some gestation periods last 17 months; at this time, presumably, the placenta can no longer adequately provide for the fetus, and the dead fetus, still developmentally immature, is expelled. Not all syndromes of prolonged gestation in cattle fit this pattern either genetically or anatomically, and sporadic examples of fetuses unable to terminate their gestation occur. These are usually grossly deformed giants with severe brain anomalies; these animals regularly die *in utero*.

Prolonged gestation occurs in sheep as a result of chemically induced teratogenesis. If ewes ingest *Veratrum californicum* on or about the fourteenth day of pregnancy, they may fail to begin labor at term; pregnancy then continues for weeks beyond term, leading ultimately to rupture of pelvic ligaments and maternal death. The fetuses damaged by the weed have cyclopian deformities and attain giant proportions. Originally, the pituitaries were thought to be absent. It appears that in most cases the gland is displaced in the malformed head.

The functional defects which prevent the various types of fetuses from making their timely contribution to their delivery can now be interpreted in the light of experimental studies. These indicate that for a fetal lamb to initiate parturition, the fetal hypothalamus, its connection to the pituitary, the pituitary, and the adrenals must be functionally competent. If any of these links is defective, prolonged gestation results. Only hypophyseal absence, however, leads to fetal dwarfism.

An additional syndrome of prolonged gestation has been recognized in ewes that are fed the African shrub *Salsola tuberculatisformis*. The mechanisms by which this plant poisoning interferes with parturition are unknown. They appear to be different from previously reported ones, in that the effects of the plant are not teratogenic, and the intoxication prevents parturition when fed during the last 50 days of gestation. The plant may inhibit fetal hypothalamic releasing factors.

Salsola tuberculatisformis is drought resistant and the disease occurs when other feed is lacking. The Karakul breed of sheep are the breed raised where the plant is common and are the breed most often affected, but other breeds are susceptible to the toxic effects of the plant. The affected ewes fail to develop normal preparturient udder enlargement, but show no other signs of intoxication.

The fetuses continue to grow and usually initiate parturition 10–20 days past term. At delivery the affected fetuses are large, lethargic, and suffer a high mortality rate. The pelts are overgrown and worthless, the hooves long, and the teeth erupted. The adrenals of fetuses are hypoplastic, the pituitaries are small, and normal granulation of the cells of the adenohypophysis is lacking.

L. Abortion and Stillbirth

It is apparent from the foregoing discussion that diseases of the conceptus may result in death with resorption, fetal mummification, abortion, or stillbirth depending on the age of the conceptus and the species involved, but not all fetal infections result in fetal death. Some viral infections appear to cause little harm to fetuses, at least during certain stages of gestation, and in some the effects are subtle. Many fetal diseases such as brucellosis of cattle and sheep, epizootic bovine abortion, *Ureaplasma diversum,* and mycotic fetal infections in cattle are chronic and may lead to the premature delivery of fetuses which are small for gestational age and diseased fetuses.

The list of bacterial and mycotic infections known to produce sporadic abortion is so long as to be nearly valueless because any bacteremia or systemic fungal infection occurring during pregnancy carries with it a great risk of bacterial or mycotic colonization of fetal chorion. The entire basis of this susceptibility is not known; however, factors within the conceptus such as isolation from the maternal immune system, decreased oxygen concentration, elevated temperature, sluggish inflammatory and immune responses, and the provision of preferred nutrients for many organisms must be important. The organisms involved in fetal infection can be the same irrespective of the period of gestation. The effects vary greatly, and overall, fetal resistance increases with age. There is some evidence which suggests that bacterial colonization of the fetal membranes does not occur as readily before embryo attachment.

Bacterial and mycotic fetal infections are most common in cattle and horses. There are, however, important differences in the pathogenesis of the infections in the two species. Except for the venereal infections early in gestation, bacterial fetal infections in cattle are almost always hematogenous in origin. Bacterial infections of equine fetuses may arise hematogenously but are more likely to infect the fetal membranes via the less protective cervix. The route of the infection can usually be determined by examination of the placenta. Transcervical infection involves the chorion adjacent to the internal os of the cervix.

The fetal effects of viral infections depend on the virus and the age of the fetus. Where the effects are known in detail, the pattern is that infections early in gestation are more likely to kill or produce serious teratologic effects; later in gestation, the effects are less severe. There are important exceptions, and the details are not known for most viral infections. Frequent mention is made throughout these volumes of abortion in relation to specific viral diseases. The fact that no mention is made of fetal disease should not be inferred to mean that the virus in question is incapable of reaching the fetus or harming it. And, alternatively, the fact that fetal disease is associated with a specific viral disease should not be construed to mean that fetal disease is produced at all stages of gestation, or that if it is, the resultant fetal lesions are the same throughout the gestation period.

Toxic, nutritional, genetic, and physical diseases are also important causes of reproductive failure and some mention is made of them elsewhere, but often the details are not available. The plants that are thought to, or known to, produce abortion or fetal disease are *Astragalus* spp. and *Oxytropis* spp. (locoweed), *Gutierrezia* spp. (broomweed), *Claviceps* spp. (ergot fungus), *Lupinus* spp. (lupines), *Conium maculatum* (poison hemlock), *Nicotiana glauca* (wild tobacco), *Solidago ciliosa* (goldenrod), *Xanthium strumarium* (cocklebur), *Trifolium* spp. (clover), and *Pinus ponderosa* (ponderosa or yellow pine) and other *Pinus* species. There must be others still unknown, and other known poisonous plants that damage or kill the conceptus as part of the disease they cause, such as the nitrate-containing *Sorghum* spp. (for details see The Hematopoietic System, Chapter 2 of this volume). In addition to rare abortion and fetal disease, the locoweeds cause blindness, neurologic disturbances, and skeletal deformities, and these are discussed in chapters on the Nervous System (Volume 1, Chapter 3), The Eye and Ear (Volume 1, Chapter 4), and Bone and Joints (Volume 1, Chapter 1). The lupines, wild tobacco, and *Conium maculatum* are primarily skeletal teratogens, and the lesions are reported in the chapter on Bones and Joints (Volume 1, Chapter 1). The role of estrogenic clovers in causing infertility is discussed under estrogenic plants in this chapter (Section IV,C,3).

Of this list, it is only the needles of the **ponderosa pine** and the like that represent an important cause of abortion. The pine is common in the rangeland throughout western Canada and the United States of America, and abortion in cattle caused by ingestion of the needles is believed to be a substantial problem, but the real incidence of the disease is unknown. Abortions occur during the last trimester. Cows abort or deliver premature weak, but viable, calves 2–20 days after ingesting pine needles. Both green and dried needles are toxic. In affected herds the abortion rate may be as high as 20–40%. The mechanism by which the pine needles bring about the premature deliveries is not known. Degenerative changes have been observed in the binucleate trophoblastic cells and the corpus luteum; these changes paralleled a decrease in the dam's serum progesterone, but whether these were the cause or effect of the abortion is not clear. The abortions tend to occur in the late autumn and winter when pregnant cows are gathered. The needles are not palatable so poisoning is often associated with scarce feed, severe weather, or boredom.

1. Infectious Causes of Abortion: General Considerations

The diagnostic rate on the causes of abortion, based on examination of samples submitted to veterinary laboratories throughout the world, varies from 5 to 50%. The proportion of abortions which are caused by infection is not known, but ~90% of those abortions in which the cause is determined are due to infection. Not all of the infectious causes can be recognized by the examination of fetoplacental submissions. The effects of systemic infec-

tions may be manifest indirectly through the dam as a result of fever, endotoxemia, or other stress.

Pregnant animals may contact infectious agents by many routes. Some agents may be carried into the reproductive tract with the semen or embryo-transfer fluids. In cattle, *Ureaplasma diversum*, bovine herpesvirus-1 and bovine virus diarrhea viruses may contaminate either of these and, being preserved by freezing and resistant to the antibiotics commonly used, they survive there. Some agents, such as *Leptospira hardjo*, may be maintained in the uterus or uterine tube over the nonpregnant period and infect the conceptus as it develops. The role of the changing hormonal influence in the reproductive tract on the multiplication of organisms may be substantial. *Campylobacter fetus* moves from the vagina to the uterus as pregnancy advances and the hormonal influence is altered. Some organisms may multiply in the conceptus for a very long period before abortion. *Ureaplasma diversum*, for example, may remain in amniotic fluid for up to 117 days before abortion occurs and *Aspergillus fumigatus* may multiply in the placenta for at least 25 days before producing abortion.

Frequently specific lesions may be present only in the placenta. As the placenta is a very large organ, and as much of it may be normal, it is important that as much of the placenta as possible be examined. If the entire placenta is not expelled, the portion retained internally should be examined as it often has the most severe lesion and is less contaminated by bacteria from the environment.

To examine the caruncle, it should be cut sagittally and the cut surface examined for infarction or suppuration. Infarction is most commonly recognized in mycotic infections but may be caused by a variety of bacteria including *Brucella* and *Salmonella*. Infarction frequently results in retention of caruncular material by the cotyledon. A variety of bacteria including *Actinomyces pyogenes* may cause a suppurative placentitis. Occasionally fibrosis of caruncles is seen and may represent normal caruncular maturation or be the result of a chronic infection. Lesions in the bovine placenta are commonly observed toward the tip of the pregnant horn and should not be confused with the normal avascular chorion in this site.

Histologic examination of hematoxylin and eosin-stained sections of the chorionic villus often reveal a blue finely granular to amorphous material in the interstitium. This may be glycosaminoglycans, mineral, bacteria, or DNA. Glycosaminoglycan deposition is normal as is mineral in the early stages of gestation. In the later stages, however, mineral is commonly observed in areas of necrosis.

The trophoblast cells should be thoroughly examined on high power as their cytoplasm may be filled by organisms. The erythrophagocytic trophoblast has been shown to harbor *Brucella* in the rough endoplasmic reticulum. *Coxiella burnetii* is also found in the trophoblast and infection by this organism can usually be distinguished from bacterial infection in that the cytoplasm of the cell appears foamy with *Coxiella* in contrast to the finely granular appearance of bacteria.

Lesions of varying intensity are usually seen in vessels of the placenta. A discrete necrotizing vasculitis of the endothelial cells in small vessels of the villi regularly occurs with infectious bovine rhinotracheitis infection. Mycotic infections cause a very severe vasculitis resulting in thrombosis and infarction of the caruncle and cotyledon. *Haemophilus somnus,* although rare as a cause of abortion, may cause a similar lesion. Chlamydial infections of the placenta may also cause a severe vasculitis but usually without thrombosis. The bacteria-packed vessels in *Salmonella* placentitis are so near the surface that they are easily mistaken for bacteria in trophoblast cells. Lesions of hemorrhage, necrosis, mineralization, and fibrosis may also be observed on the chorionic and allantoic surfaces and are commonly seen with a variety of fungi and bacteria including *U. diversum.*

The inner surface of the amnion may be severely stained with meconium, particularly in chronic progressive infections of the chorion. It is frequently the site of lesions associated with *U. diversum* infection and is characterized by multifocal to confluent areas of necrosis, mineralization, and fibrosis, with marked thickening and a mild vasculitis being common. Mycotic and mycoplasmal infections may penetrate the chorion and allantois to the amnion and result in a similar lesion.

Evidence of the significance of an isolate may be obtained only by histologic examination of the fetus and its membranes. Microscopic examination of the eyelid (surface and palpebral conjunctiva) will reveal lesions in about two-thirds of the fetuses aborted due to infectious causes. Lesions are most frequently observed on the conjunctival surface and vary from a slight infiltrate of mononuclear cells to a severe necrotizing reaction. As in other areas, the cellular infiltrate may be determined by the duration of the infection before death of the fetus. In cattle, *Yersinia* spp., *Ureaplasma,* and mycotic infections are commonly associated with a heavy focal mononuclear cell infiltrate just under the conjunctiva, whereas *A. pyogenes* infection may result in complete loss of conjunctival epithelium, and the accumulation of large numbers of bacteria, but few inflammatory cells, on the surface. Many bacteria, including *A. pyogenes* and *Listeria* continue to multiply after the death of the animal. Bovine virus diarrhea virus infection may initiate perivascular mononuclear cell infiltration in the skin and cause injury to hair follicles. The pattern of follicular hair growth when compared to the age of the fetus may determine the time of infection. Mycotic infections cause a folliculitis; however, the lesion is very variable, as to not only its presence but also its severity, and although intracorneal pustules may be seen, frequently either no lesion is observed or traces of inflammation and fungi are found only in the outer sloughing keratinized debris.

Examination of the brain, particularly in recognition of the importance of protozoan infections as a cause of abortion in ruminants, is an essential part of any fetal examination. Many viral infections manifest their effects in the brain. Although the brain often appears soft and unworthy of fixation, informative lesions are often found only in this site.

Aspiration of meconium into airways, with or without evidence of an agent, is seen in the terminal phases of many bacterial and some mycotic infections. It is probably related to the rapidity of the development of the disease and the degree of impairment of placental function. Frequently, there will be no local reaction in the lung specific to the agent, and in these the exudate present probably represents aspiration of the agent and inflammatory cells from the amnion. In *U. diversum* and parainfluenza viral infections in bovine fetuses, upper airways may remain relatively clear, but alveolar ducts become thickened and the lumina pavemented with many macrophages and few neutrophils. A severe necrotizing bronchitis may be seen with equine herpesvirus infection in horses, whereas bovine herpesvirus in cattle usually produces only multifocal necrosis.

Examination of the intestine contributes much information in some specimens. Bacteria may be seen colonizing the luminal content. The fetus swallows amniotic fluid beginning at a very early age, and bacteria and fungi can be easily seen in meconium. In one survey lesions were observed in the intestines in association with a variety of agents in a third of the bovine fetuses examined. The lesions varied considerably depending on the agent. They were mild in association with infection by *A. fumigatus.* Bovine herpesvirus-1 produced foci of necrosis. Listerial infection caused a diffuse necrotic colitis and *Bacillus* spp. induced cryptal necrosis and lymphocytic hyperplasia of Peyer's patches. *Yersinia pseudotuberculosis* caused loss of crypts and a heavy mononuclear and plasma cell infiltrate.

Bibliography

Arthur, G. H. The fetal fluids of domestic animals. *J Reprod Fertil* (Suppl.) **9**: 45–52, 1969.

Binns, W., James, L. F., and Shupe, J. L. Toxicosis of *Veratrum californicum* in ewes and its relationship to a congenital deformity in lambs. *Ann N Y Acad Sci* **111**: 571–576, 1964.

Call, J. W., and James, L. F. Pine needle abortion in cattle. *In* "Effects of Poisonous Plants on Livestock," R. F. Keeler, K. R. Van Kampen, and L. F. James (eds.), pp. 587–590. New York, Academic Press, 1978.

Chandley, A. C. *et al.* Chromosome abnormalities as a cause of infertility in mares. *J Reprod Fertil* (Suppl.) **23**: 377–383, 1975.

DeLange, M. Prolonged gestation in Karakul ewes in southwest Africa. *Proc IVth Int Cong Anim Reprod, The Hague* **3**: 590–592, 1961.

Dollahite, J. W., and Anthony, W. V. Experimental production of abortion, premature calves and retained placentas by feeding a species of perennial broomweed. *Southwest Vet* **10**: 128–131, 1957.

Drost, M., and Holm, L. W. Prolonged gestation in ewes after foetal adrenalectomy. *J Endocrinol* **40**: 293–296, 1968.

Hubbert, W. T. *et al.* Bovine abortions in five northeastern states 1960–1970. Evaluation of diagnostic laboratory data. *Cornell Vet* **63**: 291–316, 1973.

Jensen, R. *et al.* Evaluation of histopathologic and physiologic changes in cows having premature births after consuming ponderosa pine needles. *Am J Vet Res* **50:** 285–289, 1989.

Jerrett, I. V. *et al.* Diagnostic studies of the fetus, placenta and maternal blood from 265 bovine abortions. *Cornell Vet* **74:** 8–20, 1984.

Kellerman, T. S., Coetzer, J. A. W., and Naude, T. W. "Plant Poisonings and Mycotoxicoses of Livestock in Southern Africa," Capetown, South Africa, Oxford University Press, 1988.

Kennedy, P. C., Kendrick, J. W., and Stormont, C. Adenohypophyseal aplasia, an inherited defect associated with abnormal gestation in Guernsey cattle. *Cornell Vet* **47:** 160–178, 1957.

Kirkbride, C. A. Managing an outbreak of livestock abortion. 2: Diagnosis and control of bovine abortion. *Vet Med* **June:** 70–79, 1985.

Liggins, G. C., Kennedy, P. C., and Holm, L. W. Failure of initiation of parturition after electrocoagulation of the pituitary of the fetal lamb. *Am J Obstet Gynecol* **98:** 1080–1086, 1967.

McEntee, M., Brown, T., and McEntee, K. Adenomatous dysplasia of the equine allantois. *Vet Pathol* **25:** 387–389, 1988.

Miller, R. B., and Quinn, P. J. Observations on abortions in cattle: A comparison of pathological, microbiological and immunological findings in aborted fetuses and fetuses collected at abattoirs. *Can J Comp Med* **39:** 270–290, 1975.

Poitras, B. J. *et al.* The maternal-to-fetal transfer of immunoglobulins associated with placental lesions in sheep. *Can J Vet Res* **50:** 68–73, 1986.

Roche, J. F. *et al.* Reproductive wastage following artificial insemination of heifers. *Vet Rec* **109:** 401–404, 1981.

Turnbull, A. C., Anderson, A. B. M., and Wilson, G. R. Maternal urinary oestrogen excretion as evidence of a foetal role in determining gestation at labour. *Lancet* **ii:** 627–629, 1967.

Vandeplassche, M. *et al.* Dropsy of the fetal sacs in mares: Induced and spontaneous abortion. *Vet Rec* **99:** 67–69, 1976.

a. BRUCELLOSIS   Bacteria of the genus *Brucella* are small, gram-negative bacilli or coccobacilli which are strictly parasitic, prefer the intracellular habitat, and produce in animals chronic infections with persistent or recurrent bacteremias manifested typically by abortion.

Three classic species of *Brucella* were described and defined originally largely on the basis of host of origin: *B. melitensis,* goats; *B. abortus,* cattle; and *B. suis,* swine; but they are now defined by biochemical and serologic reactions. The differences between the species are slight and quantitative rather than qualitative, and the number of biotypes within each species is large. This has led to the recommendation that all should be considered a single species, *B. melitensis,* and that present species and variants with fixed properties be regarded as biotypes of the one species. This suggestion has much to commend it. The current nomenclature is confusing to the nonspecialist. However, most of the old truths still apply; the disease in cattle is caused by strains of *B. abortus;* the disease in sheep and goats by *B. melitensis;* and *B. suis* is found chiefly in swine. Cross-infections do occur and almost all domestic species are susceptible. But changes in the nomenclature will not solve the fundamental problem posed by the tendency of organisms of the genus to acquire

new characteristics and infect new hosts. The animals in which the new varieties of organisms have been discovered usually have been hosts not previously included in the reservoir system of the genus. Examples of this phenomenon are the discovery of organisms now known as *B. suis* type 2 in hares, *B. neotomae* in wood rats, *B. ovis* in rams, and *B. canis* in dogs. The generation of new features is presumably the result of evolutionary forces; why the changes are occurring so rapidly is unknown. It seems likely that what has been done in the laboratory to change speciation has been done in nature and that rapid evolutionary changes will continue to be the norm for the genus. And this in turn will continue to pose identification problems and disrupt control and eradication programs.

The biological similarities between the various species and biotypes of *Brucella* are in keeping with the remarkable similarities in the diseases produced in the different hosts by the various strains of the organism. Infections by any of these organisms of the genus *Brucella* are, initially at least, systemic, and relapsing bacteremic phases are well-established events in the persisting infections. Localization and persistence of infection may occur in many organs, perhaps to a greater extent with *Brucella suis* than with the other species. Some organs, however, notably the genitalia and placenta, are noted for the regularity with which they develop intense persistent foci of infection.

i. *Bovine Brucellosis*   Brucellosis occurs in cattle in most parts of the world. In some countries the incidence of the disease is low, because of measures taken either to prevent its entry or to eradicate it, but where the disease is endemic and uncontrolled, the incidence may approach 20–30%.

The usual source of infection for cattle is an aborted fetus or placenta, or contaminated uterine discharges, and the usual route of infection is alimentary. Infection can occur *per vaginum,* via the conjunctiva, and through the broken or unbroken skin. The relative importance of these latter routes is not known. Coital infection can occur but is uncommon, especially if genital infection in the male is long-standing, possibly because fewer organisms are excreted in semen from chronic lesions than from early lesions. Infection can be transmitted at artificial insemination if semen from infected bulls is used. Irrespective of the route of infection, the development and establishment of infection are probably comparable and depend on the age and reproductive status of the animal, its inherent resistance, and on the dose and virulence of the infecting strain of the organism. Young cattle are relatively insusceptible up to about the age of puberty. They can be infected by the usual routes and means, including the ingestion of milk in which the organisms are intermittently excreted, but they throw off the infection in the course of a few months.

Once infection is established in sexually mature animals, females especially, it tends to persist indefinitely. Some, perhaps many, recover completely, but which ones will, and when they will, cannot be predicted. The organisms extend quickly to the lymph nodes regional to the point of entry, and there they provoke an acute lymphadenitis. The inflamed glands are enlarged, often much so, hyperplastic with no clear corticomedullary distinction, and frequently bear small or large medullary hemorrhages. The sinuses are infiltrated with neutrophils and eosinophils; germinal centers and proliferative activity become obvious; and there is a slow but remarkable accumulation of plasma cells in the medullary sinuses. The changes in the regional nodes take some weeks to develop fully, and they persist for a prolonged period. There is no fibrosis or necrosis in the nodes.

The infection may be overcome in the regional nodes, but once established, it is expected to spread during the phase of acute regional lymphadenitis. Spread is chiefly hematogenous and bacteremia may persist for several months, the duration of persistence apparently depending on the susceptibility or resistance of the host. As the infection becomes chronic, bacteremia becomes intermittent, ceases in some animals, and recurs irregularly for at least 2 years in 5–10% of animals. Also, it tends to recur at parturition. It might be expected that the bacteremic episodes would result in localization and persistence of the organism in many tissues, but strangely, localization is largely restricted to the spleen, mammary glands, mammary lymph nodes, and pregnant uterus of the female, and to the lymphoid tissues, testis, and accessory glands in the male, and such localizations occur in the early bacteremic phases. The organism has little or no predilection for the kidney (although microscopic interstitial nephritis may be present), ovary, bone marrow, or mesenteric lymph nodes, and appears not to be excreted in the urine or feces. Localization does occasionally occur in synovial structures to produce a purulent tendovaginitis, arthritis, or bursitis, but whether localization and persistence occur in synovium which is healthy is not clear; some preexisting inflammatory changes may be necessary.

Infected animals, almost without exception, excrete *Brucella abortus* in the colostrum. Thereafter, excretion of the organism in milk may cease, but frequently it continues, although intermittently, throughout the period of lactation. Organisms excreted in milk are an important source of infection for children; for calves also, but they recover without significant effect. It is still not clear whether and to what extent *Brucella abortus* causes anatomic changes in the mammary glands. Lesions must be expected, but they must also be mild and difficult to distinguish from the focal inflammatory reactions commonly present in mammary glands. Dense infiltrations of the interstitial tissue by plasma cells, lymphocytes, and histiocytes, and exudation of neutrophils in acini probably constitute the usual changes. The mastitis is focal and not attended by gross changes. The cellular content of milk is increased.

Brucella abortus has special affinity for the pregnant endometrium and fetal placenta to which it spreads hematogenously during the initial or later bacteremia. Experiments using *B. abortus* in goats have shown that the organism is carried by the bloodstream to the periphery of the caruncle where, at the extremities of the maternal villi,

capillaries leak into the narrow space (hemophagous organ) adjacent to the fetal erythrophagocytic trophoblast of the chorion. Either cells of the erythrophagocytic chorionic trophoblast phagocytize the brucella organisms, or they invade; regardless, they multiply and spread via the uterine lumen and endocytosis, or by cell-to-cell transfer, to the rough endoplastic reticulum of the adjacent chorionic trophoblast cells. They multiply here and spread to the fetus following ulceration of the trophoblast and invasion of the fetal chorionic villi. The process may be similar in the bovine caruncle where hematomas appear at the ends of maternal septa late in gestation.

Gross lesions in the placenta are characteristic, but not pathognomonic; similar lesions of less or equal severity may be caused by other bacterial infections, and similar lesions, usually of greater severity, are produced by fungi. There is considerable variation in the severity of the placental lesions, and this is reflected to some extent in the course of the local infection. If the lesion is severe, abortion or premature birth is the likely outcome, and if the lesion is of minor severity, the calf may be delivered normally at term and be viable or nonviable. The intrauterine lesions apparently progress very slowly because an interval of many months may elapse between infection and abortion or normal birth. Abortion occurs most often in the seventh and eighth months of gestation. Once the infection localizes in a pregnant uterus, it almost certainly remains there and remains active until the fetus and placenta are delivered and for some time thereafter. The nonpregnant uterus is not particularly susceptible to *Brucella abortus,* and following abortion or parturition, the organism is cleared from the uterus in a few weeks, or longer in some cases.

The external appearance of an infected pregnant uterus is normal. Sometimes the placenta is normal. Typically, between the endometrium and chorion in the intercotyledonary area, there is a more or less abundant exudate which is odorless, dirty yellow, slightly viscid and slimy, and which contains grayish-yellow, pultaceous floccules of detritus. The fetal membranes and the umbilical cord are saturated with clear edema fluid, and the membranes may be 1 cm or more thick. The fetal fluids are usually normal, although occasionally fluid in the amnion is viscid.

The placental lesions are not uniform; some cotyledons may appear more or less normal and others will be extensively necrotic, whereas still others are diseased to intermediate degrees; similarly, the intercotyledonary placenta varies in the extent to which it is changed, lesions being most prominent adjacent to the cotyledons. Affected areas of intercotyledonary placenta are thickened with a yellow gelatinous fluid, opaque and tough, and the normal smooth glistening surface takes on an appearance resembling yellow to gray morocco leather with, on the surface, a patchy coagulum of inflammatory exudate and desquamated, degenerate epithelial cells. Affected cotyledons or portions of them are necrotic, soft, yellow-gray, and may be covered with the sticky, odorless, brown exudate which is usually referred to as resembling soft caramel candy.

The edematous placental stroma contains increased numbers of leukocytes, largely mononuclear but some neutrophils, the chorionic epithelial cells are stuffed with bacteria (Fig. 4.45), and many of them, with their inhabitants, desquamate into the uterochorionic space. The organisms in intact epithelial cells are coccoid, but free in the exudate, they assume a more elongate form, even while still contained within the ghosts of dead epithelial cells. Over the placentomes, the same sort of placentitis is present, but the parasitism is not so extensive in the epithelial cells covering the cotyledonary villi, except at their base, or in the epithelium lining the caruncular crypts, although many of the syncytial trophoblastic cells may be necrotic. The intervillous portions of the placenta, the so-called placental arcades, are quite severely affected, and exudate accumulates between the arcades and the expanded outer extremities of the maternal septa. There is normally some placental exudate and minor hemorrhage in these spaces, but this is greatly exaggerated in placentitis and contains infiltrated leukocytes, epithelial debris, and bacteria. The maternal portions of the placentome are not much involved except for the expanded ends of the maternal septa where these are bathed in the exudate of the placental arcades and, in consequence, become denuded and superficially necrotic. Beneath the necrotic tips of the maternal septa, there is a more or less dense infiltration of leukocytes and a prominent productive inflammation, the fibrosis extending some distance along the sides of the septa. The inflammatory enlargement of the terminal portions of the maternal septa produces an increased degree of placental interlocking, and probably contributes to retention of the placenta. Adhesions, in the usual sense of connective tissue fusion, between the placenta do not occur. The endometrium is relatively unscathed in the early infections. The zona basalis shows some increase in lymphocytes and plasma cells, and there may be scattered microscopic granulomas of epithelioid cells. Even the intercaruncular epithelium may not be overly disturbed. Later, there is severe endometritis. (For discussion of lesions in bulls caused by *B. abortus,* see The Male Genital System, Chapter 5 of this volume.)

The fetus is usually somewhat edematous with blood-tinged subcutaneous fluid. The same fluid is present in excess in the body cavities and the dorsal retroperitoneum. The normal abomasal content of a fetus is clear, translucent, thick, and viscid; in brucellosis, it often becomes very turbid, of a lemon-yellow color, and flaky. The important fetal lesion in brucellosis is a pneumonia which is present to some degree in most cases aborted in the last half of pregnancy. The lungs may appear grossly normal, but histologic examination reveals scattered microscopic foci of bronchitis and bronchopneumonia (Fig. 4.46). When severely effected, the lungs are enlarged and shaped to the thoracic contour, firm on palpation, reddened on the pleural surface or hemorrhagic, and fine yellow-white strands of fibrin are deposited on the pleura. Microscopically, there may be any stage from the minor changes mentioned through a well-developed catarrhal bronchopneumonia to the fibrinous variety. The predominant inflammatory cells are mononuclear, although many immature and mature neutrophils may be present in some areas. The septa may be edematous and the perivascular lymphatics may be infiltrated with leukocytes. By the time the placentitis has advanced to an extent that abortion is inevitable, an acute diffuse endometritis, without histo-

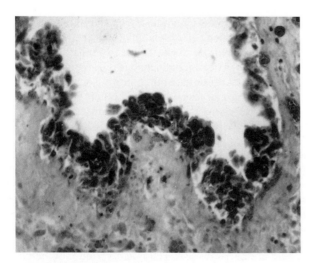

Fig. 4.45 Brucellosis. Colonies of *Brucella abortus* in cells of chorionic epithelium.

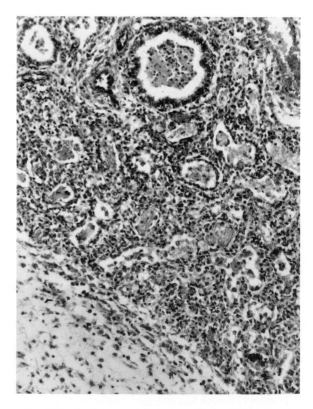

Fig. 4.46 Fetal pneumonia in brucellosis. Bovine fetus.

logic specificity, has developed. Variable lesions in the fetus include necrotizing arteritis, especially of pulmonary vessels, focal areas of necrosis, and granulomas with giant cell formation in lymph nodes, liver, spleen, and kidney.

ii. *Brucellosis in Swine* Pigs are susceptible to *B. melitensis* and slightly susceptible to *B. abortus*. The disease is usually caused by *B. suis,* and this organism presents important problems in many countries. There are some differences in the diseases produced by *B. suis* and *B. abortus,* and these depend largely on the frequency with which *B. suis* in swine produces focal granulomatous lesions with coagulation necrosis, the affinity of this organism for the skeleton and joints, and its tendency to remain in granulomatous foci in the nonpregnant endometrium.

The early stages of the pathogenesis of the infection are comparable with the early stages of bovine brucellosis. Infection can occur by the same variety of routes, but in swine, brucellosis is transmitted chiefly by coitus. Boars are as readily infected as sows, and most infected boars develop lesions in the testes or accessory genitalia from which the organisms are shed in the semen, often for life. They may also be shed in the urine from a focus of infection in the bladder. Infected females may discharge the organism from the uterus for up to 2.5 years. Suckling piglets can be infected, although they are less susceptible than weaners or adults, and although most infected piglets cast off the infection, a few of them carry the infection into adulthood.

With the development of lymphadenitis regional to the point of entry, the infection becomes bacteremic. The bacteremia may be transient or persist for many months or even some years. Localization may occur in many organs, but especially in the male and female genitalia, the skeleton including the vertebral column, synovial structures, mammary glands, lymph nodes, spleen, liver, kidney, bladder, and even in the brain.

Brucella suis can grow and multiply in phagocytes, and the typical granulomatous lesion begins with the accumulation of histiocytes and epithelioid cells. Perhaps as a response to developing hypersensitivity to the organisms, as the lesion enlarges, caseous necrosis occurs centrally, and fibrous tissue forms a capsule. The granulomas enlarge progressively, and the necrotic tissue attracts neutrophils. Giant cells are absent or scarce. Calcium salts may be deposited in the necrotic foci.

Articular lesions caused by *B. suis* are quite common. They begin as a synovitis and affect chiefly the compound and large joints of the limbs. The reaction is purulent or fibrinopurulent. Osteomyelitis in this disease is typically vertebral (or usually observed there). As with other causes of osteomyelitis, localization is typically in the vertebral epiphyses of the lumbar region. There is, however, an unusual tendency to cause diskospondylitis and destroy the intervertebral cartilages. The smaller bony lesions are typically granulomatous with dry caseation necrosis, but the necrotic cores of larger lesions may liquefy and the suppurative reaction may extend to the meninges or fistulate to produce paravertebral abscesses.

In the uterus and uterine tubes, there are often conspicuous and characteristic lesions which are not dependent on an association with pregnancy. They have been referred to as miliary uterine brucellosis and, as the name implies, there are few or very many yellowish-white nodules with an average diameter of 2–3 mm seeded in the mucosa. When the nodules are numerous, they may coalesce to form irregular plaques and then are associated with thickening of the uterine wall and stricture of the lumen. The same lesion is usually present also in the uterine tubes where obstruction results in pyosalpinx. Incised, a small quantity of caseous exudate can be expressed from the nodules. Small, red, flat, and irregular granulomas are often scattered over the surface of the supporting ligaments; grossly, they resemble fetal fat and are easily overlooked. There is some stromal fibrosis and a diffuse cellularity in the tubes and endometrium due to infiltrated plasma and lymphoid cells. In addition, well-developed and multiple hyperplastic lymphocytic nodules are evident. There are few neutrophils in the stroma, but with the mononuclear cells, they can be seen pushing their way through the epithelium to collect in abundance in the lumen of the uterus and tube and in the more superficial glands. The glands are dilated and the leukocytes are enmeshed in strands of mucin and mixed with amorphous globs of mucus. The deeper glands are cystically dilated with an attenuated epithelium, a serous or thin mucinous content, and few, if any, inflammatory cells. The epithelium lining the lumen and superficial glands is in part retained, in part desquamated, and in part shows the development of a remarkable degree of squamous metaplasia, even to the development of rete pegs and intercellular bridges. The organism is not visible in the lumen.

Abortion usually occurs between the second and third month of pregnancy, but the incidence of abortion is relatively less than in the bovine disease; there is also a high incidence of stillborn and weak piglets born at term, and probably also a high incidence of early undetected embryonic deaths. Retention of the placenta may occur. The specific uterine lesions of porcine brucellosis are described earlier. Apparently, the miliary lesions may develop during pregnancy to a superimposed diffuse catarrhal type of endometritis with patchy congestion, hemorrhage, and edema, and a small amount of creamy-pink catarrhal exudate, which contains large numbers of organisms. The fetal placenta may not show conspicuous changes, but as a rule, it is congested with small hemorrhages and patchy edema. Usually too, there is a thin layer of exudate which is slimy, grayish yellow or grayish brown like mucopus. It is more copious than the secretion normally present in the interplacental space, and smears from it show numerous free organisms and many epithelioid cells, presumably chorionic, which contain clumps of bacteria. The fetus shows subcutaneous edema, which is especially prominent about the umbilicus, and effusions into the body cavities. The edema fluid is often bloodstained. The stomach contents may appear normal or be slimy, turbid, or yellowish and may contain small flakes like curd.

iii. *Brucellosis in Sheep* A *Brucella* mutant, *B. ovis*, which was identified in the 1950s, causes a specific form of epididymitis in rams in most parts of the world in which sheep are raised. The organism also causes placentitis in pregnant ewes, but epididymitis is the more common and important manifestation of the infection and the main discussion of the disease is included with The Male Genital System (Chapter 5, Section VI,B of this volume).

There are probably as many modes of transmission as there are for the *Brucella* in general, but in this disease, the coital route is important; infected rams excrete large numbers of the organism in the semen. The infection passes through a stage of generalization, but the disease is never generalized. The organism is a humble one and not very virulent. It grows abundantly but harmlessly in the incubated egg and survives for a long time in the rat testis without provoking anything. Even the epididymitis of rams is largely an indirect effect. The epididymal lesion produced by the organism, possibly immune mediated, is modest, but it does predispose to sperm stasis and extravasation which produces the characteristic chronic granulomatous epididymitis, by which the disease is recognized. In contrast to its nonpathogenic nature in the nonpregnant ewe, the organism readily parasitizes the placenta and produces abortion; this is not common, although it may be important in individual flocks. Whereas the manifestations of infection in both ewes and rams are solely genital, the organism can be cultured in large numbers from other organs in which it does not, however, produce lesions.

The placenta is grossly edematous, being thickened to 2–5 cm by gelatinous fluid. Periarteritis and arteritis are distinctive features as in most forms of placentitis. The amnion is adherent to the chorioallantois and thickened in patches. The intercotyledonary placenta has plaquelike thickenings which may coalesce to resemble yellow-white chamois leather (Fig. 4.47A,B). Diseased cotyledons may become partially detached; they are firmer than normal, enlarged, and yellowish white. There is extensive necrosis of the epithelial elements of the cotyledons with edema and cellular infiltration of the stroma. Organisms in abundance inhabit the epithelial cells of the chorion.

Most lambs are alive at the commencement of parturition, but mummification or progressive autolysis are also common. As usual, the fetus tends to be moderately edematous, the fluid bloodstained, and that in the serous cavities may contain flecks or strands of fibrin. When present, calcified plaques on the walls, soles, or both, of the hooves and accessory digits are characteristic. The fetus shows little histologic evidence of systemic infection, even though the gastric contents may be heavily infected. Mild pneumonic alterations and lymphadenitis, especially in anatomic association to the lungs, may be present together with the development of germinal centers and plasma cells in nodes and spleen. Acute interstitial nephritis, particularly about the corticomedullary junction, and inflammatory infiltrates in portal triads are common.

The nonpregnant uterus is not susceptible to the infection. Infection of the fetus may be produced experimen-

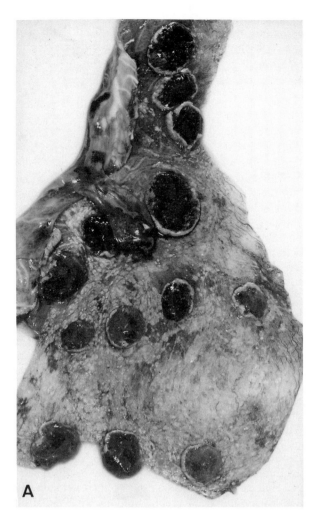

Fig. 4.47A Brucellosis. Placentitis caused by *Brucella ovis*. Sheep.

tally at any stage, but following systemic exposure, as per conjunctiva, the period of greatest susceptibility to intrauterine localization is from ~21 to 90 days of pregnancy.

Following artificial exposure, there is a lag of 2–3 weeks before bacteremia develops, and the bacteremic phase, in rams and probably in ewes, lasts ~2 weeks. Breeding ewes to rams with open epididymal infection is more likely to result in infertility than in abortion. The routes of natural exposure of pregnant ewes are not known. Placental localization is followed by slow development of the intrauterine infection, and the pregnancy may survive for 2–3 months after first being infected.

Brucella melitensis is the principal cause of brucellosis in sheep and goats, although natural infections with *B. abortus* occur occasionally. Brucellosis in goats and sheep is prevalent in countries bordering the Mediterranean, in the Near East, and in South America.

In most respects, the caprine disease resembles bovine brucellosis, but the disease in sheep is usually less pro-

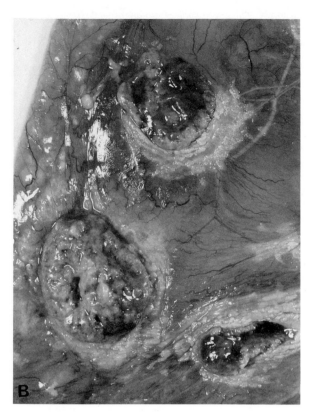

Fig. 4.47B Brucellosis. Details of pericotyledonary placentitis caused by *Brucella ovis*. Sheep.

tracted and spontaneous recovery is fairly common in a few weeks or months. In the early bacteremic phases, goats may suffer severe illness and even die, but many infections are asymptomatic. The early signs of the disease may be referable to acute mastitis with palpable nodules in the gland and a secretion which is clotted and watery. The organism is excreted in the milk, usually only for a few weeks in sheep, but often for several months or even some years in goats.

As is usual for ruminants, abortion may be the only sign observed and it tends to occur late in pregnancy. The uterine and vaginal discharges after pregnancy or parturition contain large numbers of organisms. Abortion or still-birth may terminate successive pregnancies.

iv. *Brucellosis in Dogs* *Brucella canis* was first isolated and recognized as a cause of abortion and epididymitis in dogs in 1966. The disease was identified in beagle colonies in North America at that time. Since then, it has been found in many parts of the world and in a variety of breeds. However, accurate incidence rates are not available; in the United States of America where the infection is widespread, the reported incidence rates range from 0.5 to 5%.

The disease is caused by a mucoid strain of *Brucella*, which shares biochemical features with *B. suis*. Because of the mucoid nature of the organism, it does not possess the smooth (O) antigens of *B. abortus* or *B. melitensis*. As

a consequence, the conventional brucellosis test antigens that are used for diagnosis of the disease in other domestic species are ineffective in the diagnosis of *B. canis* infections.

Transmission of the disease can be by ingestion of infected vaginal discharge after abortion or venereally by infected seminal fluids. The seminal fluids may remain infected for months. Pregnant bitches may abort after 30 days of gestation, but most abortions occur after 50 days of gestation. Testicular degeneration and epididymitis are the usual manifestations of the disease in male dogs. In response to the discomfiture of the epididymitis, the dogs frequently lick the scrotal skin, producing severe ulcerations. Persistent bacteremias are common in many dogs that show no signs of disease. Bacteria have been recovered from the blood for periods as long as a year, even though high agglutinating titers were maintained throughout the bacteremic period. The nongenital lesions consist initially of nonspecific enlargement of the retropharyngeal and mandibular lymph nodes. Later the lymph nodes throughout the body may be enlarged due to diffuse hyperplasia of lymphocytes and macrophages. The aborted fetuses may be dead or alive when expelled, but the latter usually die within a few hours or days. Pneumonia, endocarditis, and hepatitis are observed in the infected fetuses.

Focal necrosis is prominent in the chorionic villi and many of the trophoblastic epithelial cells are laden with bacteria. In the bitch, serosanguinous to gray-green vaginal discharge may be evident for 1–6 weeks following abortion and may provide a source of infection for other dogs.

v. *Brucellosis in Other Species* *Brucella abortus* or *B. suis* are regularly found in the bursal lesions described elsewhere as poll evil and fistulous withers (see Diseases of Joints, in Volume 1, Chapter 1). There are other isolated reports of suppurative skeletal or synovial lesions in horses caused by these organisms.

Whereas cats are resistant to natural infection with *Brucella* spp., dogs are relatively resistant to the classic strains of *Brucella*, but natural infections caused by all three species of the organism do occur. The majority of such infections are asymptomatic and are detected serologically. Orchitis and epididymitis have been observed. Other lesions described in dogs with brucellosis, chiefly minute granulomas in liver, kidney, and lymph nodes, are neither specific, nor necessarily related to the infection.

Bibliography

Anderson, T. D., Cheville, N. F., and Meador, V. P. Pathogenesis of placentitis in the goat inoculated with *Brucella abortus*. II. Ultrastructural studies. *Vet Pathol* **23:** 227–239, 1986.

Anderson, W. A., and Davis, C. L. Nodular splenitis in swine associated with brucellosis. *J Am Vet Med Assoc* **131:** 141–145, 1957.

Biberstein, E. L. *et al.* Epididymitis of rams. Antibody in newborn lambs. *Cornell Vet* **56:** 54–66, 1966.

Brown, I. W., Forbus, W. D., and Kerby, G. P. The reaction of the reticulo-endothelial system in experimental and naturally acquired brucellosis of swine. *Am J Pathol* **21:** 205–223, 1945.

Buddle, M. B., and Boyes, B. W. A *Brucella* mutant causing genital disease of sheep in New Zealand. *Aust Vet J* **29:** 145–153, 1953.

Cameron, H. S. Swine brucellosis. *Adv Vet Sci* **3:** 275–285, 1957.

Carmichael, L. E., and Kenney, R. M. Canine abortion caused by *Brucella canis. J Am Vet Med Assoc* **152:** 605–616, 1968.

Doyle, T. M. *Brucella abortus* infection of goats. *J Comp Pathol* **52:** 89–115, 1939.

Fitzgeorge, R. B., and Smith, H. The chemical basis of the virulence of *Brucella abortus.* VII. The production *in vitro* of organisms with an enhanced capacity to survive intracellularly in bovine phagocytes. *Br J Exp Pathol* **47:** 558–562, 1966.

Hartley, W. J. The pathology of *Brucella ovis* infection in the pregnant ewe. *N Z Vet J* **9:** 115–120, 1961.

Jones, L. M., Thomson, P. D., and Alton, G. G. Production of immunity against experimental *B. melitensis* infection in goats. *J Comp Pathol* **68:** 275–287, 1958.

Keogh, J., and Doolette, J. B. The epidemiology of ovine brucellosis in South Australia. *Aust Vet J* **34:** 412–417, 1958.

Kernkamp, H. C. H., Roepke, M. H., and Jasper, D. E. Orchitis in swine due to *Brucella suis. J Am Vet Med Assoc* **108:** 215–221, 1946.

Lowbeer, L. Skeletal and articular involvement in brucellosis of animals. *Lab Invest* **8:** 1448–1455, 1959.

Manthei, C. A., and Carter, R. W. Persistence of *Brucella abortus* infection in cattle. *Am J Vet Res* **11:** 173–180, 1950.

Meyer, M. E. Evolution and taxonomy in the genus *Brucella:* Steroid hormone induction of filterable forms with altered characteristics after reversion. *Am J Vet Res* **37;** 207–210, 1976.

Meyer, M. E. Evolution and taxonomy in the genus *Brucella:* Concepts on the origins of the contemporary species. *Am J Vet Res* **37:** 199–202, 1976.

Molello, J. A. *et al.* Placental pathology. I. Placental lesions of sheep experimentally infected with *Brucella ovis. Am J Vet Res* **24:** 897–904, 1963.

Molello, J. A. *et al.* Placental pathology. II. Placental lesions of sheep experimentally infected with *Brucella melitensis. Am J Vet Res* **24:** 905–914, 1963.

Molello, J. A. *et al.* Placental pathology. III. Placental lesions of sheep experimentally infected with *Brucella abortus. Am J Vet Res* **24:** 915–922, 1963.

Morse, E. V. Canine brucellosis—a review of the literature. *J Am Vet Med Assoc* **119:** 304–308, 1951.

Osburn, B. I. The relation of fetal age to the character of lesions in fetal lambs infected with *Brucella ovis. Pathol Vet* **5:** 395–406, 1968.

Osburn, B. I., and Kennedy, P. C. Pathologic and immunologic responses of the fetal lamb to *Brucella ovis. Pathol Vet* **3:** 110–136, 1966.

Payne, J. M. The pathogenesis of experimental brucellosis in the pregnant cow. *J Pathol Bacteriol* **78:** 447–463, 1959.

Renoux, G. Brucellosis in goats and sheep. *Adv Vet Sci* **3:** 242–273, 1957.

Smith, H. *et al.* The chemical basis of the virulence of *Brucella abortus.* I. Isolation of *B. abortus* from bovine foetal tissue. *Br J Exp Pathol* **42:** 631–637, 1961.

Smith, H. *et al.* Foetal erythritol: A cause of the localization of *Brucella abortus* in bovine contagious abortion. *Nature* **193:** 47–49, 1962.

Taul, L. K., Powell, H. S., and Baker, O. E. Canine abortion due to an unclassified gram-negative bacterium. *Vet Med Small Anim Clin* **62:** 543–544, 1967.

Thomsen, A. The eradication of bovine brucellosis in Scandinavia. *Adv Vet Sci* **3:** 197–240, 1957.

Watts, P. S. Genital infections in sheep with particular reference to *Brucella*-like organisms. *Aust Vet J* **31:** 1–6, 1955.

Wipf, L. *et al.* Pathological aspects of canine brucellosis following oral exposure. *Am J Vet Res* **13:** 366–372, 1952.

b. CAMPYLOBACTER INFECTIONS Infections caused by *Campylobacter* spp. are common and widespread in humans, cattle, sheep, swine, and chickens and also occur occasionally in dogs. In sheep and cattle, diseases caused by these organisms previously were grouped under the title of vibriosis in association with their classification in the genus *Vibrio* at that time. The principal patterns of disease are either genital, characterized by infertility or abortion, or intestinal, characterized by enteritis and diarrhea. *Campylobacter fetus* subsp. *venerealis* is a true genital infection and is an important cause of infertility in cattle and can cause abortion in cattle and sheep. *Campylobacter fetus* subsp. *fetus* is a common resident of the ovine and bovine intestine and can cause abortion in both species but does so more commonly in sheep. *Campylobacter jejuni* is a common inhabitant of the intestinal tract in sheep and cattle and may cause enteritis in humans and occasionally in animals. It frequently causes abortion in sheep and less often in cattle, although heavy losses in individual cattle herds do occur.

i. *Genital Infection of Cattle Caused by* Campylobacter fetus *subsp.* venerealis The disease in cattle due to *C. fetus* subsp. *venerealis* is a specific venereal disease, transmitted by coitus. Infected bulls may carry the organism indefinitely in the preputial cavity, although some recover spontaneously. Bulls do not become permanent carriers until they are at least 4 years of age, and most do not become infected readily until they are 5–6 years old. Epithelial crypts develop in the penile mucosa with advancing age and provide a favorable habitat for the bacteria. This also appears to be the natural habitat for these organisms, and once the infection is established in these older animals, it tends to be retained permanently as a locally innocuous surface contaminant together with the nonpathogenic *C. sputorum* subsp. *bubulus*. The organism may also survive for long periods without producing lesions on the surface of the bovine vagina, but not in the uterus. Infected cows develop an immunity and control the infection. The outstanding feature of the infection is not observed abortion but temporary sterility or repeat-breeding with prolongation of the interestrual periods. It is probable that the delayed return to estrus is the consequence of early embryonic death following fertilization and embryo attachment. Detectable abortions occur at any time but chiefly about the fourth to sixth month of gestation. The placenta is not usually retained.

The endometrial lesions in cows which are repeat-breeders are mild, and consist of lymphocytic infiltrations and nodules and scattered cystic glands. Aborted placentae are often autolyzed, indicating that fetal death occurs some time before expulsion. Placental lesions resemble those in brucellosis but are less severe. The intercotyledonary placenta is edematous, opaque, and may be leathery; there is a diffuse infiltration of cells, which are mainly macrophages. Diseased cotyledons are yellow and pultaceous; many have yellow necrotic villi at the margins, and in others, they are scattered throughout. There may be dense accumulations of polymorphs among the denuded villi and in the stroma. The degree of placentitis is quite variable and may be inconspicuous. The parasitism of the chorionic epithelium which is characteristic of *Brucella* occurs but is less evident in placentitis caused by *C. fetus* subsp. *venerealis*. Lesions in the fetus are nonspecific. There are commonly bloodstained effusions in the subcutis and body cavities with loose deposits of fibrin on the serous membranes. The normal colorless, thick, viscid mucus of the stomach content becomes yellow, very turbid, and flaky.

ii. *Genital Infection of Sheep and Cattle Caused by* Campylobacter jejuni *or* Campylobacter fetus *subsp.* fetus The chief manifestations of these campylobacter infections in sheep are late abortion, premature birth, and the birth of weak lambs. There is not usually retention of the placenta or sterility. Occasional maternal deaths occur as a result of metritis. Infection is not synonymous with abortion as infected ewes may deliver infected but clinically normal lambs.

Transmission of *C. jejuni* most often occurs by the oral route and fecal contamination of water supplies is relatively frequent. The organism is carried in the intestinal tract of poultry, and it is a commensal of cattle, sheep, and swine. The organism is also a frequent inhabitant of the intestine in dogs and cats and many rodents.

When strains of *C. fetus* subsp. *fetus* isolated from feces or bile are given orally to susceptible sheep, there is a transient bacteremia followed by localization in the gut and bile, and the infection can spread by contact between sheep. In pregnant, nonimmune ewes, localization also occurs in the uterus. With both of these *Campylobacter* spp., the infection is primarily an intestinal infection: transmission is by ingestion, and uterine localization is an accidental outcome of a brief bacteremic phase in a nonimmune sheep. The incidence of abortion will depend on the number of ewes pregnant more than 1 month and on their experience with the infecting strain. Immunity following infection or abortion usually lasts for 2 years; however, if other strains are introduced, they may cause abortion as cross-protection between strains is incomplete. This also applies to vaccination strains.

The abortion rate in natural outbreaks varies between 5 and 70% but is usually ~25%. Aborted fetuses may have only nonspecific edematous changes; however, some have rather specific hepatic lesions. Affected livers have few or many light tan areas, from 1 to 2 mm up to 1–2 cm in diameter and randomly distributed (Fig. 4.48). Frequently they have a target pattern with a more or less distinct, slighty raised white outer rim and a depressed tan inner zone. The abdomen is frequently distended with fluid, fibrin, and the enlarged liver. If the lamb has lived a while

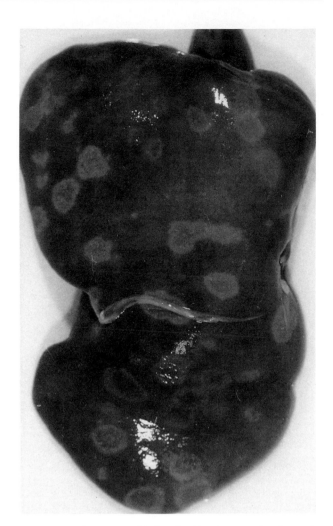

Fig. 4.48 Focal hepatitis in a fetal lamb caused by *Campylobacter fetus*.

before succumbing, these lesions must be differentiated from those of putrefactive origin and from the focal necrobacillary hepatitis of umbilical origin. Histologically, the lesion is a multifocal necrotizing hepatitis, the necrotic region represented by the tan area, and the outer rim by cellular infiltrate and necrosis (Fig. 4.49). There is frequently a bronchopneumonia with large numbers of neutrophils in major airways and alveolar ducts. Small renal cortical hemorrhages may also be present.

The cotyledons are enlarged, pale yellow to tan, dull and pultaceous, and covered with brown exudate. The chorionic stroma is edematous and infiltrated with leukocytes which are mainly histiocytes. There is often a vasculitis. There is a mild endometritis with rather more prominent inflammatory exudation in the caruncular septa. The placental lesions are more severe over the placentomes than in the intercotyledonary areas and are qualitatively similar to the lesions described in the placenta in bovine brucellosis.

Diagnosis is based on the typical hepatic lesions and

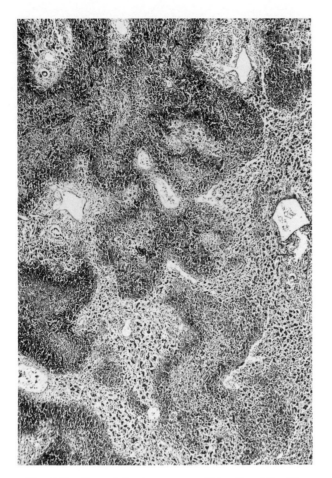

Fig. 4.49 *Campylobacter* sp. necrotizing hepatitis. Sheep fetus.

demonstration of the organism. The lesions are not patho-gnomonic. Identical morphologic changes are produced by *Flexispira rappini*. Distinctive organisms may be observed in smears of abomasal content or of affected cotyledons and show up well with special stains. Dark-field or phase contrast microscopy may be used to see the characteristic darting motility. Discrimination of the species can be made by culture or application of specific fluorescent antibody techniques.

Bibliography

Coid, C. R., and Fox, H. Short review: Campylobacters as placental pathogens. *Placenta* **4:** 295–306, 1983.

Florent, A., Vandeplassche, M., and Huysman, A. Evolution de l'infection a *Vibrio foetus* chez la genisse a la suite d'une primo-infection unique. *Rec Med Vet* **134:** 97–104, 1958.

Frank, A. H., Bryner, J. H., and O'Berry, P. A. Reproductive patterns of female cattle bred for successive gestations to *Vibrio fetus*-infected bulls. *Am J Vet Res* **25:** 988–992, 1964.

Hoerlein, A. B., and Kramer, T. Artificial stimulation of resistance to bovine vibriosis; use of bacterins. *Am J Vet Res* **25:** 371–373, 1964.

Jensen, R. *et al*. Vibrionic abortion in sheep. 1. Transmission and immunity. *Am J Vet Res* **18:** 326–328, 1957.

Jensen, R., Miller, V. A., and Molello, J. A. Placental pathology of sheep with vibriosis. *Am J Vet Res* **22:** 169–185, 1961.

Samuelson, J. D. and Winter, A. J. Bovine vibriosis: The nature of the carrier state in the bull. *J Infect Dis* **116:** 581–592, 1966.

Simon, J., and McNutt, S. H. Histopathological alterations of the bovine uterus. 1. Studies with *Vibrio fetus. Am J Vet Res* **18:** 53–65, 1957.

Storz, J., Miner, M. L., and Marriott, M. E. Early events in *Vibrio fetus* infection of ewes. *Zentralbl Veterinaermed* **11:** 475–486, 1964.

Van Donkersgoed, J. *et al*. Campylobacter jejuni abortions in two beef cattle herds in Saskatchewan. *Can Vet J* **31:** 373–377, 1990.

c. *FLEXISPIRA RAPPINI* INFECTIONS *Flexispira rappini* is a highly motile, anaerobic rod-shaped bacterium with multiple flagella at both of its tapered ends. While *F. rappini* is the generally accepted name, it has not been firmly classified. The organism is closely related to organisms in the genus *Helicobacter* but is distinguished from them by its unusual spiral grooves visible on electron-microscopic examination. Comparable organisms are frequently found in the feces of dogs and pigs and the stomachs and intestines of a variety of other species.

Flexispira rappini infections in pregnant ewes result in fetal mummification, abortion, and the birth of infected lambs, born weak. It is not considered to be the cause of epizootics of abortion and, as the lesions closely resemble those occurring in abortion due to *Campylobacter* sp. (which does cause epizootics), it is important that it be differentiated.

Naturally or experimentally infected ewes usually abort in the last trimester of pregnancy. The fetus is usually well preserved; however, if it is retained in the uterus, it decomposes rapidly, and the liver may be liquified. An infected lamb may be born alive, and occasionally only one lamb of a twin may be affected.

Gross and microscopic lesions may be completely absent but when present, are in the placenta and liver. In the placenta, they consist of a light covering of brown-to-gray exudate on the chorionic surface. In the abdomen, there may be a fibrinous peritonitis, with a thin, loosely attached layer covering the liver. The liver may be congested and enlarged or tan and soft. Multifocal to coalescing pale, brown-to-grey, clearly demarcated, irregularly shaped areas (0.5 to several centimeters across) are often visible on the surface and extend into the parenchyma.

In both the placenta and liver, lesions not visible grossly may be obvious on microscopic examination. In the placenta, foci of necrosis with accumulations of neutrophils and mononuclear cells may be present on chorionic epithelium. There is a heavy infiltration of neutrophils and macrophages in the villous stroma and a severe neutrophil-rich vasculitis may be present in the placental vessels. Capillaries in the villous stroma may be distended with organisms. In the liver, there are multiple, sometimes connecting, foci of necrotic hepatocytes infiltrated by large numbers of degenerating neutrophils, macrophages, and lymphocytes. Hepatocytes in surrounding areas are dissociated.

The condition should be suspected when these lesions are present and *Campylobacter* cannot be isolated. The organism can be seen on dark-field or phase-microscopic examination of a drop of stomach content. It stains weakly Gram negative and is more easily visualized using Giemsa or crystal violet. Special conditions are needed for culture. The organism also produces abortion in guinea pigs.

Bibliography

Bryner, J. H. *et al.* Experimental infection and abortion of pregnant guinea pigs with a unique spirillum-like bacterium isolated from aborted ovine fetuses. *Am J Vet Res* **48:** 91–95, 1987.

Kirkbride, C. A. *et al.* Ovine abortion associated with an anaerobic bacterium. *J Am Vet Med Assoc* **186:** 789–791, 1985.

Kirkbride, C. A. *et al.* Abortion in sheep caused by a nonclassified, anaerobic flagellated bacterium. *Am J Vet Res* **47:** 259–262, 1986.

d. LISTERIOSIS Infections by *Listeria monocytogenes* in ruminants are traditionally associated with cerebral localization and encephalitis (for general discussion of listeriosis see The Nervous System, Volume 1, Chapter 3), but localization occurs in the pregnant uterus and abortion or stillbirth is then the common sign. The syndromes of encephalitis and abortion may occur together in a herd or flock, but rather surprisingly, this is the exception. More commonly one or the other syndrome occurs exclusively. *Listeria monocytogenes* varies greatly antigenically and serovars 1 and 4b are commonly isolated from cattle and serovars 4b and 5 in sheep. Serovar 5 (*Listeria ivanovii*) seems particularly virulent for sheep.

Listeria monocytogenes is shed in the feces of normal carrier and clinically infected animals alike, and has been reported to survive in some soils for as long as 2 years. It is most often spread by ingestion of food or water contaminated by feces, urine, placenta, or vaginal discharges from aborting animals. Infection by the oral route, and others, has been shown experimentally to produce abortion. Silage is often the source of the organism in severe outbreaks of the infection. *Listeria monocytogenes* can multiply in poorly preserved silage with a high pH (6–7.8) but dies at a pH lower than 5.5.

Abortions in both cattle and sheep due to *Listeria* occur during the last trimester of pregnancy. The percentage of pregnant animals which abort within any group is variable; usually the abortions produced are sporadic, but sometimes 50% abort. If uterine infection develops during the early part of the last trimester, the placenta is quickly invaded by the organism, and the fetus dies as a result of a septicemia. The dead fetus is expelled in ~5 days; by this time autolytic changes mask minor gross lesions produced by the organism and all organs teem with bacteria, which continue to multiply after death of the fetus. Abortions at this stage are not usually accompanied by severe systemic disease in the dam. The placenta is typically retained as a result of a mild metritis, but the organisms and associated inflammation are quickly cleared from the nonpregnant uterus.

If the fetus is near term when the infection develops,

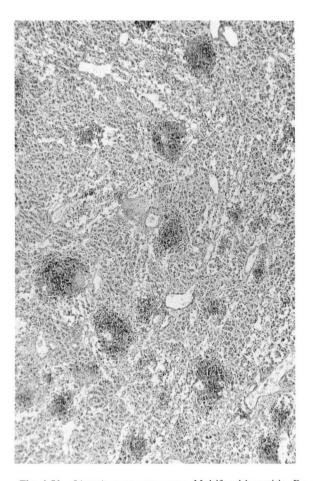

Fig. 4.51 *Listeria monocytogenes.* Multifocal hepatitis. Bovine fetus.

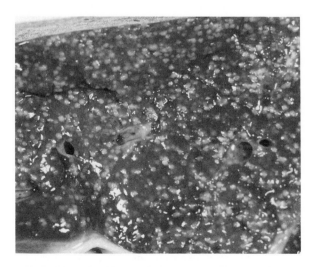

Fig. 4.50 Focal hepatitis in a fetal calf caused by *Listeria monocytogenes.*

an abnormal parturition is instituted in which dystocia is the rule, with severe metritis and septicemia being the most common complications. Lesions in the fetuses and placentae in this group are less likely to be obscured by autolytic changes. Gross lesions consist of tiny pinpoint yellow foci in the liver (Fig. 4.50). Similar foci, but usually visible only microscopically, are present in the lung, myocardium, kidney, adrenal, spleen, and brain. These foci have a central area of necrosis in which the organism can be seen on routine sections (Fig. 4.51). Surrounding, and to a lesser extent within, the area of necrosis are small numbers of degenerating neutrophils and mononuclear cells. If the fetus is near term, there may be a severe diffuse nonsuppurative cerebrospinal meningitis in which the organism can also be seen. In the bovine fetus, despite marked autolysis, the most dramatic and distinguishing lesion may be in the mucosa of the colon where a severe necrotizing enteritis with clusters of colonizing gram-positive bacteria can be seen (Fig. 4.52). The jejunum is autolyzed but is otherwise normal. Grossly and microscopically the placental lesion is severe. The necrotic tips of the villi are covered by a purulent exudate in which many bacteria are present.

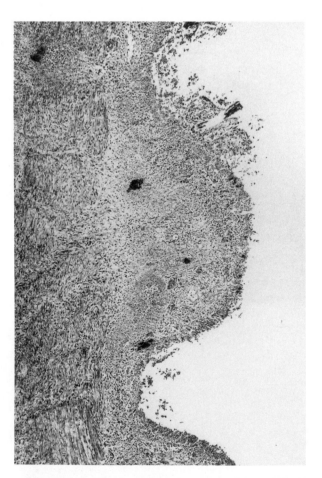

Fig. 4.52 *Listeria monocytogenes.* Necrotizing colitis. Bovine fetus. Note dark-staining colonies of bacteria.

Recovery of the organism is sometimes difficult and is probably because of the advanced autolysis or contamination of some samples. This difficulty can often be overcome by refrigerating the specimen (4°C) and reculturing after a period of storage using enriched or selective media.

Bibliography

Blenden, D. C., and Szatakowicz, R. T. Ecologic aspects of listeriosis. *J Am Vet Med Assoc* **151:** 1761–1766, 1967.

Diplock, P. T. Ovine listerial abortion. *Aust Vet J* **33:** 68–70, 1957.

Gray, M. L., and McWade, D. H. The isolation of *Listeria monocytogenes* from the bovine cervix. *J Bacteriol* **68:** 634–635, 1954.

Gray, M. L., Singh, C., and Thorp, F. Abortion and pre- or postnatal death of young due to *Listeria monocytogenes.* 3. Studies in ruminants. *Am J Vet Res* **17:** 510–516, 1956.

Gray, M. L. Experimental listeriosis in pregnant animals. *Zentralbl Veterinaermed* (Suppl.) **1:** 110–116, 1958.

Ladds, P. W., and Dennis, S. M. Sequential studies of experimentally induced ovine listerial abortion: Pathologic changes. *Am J Vet Res* **35:** 161–170, 1974.

Molello, J. A., and Jensen, R. Placental pathology. IV. Placental lesions of sheep experimentally infected with *Listeria monocytogenesis. Am J Vet Res* **25:** 441–449, 1964.

Osebold, J. W., Kendrick, J. W., and Njoku-Obi, A. Cattle abortion associated with natural *Listeria monocytogenes* infections. Abortion of cattle experimentally with *Listeria monocytogenes. J Am Vet Med Assoc* **137:** 221–226, 227–233, 1960.

e. LEPTOSPIRAL INFECTIONS Leptospires cause several disease syndromes in different species (see The Urinary System, Volume 2, Chapter 5). Maintenance hosts—that is, hosts in which the organism survives for very long periods in the kidney and genital tract (mammary glands, uterus, uterine tube, testes, and vesicular glands)—vary with the geographic region and may change with population shifts. For example, in central Canada, the maintenance host for *L. pomona* is wildlife (especially the skunk), but in the United States, it is the pig; whereas for *L. hardjo,* in both regions, it is mainly the cow. The organism is spread from a carrier cow to the fetus through the placenta, between animals by mating, by contact, or through the contaminated environment. Environments favoring maintenance of the organism include surface water with a pH around seven (or slightly higher) and moderate temperatures. The geographic region and conditions favoring an organism differ for some serovars. The organism enters the host through abraded skin or coronet and by the intestinal, genital, nasal, and ocular mucous membranes.

i. *Cattle* Leptospirosis is widespread throughout the world and is an important cause of abortion and infertility in cattle. Most often, leptospirosis in cattle is caused by *Leptospira interrogans,* serovar *pomona,* and serovar *hardjo.*

In Europe *L. interrogans* serovar *hardjo* type hardjobovis is usually the genotype of *hardjo* isolated from slaughter cattle, whereas *L. interrogans* serovar *hardjo* type hardjoprajitno (considered to be a highly virulent type of leptospire) is most commonly isolated from aborted fetuses. This latter genotype has not yet been identified in North America. Formerly it was difficult to confirm a diagnosis of leptospirosis; but with the advent of immunologically based tests to reveal and identify the organism, and routine testing of fetal cavity fluids for antibody, it has become a relatively common diagnosis (0–40% of bovine fetuses submitted), depending on the geographic location.

Infection of cattle by either *L. hardjo* or *L. pomona* results in a generalized infection followed by localization of the organisms in the renal tubules. *Leptospira hardjo* may localize in the uterine tube and descend into the uterus and placenta during pregnancy. Systemic infection may produce a febrile response and a flaccid agalactia, with the udder secretion becoming yellow and thick. The organism may also be localized in the placenta, and in some animals, produces fetal disease and abortion. Most of the abortions occur in the last trimester and often follow a delay of 6 weeks or more between the acute infection of the dam and the abortion. Fetal infection is not invariably fatal, and weak calves may be born with a substantial antibody titer. Fetuses that die *in utero* because of leptospiral infection are usually expelled in an autolyzed state.

The placenta may be edematous, but inflammatory changes are modest in spite of the fact that there may be clumps of leptospires in the chorionic epithelium. Histologically, multifocal necrotizing tubular necrosis or interstitial inflammatory infiltrates, usually with plasma cells, can be found in the kidneys of some of the fetuses. Occasionally the fetus will have a nonsuppurative meningitis (Fig. 4.53).

The organism may be demonstrated in smears of fetal kidney, lung, liver, and placenta or in fetal aqueous humor by the indirect fluorescent antibody test or by culture. The organism can also be demonstrated in fixed tissue sections of fetal kidney by the immunoperoxidase stain or in frozen sections using the fluorescent antibody technique.

ii. *Swine* Abortion in swine is often the only evidence of leptospiral infection in a herd. Several serovars of *Leptospira interrogans* regularly infect swine; *canicola, grippotyphosa, icterohaemorrhagiae, sejroe, pomona,* and *tarassovi.* The latter two serovars, which are adapted to spread in swine, are most commonly associated with fetal disease and abortion. Serologic testing has indicated that infection by serovar *bratislava,* or a closely related serovar, is also widespread; and whereas stillborn and weak liveborn piglets may have the organism and antibodies to it, the causative relationship between the clinical syndrome and infection is not firmly established.

Leptospires are passed in the urine, and can enter the body through mucous membranes or breaks in the skin. Infected swine develop bacteremia before the leptospires localize in the kidneys and uterus, where they may persist and can be shed for months. Systemic signs, if they occur at all, develop during the stage of leptospiremia. It is also during this stage that the organism invades the placenta and infects the fetus. Most of the abortions due to lep-

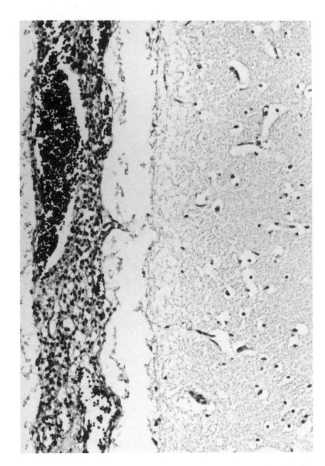

Fig. 4.53 *Leptospira* sp. Nonsuppurative meningitis. Bovine fetus.

tospirosis occur late in gestation, but stillbirths, sometimes with a few partially mummified fetuses, and births of live but weak pigs are also part of the pattern. Infected fetuses often die *in utero* and undergo autolysis, which obscures any lesions. Some aborted fetuses have elevated immunoglobulin levels. The placentas of infected pigs may be edematous, but the most severe lesions are seen in those piglets delivered alive, but sick, at or near term. Some of these will be icteric, have patchy hepatic necrosis, fluid and fibrin in the peritoneal cavity, and inflammatory changes in multiple organs. Focal aggregations of lymphocytes in the renal cortex, medulla, and pelvis are characteristic but inconstant. Leptospires can be demonstrated in the kidneys or lungs on section using a silver stain (see The Urinary System, Volume 2, Chapter 5). Special stains are much more useful on tissue sections taken from swine than from cattle; however, the fluorescent antibody test for the organism on kidney, lung, and placenta homogenates gives a more definitive answer. In addition, leptospires may frequently be demonstrated by dark-field microscopy.

iii. *Horses* Leptospires have not generally been considered to be a major cause of reproductive failure in horses; however, in Ireland, where tissues from 22 aborted equine fetuses were cultured and examined for lepto-spires, nine fetuses were found to contain organisms representing four serogroups: australis, hebdomadis, ictero-hemorrhagiae, and pomona. Serovars *bratislava* and *pomona* are considered to be of major importance in horses, and in one large group of horses tested in Ireland, most were found to carry antibody to *bratislava*. For the serovar *bratislava*, carrier horses generally do not have antibody to the infection.

Aborted foals are usually in the last trimester of pregnancy, and gross and microscopic lesions are seldom present. Gross lesions occasionally observed in aborted foals include icterus and petechial hemorrhages on mucous membranes. Histopathology may reveal dissociation of hepatocytes, multifocal hepatic necrosis, and renal tubular necrosis. Cerebral perivascular hemorrhage, thrombosis, and severe meningitis with heavy neutrophil and macrophage infiltrates have been observed. Leptospires were identified as the cause of abortion in four aborted foals with giant-cell hepatitis. High titers of fetal antibody were detected in fetal fluid by the microscopic agglutination test and the organism was identified in impression smears of kidney and placenta by immunofluorescence.

iv. Sheep A variety of serovars including *bratislava*, *ballum*, and *pomona* will cause abortion and stillbirths in sheep; however, the serovar of most importance is *hardjo*, and with this serovar there seems to be a sheep reservoir independent of the cattle reservoir. Sheep seem somewhat resistant, but have a vulnerable period 2 weeks before lambing and for 1 week after. With different serovars, sheep are usually infected from an environment contaminated by other animals. Leptospirosis is generally less common in sheep than in horses, cattle, and pigs; the grazing lands for sheep are usually arid and therefore do not support leptospires; and the drinking habits of sheep are more fastidious, in that they do not normally drink out of puddles.

Bibliography

Amatredjo, A., and Campbell, R. S. F. Bovine leptospirosis. *Vet Bull* **43:** 875–890, 1975.

Ellis, W. A. Leptospirosis. *J Small Anim Pract* **27:** 617–731, 1986

Ellis, W. A., and Michna, S. W. Bovine leptospirosis: Experimental infection of pregnant heifers with a strain belonging to the hebdomadis serogroup. *Res Vet Sci* **22:** 229–236, 1977.

Ellis, W. A. *et al.* Bovine leptospirosis: Microbiological and serological findings in aborted fetuses. *Vet Rec* **110:** 147–150, 1982.

Ellis, W. A. *et al.* Bovine leptospirosis: Serological findings in aborting cows. *Vet Rec* **110:** 178–180, 1982.

Ellis, W. A. *et al.* Leptospiral infection in aborted equine foetuses. *Equine Vet J* **15:** 321–324, 1983.

Hathaway, S. C., and Little, T. W. A. Epidemiological study of *Leptospira hardjo* infection in second calf dairy cows. *Vet Rec* **112:** 215–218, 1983.

Johnson, R. H., Allan, P. J., and Dennett, D. P. Association of *Leptospira hardjo* with abortion in heifers. *Aust Vet J* **50:** 325–326, 1974.

Kirkbride, C. A. "Laboratory Diagnosis of Livestock Abortion." 3rd Ed. Ames, Iowa, Iowa State University Press, 1990.

Logan, E. F. *et al.* Immunoglobulins in fluids of aborted and unaborted bovine fetuses. *Res Vet Sci* **31:** 161–163, 1981.

Murphy, J. C., and Jensen, R. Experimental pathogenesis of leptospiral abortion in cattle. *Am J Vet Res* **30:** 703–713, 1969.

f. *Ureaplasma diversum* Infections *Ureaplasma diversum* (formerly T mycoplasma) is an important cause of reproductive failure in cattle and ranks second only to bovine virus diarrhea virus as a cause of abortion identified by some laboratories. The true prevalence of abortion due to this organism is obscure, however, as few laboratories have the capability to grow the organism. *Ureaplasma diversum* is a bacterium in the family Mycoplasmataceae. The organism is frequently present on the mucous membranes of the nasal passages, vulva, and vagina of the cow, the sheath of bulls, and in semen and embryo transfer fluids. It remains in these sites in the animal for long periods and can be found in urine, and vaginal and nasal discharges. It may be transmitted to cows (or bulls) during breeding; and as the organism is preserved by freezing, it remains viable in diluted frozen semen.

Virulent strains of *U. diversum* may produce vulvitis, embryonic death, abortion, or the birth of dead or weak

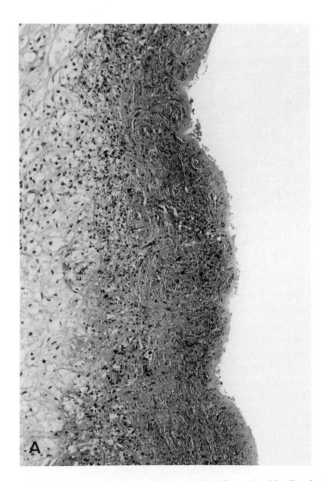

Fig. 4.54A *Ureaplasma diversum* chronic amnionitis. Bovine fetus.

calves at term. Epidemics of abortion are rare in chronically infected herds, but may occur where large groups of naive animals are congregated. Abortion usually occurs in the last trimester of gestation; or infection may manifest as premature delivery, with the birth of weak or dead calves near term. The fetal membranes may be retained. On gross examination the amnion is frequently the most severely affected portion of the placenta. It has patchy thickening with fibrosis and, in addition, may have foci of necrosis, hemorrhage, fibrin exudation, mineralization, and areas which are stained with meconium. The chorioallantois may be similarly affected, and if so, the lesion is often more severe on the allantoic surface. On histologic examination of the placenta, fibrosis and ongoing interstitial necrosis are extensive. The inflammatory cell component is mainly mononuclear with macrophages and plasma cells predominating (Fig. 4.54A). A mild arteritis is also present (Fig. 4.54B).

The fetus is usually well preserved and may be stained with meconium. The lungs are firm and, even if the calf is born alive, are poorly aerated. There is an erosive conjunctivitis with prominent goblet cell formation of the palpebral conjunctiva. Foci of lymphocytes, including plasma cells,

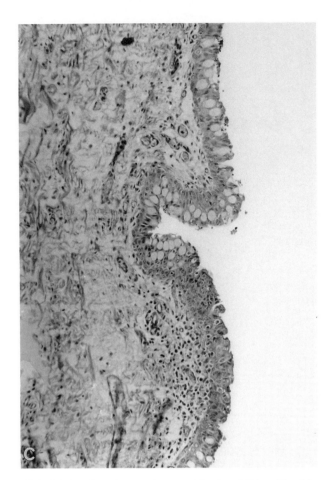

Fig. 4.54C *Ureaplasma diversum*. Conjunctivitis, with goblet cell metaplasia. Bovine fetus.

are present in the lamina propria (Fig. 4.54C). The lungs have a nonsuppurative alveolitis, and periairway lymphocytic, mononuclear cell infiltrates may be prominent (Fig. 4.54D). The lesions in the amnion, chorioallantois, and lung are characteristic but not pathognomonic and require validation by positive culture from placenta, stomach contents, or lung.

Bibliography

Doig, P. A. *et al*. Bovine granular vulvitis associated with ureaplasma infection. *Can Vet J* **20:** 89–94, 1979.

Miller, R. B. *et al*. The effects of *Ureaplasma diversum* inoculated into the amniotic cavity in cows. *Theriogenology* **21:** 367–374, 1983.

Ruhnke, H. L. *et al*. Bovine abortion and neonatal death associated with *Ureaplasma diversum*. *Theriogenology* **21:** 295–301, 1984.

g. *ACTINOMYCES PYOGENES* INFECTIONS *Actinomyces pyogenes* (*Corynebacterium pyogenes*) is widespread throughout the world as a common cause of pyogenic infection in a variety of domestic animals. The organism is commonly isolated from aborted fetuses and is fairly consistently diagnosed in from 1 to 15% of bovine fetuses

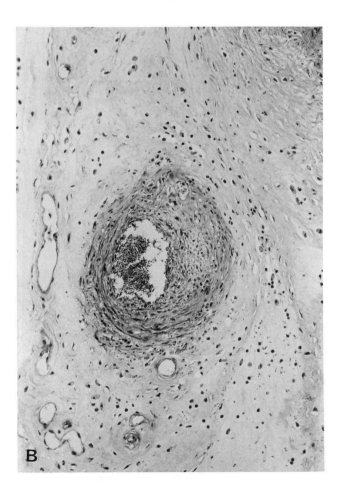

Fig. 4.54B *Ureaplasma diversum*. Placental vasculitis.

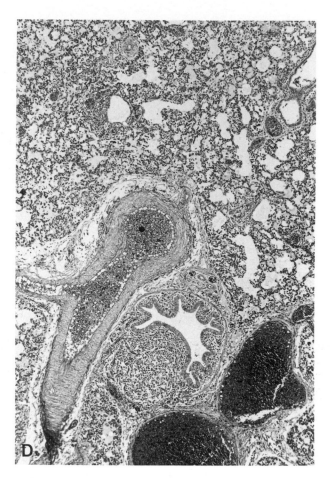

Fig. 4.54D *Ureaplasma diversum.* Nonsuppurative alveolitis and periairway lymphocyte accumulation. Bovine fetus.

Fig. 4.55A *Actinomyces pyogenes.* Suppurative placentitis. Bovine fetus.

submitted. It also causes abortion in sheep, but this is diagnosed less often. It is believed to be a primary pathogen in the cause of abortion.

The organism is a common contaminant on the mucous membranes and is found in the tonsils and vagina of healthy cattle. It is suspected the organism may penetrate the bloodstream from a mucosal surface, resulting in a transient bacteremia which, in a pregnant animal, can result in placental localization and abortion. Abortion is usually a single event in a herd or flock but is occasionally multiple. Abortion may occur at any stage of gestation but is usually in the last trimester. The dam may be ill following abortion, and some die with a suppurative endometritis, arthritis, or mastitis.

The fetus may be severely autolyzed or well preserved, and gross lesions are visible in the placenta, trachea, and lung. The placenta is often retained and when submitted may be heavily contaminated. Gross examination of portions submitted reveal marked autolysis and a suppurative placentitis with yellow to brown exudate over swollen edematous cotyledons (Fig. 4.55A). In some fetuses a characteristic hemorrhagic cast is present in the trachea (Fig. 4.55B). The lung is dark red and swollen with minute yellow foci visible on the pleural surface.

Lesions may be modest, with large numbers of bacteria but no inflammatory cells in placenta and lung or severe with a necrotizing, suppurative placentitis and an acute fibrinous bronchopneumonia. Bacteria may colonize in vessels, on the skin surface, and conjunctiva with destruction of the epithelium.

The diagnosis may be made on seeing the typical lesions and isolating the organism preferably from the placenta, lung, and stomach contents. In the absence of lesions, the diagnosis can be made with less confidence. It has been suggested that the young fetus may be unable to respond to the organism.

h. *SALMONELLA* INFECTIONS The bacterium *Salmonella* is in the family Enterobacteriaceae. It is a facultative anaerobe, a gram-negative rod, and usually motile. There are over 1300 serotypes in the genus *Salmonella*. Some of these are auxotrophic and have an affinity for just one host, whereas others are ubiquitous and found in many animals. *Salmonella typhimurium* is ubiquitous, whereas *S. abortus-ovis* is most prevalent in sheep and therefore considered auxotrophic. Many of the serotypes causing abortion in livestock are ubiquitous. Salmonellosis is of worldwide concern as a human disease and causes major

Fig. 4.55B *Actinomyces pyogenes* abortion. Hemorrhagic cast in the trachea. Bovine fetus.

losses in calves. It is also an important cause of abortion in both cattle and sheep and, in some situations, in mares.

The organism may be carried by the host without clinical disease. It is carried by many animals and birds, including cattle, sheep, pigs, horses, mice, rats, and sea gulls, and is excreted in the saliva, milk, feces, urine, and in the fluids discharged when animals abort. Animals may become infected through their feed, water, bedding, or the equipment used in feeding, handling, and cleaning. Abortions are often precipitated by a stressful event such as transport, change in diet, or feeding spoiled feed. It is not completely defined how these events precipitate the disease, but it may be partly due to disturbance of the normal intestinal flora. Whereas specific nutrients in the placenta which enhance the growth of the bacterium have not been demonstrated, *in vivo* experiments have shown marked growth enhancement following inoculation of whole placental extracts into mice.

The organism may stay in a herd for several years, and abortions may be sporadic or epizootic, usually occurring in late gestation. Abortion can occur without illness in the dam or may follow within a week of clinical signs. Cattle may suffer with a variety of clinical manifestations, and according to one report, the main manifestation exhibited

by *S. dublin* infections varied from pneumonia, to abortion and enteritis, to polyarthritis, in successive years. Signs of enteritis and abortion appear to be commonly associated. Sheep may have a mild upset with fever and anorexia shortly before aborting.

The following pathogenesis of placental infection has been proposed. The organism first localizes in the intestine, and this is followed by a brief bacteremia with localization in the lymph nodes, spleen, and lung. After a further period of growth, a second bacteremia occurs in which the placentome becomes infected. Excessive growth of the organism leads to almost complete destruction of the fetal villi, which is followed by abortion and may occur without invasion of the fetus. If *S. dublin* infection occurs within a month of term, the calf may be born alive, carrying the organism.

Following abortion the placenta is often retained, and the fetus is usually autolyzed. The chorioallantois is thickened with amber fibrin-containing fluid, the chorionic surface is diffusely gray to red, and there is a yellow exudate on the variably tan and red cotyledons. Portions of the caruncle may remain adherent to the cotyledon. There is mineralization of individual trophoblast cells, interstitium of the villi, and chorioallantoic arcade. When the chorion is severely affected, mineralization may also be extensive on the allantoic surface. Accompanying the mineral, exuberant bacterial growth expands the villi. In some villi a heavy neutrophil infiltrate is present, and if the caruncle remains attached to the cotyledon, there is a clear line of demarcation between the affected villus within the maternal crypt and the near normal maternal septa (Fig. 4.56A). Dilated capillaries, immediately under the sloughed trophoblast cells in the arcade zone, are stuffed with bacteria and in cross section resemble large rounded trophoblast cells filled with bacteria (Fig. 4.56B). In spite of the severe placentitis, there may be no lesions in the lung; however, when present, they consist of a moderate to light accumulation of neutrophils in bronchi, and bacteria may be seen colonizing airway epithelium. In the liver there can be a multifocal suppurative hepatitis.

i. *YERSINIA* INFECTIONS *Yersinia pseudotuberculosis* is a gram-negative coccobacillus in the family Enterobacteriaceae. The DNA of *Yersinia pestis*, the cause of human plague, and *Y. pseudotuberculosis* are at least 90% interrelated. Six serogroups of *Y. pseudotuberculosis* are recognized, based on the immunoreactivity of the O antigens. *Yersinia pseudotuberculosis* is the cause of ileitis, lymphadenitis, and abscessation in humans and infects a wide variety of domestic, laboratory, and native animals and birds throughout the world. The organism is frequently isolated from the feces of normal cattle, and from abortions in sheep, goats, and cattle. In sheep the organism also causes abdominal abscessation, enteritis, and inflammation in the testis and epididymis.

The pathogenesis of abortion probably follows invasion of the intestinal epithelium, transient bacteremia, and localization in the maternal caruncle, followed by passage

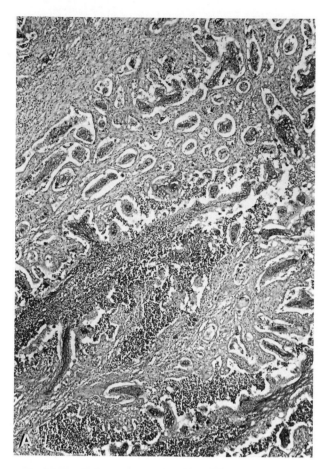

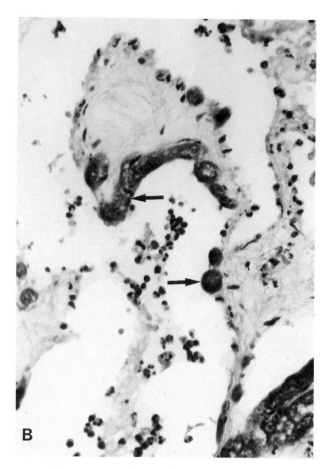

Fig. 4.56A *Salmonella* abortion. Cow. Placentome, crypt region. Acute suppurative placentitis, with complete destruction of villi, adjacent caruncle near normal.

Fig. 4.56B *Salmonella* abortion. Cow. Chorionic arcade region, absence of trophoblast cells. Bacteria fill vessels (arrows).

to the chorioallantois and fetus. It is possible that nutrients in the caruncle favor localization and growth of *Yersinia*, which is followed by villus infarction and invasion of the chorion.

Limited observations suggest that the lesions are generally similar in aborted fetuses and stillborns from sheep, goats, and cattle. The aborted conceptus is usually well preserved and contains a variety of lesions. The placentitis is largely confined to the cotyledons. Cotyledons are red or tan, thickened, and relatively devoid of exudate on the surface. Portions of the caruncle may remain attached. The intercotyledonary region is usually translucent, but a frosting of fibrin and some fibrosis may closely surround the affected cotyledons (Fig. 4.57). The thoracic and abdominal cavities contain excess amber fluid with some fibrin. In the fetal liver there may be pale tan focal areas of necrosis ranging in size from 0.1 to 1 cm. No other gross lesions are observed.

Histologically in the placenta, there is necrosis of villi and a moderate infiltration of granulocytes, macrophages, and mononuclear cells in the interstitium of the chorioallantoic arcade. In placental vessels there is fibrinoid necrosis of the media accompanied by mononuclear cell and

neutrophil infiltrates. Lesions in the attached portions of caruncle consist of thrombosis of septal vessels, hemorrhage, and necrosis, with a heavy diffuse infiltration of neutrophils and mononuclear cells. In the lamina propria of the conjunctiva there are focal infiltrations of mononuclear cells, plasma cells, and a few neutrophils (Fig. 4.58A). In the lung there are a few mononuclear cells in airways. Hepatic lesions consist of focal areas of necrosis infiltrated by granulocytes and mononuclear cells. Similar foci are visible on microscopic examination of myocardium and lymph nodes. Peyer's patches are precociously cellular. The lamina propria and submucosa of the colon is heavily populated by plasma cells and mononuclear cells, and within the lumen, colonies of the organism and scattered inflammatory cells can be seen in the meconium (Fig. 4.58B).

Extensive lesions in the uterus follow the experimental production of abortion in sheep. They consist of thrombosis of septal vessels, necrosis of septa, and heavy infiltrations of granulocytes in caruncles and intercaruncular regions. Diagnosis is based on recovery of the organism from placenta, stomach contents, or lung and the lesions described.

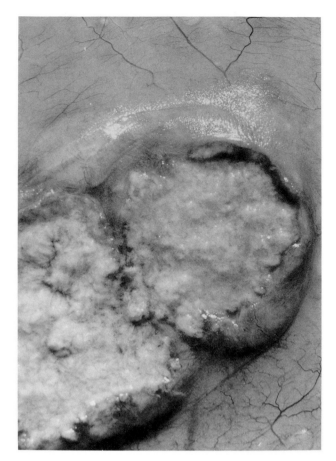

Fig. 4.57 *Yersinia pseudotuberculosis* abortion. Cow. Chronic suppurative cotyledonary placentitis and pericotyledonary fibrin.

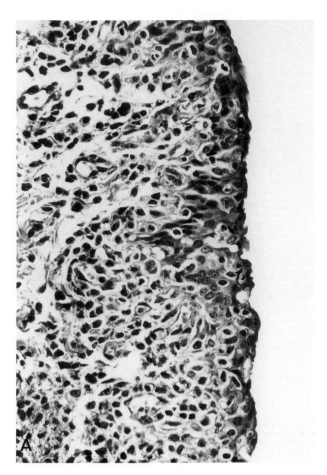

Fig. 4.58A *Yersinia pseudotuberculosis.* Nonsuppurative conjunctivitis in aborted calf.

j. HAEMOPHILUS INFECTIONS *Haemophilus somnus* is provisionally classified with the genus *Haemophilus*, but by the criteria currently in use for this genus does not belong there. The organism is a gram-negative coccobacillus. It is closely related to *Histophilus ovis* and *Haemophilus agni*.

Haemophilus somnus is best known for its association with central nervous system disease and pneumonia in cattle. There are marked differences in the ability of different isolates from the reproductive tract to produce central nervous system disease. Preliminary experiments have not shown the reverse to be true (see Bones and Joints, Volume 1, Chapter 1 and The Nervous System, Volume 1, Chapter 3). *Haemophilus somnus* is widespread throughout the world and is a common inhabitant of the vagina in cows and under certain conditions produces vaginitis, infertility, abortion, and a fatal endometritis. Usually abortions induced by *H. somnus* involve solitary animals; however, occasionally an entire herd will develop reproductive disease, including vaginitis and infertility attributable to this organism. When this occurs, predisposing factors, in the form of malnutrition, concurrent virus infection, or stress, should be considered.

The organism is commonly resident in the vagina of cows (3–76% of animals sampled) and the prepuce of bulls (more than 75% are positive). It is therefore easily spread by natural breeding, but most strains of the organism are sensitive to the antibiotics used in preparation of frozen semen. Sampling of a naturally infected herd can reveal more than 50% of calves infected by 7 months of age. The presence of the organism in the vagina does not imply clinical disease.

The pathogenesis of abortion has not been clarified. The organism shows some ability to penetrate the cervical area to infect the placenta; however, lesions in the placenta are usually in cotyledons, which suggests a hematogenous route. A bacteremia following a vaginal or respiratory infection with uterine localization is likely. Death of the fetus after infection is usually rapid, resulting in the abortion of a severely autolyzed fetus.

Lesions in the placenta are primarily confined to the cotyledon and consist of an acute nonsuppurative placentitis, which does not appear to penetrate to either the amnion or allantoic surface. The most distinctive lesion in the placenta consists of a severe fibrinoid necrosis of large and small arteries and thrombosis (Fig. 4.59). The cellular

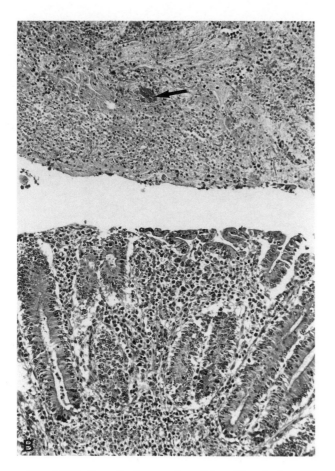

Fig. 4.58B Lymphoplasmacytic colitis, with colonies of *Y. pseudotuberculosis* in meconium (arrow).

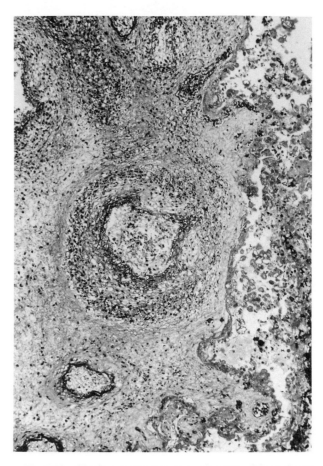

Fig. 4.59 Chorion, acute nonsuppurative vasculitis, and cotyledonary placentitis. *Haemophilus somnus*. Cow.

infiltrate in vessel walls and adjacent villi is mainly of macrophages with some neutrophils. There is sparing of veins and capillaries. Colonies of bacteria are visible next to the trophoblast cells. Lesions in the fetus are usually sparse; however, an acute fibrinous bronchopneumonia is occasionally observed. Airways contain degenerating mononuclear cells, macrophages, but few neutrophils. Bacteria are frequently observed adjacent to the epithelium of alveolar ducts, and swirling (oat-type) macrophages are numerous in these areas. Fibrin often fills lymphatics distending interlobular septa.

Bibliography

Addo, P. B., and Dennis, S. M. Experimental production of *Corynebacterium pyogenes* abortion in sheep. *Cornell Vet* **69:** 20–32, 1979.

Bullians, J. A. *Yersinia* species infection of lambs and cull cows at an abattoir. *N Z Vet J* **35:** 65–67, 1986.

Hall, G. A., and Jones, P. W. An experimental study of *Salmonella dublin* abortion in cattle. *Br Vet J* **132:** 60–65, 1976.

Hall, G. A., and Jones, P. W. A study of the pathogenesis of experimental *Salmonella dublin* abortion in cattle. *J Comp Pathol* **87:** 53–65, 1977.

Hinton, M. Salmonella abortion in cattle. *Vet Bull* **41:** 973–980, 1971.

Hinton, M. Bovine abortion associated with *Corynebacterium pyogenes*. *Vet Bull* **42:** 753–756, 1972.

Hodges, R. T., and Carman, G. Recovery of *Yersinia pseudotuberculosis* from the faeces of healthy cattle. *N Z Vet J* **33:** 175–176, 1985.

Humphrey, J. D., and Stephens, L. R. *Haemophilus somnus:* A review. *Vet Bull* **53:** 987–1004, 1983.

Jerret, I. V., and Slee, K. J. Bovine abortion associated with *Yersinia pseudotuberculosis* infection. *Vet Pathol* **26:** 181–183, 1989.

Karbe, E., and Erickson, E. D. Ovine abortion and stillbirth due to purulent placentitis caused by *Yersinia pseudotuberculosis*. *Vet Pathol* **21:** 601–606, 1984.

Kwiecien, J. M., and Little, P. B. *Haemophilus somnus* and reproductive disease in the cow: A review. *Can Vet J* **32:** 595–601, 1991.

Miller, R. B., and Barnum, D. A. Effects of *Hemophilus somnus* on the pregnant bovine reproductive tract and conceptus following cervical infusion. *Vet Pathol* **20:** 584–589, 1983.

Richardson, A. Salmonellosis in cattle. *Vet Rec* **96:** 329–331, 1975.

Schiefer, B. *et al.* The pathology of bovine abortion due to *Corynebacterium pyogenes*. *Can Vet J* **15:** 322–326, 1974.

Singh, I. P. *et al.* Some aspects of the epidemiology of *Salmonella abortus-equi* infection in equines. *Br Vet J* **127:** 378–383, 1971.

Smith, R. E. *et al.* Fetoplacental effects of *Corynebacterium pyogenes* in sheep. *Cornell Vet* **61:** 573–590, 1971.

Sorensen, G. H. Studies on the occurrence of *Peptococcus indolicus* and *Corynebacterium pyogenes* in apparently healthy cattle. *Acta Vet Scand* **17:** 15–24, 1976.

Witte, S. T., and Collins, T. C. Abortion and early neonatal death of kids attributed to intrauterine *Yersinia pseudotuberculosis* infection. *J Am Vet Med Assoc* **187:** 834, 1985.

Wray, C., and Corbel, M. J. Observations on the effect of extracts of bovine placenta on the growth of salmonella *in vitro* and *in vivo. Br Vet J* **136:** 39–43, 1980.

k. CHLAMYDIAL INFECTIONS   Chlamydiae are obligate intracellular parasites which multiply in membrane-bound vacuoles in a variety of cells. They have two distinct forms, the reticulate body, which is not infectious, and the elementary body, which is infectious and released from the cytoplasm by a mechanism not fully understood, but which involves disruption of the cell. The inclusions which can be seen on light microscopy consist of hundreds of gram-negative chlamydiae organisms bound by a membrane in a cytoplasmic vacuole. Three species in the genus *Chlamydia* are currently defined; *C. trachomatis, C. pneumoniae,* and *C. psittaci*. In humans, strains of *C. trachoma* cause disease in the eye, genital and respiratory tracts and a strain of *C. pneumoniae* causes severe respiratory disease. *Chlamydia psittaci* causes disease in animals and birds and is an important pathogen in humans. There are marked differences between different strains of *C. psittaci* and the species has been divided into nine serovars. The strains of *Chlamydia psittaci* causing reproductive failure

and intestinal infections in ruminants appear to be similar antigenically and are classified as immunotype 1. Strains causing arthritis, conjunctivitis, and encephalitis in ruminants are different and classified as immunotype 2. Strains isolated from pigs are also different from those infecting cattle and sheep. Further classification using major outer membrane proteins will attempt to relate structure to function. Antibodies developed against these components are being used to identify important functional differences between strains and will be of use in pathogenesis and immunization studies and diagnosis.

Chlamydia psittaci is an important cause of *in utero* infections in sheep and goats, resulting in abortion, stillbirth, and the birth of weak offspring. The disease in sheep has been known variously as ovine enzootic abortion, and enzootic abortion of ewes, but will be referred to here as chlamydial abortion.

The organism is transmitted orally by contamination of foodstuffs with feces or the products of abortion containing the organism. On entry into sheep and goats, *C. psittaci* may infect the intestine permanently, and from there invade the bloodstream and infect the placenta and frequently the fetus in pregnant animals. Naive animals,

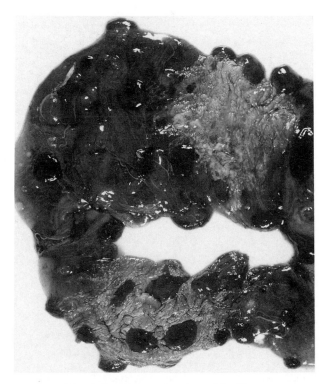

Fig. 4.60   Chlamydial placentitis. Sheep.

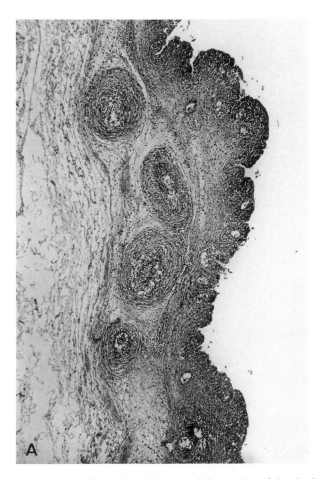

Fig. 4.61A   Placentitis with severe inflammation of the chorionic surface and deeper vasculitis. *Chlamydia psittaci*. Sheep.

including newly introduced sheep and females pregnant for the first time, are most vulnerable. The infection is chronic and abortion usually occurs in the final trimester of pregnancy, or infection may result in stillbirths or weak lambs. Extensive losses (30%) may occur on introduction of the organism to a flock, and yearly rates of 5% are common when it exists in the enzootic form.

Retention of the fetal membranes occurs in some animals and aborting females occasionally may be ill. The fetal placenta resembles that seen in bovine brucellosis, and there is patchy, almost equal involvement of cotyledons and intercotyledonary regions. Affected cotyledons are a dull-clay or dark-red color, firm, and matted with a dirty red exudate. The intercotyledonary region is red to brown, and has irregular patches of edema amidst a dry, leathery thickening of the chorioallantois (Fig. 4.60). These thickenings, often concentrated in ridges, consist of medium to large vessels, prominent because of a marked vaculitis. The vasculitis is characterized by fibrinoid necrosis of the media with a moderate to heavy infiltrate of neutrophils and some mononuclear cells (Fig. 4.61A,B). In the chorion there is superficial necrosis and characteristic inclusions can be seen in the cytoplasm of trophoblast cells.

The fetus is usually well preserved and may have few gross lesions. These consist of scattered hemorrhages in the subcutis, thymus, lymph nodes, and muscles. The liver is sometimes swollen and has pinpoint yellow foci on the surface. Histologic lesions in the fetus are modest but characteristic. There are focal areas of coagulation necrosis in the liver and spleen. These foci consist of as few as five cells, stain pale pink, and are variably surrounded by a scant infiltrate of mononuclear cells (Fig. 4.62A,B). Throughout the liver but concentrated in portal areas, there is an increase in mononuclear cells. In the lung, alveolar septa are thickened by a mononuclear cell infiltrate. A mild meningoencephalitis with vasculitis and hemorrhage has also been reported.

The situation is less clear in cattle. Most of the evidence incriminating *Chlamydia* as a cause of abortion in cattle was developed in the course of investigation of epizootic bovine abortion, the enzootic tick-borne disease that causes abortion in the California foothills. It now appears that chlamydial organisms are not the cause of epizootic bovine abortion. Cattle do carry the organism in the intestine, as do sheep, and the organism in the intestine can experimentally produce abortion. It is reported that abor-

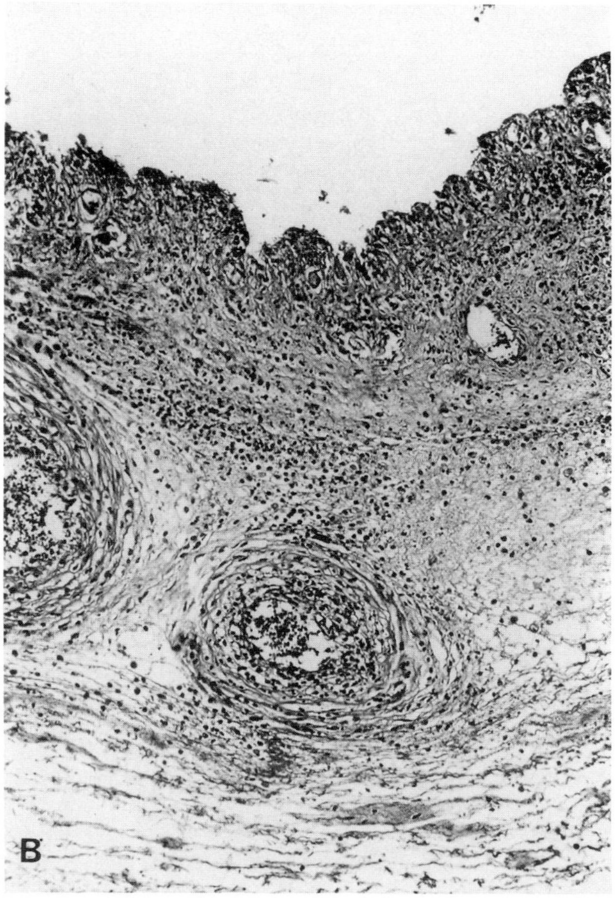

Fig. 4.61B Detail of placentitis and vasculitis in Fig. 4.61A.

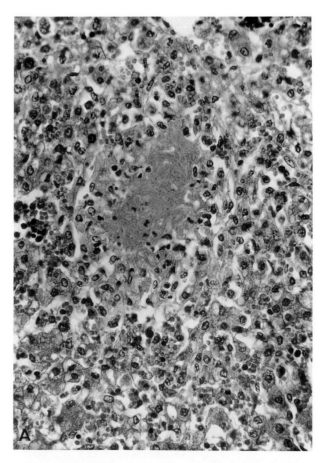

Fig. 4.62A *Chlamydia psittaci* abortion. Focal coagulation necrosis in liver.

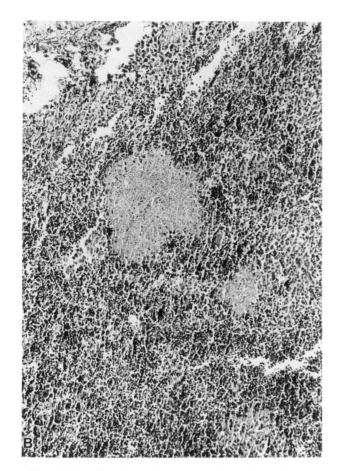

Fig. 4.62B Focal coagulation necrosis in spleen. Sheep.

tions in cattle are usually sporadic, occurring throughout the year with occasional outbreaks in isolated herds.

In order to establish *Chlamydia* as the cause of an abortion in cattle or sheep, the organism should be demonstrated in a smear (preferably of placenta) stained by a modified Ziehl–Neelsen, Gimenez, or Giemsa method. The organism can also be demonstrated in sections from the diseased placenta using a specific fluorescence antibody test or immunoperoxidase procedure. The presence of specific antibody in the fetus is confirmative, but antibody is not always present.

Bibliography

Blewett, D. A. *et al.* Ovine enzootic abortion: The acquisition of infection and consequent abortion within a single lambing season. *Vet Rec* **111:** 499–501, 1982.

Boulanger, P., and Bannister, G. L. Abortion produced experimentally in cattle with an agent of the psittacosis–lymphogranuloma–venereum group of viruses. *Can J Comp Med* **23:** 259–265, 1959.

McEwen, A. D., and Foggie, A. Enzootic abortion in ewes. A note on the susceptibility of the pregnant bovine to infection and abortion. *Vet Rec* **67:** 373, 1955.

Puy, H. *et al.* Immunological specificity of monoclonal antibodies to *Chlamydia psittaci* ovine abortion strain. *Immunol Lett* **23:** 217–222, 1989/1990.

Rodolakis, A. *et al. Chlamydia psittaci* experimental abortion in goats. *Am J Vet Res* **45:** 2086–2089, 1984.

Schoop, G., and Kauker, E. Infektion eines Rinderbestandes durch ein Virus der Psittakosis–Lymphogranuloma–Gruppe gehaufte Aborte im Verlauf der Erkrankungen. *Dtsch Tierartzl Wochenschr* **63:** 233–235, 1956.

Stamp, J. T. *et al.* Enzootic abortion in ewes. 1. Transmission of the disease. *Vet Rec* **62:** 251–255, 1950.

Storz, J., and McKercher, D. G. Etiological studies on epizootic bovine abortion. *Zentralbl Veterinaermed* **9:** 411–427, 520–541, 1962.

Studdert, M. J. Bedsonia abortion of sheep. II. Pathology and pathogenesis with observations on the normal ovine placenta. *Res Vet Sci* **9:** 57–64, 1968.

Su, H. *et al. Chlamydia trachomatis*–host cell interactions: Role of the chlamydial major outer membrane protein as an adhesin. *Infect Immun* **58:** 1017–1025, 1990.

l. *COXIELLA* INFECTIONS *Coxiella burnetii* is in the family Rickettsiaceae and is the only member of the genus *Coxiella*. The organism is well known as the cause of Q (Query) fever in humans. It is an intracellular parasite. Dairy cows, goats, and sheep are the common reservoirs of the organism, and some forms of wildlife are also reservoirs. Human and domestic animal infection occurs when

Fig. 4.63 *Coxiella burnetii* abortion. Suppurative intercotyledonitis. Goat.

contaminated dust or fluids are inhaled; however, massive contamination of a paddock or pasture facilitates oral transmission. Infection apparently persists indefinitely in sheep and cattle, and organisms are shed at parturition and in the milk. Abortion usually follows initial exposure and may be unusual in endemically infected flocks.

Abortion tends to occur late in the gestation period, and weak lambs and kids may be born during an outbreak. The aborted fetus may be well preserved or autolyzed. Gross lesions are confined to the placenta, which is thickened and leathery with multifocal to confluent areas of mineralization. The exudate is copious, off-white, and most obvious in the intercotyledonary region (Fig. 4.63). If the cotyledon is involved, the early lesion appears as a white outer ring with flecks of white scattered in the central region. Inflammation in the placenta is an acute diffuse suppurative placentitis with heavy neutrophil infiltration and extensive necrosis of cotyledonary villi and intercotyledonary epithelium (Fig. 4.64A). In contrast, the cellular infiltrate in the interstitium of the chorioallantoic arcade is largely mononuclear with a predominance of plasma cells. The lesion, which on gross examination appears to be largely confined to the intercotyledonary region, is

frequently found histologically to be also extensive in the cotyledons, especially at the periphery. The vasculitis which develops with chlamydial infections is not usually a feature of placentitis due to *Coxiella*. Smears of the placental exudate contain large numbers of organisms which can be stained by the modified Ziehl–Neelsen or Macchiavello's stain. Invasion of the cells of the chorion by the organism produces microcolonies which distend the cytoplasm of the infected cells (Fig. 4.64B). These microcolonies can be seen in conventionally stained sections, but must be distinguished from similar colonies that chlamydial organisms, and sometimes bacteria, form. With hematoxylin and eosin stains, chlamydial inclusions tend to stain poorly and appear more homogeneous, whereas with *Coxiella*, cells containing the organism frequently have a characteristic foamy appearance with multiple unstained vacuoles within a pale blue cytoplasm. The nucleus, if visible, is usually pushed against the wall of the cell and assumes a crescent shape.

The histologic lesions in affected fetuses are usually modest. A granulomatous hepatitis and nonsuppurative pneumonia with occasional focal lymphoid accumulations around bronchioles has been described. A few lympho-

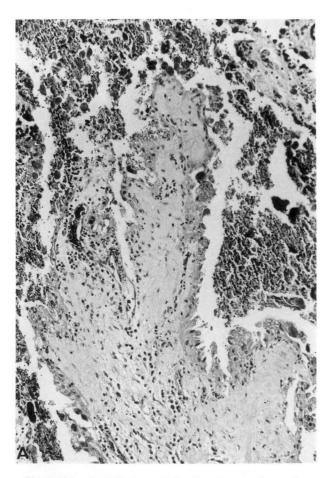

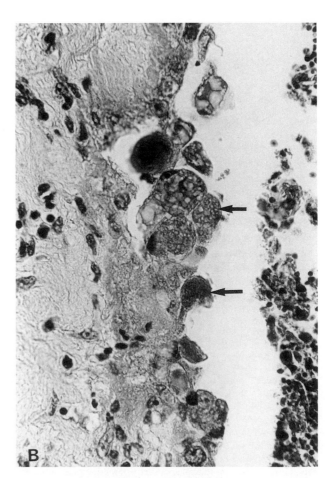

Fig. 4.64A *Coxiella burnetii* abortion. Suppuration and necrosis of placental villi, plasma cells in the interstitium. Sheep.

Fig. 4.64B *Coxiella burnetii* within trophoblast cells (arrows). Sheep.

cytes and macrophages may infiltrate the renal medulla and surround the portal vessels of the liver.

Bovine fetuses may also be aborted due to coxiella infections. The lesion in the placenta is characteristic and similar to that observed in sheep and goats. The simple presence of *Coxiella* in a placenta is not sufficient evidence to diagnose it as the cause of abortion. Animals may carry the organism for a prolonged period, and although it appears that they abort only once, there may be large numbers of organisms in the placenta in subsequent pregnancies. Assessment of the extent of the placental lesion may be helpful.

Bibliography

Crowther, R. W., and Spicer, A. J. Abortion in sheep and goats in Cyprus caused by *Coxiella burnetii*. *Vet Rec* **99:** 29–30, 1976.

Marmion, B. P., and Watson, W. A. Q fever and ovine abortion. *J Comp Pathol* **71:** 360–369, 1961.

Moore, J. D. *et al.* Pathology and diagnosis of *Coxiella burnetii* infection in a goat herd. *Vet Pathol* **28:** 82–84, 1991.

Palmer, N. C. *et al.* Placentitis and abortion in goats and sheep in Ontario caused by *Coxiella burnetii*. *Can Vet J* **24:** 60–61, 1983.

Roberts, W., Grist, N. R., and Giroud, P. Human abortion associated with infection by ovine abortion agent. *Br Med J* **4:** 37, 1967.

Tunnicliff, E. A. Ovine virus abortion. *J Am Vet Med Assoc* **136:** 132–134, 1960.

Waldhalm, D. G. *et al.* Abortion associated with *Coxiella burnetii* infection in dairy goats. *J Am Vet Med Assoc* **173:** 1580–1581, 1978.

m. MISCELLANEOUS INFECTIONS IN MARES That group of bacterial infection in foals producing the syndromes variously known as joint ill, navel ill, foal septicemia, and pyosepticemia neonatorum can also produce abortions and stillbirths. The implications of this are that these diseases of the newborn can be continuations of intrauterine disease. The organisms involved, in approximately the order of frequency, are β-hemolytic streptococci, *E. coli, Pseudomonas* spp., *Staphylococcus aureus, Klebsiella pneumoniae,* and *Actinobacillus equuli.* These organisms may occasionally reach the chorion by way of the bloodstream of the dam, but most ascend through a patent cervix. This produces an inflammatory thickening of the placenta adjacent to the internal os of the cervix (Fig. 4.65A). Fungal infections frequently have the same pathogenesis and produce similar placental lesion with a sticky brown exudate covering the surface of the affected areas of the chorion (Fig. 4.65B). *Aspergillus fumigatus* is the most commonly isolated fungus, but *Absidia corymbifera* and other *Aspergillus* spp. have been recovered.

Actinobacillus equuli is an important cause of neonatal death of foals in most countries. Foals may become infected *in utero* and be aborted, as discussed, or they may die of septicemia in the early neonatal period, or they may survive for several days and develop small abscesses in many tissues. It is probable that some of these infections

Fig. 4.65A Placental thickening following transcervical infection. *Streptococcus* sp. Mare.

are acquired during or after parturition. Aborted foals and those dying acutely of septicemia do not have distinctive lesions. After a course of 3–4 days or more, evidence of localization of the infection with miliary microabscessations and polyarthritis develops. The microabscesses, which are embolic in type, may be found in many organs, and they are readily visible to the naked eye in the renal cortices. The renal abscesses are small, being only 1–3 mm in diameter, and they are numerous. Histologically, bacterial colonies are obvious in the glomeruli and intertubular capillaries and are enclosed in foci of intense suppuration. The arthritis is of fibrinopurulent type.

There is little information on the epidemiology and pathogenesis of actinobacillosis. Mares which are caused to abort by this infection are not ill, and the organism does not persist for long in the uterus after abortion. However, the same mare may abort successive pregnancies, a fact which suggests that the uterus can be reinfected from some endogenous asymptomatic focus. *Actinobacillus equuli* can occasionally be found in the intestine and tissues of healthy animals, and occasionally also as an opportunist in lesions such as verminous thromboses. It is, however, seldom of pathologic significance in adult animals, although septicemic infections have been observed.

Bibliography

Platte, H. Etiological aspects of perinatal mortality in the thoroughbred. *Equine Vet J* **5:** 116–120, 1973.

Platte, H. Infections of the horse fetus. *J Reprod Fertil* (Suppl.) **23:** 605–610, 1975.

Fig. 4.66 Mycotic abortion. Ox. Mycotic dermatitis in a fetal calf.

Fig. 4.65B Mycotic placentitis following transcervical infection. Mare.

Prickett, M. E. Abortion and placental lesions in the mare. *J Am Vet Med Assoc* **157:** 1465–1470, 1970.

Swerczek, T. W. Equine fetal diseases. *In* "Current Therapy in Theriogenology 2," D. A. Morrow, (ed.), pp. 699–704. Philadelphia, Pennsylvania, W. B. Saunders, 1986.

n. MYCOTIC ABORTION IN CATTLE Sporadic cases of bovine abortion occur due to infection with species of the rather ubiquitous genera, *Aspergillus, Mortierella, Absidia, Mucor,* and *Rhizopus,* their relative frequency in about this order, but the order varies from area to area. The portal of entry is not known, but that the initial development of lesions is in placentomes indicates hematogenous arrival. Extension from either respiratory or rumen infections is possible. With the exception of *Rhizopus,* these organisms are well-known, secondary pathogens. Abortion occurs late in gestation, between the sixth and eighth month, and the placenta is often firmly retained.

The fetus may appear normal, but often there are characteristic cutaneous lesions in the form of irregular elevated plaques resembling ichthyosis or extensive ringworm. These lesions are seen most commonly about the

periorbit, occiput, shoulders, back, and sides (Fig. 4.66). Affected areas are slightly elevated, grayish, irregular in outline, and tend to coalesce. Histologically, the infection is seen to be superficial, involving the epidermis mainly, with parakeratosis, and edema and inflammatory-cell infiltration of the underlying dermis. Sometimes the hair follicles and the dermis are invaded by the fungus, but with the exception of an occasional case of bronchopneumonia or encephalitis, there are no lesions in the internal organs, although the fungus can be isolated from the stomach contents (Fig. 4.67). The placental lesions are remarkable. The gross appearance resembles that seen in brucellosis but is often much more severe. The allantois–chorion is then leathery with extensive superficial necrosis. The placentomes are greatly enlarged and necrotic with swollen margins, and are firmly incarcerated (Fig. 4.68A). The infecting organisms are readily demonstrated in the necrotic tissue and typically extend along the blood vessels to produce necrotizing vasculitis.

Lesser degrees of placentitis are probably quite common with the lesions restricted to placentomes as areas of hemorrhage, softening, and necrosis (Fig. 4.68B). The infection appears to begin in the placentomes with later spread to the interplacental space between the cotyledons, where a rich growth of hyphae occurs. With placentitis of lesser degree, there usually are no cutaneous lesions in the fetus, although the fungi can be found in the gastric contents. The endometrial lesions are less severe than

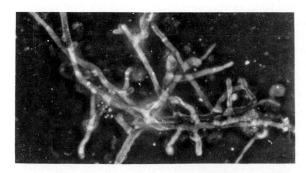

Fig. 4.67 Mycotic abortion. Ox. *Aspergillus* sp. in smear from fetal stomach (phase-contrast microscopy.)

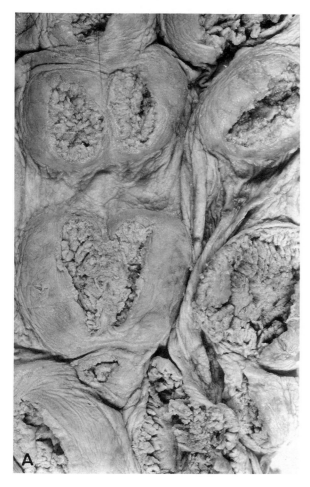

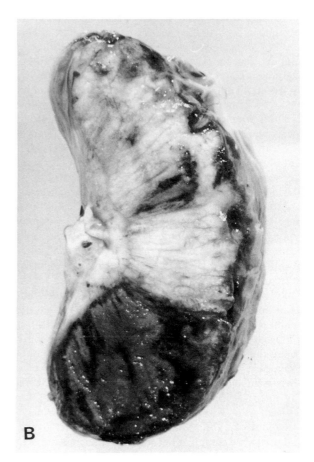

Fig. 4.68B Mycotic abortion. Ox. Infarction of placentome.

Fig. 4.68A Mycotic abortion. Ox. Mycotic placentitis.

those in the placenta. Secondary infections may follow retention of the placenta. The majority of cows recover sufficiently after abortion to allow subsequent pregnancies to be carried to term, but in some, endometrial destruction is severe. *Mortierella wolfii* may produce a fatal mycotic pneumonia in the dam following abortion.

Bibliography

Cordes, D. O., Carter, M. E., and di Menna, M. E. Mycotic pneumonia and placentitis caused by *Mortierella wolfii*. *Vet Pathol* **9:** 190–201, 1972.

Eustis, S. L. *et al.* Porcine abortions associated with fungi, actinomycetes and *Rhodococcus* sp. *Vet Pathol* **18:** 608–613, 1981.

Hill, M. W. M. *et al.* Pathogenesis of experimental bovine mycotic placentitis produced by *Aspergillus fumigatus*. *Vet Pathol* **8:** 175–192, 1971.

Hillman, R. B., and McEntee, K. Experimental studies in bovine mycotic placentitis. *Cornell Vet* **59:** 289–302, 1968.

Kirkbride, C. A. Diagnosis of mycotic abortion in cattle. *Proc 75th Mtg U.S. Anim Hlth Assoc,* Oklahoma City, Oklahoma 1971.

o. PROTOZOAN INFECTIONS

i. *Toxoplasmosis* *Toxoplasma gondii* infects a large range of hosts and has a worldwide distribution. Although *T. gondii* infections of the fetus are relatively common, the disease toxoplasmosis is still somewhat of an event. The present discussion is limited to genital manifestations. A more complete discussion of the disease is given with coccidial disease of the intestine (see The Alimentary System, Volume 2, Chapter 1).

The intrauterine transmission of infection has long been established in women and in a number of domestic species; but until New Zealand investigators demonstrated that the organism was an important cause of abortion in ewes, toxoplasmosis was not recognized as playing a significant part in the general scheme of abortifacient infections. In sheep and goats, infection during pregnancy frequently results in embryonic death, mummification, abortion, stillbirth, and neonatal death. Most infected fetuses are expelled close to the expected lambing period. Infection early in gestation may result in severe fetal injury, whereas ewes infected from 110 days onward usually lamb normally. The lamb may be congenitally infected but appear healthy. The size of the infecting dose and the route of infection may produce different effects on the fetus; but whereas strains of the organism differ, the differences do not appear to influence the pattern of disease. Ewes or does are unlikely to abort twice; however, stress with recrudescence of encysted *T. gondii* followed by a series of abortions is reported to have been related to the administration of a vaccine to prevent endotoxemia in sheep.

Animals that are particularly at risk are those moved during pregnancy to areas heavily contaminated with cat feces. Contamination of grain or hay with feces containing oocysts is a common route of transmission and occurs most often when ewes are brought in to be housed in barns. Under such circumstances, abortion rates can be high. Straw bedding contaminated with cat feces when spread on pasture can lead to infection of the dams and fetal disease. Only a small number of oocysts is necessary to produce fetal infection in ewes and does. Sheep-to-sheep infection is unlikely even when infected males breed noninfected females. Placentas and fetuses containing cysts may be consumed and thereby infect rodents, birds, and cats. If susceptible cats eat infected rodents and birds, they will excrete large numbers of oocysts, which can survive for long periods in the environment.

Following experimental inoculation of *T. gondii* into pregnant ewes, parasitemia occurs within 3–4 days; this coincides with a transient rise in the ewe's temperature, but little else is seen. During parasitemia the organism first passes to the caruncle, then to the trophoblast and fetus, arriving there 8–14 days following infection in the ewe. The organism multiplies in the caruncular septa, producing foci of necrosis. Infiltration of maternal neutrophils and mononuclear cells is extensive, and encysted organisms can frequently be found in the adjacent endometrium.

In the aborted fetal lamb or kid, gross lesions are seen in about half of the placentas and are almost invariably confined to the cotyledons. The usual lesion is a focal, 1-2 mm yellow-to-white area seen anywhere along the villus, from the tip to the base (Fig. 4.69). Lesions are more easily observed by placing the cotyledon in saline. In trophoblastic epithelium and villous stroma there are foci of necrosis with mineralization and a light to moderate mononuclear cell infiltrate. Degenerating *Toxoplasma* cysts and tachyzoites may rarely be seen in the trophoblastic cells. These are most easily distinguished away from the lesion as tissue necrosis also disturbs the conformation of the parasite. On tissue sections of fresh material, the tachyzoite is oval to rounded (2–4 μm in diameter) with a central nucleus.

Mild nonsuppurative encephalitis is the common lesion in fetuses aborted due to *T. gondii*; it is present in ~95% of brains of infected fetuses. These lesions consist of a few widely scattered foci of necrosis (some of which are mineralized), with infiltrates of glial and mononuclear cells. Focal areas of leukoencephalomalacia without cellular infiltrate may also be seen. The brain (particularly gray matter) is the most common site in which to find cysts of *Toxoplasma*. They are most profitably sought at the periphery of the lesion or beyond. In brain, the cysts tend to be round and up to 100 μm in diameter. Lesions consisting of multifocal necrotizing nonsuppurative hepatitis and pneumonia, and nonsuppurative myositis, myocarditis, and nephritis may also be present. In muscle,

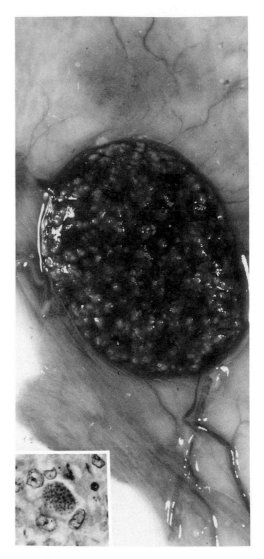

Fig. 4.69 Placentitis in toxoplasmosis. Sheep. Focal necrosis in cotyledon and normal intercotyledonary placenta. Inset: details of the organisms in cotyledon. (Courtesy of W. J. Hartley.)

cysts are usually spindle shaped. Lymph nodes in fetuses which have become immunocompetent are hyperplastic.

Toxoplasmosis in pigs is not recognized as a major cause of reproductive failure, but occasional outbreaks have been reported. Embryonic deaths may be associated with fever and illness in the sow rather than *in utero* infection. However, when sows are infected by the oral route in the late first and second trimesters, *T. gondii* is capable of infecting the conceptus, causing intrauterine death resulting in absorption, abortion, autolysis, and mummification. Gross lesions are rare. When present they consist of multiple white foci 1–3 mm in diameter in the heart, liver, and adrenals. Histologically, necrotizing lesions may be present in the endometrium, placenta, brain, spinal cord, heart, lung, liver, kidneys, skeletal muscle, tongue, and eye. Cysts and tachyzoites may be best observed in the brain and spinal cord and are less frequent in the placenta. Some fetuses, however, may be completely devoid of lesions, cysts, or tachyzoites. Congenitally acquired infections may not produce signs until some time after birth, when piglets manifest the infection with hepatitis and lymphadenitis.

The diagnosis of abortion due to *T. gondii* may be accomplished with varying degrees of certainty by a variety of methods that include the following: observing lesions and cysts, tissue examination for the detection of cysts using immunologic or immunohistochemical tests, testing for fetal antibody, and animal inoculation. A provisional diagnosis may be made on gross examination of the placenta, and histologic examination of placenta and fetal brain where, if typical lesions, cysts, and tachyzoites are observed, the diagnosis is probably reliable. Convenient, specific immunologically based methods are available for the detection of *T. gondii* antigens in tissue. These include the fluorescent antibody test, immunoperoxidase staining, avidin–biotin complexing, and the polymerase chain reaction. The polymerase chain reaction is highly sensitive and, as it can be used to identify specific fragments of DNA, may be used on severely autolyzed samples.

Cattle rarely if ever abort due to toxoplasmosis. Although the organism may be present in cattle, it does not cause disease.

Bibliography

Arthur, M. J., and Blewett, D. A. IFAT detection of IgG specific to *Toxoplasma* in thoracic fluids from aborted lambs; Evaluation on routine diagnostic submissions. *Vet Rec* **122:** 29–31, 1988.

Blewett, D. A., and Watson, W. A. The epidemiology of ovine toxoplasmosis. II. Possible sources of infection in outbreaks of clinical disease. *Br Vet J* **139:** 546–555, 1983.

Buxton, D. Ovine toxoplasmosis: A review. *J R Soc Med* **83:** 509–511, 1990.

Buxton, D., and Finlayson, J. Experimental infection of pregnant sheep with *Toxoplasma gondii;* pathological and immunological observations on the placenta and foetus. *J Comp Pathol* **96:** 320–333, 1986.

Buxton, D. *et al.* Perinatal changes in lambs infected with *Toxoplasma gondii. Res Vet Sci* **32:** 170–176, 1982.

Dubey, J. P. *Toxoplasma gondii* cysts in placentas of experimentally infected sheep. *Am J Vet Res* **48:** 352–353, 1987.

Dubey, J. P. Status of toxoplasmosis in sheep and goats in the United States. *J Am Vet Med Assoc* **196:** 259–262, 1990.

Dubey, J. P., and Beattie, C. P. "Toxoplasmosis of Animals and Man," Boca Raton, Florida, CRC Press, 1988.

Dubey, J. P., and Kirkbride, C. A. Toxoplasmosis and other causes of abortions in sheep from north central United States. *J Am Vet Med Assoc* **196:** 287–190, 1990.

Dubey, J. P., and Livingston, C. W. *Sarcocystis capracanis* and *Toxoplasma gondii* infections in range goats from Texas. *Am J Vet Res* **47:** 523–524, 1986.

Dubey J. P. *et al.* Serologic diagnosis of toxoplasmosis in experimentally infected pregnant goats and transplacentally infected kids. *Am J Vet Res* **46:** 1137–1140, 1985.

Dubey, J. P. *et al. Toxoplasma gondii*-induced abortion in dairy goats. *J Am Vet Med Assoc* **188:** 159–162, 1986.

Dubey, J. P. *et al.* Placental transfer of specific antibodies during ovine congenital toxoplasmosis. *Am J Vet Res* **48:** 474–476, 1987.

Dubey, J. P. *et al.* Serologic and histologic diagnosis of toxoplasmic abortions in sheep in Oregon. *J Am Vet Med Assoc* **196:** 291–294, 1990.

Dubey, J. P. *et al.* Lesions in fetal pigs with transplacentally-induced toxoplasmosis. *Vet Pathol* **27:** 411–418, 1990.

Frenkel, J. K. Toxoplasmosis: Parasite life cycle, pathology and immunology. *In* "The Coccidia," D. M. Hammond and P. L. Long (eds.), pp. 343–410. Baltimore, Maryland, University Park Press, 1973.

Greig, A. Toxoplasmosis in sheep. *Vet Ann* **30:** 85–91, 1990.

Hartley, W. J., and Bridge, P. S. A case of suspected congenital *Toxoplasma* encephalomyelitis in a lamb associated with a spinal cord anomaly. *Br Vet J* **131:** 380–384, 1975.

Hunter, D. *et al.* Examination of ovine foetal fluid for antibodies to *Toxoplasma gondii* by the Dye Test and an indirect immunofluorescence test specific for IgM. *Br Vet J* **138:** 29–34, 1982.

Munday, B. L. *et al.* Diagnosis of congenital toxoplasmosis in ovine foetuses. *Aust Vet J* **64:** 292, 1987.

Obendorf, D. L., Statham, P., and Munday, B. L. Resistance to *Toxoplasma* abortion in female goats previously exposed to *Toxoplasma* infection. *Aust Vet J* **67:** 233–234, 1990.

Poitras, B. J. *et al.* The maternal to fetal transfer of immunoglobulins associated with placenta lesions in sheep. *Can J Vet Res* **50:** 68–73, 1986.

Poncelet, J. L. Abortions in sheep following vaccination. *Bulletin de GTV Paris, France; Groupements Techniques Veterinaires* **6:** 41–42, 1989.

Uggla, A., Sjoland, L., and Dubey, J. P. Immunohistochemical diagnosis of toxoplasmosis in fetuses and fetal membranes of sheep. *Am J Vet Res* **48:** 348–351, 1987.

Wheeler, R. *et al.* Diagnosis of ovine toxoplasmosis using PCR. *Vet Rec* **126:** 249, 1990.

ii. Neosporal Infections *Neospora caninum* is a newly recognized parasite belonging to the class Apicomplexa. It differs morphologically and antigenically from toxoplasma and sarcocystis. The organism has been isolated from dogs and cats and causes severe neuromuscular paralytic disease. Experimentally induced infections in pregnant dogs and cats resulted in *in utero* death as well as birth of infected offspring, some of which died, while others remained normal. Orally infected nonpregnant cats which subsequently became pregnant gave birth to infected, but clinically normal, kittens.

A *N. caninum*-like organism has been identified in the placenta and fetus in bovine abortions and in newborn calves with ascending paralysis. Investigations in California, using the avidin–biotin–immunoperoxidase technique, identified protozoa as the cause of abortion in 95 of 391 (24%) bovine abortions submitted. Ninety-three of these were *N. caninum*-like and one was a *Sarcocystis* species. All fetuses had a nonsuppurative necrotizing encephalitis.

The organism has also been identified as a cause of abortion in sheep. Lesions appear to be similar in bovine and ovine aborted fetuses and in clinically affected bovine, ovine, canine, and feline neonates.

Bovine fetuses are usually expelled in the second or

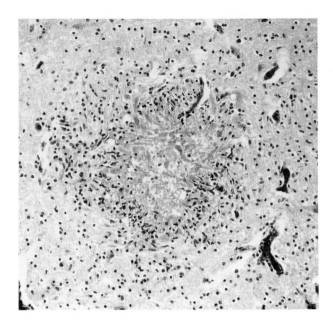

Fig. 4.70 *Neospora*-like organism in bovine abortion. Focal necrosis surrounded by a narrow zone of glial cells in brain. (Figs. 4.70–4.72, courtesy of B. C. Barr. Reprinted from B. C. Barr et al., *Vet Pathol* **27:** 354–361, 1990.)

third trimester and are autolyzing. In the placenta, the villi are necrotic, and the intercotyledonary region is normal. Clusters of *N. caninum*-like zoites are only occasionally observed in poorly defined cysts within trophoblasts. Lesions are prominent throughout the brain and, in particular, in the brain stem. They consist of multifocal encephalitis with scattered foci of gliosis or necrosis, which are occasionally mineralized and surrounded by mononuclear cells (Fig. 4.70). Clusters of zoites are present in areas of inflammation, but also randomly distributed (Fig. 4.71A,B,C,D). There are also scattered foci of necrosis and mononuclear cell infiltrates in skeletal muscle and heart where intracellular zoites are seen in myocytes, Pur-

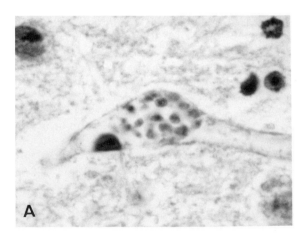

Fig. 4.71A *Neospora*-like organisms in brain. Clusters of zoites in vessel.

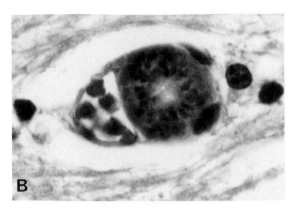

4.71B Clusters of zoites in endothelial cell.

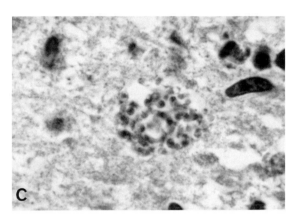

4.71C Clusters of zoites lying free in parenchyma.

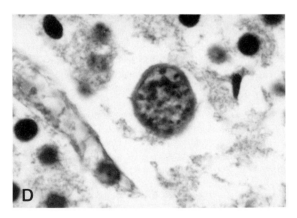

Fig. 4.71D Clusters of zoites in parenchyma surrounded by cystlike wall.

kinje fibers, and endothelial cells. Less-consistent lesions include multifocal hepatic necrosis (Fig. 4.72), focal non-suppurative interstitial nephritis, interstitial pneumonia, and adrenal adenitis.

There is a strong correlation between the presence of infection and characteristic lesions in the brain and spinal cord; however, free protozoa are difficult to recognize and

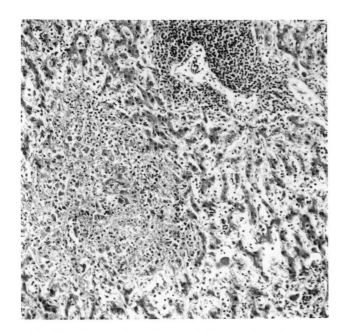

Fig. 4.72 Focus of hepatic necrosis (below, left) and mononuclear cell infiltrate in portal area. Bovine fetus. Abortion due to *Neospora*-like organism.

are most easily identified using specific avidin–biotin–immunoperoxidase procedures. *Toxoplasma gondii*, *Sarcocystis* species, and *N. caninum* may need to be differentiated.

Bibliography

Anderson, M. L. *et al.* A survey of causes of bovine abortion occurring in the San Joaquin Valley, California. *J Vet Diagn Invest* **2:** 283–287, 1990.

Barr, B. C. *et al.* *Neospora*-like protozoal infections associated with bovine abortions. *Vet Pathol* **28:** 110–116, 1991.

Barr, B. C. *et al.* Neospora-like encephalomyelitis in a calf: Pathology, ultrastructure, and immunoreactivity. *J Vet Diagn Invest* **3:** 39–46, 1991

Dubey, J. P., and Lindsay, D. S. Transplacental *Neospora caninum* infection in cats. *J Parasitol* **75:** 765–771, 1989.

Dubey, J. P., and Lindsay, D. S. Transplacental *Neospora caninum* infection in dogs. *Am J Vet Res* **50:** 1578–1579, 1989.

Dubey, J. P., and Lindsay, D. S. *Neospora caninum* induced abortion in sheep. *J Vet Diagn Invest* **2:** 230–233, 1990.

Dubey, J. P. *et al.* *Neospora caninum*-associated myocarditis and encephalitis in an aborted calf. *J Vet Diagn Invest* **2:** 66–69, 1990.

Dubey, J. P. *et al.* Fatal congenital *Neospora caninum* infection in a lamb. *J Parasitol* **76:** 127–130, 1990.

Lindsay, D. S., and Dubey, J. P. Immunohistochemical diagnosis of *Neospora caninum* in tissue sections. *Am J Vet Res* **50:** 1981–1983, 1989.

Shivaprasad, H. L., Ely, R., and Dubey, J. P. A *Neospora*-like protozoon found in an aborted bovine placenta. *Vet Parasitol* **34:** 145–148, 1989.

Thilsted, J. P., and Dubey, J. P. Neosporosis-like abortions in a herd of dairy cattle. *J Vet Diagn Invest* **1:** 205–209, 1989.

iii. Sarcocystis *Infections Sarcocystis* spp. cause abortion, stillbirth, and neonatal deaths in cattle, goats, sheep, and pigs. In contrast to some of the other protozoa, they are usually specific for their intermediate host and may be named using the intermediate and definitive host in the species name; for example, *Sarcocystis bovicanis* (syn. *S. cruzi*), *Sarcocystis capracanis* (syn. *S. tenella*). Transmission, pathogenesis, lesions, and diagnosis are similar in all intermediates, and only the lesions observed in cattle will be described.

The animal becomes infected by ingesting food or water contaminated by feces containing oocysts, which are immediately infective to the intermediate host. Definitive hosts include the dog, coyote, wolf, red fox, and raccoon for the pathogenic organism *Sarcocystis bovicanis*.

Infection may result in an acute necrotizing endometritis in the dam followed by multiple foci of necrosis in the soft tissues of the fetus. Why the organism is sometimes visible only in the dam, and at other times also in the fetus is not known. In the dam, the zoite-containing cysts may be observed in the endometrium or caruncle following abortion. If infection spreads to the fetus, extensive lesions are evident on microscopic examination, in spite of autolysis and near absence of gross lesions. In the brain

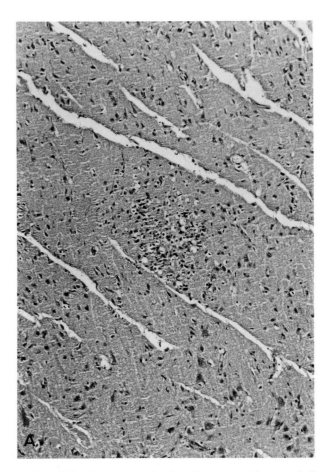

Fig. 4.73A *Sarcocystis* abortion. Nonsuppurative encephalitis. Bovine fetus.

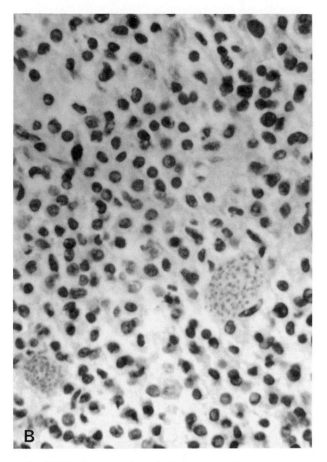

Fig. 4.73B Sarcocyst in endothelium of small vessel, Peyer's patch. Bovine fetus.

and meninges, there are multifocal areas of necrosis surrounded by undifferentiated mononuclear cells and lymphocytes (Fig. 4.73A). Cardiac muscle, kidney, liver, lung, and chorioallantoic membranes may have similar lesions, some of which are mineralized and undergoing fibrosis. Endothelial cells in several soft tissues may contain thin-walled cysts which bulge into the lumen (Fig. 4.73B) of the frequently thrombosed vessel. The cysts, packed with elongate zoites, though often difficult to find, are most easily distinguished in nervous tissue, maternal caruncle, and endometrium away from the site of inflammation.

Diagnosis is based on finding typical lesions of a nonsuppurative meningoencephalitis, multifocal necrosis in soft tissues, and the presence of cysts containing zoites with a tropism for endothelium. If the organism cannot be found on routine sectioning, it can readily be located using specific fluorescent antibody, immunogold, or avidin–biotin complex techniques.

Bibliography

Barnett, D. *et al.* Practicable diagnosis of acute bovine sarcocystosis causally related to bovine abortion. *Proceedings, 20th Annu Meet, Am Assoc Vet Lab Diag:* 131–138, 1977.

Corner, A. H. *et al.* Dalmeny disease. An infection of cattle presumed to be caused by an unidentified protozoon. *Can Vet J* **4:** 252–264, 1963.

Dubey, J. P. Abortion and death in goats inoculated with *Sarcocystis* sporocysts from coyote feces. *J Am Vet Med Assoc* **178:** 700–703, 1981.

Erber, M., and Geisel, O. Untersuchungen zur klinik und pathologie der *Sarcocystis suicanis*-infektion beim schwein. *Berl Munch Tierarztl Wochenschr* **92:** 197–202, 1979.

Erber, M., Meyer, J., and Boch, J. Aborte beim schwein durch Sarkosporidien (*Sarcocystis suicanis*). *Berl Munch Tierarztl Wochenschr* **91:** 393–395, 1978.

Hong, C. B. *et al.* Sarcocystosis in an aborted bovine fetus. *J Am Vet Med Assoc* **181:** 585–588, 1982.

Jolley, W. R. *et al.* Encephalitic sarcocystosis in a newborn calf. *Am J Vet Res* **44:** 1908–1911, 1983.

McCausland, I. P. *et al.* Multiple apparent *Sarcocystis* abortion in four bovine herds. *Cornell Vet* **74:** 146–154, 1984.

iv. *Genital Trichomoniasis* This is a specific contagious venereal disease of cattle caused by the flagellated protozoan *Tritrichomonas foetus* and transmitted at coitus.

Infection in the bull remains in the preputial cavity and must be considered, in the absence of effective treatment, as permanent. In early infections, there is a balanoposthitis of moderate severity with preputial swelling and a slight purulent discharge. As the infection becomes chronic, the inflammatory reaction disappears and the organisms become fewer in number. There is a tendency for them to concentrate on the glans penis and adjacent areas of the prepuce, but they are quite difficult to find in preputial washings and may be detected only by test-mating suspected bulls to susceptible heifers.

Females are not readily infected, if at all, except by service or experimentally by placing a culture of the organism in the vagina. A few days after infection, an acute vaginitis with swelling of the vulva develops, and there is a moderate amount of mucoid floccular discharge in the vagina. The protozoa may be easy or impossible to find in this exudate. The vaginitis resolves shortly, and the infection localizes in the uterus and cervix. Immediately prior to the estrus which follows the infective service, large numbers of organisms can usually be found in exudate aspirated from the cervix, but as estrus advances, the number of protozoa is greatly reduced.

The manifestations of established trichomoniasis in females are cervicitis and endometritis which result in repeat breeding, abortion, or pyometra. The inflammatory changes in the endometrium and cervix are relatively mild and nonspecific, although the exudate, mucopurulent in character, may be rather copious. Discharge of exudate into the vagina may be more or less continuous or intermittent, and the numbers and activity of the organisms in the discharge vary considerably over short periods. The discharge may not be apparent at the vulva.

The pattern of repeat breeding in trichomoniasis is the same as that in venereal campylobacter infections, return to service occurring at irregular intervals, which indicates that fertilization and implantation are followed by embryonic death. When the embryo or fetus dies, it may be resorbed, aborted, or retained with the development of pyometra. Abortions due to trichomoniasis may occur at any time but mainly in the first half of pregnancy. There are no specific gross fetal lesions, but large numbers of protozoa may be found in the fetal fluids and stomach. The placenta is not severely altered as in brucellosis; it may be covered by a white or yellowish flocculent exudate in small amounts, thickened and slightly tough, and hemorrhage without much necrosis may be evident on the cotyledons.

On histologic examination, the placenta is edematous with a light diffuse infiltrate of mononuclear cells and mild, spotty necrosis of chorionic epithelium. Many trichomonads may be present in the stroma of the chorion. In the lung, there may be no changes, or there can be a marked bronchopneumonia with heavy infiltrates of neutrophils and mononuclear cells. Many large multinucleated giant cells may be present; some of these contain trichomonads and meconium. Occasionally there is an alveolitis with thickening of alveolar septa by neutrophils and macrophages. Trichomonads are present within airways and scattered throughout the alveoli. The typical trichomonad morphology can be seen on routine staining; however, the flagella and undulating membrane cannot. These can be identified by the use of Bodian's silver protargol stain. The organism can be readily identified in abomasal and intestinal content and, on some occasions, in the lungs and placenta.

Pyometra is one of the remarkable changes of trichomoniasis, but it is a relatively uncommon complication. Its pathogenesis follows the scheme outlined earlier, and its character is remarkable only for the copiousness of the exudate which is present. Any volume up to a gallon or more may be present, and the exudate may be watery with floccules, colostrumlike, or brownish and sticky. It is without odor and swarms with organisms.

Bibliography

Clark, B. L., Dufty, J. H., and Parsonson, I. M. The frequency of infertility and abortion in cows infected with *Tritrichomonas foetus* var. *brisbane. Aust Vet J* **63**: 31–32, 1986.

Parsonson, I. M., Clark, B. L., and Dufty, J. H. Early pathogenesis and pathology of *Tritrichomonas foetus* infection in virgin heifers. *J Comp Pathol* **86**: 59–66, 1976.

Rhyan, J. C., Stackhouse, L. L., and Quinn, W. J. Fetal and placental lesions in bovine abortion due to *Tritrichomonas foetus. Vet Pathol* **25**: 350–355, 1988.

Skirrow, S. Identification of trichomonad-carrier cows. *J Am Vet Med Assoc* **191**: 553–554, 1987.

2. Viral Infections of the Pregnant Uterus

a. HOG CHOLERA   This disease is discussed in detail with The Cardiovascular System (Chapter 1 of this volume).

Occasionally, for what is usually a hyperacute or acute disease, pigs will only be mildly affected, and during such an inapparent viremic phase in a pregnant sow, the virus may be carried to the fetus and cause embryonic death, malformations, mummification, and stillbirth, or birth of live clinically normal pigs or pigs with tremors. Persistently infected piglets may be born at term and appear clinically normal or weak but, in either circumstance, eventually succumb to the effects of the virus. At one time the use of vaccines containing hog cholera virus of low virulence was the cause of severe losses due to *in utero* infections in piglets.

After exposure of a pregnant sow, the virus multiplies in the tonsils and spreads to other lymphoid organs, eventually reaching the fetus between 13 and 18 days. During the viremic stage, the virus crosses the uteroplacental interface, infecting all or some fetuses. In the latter situation, the virus may continue to pass from fetus to fetus, eventually infecting many.

Lesions produced in fetuses are very variable and depend on the virulence of the infecting virus, the stage of gestation when the infection occurs, and an apparent variation in the development time of immune competence to the virus among fetal piglets. Generally speaking, early infections result in embryonic or fetal death or persistent infections. Early to midgestation infections result in malformations including pulmonary hypoplasia, pulmonary artery malformation, micrognathia, arthrogryposis, and central nervous system malformations (including cerebellar hypoplasia, microcephaly, and defective myelination in brain and spinal cord). Infections during the last trimester may produce no abnormalities or result in mummification or stillbirths.

Persistent infections in piglets most commonly follow exposure to the virus from 22 to 70 days of gestation and, rarely, up to 100 days. These fetuses may or may not die *in utero* and often show retarded development and runting. Many show no signs until some time after birth, and persistently infected pigs which lived to 11 months have been reported. The latter commonly have marked depletion of lymphocytes especially in the thymic cortex, and B cell-dependent regions of lymph nodes and spleen. Lesions are also observed in heart (degeneration of endothelium and valvular fibrosis), liver (portal fibrosis), intestine (villous atrophy and ulceration), lungs (interstitial pneumonia), skin (necrosis and degeneration of epithelial cells), and central nervous system (neuron degeneration and hydrocephalus). Many persistently infected piglets are born alive with a congenital tremor associated with central nervous system dysgenesis, cerebellar hypoplasia, and hypomyelination (see The Nervous System, Volume 1, Chapter 3). Piglets with tremor which survive for long periods have reduced clinical signs but continue to excrete the virus.

Many fetal piglets show some ability to respond immunologically to hog cholera virus between 70 and 90 days of gestation with an immunoglobulin of low specificity and avidity. Fetuses collected from sows infected during this period may show general lymphoid hyperplasia and development of primary follicles. Kidney and brain may have mild lesions with scattered focal mononuclear cell infiltrates, and spleen and kidney may have foci of necrosis and hemorrhage.

Bibliography

Benten, von K., *et al.* Experimental transplacental transmission of hog cholera virus in pigs. III. Histopathological findings in the fetus. *Zentbl Veterinarmed* **27:** 714–724, 1980.

Collett, M. S., Moennig, V., and Horzinek, M. C. Recent advances in Pestivirus research. *J Gen Virol* **70:** 253–266, 1989.

Harkness, J. W. Classical swine fever and its diagnosis: A current view. *Vet Rec* **116:** 288–293, 1985.

Johnson, K. P. *et al.* Multiple fetal malformations due to persistent viral infection. I. Abortion, intrauterine death, and gross abnormalities in fetal swine infected with hog cholera vaccine virus. *Lab Invest* **30:** 508–617, 1974.

b. BOVINE VIRUS DIARRHEA   This disease is described in detail with The Alimentary System (Volume 2, Chapter 1).

Bovine virus diarrhea virus is closely related to Border disease virus of sheep and less closely to hog cholera virus. These three viruses have been shown to cross-infect naturally between species and are rapidly transferred to the fetus from a tolerant persistently infected dam or a dam which is not immune when initially exposed to infection.

Naturally occurring bovine virus diarrhea virus infections in pregnant sows have been reported to pass to the fetus, causing stillbirth and neonatal death similar to chronic hog cholera. Piglets infected *in utero* can either carry the virus (without antibody), develop antibody later, or become clinically affected and develop intestinal lesions, depending on the stage of gestation when infection occurs. On necropsy, infected piglets may also have a nonsuppurative meningitis and choroiditis. Swine fetuses dying from *in utero* infections with this virus may be resorbed.

Reproductive disease in cattle induced by bovine virus diarrhea virus includes oophoritis, fertilization failure, embryonic death, absorption or abortion, mummification, stillbirth, birth of calves small for gestational age or with congenital defects, and weak calves born persistently infected with noncytopathic strains of the virus. Abortion storms can occur, and occasionally severe losses may follow superovulation and embryo transfer. This may be related to the creation of a population of fetuses of the same gestational age (and therefore equally susceptible to the infection) or contamination of transfer fluids by the virus.

Fertilization failure may occur when the virus is introduced *in utero* into animals which are seronegative. This may be more important than embryo death, since the virus is not believed to readily cross the placenta during the first two months of gestation. Acutely infected bulls shed the virus for at least 14 days in their semen but may have abnormal semen for several months. Spread of the virus is more likely to occur with natural service than with artificial insemination.

Bovine fetuses exposed to bovine virus diarrhea virus

between 40 and 120 days gestation may become persistently infected and, if born alive, constitute the main source of virus for other animals. Transmission of virus may occur through the usual routes of excretion, or through the placenta to their own fetus and, if male, through their semen to susceptible females. Fetuses dying from the infection may remain in the uterus for up to 50 days after death, and recovery of the virus may be difficult. Fetuses exposed to the virus between 120 and 150 days of gestation (during the period of rapid brain growth) frequently develop central nervous system lesions (see The Nervous System, Volume 1, Chapter 3).

Cataracts and retinal atrophy may be present, and the latter is presumed to follow a focal-to-diffuse necrotizing, nonsuppurative retinitis and usually accompanies the cerebellar lesions. Microphthalmia and optic neuritis with subsequent atrophy also occur. Other lesions associated with bovine virus diarrhea (BVD) virus infections in the fetus include thymic cortical atrophy with only epithelial stroma and macrophages persisting in the cortex, generalized vasculitis, multifocal myocarditis, peribronchiolar lymphoid hyperplasia, pulmonary hypoplasia, multifocal perivascular dermatitis with hypotrichosis, and growth arrest lines in long bones. Cellular accumulations are usually mononuclear, lymphocytic, and rich in plasma cells. Plasmablasts, plasma cells, and lymphoblasts may be prominent in lymphoid tissues.

c. BORDER DISEASE   Border disease (see The Nervous System, Volume 1, Chapter 3) is the result of an *in utero* infection of pregnant sheep and goats by the *Pestivirus*, border disease virus. Infection can result in embryonic or fetal death, abortion, mummification, dysmorphogenesis, early postnatal death, and birth of weak or clinically normal young. The clinically normal kids or lambs can, however, be persistently infected. Infection in the nonpregnant adult animal may result in fever and leukopenia, but clinical signs are usually absent.

There are a wide variety of fetal tissues affected by border disease virus infection and the name hairy shaker was coined because of the frequently observed hairy appearance of the wool and the tremor associated with hypomyelinogenesis. Abnormalities induced in the fetus depend mainly on the stage of gestation, and reflect the stepwise development of fetal immune competence to the virus. Strains of the virus also vary in virulence, as does the susceptibility of breeds of sheep to the effects of the agent. The ovine and caprine fetuses become immunocompetent to this virus by about 90 days, and most injury follows infections occurring before this time.

Intrauterine growth retardation commonly occurs with border disease in goats and sheep and was reported to be of similar degree except for lung and brain to that seen in fetal starvation. The lung and brain are more severely affected. Growth arrest lines are commonly observed in the long bones.

Placentitis is particularly evident when infection occurs during the first trimester of gestation. Multiple white, 1-

to 2-mm foci of necrosis in the caruncles at the base of the crypts appear early in the infection. These may fuse into a continuous band of necrosis, which may become mineralized and infiltrated with connective tissue and a few macrophages, or the entire base of the caruncle may be infarcted, necrotic, and infiltrated with neutrophils. The lesion is believed to be associated with endothelial swelling and thrombosis of capillaries in the endometrium and adjacent caruncle. The trophoblast usually does not undergo necrosis but may atrophy.

Bibliography

Anderson, C. A. et al. Border disease: Virus-induced decrease in thyroid hormone levels with associated hypomyelination. Lab Invest 57: 168–175, 1987.

Anderson, C. A. et al. Tropism of border disease virus for oligodendrocytes in ovine fetal brain cell cultures. Am J Vet Res 48: 822–827, 1987.

Anderson, C. A. et al. Experimentally induced ovine border disease: Extensive hypomyelination with minimal viral antigen in neonatal spinal cord. Am J Vet Res 48: 499–503, 1987.

Archbald, L. F. et al. Effects of intrauterine inoculation of bovine viral diarrhea–mucosal disease virus on uterine tubes and uterus of nonpregnant cows. Am J Vet Res 34: 1133–1135, 1973.

Baker, J. C. Bovine viral diarrhea virus: A review. J Am Vet Med Assoc 190: 1449–1458, 1987.

Barlow, R. M. et al. Experiments in border disease VII. The disease in goats. J Comp Pathol 85: 291–297, 1975.

Barlow, R. M. et al. Mechanisms of natural transmission of border disease. J Comp Pathol 90: 57–63, 1980.

Brown, T. T. et al. Virus induced congenital anomalies of the bovine fetus. Cornell Vet 63: 561–578, 1973.

Brown, T. T. et al. Pathogenetic studies of infection of the bovine fetus with bovine viral diarrhea virus. Vet Pathol 11: 486–505, 1974.

Carlsson, U. Border disease in sheep caused by transmission of virus from cattle persistently infected with bovine virus diarrhoea virus. Vet Rec 128: 145–147, 1991.

Carlsson, U. et al. Bovine virus diarrhoea virus, a cause of early pregnancy failure in the cow. J Vet Med (A) 36: 15–23, 1989.

Casaro, A. P. E., Kendrick, J. W., and Kennedy, P. C. Response of the bovine fetus to bovine viral diarrhea mucosal disease virus. Am J Vet Res 32: 1543–1562, 1971.

Cay, B. et al. Comparative analysis of monoclonal antibodies against Pestivirus: Report of an international workshop. Vet Microbiol 20: 123–129, 1989.

Done, J. T. et al. Bovine virus diarrhoea–mucosal disease virus: Pathogenicity for the fetal calf following maternal infection. Vet Rec 106: 473–479, 1980.

Duffell, S. J., and Harkness, J. W. Bovine virus diarrhoea–mucosal disease infection in cattle. Vet Rec 117: 240–245, 1985.

Fernandez, A. et al. Viral antigen distribution in the central nervous system of cattle persistently infected with bovine viral diarrhea virus. Vet Pathol 26: 26–32, 1989.

Gardiner, A. C. The distribution and significance of border disease viral antigen in infected lambs and foetuses. J Comp Pathol 90: 513–518, 1980.

Gardiner, A. C., Nettleton, A. C., and Barlow, R. M. Virology and immunology of a spontaneous and experimental mucosal disease-like syndrome in sheep recovered from clinical border disease. J Comp Pathol 93: 463–469, 1983.

Howard, C. J., Brownlie, J., and Thomas, L. H. Prevalence of bovine virus diarrhoea virus viraemia in cattle in the U.K. Vet Rec 119: 628–629, 1986.

Kahrs, R. F. Effects of bovine viral diarrhea on the developing fetus. J Am Vet Med Assoc 163: 877–878, 1973.

Moennig, V. Pestiviruses: A review. Vet Microbiol 23: 35–54, 1990.

Ohmann, H. B. Experimental fetal infection with bovine viral diarrhea virus. II. Morphological reactions and distribution of viral antigen. Can J Comp Med 46: 363–369, 1982.

Orr, M. B. Pathology of the skin and fleece. In "Border Disease of Sheep: A Virus-Induced Teratogenic Disorder," R. M. Barlow and D. S. P. Patterson (eds.), pp. 13–21. Berlin and Hamburg, Verlag Paul Parey, 1982.

Parsonson, I. M. et al. The effects of bovine viral diarrhoea–mucosal disease (BVD) virus on the ovine foetus. Vet Microbiol 4: 279–292, 1979.

Roeder, P. L., Jeffrey, M., and Cranwell, M. P. Pestivirus fetopathogenicity in cattle: Changing sequelae with fetal maturation. Vet Rec 118: 44–48, 1986.

Stewart, W. C. et al. Bovine viral diarrhea infection in pregnant swine. Am J Vet Res 41: 459–462, 1980.

Terpstra, C. Border disease: A congenital infection of small ruminants. Prog Vet Microbiol Immunol 1: 175–198, 1985.

Welhelmsen, C. L. et al. Lesions and localization of viral antigen in tissues of cattle with experimentally induced or naturally acquired mucosal disease, or with naturally acquired chronic bovine viral diarrhea. Am J Vet Res 52: 269–275, 1991.

d. EQUINE VIRAL ARTERITIS Equine viral arteritis is a disease of horses (see Section VI,A of The Vascular System in The Cardiovascular System, Chapter 1 of this volume) caused by a virus in the family Togaviridae. Strains of the virus vary greatly in virulence: consequences of infection range from subclinical to mild fever; ocular and nasal discharges; edema around the eyes, legs, and scrotum; skin rash; and in severe forms, diarrhea, respiratory distress, and rarely death. Subclinical infections are common especially in standardbred horses. Forty to 80% of pregnant animals abort or deliver a stillborn foal.

Aborted fetuses may be fresh or autolyzed, but gross and microscopic lesions are rarely observed as the virus seldom produces an active infection in the fetus. A moderate arteritis with necrosis and mononuclear cell infiltrate has been reported in the myocardium of a few fetuses. In the uterus of the mare, however, there may be an acute multifocal necrotizing myometritis associated with replication of the virus in smooth muscle cells. The changes observed on electron-microscopic examination of muscle consisted of accumulations of ribosomes near the membrane of the smooth muscle cell. These are similar to those produced in endothelial cells by the virus. It is believed that the uterine infection may be the critical process contributing to abortion.

Bibliography

Coignoul, F. L., and Cheville, N. F. Pathology of maternal genital tract, placenta, and fetus in equine viral arteritis. Vet Pathol 21: 333–340, 1984.

Cole, J. R. et al. Transmissibility and abortigenic effect of equine viral arteritis in mares. *J Am Vet Med Assoc* **189**: 769–771, 1986.

Doll, E. R. *et al.* Isolation of a filterable agent causing arteritis of horses and abortion in mares. Its differentiation from the equine abortion (influenza) virus. *Cornell Vet* **47**: 3–41, 1957.

Golnik, W., Michalska, Z., and Michalak, T. Natural equine viral arteritis in foals. *Schweiz Archiv Tierheilk* **123**: 523–533, 1981.

Timoney, P. J., and McCollum, W. H. Equine viral arteritis. *Can Vet J* **28**: 693–695, 1987.

e. PORCINE PARVOVIRUS Porcine parvovirus infection is an important and common cause of reproductive failure in pigs and manifests as embryonic death with reabsorption, mummification, stillbirths, and reduced litter size. Farrow-to-finish farms are affected nearly twice as often as farms that sell feeder pigs, probably due to the maintenance of a continual reservoir of infection in the finishing pigs. The virus is most commonly spread by the ingestion of contaminated feces. A viremia develops in susceptible pigs and the virus will adhere to the zona pellucida of embryos. In pregnant pigs the virus enters the placenta and passes to the embryo or fetus approximately 23–32 days following oral inoculation. Up to a fetal age of 70 days, the virus may cause death of the embryo with resorption or of the fetus with mummification; however, the fetus becomes increasingly resistant to the effects of the virus after 36 days and is immunocompetent by about 70 days. Death of the fetus may be due to the total effect of the virus on rapidly dividing cells, including neurons and capillary endothelium of the cortical laminae in cerebellum and cerebrum, and cells in the lung, liver, pancreas, and chorioallantois. Movement of virus from one fetus to another may be affected by development of maternal antibody, increasing resistance of fetuses with advancing age, and the variability of the circulation and contact between the placental membranes of adjacent fetuses.

When the virus is experimentally introduced by the uterine route it causes inflammation in the ovaries and degeneration of uterine epithelium, both of which may contribute to reproductive failure, but does not otherwise cause clinical signs. Uterine changes may include marked variation in surface epithelial cell height, epithelial cell basophilia, degeneration, and ulceration. A light infiltration of neutrophils, eosinophils, and lymphocytes may be present throughout the endometrium. Corpora lutea contain foci of mononuclear cells, many of which are lymphoplasmacytic. In the fetus, lesions are most commonly observed in older fetuses and may be present in kidney, liver, brain, and fetal placenta. In the cerebrum, which is the most commonly affected site, perivenous accumulations of adventitial cells, plasma cells, and other mononuclear cells occur in white and gray matter and in meninges. Similar accumulations are observed in the renal interstitium, hepatic portal regions, and in the chorioallantois.

Viral antigen is demonstrable by the fluorescent antibody test in affected tissues of embryos, and mummified and stillborn fetuses; the lung is the organ of choice as it is easily collected even from mummified fetuses and has a minimum of autofluorescence.

Bibliography

Cutlip, R. D., and Mengeling, W. L. Experimentally induced infection of neonatal swine with porcine parvovirus. *Am J Vet Res* **36**: 1179–1182, 1975.

Cutlip, R. D., and Mengeling, W. L. Pathogenesis of *in utero* infection: Experimental infection of eight- and ten-week-old porcine fetuses with porcine parvovirus. *Am J Vet Res* **36**: 1751–1754, 1975.

Hohdatsu, T. *et al.* Detection of antibodies against porcine parvovirus in swine sera by enzyme-linked immunosorbent assay. *Vet Microbiol* **17**: 11–19, 1988.

Johnson, R. H. *et al.* Observations on the epidemiology of porcine parvovirus. *Aust Vet J* **52**: 80–84, 1976.

Joo, H. S. *et al.* Pathogenesis of porcine parvovirus infection: Pathology and immunofluorescence in the foetus. *J Comp Pathol* **87**: 383–391, 1977.

Krell, P. J., Salas, T., and Johnson, R. P. Mapping of porcine parvovirus DNA and development of a diagnostic DNA probe. *Vet Microbiol* **17**: 29–43, 1988.

Mengeling, W. L., and Cutlip, R. C. Pathogenesis of *in utero* infection: Experimental infection of five-week-old porcine fetuses with porcine parvovirus. *Am J Vet Res* **36**: 1173–1175, 1975.

Mengeling, W. L., and Paul, P. S. Reproductive performance of gilts exposed to porcine parvovirus at 56 or 70 days of gestation. *Am J Vet Res* **42**: 2074–2076, 1981.

Meyers, P. J. *et al.* Hormonal changes in sows after induced porcine parvovirus infection in early pregnancy. *Am J Vet Res* **48**: 621–626, 1987.

Narita, M. *et al.* Histopathological changes of the brain in swine foetuses naturally infected with porcine parvovirus. *Natl Inst Anim Health Q* **15**: 24–28, 1975.

Redman, D. R., Bohl, E. H., and Ferguson, L. C. Porcine parvovirus: Natural and experimental infections of the porcine fetus and prevalence in mature swine. *Infect Immun* **10**: 718–723, 1974.

Sorensen, K. J., Askaa, J., and Dalsgaard, K. Assay for antibody in pig fetuses infected with porcine parvovirus. *Acta Vet Scand* **21**: 312–317, 1980.

Thacker, B. J. *et al.* Clinical, virologic, and histopathologic observations of induced porcine parvovirus infection in boars. *Am J Vet Res* **48**: 763–767, 1987.

Wrathall, A. E., and Mengeling, W. L. Effect of porcine parvovirus on development of fertilized pig eggs *in vitro*. *Br Vet J* **135**: 249–254, 1979.

f. BOVINE PARVOVIRUS Bovine parvovirus is widely distributed in cattle. It is reported to cause enteritis and diarrhea in young calves and abortion and birth of weak calves when infections occur in naive pregnant cows. The organism is not considered a major cause of abortion, but this may be because it is not looked for consistently in submissions. From 10 to 100% of the cattle in a population will have antibody to parvovirus. Work in the United States suggests when reproductive problems including repeat breeding, embryonic mortality, and abortion are considered, animals with antibody to bovine parvovirus had about three times the number of these problems as did seronegative animals.

Bovine parvovirus is excreted in the feces of affected and normal animals, and the oral route may be important in transmission. Inoculation of parvovirus into pregnant cows can cause fetal death and abortion in the early stages of gestation. Some experimentally infected calves may be born with cerebellar hypoplasia due to necrosis of the cells of the external granular layer. Cerebellar hypoplasia is believed to follow infection during the period of rapid cell multiplication (107–150 days gestation) in the cortex. Intranuclear inclusions are seen in external granular cells, hepatocytes, adrenal cortical cells, and intestinal crypt cells associated with areas of necrosis. Bovine fetuses are capable of forming IgM antibody to the virus by 93 days gestation, and infections in the third trimester usually result only in antibody formation with no abortion.

g. MINUTE VIRUS OF CANINES   Minute virus of canines (MVC) is an incompletely characterized autonomous parvovirus of dogs. It has been isolated from normal dog feces and from pups with diarrhea. The antigenic and genomic features are different from canine parvovirus-2 and it is different from other parvoviruses. The dog is the only known host of the virus, and the cultural requirements are remarkably strict. Only a single cell line has been reported to support its growth.

The epidemiology of the infection is undetermined, but hemagglutination-inhibition tests suggest many adult dogs have had exposure to the virus. Serologic evidence has suggested that the virus has caused fetal death and resorption in dogs. Experimental transplacental infection of fetal pups can occur, the outcome depending on the stage of pregnancy. Infection early in pregnancy was more likely to be lethal to embryos and fetuses than it was later. Virus was demonstrated in fetal lung and intestinal villi immediately after the death of fetuses, but could not be demonstrated in fetal or uterine tissues in the autolyzed conceptuses or in fetuses undergoing resorption 2 weeks or more after their deaths.

The gross lesions observed in the experimental disease were not specific. Histopathologic changes were most obvious in the lung; they consisted of interstitial pneumonia with basophilic intranuclear inclusion bodies in alveolar and bronchial epithelium. Other lesions reported were necrosis of the intestinal villi and myocarditis. The inability to isolate the virus or demonstrate viral antigens in tissues by immunostaining except at or near the time of death of the conceptus makes this a difficult diagnosis to prove.

Bibliography

Carmichael, L. E., Schlafer, D. H., and Hashimoto, A. Pathogenicity of minute virus of canines (MVC) for the canine fetus. *Cornell Vet* **81:** 151–171, 1991.
Kahrs, R. F. Parvoviruses. *In* ''Viral Diseases of Cattle,'' pp. 183–188. Ames, Iowa, Iowa State University Press, 1981.
Storz, J. *et al.* Parvovirus infection of the bovine fetus: Distribution of infection, antibody response, and age-related susceptibility. *Am J Vet Res* **39:** 1099–1102, 1978.

h. BLUETONGUE VIRUS   The disease is described in detail with The Alimentary System (Volume 2, Chapter 1).

The consequences of infection during pregnancy depend largely on the stage of gestation when infection occurs and on the virulence of the virus. Infections in sheep up to ~50 days of pregnancy may result in death of the conceptus and absorption or abortion. Between 50 and 80 days of gestation, a necrotizing encephalopathy occurs (see The Nervous System, Volume 1, Chapter 3); in the last third of gestation, as immune competence increases, there may be no obvious disease. On the other hand, infections of near-term and newborn lambs can result in a sluggish immune response, possibly associated with the rise in fetal cortisol at parturition and the ingestion of colostrum, both of which may suppress the active immune response.

Lesions in bovine fetuses develop similarly; and up to 70 days of gestation, infection may cause *in utero* death and resorption or abortion of the conceptus. Between 70 and 130 days of gestation, serotypes 10, 11, 13, and 17 can cause hydranencephaly and death of the fetus.

i. EPIZOOTIC HEMORRHAGIC DISEASE   Epizootic hemorrhagic disease virus is an *Arbovirus* in the family Reoviridae, genus *Orbivirus*. It exists as at least two serotypes, 1 and 2, and is present in Africa, Australia, North America, and Japan. Ibaraki disease virus, isolated first in Japan, is a serotype-2 strain. Epizootic hemorrhagic disease virus is spread by *Culicoides* spp. and causes a fatal hemorrhagic disease of white-tailed deer. In cattle, the Ibaraki strain can cause severe ulceration of the mucosa of the mouth, tongue, and esophagus with necrosis of esophageal musculature and marked pain on swallowing. Many animals die. The disease in many respects is similar to bluetongue virus infection in sheep. However, epizootic hemorrhagic disease virus has a very low pathogenicity for sheep. Experimental inoculation of bluetongue virus causes abortion and mummification in white-tailed deer. Epizootic hemorrhagic disease virus, because of its similarity to bluetongue virus, is also believed to be a cause of abortion and congenital anomalies in cattle, but this has not been proven.

Bibliography

Enright, F. M., and Osburn, B. I. Ontogeny of host responses in ovine fetuses infected with bluetongue virus. *Am J Vet Res* **41:** 224–229, 1980.
Hoff, G. L., and Trainer, D. O. Bluetongue and epizootic hemorrhagic disease viruses: Their relationship to wildlife species. *In* ''Advances in Veterinary Science and Comparative Medicine,'' C. A. Brandly and C. E. Cornelius (eds.), pp. 111–132. New York, Academic Press, 1978.
Huismans, H., Bremer, C. W., and Barber, T. L. The nucleic acid and proteins of epizootic haemorrhagic disease virus. *Onderstepoort J Vet Res* **46:** 95–104, 1979.
Luedke, A. J., Jochim, M. M., and Jones, R. H. Bluetongue in cattle: Effects of *Culicoides variipennis*-transmitted bluetongue virus on pregnant heifers and their calves. *Am J Vet Res* **38:** 1687–1695, 1977.

MacLachlan, N. J. *et al.* Bluetongue virus-induced encephalopathy in fetal cattle. *Vet Pathol* **22:** 415–417, 1985.

Mahrt, C. R., and Osburn, B. I. Experimental bluetongue virus infection of sheep; effect of vaccination: Pathologic, immunofluorescent, and ultrastructural studies. *Am J Vet Res* **47:** 1198–1203, 1986.

Omori, T. Ibaraki disease: A bovine epizootic disease resembling bluetongue. *Natl Inst Anim Health Q* **10:** 45–55, 1970.

Richardson, C. *et al.* Observations on transplacental infection with bluetongue virus in sheep. *Am J Vet Res* **46:** 1912–1922, 1985.

Uren, M. F. Clinical and pathological responses of sheep and cattle to experimental infection with five different viruses of the epizootic haemorrhagic disease of deer serogroup. *Aust Vet J* **63:** 199–200, 1986.

j. CANINE HERPESVIRUS   In adults or weaned puppies, canine herpesvirus may produce a mild upper respiratory infection, but inapparent infections are common. Infection of the neonate is possible by various routes. It may be transmitted from maternal lesions of genital herpes to the pup as it traverses the birth canal, from initial maternal infection, or by recrudescence of latent virus. Viral recrudescence can be induced in recovered animals by corticosteroids. There is considerable homology between endonuclease fragments of the canine virus and feline herpesvirus-1.

Canine herpesvirus infection is regularly fatal for newborn puppies. Resistance to the disease is sharply age related; pups exposed after 2 weeks of age will not develop severe illness. The disease has an incubation period of from 3 to 7 days, after which the affected puppies rapidly sicken and die, usually within 2 days. The clinical signs lack specificity. Affected animals usually are not febrile; most vomit and refuse food. Their breathing becomes shallow and rapid, and shortly before death, the pups give evidence of abdominal pain.

At autopsy, usually, there is a pleural and peritoneal effusion which may be blood-tinged. Petechial and ecchymotic hemorrhages are scattered throughout the subserosal tissue, usually representing the most impressive gross feature of the disease. Large hemorrhages are regularly seen in the kidney (Fig. 4.74). The lungs are wet, and

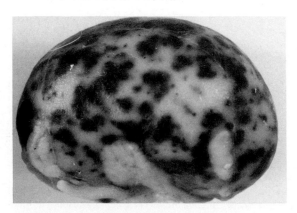

Fig. 4.74 Renal hemorrhages in newborn pup. Canine herpesvirus infection. (Courtesy of D. H. Percy.)

frothy fluid fills the bronchi and bronchioles. The lymph nodes are enlarged and reddened, and splenomegaly is present. Tiny red foci, determined microscopically to be foci of necrosis, are occasionally seen in the liver. Foci of necrosis are characteristic of the disease and may occur in any organ (Fig. 4.75A). Lung, heart, kidney, intestine, pancreas, adrenal, and spleen are common sites. The inflammatory response associated with these disseminated focal areas of necrosis is slight or absent. Viral inclusion bodies are not numerous but can best be seen adjacent to the areas of necrosis in the liver or lung, or in the renal glomeruli. The inclusion bodies are located within the nucleus. Most are basophilic, but some are faintly acidophilic.

Discrete microscopic lesions are present in the brains of the pup as early as 72 hr after experimental infection. These increase in severity as the disease progresses. They consist at death of a nonsuppurative meningoencephalomyelitis, characterized by destruction of gray and white matter (Fig. 4.75B). The gray is somewhat more severely involved. There are focal and segmental areas of destruction of the cerebral cortices with concurrent microgliosis. Vascular changes vary from endothelial swelling to mononuclear cuffing. Diffuse pial meningitis with infiltrates of mononuclear cells and neutrophils is frequent, as is ganglioneuritis of the trigeminal nerve ganglia.

Canine herpesvirus has been associated with abortion, stillbirth, and infertility. Depending on the time of gestation, inoculation of pregnant bitches with the virus results in death and expulsion of fetuses, mummification, or premature delivery of live puppies. Foci of necrosis with intranuclear inclusions are variably observed in liver, spleen, kidney, placental labyrinth, and heart. Endothelium and trophoblast are commonly involved. Virus can be recovered from many fetuses.

k. PORCINE HERPESVIRUS-1   Porcine herpesvirus-1 (pseudorabies virus, Aujeszky's disease virus) is capable of causing reduced fertility in boars and sows, fetal resorption, mummification, abortion, and stillbirth in sows and a fatal meningoencephalitis in young pigs, sheep, cattle, and dogs (for details see The Nervous System, Volume 1, Chapter 3). Strains of the virus vary in their ability to invade the placenta and fetus; however, abortion without invasion of the conceptus may occur secondary to fever and nervous disease in the sow. When the virus does invade the placenta and fetus, characteristic gross and microscopic lesions are produced. There is a strong correlation between the ability of strains to cause syncytia formation in cultures of cells and their ability to produce disease in pigs and cattle.

Abortion usually occurs ~10 days after the onset of clinical illness in the sow. If the placenta and fetus are invaded, lesions consist of multifocal coagulative necrosis of chorionic villi and focal coagulative necrosis in many fetal organs including liver, adrenal, and spleen. Intranuclear inclusions have been observed in trophoblast and interstitial cells and hepatocytes around areas of necrosis.

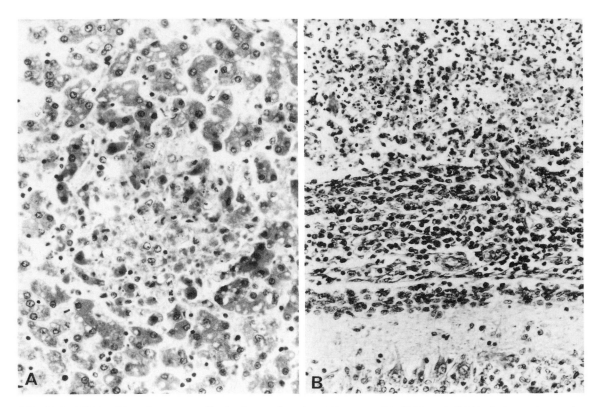

Fig. 4.75 (A) Focal hepatic necrosis. Canine herpesvirus. Pup. (Courtesy of D. H. Percy.) (B) Meningitis and necrosis of cerebellum. Canine herpesvirus. Pup. (Courtesy of D. H. Percy.)

l. PORCINE CYTOMEGALOVIRUS Porcine cytomegalovirus, Suid herpesvirus 2, is in the family Herpesviridae, and subfamily Betaherpesvirinae (cell-associated virus). Infection by this virus is normally confined to pigs and was first described in England in 1955 as inclusion-body rhinitis and subsequently, as a transplacental infection of fetal pigs, in 1961 in Australia.

In neonates, infection causes a mild to severe nonsuppurative necrotizing rhinitis characterized by the presence of large basophilic intranuclear inclusions in the epithelium of the mucous glands and ducts in the nasal mucosa (for details see The Respiratory System, Volume 2, Chapter 6). The virus usually affects young pigs (younger than 4 weeks), but in naive herds may cause death in pigs of 4–12 weeks. The virus occasionally causes severe systemic disease in mature pigs. When infection occurs in pregnant sows, they may deliver small litters, mummified fetuses, or stillborn or weak piglets either before or on the due date. Pigs infected *in utero* may be born alive but small for gestational age, and may develop systemic signs and die. Subsequent reproductive performance in infected sows may be reduced and manifested as decreased conception rates and decreased litter size.

The virus seems to pass through the placenta without inducing injury. Experimentally the virus has not been isolated from the placenta, even when it can be isolated from the fetus that has just died. Neither has antibody been detected in piglets which become infected *in utero*.

Fetuses die 4–6 weeks after inoculation of the sow, and the time is independent of the stage of gestation. Casualties of infection are scattered randomly throughout the uterus in various stages of decomposition and mummification, much as with porcine parvovirus infection. Following experimental infection, no gross lesions were seen on necropsy of the autolyzed fetuses. However, on microscopic examination, there was cytomegaly, and intranuclear inclusions were observed in liver and lung.

In natural outbreaks of infection, pigs born alive with a congenital infection usually died within a week. Gross lesions in these consist of pulmonary edema and congestion, hydrothorax and hydropericardium, and multiple firm, gray foci in the most ventral portions of the lungs. Mediastinal lymph nodes are enlarged, and petechial hemorrhages are visible on the heart, lungs, intestine, and kidneys. Large basophilic intranuclear inclusions are present in capillary endothelial cells and macrophages. Similar inclusions are present in capillaries of the renal medulla and tubular epithelium. Two types of inclusions have been observed: the usual large basophilic type found in pulmonary macrophages and glands and ducts of the nasal mucosa and renal tubular epithelium, and a small herpeslike inclusion found mainly in reticuloendothelial cells. A nonsuppurative meningoencephalitis has been reported in some pigs and was characterized by perivascular lymphocytic infiltrates in the choroid plexus and vessels outside the cerebral cortex.

Experimental infections of piglets *in utero* or newborns younger than 2 weeks of age has shown them to be very susceptible and to develop a severe debilitating disease with death a common sequel. This severe infection has been related to viral infection in the cells of the reticuloendothelial system. This results in disseminated hemorrhage and edema in the lung, kidney, adrenal, liver, and lymph nodes, all of which is associated with viral destruction of macrophages and capillary endothelial cells. In the less severe and nonfatal forms of the disease there is invasion of the mucus-producing cells of the nasal mucosa, the cells of the renal tubules and, less frequently, the epithelium of the salivary glands, Harderian and lacrimal glands, seminiferous epithelium, epithelium of ductus epididymis, mucous glands of the esophagus, duodenum, jejunum, and hepatocytes. Invasion of macrophages and endothelium or epithelium is not completely exclusive, and both occur often.

The cytomegalovirus inclusions, when correlated with appropriate signs, are diagnostic of this infection in pigs.

m. BOVINE HERPESVIRUS-1   Bovine herpesviruses cause a wide variety of diseases in several species, including cattle, sheep, goats, wildebeest, and other wild ruminants. They are divided into six different groups designated bovine herpesvirus 1–6. The infections of the pregnant uterus caused by bovine herpesviruses which will be discussed in this section include bovine herpesvirus-1 (infectious bovine rhinotracheitis/infectious pustular vulvovaginitis), and bovine herpesvirus-4 (a cytomegalovirus). Bovine herpesvirus-5 (bovine encephalitis herpesvirus) has also been reported to cause fetal disease in cattle, but details are not available.

Bovine herpesvirus-1 is in the family Herpesviridae and the subfamily Alphaherpesvirinae. It has been associated with two more or less distinct disease syndromes: a respiratory disease, infectious bovine rhinotracheitis (IBR), and a genital disease, infectious pustular vulvovaginitis (IPV). There are conflicting reports as to whether genital and respiratory isolates can be distinguished by restriction endonuclease DNA analysis. Some isolates of the IBR subgroup are more virulent than most strains of either respiratory or genital origin, and these variants, in addition to severe upper respiratory disease, can cause endometritis, oophoritis, mastitis, dermatitis, and fatal diarrhea in young calves. Both respiratory and genital strains can cause fetal disease and abortion. The bovine herpesvirus that is encephalitogenic is reported to be antigenically indistinguishable from BHV-1, but by restriction endonuclease DNA fingerprints, it is totally different from BHV-1. The encephalitogenic virus is being designated as BHV-5. It too has been recovered from an aborted bovine fetus. Infectious pustular vulvovaginitis virus primarily causes inflammation of the penis and sheath in bulls (see The Male Genital System, Chapter 5 of this volume) and inflammation in the vagina and vulva in cows and abortion in pregnant cows.

Bovine herpesvirus-1 is distributed worldwide. Cattle-to-cattle transmission is the principal method of spread, and virus is easily dispatched from naturally infected or vaccinated cattle to susceptible animals through respiratory, ocular, or vaginal discharges. All BHV-1 strains, including thermosensitive and thymidine kinase-negative strains, are capable of becoming latent and therefore undergoing recrudescence and shedding from mucous membranes. Virus is more abundant in excretions from primary than from recrudescent infections; however, transmission may occur from either. Live-vaccine viruses may become latent, and none of the vaccines can prevent latency by a superinfecting challenge virus, but if the antibody titer is high in the vaccinate, the incidence of latent superinfection and amount of virus excreted may be lowered. Recrudescence may be induced by superinfections, corticosteroids, transport, parturition, and other stressful situations. Semen may also be an important source of the virus, and embryo transfer fluids, less so.

Sheep, goats, and pigs often have antibody titers to BHV-1 and may occasionally show clinical signs of disease. Infection may manifest as an upper respiratory disease which usually goes unnoticed; however, sheep may develop pneumonia. Abortions from which BHV-1 has been recovered have been described in sheep and pigs. Sheep, goats, pigs, and wild ruminants are rarely the source of BHV-1 for other animals. It is more common for cattle to transmit the virus to them, and disease is an infrequent event.

In pregnant animals when the virus enters the mucosa of the respiratory or genital tract, it multiplies at that site and is carried to the rest of the body in infected leukocytes and in blood to the uterus. It has been postulated that the virus, on reaching the caruncle, passes though the trophoblast to the interstitium of the villus, thereby infecting endothelium, mesenchyme, and then trophoblast. Death is due to tissue destruction in the fetus and the placenta. Early embryos are also susceptible to IBR virus infection.

Abortion rates in a herd of cattle may reach 25%; however, the prevalence of abortion in a region declines with a reduction in the naive population. Most cows infected with BHV-1 do not abort until 3–6 weeks following the initial infection, and a significantly greater number of cows abort between 5 and 8 months of gestation than earlier. It is speculated that the virus can stay in the placenta for a prolonged period before invading the fetus. As the virus invades, the fetus dies quickly with no preparation for delivery. Expulsion occurs 3–5 days following fetal death, and the carcass is in a state of advanced autolysis. Rarely fetal death coincides with the time of normal delivery, and the fetus is discharged well preserved.

Gross lesions in the fetus are usually absent or masked by autolysis, but when visible include white to tan 1–3 mm diameter foci of necrosis under the liver capsule, and more rarely on the surface of the lung. Perivascular renal hemorrhage may be present in addition to focal hemorrhages at the corticomedullary junction in the kidney (Fig. 4.76).

Fig. 4.76 Bovine herpesvirus-1 infection. Focal renal hemorrhage and necrosis. Day-old calf.

On microscopic examination, foci of necrosis with minimal cellular infiltrate can be seen in many tissues, including liver, adrenal, kidney, intestine, lymph node, lung, and spleen (Fig. 4.77A,B). As autolysis is advanced, inclusion bodies are difficult to distinguish, but are most profitably sought in the adrenal within the more normal cells surrounding a focus of necrosis. A necrotizing vasculitis in the small vessels of placental villi is consistently present (Fig. 4.78).

To confirm the diagnosis, the immunoperoxidase test using monoclonal antibodies for BHV-1 antigen is useful because it is specific, highly sensitive, can be done on fixed tissue sections, and requires no tissue cultures. Results are also available quickly. Fluorescent antibody techniques on frozen tissue sections of kidney, liver, or placenta, and tissue cultures of these samples, can also be used for detection of virus. To identify antibodies to BHV-1 the enzyme-linked immunosorbent assay (ELISA) has a higher validity in differentiating a negative from a positive sample than the serum-neutralization test. Since cows abort such a long time after initial infection with BHV-1, most animals have very low titers to the virus at the time of abortion. On an individual basis, therefore, antibody titers are of little value in diagnosis. Also, as the fetus dies quickly, there is usually no evidence of fetal antibody production.

n. BOVINE HERPESVIRUS-4 Bovine cytomegalovirus is one of several herpesviruses in the bovine herpesvirus-4 group, in the family Herpesviridae and the subfamily Gammavirinae (cell-associated herpesviruses). Viruses in the BHV-4 group include bovine cytomegalovirus, noninfectious bovine rhinotracheitis herpesvirus, orphan herpesvirus, DN-599, and Movar 33/63. These viral isolates are closely related antigenically and may represent strains of the same virus. They have been isolated from cattle in the United States, Africa, and Europe in association with

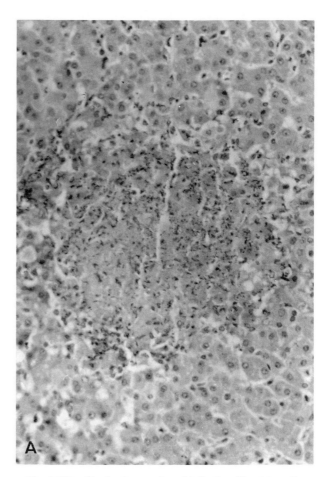

Fig. 4.77A Bovine herpesvirus-1 infection. Focal hepatic necrosis.

a wide variety of clinical conditions including pneumonia, enteritis, metritis, mammillitis, and from cattle with the disease syndrome epivag (see the following and The Male Genital System, Chapter 5 of this volume).

Bovine cytomegalovirus (BHV-4) is considered an important cause of abortion in cattle; however, experimental evidence confirming a causal relationship is incomplete. The virus is less virulent than BHV-1, and concurrent infections of BHV-4 with other pathogens are frequent (up to 75%). It has been suggested that the virus is immunosuppressive and thereby intensifies the effects of other agents. Bovine cytomegalovirus has been isolated together with bovine virus diarrhea virus and Border disease virus (from abortions) and *Mycobacterium paratuberculosis* in animals with Johne's disease.

As the virus is carried within mononuclear blood cells, grows on the mucosal surface, and is a poor immunogen, it elicits a low level of neutralizing antibody, which is also of low avidity. The virus is transmitted from cow to cow in oral and nasal secretions and, during viremia, is carried to the placenta, where it multiplies and invades the fetus. The virus may remain latent in mononuclear blood cells and in the trigeminal nerve ganglion. Recrudescence can

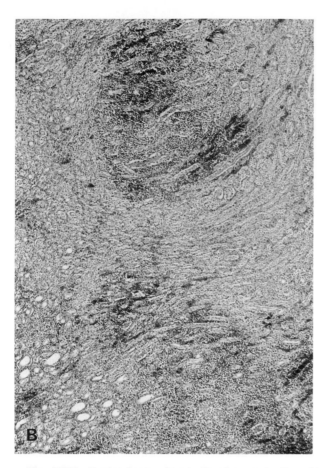

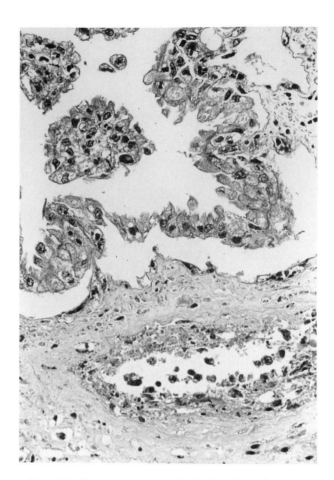

Fig. 4.77B Bovine herpesvirus-1. Renal necrosis. Day-old calf.

Fig. 4.78 Bovine herpesvirus-1 infection. Necrotizing vasculitis. Placenta, aborted bovine fetus.

be induced with corticosteroids. Herds tested for BHV-4 antibodies vary greatly in the proportion of cattle with antibody to the virus.

Mummification and the birth of weak pigs has also been reported in association with BHV-4 infections in sows. Fetal pigs are capable of developing antibody to the virus by 74 days of gestation. Recrudescence induced by corticosteroids has been reported in sheep and pigs with spread of virus to other animals.

Descriptions of the lesions seen in the aborted bovine fetus are limited to one report. No gross lesions were observed in the fetus, but on histopathologic examination, there was thickening of alveolar septa in the lungs, where many large intranuclear inclusions were observed in large alveolar cells. Similar inclusions were also seen in cells of bile duct epithelium, myocardium, spleen, and epithelium of renal tubules. The virus was not isolated from this animal, but the inclusions seen on light and electron microscopy were typical of cytomegaloviruses.

o. EQUINE HERPESVIRUS-1   Equine herpesvirus-1 (EHV-1) is in the family Herpesviridae and subfamily Alphaherpesvirinae. It is widespread throughout the world and is a major cause of abortion in pregnant mares and an infre-

quent cause of neurologic disease in horses. Another serologically related herpesvirus is a common respiratory pathogen of horses. This virus is currently designated equine herpesvirus-4, but before this, the two viruses were regarded as a single virus, equine rhinopneumonitis virus, the cause of both the common respiratory disease and abortion storms. Subsequently the viruses were separated into two subtypes on the basis of epidemiology and restriction endonuclease cleavage patterns. Subtype 1 was associated with abortion, and subtype 2, with respiratory disease. This taxonomic confusion has left a bewildering train of synonyms. We shall refer to equine herpesvirus-1 as the equine abortion virus and equine herpesvirus-4 as the rhinopneumonitis virus, recognizing that all the data do not support this distinction. Both viruses are reported to cause respiratory disease, but EHV-4, rhinopneumonitis virus, is the common cause, and both viruses can cause abortion, but EHV-1, the equine abortion virus, is the important cause of both single and multiple abortions in mares. EHV-1 is the only one causing neurologic disease (see the Nervous System). EHV-4 is commonly isolated from mild upper respiratory disease in horses and much less so from abortion. Respiratory disease due to EHV-4 is common in some areas where abortion is rare.

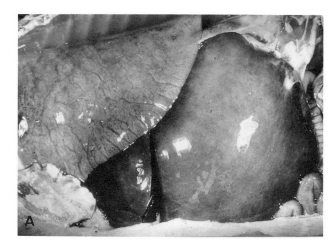

Fig. 4.79A Equine herpesvirus-1 infection. Pulmonary edema in an aborted equine fetus.

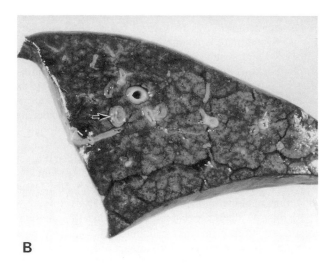

Fig. 4.79B Equine herpesvirus-1. Fibrin casts in bronchi.

Equine herpesvirus-1, equine abortion virus, is very widespread, and first exposure usually occurs before the foal is 1 year old. A mild to severe upper respiratory disease is produced, and secondary bacterial infection, frequently with *Streptococcus zooepidemicus* or other streptococci, is common. As the virus is so widespread and contact so likely at an early age, abortion usually occurs in an animal which has had previous experience

with the virus. As with herpesviruses generally, the level of immunity is low and the virus is transported in leukocytes through the bloodstream to the placenta and hence the fetus. Death of the fetus does not occur until the onset of the usually prompt and uncomplicated abortion. The dam shows no premonitory signs and the fetus is aborted in a fresh state. The time from exposure—whether by

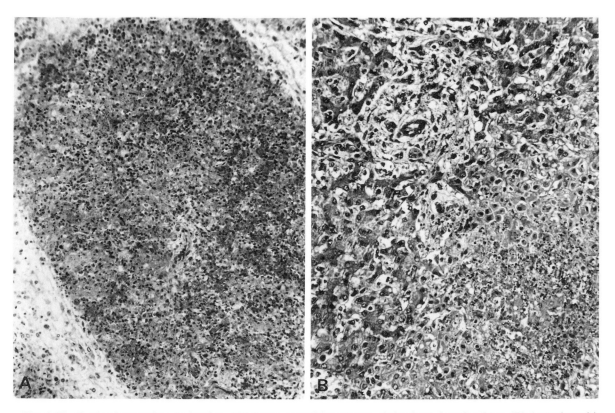

Fig. 4.80 Equine herpesvirus-1 abortion. (A) Acute necrotizing pneumonia in aborted equine fetus. (B) Acute hepatitis with focal necrosis. Equine fetus.

recrudescence of latent virus or reinfection—until abortion varies from 9 days to several months. Respiratory disease in the mare is usually not observed following the infection which results in abortion. It is reported that 95% of the abortions due to EHV-1 occur in the last 3 months of pregnancy and naturally acquired infection has not been observed to produce abortion before 5 months of gestation. Exactly where the virus is, and in what state, during the protracted incubation period has not been determined.

The aborted fetuses may show characteristic and diagnostic lesions which are variable in their development and may be modest. Edema of the subcutis and fascia and accumulated amber fluid in the body cavities are common. There may be slight general icteric discoloration and meconium staining of the foot pads and amnion. The most consistent gross lesion is severe edema of the lungs. They are heavy and rubbery, show the impressions of the ribs, and exhibit a pitting response to pressure (Fig. 4.79A). There is also edema of the interlobular septa. The lungs may be darker or lighter than normal, and tan to white foci of necrosis (2–4 mm in diameter) and petechial hemorrhages may be visible on the surface. Casts of fibrin are occasionally present in the bronchi (Fig. 4.79B) and are rarely in the trachea. Beneath the capsule of the liver there are, in ~50% of aborted fetuses, gray to white foci of necrosis varying in size from minute up to 5.0 mm in diameter. Such foci may be few or numerous. The spleen is usually enlarged with petechial hemorrhages on the capsule and unusual prominence of the follicles on cut surface. Petechial or ecchymotic hemorrhages may occur anywhere, but chiefly in the upper respiratory mucosae. Occasionally, there is hemorrhagic necrosis of the renal cortices.

Histologically the pulmonary interlobular septa are edematous and are infiltrated with mononuclear inflammatory cells. The edema and spotty necrosis and hemorrhage involve the whole organ and there is a fibrinous alveolar exudation and necrosis of bronchial and alveolar epithelial cells (Fig. 4.80A). The acidophilic inclusion bodies found in the nuclei of the bronchial and alveolar epithelium are specific. The foci of hepatic necrosis are not as common as the changes in the lungs, they are often minute and may be missed in a section. Acidophilic inclusion bodies also form in the nuclei of hepatic parenchymal cells, but they are not constant and are never numerous. If present they can usually be found around the areas of focal necrosis. There is edema of the liver, and leukocytic infiltration in the necrotic foci and portal triads is common (Fig. 4.80B). Rarely there is a diffuse hepatitis without focal necrosis. Necrosis of germinal centers occurs in the enlarged splenic follicles and other lymphocytic tissues, including thymus. Intranuclear inclusion bodies may be found in the primitive reticular cells in such foci. There are hemorrhages in the splenic pulp and about the malpighian corpuscles. The placenta is normal.

Foals infected with this virus *in utero* may be born alive

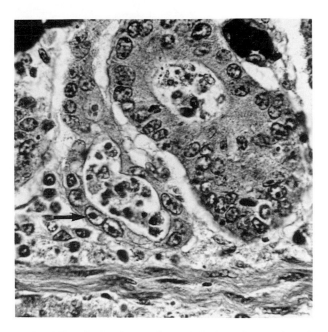

Fig. 4.81 Equine herpesvirus-1 infection. Intranuclear inclusions (arrow) and necrosis in intestinal crypts. Two-day-old foal.

at, or near, term. Whether any survive is not known. Many of them die in the first few days with severe interstitial pneumonia and secondary bacterial septicemia. Focal hepatic necroses are, as a rule, not present in these animals; however, focal necrosis of crypt epithelium with hemorrhage in the intestine is sometimes observed (Fig. 4.81).

The diagnosis can be based on the typical microscopic lesions, including the presence of inclusions. The demonstration of EHV-1 on cell cultures from samples of lung, liver, spleen, or thymus or by the immunoperoxidase technique on placenta is definitive. Identification of the specific virus may be accomplished using monoclonal antibodies. Serological examination for antibody is of little use in the diagnosis of abortion as most animals are exposed to the virus several times in their life and abortion may occur too long after the last exposure, making interpretation impossible, even on the basis of paired samples.

Bibliography

Allen, G. P. *et al.* Molecular epidemiologic studies of equine herpesvirus 1 infections by restriction endonuclease fingerprinting of viral DNA. *Am J Vet Res* **44:** 263–271, 1983.

Baker, J. C., Rust, S. T., and Walker, R. D. Transmission of a vaccinal strain of infectious bovine rhinotracheitis virus from intranasally vaccinated steers commingled with nonvaccinated steers. *Am J Vet Res* **50:** 814–816, 1989.

Bryans, J. T., and Allen, G. P. Herpesviral diseases of the horse. *In* "Herpesvirus Diseases of Cattle, Horses and Pigs," G. Wittmann (ed.), pp. 176–229, Boston, Kluvier Academic Publ., 1989.

Campbell, T. M., and Studdert, M. J. Equine herpesvirus type 1 (EHV1). *Vet Bull* **53:** 135–146, 1983.

Carmichael, L. E. Herpesvirus canis: Aspects of pathogenesis and immune response. *J Am Vet Med Assoc* **156:** 1714–1721, 1970.

Cornwell, H. J. C. *et al.* Neonatal disease in the dog associated with a herpes-like virus. *Vet Rec* **79:** 661–662, 1966.

Crabb, G. S., and Studdert, M. J. Comparative studies of proteins of equine herpesvirus 4 and 1 and asinine herpes 3: Antibody response of the natural host. *J Gen Virol* **71:** 2033–2041, 1990.

Dimock, W. W., Edwards, P. R., and Bruner, D. W. Infections observed in equine fetuses and foals. *Cornell Vet* **37:** 89–99, 1947.

Doll, E. R., and Bryans, J. T. Incubation periods for abortion in equine viral rhinopneumonitis. *J Am Vet Med Assoc* **141:** 351–354, 1962.

Donaldson A. I. *et al.* Experimental Aujeszky's disease in pigs: Excretion, survival and transmission of the virus. *Vet Rec* **113:** 490–494, 1983.

Edington, N. *et al.* Experimental transplacental transmission of porcine cytomegalovirus. *J Hyg Camb* **78:** 243–251, 1977.

Edington, N., Plowright, W., and Watt, R. G. Generalized porcine cytomegalic inclusion disease: Distribution of cytomegalic cells and virus. *J Comp Pathol* **86:** 191–202, 1976.

Jackson, T. A. *et al.* Equine herpesvirus 1 infection of horses: Studies on the experimentally induced neurologic disease. *Am J Vet Res* **38:** 709–719, 1977.

Hsu F. S. *et al.* Placental lesions caused by pseudorabies virus in pregnant sows. *J Am Vet Med Assoc* **177:** 636–641, 1981.

Kendrick, J. W., and McEntee, K. The effect of artificial insemination with semen contaminated with IBR-IPV virus. *Cornell Vet* **57:** 3–11, 1967.

Kennedy, P. C., and Richards, W. P. C. The pathology of abortion caused by the virus of infectious bovine rhinotracheitis. *Pathol Vet* **1:** 7–17, 1964.

Kluge, J. P., and Mare, C. J. Swine pseudorabies: Abortion, clinical disease and lesions in pregnant gilts infected with pseudorabies virus (Aujeszky's disease). *Am J Vet Res* **35:** 911–915, 1974.

Molello, J. A. *et al.* Placental pathology. V. Placental lesions of cattle experimentally infected with infectious bovine rhinotracheitis virus. *Am J Vet Res* **27:** 907–915, 1966.

Percy, D. H., Olander, H. J., and Carmichael, L. E. Encephalitis in the newborn pup due to a canine herpesvirus. *Pathol Vet* **5:** 135–145, 1968.

Poste, G., and King, N. Isolation of a herpesvirus from the canine genital tract: Association with infertility, abortion and stillbirths. *Vet Rec* **88:** 229–233, 1971.

Schiefer, B. Bovine abortion associated with cytomegalovirus infection. *Zentralbl Veterinaermed* **21:** 145–151, 1974.

Splitter, G. A., *et al.* Bovine herpesvirus-1: Interactions between animal and virus. *In* "Comparative Pathobiology of Viral Diseases," Vol I., G. Olsen, S. Krakowa, and J. R. Blakeslee (eds.), pp. 57–88. Boca Raton, Florida, CRC Press, 1985.

Stewart, S. E. *et al.* Herpes-like virus isolated from neonatal and fetal dogs. *Science* **148:** 1341–1343, 1965.

Studdert, M. J., Simpson, T., and Roizman, B. Differentiation of respiratory and abortigenic isolates of equine herpesvirus 1 by restriction endonucleases. *Science* **214:** 562–564, 1981.

Waldvogel, A. *et al.* Caprine herpesvirus infection in Switzerland: Some aspects of its pathogenicity. *Zentralbl Veterinaermed* (*B*) **28:** 612–623.

Whitwell, K. Studies on the pathogenesis of equine herpesvirus 1 (EHV-1). *In* "Animal Health Trust Scientific Report 1988-1989," pp. 9–10, 1990.

Wohlgemuth, K. *et al.* Pseudorabies virus associated with abortion in swine. *J Am Vet Med Assoc* **172:** 478–479, 1978.

Wyler, R., Engels, M., and Schwyzer, M. Infectious bovine rhinotracheitis/vulvovaginitis (BHV-1). *In* "Herpesvirus Diseases of Cattle, Horses, and Pigs," G. Wittman (ed), pp. 1–72. Boston, Kluvier Academic Publ., 1989.

p. VIRAL INFECTION OF THE FETUS   A variety of viral infections can produce fetal disease. The viruses, the species affected, and the organs primarily damaged are listed in Table 4.1. Most of these viruses produce diseases in adult animals, and these are discussed in detail elsewhere in these volumes. The list is necessarily incomplete; not all viral infections of domestic animals have been examined to determine their ability to produce infection in the fetus. Many of the findings have been unexpected, that a virus known for many years to produce disease in sheep, bluetongue, could produce fetal disease in another, hydranencephaly in bovine fetuses. Akabane virus, a virus not known to be pathogenic for any species, has proven to be an important teratogen in several species in widely separated parts of the world.

Most of the viruses listed in Table 4.1 are also known as teratogens, but whether a virus produces death of the conceptus, a teratogenic effect, or little or no effect depends on the age of the conceptus when infected and the virus involved. The vulnerability of the conceptus is usually inversely proportional to its age. Organogenesis, tissue differentiation, and maturation are successive stages during which the effects of viral infection are expected to be progressively less severe. These are only broad generalities influenced by different maturation speeds in different organs, the development of fetal immune competence, the viral strains involved, and their changing capability to infect the embryo or fetus during gestation.

Some of the viruses listed in Table 4.1 are minor causes of reproduction failure, but some are major causes of fetal disease. Porcine parvovirus is an important cause of fetal loss in swine. This disease can be diagnosed relatively easily. The importance of several other diseases that pose diagnostic problems, such as bovine virus diarrhea virus or bluetongue virus infection in bovine fetuses, is less clear.

At least one of the diseases listed is of such severity that few infected dams survive (rinderpest), and abortion and disease of the fetus is a minor part of the impact of the disease. But in other such diseases it has been shown that although reproduction failure is a relatively insignificant part of the disease, intrauterine fetal infection can produce persistently infected young (hog cholera, bovine virus diarrhea), which shed virus and play an important role in the maintenance and spread of the virus. Which viral infections have this vertical pattern of spread is not known; most apparently do not, but the possibility must be considered that any virus that can infect the fetus may also persist in the newborn.

Not all viral infections are important causes of fetal

TABLE 4.1

Viruses Known to Infect Fetuses of Domestic Species and the Reported Sequelae

Virus		Species affected	Pathogenicity	
Family	Specific name		Fetal organs affected	Results
Togaviridae	Hog cholera	Swine	CNS[a]	Growth retardation, malformation
	Bovine virus diarrhea	Cattle	CNS, eye, skin, skeleton	Mummification, abortion, growth retardation, malformation
	Border disease	Sheep, goats	CNS, skeleton, skin, thyroid	Growth retardation, malformation
	Equine viral arteritis	Horses	Multiple	Abortion
Flaviviridae	Japanese B encephalitis	Swine	CNS	Growth retardation, malformation
	Wesselsbron disease	Sheep, cattle	CNS, multiple	Mummification, abortion, malformation
Parvoviridae	Feline panleukopenia	Cats	CNS	Malformation
	Porcine parvovirus	Swine	CNS, multiple	Mummification, malformation
	Bovine parvovirus	Cattle	CNS, multiple	Abortion
	Minute virus of canines	Dogs	Multiple	Fetal death, resorption
Reoviridae	Bluetongue	Sheep, cattle	CNS	Fetal death, malformation
	Epizootic hemorrhagic disease	Cattle	CNS	Abortion, malformation
	Chuzan	Cattle	CNS	Malformation
Bunyaviridae	Akabane	Cattle, sheep, goats	CNS, muscle, skeleton	Abortion, malformation
	Cache Valley	Sheep	CNS, muscle	Malformation
	Rift Valley fever	Cattle, sheep	Liver	Abortion
	Aino	Cattle	CNS	Abortion, malformation
	Nairobi sheep disease	Sheep	Multiple	Abortion
Herpesviridae	Equine herpesvirus-1 (equine viral abortion)	Horses	Multiple	Abortion
	Canine herpesvirus	Dogs	Multiple, CNS, eye	Abortion, fetal death
	Bovine herpesvirus-1 (infectious bovine rhinotracheitis)	Cattle, sheep, swine	Multiple	Abortion
	Bovine herpesvirus-4 (bovine cytomegalovirus)	Cattle, swine	Multiple	Abortion, fetal death
	Bovine herpesvirus-5 (bovine encephalitis)	Cattle	Multiple	Abortion
	Caprine herpesvirus	Goats	Multiple	Abortion
	Aujeszky's disease (pseudorabies)	Swine	Multiple	Fetal death
	Porcine cytomegalovirus (inclusion-body rhinitis)	Swine	Multiple	Fetal death, growth retardation
Paramyxoviridae	Rinderpest	Cattle	Multiple	Abortion
Iridoviridae	African swine fever	Swine	Multiple	Fetal death
Picornaviridae	Encephalomyocarditis	Swine	Heart	Fetal death

[a] CNS, central nervous system.

disease. Some infections are so common that most animals will have become immune before reaching breeding age. This is thought to be important in limiting the role of parainfluenza 3 infection in fetal disease. Those virus infections that do not usually produce viremias, such as the feline herpesvirus, feline rhinotracheitis virus, rarely produce fetal infections. Other factors, such as ability to cross or infect the placenta, must be involved, but few studies of these features have been made. There are also subtle differences between strains that influence the fetal pathogenicity. Some strains of the hog cholera virus pro-

duce congenital tremors in pigs, whereas others do not, and similar strain differences appear to exist between various isolates of bovine virus diarrhea and Border disease. The basis of these differences is not known.

q. EPIZOOTIC BOVINE ABORTION This disease is a tickborne infection of cattle that produces chronic fetal disease and abortion. The name is a misnomer, the disease being provincial and endemic. Its distribution is apparently limited by that of the vector, the argasid tick *Ornithodoros coriaceus*. The precise geographic range of the tick is not known. It is known to inhabit brushy foothills of California and adjacent areas of Nevada, Oregon, and northern Mexico. The most common hosts of the tick are cattle and deer. It is presumed that the tick is transmitting a virus from the deer to cattle, as the infection remains endemic in ranges on which no cattle have been grazed. This hypothetical virus remain elusive however, and the cause of the disease is unknown. Previously it had been thought that the disease was caused by chlamydial organisms, then known as members of the psittacosis–lymphogranuloma group. The tick carries a spirochete, *Borrelia coriaceae,* and this organism has been proposed as the cause of the disease, but this has not been proved, and aborted fetuses do not develop antibodies to the spirochete.

Cattle being exposed to the ticks for the first time are primarily at risk. The infection produces no recognizable clinical signs in cattle of any age, but if the animals are pregnant, the infection is passed to their fetuses where a chronic disease develops. There is a 3-month or longer incubation period between the exposure of the dam to ticks and the abortion of the fetus. Not all diseased fetuses are aborted. Weak calves are often associated with outbreaks of abortion in this disease. Cows that abort due to the disease are not usually ill, and the placenta is shed without difficulty. They conceive again when bred, and in most cases, do not abort again, even if exposed to tick-infested range. If there is little movement of animals, abortions may be rare in endemic areas. The disease becomes a major cause of abortion in a herd only when pregnant animals are moved into an area where the infection is endemic.

The fetuses affected by the disease are most often aborted during the last trimester. Most trigger their own deliveries and die either during delivery or shortly thereafter. Only rarely do they die *in utero* and undergo autolysis. Many affected fetuses have a characteristic abdominal distension due to ascites which is a striking, but inconstant, lesion. Lymph nodes throughout the carcass are enlarged, usually impressively so. The normal superficial cervical lymph node of a bovine term fetus weighs between 3.5 and 7.0 g. Fetuses aborted in this disease have cervical lymph nodes which weigh 16 g or more. The spleens of affected fetuses are similarly enlarged; the thymuses in contrast are slightly reduced in size.

Petechial hemorrhages are regularly seen in the conjunctival and oral mucosa as well as in the mucosa of the trachea. The thymuses are often embedded in massive hemorrhages and edema. These hemorrhages are, at least in part, traumatic, developing during parturition, the protected portion of the thymus within the thoracic cavity being spared. The enlarged, coarsely nodular liver is, when present, an impressive gross change in fetuses affected by enzootic bovine abortion (Fig. 4.82). However, the lesion is not present in all diseased fetuses, and similar hepatic congestion with ascites occurs in fetuses with cardiac anomalies or myocardial degeneration. Many aborted fetuses breathe, and the lungs of these are partially aer-

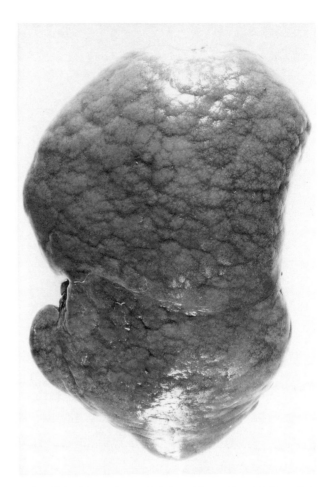

Fig. 4.82 Congestive nodularity of liver. Aborted fetus. Epizootic bovine abortion.

ated. Small gray foci of inflammation can be seen in a variety of tissue, but they can be seen most easily in organs such as kidney and heart, which provide a dark background.

The changes that develop in the lymphoid organs are the most specific and diagnostic. The enlargement of lymph nodes and spleen consists of remarkable lymphoid and mononuclear cell hyperplasia. Well-defined secondary follicles develop in the cortical and paracortical regions of the lymph nodes. The sinuses are stuffed with multinuclear giant cells and macrophages, which form sheets in the medullary areas (Fig. 4.83A). The spleens show a similar hyperplastic response involving lymphocytes of the follicles and periarterial sheaths; there is as well a widespread infiltration of macrophages. Foci of acute necrosis are often superimposed on the chronic proliferative change (Fig. 4.83B)

The most distinctive fetal lesion is a remarkable thymic inflammatory change. The cortical mantle of thymocytes is greatly reduced, and macrophages diffusely infiltrate both the medulla and the septa of the gland (Fig. 4.84A). The central veins of the liver are distended and the liver

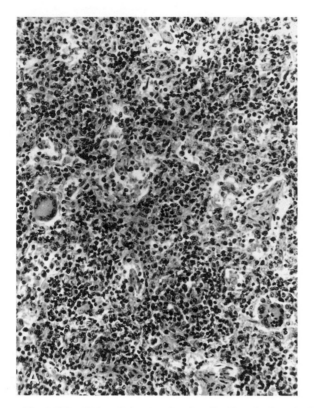

Fig. 4.83A Epizootic bovine abortion. Proliferative lymphadenitis with giant-cell formation in aborted calf.

Fig. 4.84A Epizootic bovine abortion. Fetus. Thymitis with distortion of thymic architecture by infiltrating macrophages.

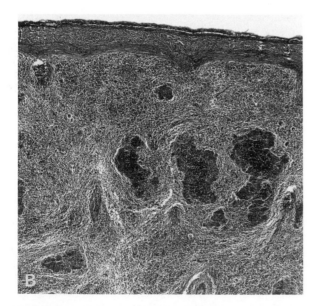

Fig. 4.83B Focal areas of acute necrotizing splenitis. Fetus.

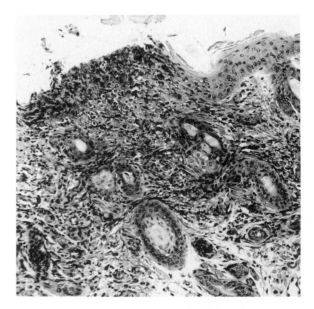

Fig. 4.84B Focal necrotizing dermatitis.

plates are thinned. A mononuclear cell infiltration is present around portal vessels. Foci may be 100 μm in diameter and are often granulomatous in appearance.

Affected fetuses may have inflammatory vascular lesions involving small- and large-caliber vessels in any, or

all, organs (Fig. 4.84B). Lesions that are particularly useful in establishing the diagnosis are present regularly in the lung and brain. The alveolar walls of the lung are thickened, and granulomatous inflammatory foci are present in the septa. In the brain, foci of vasculitis are scattered

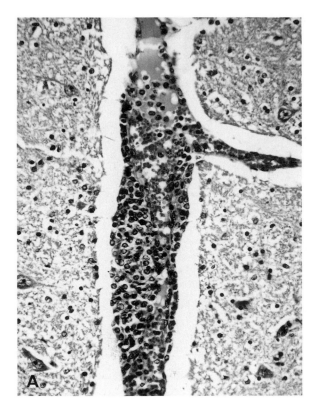

Fig. 4.85A Epizootic bovine abortion. Fetus. Cerebral vasculitis.

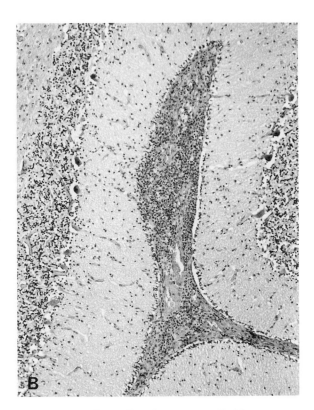

Fig. 4.85B Granulomatous meningitis.

throughout, and the meninges are thickened by granulomatous inflammation (Fig. 4.85A,B).

Lesions do develop in the placenta but are usually mild and involve the loose connective tissue. They contrast with the acute severe inflammatory reaction of the chorionic surface seen in bacterial and chlamydial infections.

The fetal disease is a very chronic one. The entire spectrum of lesions requires at least 3 months to develop. Since the lesions gain their specificity only during the latter part of this prolonged incubation period, the disease can be confidently diagnosed only in those animals exposed before the last trimester.

Bibliography

Johnson, R. C. *et al. Borrelia coriaceae* sp. nov.: Putative agent of epizootic bovine abortion. *Intl J Syst Bacteriol* 37: 72–74, 1987.

Kennedy, P. C. *et al.* Epizootic bovine abortion: Histogenesis of the fetal lesions. *Am J Vet Res* 44: 1040–1048, 1983.

Kimsey, P. B. *et al.* Studies on the pathogenesis of epizootic bovine abortion 44: 1266–1271, 1983.

Lane, R. S. *et al.* Isolation of a spirochete from the soft tick, *Ornithodoros coriaceus:* A possible agent of epizootic bovine abortion. *Science* 230: 85–87, 1985.

LeFebvre, R. B., and Perng, G. C. Genetic and antigenic characterization of *Borrelia coriaceae*, putative agent of epizootic bovine abortion. *J Clin Microbiol* 27: 389–393, 1989.

Osebold, J. W. *et al.* Congenital spirochetosis in calves: Association with epizootic bovine abortion. *J Am Vet Med Assoc* 188: 371–376, 1986.

Osebold, J. W. *et al.* Histopathologic changes in bovine fetuses after repeated reintroduction of a spirochete-like agent into pregnant heifers: Association with epizootic bovine abortion. *Am J Vet Res* 48: 627–633, 1987.

Schmidtmann, E. T. *et al.* Experimental and epizootiologic evidence associating *Ornithodoros coriaceus* Koch (Acari–Argasidae) with the exposure of cattle to epizootic bovine abortion in California. *J Med Entomol* 13: 292–299, 1976.

Spezialetti, R., and Osebold, J. W. Lymphocyte blastogenesis and cellular cytotoxicity in a congenital infection of bovine fetuses related to epizootic bovine abortion. *Res Vet Sci* 46: 160–167, 1989.

3. Miscellaneous Lesions of the Postpartum Uterus

Serosal cysts are observed occasionally in the aged pluriparous bitch and very rarely in the cow. They are thin-walled cysts containing a clear watery fluid and vary in size from a few millimeters to several centimeters. They appear to be retention cysts formed during involution of the uterus from pinched-off segments of serosal epithelium.

Endometrial cysts, adjacent to caruncles, develop in some cows and ewes during the process of uterine involution (Fig. 4.86). Adhesion of the caruncular stalk to adjacent glandular tissue causes blockage of the underlying glands. The retention cysts enlarge progressively with age, and care must be taken to differentiate them from estrogen-linked cystic hyperplasia.

Fig. 4.86   Endometrial cysts. Cow.

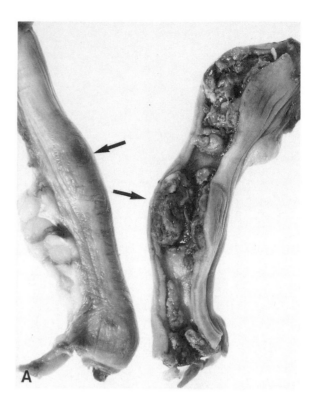

Fig. 4.87A   Subinvolution of placental sites. Bitch. Fusiform enlargement of uterine horns at sites of failure of involution (arrows).

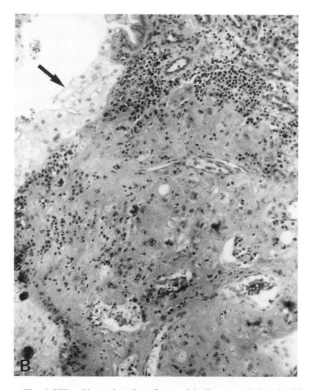

Fig. 4.87B   Vacuolated surface epithelium overlying decidual cells, necrotic debris, and inflammatory cells.

Subinvolution of **placental sites** in the bitch is manifested clinically by a prolonged, blood-tinged vaginal discharge. In the normal bitch, overt uterine bleeding usually ceases within 7–10 days following whelping, but the affected bitch may bleed for several weeks or months. In some cases, the blood loss causes severe anemia and occasionally death. Ellipsoidal enlargements, located in the areas of previous placental attachment, are evident in the uterine cornua (Fig. 4.87A). The endometrium in the affected areas is hemorrhagic, irregularly thickened, and gray to brown. The endometrium between the enlargements is normal in appearance. The surface epithelium is often detached, but when present, it has heavily vacuolated cytoplasm indicating progestational stimulation (Fig. 4.87B). Corpora lutea are invariably present, but progesterone levels have been low in the few cases in which the determination was made. The uterine mass adjacent to the uterine lumen is composed of an admixture of amorphous eosinophilic debris, thrombotic material, degenerating placental site tissue, and regenerating endometrium (Fig. 4.88A). The deeper

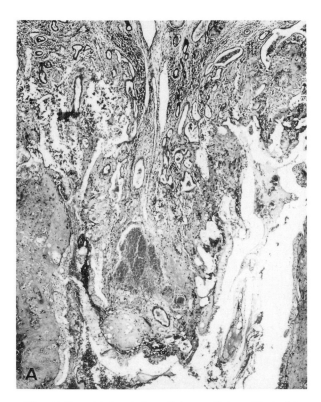

Fig. 4.88A Subinvolution of placental sites. Bitch. Hemorrhage and necrotic debris in pedunculated endometrium.

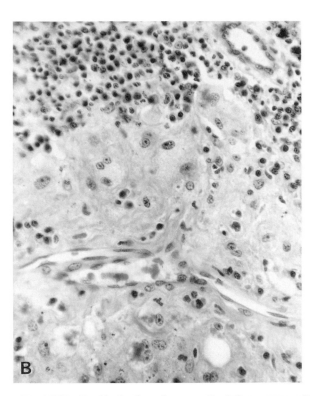

Fig. 4.88B Decidual cells and surrounding inflammatory cells deep in the placental site.

tissue is permeated with numerous irregular-shaped cells with large nuclei and abundant vacuolated cytoplasm (Fig. 4.88B). They are interpreted as syncytial trophoblasts or alternatively as decidual cells, and are far more numerous than in the normal gravid uterus. Some degree of invasion of the myometrium by these cells is not unusual, and in some cases there is perforation of the serosa, allowing uterine contents to escape into the peritoneal cavity. The condition appears to be more prevalent in young bitches, but the cause has not been established.

VI. The Cervix

Cysts of the cervix occur in cows, and probably all are retention cysts formed by fusion of the rugae. Loss of the original epithelium of the rugae is first necessary and may be incidental to the lacerations of parturition, artificial insemination, or inflammation. The cysts are usually small and not significant. Larger ones may cause partial occlusion of the cervical canal, but this is seldom of importance.

Stenosis of the cervix is extraordinary. It is acquired rather than congenital and consists of fusion across epithelial surfaces and scarification. It may follow severe laceration or long-standing inflammation.

Cervicitis is not in itself an entity but, instead, is an extension of an endometritis or vaginitis. The mucus-secreting epithelium provides good defense against bacte-

rial invasion and, if the epithelium breaks down, it exposes densely arranged connective and muscular tissues which are not especially sensitive to the actions of bacteria. In the cow, the epithelium lining the external os and the adjacent few rugae is of the simple vaginal type, and this is more susceptible to irritation than the mucus-secreting surface anterior to it. Most inflammations then are superficial, but there are some exceptions such as necrobacillosis. The circumstances in which cervicitis occurs have been mentioned under diseases of the uterus or will be mentioned under diseases of the vagina. The usual form of bovine simple cervicitis is seen as a swelling of the caudal annular rugae, which are edematous and hyperemic. They soon protrude through the external os into the vagina, and a thin mucopurulent exudate accumulates between the folds and collects in the vagina. Histologically, there is epithelial degeneration and desquamation and an infiltration by inflammatory cells, which are predominantly mononuclear. Neutrophils penetrate the epithelium and become mixed with mucus, or in more acute cases, there may be frank suppuration. Cervicitis is not usually more severe than this, although in older cows, the prolapse and tumefaction of the rings may be grossly obvious. Although degrees of prolapse result from inflammation, they may also predispose to it. Slight but progressive degrees of eversion of the cervical rings occur with succeeding pregnancies to expose portions of the cervical mucosa to the contaminated environment of the vagina. Chronic cervicitis may lead in time to enlargement and induration of the cervix with some stenosis, but enlargement is not a criterion of inflammatory sclerosis. Discrete cervical abscesses or suppurative fistulous tracts occasionally result from accidental injury acquired during uterine irrigation or artificial insemination.

VII. The Vagina and Vulva

Cysts in the vagina or vulva are not important in themselves, but they do have some diagnostic significance. They occur as **cystic dilations in Gartner's ducts and Bartholin's glands** in cattle. Gartner's ducts are remnants of the embryonic Wolffian ducts; they lie, one on each side of the floor of the vagina, beneath the epithelium. They are invariably present to some degree in cattle although, when normal, they are detectable only microscopically as discontinuous ducts lined by a simple epithelium and found in the anterior vagina and disappearing caudally. They become cystic in cows poisoned with highly chlorinated naphthalenes, in cows with ovarian follicular cysts, and occasionally following acute vaginitis. With lesser degrees of dilation, the cysts are more readily palpable than visible, but in well-developed instances, they are clearly visible through the elevated and thinned vaginal wall (Fig. 4.16). The cysts may be isolated, or they may have a string-of-beads disposition, or the whole duct may be present, dilated to 1–2 cm and tortuous. Cysts of Bartholin's glands occur under the same stimuli, but chiefly as a consequence of inflammation. The glands lie one on each side of the floor of the vestibule and normally are about 3 × 1.5 cm in dimensions. They become visible when enlarged, especially if the vestibular mucosa is slightly everted. These glands are very sensitive to estrogens, responding with the elaboration of a thin mucus and hyperplasia of the ductal epithelium. Gross exaggeration of responses in hyperestrogenism accounts for the cystic development. Large retention cysts follow inflammatory stricture of the short excretory ducts. Abscessation may follow localization of infection in the cysts.

Ruptures of the vagina and vulva are quite frequently acquired as parturient injuries. The mucosa alone may be ruptured, or the entire thickness of the vaginal wall. Hemorrhages occur into the vagina or the perivaginal wall from fetal pressure or vascular disruption. Pelvic fat may herniate into the vagina; it is often mistaken for a neoplasm. Defects which extend deeper than the mucosa heal with cicatrization which may, in turn, result in partial stricture. Otherwise, the outcome depends on whether the lacerations become infected and, if so, with what. Occasionally, diffuse phlegmon, abscess, gangrene, or peritonitis are sequelae.

A. Vulval Tumefaction (Vulvovaginitis) of Swine

Tumefaction of the vulva is a physiological response to estrogens. It may be of exceptional development or persistence in hyperestrogenism.

Vulval tumefaction in swine is due to mycotoxins. A distinctive syndrome characterized by vulval hyperemia and edema occurs in swine associated with the feeding of moldy grains. The mycotoxin that causes this estrogenic effect is zearalenone or F-2, which is produced by at least four different *Fusarium* species, *Fusarium graminearum* being the most common.

Young gilts are chiefly affected, older animals being more resistant. There is remarkable edematous swelling of the vulva and vagina, which may be severe enough to lead to eversion and prolapse of the mucosa. Prolapse of the vagina may occur in up to 30% of cases and secondary rectal prolapse occurs in a lesser percentage. The uteri of affected animals may be enlarged by endometrial gland hyperplasia and edema. The ovaries become inactive and atrophic. Gilts may also show mammary gland enlargement with squamous metaplasia of the ductal epithelium. Decreased libido, testicular atrophy, and balanoposthitis have been reported in young boars.

Strains of *Fusarium* vary in toxigenic potency, not only in regard to zearalenone but also to trichothecene toxins which, if produced in quantity, may broaden the clinical picture to include anorexia and vomition.

B. Inflammatory Diseases of the Vagina and Vulva

The mucous membrane of the vagina and vulva shares, with mucous membranes in general, a sensitivity to irritants. Although the vagina is of Müllerian duct origin, the original epithelium is replaced by stratified squamous epithelium from the urogenital sinus. This epithelium proliferates and matures under the influence of estrogen and is then more resistant to infection. This enhanced resistance may be due to mechanical factors in the thickened keratinized epithelium and to local production of lactic acid from the glycogen which is deposited in the epithelium under the influence of estrogens. There are probably additional ingredients in the recipe for resistance. We shall be concerned here only with the specific types of vaginitis and vulvitis, although both are more commonly nonspecific and of simple catarrhal type.

1. Granular Venereal Disease

Papular eruptions of the vulval mucosa are common in most domestic species but are best known in the bovine species by this name or as nodular venereal disease, granular vaginitis, etc. The term vaginitis is a misnomer, as the papules are strictly limited to the vulval mucosa and are not found in the vagina, although in acute cases there may be an associated nonspecific vaginitis. Vulval granules may be found in any herd, affecting animals of any age, but are usually more prominent in heifers bred naturally. They are much less common in pregnancy and are almost never present about the time of parturition. In severe cases, the papules may be found on all aspects of the vulval mucosa, but usually they are clustered in the ventral commissure about the clitoris as pale or pink elevations a few millimeters in diameter and covered by a normal intact vulval mucosa. When numerous, they are likely also to be

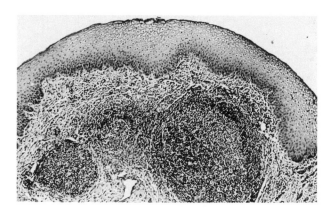

Fig. 4.89 Granular venereal disease. Cow. Lymphoid follicle in submucosa.

more active, larger, often coalescent, congested and red with a covering catarrhal vulvitis and vulval swelling. The overlying epithelium is then easily injured and bleeding occurs freely from the papules. The resting papules are composed of respectably organized lymphoid follicles (Fig. 4.89). When the vulval mucosa is irritated, these become congested, with small intrafollicular hemorrhages and edema, and hyperplastic, the mitotic frequency being quite high. The same lesions occur on the penis and prepuce of the bull and, as do those of the vulva, often persist for many months. These papules have been produced experimentally with ureaplasmas and occur commonly in herds with natural *Ureaplasma* infections but are not specific for this organism. The development of subepithelial lymphocytic foci is a characteristic response of mucous membranes to mild persistent or recurrent irritation and the simplest explanation of granular venereal disease is that it is the result of mild inflammation of the vulval mucosa.

2. Infectious Bovine Cervicovaginitis and Epididymitis (Epivag)

This is a specific infectious disease that has been an important cause of infertility in east and southern Africa. A slow-growing cytomegalovirus of the bovine herpesvirus 4 group has been recovered from infected animals. In limited studies, these isolates have not reproduced the full spectrum of clinical signs, so there remains some uncertainty about their role in the disease. The slow-growing herpes strains are not serologically related to the bovine herpesvirus-1 virus of infectious bovine rhinotracheitis.

Natural transmission is solely by coitus. Experimental transmission is easy if infective discharges are placed in the vagina or prepuce. After incubating for a few days, diffuse purplish inflammatory streaks or patches develop in the anterior vagina to be followed by the development of a copious tenacious creamy discharge in which there are large numbers of neutrophils but few or no organisms. The infection spreads readily from the vagina to the cervix and uterus, also with the production there of the same copious pus. About 25% of infected females are sterile because of the development of chronic salpingitis with

hydrosalpinx and bursal adhesions. In the bull, there is early but slight palpable enlargement of the spermatic cord and epididymis. The disease usually commences in one epididymis but later involves both organs and spreads from the tail of the epididymis to the head of it and ultimately to the testicle. The lesion is apparently an interstitial epididymitis with the production of excessive peritubular fibrosis and tubular obstruction. Testicular changes are probably secondary to those in the epididymis and to obliterative adhesions in the cavity of the tunica vaginalis. Similar productive inflammatory lesions occur in the ampullae and the seminal vesicles.

3. Infectious Pustular Vulvovaginitis of Cattle

We are assuming for this description of infectious pustular vulvovaginitis that it is the disease which for many decades has been variously termed vesicular venereal disease, vesicular vaginitis, coital exanthema, coital vesicular exanthema, and *Blaschenausschlag*. The assumption is probably valid, but proof is lacking and may no longer be obtainable. Infectious pustular vulvovaginitis is caused by the same herpesvirus (bovine herpesvirus-1) that causes infectious bovine rhinotracheitis (see The Respiratory System, Volume 2, Chapter 6), or at least is caused by a subtype of that virus, which is serologically indistinguishable from it. As a rule, nasal and vaginal infections behave epidemiologically as distinct diseases, although occasionally the syndromes occur together in individual animals. The infection can be transmitted to sheep and goats, producing a vaginitis.

Infectious pustular vulvovaginitis is highly contagious. It is frequently transmitted by coitus, but it can also be transmitted by other mechanical means and is contagious by close contact. It may involve individual or few animals in a herd, but frequently spreads rapidly to involve all exposed females in a few days. The disease subsides in about 10 days, leaving immunity which is fragile and transient. Reinfection can occur, but early reinfection produces only a mild disease.

The incubation period is 1–3 days but may be as brief as 12 hr. The lesions are restricted to the genital tract, but a viremic phase probably occurs because there is early fever and leukopenia. Initially, there is hyperemia of the vaginal and vulval mucosa with focal hemorrhages in the lymphocytic follicles of the submucosa. The severity of the vulvovaginitis increases rapidly, and edema of the vulva and mucopurulent vaginal discharge develop. The focal lesions replace the hemorrhages over the lymphoid follicles and consist of small (2–3 mm) pocklike foci, slightly elevated, pale, soft, and friable (Fig. 4.90). The focal lesions, being related to the lymphoid follicles, may be in short linear arrangements. The epithelium in the focal lesions erodes or ulcerates so that in a few days the foci are flat, gray, semitransparent plaques the size of the original lesions.

The virus is epitheliotropic, the initial and most severe alterations occurring in the epithelium of the vagina and vulva. There is ballooning degeneration of the epithelial

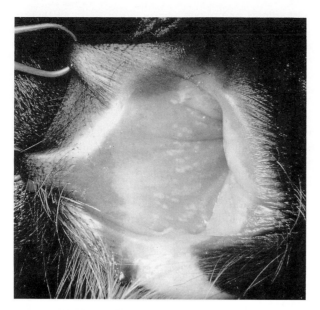

Fig. 4.90 Vulva. Cow. White plaques of epithelial necrosis caused by bovine herpesvirus-1 (infectious pustular vulvovaginitis). (Courtesy of *Cornell Veterinarian.*)

cells, and at about 24 hr, intranuclear inclusions can be found in the epithelium (Fig. 4.91A). The inclusions are lightly acidophilic or amphophilic and large; they can be found for 3–4 days, by which time the lesion has reached its zenith and is beginning to resolve. The parasitized cells undergo necrosis, and epithelial disruption and ulceration occur accompanied by an intense infiltration by neutrophils (Fig. 4.91B). Vesicles and true pustules do not form. Acute inflammation occurs in the lamina propria, with hyperemia and edema and the exudation of numerous plasma cells and lymphocytes. Many of the small vessels are occluded by adventitial and endothelial swelling. The lymphocytic follicles are remarkably hyperplastic and edematous, and their outlines are obscured by the infiltrating cells in the lamina propria. Resolution occurs in about 8 days, with hyperplastic lymphoid follicles and slight epithelial thickening as residues.

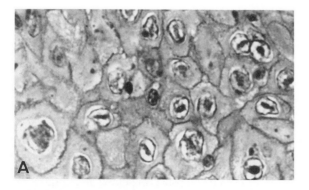

Fig. 4.91A Intranuclear inclusions in vaginal epithelium in bovine herpesvirus-1 infection.

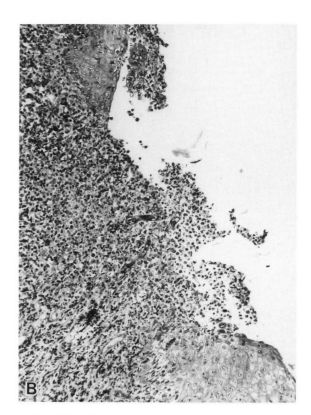

Fig. 4.91B Histologic detail of Fig. 4.90. Ballooning degeneration and ulceration of epithelium.

Although most cows which are served naturally by infected bulls do not appear to experience infertility, susceptible heifers, which are inseminated with semen containing virus, fail to conceive. Intranuclear inclusions may be found in the epithelial cells of the zona compacta within 48 hr but are absent by 72 hr after exposure. The surface epithelium and the underlying connective tissue become necrotic and infiltrated with neutrophils. There is pronounced edema of the lamina propria and infiltration of mononuclear-type cells, especially lymphocytes. The oviducts are involved but to a less severe degree.

The virus can produce similar lesions on the mucous membrane of the penis of infected bulls. Since recrudescence with viral shedding is a feature of this, as well as other, herpetic diseases, animals with inapparent infections can also transmit the disease.

A herpesvirus identified as equine herpesvirus-3 produces a comparable genital disease in horses, equine coital exanthema. It is discussed with diseases of the penis in The Male Genital System (Chapter 5 of this volume).

4. Necrotic Vaginitis and Vulvitis

Necrotic vaginitis is a deep diphtheritic inflammation of the vaginal mucosa; it occurs in two fairly distinct syndromes, being either a vulvovaginitis or a cervicovaginitis. Necrotic vulvovaginitis is uncommon but may involve a number of cows in a herd. It is primarily due to trauma with contamination, often the result of bite wounds by

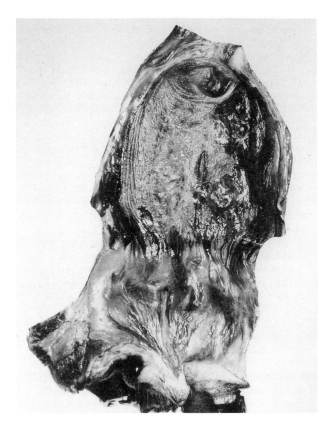

Fig. 4.92 Necrotic vaginitis of anterior vagina secondary to dystocia. Cow. (Courtesy of C. A. V. Barker.)

pigs or dogs. Necrotic cervicovaginitis (Fig. 4.92) is a complication of parturition and is observed chiefly in ewes and cows as a consequence of dystocia. A number of influences, mainly prolonged pressure necrosis, laceration, and abrasion, usually acting in combination, are responsible for the lesion. Necrotic cervicovaginitis is frequently fatal either in association with direct extension of the inflammation to the peritoneum or in association with the uterine complications of prolonged dystocia and fetal emphysema.

5. Dourine

Dourine of horses and their relatives is caused by *Trypanosoma equiperdum* and is primarily a venereal disease. The other trypanosomiases of domestic animals are discussed with The Hematopoietic System (Section II,F,4 of The Erythron in Chapter 2 of this volume), they being essentially hemic infections with arthropod vectors. Dourine differs from other trypanosomiases in that the organisms are in the blood only intermittently or in very virulent infections, and transmission is effected not by bloodsucking insects but by contact of infected mucous membranes. Natural transmission is by coitus so that the disease is almost exclusively one of stallions and breeding mares. Rare cases occur in unbred animals and in young foals. The disease prevails in the Balkans, much of Africa, Asia,

and South America. It has been eradicated from most of Europe and North America.

Strains of *Trypanosoma equiperdum* vary considerably in their virulence although not notably in their capacity to infect. Irrespective of the route of experimental infection in horses, the trypanosomes demonstrate some predilection for genital mucosae. Following natural infection, the organisms are frequently numerous in the vaginal discharges and male urethra. There are periods, however, sometimes of several weeks or months, in which the organisms are not present in these sites, so that infected animals are not always infective for others at breeding. The organisms penetrate intact mucosa at the site of implantation and proliferate in the submucosal lymph spaces. The incubation period may be several weeks or months, during which trypanosomes are present in the genital discharges, so that there is a possibility that they may proliferate in the lumen of the genital tract for long periods before invading the tissues. From the initial lesions, the organisms are disseminated in the blood to other parts of the body, and edematous swellings occur where they localize. Infection of the blood may be intermittent or continuous, and there may be few organisms or many, depending on the virulence of the strain and the susceptibility of the host. Strains of low virulence may not be demonstrated in the blood by direct smears, but may be demonstrated by centrifuging plasma or directly by transfusing blood in large volumes. In these mild infections, it may also be possible to demonstrate the organisms in genital washes or discharges, but diagnostic problems are simplified by an efficient complement-fixation test.

The signs and course of the infections vary considerably depending on the virulence of the organism and on the susceptibility of the host. The resistance of the host can be greatly modified by climatic conditions, physical condition, intercurrent disease, nutritional status, and so forth, and the South African varieties of the disease can be asymptomatic in animals that are well husbanded. Animals affected with this insidious type of the disease do act as carriers and reservoirs of infection. When the infection is attended by clinical signs, the course is variable; it may be severe and fatal in a few weeks, fatal after a chronic or intermittent course of from several months to 2 years, or clinical recovery may occur.

The signs of dourine can be divided into genital, cutaneous, nervous, and general manifestations, which occur separately or concurrently. The initial signs are usually genital but may be nervous or cutaneous. The incubation period of the genital signs varies from several days to several months during which the organisms are present in genital discharges or washings. The external genitalia are swollen and doughy, but characteristically the swellings are neither hot nor painful. The degree of swelling varies considerably from case to case and periodically in an established case, with a tendency to be permanent due to induration in chronic infections. The swelling, when severe, extends to the perineum and ventral abdominal wall. The lymphocytic follicles of the mucosa of the female genitalia become hyperplastic and ulcerate, and there may at this stage be a copious fluid discharge. In males, the swelling involves the glans penis as well as the prepuce and may cause prolapse of the urethra and penis. In virulent infections, flat circular ulcers develop in the glans penis. Healed ulcers in both males and females remain often as depigmented scars, and pigmentary atrophy of the skin and genital mucosa may also occur in the absence of ulceration, the loss of pigment being when present a very characteristic sign of this disease.

Cutaneous lesions may never occur in the mild disease. In virulent infections, edematous urticarialike plaques, usually circular but sometimes linear or in rings, occur in the skin, especially on the sides of the body and croup. The swellings may be up to 15 cm in diameter and painless and free of itch. They disappear in a few days and new ones form. There is usually no residuum, but sometimes there are local disturbances of sweating and pigmentation.

Nervous manifestations develop late in the course and usually lead to death. There is acute hyperesthesia initially, which may be generalized or localized to the distribution of particular nerves. Later, there is diminished sensitivity or even anesthesia, and this is accompanied by paresis or paralysis of individual motor nerves. The pareses are either unilateral in distribution or asymmetrical in severity and involve most commonly the facial nerves and the motor nerves of the hind limbs. Decubitus follows.

The general manifestations are chiefly of continued or remittent fever, emaciation, and severe anemia. Noninflammatory synovial effusions, superficial lymphadenopathy, uveitis, and optic atrophy are described.

Apart from changes in the peripheral nerves which are responsible for the paralytic phenomena, pathologic changes additional to those which are clinically observable are not reported. Degenerative changes occur chiefly in the lumbar and fifth and seventh cranial nerves, being most severe in the roots and involving also the ganglia. The neuropathy is probably preceded by edematous infiltrations, and these may still be present at death. The large nerve trunks are transformed into fibrous cords, which are fused with the surrounding muscular fascia. Microscopically, there is edema, mononuclear infiltration, and fibrosis of the perineurium. There is slight endoneural reactivity in fascicles in which there is extensive fiber degeneration. Sclerosing changes are present also in the ganglia.

Bibliography

Afshar, A., Stuart, P., and Huck, R. A. Granular vulvovaginitis (nodular venereal disease) of cattle associated with *Mycoplasma bovigenitalium*. *Vet Rec* **78:** 512–519, 1966.

Al-Bassam, M. A., Thomson, R. G., and O'Donnell, L. Involution abnormalities in the postpartum uterus of the bitch. *Vet Pathol* **18:** 208–218, 1981.

Anderson, J., Plowright, W., and Purchase, H. S. Pathological and seminal changes in bulls affected with a specific venereal infection. *J Comp Pathol* **61:** 219–230, 1951.

Baker, J. A., McEntee, K., and Gillespie, J. H. Effects of infectious bovine rhinotracheitis—infectious pustular vulvovaginitis (IBR-IPV) virus in newborn calves. *Cornell Vet* **50:** 156–170, 1960.

Barrowman, P. R., and Vuuren, M. van. The prevalence of dourine in southern Africa. *J South Afr Vet Assoc* **47:** 83–85, 1976.

Bristol, F. N., and Djurikovic, S. Hyperestrogenism in female swine as a result of feeding moldy corn. *Can Vet J* **12:** 132–135, 1971.

Bryans, J. T., and Allen, G. P. *In vitro* and *in vivo* studies of equine "coital" exanthema. *Proc 3rd Int Conf Equine Infect Dis* 322–336, 1973.

DeKock, G., Robinson, E. M., and Parkin, B. S. Some observations on dourine. *J South Afr Vet Med Assoc* **10:** 1–11, 1947.

Gillespie, J. H. *et al.* Comparison of infectious pustular vulvovaginitis virus with infectious bovine rhinotracheitis virus. *Cornell Vet* **49:** 288–297, 1959.

Grieg, A. S. *et al.* Cultivation in tissue culture of an infectious agent from coital exanthema of cattle. A preliminary report. *Can J Comp Med* **22:** 119–122, 1958.

Hellig, H. Investigation into a natural outbreak of infectious pustular vulvovaginitis (IPV) in cattle in South Africa. *J South Afr Vet Med Assoc* **36:** 219–227, 1965.

Hudson, J. R. A specific venereal disease of cattle characterized by epididymitis in bulls and vaginitis in cows and heifers. *Proc 14th Int Vet Congr, Lond* **3:** 487–491, 1949.

Kendrick, J. W., Gillespie, J. H., and McEntee, K. Infectious pustular vulvovaginitis of cattle. *Cornell Vet* **48:** 458–495, 1958.

Koen, J. S., and Smith, H. C. An unusual case of genital involvement in swine associated with eating moldy corn. *Vet Med* **40:** 131–133, 1945.

Kurtz, H. J. *et al.* Histologic changes in the genital tracts of swine fed estrogenic mycotoxin. *Am J Vet Res* **30:** 551–556, 1969.

Mare, C. J., and Van Rensberg, S. J. The isolation of viruses associated with infertility in cattle: A preliminary report. *J S Afr Vet Med* **32:** 201–210, 1961.

McErlean, B. A. Vulvovaginitis of swine. *Vet Rec* **64:** 539–540, 1952.

Mirocha, C. J., Christensen, C. M., and Nelson, G. H. Estrogenic metabolite produced by *Fusarium graminearum* in stored grain. *Appl Microbiol* **15:** 497–503, 1967.

Moulton, J. E., Coleman, J. L., and Gee, M. K. Pathogenesis of *Trypanosoma equiperdum* in rabbits. *Am J Vet Res* **36:** 357–366, 1975.

Parkin, B. S. The demonstration and transmission of the South African strain of *Trypanosoma equiperdum* of horses. *Onderstepoort J Vet Sci* **23:** 41–58, 1947.

Stob, M. *et al.* Isolation of an anabolic, uterotrophic compound from corn infected with *Gibberella zeae*. *Nature* **196:** 1318, 1962.

Theodoridis, A. Preliminary characterization of viruses isolated from cases of epididymitis and vaginitis in cattle. *Onderstepoort J Vet Res* **45:** 187–195, 1978.

Van Rensberg, S. W. J. Bovine sterility caused by infectious disease in South Africa. *Br Vet J* **109:** 226–233, 1953.

Wyllie, T. D., and Morehouse, L. G. (eds.). "Mycotoxic Fungi, Mycotoxins, Mycotoxicoses." New York, Marcel Dekker, 1977.

C. Neoplastic Diseases of the Tubular Genitalia

Isolated reports have ascribed parentage of both benign and malignant tumors to all histologic components of the tubular genitalia, sometimes with doubtful credentials. Some of the least uncommon tumors warrant brief discussion.

Tumors of the vulva could be discussed with tumors of the skin. Squamous cell carcinoma is well known in the mare and cow, and, indeed, a high incidence in cows is reported from Kenya. These bovine cases are analogous to the orbital tumors of cattle, thought to be initiated in a solar dermatosis, their incidence negatively correlated with the degree of local epithelial pigmentation.

1. Leiomyoma

The leiomyoma (clinical fibroid) is the most common tumor of the tubular genitalia of the bitch but is rarely met with in other species. Its origin is from the smooth muscle in the wall of the uterus, cervix, or vagina, and, although it may be solitary, usually there are multiple foci in each division. The tumor in the bitch is not malignant or wholly autonomous, being, in some unknown manner, endocrine dependent. It rarely occurs in the bitch earlier than middle age and is frequently associated with ovarian follicular cysts or estrogen-secreting tumors, and often also with endometrial hyperplasia, mammary hyperplasia, and mammary neoplasia. Bitches castrated early in life are exempt, and established tumors regress following castration. This tumor has been provoked in guinea pigs by continuous low-level doses of estrogen but, for spontaneous leiomyomas as well as endometrial and mammary neoplasms, there is as yet no precise knowledge of the role of estrogen.

The genital leiomyoma may grow to be as large as 10–12 cm in diameter but is not invasive. The smaller tumors are fleshy but, as they enlarge, become firm or hard (hence the clinical term fibroid) due to the connective tissue stroma. On cut surface, they have a watered-silk appearance, and the color, whether more fleshy or white, depends on relative amounts of muscle and connective tissue. The tumor is not encapsulated but is well demarcated and easily shelled out. In almost all cases, the tumors project as globose or elliptical masses or as bulbous polyps into the lumen of the vagina, uterus, or cervix, but some project outward and a few are found in the mesometrium (Fig. 4.93).

Degenerative changes occur in the larger tumors and are of two types, being either progressive replacement of myoma tissue by fibrous tissue of the stroma, or edema and liquefaction with cyst formation. In the first type, the fibrosis may progress through hyalinization to mineralization. The second type of degenerative change afflicts the vaginal leiomyomas especially, probably because these more often become pedunculated and susceptible to circulatory embarrassment.

Histologically, the tumor is composed of whorling bundles of smooth muscle cells with abundant stroma but scant intercellular connective tissue. The organization does not depart much from normal, and often the presence of neoplasm is best appreciated by the naked-eye appearance of the tissue (Fig. 4.94A). The overlying epithelium

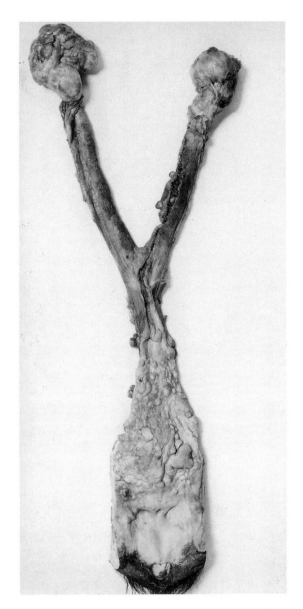

Fig. 4.93 Vaginal leiomyomas associated with follicular and luteal cysts of ovaries. Bitch.

is intact. Malignant transformation has not been observed. In species other than the bitch, genital leiomyomas are uncommon, usually solitary, and not related in incidence to endocrine disturbances (Fig. 4.94B).

2. Transmissible Venereal Tumor of Dogs

This tumor is a contagious neoplasm, transmitted most commonly by coitus, and occurs on the external genitalia of either sex. It is also known as Sticker's sarcoma. The noncommittal name of the tumor, transmissible venereal tumor, is preferred to such terms as lymphosarcoma, histiocytoma, sarcoma, etc., which, suitably prefaced to indicate its transmissibility, have been used. Because the host's tissues do not contribute to the growth, the histo-

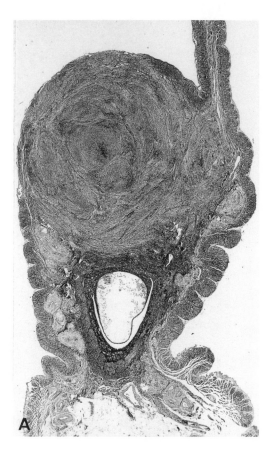

Fig. 4.94A Uterine leiomyoma. Bitch.

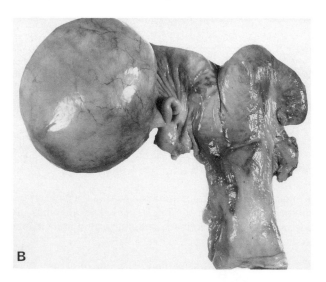

Fig. 4.94B Uterine leiomyoma. Cow.

genesis is obscure and likely to remain so. This tumor differs from other infectious tumors in that the infecting cells are transplanted and grow like a graft. This was the first neoplasm in the history of pathology to be transmitted experimentally; this was accomplished by the Russian veterinarian Novinsky in 1863.

The tumor has been reported from Europe and Asia as well as from the Americas. There are marked fluctuations in its prevalence in endemic areas over the course of a few years. The tumor tends to be common where dogs are allowed to run free and is rare where dogs are controlled. The tumor is not confined to the genitalia, but in some few cases is found in other cutaneous locations. Metastasis sometimes occurs to regional lymph nodes, but spontaneous regression usually occurs in less than 6 months.

It seems unlikely that a tumor of such character would originate in multiple geographic centers; more probably its wide distribution, especially in tropical and subtropical areas, is due to a manner of spread akin to that of infectious disease. This is an assumption impossible to prove, but it is supported by the fairly constant karyotypes of the tumor cells to the extent that these have been examined. The diploid number of chromosomes in dogs is 78, and, normally, they are acrocentric except for X and Y. The chromosome complement of tumor cells originating in separated areas of Japan and North America is closely similar, most tumor cells containing 59 chromosomes; the minor variations in chromosome numbers in some cells and in chromosome morphology are within the variability accepted for stem-lines of transplantable tumors and long-term cell cultures.

In the vagina, the tumor takes origin in the submucosa, usually of the dorsal wall, as one or more papular or papillary proliferations. Expansive growth also occurs mainly in the submucosa, and the overlying epithelium becomes stretched and attenuated. With rupture or penetration of the epithelium, the tumor projects into the vagina as an irregular, ulcerated, and friable mass, which may protrude from the vulva. Histologically, the tumor varies depending on the stage of growth or regression. During the early stage of growth, the tumor is composed of round, oval, or polyhedral cells with indistinct boundaries and a poorly stained or clear cytoplasm. The nuclei are large in proportion to cell size with a single, well-defined nucleolus and many chromatin granules. Variability in the size of the cells is rather characteristic, and mitoses are frequent. Electron-microscopic studies have shown that most cells in growing tumors are round cells with microvilli, whereas regressing tumors have spindle-shaped cells suggestive of tumor cell differentiation toward fibroblastic cells. The number of infiltrating lymphocytes increases as growth of the tumor slows and the tumor begins to regress. Most of the infiltrating lymphocytes are T cells.

In the skin also, the tumor seems to arise in the subcutis, and the epidermis is not usually penetrated. It is suggested that natural cutaneous implantation occurs in bite or other wounds. In any locale, the tumor shows a marked tendency to break down after a few months of rapid progression, and on incision the necrotic tissue resembles pus.

It is not known how this tumor can be transplanted with living cells across major histocompatibility barriers and survive for a substantial time before regressing. Infected dogs do develop antibodies, but in dogs with growing tumors, some of these antibodies are blocked by a tumor-associated antigen. This suggests that the expression of this tumor-associated antigen may play a role in blocking systemic immune responses and facilitate tumor growth. Unlike tumor-associated antigen, which is expressed in progressive tumors, class I and class II major histocompatibility antigens are expressed only when regression begins. Dogs are resistant to challenge after natural regression of the tumor. Secondary tumor challenges begin to regress on day 9 and disappear after about 2 weeks. The secondary tumors are rapidly infiltrated by lymphocytes, most of which are T cells, and undergo degeneration. The tumor is transmissible not only between dogs but also to the fox, coyote, and jackal.

The life history of the tumors as outlined and especially the limited metastatic capability are not constant, and greater contagiousness and more virulent expressions may be expected in immunosuppressed individuals and in a canine population of suboptimal physiologic status. Such appears to be the case, for example, in the promiscuous, scavenging, malnourished, feral populations. There, metastases are common in both sexes as sequelae to the primary lesions. In the female, extension occurs to the uterus, cervix, and uterine tubes. In both sexes, there is early spread to inguinal nodes, which become large and firm. Cutaneous involvement is also common with few or many lesions, some up to 6 cm in diameter, often ulcerated and hemorrhagic. Buccal lesions and large periorbital tumors are also described. Ocular lesions are common in this disease, but, apart from the sites mentioned, conventional metastatic patterns in internal viscera do not occur.

3. Fibropapilloma of the Vulva

This is probably the most common tumor of the bovine vulva and affects young animals primarily. It is, briefly, a wart growing on a mucous membrane, the nature of the host tissue apparently determining the reaction to the virus of bovine verrucae (bovine papillomavirus). This virus infecting the keratinized bovine skin provokes the typical papilloma of relatively scant connective tissue core and abundant epidermal overgrowth. Transplanted instead to the penile or vulval mucosa, the contribution of the two moieties is reversed, the bulk of the tumor being of connective tissue with (barring accidents) just enough epithelium to cover it. Histologically, the bulk of the tumor consists of interlacing bundles of fibrocytes. In the younger tumors, there may be many mitotic figures, and such cases have often been misdiagnosed as fibrosarcomas. Many of the plump spindle cells have large nuclei with bizarre nucleoli and sometimes pale, eosinophilic, inclusionlike intranuclear structures. Collagen formation from the stroma is progressive with duration. Surface ulceration is followed by superficial inflammatory infiltration.

The vulval tumors are usually sessile, rounded growths when young but become progressively more cauliflowerlike. Those attached to the penis are often pedunculated. The natural history of the genital fibropapilloma is the same as that of the common cutaneous papilloma, spontaneous regression occurring in 1–6 months or so. Within

this period, surgical excision may be followed by recurrence.

4. Transmissible Genital Papilloma of the Pig

Ordinary cutaneous warts are not described in pigs. A transmissible papilloma occurs in the preputial diverticulum of the boar. It is transmissible to the lightly scarified vulval mucosa with an experimental incubation period of about 8 weeks. The lesions are 1–3 cm in size and papular, some of the larger lesions being papillary. The clinical course is not many weeks, after which the lesions begin to regress and are eventually sloughed. Recovered animals are immune to reinfection.

The lesion histologically is typically papillomatous with extensive uneven acanthosis and epithelial overgrowth, but with none of the abundant mesodermal reaction that occurs in the bovine disease previously outlined. The broadened malpighian zone contains scattered cells with intracytoplasmic inclusion material. The inclusions are large, spherical, homogeneous with acidophilic stains, and often surrounded by a halo. An inflammatory response with mononuclear cell infiltration occurs in the underlying dermis.

5. Carcinoma of the Endometrium and Cervix

These are rare neoplasms in domestic animals (Fig. 4.95) and therein lies their interest. The rarity is real and not, as has been implied, apparent because of inadequate

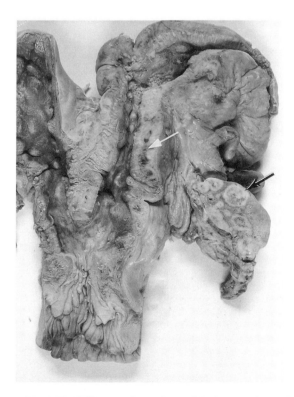

Fig. 4.95 Diffuse uterine endometrial adenocarcinoma (white arrow) with metastasis to right ovary (black arrow). Cow.

postmortem examination. The most popular current opinion on the pathogenesis of the endometrial carcinomas in women relates them to prolonged estrogenism, with cystic endometrial hyperplasia of the same cause as the precancerous lesion. In domestic animals, on the other hand, cystic endometrial hyperplasia is common, but carcinoma is rare and, when it does occur, does not seem to require prior hormonal conditioning of the endometrium.

Carcinoma of the endometrium appears to occur more frequently in cattle than in other domestic animals. The tumors may be single or multiple, hard, nodular masses of varying size in the uterine wall. Umbilication of the serosal surface is a characteristic feature of the neoplasm. The regional lymph nodes and the lungs are the usual sites of metastatic lesions.

6. Lymphosarcoma

Involvement of the uterus is not uncommon in the multicentric form of bovine lymphosarcoma but is distinctly so in other species. As elsewhere, there are two anatomic forms of uterine involvement, diffuse and nodular. In either event, the initial deposition seems to occur in the endometrium. The diffuse lesion involves both horns, body, and occasionally the cervix in a more or less uniform thickening, although the mucosa may be thrown into broad folds, and the atrophic caruncles may still be easily recognized. The thickened wall loses its elasticity and on cut surface is rather structureless with an appearance like firm

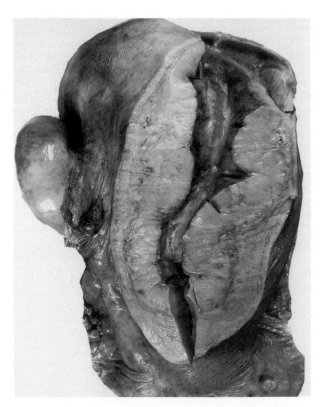

Fig. 4.96 Diffuse lymphosarcomatous invasion of uterine wall. Cow.

fat (Fig. 4.96). Gentle pressure may expel small quantities of cloudy, highly cellular fluid. There is patchy ulceration of the endometrial mucosa. There is nothing peculiar to the nodular form of the disease. These nodules, few or many, may attain large size and cause corresponding deformity. Central liquefactive necrosis is common in them. Microscopically, there is, with either form of the disease, gradual replacement of the normal uterine structures by infiltrating tumor cells. Lesions of corresponding type occur in the vagina, but do so less commonly.

7. Metastatic Tumors

With the exception of lymphosarcoma, secondary neoplastic diseases of the tubular genitalia are rather rare. Serosal implantations occur in peritoneal carcinomatosis. Hematogenous metastases probably have a predilection for the endometrium.

Bibliography

Burdin, M. L. Squamous cell carcinoma of the vulva of cattle in Kenya. *Res Vet Sci* **5:** 497–505, 1964.
Cockrill, J. M., and Beasley, J. N. Ultrastructural characteristics of canine transmissible venereal tumor at various stages of growth and regression. *Am J Vet Res* **36:** 679–681, 1975.
Cotchin, E. Spontaneous tumors of the uterus and ovaries in animals. *In* "Pathology of the Female Genital Tract," A. Blaustein (ed.), pp. 822–861. New York, Springer-Verlag, 1977.
Craig, J. M. The pathology of the female reproductive tract. *Am J Pathol* **94:** 385–437, 1979.
Higgins, D. A. Observations on the canine transmissible venereal tumour as seen in the Bahamas. *Vet Rec* **79:** 67–71, 1966.
Mizuno, T. *et al.* Role of lymphocytes in dogs experimentally rechallenged with canine transmissible sarcoma. *Jpn J Vet Sci* **51:** 86–95, 1989.
Parish, W. E. A transmissible genital papilloma of the pig resembling condyloma acuminatum of man. *J Pathol Bacteriol* **81:** 331–345, 1961.
Prier, J. E. Chromosome pattern of canine transmissible sarcoma cells in culture. *Nature* **212:** 724–726, 1966.
Yang, T. J. Immunobiology of a spontaneously regressive tumor, the canine transmissible venereal sarcoma. *Anticancer Res* **8:** 93–96, 1988.

General

Kurman, R. J. (ed.). "Blaustein's Pathology of the Female Genital Tract," 3rd Ed. New York, Springer-Verlag, 1987.
McEntee, K. "Reproductive Pathology of Domestic Mammals." San Diego, California, Academic Press, 1990.

VIII. The Mammary Glands

The mammary glands are modified cutaneous glands with the important roles of providing nourishment to immature neonates and providing to the newborn of many species passive immunologic protection against pathogenic microbes. The mammary glands may, however, provide a source of infection by microorganisms such as *Mycobacterium paratuberculosis* and the Brucellae, which arrive in the gland hematogenously and provoke minimal pathologic change, or *Pasteurella* spp. and *Streptococcus* spp. during the course of galactogenic infection and inflammation. A number of helminth parasites have adapted their life cycles in ways that ensure their transfer to neonates in milk.

Toxins of various sources may be excreted in milk. Those of most veterinary concern are principally of plant origin and are not usually accompanied by pathological changes in the gland. They may, as a result of systemic or distant, rather than local, action on the gland, lead to involution and agalactia as exemplified by the agalactia of sows ingesting ergot of rye, *Claviceps purpurea,* which leads to inhibition of prolactin secretion by the anterior pituitary gland. The avocado plant, *Persea americana,* is an exception in which selective toxic injury to the lactating gland occurs.

The major pathological issues are those of inflammatory and neoplastic disease; there are others of minor or surgical importance. There are important diseases affecting the skin of the mammary glands, and neoplastic diseases, of importance in dogs and cats. These are discussed with diseases of The Skin and Appendages (Volume 1, Chapter 5).

A. Developmental Considerations

The weight of evidence supports the derivation of mammary glands as specializations from sweat glands. The myoepithelial cells are thought to be of epithelial origin, and the relationship to sweat glands possibly accounts for the structural and behavioral similarities, and the participation of myoepithelial cells, of mammary and sweat-gland carcinomas in the bitch.

The glands form on the mammary line or ridge detectable in the ventrolateral ectoderm of the embryo. The ectodermal cells of the ridge, which are destined to develop into the mammary glands, congregate in specific areas, of number and location appropriate to the species, to form the mammary buds. From the buds, primary sprouts push into the mesenchyme, the number of sprouts from each bud determining the number of openings which will eventually develop in each teat. Only one primary sprout develops from each bud in cattle and is destined to form the streak canal and teat cistern and, proximally, the gland cistern. Secondary sprouts develop from the primary sprout to form the early milk ducts. In species such as cattle, sheep, and goats with one teat orifice, the gland is a single large secretory structure, but ectopic mammary tissue may form in the wall of teat canal (Fig. 4.97). There are usually 2 streak canals in horses and pigs, 3–7 in the cat, and 8–14 in the dog. In species with multiple teat orifices, a mammary gland is in reality a composite of glands, the number corresponding to the number of streak canals, and each autonomous and separate from its neighbors.

Supernumerary teats or nipples are common, especially in cattle in which the incidence is approximately 30% and probably subject to a genetic control mechanism. They occur in males as well as females (except in male horses,

Fig. 4.97 Ectopic mammary tissue in wall of teat canal. Cow.

which do not develop nipples), and are located in relation to the embryonic mammary ridge although some may be displaced laterally. In many cases the supernumerary teat will have functional mammary tissue associated with it.

Mammary gland development in males is generally similar to that in females in the embryonic and fetal stages. In those species such as rats, mice, and horses, the males of which do not have teats or nipples, the primary sprouts separate from the surface epithelium of the mammary bud under the influence of androgenic hormone and regress. The male mammary gland is susceptible to hormonal stimulation but is not as sensitive as the female. Alveolar structures are not present and the enlargement which occurs under the influence of estrogen is due to cystic ductal hyperplasia. This is seen most often in dogs with estrogen-secreting tumors of the testes.

B. Inflammatory Disease of the Mammary Glands

Mastitis is inflammation of the mammary gland. Whereas injury of any type or cause is expected to produce an inflammatory response, for all practical purposes the disease is a response to invasion of the gland by microbes. The route of invasion may be hematogenous, as in tuberculosis and brucellosis, it may be percutaneous as a complication of local dermatologic conditions of the skin or teat, but the usual route of invasion is through the teat canal. The pathogenesis and characteristics of the disease, which are intensively studied only in cows, are dependent on factors in microbial ecology and the nature of the pathogen as well as on host factors.

It appears that the major and critical barrier to infection is some component or components of the teat canal. The establishment of a bacterial population in the teat cistern is usually followed by inflammation, but the application of even large numbers of bacteria to the teat orifice results in infection in only a few cases. The length of the teat canal seems not to be important. Its patency may be, but there is no good measure of this. There is contradictory evidence on the relationship between ease of milking and susceptibility to mastitis, but ease of milking is not a criterion of patency of the teat orifice. There is little doubt that structural factors of the teat orifice and canal are important in regulating the entry of organisms in some cases.

Whether other factors are also involved is not known, but it seems likely that chemical factors within the teat canal are of primary importance in determining whether the infecting organisms will be repulsed or permitted to proliferate and grow through the canal into the cistern. A purely mechanical progression is unlikely, although reflux can occur in improper milking procedure.

Resistance to mastitis caused by *Streptococcus agalactiae* is rather strong in some cows so that it is possible to compare cows which are highly susceptible with those that are highly resistant. The difference between resistant and susceptible animals is apparent only if the infecting organism is applied to the teat orifice or within the teat canal; disposition of the organism, even in small numbers, within the sinus rather regularly results in infection— apparently the inherited mechanisms of resistance have been bypassed. Resistance is not an all-or-none phenomenon. It declines progressively with age and is, no doubt, subject to the general influence of body health, but at any single exposure to the teat orifice or canal, it is all or none. In a susceptible animal, exposure of the canal to even small numbers of the organism is followed by their multiplication and entry into the cistern, whereas, in resistant animals, the inoculated organisms multiply briefly if at all in the teat canal, and disappear in the course of two to three milkings. But resistance to this sort of exposure, which most nearly parallels natural exposure, can be removed by removing the smegma of the teat canal. The teat canal is lined by a stratified squamous epithelium which is surfaced by a layer of keratinlike material and then an innermost layer of smegma—a waxy material composed probably of epithelial debris and milk solids. Although the lining of the teat canal is an invagination from the normal surface epithelium, its chemical composition, as judged by staining reactions, is quite different. It is possible that there is, in the smegma of the teat orifice of resistant cows, some substance which is actually inhibitory to the growth of *S. agalactiae*. The phenomenon of resistance can be demonstrated to operate also in infections by *Staphylococcus aureus,* but with two differences. There is no increasing-with-age susceptibility to staphylococcal infections, and cows which are resistant to *S. agalactiae* are not necessarily resistant to *S. aureus* and vice versa; apparently, the ingredients of resistance to the two infections are not the same.

There are humoral and cellular factors in normal milk which are inhibitory to bacterial growth and which may be increased in inflammation.

Lactoferrin is an iron-binding protein present in secretions and in neutrophils, which is inhibitory to the multiplication of bacteria with high iron requirements. The high concentration of citrate in normal milk prevents the action of lactoferrin, but it is effective in the secretion of the nonlactating gland.

Lysozyme is locally synthesized in the gland and although the concentration in milk is low, there is some correlation between milk titer and susceptibility to infection. The titer may be increased during inflammation.

Lactoperoxidase is synthesized by mammary epithelium and is potentially inhibitory of *Staphylococcus aureus* and most streptococci, and is a source of hydrogen peroxide available as a byproduct of bacterial fermentation of milk carbohydrates.

There are differences between species in the nature of immunoglobulins in colostrum and milk. In the cow, the immunoglobulins are predominantly of the IgG class and are selectively transferred from serum to milk. The levels are low in normal glands, but permeability changes in inflammation allow the titer to be much increased. The same class of antibody is identified with the concentration of plasma cells in the distal teat cistern and proximal streak canal and may have an effect against organisms resident in the streak canal. The major role of these antibodies is to promote opsonization of microorganisms. Immunoglobulin A is also present in milk and may have a defensive effect on bacterial adherence to epithelial surfaces. Gramnegative bacteria can be killed by the complement– antibody system. The activity of this system in milk is normally low but can be greatly increased in inflammation as a result of permeability changes in mammary vasculature. There is variation between animals in the level of bacteriocidal activity in serum, but this appears to be the main mechanism limiting the duration of acute coliform infections.

Phagocytosis is the most important component of defence against gram-positive bacteria which have gained entry to the gland. Neutrophils enter the secretion in large numbers during mastitis, but the efficiency of phagocytosis and killing of engulfed bacteria is less in milk than in serum. The reduced efficiency may be due to depletion of energy substrates or lysosomal exhaustion following ingestion of lipid droplets or other solid materials in milk. Nonetheless, leukocytes do have a very significant role in controlling infection and in preventing superimposed infections in inflamed glands.

1. Bovine Mastitis

From practical considerations, bovine mastitis is a response to ascending infection of the gland by way of the teat canal. In spite of diligent efforts to limit infective opportunities and to treat established infections, the incidence of the disease and of reinfection is, overall, not much changed. The problems appear to reside in limited defense against colonization by ascending infection and in clearing mechanisms of limited effectiveness.

Potential mammary pathogens are ubiquitous. Modern techniques of microbial classification have identified more than a hundred species, subspecies, and serovars isolated from the mammary gland. *Streptococcus agalactiae* and perhaps some types of *Staphylococcus aureus* are obligate parasites of the gland and inevitable pathogens, but the great majority of infections are opportunistic. It is perhaps useful to view the mammary gland and its microbiota in the same way as the bacterial colonization of other epithelial surfaces is viewed, whereby the local epithelial microenvironment is available for colonization by competent microbes. In this view, the consistent presence of *Coryne-*

bacterium bovis and coagulase-negative staphylococci as well as other ill-characterized microaerophilic and anaerobic organisms reflects the normal ecology on which from time to time populations of other microorganisms may be superimposed.

Mastitis, excepting that caused by obligate pathogens, might best be regarded as an inflammatory disease rather than an infectious disease and is one expressed by tissue injury when the microbial balance is upset and becomes dominated by organisms which are injurious by virtue of tissue invasiveness or toxigenicity. The pathologic process may range from transient to persistent and from mild or subclinical to peracute and fatal.

Notwithstanding the large number of microbial species which may be isolated from the diseased mammary gland, the epidemiologic picture of bovine mastitis is dominated by the streptococci, staphylococci, and coliforms. Some infections, such as by *Cryptococcus* and the atypical mycobacteria, are usually iatrogenic. Some infections, such as by *Pseudomonas* and *Prototheca*, reflect heavy environmental contamination. Once the inflammatory process is established, additional microbial species, particularly the anaerobes, may participate by permission of a changed intramammary environment, allowing growth of resident organisms or altered competence of the teat canal, allowing entry of additional species.

a. STREPTOCOCCAL MASTITIS   Bacteria of the family Streptococcaceae are important causes of bovine mastitis. *Streptococcus agalactiae, S. dysgalactiae,* and *S. uberis* are the most frequently implicated. Sporadic disease has been associated with *S. equi* and its subspecies *zooepidemicus, S. bovis, S. pyogenes, S. pneumoniae, Enterococcus faecalis, E. faecium,* and *E. durans.*

Bovine mastitis due to *Streptococcus agalactiae* was by far the most common and important of the forms of bovine mastitis until it was subdued by therapy. The causative agent has the mammary gland of the cow and goat as its natural and sole habitat; resistance to the extramammary environment is low although the organism may survive for a month or more on fomites, but environmental factors are important in the transfer of infection from cow to cow.

The only significant portal of entry of *S. agalactiae* into the mammary gland is through the teat canal. Mastitis is always a possible complication of traumatic teat injuries, but infection in such cases is more often due to environmental organisms than to *S. agalactiae.*

Streptococcal mastitis is usually permanent. The organism is capable of maintaining its numbers in the cisterns in opposition to inflammatory products and the irrigative force of milking. How it remains in equilibrium with its host in the infection phase is not clear. At a variable period following the establishment of infection, the bacterial population may suddenly increase, and the tissue become invaded for a brief period producing a clinical crisis. This proliferation of bacteria within the cisterns precedes tissue penetration and is associated with altered virulence of

the organism as measured by mouse-pathogenicity tests. Whether there is also some change in the host system is not known, but change ought perhaps to be expected, because the detrimental effects of improper machine milking, in terms of height of vacuum, rate of pulsation, period of application, etc., are expressions, largely, of flare-ups of established quiescent infections. Repeated invasions usher in a series of inflammatory and reparative reactions, which if unchecked therapeutically, culminate in fibrosis and involution of the affected quarter.

There are varying degrees of severity of the clinical reaction of streptococcal mastitis, the more acute cases being associated with systemic disturbances and the less acute ones, not. The anatomic basis lies not in the acuteness of the inflammation, but in the amount of mammary tissue involved. A greater or lesser amount of normal uninflamed tissue is always present, the inflammatory reaction being confined to those areas in which the organisms manage to penetrate the duct epithelium. It appears that the streptococci are unable to penetrate the interstitial tissues to a significant degree, although sometimes they are found in leukocytes in the lymphatics and supramammary lymph node, and that the inflammatory reaction is due to diffusing toxic products of the organisms. This period of epithelial penetration is brief, being of only a few hours' duration, and the organism in tissue is rapidly destroyed by leukocytes. This destruction of the organism may at times sterilize the udder or so diminish numbers that bacterial cultures of milk taken during a crisis may be negative, a finding that applies also to other acute mammary infections.

The first response to the penetration of streptococci is a remarkable interstitial edema and an extensive migration of neutrophils into the interlobular tissue and secretory acini. The stromal lymphatics are widely dilated and contain numerous leukocytes, which are delivered to the regional lymph node. The acinar epithelium becomes vacuolated and desquamated, or in places heaped up and ragged, over accumulations of macrophages and fibroblasts which make their appearance very early in the course of the reaction. Streptococci are very numerous at this stage, both within the ducts and acini and in and under the epithelium. The acute exudative reaction gives way to two processes which proceed together—pathologic fibrosis and involution. A few organisms still persist in the larger ducts and the neutrophil reaction is reduced, but macrophages and fibroblasts continue to increase in number (Fig. 4.98) and eventually obliterate many of the acini. Lymphocytic foci begin to develop in the interstitial tissue. Other acini become dilated and rounded and contain a stringy coagulum with intact cells and cellular debris which constitutes the first indications of the involution that follows interruption of secretion and acinar stagnation.

The involution affects lobules directly involved in the exudative inflammation as well as those surrounded by interacinar tissue to which the fibroplasia has extended. Stagnation of secretion occurs also in many of the smaller ducts. It is pertinent to note that this is the stage which on

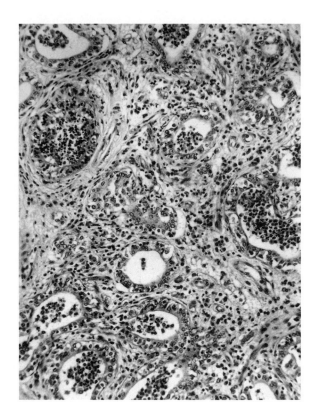

Fig. 4.98 Acute streptococcal mastitis. Cow.

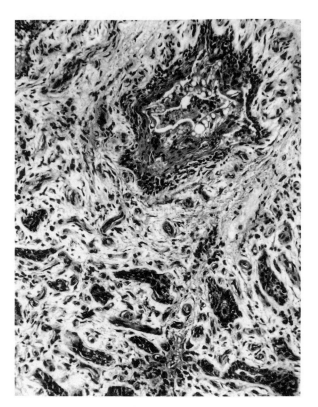

Fig. 4.99 Chronic streptococcal mastitis with fibrosis, glandular atrophy, and squamous metaplasia of duct. Cow.

palpation is swollen, firm, and painful and often referred to as fibrosed. But there is as yet little fibrosis, the induration being due instead to retained secretion. These nodular or diffuse indurations are firm and liverlike in consistency but of normal color and are easily cut in contrast to the normal slack elasticity of mammary tissue, which makes it difficult to cut. The processes of fibrosis and involution continue until the end stages when some lobules show normally involuted tissue, some are obliterated by fibrosis (Fig. 4.99), and others show a varying balance between both processes. Such a gland is atrophic, dry and, because of the fibrosis, poorly elastic. Some lobules escape for a time, often a long time, only to be involved ultimately in a flare-up or by extension of the disease process. It is evident then that the many lobules involved are often out of phase with each other, but that in each, the succession of events proceeds rapidly.

Adjacent to the spots where the streptococci invade the epithelium of the smaller ducts, granulation tissue develops rapidly and exuberantly and protrudes into the duct lumen, which is dilated and contains stagnant milk, exudate, and numerous organisms. These polypoid proliferations may completely obstruct the duct or, by fusing with the opposite surface, produce loculations resembling abscess cavities, which are clearly visible macroscopically. There is concomitant proliferation of periductal fibrous tissue spreading centrifugally to involve and obliterate large amounts of lobular tissue. Such granulation tissue in time cicatrizes and the duct epithelium is restored. Similar

changes occur in the larger ducts and teat cisterns at points of bacterial penetration, but they are less exuberant and the epithelium is shortly transformed to a squamous, sometimes keratinizing, type, which is later desquamated and reconstituted as the acute phase subsides to the chronic one of involution and as fibrosis supervenes. Involution has been regarded as a defense mechanism, but this misconstrues its purely passive development as a consequence of stagnation. Such involuted lobules are capable of function at subsequent lactations. Whether there is any replacement of destroyed lobules by regeneration from persistent ducts has not been established, but it cannot amount to much.

The gross appearances of streptococcal mastitis vary according to the stage of the disease. Usually more than one quarter is involved, but not uniformly, with most of the change taking place in the distal portion of the gland about the cisterns and large ducts. The cisterns and ducts are filled with secretion in the early stage of the disease, the secretion being serous and floccular, or distinctly purulent. The mucous membrane of the cistern may not be appreciably altered or it may be hyperemic and granular. The glandular tissue is swollen and turgid. It cuts easily with a knife and on the cut surface the lobulations are distinct because the swollen lobules protrude. It is not possible to recognize the interstitial edema. The affected lobular tissue is grayish in color and can be distinguished from the milky whiteness of normal lactating tissue, but not readily from involuted parenchyma.

In the later stages of the disease, the most obvious changes occur in the cisterns and large ducts as a moderate thickening of the epithelium and small rounded polypoid proliferations into the lumen of the cistern and large ducts. Fibrosis is recognizable in these cases surrounding the ducts, surrounding lobules and obliterating some, and diminishing in severity toward the base of the gland.

Streptococcus uberis is usually associated with a mild and chronic disease, but the other streptococcal species listed usually produce acute, but transient, inflammation, often accompanied by systemic signs of illness. The incidence of infection with any of these organisms tends to be sporadic, but occasionally there are severe outbreaks in individual herds. Any of the first three species may be present with *Actinomyces pyogenes* in the severe summer mastitis of dry cows and heifers.

Very little is known of the pathogenesis of these infections, although predisposing teat injuries are often incriminated; and little is known of the history of the lesions, although apparently it is similar to that of the lesions of *S. agalactiae* infection. With the exception of *S. uberis,* these infections do not persist as endemic infections.

Mastitis caused by *S. pyogenes* is of some importance because the organism is a human pathogen; the secretion from infected quarters may contain enormous numbers of these organisms, but they proliferate barely, if at all, in drawn samples of milk. The mammary infections may arise from close association with infected humans and outbreaks of upper airway infection have occurred in humans consuming milk from infected cows.

b. STAPHYLOCOCCAL MASTITIS   Staphylococcal mastitis is predominantly an infection of the younger age groups, and there is no increase in susceptibility with age.

Pathogenic strains of staphylococci are always of human or animal origin and persist as permanent inhabitants of the skin and mucous membranes, although they are capable of more than average resistance to the general environment. It is generally accepted that the infection of the udder is contagious, and it seems probable that entry through the teat canal is, in general, subject to the same sort of local controls as is the entry of *Streptococcus agalactiae.*

Pathogenic staphylococci can be broadly grouped as to whether they are catalase positive or catalase negative. The ability of a strain to produce catalase and hemolysis is the best single criterion of pathogenicity for animals. The coagulase-positive staphylococci include the important *Staphylococcus aureus* as well as *S. intermedius* and *S. hyicus.* The coagulase-negative staphylococci comprise a large number of species of which *S. epidermidis* is the species most frequently isolated from milk. These species are of low pathogenicity and may perhaps be best regarded as part of the normal flora. They are commonly recovered from normal milk samples and from cases of subclinical mastitis expressed as elevated somatic cell counts in drawn samples.

Mastitis paralleling the natural disease can be produced by infusing the udder with the α toxin; the β toxin is nonirritant. Strains of the organism, not separable by other criteria, differ in their toxigenicity, but this is not necessarily related to pathogenicity; one strain may produce gangrenous mastitis on some occasions and only a mild disease on other occasions.

Clinically, staphylococcal mastitis may be peracute and fulminating or milder and more chronic; the latter is the more common. The acute forms of the disease, typically, occur shortly after parturition and tend to produce gangrene of affected quarters and a high mortality. The affected quarters are swollen and tense, hot and firm, and very painful. There is almost complete stagnation of secretion and only a few milliliters of brownish bloodstained or straw-colored watery fluid can be expressed from the teat. Uninfected quarters of the same udder are also swollen and tense and the secretion is reduced but otherwise normal, an effect probably due to the diffusion of toxins through the vascular bed of the gland.

Gangrene usually affects first the teat and adjacent portions of the udder and may not be more extensive, or it may extend even to involve the whole quarter. The tissues become blue and eventually black and are softer, insensitive, and cold. There is pitting edema of the inguinal area, flank, and venter, and in a day or so the necrotic skin begins to exude serum and to slough, and crepitating gas bubbles develop beneath it. The changes are those of ordinary moist gangrene, remarkable only for the abundance of fluid exudation, and are attributable to the direct action of toxin on the acinar tissue and to venous thrombosis. The amount of tissue involved in the gangrenous process is quite variable and groups of necrotic lobules adjoin others which are near normal. Natural separation of the gangrenous areas begins about a week after the onset but proceeds slowly with the development of a suppurative surface and fistulae.

The acute, nongangrenous, and mild forms of the disease progress more closely along the lines already described for streptococcal mastitis, but with some important differences which will be mentioned later. It appears that early events are comparable to those of streptococcal infection and that the bacterial population, resident in the ducts and cisterns during the infection phase, multiplies rapidly (Fig. 4.100A) and then penetrates the duct wall to the interacinar tissue. If the penetration is massive and the organisms are highly toxigenic, the acute and gangrenous forms of the disease occur. But the invasion, even if massive, is massive only in patches and intervening areas remain normal for some time (Fig. 4.100B).

The differences in the manner of progression of streptococcal and the milder staphylococcal mastitis depend on different toxigenicity of the genera and the ability of the staphylococci to invade more deeply into the interacinar tissue and to establish themselves there as persistent foci

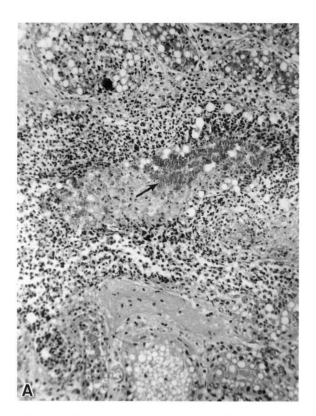

Fig. 4.100A Acute staphylococcal mastitis. Cow. Note colony of bacteria in duct at center (arrow).

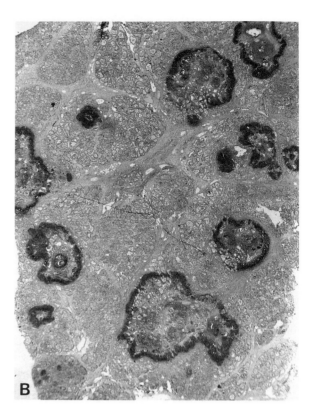

Fig. 4.100B Acute staphylococcal mastitis with patchy zones of early involvement. Cow.

of infection to provoke the granulomatous reaction known as botryomycosis (Fig. 4.101A,B). The initial reaction is necrotizing and these necrotic foci are soon surrounded by an intense leukocytic response, and fibroplasia develops rapidly to wall off the irritant foci, thus obliterating large portions of the normal mammary structure. Each granulomatous focus may be no more than 1–2 cm in diameter, but they may be numerous and involve a large proportion of the gland with, between them, a residuum of involuted lobules which is surrounded by septa greatly thickened by confluence of fibrosis. The thickened septa and the granulomas are readily visible to the naked eye and small amounts of pus can be expressed from them. Microscopically, the granulomas closely resemble those of actinomycosis and have been so called, but the coccal organisms are easily visible in the microabscesses. Walled off by connective tissue, the bacteria are not exposed to the action of antibiotics. The infection is always a localized one, and even in peracute cases, bacteremia does not occur.

The organ changes in glands shedding nonhemolytic, coagulase-negative staphylococci of low pathogenicity without clinical signs of infection are mild progressive fibrosis and lobular atrophy. Hemolytic coagulase-positive staphylococci produced lobular lesions of greater extent and rate of progression. The different stages of the inflammatory reaction in different lobules is quite like that in the streptococcal disease, and the patchy progression

corresponds to the irregular distribution of cultivable organisms in affected glands. It is proper to emphasize the importance in this mild disease of partial or complete occlusion of small collecting ducts or intralobular ducts by exudate or granulation tissue as mechanisms inducing acinar atrophy. Occluded ducts also produce a locus for bacterial persistence and growth leading to small abscesses and obliterative granulomas.

Staphylococcus epidermidis is being recognized as an intramammary pathogen, but not a serious one. Infections tend to be eliminated spontaneously.

c. COLIFORM MASTITIS   Coliform bacteria are more common and important than their treatment as a miscellany suggests, but very little is known of the pathogenesis and lesions. The bacteria usually grouped in this category include *Escherichia coli* and species of the genera *Enterobacter, Klebsiella, Citrobacter, Serratia,* and *Proteus.* These organisms are part of the environmental flora, and they do not have a particular predilection for the mammary gland; mammary infection may be correlated with the level of environmental exposure. Each usually produces a clinically acute form of the disease with systemic reaction and, especially in the case of *E. coli,* septicemia. Acute tendovaginitis may also occur. Both *E. coli* and *E. aerogenes* may be associated with clinical flare-ups of the milder progressive type of disease with a history closely resembling that caused by *Streptococcus agalactiae.* With

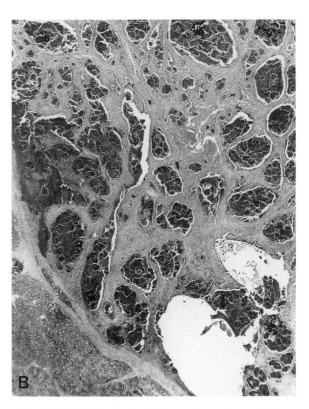

Fig. 4.101A Granulomatous nodules and fibrosis in chronic staphylococcal mastitis of botryomycotic type. Cow.

Fig. 4.101B Chronic staphylococcal granulomas. Mammary gland. Cow.

these infections also, negative cultures may be obtained during the acute phase.

Coliform mastitis is a serious problems in herds in which the more common mammary infections are suppressed or eliminated. Coliform mastitis can be viewed as a disease of a normal lactating gland, the microbe being ineffectual in initiating infection in glands which, as a result of prior irritation, produce large numbers of neutrophils in the secretion. Coliform bacteria are ubiquitous in the dairy environment and microbial density may be enormously increased by well-intentioned husbandry procedures.

In accord with convention, the mammary changes produced by coliform organisms are attributed to the action of endotoxins. Cellular susceptibility to the actions of endotoxins are of course important, but the tissue changes which dominate the response are the result of endotoxic injury to the microvasculature of the alveolar walls and mammary interstitium. Acute infections which provoke massive neutrophil release may be self-cleansing. Chronic infections do, however, occur, characterized by irregular exacerbations, but the mechanisms of persistent infection are unclear and may be related to the adequacy of host responses and the sensitivity of the bacteria to the lytic effects of the antibody/complement mechanism.

The pathology of the coliform infections has not been studied in detail. Indeed, there is even some debate on the pathogenesis of those infections produced by the entero-

bacteria, a hematogenous route of infection frequently being postulated because mastitis often coexists with puerperal infection of the uterus. It appears much more likely that infection is galactogenic because the cisterns and ducts bear the brunt of the injury and contain numerous bacteria, whereas neither bacteria nor significant leukocytosis can be demontrated in the interstitial tissue in the peracute infections which are the fatal ones.

The infection is often limited to one quarter and, being of short course, the inflammation may be predominantly serous with much edema or it may be severely hemorrhagic (Fig. 4.102A). The secretion in the cistern is scanty and may be watery, or cloudy and bloodstained; it contains floccules of fibrin and coagulated casein. Similar floccules can often be seen as plugs in the milk ducts (Fig. 4.102B). There is usually severe edema of the subcutis.

Microscopically, the inflammatory reaction is seen to be centered on the ducts. The lining of the larger ducts is destroyed and replaced by fibrinocellular exudate. The lining of the smaller, intralobular ducts is also destroyed and in these there are plugs of necrotic detritus (Fig. 4.103A). The acini are filled with a serous fluid in which there are vacuolated desquamated epithelial cells, but leukocytes are few or absent, either in the alveoli or in the septa. The septal tissues, especially the interlobular septa, are widened by edema fluid, and the lymphatics, which are greatly dilated, contain many plugs of fibrin. In areas of hemorrhagic inflammation, the septal vessels are suffused

Fig. 4.102A Coliform mastitis. Cow. Typical hemorrhagic consolidation, sharply demarcated.

Fig. 4.102B Hemorrhagic inflammation of gland cistern with plugs of exudate in ducts.

with blood and tortuous, and there is extensive hemorrhage into the stroma (Fig. 4.103B) and into the acini.

If the course of the peracute inflammation is prolonged beyond 1–2 days, extensive necrosis of tissue may occur. Gangrene is not produced by coliform bacilli alone. The necrosis is evident by the altered stainability of the tissues; the histoarchitecture of the tissue is retained for some time. If the animal survives, the necrotic tissue, usually most of one quarter, is sequestrated.

d. SUMMER MASTITIS *Actinomyces* (formerly *Corynebacterium*) *pyogenes* causes sporadic cases of acute mastitis as a result of penetrating injury and is associated frequently with a form of mammary disease known in the United Kingdom as summer mastitis and in Europe as Holstein udder plague. This is usually a mixed bacterial infection affecting immature and nonlactating glands of animals at pasture. Whereas *A. pyogenes* can cause sporadic cases of suppurative mastitis and can be experimentally transmitted via the teat canal, it cannot always be recovered from the exudates of summer mastitis. It is

more usual to recover a mixture of organisms, which may include *A. pyogenes, S. dysgalactiae, Peptococcus indolicus, Bacteroides melaninogenicus, Fusobacterium necrophorum,* microaerophilic and unidentified obligate anaerobes. Infection is acquired via the teat canal probably as a result of contamination by flies attracted to preexisting teat lesions such as impetigo, papular stomatitis, trauma, and photosensitization.

The sequence of events in actinomycotic mastitis has not been studied, but the lesion is clearly a necrotizing suppurative galactophoritis with slight primary involvement of the acinar tissue. Grossly visible abscesses form where the exudate remains stagnant in the ducts. The smaller abscesses occur in the intralobular ducts, and these are associated with desquamative changes and the exudation of a few leukocytes into the acini, and with rapid fibroplasia and infiltration of leukocytes, including many plasma cells, in the septa. The smaller intralobular abscesses may heal with obliteration of the ducts and scarification of the lobule. The large abscesses are centered on the larger ducts, the walls of which are remodeled

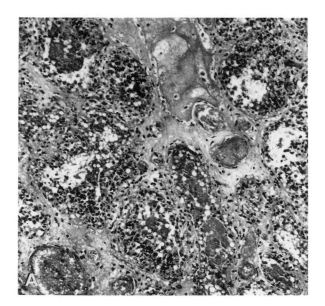

Fig. 4.103A Coliform mastitis. Cow. Necrosis and desquamation in acini and fibrin plugs in lymphatics.

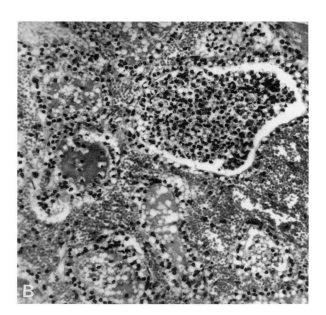

Fig. 4.103B Interstitial hemorrhage in acute disease.

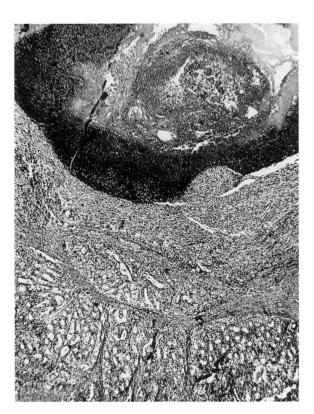

Fig. 4.104 Intraductal abscess in corynebacterial mastitis. Cow.

by exuberant granulation tissue although still lined in part by hyperplastic and squamous epithelium (Fig. 4.104). Bacteria are numerous in the abscesses and in the secretions in other parts of the duct system. The walls of the teat canal and cistern are thickened by granulation tissue and the mucosa is ragged. The cistern is narrowed by the fibrosis and the teat canal may be stenotic or atretic.

e. MYCOPLASMAL MASTITIS   The mycoplasmas are difficult organisms to handle, and much sorting out of them remains to be done in terms of their classification, ecology,

and pathogenic capabilities. They have leaped to prominence in the last few years as causes of bovine mastitis. In herds where sincere attempts are made to control mammary infections, their presence is suggested by high attack rates, rapid spread to other quarters and within the herd, unrewarded efforts to culture bacteria, and very poor response to therapy. The diseases produced by *Mycoplasma* spp. and *Acholeplasma* are not clearly separable pathologically or clinically. The following discussion relates in particular to *Mycoplasma bovis* which, being the most frequent of these infections, is here used as the type infection, although infections by *M. canadense, M. bovigenitalium,* and *M. californicum* are also common.

The disease is characterized by sudden onset of agalactia. The mammary gland is initially swollen, firm, and painless. Ordinarily there are no systemic signs, but arthritis may complicate some outbreaks. When clinical signs are more extravagant, coexisting mammary infections should be suspected. The disease spreads rapidly in a herd, and animals in full lactation tend to be affected simultaneously in all glands. The milk as drawn appears normal but rapidly forms a floccular deposit and clear supernatant on standing. The presence of pus or blood probably indicates coexistent bacterial infection, although the mycoplasmas do cause massive neutrophil emigration.

Affected quarters are, in the active stages, swollen and firm but later become slack as rapid involution occurs.

The altered secretion and glandular enlargement may persist for several weeks. Clinical recovery may occur without return to full normal function, and such animals may continue to excrete the organism intermittently for more than a year. It is not clear whether clinical relapses occur or whether over this long period there is cumulative mammary injury.

The route of infection of the mammary gland is not clear. The disease can be produced experimentally, usually in an acute form, by applying the organisms on the teat or within the canal, as well as by intravenous injection. The mode of spread suggests a galactopoietic infection, but the occurrence of arthritis suggests that systemic dissemination occurs in some cases.

The pathological changes in the gland appear fairly distinguishable from other causative types of mastitis, and the available descriptions of tissue changes have much in common. The distinguishing features appear to be the abundant exudation of granulocytes in the early stages, and the hyperplasia of alveolar as well as ductal epithelium in later stages.

The exudation of granulocytes begins within a few hours of experimental infection as perivascular accumulations in walls of the cistern, in the lobular interstitium, which is edematous, and mixed with the alveolar secretion. Initially patchy, the reaction extends quickly to involve most of the parenchyma, and, after several days, the reaction begins to subside, mononuclear cells, including plasma cells and macrophages, appear in the exudate, and dense focal lymphocytic accumulations develop in intralobular and periductal connective tissue. The lymphocytic infiltrations persist for many months. The epithelial cells of alveoli and ducts develop large fat vacuoles, which are discharged into the lumen. The alveolar epithelium shortly becomes hyperplastic, the rather undifferentiated cells being present in several layers. In the course of some weeks, fibrosis of intralobular septa is accompanied by alveolar atrophy. Hyperplastic and metaplastic alterations also occur in the ductal epithelium, around which is progressive fibroplasia and lymphocytic infiltration. There is extensive but discontinuous erosion of the epithelium, and similar changes occur in the epithelium of the milk cistern. Exuberant granulation tissue forms, especially at the sites of epithelial erosion, producing polypous protrusions into the ducts and cistern. Retention of exudate occurs in small ducts. The exudate tends to calcify and serve as a nidus for discrete granulomas.

f. BOVINE TUBERCULOUS MASTITIS   Various figures have been quoted for the incidence of mastitis in bovine tuberculosis, but no satisfactory estimation has so far been made. The most significant generalization is made by Nieberle to the effect that involvement of the mammary glands is second only to that of the lungs. Its public health importance, however, depends not only on Nieberle's generalization, but also on one by Stamp, who states that in all cases, irrespective of age or activity of the lesion,

the ducts are involved and, therefore, that every case of mammary tuberculosis is an open case.

There are three major anatomical forms of mammary tuberculosis; disseminated miliary tuberculosis, chronic organ tuberculosis, and caseous tuberculous mastitis. In the great majority of cases, the infection is blood borne, with subsequent spread along the ducts (Fig. 4.105). The anatomic form of the lesions is deductively explained on the basis of degrees of resistance and hypersensitivity but these are essentially imprecise (see Tuberculosis, Diseases of Lung, Volume 2, Chapter 6). The disease may occur with outstanding duct involvement following infusion with material contaminated with *Mycobacterium tuberculosis*.

Disseminated miliary tuberculosis develops as part of an early generalization and is not common even in a highly infected community. The lesions are rather like those of miliary tuberculosis elsewhere, occurring as nodules up to 1 cm or so in diameter which project above the cut surface. The tubercles are not evenly distributed, but rather tend to occur in groups in the deeper tissues. Some are caseous, other are caseous and calcified, and all are of typical tubercle structure, often with very heavy capsules. The foci begin to develop in the interacinar connective tissue and remain localized to lobules which are completely destroyed (Fig. 4.106), but there is no direct progression between adjacent lobules or encroachment on the interlobular connective tissue. The reaction is typically tuberculous, and often there is more than one tubercle

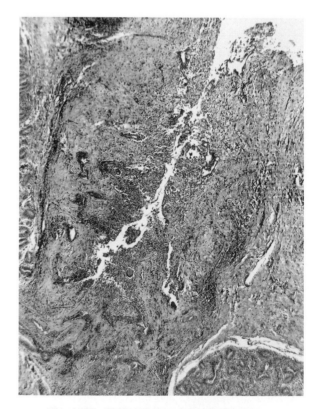

Fig. 4.105 Tuberculous galactophoritis. Cow.

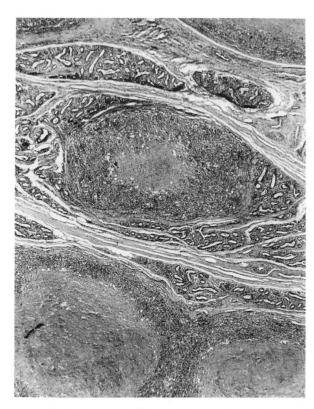

Fig. 4.106 Miliary tuberculous mastitis. Cow.

in each lobule affected. The interlobular ducts are also extensively involved, and the dilated lumen is filled with cellular exudate. The wall itself is greatly increased in thickness by tuberculous granulation tissue and epithelial hyperplasia. The supramammary lymph nodes in such cases usually contain many typical tubercles.

Chronic organ tuberculosis is the most common form, representing 80–90% of cases. It is rather different from the ordinary tuberculous lesion, and the regional lymph nodes are not usually involved, although some may show numerous microscopic foci of infection with few organisms. The mammary tissue is very firm and cuts readily. The lobulated structure is exaggerated and, when cut, projects above the surface of the organ, giving a smoothly bumpy appearance, surrounded by the indentation of the interlobular tissue. The lobules which are affected, and most of them tend to be, vary from grayish red to white and the surface is dry. The lesions begin as one or more foci of granulation tissue within the lobules, and they expand and coalesce to involve the entire lobule. A few of them break down and caseate. Histologically, the lobular outlines are retained and the interlobular septa are not involved. The process begins in the intralobular septa, and the diffuse tuberculous granulation tissue overruns and obliterates the acinar tissue. Typical tubercules do not form although the types of cells involved in the reaction are as usual. The intra- and interlobular ducts are always involved and the walls are greatly thickened by granulation

tissue. The surface tends to caseate, although some epithelium, altered to a squamous type, may persist. Caseous exudate collects in the sinuses. In this form of the disease, intramammary spread is entirely by way of the ducts, and the reaction is typical of any chronic organ tuberculosis and of exogenous or endogenous reinfection in an animal highly resistant by virtue of previous exposure.

Caseous tuberculous mastitis causes great enlargement of affected glands, and they are not nodular. The caseous areas are large and irregular with the dry yellowish caseous appearance and hyperemic margin suggestive of ischemic infarction. Other portions of the gland may show areas of chronic organ tuberculosis, often with commencing caseation and transformation to the diffuse type. Confluence is the rule and there is no limitation of spread by interlobular boundaries. The histologic change is characterized by inflammatory exudation of fibrin and numerous leukocytes. The caseated areas are surrounded by a zone of hyperemic granulation tissue of the sort that occurs in chronic organ tuberculosis, and hemorrhage in it occurs frequently. This form of the disease appears to develop from chronic organ tuberculosis, or possibly as part of an early generalization of the disease, in animals whose resistance is lowered by nonspecific factors or in which a high degree of sensitivity to the organism has developed.

There is, in addition, a form of mammary tuberculosis in which the process centers on the interlobular ducts with subsequent spread to the larger ducts. In all cases of tuberculous mastitis, there is duct involvement, but in the types already described, duct involvement is by spread from initial foci in the intralobular tissues. In this last form, so-called **tuberculous galactophoritis**, there may be no discoverable acinar lesion, or, if the lesions are present, they are more recent than, and obviously derived from, the duct infection. The glandular tissue in such cases retains its lobular character. The ducts are dilated and filled with an exudate of mixed inflammatory cells. The duct epithelium undergoes a patchy squamous or keratinizing metaplasia and the walls are greatly thickened by granulation tissue and infiltrated mononuclear cells but have few or no tubercles. The infectious process apparently begins in the walls of the larger interlobular ducts and extends up and down the duct system. Proximal progression leads to involvement of the smaller interlobular ducts and their obstruction, and this is followed by involution of the associated lobules. Tubercles occur in the supramammary lymph gland in this form of the disease.

Mammary tuberculosis usually develops insidiously without signs of acute inflammation except in a few cases of the diffuse caseous type. The gland progressively increases in size and firmness. Rather importantly, the milk may be physically normal for a long period after the onset of the disease even though it contains the bacilli. In the later stages of the disease there is a reduction in the amount of the secretion, and it is converted to a thin wheylike fluid which contains floccules of caseous exudate and very large numbers of bacilli. Because of the predominance of duct involvement, all cases are open and bacilli may be passed in the milk before gross lesions of tuberculosis are evident.

g. MISCELLANEOUS INFECTIONS Included within this grouping are *Candida* spp., *Nocardia asteroides*, *Mycobacterium* spp., and *Cryptococcus neoformans*. With the exception of *Nocardia asteroides*, the infections develop only after infusion of the udder with compounds, usually penicillin, suspended in oil and tend, therefore, to involve a number of animals in individual herds.

Mastitis caused by **N. asteroides** (primarily a pathogen of humans and dog) is sporadic, although occasionally it reaches significant proportions in a herd. The source of infection is predominantly the soil, and under unknown circumstances, entry is gained to the udder via the teat canal. Although perhaps best regarded as a sporadic infection of individual quarters, it is possible the outbreaks in herds could have resulted from repeated mammary infusions for the control of mastitis.

Nocardia asteroides may be present in mammary secretions for many months without causing clinical signs of mastitis. In other cases, the response to infection is a mild chronic involvement of the mammary gland, whereas in yet other cases, the reaction is of acute onset and severe clinical course. Severe reactions seem particularly likely to occur in infections acquired near the time of parturition. The severe reactions are characterized by fever that may last for several weeks with other signs of systemic illness, and spread of infection to supramammary nodes and lungs has been demonstrated in some of these cases. Affected quarters enlarge rapidly and in one week may be very firm from fibrosis. In some cases, there is a palpable nodularity, the nodules being 2–5 cm in diameter. These may be small sinus tracts discharging onto the skin, or larger defects that qualify as ruptures. The acute infections may be purulent in small foci, but larger abscesses suggest a mixed infection with *Pseudomonas aeruginosa, Actinomyces pyogenes,* or other organisms.

The mastitis is essentially lobular, with the ducts bearing the brunt of the injury. The lining of the ducts and sinuses is thick and corrugated. The drawn milk may be near normal or grossly altered, and in the early stages of infection may contain small white granules up to 1 mm in size, which are tangled masses of bacteria. Similar colonies may be found in the discharge of sinus tracts. The glandular lesions remain fairly restricted to lobules without much coalescence across interlobular septa except in virulent infections or those with much suppuration. The cut surface of the tissues shows extensive fibrosis and a lumpy lobular appearance which may be beset with small abscesses. The process from there is typically granulomatous without club formation (Fig. 4.107A,B).

Cryptococcal mastitis caused by *Cryptococcus neoformans* has, in most reported instances, followed repeated udder infusion, usually with aqueous or oily penicillin. There is usually involvement of more than one quarter with, in the acute stages, rather severe swelling and an increased firmness of the gland, and a subcutaneous edema which, when chronic, produces tight adhesion of the skin to the gland. Changes in the milk occur after 2–3 weeks, when it becomes watery and flaky, although large numbers of organisms are excreted during the acute phase. The diseased tissue is abnormally fleshy, firm, and grayish white with spotty hemorrhages. Lobulation is exaggerated in the acute stages, and the cut surface is very slimy with viscid grayish secretion that makes the gland hard to handle. Later the tissues become involuted and gray and contain numerous small granulomas.

The histologic picture is variable. In severely affected areas there tends to be complete liquefaction affecting epithelial cells, and the acini, in consequence, become distended and confined only by the skeleton of connective tissue (Fig. 4.108). The organism is present in very large numbers, and sometimes even the connective tissue is lysed to produce multilocular cystic acini. Thus far the inflammatory reaction is minor, but there develops a massive histiocytic response. Large vacuolated histiocytes infiltrate the acini and ductules and crowd beneath remaining epithelium, lifting it from its basement membrane. These cells engulf large numbers of yeasts, and the histolytic effect of the organisms is no longer so obvious. More chronically, there develops extensive inter- and intralobular fibrosis and a variable number of more typical granulomas. The corresponding lymph node is greatly enlarged and the lesions in it parallel those in the udder (Fig. 4.109). The infection may metastasize to the lungs.

Mastitis caused by **Mycobacterium** spp. other than *M. tuberculosis* also follows mammary infusion measures. These infections too cause a very marked enlargement of the gland, but there is no systemic reaction, nodularity, or subcutaneous edema. The basic tissue reaction is granulomatous and well restricted to lobules. Giant cells are common, but the main infiltrating cells are histiocytes and lymphocytes organized in granulomatous formations, which are often centered on oil droplets. Many of the lobules are completely destroyed by granulation tissue, and there is considerable thickening and lymphocytic infiltration in the interlobular connective tissue. In the early stages, a ring of neutrophils may surround the oil droplets, which contain large numbers of bacilli. Neither the organisms alone nor the oil alone is capable of provoking a granulomatous mastitis.

Candida spp. infection has usually occurred under the same conditions of infusion. Mastitis caused by *Trichosporon* species is also reported. The tissue changes were not examined in either of these infections, but the infection produced by *Candida* spp. is often acute and transient, the infecting organism disappearing in about 3 weeks. Candidiasis is more likely to occur in animals being fed brewers' grains.

2. Mastitis in Sheep and Goats

Mastitis in the ewe and goat is usually caused by *Staphylococcus aureus* or *Pasteurella haemolytica. Corynebacterium pseudotuberculosis* and *A. pyogenes* cause cold abscesses and probably enter via skin wounds. Mastitis in

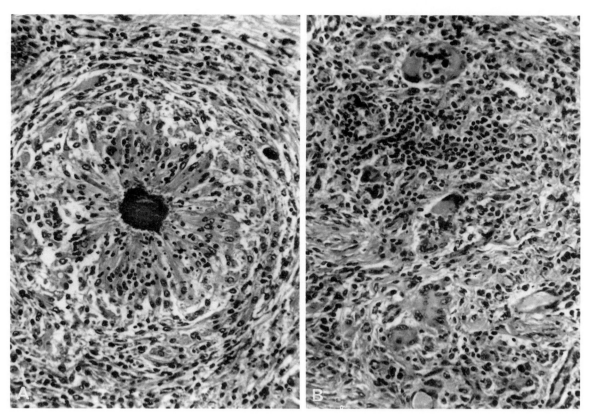

Fig. 4.107 Nocardial mastitis. Cow. (A) Granuloma formation around bacterial colony. (B) Giant cell reaction in chronic stage.

the ewe and doe is important where milk is used for human consumption either directly or in products such as cheese, and because the mortality rate can be as high as 50%. Except for infection by *Mycoplasma agalactiae,* clinical mastitis usually affects only one gland. The range of organisms causing bacterial disease includes the species pathogenic in the cow.

Staphylococcus aureus causes a galactogenic infection occurring very early in lactation with a course and consequences paralleling the disease in the cow. The morbidity, however, is higher and may approach 25% of a flock. *Pasteurella haemolytica* causes mastitis in sheep on summer range at or near the end of lactation. It too is probably galactogenic and affects ~5% of a flock. There is an early acute phase with systemic reaction, but in the course of 48 hr, the systemic reaction has subsided. The affected gland, usually one side, is greatly enlarged and tense, and the secretion becomes watery and contains flakes. Later there may be developing bluish discoloration and widespread necrosis of tissue. Animals that do not die develop abscesses in the course of a week or so with rupture and fistulation, and eventually most of the affected gland sloughs. Pneumonia in lambs has been reported to accompany this form of mastitis.

Contagious agalactia is primarily a disease of goats, sheep being slightly less susceptible, and is caused by *Mycoplasma agalactiae.* The disease is endemic in countries bordering on the Mediterranean Sea.

The name, contagious agalactia, merely emphasizes one, although perhaps the most consistent, feature of the disease, namely mastitis. The disease is initially septicemic, and often fatal in this phase. If the infected animal survives the acute stage, there may develop signs of localization in the eyes and periarticular tissues and, in the case of a lactating female, in the mammary gland. These are the sites in which localization is obvious, but it might be anticipated that *M. agalactiae* shares the affinity of this class of microorganisms for connective tissues and lymphatics and that it localizes in other less obvious places. The organisms are eliminated in the secretions and discharges. Infection is probably by ingestion, but it can be galactogenic and per conjunctivum.

Lactating females and kids are particularly susceptible to the infection and likely to succumb in the septicemic phase of the disease. Pregnant females may abort or deliver live but infected fetuses. There are no specific lesions in the septicemic phase, but the organism can be cultured from all tissues and from the blood. In infections of lesser severity or following recovery from the peracute disease, the febrile reaction subsides, and lameness, keratoconjunctivitis, and alterations of the mammary secretion occur. This complete triad is not always present, and even

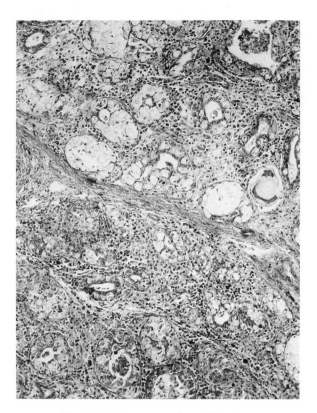

Fig. 4.108 Cryptococcal mastitis. Cow.

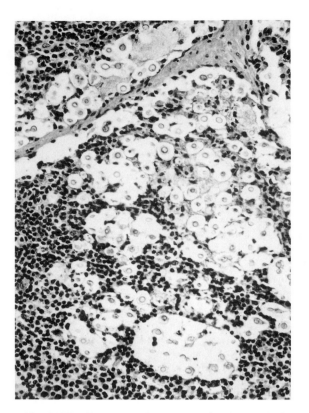

Fig. 4.109 Cryptococcal organisms forming typical soap-bubble lesion in supramammary node.

when there are signs of localization the animal may continue to decline and die in 2 weeks or so.

The mastitis apparently is an inflammation primarily of the interstitial tissue with secondary changes in the acini. Mild inflammations, although they cause cessation of lactation, may be followed by full functional recovery, but the more severe reactions may lead to progressive fibrosis with parenchymal atrophy. The organisms are eliminated in the milk and in other discharges for many months in animals which survive the initial stages of the disease. Lameness is indicative of arthritis and periarthritis and this is more common at the carpal and tarsal joints than at others. Inflammatory edema thickens the periarticular tissues. The synovial membranes may be normal or they may be hyperemic or ulcerated with an increased volume of turbid fluid which may be bloodstained. The ocular lesions, which are present in ~50% of cases, consist of mucopurulent conjunctivitis and keratitis which may be complicated by ulceration and staphyloma.

Mastitis affecting both glands of ewes is a common manifestation of infection by the **lentiviruses of maedi–visna and progressive pneumonia** (see The Respiratory System, Volume 2, Chapter 6). These infections are grouped with the slow-virus diseases because of the long interval between infection and the onset of clinical neurologic or pulmonary disease. The mammary disease can be of much earlier onset and of much higher incidence than other manifestations of these infections, although the ex-

pression of the various syndromes may be strongly influenced by breed and genotype of the hosts. The presenting indications of lentivirus mastitis are reduced milk yield, poor preweaning growth of lambs, and progressive atrophy and induration of both glands.

The virus is present in the mammary gland shortly after exposure. The severity of the lesions increases with the duration of infection. The virus replicates in circulating monocytes and more freely in tissue macrophages. The presence of macrophages in the mammary interstitium encourages a local inflammatory response consisting largely of lymphocytes and plasma cells. The infiltrations are diffuse in the interstitial and periductal connective tissue, and lymphatic nodules are formed. Fibrosis is not a feature of the response. Degeneration and loss of acinar and ductular epithelium which is present in advanced lesions may be a consequence of the infiltration and not a direct virus effect.

The **virus of caprine arthritis encephalitis** is closely related to that of ovine progressive pneumonia and the pathogeneses of the infections in the separate species have very much in common. Mastitis does occur in the caprine disease and has similar morphologic features to that in the ewe but is not a prominent component of the syndrome.

The toxic principle of the avocado, *Persea americana*, is not known and only some varieties of the plant are toxic. Mammary injury is reported in lactating goats. The gland becomes swollen, firm, and congested, especially in dorsal parts, and there may be subcutaneous edema affecting the gland and adjacent body wall. Histologically, there is widespread degeneration and necrosis of secretory epithelium, more severe at the centers of the lobules, and patchy loss of duct epithelium. Cellular inflammatory response is minimal. Restitution of function may be incomplete.

3. Mastitis in the Dog and Cat

In the dog and cat, inflammations of the breast are uncommon and nonspecific. The most important is nonspecific acute mastitis, which is usually confined to early lactating or pseudopregnant periods. The exciting agents are staphylococci and streptococci, the former being the more common, and both are believed to gain entrance via fissures in the nipples and adjacent skin and to spread by way of both the ducts and the lymphatics. The staphylococci may cause gangrene, but they tend to localize in the course of a few days and form abscesses, whereas the streptococci tend to produce a more diffuse and spreading purulent inflammation. The acutely affected glands are large, firm, and edematous, and the overlying skin becomes taut and shiny. Only a small amount of grayish secretion can be expressed; it may be blood tinged or contain pus. The chronic cases of mastitis in these species cannot be satisfactorily distinguished from neoplasms until examined cytologically. Indeed, mastitis is often superimposed on cystic mammary hyperplasia and on mammary neoplasia, especially those tumors which involve the ducts.

4. Mastitis in Swine

Chronic granulomatous mastitis, affecting one or more glands, is occasionally observed in old sows. The lesions are of botryomycotic type, being suppurative granulomas, and most of them are produced by *Staphylococcus aureus*. *Actinomyces bovis* was an important cause of this form of mastitis some decades ago but presently appears to be of very low incidence. It is true for sows as for cows that the designation actinomycosis of the udder refers more usually to staphylococci than to actinomycotic infection. In either event, it appears that the infections are percutaneous, entry being gained through wounds of the skin.

Mastitis caused by coliform bacteria occurs as a very acute infection shortly after parturition. It usually affects more than one gland and produces lesions which are similar to those of cows with coliform mastitis. Hemorrhage mastitis is very rare in sows, and the usual gross appearance is of slight to moderate reddening of the glands with severe edema (Fig. 4.110). Affected mammae are firmer than normal. The severe systemic reaction and the occasional presence of clinical synovitis indicate that septicemia may occur. The pathogenesis is unknown, but ascending infections from environmental contamination may be important.

The important syndrome in sows is that known as the **mastitis–metritis–agalactia syndrome.** It occurs in sows 12–48 hr after parturition and presents as lethargy, fever, swelling and firmness of the mammary glands, and

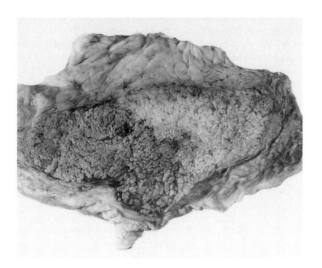

Fig. 4.110 Acute coliform mastitis. Sow. Zone of inflammation (left) is demarcated from less-affected tissue.

agalactia. There is some overlying hyperemia, and edema may be extensive. There are probably several causes for the presence of agalactia in postpartum sows, and acute infectious mastitis such as that due to coliforms is one. The mastitis–metritis–agalactia syndrome is most common in intensive units, and hormonal dysfunction stimulated by sudden changes in management may be precipitating. Metritis is seldom a significant part of the syndrome.

Bibliography

Arnold, J. P. Anatomy and physiology of the bovine teat. *J Am Vet Med Assoc* **116:** 112–116, 1950.

Boughton, E. *Mycoplasma bovis* mastitis. *Vet Bull* **49:** 377–387, 1979.

Brooker, B. E. Pseudopod formation and phagocytosis of milk components by epithelial cells of the bovine mammary gland. *Cell Tissue Res* **229:** 639–650, 1983.

Collins, R. A., Parsons, K. R., and Bland, A. P. Antibody-containing cells and specialized epithelial cells in the bovine teat. *Res Vet Sci* **41:** 50–55, 1986.

Craigmill, A. L. *et al.* Pathological changes in the mammary gland and biochemical changes in milk of the goat following oral dosing with leaf of the avocado *Persea americana. Aust Vet J* **66:** 206–211, 1989.

Deng, P. *et al.* Ultrastructure and frequency of mastitis caused by ovine progressive pneumonia virus infection in sheep. *Vet Pathol* **23:** 184–189, 1986.

Derbyshire, J. B. The experimental production of staphylococcal mastitis in the goat. *J Comp Pathol* **68:** 232–241, 1958.

Du Preez, J. H. The role of anerobic bacteria in bovine mastitis: A review. *J South Afr Vet Assoc* **60:** 159–168, 1989.

Frost, A. J., Hill, A. W., and Brooker, B. E. The early pathogenesis of bovine mastitis due to *Escherichia coli. Proc R Soc Lond (Biol)* **209:** 431–439, 1980.

Houwers, D. J. *et al.* Incidence of indurative lymphocytic mastitis in a flock of sheep infected with maedi-visna virus. *Vet Rec* **122:** 435–437, 1988.

Jones, T. O. *Escherichia coli* mastitis in dairy cattle—A review of literature. *Vet Bull* **60:** 205–231, 1990.

Kremer, W. D. J., Noordhuizen-Stassen, E. N., and Lohius, J. A. C. M. Host defence and bovine coliform mastitis. *Vet Q* **12:** 103–112, 1990.

Madsen, M., Sorenson, G. H., and Aalbaek, B. Summer mastitis in heifers: A bacteriological examination of secretions from clinical cases of summer mastitis in Denmark. *Vet Microbiol* **22:** 319–328, 1990.

Nickerson, S. C. Immune mechanisms of the bovine udder: An overview. *J Am Vet Med Assoc* **187:** 41–45, 1985.

Oz, H. H., Farnsworth, R. J., and Larson, V. L. Environmental mastitis. *Vet Bull* **55:** 829–840, 1985.

Panter, K. E., and James, L. F. Natural plant toxicants in milk. A review. *J Anim Sci* **68:** 892–904, 1990.

Pattison, I. H. The progressive pathology of bacterial mastitis. *Vet Rec* **70:** 114–117, 1958.

Rainard, P., and Poutrel, B. Effect of naturally occurring intramammary infections by minor pathogens on new infections by major pathogens in cattle. *Am J Vet Res* **49:** 327–329, 1988.

Watts, J. L. Etiological agents of bovine mastitis. *Vet Microbiol* **16:** 41–66, 1988.

CHAPTER 5

The Male Genital System

P. W. LADDS
James Cook University, Australia

I. General Considerations

The male genital system resists the sort of integrated generalization which is in pathology a measure of the knowledge and understanding of the disease processes that affect a system. The embryologic elements are disparate and are required to interact at the right place and time, but anatomically they are widely separated, and many of their diseases are linked with those of the urinary apparatus with which they share developmental pathways. The structures and functions of the system are sequential and the development to maturity depends on local inductive mechanisms and the influences of hormones. The expression of biological purpose, the ejaculation of sufficient numbers of competent spermatozoa, is subject to hormonal and nervous integration.

Abnormalities of the male genital system may be expressed in the categories familiar to pathologists, such as anatomic maldevelopments, degenerative alterations, inflammatory processes, or neoplastic transformations. Most of these changes are well described and reasonably understood. What may be the most important changes are those which, in the absence of ordinary pathologic processes, impair the production of spermatozoa in the numbers and of the quality required for fertilization. These are subtle and not well understood. After puberty there are age-related changes in the number of spermatozoa produced by normal males of each species. Departures from normal tend to include both reduction of the number of sperm produced and reduction of the fertilizing capacity of those which are produced.

The evaluation of ejaculated spermatozoa is highly developed in its techniques and its dogma, substantiated for some species by the analysis of mass data derived from artificial insemination results. The evaluation of ejaculated semen has considerable practical importance, but this is not usually within the domain of the pathologist and is not further considered here; our interests are the organs and tissues of the male genitalia.

The germ cells, or gonocytes, occupy a basal position in the seminiferous tubules of the testicle early in organogenesis, and from there produce the several developmental stages which lead to mature spermatozoa. The primordial germ cells are believed to originate in the yolk-sac endoderm and migrate from there to the gonadal ridges of the embryo. This scheme is generally accepted, but it is based on static morphologic evidence and is very incomplete. What is clear is that the primordial germ cells have the unique capability of both mitotic and meiotic division, which might well influence the sensitivity of the seminiferous epithelium to adverse influences.

There is an inductive relationship between the primitive germ cells and the primordium of the gonad. The germ cells need the environment of the gonadal ridge in order to survive and mature, and the rudimentary gonad requires the presence of germ cells if it is progressively to mature. In the developing testis of animals there is massive loss of gonocytes in late fetal or early neonatal periods. The timing of episodes of degeneration is probably species specific, and its extent is difficult to quantify, but the possibility exists that excessive gonocyte loss may be the basis for some patterns of gonadal hypoplasia. The reasons for gonocyte loss are not known, but analogies are available from other tissues in which organogenesis is compounded of cellular growth and differentiation on the one hand and cell death on the other, possibly as expressions of local processes of genetic selection.

The mature testis yields two major products. The androgenic hormone, testosterone, is secreted in a relatively straightforward manner by the Leydig cells under appropriate stimulation by luteinizing hormone of the pituitary. Testosterone is secreted into the interstitial space of the testis, which allows free access to the lymphatics and capillaries, and the hormone is carried throughout the body, completing the secondary masculinization. The second product, the spermatozoon, is generated by a process that is anything but straightforward, but it is orderly, and this gives a morphologic predictability to the seminiferous tubules. This predictability is unexpected, considering the remarkable dynamism of the organ, the usual figure given for sperm production being ~300–500 sperm per gram of testis per second. This production is carried out in three phases. The first phase consists of increasing the number of potential spermatozoa by mitotic division of activated spermatogonia up to the stage of the primary spermatocyte. The number of cell divisions varies with species. In the second phase, meiotic divisions of the primary and secondary spermatocytes reshuffle the genetic information and produce early spermatids with haploid chromosome numbers. In the third phase of spermatogenesis, the early spermatids undergo cytodifferentiation to produce spermatozoa.

The time from the start of mitotic division of a single type A spermatogonium, or activated spermatogonium, until the release of spermatozoa generated from this division also varies between species, but the time taken in spermatogenesis is remarkably constant within a species. The time for completion of spermatogenesis in the bull is 56 days; in the ram, 40 days; and in the boar, 32 days.

Not only is the time needed for spermatogenesis precise, but the timing between successive generations is also precise. It is governed by the period of quiescence for all regenerated type A spermatogonia, which is constant and species specific. In bulls it is 14 days. This is one quarter of the time required for spermatogenesis so that four successive clones of spermatogenic cells are developing at the same time. Each generation displaces the previous one toward the luminal position. A section through a seminiferous tubule will show spermatogenesis at four distinct stages in the progression to spermatozoa.

Since the interval between activation of the quiescent type A spermatogonia is constant and the rate of maturation of cells in spermatogenesis is constant, it follows that each radial segment will have a characteristic set of cell associations. Spermatogenesis has a remarkable timing organization, but how it is achieved is not known.

Spermatogenesis is dependent on hormonal support. Both testosterone and follicle-stimulating hormone are required for sperm production, but their role is apparently only permissive. Without these hormones spermatogenesis ceases, but they do not influence the rate at which spermatogenesis proceeds, nor do they alter the quiescent period of the spermatogenic cycle.

In addition to the phenomenon of gonocyte death, degeneration or aberrant development of gametes occurs with a regularity which gives a flavor of normalcy to the degenerative process. Quite separately from the regression which occurs seasonally in some species, peaks of degeneration of developing gametes occur at stages of the spermatogenic process which appear to be species specific. The stages most sensitive to degeneration are the early spermatogonal stage, the stage of reduction division, and the stage of chromatin condensation to form the spermatid nucleus. Degenerate cells may be recognized singly or in groups (possibly clones) and the calculated losses of cells have been as high as 35% of the germ cells present in the testis.

The production of spermatozoa which are structurally abnormal is a universal phenomenon. The nature of the morphologic abnormalities and the relative numbers of normal to abnormal spermatozoa vary widely; many of the abnormal gametes degenerate in the tubule, and it is doubtful whether those ejaculated are capable of fertilizing an ovum.

Degeneration of gametogenic stages may represent a process of natural selection. The process of selection is susceptible to error as exemplified by chromosomal aberrations. The degenerative phenomena, including chromosomal abnormalities, may be expressions of genetic influences or of the whole range of noxious influences which cause disease.

The role of the fetal testes in influencing development of the remaining genitalia in the male is becoming clearer. Once the so-called chromosomal sex has been determined at fertilization, presence of the Y chromosome, and specifically the testis determining gene, direct subsequent events. Important in this regard is synthesis, in Sertoli cells, of Müllerian inhibiting hormone, which causes regression of the Müllerian ducts in the developing male fetus. This programmed cell death is considered analogous to that which occurs during morphogenesis in insects and amphibians. Müllerian duct regression is evident by day 39 in the fetal dog. Testosterone, from Leydig cells in the fetal testes, acts both directly and indirectly by conversion to dihydrotestosterone to bring about differentiation of the mesonephric (Wolffian) ducts and external genitalia. Pseudohermaphroditism (discussed in Chapter 4, Section I,A) in a genetic male may thus result either from defective secretion of these hormones or from unresponsiveness of target cells.

Blood vessels do not enter the seminiferous tubules. The complex morphologic and physiological events within the tubule must require an environment very different from that in other tissues and, indeed, separate local environments are probably required to separate sequential events within the epithelium. Attention was drawn to the existence of a blood–testis barrier by remarkable differences in the composition of blood, lymph, and rete testis fluid, and a structural basis has been sought for it. Two structural components of the barrier have been identified, namely, the contractile myoid cells in the wall of the tubule and a series of tight junctions between Sertoli cells, which place spermatocytes, spermatids, and spermatozoa in an adluminal compartment. It appears not to be fully developed until puberty, and indeed studies in young bulls suggest that the tubular basal lamina may function as an early barrier monitoring the passage of substances between tubular epithelium and the intertubular compartment. At 16 weeks the basal lamina is composed of 25–30 lamellae and has a maximal thickness of 3 μm. It subsequently decreases in thickness, coincident with the development of Sertoli cells with their characteristic tight junctions.

Alterations to the barrier function are important in the pathogenesis of some types of testicular injury. The products of germinal cell activity, especially those segregated on the luminal side of the tight junctions between Sertoli cells, are not recognized as self by the immune mechanisms of the body, and require the protection of the blood–testis barrier. There are opportunities in many diseases for antigenic components of semen to gain access from the long tubular ducts to the blood and to induce humoral and cell-mediated responses.

Recent studies on immunology of the mammalian testis suggest that it is an immunologically suppressed organ, so that responses to antigens that are clearly outside the blood–testis barrier may be distinctly less apparent than in other tissues. Both allografts and xenografts in rodent testes, for example, survive longer than elsewhere, and it has been demonstrated that this survival is not due to absence from the testes of effective immune or inflammatory mechanisms. The reason for this inhibition or downregulation of testicular immune response is not clear, but local suppression, mediated by the gonadal hormones inhibin and activin, is suggested. Inhibin, which is believed to be produced solely by Sertoli cells, is in high concentration in testicular interstitial fluid; at least *in vitro*, (bovine) inhibin dramatically potentiates the response of rat thymocytes to phytohemagglutinin. Activin, however, which may be secreted by Leydig cells, inhibits the actions of both phytohemagglutinin and inhibin. Possibly the mechanism involves local activation of testis antigen-specific T suppressor cell functions, and it is relevant that neonatal thymectomy in mice results in subsequent allergic orchitis. Quite pronounced changes on Leydig cell populations in the pre-pubertal bull have been demonstrated, and perhaps such changes relate to the special immunologic status of the testis. Between 4 and 8 weeks of age, fetal Leydig cells are replaced by postnatal ones, many of which then degenerate during the 16- to 30-week period. Leydig cells that survive this degenerative phase constitute the long-lasting adult population.

So-called immunologic infertility has been recognized in several domestic species, and it is of interest that in more than 50% of infertile men with antisperm antibodies, no clinical cause for these antibodies can be demonstrated, thereby suggesting a defect in the normal immunoregulation of the testis.

Other immunologic studies, particularly in ruminants, have clearly demonstrated the existence of both secretory and cell-mediated immunity at epithelial surfaces elsewhere in the genitalia. Specific antibodies and elevated immunoglobulin levels occur in semen and preputial washings in some venereal infections. These are derived from serum but probably also by selective transfer, in the case of immunoglobulin A (IgA), across epithelium, especially in the accessory sex glands. Immunoglobulin-containing cells are present in many subepithelial locations in normal genitalia and are greatly increased in genital infections; they would appear to be the source of much secreted immunoglobulin.

In addition to immunoglobulin-containing cells, T lymphocytes are present in male genital epithelium, and it has been suggested that an observed higher prevalence in normal epididymal epithelium of T suppressor cells than of T helper cells is an integral part of the blood–testis barrier.

The pronephros is a cervical kidney, the mesonephros is a thoracic kidney, and the metanephros, or definitive kidney, is lumbar in its location. The developmental associations between the mesonephros and the organization of the urinary and genital systems have been known in some detail for a long time. More recent evidence confirms the mesonephric origin of the granulosa cells of the ovary and the Sertoli cells of the testis. Masses of cells, the rete blastema, migrating from a mesonephric glomerulus (or glomeruli in some species), accumulate with primordial germ cells in the rudimentary gonad and, apparently under the influence of a developing vascular network from the developing tunica albuginea, begin to organize into seminiferous cords. The seminiferous cords develop centripetally and recruit mesonephric tubules, one or more depending on the species, to form the tubuli recti, rete cords, and efferent ducts of the testis.

Additional details of embryologic development help explain gonadal dysgenesis, segmental defects in the mesonephric (Wolffian) ducts, and the relatively common failure, in some species, to develop patent connections between all efferent ducts of the testis and the epididymal tubule. Under appropriate androgen support, the mesonephric ducts develop to form the epididymides and ductus deferentes. In the absence of such hormonal influence they are vestigial and discontinuous, as they are in the normal female. The epididymis functions to aid in the maturation of sperm in the head of the gland and to store it in the tail. These sequential functions require continued hormonal support.

The accessory sex glands include the ampullae, the seminal vesicles, the prostate, and the bulbourethral glands. There are species differences in the accessory glands which are present. Pathologic changes in them are limited in their variety; some will reflect the origins of the glands, the ampullae, and seminal vesicles, from the mesonephric duct, and the prostate and bulbourethral glands from the urogenital sinus.

Bibliography

Campero, C. M. *et al.* Immunoglobulin-containing cells in normal and inflamed accessory sex glands of bulls. *Aust Vet J* **66:** 137–141, 1989.

Campero, C. M. *et al.* Immunopathology of experimental *Bru-*

cella abortus strain 19 infection of the genitalia of bulls. *Vet Immunol Immunopathol* **24**: 235–246, 1990.

Foster, R. A. *et al.* Immunoglobulins and immunoglobulin-containing cells in the reproductive tract of normal rams. *Aust Vet J* **65**: 16–20, 1988.

Foster, R. A. *et al.* Immunoglobulins and immunoglobulin-containing cells in the reproductive tracts of rams naturally infected with *Brucella ovis. Aust Vet J* **65**: 37–40, 1988.

Hedger, M. P. The testes: An "immunologically suppressed" tissue? *Reprod Fertil Dev* **1**: 75–79, 1989.

Johnson, C. A. The role of the fetal testicle in sexual differentiation. *Compend Contin Educ* **5**: 129–132, 1983.

Marshall, L. S. *et al.* Persistent Müllerian duct syndrome in miniature schnauzers. *J Am Vet Med Assoc* **181**: 798–801, 1982.

McEntee, K. "Reproductive Pathology of Domestic Mammals." Academic Press, New York, 1990.

McLaren, A. The making of male mice. *Nature* **351**: 96, 1991.

Rosenthal, R. C. *et al.* Detection of canine antisperm antibodies by indirect fluorescence and gelatin agglutination. *Am J Vet Res* **45**: 370–374, 1984.

Suri, A. K. *et al.* Effect of infection of the genital tract on the concentration of IgG and albumin in bull serum and semen. *Vet Immunol Immunopathol* **13**: 273–278, 1986.

Wrobel, K.-H. *et al.* Postnatal development of the tubular lamina propria and the intertubular tissue in the bovine testis. *Cell Tissue Res* **252**: 639–653, 1988.

II. Scrotum

The scrotum evolves as an outpouching of perineal skin and is lined by evaginations of the peritoneum. The theory that the development of the scrotum is directed solely to providing a superficial cool environment for the testicle is too limited. The storage receptacles for mature sperm in most mammals also need to be located in sites in which the temperature is less than core temperature. Furthermore, since maturation of sperm is a slow process, a fact reflected in the extraordinary length of the epididymal tubule, the scrotum may evolve to accommodate the epididymis. Finally, the testicular artery, which is quite separate from the blood supply to the epididymis, may have, by virtue of its length, physiological consequences with evolutionary advantage, which are assisted by the development of an abdominal evagination.

The scrotal skin is thinner than skin elsewhere and is well covered with hair only in the cat, but in all species it is well supplied with sweat glands. The contractility of the scrotum is a property of the dermal dartos, which consists of fibroelastic tissue and smooth muscle. The structure of the scrotum is adapted to its function of maintaining the testis at a temperature less than the intra-abdominal; it is provided with abundant glands and has little or no subcutaneous fat or connective tissue. The thermoregulatory function is shared with the cremaster muscle, which regulates the proximity of the testicle to the abdominal wall, and with the pampiniform plexus of veins, which ensures maximal contact of cooled venous circulation with the warmer arterial blood. If the testes of the common domestic animals are maintained at or above normal body temperature, degeneration or arrest of development of the seminiferous epithelium occurs, although the impairment of function of the interstitial cells is minimal or absent. Experimental shortening of the scrotum of ram lambs results ultimately in mature rams which have small testes and epididymides, low sperm production, and in most cases, no epididymal sperm reserves.

The formation of the scrotum from the paired genital swellings of the embryo, and its full development to puberty, is dependent on gonadotrophin-induced androgens, and is dependent therefore on the development of the gonad from the indifferent stage into a testis. Anomalies

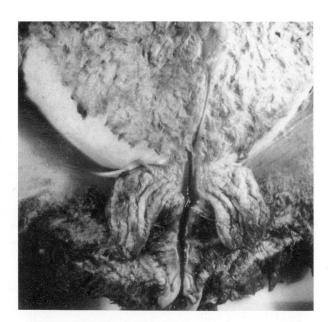

Fig. 5.1 Hypospadias with bifurcation of the scrotum. Ram.

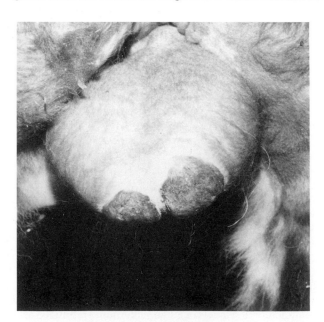

Fig. 5.2 Scrotal frostbite. Bull. (Courtesy of W. F. Cates.)

of scrotal development are simple in their expression in spite of the complicated steps in the development of the paired primordia. Apparent absence of the scrotum has been observed in cryptorchidism. Various degrees of bifurcation of the scrotum, and scrotal clefts, represent failures of fusion (Fig. 5.1). The defects may be local and accompanied by hypospadias or be a reflection of gonadal dysgenesis in intersex states.

Range bulls exposed to extreme cold may develop scrotal frostbite. A higher incidence of lesions in old bulls is attributed to their more pendulous scrotums. Lesions consist essentially of necrosis of skin on the ventral aspect of the scrotum and vary from mild (about 3 cm diameter) to severe, in which more than 75% of the scrotum is affected (Fig. 5.2). Associated changes include scrotal swelling, tunic adhesions, and decreased semen quality.

Because of the delicacy of scrotal skin, it is especially vulnerable to inflammation. The expressions of inflammation are influenced by the abundant arteriovenous anastomoses which are present, and the effects include impairment of the function of the tunica dartos. The causes of **scrotal dermatitis** are frequently nonspecific environmental irritants. Important specific causes in ruminants include *Dermatophilus congolensis*, *Besnoitia besnoiti*, fungi, and various ectoparasites, especially *Chorioptes* (Fig. 5.3A). These infections are discussed with Diseases of the Skin (Volume 1, Chapter 5).

Although the scrotal skin of bulls seems especially susceptible to infection by *Dermatophilus congolensis*, possibly because of the role of ticks in transmission, the infection is not necessarily associated with testicular degeneration. Severe, prolonged infection may, however, result in thickening of the skin to 1 cm, with advanced testicular degeneration and sterility. Infection of scrotal skin of bulls and goats by *Besnoitia besnoiti* may also lead to severe testicular degeneration. This infection is not limited to skin and the small nodules of infection containing white gritty material may be observed grossly in the subcutis of the scrotum as well as in the tunics, the parenchyma of the testis, and in the epididymis; vascular damage is considered important in the development of lesions, and large numbers of *Besnoitia* cysts may be present in intima of vessels of the pampiniform plexus. A feature of the testicular lesions is extensive intratubular mineralization.

Infestation by *Chorioptes* is common on the scrotum of rams. Indeed, the scrotum appears to be a preferred site

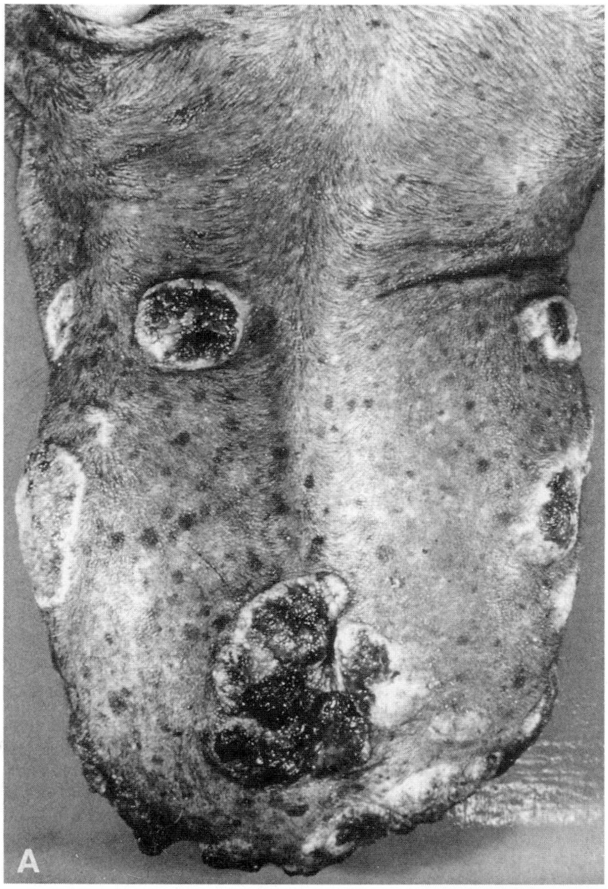

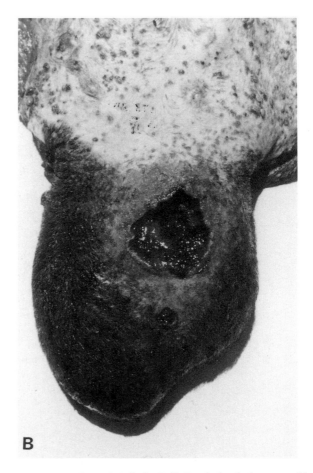

Fig. 5.3A Mycetomas of the scrotum. Bull. Fungal organisms had metastasized to the regional lymph nodes.

Fig. 5.3B Scrotal myiasis. Bull. Exudative lesion caused by infestation with the screwworm *Chrysomyia bezziana*. (Courtesy of J. D. Humphrey.)

for the mite. The lesions can be mild but are frequently sufficiently severe to influence the size and function of the testis. In such cases there is marked thickening of skin and matting of overlying hair and wool by excessive secretions. *Linognathus pedalis,* the foot louse of sheep, and *Haematopinus eurysternus* in cattle can be important causes of scrotal dermatitis. The flea *Tunga penetrans* may also cause scrotal dermatitis in the boar by burrowing into the skin. Scrotal myiasis is caused occasionally by screwworm and other fly larvae (Fig. 5.3B).

Any of the **neoplastic diseases** of the skin may arise in the scrotum, but the scrotum is not a particularly susceptible site. Melanotic tumors and mast cell tumors are not uncommon in the scrotal skin of the dog. Varicose tumors of blood vessels in the scrotal skin, often referred to as hemangiomas, occur in dogs and boars. In dogs, the lesions commence as focal pigmented areas and progress to a small plaque and then to a capillary, then cavernous hemangioma. Microscopically, small arterioles are seen between cavernous spaces, which may extend quite deeply into underlying muscle. The lesions may become ulcerated, and severe hemorrhage of the arterial type may occur. In boars, the multiple warty lesions are unsightly and may occasionally hemorrhage but do not appear to affect fertility (Fig. 5.3C). Lesions are benign and, as in dogs, they may not be true neoplasms. Age, altered hemodynamic forces in the capillary bed of the scrotal dermis, and genetic predisposition are factors which might be involved in the development of these lesions.

Varicose dilatations of scrotal veins occur in older bulls.

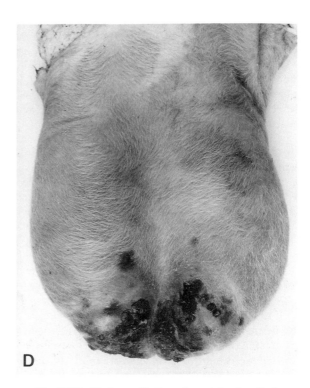

Fig. 5.3D Varicose dilation of scrotal veins. Bull.

They appear as flattened and irregular thickenings of the scrotal skin. The overlying epithelium is usually normal unless it ulcerates, when oozing of blood occurs (Fig. 5.3D). The affected venous channels are irregular in shape, widely dilated, and have thickened walls. Thrombosis does not occur unless the lesion is traumatized.

Bibliography

Cooper, J. E. An outbreak of *Tunga penetrans* in a pig herd. *Vet Rec* **80:** 365–366, 1967.

Probert, A. D., and Davies, A. S. A study of short scrotum, castrated, and entire rams. *Proc NZ Soc Anim Prod* **46:** 55–58, 1986.

Sekoni, V. O. Terminal sterility in a Friesian bull naturally infected with chronic scrotal cutaneous streptothricosis (Kirchi). *Theriogenology* **20:** 27–36, 1983.

III. Tunica Vaginalis

The tunica vaginalis is an extremely thin mesothelial layer continuous with and with the same structure as the peritoneum. Considerations of it cannot be separated from the scrotum, which is lined by the parietal layer of the tunic or from the capsule of the testicle, of which it forms the outer layer. The parietal and visceral layers are separated by the cavity of the tunica vaginalis.

The cavity of the tunica vaginalis communicates with the peritoneal cavity and is susceptible to the accumulation of fluid, hydrocele, in conditions leading to ascites, anasarca, or local lymphedema. Hydrocele may lead to quite severe testicular degeneration on the involved side

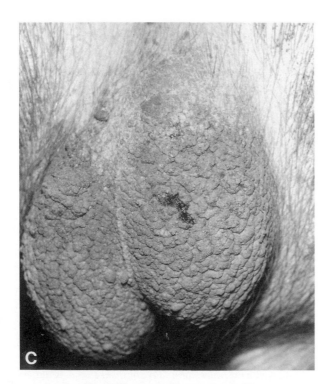

Fig. 5.3C Scrotal hemangioma. Boar.

and to lesser damage on the opposite side, mostly in response to impaired thermoregulation. Lymphatic drainage from the cavity of the tunica vaginalis is through scrotal lymphatics beneath the parietal tunic, and this accounts for the spread of inflammatory processes from the cavity of the tunica to the scrotum. Hematocele, the accumulation of much blood in the vaginal cavity, is mostly the result of trauma. It is not uncommon in boars housed together.

The visceral tunic is an integral part of the testicular capsule, the other two layers of it being the tunica albuginea and tunica vasculosa. The tunica albuginea is composed of inelastic collagen and fibrous tissue, and smooth muscle through which the tunica vasculosa is normally visible. In spite of the presence of smooth muscle, the capsule is not contractile to a significant degree; it probably contributes to the flow of semen by maintaining intratesticular pressure. The tunic is thickened and opaque in testicular hypoplasia and atrophy.

Inflammatory changes in the tunica vaginalis may be part of disseminated infection, with lesions typical of the infection, such as feline infectious peritonitis, tuberculosis, and caseous lymphadenitis (Fig. 5.4). Suppurative inflammation occurs as an extension of scrotal injury or from sporadic epididymitis, such as that following reflux of urine.

Severe periorchitis is a complication of epididymitis irrespective of cause, but is of notable frequency where infections with *Trypanosoma brucei* are endemic and in rams infected with *Brucella ovis* or *Actinobacillus seminis*. Ectopic parasites, such as equine strongyles, can be found on the tunics. *Setaria labiatopapillosa* causes a granulomatous periorchitis in bulls, and lesions occur particularly after worms are killed by anthelmintic treatment. A high occurrence of eosinophilic, then granulomatous, periorchitis in buffaloes is also attributed to *Setaria* spp. Cysts of taenid larvae, mostly *Cysticercus tenuicollis*, may be found in the scrotal cavity of rams, unassociated with inflammation. In dogs, peritoneal cestodiasis may extend to involve the vaginal tunics; saccular dilatations within the tunics, thickening of the spermatic cord, and pyogranulomatous inflammation involving the tunica albuginea and extending into the testes are in response to larvae of *Mesocestoides* spp.

Adhesions between the visceral and parietal layers of the tunica vaginalis are common, especially in older animals. Adhesions are at first fibrinous and indicate acute periorchitis but later become fibrous. Fine, threadlike, fibrous adhesions located at the epididymal tail in bulls are normal. Although adhesions may result from infectious periorchitis and may be associated with epididymitis, it seems probable that most are the result of trauma. Most adhesions are focal and solitary and are not accompanied

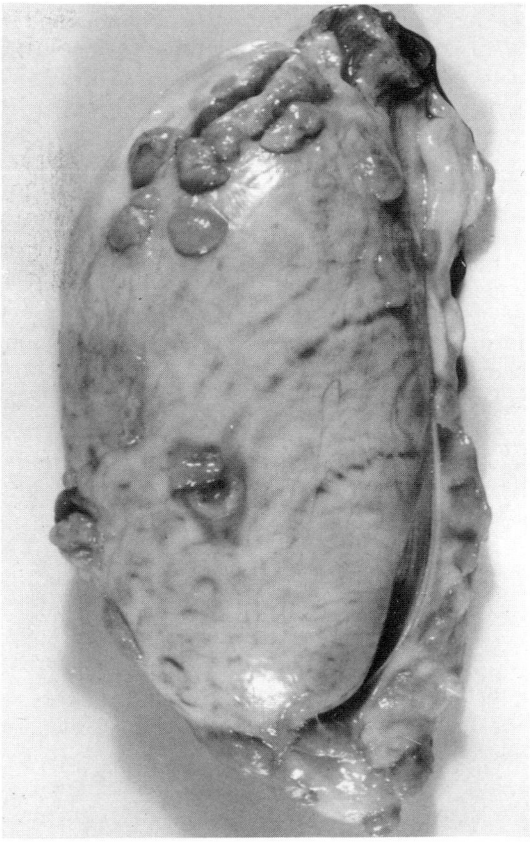

Fig. 5.4 Tuberculous periorchitis. Bull.

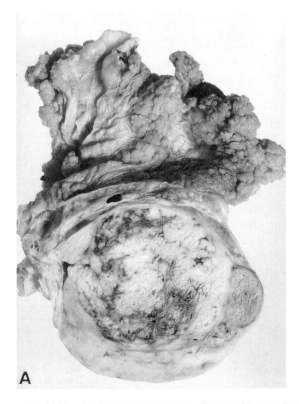

Fig. 5.5A Papillary mesothelioma of mesorchium and tunica vaginalis. Dog.

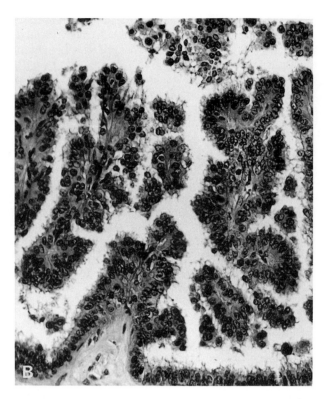

Fig. 5.5B Histologic appearance of papillary mesothelioma shown in Fig. 5.5A.

by lesions in the testis or epididymis; they are probably of little consequence.

Cystic lesions of the tunica vaginalis may be loculations within adhesions. Very small cysts, probably of Müllerian duct origin, are common adjacent to the head of the epididymis in the horse, and are referred to as appendix testis or Morgagni's hydatid.

Neoplastic disease involving the tunica vaginalis is rare, even as secondary involvement from the testis or scrotum. Mesotheliomas with primary involvement of the tunica vaginalis occur in the dog and bull (Fig. 5.5 A,B). Extension of the tumor into the peritoneal cavity has been observed in the dog. Ultrastructural or immunohistologic study may be necessary to differentiate such mesotheliomas from adenocarcinoma. Origin of intrascrotal mesothelioma from paramesonephric duct remnant has been suggested. There is a report of a lipoma occupying the right scrotal cavity in a ram, causing compression of, but apparently not directly involving, the testis.

Bibliography

Akusu, M. O. *et al.* Genital besnoitiosis in bulls: report of five cases. *Nigerian Vet J* **11:** 69–71, 1982.

Cihak, R. W., Roen, D. R., and Klassen, J. Malignant mesothelioma of the tunica vaginalis in a dog. *J Comp Pathol* **96:** 459–462, 1986.

Powe, T. A., and Powers, R. D. Periorchitis after tetramisole treatment in bulls implanted with *Setaria labiatopapillosa. J Am Vet Med Assoc* **186:** 588–589, 1985.

Sutton, R. H. Mesothelioma in the tunica vaginalis of a bull. *J Comp Path* **99:** 77–82, 1988.

Tageldin, M. H. Testicular lipoma in a desert ram. *Sudan J Vet Res* **2:** 92–96, 1980.

Zeman, D. H., Cheney, J. M., and Waldrup, K. A. Scrotal cestodiasis in a dog. *Cornell Vet* **78:** 273–279, 1988.

IV. Penis and Prepuce

It is convenient and, for the pathologist, logical to consider these together. There are a number of developmental penile and preputial abnormalities which are of surgical importance, but which do not require the contributions of the pathologist; those which do, tend to involve simultaneously the penis and prepuce.

Penile and preputial hypoplasia occur mainly as a result of early castration or in intersex states associated with other defects of the urogenital system. Partial or complete lack of the sigmoid flexure of the penis has been observed in rams and bulls. In the bull, a form of hypoplasia described as congenital short penis, possibly associated with shortening of the retractor penis muscle, has been described. Hypoplasia of the glans penis only has been reported. Other rare anomalies in the bull include abnormal insertion of the retractor penis muscle, with stretching of the skin anterior to the testis during erection, partial or complete duplication of the penis, supernumerary ectopic penis, and detached urethral process in which the free end of the penis has a bifid appearance resembling diphallia. Congenital dilation of the penile urethra has been described in the goat. Directional deviations of the erect penis may be attributed to congenital curvature of the os penis in the dog, to asymmetrical development of the corpus cavernosum penis in the horse and donkey, and to persistence of the penile frenulum in the bull, buck, boar, and dog (Fig. 5.6), or malfunction of the apical ligament in the ram and bull.

Erection of the penis requires participation of the local circulatory apparatus. Contraction of the ischiocavernosus muscles allows arterial input to the corpus cavernosum and spongiosum, but occludes venous drainage. Abnormalities in the vascular function may lead to hemorrhage from the corpus spongiosum to the urethra. Forced deviation of an erect penis may exaggerate the vascular pressures and the structural tensions of the penis. Several important vascular defects of the penis which cause erection failure and impotence have been studied extensively in the bull and boar, but their precise diagnosis is not possible at necropsy unless prior study of abnormal angioarchitecture of the penis has been achieved by radiography or injection of plastic into vessels to form casts. These defects in the bull include rupture of the corpus cavernosum penis with subsequent hematoma formation, mostly at the distal bend of the sigmoid flexure (Fig. 5.7), but sometimes proximal to it, and vascular shunts from the corpus cavernosum to the corpus spongiosum of the penis or to peripenile vasculature, thereby preventing effective erection. Rupture of the corpus cavernosum has also been

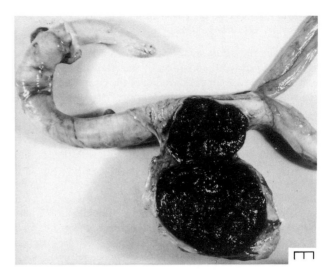

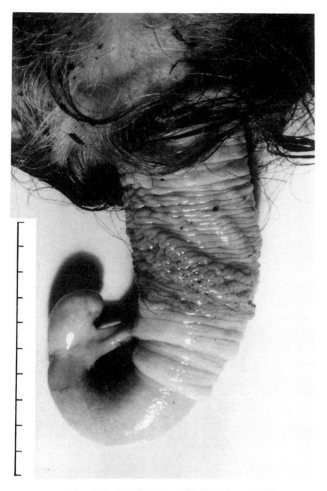

Fig. 5.7   Hematoma of penis with rupture of tunica albuginea opposite point of attachment of retractor penis muscle (above, right). Bull.

Fig. 5.6   Persistent penile frenulum. Bull.

used for artificial insemination. The precise cause is unknown.

Paraphimosis, or inability to retract the penis, is a particular problem in the stallion. Trauma is the most common cause, but inflammation of other cause, neoplasia, or primary penile paralysis may be involved. Penile paralysis occurs in severely debilitated stallions, after administration of certain phenothiazine-derivative tranquilizers, or as a result of local neurologic disorders.

In the dog, fracture of the os penis with resultant callus formation may lead to obstruction of the urethra. Calcification and ossification of the penis caudal to the os penis is a frequent (usually radiographic) finding and its prevalence increases with age.

observed in the stallion, but in this species hemorrhage from vessels outside an intact tunica albuginea is considered more frequent; trauma is the likely cause. Whereas in the bull these defects seem mostly to have an acquired origin, shunts in the boar from the corpus cavernosum penis to neighboring veins appear to be developmental in origin and may be inherited.

Surgical conditions of the prepuce include varicosities of preputial veins in stallions, which may lead to thrombosis, edema, and inflammation. Eversion of the preputial mucosa as a temporary event is common in bulls of the *Bos indicus* species and does occur in polled breeds of *Bos taurus*. The eversion is permitted by inadequate muscular arrangements in the prepuce, and its importance lies in the injuries to which the everted epithelium is exposed. Trauma and desiccation lead to edema and inflammatory change (Fig. 5.8). A preputial diverticulum is anatomically normal in the boar but abnormal in the bull and buffalo. Deflection of the penis into the diverticulum results in accumulation of debris, urine, and semen, which predispose to local infection and inflammation.

Tearing of preputial mucosa, typically at the ventral preputial fornix, is a common injury which occurs in bulls

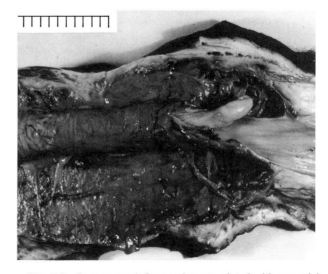

Fig. 5.8   Gangrenous inflammation associated with preputial prolapse. Bull.

A. Inflammation of the Penis and Prepuce (Balanoposthitis)

Inflammation of the glans penis **(balanitis)** is frequently accompanied by inflammation of the prepuce **(posthitis).** Fundamental to an understanding of the causes and pathogenesis of balanoposthitis is an appreciation of the large variety and number of organisms resident in the preputial cavity, and of the existence of local immunity in this location. Viruses such as parainfluenza-3 in bulls and herpesviruses in many species, nonpathogenic and potentially pathogenic bacteria such as *Corynebacterium renale* and *Haemophilus somnus,* fungi, mycoplasmas and ureaplasmas, chlamydia, and protozoa are readily isolated from preputial cavities which are normal on clinical and pathologic examination. Studies on *Mycoplasma* species and *Ureaplasma diversum* in bulls suggest that the prepuce and distal urethra are areas these organisms mostly colonize. Reports on whether these organisms cause lesions in the prepuce are conflicting. Lymphoplasmacytic infiltrations of the preputial mucosa, and sometimes distinct lymphoid follicles, probably are responses to the presence of these organisms, but these histologic lesions are sufficiently frequent to be considered normal. Immunoglobulins are present in preputial washings, and it seems logical that these are derived from plasma cells of the prepuce. In the bull, the prevalence of these plasma cells tends to increase with age. Possibly, small (T) lymphocytes, frequently seen in preputial and penile epithelium, participate in cell-mediated immune reactions at the epithelial surface. Mucous glands are absent from the epithelium of the penis and prepuce, thus limiting the analogies which can be drawn with inflammatory processes on mucous membranes.

Balanoposthitis in the bull can be caused by the bovine herpesvirus-1. This virus causes both respiratory disease, infectious bovine rhinotracheitis (IBR), and genital disease, infectious pustular vulvovaginitis (IPV). The genital disease in the bull is characterized clinically by a thin purulent preputial discharge. Simultaneous occurrence of the respiratory and genital forms of the disease is rare. In the acute stage of balanoposthitis, 2–3 days postinfection, numerous small gray-white opaque foci of necrosis are present (Fig. 5.9A,B). These areas of necrosis may form confluent and flat efflorescences. In severe cases, edematous swelling of the penis and prepuce may occur at this time.

The foci of necrotic mucosa, which exist for 1–2 days only, subsequently become indistinct, the surface sloughs, and sharp ulcers or erosions remain, especially in the area of the glans. Many ulcers are surrounded by a distinct zone of hyperemia. Healing commences after 6–8 days and in uncomplicated cases is complete in 2 weeks. For a variable period thereafter, however, lymphoid follicles are present as pale to dark red nodules, arranged separately or in clusters, and a seromucoid exudation is present.

The microscopic lesion is one of epithelial necrosis with neutrophilic accumulation, lymphocytic infiltration of sur-

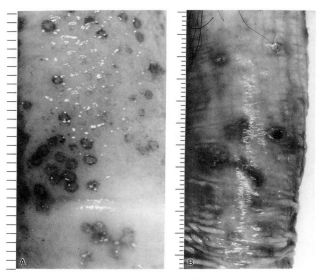

Fig. 5.9 (A) Bovine herpesvirus-1 (IBR) infection producing ulceration of epithelium. Earlier stages have distinct zones of hyperemia. (B) Bovine herpesvirus-1 (IBR) infection producing ulceration of epithelium. Penis. Bull.

rounding stroma, and the transient appearance of intranuclear inclusion bodies in degenerating epithelial cells. The residual changes are nonspecific. As with other herpesvirus infections, latency occurs, and viral production can be reactivated by natural mechanisms or in response to corticosteroids. Although orchitis has been observed in bulls in association with this herpetic infection, it is exceptional.

Herpesvirus infection of male **goats** may also cause similar lesions, which may, however, progress to an extensive suppurative and necrotizing balanoposthitis involving the glans, fornix, and entire urethral process.

The lesions and pathogenesis of **equine coital exanthema,** caused by equine herpesvirus type 3, are in many respects comparable to those of IBR in cattle, although the penile and preputial pustules and subsequent ulcers are somewhat larger, being up to about 15 mm in diameter (Fig. 5.10). Lesions in the stallion tend to involve the body of the penis more frequently than the glans. They first appear 2–5 days postinfection as watery blisters but rapidly progress to circumscribed yellow pustules with raised borders and depressed centers. Resolution usually occurs within a few weeks leaving depigmented spots, but it is unclear whether the virus remains in the penis and prepuce in a latent form.

In dogs, the genital lesions of **canine herpesvirus** infection consist of hyperemia, petechial hemorrhages, and the development of lymphoid nodules, especially over the base of the penis and the preputial reflection. There is an associated serous discharge from the preputial orifice. Lesions appear about 3 days after experimental infection, but are self-limiting, and regress 4–5 days subsequently, with no apparent sequelae. Simultaneous conjunctivitis may sometimes occur. Recurrence of lesions, an im-

Fig. 5.10 Healing ulcers of penis due to equine herpesvirus-3 (coital exanthema). Stallion.

portant feature of the disease in bitches, seems to be less common in the male. Pustule formation and ulceration do not appear to be a feature of genital herpesvirus infection in the male dog.

A mild but chronic balanoposthitis, not associated with herpesvirus infection and presumed to be bacterial in origin, is common in dogs. Depending on duration, there is intense hyperemia of the epithelium, a mucopurulent exudate, and sometimes ulceration. Lymphoid follicles become enlarged and prominent. Hemolytic strains of *Escherichia coli* are most frequently isolated from such cases, other common isolates being *Proteus vulgaris* and hemolytic streptococci.

Multiple pale, dome-shaped papules up to ~3 mm in diameter on hairless parts of the equine prepuce, and on the muzzle, may result from infection with the poxvirus of **molluscum contagiosum** (for details see The Skin and Appendages, Volume 1, Chapter 5). Histologically, discrete epithelial proliferations with characteristic molluscum bodies are present.

In sheep, **ulcerative posthitis** of wethers particularly, but also of rams, is common and important. Vulvitis may occur in ewes in affected flocks. Differences in susceptibility between wethers and rams is probably attributable to incomplete development of the prepuce and penis in the former, and the tendency of the wether to urinate within its sheath.

Development of the primary lesion depends on the occurrence of a transmissible, urea-hydrolyzing bacterium, now considered to be *Corynebacterium renale,* and on the excretion of urine rich in urea. In one outbreak, *Rhodococcus equi* and *C. hofmanni,* both of which also produce urease, were isolated from lesions. Other factors in urine, possibly hormonal in origin, may also be involved. High-protein and especially leguminous diets predispose to the disease and the incidence is lowest during the summer. It is presumed that wethers are infected by the transmission of material on contaminated bedding, herbage, or by flies. Venereal transmission from rams to ewes also occurs.

The lesion in wethers begins as a small yellowish area of epidermal necrosis, usually on the dorsal part of the bare tip of the prepuce. This area ulcerates, and slow expansion of the lesion may result in occlusion of the preputial orifice. Secondary lesions occur when the prepuce becomes swollen due to the accumulation of urine or pus. At this stage, there is more or less extensive internal ulceration of the prepuce which may slough, destruction of the urethral process, and ulceration of the glans penis (Fig. 5.11). The common name, sheath rot, describes the condition as seen at this time. Secondary lesions are uncommon in rams.

A severe **ulcerative balanitis,** distinct from this condition, occurs in Border Leicester rams, in association with vulvovaginitis in ewes mated to them. Morbidity rate can be quite high. Acute deep ulcerations up to several centimeters in diameter occur on the ventral surface of the collum glandis. Excessive granulation tissue and varying hemorrhage become apparent in ulcerated areas and the lesion may extend well above the surface of the penis. Much necrotic and purulent material covers the glandis and urethral process, which is markedly thickened. Adhesions of the penis and prepuce may result. Ewes in contact with affected rams have shallow ulcers on the ventral commissure of the labia and posterior vagina. The causative agent has not been identified and it is not clear why Border Leicester rams are predisposed.

Traumatic injury by the cockleburr (*Xanthium strumarium*) may cause outbreaks of severe acroposthitis in rams, with subsequent suppurative vaginitis in ewes mated to them.

Outbreaks of **ulcerative posthitis** in bulls occur occasionally, but these are less severe than those in sheep, and it is unclear how closely this condition parallels ovine posthitis in regard to cause and pathogenesis. Lesions in bulls consist of mild to severe erosions and ulcerations confined to the preputial orifice and anterior ventral area of the prepuce. There is associated edema. Lesions begin as small raised areas of necrosis which are easily dislodged to leave shallow erosions. The ulcers which subsequently develop in these locations are irregular in shape, may be several centimeters in diameter, are sharply defined, and frequently bleed. The histologic appearance of lesions is nonspecific, being one of superficial necrosis, reactive

Fig. 5.11 Ulcerative posthitis with secondary balanitis. Wether. (Courtesy of W. J. Hartley.)

epithelial hyperplasia, underlying granulation tissue, and a mixed cellular infiltration. Causal organisms cannot be demonstrated histologically in early lesions, preceding ulceration. In more severe forms of this condition, there is increasing edema, abscessation, and myiasis. Deformity of the preputial orifice and phimosis may occur following healing of ulcers.

Corynebacterium renale, an inhabitant of the prepuce of apparently healthy bulls, can frequently be isolated from bulls with ulcerative posthitis, and as animals on a high plane of nutrition are predisposed, a parallel of this condition with ovine posthitis is suggested. However, it is not clear why steers are infrequently affected.

In bulls, the larvae of *Strongyloides papillosus* may cause severe balanoposthitis. Traumatic injury with subsequent infection by such organisms as *Actinomyces (Corynebacterium) pyogenes, Escherichia coli,* streptococci, and staphylococci is also common in bulls. Tuberculous balanoposthitis, with characteristic granulomas involving the penis or prepuce, is seen occasionally in bulls from infected herds, sometimes as the only lesion. Genital transmission is possible. The presence of *Campylobacter fetus* subsp. *venerealis* in the preputial sac is not associated with gross or histologic changes, although the recognized higher susceptibility of older bulls has been associated with an increase in size and number of crypts in the epithelium of the penis. It seems unlikely that *Campylobacter* spp. can colonize the preputial cavity of male goats; three species (*C. fetus* subsp. *venerealis, C. fetus* subsp. *fetus,* and *C. jejuni*) did not persist following experimental intrapreputial inoculation.

Preputial infection in bulls by *Tritrichomonas foetus* is not accompanied by specific lesions. Microscopic examination of infected penile and preputial epithelium may reveal cells and debris in crypts, and a slight but mixed cellular infiltrate. Trichomonads cannot be demonstrated in histologic sections.

Balanoposthitis is an uncommon lesion in the stallion but occurs in **dourine,** caused by infection with *Trypanosoma equiperdum* (see also The Female Genital System, Chapter 4 of this volume) and in **cutaneous habronemiasis.** Primary infection in dourine occurs in the genitalia following transmission by coitus. There is marked edema of the prepuce and adjacent areas, and phimosis may result. In many cases yellow-red nodules up to ~8 mm in diameter may subsequently appear on the penis, especially near the urethral orifice. These later transform to shallow ulcers, which after healing may leave distinct nonpigmented foci on the skin of the penis, prepuce, scrotum, and perineum. In cutaneous habronemiasis (for detailed discussion see The Skin and Appendages, Volume 1, Chapter 5), also known as genital bursatti and summer sores, the verminous lesions may also be present on other parts of the body. The lesions may affect the penis only, about the urethral meatus, or may at the same time involve the preputial mucosa at about the point of its contact with the end of the retracted penis. The lesions are elevated, ulcerated, fungating, and very liable to hemorrhage; they consist of

exuberant fibrous tissue enclosing few or many larvae of *Habronema* spp. (Fig. 5.12) and may be sufficiently large to cause paraphimosis. There is a very dense infiltration by eosinophils, and these accumulate as small eosinophil cell abscesses about immobilized and dead larvae. They may be discernible with the unaided eye as small yellow foci on the cut surface. The lesion develops during the summer months. Squamous cell carcinomas are very often infected by these larvae, and all summer-sore lesions should be examined with this in mind.

Preputial diverticulitis occurs in swine, and the characteristic plaques and ulcers occur in the preputial diverticula of 25–40% or more of castrated and entire male pigs. The lesions commence as foci of hyperkeratosis and progress to distinct white-gray plaques 2–4 mm in diameter with obvious para- and dyskeratosis. As a result of central necrosis, sloughing of epithelium, and neutrophilic infiltration, they form discrete yellow-brown ulcers up to about 10 mm in diameter with raised borders (Fig. 5.13A,B). Such lesions may become confluent, and hemorrhage from them is common. Ulceration is seen most often in older pigs. Calculi may be present in the diverticulum. The precise cause of preputial diverticulitis in pigs is unclear, but accumulation and decomposition of urine are considered of primary importance. *Eubacterium suis,* a cause of pyelonephritis and cystitis in sows, may be isolated from the prepuce of boars with diverticulitis and healthy boars; the likelihood of venereal transfer exists.

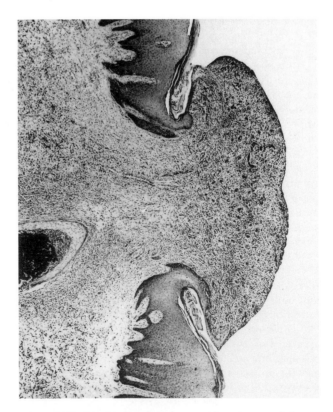

Fig. 5.12 Preputial granuloma caused by larvae of *Habronema* sp. Horse.

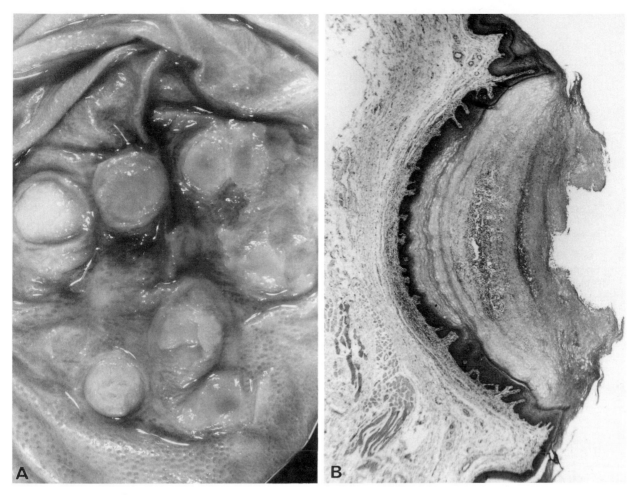

Fig. 5.13 (A) Plaque and ulcer of the preputial diverticulum. (B) Histologic appearance of plaque in preputial diverticulum. Boar.

Other bacteria frequently present in the diverticulum include *Proteus* spp., and staphylococci. There appears to be no correlation between types of bacteria present and boar performance. An age-related change in bacterial flora of the diverticulum has been demonstrated, *E. suis* being generally more common in pigs older than 4 months of age and in adult boars than in younger animals. Spirochetes also colonize the preputial diverticulum and may contribute to inflammation. Miscellaneous conditions causing variable inflammation of the penis and prepuce include intrapreputial foreign bodies, and hair ring in the bull, buck, and cat, in which in-drawn preputial hairs encircle and may constrict the penis.

B. Neoplasms of the Penis and Prepuce

The important primary tumors are transmissible fibropapilloma in the bull, squamous papilloma and squamous cell carcinoma in the horse, and transmissible venereal tumor of dogs. There is a report of primary chondrosarcoma of the os penis of a dog. Additional details on bovine fibropapillomatosis and canine venereal tumor are given in The Female Genital System (Chapter 4 of this volume).

Fibropapillomas in bulls caused by bovine fibropapilloma virus type 2 occur on the glans penis. They are most common in young bulls, 1 to 2 years of age, and may not be noticed until after service, when hemorrhage occurs. Mostly they are multiple and up to several centimeters in diameter (Fig. 5.14). Tumors are pink or gray-white on section and are composed mainly of fibrous tissue with an epithelial covering of variable thickness (Fig. 5.15A,B). Histologically, there are differences between tumors in young and old bulls, those in young bulls being more cellular with frequent mitoses.

Although benign, large fibropapillomas in bulls may interfere with free movement of the penis through the preputial orifice, and retraction of the glans may be prevented. In such cases, preputial hairs may engage the prolapsed penis and cause strangulation. Alternatively, continued growth of the mass within the preputial cavity may prevent penile protrusion. The urethra may then be

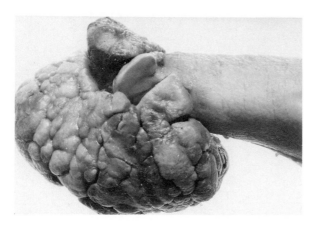

Fig. 5.14 Transmissible fibropapilloma. Bull.

so compressed that rupture occurs and extensive cellulitis results from infiltration of urine in the peripenile tissues.

A **transmissible genital papilloma** has been reported in swine. Both the natural and experimental neoplasms regress with time. Papilloma virus antigen is present in the tumor, but the virus has not been recovered. The tumors are small, round, and project 3 cm from the mucosal surface. Histologically, they are characterized by excessive thickening of the stratum spinosum with exaggeration of the rete pegs. There is little proliferation of the underlying connective tissue. The hyperplastic epithelium is covered by a thin layer of keratinizing cells.

Squamous papilloma in all species is a benign, keratinizing epithelial papilliform tumor with little fibrous stroma; it is usually small and is frequently associated with lymphoplasmacytic infiltration. Squamous papilloma is most common in the horse, however, and is the benign counterpart of squamous cell carcinoma.

Squamous cell carcinoma of the penis and prepuce has been described in the horse, dog, and bull, but it is possible the bovine cases were fibropapillomas. In the horse, carcinomas occur with about equal frequency in stallions and geldings and arise mostly from the glans. Superficial ulceration and necrosis of large tumors is common (Fig. 5.16). Histologically, the penile tumor is usually well differentiated, and keratinization is almost always present. Infiltration of inflammatory cells, especially eosinophils, around the tumor is common, and foci of necrosis and calcification are frequent. Extension of the tumor occurs either by infiltration into the corpus cavernosum or by metastasis, especially to the inguinal lymph nodes and less frequently to other organs such as lung or liver.

In the dog, squamous cell carcinoma of the penis and prepuce is of similar appearance to that in the horse but may be nonkeratinizing. Approximately 25% of reported cases had metastasized, chiefly to the inguinal lymph nodes.

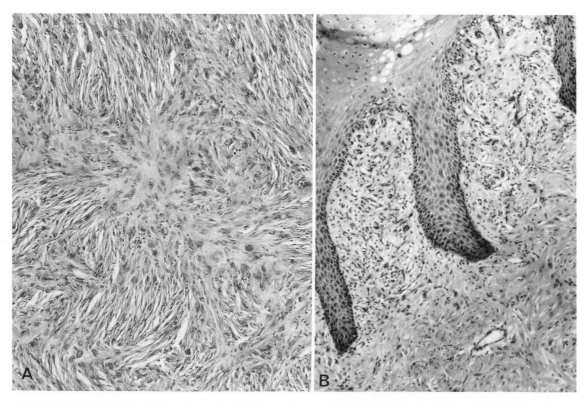

Fig. 5.15 (A) Histologic details of transmissible fibropapilloma. (B) Histologic details of transmissible fibropapilloma. Bull.

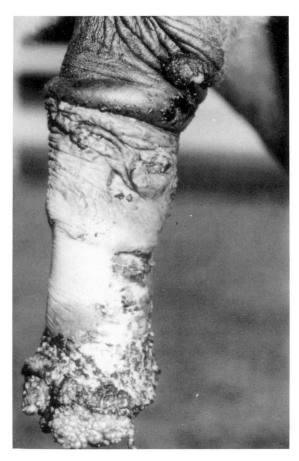

Fig. 5.16 Squamous cell carcinoma. Penis. Horse. (Courtesy of K. G. Johnston.)

Transmissible venereal tumors in the male dog mostly arise on the penis or within the prepuce. They may be single or multiple, sessile or pedunculated, nodular or papillary, soft to firm consistency, and up to ~15 cm in diameter (Fig. 5.17). Histologically the usual pattern is

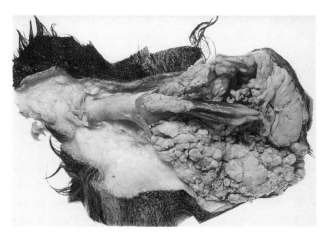

Fig. 5.17 Transmissible venereal tumor on penis and prepuce. Dog.

one of solid sheets of uniform round, ovoid, or polyhedral cells with large round vesicular nuclei. Mitoses are common. Metastasis is uncommon but may involve the superficial inguinal lymph nodes and various organs. Most cutaneous metastases are considered to result from trauma and mechanical implantation (for more detail, see The Female Genital System, Chapter 4 of this volume).

Secondary tumors of the penis or prepuce include squamous cell carcinoma, mast cell tumor, malignant melanoma, hemangiosarcoma, and lymphosarcoma, in which cell infiltration may be either diffuse or nodular.

Bibliography

Ashdown, R. R., and Glossop, C. E. Impotence in the bull: (3) Rupture of the corpus cavernosum penis proximal to the sigmoid flexure. *Vet Rec* **113:** 30–37, 1983.

Bonte, P. *et al.* Primary erection failure in adult Belgian Pietrain: Clinical symptoms and post-mortem findings. *Vlaams Diergeneesk Tijdsch* **55:** 103–115, 1986.

Doherty, M. L. Outbreak of posthitis in grazing wethers in Scotland. *Vet Rec* **116:** 372–373, 1985.

Elliot, G. Preputial diverticulitis, vaginitis, cystitis and pyelonephritis, a social problem. *Pig Vet Soc Proc* **19:** 43–48, 1987.

Ellis, B., and Redwood, R. W. Attempted experimental infection of the preputial cavity of three male goats with *Campylobacter* species. *Vet Rec* **123:** 568, 1988.

Fish, N. A., Rosendal, S., and Miller, R. B. The distribution of mycoplasmas and ureaplasmas in the genital tract of normal artificial insemination bulls. *Can Vet J* **26:** 13–15, 1985.

Jones, J. E. T., and Dagnall, G. J. R. The carriage of *Corynebacterium suis* in male pigs. *J Hyg* **93:** 381–388, 1984.

Makady, F. M., Yousseff, H. A., and Mahmoud, A. Z. Some congenital anomalies of the penis of the goat (dilation of penile urethra, persistent penile frenulum). *Assiut Vet Med J* **18:** 168–172, 1987.

Parish, W. E. An immunological study of the transmissible genital papilloma of the pig. *J Pathol Bacteriol* **83:** 429–442, 1962.

Parker, W. G. *et al.* Avulsion of the bovine prepuce from its attachment to the penile integument during semen collection with an artificial vagina. *Theriogenology* **28:** 237–256, 1987.

Patnaik, A. K., Matthiesen, D. T., and Zarvie, D. A. Two cases of canine penile neoplasm: Squamous cell carcinoma and mesenchymal chondrosarcoma. *J Am Anim Hosp Assoc* **24:** 403–406, 1987.

Tarigan, S., Ladds, P. W., and Foster, R. A. Genital pathology of feral male goats. *Aust Vet J* **67:** 286–290, 1990.

Weiringa, W., and Mouwen, J. M. V. M. Ulceration of the preputial diverticulum in swine. *Tijdschr Diergeneeskd* **108:** 751–760, 1983.

Wolfe, D. F. *et al.* Failure of penile erection due to vascular shunt from corpus cavernosum penis to the corpus spongiosum penis in a bull. *J Am Vet Med Assoc* **184:** 1511–1572, 1984.

Wolfe, D. F., Carson, R. L., and Hanrahan, L. A. Supernumerary ectopic penis in a bull. *J Am Vet Med Assoc* **191:** 539, 1987.

V. Testes

Perhaps more so than in other genital or, indeed, extragenital tissues, meaningful study of the testis and epididymis demands prompt examination, and use of special fixatives. Autolysis is apparent after only a few hours at room temperature and after 7–8 hr refrigeration. Use of special fixatives such as Bouin's solution is essential for good preservation of developing germ cells in the testis.

A. Anomalies of Development

1. Testicular Hypoplasia

Hypoplasia of the testis occurs in all species of domestic animals but has been most extensively studied in the bull. Testicular hypoplasia also occurs in association with cryptorchidism and intersex states (Fig. 5.18), but of particular importance is uncomplicated testicular hypoplasia, in which hypoplasia of the intrascrotal testis occurs as a distinct entity. Differentiation of testicular hypoplasia from immaturity due to adverse environmental factors or disease, and from degeneration, is difficult, and in the absence of appropriate history, sometimes impossible. For this reason data on the occurrence of testicular hypoplasia per se are often misleading. A particular problem in defining testicular hypoplasia arises in determining whether its cause(s) was operative prenatally or during the period between birth and puberty.

In bulls, extensive and detailed studies indicate that occurrence of testicular hypoplasia in most bull populations is probably in the order of 0.5–1%. However, it may sometimes be much higher, as in the Swedish Highland breed, in which an occurrence of 30% was recorded. Studies in this breed revealed a hereditary basis for hypoplasia due to an autosomal recessive gene with incomplete penetrance. In addition to this classic form of testicular hypoplasia, several other forms have been identified in Swedish breeds. These include a low germ-cell resistance form in the Swedish Red and White breed, and an arrested spermatogenesis form in the Swedish Friesian breed. In the classic form, and in the Swedish Red and White breed, hypoplasia affects the gonads of both sexes. A high occurrence of testicular hypoplasia has also been observed in some breeds of *Bos indicus*, but later puberty in these animals confounds diagnosis. Further studies are necessary before a satisfactory classification of testicular hypoplasia in bulls can be established.

The hypoplastic testis, often not observed until after puberty, may be as small as one quarter of normal size and is freely movable in the scrotum. Hypoplasia may be

uni- or bilateral. Although reports indicate that unilateral hypoplasia, especially of the left testis, is much more prevalent than the bilateral condition, this may to some extent reflect its easier recognition (Fig. 5.19). Bilateral cases of mild testicular hypoplasia may escape precise clinical diagnosis unless procedures such as scrotal circumference measurement are used. Consistency of the hypoplastic testis is closer to normal than that of the degenerate testicle and the cut surface usually tends to bulge on sectioning.

Histologically, testicular hypoplasia in bulls is arbitrarily divided into mild, intermediate, and severe degrees, although the picture derived from a large series provides a continuous spectrum. Familiarity with normal testicular histology is necessary and it is important to note that in affected testes, other than those with severe hypoplasia, hypoplastic tubules are usually intermingled with normal seminiferous tubules (Fig. 5.20A,B). The measurement of scrotal circumference, when used systematically, affords a useful clinical technique for judging the prevalence of hypoplastic tubules, the circumference being reduced with increased numbers of hypoplastic tubules.

In the severe form of hypoplasia, most or all of the seminiferous tubules are of small diameter and are lined by Sertoli cells only, or Sertoli cells and perhaps a basal layer of stem cells or spermatogonia which do not show

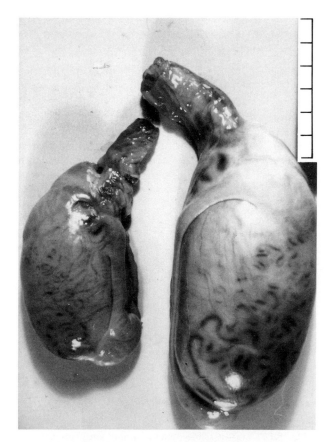

Fig. 5.19 Unilateral testicular hypoplasia. Bull.

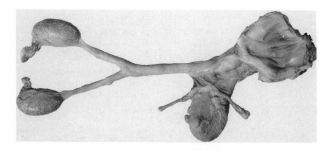

Fig. 5.18 Male pseudohermaphrodite. The tubular genitalia are predominantly female; both gonads are testes. Ox.

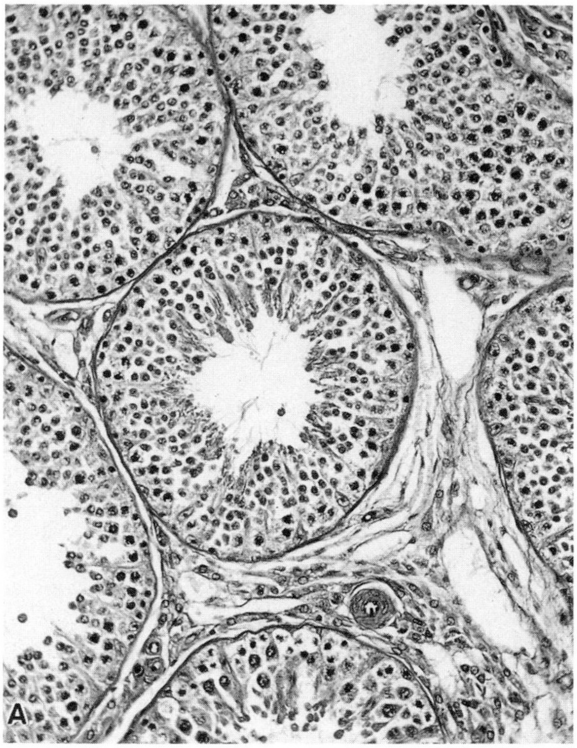

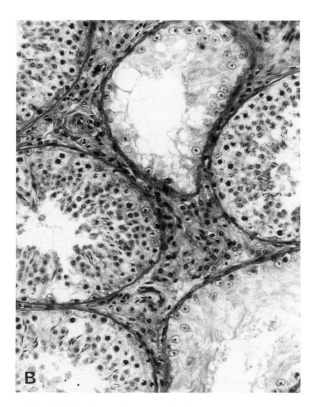

Fig. 5.20A   Normal testis. Bull.

mitotic activity (Fig. 5.21). The basement membranes are thickened and hyaline, and there is an increase in peritubular connective tissue. The interstitial cells of Leydig appear to be increased in number. Quantitative histologic studies, however, have revealed that the actual volume occupied by the Leydig cells may be less than that in normal bulls. In other quantitative histologic studies of severe testicular hypoplasia in bulls, concurrent degeneration of Sertoli and Leydig cells was observed, and it was postulated that maintenance and regulation of the latter depended on viability of the former.

In the intermediate form, 50% or more of the tubules are hypoplastic, whereas in the remaining tubules there are varying degrees of spermatogenic activity and occasional spermatozoa. In the majority of active tubules however, differentiation proceeds only to the spermatocyte stage, and subsequent degeneration of cells results in obvious vacuolation of the germinal epithelium. These active tubules may be of normal diameter. In addition to vacuolation, a variable number of multinucleate giant cells, resulting from division, but not separation of germinal cells, are present (Fig. 5.22). Additionally, distinct intratubular concrements surrounded by a corona of cells are occasionally observed and are a further indication of disordered spermatogenesis.

In the mild form of testicular hypoplasia, occasional tubules are of the Sertoli-cell-only type, but in the majority of tubules there is active spermatogenesis to the spermato-

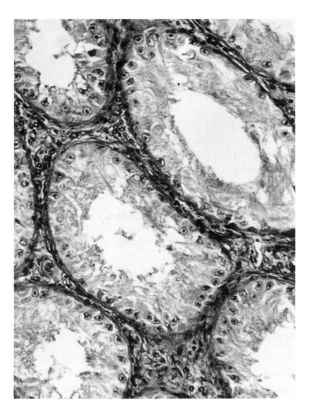

Fig. 5.20B   Hypoplastic seminiferous tubules in otherwise normal testicle. Bull.

Fig. 5.21   Severe testicular hypoplasia with seminiferous tubules lined almost entirely by Sertoli cells. Bull.

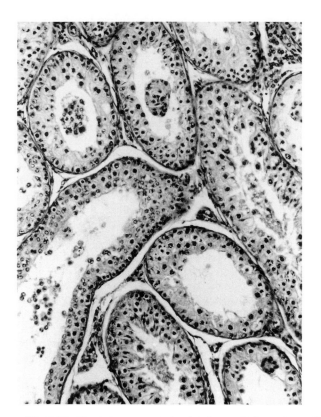

Fig. 5.22 Testicular hypoplasia with giant-cell formation and intraluminal cellular debris. Bull.

cyte stage or beyond. Intratubular giant cells are again a feature and differentiation of this form of hypoplasia from testicular degeneration presents a problem that is difficult to resolve, especially as degeneration is superimposed on hypoplasia in postpubertal bulls. Although infiltration of inflammatory cells usually is not a feature of testicular hypoplasia, lymphocytic infiltrates are observed in hypoplastic testes of Swedish Red and White bulls.

The causes of testicular hypoplasia have not been identified, but available literature suggests involvement of a number of diverse factors, perhaps operating through a common pathway. Possibilities include a deficiency or abnormality of germinal cells, including their failure to reach the gonads, failure of germ cells to proliferate in the gonads, and excessive gonocyte death and transplacental infections or intoxications. Alternative possibilities include zinc deficiency; endocrinologic deficiency involving the hypophysis, the hypophyseal–hypothalamic pathway, or the testis itself; chromosomal aberrations; and impaired descent of the testis. Immunologic castration of calves, with subsequent testicular hypoplasia at puberty, has been achieved by injection of luteinizing hormone-releasing hormone, thereby stimulating the production of antibodies directed against the endogenous releasing hormone.

Endocrinologic disturbances associated with hypogonadism in humans have been extensively studied, and it is established that deficiencies of luteinizing hormone-

releasing hormone, follicle-stimulating hormone, and luteinizing hormone, decreased testosterone, overproduction of estrogen and prolactin, and lack of receptors in the target organs can all result in small testes. In bulls, cytogenetic–chromosomal abnormalities associated with hypoplastic testes include translocations; sticky chromosomes and multiple spindle formations in germinal cells with impaired division; chromosomal mosaics and chimerism; and a condition characterized by an XXY chromosome configuration resembling Klinefelter's syndrome in humans (Fig. 5.23). Involvement of these factors in causing unilateral testicular hypoplasia are less clear, however, and in such cases the possibility of a local endocrinologic or vascular disturbance, or testicular maldescent, is more attractive.

An association between hypoplasia and anomalous branching of the testicular artery in the same testis has been demonstrated in one breed (Sanga Nguni) of *Bos indicus* cattle in which inherited testicular hypoplasia is recognized.

The occurrence of testicular hypoplasia in rams and probably bucks is comparable to that in bulls and similar comments concerning its cause and pathogenesis apply (Fig. 5.24). In unilateral cases, however, the right testis may be more often involved. Evidence for inheritance is good in regard to unilateral testicular hypoplasia in rams, which occurs simultaneously with monorchidism and cryptorchidism in affected flocks. Impaired testicular de-

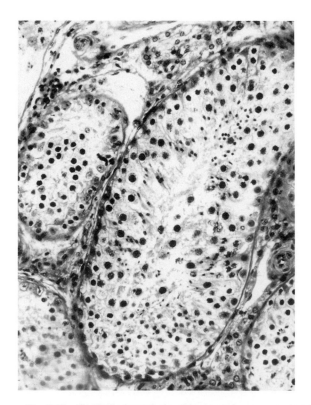

Fig. 5.23 Seminiferous tubule with arrested spermatogenesis due to sticky chromosomes. Bull.

Fig. 5.24 Testicular hypoplasia. Rams. Normal testes lower right.

scent is the immediate likely cause. Chromosomal translocation syndromes have also been observed in rams. Testicular hypoplasia in rams has been associated with diets deficient in zinc. Experimentally it has been shown in lambs, and also in calves and rats, that dietary levels of zinc adequate for normal growth are inadequate for normal testicular development and function.

In goats, several independent studies have recognized a distinct segmental hypoplasia involving groups of seminiferous tubules which microscopically appear as a discrete pale zone.

In boars also, there is strong evidence of inheritance of testicular hypoplasia, and translocation syndromes and mosaicism with testicular hypoplasia have been described.

No detailed studies of testicular hypoplasia in stallions have been reported, although in one clinical study on almost 1000 young stallions it was observed, often in association with epididymal hypoplasia, in more than 3% of animals. The right testis was most frequently involved.

In dogs, testicular hypoplasia has also been observed in the gray collie syndrome in association with gray-silver hair coloration, cyclic hematopoiesis, lameness, and other signs (see The Hematopoietic System, Chapter 2 of this volume).

Marked testicular hypoplasia with partial or complete arrest of spermatogenesis occurs commonly in mammalian hybrids. Examples include dog–coyote hybrids, equine hybrids resulting from crossing stallions with female donkeys or crossing donkeys with mares, and various bovine hybrids produced by crossing swamp buffalo with water buffalo or by crossing domestic cattle (*Bos taurus, Bos indicus*) with other species such a Bali cattle (*Bos sondaicus*), the mithun (*Bos frontalis*), or the American bison (*Bos bison*).

Since normal meiotic division requires pairing of homologous chromosomes, true pairing may be impossible in hybrids if the maternal and paternal chromosomes are different in size and number. This can result in spermatic arrest in which spermatogenesis is blocked at near the pachytene stage. Even if the block is incomplete, as would

be the case in which maternal and paternal chromosomes were a close match, the reduction in sperm production can render the male hybrid sterile, whereas the female hybrid may successfully produce a few fertile ova and can conceive. Cytogenetic and histologic studies of the bovine hybrids have revealed considerable variation in the extent and stage of spermatogenic arrest, and it may be that other more subtle factors contribute to the male infertility of these hybrids.

2. Testicular Hypoplasia Due to Chromosomal Anomalies

The normal sex chromosome constitution in mammalian species is XY in the male and XX in the female. One of the female's two X chromosomes is inactivated soon after fertilization. This is thought to be a dosage-compensation device, as the gene complement of the large X chromosome is much greater than that of the small Y chromosome. The early embryonic inactivation is random in that either the paternal X chromosome or the maternal X chromosome may be inactivated, forming the sex chromatin body which can be seen adjacent to the nuclear membrane in many somatic cells of normal females. The inactivation once made determines the X inactivation for the descendants of each cell. Therefore the normal female represents a functional mosaic with approximately half of the cells expressing genes of the paternal X chromosome and half expressing genes of the maternal X chromosome. Although only a single X chromosome is active in somatic cells in the female, the loss of an X chromosome, as the result of nondisjunction during cell division, creates a sex chromosome monosomy, causing gonadal dysgenesis with severely hypoplastic ovaries (see The Female Reproductive System, Chapter 4 of this volume). Nondisjunction can also result in aneuploidy with more than the normal chromosomal complement. Polysomies of sex chromosomes, with XXY or a variation of this constitution, produce testicular hypoplasia. The syndrome is identified as Klinefelter's syndrome in humans, in which the condition was first recognized. The features of the syndrome are small testes and prostate, aspermia, frequent breast enlargement, and typically an XXY karyotype with positive sex chromatin.

The Y chromosome is male determining in mammals and, if present, the gonad formed is recognizably testicular, but for testicular development to be normal, the individual must have only a single X chromosome. In conditions in which there is polysomy of sex chromosomes, as in Klinefelter's syndrome, the extra X chromosome (or chromosomes) is inactivated as in female cells, but in spite of this inactivation, the extra X chromosome usually produces testicular hypoplasia and sterility.

Trisomy of sex chromosomes arises as the result of nondisjunction, or failure of separation, during meiosis. This produces sperm or ova with abnormal numbers of sex chromosomes. The fertilization of such abnormal ova or the fertilization by such sperm can, as one consequence, produce an individual with an XXY sex chromosome constitution. Not all individuals with testicular hypoplasia caused by an extra X chromosome have the classic XXY aneuploidy. Variants consist of X polyploidy beyond trisomy, as the result of meiotic nondisjunction; chimerism (two populations of cells each from a different source), as from fusion of littermates; and mosaicism (two distinct cell populations from a single source), as the result of nondisjunction during mitosis at an early stage of embryo development, giving rise to XY/XXY individuals.

Sex chromosome anomalies are common in humans. The estimated incidence of Klinefelter's syndrome is 1 in 500 male births. The syndrome has been identified in stallions, bulls, rams, dogs, and cats. The number of reported cases in domestic animals is very low, but the studies of the condition in cats indicate that the incidence in domestic species is much higher than the limited number of reports would suggest.

Cats have been overrepresented in the reported cases of XXY patterns of testicular disease because coat color in tricolor, tortoiseshell, and calico cats provides a marker system for recognition of affected animals. Although cat fanciers make distinctions, the genetic mechanisms which produce tricolor, tortoiseshell, and calico cats are similar. Tricolored cats should be female; male tricolored cats are prime candidates to have XXY testicular hypoplasia. The gene for orange is X linked, and is dominant or epistatic over the gene for black. Therefore normal male cats having only a single X chromosome can be either orange or black but not both. Normal females have two X chromosomes and can have both orange and black coloration. Male cats with an XXY sex chromosome complement, or males which are whole body chimerics as the result of embryonic fusion of littermates, can have the expression of two or more X chromosomes, allowing the production of tortoiseshell or calico patterns. Most, but not all, cats with XXY chromosomal complement have hypoplastic testicles and are sterile. The testicular hypoplasia can be associated with aplasia of epididymides and vasa deferentia. Rarely, tricolored male cats have normal testicles and normal karyotypes, the animals being whole-body chimerics resulting from the embryonic fusion of male littermates.

The expression of the defect varies. In severely affected animals, the testicles are small, the seminiferous tubules, although well formed, are lined only by Sertoli cells, and no spermatogenesis is present. Leydig cells are usually present, and studies in humans have shown that total Leydig cell volume per testis in Klinefelter's syndrome is comparable to that of control testes. The actual number of Leydig cells in testes in this syndrome is, however, reduced, and ultrastructural studies reveal two populations of Leydig cells—hypertrophied normal cells, and abnormal cells which are immature, abnormally differentiated, or multivacuolated. Bulls may be eunuchoid and in some the scrotums have failed to form and the hypoplastic testicles are located subcutaneously.

3. Cryptorchidism

Incomplete descent of the testes is known as cryptorchidism. Testicular descent, the process by which a gonad descends from the dorsal abdominal wall into the scrotum, occurs only within the mammalian species. Many mammals, however, retain the testes in the abdominal cavity. Formation of a scrotum and descent of the testis are regarded as late evolutionary developments. The occurrence of cryptorchidism in most species seems to be about 1% but is sometimes as high as 10%. Most cases are unilateral, the left testis being retained more often than the right. Bilateral cryptorchids are always sterile, but fertility of unilateral cases is impaired to a varying degree. Rarely, retained testes may contain teratomas, but cryptorchid testicles have a high risk of developing other primary testicular neoplasms.

The cryptorchid testis may be located at any point along its migration path, such as near the kidney, in the inguinal canal, or subcutaneously at the external inguinal ring. Sometimes the epididymis may be located in the inguinal canal, while the testis is in the abdomen. The gross and microscopic appearance of the cryptorchid testis will depend on its location and the age of the affected animal. Marked fibrosis, especially of the tunics, is a feature of cryptorchid testes in older animals. Affected testes are small and firm to hard, and histologically resemble the severe form of hypoplasia. Intratubular concrements similar to those seen in testicular hypoplasia may be present. In unilateral cases, compensatory hypertrophy of the normal testis may occur.

In domestic species other than the horse, cryptorchidism is generally regarded as being inherited as a single, autosomal sex-linked character which is probably recessive. Difficulty arises, however, in distinguishing a recessive gene from a dominant one with incomplete penetrance. Cryptorchidism in horses is considered to be inherited in a dominant manner and has been studied extensively. Probably about 10% of cases are bilateral. Reports on whether retention occurs more commonly in the inguinal canal than in the abdomen vary. The overall incidence of cryptorchidism appears to be marginally greater on the left, and a greater proportion of left cryptorchids have intra-abdominal testes; on the right side, however, there is a tendency for inguinal retention. This might be related to the retarded descent of the left testes as compared with the right. It has been found that whereas about 50% of right-sided unilateral abdominal cases have the epididymal tail descended, only 20% of left-sided cases do. The abdominal testis does not normally increase in size with age; it undergoes marked hypertrophy, however, if the (opposite) scrotal testis is removed.

The striking change in size of the equine fetal testis may have some bearing on the occurrence of cryptorchidism. Whereas in other species, testis migration from near the kidney to the inguinal ring is obvious at an early fetal age, in the horse, remarkable enlargement of both the fetal ovary and testis occurs as the result of hypertrophy and hyperplasia of the interstitial cells between 100 and 250 days of gestation. During this time the cranial pole of the gonad remains in contact with the kidney. After ~250 days of gestation the interstitial cells undergo degeneration and the size of the gonads is greatly reduced, facilitating the testicle's passage through the inguinal canal.

In addition to those changes, lesions associated with cryptorchidism in horses can include orchitis and testicular mineralization, and the presence in the testicle of migratory strongyle larvae. Migration tracts commonly observed in the cryptorchid testis and epididymis, especially if intra-abdominal, are those of aberrant fifth-stage larvae of *Strongylus edentatus* (Fig. 5.25). They consist of channels about 2 mm wide containing blood and larvae. Hemorrhage may extend to adjacent parenchyma. Larvae enter the mesorchium and migrate under the peritoneum down the spermatic cord.

The frequency of testicular tumors, and especially Sertoli cell tumors, is more than 10 times greater in cryptorchid dogs than in normal dogs. Retained testes that are enlarged by neoplastic growth are especially prone to torsion. There is evidence that certain breeds, particularly poodles, are predisposed to cryptorchidism. A distinct difference between uni- and bilateral abdominal testes has been observed in cryptorchid dogs, the epididymides in bilateral cases being more immature and attached to the testes in the head regions only. Higher testosterone levels

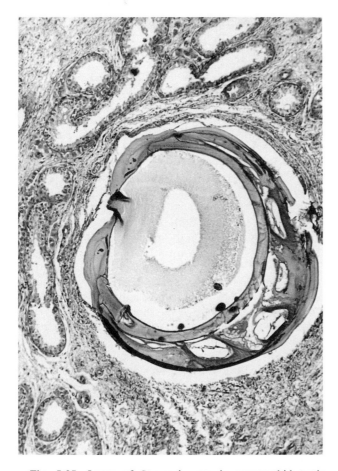

Fig. 5.25 Larva of *Strongylus* sp. in cryptorchid testis. Horse.

originating from the descended contralateral testis in unilateral cases may be responsible for this difference.

The pathogenesis of most cases of cryptorchidism is unresolved, but the process is, in part, under hormonal control. Normal testicular descent requires both testosterone and Müllerian inhibiting hormone. Animals with testicles in which hormone production is abnormal or individuals in which normal hormone tissue sensitivity is lacking have a greatly increased chance of having failure of testicular descent. This includes, among others, animals with sex chromosome abnormalities, individuals with androgen insensitivity, and dogs with persistent Müllerian duct syndrome. During normal descent of the testis, the gubernaculum acts mainly as a guide, and its aberrant insertion may account for some cases of testicular ectopia. The vaginal process facilitates testicular passage through the inguinal canal, but defects in these structures do not appear to be related to cryptorchidism. Fibrosis and shrinkage of the gubernaculum is considered to cause traction on the testis and aid descent. In the neonatal period, there is fibrosis and constriction of the internal inguinal ring, which has the effect of trapping the testis in its position at that time.

Normal descent of the testis is coordinated with epididymal differentiation, and may be in part controlled by it. The epididymis precedes the testis during descent, and its differentiation is completed when testicular descent ends. Epididymal differentiation is stimulated by local testosterone concentration, and is retarded in some cryptorchid testicles, but whether this causes maldescent or is caused by it is not clear.

Studies in dogs orchidectomized as fetuses or neonates then treated with testosterone have, however, shown that both testosterone and an unidentified nonandrogenic testicular factor play a role in outgrowth of the gubernaculum and therefore the early events of normal testicular descent; also testosterone was found necessary to induce gubernaculum regression in the second phase of testicular descent. It therefore seems clear that early impairment of endocrinologic events is important in testes destined to be cryptorchid. There is evidence to support the view that Müllerian inhibiting hormone initiates transabdominal migration, whereas testosterone mediates inguinal scrotal descent. Irrespective of the role of endocrinologic disturbances in causing cryptorchidism, it is also clear that retained testes are defective in endocrine secretion. Testosterone secretion by Leydig cells in surgically induced cryptorchidism is significantly reduced. Also, morphologic studies on cryptorchid and descended testes in young unilateral cryptorchid goats have revealed progressive degeneration of Sertoli cells in intra-abdominal testes, which were initially comparable to those in controls. Many organelles involved in secretion were no longer recognizable ultrastructurally in affected testes from goats 12–15 months old.

Failure to confirm any endocrinologic irregularities in boys with cryptorchid testicles and disappointing responses to hormonal treatment of cryptorchidism in some individuals has given rise to the hypothesis that abnormalities of the anatomy of the genitofemoral nerve may also cause cryptorchidism.

4. Other Anomalies of the Testis

Complete **agenesis** or lack of one or both testes occurs occasionally, usually in association with anomalous development of other organs. **Fusion** of both testes has been observed in the boar. Testicular **ectopia** implies that the testis is in an abnormal location and away from the normal route of descent; in this respect it differs from cryptorchidism. Ectopic testes are rare in dogs but may be located in the perineal or crural regions, in the abdomen, or adjacent to the prepuce but away from the inguinal canal. Both testes may occur on one side of the scrotum.

Whereas ectopic testes are complete and normally attached to their epididymides, **heterotopia** of testicular tissue occurs as multiple nodules which lack auxillary structures. Testicular heterotopia occurs in pigs in which discrete pale nodules up to ~5 cm in diameter are located anywhere on the parietal or visceral peritoneum. The nodules suggest metastatic neoplasms, but histologically resemble the testes of cryptorchids. The cause of testicular heterotopia is not known, but it is considered to result from defective migration of primordial germ cells in the embryo. Dispersion of cells of the germinal epithelium due to injury is an alternative, though less likely, possibility.

Polyorchidism or the presence of supernumerary testes has been observed in the horse, calf, and pig. Three testes and attached epididymides are present in such cases and the duplicated testis may be either scrotal or abdominal.

Accessory adrenal cortical tissue in the testes or epididymides has been observed in the stallion, ram, dog, and cat (Fig. 5.26). In horses, these are frequently seen as small yellow nodules between the testis and epididymal head or in the mediastinum testis. The masses are composed of irregular formations of adrenal cortical cells. The adrenal cortical cells may resemble the interstitial cells of Leydig but have a more heavily vacuolated cytoplasm, especially in the immature testis.

Heterotopic Leydig cells are observed in the cat in the tunica albuginea and stroma of the mediastinum testis; they may be important in supplying luminal androgen to the head of the epididymis.

Cystic rete testis due to an apparently congenital lack of communication between efferent duct and the epididymis has been described in a cat with a large, centrally located cyst in the affected testis.

B. Circulatory Disturbances in the Testis

The spermatic artery is very long in species with scrotal testes, much of the length being contributed by the coiled portions immediately beyond the inguinal ring. The degree of coiling varies between species, being quite pronounced in ungulates, in which the spermatic artery is several meters long. The extreme length and small diameter of the artery suggest that fine regulation of blood flow is possible, but that reactive hyperemia is unlikely to occur. Testicular

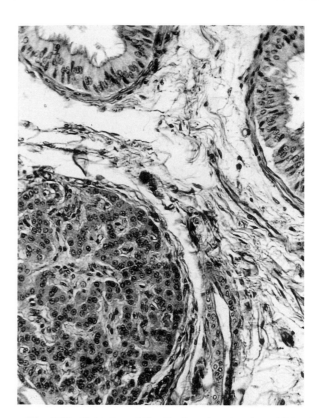

Fig. 5.26 Ectopic nodule of adrenal cortex in epididymis. Ram.

blood flow is low in relation to metabolic needs, and hypoxia develops quickly if metabolic demand is increased, such as by raising testicular temperature, or if blood flow is impaired. By the time the artery penetrates the tunica albuginea, pulsatile flow is almost eliminated and the structure of the vessel changes. The diameter enlarges, the wall becomes thinner and the elastic fibers are reduced; blood flow is then susceptible to increased intratesticular pressure.

The interstitial tissue of the testis is richly supplied with lymphatics. In addition to the usual role in fluid exchange, they probably provide a local transport mechanism between Leydig cells and the tubules. Edema of the testis occurs after trauma and is a frequent early change in orchitis (Fig. 5.27A). Macroscopically it is evidenced by enlargement, to perhaps twice normal size, and serous fluid drips from the cut surface. Histologically there is distinct separation and dilation of tubules, diffuse vacuolation of germinal epithelium, and dilation of lymphatics. Following obstruction of testicular lymphatics in rats, marked edema of the testis and degeneration occur within 3 days. Degeneration increases in severity for 3 months, but subsequently recovers completely.

Hyaline degeneration of the walls of arterioles is associated with wedge-shaped areas of fibrosis in the testes of old bulls, but the precise cause of this lesion is not clear. In old dogs, hyaline degeneration of both arterioles and arteries accompanies testicular degeneration. Initial le-

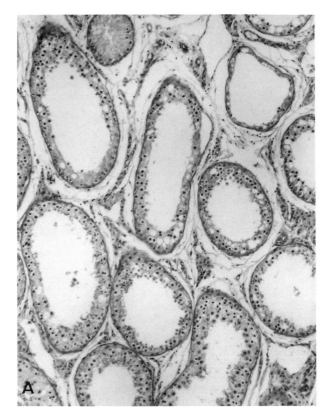

Fig. 5.27A Testicular edema. Bull. Probably traumatic. Note separation of tubules, dilated lymphatics, and early degeneration.

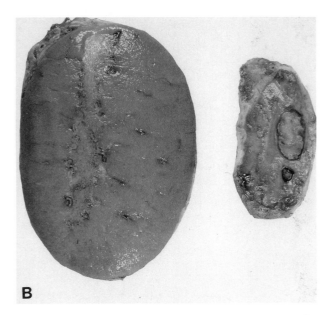

Fig. 5.27B Infarction of left testis of bull due to compression of spermatic cord while removing tissue for biopsy.

with marked interstitial epididymitis, orchitis, and testicular degeneration occurs in malignant catarrhal fever in buffaloes in association with generalized vasculitis (Fig. 5.28A,B).

Occlusion of the testicular artery with resulting isch-

sions occur in the tunic and parenchyma, but larger vessels in the spermatic cord are subsequently involved. Focal areas of infarction may occur as a result of severe vascular lesions. Atheromatous lesions are rarely observed.

Thrombosis of testicular arteries, usually of unknown cause, is seen occasionally in bulls (Fig. 5.27B). Such thrombi are present in both the parenchyma and tunics and may be partially mineralized and sometimes occluding. The mild degenerative changes in germinal epithelium in association with thrombi emphasize the good collateral circulation. In experimental *Trypanosoma vivax* infection in sheep, thrombosis of testicular vessels probably contributes to testicular degeneration caused mainly by pyrexia; subsequent infarction and necrosis may sometimes occur. Occluding venous thrombosis has been observed in the testes of rams which were, or were not, affected by varicocele; testicular degeneration associated with such thrombi was mild.

Inflammation of the testicular artery occurs frequently in the horse. The known causes are migrating strongyle larvae and the equine arteritis virus, but the cause is not established in many cases. The usual morphologic manifestations in the testicular tissue include focal infiltration of lymphocytes and degeneration of seminiferous tubules adjacent to the inflamed arteries and arterioles. The inflammatory reaction is rarely of such a severe degree as to cause thrombosis and infarction. A striking vasculitis

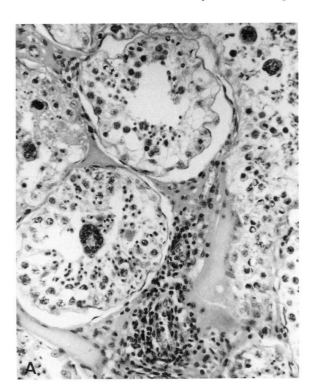

Fig. 5.28A Interstitial orchitis and acute degeneration of seminiferous tubules with giant-cell formation in malignant catarrhal fever. Buffalo.

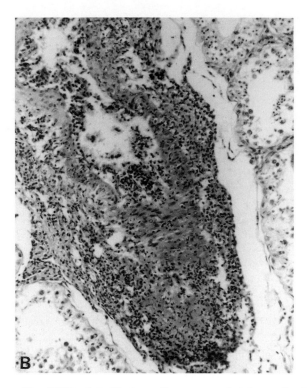

Fig. 5.28B Arteritis in malignant catarrhal fever. Testis. Buffalo.

emia of the testis may result from torsion (see the following), contusion of the spermatic cord, or the use of an emasculatome for castrating young lambs and calves. Experimentally, destruction of germinal epithelium can be demonstrated after ischemia of more than 1 hr. Necrosis of testicular parenchyma follows after 4–6 hr, although sperm are relatively resistant to lysis and may retain their staining characteristics for weeks or months. Following ligation of the spermatic artery in prepubertal rams, however, revascularization of the tunica albuginea occurs by penetration of capillaries from the epididymal arteries. Islets of seminiferous tubules associated with the tunic (Moskoff's islets) may thus survive, and considerable regeneration occurs during the ensuing months. However, as the regenerated tubules lack a normal drainage system, there is spermiostasis and ultimate degeneration.

Bibliography

Baumans, V., Dijkstra, G., and Wensing, C. J. G. The role of non-androgenic testicular factor in the process of testicular descent in the dog. *Int J Androl* **6:** 541–552, 1983.

Cox, J. E. Factors affecting testis weight in normal and cryptorchid horses. *J Reprod Fertil* (Suppl.) **32:** 129–134, 1982.

Cox, J. E., Edwards, G. B., and Neal, P. A. An analysis of 500 cases of equine cryptorchidism. *Equine Vet J* **11:** 113–116, 1979.

Ezeasor, D. N., and Singh, A. Morphologic features of Sertoli cells in the intra-abdominal testes of cryptorchid dwarf goats. *Am J Vet Res* **48:** 1736–1745, 1987.

Farner, J. H. *et al.* Impaired testosterone biosynthesis in cryptorchidism. *Fertil Steril* **44:** 125–132, 1985.

Fujikake, N. *et al.* Relationship between development of the gubernaculum and testicular descent in the rat fetus: Macroscopic and light and electron microscopic detection. *Jpn J Vet Sci* **51:**416–427, 1989.

Gelberg, H. B., and McEntee, K. Cystic rete testis in a cat and fox. *Vet Pathol* **20:** 634–636, 1983.

Genetsky, R. M. *et al.* Equine cryptorchidism: Pathogenesis, diagnosis, and treatment. *Compend Contin Educ* **6:** 5577–5582, 1984.

Gimbo, A., Zanghi, A., and Gianetto, S. Ram testicular hypoplasia. Anatomical and histopathological observations. *Schweiz Arch Tierheilkd* **129:** 481–491, 1987.

Hayes, H. M. Epidemiological features of 5009 cases of equine cryptorchidism. *Equine Vet J* **18:** 467–471, 1986.

Hutson, J. M. *et al.* Testicular descent: new insights into its hormonal control. *In* "Oxford Reviews of Reproductive Biology," Vol. 12, S. R. Milligan, (ed.), pp. 1–56. Clarendon Press, Oxford, 1990.

König, H. *et al.* Testicular hypoplasia (lack of spermatogonia) and left-sided epididymal and deferential aplasia in a tricolor male cat with the 39/XXY karyotype. *Deutsche Tierärztl Wschr* **90:** 341–343, 1983.

Madrid, N. *et al.* Scrotal circumference, seminal characteristics, and testicular lesions of yearling Angus bulls. *Am J Vet Res* **49:** 579–585, 1988.

McCool, C. J. Spermatogenesis in Bali cattle (*Bos sondaicus*) and hybrids with *Bos indicus* and *Bos taurus*. *Res Vet Sci* **48:** 288–294, 1990.

Nistal, M., Santamaria, L., and Paniagua, R. Quantitative and ultrastructural study of Leydig cells in Klinefelter's syndrome. *J Pathol* **146:** 323–331, 1985.

Santschi, E. M., Juzwiak, J. S., and Slone, D. E. Monorchidism in three colts. *J Am Vet Med Assoc* **194:** 265–266, 1989.

Sponenberg, D. P., Smith, M. C., and Johnson, R. J., Jr. Unilateral testicular hypoplasia in a goat. *Vet Pathol* **20:** 503–506, 1983.

Veeramachaneni, D. N. R. *et al.* Pathophysiology of small testes in beef bulls: Relationship between scrotal circumference, histopathologic features of testes and epididymides, seminal characteristics, and endocrine profiles. *Am J Vet Res* **47:** 1988–1999, 1986.

Wensing, C. J. G., and Colenbrander, B. Cryptorchidism and inguinal hernia. *Proc K Ned Acad Wet* (C) **76:**489–494, 1973.

Wilson, D. G., and Nixon, A. J. Case of equine cryptorchidism resulting from persistence of the suspensory ligament of the gonad. *Equine Vet J* **18:** 412–413, 1986.

Winter, H. *et al.* Mithun cross siri hybrids: cyto- and immunogenetic examinations and characterisation of abnormal spermatogenesis. *Res Vet Sci* **45:** 86–100, 1988.

Wrobel, K.-H., and Hees, H. Heterotopic Leydig cells in the cat. *Anat Histol Embryol* **16:** 289–292, 1987.

C. Testicular Degeneration

It is well recognized that the germinal epithelium is extremely sensitive to a great variety of adverse influences and that testicular degeneration or atrophy is the most frequent cause of reduced fertility in male animals. Within the seminiferous tubules, the dividing primary spermatocytes especially, then other differentiating germinal cells up to the spermatid stage, are most susceptible to injury, whereas the (A type) spermatogonia, or stem cells, and the nongerminal Sertoli cells are comparatively resistant, thereby allowing for considerable regeneration to occur. The Sertoli cells are most resistant and are often the only cells remaining after prolonged testicular insult. It is currently assumed that they do not divide and that their number does not vary in the adult. Therefore a Sertoli cell index, the ratio of the number of Sertoli cells to germinal cells, may sometimes be useful as a measure of the degree of testicular degeneration.

Irrespective of the cause of disturbed spermatogenesis, the reactions of the seminiferous epithelium are essentially similar, variation occurring only in extent and degree. This limited repertoire of reactivity extends to impaired testicular development in young animals so that differentiation between degeneration and hypoplasia is frequently difficult, and sometimes impossible, especially as hypoplastic testes are predisposed to degeneration.

Macroscopically, the testis undergoing degeneration may at first be enlarged by edema but is usually reduced in size. In early or rapidly progressing degeneration, the testis is soft and flabby, lacks turgor, and the cut surface does not bulge. Distinct wrinkling of the tunica albuginea may be apparent. The decrease in parenchyma is not paralleled by a decrease in stroma, and the end result of degeneration is a small testis of firm consistency. Since the epididymis is usually less affected than the testis, it will ultimately appear to be disproportionally large (Fig. 5.29A). With continued degeneration and fibrosis, the testis becomes increasingly hard, and variable mineralization may occur (Fig. 5.29B). The cut surface of such a testis has a coarse granular appearance.

Testicular degeneration may be uni- or bilateral, depending on whether the causes are local or general. The degenerative processes may not involve the whole testis uniformly. In old bulls, for example, a pattern of ventral degeneration of the testis is common (Fig. 5.30), but in epididymal aplasia, degeneration initially involves the dorsal portion of the testis as a result of sperm obstruction, accumulation, and later extravasation in that area.

The microscopic changes vary with the severity and stage of the degeneration. In the early stages, there is failure of maturation of spermatozoa and degeneration of spermatids; many spermatids are necrotic and others produce characteristic spermatidic multinuclear giant cells. The degenerative processes do not involve the testes uniformly. When the degeneration is more advanced, the affected areas are more extensive and degenerative changes appear in the precursors of spermatids, changes which are characterized by cytoplasmic vacuolation and nuclear pyknosis and, with progression, the tubules may be denuded to the basement membranes with loss ultimately of even resistant Sertoli cells. The tubules then collapse and there is wavelike hyaline thickening of the basement membrane (Figs. 5.31A,B).

Although increased basement membrane thickness is also a feature of testicular hypoplasia, the more pronounced shrinkage and collapse of tubules in testicular degeneration results in buckling of the basement membrane. This is a particularly useful feature in the differentiation of hypoplasia and degeneration and is best demonstrated in sections stained by the periodic acid–Schiff (PAS) method. Use of a trichrome stain is also helpful in defining the extent of fibrosis and outlining areas of spermiostasis and mineralization. The PAS stain clearly accentuates the basement membrane change. In ultrastructural studies of the testes of normal bulls and bulls with atrophy or hypoplasia, mean thicknesses of the actual basal lamina were approximately 0.7, 1.5, and 1.0 μm, respectively.

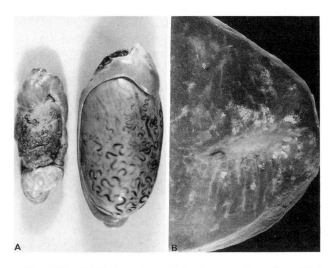

Fig. 5.29 (A) Unilateral testicular degeneration. The epididymis of the atrophic testis appears disproportionately large. (B) Mineralization of seminiferous tubules. Bull.

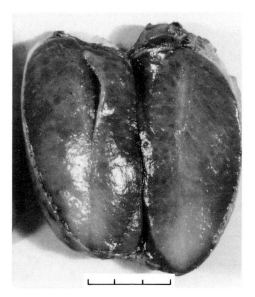

Fig. 5.30 Advanced testicular degeneration with diffuse fibrosis that is most severe ventrally. Bull.

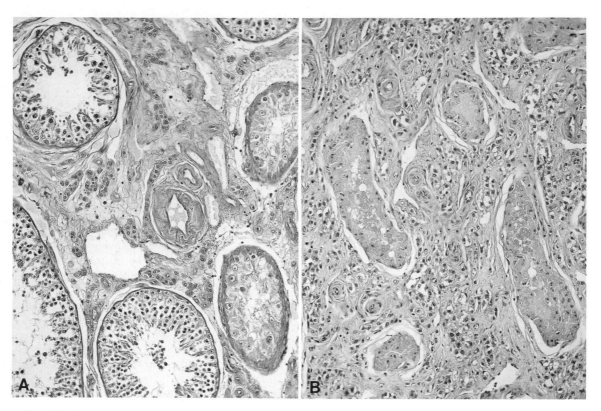

Fig. 5.31 Testicular degeneration. Bull. (A) Tubules are at varying stages in the process. Hyalinized vessels at center. (B) More advanced degeneration with hyalinized tubules. Depletion of both germinal and Sertoli cells.

The detection of giant cells in degenerated testes and perhaps in the semen is of some importance, and such cells have been the subject of several experiments in rodents. Following once-only scrotal heating, two types of giant cells, mononuclear giant cells probably derived from pachytene spermatocytes which fail to differentiate further, and multinucleate cells considered to be derived from coalescence of identical spermatids, are observed histologically. Even extremely brief heating, of several minutes only, induces giant-cell formation. Such giant cells are seen as early as 6 hr and as late as 7 weeks postheating but seem to be most prevalent at about 1 week. The fate of these cells is unclear. They may disintegrate or pass out of the testis. In addition to giant cell formation, minor increases in testicular temperature in sheep produce a marked accumulation of B type spermatogonia; histologically this can be demonstrated by the presence of many cells in which mitosis is incomplete.

The described changes in degeneration are frequently confounded by granuloma formation which results from contact between degenerate sperm and the intertubular connective tissue. The frequent presence of lymphocytes and plasma cells in such lesions probably indicates an immune response to sperm, but complicates the differentiation of simple degeneration from interstitial orchitis. Osseous metaplasia sometimes occurs in tubules affected by sperm stasis for a long time.

The causes of testicular degeneration are many and varied. Accidental or therapeutic ionizing radiation has long been recognized as an important cause of testicular atrophy; dividing germinal cells are primarily affected, but there is evidence that, in continuous low-dose γ-radiation, Sertoli cell function is also impaired. Ionizing radiation and cytotoxic chemicals have both been shown to be more damaging to germ cell populations in developing animals than in mature ones. The A_1 spermatogonium is the most sensitive cell, but there are large differences in sensitivity between species.

Because a temperature differential of several degrees between the intrascrotal testis and the rest of the body is a prerequisite for normal spermatogenesis in domestic animals, thermal degeneration occurs if the testis is subjected to temperatures equal to or in excess of normal body temperature. Thermal damage occurs in cryptorchid or ectopic testes, and with excessive scrotal fat and other local influences such as scrotal dermatitis, edema, hydrocele, and periorchitis. In rams, and possibly bulls, exposed to high environmental temperatures under range conditions, thermal degeneration may temporarily decrease fertility.

Localized or systemic infections are also common causes of testicular degeneration, but in such cases it is not always possible to separate the effects of fever from those of toxemia. Examples of localized conditions include epididymitis–orchitis caused by brucellae or other organisms, and periorchitis perhaps occurring as an extension of peritonitis. Tunic adhesions formed during the course of healing in periorchitis may permanently compromise thermoregulation of the testis. In systemic disease, testicular degeneration results from marked pyrexia in babesiosis, anaplasmosis, trypanosomiasis, and others. Testicular degeneration and mineralization in besnoitiosis (see the preceding) results from parasitism of both the scrotum and spermatic cord.

General or specific nutritional deficiencies or excesses leading to testicular degeneration include malnutrition, perhaps superimposed on chronic disease, and vitamin A deficiency. The effect on the gonads of vitamin A deficiency is probably indirect and due to suppressed release of gonadotrophic hormone from the pituitary gland. Severe testicular degeneration, accompanied by night blindness, may occur in rams kept for prolonged periods on dry natural pastures or otherwise denied access to green feed. In cats, testicular degeneration occurs after administration of excess vitamin A. Cats also lack the ability to convert (in the liver) sufficient quantities of linoleate to arachidonate, so that male cats fed experimentally for prolonged periods on a diet deficient in essential fatty acids develop extensive testicular degeneration.

With the probable exception of zinc deficiency, which specifically retards testicular development and function, and which may occur in ruminants on pastures in some areas, there are no known nutritional factors which specifically affect spermatogenesis in mature domestic animals; diets which are adequate for growth and maintenance are adequate for full fertility.

Various circulatory disturbances, and most conspicuously the partial or complete occlusion of testicular vessels in torsion of the spermatic cord, are important causes of testicular degeneration. Incomplete infarction after torsion, or use of an emasculatome for castrating young lambs and calves, may leave a zone of viable Leydig cells adjacent to the tunic. Steroid production from these may be sufficient to maintain male characteristics.

Obstructive lesions, particularly involving the epididymal head, and malformations of the efferent tubules, result in sperm accumulation and degeneration due to back pressure. Obstructive lesions occurring in the body or tail of the epididymis are less likely to cause testicular degeneration because fluids and disintegrating spermatozoa can be absorbed in the head of the epididymis.

Degeneration of the germinal epithelium may occur with or without an inflammatory response. Noxious agents, including a number of chemicals, metals and rare earth salts, and ionizing radiation are capable of causing testicular degeneration in a variety of mammalian species. Alkylating agents, including tretamine (*N,N'*, *N''*-triethylenemalamine), busulfan (Myleran, 1,4-dimethanesulfonoxy-butane), and isopropyl methanesulfonate, have an adverse effect on spermatogenesis, but fertility usually returns following cessation of treatment. Daily subcutaneous injections of certain metallic and rare earth

salts in rodents causes spermatogenic arrest at the primary spermatocyte or spermatogonial stages and destruction of spermatozoa in the epididymis. Spermatogenesis returns to normal after the administration of the salt is stopped. The intratesticular injection of these substances injures all elements of the seminiferous tubules, as well as the interstitial cells, with permanent damage to the tubules. Repeated administration of lead acetate to growing rats causes impaired spermatogenesis and concurrent damage to Leydig cells. Such changes are exacerbated by a protein-deficient diet. The subcutaneous injection of cadmium chloride in laboratory rodents damages the vascular endothelium with resulting thrombosis and necrosis of the testis. The administration of zinc protects against the necrotizing effects of cadmium. Elevated levels of dietary cobalt in rats cause cyanosis and engorgement of testicular vasculature after about a month, followed by degeneration and necrosis of both germinal epithelium and Sertoli cells.

The oral administration of highly chlorinated naphthalene causes testicular degeneration which is reversible in the bull. In the rabbit, the intravenous injection of amphotericin B (Fungizone) decreases the rate of migration of spermatozoa from the Sertoli cells toward the lumen of the seminiferous tubules, and interferes with the release of spermatozoa from the tubular epithelium. In humans, gentamicin has been reported to interfere with meiosis and cause an increased number of normal and abnormal primary spermatocytes. Other agents causing testicular degeneration include melatonin, carbamate pesticides, ethylene dibromide, nitrofurans and several related pharmaceutical compounds, and the environmental contaminants photomirex (8-monohydromirex) and mirex. In humans, use of several cytostatic anticancer drugs such as cyclophosphamide and chlorambucil cause testicular atrophy, which is usually reversible after cessation of therapy.

There are many mechanisms of chemically induced testicular degeneration. Whereas cadmium chloride induces vascular damage, organic mercurials reduce the uptake of amino acids by spermatogonia and spermatids. Testicular damage caused by busulfan and related drugs involves destruction of spermatogonia and therefore mimics radiation damage. Nitrofurans also affect germinal cells early, and prevent them developing beyond the primary spermatocyte stage. Other drugs, such as ORF 1616 (a dinitropyrrole) and the glycol ethers (used extensively as solvents and emulsifiers), damage the spermatocytes and spermatids, with no apparent changes in spermatogonia. A further possibility is primary damage to Sertoli cells; esters of *o*-phthalic acid (used as plasticizers in polyvinyl chloride products, and therefore now widely distributed in the environment) cause testicular atrophy in rats, and following oral administration, the initial change, a degenerative vacuolation in Sertoli cells, precedes changes in germinal cells. The Sertoli cell is also regarded as the initial target in testicular atrophy caused by the neurotoxic hexacarbon 2,5-hexanedione, thought to be the principal toxic agent in *N*-hexane poisoning.

Toxic plants such as locoweed (*Astragalus lentigino-*

sus) and mycotoxins on certain pastures are known to cause testicular degeneration in ruminants. In locoweed toxicity, impaired spermatogenesis is indicated by decreased numbers of mature sperm in the testes and epididymis, and pronounced vacuolation of spermatogonia, primary spermatocytes, and Sertoli cells, as well as epithelium of the epididymis and vas deferens. Degeneration is transient, however, and regeneration occurs following cessation of feeding locoweed. Altered reproductive function has also been noted in stallions fed alfalfa hay contaminated with *Senecio vulgaris;* the change has been associated with a decrease in luteinizing hormone/human chorionic gonadotrophin receptors in the testes.

Hormonal factors may be implicated in testicular degeneration in a variety of ways. Spermatogenesis and steroidogenesis are controlled by follicle-stimulating hormone and luteinizing hormone and therefore indirectly by releasing hormones involving the hypothalamic–hypophyseal pathway. Androgens are secreted by Leydig cells under the control of luteinizing hormone, whereas follicle-stimulating hormone regulates spermatogenesis, largely by promoting the production of androgen-binding protein by Sertoli cells. Under particular circumstances, estrogenic hormones are produced by the Sertoli cells. The administration, or imbalanced production, of one or more of these hormones may thus cause testicular degeneration. Genital atrophy occurs frequently in association with neoplasms of the pituitary gland and hypothalamus. Excess steroid production by Sertoli and Leydig cell tumors induces degeneration of the seminiferous epithelium in dogs. In hypothyroidism, experimentally induced by administration of thiourea to male goats, there is a reversible atrophy of both the testes and accessory sex glands.

In addition to these more obvious factors, gradual testicular degeneration is associated with diabetes, and increasing age. The pathogenesis of diabetic testicular atrophy has not been elucidated. Whereas age-associated degeneration may largely be secondary to degenerative vascular lesions within the testis, it is, at least in bulls, characterized by a diffuse increase in intertubular stroma and a decreased proportion of parenchyma. Other age-related microscopic changes include increased thickness of tunics and tubule basement membrane, increased proportion of degenerated tubules, and hyperplasia of Leydig cells, which contain increased lipofuscin.

Regeneration of germinal epithelium depends on persistence of spermatogonia and Sertoli cells and elimination of the injurious agent. If spermatogonia have disappeared or the basement membrane becomes hyalinized (Fig. 5.31B), regeneration is no longer possible. Otherwise regeneration may be complete. Time required for regeneration is quite variable, however, and return to normal fertility may be delayed for some time after other signs of clinical recovery. Initial signs of regeneration after testicular degeneration induced by x-irradiation are seen at 10 weeks, and regeneration is complete at about 30 weeks. Following thermal degeneration of the testes in rams, recovery may take 3–6 months.

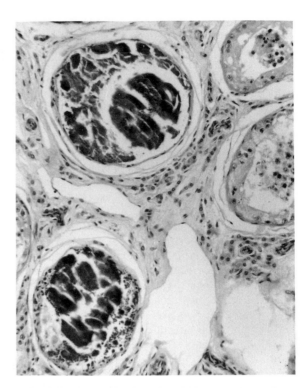

Fig. 5.32 Mineralization of seminiferous tubules. Bull.

Fibrosis and calcinosis of the testis occur as a result of degeneration or inflammation, but may also be of insidious or spontaneous onset. The degree of fibrosis varies from a mild diffuse change with increased consistency to extreme fibrosis where the testis is small and hard with an obvious, grossly visible, increase in stroma. Ventral testicular fibrosis in old bulls (Fig. 5.30) is attributed to degenerative vascular changes (see preceding sections). As the germinal epithelium is avascular, progressive fibrosis may also be consequent to basement membrane changes restricting diffusion from underlying vessels.

Testicular calcinosis frequently accompanies fibrosis, but may also occur independently (Figs. 5.29B, 5.32). In mature and old bulls, an occurrence of calcinosis in the order of 10–30% can be expected, although in most testes, only occasional isolated tubules will be involved. These are of little significance. The occurrence of calcinosis in bucks is comparable to that in bulls. In rams, it is considerably less frequent. Calcinosis is usually bilateral and intratubular (Fig. 5.33A,B) and follows spermiostasis and epithelial degeneration. In acute degenerations, however, calcium salts may also be deposited in the stroma. Sometimes mineralization of the already thickened basement membrane occurs. More extensive degrees of calcinosis, occasionally progressing to complete testicular involvement, are associated with obvious degeneration or inflammation (Fig. 5.34).

Bibliography

Bargai, V., Nobel, T., and Pearl, S. Radiographic changes in testes of bulls infected with besnoitiosis. A correlated radiologic–pathologic study. *Vet Radiol* **25:** 235–239, 1984.

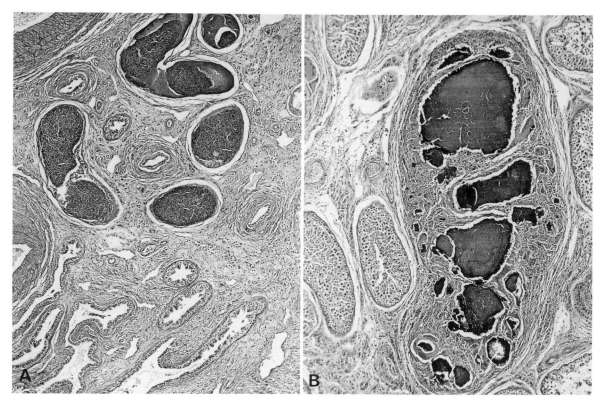

Fig. 5.33   (A) Sperm stasis near mediastinum testis. (B) Spermatic granuloma in focus of sperm stasis in testis. Bull.

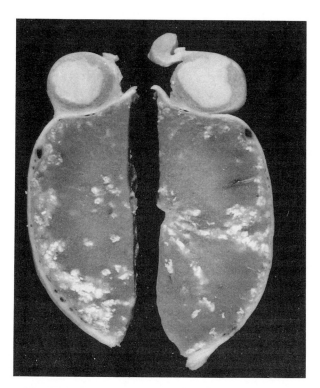

Fig. 5.34   Spermatic granuloma in epididymis. Mineralization of testis. Goat. (Courtesy of R. A. Foster.)

Chapin, R. E., Morgan, K. T., and Bus, J. S. The morphogenesis of testicular degeneration induced in rats by orally administered 2,5-hexanedione. *Exp Mol Pathol* **38:** 149–169, 1983.

Corrier, D. E. *et al.* Testicular degeneration and necrosis induced by dietary cobalt (in rats). *Vet Pathol* **22:** 610–616, 1985.

Creasy, D. M., and Foster, P. M. D. The morphological development of glycol ether-induced testicular atrophy in the rat. *Exp Mol Pathol* **40:** 169–176, 1984.

Creasy, D. M., and Foster, P. M. D. Male reproductive system. *In* "Handbook of Toxicologic Pathology," W. M. Haschek and C. G. Rousseaux (eds.), pp. 829–889. New York, Academic Press, 1991.

Elcock, L. H., and Schoning, P. Age-related changes in the cat testis and epididymis. *Am J Vet Res* **45:** 2380–2384, 1984.

Evans, J. W. *et al.* Relationship of age and season and consumption of *Senecio vulgaris* to LH/hCG receptors in the stallion testis. *J Reprod Fertil* (Suppl.) **35:** 59–65, 1987.

Kamtchoning, P. *et al.* Effect of continuous low dose γ-irradiation on rat Sertoli cell function. *Reprod Nutr Dev* **28:** 1009–1017, 1988.

MacDonald, M. L. *et al.* Effects of linoleate and arachidonate deficiencies on reproduction and spermatogenesis in the cat. *J Nutr* **114:** 719–726, 1984.

Panter, K. E., and Hartley, W. J. Transient testicular degeneration in rams fed locoweed (*Astragalus lentiginosus*). *Vet Hum Toxicol* **31:** 42–46, 1989.

Pitts, W. J. *et al.* Effects of zinc deficiency and restricted feeding from 2 to 5 months of age on reproduction in Holstein bulls. *J Dairy Sci* **49:** 995–1000, 1966.

Reddi, M., and Rajan, A. Pathology of the reproductive organs in experimental hypothyroidism in male goats. *Indian Vet J* **62:** 837–842, 1985.

Saxena, D. K. *et al*. The effect of lead exposure on the testis of growing rats. *Exp Pathol* **31:** 249–252, 1987.

Saxena, D. K. *et al*. Lead-induced testicular changes in protein-malnourished rats. *Folia Histochem Cytobiol* **27:** 57–62, 1989.

Veeramachaneni, D. N. R. *et al*. Changes in basal lamina of seminiferous tubules associated with deranged spermatogenesis in the bull. *Am J Vet Res* **48:** 243–245, 1987.

D. Orchitis

Orchitis may be interstitial (intertubular), intratubular, or necrotizing. Ease of separation of these types varies with the stage of the lesion. With the possible exception of mild interstitial orchitis, which may be confused with testicular degeneration, orchitis is an uncommon lesion in domestic animals, particularly if orchitis consequent to trauma is excluded. Most often it arises by hematogenous infection; spread of infection from neighboring organs through the genitourinary passage is less important. Frequently, inflammation of the tunica vaginalis accompanies orchitis and may precede it.

Interstitial orchitis may not be recognized macroscopically, but histologically it is characterized by mononuclear infiltration of intertubular stroma, with concurrent or subsequent fibrosis (Fig. 5.35A). In bulls, small mononuclear infiltrates are frequently observed adjacent to seminiferous or rete tubules or efferent ducts of otherwise normal testes. Such foci may be of infectious or immune origin,

the latter being in response to antigen leaking from a damaged tubule. In stallions, interstitial lymphocytic foci, often perivascular, are particularly common and occur in areas of tubule degeneration and vasculitis (Fig. 5.35B).

Macroscopically in **intratubular orchitis,** solitary or multiple white-yellow foci of up to about 1 cm are seen on section. Histologically, the tubule outline is retained in the affected area, but the seminiferous epithelium is obliterated and replaced centrally by neutrophils and detritus. Peripheral to this are numerous mononuclear or giant cells indicating granulomatous orchitis, the pathogenesis of which is comparable to spermatic granuloma formation in the epididymis. Sertoli cell hyperplasia and calcification may accompany these changes. Probably intratubular orchitis results from ascending infection.

Necrotizing orchitis is characteristic of brucellosis but may result from other infections, or conditions causing severe trauma or ischemia of the testis (Fig. 5.36A,B). Sometimes severe chronic periorchitis may completely obliterate blood supply to the testis, which eventually becomes a necrotic mass encased within the markedly thickened tunics (Fig. 5.37). On sectioning, necrotic areas are dry, yellow, often laminated and only slightly calcified. The histologic picture is ultimately one of coagulation necrosis bordered by fibrosis and mononuclear cell infiltration. Abscessation and fistulation through the scrotum may accompany necrotizing or other forms of orchitis.

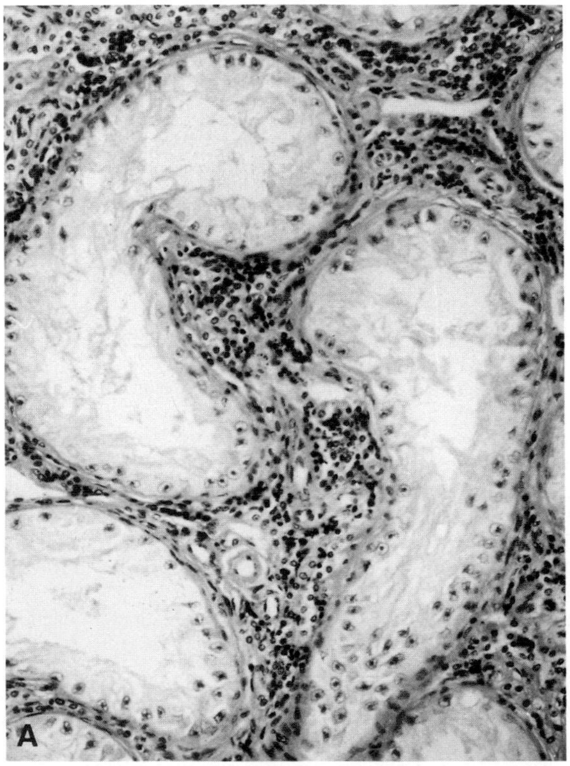

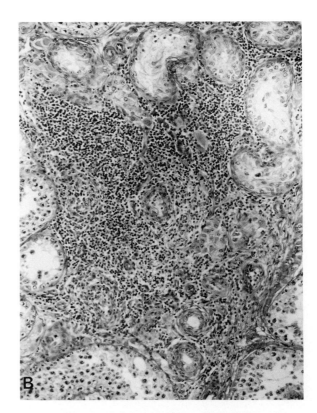

Fig. 5.35A Interstitial orchitis and degeneration of seminiferous tubules. Bull.

Fig. 5.35B Focal interstitial accumulation of lymphocytes. Horse. Tubular degeneration above.

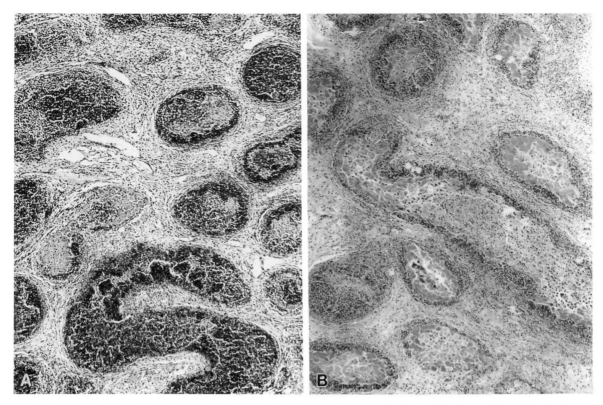

Fig. 5.36 *Brucella* orchitis. (A) Reaction predominantly intratubular. (B) Necrotizing bacterial orchitis. Bull.

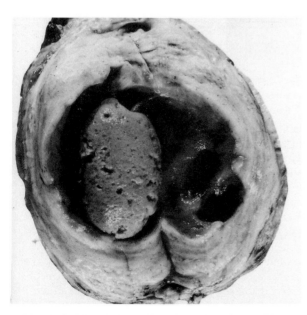

Fig. 5.37 Chronic *Brucella* orchitis. Bull. Severe fibrous periorchitis with infarction of testis.

Many infectious agents have been isolated from testes or semen of animals with orchitis, but the significance of many and perhaps most isolates is unclear. In bulls, a number of viruses have been isolated from testes or semen, but pathologic changes in the testes have been observed in association with only some of them. In some cases, such as in persistent infection with bovine virus diarrhea virus, sperm defects, but no distinct histologic lesions in the testis, are described. A severe interstitial orchitis and testicular degeneration associated with inflammation of spermatic arteries occurs in malignant catarrhal fever in buffaloes (Fig. 5.28A,B). In experimental bluetongue virus infection in bulls, an interstitial orchitis is also associated with arteritis. Clinical orchitis and aspermatogenesis have been observed in association with enterovirus infection, but pathologic changes are not described. A focal nodular orchitis may be observed in lumpy-skin disease. There appears to be no viral orchitis in bulls or other domestic animals comparable to mumps in humans, in which the full extent of damage, resulting in progressive tubular hyalinization and sclerosis, may not become apparent until a considerable time after the acute orchitis which is characterized by edema, mononuclear and sometimes neutrophilic infiltration, and germinal cell degeneration.

Reproductive disorders occur in boars with porcine herpesvirus infection (pseudorabies, Aujeszky's disease), and there may be edema of the scrotal region, but results of experimental infection have varied. It appears the virus, at least following intratesticular inoculation, does not replicate in germinal epithelium of the seminiferous tubules, but does in the serosa, causing an exudative periorchitis. Intratesticular inoculation of boars with porcine parvovirus caused testicular degeneration comparable to that in controls given culture medium only via the same route; no lesion occurred after intramuscular inoculation.

Pathologic changes in bacterial orchitis in bulls are mostly non-specific. Lesions in **brucellosis** are, however, characteristic. In areas where this disease is not controlled, the most common infectious cause of orchitis is *Brucella abortus,* and vaccine strain 19 of this organism is capable of producing the lesion. In most instances, the orchitis is acute and the lesion is irreversible. It may be unilateral but, even so, affected animals are sterile because of the admixing of inflammatory products with the semen of the opposite testis and because the latter undergoes thermal degeneration. The swelling in the scrotum develops quickly and is hot and doughy; it is due largely to inflammatory changes in the tunics and to a lesser extent in the epididymis. Swelling of the testis is never very obvious, being limited by the toughness of the tunica albuginea; this effect predisposes to pressure necrosis within the organ. The cavity of the tunica vaginalis is distended with a fibrinopurulent exudate flecked with hemorrhage, and on the surface of both parietal and visceral tunics, a thick layer of moist yellow fibrin is deposited (Fig. 5.38). At this stage, the testis and epididymis may be grossly unchanged; but, shortly, scattered yellow flecks and foci of necrosis occur in the testicular parenchyma; they progress and coalesce to produce total testicular necrosis, which becomes sequestered by inflammation and thickening of the tunica. Sometimes the necrotic parenchyma liquefies into pus, and the organ then is a pus-filled cavity surrounded by a thick connective tissue capsule. Perforation may occur but is unusual. Occasionally, the necrotic foci may not expand and coalesce but rather remain as areas of dry necrosis with little or no liquefaction, to be rapidly surrounded by large amounts of fibrous tissue. These necrotic foci are usually multiple and do cause enlargement of the organ, but the size may ultimately be reduced by scarification.

Microscopically, the inflammatory involvement of the regional tissues is the same as a fibrinopurulent exudation on any serous membranes, and healing always results in adhesions between the parietal and visceral layers. These adhesions may become extremely dense. Within the testes, the infection appears to progress along the lumen of the seminiferous tubules (Fig. 5.36A). The seminal epithelium becomes necrotic and desquamates, and large numbers of the organisms are visible in the necrotic cells and in the lumen. Use of the modified Ziehl–Neelsen stain facilitates recognition of bacteria present. At this early stage, a variety of leukocytes invade the interstitial tissues and form cuffs about the tubules. With progression of the disease, the walls of the tubules and the interstitial tissues become necrotic. There is, rather regularly, a focal necrotizing epididymitis complicated by the development of spermatic granulomas. In those instances in which the testicular lesion remains focal, the reaction to the initial necrosis becomes abetted by a tuberculoid granulomatous response to the dead sperm.

Tuberculous orchitis in bulls is an uncommon lesion,

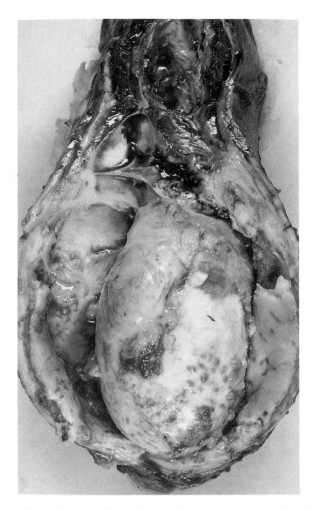

Fig. 5.38 *Brucella* orchitis. Bull. Tunic reflected to show fibrinous exudate adherent to tunic, epididymis, and testis.

even in areas of endemic infection. The testis is less frequently involved than are the epididymis or tunics. Similarity of the inflammatory response to sperm and tubercle bacilli confounds interpretation of histologic lesions in the male genitalia.

Involvement of the testis may take one of two anatomic forms, being either miliary tuberculosis or chronic testicular tuberculosis. In the former, small or large caseous and calcified foci are irregularly scattered throughout the testes but may spare the epididymis entirely. In chronic testicular tuberculosis (Fig. 5.39), the cut surface of the enlarged testis reveals broad bands of caseous necrosis radiating out from the rete testis, and the epididymis is usually involved. This typical picture is due to the extension of the chronic tuberculous process within the seminiferous tubules (Fig. 5.40). The path of infection in such cases is probably intratubular from a primary epididymal lesion.

Other bacteria causing orchitis in bulls, sometimes in association with overt abscessation, include streptococci, staphylococci, *Actinomyces (Cornyebacterium) pyo-*

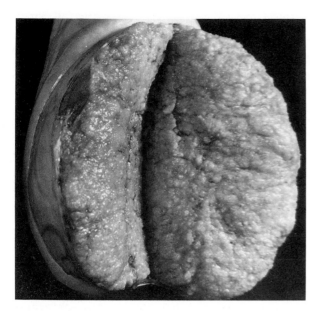

Fig. 5.39 Chronic tuberculous orchitis. Bull. (Courtesy of C. A. V. Barker.)

genes, *Escherichia coli*, *Haemophilus* spp., and *Salmonella* spp. *Actinomyces bovis*, *Actinobacillus* sp., and *Nocardia farcinica* may also cause bovine orchitis. In nocardiosis, lesions, which may also be present in other organs, are at first nodular but ultimately transform the whole testis into an abscess, the capsule of which is the tunica vaginalis.

It seems likely that infection of bulls with *Chlamydia psittaci* causes orchitis, and in field cases, focal granulomatous lesions have been observed. Such lesions have not been reproduced experimentally, however. The spontaneous occurrence of orchitis and epididymitis has been described in the bull in association with high counts of *Mycoplasma* sp., but the frequent presence of this organism in normal genitalia makes interpretation of this observation difficult.

In boars, enteroviruses and parvovirus have been isolated from semen, but viral orchitis is not described. Orchitis caused by *Brucella suis* is characterized more by multiple abscessation than by confluent necrosis. In some cases there is an associated fibrinopurulent, perhaps bloodstained inflammation of the tunica. Abscessation develops in the epididymis as well as in the testis, and these lesions are characterized by central caseation surrounded by a zone of epithelioid cells, and these in turn, by a broad connective tissue capsule infiltrated by leukocytes.

In some tropical countries, orchitis caused by infection with *Pseudomonas pseudomallei* occurs in boars and other small domestic animals, sometimes associated with lesions in seminal vesicles and prostate and other organs. Extreme enlargement of the testis due to accumulation of purulent exudate may occur. Microscopically, the main lesion is multifocal caseation necrosis with marked mononuclear cell infiltration and encapsulation by much fibrous

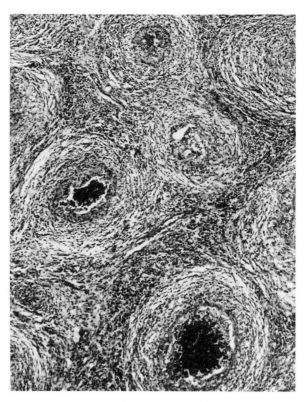

Fig. 5.40 Tuberculous orchitis. Bull.

connective tissue. Severe testicular degeneration accompanies orchitis. Other organisms isolated from orchitis in boars include *A. pyogenes*, *Streptococcus zooepidemicus*, and *Streptococcus equisimilis*.

In stallions, a mild interstitial orchitis is common and may be associated with vasculitis (Fig. 5.35). Such lesions may sometimes be part of a generalized vascular involvement in equine viral arteritis, and in equine infectious anemia in which infarcts may occur. Orchitis occurring as part of systemic disease may also be observed in the form of typical nodules in glanders, and as an acute suppurative, sometimes abscess-forming orchitis in infection with *Salmonella abortus-equi*. Experimental infection of stallions with the causal organism of contagious equine metritis revealed no reaction to the organism or its presence in the testis or accessory sex glands. Focal lesions due to the larvae of *Strongylus edentatus* may be seen in the testis, tunics, and epididymis, especially of young horses (Fig. 5.25).

Nodular orchitis occurs in sheep pox and a chronic interstitial orchitis, but not epididymitis, has been observed in rams infected with the maedi–visna virus. Sporadic testicular abscesses are caused by *Actinomyces pyogenes* and *Corynebacterium psuedotuberculosis*. In bucks, orchitis, which in many respects resembles brucellosis lesions in bulls, may result from infection with *Brucella melitensis*. Orchitis in rams occurs mostly in association with epididymitis.

In dogs, orchitis is not uncommon and is usually accompanied by epididymitis. Intranuclear and cytoplasmic inclusions are found in the Sertoli cells in mature dogs with distemper. The majority of seminiferous tubules degenerate, and inflammation occurs in a few tubules. Penetrating wounds of the scrotum may occasionally be implicated in the pathogenesis of epididymo-orchitis in the dog, but the commonest route is by reflux along the vas deferens from the bladder, urethra, or prostate of infection chiefly by *Escherichia coli, Proteus vulgaris,* and other miscellaneous organisms. An acute inflammatory response in either the epididymis or testis is usually suppurative with the formation of one or more abscesses (Fig. 5.41). The tunica vaginalis may be involved by extension of the inflammatory process and fistulation through the scrotal skin to the exterior may occur. The acute inflammations are usually centered on the ducts with the usual degenerative and desquamative changes in the epithelium, and edema and mononuclear cell infiltrates develop in the surrounding stroma. Healing occurs with dense cicatrization, which in the epididymis will cause some tubular obstruction with spermatocele formation. The subacute and chronic inflammations present no special features. The affected testis is usually firm, small, and irregular, although the epididymis may be enlarged and woody. Of the infiltrating cells, lymphocytes and plasma cells now predominate and fibrosis is well developed. Neutrophils may predominate in the lumen of the epididymis.

Other important bacterial causes of orchitis in dogs are *Brucella canis* and *Pseudomonas pseudomallei,* both of

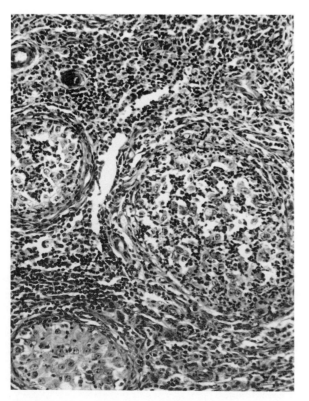

Fig. 5.41   Acute orchitis caused by *Escherichia coli*. Dog.

which are associated with epididymitis (see the following). A familial occurrence of interstitial, lymphocytic orchitis, associated with testicular atrophy and reduced fertility, has been observed in inbred beagle dogs with lymphocytic thyroiditis; immune factors have been implicated.

Bibliography

Dezoe, B. L. Histopathologic changes in male swine with experimental brucellosis. *Am J Vet Res* **29:** 1215–1220, 1968.

Fritz, T. E. *et al.* Pathology and familial incidence of orchitis and its relation to thyroiditis in an enclosed beagle colony. *Exp Mol Pathol* **24:** 142–158, 1976.

Hall, L. B. *et al.* Testicular changes observed in boars following experimental inoculation with pseudorabies (Aujeszky's) virus. *Can J Comp Med* **48:** 303–307, 1984.

Ikede, B. O. Genital lesions in experimental chronic *Trypanosoma brucei* infection in rams. *Res Vet Sci* **26:** 145–151, 1979.

Miry, C. *et al.* Effect of intratesticular inoculation with Aujeszky's disease virus on genital organs of boars. *Vet Microbiol* **14:** 355–363, 1987.

Palfi, V., Glavits, R., and Hajtos, I. Testicular lesions in rams infected by maedi–visna virus. *Acta Vet Hung* **37:** 97–102, 1989.

Revell, S. G. *et al.* Some observations on the semen of bulls persistently infected with bovine virus diarrhoea virus. *Vet Rec* **123:** 122–125, 1988.

Thacker, B. J. *et al.* Clinical, virologic, and histopathologic observations of induced porcine parvovirus infection in boars. *Am J Vet Res* **48:** 763–766, 1987.

E. Neoplastic Diseases of the Testis and Epididymis

Primary epididymal neoplasms are extremely rare in domestic animals, and secondary neoplastic involvement of the epididymis is rare, if spread from a neoplasm in the contiguous testis is excluded. Therefore, of particular concern here are the primary testicular tumors, especially in dogs. Lymphosarcoma may sometimes involve the testis, and there are reports of testicular fibroma, leiomyoma (in the stallion and ram), lipoma, hemangioma, sarcoma, carcinoma, and granulosa cell tumor. Proliferation of tunica mesothelium may encroach on the testes.

Primary testicular tumors are quite common in older dogs, somewhat less so in older bulls, and distinctly unusual in other species. Undoubtedly, this represents largely a species difference in susceptibility with, in addition, factors associated with aging. The primary tumors are of three main types and derived from the three specialized testicular elements, the interstitial cells of Leydig, the sustentacular cells of Sertoli, and the spermatic germinal epithelium. Testicular tumors are seen mostly in mature and old animals; the occurrence of Leydig cell tumors in dogs is especially age associated. Reports on the incidence of the three primary testicular neoplasms in dogs are conflicting and probably permit the generalization that the different types occur with approximately equal frequency.

Most testicular tumors in domestic animals can be diagnosed macroscopically. In general, interstitial cell tumors are tan to yellow-orange, discrete, and soft; Sertoli cell tumors are mostly firm, lobulated, and white; seminomas tend to be homogeneous pale to gray, soft and with a glistening surface resembling lymphosarcoma.

Combinations of the common types of tumor occur in about 25% of canine testicular neoplasia. Any combination is possible and, although the neoplastic nodules are usually separate, they may intermix. There are no implications of common causation but rather of multiplicity of types as probable fortuitous developments in an age group predisposed to testicular neoplasms. All primary testicular tumors occur in older dogs, but the mean age of dogs in which Sertoli cell tumors are diagnosed is slightly less than that for the other tumor types. Certain breeds, notably boxers, have a higher occurrence of all three primary tumors, and the mean age at time of tumor recognition appears to be younger in boxers than in other breeds. There is also evidence of a relatively high occurrence of seminomas in German shepherds.

The relationship between cryptorchidism and testicular neoplasia is well recognized, cryptorchid dogs being at least 10 times more likely to develop primary testicular neoplasms than are normal dogs. Neoplasms appear to develop somewhat earlier in cryptorchid than in scrotal testes. Canine testicular tumors are found more frequently in the right than in the left testis, and this is also true for the cryptorchid testis.

An appropriate clinical history, especially one of feminization with gynecomastia in dogs with Sertoli cell tumors, is of some help in the diagnosis and differentiation of testicular tumors. Sertoli cell tumors (and particularly those in an extrascrotal location) are associated with estrogen secretion. Blood dyscrasias, characterized by anemia, leukopenia, and pancytopenia, may occur in dogs with Sertoli cell tumors and are attributed to estrogen-induced bone marrow toxicosis. Irrespective of particular tumor type, however, clinical signs attributable to endocrinologic disturbances are observed in only a minority of affected dogs.

Sertoli cell tumors and seminomas of the dog, although potentially malignant, seldom metastasize. Extension of these neoplasms is most likely to occur to the epididymis or by local infiltration of lymphatics and the pampiniform venous plexus of the spermatic cord. Examination for tumor emboli spread should include the deep inguinal lymph nodes and cross sections of spermatic cord adjacent to the point of excision. Feminization due to metastatic lesions of Sertoli cell tumor is well known.

1. Interstitial (Leydig) Cell Tumors

These are often referred to as interstitial cell adenomas, which is appropriate enough. Lesions which are less than 1 cm in diameter are usually regarded as hyperplasia, although this distinction may have little biologic significance.

The interstitial cell tumor in the dog is found chiefly in the older animals. It also occurs in the bovine testicle, is found in the older age groups, and, by current figures, mainly in the Guernsey breed. In horses the tumor devel-

ops almost exclusively in cryptorchid testicles. There is one report of a large Leydig cell tumor associated with aspermia in a boar. This tumor appears to be as yet undescribed in the ram or buck.

Steroid production by interstitial cell tumors has not been studied adequately in domestic animals. Androgens are produced by normal Leydig cells, and some interstitial cell tumors of dogs have been shown to produce excess androgen, but most tumors do not cause signs of hyperandrogenism. Interstitial cell tumors of humans produce either androgenic or estrogenic excess. Signs of hyperestrogenism have been observed in a few dogs with Leydig cell tumors and the condition has been corrected by removal of the neoplastic gland, but in dogs, the more common association is with perianal gland neoplasia, tail-gland hyperplasia, and prostatic enlargement, which suggests that secretion of excess androgens is more frequent.

Leydig cell tumors in stallions have been seen most often in undescended testes. They contain two cell types; the first is essentially a hypertrophic Leydig cell, but the other is a pleomorphic fusiform cell with fibrillar, vacuolated cytoplasm and indistinct borders. Hormone determinations have not been reported on horses with these tumors, but viciousness, which was corrected by castration, has been observed.

A high incidence of telangiectasis of the liver, thyroid C (parafollicular) cell tumors, and infertility has been observed in Guernsey bulls with interstitial cell tumors, but it is not known whether the lesions are in any way interdependent or possess a common denominator.

Nodular hyperplasia is probably a preneoplastic change in some instances of interstitial cell tumor of the dog. The hyperplastic nodules occur principally in testes which have undergone senile atrophy and, although they may be macroscopically visible, the nodules are small, nonencapsulated, and consist of an increased number of interstitial cells in the intertubular stroma. The apparent diffuse hyperplasia of interstitial cells in cryptorchid and hypoplastic testes has no significance for the development of tumors. In both nodular and diffuse hyperplasia, the hyperplastic cells are regular in form and size with increased acidophilia of the cytoplasm and without mitoses.

Interstitial cell tumors in the dog are often multiple, but may be solitary and unilateral or bilateral (Fig. 5.42). The common size is from 1 mm or so to 2 cm or so; only exceptionally are they large enough to increase the size of the organ, but they may lend the organ an irregularity in contour, the rounded bulge of the tumor being visible in an otherwise small, soft, atrophic testis. On cut surface, the tumors are well demarcated but only lightly encapsulated, spheroidal, and yellowish or some variant thereof, like corpora lutea. The tumor is predisposed to hemorrhage, which causes dark discoloration, and to cyst formation in areas. Even a large tumor may be obviously composed of confluent nodules indicating multicentricity of origin, and this is readily apparent in the form of discrete tumors when they are small. The consistency is soft, there being little stroma, and the cut surface is slightly greasy.

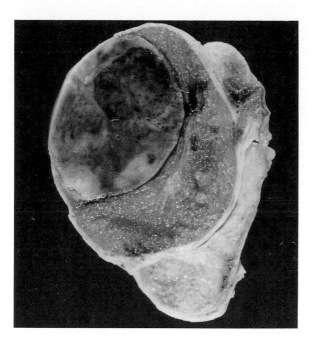

Fig. 5.42 Interstitial cell tumor. Dog.

Histologically, classification of Leydig cell tumors into solid diffuse or cystic–vascular types, and an intermediate pseudoadenomatous type, has been proposed, but such classification appears to have little relationship to biological behavior of the tumor.

Typically, the component cells in the dog are rather respectable interstitial cells, being round or polyhedral, with abundant cytoplasm which may be granular or vacuolar and which often contains yellowish lipochrome pigment (Fig. 5.43A,B). The neoplastic cells in the bull are not vacuolated and contain very little lipid. Sometimes, and in some tumors, the cells have a more mesenchymal appearance, being spindle shaped with indistinct cytoplasmic outline and a streaming arrangement. It is in such tumors especially that necrosis and cyst formation occur. The nuclei are regular in size and staining affinity, and mitoses are rare. The stroma is scant, supporting a capillary network, but occasionally suffices to give the tumor an organoid or pseudoalveolar appearance.

Intranuclear cytoplasmic invaginations, appearing as pseudoinclusions in the nucleus, have been observed consistently in up to 15% of tumor cells in canine Leydig cell tumors. The inclusions, which are strongly PAS positive and which are composed of smooth and rough endoplasmic reticulum, vesicles and lipid vacuoles, myelin figures, and disrupted membranous profiles, have not been demonstrated in other testicular tumors. Nuclei containing these invaginations are enlarged. Except where invaginations are present, the ultrastructural appearances of neoplastic and normal Leydig cells in the dog are similar. The tumors grow slowly and expansively with surrounding compression atrophy and often with a thin condensed capsule. They are not notably invasive. The great majority are

benign, and metastatic spread is an exceptional development. In bulls with Leydig cell tumors, semen production and fertility may be reduced.

2. Sertoli Cell Tumor

The Sertoli cell tumor is rare in domestic species other than the dog, but has been observed in the bull, horse, ram, and cat. These tumors often cause enlargement of the affected testis, and there is, sometimes, the development of a feminization syndrome in the host. The feminizing effect of these tumors is due to their high content of estrogen. The degree of differentiation of the neoplastic cells will influence the content of estrogens but, usually, it is the larger tumors which are responsible for the feminizing syndromes, a fact which suggests that the bulk of the tumor determines in large measure the quantity of estrogen elaborated. The attractiveness of affected dogs to other male dogs is well known, but other evidences of estrogenism are manifest as reduction of libido, female distribution of body fat, cutaneous and pilosebaceous atrophy leading to symmetrical alopecia, atrophy of testes and penis, an estrogenic form of mammary development, swelling of the prepuce, and hyperplasia or squamous metaplasia of the prostate, which may be accompanied by perineal hernia. Metaplasia can progress to the stage where quite large keratin accumulations fill glandular lumina. Enlargement of the seminal colliculus with partial obstruction of the urethral lumen may also accompany prostatic changes in these dogs.

As well as feminization, excess endogenous estrogen in dogs with Sertoli cell tumors may cause depression of bone marrow activity with resultant clinical signs of hemorrhage caused by thrombocytopenia, anemia caused by blood loss or diminished erythrocyte production, and infection and fever associated with granulocytopenia. Recovery may follow castration and supportive therapy.

The gross appearance of a Sertoli cell tumor can be quite distinctive. The larger tumors are somewhat irregularly ovoid, lobulated, and enclosed in a tense tunica albuginea (Fig. 5.44). The cut surface bulges and usually is whitish in color and quite firm or even hard, although sometimes it may be discolored or cystic. The firmness of the tumor is due to the abundance of its stroma, something the other two common types of tumor have in small amounts only.

Although metastasis of Sertoli cell tumor even to the regional lymph node is unusual, local extension of neoplastic tissue into the testicular vein and associated lymphatics may result in hydrocele with massive swelling of the scrotum.

Histologically, Sertoli cell tumors may be of two types, either intratubular (with or without invasion as determined by penetration of the basement membrane), or diffuse. There seems to be little correlation, however, between histologic type and metastasis. Stroma is always plentiful, and it may be hyalinized. The stromal tissues are commonly arranged to provide a pseudotubular pattern in which the neoplastic cells tend to palisade. Such arrangements to some degree can be found in most Sertoli cell

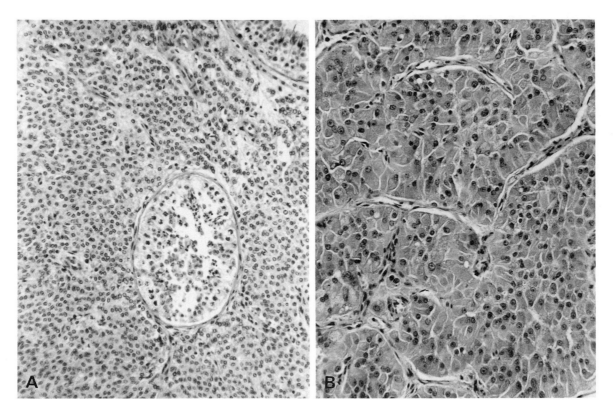

Fig. 5.43 Histologic appearance of an interstitial cell tumor. (A) The tumor cells surround a degenerating tubule. (B) Histologic appearance of an interstitial cell tumor. Dog.

tumors (Fig. 5.45A). In the early and well-differentiated tumors, the cells resemble normal Sertoli cells, being rather elongate with foamy acidophilic cytoplasm and small, basally situated, dark-staining nuclei. In less-differentiated varieties, the cells are still elongate and possess eosinophilic cytoplasm, but the nuclei are elongate and pleomorphic and are no longer basally located against the trabecular pole (Fig. 5.45B). In the less common form of the tumor, the cells show little or no tendency to palisade but are discrete and spherical with well-defined eosinophilic cytoplasm and some nuclear irregularity. Mitoses are always sparse, except where malignancy is obvious. Lipids are demonstrable in the neoplastic cells as large droplets and globules in more-differentiated tumors and as fine droplets in the least-differentiated tumors.

Ultrastructural examination of canine Sertoli cell tumors reveals that the characteristic specific intercellular junctions and crystals of Charcot–Bottcher do not occur in tumor cells. Nevertheless, a number of common features do persist so that these and the abundant intracytoplasmic organelles permit differentiation from Leydig cell tumors and seminomas (Fig. 5.46A,B). Prominent intercellular gap junctions have been observed in Sertoli cell tumor in the dog and it is possible that these were induced by the hormonal activity of the tumor.

Early age of slaughter of most food animals precludes studies on true tumor occurrence with increasing age. In one study, however, in which bulls and buffaloes were kept until 9–14 years for draft purposes, 20 of 161 testes examined contained neoplasms; all but one were Sertoli cell tumors. Seven tumors were in undescended testes. In contrast to Sertoli cell tumors in the dog, those in cattle have been observed in newborn or young calves with sufficient frequency to suggest that impaired embryogenesis, possibly of genetic origin, might have some role in causation. In support of this suggestion is the presence, in some bovine Sertoli cell tumors, of laminated intratubular concrements resembling those seen in bovine testicular hypoplasia and cryptorchidism. Moreover the simultaneous occurrence of Sertoli cell tumor and epididymal aplasia has been observed. One described bovine Sertoli cell tumor was in a testis of an animal in which castration by the Burdizzo method had been attempted 5 years previously. In general, the gross and microscopic appearance of Sertoli cell tumors in the bull resemble those in the dog (Fig. 5.47A,B). Metastasis of Sertoli cell tumor in the bull has not been observed, and there is no clear evidence of hyperestrogenism.

Occasional Sertoli cell tumors in rams, comparable to those in the bull, have been observed. One case of Sertoli cell tumor in a stallion involved the single descended testis; a ductal pattern of neoplastic Sertoli cells, which contained clusters of distinct hyaline bodies, was evident.

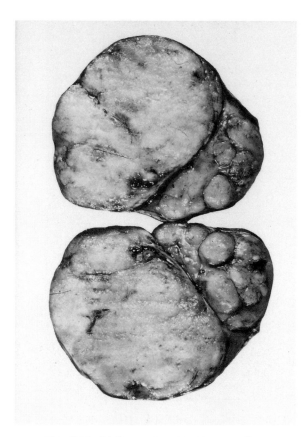

Fig. 5.44 Malignant Sertoli cell tumor. Dog.

3. Seminoma

Seminomas are common in canine testes and have also been observed in the stallion (and mule), ram, buck, and bull. They occur in older animals and are disproportionately common in cryptorchid testes. They arise from cells of the spermatogenic series, presumably from basal spermatogonia, and there are usually multiple foci of origin in the affected testes. These tumors do not produce hormones. They are not often malignant, but are probably more so than either of the other two types. They tend, however, to be locally invasive.

Seminomas often attain a size of 6 cm or more before removal, and by then the stretched tunica albuginea encloses neoplastic tissues only. Sudden enlargement of the testis and pain caused by hemorrhage and necrosis in the tumor are often the presenting signs in dogs. In all species the sectional surface is coarsely lobulated by a few fine trabeculae, the color is usually white or grayish white, and the texture is soft to moderately firm (Fig. 5.48A). If lightly squeezed, a milky fluid may exude from the cut surface. The texture and color closely resemble those of neoplastic lymphoid tissue.

Microscopically, as with Sertoli cells tumors, intratubular and diffuse types are recognized. The earliest development of the tumor is intratubular, and even in some large specimens, intratubular growth is still evident and undoubtedly comprises one method of spread (Fig. 5.48B). Rupture of the tubules soon occurs and the growth becomes confluent, forming broad sheets of closely packed cells with scant supporting stroma. There is but slight cytologic variation from tumor to tumor. The cells are

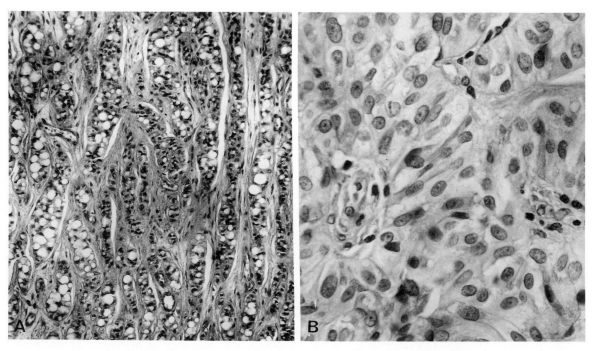

Fig. 5.45 (A and B) Histologic appearance of Sertoli cell tumor. Dog.

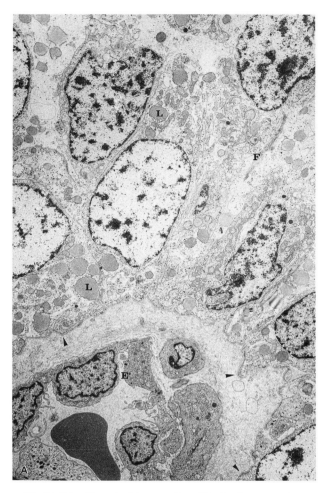

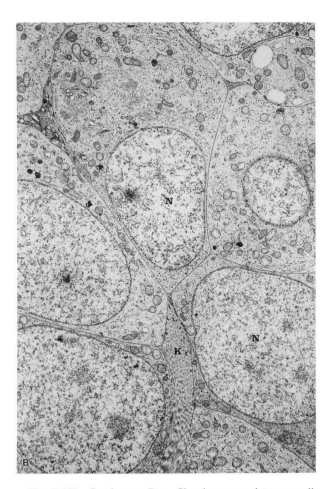

Fig. 5.46A Sertoli cell tumor. Dog. Vertically arranged tumor cells on a basement membrane (arrowheads) with lipid droplets (L) and filamentous material (F). Endothelial cells (E) of capillary, bottom left. ×3000. (Courtesy of D. von Bomhard and *Journal of Comparative Pathology*.)

Fig. 5.46B Seminoma. Dog. Closely arranged tumor cells with oval nuclei (N) and few organelles. Intercellular space filled with collagenous fibrils (K) which are not limited by a basement membrane. ×3000. (Courtesy of D. von Bomhard and *Journal of Comparative Pathology*.)

large and polyhedral and fairly discrete, with a rim of visible cytoplasm which may be basophilic or eosinophilic (Fig. 5.48C). The nuclei are large and rounded and usually strongly chromatic with one or more large acidophilic nucleoli. In some tumors, the nuclei are larger and vesiculate but still regular in shape, and the cells are closely packed with scant or invisible cytoplasm. In many seminomas, it is possible to find scattered mono- or multinucleate giant cells with abundant granular acidophilic cytoplasm. Focal or diffuse accumulations of lymphocytes occur in most seminomas and are a useful distinguishing feature. Whereas atrophy of tubules at the edge of the tumor is normally evident, occasional foci of marked intratubular spermatogonial proliferation and early seminoma formation in this location are sometimes seen.

When examined by electron microscopy, tumor cells in canine seminoma resemble normal germinal epithelium and are characterized by a relative scarcity of cytoplasmic organelles, oval nuclei, straight cell borders, and distinct

Golgi complex (Fig. 5.46B). Intercellular bridges, as seen in normal germinal cells, are present in seminomas. In all cases the cells are markedly distinct from cells of Leydig or Sertoli cell tumors.

Presumably because abdominal location of seminomas in retained testes precludes their early clinical recognition, they may become quite large and metastasize widely. In the stallion, such tumors up to 9 kg have been found, and metastasis to most parts of the abdominal cavity and also the thoracic cavity has been observed.

Intratubular seminomas have been reported in mature and aged rams with testicular degeneration. The proliferating spermatogonia are confined to the seminiferous tubules and do not attain sufficient size to be recognized on gross examination. It appears that this may be a dysplastic rather than a neoplastic process. A similar dysplastic process in otherwise degenerate testes is also observed in dogs. Several such foci occur in the organ, the affected tubules being slightly dilated and the abnormal cells distinctively large, rounded, or polyhedral, usually multinucleate, and

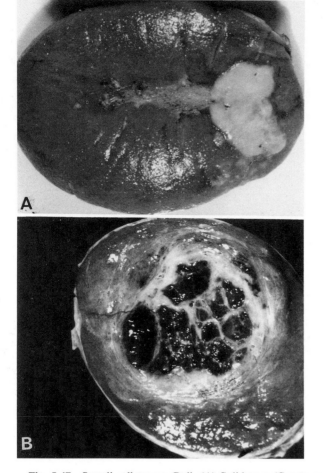

Fig. 5.47 Sertoli cell tumor. Bull. (A) Solid type.(Courtesy of *Journal of Comparative Pathology*.) (B) Cystic type. (Courtesy of *Journal of Comparative Pathology*.)

with pigments in the cytoplasm. Rarely, a highly malignant form of seminoma occurs in the ram, producing overall enlargement of the testis, and hemorrhage, necrosis, and massive neoplastic infiltration of the entire testis.

4. Teratoma

Teratoma of the testis is a tumor which is virtually unknown in domesticated mammals other than the horse, in which it is the most frequently reported testicular tumor. Equine teratoma is a benign tumor found in the scrotal, or more often, cryptorchid testis of young animals. It is probable that its presence in a fetal testis would prevent normal descent. There is a report of testicular teratoma which caused partial obstruction of the colon in a neonatal foal.

Macroscopically teratomas are single or multiple and quite varied in color and texture. They are usually less than 10 cm in diameter but may be greater than 25 cm. A cystic and/or multilocular structure is common and, on section, hair and mucoid or sebaceous like secretions are often seen, hence frequent use of the term dermoid cyst.

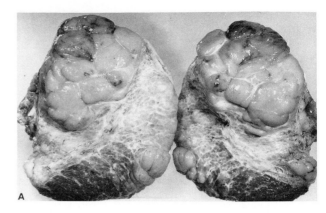

Fig. 5.48A Multiple nodules of seminoma in fibrotic testis. Horse.

Yellow-white solid masses with fibrous, adipose, cartilaginous, and bony tissue are also frequent.

Histologically, structures derived from all embryonic germ layers including ectodermal (dermoid cysts, hair, teeth), neuroectodermal (nervous tissue, melanoblasts), entodermal (salivary gland, respiratory), or mesodermal (fibrous or adipose tissue, bone, muscle) may be present. Nervous tissue is almost always present, and adipose tissue is also very common. Testis adjacent to the teratoma mass may show reduced spermatogenesis, with varying degrees of tubular atrophy. The histogenesis of gonadal teratoma is discussed with The Female Genital System (Chapter 4 of this volume).

5. Other Rare Primary Tumors

Embryonal carcinoma is rare in animals but is of significant frequency among the germinal tumors of humans, in which it is highly malignant. It is probably best regarded as belonging, histogenetically, with the teratomas but not displaying the tissue differentiations which identify the teratoma. The cells are of indifferent embryonic types. Trophoblastic differentiation and demonstrable alpha-fetoprotein in the epithelial cells support the diagnosis.

The **gonadoblastoma** may occur in testes and ovaries which are normally differentiated, but the majority of such tumors in humans occur in young patients who are phenotypically female but have a male chromosome complement and anatomic stigmata of intersexuality. These rare tumors are of solid appearance and are comprised microscopically of a varying but intimate admixture of germ cells, small epithelial cells of Sertoli appearance, cells resembling granulosa cells (which may be organized to form Call–Exner-like bodies), and clusters of acidophilic interstitial cells.

Adenomas and **adenocarcinomas** of presumptive origin from the **rete testis** are described in dogs and horses. These are tubulopapillary structures with scant supporting stroma. The papillae and anastomosing tubules are lined

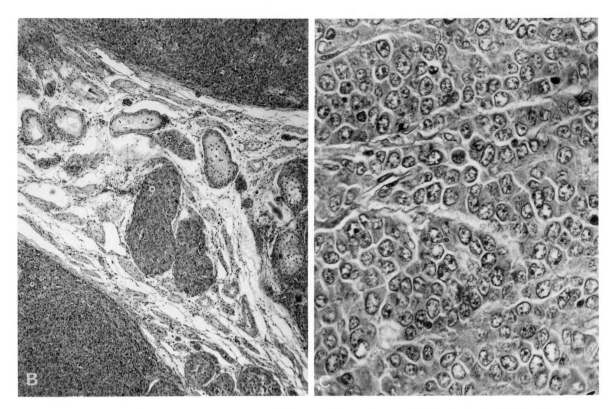

Fig. 5.48 Fibrotic testis. (B) Horse. Intratubular growth in seminoma. (C) Histologic detail of seminoma.

by small, closely packed cells with scant cytoplasm in either single or multiple layers. Evidence of transformation from normal to neoplastic rete epithelium is a useful criterion for diagnosis, and it is desirable to exclude other teratomatous structures or a primary tumor elsewhere.

Bibliography

Cullen, J. M. *et al.* A mixed germ cell–sex cord–stromal neoplasm of the testis in a stallion. *Vet Pathol* **24:** 575–577, 1987.

Dass, L. L. *et al.* A rare case of Sertoli cell tumour in a bullock. *Indian Vet J* **64:** 740–741, 1987.

Düe, W. *et al.* Immunohistochemical determination of oestrogen receptor, progesterone receptor, and intermediate filaments in Leydig cell tumours, Leydig cell hyperplasia, and normal Leydig cells of the human testis. *J Pathol* **157:** 225–234, 1989.

Foster, R. A., Ladds, P. W., and Hoffmann, D. Testicular leiomyoma in a ram. *Vet Pathol* **26:** 184–185, 1989.

Galofaro, V., and Di Guardo, G. Spontaneous seminoma in a mule. *Equine Vet J* **18:** 218–219, 1986.

Gelberg, H. B., and McEntee, K. Equine testicular interstitial cell tumors. *Vet Pathol* **24:** 231–234, 1987.

Houszka, M., and Dubiel, A. Leydigoma (interstitial cell tumour) of the testicle in a boar. *Medycyna Weterynaryjna* **41:** 498–500, 1985.

Jacobs, G. *et al.* Colliculus seminalis as a cause of a urethral filling defect in two dogs with Sertoli cell testicular neoplasms. *J Am Vet Med Assoc* **192:** 1748–1750, 1988.

Johnson, R. C., and Steinberg, H. Leiomyoma of the tunica albuginea in a horse. *J Comp Pathol* **100:** 465–468, 1989.

Morgan, R. V. Blood dyscrasias associated with testicular tumors in the dog. *J Am Anim Hosp Assoc* **18:** 970–975, 1982.

Parks, A. H. *et al.* Partial obstruction of the small colon associated with an abdominal testicular teratoma in a foal. *Equine Vet J* **18:** 342–343, 1986.

Prange, H. *et al.* Pathology of testicular tumors of the dog. I. Epidemiology and comparative epidemiological aspects. *Arch Exp Vet Med* **40:** 555–565, 1986.

Prange, H., Kosmehl, H., and Katenkamp, D. Pathology of testicular tumors of the dog. 2. Morphology and comparative morphological aspects. *Arch Exp Vet Med* **41:** 366–368, 1987.

Rahaley, R. S. *et al.* Sertoli cell tumor in a horse. *Equine Vet J* **15:** 68–70, 1983.

Reifinger, M. Statistical investigations on the occurrence of testicular neoplasms in domestic animals. *J Vet Med (A)* **35:** 63–72, 1988.

Sanford, S. E., Miller, R. B., and Hoover, D. M. A light- and electron-microscopical study of intranuclear cytoplasmic invaginations in interstitial cell tumors of dogs. *J Comp Pathol* **97:** 629–635, 1987.

Schönbauer, M. Papillary carcinoma of the testis from a zebra. *Wien Tierärztl Monatschr* **69:** 95, 1982.

Schönbauer, M., and Schönbauer-Längle, A. Seminoma in the horse. A retrospective study. *Zentralbl Veterinarmed (A)* **30:** 189–198, 1983.

Scott, D. W., and Reimers, T. J. Tail gland and perianal gland hyperplasia associated with testicular neoplasia and hypertestosteronemia in a dog. *Can Pract* **13:** 15–17, 1986.

Thilander, G., Lindberg, R., and Plöen, L. Ultrastructural features of the neoplastic Sertoli cell in a dog. *Acta Vet Scand* **28:** 445–446, 1987.

Trigo, F. J., Miller, R. A., and Torbeck, R. C. Metastatic equine seminoma: Report of two cases. *Vet Pathol* **21:** 259–260, 1984.

Turk, J. R., Turk, M. A., and Gallina, A. M. A canine testicular tumor resembling gonadoblastoma. *Vet Pathol* **18:** 201–207, 1981.

Valentine, B. A., and Weinstock, D. Metastatic testicular embryonal carcinoma in a horse. *Vet Pathol* **23:** 92–96, 1986.

Weaver, A. D. Survey with follow-up of 67 dogs with testicular Sertoli cell tumours. *Vet Rec* **113:** 105–107, 1983.

Zanwar, S. G., Sardeshpande, P. D., and Deshpande, B. R. Testicular neoplasia in the bovines. *Indian J Anim Reprod* **3:** 31–34, 1983.

VI. Epididymis

A. Anomalies of Development

The epididymis, vas deferens, ampulla, and the seminal vesicle are derived from the mesonephric duct; the tubuli recti, the rete testis, and the efferent ducts of the testis are derived from the mesonephros by way of the gonadal blastema, which migrates from the mesonephros to the gonadal ridge. The separate origins for these excretory ducts may not be appropriate to all species but assist understanding of congenital obstructions in the excretory ducts. Obstruction of a duct leads to impaction with sperm, local dilation of the duct (spermatocele), and extravasation of sperm to produce a spermatic granuloma.

The spermatic granulomas which are rather common in the head of the epididymis of the ram and goat, and less so in other species, are probably, in the absence of other epididymal lesions, the result of defective development of testicular ducts rather than of the epididymal tubule. The majority of such defects are unilateral, affecting the right side, but some are bilateral. The condition appears to be inherited. In each testicle, the number of efferent ducts ranges from ~10 to 20 depending on the species, and ducts which end blindly are rather common. Individual blind ducts may be obliterated quickly and not be of significance. Expanding spermatoceles and spermatic granulomas may compromise all of the efferent ducts of an organ.

Segmental aplasia of the epididymis is segmented aplasia of the mesonephric duct and therefore is unlikely to be bilateral. Segmental aplasia most frequently involves the body and tail of the epididymis and, concurrently, the corresponding ampulla and seminal vesicles in about one third of cases. It has been best studied in the bull in which its inheritance seems likely, but has also been observed occasionally in the ram, buck, boar, and dog. In the stallion, there is a report of unilateral aplasia of the epididymal tail, but otherwise epididymal aplasia appears to have been observed only in association with anomalies of the testis and urinary organs, as was also the case in one tortoise-shell cat.

In bulls, segmental aplasia of the epididymis most often involves the body or tail (Fig. 5.49) and the right organ is most often affected. After puberty, aplasia results in spermiostasis and increased pressure, leading to spermatocele formation and testicular degeneration, the severity

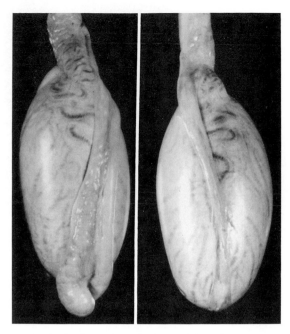

Fig. 5.49 Unilateral aplasia of epididymis. Bull.

of which is proportional to proximity of the aplastic segment to the testis. Degenerative changes are most pronounced in the proximal testis and diminish in severity toward the distal pole. At least in young postpubertal bulls with unilateral epididymal aplasia, the testis on the affected side is more turgid and resilient than the opposite, normal testis and tends to be heavier, presumably because of sperm accumulation.

Microscopic testicular changes in epididymal aplasia include marked dilation of rete tubules, visible grossly, spermiophagy, and epithelial proliferations in the efferent ducts and the rete testis (Fig. 5.50A,B). Elsewhere in testicular parenchyma, a granulomatous reaction may be centered on tubules as a reaction to sperm or other seminiferous material (Fig. 5.50C). In bulls, hypoplasia of the epididymis may accompany segmental aplasia or spermatocele, presumably as a consequence of impaired sperm transport. The occurrence of bilateral epididymal hypoplasia, unassociated with blockage of sperm, is, however, described. Affected bulls had decreased sperm motility.

Several congenital cystic conditions of the epididymis, namely paradidymis, blind efferent ductules, and aberrant ductules, are considered to result from cystic dilation of remnants of mesonephric tubules. They are observed with varying frequency in most and perhaps all domestic species and are usually of no clinical significance. In paradidymis externus, isolated cysts up to several millimeters in diameter and of rounded to oblong shape are located in the spermatic cord just proximal to the head of the epididymis. The cysts have no apparent connection with the epididymis. In paradidymis internus, similar isolated tubules are located entirely within the epididymal stroma. Unlike paradidymis, blind efferent and aberrant ductules do retain

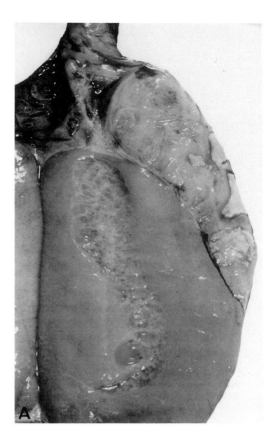

Fig. 5.50A Spermatic granuloma in head of epididymis. Ram. Note dilation of rete tubules.

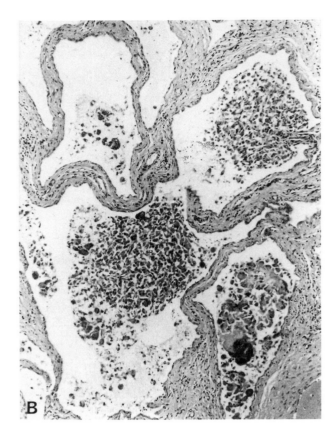

Fig. 5.50B Dilated rete tubules. Bull. Dilation caused by epididymal aplasia. Tubules partly filled with macrophages.

communication with the rete testis or epididymal duct, respectively, and would normally have formed an efferent duct. Microscopically, paradidymis, blind and aberrant ductules are lined by ciliated epithelium, and cystic enlargement results from secretions. Larger epididymal cysts up to about 4 cm in diameter and containing clear or milky fluid (congenital retention cysts, galactocele) or sperm, may thus develop, especially in the head of the epididymis. It is possible that cysts located between the head of the epididymis and the testis, and containing clear serous fluid may be the appendix epididymis, and therefore represent the blind cranial segment of the mesonephric duct rather than dilated mesonephric tubules.

The higher occurrence of cysts in neonatal calves than in bulls suggests that regression of cysts is common. Spermatocele formation in blind and aberrant ductules may, however, result in spermatic granuloma (Fig. 5.51) and, based on the high occurrence of spermatic granulomas in the head of the epididymis of polled goats, it appears that the condition is inherited. It also seems possible, however, that pressure from large retention cysts containing no sperm might cause spermiostasis and perhaps eventual granuloma formation in adjacent tubules (Fig. 5.52).

Ectopic rests of adrenal cortical tissue may be found in the epididymis, testis, or spermatic cord, especially in horses (Fig. 5.26).

Bibliography

Bader, H. *et al.* A contribution to the segmental aplasia and hypoplasia of the epididymis of the bull. *Deutsch Tierarztl Wschr* **90:** 448–456, 1983.

Hemeida, N. A., and McEntee, K. Spermiostasis of the ductuli efferentes of the bovine epididymis. *Assiut Vet Med J* **13:** 317–324, 1984.

Ladds, P. W., Briggs, G. D., and Foster, R. A. Epididymal aplasia in two rams. *Aust Vet J* **67:** 457–458, 1990.

B. Epididymitis

In most species, epididymitis occurs more frequently than orchitis, although they are often concurrent and associated with inflammation of the accessory sex glands. Also, in contrast to orchitis, which is often hematogenous in origin, sporadic epididymitis arises chiefly by spread of infection in the genitourinary passages, even though existence of a protective local immune mechanism in these passages has been confirmed.

Once initiated, the course of epididymitis is variable. The acute stage with edematous enlargement may be followed by abscess formation, sometimes with perforation, periorchitis, and peritonitis, and increasing fibrosis.

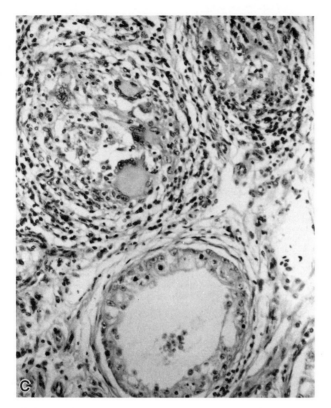

Fig. 5.50C Granulomatous orchitis associated with epididymal aplasia. Bull.

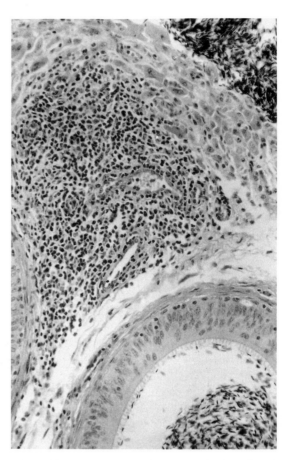

Fig. 5.52 Spermatic granuloma in head of epididymis. Bull. Extravasated sperm (upper right) are surrounded by a zone of macrophages.

Spermatic granulomas occur after extravasation of spermatozoa (Fig. 5.53).

Macroscopically, irregularity of epididymal size and shape is apparent especially in unilateral cases in comparison with the contralateral side. Fibrinous or fibrous adhesions may be present between affected epididymis and adjacent tunics. Consistency will depend on duration of inflammation and the development of spermatic granulomas. In chronic epididymitis, the marked increase in fibrous tissue both within the epididymis and between the epididymis and tunics results in atrophy of epithelial elements, and the recognizable epididymis becomes hard and nodular, perhaps enclosing dilated remnants of epididymal duct (Fig. 5.54). Distinct hard sperm stones, the end products of spermiostasis, may be present within such ducts. Concurrent testicular atrophy will also result in the epididymis appearing disproportionately large in relation to the testis.

Histologically, affected epididymal ducts contain fibrin, neutrophils, spermatozoa in various stages of disintegration, damaged epithelium, macrophages, and multinucleate giant cells, many of which contain spermatozoa. Other features of spermatic granuloma, with prominent intersti-

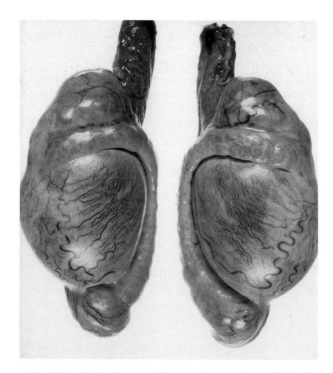

Fig. 5.51 Bilateral spermatic granulomas in head of epididymis. Goat.

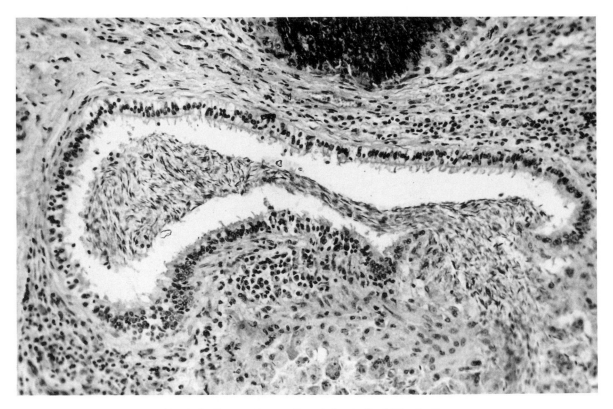

Fig. 5.53 Spermatic granuloma in head of epididymis. Ram.

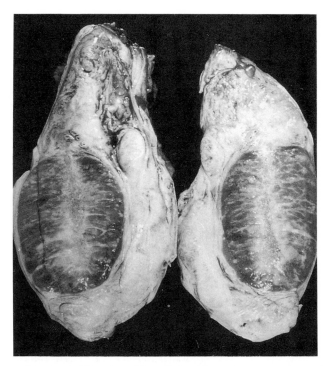

Fig. 5.54 Chronic epididymitis with testicular atrophy and fibrosis. Bull.

tial accumulation of mononuclear cells, become apparent. Epithelial hyperplasia, with the development of intraepithelial lumina (Fig. 5.55), has been observed in most domestic species in association with epididymitis, but such lumina also occur in noninflammatory lesions. In chronic epididymitis, squamous metaplasia of epithelium may also occur in affected ducts, in association with progressive fibrosis, and there may be hypertrophy of smooth muscle surrounding ducts.

Epididymitis may be caused by spermatic granulomas resulting from congenital ductal anomalies, trauma, and infections. There is evidence that reflux of sterile urine via the vas deferens to the epididymal tail may cause epididymitis.

Infectious and immunologic causes of epididymitis in **bulls** are largely the same as those for orchitis, but bulls develop a specific infectious epididymitis, so-called **epididymitis–vaginitis** (epivag), first described in Kenya and later in South Africa. Lesions in the bull consist of initial soft swelling of the epididymis with subsequent enlargement and fibrosis. Associated lesions are abscess formation, tunic adhesions, ampullitis and seminal vesiculitis, and testicular degeneration. Sometimes, however, the seminal vesicles only are affected. The possible role of a herpes virus in the pathogenesis of epivag is discussed with The Female Genital System (Chapter 4 of this volume).

Brucella abortus rarely causes epididymitis in the ab-

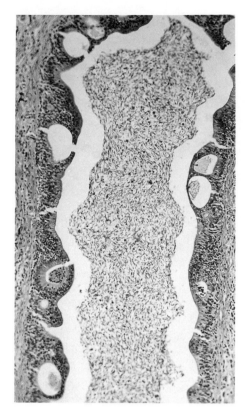

Fig. 5.55 Epididymitis caused by *Brucella ovis*. Ram. Hyperplasia of epithelium and epithelial lumina formation.

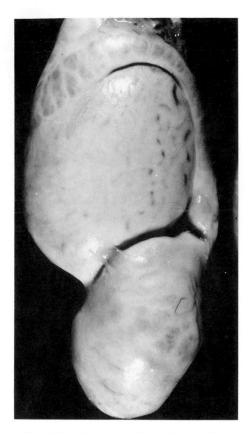

Fig. 5.56 Chronic epididymitis of tail caused by *Brucella ovis*. Ram.

sence of orchitis. *Actinobacillus seminis,* a frequent cause of epididymitis in rams, has been isolated from the semen of a bull with bilateral epididymitis. Epididymitis has been observed in bulls with seminal vesiculitis induced by inoculation of *Mycoplasma bovigenitalium,* but the roles of mycoplasmas and *Chlamydia psittaci* in causing epididymitis await clarification. There is one report of severe spontaneous epididymo-orchitis caused by chronic mycoplasma infection. Epididymitis may accompany orchitis, periorchitis, and testicular degeneration in cattle, sheep, goats, horses, and dogs infected with *Trypanosoma brucei.*

Epididymitis is of particular importance in **rams** in which it is a frequent and serious cause of reduced fertility. Although one of two organisms, *Brucella ovis* or *Actinobacillus seminis,* are usually associated with the condition, many other organisms such as *Pasteurella haemolytica, Escherichia coli,* and *Actinomyces (Corynebacterium) pyogenes* have been isolated from cases of epididymitis. Whereas *Brucella ovis* is a cause of epididymitis in mature rams, other gram-negative pleomorphic organisms are found more commonly in epididymitis in virgin rams, suggesting the existence of two separate disease entities, dependent on sexual experience of the animal. Alternatively, slow development of *Brucella ovis* lesions may be partially responsible for this apparent age difference.

Ovine epididymitis caused by *Brucella ovis* is an im-

portant cause of reduced fertility in many countries and has been extensively studied. Progression of the disease is very slow, from a local infection persisting in the exposed mucous surface for 1 month, to a regional one involving adjacent lymph nodes, leading to a bacteremia. The bacteremic stage appears to subside after about 2 months, but organisms localize in the genital tract, spleen, kidney, and liver, where they persist for an indefinite period. Following experimental infection, neither gross nor microscopic lesions are seen in organs other than genitalia.

In approximately 90% of epididymides with lesions caused by *Brucella ovis,* the epididymal tail is involved, and lesions in this location probably occur in all epididymides infected with this organism (Fig. 5.56). Initial localization of the bacteria produces edema, and lymphocytic and macrophage infiltration. Later, neutrophils are added when sperm enter the interstitium. Early epithelial changes include hyperplasia and hydropic degeneration, with the formation of intraepithelial lumina (Fig. 5.55). At the same time there is increasing fibrosis in interstitial areas. The combination of fibrosis and epithelial hyperplasia obstructs the lumen and causes sperm stasis. These changes develop over many months, and large numbers of organisms are excreted in the ejaculate. Subsequent events depend on the extravasation of sperm and formation of spermatic granuloma. The tail of the epididymis in these cases may be enlarged four to five times, and the lesion is often bilateral. If the extravasated sperm enter the cavity of the tunica vaginalis, adhesions will result, and testicular degeneration increases. Unlike brucellosis in the bull, there is no primary orchitis. Lesions in the vas deferens similar to those in the epididymis may occur, but are not associated with sperm stasis or leakage. There is pronounced epithelial hyperplasia, with thickening and folding of the wall, and the lamina propria is densely infiltrated with lymphocytes, plasma cells, and histiocytes. Unfortunately for control purposes, many rams do not develop detectable gross lesions or they develop them only late in the course of the disease. Identification of *B. ovis* in histologic sections is difficult but can be aided by application of immunolabeling procedures.

Actinobacillus seminis and related strains of the so-called gram-negative pleomorphic organisms also produce an epididymitis. These bacteria consist principally of *Haemophilus somnus* (*Histophilus ovis*) and *Actinobacillus* spp. The species identification of *Actinobacillus* responsible for epididymitis in young rams is indefinite, but it is suggested that *A. seminis* is the bacterium involved. Both *H. somnus* and *A. seminis* are temporarily resident in the prepuce and become opportunistic pathogens by ascending infection, under the appropriate conditions such as those resulting at puberty from elevated levels of luteinizing hormone and follicle-stimulating hormone releasing hormones. It is suggested that the pathogenesis of *E. coli* epididymitis in rams is similar. Typically, these are acute infections which occur mostly in young rams, and there may be a severe and diffuse periorchitis. In epididymitis caused by *A. seminis*, there is abscessation of one or both epididymides, and these may fistulate through the scrotal wall (Fig. 5.57). Histologically, the initial epididymal lesion is similar to that of *Brucella ovis*, being characterized by intraepithelial lumina. In the chronic form of the disease seen in older rams, the epididymides are enlarged and fibrotic, and the testes are atrophic. The seminal vesicles or prostate may also be affected. Experimental inoculation of rams by various routes with *Actinobacillus seminis* has shown that part or all of the genital tract may become infected, but that the epididymis is most constantly involved.

Lesions in the epididymis caused by *H. somnus* are similar. The disease occurs sporadically, is sometimes accompanied by a febrile response, and the most frequent lesion is a large multilobulated abscess affecting one epididymis. The abscess cavities contain much fluid and green-yellow flocculent pus, often with lumps of necrotic material. Fibrous adhesions of the tunics also occur. In early cases with epididymal abscessation, the vas deferens is usually patent and contains pus and spermatic fluid, but in older cases that part of the epididymis adjacent to the vas deferens is usually empty and has grossly fibrosed walls. Microscopically, the early response is a severe suppurative one involving the walls of epididymal tubules. In appropriately stained sections, numerous bacteria can be demonstrated in intraluminal necrotic debris. Subsequently, infection may extend to the stroma, so that in

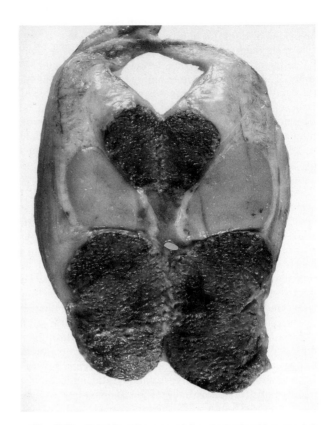

Fig. 5.57 Epididymitis caused by *Actinobacillus seminis*. Ram. (Courtesy of K. L. Hughes.)

chronic cases there is extensive interstitial epididymitis with much fibrosis. As with brucellosis, the testis is only involved secondarily. There is one report of severe suppurative epididymitis and orchitis caused by *H. somnus* in a 3-week-old calf.

Epididymitis is less important in **bucks** than in rams; *Actinobacillus seminis* and *Staphylococcus aureus* have been isolated from lesions, and there is a report of possible *B. ovis* epididymitis in an Angora goat. Epididymitis may also be secondary to sperm stasis, and in such cases *Escherichia coli* and *Pseudomonas* spp. may be isolated.

In **boars,** the macroscopic lesions caused by *Brucella suis* infection are quite variable. Single or multiple abscesses are frequent in the epididymis but less so in the testes, which may be enlarged or atrophic. Enlargement of the seminal vesicles due to localization of *B. suis* may also occur and abscessation, perhaps seen only microscopically, occurs in the seminal vesicles, prostate, and bulbourethral glands.

Epididymitis occurs in sexually mature dogs with canine distemper. The histologic appearance is similar to that seen in epididymitis of other causes, but, in addition, cytoplasmic and intranuclear inclusions are present in the epithelial cells. Some care is necessary to differentiate the specific inclusions from the eosinophilic cytoplasmic bodies normally present in the head of the epididymis and from the intranuclear bodies which are present in normal dogs; the fluorescent-antibody technique will differentiate inclusions of doubtful nature.

In male **dogs** infected with *Brucella canis* there is epididymitis, prostatitis, scrotal dermatitis, and testicular atrophy. These changes can be unilateral. Infected animals are bacteremic and the organism can persist in certain tissues, particularly the prostate, for many months, but recovery can eventually occur and recovered dogs are immune to reinfection. Venereal transmission to females by infected males can occur, although the organism is not consistently isolated from semen. Scrotal swelling is apparent 1–2 weeks after experimental intravenous inoculation, or 3–5 weeks after oral infection. Such swelling is caused by accumulation of fibrinopurulent exudate in the cavity of the tunica vaginalis. Scrotal ulceration is the result of persistent licking of the scrotum caused by pain of the epididymitis. Testicular abscesses are seldom if ever observed, but testicular necrosis (as observed in the bull) may occur rarely, and is accompanied by marked fibrous thickening of tunics. Microscopically in such cases, there is coagulation necrosis with a predominantly mononuclear, reparative response in peripheral areas where a necrotizing vasculitis and associated thrombosis may also occur. More frequently, however, histologic findings are those of interstitial epididymitis and prostatitis, and testicular atrophy. Lymphocytic infiltration of epididymal stroma is variable. Fibrosis may be extensive but, in contrast to brucellosis in other species, obliteration or stricture of ducts is unusual. Lymphocytes, neutrophils, and macrophages are present in epididymal ducts. In chronic cases there is marked enlargement of the epididymis, especially the tail, with possible spread of inflammation to the vas deferens. As well as containing inflammatory cells and an increased percentage of abnormal sperm, the ejaculate of male dogs with chronic brucellosis contains sperm agglutinins, and it is suggested that the resulting infertility is, in part, mediated by isoimmune reactions resulting from the heightened nonspecific phagocytic activity of inflammatory cells attracted to the sites of bacterial growth in the epididymis. *Brucella suis* may also cause a spontaneous granulomatous epididymitis and prostatitis in the dog.

Following infection of dogs with *Pseudomonas pseudomallei,* epididymitis, orchitis, and scrotal edema may occur in association with pyrexia, depression, swelling of one or more limbs, and lameness. Macroscopically, the epididymides are enlarged to perhaps three or four times normal size, are firm and hemorrhagic, and may contain small abscesses or necrotic foci. The testis and vas deferens may also be involved.

In the **stallion,** migrating strongyle larvae are considered a likely cause of epididymitis and spermatic granuloma of the epididymis, especially the head (Fig. 5.58); adenomyosis (see the following) has been shown to be a further cause.

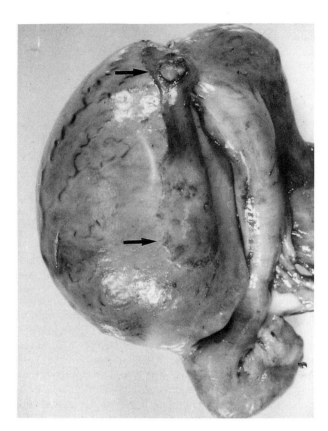

Fig. 5.58 Granulomatous lesion due to migration of strongyle larva. (arrows.) Testis. Horse.

Bibliography

Anderson, J., Plowright, W., and Purchase, H. S. Pathological and seminal changes in bulls affected with a specific venereal infection. *J Comp Pathol* **61:** 219–230, 1951. (Epivag)

Barr, S. C. *et al. Brucella suis* biotype 1 infection in a dog. *J Am Vet Med Assoc* **189:** 686–687, 1986.

Blue, M. G., and McEntee K. Epididymal spermatic granuloma in a stallion. *Equine Vet J* **17:** 246–251, 1985.

Burgess, G. W. Ovine contagious epididymitis: A review. *Vet Microbiol* **7:** 551–575, 1983.

Constable, P. D., and Webber, J. J. *Escherichia coli* epididymitis in rams. *Aust Vet J* **64:** 123, 1987.

Fodor, L., Hajtos, I., and Glavits, R. Purulent epididymitis in a sucking lamb caused by *Pasteurella haemolytica. Acta Vet Hung* **35:** 427–431, 1987.

Foster, R. A. *et al.* Identification of *Brucella ovis* in formalin-fixed paraffin-embedded genital tissues of naturally infected rams by the indirect peroxidase–antiperoxidase technique. *Aust Vet J* **65:** 324–326, 1988.

George, L., and Carmichael, L. Antisperm responses in male dogs with chronic *Brucella canis* infections. *Am J Vet Res* **45:** 274–281, 1984.

Hajtos, I. *et al.* Ovine suppurative epididymo-orchitis caused by *Histophilus ovis. J Vet Med (B)* **33:** 528–536, 1986.

Healy, M. C. *et al.* Comparison and partial characterization of the protein profiles and outer membrane antigens of *Actinobacillus* species isolated from ram lambs with epididymitis. *Am J Vet Res* **49:** 1824–1831, 1988.

Holzmann, A. *et al.* Orchiepididymitis in a bull with *Mycoplasma* in its ejaculate. *Zentralbl Veterinarmed (A)* **30:** 760–766, 1983.

Jansen, B. C. The epidemiology of bacterial infection of the genitalia in rams. *Onderstepoort J Vet Res* **50:** 275–282, 1983.

Jansen, B. C., Hayes, M., and Knoetze, P. C. The reaction of ovine neutrophils to *Histophilus ovis* in relation to genital infection in rams. *Onderstepoort J Vet Res* **50:** 125–132, 1983.

Metz, A. L., Haggard, D. L., and Hakomaki, M. R. Chronic suppurative orchiepididymitis associated with *Haemophilus somnus* in a calf. *J Am Vet Med Assoc* **184:** 1507–1508, 1984.

Vermeulen, S. O. *et al. Brucella ovis* as a possible cause of epididymitis in an Angora goat. *J South Afr Vet Assoc* **54:** 177–179, 1988.

Walker, R. L. *et al.* Association of age of ram with distribution of epididymal lesions and etiologic agent. *J Am Vet Med Assoc* **188:** 393–396, 1986.

Webb, R. F. Clinical findings and pathological changes in *Histophilus ovis* infections of sheep. *Res Vet Sci* **35:** 30–34, 1983.

C. Miscellaneous Lesions of the Epididymis

A **spermatocele** is defined as a cystic dilation of the epididymal duct with the accumulation of sperm in the cyst. The term is often used, however, to describe grossly visible pockets of sperm in the epididymis, although these may long have left the confines of the dilated duct. The development of a spermatocele follows congenital or acquired occlusion of the duct and the onset of spermatogenesis by the testis. Most spermatoceles progress to the development of spermatic granulomas. Impaction and inspissation of sperm in a cystic dilation is followed by atrophy of the surrounding ductal epithelium with fragmentation of the basement membrane or, in some instances, spontaneous rupture of the spermatocele occurs. In either event, extravasation of sperm occurs, and when the sperm contact stromal tissue, there is a characteristic granulomatous response to them (Fig. 5.52).

Lesions usually referred to as intraepithelial cysts or vacuoles but perhaps more precisely as **intraepithelial lumina** (as they have no distinct limiting membrane) are frequently associated with epithelial hyperplasia in the epididymis. Such lumina may be observed in the head, body, or tail and, if large, will result in the appearance of distinct epithelial bridges across affected ducts. The lumina probably form during regeneration following hydropic degeneration of varying cause. Causes include epididymitis as a nutritional deficiency. They also develop following estrogenic stimulation in immature mice, but they have been observed in apparently healthy dogs.

Adenomyosis of the epididymis is a condition characterized by invasion of the muscular layers and surrounding stroma by the epithelium and, insofar as sperm may become sequestered in these ductal outgrowths, spermatic granuloma may result. Adenomyosis occurs in older animals, particularly bulls and dogs, and there is evidence to suggest that chronic estrogenic stimulation is involved, such as in dogs with Sertoli cell tumors. Adenomyosis, and associated spermatic granuloma formation, have been observed in the epididymal tail of bulls implanted with the estrogenlike anabolic agent, zeranol; the structure of accessory sex glands was also altered. Spermatic granulomas secondary to adenomyosis will have wider distribution than those resulting from aberrant ducts.

Male English springer spaniel dogs affected with the lysosomal storage disease αL-**fucosidosis** are infertile, and light microscopy reveals severely vacuolated epididymal epithelium. On electron-microscopic examination, membrane-bound granular or fibrillar material is in epididymal epithelium, smooth muscle and myoid cells, and Sertoli cells. Some reduction in germinal cells is apparent in the testes.

Bibliography

Deschamps, J. C. *et al.* Effects of zeranol on reproduction in beef bulls: Scrotal circumference, serving ability, semen characteristics and pathologic changes of the reproductive organs. *Am J Vet Res* **48:** 137–147, 1987.

Taylor, R. M., Martin, I. C. A., and Farrow, B. R. H. Reproductive abnormalities in canine fucosidosis. *J Comp Pathol* **100:** 369–380, 1989.

VII. Spermatic Cord

A. Anomalies of the Spermatic Cord

Segmental aplasia of the vas deferens may occur independently or accompany aplasia of other mesonephric duct derivatives such as the seminal vesicle or epididymis. The condition has been observed occasionally in the bull and also in the dog and boar. Where the vas deferens persists in cases of epididymal aplasia, its lumen is narrow, and intraepithelial lumina, similar to those seen in the epididymis, may be present.

B. Circulatory Disturbances of the Spermatic Cord

A **varicocele** is a dilation and tortuosity of the veins of the pampiniform plexus and the cremasteric veins. It is observed occasionally in stallions, rarely in the bull, and in about 1–2% of rams, in which occurrence increases with age. Varicoceles appear as dark red nodules, 1–3 cm or more in diameter, enclosed in fascia of the spermatic cord proximal to the testis (Fig. 5.59). Dissection of varicoceles may reveal large organizing laminated thrombi. The etiology of varicocele is unknown, but (in humans) deficiency of valves in veins draining the testis is considered likely. Deficiency of elastic and fibrous tissue in surrounding fascia is another possibility.

Varicocele in the ram is bilateral, or unilateral with no apparent predisposition of side. Large varicoceles are always thrombosed and are associated with overt testicular degeneration, occasional thrombosis of testicular vessels, and reduced testis : bodyweight ratios, the latter possibly resulting from compromised thermoregulation in the testis. In adolescent varicocele in humans, testicular degeneration is evidenced by premature germ cell sloughing, decreased numbers of late-stage forms of germ cells (secondary spermatocytes, spermatids, and spermatozoa) and some degree of tubular sclerosis.

Detailed histologic studies have revealed an apparently independent sclerosis involving almost 50% of both arter-

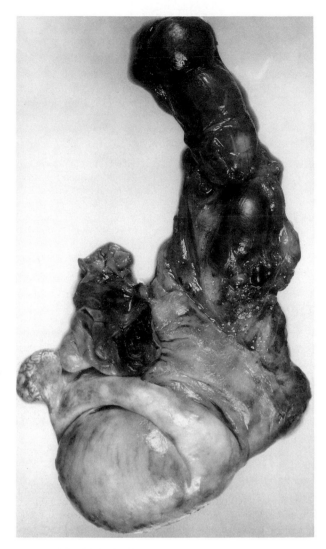

Fig. 5.59 Varicocele with thrombosis. Horse.

ies and veins in the spermatic cord of rams. The arterial changes, which tend to be bilateral, consist of fibroplasia of the tunica intima. A variable degee of intimal and or medial mineralization may also be seen in branches of the testicular artery in the spermatic cords of rams and bucks. An apparent hypoplasia of veins of the spermatic cord has been observed in bulls associated with thickening and degeneration of the testicular artery.

Although studies in the bull have not demonstrated a direct connection between venous and arterial blood vessels of the pampiniform plexus, functional arteriovenous anastomoses through small vessels do occur in normal bulls, boars, and rams and may represent a further means of thermoregulation of the testis. A large anastomosis of the spermatic artery and vein has been observed in the stallion associated with an abdominal testis.

Torsion of the spermatic cord, mechanically permissible if there is a broad mesorchium, occurs in dogs, pigs, and occasionally horses, and may involve a cryptorchid testis

or one in the inguinal position. Neoplastic enlargement of a cryptorchid testis also appears to favor the development of torsion, as may loose attachment of the epididymal tail to the testis.

A normal descended testis is seldom involved in torsion. Depending on the degree of torsion, there may be venous occlusion with congestion, hemorrhage, and edema, or occlusion of both artery and vein with infarction and necrosis. This is usually a very painful condition in the dog. In acute torsion (most easily recognized at laparotomy), the marked venous engorgement is apparent as enlargement and darkening of the testis. Frequently, however, examination reveals focal accumulation of hemosiderin and bilirubin, and erythrophagocytosis as evidence of older hemorrhages often within the deeper parts of the affected testis, thus indicating that torsion and the resultant congestion and extravasation may be a gradual or repetitive process.

Incomplete torsion of cryptorchid testes is sometimes observed in pigs in abattoirs, and it is considered that the slaughtering process itself may be responsible.

C. Inflammation of the Spermatic Cord

Referred to as **funiculitis,** inflammation of the spermatic cord follows open castration. It may be acute and necrotizing, as is often seen in the pig, in which species there is ample opportunity for contamination, or it may be chronic as in the typical scirrhous cord of horses and cattle. The nature of the infecting organisms determines the nature of the reaction. In the pig, it is often a necrotizing purulent response and there may be very little granulation tissue, and the cutaneous incision may not heal. In this species too, ascending infection provoking a diffuse peritonitis, and tetanus, are common complications. The classic scirrhous cord of the gelding is a pyogenic infection, usually by staphylococci, which produces typical botryomycosis, an exuberant granulation tissue enclosing multiple small abscesses and containing many fistulae.

Verminous granulomas caused by wandering larvae of *Strongylus* spp. are occasionally observed in the spermatic cord and testes of horses.

Bibliography

Amselgruber, W., and Sinowatz, F. Relationship of the testicular artery to the veins of the pampiniform plexus in the bull. *Anat Histol Embryol* **16:** 363–370, 1987.

David, J. S. E., and McCullagh, K. G. A case of spermatic arteriovenous anastomosis in the horse. *Equine Vet J* **10:** 94–96, 1978.

Ezzi, A. *et al.* Pathology of varicocele in the ram. *Aust Vet J* **65:** 11–15, 1988.

Hees, H. *et al.* Vascular morphology of the bovine spermatic cord and testis 1. Light and scanning electronmicroscopic studies on the testicular artery and pampiniform plexus. *Cell Tissue Res* **237:** 31–38, 1984.

Noordhuizen-Stassen, E. N. *et al.* Functional arterio-venous anastomoses between the testicular artery and the pampiniform plexus in the spermatic cord of rams. *J Reprod Fertil* **75:** 193–201, 1985.

Panebianco, A., Zanghi, A., and Catone, G. Some angiopathies of the spermatic cord in sheep and goats. *Clin Vet* **108:** 441–449, 1985.

Pascoe, J. R. Torsion of the spermatic cord in a horse. *J Am Vet Med Assoc* **178:** 242–245, 1981.

VIII. Seminal Vesicle and Ampulla

A. Anomalies of Development

The chief congenital anomaly affecting the seminal vesicles and ampullae is **segmental aplasia** of the mesonephric duct. This condition appears to occur most frequently, and has been most studied in the bull, in which it seems to be heritable. Aplasia or hypoplasia of this duct may affect any or all of the sex organs derived from it so that the vas deferens and epididymis may be simultaneously affected. Aplasia is mostly unilateral, and on the affected side, the ampulla is completely absent or rudimentary, and the seminal vesicle is markedly hypoplastic and sometimes cystic. Seminal vesicle hypoplasia of varying severity may occur independent of aplasia of the ampullae or vas deferens. Other anomalies of the seminal vesicles and ampullae of the bull include fusion (Fig. 5.60A) and appendages of the ampullae. There is significant variation in the relationship of ampullae to the seminal vesicles in normal bulls; the ampullae are entirely dorsal to the seminal vesicles in 40% of bulls, are entirely ventral to the seminal vesicles in 40% of bulls, and in the remaining 20% of animals are intermediate. Realization of these anomalies and anatomic variations is important in rectal examination of bulls. Although the anomalies per se seem incidental, and unrelated to diminished fertility, evidence suggests that anomalous glands are predisposed to inflammation.

Cystic dilation of the lumina of occasional lobules, or a general dilation of duct lumina in one, and sometimes both, seminal vesicles, occurs in bulls, horses, and swine.

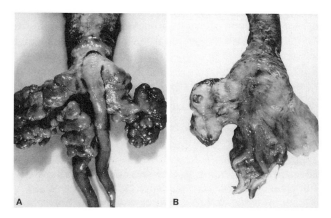

Fig. 5.60 (A) Fusion of seminal vesicles with displaced ampulla. (B) Chronic seminal vesiculitis with scarring. Bull.

B. Inflammation of Seminal Vesicles

The discussion here relates in particular to seminal vesiculitis, but it is frequently accompanied by reaction of similar types in ampullae. Seminal vesiculitis is a common lesion in the bull, and is seen rarely in the stallion and boar (Fig. 5.60B). Macroscopically apparent seminal vesiculitis is quite rare in the ram, but histologic lesions accompany the frequent localization of *Brucella ovis, Actinobacillus seminis,* and *Haemophilus somnus* (*Histophilus ovis*) in the seminal vesicles and ampulla in association with epididymitis caused by these organisms. Typically, seminal vesiculitis is characterized clinically by palpable changes in the seminal vesicles on rectal examination, and by changes in the semen, particularly the presence of pus. Young bulls, younger than 2 years of age, are frequently affected.

Two forms of seminal vesiculitis in the bull are recognized, a chronic interstitial form characterized by a considerable increase in size, excessive fibrosis, firm consistency, and loss of lobulation, and a predominantly degenerative form characterized by a slight or no change in size, and only a slight increase in consistency.

Microscopically, the chronic interstitial form is characterized by fibrosis and cellular infiltration of stroma with lymphocytes, plasma cells, histiocytes, neutrophils, and occasional eosinophils. The epithelium is usually normal but is metaplastic in some areas (Fig. 5.61A). Acini contain a small amount of finely granular eosinophilic material, occasional neutrophils, and desquamated epithelium. In the predominantly degenerative form of seminal vesiculitis, only a slight increase in stroma is observed, but leukocytic infiltration is often present. In addition, neutrophils, sloughed epithelium, and intensely basophilic material are present in the acini in clumps or strands (Fig. 5.61B).

Numerous infectious agents may be isolated from **bovine seminal vesiculitis,** but the precise roles of most have yet to be determined. In chronic seminal vesiculitis, bacteriologic investigations are often negative. *Actinomyces* (*Corynebacterium*) *pyogenes* appears to be the most common isolate, and lesions in general are those of chronic interstitial seminal vesiculitis. Large abscesses may be present also. These may involve adjacent tissue, and adhesions and fistulae form with the rectum or bladder. Staphylococci and streptococci generally do not form large abscesses, but more often, small focal or diffuse suppurative inflammation. Seminal vesiculitis caused by *Brucella abortus* is fibrinopurulent, progressing to necrosis, suppuration, and dystrophic calcification; granuloma formation is rarely observed. Tuberculous seminal vesiculitis causes a great increase in the size of seminal vesicles with characteristic nodulation and caseation.

Several viruses, including the virus of infectious bovine rhinotracheitis–infectious pustular vulvovaginitis (bovine herpesvirus-1), may infect the accessory sex glands of the bull. Clinical signs of seminal vesiculitis in such cases are reported to be mild and transitory. *Mycoplasma bovigeni-*

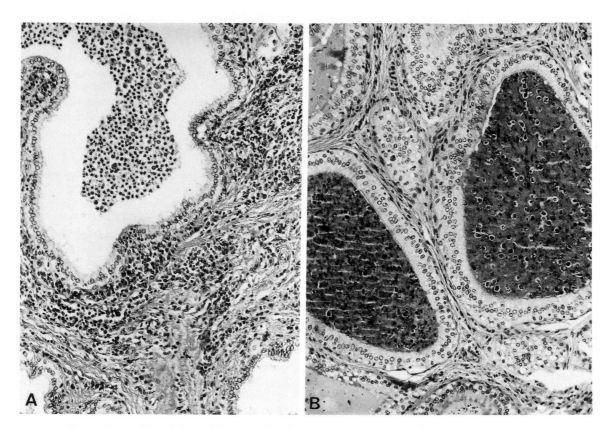

Fig. 5.61 (A) Chronic interstitial seminal vesiculitis. Bull. (B) Acini filled with basophilic detritus.

talium and ureaplasmas have been isolated from bulls with seminal vesiculitis, and unequivocal lesions have resulted from experimental infection with these organisms and *Mycoplasma bovis*. Apart from a slight to pronounced infiltration of eosinophils into these lesions, they are nonspecific, and it is unclear to what extent mycoplasmas cause spontaneous seminal vesiculitis. *Mycoplasma bovigenitalium* can be isolated from the vesicular gland secretions from a large proportion of normal pubescent bulls. As with the mycoplasmas, *Chlamydia psittaci* has been isolated from the semen and epididymides of bulls, and its possible role in seminal vesiculitis has yet to be clarified. The seminal vesicles may also be an important site for localization of *Leptospira* spp. *Leptospira interrogans* serovar *hardjo* has been isolated with equal frequency from seminal vesicles and kidneys of naturally infected bulls, thereby suggesting the possibility of venereal transmission.

The pathogenesis of bovine seminal vesiculitis is as yet unresolved. Hematogenous infection of the glands from a primary point of infection elsewhere seems probable in most cases, but ascending infection from the urethra and infection via peritoneal fluid, perhaps following navel infections during calfhood, are also possible. Establishment of infection may be aided by stress, reflux of urine or semen into the seminal vesicles, and by congenital defects in the duct system of the pelvic organs.

Seminal vesiculitis is of lesser importance in **sheep and**

goats, and **pigs,** than in the bull, but occurs in brucellosis caused by *Brucella melitensis* and *Brucella suis,* respectively. The characteristic lesion is chronic, and resembles that seen in the bull. Abscessation of the seminal vesicles and prostate may accompany orchitis in boars caused by *Pseudomonas pseudomallei*. Abscesses in the seminal vesicles of barrows may be the result of infection at time of castration. Purulent, abscess-forming seminal vesiculitis in stallions has been described but appears to be uncommon. In the **stallion,** following experimental infection with virus of equine arteritis, virus is shed in semen for months and the ampulla and bulbourethral glands seem to be predilection sites. It is unclear whether there are lesions in infected tissues, or associated sperm abnormalities.

Bibliography

Blanchard, T. L. *et al.* Bilateral seminal vesiculitis and ampullitis in a stallion. *J Am Vet Med Assoc* **192:** 525–526, 1988.

Campero, C. M., Bagshaw, P. A., and Ladds, P. W. Lesions of presumed congenital origin in the accessory sex glands of bulls. *Aust Vet J* **66:** 80–85, 1989.

Dargatz, D. A., Mortimer, R. G., and Ball, C. Vesicular adenitis of bulls: A review. *Theriogenology* **28:** 513–521, 1987.

Ellis, W. A., Cassells, J. A., and Doyle, J. Genital leptospirosis in bulls. *Vet Rec* **118:** 333, 1986.

Foster, R. *et al.* Pathology of the accessory sex glands of rams infected with *Brucella ovis*. *Aust Vet J* **64:** 248–250, 1987.

Linhart, R. D., and Parker, W. G. Seminal vesiculitis in bulls. *Compend Contin Educ* **10:** 1428–1432, 1448, 1988.

Neu, S. M., Timoney, P. J., and McCallum, W. H. Persistent infection of the reproductive tract in stallions experimentally infected with equine arteritis virus. *Equine Inf Dis V: Proc Vth Int Conf* D. G. Powell (ed.), 149–154, 1988.

IX. Prostate and Bulbourethral Gland

These accessory sex glands are derived from urogenital sinus epithelium and their responses to some adverse influences are similar. Studies on the accessory sex glands of normal bulls and rams have demonstrated that a protective local secretory immune system is present and especially involves the bulbourethral glands and prostate which both have a predominance of IgA-containing cells in subepithelial locations.

A. Anomalies of Development

Anomalies of the bulbourethral gland include congenital retention cysts in bulls and cats, and aplasia, hypoplasia, and fusion in bulls. Melanosis of the bulbourethral glands has been observed in the bull and in swine. Malformations of the prostate are infrequent and are usually seen together with anomalies of the remaining genitalia, especially in intersex states and in cryptorchidism. In some dogs, the dorsal median groove of the prostate may be indistinct so that the bilobed nature of the gland is not clearly evident. Other prostatic anomalies include prostatic appendage in the bull, with polyplike protrusion of prostatic tissue into the urethral lumen, and congenital retention cysts in the bull and cryptorchid boar.

Prostatic cysts in the dog may be congenital or be secondary to hyperplasia, neoplasia, or inflammation (see the following). Classification of prostatic cysts in the dog into four types—multiple cysts associated with prostatic hypertrophy, retention cysts, paraprostatic cysts, and cysts associated with squamous metaplasia—has been proposed. Although their precise origin is still unclear, paraprostatic cysts probably result from anomalous development. Most cysts appear to be attached to the prostate at a small localized area but without a definite stalk (Fig. 5.62A). The smaller cysts are up to ~7 cm in diameter and have a wall 2–5 mm thick, whereas larger cysts may be up to 24 cm long and 14 cm in diameter and contain much collagen and even bone in their walls. Both types of cysts are lined by epithelium which appears to be secretory. An accumulation of fibrin on the inner aspect of the larger cysts, as well as cauliflowerlike bony lesions extending into the lumen, is common. Cysts may rarely become infected and rupture, but their content is almost always sterile and is devoid of pus, urine, and spermatozoa. It has been suggested that such cysts arise from vestiges of the Müllerian duct, the uterus masculinus, but most reported cysts appear to have arisen from one or the other prostatic lobe. Enlarged remnants of ducts which form prostatic or paraprostatic cysts may be expected to arise

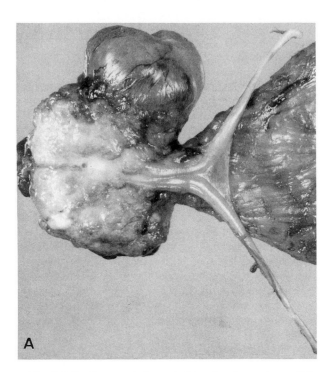

Fig. 5.62A Paraprostatic cyst distorting the contour of the prostate. Dog.

in the dorsal aspect of the prostate and have their origin in the midline. Where such cysts are bilobed and lined by simple columnar to pseudostratified columnar epithelium that resembles endometrium, their designation as cystic uterus masculinus is justified. Cystic enlargement, which occurs mostly in old dogs, appears to be stimulated by estrogenism, as in response to a Sertoli cell tumor, and may cause constipation, dysuria, or anuria.

A **uterus masculinus** is also frequently observed in males of other species (Fig. 5.62B). As in the dog, it may be cystic or communicate with the urethra.

B. Inflammation of the Prostate and Bulbourethral Gland

Inflammation of the bulbourethral gland in the bull often accompanies seminal vesiculitis, which it resembles in both causation and tissue response. Concretions may result from chronic inflammation.

Prostatitis is common in the dog. It is often a disease of older dogs in which hyperplastic prostatic changes are present. The inflammatory changes contribute to the enlargement in a significant percentage of the cases, but the disease may occur in younger dogs with normal prostates. The infecting agents are urinary pathogens, *Escherichia coli*, *Proteus vulgaris*, streptococci, and staphylococci, which invade via the prostatic urethra.

Prostatitis is often acute with systemic signs of illness, and about two thirds of affected dogs have a history of urinary tract signs, which include gross blood and or pus in the urine, urethral discharge, incontinence, or dysuria.

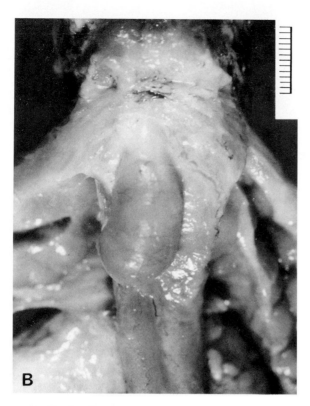

Fig. 5.62B Cystic uterus masculinus overlying ampullae. Bull.

Except for dramatic elevations of neutrophil counts with prostatic abscesses, clinical pathology data are mostly equivocal so that cytologic evaluation and culture of prostatic fluid collected by prostatic massage or ejaculation is needed to assist in the differential diagnosis of prostatic disease in the dog.

Acute prostatitis is a diffuse or focal suppurative inflammation with a tendency to abscess formation. The abscesses may be minute and multiple, or large with confluent areas of necrosis. The larger abscesses fluctuate on capsular palpation. Such abscesses may lead to metastatic sepsis, septicemia, peritonitis, and death, especially if there is sudden rupture.

In diffuse inflammations, the gland is often asymmetrically enlarged, congested, and edematous with a soggy consistency, and pressure causes welling of pus over the cut surface. Depending on the duration and severity of the inflammation, the reaction may be centered on the acini with ordinary catarrhal changes and only slight stromal involvement, or the reaction with neutrophilic infiltration may pervade all structures (Fig. 5.63A). Localization with destruction of acini proceeds to abscess formation and these may become confluent, converting the whole gland into a multiloculate abscess cavity, but this is unusual. The acute inflammations may resolve and leave only extensive scarring, or they may become chronic, especially if ducts are obstructed, and the infection then tends to persist in small walled-off foci.

Chronic prostatitis in the dog is also a common lesion (Fig. 5.63B). The prostatic epithelium is atrophic and its cytoplasm loses its characteristic eosinophilic staining quality. The lumina of the gland contain a variable number of inflammatory cells, neutrophils, and macrophages, and debris. The inflammation is apt to involve the gland segmentally and may spare large portions. Aggregations of lymphocytes in the fibromuscular stroma, especially about the ducts, is a very common change.

Prostatitis is a constant feature of *Brucella canis* infection. Typically, involvement is extensive but lobular in distribution, and consists of a generalized lymphocytic infiltration with destruction of adjacent epithelium and associated fibrosis.

C. Metaplasia and Hyperplasia of the Prostate and Bulbourethral Gland

These changes primarily concern the canine prostate, but a reversible enlargement of the bulbourethral glands with squamous metaplasia, hyperplasia, and cystic dilation, occurs in wethers ingesting strains of certain clovers (*Trifolium pratense, T. repens, T. subterraneum*) possessing high estrogenic potency (Fig. 5.64).

Squamous metaplasia of the epithelium of the prostate gland in the male dog may occur spontaneously in association with neoplasia of the testes, particularly Sertoli cell tumors, or following the administration of estrogens. The metaplastic change may involve acini in all parts of the gland as well as in the prostatic urethra, uterus masculinus, and ducts.

Affected epithelium is converted to a stratified squamous type from the surface of which squames are shed into the lumen (Fig. 5.65). Neutrophils and macrophages are often numerous in lumina containing squames which have become fused into amorphous masses. Squamous metaplasia is associated with an increased proliferation of basal cells. Although squamous metaplasia of the canine prostate predisposes it to inflammation, there is no evidence that it is a preneoplastic change. Squamous metaplasia of the prostate, comparable to that in the dog, has also been observed in swine. In cats following estrogen administration, prostatic enlargement with epithelial hyperplasia and cystic dilation of glands occurs, but metaplastic cornification is restricted to urethral epithelium.

In wethers it has been demonstrated that a few weeks grazing on potent green clover pastures provides adequate estrogenic stimulus for the induction of profound and extensive metaplastic changes in the accessory sex organs, particularly the bulbourethral glands. Entire male sheep are not affected.

The bulbourethral glands become enlarged and firmer than normal and are marbled on the cut surface. They may also contain cysts, and in a small percentage these cysts may become relatively enormous and cause a fluctuating swelling of the perineal region (Fig. 5.64). The large cavity fills with urine and cellular detritus and communicates with the lumen of the urethra through the dilated pores of

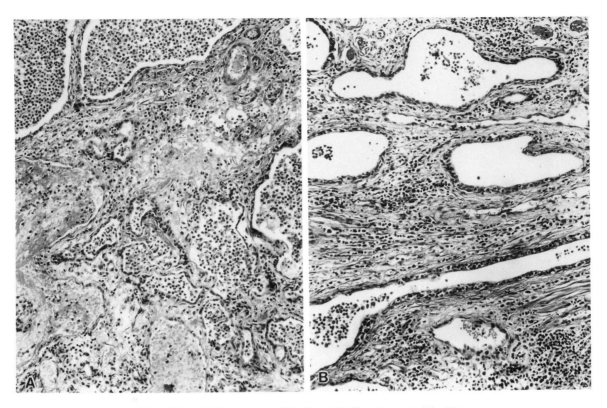

Fig. 5.63 (A) Acute prostatitis. Dog. (B) Chronic prostatitis. Dog.

bulbourethral glands in the dorsal wall of the pelvic urethra. The epithelial hyperplasia, together with the formation of cellular casts, may produce urethral obstruction
with death. Prolapse of the rectum is not uncommon.
Cysts of the uterus masculinus occur and castrate males
show mammary development. (For effect on females, see
The Female Genital System, Chapter 4 of this volume).
The prostatic changes consist of squamous metaplasia
with conversion of the acini to keratinaceous cysts. There
is also an increase in the stroma of the prostate, and the
changes are grossly visible as chalky or yellowish streaks
and flecks. In addition, the prostatic urethra may show
metaplastic changes. Comparable changes after stilbestrol
administration have been observed in the bulbourethral
glands of feedlot lambs, and in both the bulbourethral
glands and seminal vesicles of bull calves. Marked squamous metaplasia of glandular genital organs, but particularly the seminal vesicles of bulls, occurs following ingestion of chlorinated naphthalenes.

Hyperplasia–hypertrophy of the prostate is observed
occasionally in the bull but is of common occurrence only
in the dog; the organ is more or less diffusely involved,
although in some instances normal or atrophic lobules may
be interspersed in small numbers with hyperplastic ones.
Some degree of hyperplasia is often evident in dogs 4–5
years of age, and the incidence and degree increase with
advancing years such that 80% or more of mature or old
dogs may have enlarged prostates. Enlargement of the
prostate is frequently associated with constipation, pre-

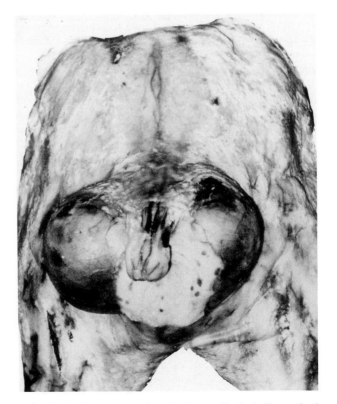

Fig. 5.64 Phytoestrogenism in sheep. Cystic bulbourethral
gland in wether.

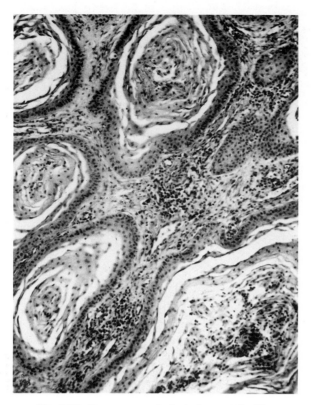

Fig. 5.65 Squamous metaplasia in prostate. Dog.

sumably caused by pressure on the rectum. Less common, but more important, is interference with micturition. It is not agreed just why urinary retention occurs. Compression with stenosis of the prostatic urethra has been advanced in explanation but is unsatisfactory because, at least in postmortem specimens, it is impossible to demonstrate any compression of the urethra, and because in the living animal, incontinence is common and slight manual pressure on the bladder will serve to eliminate urine, nor is the urethra completely surrounded by the prostate gland. It is possible that pressure of the enlarged gland on the sacral parasympathetic outflow causes paresis of the bladder. It is also possible that attenuation of the urethral lumen by longitudinal stretching may contribute in those cases in which the prostatic enlargement is sufficient to cause its displacement forward out of the pelvic cavity. Acute infections of the urinary tract and hydronephrosis often complicate the urinary obstruction.

In point of causation, there is little doubt that hormonal imbalances are important. Hyperplasia does not occur in eunuchoid or castrate dogs and, once established, castration is a useful therapeutic measure. There are rare exceptions in which an adrenocortical tumor in a castrate dog has been associated with hyperplasia. An altered androgen : estrogen ratio seems to underlie prostatic hyperplasia in dogs. Two phases in the development of lesions are suggested. In the first phase in young dogs, actual epithelial hyperplasia is associated with normal interstitial cells and normal androgen production but with the secretion in

the testes of an as yet unidentified estrogenlike molecule. In the second, cystic, phase there is decreased androgen production and decreased activity of interstitial cells but continued production of the estrogenic substance. Importantly, however, although levels of circulating testosterone may decrease in old dogs, actual concentrations of dihydrotestosterone in hypertrophied or hyperplastic tissue are increased. Also it has been demonstrated that in such tissue in aged beagles, nuclear androgen receptors are elevated. Therefore estrogens appear to act synergistically with androgens to potentiate hyperplasia and act directly on the prostate. It is unclear in spontaneous prostatic hyperplasia, however, whether and to what extent estrogen induces accompanying stromal changes.

The hyperplastic gland is almost invariably larger than normal, being up to four times its normal size, and the surface is irregularly nodular (Fig. 5.66A), in some cases obscuring the normal bilobed appearance. Palpably fluctuating cysts and venous and lymphatic ectasias may be present beneath the capsule. The appearance of the sectional surface varies depending on the degree of acinar and stromal hyperplasia and on the presence and size of cysts. The lobules vary much in size and may be poorly defined, or well defined if there is a prominent increase in size of the interlobular trabeculae.

The glandular elements are often spongy, and close examination reveals numerous small cysts containing a milky fluid. The larger readily visible cysts are irregularly distributed, but the largest ones tend to be located beneath the capsule. The prostatic urethra may be somewhat distorted with small yellow elevations of the mucosa corresponding to the papillae of the excretory ducts. Microscopically, the structure is very diverse and consists of adenomatoid hyperplasia, stromal hyperplasia, and condensation, and cystic formations variously interspersed (Fig. 5.66B). The adenomatoid or acinar hyperplasia consists of epithelial hyperplasia without hypertrophy, leading, as in the thyroid gland, to papillary proliferations into the lumen, but always in a single layer on an intact basement membrane. Within the same lobule, some acini may be cystic, and these are of irregular size, rounded, and lined by an epithelium which has become pressed and flattened. There may be secretion in the lumen but usually not. The interlobular connective tissue is frequently and irregularly increased in amount and condensation, and may exist as broad sheets with some extension into the intralobular stroma. In most instances, this interstitial tissue contains accumulations of mononuclear inflammatory cells, but the infiltrations do not involve the epithelium or acinar lumen.

There is no clear distinction between a normal prostate and one which is in the early stages of pathologic hyperplasia, although arbitrary decisions may be made on the weight or size of the gland relative to body weight. The relative weight corrected for age is fairly consistent except in Scottish terriers, in which the relative weight of the prostate is about four times that of any other breed. Although estrogens in some cases are therapeutically effec-

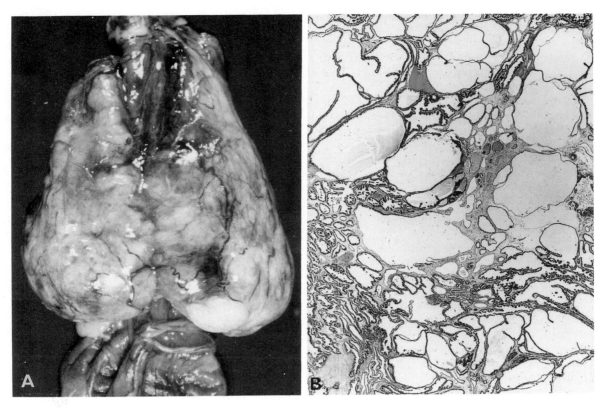

Fig. 5.66 (A) Prostatic hyperplasia. (B) Cystic hyperplasia of prostate. Dog.

tive in causing temporary regression of an enlarged prostate, these hormones in suitable dosage cause rapid enlargement of the gland.

Hypertrophy induced by estrogen per se is distinguishable by virtue of squamous metaplasia in the gland, and frequently in the uterus masculinus also.

D. Neoplastic Diseases of the Prostate and Bulbourethral Gland

Neoplasia of the accessory sex glands is rare in all domestic animals. Hemangiosarcoma has, however, been observed in the bulbourethral gland of the goat. Major interest centers on adenocarcinoma of the canine prostate. Although also uncommon, in several studies this neoplasm was diagnosed as the cause of prostatic enlargement in a significant percentage of cases. Neoplasia of the canine prostate is considered not to result from hyperplasia. The cause of prostatic carcinoma is unknown but, as with hyperplasia, it seems clear that the neoplasm develops in an environment of hormonal imbalance, probably with other causative factors also being involved. Prostatic neoplasia is not causally related to testicular neoplasia.

Prostatic adenocarcinoma is seen mostly in old dogs, especially those older than 10 years. Clinical signs are those of prostatic disease generally, but dogs with adenocarcinoma seem more likely to exhibit emaciation and rear limb locomotory disturbances, apparently due to metasta-

sis of the tumor to the lower lumbar vertebrae, bones of the pelvis and perhaps long bones of the rear limbs. In several published reports on prostatic adenocarcinoma, bone metastasis occurred in more than one third of cases and was always accompanied and probably preceded by visceral metastasis. Bone metastasis may involve other bones in addition to the pelvis and lumbosacral spine, such as the ribs, scapula, or digits. Bone metastasis may result from spread of neoplastic cells in the vertebral venous plexus, systemic circulation, or by direct extension. Macroscopically, enlargement of the prostate with adenocarcinoma is more likely to be asymmetric and irregular than is the case in benign hyperplasia.

A severe fibrosing reaction to invading tumor cells is almost always evident, and the neoplasm is usually quite hard, sometimes containing foci of ossification. The capsule of the prostate and adjacent organs are often invaded by the neoplasm, which may be adherent to the pubis or pelvis, probably more so than in benign prostatic hyperplasia. Cyst formation in association with the neoplasm is common. Neoplastic tissue may extend into the sublumbar area and metastasis to iliac, lumbar, and pelvic lymph nodes, bone, kidneys, bladder, lungs, liver, heart, mesentery, and omentum occurs. Neoplastic cells may also be observed in the urine and occasionally in blood smears. Prostatic adenocarcinoma metastatic to the regional lymph nodes and urinary bladder, but not the lungs, has been associated with hypertrophic osteopathy. Histologic confirmation of the prostatic origin of metastases in bone and other tissues may be aided by immunocytochemistry. However, whereas both normal and hyperplastic canine prostatic epithelium label strongly with antibodies to both canine and human prostatic antigen, most reported canine prostatic adenocarcinomas, unlike those in human studies, were negative.

Microscopically it is important to note that most prostates with carcinoma also contain areas of hyperplasia. A variety of patterns may be seen in prostatic carcinoma, but a classification into two broad types of adenocarcinoma and undifferentiated adenocarcinoma has been suggested. Subclassification of adenocarcinoma into intraalveolar proliferative or small acinar types, and undifferentiated adenocarcinoma into syncytial or discrete epithelial types, is proposed.

The intra-alveolar proliferative type appears to be most common and consists of multiple foci of irregularly shaped alveoli that contain fronds of neoplastic cells radiating from a dense basal layer toward a central lumen. The cells comprising the fronds are not located on a basement membrane and this feature is important in differentiating adenocarcinoma from benign hyperplasia. In the second (small acinar) type, acini are embedded in excessive proliferating stroma and close attention to the cytology and pleomorphism of acinar cells is necessary in order to differentiate this form from acinar atrophy.

In the third (syncytial) form, the neoplasm is composed of pleomorphic, often fusiform cells, frequently arranged in sheets and resembling a sarcoma; such cells are intermingled with neoplastic acini. Singular, discrete, round to polygonal, often signet-ring-shaped epithelial cells embedded in stroma are characteristic of the fourth (discrete epithelial) type of adenocarcinoma.

Other primary tumors of the prostate are sarcoma (fibrosarcoma, leiomyosarcoma) and squamous cell carcinoma.

Secondary tumors, especially tumors of the lymphoreticular system, are not too uncommon although uncommonly looked for. It is sometimes difficult to differentiate lymphomatous infiltration from the common senile infiltration of lymphocytes. Carcinomas of the neck of the bladder may invade the prostate, as may similar transitional cell carcinomas of the prostatic urethra and ducts; indeed, in some series the transitional cell carcinoma is more numerous than is adenocarcinoma.

E. Miscellaneous Conditions of Accessory Sex Glands

Atrophy of the accessory sex glands occurs following castration, in advanced age, and sometimes with chronic inflammation. The glands become small, dense, and of tough consistency.

In the dog, the atrophic prostate may be reduced to one-half or one-fourth its normal size. The response to androgen deprivation caused by castration is sudden; ultrastructural studies have revealed that within 3 days many prostatic acinar cells were atrophic and contained large lipid droplets. There was, however, a concurrent proliferation of basal cells. Microscopically after castration, acinar and ductal epithelial cells decrease in size, become less differentiated and more basophilic, and within 3 months epithelium is flat, and acinar lumina small or nonexistent. The stroma concurrently becomes more conspicuous, and smooth muscle of the capsule and trabeculae disappears and is replaced by dense fibrous tissue. Senile atrophy occurs normally in dogs older than 11 years of age. Atrophy may also result from distemper, diabetes mellitus, and other systemic diseases.

Concretions in the seminal vesicles, bulbourethral glands, or prostate are attributed to precipitation of retained secretions or to chronic inflammation. Corpora amylacea are extremely rare in the accessory sex glands of domestic animals. In bulls, vesicular concretions are up to 1.5 cm or more in diameter, irregular, friable, rough externally, and distinctly laminated on section. Histologically, the concretions consist of amorphous eosinophilic debris with occasional clumps of enmeshed nuclear material. They are usually found in inflamed seminal vesicles and the lining of the concretion cavity varies from normal to pseudostratified. Concretions are composed of organic components, phosphates, carbonates, and spermatozoa—so-called semen stones. Microconcretions are common in the seminal vesicles of rams, unassociated with inflammation.

In dogs, the occurrence of prostatic concrements or calculi is extremely rare, and indeed, their existence is debated; it is suggested that such calculi may be confused with urinary calculi. Described calculi are 1–5 mm in size, hard, white, and spherical, and consist of phosphates and carbonates of calcium as well as urates and oxalates. These form around a crystallization point of organic material, mostly desquamated gland epithelium. The presence of calculi seldom appears to elicit clinical signs. Prostatic calculi have also been observed in sheep.

Metaplastic ossification of prostatic stroma occurs in the dog in association with neoplasia and other chronic conditions. Foci of mineralized bone could, on radiographic examination, be mistaken for calculi.

Intraductal foreign (plant) material has been observed in the bulbourethral gland of the bull associated with chronic adenitis.

Bibliography

Atilola, M. A. O., and Pennock, P. W. Cystic uterus masculinus in the dog. Six case history reports. *Vet Radiol* 27: 8–14, 1986.

Berry, S. J., and Isaacs, J. T. Comparative aspects of prostatic growth and androgen metabolism with aging in the dog *versus* the rat. *Endocrinology* 114: 511–520, 1984.

Brendler, C. B. *et al.* Spontaneous benign prostate hyperplasia in the beagle. *J Clin Invest* 71: 1114–1123, 1983.

Campero, C. M., Ladds, P. W., and Thomas, A. D. Pathological findings in the bulbourethral glands of bulls. *Aust Vet J* 65: 241–244, 1988.

Durham, S. K., and Dietze, A. E. Prostatic adenocarcinoma with and without metastasis to bone in dogs. *J Am Vet Med Assoc* 188: 1432–1436, 1986.

Ewing, L. L. *et al.* Testicular androgen and estrogen secretion and benign prostatic hyperplasia in the beagle. *Endocrinology* **114:** 1308–1314, 1984.

Hargis, A. M., and Miller, L. M. Prostatic carcinoma in dogs. *Compend Contin Educ* **5:** 647–653, 1983.

Leib, M. S. *et al.* Squamous cell carcinoma of the prostate gland. *J Am Anim Hosp Assoc* **22:** 509–514, 1986.

McEntee, M., Isaacs, W., and Smith, C. Adenocarcinoma of the canine prostate; immunohistochemical examination for secretory antigens. *Prostate* **11:** 163–170, 1987.

Olson, P. W. *et al.* Disorders of the canine prostate gland: Pathogenesis diagnosis, and medical therapy. *Compend Contin Educ* **9:** 613–624, 1987.

Rendano, V. T., and Slauson, D. O. Hypertrophic osteopathy in a dog with prostatic adenocarcinoma and without thoracic metastasis. *J Am Anim Hosp Assoc* **18:** 905–909, 1982.

Sarma, D. K. *et al.* Isolation of chlamydia from a pig with lesions in the urethra and prostate gland. *Vet Rec* **11:** 525, 1983.

Sinowatz, F. Early changes in the dog prostate after castration: An ultrastructural study. *Acta Anat* **120:** 103–107, 1984.

Cumulative Index

Boldface, major discussion; f, figure; t, table; italic, volume number

K

on kidney, *3*:292
on osteoclast, *1*:9
fetus and, *1*:70
meat diet and, *1*:77
metabolism of, *1*:66, *3*:291
calcium and, *3*:292
parathyroid hormone and, *3*:292
phosphorus and, *3*:292
in milk, *1*:81
osteomalacia and, *1*:66
Vitamin D poisoning, **1:81–83**, *1*:82f. *See also* Calcinogenic plants, intoxication by
action of vitamin A and, *1*:82
basophilic osteoid and, *1*:82
effects on teeth, *1*:83
mineral deposition in cornea and, *1*:465
subendocardial mineralization and, *3*:23
vascular mineralization and, *3*:56
Vitamin E
deficiency of
anemia and, *3*:174
effects on skin, *1*:600
intestinal lipofuscinosis and, *2*:87, *2*:87f
iron toxicity and, *3*:174
mulberry heart disease and, *3*:33
myocardial necrosis and, *3*:28, *3*:28f
nutritional myopathy and, *1*:228
dystrophy of pigmented epithelium and, *1*:502
equine degenerative myeloencephalopathy and, *1*:356
free radicals and, *1*:229
Vitamin E-responsive dermatosis in goats, *1*:600
Vitamin E-selenium deficiency, microangiopathic hemolytic anemia and, *3*:205
Vitamin K₃. *See* Menadione
Vitamin K, absorption of, hepatobiliary disease and, *3*:265
Vitamin K antagonists, poisoning by, *3*:264–265
Vitamin K-dependent factors, deficiency of, *3*:264
Vitiligo, *1*:578
Vizsla, sebaceous adenitis and, *1*:573
Vogeloides massinoi, in bronchi of cats, *2*:683
Vogeloides ramanujacharii, in lungs of cats, *2*:683
Vogt-Koyanagi-Harada syndrome, *1*:484, *1*:578, *1*:627
Volvulus of large colon in horses, *2*:100
Vomiting and wasting disease in pigs, hemagglutinating encephalomyelitis virus and, *1*:409, *2*:185
von Brunn's nests, *2*:522, *2*:535, *2*:536
von Ebner, incremental lines of, in dentin, *2*:4
von Willebrand factor, *3*:257, **3:258**

von Willebrand's disease, *3*:258, **3:262–263**
VTEC. *See Escherichia coli*, verotoxigenic
Vulva. *See also* Vagina and vulva
fibropapilloma of, *3*:453
squamous cell carcinoma of, *3*:451
Vulva and vestibule
embryology of, *3*:352
hypoplasia of, *3*:352
Vulval tumefaction of swine, zearalenone and, *3*:446
Vulvar-fold dermatitis. *See* Intertrigo
Vulvitis, granular, *Ureaplasma diversum* and, *3*:408
Vulvovaginitis
associated with ulcerative balanitis of Border Leicester rams, *3*:481
associated with ulcerative posthitis of sheep, *3*:481
of swine. *See* Vulval tumefaction of swine

W

Waardenburg-Klein syndrome, *1*:577
Walchia americana, trombiculidiasis and, *1*:690
Walkabout in horses, pyrrolizidine alkaloid poisoning and, *2*:394
Walking dandruff. *See* Cheyletiellosis
Walking disease in horses, pyrrolizidine alkaloid poisoning and, *2*:394
Wallaby, galactose-induced cataract and, *1*:474
Wallerian degeneration, **1:297**, *1*:302
bovine ephemeral fever and, *3*:94
in optic nerve, *1*:352
Wandering Jew. *See Tradescantia fluminensis*
Wangiella sp., subcutaneous phaeohyphomycosis and, *1*:668
Warble flies, *1*:676
Warbles, *1*:676
Warfarin, disorders of hemostasis and, *3*:265
Water diuresis, *2*:451
Waterhouse-Friderichsen syndrome, disseminated intravascular coagulation and, *3*:63
Water intoxication, polioencephalomalacia of ruminants and, *1*:343
Water-line disease in Labrador retriever. *See* Labrador retriever, seborrhea, primary and
Water-rot in sheep. *See* Fleece-rot
Watery mouth in lambs, *Escherichia coli* and, *2*:213
Wattles, *1*:554
Weaner colitis in sheep, *Campylobacter* and, *2*:234
Weather stain in sheep. *See* Fleece-rot

Weave mouth, *2*:7
Weaver syndrome in Brown Swiss cattle. *See* Myeloencephalopathy in Brown Swiss cattle
Weddellite, urinary calculi and, *2*:529, *2*:530
Weed. *See* Sporadic lymphangitis of horses
Weibel-Palade body, *3*:50
hemangiosarcoma and, *3*:99
Weimaraner dog
hyperestrogenism and, *1*:606
hypertrophic osteodystrophy and, *1*:109
hypomyelinogenesis in, *1*:370
myelodysplastic and dysraphic lesions in, *1*:274
oral malignancies and, *2*:22
persistent infections in, *3*:107
scoliosis in, *1*:274
sterile pyogranuloma syndrome and, *1*:703
thymic atrophy and, *3*:223
tricuspid insufficiency and, *3*:9t
Werdnig-Hoffmann disease, *1*:363
Wesselsbron disease, *2*:366–367
fetal infections and, *3*:440t
prolonged gestation and, *3*:392
Wesselsbron disease virus
hydranencephaly and, *1*:281
nervous system defects and, **1:287**
West Highland white terrier
atopy and, *1*:610
canine craniomandibular osteopathy and, *1*:112
congenital inguinal hernia in, *2*:97
copper toxicity and, *2*:400
ectopic ureter and, *2*:523
epidermal dysplasia with *Malassezia pachydermatis* infection and, *1*:575
globoid cell leukodystrophy in, *1*:372
keratoconjunctivitis sicca and, *1*:468
seborrhea, primary and, *1*:571
West Nile virus, *1*:412
cerebral anomalies and, *1*:287
Wheat diet in dogs, villus atrophy with increased epithelial loss and, *2*:108
Wheaten terrier. *See* Soft-coated Wheaten terrier
Whewellite, urinary calculi and, *2*:529
Whey feeding, esophagogastric ulcer in pigs and, *2*:68
Whippet
color dilution alopecia in, *1*:559
regional hypotrichosis and, *1*:558
Whipworms. *See Trichuris*; Trichurosis
White clover. *See Trifolium repens*
White heifer disease, *3*:355
White muscle disease. *See* Nutritional myopathy
White-spotted kidney of calves
Escherichia coli and, *1*:169, *2*:213

ISBN 0-12-391607-0

9 780123 916075

90018